NUMBERS EXPRESSED IN SCIENTIFIC NOTATION

$$
\begin{aligned}
1\ 000\ 000 &= 10 \times 10 \times 10 \times 10 \times 10 \times 10 &&= 10^6 \\
100\ 000 &= 10 \times 10 \times 10 \times 10 \times 10 &&= 10^5 \\
10\ 000 &= 10 \times 10 \times 10 \times 10 &&= 10^4 \\
1000 &= 10 \times 10 \times 10 &&= 10^3 \\
100 &= 10 \times 10 &&= 10^2 \\
10 &= 10 &&= 10^1 \\
1 &= 1 &&= 10^0 \\
0.1 &= 1/10 &&= 10^{-1} \\
0.01 &= 1/100 = 1/10^2 &&= 10^{-2} \\
0.001 &= 1/1000 = 1/10^3 &&= 10^{-3} \\
0.0001 &= 1/10\ 000 = 1/10^4 &&= 10^{-4} \\
0.000\ 01 &= 1/100\ 000 = 1/10^5 &&= 10^{-5} \\
0.000\ 001 &= 1/1\ 000\ 000 = 1/10^6 &&= 10^{-6}
\end{aligned}
$$

PHYSICAL DATA

Speed of light in a vacuum $= 2.9979 \times 10^8$ m/s

Speed of sound (20°C, 1 atm) $= 343$ m/s

Standard atmospheric pressure $= 1.01 \times 10^5$ Pa

1 light-year $= 9.461 \times 10^{12}$ km

1 astronomical unit (A.U.),
(average Earth-Sun distance) $= 1.50 \times 10^{11}$ m

Average Earth-Moon distance $= 3.84 \times 10^8$ m

Equatorial radius of the Sun $= 6.96 \times 10^8$ m

Equatorial radius of Jupiter $= 7.14 \times 10^7$ m

Equatorial radius of the Earth $= 6.37 \times 10^6$ m

Equatorial radius of the Moon $= 1.74 \times 10^6$ m

Average radius of hydrogen atom $= 5 \times 10^{-11}$ m

Mass of the Sun $= 1.99 \times 10^{30}$ kg

Mass of Jupiter $= 1.90 \times 10^{27}$ kg

Mass of the Earth $= 5.98 \times 10^{24}$ kg

Mass of the Moon $= 7.36 \times 10^{22}$ kg

Proton mass $= 1.6726 \times 10^{-27}$ kg

Neutron mass $= 1.6749 \times 10^{-27}$ kg

Electron mass $= 9.1 \times 10^{-31}$ kg

Electron charge $= 1.602 \times 10^{-19}$ C

STANDARD ABBREVIATIONS

A	ampere	g	gram	M	molarity
amu	atomic mass unit	h	hour	min	minute
atm	atmosphere	hp	horsepower	mph	mile per hour
Btu	British thermal unit	Hz	Hertz	N	newton
C	coulomb	in.	inch	Pa	pascal
°C	degree Celsius	J	joule	psi	pound per square inch
cal	calorie	K	kelvin	s	second
eV	electron volt	kg	kilogram	V	volt
°F	degree Fahrenheit	lb	pound	W	watt
ft	foot	m	meter	Ω	ohm

Media for *Conceptual Physical Science*, **Fourth Edition** by Paul G. Hewitt,

Prologue: The Nature of Science

1 Patterns of Motion and Equilibrium

INTERACTIVE FIGURES
- 1.21
- 1.23
- 1.24

VIDEOS
- Newton's Law of Inertia
- The Old Tablecloth Trick
- Toilet Paper Roll
- Inertia of a Cylinder
- Inertia of an Anvil
- Definition of a Newton
- Definition of Speed
- Average Speed
- Velocity
- Changing Velocity
- Definition of Acceleration
- Numerical Example of Acceleration
- Free Fall: How Fast?
- Free Fall: How Far?
- Free-Fall Acceleration Explained

2 Newton's Laws of Motion

INTERACTIVE FIGURES
- 2.6
- 2.9
- 2.12
- 2.21
- 2.24
- 2.25
- 2.26
- 2.27

TUTORIALS
- Parachuting and Newton's Second Law
- Newton's Third Law

VIDEOS
- Newton's Second Law
- Force Causes Acceleration
- Friction
- Falling and Air Resistance
- Forces and Interactions
- Action and Reaction on Different Masses
- Action and Reaction on Rifle and Bullet

3 M

INTERACTIVE FIGURES
- 3.10
- 3.12
- 3.13
- 3.22
- 3.26

TUTORIALS
- Momentum and
- Energy

VIDEOS
- Definition of Mo
- Changing Mome
- Decreasing Mor
- Bowling Ball an
- Conservation of Example
- Machines: Pulle

10 Waves and Sound

INTERACTIVE FIGURES
- 10.1
- 10.5
- 10.6
- 10.27
- 10.28
- 10.30
- 10.31

TUTORIALS
- Waves and Vibrations
- Doppler Effect

VIDEOS
- Transverse vs. Longitudinal Waves
- Refraction of Sound
- Resonance
- Resonance and Bridges
- Interference and Beats
- Doppler Effect

11 Light

INTERACTIVE FIGURES
- 11.2
- 11.3
- 11.10
- 11.19
- 11.33
- 11.34
- 11.41
- 11.58

TUTORIALS
- Light and Spectroscopy
- Color

VIDEOS
- Light and Transparent Materials
- Image Formation in a Mirror
- Model of Refraction
- Colored Shadows
- Yellow-Green Peak of Sunlight
- Why the Sky Is Blue and the Sunset Is Red
- The Rainbow
- Soap Bubble Interference
- Polarization and 3D Viewing

12 Atoms and the Periodic Table

INTERACTIVE FIGURE
- 12.22

TUTORIALS
- Atomic Structure
- Atoms and Isotopes
- Bohr's Shell Model

VIDEOS
- Evidence for Atoms
- Atoms Are Recyclable
- Electron Waves

13 The

IN
- 13.2
- 13.3
- 13.4
- 13.11
- 13.15
- 13.22
- 13.28
- 13.34
- 13.35

TUTORIAL
- Nuclear Physic

VIDEOS
- Radioactive De
- Half-Life
- Plutonium

20 Rocks and Minerals

TUTORIAL
- The Rock Cycle Activity

VIDEO
- The Rock Cycle

21 Geologic Time—Reading the Rock Record

INTERACTIVE FIGURES
- 21.6
- 21.8
- 21.21

TUTORIAL
- Formation of an Angular Unconformity

22 Plate Tectonics and Earth's Interior

INTERACTIVE FIGURE
- 22.3

TUTORIALS
- Mantle Convection and Seafloor Spreading
- Plate Boundaries and Plate Tectonics

VIDEOS
- The Mantle and Crust
- Plate Tectonics
- Folds and Faults in Earth's Crust
- The San Andreas Fault Story

23 Sh

TUTORIALS
- The Hydrologic
- Shaping Planet

VIDEO
- The Hydrologic

John Suchocki, and Leslie A. Hewitt

mentum and Energy

Collisions

nentum
ntum
nentum Over a Short Time
Conservation of Momentum
Momentum: Numerical

vs

4 Gravity, Projectiles, and Satellites

INTERACTIVE FIGURES
• 4.6
• 4.14
• 4.15
• 4.19
• 4.21
• 4.28
• 4.32
• 4.35
TUTORIALS
• Motion and Gravity
• Orbits and Kepler's Laws
• Projectile Motion
VIDEOS
• von Jolly's Method of Measuring the Attraction Between Two Masses
• Inverse-Square Law
• Weight and Weightlessness
• Apparent Weightlessness
• Discovery of Neptune
• Projectile Motion Demo
• More Projectile Motion
• Circular Orbits

5 Fluid Mechanics

VIDEOS
• Dam and Water
• Buoyancy
• Archimedes' Principle
• Flotation
• Air Has Weight
• Air Is Matter
• Air Has Pressure
• Buoyancy of Air

6 Thermal Energy Thermodynam

VIDEOS
• Low Temperatures with Liquid Nitro
• How a Thermostat Works

Atomic Nucleus and Radioactivity

ERACTIVE FIGURES

ay

14 Elements of Chemistry

INTERACTIVE FIGURE
• 14.9
TUTORIAL
• What Is Chemistry?
VIDEOS
• Oxygen Bubble Burst
• Fire Water

15 How Atoms Bond and Molecules Attract

INTERACTIVE FIGURES
• 15.18
• 15.35
TUTORIALS
• Intermolecular Forces
• Covalent Bonds
• Bonds and Bond Polarity
• Polar Attraction
VIDEOS
• Ionic Bonds
• Covalent Bonds

16 Mixtures

INTERACTIVE FIGURE
• 16.32
TUTORIAL
• Solubility
VIDEOS
• Mixtures Can Be Separated by Phys Means
• Solutions
• Solubility Changes with Temperature Pressure
• Soap Works by Being Both Polar and Nonpolar

ping Earth's Surface

ycle Activity
y Surfaces

ycle

24 The Oceans, Atmosphere, and Climatic Effects

INTERACTIVE FIGURES
• 24.8
• 24.9
TUTORIALS
• Vertical Structure of the Atmosphere
• Seasons
• The Coriolis Effect

25 Driving Forces of Weather

INTERACTIVE FIGURE
25.12
TUTORIALS
• Surface Temperature of Planets
• Rain Shadow Activity
• Cold Front Activity
• Warm Front Activity
VIDEO
• Adiabatic Process

26 The Solar Sys

INTERACTIVE FIGURE
• 26.19
TUTORIALS
• Formation of the Solar System
• The Sun
VIDEOS
• History of the Solar System
• Orbits in the Solar System

Unfold to explore how The Physical Science Place Web site makes learning easy and fun!

The observer sees the colored components of white light as a smooth, continuous spectrum.

Reset

Main

This media grid provides you with an at-a-glance, chapter-by-chapter overview of how the text and media work together to illustrate key concepts in Physics, Chemistry, Earth Science, and Astronomy. The grid outlines the extensive library of Interactive Figures, Tutorials, and Videos available for use with *Conceptual Physical Science,* Fourth Edition. In addition to these resources, The Physical Science Place provides self-assessment quizzes and other self-study tools for each chapter.

and
cs

7 Heat Transfer and Change of Phase

INTERACTIVE FIGURE
- 7.13

VIDEOS
- The Secret to Walking on Hot Coals
- Air Is a Poor Conductor
- Condensation Is a Warming Process
- Boiling Is a Cooling Process
- Pressure Cooker: Boiling and Freezing at the Same Time

8 Static and Current Electricity

INTERACTIVE FIGURES
- 8.2
- 8.3
- 8.9
- 8.10
- 8.29
- 8.30

TUTORIALS
- Electrostatics
- Electricity and Circuits

VIDEOS
- Electric Potential
- Van de Graaff Generator
- Caution on Handling Electric Wires
- Birds and High-Voltage Wires
- Alternating Current
- Ohm's Law
- Electric Circuits

9 Magnetism and Electromagnet Induction

INTERACTIVE FIGURES
- 9.3
- 9.8
- 9.17
- 9.21
- 9.24
- 9.28

TUTORIAL
- Magnetic Fields

VIDEOS
- Oersted's Discovery
- Magnetic Forces on Current-Carrying Wires
- Applications of Induction
- Faraday's Law

17 How Chemicals React

INTERACTIVE FIGURES
- 17.7
- 17.8
- 17.9
- 17.10

TUTORIALS
- Chemical Reactions and Equations
- Equilibrium

cal

and

18 Two Classes of Chemical Reactions

INTERACTIVE FIGURE
- 18.27

TUTORIALS
- Nature of Acids and Bases
- Strong and Weak Acids and Bases
- The pH Scale

VIDEOS
- Some Acids and Bases Are Stronger Than Others
- Rainwater Is Acidic and Ocean Water Is Basic
- Beach Sand Composition

19 Organic Compounds

TUTORIALS
- Introduction to Organic Molecules
- Organic Molecules and Isomers
- Functional Groups
- Polymers
- Polymers from Monomers

em

27 Stars and Galaxies

TUTORIALS
- Stellar Evolution
- Black Holes

VIDEO
- Lives of Stars

28 The Structure of Space and Time

INTERACTIVE FIGURE
- 28.1

TUTORIALS
- Hubble's Law
- Fate of the Universe

VIDEO
- From the Big Bang to Galaxies

Appendices

TUTORIALS
- Significant Figures
- Vectors

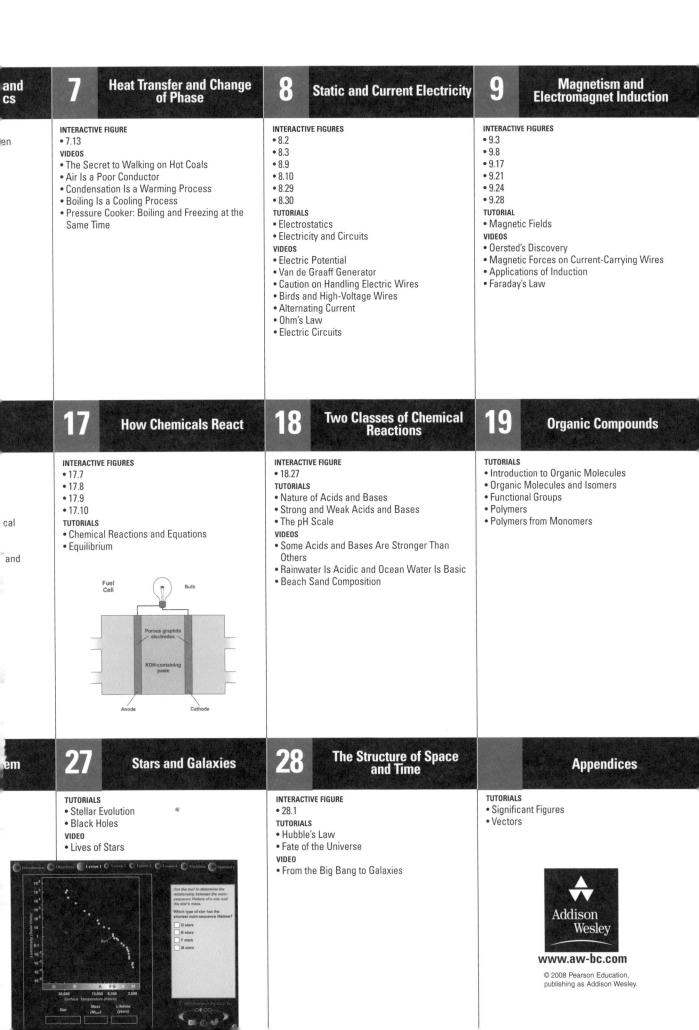

Addison Wesley

www.aw-bc.com

© 2008 Pearson Education,
publishing as Addison Wesley.

Log on.

Tune in.

Succeed.

Your steps to success.

STEP 1: Register

All you need to get started is a valid email address and the access code below. To register, simply:

1. Go to **www.physicsplace.com**
2. Click the appropriate book cover.
   Cover must match the textbook edition being used for your class.
3. Click "**Register**" under "**First-Time User?**"
4. Leave "**No, I Am a New User**" selected.
5. Using a coin, scratch off the silver coating below to reveal your access code.
   Do not use a knife or other sharp object, which can damage the code.
6. Enter your access code in lowercase or uppercase, without the dashes.
7. Follow the on-screen instructions to complete registration.
   During registration, you will establish a personal login name and password to use for logging into the website. You will also be sent a registration confirmation email that contains your login name and password.

Your Access Code is:

Note: If there is no silver foil covering the access code, it may already have been redeemed, and therefore may no longer be valid. In that case, you can purchase access online using a major credit card. To do so, go to **www.physicsplace.com**, click the cover of your textbook, click "**Buy Now**", and follow the on-screen instructions.

STEP 2: Log in

1. Go to **www.physicsplace.com** and click the appropriate book cover.
2. Under "**Established User?**" enter the login name and password that you created during registration. *If unsure of this information, refer to your registration confirmation email.*
3. Click "**Log In**".

STEP 3: (Optional) Join a class

Instructors have the option of creating an online class for you to use with this website. If your instructor decides to do this, you'll need to complete the following steps using the Class ID your instructor provides you. By "joining a class," you enable your instructor to view the scored results of your work on the website in his or her online gradebook.

To join a class:

1. Log into the website. For instructions, see "STEP 2: Log in."
2. Click "**Join a Class**" near the top right.
3. Enter your instructor's "**Class ID**" and then click "**Next**".
4. At the Confirm Class page you will see your instructor's name and class information. If this information is correct, click "**Next**".
5. Click "**Enter Class Now**" from the Class Confirmation page.
- *To confirm your enrollment in the class, check for your instructor and class name at the top right of the page. You will be sent a class enrollment confirmation email.*
- *As you complete activities on the website from now through the class end date, your results will post to your instructor's gradebook, in addition to appearing in your personal view of the Results Reporter.*

To log into the class later, follow the instructions under "STEP 2: Log in."

Got technical questions?

Customer Technical Support: To obtain support, please visit us online anytime at http://247.aw.com where you can search our knowledgebase for common solutions, view product alerts, and review all options for additional assistance.

SITE REQUIREMENTS

For the latest updates on Site Requirements, go to **www.physicsplace.com**, choose your text cover, and click Site Reqs.

WINDOWS
OS: Windows 2000, XP
Resolution: 1024 x 768
Plugins: Latest Version of Flash/QuickTime/Shockwave (as needed)
Browsers: Internet Explorer 6.0; Firefox 1.0

MACINTOSH
OS: 10.2.4, 10.3.2
Resolution: 1024 x 768
Plugins: Latest Version of Flash/QuickTime/Shockwave (as needed)
Browsers: Firefox 1.0; Safari 1.3

Internet Connection: 56k modem minimum

Register and log in

Join a class

Important: Please read the Subscription and End-User License agreement, accessible from the book website's login page, before using *The Physics Place* website. By using the website, you indicate that you have read, understood, and accepted the terms of this agreement.

Conceptual

Physical Science

Fourth Edition

Paul G. Hewitt

City College of San Francisco

John Suchocki

Saint Michael's College

Leslie A. Hewitt

PEARSON

Addison
Wesley

San Francisco Boston New York
Cape Town Hong Kong London Madrid Mexico City
Montreal Munich Paris Singapore Sydney Tokyo Toronto

Publisher: Jim Smith
Associate Project Editor: Chandrika Madhavan
Assistant Editor and Media Producer: Ashley Taylor Anderson
Editorial Manager: Laura Kenney
Development Manager: Michael Gillespie
Executive Marketing Manager: Scott Dustan
Managing Editor: Corinne Benson
Senior Production Supervisor: Nancy Tabor
Production Service: Progressive Publishing Alternatives
Illustrations: Dartmouth Publishing, Inc.
Text Design: Seventeenth Street Studios
Cover Design: Richard Leeds, BigWig Design
Manufacturing Buyer: Jeffrey Sargent
Photo Research: Truitt and Marshall
Director, Image Resource Center: Melinda Patelli
Manager, Rights and Permissions: Zina Arabia
Image Permission Coordinator: Elaine Soares
Cover Printer: Phoenix Color Corporation
Text Printer and Binder: Quebecor World, Dubuque

Cover Image: Photo of Comet McNaught (c/2006 P1) by Robert McNaught
Photo Credits: see page P-1

Library of Congress Cataloging-in-Publication Data

Hewitt, Paul G.
    Conceptual physical science/Paul G. Hewitt, John Suchocki, Leslie A. Hewitt.—4th ed.
        p. cm.
    Includes index.
    ISBN-13: 978-0-321-51695-4
    ISBN-10: 0-321-51695-8
1. Physical science. I. Suchocki, John. II. Hewitt, Leslie A. III. Title.
    Q158.5.H48 2007
    500.2—dc22

                                                                                2007036288

ISBN-13: 978-0-321-51695-4
ISBN-10: 0-321-51695-8

PEARSON

Addison
Wesley

www.aw-bc.com

3 4 5 6 7 8 9 10—QWD—10 09

■ **To the Memory of Carl Sagan**
   for inspiring us all to be candles in the dark

BRIEF TABLE OF CONTENTS

■ **Prologue: The Nature of Science**   1

PART ONE

physics   13

■ **1**   Patterns of Motion and Equilibrium   15

■ **2**   Newton's Laws of Motion   37

■ **3**   Momentum and Energy   59

■ **4**   Gravity, Projectiles, and Satellites   87

■ **5**   Fluid Mechanics   115

■ **6**   Thermal Energy and Thermodynamics   141

■ **7**   Heat Transfer and Change of Phase   159

■ **8**   Static and Current Electricity   181

■ **9**   Magnetism and Electromagnet Induction   209

■ **10**   Waves and Sound   231

■ **11**   Light   255

PART TWO

chemistry   283

■ **12**   Atoms and the Periodic Table   285

■ **13**   The Atomic Nucleus and Radioactivity   311

■ **14**   Elements of Chemistry   337

■ **15**   How Atoms Bond and Molecules Attract   353

■ **16**   Mixtures   381

■ **17**   How Chemicals React   411

■ **18**   Two Classes of Chemical Reactions   437

■ **19**   Organic Compounds   471

PART THREE

earth science   503

■ **20**   Rocks and Minerals   505

■ **21**   Geologic Time—Reading the Rock Record   539

■ **22**   Plate Tectonics and Earth's Interior   565

■ **23**   Shaping Earth's Surface   599

■ **24**   The Oceans, Atmosphere, and Climatic Effects   629

■ **25**   Driving Forces of Weather   663

PART FOUR

astronomy   687

■ **26**   The Solar System   689

■ **27**   Stars and Galaxies   719

■ **28**   The Structure of Space and Time   749

Appendices

Appendix A: On Measurement and Unit Conversion   A-1

Appendix B: Linear and Rotational Motion   B-1

Appendix C: Vectors   C-1

Appendix D: Exponential Growth and Doubling Time   D-1

Appendix E: Special Relativity   E-1

■ **Glossary**   G-1

■ **Photo Credits**   P-1

■ **Index**   I-1

TABLE OF CONTENTS

■ **Prologue: The Nature of Science** 1

   A Brief History of Advances in Science 2

   Mathematics and Conceptual
Physical Science 2

   Scientific Methods 3

   The Scientific Attitude 3

   Science Has Limitations 6

   Science, Art, and Religion 7

   Technology—The Practical Use
of Science 8

   The Physical Sciences: Physics, Chemistry,
Earth Science, and Astronomy 9

   In Perspective 10

PART ONE

physics 13

■1 Patterns of Motion and Equilibrium 15

1.1 Aristotle on Motion 16

1.2 Galileo's Concept of Inertia 17

1.3 Mass—A Measure of Inertia 18

   One Kilogram Weighs 9.8 N 20

1.4 Net Force 20

1.5 The Equilibrium Rule 22

1.6 Support Force 23

1.7 Dynamic Equilibrium 23

1.8 The Force of Friction 24

1.9 Speed and Velocity 25

   Speed 25

   Instantaneous Speed 25

   Average Speed 26

   Velocity 27

   Motion is Relative 27

1.10 Acceleration 27

■2 Newton's Laws of Motion 37

2.1 Newton's First Law of Motion 38

   The Moving Earth 39

2.2 Newton's Second Law of Motion 40

   When Acceleration is g—Free Fall 41

   *When Acceleration of Fall is Less Than g—
Non-Free Fall* 43

2.3 Forces and Interactions 45

2.4 Newton's Third Law of Motion 46

   Simple Rule to Identify Action and Reaction 46

   Action and Reaction on Different Masses 48

   Defining Your System 50

2.5 Summary of Newton's Three Laws 52

■3 Momentum and Energy 59

3.1 Momentum and Impulse 60

3.2 Impulse Changes Momentum 61

   Case 1: Increasing Momentum 61

   Case 2: Decreasing Momentum Over a Long Time 61

   Case 3: Decreasing Momentum Over a Short Time 62

   Bouncing 63

3.3 Conservation of Momentum 65

   Collisions 66

3.4 Energy and Work 68

   Potential Energy 69

   Kinetic Energy 70

3.5 Work–Energy Theorem 71

   Kinetic Energy and Momentum Compared 73

3.6 Conservation of Energy 74

3.7 Power 75

3.8 Machines 76

3.9 Efficiency 77

3.10 Sources of Energy 78

4 Gravity, Projectiles, and Satellites   87

4.1   The Universal Law of Gravity   88
       The Universal Gravitational Constant, G   *89*
4.2   Gravity and Distance: The Inverse-Square Law   90
4.3   Weight and Weightlessness   92
4.4   Universal Gravitation   93
4.5   Projectile Motion   94
       Projectiles Launched Horizontally   *95*
       Projectiles Launched at an Angle   *96*
4.6   Fast-Moving Projectiles—Satellites   101
4.7   Circular Satellite Orbits   103
4.8   Elliptical Orbits   104
4.9   Escape Speed   106

5 Fluid Mechanics   115

5.1   Density   116
5.2   Pressure   117
       Pressure in a Liquid   *118*
5.3   Buoyancy in a Liquid   119
5.4   Archimedes' Principle   120
       Flotation   *122*
5.5   Pressure in a Gas   124
       Boyle's Law   *124*
5.6   Atmospheric Pressure   125
       Barometers   *127*
5.7   Pascal's Principle   129
5.8   Buoyancy in a Gas   130
5.9   Bernoulli's Principle   131
       Application of Bernoulli's Principle   *132*

6 Thermal Energy and Thermodynamics   141

6.1   Temperature   142
6.2   Absolute Zero   143
6.3   Heat   144
6.4   Quantity of Heat   145
6.5   The Laws of Thermodynamics   145
6.6   Entropy   146
6.7   Specific Heat Capacity   147
       The High Specific Heat Capacity of Water   *148*
6.8   Thermal Expansion   150
6.9   Expansion of Water   151

7 Heat Transfer and Change of Phase   159

7.1   Conduction   160
7.2   Convection   161
7.3   Radiation   163
       Emission of Radiant Energy   *163*
       Absorption of Radiant Energy   *164*
       Reflection of Radiant Energy   *165*
7.4   Newton's Law of Cooling   166
7.5   Global Warming and the Greenhouse Effect   167
7.6   Heat Transfer and Change of Phase   168
       Evaporation   *169*
       Condensation   *170*
7.7   Boiling   171
7.8   Melting and Freezing   173
7.9   Energy and Change of Phase   174

8 Static and Current Electricity   181

8.1   Electric Charge   182
       Conservation of Charge   *183*
8.2   Coulomb's Law   184
       Charge Polarization   *186*
8.3   Electric Field   186
8.4   Electric Potential   189
8.5   Voltage Sources   190
8.6   Electric Current   191
       Direct Current and Alternating Current   *193*
8.7   Electric Resistance   193
       Superconductors   *194*

8.8 Ohm's Law 194
 Electric Shock 195
8.9 Electric Circuits 197
 Series Circuits 197
 Parallel Circuits 198
 Parallel Circuits and Overloading 199
 Safety Fuses 200
8.10 Electric Power 200

9 Magnetism and Electromagnet Induction 209

9.1 Magnetic Poles 210
9.2 Magnetic Fields 211
9.3 Magnetic Domains 212
9.4 Electric Currents and Magnetic Fields 213
 Electromagnets 214
 Superconducting Electromagnets 215
9.5 Magnetic Forces on Moving Charges 215
 Magnetic Force on Current-Carrying Wires 216
 Electric Meters 216
 Electric Motors 217
9.6 Electromagnetic Induction 218
 Faraday's Law 219
9.7 Generators and Alternating Current 220
9.8 Power Production 221
9.9 The Transformer—Boosting or Lowering Voltage 222
9.10 Field Induction 223

10 Waves and Sound 231

10.1 Vibrations and Waves 232
10.2 Wave Motion 233
 Wave Speed 233
10.3 Transverse and Longitudinal Waves 234
10.4 Sound Waves 235
 Speed of Sound 236

10.5 Reflection and Refraction of Sound 237
10.6 Forced Vibrations and Resonance 239
10.7 Interference 240
 Beats 242
 Standing Waves 243
10.8 Doppler Effect 244
10.9 Bow Waves and the Sonic Boom 245
10.10 Musical Sounds 247

11 Light 255

11.1 Electromagnetic Spectrum 256
11.2 Transparent and Opaque Materials 257
11.3 Reflection 260
 Law of Reflection 260
 Diffuse Reflection 262
11.4 Refraction 262
11.5 Color 265
 Selective Reflection 266
 Selective Transmission 266
 Mixed Colored Lights 267
 Complementary Colors 268
 Mixed Colored Pigments 269
 Why the Sky is Blue 269
 Why Sunsets are Red 270
 Why Clouds are White 271
11.6 Dispersion 272
 Rainbows 272
11.7 Polarization 274

PART TWO

chemistry 283

12 Atoms and the Periodic Table 285

12.1 Atoms are Ancient and Empty 286
12.2 The Elements 287
12.3 Protons and Neutrons 288
 Isotopes and Atomic Mass 290

12.4 The Periodic Table   292

   Periods and Groups   *293*

12.5 Physical and Conceptual Models   296

12.6 Identifying Atoms Using the
     Spectroscope   298

12.7 The Quantum Hypothesis   300

12.8 Electron Waves   302

12.9 The Shell Model   304

▪13 The Atomic Nucleus and Radioactivity   311

13.1 Radioactivity   312

   Alpha, Beta, and Gamma Rays   *312*

   Radiation Dosage   *314*

   Radioactive Tracers   *316*

13.2 The Atomic Nucleus and the
     Strong Nuclear Force   316

13.3 Half-Life and Transmutation   318

   Natural Transmutation   *319*

   Artificial Transmutation   *321*

13.4 Radiometric Dating   321

13.5 Nuclear Fission   323

   Nuclear Fission Reactors   *325*

   The Breeder Reactor   *326*

13.6 Mass–Energy Equivalence:
     $E = mc^2$ 327

13.7 Nuclear Fusion   330

   Controlling Fusion   *331*

▪14 Elements of Chemistry   337

14.1 Chemistry: The Central Science   338

14.2 The Submicroscopic World   340

14.3 Physical and Chemical Properties   341

14.4 Determining Physical and
     Chemical Changes   343

14.5 Elements to Compounds   345

14.6 Naming Compounds   347

▪15 How Atoms Bond and Molecules Attract   353

15.1 Electron-Dot Structures   354

15.2 The Formation of Ions   356

   Molecules Can Form Ions   *358*

15.3 Ionic Bonds   359

15.4 Metallic Bonds   361

15.5 Covalent Bonds   362

15.6 Polar Covalent Bonds   365

15.7 Molecular Polarity   368

15.8 Molecular Attractions   371

   Ions and Dipoles   *372*

   Induced Dipoles   *373*

▪16 Mixtures   381

16.1 Most Materials are Mixtures   382

   *Mixtures Can Be Separated by
   Physical Means*   *383*

16.2 The Chemist's Classification
     of Matter   384

16.3 Solutions   386

16.4 Solubility   391

   Solubility Changes with Temperature   *392*

   Solubility of Gases   *393*

16.5 Soaps and Detergents   394

16.6 Softening Hard Water   396

16.7 Purifying the Water We Drink   397

   Desalination   *399*

   Bottled Water   *402*

16.8 Wastewater Treatment   403

▪17 How Chemicals React   411

17.1 Chemical Equations   412

17.2 Counting Atoms and Molecules by Mass   414

   Converting Between Grams and Moles   *416*

17.3 Reaction Rates   418

17.4   Catalysts   423

17.5   Energy and Chemical Reactions   425

*An Exothermic Reaction Involves a Net
Release of Energy*   426

*An Endothermic Reaction Involves a Net
Absorption of Energy*   428

17.6   Chemical Reactions Are Driven
by Entropy   429

18 Two Classes of Chemical Reactions   437

18.1   Acids Donate Protons; Bases
Accept Them   438

*A Salt Is the Ionic Product of an Acid–Base
Reaction*   441

18.2   Relative Strengths of Acids
and Bases   442

18.3   Acidic, Basic, and Neutral
Solutions   445

The pH Scale Is Used to Describe Acidity   447

18.4   Acidic Rain and Basic Oceans   448

18.5   Losing and Gaining Electrons   452

18.6   Harnessing the Energy of Flowing
Electrons   454

Batteries   455

Fuel Cells   457

18.7   Electrolysis   459

18.8   Corrosion and Combustion   461

19 Organic Compounds   471

19.1   Hydrocarbons   472

19.2   Unsaturated Hydrocarbons   476

19.3   Functional Groups   478

19.4   Alcohols, Phenols, and Ethers   480

19.5   Amines and Alkaloids   484

19.6   Carbonyl Compounds   485

19.7   Polymers   489

Addition Polymers   492

Condensation Polymers   494

PART THREE

earth science   503

20 Rocks and Minerals   505

20.1   The Geosphere Is Made Up of
Rocks and Minerals   506

20.2   Minerals   508

20.3   Mineral Properties   509

Crystal Form   509

Hardness   510

Cleavage and Fracture   511

Color   512

Density   512

20.4   Classification of Rock-Forming Minerals   513

20.5   The Formation of Minerals and Rock   514

Crystallization in Magma   515

Crystallization in Water Solutions   516

20.6   Rock Types   517

20.7   Igneous Rocks   518

Generation of Magma   518

*Three Types of Magma, Three Major
Igneous Rocks*   520

Igneous Rocks at Earth's Surface   521

Igneous Rocks Beneath Earth's Surface   523

20.8   Sedimentary Rocks   523

The Formation of Sedimentary Rock   524

Classifying Sedimentary Rocks   526

Chemical Sedimentary Rocks   527

Fossils: Clues to Life in the Past   528

20.9   Metamorphic Rocks   528

Types of Metamorphism: Contact and Regional   530

Classifying Metamorphic Rocks   531

20.10 The Rock Cycle   532

21 Geologic Time—Reading the Rock Record   539

21.1   The Rock Record—Relative Dating   540

Gaps in the Rock Record   543

21.2   Radiometric Dating   544

21.3   Geologic Time   546

21.4   Precambrian Time (4500 to
       543 Million Years Ago)   547

21.5   The Paleozoic Era (543 to
       248 Million Years Ago)   549

       *The Cambrian Period (543 to 490 Million
       Years Ago)*   *549*

       *The Ordovician Period (490 to 443 Million
       Years Ago)*   *550*

       *The Silurian Period (443 to 417 Million
       Years Ago)*   *550*

       *The Denovian Period (417 to 354 Million
       Years Ago)*   *551*

       *The Carboniferous Period (354 to 290
       Million Years Ago)*   *551*

       *The Permian Period (290 to 248 Million
       Years Ago)*   *552*

21.6   The Mesozoic Era (248 to 65 Million
       Years Ago)   554

       The Cretaceous Extinction   *555*

21.7   The Cenozoic Era (65 Million Years
       Ago to the Present)   556

       Cenozoic Life   *558*

21.8   Earth History in a Capsule   559

22   Plate Tectonics and
     Earth's Interior   565

22.1   Seismic Waves   566

22.2   Earth's Internal Layers   567

       The Core   *568*

       The Mantle   *569*

       The Crustal Surface   *570*

22.3   Continental Drift—An Idea Before
       Its Time   571

22.4   Acceptance of Continental Drift   573

22.5   The Theory of Plate Tectonics   576

       Divergent Plate Boundaries   *577*

       Convergent Plate Boundaries   *579*

       Transform Plate Boundaries   *583*

22.6   Continental Evidence for Plate
       Tectonics   584

       Folds   *585*

       Faults   *585*

       Earthquakes   *587*

22.7   The Theory That Explains the
       Geosphere   592

23   Shaping Earth's
     Surface   599

23.1   The Hydrologic Cycle   600

23.2   Groundwater   601

       The Water Table   *603*

       Aquifers and Springs   *604*

       Groundwater Movement   *606*

23.3   The Work of Groundwater   607

       Land Subsidence   *607*

       Carbonate Dissolution   *608*

23.4   Surface Water and Drainage
       Systems   609

       Stream Flow Geometry   *610*

       Drainage Basins and Networks   *612*

23.5   The Work of Surface Water   613

       Erosion and Transport of Sediment   *613*

       Erosional and Depositional Environments   *614*

       Stream Valleys and Floodplains   *614*

       Deltas: The End of the Line for a River   *617*

23.6   Glaciers and Glaciation   618

       Glacier Formation and Movement   *618*

       Glacial Mass Balance   *619*

23.7   The Work of Glaciers   621

       Glacial Erosion and Erosional Landforms   *621*

       *Glacial Sedimentation and Depositional
       Landforms*   *622*

23.8   The Work of Air   624

24   The Oceans,
     Atmosphere, and
     Climatic Effects   629

24.1   Earth's Atmosphere and Oceans   630

       *Evolution of Earth's Atmosphere
       and Oceans*   *630*

24.2   Components of Earth's Oceans   632
   The Ocean Floor   632
   Seawater   634
24.3   Ocean Waves, Tides, and Shorelines   635
   Wave Refraction   636
   The Work of Ocean Waves   638
   Along the Coast   638
   Tides   640
24.4   Components of Earth's Atmosphere   642
   Vertical Structure of the Atmosphere   643
24.5   Solar Energy   645
   The Seasons   645
   Terrestrial Radiation   646
   The Greenhouse Effect and Global Warming   646
24.6   Driving Forces of Air Motion   648
   The Temperature-Pressure Relationship   648
   Large-Scale Air Movement   650
24.7   Global Circulation Patterns   651
   Upper Atmospheric Circulation   653
   Oceanic Circulation   653
   Surface Currents   654
   Deep-Water Currents   657

25 Driving Forces of Weather   663

25.1   Atmospheric Moisture   664
   Temperature Changes and Condensation   666
25.2   Weather Variables   667
   Adiabatic Processes in Air   667
   Atmospheric Stability   669
25.3   Cloud Development   670
   High Clouds   671
   Middle Clouds   672
   Low Clouds   672
   Clouds With Vertical Development   673
   Precipitation Formation   673
25.4   Air Masses, Fronts, and Storms   674
   Atmospheric Lifting   675
   Convectional Lifting   675
   Orographic Lifting   675
   Frontal Lifting   676
   Midlatitude Cyclones   678
25.5   Violent Weather   679
   Thunderstorms   680
   Tornados   681
   Hurricanes   681
25.6   The Weather—The Number One Topic of Conversation   682

PART FOUR
astronomy   687

26 The Solar System   689

26.1   The Solar System and Its Formation   690
   Nebular Theory   692
26.2   The Sun   693
26.3   The Inner Planets   695
   Mercury   695
   Venus   696
   Earth   697
   Mars   698
26.4   The Outer Planets   699
   Jupiter   699
   Saturn   700
   Uranus   702
   Neptune   702
26.5   Earth's Moon   703
   The Phases of the Moon   704
   Why One Side Always Faces Us   706
   Eclipses   707
26.6   Failed Planet Formation   709
   The Asteroid Belt and Meteors   709
   The Kuiper Belt and Dwarf Planets   710
   The Oort Cloud and Comets   711

27 Stars and Galaxies   719

27.1   Observing the Night Sky   720
27.2   The Brightness and Color of Stars   722

	Radiation Curves of Stars	723
27.3	The Hertzsprung-Russell Diagram	724
27.4	The Life Cycles of Stars	727
	Novae and Supernovae	729
27.5	Black Holes	732
	Black Hole Geometry	733
27.6	Galaxies	735
	Elliptical, Spiral, and Irregular Galaxies	737
	Active Galaxies	738
	Clusters and Superclusters	740

28 The Structure of Space and Time 749

28.1	The Big Bang	750
	Cosmic Background Radiation	753
	The Abundance of Hydrogen and Helium	754
28.2	Cosmic Inflation	755
28.3	General Relativity	757
	Tests of General Relativity	760
28.4	Dark Matter	761
	Galaxy Formation	763
28.5	Dark Energy	763
28.6	The Fate of the Universe	764

Appendix A: On Measurement and Unit Conversion	**A-1**
Appendix B: Linear and Rotational Motion	**B-1**
Appendix C: Vectors	**C-1**
Appendix D: Exponential Growth and Doubling Time	**D-1**
Appendix E: Special Relativity	**E-1**
Glossary	**G-1**
Photo Credits	**P-1**
Index	**I-1**

This is a very personal book, a family undertaking shown in the many photographs throughout. Author Paul is seen with his wife Lillian on page 51, and Lil appears again on pages 160 and 236, and with her pet conure, Sneezlee, on page 269. Lil's mom Siu Bik and dad Wai Tsan Lee are on pages 174 and 213, and Lil's niece Allison Lee Wong and nephew Erik Lee Wong are on page 171. Paul's grown children begin with author Leslie in her student days on page 307, and son Paul on pages 146 and 162. Son Paul's lovely wife Ludmila shows crossed Polaroids on page 277, and their children Alexander and Grace Hewitt open the physics chapters on page 13. Grace alone opens the astronomy chapters on page 687. Alexander and Grace team up with grandchildren Megan and Emily Abrams for the series of group photos on page 269. Author Paul's first grandchild, Manuel Hewitt, swings on page 251.

Paul's sister (and John's mom) Marjorie Hewitt Suchocki (pronounced Su-hock-ee, with a silent *c*), a prominent theologian, is shown reflectively on page 261. Paul's brother Steve shows Newton's third law with his daughter Gretchen on page 56. Paul's other brother Dave with his wife Barbara pump water on page 128. Their son Davey shows some simple electronics on page 200.

John, who in his "other life" is John Andrew, singer and songwriter, plays his guitar on page 220. He is shown again walking barefoot on red-hot coals on page 159. His wife Tracy is shown with son Ian on page 287, and with son Evan on page 344. Daughter Maitreya is seen on pages 283 and 471. John's nephew Graham Orr on page 382 is seen at ages 7 and 21, demonstrating how water is essential for growth. The Suchocki dog, Sam, pants on page 169. The "just-married" John and Tracy are flanked by John's sisters Cathy and Joan on page 248. (Tracy's wedding ring is figured prominently on page 337.) Sister Joan is riding her horse on page 25. Cousin George Webster is seen with his scanning electron microscope on page 302. Several dear friends from John's years of teaching in Hawaii include Rinchen Trashi on page 299 and Kai Dodge and Maile Ventura on page 467. The Suchocki's Vermont friend, Nikki Jiraff, is seen carbonating water on page 402.

Leslie is seen at age 16 illustrating the wonderful idea that we're all made of stardust on page 307. Leslie's husband Bob Abrams is shown on pages 618 and 622. The late Millie Hewitt, Leslie's mom, illustrates the cooling effect of rapid evaporation on page 162. Leslie's children, Megan and Emily, open the Earth science chapters on page 503. And dear to Paul, John, and Leslie, our late friend Charlie Spiegel is shown on page 259.

Contributions to the physics chapters include renown physicist Ken Ford, who shows his passion for flying on page 242. Marshall Ellenstein, a contributor, editor, and producer of Paul's DVDs on physics, walks barefoot on broken pieces of glass on page 136. Diane Reindeau, shown on page 233, is another physics contributor.

Physics professor friends include Tsing Bardin illustrating liquid pressure on page 118, Bob Greenler displaying a colorful giant bubble on page 255, Ron Hipschman freezing water on page 173, Peter Hopkinson with his zany mirror on page 278, David Housden with an impressive circuit display on page 199, John Hubisz and entropy on page 147, Chelcie Liu with his novel race tracks in Figure B.3 in Appendix B, Jennie McKelvie making waves on page 241, Fred Myers showing magnetic force on page 212, Sheron Snyder generating light on page 225, Jim Stith turning his impressive Wimhurst generator on page 190, Roy Unruh with an electric car on page 201, and Lynda Williams singing her heart out on page 247.

Paul's dear personal friends include Burl Grey on page 22, who stimulated Paul's love of physics a half century ago, and Will Maynez showing the airtrack he built for CCSF on page 67. Tim Gardner shows air pressure on page 132 and induction on page 227. Friend from teen years, Paul Ryan, sweeps his finger through molten lead on page 175. Friend from college days, Howie Brand, illustrates impulse and changes in momentum on page 64. Tenny Lim, former student and now a design engineer for Jet Propulsion Labs, puts energy into her bow on page 70. Another former physics student, Helen Yan, now an orbit analyst for Lockheed Martin Corporation, poses with a black and white box on page 166. Jean Curtis demonstrates magnetic levitation on page 220. Science author Suzanne Lyons with children Tristan and Simone illustrates complementary colors on page 279. Lori Patterson is electrified on page 188, while her son Ryan resonates on page 240. Tammy and Larry Tunison demonstrate radiation safety on page 315. Dave Vasquez with his family are barely seen in the solar-powered train on page 79. Chiu Man Wu is on page 169, with daughter Andrea on page 258. Little Miriam Dijamco and Michelle Anna Wong make touching music on page 231. Former student Cassy Cosme safely breaks bricks with her bare hand on page 63. On page 74 Paul also safely breaks a cement block above Pablo Robinson while he lies between beds of sharp nails. Pablo also shows color addition on page 268. His wife, bio-tech author Ellyn Daugherty, shows thermal expansion on page 156.

These photographs are of people very dear to the authors, which all the more makes *Conceptual Physical Science* our labor of love.

Physical Science is about the rules of the physical world—physics, chemistry, geology, and astronomy. Just as you can't enjoy a ball game, computer game, or party game until you know its rules, so it is with nature. Nature's rules are beautifully elegant and can be neatly described mathematically. That's why many physical science texts are treated as applied mathematics. But too much emphasis on computation misses something essential—*comprehension*—a gut feeling for the concepts. This book is *conceptual*, focusing on concepts in down-to-earth English rather than in mathematical language. You'll see the mathematical structure in frequent equations, but you'll find them *guides to thinking* rather than recipes for computation.

We enjoy physical science, and you will too—because you'll understand it. Just as a person who knows the rules of botany best appreciates plants, and a person who knows the intricacies of music best appreciates music, you'll better appreciate the physical world about you when you learn its rules.

Enjoy your physical science!

Paul G. Hewitt

John Suchocki

Leslie A. Hewitt

TO THE INSTRUCTOR

This fourth edition of Conceptual Physical Science with its important ancillaries provides your students an enjoyable and readable introduction to physics, chemistry, Earth science, and astronomy—all melded in a captivating manner. The 28 chapters are divided into four main parts—Physics, Chemistry, Earth Science, and Astronomy. We begin with physics because we see it as the most basic of the sciences. It reaches up to chemistry, which in turn reaches to Earth science, astronomy, and ultimately, beyond this book, serves as a foundation for the life sciences.

For the nonscience student, this book provides a base from which to view nature more perceptively—to see that a surprisingly few relationships make up its rules. For the science student, it is this and a springboard to involvement in other sciences such as biology and health-related fields.

Physics begins with static equilibrium so that your students start with forces and vectors rather than velocity and acceleration. After success with simple forces, the chapter touches on kinematics and proceeds to Newton's laws of motion. The pace picks up with the conventional order of mechanics topics followed by heat, thermodynamics, electricity and magnetism, sound, and light.

As with previous editions, the focus of physics is on qualitative comprehension. Mathematical expression follows in the problem sets. We minimize mathematical language and mathematical problems that are roadblocks to many students. Although a flip though the pages will show that the equations are there, they are presented as guides to thinking rather than as recipes for algebraic manipulation. Most of their derivations are addressed in the footnotes. The treatment of physics is followed with the realm of the atom—a bridge to chemistry.

The chemistry chapters in this edition have been reworked to streamline the flow of concepts. For example, following the rule of "keep it simple," details of the quantum hypothesis and atomic orbitals are omitted. This allows condensation of the two previous third-edition chapters on atoms into a single chapter. Further consolidation is achieved by combination of chemical bonding and molecular attractions into a single chapter entitled "How Atoms Bond and Molecules Attract," the theme being the macroscopic consequences of atomic and molecular "stickiness." Thus, the stage is set for the following chapter on mixtures, which is a consolidation of two chapters in the third edition. Furthermore, the section on balancing chemical equations is reunited with the reactions chapter, which, as in the third edition, focuses on the mechanics of chemical reactions. A new section on the role of entropy now helps emphasize the driving forces behind chemical reactions. Also, for this fourth edition, the discussions of acid-base and oxidation-reductions reactions now comprise a single chapter. As with previous editions, chemistry is related to the students' familiar world—the fluorine in their toothpaste, the Teflon on their frying pans, and the flavors produced by various organic molecules. The environmental aspects of chemistry are also highlighted.

Part three, Earth science, has been reorganized and refocused in many ways, giving attention to all areas of Earth science. The first three chapters are focused on the geosphere. The two chapters on minerals and rocks from the third edition have been combined into a single chapter, which provides an overview of Earth's composition and structure and acts as an introduction to plate tectonics. Foundation material from "Rocks and Minerals" is then used to examine Earth over geologic time. The chapter on Earth's interior and plate tectonics has been supplemented, refocused, and edited for clarity. The interrelated topics on surface water and surface processes from the third edition have been combined into one chapter entitled "Shaping Earth's Surface." The last two Earth science chapters, which focus on the hydrosphere and atmosphere, have been substantially supplemented with new material added to most sections. There is now a stronger emphasis on the role of density in atmospheric and oceanic circulation and more material on tides, with a more rigorous discussion of surface currents, gyres, and upwelling and downwelling. New material on atmospheric moisture, saturation vapor pressure, evaporation and condensation, adiabatic processes and the ideal gas law, atmospheric stability, cloud and precipitation formation, midlatitude cyclones, and latent heat and hurricanes add depth to our study of the atmosphere.

The applications of physics, chemistry, and the Earth sciences applied to other massive bodies in the universe culminate in part four—Astronomy. Most revision from the third to fourth edition occurs in these chapters; hence the astronomy-oriented cover of this edition and dedication to one of our heroes—the astronomer and popularizer of science, Carl Sagan. Of all the physical sciences, astronomy and cosmology are arguably undergoing the most rapid growth. Many recent discoveries are featured in this edition, illustrating how science is more than a growing body of knowledge and how it is an arena in which humans actively and systematically reach out to learn more about our place in the universe. The first astronomy chapter delves into our current understandings of our solar system, and it is followed by a chapter on stars, galaxies, and the universe as a whole. In the third chapter of part four, we explore the frontiers of cosmology, covering topics such as the Big Bang, cosmic inflation, dark matter, dark energy, and potential fates of the universe. Also included are discussions on general relativity from the third edition. For brevity, discussions on special relativity are condensed and occur in Appendix E.

■ Pedagogy

At the end of each chapter are Summary of Terms, Review Questions, Activities, Exercises, and in many chapters, Problems. All of the important ideas from each chapter are framed in the relatively easy-to-answer Review Questions, grouped by chapter sections. They are, as the name implies, a review of chapter material. Their purpose is simply to provide a structured way to review the chapter. They are not meant to challenge the student's intellect, for in the vast majority of cases, the answers can simply be looked up. The Exercises, on the other hand, play a different role. Some of these are designed to prompt the application of physical science into everyday situations, while others are more sophisticated and call for considerable critical thinking. A new feature of this edition is the symbols that indicate the level of difficulty, which assists you in creating student assignments. Also new to this edition is a Readiness Assurance Test (RAT) at the end of every chapter. Students take this for self assessment. We recommend that they not move on to new material until they can answer at least 7 of the 10 RAT questions correctly in a single sitting.

The Problems are mainly simple computations that aid in learning concepts. Note the inclusion of "conceptual" problems, where symbols for concepts rather than numerical values are given (identified by the black diamond symbol). There are fewer Problems than Exercises to decrease the likelihood of students focusing on problems rather than conceptual reasoning. Exercises call for critical thinking. Although building confidence in math is a worthy goal, it is not the focus of this book.

Students can find the answers to the odd-numbered Exercises and Problems in the back part of the *Practice Book for Conceptual Physical Science.* Complete answers to all Exercises and Problems are in the *Instructor Manual for Conceptual Physical Science.*

Units of measurement are not emphasized in this text. When used, they are almost exclusively expressed in SI (exceptions include such units as calories, grams per centimeter cubed, and light years). Mathematical derivations are avoided in the main body of the text and appear in footnotes or in the appendixes.

◾ Ancillary Materials

More than enough material is included for a one-year course, which allows for a variety of course designs to fit your taste. These are suggested in the *Instructor Manual,* which you'll find to be different from most instructors' manuals. It contains many lecture ideas and topics not treated in the textbook as well as teaching tips and suggested step-by-step lectures and demonstrations. It has full-page answers to all Review Questions, Exercises, and Problems in the text.

Answers to the odd-numbered Exercises and Problems are available to students in the student supplement, *Conceptual Physical Science Practice Book.* This very important book, our most creative work, guides your students to a sometimes computational way of developing concepts. It spans a wide use of analogies and intriguing situations, all with a user-friendly tone.

The *Computerized Test Bank for Conceptual Physical Science* has more than 2400 multiple choice questions as well as short answer and essay questions. The questions are categorized according to level of difficulty. The *Computerized Test Bank* allows you to edit questions, add questions, and create multiple test versions and is included as part of the Media Manager CD-ROM package.

The *Laboratory Manual for Conceptual Physical Science* is written by the authors and Dean Baird. In addition to interesting laboratory experiments, it includes a range of activities similar to the activities in the textbook. These guide students to experience phenomena before they quantify the same phenomena in a follow-up laboratory experiment. Answers to questions in the lab manual are in the *Instructor Manual.*

We encourage you to explore the rich media available to you and your students through *The Physical Science Place* (**www.physicsplace.com**), the popular Web site that features award-winning, self-paced interactive figures™ and tutorials that students love. This site also provides a library of videos created by the authors and other well-known sources, chapter-specific self-study quizzes, an interactive glossary and flashcard deck, relevant web links, an interactive periodic table, and lots more! Instructors can now also track students' completion of select tutorials and all quizzes using the gradebook feature. Access to the Web site is provided with every new book.

Another valuable media resource available to you is the *Media Manager for Conceptual Physical Science.* This cross-platform CD-ROM set provides instructors with the largest library available of purpose-built, in-class presentation materials, including all the images from the book in high-resolution JPEG format; interactive figures™ and videos from the Web site; PowerPoint® lecture outlines and clicker questions in PRS-enabled format for each chapter, all of which are written by the authors; Hewitt's acclaimed next-time questions in PDF format; and electronic, editable versions of the *Computerized Test Bank* and *Instructor Manual* in Word format. The Media Manager provides you with everything you need to prepare for dynamic, engaging lectures in no time.

Lastly, as a supplement for more on algebraic problem solving in physics, consider *Problem Solving in Conceptual Physics*, by Hewitt and Wolf, Addison-Wesley ISBN 0-8053-9377-3.

Go to it! Your conceptual physical science course really can be the most interesting, informative, and worthwhile science course available to your students.

■ Acknowledgments

Great thanks go to Phil Wolf for authoring many of the new physics problems. For contributions to problem solutions, thanks go to David Housden and Evan Jones. For general physics input to this and previous editions, we remain grateful to Dean Baird, Tsing Bardin, Howie Brand, George Curtis, Paul Doherty, Marshall Ellenstein, Ken Ford, John Hubisz, Dan Johnson, Tenny Lim, Iain McInnes, Fred Myers, Diane Reindeau, Kenn Sherey, Chuck Stone, Larry Weinstein, David Williamson, and Dean Zollman.

For development of chemistry chapters, thanks go to Bill Baker, Martin Brook, Pamela Burnley, Steven Burns, Thomas Chasteen, Randy Criss, Keith Dunn, Michael Epstein, Bambi Failey, Brian Goodman, Jeremiah Jerrett, Frank Lambert, Holly Lawson, Jim McClintock, Christos Valiotos, and Will Wickun.

For assistance with the writing and presentation of the Earth science material, we give heartfelt thanks to Bob Abrams. We are also thankful for contributions from Mary Brown, Ann Bykerk-Kauffman, Oswaldo Garcia, Newell Garfield, Karen Grove, Trayle Kulshan, Jan Null, Katryn Weiss, Lisa White, and Mike Young.

For astronomy, we're grateful for permission to use many of the graphics that appear in the textbook *The Cosmic Perspective*, 4th edition. Much appreciation to Jeffrey Bennett, Megan Donahue, Nicholas Schneider, and Mark Volt. Also, for previous reviews of the astronomy chapters, we remain grateful to Richard Crowe, Bjorn Davidson, Stacy McGaugh, Michelle Mizuno-Wiedner, Neil de Grass Tyson, Joe Wesney, Lynda Williams, and Erick Zackrisson.

For editorial help in all stages of production, we are especially grateful to spouses Lillian Lee Hewitt, Tracy Suchocki, and Bob Abrams.

For their dedication to this edition, we praise the staff at Addison-Wesley in San Francisco. We are especially thankful to Ashley Taylor Anderson, Adam Black, Chandrika Madhavan, and Jim Smith. We're grateful to Crystal Clifton, Sylvia Rebert, and the production team at Progressive Publishing Alternatives for their patience with our last-minute changes. Thanks to you all!

prologue

the nature of science

A BRIEF HISTORY OF ADVANCES IN SCIENCE

MATHEMATICS AND CONCEPTUAL PHYSICAL SCIENCE

SCIENTIFIC METHODS

THE SCIENTIFIC ATTITUDE

SCIENCE HAS LIMITATIONS

SCIENCE, ART, AND RELIGION

TECHNOLOGY—THE PRACTICAL USE OF SCIENCE

THE PHYSICAL SCIENCES: PHYSICS, CHEMISTRY, EARTH SCIENCE, AND ASTRONOMY

IN PERSPECTIVE

■ *Science* is the product of human curiosity about how the world works—an organized body of knowledge that describes the order within nature and the causes of that order. **Science** is an ongoing human activity that represents the collective efforts, findings, and wisdom of the human race, an activity that is dedicated to gathering knowledge about the world and organizing and condensing it into testable laws and theories. In our study of science, we are learning about the rules of nature—how one thing is connected to another and how patterns underlie all we see in our surroundings. Any activity, whether a sports game, computer game, or the game of life, is meaningful only if we understand its rules. Learning about nature's rules is relevant with a capital R!

The beginnings of science go back before recorded history, when people first discovered repeating patterns in nature. They noted star patterns in the night sky, patterns in the weather, and patterns in animal migration. From these patterns, people learned to make predictions that gave them some control over their surroundings. Science is based on rational thinking about the physical world.

A Brief History of Advances in Science

Science made great headway in Greece in the 4th and 3rd centuries BC and spread throughout the Mediterranean world. Scientific advance came to a near halt in Europe when the Roman Empire fell in the 5th century AD. Barbarian hordes destroyed almost everything in their paths as they overran Europe. Reason gave way to religion, which ushered in what came to be known as the Dark Ages. During this time, the Chinese and Polynesians were charting the stars and the planets. Before the advent of Islam, Arab nations developed mathematics and learned about the production of glass, paper, metals, and various chemicals. Greek science was reintroduced to Europe by Islamic influences that penetrated into Spain during the 10th, 11th, and 12th centuries. Universities emerged in Europe in the 13th century, and the introduction of gunpowder changed the social and political structure of Europe in the 14th century. The 15th century saw art and science beautifully blended by Leonardo da Vinci. Scientific thought was furthered in the 16th century with the advent of the printing press.

The 16th-century Polish astronomer Nicolaus Copernicus caused great controversy when he published a book proposing that the Sun is stationary and that Earth revolves around the Sun. These ideas conflicted with the popular view that Earth was the center of the universe. They also conflicted with Church teachings and were banned for 200 years. The Italian physicist Galileo Galilei was arrested for popularizing the Copernican theory and for his other contributions to scientific thought. Yet a century later, those who advocated Copernican ideas were accepted.

These cycles occur age after age. In the early 1800s, geologists met with violent condemnation because they differed with the account of creation in the book of Genesis. Later in the same century, geology was accepted, but theories of evolution were condemned and the teaching of them was forbidden. Every age has its groups of intellectual rebels who are scoffed at, condemned, and sometimes even persecuted at the time but who later seem beneficial and often essential to the elevation of human conditions. "At every crossway on the road that leads to the future, each progressive spirit is opposed by a thousand men appointed to guard the past."*

- In pre-Copernican times the Sun and Moon were viewed as planets. Their planetary status was removed when Copernicus substituted the Sun for Earth's central position. Only then was Earth regarded as a planet among others. More than 200 years later, in 1781, telescope observers added Uranus to the list of planets. Neptune was added in 1846. Pluto was added in 1930—and removed in 2006.

Mathematics and Conceptual Physical Science

Science and human conditions advanced dramatically after science and mathematics became integrated some four centuries ago. When the ideas of science are expressed in mathematical terms, they are unambiguous. The equations of science provide compact expressions of relationships between concepts. They don't have the multiple meanings that so often confuse the discussion of ideas expressed in common language. When findings in nature are expressed mathematically, they are easier to verify or to disprove by experiment.

* From Count Maurice Maeterlinck's "Our Social Duty."

The mathematical structure of physics is evident in the many equations you will encounter throughout this book. The equations are guides to thinking that show the connections between concepts in nature. The methods of mathematics and experimentation led to enormous success in science.*

> Science is a way to teach how something gets to be known, what is not known, to what extent things are known (for nothing is known absolutely), how to handle doubt and uncertainty, what the rules of evidence are, how to think about things so that judgments can be made, and how to distinguish truth from fraud and from show.
> —Richard Feynman

■ Scientific Methods

There is no *one* scientific method. But there are common features in the way scientists do their work. Although no cookbook description of the **scientific method** is really adequate, some or all of the following steps are likely to be found in the way most scientists carry out their work.

1. *Observe.* Closely observe the physical world around you. Recognize a question or a puzzle—such as an unexplained observation.

2. *Question.* Make an educated guess—a **hypothesis**—to answer the question.

3. *Predict.* Predict consequences that can be observed if the hypothesis is correct. The consequences should be *absent* if the hypothesis is not correct.

4. *Test predictions.* Do experiments to see if the consequences you predicted are present.

5. *Draw a conclusion.* Formulate the simplest general rule that organizes the hypothesis, predicted effects, and experimental findings.

Although these steps are appealing, much progress in science has come from trial and error, experimentation without hypotheses, or just plain accidental discovery by a well-prepared mind. The success of science rests more on an attitude common to scientists than on a particular method. This attitude is one of inquiry, experimentation, and humility—that is, a willingness to admit error.

■ The Scientific Attitude

It is common to think of a fact as something that is unchanging and absolute. But in science, a **fact** is generally a close agreement by competent observers who make a series of observations about the same phenomenon. For example, although it was once a fact that the universe is unchanging and permanent, today it is a fact that the universe is expanding and evolving. A scientific hypothesis, on the other hand, is an educated guess that is only presumed to be factual until supported by experiment. When a hypothesis has been tested over and over again and has not been contradicted, it may become known as a **law** or *principle*.

If a scientist finds evidence that contradicts a hypothesis, law, or principle, the scientific spirit requires that the hypothesis be changed or abandoned (unless the contradicting evidence, upon testing, turns out to be wrong—which sometimes happens). For example, the greatly respected Greek philosopher Aristotle (384–322 BC) claimed that an object falls at a speed proportional to its weight. This idea was held to be true for nearly 2000 years because of Aristotle's compelling authority. Galileo allegedly showed the falseness of Aristotle's claim with

---

* We distinguish between the mathematical structure of science and the practice of mathematical problem solving—the focus of most nonconceptual courses. Note that there are fewer mathematical problems than exercises at the ends of the chapters in this book. The focus is on comprehension before computation.

Experiment, not philosophical discussion, decides what is correct in science.

Facts are revisable data about the world. Theories interpret facts.

Before a theory is accepted, it must be tested by experiment and make one or more new predictions— different from those made by previous theories.

one experiment—demonstrating that heavy and light objects dropped from the Leaning Tower of Pisa fell at nearly equal speeds. In the scientific spirit, a single verifiable experiment to the contrary outweighs any authority, regardless of reputation or the number of followers or advocates. In modern science, argument by appeal to authority has little value.*

Scientists must accept their experimental findings even when they would like them to be different. They must strive to distinguish between what they see and what they wish to see, for scientists, like most people, have a vast capacity for fooling themselves.** People have always tended to adopt general rules, beliefs, creeds, ideas, and hypotheses without thoroughly questioning their validity and to retain them long after they have been shown to be meaningless, false, or at least questionable. The most widespread assumptions are often the least questioned. Most often, when an idea is adopted, particular attention is given to cases that seem to support it, while cases that seem to refute it are distorted, belittled, or ignored.

Scientists use the word *theory* in a way that differs from its usage in everyday speech. In everyday speech, a theory is no different from a hypothesis—a supposition that has not been verified. A scientific **theory**, on the other hand, is a synthesis of a large body of information that encompasses well-tested and verified hypotheses about certain aspects of the natural world. Physicists, for example, speak of the quark theory of the atomic nucleus, chemists speak of the theory of metallic bonding in metals, and biologists speak of the cell theory.

The theories of science are not fixed; rather, they undergo change. Scientific theories evolve as they go through stages of redefinition and refinement. During the past hundred years, for example, the theory of the atom has been repeatedly refined as new evidence on atomic behavior has been gathered. Similarly, chemists have refined their view of the way molecules bond together, and biologists have refined the cell theory. The refinement of theories is a strength of science, not a weakness. Many people feel that it is a sign of weakness to change their minds. Competent scientists must be experts at changing their minds. They change their minds, however, only when confronted with solid experimental evidence or when a conceptually simpler hypothesis forces them to a new point of view. More important than defending beliefs is improving them. Better hypotheses are made by those who are honest in the face of experimental evidence.

Away from their profession, scientists are inherently no more honest or ethical than most other people. But in their profession, they work in an arena that places a high premium on honesty. The cardinal rule in science is that all hypotheses must be testable—they must be susceptible, at least in principle, to being shown to be *wrong*. Speculations that cannot be tested are regarded as "unscientific." This has the long-run effect of compelling honesty—findings widely publicized among fellow scientists are generally subjected to further testing. Sooner or later, mistakes (and deception) are found out; wishful thinking is exposed. A discredited scientist does not get a second chance in the community of scientists. The penalty for fraud is professional excommunication. Honesty, so important to the progress of science, thus becomes a matter of self-interest to scientists. There is relatively little bluffing in a game in which all bets are called. In fields of study where right and wrong are not so easily established, the pressure to be honest is considerably less.

---

* But appeal to *beauty* has value in science. More than one experimental result in modern times has contradicted a lovely theory that, upon further investigation, proved to be wrong. This has bolstered scientists' faith that the ultimately correct description of nature involves conciseness of expression and economy of concepts—a combination that deserves to be called beautiful.

** In your education it is not enough to be aware that other people may try to fool you; it is more important to be aware of your own tendency to fool yourself.

In science, it is more important to have a means of proving an idea wrong than to have a means of proving it right. This is a major factor that distinguishes science from nonscience. At first this may seem strange, for when we wonder about most things, we concern ourselves with ways of finding out whether they are true. Scientific hypotheses are different. In fact, if you want to distinguish whether a hypothesis is scientific, look to see if there is a test for proving it wrong. If there is no test for its possible wrongness, then the hypothesis is not scientific. Albert Einstein put it well when he stated, "No number of experiments can prove me right; a single experiment can prove me wrong."

Consider the biologist Charles Darwin's hypothesis that life forms evolve from simpler to more complex forms. This could be proven wrong if paleontologists were to find that more complex forms of life appeared before their simpler counterparts. Einstein hypothesized that light is bent by gravity. This might be proven wrong if starlight that grazed the Sun and could be seen during a solar eclipse were undeflected from its normal path. As it turns out, less complex life forms are found to precede their more complex counterparts and starlight is found to bend as it passes close to the Sun, which support the claims. If and when a hypothesis or scientific claim is confirmed, it is regarded as useful and as a stepping-stone to additional knowledge.

Consider the hypothesis "The alignment of planets in the sky determines the best time for making decisions." Many people believe it, but this hypothesis is not scientific. It cannot be proven wrong, nor can it be proven right. It is *speculation*. Likewise, the hypothesis "Intelligent life exists on other planets somewhere in the universe" is not scientific. Although it can be proven correct by the verification of a single instance of intelligent life existing elsewhere in the universe, there is no way to prove it wrong if no intelligent life is ever found. If we searched the far reaches of the universe for eons and found no life, then that would not prove that it doesn't exist "around the next corner." A hypothesis that is capable of being proven right but not capable of being proven wrong is not a scientific hypothesis. Many such statements are quite reasonable and useful, but they lie outside the domain of science.

The essence of science is expressed in two questions: How would we know? What evidence would prove this idea wrong? Assertions without evidence are unscientific and can be dismissed without evidence.

CHECK POINT

Which of these statements is a scientific hypothesis?
(a) Atoms are the smallest particles of matter that exist.
(b) Space is permeated with an essence that is undetectable.
(c) Albert Einstein was the greatest physicist of the 20th century.

Was this your answer?

Only statement (a) is scientific, because there is a test for falseness. The statement not only is *capable* of being proven wrong, but *has* been proven wrong. Statement (b) has no test for possible wrongness and is therefore unscientific. Likewise for any principle or concept for which there is no means, procedure, or test whereby it can be shown to be wrong (if it is wrong). Some pseudoscientists and other pretenders of knowledge will not even consider a test for the possible wrongness of their statements. Statement (c) is an assertion that has no test for possible wrongness. If Einstein was not the greatest physicist, how could we know? Note that because the name Einstein is generally held in high esteem, it is a favorite of pseudoscientists. So we should not be surprised that the name of Einstein, like that of Jesus or of any other highly respected person, is cited often by charlatans who wish to bring respect to themselves and their points of view. In all fields, it is prudent to be skeptical of those who wish to credit themselves by calling upon the authority of others.

We each need a *knowledge filter* to tell the difference between what is true and what only pretends to be true. The best knowledge filter ever invented for explaining the physical world is science.

Science Has Limitations

Science deals only with hypotheses that are testable. Its domain is therefore restricted to the observable natural world. Although scientific methods can be used to debunk various paranormal claims, they have no way of accounting for testimonies involving the supernatural. The term *supernatural* literally means "above nature." Science works within nature, not above it. Likewise, science is unable to answer philosophical questions, such as "What is the purpose of life?" or religious questions, such as "What is the nature of the human spirit?" Though these questions are valid and may have great importance to us, they rely on subjective personal experience and do not lead to testable hypotheses. They lie outside the realm of science.

Science and Society

Pseudoscience

For a claim to qualify as scientific, it must meet certain standards. For example, the claim must be reproducible by others who have no stake in whether the claim is true or false. The data and subsequent interpretations are open to scrutiny in a social environment where it's okay to have made an honest mistake, but not okay to have been dishonest or deceiving. Claims that are presented as scientific but do not meet these standards are what we call **pseudoscience**, which literally means "fake science." In the realm of pseudoscience, skepticism and tests for possible wrongness are downplayed or flatly ignored.

Examples of pseudoscience abound. Astrology is an ancient belief system that supposes that a person's future is determined by the positions and movements of planets and other celestial bodies. Astrology mimics science in that astrological predictions are based on careful astronomical observations. Yet astrology is not a science because there is no validity to the claim that the positions of celestial objects influence the events of a person's life. After all, the gravitational force exerted by celestial bodies on a person is smaller than the gravitational force exerted by

objects making up the earthly environment: trees, chairs, other people, bars of soap, and so on. Further, the predictions of astrology are not borne out; there just is no evidence that astrology works.

For more examples of pseudoscience, turn on the television. You can find advertisements for a plethora of pseudoscientific products. Watch out for remedies to ailments such as baldness, obesity, and cancer; for air-purifying mechanisms; and for "germ-fighting" cleaning products in particular. Although many such products operate on solid science, others are pure pseudoscience. Buyer beware!

Humans are very good at denial, which may explain why pseudoscience is such a thriving enterprise. Many pseudoscientists do not recognize their efforts as pseudoscience. A practitioner of "absent healing," for example, may truly believe in her ability to cure people she will never meet except through e-mail and credit card exchanges.

She may even find anecdotal evidence to support her contentions. The *placebo effect*, discussed in Section 8.2, can mask the ineffectiveness of various healing modalities. In terms of the human body, what people believe *will* happen often *can* happen because of the physical connection between the mind and body.

That said, consider the enormous downside of pseudoscientific practices. Today more than 20,000 astrologers are practicing in the United States. Do people listen to these astrologers just for the fun of it? Or do they base important decisions on astrology? You might lose money by listening to pseudoscientific entrepreneurs; worse, you could become ill. Delusional thinking, in general, carries risk.

Meanwhile, the results of science literacy tests given to the general public show that most Americans lack a basic understanding of basic concepts of science. Some 63% of American adults are unaware that the mass extinction of the dinosaurs occurred long before the first human evolved; 75% do not know that antibiotics kill bacteria but not viruses; 57% do not know that electrons are smaller than atoms. What we find is a rift—a growing divide—between those who have a realistic sense of the capabilities of science and those who do not understand the nature of science, its core concepts, or, worse, feel that scientific knowledge is too complex for them to understand. Science is a powerful method for understanding the physical world, and a whole lot more reliable than pseudoscience as a means for bettering the human condition.

Science, Art, and Religion

The search for a deeper understanding of the world around us has taken different forms, including science, art, and religion. Science is a system by which we discover and record physical phenomena and think about possible explanations for such phenomena. The arts are concerned with personal interpretation and creative expression. Religion addresses the source, purpose, and meaning of it all. Simply put, science asks *how*, art asks *who*, and religion asks *why*.

Science and the arts have certain things in common. In the art of literature, we find out about what is possible in human experience. We can learn about emotions such as rage and love, even if we haven't yet experienced them. The arts describe these experiences and suggest what may be possible for us. Similarly, a knowledge of science tells us what is possible in nature. Scientific knowledge helps us predict possibilities in nature even before we experience them. It provides us with a way of connecting things, of seeing relationships between and among them, and of making sense of the great variety of natural events around us. While art broadens our understanding of ourselves, science broadens our understanding of our environment.

Science and religion have similarities also. For example, both are motivated by curiosity for the natural. Both have great impact on society. Science, for example, leads to useful technological innovations, while religion provides a foothold for many social services. Science and religion, however, are basically different. Science is concerned with understanding the physical universe, while religion is concerned with spiritual matters, such as belief and faith. While scientific truth is a matter of public scrutiny, religion is a deeply personal matter. In these respects, science and religion are as different as apples and oranges and do not contradict each other. Science, art, and religion can work very well together, which is why we should never feel forced into choosing one over the other.

That science and religion can work very well together deserves special emphasis. When we study the nature of light later in this book, we treat light first as a wave and then as a particle. At first, waves and particles may appear contradictory. You might believe that light can be only one or the other, and that you must choose between them. What scientists have discovered, however, is that light waves and light particles *complement* each other, and that when these two ideas are taken together, they provide a deeper understanding of light. In a similar way, it is mainly people who are either uninformed or misinformed about the deeper natures of both science and religion who feel that they must choose between believing in religion and believing in science. Unless one has a shallow understanding of either or both, there is no contradiction in being religious in one's belief system and being scientific in one's understanding of the natural world.*

Many people are troubled about not knowing the answers to religious and philosophical questions. Some avoid uncertainty by uncritically accepting almost any comforting answer. An important message in science, however, is that uncertainty is acceptable. For example, in Chapter 15 you'll learn that it is not possible to know with certainty both the momentum and position of an electron in an atom. The more you know about one, the less you can know

> Art is about cosmic beauty. Science is about cosmic order. Religion is about cosmic purpose.

---

* Of course, this does not apply to certain religious extremists who steadfastly assert that one cannot embrace both science and their brand of religion.

> The belief that there is only one truth and that oneself is in possession of it seems to me the deepest root of all the evil that is in the world.
> —Max Born

about the other. Uncertainty is a part of the scientific process. It's okay not to know the answers to fundamental questions. Why are apples gravitationally attracted to Earth? Why do electrons repel one another? Why do magnets interact with other magnets? Why does energy have mass? At the deepest level, scientists don't know the answers to these questions—at least not yet. We know a lot about where we are, but nothing really about *why* we are. It's okay not to know the answers to such religious questions. Given a choice between a closed mind with comforting answers and an open and exploring mind without answers, most scientists choose the latter. Scientists in general are comfortable with not knowing.

CHECK POINT

Which of the following activities involves the utmost human expression of passion, talent, and intelligence? (a) painting and sculpture (b) literature (c) music (d) religion (e) science

Was this your answer?
All of them. In this book, we focus on science, which is an enchanting human activity shared by a wide variety of people. With present-day tools and know-how, scientists are reaching further and finding out more about themselves and their environment than people in the past were ever able to do. The more you know about science, the more passionate you feel toward your surroundings. There is science in everything you see, hear, smell, taste, and touch!

Technology—The Practical Use of Science

Science and technology are also different from each other. Science is concerned with gathering knowledge and organizing it. Technology lets humans use that knowledge for practical purposes, and it provides the instruments scientists need to conduct their investigations.

Technology is a double-edged sword. It can be both helpful and harmful. We have the technology, for example, to extract fossil fuels from the ground and then burn the fossil fuels to produce energy. Energy production from fossil fuels has benefited society in countless ways. On the flip side, the burning of fossil fuels damages the environment. It is tempting to blame technology itself for such problems as pollution, resource depletion, and even overpopulation. These problems, however, are not the fault of technology any more than a stabbing is the fault of the knife. It is humans who use the technology, and humans who are responsible for how it is used.

Remarkably, we already possess the technology to solve many environmental problems. The 21st century will likely see a switch from fossil fuels to more sustainable energy sources. We recycle waste products in new and better ways. In some parts of the world, progress is being made toward limiting human population growth, a serious threat that worsens almost every problem faced by humans today. Difficulty in solving today's problems results more from social inertia than from failing technology. Technology is our tool. What we do with this tool is up to us. The promise of technology is a cleaner and healthier world. Wise applications of technology *can* improve conditions on planet Earth.

Risk Assessment

The numerous benefits of technology are paired with risks. X-rays, for example, continue to be used for medical diagnosis despite their potential for causing cancer. But when the risks of a technology are perceived to outweigh its benefits, it should be used very sparingly or not at all.

Risk can vary for different groups. Aspirin is useful for adults, but for young children it can cause a potentially lethal condition known as *Reye's syndrome*. Dumping raw sewage into the local river may pose little risk for a town located upstream, but for towns downstream the untreated sewage is a health hazard. Similarly, storing radioactive wastes underground may pose little risk for us today, but for future generations the risks of such storage are greater if there is leakage into groundwater. Technologies involving different risks for different people, as well as differing benefits, raise questions that are often hotly debated. Which medications should be sold to the general public over the counter and how should they be labeled? Should food be irradiated in order to put an end to food poisoning, which kills more than 5000 Americans each year? The risks to all members of society need consideration when public policies are decided.

The risks of technology are not always immediately apparent. No one fully realized the dangers of combustion products when petroleum was selected as the fuel of choice for automobiles early in the last century. From the hindsight of 20/20 vision, alcohols from biomass would have been a superior choice environmentally, but they were banned by the prohibition movements of the day.

Because we are now more aware of the environmental costs of fossil-fuel combustion, biomass fuels are making a slow comeback. An awareness of both the short-term risks and the long-term risks of a technology is crucial.

People seem to have a hard time accepting the impossibility of zero risk. Airplanes cannot be made perfectly safe. Processed foods cannot be rendered completely free of toxicity, for all foods are toxic to some degree. You cannot go to the beach without risking skin cancer, no matter how much sunscreen you apply. You cannot avoid radioactivity, for it's in the air you breathe and the foods you eat, and it has been that way since before humans first walked on Earth. Even the cleanest rain contains radioactive carbon-14, as do our bodies. Between each heartbeat in each human body, there have always been about 10,000 naturally occurring radioactive decays. You might hide yourself in the hills, eat the most natural foods, practice obsessive hygiene, and still die from cancer caused by the radioactivity emanating from your own body. The probability of eventual death is 100%. Nobody is exempt.

Science helps determine the most probable. As the tools of science improve, then assessment of the most probable gets closer to being on target. Acceptance of risk, on the other hand, is a societal issue. Placing zero risk as a societal goal is not only impractical but selfish. Any society striving toward a policy of zero risk would consume its present and future economic resources. Isn't it more noble to accept nonzero risk and to minimize risk as much as possible within the limits of practicality? A society that accepts no risks receives no benefits.

■ The Physical Sciences: Physics, Chemistry, Earth Science, and Astronomy

Science is the present-day equivalent of what used to be called *natural philosophy*. Natural philosophy was the study of unanswered questions about nature. As the answers were found, they became part of what is now called science. The study of science today branches into the study of living things and nonliving things: the life sciences and the physical sciences. The life sciences branch into such areas as molecular biology, microbiology, and ecology. The *physical sciences* branch into such areas as physics, chemistry, the Earth sciences, and astronomy.

A few words of explanation about each of the major divisions of science: Physics is the study of such concepts as motion, force, energy, matter, heat, sound, light, and the components of atoms. Chemistry builds on physics by telling us how matter is put together, how atoms combine to form molecules, and how the molecules combine to make the materials around us. Physics and chemistry, applied to Earth and its processes, make up Earth science—geology, meteorology, and oceanography. When we apply physics, chemistry, and geology to other planets and to the stars, we are speaking about astronomy. Biology is more complex than physical science, for it involves matter that is alive. Underlying biology is chemistry, and underlying chemistry is physics. So physics is basic to both physical science and life science. That is why we begin with physics, then follow with chemistry, then investigate Earth science and conclude with astronomy. All are treated conceptually, with the twin goals of enjoyment and understanding.

■ In Perspective

Just as you can't enjoy a ball game, computer game, or party game until you know its rules, so it is with nature. Because science helps us learn the rules of nature, it also helps us appreciate nature. You may see beauty in a structure such as the Golden Gate Bridge, but you'll see more beauty in that structure when you understand how all the forces that act on it balance. Similarly, when you look at the stars, your sense of their beauty is enhanced if you know how stars are born from mere clouds of gas and dust—with a little help from the laws of physics, of course. And how much richer it is, when you look at the myriad objects in your environment, to know that they are all composed of atoms—amazing, ancient, invisible systems of particles regulated by an eminently knowable set of laws.

If the complexity of science intimidates you, bear this in mind: All the branches of science rest upon a relatively small number of basic rules. Learn these underlying rules (physical laws), and you have a tool kit to bring to any phenomenon you wish to understand.

Go to it—we live in a time of rapid and fascinating scientific discovery!

SUMMARY OF TERMS

Science The collective findings of humans about nature, and a process of gathering and organizing knowledge about nature.

Scientific method Principles and procedures for the systematic pursuit of knowledge involving recognition and formulation of a problem, collection of data through observation and experiment, and formulation and testing of hypotheses.

Hypothesis An educated guess; a reasonable explanation of an observation or experimental result that is not fully accepted as factual until tested over and over again by experiment.

Fact A phenomenon about which competent observers who have made a series of observations agree.

Law A general hypothesis or statement about the relationship of natural quantities that has been tested over and over again and has not been contradicted. Also known as a *principle*.

Theory A synthesis of a large body of information that encompasses well-tested and verified hypotheses about certain aspects of the natural world.

Pseudoscience Fake science that pretends to be real science.

REVIEW QUESTIONS

1. Briefly, what is science?

A Brief History of Advances in Science

2. Throughout the ages, what has been the general reaction to new ideas about established "truths"?

Mathematics and Conceptual Physical Science

3. What is the role of equations in this course?

Scientific Methods

4. Outline the steps of the classic scientific method.

The Scientific Attitude

5. Distinguish between a scientific fact, a hypothesis, a law, and a theory.
6. In daily life, people are often praised for maintaining some particular point of view, for the "courage of their convictions." A change of mind is seen as a sign of weakness. How is this different in science?
7. What is the test for whether a hypothesis is scientific or not?
8. In daily life, we see many cases of people who are caught misrepresenting things and who soon thereafter are excused and accepted by their contemporaries. How is this different in science?

Science Has Limitations

9. What is meant by the term *supernatural*?

Science, Art, and Religion

10. Why are students of the arts encouraged to learn about science and science students encouraged to learn about the arts?

11. Why do many people believe they must choose between science and religion?
12. Psychological comfort is a benefit of having solid answers to religious questions. What benefit accompanies a position of not knowing the answers?

Technology—The Practical Use of Science

13. Clearly distinguish between science and technology.

The Physical Sciences: Physics, Chemistry, Earth Science, and Astronomy

14. Why is physics considered to be the basic science?

In Perspective

15. What is the importance to you in learning nature's rules?

EXERCISES

1. Which of the following are scientific hypotheses? (a) Chlorophyll makes grass green. (b) Earth rotates about its axis because living things need an alternation of light and darkness. (c) Tides are caused by the Moon.
2. In answer to the question, "When a plant grows, where does the material come from?", Aristotle hypothesized by logic that all material came from the soil. Do you consider his hypothesis correct, incorrect, or partially correct? What experiments do you propose to support your choice?
3. The great philosopher and mathematician Bertrand Russell (1872–1970) wrote about ideas in the early part of his life that he rejected in the latter part of his life. Do you see this as a sign of weakness or as a sign of strength in Bertrand Russell? (Do you speculate that your present ideas about the world around you will change as you learn and experience more, or do you speculate that further knowledge and experience will solidify your present understanding?)
4. Bertrand Russell wrote, "I think we must retain the belief that scientific knowledge is one of the glories of man. I will not maintain that knowledge can never do harm. I think such general propositions can almost always be refuted by well-chosen examples. What I will maintain—and maintain vigorously—is that knowledge is very much more often useful than harmful and that fear of knowledge is very much more often harmful than useful." Think of examples to support this statement.

EXPLORING FURTHER

Bodanis, David. *E = mc²: A Biography of the World's Most Famous Equation.* New York: Berkley Publishing Group, 2002. This is an engaging book about the people of science.

Bryson, Bill. *A Short History of Nearly Everything.* New York: Broadway Books, 2003. Another engaging book about the people of science, and from time to time, their zany antics.

Feynman, Richard P. *Surely You're Joking, Mr. Feynman.* New York: Norton, 1986. This is such tasty reading that I (PGH) allowed myself only one chapter per reading!

Gleick, James. *Genius—The Life and Science of Richard Feynman.* New York: Pantheon Books, 1992. More in depth about perhaps the most colorful physicist of the 20th century.

part one
physics

patterns of motion and equilibrium

1

1.1 ARISTOTLE ON MOTION

1.2 GALILEO'S CONCEPT OF INERTIA

1.3 MASS—A MEASURE OF INERTIA

1.4 NET FORCE

1.5 THE EQUILIBRIUM RULE

1.6 SUPPORT FORCE

1.7 DYNAMIC EQUILIBRIUM

1.8 THE FORCE OF FRICTION

1.9 SPEED AND VELOCITY

1.10 ACCELERATION

■ More than 2000 years ago Greek scientists understood some of the physics we understand today. They had a good grasp of the physics of floating objects and some of the properties of light. But they were confused about motion. One of the first to study motion seriously was Aristotle, the most outstanding philosopher-scientist in ancient Greece. Aristotle attempted to clarify motion by classification.

fyi

- Rather than reading chapters in this book slowly, try reading quickly and more than once. You'll better learn physics by going over the same material several times. With each time, it makes more sense. Don't worry if you don't understand things right away—just keep on reading.

FIGURE 1.1

Galileo's famous demonstration.

FIGURE 1.2

Does a force keep the cannon-ball moving after it leaves the cannon?

1.1   Aristotle on Motion

Aristotle divided motion into two classes: *natural motion* and *violent motion*. Natural motion had to do with the nature of bodies. Light things like smoke rose, and heavy things like dropped boulders fell. The motion of stars across the night sky were natural. Violent motion, on the other hand, resulted from pushing or pulling forces. Objects whose motions were unnatural were either pushed or pulled. Aristotle believed that natural laws could be understood by logical reasoning.

Two assertions of Aristotle held sway for some 2000 years. One was that heavy objects necessarily fall faster than lighter objects. The other was that moving objects must necessarily have forces exerted on them to keep them moving.

These ideas were completely turned around in the 16th century by Galileo, who held that experiment was superior to logic in uncovering natural laws. Galileo demolished the idea that heavy things fall faster than lighter things in his famous Leaning Tower of Pisa experiment, where he allegedly dropped objects of different weights and showed that—except for the effects of air resistance—they fell to the ground together.

CHECK POINT

Isn't it common sense to think of Earth as in its proper place, and that a force to move it is inconceivable, as Aristotle held, and that the Earth *is* at rest in this universe? (*Think and formulate your own answer. Then check your thinking below.*)

Was this your answer?

Common sense is relative to one's time and place. Aristotle's views were logical and consistent with everyday observations. So unless you become familiar with the physics to follow in this book, Aristotle's views about motion *do* make common sense (and are held by many uneducated people today). But as you acquire new information about nature's rules, you'll likely find your common sense progressing beyond Aristotelian thinking.

Aristotle (384–322 BC)

Aristotle was the foremost philosopher, scientist, and educator of his time. Born in Greece, he was the son of a physician who personally served the king of Macedonia. At age 17, he entered the Academy of Plato, where he worked and studied for 20 years until Plato's death. He then became the tutor of young Alexander the Great. Eight years later, he formed his own school. Aristotle's aim was to systematize existing knowledge, just as Euclid had systematized geometry. Aristotle made critical observations; collected specimens; and gathered, summarized, and classified almost all of the existing knowledge of the physical world. His systematic approach became the method from which Western science later arose. After his death, his voluminous notebooks were preserved in caves near his home and were later sold to the library at Alexandria. Scholarly activity ceased in most of Europe through the Dark Ages, and the works of Aristotle were forgotten and lost in the scholarship that contin-

ued in the Byzantine and Islamic empires. Several of his texts were reintroduced to Europe during the 11th and 12th centuries and were translated into Latin. The Church, the dominant political and cultural force in Western Europe, at first prohibited the works of Aristotle and then accepted and incorporated them into Christian doctrine.

■ 1.2 Galileo's Concept of Inertia

Galileo tested his revolutionary idea by *experiment*. On inclined planes he showed that moving things, once moving, continued in motion *without* the application of forces. In the simplest sense, a **force** is a push or a pull. Although a force is needed to start an object moving, Galileo showed that once moving, no force is needed to keep it moving—except for the force needed to overcome friction (more about friction in Section 1.8). When friction is absent, a moving object needs no force to keep it moving. Galileo reasoned that a ball moving horizontally would move forever if friction were entirely absent. A ball would move of itself—of its own **inertia**.

This was the beginning of modern science. Experiment, not philosophical speculation, is the test of truth.

Slope downward—
Speed increases

Slope upward—
Speed decreases

No slope—
Does speed change?

FIGURE 1.3

The motion of balls on various planes.

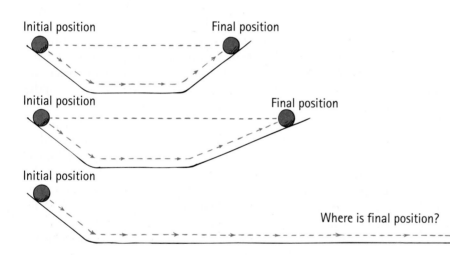

Initial position   Final position

Initial position   Final position

Initial position

Where is final position?

FIGURE 1.4

A ball rolling down an incline on the left tends to roll up to its initial height on the right. The ball must roll a greater distance as the angle of incline on the right is reduced.

Galileo Galilei (1564–1642)

Galileo was born in Pisa, Italy, in the same year Shakespeare was born and Michelangelo died. He studied medicine at the University of Pisa and then changed to mathematics. He developed an early interest in motion and was soon at odds with others around him, who held to Aristotelian ideas on falling bodies. He left Pisa to teach at the University of Padua and became an advocate of the new theory of the solar system advanced by the Polish astronomer Copernicus. Galileo was one of the first to build a telescope, and

the first to direct it to the nighttime sky and discover mountains on the Moon and the moons of Jupiter. Because he published his findings in Italian instead of in Latin, which was expected of so reputable a scholar, and because of the recent invention of the printing press, his ideas reached many people. He soon ran afoul of the Church and was warned not to teach and not to hold to Copernican views. He restrained himself publicly for nearly 15 years. Then he defiantly published his observations and conclusions, which were counter to Church doctrine. The outcome was a

trial in which he was found guilty, and he was forced to renounce his discoveries. By then an old man broken in health and spirit, he was sentenced to perpetual house arrest. Nevertheless, he completed his studies on motion, and his writings were smuggled out of Italy and published in Holland. His eyes had been damaged earlier by viewing the Sun through a telescope, which led to blindness at age 74. He died four years later.

fyi

- Galileo and William Shakespeare were born in the same year, 1564. In 1632 Galileo published his first mathematical treatment of motion—13 years after the Pilgrims landed at Plymouth Rock.

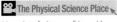
The Physical Science Place
Newton's Law of Inertia
The Old Tablecloth Trick
Toilet Paper Roll
Inertia of a Cylinder
Inertia of an Anvil
Definition of a Newton

FIGURE 1.5

An anvil in outer space—beyond the Sun for example—may be weightless, but it still has mass.

CHECKPOINT

A ball rolling along a level surface slowly comes to a stop. How would Aristotle explain this behavior? How would Galileo explain it? How would you explain it?

Were these your answers?
As mentioned, think about the Checkpoint questions throughout this book *before* reading the answers. When you first formulate your own answers, you'll find yourself learning more—much more!

Aristotle would probably say that the ball stops because it seeks its natural state of rest. Galileo would probably say that friction overcomes the ball's natural tendency to continue rolling—that friction overcomes the ball's *inertia*, and brings it to a stop. Only you can answer the last question!

■ 1.3   Mass—A Measure of Inertia

Every material object possesses inertia; how much depends on its amount of matter—the more matter, the more inertia. In speaking of how much matter something has, we use the term *mass*—the greater the mass of an object, the greater the amount of matter and the greater its inertia. **Mass** is a measure of the inertia of a material object.

Loosely speaking, mass corresponds to our intuitive notion of **weight**. We say something has a lot of matter if it is heavy. That's because we are accustomed to measuring matter by gravitational attraction to Earth. But mass is more fundamental than weight; it is a fundamental quantity that completely escapes the notice of most people. There are times, however, when weight corresponds to our unconscious notion of inertia. For example, if you are trying to determine which of two small objects is heavier, you might shake them back and forth in your hands or move them in some way instead of lifting them. In doing so, you are judging which of the two is more difficult to get moving, seeing which is the more resistant to a *change* in motion. You are really comparing the inertias of the objects.

It is easy to confuse the ideas of mass and weight. We define each as follows:

Mass: The quantity of matter in an object. It is also the measure of the inertia or sluggishness that an object exhibits in response to any effort made to start it, stop it, or change its state of motion in any way.

Weight: The force upon an object due to gravity.

The standard unit of mass is the **kilogram**, abbreviated kg. Weight is measured in units of force (such as pounds). The scientific unit of force is the **newton**, abbreviated N, which we'll use in this book. The abbreviation is written with a capital letter because the unit is named after a person.

Mass and weight are directly proportional to each other.* If the mass of an object is doubled, its weight is also doubled; if the mass is halved, the weight is

---

* *Directly proportional* means directly related. If you change one, the other changes proportionally. The constant of proportionality is g, the acceleration due to gravity. As we shall soon see, weight = mg, (or mass × acceleration due to gravity), so $9.8 \text{ N} = (1 \text{ kg})(9.8 \text{ m/s}^2)$. In Chapter 5 we'll extend our definition of weight to be the gravitational force of a body pressing against a support (for example, against a weighing scale).

halved. Because of this, mass and weight are often interchanged. Also, mass and weight are sometimes confused because it is customary to measure the quantity of matter in things (their mass) by their gravitational attraction to Earth (their weight). But mass doesn't depend on gravity. Gravity on the Moon, for example, is much less than it is on Earth. Whereas your weight on the surface of the Moon would be much less than it is on Earth, your mass would be the same in both locations.

Don't confuse mass and **volume**. When we think of a massive object, we often think of a big object. An object's size, however, is not necessarily a good way to judge its mass. Which is easier to get moving: a car battery or a king-size pillow? So we find that mass is neither weight nor volume.

A nice demonstration that distinguishes mass from weight is the massive ball suspended on the string shown in Figure 1.7. The top string breaks when the lower string is pulled with a gradual increase in force, but the bottom string breaks when the string is jerked. Which of these cases illustrates the weight of the ball, and which illustrates the mass of the ball? Note that only the top string bears the weight of the ball. So when the lower string is gradually pulled, the tension supplied by the pull is transmitted to the top string. So total tension in the top string is pull plus the weight of the ball. The top string breaks when the breaking point is reached. But when the bottom string is jerked, the mass of the ball—its tendency to remain at rest—is responsible for breakage of the bottom string.

FIGURE 1.6

The astronaut in space finds it just as difficult to shake the "weightless" anvil as on Earth. If the anvil is more massive than the astronaut, which shakes more—the anvil or the astronaut?

FIGURE 1.7

Why will a slow continuous increase in downward force break the string above the massive ball, whereas a sudden increase in downward force breaks the lower string?

CHECK POINT

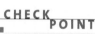

1. Does a 2-kg iron block have twice as much *inertia* as a 1-kg iron block? Twice as much *mass*? Twice as much *volume*? Twice as much *weight* when weighed in the same location?
2. Does a 2-kg iron block have twice as much *inertia* as a 1-kg bunch of bananas? Twice as much *mass*? Twice as much *volume*? Twice as much *weight* when weighed in the same location?
3. How does the mass of a bar of gold vary with location?

Were these your answers?
1. The answer is yes to all questions. A 2-kg block of iron has twice as many iron atoms, and therefore twice the amount of matter, mass, and weight. The blocks consist of the same material, so the 2-kg block also has twice the volume.
2. Two kilograms of *anything* has twice the inertia and twice the mass of 1 kg of anything else. Because mass and weight are proportional in the same location, 2 kg of anything will weigh twice as much as 1 kg of anything. Except for volume, the answer to all the questions is yes. Volume and mass are proportional only when the materials are identical—when they have the same *density*. (Density is mass/volume, as we'll discuss in Chapter 5). Iron is much more dense than bananas, so 2 kg of iron must occupy less volume than 1 kg of bananas.
3. Not at all! It consists of the same number of atoms no matter what the location. Although its weight may vary with location, it has the same mass everywhere. This is why mass is preferred to weight in scientific studies.

FIGURE 1.8

Why does the blow of the hammer not harm her?

FIGURE 1.9

One kilogram of nails weighs 9.8 N, which is equal to 2.2 lb.

ONE KILOGRAM WEIGHS 9.8 N

A 1-kg bag of any material at Earth's surface has a weight of 9.8 N. Away from Earth's surface where the force of gravity is less (on the Moon, for example), the bag would weigh less.

Except in cases where precision is needed, we will round off 9.8 and call it 10. So 1 kg of something on Earth's surface weighs about 10 N. If you know the mass in kilograms and want weight in newtons, multiply the number of kilograms by 10. Or, if you know the weight in newtons, divide by 10 and you'll have the mass in kilograms. As previously mentioned, weight and mass are proportional to each other.

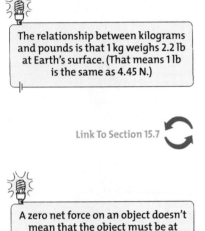

The relationship between kilograms and pounds is that 1 kg weighs 2.2 lb at Earth's surface. (That means 1 lb is the same as 4.45 N.)

Link To Section 15.7

A zero net force on an object doesn't mean that the object must be at rest, but that its state of motion remains unchanged. It can be at rest or moving uniformly in a straight line.

1.4 Net Force

In simplest terms, a force is a push or a pull. Objects don't speed up, slow down, or change direction unless a force acts. When we say "force," we imply the total force, or *net* force, acting on an object. Often more than one force acts. For example, when you throw a baseball, the force of gravity, air friction, and the pushing force you apply with your muscles all act on the ball. The **net force** on the ball is the combination of all these forces. It is the net force that changes an object's state of motion.

For example, suppose you pull on a box with a force of 5 N (slightly more than 1 lb). If your friend also pulls with 5 N in the same direction, the net force on the box is 10 N. If your friend pulls on the box with the same magnitude of force as you in the opposite direction, the net force on it is zero. Now if you increase your pull to 10 N and your friend pulls oppositely with 5 N, the net force is 5 N in the direction of your pull. This is shown in Figure 1.10.

The forces in Figure 1.10 are shown by arrows. Forces are vector quantities. A **vector quantity** has both magnitude (how much) and direction (which way). When an arrow represents a vector quantity, the arrow's length represents magnitude and its direction shows the direction of the quantity. Such an arrow is called a **vector**. (You'll find more on vectors in the next chapter, in Appendix C, and in the *Conceptual Physical Science Practice Book*.)

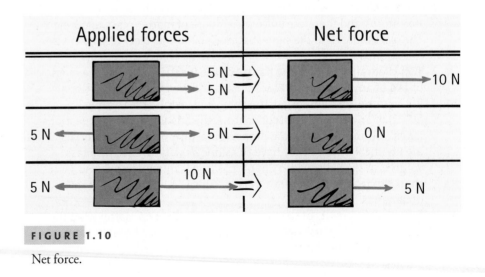

FIGURE 1.10

Net force.

Personal Essay

When I was in high school, my counselor advised me not to enroll in science and math classes but instead focus on what seemed to be my gift for art. I took this advice. I was then interested in drawing comic strips and in boxing, neither of which earned me much success. After a stint in the Army, I tried my luck at sign painting, and the cold Boston winters drove me south to Miami, Florida. There, at age 26, I got a job painting billboards and met an intellectual friend, Burl Grey. Like me, Burl had never studied physics in high school. But he was passionate about science in general, and shared his passion with many questions as we painted together.

I remember Burl asking me about the tensions in the ropes that held up the scaffold we were on. The scaffold was simply a heavy horizontal plank suspended by a pair of ropes. Burl twanged the rope nearest his end of the scaffold and asked me to do the same with mine. He was comparing the tensions in both ropes—to determine which was greater. Burl was heavier than I was, and he guessed the tension in his rope was greater. Like a more tightly stretched guitar string, the rope with greater tension twangs at a higher pitch. The finding that Burl's rope had a higher pitch seemed reasonable because his rope supported more of the load.

When I walked toward Burl to borrow one of his brushes, he asked if tensions in the ropes changed. Did tension in his rope increase as I moved closer? We agreed that it should have, because even more of the load was then supported by Burl's rope. How about my rope? Would its tension decrease? We agreed that it would, for it would be supporting less of the total load. I was unaware at the time that I was discussing physics.

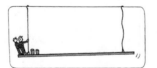

Burl and I used exaggeration to bolster our reasoning (just as physicists do). If we both stood at an extreme end of the scaffold and leaned outward, it was easy to imagine the opposite end of the scaffold rising like the end of a seesaw, with the opposite rope going limp. Then there would be no tension in that rope. We then reasoned that the tension in my rope would gradually decrease as I walked toward Burl. It was fun posing such questions and seeing if we could answer them.

A question that we couldn't answer was whether the decrease in tension in my rope when I walked away from it would be *exactly* compensated by a tension increase in Burl's rope. For example, if my rope underwent a decrease of 50 N,

would Burl's rope gain 50 N? (We talked pounds back then, but here we use the scientific unit of force, the *newton*—abbreviated N.) Would the gain be *exactly* 50 N? And if so, would this be a grand coincidence? I didn't know the answer until more than a year later, when Burl's stimulation resulted in my leaving full-time painting and going to college to learn more about science.*

There I learned that any object at rest, such as the sign-painting scaffold I worked on with Burl, is said to be in equilibrium. That is, all the forces that act on it balance to zero. So the sum of the upward forces supplied by the supporting ropes indeed do add up to our weights plus the weight of the scaffold. A 50-N loss in one would be accompanied by a 50-N gain in the other.

I tell this true story to make the point that one's thinking is very different when there is a rule to guide it. Now, when I look at any motionless object, I know immediately that all the forces acting on it cancel out. We see nature differently when we know its rules. It makes nature simpler and easier to understand. Without the rules of physics, we tend to be superstitious and see magic where there is none. Quite wonderfully, everything is beautifully connected to everything else by a surprisingly small number of rules. Physics is the study of nature's rules.

* I am indebted to Burl Grey for the stimulation he provided, for when I continued with formal education, it was with enthusiasm. I lost contact with Burl for 40 years. A student in my class at the Exploratorium in San Francisco, Jayson Wechter, who was a private detective, located him in 1998 and put us back in contact. Friendship renewed, we continue in our spirited conversations. It was via Burl that I met my teaching role model, futurist Jacque Fresco, the most talented teacher I've ever met. Now in his 90s, he continues to inspire people toward a positive future through his books, TV documentaries, and most recently by the movie that features his vision, "Future By Design."

FIGURE 1.11

Burl Grey, who first introduced the author to tension forces, suspends a 2 lb. bag of flour from a spring scale, showing its weight and the tension in the string of about 9 N.

■ 1.5 The Equilibrium Rule

If you tie a string around a 2-lb bag of flour and suspend it on a weighing scale (Figure 1.11), a spring in the scale stretches until the scale reads 2 lb. The stretched spring is under a "stretching force" called *tension*. A scale in a science lab is likely calibrated to read the same force as 9 N. Both pounds and newtons are units of weight, which in turn, are units of *force*. The bag of flour is attracted to Earth with a gravitational force of 2 lb—or equivalently, 9 N. Suspend twice as much flour from the scale and the reading will be 18 N.

Two forces are acting on the bag of flour—tension force acting upward and weight acting downward. The two forces on the bag are equal in magnitude and opposite in direction, and they cancel to zero. Hence the bag remains at rest.

When the net force on something is zero, we say that the object is in *mechanical equilibrium.** In mathematical notation, the **equilibrium rule** is

$$\Sigma \boldsymbol{F} = 0$$

The symbol Σ stands for "the vector sum of" and $\boldsymbol{F}$ stands for "forces." For a suspended object at rest, like the bag of flour, the rule states that the forces acting upward on the body must be balanced by other forces acting downward to make the vector sum equal zero. (Vector quantities take direction into account, so if upward forces are positive, downward ones are negative; the resulting sum is equal to zero.)

In Figure 1.12 we see the forces of interest to Burl and Paul on their sign-painting scaffold. The sum of the upward tensions is equal to the sum of their weights plus the weight of the scaffold. Note how the magnitudes of the two upward vectors equal the magnitude of the three downward vectors. Net force on the scaffold is zero, so we say it is in mechanical equilibrium.

FIGURE 1.12

The sum of the upward vectors equals the sum of the downward vectors. $\Sigma \boldsymbol{F} = 0$, and the scaffold is in equilibrium.

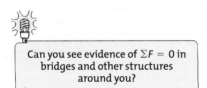

Can you see evidence of $\Sigma F = 0$ in bridges and other structures around you?

CHECK POINT

If you hang from a trapeze at rest, what is the tension in each of the two supporting vertical ropes?

Was this your answer?

The tension would be half your weight in each rope. In this way, $\Sigma F = 0$.

---

* We'll see in Appendix B that another condition for mechanical equilibrium is that the net torque equals zero.

■ 1.6 Support Force

Consider a book lying at rest on a table. It is in equilibrium. What forces act on the book? One force is that due to gravity—the *weight* of the book. Because the book is in equilibrium, there must be another force acting on it to produce a net force of zero—an upward force opposite to the force of gravity. The table exerts this upward force, called the **support force**. This upward support force, often called the *normal force*, must equal the weight of the book.* If we designate the upward force as positive, then the downward force (weight) is negative, and the sum of the two is zero. The net force on the book is zero. Stating it another way, $\Sigma F = 0$.

To better understand that the table pushes up on the book, compare the case of compressing a spring (Figure 1.13). If you push the spring down, you can feel the spring pushing up on your hand. Similarly, the book lying on the table compresses atoms in the table, which behave like microscopic springs. The weight of the book squeezes downward on the atoms, and they squeeze upward on the book. In this way, the compressed atoms produce the support force.

When you step on a bathroom scale, two forces act on the scale. One is the downward pull of gravity, your weight, and the other is the upward support force of the floor. These forces compress a spring that is calibrated to show your weight (Figure 1.14). In effect, the scale shows the support force. When you weigh yourself on a bathroom scale at rest, the support force and your weight have the same magnitude.

FIGURE 1.13

The table pushes up on the book with as much force as the downward force of gravity on the book. The spring pushes up on your hand with as much force as you exert to push down on the spring.

CHECK POINT

Suppose you stand on two bathroom scales with your weight evenly divided between the two scales. What will each scale read? How about if you stand with more of your weight on one foot than the other?

Were these your answers?

The reading on both scales adds up to your weight. This is because the sum of the scale readings, which equals the supporting normal force by the floor, must counteract your weight so the net force on you will be zero. That is, the vector sum $\Sigma F = 0$. If you stand equally on each scale, each will read half your weight. If you lean more on one scale than the other, more than half your weight will be read on that scale but less on the other, so they will still add up to your weight. For example, if one scale reads two-thirds your weight, the other scale will read one-third your weight. In whatever case, $\Sigma F = 0$. Get it?

FIGURE 1.14

The upward support is as much as the downward gravitational force.

■ 1.7 Dynamic Equilibrium

When an object isn't moving, the forces on it add up to zero—it's in equilibrium. More specifically, we say it's in *static equilibrium*. But the state of rest is only one form of equilibrium. An object moving at constant speed in a straight-line path is also in equilibrium. We say it's in *dynamic equilibrium*.

---

* This force acts at right angles to the surface. When we say "normal to," we are saying "at right angles to," which is why this force is called a normal force.

When the push on the desk is as great as the force of friction between it and the floor, the net force on the desk is zero and it slides at an unchanging speed.

fyi

- In Chapter 6 we'll discuss thermal equilibrium, and in Appendix B we'll discuss rotational equilibrium.

FIGURE 1.16

Friction results from the mutual contact of irregularities in the surfaces of sliding objects. Even surfaces that appear to be smooth have irregular surfaces when viewed at the microscopic level.

Once in motion, if there is no net force to change the state of motion, it moves at an unchanging speed and is in dynamic equilibrium. Whether equilibrium is static or dynamic, $\Sigma F = 0$.

Interestingly, an object under the influence of only one force cannot be in static or dynamic equilibrium. Net force couldn't be zero. Only when there is no force at all, or when two or more forces combine to zero, can an object be in equilibrium. We can test whether something is in equilibrium by noting whether it undergoes changes in motion.

Consider pushing a desk across a schoolroom floor. If it moves steadily at constant speed, with no change in its motion, it is in equilibrium. This tells us that more than one horizontal force acts on the desk—likely the force of friction between the desk and the floor. The fact that the net force on it equals zero means that the force of friction must be equal in magnitude and act opposite to our pushing force.

1.8 The Force of Friction

friction is the resistive force that opposes the motion or attempted motion of an object past another with which it is in contact. It occurs when one object rubs against something else.* Friction occurs for solids, liquids, and gases. An important rule of friction is that it always acts in a direction to oppose motion. If you push a solid block along a floor to the right, the force of friction on the block will be to the left. A boat propelled to the east by its motor experiences water friction to the west. When an object falls downward through the air, the force of friction, **air resistance**, acts upward. Again, for emphasis: friction always acts in a direction to oppose motion.

CHECK POINT

You push on a piece of furniture and it slides at constant speed across the living room floor. In other words, it is in equilibrium. Two horizontal forces act on it. One is your push and the other is the force of friction that acts in the opposite direction. Which force is greater?

Was this your answer?

Neither, for both forces have the same magnitude. If you call your push positive, then the friction is negative. Because the pushed furniture is in equilibrium, can you see that the two forces combine to equal zero?

The amount of friction between two surfaces depends on the kinds of material and how much they are pressed together. Friction is due to tiny surface bumps and also to the "stickiness" of the atoms on the surfaces of the two materials (Figure 1.16). Friction between a sliding desk you're pushing and a smooth linoleum floor is less than between the desk and a rough floor. And if the surface is inclined, friction is less because it doesn't press as much on the inclined surface (we won't treat inclined surfaces in this chapter).

So we see that when you push horizontally on a piece of furniture and it slides across the floor, both your force and the opposite force of friction affect the

---

* Unlike most concepts in physics, friction is a very complicated phenomenon. The findings are empirical (gained from a wide range of experiments) and the predictions are approximate (also based on experiment).

motion. When you push hard enough on the sliding furniture to match the friction, the net force on it is zero, and it slides at constant velocity. Notice that we are talking about what we recently learned—that no change in motion occurs when $\Sigma F = 0$.

CHECK POINT

1. Suppose you exert a 50-N horizontal force on a heavy desk resting motionless on your schoolroom floor. The fact that it remains at rest indicates that 50 N isn't great enough to make it slide. How does the force of friction between the desk and floor compare with your push?
2. You push harder—say, 55 N—and the desk still doesn't slide. How much friction acts on it?
3. You push still harder and the desk moves. Once it is in motion, you push with 60 N, which is just sufficient to keep it sliding at constant velocity. How much friction acts on the desk?
4. What net force does a sliding desk experience when you exert a force of 65 N and friction between the desk and the floor is 60 N?

Were these your answers?
1. The force of friction is 50 N in the opposite direction. Friction opposes the motion that would occur otherwise. The fact that the desk is at rest is evidence that $\Sigma F = 0$.
2. Friction increases to 55 N, and again $\Sigma F = 0$.
3. The force of friction is 60 N, because when moving at constant velocity, $\Sigma F = 0$.
4. The net force is 5 N, because $\Sigma F = 65 \text{ N} - 60 \text{ N}$. In this case the desk picks up speed. As we will see, it *accelerates*.

FIGURE 1.17

The greater the distance traveled each second, the faster the horse gallops.

1.9 Speed and Velocity

SPEED

Before the time of Galileo, people described moving things as simply "slow" or "fast." Such descriptions were vague. Galileo was the first to measure speed by comparing the distance covered with the *time* it takes to move that distance. He defined **speed** as the distance covered per amount of travel time.

$$\text{Speed} = \frac{\text{distance covered}}{\text{travel time}}$$

For example, if a bicyclist covers 20 kilometers in 1 hour, her speed is 20 km/h. Or, if she runs 6 meters in 1 second, her speed is 6 m/s.

Any combination of units for distance and time can be used for speed—kilometers per hour (km/h), centimeters per day (the speed of a sick snail), or whatever is useful and convenient. The slash symbol (/) is read as "per," and means "divided by." In physics the preferred unit of speed is meters per second (m/s). Table 1.1 compares some speeds in different units.

INSTANTANEOUS SPEED

Moving things often have variations in speed. A car, for example, may travel along a street at 50 km/h, slow to 0 km/h at a red light, and speed up to only 30 km/h because of traffic. At any instant you can tell the speed of the car by looking at its speedometer. The speed at any instant is the *instantaneous speed*.

The Physical Science Place

Definition of Speed
Average Speed
Velocity
Changing Velocity

FIGURE 1.18

A common automobile speedometer. Note that speed is shown in units of km/h and mi/h.

If you get a traffic ticket for speeding, is the speed written on your ticket your *instantaneous speed* or your *average speed*?

TABLE 1.1   APPROXIMATE SPEEDS IN DIFFERENT UNITS

12 mi/h = 20 km/h = 6 m/s (bowling ball)
25 mi/h = 40 km/h = 11 m/s (very good sprinter)
37 mi/h = 60 km/h = 17 m/s (sprinting rabbit)
50 mi/h = 80 km/h = 22 m/s (tsunami)
62 mi/h = 100 km/h = 28 m/s (sprinting cheetah)
75 mi/h = 120 km/h = 33 m/s (batted softball)
100 mi/h = 160 km/h = 44 m/s (batted baseball)

AVERAGE SPEED

In planning a trip by car, the driver often wants to know the travel time. The driver is concerned with the *average speed* for the trip. How is average speed defined?

$$\text{Average speed} = \frac{\text{total distance covered}}{\text{travel time}}$$

Average speed can be calculated rather easily. For example, if you drive a distance of 80 km in 1 h, your average speed is 80 km/h. Likewise, if you travel 320 km in 4 h,

$$\text{Average speed} = \frac{\text{total distance covered}}{\text{travel time}} = \frac{320 \text{ km}}{4 \text{ h}} = 80 \text{ km/h}$$

Note that, when a distance in kilometers (km) is divided by a time in hours (h), the answer is in kilometers per hour (km/h).

Because average speed is the entire distance covered divided by the total time of travel, it doesn't indicate the various instantaneous speeds that may have occurred along the way. On most trips, the instantaneous speed is often different from the average speed.

If we know average speed and travel time, distance traveled is easy to find. A simple rearrangement of the definition above gives

$$\text{Total distance covered} = \text{average speed} \times \text{travel time}$$

For example, if your average speed on a 4-h trip is 80 km/h, then you cover a total distance of 320 km.

CHECK POINT

1. What is the average speed of a horse that gallops 100 m in 8 s? How about if it gallops 50 m in 4 s?
2. If a car travels with an average speed of 60 km/h for an hour, it will cover a distance of 60 km. (a) How far would the car travel if it moved at this rate for 4 h? (b) For 10 h?

Were these your answers?
*(Are you reading this before you have reasoned answers in your mind? As mentioned earlier, **think** before you read the answers. You'll not only learn more; you'll enjoy learning more.)*
1. In both cases the answer is 12.5 m/s:

$$\text{Average speed} = \frac{\text{total distance covered}}{\text{travel time}} = \frac{100 \text{ meters}}{8 \text{ seconds}} = \frac{50 \text{ meters}}{4 \text{ seconds}} = 12.5 \text{ m/s}$$

2. The distance traveled is the average speed × time of travel, so
   a. Distance = 60 km/h × 4 h = 240 km
   b. Distance = 60 km/h × 10 h = 600 km

VELOCITY

When we know both the speed and direction of an object, we know its **velocity**. For example, if a vehicle travels at 60 km/h, we know its speed. But, if we say it moves at 60 km/h to the north, we specify its *velocity*. Speed is a description of how fast; velocity is a description of how fast *and* in what direction. As previously mentioned, a quantity such as velocity that specifies direction as well as magnitude is called a *vector quantity*. Velocity is a vector quantity. (Vectors are treated in Appendix C, and are nicely developed in the *Conceptual Physical Science Practice Book*.)

Constant speed means steady speed, neither speeding up nor slowing down. Constant velocity, on the other hand, means both constant speed *and* constant direction. Constant direction is a straight line—the object's path doesn't curve. So, constant velocity means motion in a straight line at constant speed—motion with no acceleration.

FIGURE 1.19

Although the car can maintain a constant speed along the circular track, it cannot maintain a constant velocity. Why?

CHECK POINT
"She moves at a constant speed in a constant direction." Say the same sentence in fewer words.

Was this your answer?
   "She moves at constant velocity."

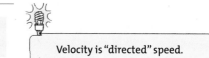

Velocity is "directed" speed.

MOTION IS RELATIVE

Everything is always moving. Even when you think you're standing still, you're actually speeding through space. You're moving relative to the Sun and stars—though you are at rest relative to Earth. At this moment, your speed relative to the Sun is about 100,000 km/h, and that speed is even faster relative to the center of our galaxy.

When we discuss the speed or velocity of something, we mean speed or velocity relative to something else. For example, when we say a space shuttle travels at 30,000 km/h, we mean relative to Earth below. Or when we say a racing car reaches a speed of 300 km/h, we mean relative to the track. Unless stated otherwise, all speeds discussed in this book are relative to the surface of Earth. Motion is relative.

FIGURE 1.20

Although you may be at rest relative to Earth's surface, you're moving about 100,000 km/h relative to the Sun.

CHECK POINT
A hungry mosquito sees you resting in a hammock in a 3-m/s breeze. How fast and in what direction should the mosquito fly in order to hover above you for lunch?

Was this your answer?
   The mosquito should fly toward you into the breeze. When just above you, it should fly at 3 m/s in order to hover at rest. Unless its grip on your skin is strong enough after landing, it must continue flying at 3 m/s to keep from being blown off. That's why a breeze is an effective deterrent to mosquito bites.

■ 1.10 Acceleration

Most moving things undergo variations in their motion. We say they undergo *acceleration*. The first to formulate the concept of acceleration was Galileo, who developed the concept in his experiments with inclined planes.

The Physical Science Place
Definition of Acceleration
Numerical Example of Acceleration
Free Fall: How Fast?
Free Fall: How Far?
Free-Fall Acceleration Explained

FIGURE 1.21

INTERACTIVE FIGURE

A ball gains the same amount of speed in equal intervals of time. It undergoes constant acceleration.

Can you see that a car has three controls that change velocity—the gas pedal (accelerator), the brakes, and the steering wheel?

FIGURE 1.22

We say that a body undergoes acceleration when there is a *change* in its state of motion.

He found that balls rolling down inclines rolled faster and faster. Their velocity changed as they rolled. Further, the balls gained the same amount of velocity in equal time intervals.

Galileo defined the rate of change of velocity as **acceleration.***

$$\text{Acceleration} = \frac{\text{change of velocity}}{\text{time interval}}$$

Acceleration is experienced when you're in a moving car or bus. When the bus driver steps on the gas pedal, the vehicle gains speed. We say that the bus accelerates. Thus, we can see why the gas pedal is called the "accelerator"! When the brakes are applied, the vehicle slows. This is also acceleration, because the velocity of the vehicle is changing. When something slows down, we often call this *deceleration*.

Consider driving in a car that steadily increases in speed. Suppose that in 1 s, you steadily increase your velocity from 30 km/h to 35 km/h. In the next second, you go from 35 km/h to 40 km/h, and so on. You change your velocity by 5 km/h each second. We see that

$$\text{Acceleration} = \frac{\text{change of velocity}}{\text{time interval}} = \frac{5 \text{ km/h}}{1 \text{ s}} = 5 \text{ km/h} \cdot \text{s}$$

In this example the acceleration is 5 km/h-second (abbreviated as 5 km/h · s).** Note that the unit for time appears twice: once for the unit of velocity and again for the interval of time in which the velocity is changing. Also note that acceleration is not just the change in velocity; it is the change in velocity per second. If either speed or direction changes, or if both change, then velocity changes.

When a car makes a turn, even if its speed does not change, it is accelerating. Can you see why? Acceleration occurs because the car's direction is changing. Acceleration refers to a change in velocity. So acceleration involves a change in speed, a change in direction, or a change in both speed *and* direction. Figure 1.22 illustrates this.

---

* The Greek letter Δ (delta) is often used as a symbol for "change in" or "difference in." In "delta" notation, $a = \frac{\Delta v}{\Delta t}$, where Δv is the change in velocity and Δt is the change in time (the time interval). From this we see that $v = at$. See further development of linear motion in Appendix B.

** When we divide $\frac{\text{km}}{\text{h}}$ by s $\left(\frac{\text{km}}{\text{h}} \div \text{s}\right)$, we can express this as $\frac{\text{km}}{\text{h}} \times \frac{1}{\text{s}} = \frac{\text{km}}{\text{h} \cdot \text{s}}$ (some textbooks express this as km/h/s). Or when we divide $\frac{\text{m}}{\text{s}}$ by s $\left(\frac{\text{m}}{\text{s}} \div \text{s}\right)$, we can express this as $\frac{\text{m}}{\text{s}} \times \frac{1}{\text{s}} = \frac{\text{m}}{\text{s} \cdot \text{s}} = \frac{\text{m}}{\text{s}^2}$ (which can also be written as (m/s)/s, or ms^{-2}.

TABLE 1.2   FREE-FALL VELOCITY ACQUIRED AND DISTANCE FALLEN

Time of Fall (s)	Velocity Acquired (m/s)	Distance Fallen (m)
0	0	0
1	10	5
2	20	20
3	30	45
4	40	80
5	50	125

Hold a stone above your head (not directly above your head!) and drop it. It accelerates during its fall. When the only force that acts on a falling object is that due to gravity, when air resistance doesn't affect its motion, we say the object is in **free fall**. All freely falling objects in the same vicinity have the same acceleration. At Earth's surface an object in free fall gains speed at the rate of 10 m/s each second, as shown in Table 1.2.

$$\text{Acceleration} = \frac{\text{change in speed}}{\text{time interval}} = \frac{10 \text{ m/s}}{1 \text{ s}} = 10 \text{ m/s} \cdot \text{s} = 10 \text{ m/s}^2$$

We read the acceleration of free fall as 10 meters per second squared. (More precisely, 9.8 m/s^2.) This is the same as saying that acceleration is 10 meters per second per second. Note again that the unit of time, the second, appears twice. It appears once for the unit of velocity and again for the time during which the velocity changes.

> "When you're over the hill, that's when you pick up speed."
> —Quincy Jones

CHECK POINT

In 2.0 s a car increases its speed from 60 km/h to 65 km/h while a bicycle goes from rest to 5 km/h. Which has the greater acceleration?

Was this your answer?
Both have the same acceleration because both gain the same amount of speed in the same time. Both accelerate at 2.5 km/h · s.

In Figure 1.23, we imagine a freely falling boulder with a speedometer attached to it. As the boulder falls, the speedometer shows that the boulder goes 10 m/s faster each second. This 10 m/s gain each second is the boulder's acceleration. Velocity acquired and distance fallen* are shown in Table 1.2.

> Why do all freely falling objects fall with equal acceleration? The answer to this question awaits you in Chapter 3.

*Distance fallen from rest: $d = \text{average velocity} \times \text{time}$

$$d = \frac{\text{initial velocity} + \text{final velocity}}{2} \times \text{time}$$

$$d = \frac{0 + gt}{2} \times t$$

$$d = \frac{1}{2}gt^2$$

(See Appendix B for further explanation.)

FIGURE 1.23

INTERACTIVE FIGURE

Imagine that a falling boulder is equipped with a speedometer. In each succeeding second of fall, you'd find the boulder's speed increasing by the same amount: 10 m/s. Sketch in the missing speedometer needle at $t = 3$ s, $t = 4$ s, and $t = 5$ s.

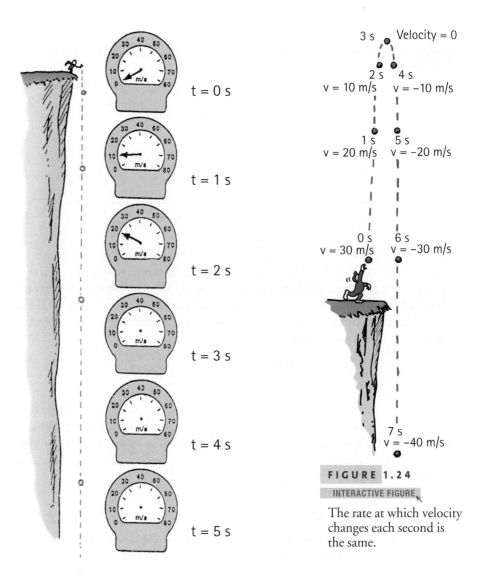

t = 0 s

t = 1 s

t = 2 s

t = 3 s

t = 4 s

t = 5 s

3 s   Velocity = 0

2 s   4 s
v = 10 m/s   v = −10 m/s

1 s   5 s
v = 20 m/s   v = −20 m/s

0 s   6 s
v = 30 m/s   v = −30 m/s

7 s
v = −40 m/s

FIGURE 1.24

INTERACTIVE FIGURE

The rate at which velocity changes each second is the same.

(The acceleration of free fall is further developed in Appendix B and in the *Conceptual Physical Science Practice Book*.) We see that the distance of free fall from rest is directly proportional to the square of the time of fall. In equation form,

$$d = \frac{1}{2}gt^2$$

Up-and-down motion is shown in Figure 1.24. The ball leaves the thrower's hand at 30 m/s. Call this the initial velocity. The figure uses the convention of up being + and down being −. The minus sign of downward values of velocity indicate a downward direction. More important, notice that the 1-s interval positions correspond to 10-m/s velocity changes.

Aristotle used logic to establish his ideas of motion, whereas Galileo used experiment. Galileo showed that experiments are superior to logic in testing knowledge. Galileo was concerned with *how* things move rather than *why* they move. The path was paved for Isaac Newton to make further connections of concepts of motion.

Hang Time

Some athletes and dancers have great jumping ability. Leaping straight up, they seem to "hang in the air," apparently defying gravity. Ask your friends to estimate the **hang time** of the great jumpers—the time a jumper is airborne with his or her feet off the ground. They may estimate 2 or 3 s. But, surprisingly, the hang time of the greatest jumpers is almost always less than 1 s. The perception of a longer time is one of many illusions we have about nature.

People often have a related illusion about the vertical height a human can jump. Most of your classmates probably cannot jump higher than 0.5 m. They can easily step over a 0.5-m fence, but in doing so, their bodies rise only slightly. The height of the barrier is different from the height a jumper's "center of gravity" rises. Many people can leap over a 1-m fence, but only rarely does anybody raise the "center of gravity" of his or her body by 1 m. Even basketball star Michael Jordan in a standing jump during his prime couldn't raise his body 1.25 m high, although he could easily reach considerably above the basket, which is more than 3 m high.

Jumping ability is best measured by a standing vertical jump. Stand facing a wall with feet flat on the floor and arms extended upward. Make a mark on the wall at the top of your reach. Then make your jump, and, at the point you are able to reach, make another mark. The distance between these two marks measures your vertical leap. If it's more than 0.6 m (2 ft), you're exceptional.

Here's the physics. When you leap upward, jumping force is applied only while your feet are still making contact with the ground. The greater the force, the greater your launch speed and the higher your jump. When your feet leave the ground, your upward speed immediately decreases at the steady rate of g, which is 10 m/s². At the top of your jump, your upward speed decreases to zero. Then you begin to fall, gaining speed at exactly the same rate, g. If you land as you took off, upright with legs extended, then your time rising equals your time falling; hang time is time up plus time down. While you are airborne, no amount of leg or arm pumping or other bodily motions can change your hang time.

As will be shown in Appendix B, the relationship between time up or down and vertical height is given by

$$d = \frac{1}{2}gt^2$$

If the vertical height d is known, we can rearrange this expression to read

$$t = \sqrt{\frac{2d}{g}}$$

Quite interestingly, no basketball player on record has exceeded 1.25 m in a vertical standing jump. For the corresponding hang time, let's use 1.25 m for d, and the more precise value of 9.8 m/s² for g. Solving for t, half the hang time (one way), we get

$$t = \sqrt{\frac{2d}{g}} = \sqrt{\frac{2(1.25)\ \text{m}}{9.8\ \text{m/s}^2}} = 0.50\ \text{s}$$

Double this amount (because this is the time for one direction of an up-and-down round trip) and we see that such record-breaking hang time is 1 s.

We're discussing vertical motion here. How about running jumps? We'll see in Chapter 4 that hang time depends only on the jumper's vertical speed at launch. While the jumper is airborne, his or her horizontal speed remains constant while the vertical speed undergoes acceleration. Intriguing physics!

SUMMARY OF TERMS

Inertia The property by which objects resist changes in motion.

Mass The quantity of matter in an object. More specifically, it is the measure of the inertia or sluggishness that an object exhibits in response to any effort made to start it, stop it, deflect it, or change in any way its state of motion.

Weight Simply stated, the force due to gravity on an object. More specifically, the gravitational force with which a body presses against a supporting surface.

Kilogram The unit of mass. One kilogram (symbol kg) is the mass of 1 liter (l) of water at 4°C.

Force Simply stated, a push or a pull.

Newton The scientific unit of force.

Volume Quantity of space an object occupies.

Net force The combination of all forces that act on an object.

Vector quantity A quantity whose description requires both magnitude and direction.

Vector An arrow to represent the magnitude and direction of a quantity.

Equilibrium rule The vector sum of forces acting on a non-accelerating object equals zero: $\Sigma \boldsymbol{F} = 0$.

Support force The force that supports an object against gravity, often called the *normal force*.

Friction The resistive force that opposes the motion or attempted motion of an object past another with which it is in contact, or through a fluid.

Air resistance The force of friction acting on an object due to its motion in air.

Speed The distance traveled per time.

Velocity The speed of an object and specification of its direction of motion.

Acceleration The rate at which velocity changes with time; the change in velocity may be in magnitude or direction or both, usually measured in units of m/s^2.

Free fall Falling only under the influence of gravity—falling without air resistance.

Hang time The time that one's feet are off the ground during a vertical jump.

REVIEW QUESTIONS

*Each chapter in this book concludes with a set of review questions and exercises, and some chapters include problems. The **Review Questions** are designed to help you comprehend ideas and catch the essentials of the chapter material. You'll notice that answers to the questions can be found within the chapters. The **Exercises** stress thinking rather than mere recall of information and call for an understanding of the definitions, principles, and relationships of the chapter material. In many cases the intention of particular exercises is to help you apply the ideas of physics to familiar situations. Unless you cover only a few chapters in your course, you will likely be expected to tackle only a few exercises for each chapter. Answers should be in complete sentences, with an explanation or sketches when applicable. The large number of exercises is to allow your instructor a wide choice of assignments. **Problems** are math-based exercises where calculations aid concept understanding.*

1.1 Aristotle on Motion

1. What did Aristotle believe about the relative speeds of fall for heavy and light objects?
2. Did Aristotle believe that forces are necessary to keep moving objects moving, or did he believe that once moving, they'd move of themselves?

1.2 Galileo's Concept of Inertia

3. What idea of Aristotle did Galileo discredit with his inclined-plane experiments?
4. Which dominated Galileo's way of extending knowledge: philosophical discussion or experiment?
5. What name is given to the property by which objects resist changes in motion?

1.3 Mass—A Measure of Inertia

6. Which depends on location, weight or mass?
7. Where is your weight greater, on Earth or on the Moon? How about your mass?
8. What are the units of measurement for weight and for mass?
9. One kilogram weighs 9.8 N on Earth. Would it weigh more or less on the Moon?

1.4 Net Force

10. What is the net force on a box pushed to the right with 50 N of force, while being pushed to the left with 20 N of force?
11. What two quantities are necessary for a vector quantity?

1.5 The Equilibrium Rule

12. Name the force that occurs in a rope when both ends are pulled in opposite directions.
13. How much tension is there in a vertical rope that holds a 20-N bag of apples at rest?
14. What does $\Sigma F = 0$ mean?

1.6 Support Force

15. Why is the support force on an object often called the normal force?
16. When you weigh yourself, how does the support force of the scale acting on you compare with the gravitational force between you and Earth?

1.7 Dynamic Equilibrium

17. A bowling ball sits at rest. Another ball rolls down a lane at constant speed. Which, if either, is in equilibrium? Defend your answer.
18. If we push an object at constant velocity, how do we know how much friction acts on the object compared to our pushing force?

1.8 The Force of Friction

19. How does the direction of a friction force compare with the velocity of a sliding object?
20. If you push to the right on a heavy piece of furniture and it slides, what is the direction of friction on the furniture?
21. Suppose you push to the right on a heavy piece of furniture, but not hard enough to make it slide. Does a friction force act on the furniture?

1.9 Speed and Velocity

22. Distinguish between speed and velocity.
23. Why do we say velocity is a vector and speed is not?
24. Does the speedometer on a vehicle show average speed or instantaneous speed?

25. How can you be at rest and moving at 100,000 km/h at the same time?

1.10 Acceleration

26. Distinguish between velocity and acceleration.

27. What is the acceleration of an object that moves at constant velocity? What is the net force on the object in this case?

28. What is the acceleration of an object in free fall at Earth's surface?

EXERCISES

● BEGINNER ■ INTERMEDIATE ◆ EXPERT

Please do not be intimidated by the large number of exercises in this book. If your course work is to cover many chapters, your instructor will likely assign only a few exercises from each.

1. ● Asteroids have been moving through space for billions of years. What keeps them moving?

2. ● A space probe is carried by a rocket into outer space. What keeps the probe moving after the rocket no longer pushes it?

3. ● A bowling ball rolling along a lane gradually slows as it rolls. How would Aristotle interpret this observation? How would Galileo interpret it?

4. ● What Aristotelian idea did Galileo discredit in his fabled Leaning Tower of Pisa experiment? With his inclined-plane experiments?

5. ● When a ball rolls down an inclined plane, it gains speed because of gravity. When rolling up an inclined plane, it loses speed because of gravity. Why doesn't gravity play a role when it rolls on a horizontal surface?

6. ● What physical quantity is a measure of how much inertia an object has?

7. ● Which has more mass, a 2-kg fluffy pillow or a 3-kg small piece of iron? More volume? Why are your answers different?

8. ● Does a dieting person more accurately lose mass or lose weight?

9. ● A favorite class demonstration by Paul Hewitt is lying on his back with a blacksmith's anvil placed on his chest. When an assistant whacks the anvil with a strong sledge-hammer blow, Hewitt is not injured. How is the physics here similar to that illustrated in Figure 1.8?

10. ● What is your own mass in kilograms? Your weight in newtons?

11. ● Gravitational force on the Moon is only 1/6 that of the gravitational force on Earth. What would be the weight of a 10-kg object on the Moon and on Earth? What would its mass be on the Moon and on Earth?

12. ● The sketch shows a painter's scaffold in mechanical equilibrium. The person in the middle weighs 250 N, and the tensions in each rope are 200 N. What is the weight of the scaffold?

13. ● A different scaffold that weighs 300 N supports two painters, one weighing 250 N and the other weighing 300 N. The reading on the left scale is 400 N. What should the reading on the right scale be?

14. ■ Nellie Newton hangs at rest from the ends of the rope as shown. How does the reading on the scale compare to her weight?

15. ■ Harry the painter swings year after year from his bosun's chair. His weight is 500 N, and the rope, unknown to him, has a breaking point of 300 N. Why doesn't the rope break when he is supported as shown to the left? One day, Harry was painting near a flagpole, and, for a change, he tied the free end of the rope to the flagpole instead of to his chair, as shown to the right. Why did Harry end up taking his vacation early?

16. ● A hockey puck slides across the ice at a constant velocity. Is it in mechanical equilibrium? Why or why not?

17. ● If you push horizontally on a crate that contains your new desk and it slides across the floor, slightly gaining speed, how does the friction acting on the crate compare with your push?

18. ● When you place a heavy book on a table, the table pushes up on the book. Why doesn't this upward push cause the book to rise from the table?

19. ■ An empty jug of weight W rests on a table. What is the support force exerted on the jug by the table? What is the support force when water of weight w is poured into the jug?

20. ● In order to slide a heavy cabinet across the floor at constant speed, you exert a horizontal force of 600 N. Is the force of friction between the cabinet and the floor greater than, less than, or equal to 600 N? Defend your answer.

21. ● Consider your desk at rest on your bedroom floor. As you and your friend start to lift it, does the support force on the desk provided by the floor increase, decrease, or remain unchanged? What happens to the support force on the feet of you and your friend?

22. ● Correct your friend who says, "The dragster rounded the curve at a constant velocity of 100 km/h."

23. ● What is the impact speed when a car moving at 100 km/h bumps into the rear of another car traveling in the same direction at 98 km/h?

24. ● You're in a car traveling at some specified speed limit. You see another car moving at the same speed toward you. How fast is the car approaching you, compared with the speed limit?

25. ● Harry Hotshot can paddle a canoe in still water at 8 km/h. How successful will he be at canoeing upstream in a river that flows at 8 km/h?

26. ● Suppose that a freely falling object were somehow equipped with a speedometer. By how much would its speed readings increase with each second of fall?

27. ● Suppose that the freely falling object in the preceding exercise was also equipped with an odometer. What equation is most appropriate for readings of distance fallen each second? Do the readings indicate equal or unequal distances of fall for successive seconds? Explain.

28. ● When a ballplayer throws a ball straight up, by how much does the speed of the ball decrease each second while it is ascending? In the absence of air resistance, by how much does its speed increase each second while it is descending? How much time is required for its ascent? How much time is required for its descent?

29. ■ Someone standing at the edge of a cliff (as in Figure 1.24) throws a ball straight up at a certain speed and another ball straight down with the same initial speed. If air resistance is negligible, which ball has the greater speed when it strikes the ground below?

30. ● What is the acceleration of a car that moves at a steady velocity of 100 km/h for 100 s? Explain your answer, and state why this question is an exercise in careful reading as well as in physics.

31. ● Jacob says acceleration is how fast you go. Katelyn says acceleration is how fast you get fast. They look to you for confirmation. Who's correct?

32. ● For a freely falling object dropped from rest, what is its acceleration at the end of the 5th second of fall? At the end of the 10th second? Defend your answer (and distinguish between velocity and acceleration).

33. ■ Two balls, A and B, are released simultaneously from rest at the left end of the equal-length tracks A and B, as shown. Which ball will reach the end of its track first?

34. ■ Refer to the tracks above.
(a) Does ball B roll faster along the lower part of track B than ball A rolls along track A?
(b) Is the speed gained by ball B going down the extra dip the same as the speed it loses going up near the right-hand end—and doesn't this mean that the speed of balls A and B will be the same at the ends of both tracks?
(c) On track B, won't the average speed dipping down and up be greater than the average speed of ball A during the same time?
(d) So, overall, does ball A or ball B have the greater average speed? (Do you wish to change your answer to the previous exercise?)

PROBLEMS

● BEGINNER   ■ INTERMEDIATE   ◆ EXPERT

1. ● Find the net force produced by a 30-N force and a 20-N force in each of the following cases:
(a) Both forces act in the same direction.
(b) The two act in opposite directions.

2. ● A horizontal force of 100 N is required to push a piece of furniture across a floor at a constant velocity.
(a) What is the net force acting on the furniture?
(b) How much is the friction force that acts on the sliding furniture?
(c) How much friction acts on the furniture when it is at rest on a horizontal surface?

3. ● Calculate the average speed of a tennis ball that travels the full length of the court, 24 m, in 0.5 s.

4. ● Calculate the average speed (in kilometers per hour) of Larry, who runs to the store 4 km away in 30 min.

5. ● Calculate the acceleration of a ball that starts from rest and rolls down a ramp and gains a speed of 25 m/s in 5 s.

6. ● Extend Table 1.2 (which gives values from 0 to 5 s) to 0 to 10 s, assuming no air resistance.

7. ■ Lillian rides her bicycle along a straight road at an average velocity v.
(a) Write an equation showing how far she travels in time t.
(b) If Lillian's average speed is 7.5 m/s for a time of 5.0 min, show that she travels a distance of 2300 m.

8. ■ A race car races on a circular racetrack of radius r.
(a) Write an equation for the car's average speed when it travels a complete lap in time t.
(b) The radius of the track is 400 m and the time to make a lap is 40 s. Show that the average speed around the track is 63 m/s.

9. ■ A ball is thrown straight up with an initial speed of 30 m/s.
(a) Show that the time it takes to reach the top of its trajectory will be 3 seconds.
(b) Show that it will reach a height of 45 m (neglecting air resistance).

10. ■ A ball is thrown straight up with enough speed so that it is in the air for several seconds.
(a) What is the velocity of the ball when it reaches its highest point?
(b) What is its velocity 1 s before it reaches its highest point?
(c) What is the change in its velocity during this 1-s interval?
(d) What is its velocity 1 s after it reaches its highest point?
(e) What is the change in velocity during this 1-s interval?
(f) What is the change in velocity during the 2-s interval from 1 s before the highest point to 1 s after the highest point? (Caution: we are asking for velocity, not speed.)

(g) What is the acceleration of the ball during any of these time intervals and at the moment the ball has zero velocity?

11. ◆ Starting from rest, the change in an object's velocity $= at$. That is, $v_f - v_0 = at$, or $t = \dfrac{v_f - v_0}{a}$. The distance traveled by an object is given by $d = v_{ave}t$, where $v_{ave} = \dfrac{v_f + v_0}{2}$. Begin with $d = v_{ave}t$ and with appropriate substitutions show that $d = \dfrac{v_f^2 - v_0^2}{2a}$.

Note that this equation does not include time, so it's a good one to use when time is not given in a problem!

12. ◆ An electrically charged particle accelerates uniformly from rest to speed v while traveling a distance x.

(a) Show that the acceleration of the particle is $a = \dfrac{v^2}{2x}$.

(b) If the particle starts from rest and reaches a speed of 1.8×10^7 m/s over a distance of 0.10 m, show that its acceleration is 1.6×10^{15} m/s^2.

ACTIVE EXPLORATIONS

1. Grandma is interested in your educational progress. Like most grandmothers, she likely has little science background and may be mathematically challenged. Write a letter to Grandma, without using equations, and explain to her the difference between velocity and acceleration. Tell her why some of your classmates confuse the two, and state some examples that clear the confusion.
2. Stand flatfooted next to a wall and make a mark at the highest point you can reach. Then jump vertically and

mark this highest point. The distance between the marks is your vertical jumping distance. Use this to calculate your hang time.
3. By any method you choose, determine both your walking speed and your running speed.
4. Go further than the previous activity, and try walking across a room with a constant acceleration. (Not so easy!)

READINESS ASSURANCE TEST (RAT)

If you have a good handle on this chapter, if you really do, then you should be able to score at least 7 out of 10 on this RAT. If you score less than 7, you need to study further before moving on.

Choose the BEST answer to the following.

1. The concept of inertia mostly involves
   (a) mass.
   (b) weight.
   (c) volume.
   (d) density.
2. The mass of 1 kg of iron on Earth
   (a) is less on the Moon.
   (b) is the same on the Moon.
   (c) is greater on the Moon.
   (d) weighs the same everywhere.
3. The weight of 1 kg of iron on Earth
   (a) is less on the Moon.
   (b) is the same on the Moon.
   (c) is greater on the Moon.
   (d) is the same everywhere.
4. When we say that 1 kg weighs 9.8 N, we mean that
   (a) 1 kg is 1 N.
   (b) it's true at Earth's surface.
   (c) it's true everywhere.
   (d) mass and weight are one and the same.
5. The equilibrium rule, $\Sigma F = 0$, applies to
   (a) objects or systems at rest.
   (b) objects or systems in uniform motion in a straight line.
   (c) both of these.
   (d) neither of these.

6. When you stand on two bathroom scales, one foot on each scale with weight evenly distributed, each scale will read
   (a) your weight.
   (b) half your weight.
   (c) zero.
   (d) actually more than your weight.
7. The difference between speed and velocity most involves
   (a) acceleration.
   (b) amount.
   (c) direction.
   (d) all of these.
8. When a ball increases in speed by the same amount each second, its acceleration
   (a) also increases each second.
   (b) decreases each second.
   (c) is constant.
   (d) none of these.
9. If a falling object gains 10 m/s each second it falls, its acceleration is
   (a) 10 m/s.
   (b) 10 m/s per second.
   (c) both of these.
   (d) neither of these.
10. A freely falling object has a speed of 30 m/s at one instant. Exactly 1 s later its speed will be
    (a) the same.
    (b) 35 m/s.
    (c) more than 35 m/s.
    (d) 60 m/s.

Answers to RAT
1a, 2b, 3a, 4b, 5c, 6b, 7c, 8c, 9b, 10c

CHAPTER 1 ONLINE RESOURCES

The Physical Science Place

Interactive Figures

- 1.21, 1.23, 1.24

Videos

- Newton's Law of Inertia
- The Old Tablecloth Trick
- Toilet Paper Roll
- Inertia of a Cylinder
- Inertia of an Anvil
- Definition of a Newton
- Definition of Speed
- Average Speed

- Velocity
- Changing Velocity
- Definition of Acceleration
- Numerical Example of Acceleration
- Free Fall: How Fast?
- Free Fall: How Far?
- Free-Fall Acceleration Explained

Quiz

Flashcards

Links

newton's laws of motion

2.1 NEWTON'S FIRST LAW OF MOTION

2.2 NEWTON'S SECOND LAW OF MOTION

2.3 FORCES AND INTERACTIONS

2.4 NEWTON'S THIRD LAW OF MOTION

2.5 SUMMARY OF NEWTON'S THREE LAWS

■ Galileo's work set the stage for Isaac Newton, who was born shortly after Galileo's death in 1642. By the time Newton was 23, he had developed his famous three laws of motion that completed the overthrow of Aristotelian physics. These three laws first appeared in one of the most famous books of all time, Newton's *Philosophiae Naturalis Principia Mathematica,** often simply known as the *Principia.* The first law is a restatement of Galileo's concept of inertia; the second law relates acceleration to its cause—force; and the third is the law of action and reaction.

Newton's three laws of motion are the foundation of present-day mechanics. It was Newton's laws that got humans to the Moon.

* The Latin title means "Mathematical Principles of Natural Philosophy." See Newton's biography on page 53.

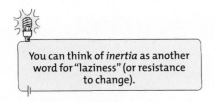

2.1 Newton's First Law of Motion

Newton's first law of motion, usually called the *law of inertia*, is a restatement of Galileo's idea.

Every object continues in a state of rest or of uniform speed in a straight line unless acted on by a nonzero force.

The key word in this law is *continues*: an object *continues* to do whatever it happens to be doing unless a force is exerted upon it. If the object is at rest, it *continues* in a state of rest. This is nicely demonstrated when a tablecloth is skillfully whipped from beneath dishes sitting on a tabletop, leaving the dishes in their initial state of rest.* On the other hand, if an object is moving, it *continues* to move without changing its speed or direction, as evidenced by space probes that continually move in outer space. This property of objects to resist changes in motion is called **inertia**.

FIGURE 2.1

Inertia in action.

CHECK POINT

When a space shuttle travels in a nearly circular orbit around Earth, is a force required to maintain its high speed? If suddenly the force of gravity were cut off, what type of path would the shuttle follow?

Was this your answer?

No force in the direction of the shuttle's motion exists. The shuttle "coasts" by its own inertia. The only force acting on it is the force of gravity, which acts at right angles to its motion (toward Earth's center). We'll see later that this right-angled force holds the shuttle in a circular path. If it were cut off, the shuttle would move in a straight-line path at constant speed (constant velocity).

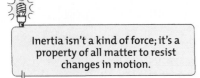

Why will the coin drop into the glass when a force accelerates the card?

Why is it that a slow continuous increase in the downward force breaks the string above the massive ball, but a sudden increase breaks the lower string?

Why does the downward motion and sudden stop of the hammer tighten the hammerhead?

FIGURE 2.2

Examples of inertia.

FIGURE 2.3

Rapid deceleration is sensed by the driver, who lurches forward—inertia in action!

* Close inspection shows that brief friction between the dishes and the fast-moving tablecloth starts the dishes moving, but then friction between the dishes and table stops the dishes before they slide very far. If you try this, use unbreakable dishes!

THE MOVING EARTH

As mentioned in the Prologue, the 16th-century Polish astronomer Coperni-cus caused great controversy when he published a book proposing that Earth revolves around the Sun.* This idea conflicted with the popular view that Earth was the center of the universe. Copernicus's concept of a Sun-centered solar system was the result of years of studying the motion of the planets. He had kept his theory from the public—for two reasons. The first reason was that he feared persecution; a theory so completely different from common opinion would surely be taken as an attack on established order. The second reason was reservations about it himself; he could not reconcile the idea of a moving Earth with the prevailing ideas of motion. The concept of inertia was unknown to him and others of his time. In the final days of his life, at the urging of close friends, he sent his manuscript, *De Revolutionibus Orbium Coelestium,*** to the printer. The first copy of his famous exposition reached him on the day he died—May 24, 1543.

Nicolaus Copernicus (1473–1543)

The idea of a moving Earth was much debated. Europeans thought like Aristotle, and the existence of a force big enough to keep Earth moving was beyond their imagination. They had no idea of the concept of inertia. One of the arguments against a moving Earth was the following: Consider a bird sit-ting at rest on a branch of a tall tree. On the ground below is a fat, juicy worm. The bird sees the worm and drops vertically below and catches it. It was argued that this would be impossible if Earth were moving. A moving Earth would have to travel at an enormous speed to circle the Sun in one year. While the bird would be in the air descending from its branch to the ground below, the worm would be swept far away along with the moving Earth. It seemed that catching a worm on a moving Earth would be an impossible task. The fact that birds do catch worms from tree branches seemed to be clear evi-dence that Earth must be at rest.

FIGURE 2.4

Can the bird drop down and catch the worm if Earth moves at 30 km/s?

Can you see the error in this argument? The concept of inertia is missing. You see, not only is Earth moving at a great speed, but so are the tree, the branch of the tree, the bird that sits on it, the worm below, and even the air in between. Things in motion remain in motion if no unbalanced forces are acting upon them. So when the bird drops from the branch, its initial sideways motion remains unchanged. It catches the worm quite unaffected by the motion of its total environment.

We live on a moving Earth. If you stand next to a wall and jump up so that your feet are no longer in contact with the floor, does the moving wall slam into you? Why not? It doesn't because you are also traveling at the same speed, before, during, and after your jump. The speed of Earth relative to the Sun is not the speed of the wall relative to you.

Four hundred years ago, people had difficulty with ideas like these. One reason is that they didn't yet travel in high-speed vehicles. Rather, they experienced slow, bumpy rides in horse-drawn carts. People were less aware of the effects of inertia. Today we flip a coin in a high-speed car, bus, or plane and catch the vertically moving coin as we would if the vehicle were at rest. We see evidence for the law of inertia when the horizontal motion of the coin prior, during, and after the catch is the same. The coin always keeps up with us.

FIGURE 2.5

When you flip a coin in a high-speed airplane, it behaves as if the airplane were at rest. The coin keeps up with you— inertia in action!

* Copernicus was certainly not the first to think of a Sun-centered solar system. In the fifth century, for example, the Indian astronomer Aryabhata taught that Earth circles the Sun, not the other way around (as the rest of the world believed).
** The Latin title means "On the Revolutions of Heavenly Spheres," by Nicolaus Copernicus.

The Physical Science Place
Parachuting and Newton's Second Law

The Physical Science Place
Newton's Second Law
Force Causes Acceleration
Friction
Falling and Air Resistance

∎ 2.2   Newton's Second Law of Motion

Isaac Newton was the first to realize the connection between force and mass in producing acceleration, which is one of the most central rules of nature. He expressed it in his *second law of motion*. **Newton's second law of motion** states the following:

> **The acceleration produced by a net force on an object is directly proportional to the net force, is in the same direction as the net force, and is inversely proportional to the mass of the object.**

Or, in shorter notation,

$$\text{Acceleration} \sim \frac{\text{net force}}{\text{mass}}$$

By using consistent units such as newtons (N) for force, kilograms (kg) for mass, and meters per second squared (m/s^2) for acceleration, we produce the exact equation

$$\text{Acceleration} = \frac{\text{net force}}{\text{mass}}$$

In briefest form, where a is acceleration, F is net force, and m is mass:

$$a = \frac{F}{m}$$

Acceleration equals the net force divided by the mass. If the net force acting on an object is doubled, the object's acceleration will be doubled. Suppose instead that the mass is doubled. Then the acceleration will be halved. If both the net force and the mass are doubled, then the acceleration will be unchanged. (These relations are nicely developed in the *Conceptual Physical Science Practice Book*.)

FIGURE 2.6
INTERACTIVE FIGURE
Acceleration depends on both the amount of push and the mass being pushed.

Force *changes* motion, it doesn't *cause* motion.

Force of hand accelerates the brick

Twice as much force produces twice as much acceleration

Twice the force on twice the mass gives the same acceleration

FIGURE 2.7
Acceleration is directly proportional to force.

Force of hand accelerates the brick

The same force accelerates 2 bricks 1/2 as much

3 bricks, 1/3 as much acceleration

FIGURE 2.8
Acceleration is inversely proportional to mass.

CHECK POINT

1. In the previous chapter we defined acceleration to be the time rate of change of velocity; that is, a = (change in v)/time. Are we now saying that acceleration is instead the ratio of force to mass—that is, $a = F/m$? Which is it?

2. A jumbo jet cruises at constant velocity of 1000 km/h when the thrusting force of its engines is a constant 100,000 N. What is the acceleration of the jet? What is the force of air resistance on the jet?

3. Suppose you apply the same amount of force to two separate carts, one cart with a mass of 1 kg and the other with a mass of 2 kg. Which cart will accelerate more, and how much greater will the acceleration be?

Were these your answers?

1. Both are correct. Acceleration is *defined* as the time rate of change of velocity and is *produced by* a force. How much force/mass (usually the cause) determines the rate change in velocity/time (usually the effect). So we must first define acceleration and then define the terms that produce acceleration.

2. The acceleration is zero, as evidenced by the constant velocity. Because the acceleration is zero, it follows from Newton's second law that the net force is zero, which means that the force of air resistance must just equal the thrusting force of 100,000 N and act in the opposite direction. So the air resistance on the jet is 100,000 N. This is in accord with $\Sigma F = 0$. (Note that we don't need to know the velocity of the jet to answer this question, but only that it is constant—our clue that acceleration, and therefore net force, is zero.)

3. The 1-kg cart will have more acceleration—twice as much, in fact—because it has half as much mass, which means half as much resistance to a change in motion.

WHEN ACCELERATION IS g—FREE FALL

Although Galileo founded the concepts of both inertia and acceleration and was the first to measure the acceleration of falling objects, he was unable to explain why objects of various masses fall with equal accelerations. Newton's second law provides the explanation.

We know that a falling object accelerates toward Earth because of the gravitational force of attraction between the object and Earth. As mentioned earlier, when the force of gravity is the only force—that is, when air resistance is negligible—we say that the object is in a state of **free fall**. An object in free fall accelerates toward Earth at 10 m/s² (or, more precisely, at 9.8 m/s²).

The greater the mass of an object, the stronger is the gravitational pull between it and Earth. The double brick in Figure 2.9, for example, has twice the gravitational attraction of the single brick. Why, then, doesn't the double brick fall twice as fast (as Aristotle supposed it would)? The answer is evident in Newton's second law: the acceleration of an object depends not only on the force (weight, in this case), but on the object's resistance to motion—its inertia. Whereas a force produces an acceleration, inertia is a *resistance* to acceleration. So twice the force exerted on twice the inertia produces the same acceleration as half the force exerted on half the inertia. Both accelerate equally. The acceleration due to gravity is symbolized by g. We use the symbol g, rather than a, to denote that acceleration is due to gravity alone.

The ratio of weight to mass for freely falling objects equals the constant g. This is similar to the constant ratio of circumference to diameter for circles,

Here's directly proportional.

Here's inversely proportional.

When one thing is **inversely proportional** to another, then as one gets bigger, the other gets smaller.

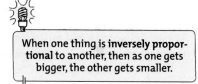

$$\frac{F}{m} = g \qquad \frac{2\,F}{2\,m} = g$$

FIGURE 2.9

INTERACTIVE FIGURE

The ratio of weight (F) to mass (m) is the same for all objects in the same locality; hence, their accelerations are the same in the absence of air resistance.

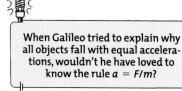
When Galileo tried to explain why all objects fall with equal accelerations, wouldn't he have loved to know the rule $a = F/m$?

FIGURING PHYSICAL SCIENCE

■ Problem Solving

If we know the mass of an object in kilograms (kg) and its acceleration in meters per second per second (m/s^2), then the force will be expressed in newtons (N). One newton is the force needed to give a mass of 1 kg an acceleration of 1 m/s^2. We can arrange Newton's second law to read

Force = mass × acceleration

$$1\,N = (1\,kg) \times (1\,m/s^2)$$

We can see that

$$1\,N = 1\,kg \cdot m/s^2$$

The dot between kg and m/s^2 means that the units are multiplied together.

If we know two of the quantities in Newton's second law, we can calculate the third.

SAMPLE PROBLEM 1

How much force, or thrust, must a 20,000-kg jet plane develop to achieve an acceleration of 1.5 m/s^2?

SOLUTION:

Using the equation

force = mass × acceleration

we can calculate the force:

$$F = ma$$
$$= (20,000\,kg) \times (1.5\,m/s^2)$$
$$= 30,000\,kg \cdot m/s^2$$
$$= 30,000\,N$$

Suppose we know the force and the mass, and we want to find the acceleration. For example, what acceleration is produced by a force of 2000 N applied to a 1000-kg automobile? Using Newton's second law, we find that

$$a = \frac{F}{m} = \frac{2000\,N}{1000\,kg}$$
$$= \frac{2000\,kg \cdot m/s^2}{1000\,kg} = 2\,m/s^2$$

If the force is 4000 N, the acceleration is

$$a = \frac{F}{m} = \frac{4000\,N}{1000\,kg}$$
$$= \frac{4000\,kg \cdot m/s^2}{1000\,kg} = 4\,m/s^2$$

Doubling the force on the same mass simply doubles the acceleration.

Physics problems are typically more complicated than these.

SAMPLE PROBLEM 2

Here is a more conceptual problem. It is conceptual because it deals not in numbers, but in concepts directly. The focus is showing symbols for concepts, rather than their numerical values. In the sample problem below, force is F, mass is m, and acceleration is a. This way you build a habit of first thinking in terms of concepts and the symbols that represent them. Part (b) follows up and brings in the numbers after you've done the physics.

A force F acts in the forward direction on a carton of chocolates of mass m. A friction force f opposes this motion.
(a) Use Newton's second law and show that the acceleration of the carton is

$$\frac{F-f}{m}.$$

(b) If the carton's mass is 4.0 kg, the applied force is 12.0 N, and the friction force is 6.0 N, show that the carton's acceleration is 1.5 m/s^2.

SOLUTION:

(a) We're asked to find the acceleration. From Newton's second law we know that $a = \frac{F_{net}}{m}$. Here the net force is $F - f$. So the solution is $a = \frac{F-f}{m}$ (where all quantities represented are known values). Notice that this answer applies to all situations in which a steady applied force is opposed by a steady frictional force. It covers many possibilities.
(b) Here we simply substitute the numerical values given:

$$a = \frac{F-f}{m} = \frac{12.0\,N - 6.0\,N}{4.0\,kg}$$
$$= 1.5\frac{N}{kg} = 1.5\,m/s^2.$$

(The units N/kg are equivalent to m/s^2.) Note that the answer, about 15 percent of g, is "reasonable." For information on units of measurement and significant figures, see Appendix A.

which equals the constant π. The ratio of weight to mass is identical for both heavy and light objects, just as the ratio of circumference to diameter is the same for both large and small circles (Figure 2.10).

We now understand that the acceleration of free fall is independent of an object's mass. A boulder 100 times as massive as a pebble falls at the same acceleration as the pebble because although the force on the boulder (its weight) is 100 times the force (or weight) on the pebble, its resistance to a change in motion (mass) is 100 times that of the pebble. The greater force offsets the correspondingly greater mass.

The speed of a vertically thrown ball at the top of its path is zero. Is the acceleration there zero also? (Answer begins with an N.)

CHECK POINT

In a vacuum, a coin and a feather fall equally, side by side. Would it be correct to say that *equal forces of gravity* act on both the coin and the feather in a vacuum?

Was this your answer?

No, no, no—a thousand times no! These objects accelerate equally not because the forces of gravity on them are equal, but because the *ratios* of their weights to masses are equal. Although air resistance is not present in a vacuum, gravity is. (You'd know this if you placed your hand into a vacuum chamber and a cement truck rolled over it!) If you answered yes to this question, let this be a signal to be more careful when you think physics!

WHEN ACCELERATION OF FALL IS LESS THAN *g*—NON–FREE FALL

Most often, air resistance is not negligible for falling objects. Then the acceleration of fall is less. Air resistance depends primarily on two things: speed and surface area. When a skydiver steps from a high-flying plane, the air resistance on the skydiver's body builds up as the falling speed increases. The result is reduced acceleration. The acceleration can be reduced further by increasing surface area. A diver does this by orienting his or her body so more air is encountered—by spreading out like a flying squirrel. So air resistance depends on speed and the frontal area encountered by the air.

For free fall, the downward net force is weight—only weight. But when air is present, the downward net force = weight – air resistance. Can you see that the presence of air resistance reduces net force? And that less net force means less acceleration? So as a diver falls faster and faster, the acceleration of fall becomes less and less.* What happens to the net force if air resistance builds up to equal weight? The answer is that net force becomes zero. Here we see $\Sigma F = 0$ again! Then acceleration becomes zero. Does this mean the diver comes to a stop? No! What it means is that the diver no longer gains speed. Acceleration terminates—it no longer occurs. We say the diver has reached **terminal speed**. If we are concerned with direction—down, for falling objects—we say the diver has reached **terminal velocity**.

Terminal speed for a human skydiver varies from about 150 to 200 km/h, depending on weight, size, and orientation of the body. A heavier person has to fall faster for air resistance to balance weight.** The greater weight is more effective in "plowing through" air, resulting in a higher terminal speed for a heavier person. Increasing frontal area reduces terminal speed. That's where a parachute is useful. A parachute increases frontal area, which greatly increases air resistance, reducing the terminal speed to a safe 15 to 25 km/h.

---

* In mathematical notation,

$$a = \frac{F_{\text{net}}}{m} = \frac{mg - R}{m}$$

where *mg* is the weight and *R* is the air resistance. Note that when $R = mg$, $a = 0$; then, with no acceleration, the object falls at constant velocity. With elementary algebra we proceed another step and get

$$a = \frac{F_{\text{net}}}{m} = \frac{mg - R}{m} = g - \frac{R}{m}$$

We see that the acceleration *a* will always be less than *g* if air resistance *R* impedes falling. Only when $R = 0$ does $a = g$.
** A skydiver's air resistance is proportional to speed squared.

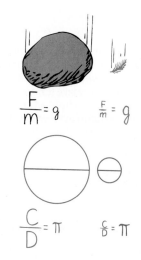

$$\frac{F}{m} = g \qquad \frac{F}{m} = g$$

$$\frac{C}{D} = \pi \qquad \frac{C}{D} = \pi$$

FIGURE 2.10

The ratio of weight (*F*) to mass (*m*) is the same for the large rock and the small feather; similarly, the ratio of circumference (*C*) to diameter (*D*) is the same for the large circle and the small circle.

In free fall, only a single force acts—the force of gravity. Whenever the force of air resistance also occurs, the falling object is not in free fall.

FIGURE 2.11

In a vacuum, a feather and a coin fall at an equal acceleration.

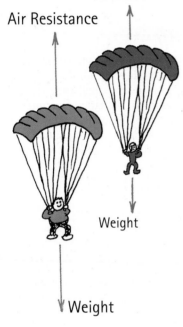

fyi

- Skydivers and flying squirrels are not alone in increasing their surface areas when falling. When the paradise tree snake (*Chrysopelea paradisi*) jumps from a tree branch, it doubles its width by flattening itself. It acquires a slightly concave shape and maneuvers itself by undulating in a graceful S shape, traveling more than 20 m in a single leap.

Air Resistance

Air Resistance

Weight

Weight

FIGURE 2.12

INTERACTIVE FIGURE

The heavier parachutist must fall faster than the lighter parachutist for air resistance to cancel her greater weight.

fyi

- Depending on the size and weight of packages dropped from airplanes, 160 km/h (100 miles per hour) is a typical terminal speed. That's about how fast a pitched baseball travels, or almost as fast as a tennis ball is served. Objects such as bags of rice and flour can survive this terminal speed, so parachutes are seldom used. In fact, parachutes are not used when dropping food supplies to citizens in the midst of an army that would confiscate the supplies.

CHECK POINT

A skydiver jumps from a high-flying helicopter. As she falls faster and faster through the air, does her acceleration increase, decrease, or remain the same?

Was this your answer?

Acceleration decreases because the net force on the skydiver decreases. Net force is equal to her weight minus her air resistance, and because air resistance increases with increasing speed, net force and hence acceleration decrease. By Newton's second law,

$$a = \frac{F_{\text{net}}}{m} = \frac{mg - R}{m}$$

where mg is her weight and R is the air resistance she encounters. As R increases, both net force and a decrease. Note that if she falls fast enough so that $R = mg$, $a = 0$, so with no acceleration she falls at constant speed.

Consider the interesting demonstration of the falling coin and feather in the glass tube (Figure 2.11). When air is inside, we see that the feather falls more slowly due to air resistance. The feather's weight is very small, so it reaches terminal speed very quickly. Can you see that it doesn't have to fall very far or fast before air resistance builds up to equal its small weight? The coin, on the other hand, doesn't have enough time to fall fast enough for air resistance to build up to equal its weight.

CHECK POINT

Consider two parachutists, a heavy person and a light person, who jump from the same altitude with parachutes of the same size.
1. Which person reaches terminal speed first?
2. Which person has the greater terminal speed?
3. Which person reaches the ground first?
4. If there were no air resistance, as on the Moon, how would your answers to these questions differ?

Were these your answers?

To answer these questions, think of a coin and a feather falling in air.
1. Just as a feather reaches terminal speed very quickly, the lighter person reaches terminal speed first.
2. Just as a coin falls faster than a feather through air, the heavier person falls faster and reaches a higher terminal speed.
3. Just like the race between a falling coin and feather, the heavier person falls faster and reaches the ground first.
4. If there were no air resistance there would be no terminal speed at all. Both would be in free fall and hit the ground at the same time.

When Galileo allegedly dropped objects of different weights from the Leaning Tower of Pisa, they didn't actually hit at the same time. They almost did, but because of air resistance, the heavier one hit a split second before the other. But this contradicted the much longer time difference expected by the followers of Aristotle. The behavior of falling objects was never really understood until Newton announced his second law of motion.

■ 2.3   Forces and Interactions

So far, we've treated force in its simplest sense—as a push or pull. In a broader sense, a force is not a thing in itself but makes up an **interaction** between one thing and another. If you push on a wall with your fingers, more is happening than you pushing on the wall. You're interacting with the wall, and the wall is also pushing on you. The fact that your fingers and the wall push on each other is evident in your bent fingers (Figure 2.14). These two forces are equal in magnitude (amount) and opposite in direction. This **force pair** constitutes a single interaction. In fact, you can't push on the wall unless the wall pushes back. A pair of forces is involved: your push on the wall and the wall's push back on you.*

In Figure 2.15, we see a boxer's fist hitting a massive punching bag. The fist hits the bag (and dents it) while the bag hits back on the fist (and stops its motion). This force pair is fairly large. But what if the boxer were hitting a piece of tissue paper? The boxer's fist can exert only as much force on the tissue paper as the tissue paper can exert on the boxer's fist. Furthermore, the fist can't exert any force at all unless what is being hit exerts the same amount of reaction force. An interaction requires a *pair* of forces acting on *two* objects.

When a hammer hits a stake and drives it into the ground, the stake exerts an equal amount of force on the hammer that brings it to an abrupt halt. And when you pull on a cart and it accelerates, the cart pulls back on you, as evidenced perhaps by the tightening of the rope wrapped around your hand. One thing interacts with another; the hammer interacts with the stake, and you interact with the cart.

FIGURE 2.13

A stroboscopic study of a golf ball (left) and a Styrofoam ball (right) falling in air. The air resistance is negligible for the heavier golf ball, and its acceleration is nearly equal to *g*. Air resistance is not negligible for the lighter Styrofoam ball, which reaches its terminal velocity sooner.

The Physical Science Place
Forces and Interactions

FIGURE 2.14

When you lean against a wall, you exert a force on the wall. The wall simultaneously exerts an equal and opposite force on you. Hence you don't topple over.

Can a boxer hurt his hand when punching a piece of tissue paper?

FIGURE 2.15

He can hit the massive bag with considerable force. But with the same punch he can exert only a tiny force on the tissue paper in midair.

* We tend to think of only living things pushing and pulling. But inanimate things can do likewise. So please don't be troubled about the idea of the inanimate wall pushing on you. It does, just as another person leaning against you would.

FIGURE 2.16

In the interaction between the hammer and the stake, each exerts the same amount of force on the other.

Which exerts the force and which receives the force? Isaac Newton's answer to this was that neither force has to be identified as "exerter" or "receiver," and he concluded that both objects must be treated equally. For example, when the hammer exerts a force on the stake, it is brought to a halt by the force the stake exerts on the hammer. Both forces are equal and oppositely directed. When you pull the cart, the cart simultaneously pulls on you. This pair of forces, your pull on the cart and the cart's pull on you, make up the single interaction between you and the cart. Such observations led Newton to his third law of motion.

2.4 Newton's Third Law of Motion

Newton's third law of motion states the following:

Whenever one object exerts a force on a second object, the second object exerts an equal and opposite force on the first.

We can call one force the *action force*, and the other the *reaction force*. Then we can express Newton's third law in the following form:

To every action there is always an opposed equal reaction.

The Physical Science Place
Newton's Third Law

It doesn't matter which force we call *action* and which we call *reaction*. The important thing is that they are co-parts of a single interaction and that neither force exists without the other. Action and reaction forces are equal in strength and opposite in direction. They occur in pairs and make up one interaction between two things.

When walking, you interact with the floor. Your push against the floor is coupled to the floor's push against you. The pair of forces occurs simultaneously. Likewise, the tires of a car push against the road while the road pushes back on the tires—the tires and the road push against each other. In swimming, you interact with the water that you push backward, while the water pushes you forward—you and the water push against each other. The reaction forces account for our motion in these cases. These forces depend on friction; a person or car on ice, for example, may not be able to exert the action force required to produce the needed reaction force. Neither force exists without the other.

When pushing my fingers together I see the same discoloration on each of them. Aha—evidence that each experiences the same amount of force!

SIMPLE RULE TO IDENTIFY ACTION AND REACTION

There is a simple rule for identifying action and reaction forces. First, identify the interaction—one thing (object A) interacts with another (object B). Then, action and reaction forces can be stated in the following form:

Action: Object A exerts a force on object B.

Reaction: Object B exerts a force on object A.

The rule is easy to remember. If action is A acting on B, reaction is B acting on A. We see that A and B are simply switched around. Consider the case of your hand pushing on the wall. The interaction is between your hand and the wall. We'll say the action is your hand (object A) exerting a force on the wall (object B). Then the reaction is the wall exerting a force on the your hand.

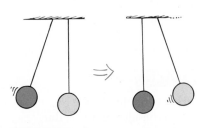

FIGURE 2.17

The impact forces between the blue ball and the yellow ball move the yellow ball and stop the blue ball.

Action: tire pushes on road   Reaction: road pushes on tire

Action: rocket pushes on gas   Reaction: gas pushes on rocket

Action: man pulls on spring   Reaction: spring pulls on man

↓ Action: Earth pulls on ball

↑ Reaction: ball pulls on Earth

FIGURE 2.18

Action and reaction forces. Note that when action is "A exerts force on B," the reaction is then simply "B exerts force on A."

Know that an action force and its reaction force always act on *different* objects. Two external forces acting on the same object, even if they are equal and opposite in direction, *cannot* be an action–reaction pair. That's the law!

FIGURE 2.19

Earth is pulled up by the boulder with just as much force as the boulder is pulled downward by Earth.

Hands-On Physical Science

Below we see two vectors on the sketch of the hand pushing the wall. The wall also pushes back on the hand.

Note that the others show only the action force. Draw appropriate vectors showing the reaction forces. Can you

specify the action–reaction pair in each case?

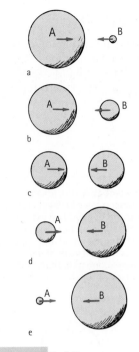

FIGURE 2.20

Which falls toward the other, planet A or planet B? Do the accelerations of each relate to their relative masses?

The Physical Science Place

Action and Reaction on
Different Masses
Action and Reaction on
Rifle and Bullet

CHECK POINT

■

1. A car accelerates along a road. Identify the force that moves the car.
2. Identify the action and reaction forces for the case of an object in free fall (no air resistance).

Were these your answers?

1. It is the road that pushes the car along. Really! Except for air resistance, only the road provides a horizontal force on the car. How does it do this? The rotating tires of the car push back on the road (action). The road simultaneously pushes forward on the tires (reaction). How about that!
2. To identify a pair of action–reaction forces in any situation, first identify the pair of interacting objects. In this case Earth interacts with the falling object via the force of gravity. So Earth pulls the falling object downward (call it *action*). Then *reaction* is the falling object pulling Earth upward. It is only because of Earth's enormous mass that you don't notice its upward acceleration.

ACTION AND REACTION ON DIFFERENT MASSES

Quite interestingly, a falling object pulls upward on Earth with as much force as Earth pulls downward on it. The resulting acceleration of a falling object is evident, while the upward acceleration of Earth is too small to detect.

Consider the exaggerated examples of two planetary bodies in parts (a) through (e) in Figure 2.20. The forces between planets A and B are equal in magnitude and oppositely directed in each case. If the acceleration of planet A is unnoticeable in part (a), then it is more noticeable in part (b), where the difference between the masses is less extreme. In part (c), where both bodies have equal mass, acceleration of planet A is as evident as it is for planet B. Continuing, we see that the acceleration of planet A becomes even more evident in part (d) and even more so in part (e). So, strictly speaking, when you step off the curb, the street rises ever so slightly to meet you.

When a cannon is fired, an interaction occurs between the cannon and the cannonball. The sudden force that the cannon exerts on the cannonball is exactly equal and opposite to the force the cannonball exerts on the cannon. This is why the cannon recoils (kicks). But the effects of these equal forces are very different. This is because the forces act on different masses. Recall Newton's second law,

$$a = \frac{F}{m}$$

Let *F* represent both the action and reaction forces, $\mathcal{M}$ the mass of the cannon, and *m* the mass of the cannonball. Different-sized symbols are

FIGURE 2.21

INTERACTIVE FIGURE

The force exerted against the recoiling cannon is just as great as the force that drives the cannonball along the barrel. Why, then, does the cannonball undergo more acceleration than the cannon?

used to indicate the relative masses and resulting accelerations. Then the acceleration of the cannonball and cannon can be represented in the following way.

$$\text{cannonball: } \frac{F}{m} = a$$

$$\text{cannon: } \frac{F}{m} = a$$

Thus we see why the change in velocity of the cannonball is so large compared with the change in velocity of the cannon. A given force exerted on a small mass produces a large acceleration, while the same force exerted on a large mass produces a small acceleration.

We can extend the idea of a cannon recoiling from the ball it fires to understanding rocket propulsion. Consider an inflated balloon recoiling when air is expelled (Figure 2.22). If the air is expelled downward, the balloon accelerates upward. The same principle applies to a rocket, which continually "recoils" from the ejected exhaust gas. Each molecule of exhaust gas is like a tiny cannonball shot from the rocket (Figure 2.23).

A common misconception is that a rocket is propelled by the impact of exhaust gases against the atmosphere. In fact, before the advent of rockets, it was commonly thought that sending a rocket to the Moon was impossible. Why? Because there is no air above Earth's atmosphere for the rocket to push against. But this is like saying a cannon wouldn't recoil unless the cannonball had air to push against. Not true! Both the rocket and recoiling cannon accelerate because of the reaction forces exerted by the material they fire—not because of any pushes on the air. In fact, a rocket operates better above the atmosphere where there is no air resistance.

CHECK POINT

1. Which pulls harder, the Moon on Earth, or Earth on the Moon?
2. A high-speed bus and an unfortunate bug have a head-on collision. The force of the bus on the bug splatters it all over the windshield. Is the corresponding force of the bug on the bus greater, less, or the same? Is the resulting deceleration of the bus greater than, less than, or the same as that of the bug?

Were these your answers?

1. Each pull is the same in magnitude. This is like asking which distance is greater, from New York to San Francisco or from San Francisco to New York. So we see that Earth and the Moon simultaneously pull on each other, each with the *same* amount of force.
2. The magnitudes of the forces are the same, for they constitute an action–reaction force pair that makes up the interaction between the bus and the bug. The accelerations, however, are very different because the masses are different! The bug undergoes an enormous and lethal deceleration, while the bus undergoes a very tiny deceleration—so tiny that the very slight slowing of the bus is unnoticed by its passengers. But if the bug were more massive, as massive as another bus, for example, the slowing down would be quite apparent.

FIGURE 2.22

The balloon recoils from the escaping air and climbs upward.

fyi

- Gases and fragments shoot out in all directions when a firecracker explodes. When fuel in a rocket burns, a slower explosion, exhaust gases shoot out in one direction.

FIGURE 2.23

The rocket recoils from the "molecular cannonballs" it fires and rises.

Hands-On Physical Science

▪ Tug-of-War

Perform a tug-of-war between guys and gals. Do it on a polished floor that's somewhat slippery, with guys wearing socks and gals wearing rubber-soled shoes. Who will surely win, and why? (Hint: Who wins a tug-of-war, those who pull harder on the rope or those who push harder against the floor?)

FIGURE 2.24

INTERACTIVE FIGURE

A force acts on the orange system and it accelerates to the right.

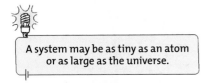

A system may be as tiny as an atom or as large as the universe.

DEFINING YOUR SYSTEM

An interesting question often arises: if action and reaction forces are equal and opposite, why don't they cancel to zero? To answer this question we must consider the *system* involved. Consider, for example, a system consisting of a single orange (Figure 2.24). The dashed line surrounding the orange encloses and defines the system. The vector that pokes outside the dashed line represents an external force on the system. The system accelerates in accord with Newton's second law. In Figure 2.25 we see that this force is provided by an apple, which doesn't change our analysis. The apple is outside the system. The fact that the orange simultaneously exerts a force on the apple, which is external to the system, may affect the apple (another system), but not the orange. You can't cancel a force on the orange with a force on the apple. So in this case, the action and reaction forces don't cancel.

Now let's consider a larger system, enclosing both the orange and the apple. We see the system bounded by the dashed line in Figure 2.26. Notice that the force pair is *internal* to the orange–apple system. These forces *do* cancel each other. They play no role in accelerating the system. A force external to the system is needed for acceleration. That's where friction with the floor comes into play (Figure 2.27). When the apple pushes against the floor, the floor simultaneously pushes on the apple—an external force on the system. The system accelerates to the right.

FIGURE 2.25

INTERACTIVE FIGURE

The force on the orange, provided by the apple, is not canceled by the reaction force on the apple. The orange still accelerates.

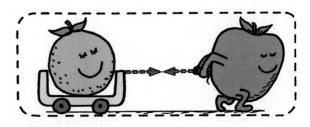

FIGURE 2.26

INTERACTIVE FIGURE

In the larger system of orange + apple, action and reaction forces are internal and do cancel. If these are the only horizontal forces, with no external force, no net acceleration of the system occurs.

Inside a baseball, trillions of interatomic forces are at play. They hold the ball together, but they play no role in accelerating the ball. Although every one of the interatomic forces is part of an action–reaction pair within the ball, they combine to zero, no matter how many of them there are. A force external to the ball, such as batting it, is needed to accelerate it.

If this is confusing, it may be well to note that Newton had difficulties with the third law himself.

FIGURE 2.27

INTERACTIVE FIGURE

An external horizontal force occurs when the floor pushes on the apple (reaction to the apple's push on the floor). The orange–apple system accelerates.

CHECK POINT

1. On a cold, rainy day, your car battery is dead, and you must push the car to move it and get it started. Why can't you move the car by remaining comfortably inside and pushing against the dashboard?
2. Does a fast-moving baseball possess force?

Were these your answers?

1. In this case, the system to be accelerated is the car. If you remain inside and push on the dashboard, the force pair you produce acts and reacts within the system. These forces cancel out, as far as any motion of the car is concerned. To accelerate the car, there must be an interaction between the car and something external—for example, you on the outside pushing against the road.
2. No, a force is not something an object *has*, like mass; it is part of an interaction between one object and another. A speeding baseball may possess the capability of exerting a force on another object when interaction occurs, but it does not possess force as a thing in itself. As we will see in the following chapters, moving things possess momentum and kinetic energy.

Using Newton's third law, we can understand how a helicopter gets its lifting force. The whirling blades are shaped to force air particles down (action), and the air forces the blades up (reaction). This upward reaction force is called *lift*. When lift equals the weight of the craft, the helicopter hovers in midair. When lift is greater, the helicopter climbs upward.

This is true for birds and airplanes. Birds fly by pushing air downward. The air simultaneously pushes the bird upward. When the bird is soaring, the wing must be shaped so that moving air particles are deflected downward. Slightly tilted wings that deflect oncoming air downward produce lift on an airplane. Air that is pushed downward continuously maintains lift. This supply of air is obtained by the forward motion of the aircraft, which results from propellers or jets that push air backward. When the propellers or jets push air backward, the air simultaneously pushes the propellers or jets forward. We will learn in Chapter 5 that the curved surface of a wing is an airfoil, which enhances the lifting force.

FIGURE 2.28

Ducks fly in a V formation because air pushed downward at the tips of their wings swirls upward, creating an updraft that is strongest off to the side of the bird. A trailing bird gets added lift by positioning itself in this updraft, pushes air downward and creates another updraft for the next bird, and so on. The result is a flock flying in a V formation.

We see Newton's third law in action everywhere. A fish propels water backward with its fins, and the water propels the fish forward. The wind caresses the branches of a tree, and the branches caress back on the wind to produce whistling

FIGURE 2.29

You cannot touch without being touched—Newton's third law.

sounds. Forces are interactions between different things. Every contact requires at least a twoness; there is no way that an object can exert a force on nothing. Forces, whether large shoves or slight nudges, always occur in pairs, each opposite to the other. Thus, we cannot touch without being touched.

■ 2.5 Summary of Newton's Three Laws

Newton's first law, the law of inertia: An object at rest tends to remain at rest; an object in motion tends to remain in motion at constant speed along a straight-line path. This property of objects to resist change in motion is called *inertia*. Mass is a measure of inertia. Objects undergo changes in motion only in the presence of a net force.

Newton's second law, the law of acceleration: When a net force acts on an object, the object accelerates. The acceleration is directly proportional to the net force and inversely proportional to the mass. Symbolically, $a \sim F/m$. Acceleration is always in the direction of the net force. When an object falls in a vacuum, the net force is simply the weight, and the acceleration is g (the symbol g denotes that acceleration is due to gravity alone.) When an object falls in air, the net force is equal to the weight minus the force of air resistance, and the acceleration is less than g. If and when the force of air resistance equals the weight of a falling object, acceleration terminates, and the object falls at constant speed (called the *terminal speed*).

Newton's third law, the law of action–reaction: Whenever one object exerts a force on a second object, the second object exerts an equal and opposite force on the first. Forces occur in pairs: one is an action and the other is a reaction, which together constitute the interaction between one object and the other. Action and reaction always act on different objects. Neither force exists without the other.

There has been a lot of new and exciting physics since the time of Isaac Newton. Nevertheless, and quite interestingly, as mentioned at the beginning of the chapter, it was primarily Newton's laws that got us to the Moon. Isaac Newton truly changed our way of viewing the world.

> A scientist has a compelling need to understand what makes the world tick.

Hands-On Activity

If you drop a sheet of paper and a book side by side, the book falls faster than the paper. Why? The book falls faster because of its greater weight compared to the air resistance it encounters. If you place the paper against the lower surface of the raised book and again drop them at the same time, it will be no surprise that they hit the surface below at the same time. The book simply pushes the paper with it as it falls. Now, repeat this, only with the paper on *top* of the book, not sticking over its edge. How will the accelerations of the book and paper compare? Will they separate and fall differently? Will they have the same acceleration? Try it and see! Then explain what happens.

Isaac Newton (1642–1727)

On Christmas Day in the year 1642, the year that Galileo died, Isaac Newton was prematurely born and barely survived. Newton's birthplace was his mother's farmhouse in Woolsthorpe, England. His father died several months before his birth, and he grew up under the care of his mother and grandmother. As a child he showed no particular signs of brightness, and at age $14\frac{1}{2}$ he was taken out of school to work on his mother's farm. As a farmer he was a failure, preferring to read books he borrowed from a neighboring pharmacist. An uncle sensed the scholarly potential in young Isaac and prompted him to study at the University of Cambridge, which he did for five years, graduating without particular distinction.

A plague swept through England, and Newton retreated to his mother's farm—this time to continue his studies. At the farm, at ages 23 and 24, he laid the foundations for the work that was to make him immortal. Seeing an apple fall to the ground led him to consider the force of gravity extending to the Moon and beyond. He formulated the law of universal gravitation. He invented calculus, a very important mathematical tool in science. He extended Galileo's work and developed the three fundamental laws of motion. He also formulated a theory of the nature of light and showed with prisms that white light is composed of all colors of the rainbow. It was his experiments with prisms that first made him famous.

When the plague subsided, Newton returned to Cambridge and soon established a reputation for himself as a first-

rate mathematician. His mathematics teacher resigned in his favor and Newton was appointed the Lucasian professor of mathematics. He held this post for 28 years. In 1672 he was elected to the Royal Society, where he exhibited the world's first reflector telescope. It can still be seen, preserved at the library of the Royal Society in London with the inscription: "The first reflecting telescope, invented by Sir Isaac Newton, and made with his own hands."

It wasn't until Newton was 42 that he began to write what is generally acknowledged as the greatest scientific book ever written, the *Philosophiae Naturalis Principia Mathematica*. He wrote the work in Latin and completed it in 18 months. It appeared in print in 1687 and wasn't printed in English until 1729, two years after his death. When asked how he was able to make so many discoveries, Newton replied that he solved his problems by continually thinking very long and hard about them—and not by sudden insight.

At age 46 he was elected a member of Parliament. He attended the sessions in Parliament for two years and never gave a speech. One day he rose and the house fell silent to hear the great man. Newton's "speech" was

very brief; he simply requested that a window be closed because of a draft.

A further turn from his work in science was his appointment as warden and then as master of the mint. Newton resigned his professorship and directed his efforts toward greatly improving the workings of the mint, to the dismay of counterfeiters who flourished at that time. He maintained his membership in the Royal Society and was elected president, then re-elected each year for the rest of his life. At age 62, he wrote *Opticks*, which summarized his work on light. Nine years later he wrote a second edition to his *Principia*.

Although Newton's hair turned gray at age 30, it remained full, long, and wavy all his life. Unlike others in his time, he did not wear a wig. He was a modest man, very sensitive to criticism, and he never married. He remained healthy in body and mind into old age. At age 80, he still had all his teeth, his eyesight and hearing were sharp, and his mind was alert. In his lifetime he was regarded by his countrymen as the greatest scientist who ever lived. In 1705 he was knighted by Queen Anne. Newton died at age 85 and was buried in Westminster Abbey along with England's kings and heroes.

Newton "opened up" the universe, showing that the same natural laws that act on Earth govern the larger cosmos as well. For humankind this led to increased humility, but also to hope and inspiration because of the evidence of a rational order. Newton ushered in the Age of Reason. His ideas and insights truly changed the world and elevated the human condition.

SUMMARY OF TERMS

Newton's first law of motion Every object continues in a state of rest, or in a state of motion in a straight line at constant speed, unless acted upon by a net force.

Inertia The property by which objects resist changes in motion.

Newton's second law of motion The acceleration produced by a net force on an object is directly proportional to the net force, is in the same direction as the net force, and is inversely proportional to the mass of the object.

Free fall Falling only under the influence of gravity—falling without air resistance.

Terminal speed The speed at which the acceleration of a falling object terminates when air resistance balances its weight.

Terminal velocity Terminal speed when direction is specified or implied.

Interaction Mutual action between objects during which each object exerts an equal and opposite force on the other.

Force pair The action and reaction pair of forces that occur in an interaction.

Newton's third law of motion Whenever one object exerts a force on a second object, the second object exerts an equal and opposite force on the first.

REVIEW QUESTIONS

2.1 Newton's First Law of Motion

1. State the law of inertia.
2. Is inertia a property of matter or a force of some kind?
3. What concept was missing from people's minds in the 16th century when they couldn't believe Earth was moving?
4. When a bird lets go of a branch and drops to the ground below, why doesn't the moving Earth sweep away from the dropping bird?
5. What kind of path would the planets follow if suddenly their attraction to the Sun no longer existed?

2.2 Newton's Second Law of Motion

6. State Newton's second law.
7. Is acceleration directly proportional to force, or is it inversely proportional to force? Give an example.
8. Is acceleration directly proportional to mass, or is it inversely proportional to mass? Give an example.
9. If the mass of a sliding block is tripled at the same time that the net force on it is tripled, how does the resulting acceleration compare to the original acceleration?
10. What is the net force that acts on a 10-N freely falling object?
11. Why doesn't a heavy object accelerate more than a light object when both are freely falling?
12. What is the net force that acts on a 10-N falling object when it encounters 4 N of air resistance? 10 N of air resistance?
13. What two principal factors affect the force of air resistance on a falling object?
14. What is the acceleration of a falling object that has reached its terminal velocity?
15. If two objects of the same size fall through air at different speeds, which encounters the greater air resistance?
16. Why does a heavy parachutist fall faster than a lighter parachutist who wears the same size parachute?

2.3 Forces and Interactions

17. Previously, we said that a force was a push or pull; now we say it is an interaction. Which is it? A push or pull, or an interaction? And what does it mean to say *interaction*?
18. How many forces are required for a single interaction?
19. When you push against a wall with your fingers, they bend because they experience a force. Identify this force.
20. A boxer can hit a heavy bag with great force. Why can't he hit a sheet of newspaper in midair with the same amount of force?

2.4 Newton's Third Law of Motion

21. State Newton's third law.
22. Consider hitting a baseball with a bat. If we call the force on the bat against the ball the action force, identify the reaction force.
23. If the forces that act on a cannonball and the recoiling cannon from which it is fired are equal in magnitude, why do the cannonball and cannon have very different accelerations?
24. Is it correct to say that action and reaction forces always act on different bodies? Defend your answer.
25. If body A and body B are both within a system, can forces between them affect the acceleration of the system?
26. What is necessary, forcewise, to accelerate a system?
27. When do action–reaction pairs of forces cancel each other and when do they not?
28. How does a helicopter get its lifting force?
29. What law of physics is inferred when we say you cannot touch without being touched?

2.5 Summary of Newton's Three Laws

30. Briefly summarize Newton's three laws of motion.

EXERCISES

● **BEGINNER**   ■ **INTERMEDIATE**   ◆ **EXPERT**

Again, please do not be intimidated by the large number of exercises and problems in this and other meatier chapters. If your course work is to cover many chapters, your instructor will likely assign only a few exercises and/or problems from each.

1. ● In the orbiting space shuttle, you are handed two identical closed boxes, one filled with sand and the other filled with feathers. How can you tell which is which without opening the boxes?

2. ● Your empty hand is not hurt when it bangs lightly against a wall. Why does your hand hurt if it is carrying a heavy load? Which of Newton's laws is most applicable here?

3. ● Why is a massive cleaver more effective for chopping vegetables than a lighter knife of the same sharpness?

4. ■ Each of the vertebrae forming your spine is separated from its neighbors by disks of elastic tissue. What happens, then, when you jump heavily on your feet from an elevated position? Can you think of a reason why you are a little shorter in the evening than you are in the morning? (Hint: Think about the hammerhead in Figure 2.2.)

5. ■ Before the time of Galileo and Newton, many learned scholars thought that a stone dropped from the top of a tall mast on a moving ship would fall vertically and hit the deck behind the mast by a distance equal to how far the ship had moved forward while the stone was falling. In light of your understanding of Newton's laws, what do you think about this idea?

6. ● As you stand on a floor, does the floor exert an upward force against your feet? How much force does it exert? Why are you not moved upward by this force?

7. ● A race car travels along a raceway at a constant velocity of 200 km/h. What horizontal forces act, and what is the net force acting on the car?

8. ● To pull a wagon across a lawn at a constant velocity, you must exert a steady force. Reconcile this fact with Newton's first law, which states that motion with a constant velocity indicates no force.

9. ● When your car moves along the highway at a constant velocity, the net force on it is zero. Why, then, do you continue running your engine?

10. ■ A rocket becomes progressively easier to accelerate as it travels through space. Why is this so? (Hint: About 90 percent of the mass of a newly launched rocket is fuel.)

11. ● When you toss a coin upward, what happens to its velocity while ascending? What happens to its acceleration? (Neglect air resistance.)

12. ● As you leap upward from the ground, how does the force that you exert on the ground compare with your weight?

13. ● A common saying goes, "It's not the fall that hurts you; it's the sudden stop." Translate this into Newton's laws of motion.

14. ■ On which of these hills does the ball roll down with increasing speed and decreasing acceleration along the path? (Use this example if you wish to explain to someone the difference between speed and acceleration.)

15. ■ If you drop an object, its acceleration toward the ground is 10 m/s^2. If you throw it down instead, would its acceleration after throwing be greater than 10 m/s^2? Ignore air resistance. Why or why not?

16. ■ Can you think of a reason why the acceleration of the object thrown downward through the air in the preceding exercise would actually be less than 10 m/s^2?

17. ■ If it were not for air resistance, would it be dangerous to go outdoors on rainy days? Defend your answer.

18. ■ What is the acceleration of a stone at the top of its trajectory when it has been thrown straight upward? (Is your answer consistent with Newton's second law?)

19. ■ Two 100-N weights are attached to a spring scale as shown. Does the scale read 0 N, 100 N, or 200 N, or does it give some other reading? (Hint: Would it read any differently if one of the ropes were tied to the wall instead of to the hanging 100-N weight?)

20. ■ You hold an apple over your head. (a) Identify all the forces acting on the apple and their reaction forces. (b) When you drop the apple, identify all the forces acting on it as it falls and the corresponding reaction forces.

21. ● What is the net force on an apple that weighs 1 N when you hold it at rest above your head? What is the net force on it when you release it?

22. ● Aristotle claimed that the speed of a falling object depends on its weight. We now know that objects in free fall, whatever their weights, undergo the same gain in speed. Why does weight not affect acceleration?

23. ● Does a stick of dynamite contain force? Defend your answer.

24. ● Can a dog wag its tail without the tail in turn "wagging the dog"? (Consider a dog with a relatively massive tail.)

25. ● When the athlete holds the barbell overhead, the reaction force is the weight of the barbell on his hand. How does this force vary for the case in which the barbell is accelerated upward? Downward?

26. ● Why can you exert greater force on the pedals of a bicycle if you pull up on the handlebars?

27. ■ The strong man will push apart the two initially stationary freight cars of equal mass before he himself drops straight to the ground. Is it possible for him to give either of the cars a greater speed than the other? Why or why not?

28. ■ Suppose two carts, one twice as massive as the other, fly apart when the compressed spring that joins them is released. How fast does the heavier cart roll compared with the lighter cart?

29. ■ If you exert a horizontal force of 200 N to slide a desk across an office floor at a constant velocity, how much friction does the floor exert on the desk? Is the force of friction equal and oppositely directed to your 200-N push? Does the force of friction make up the reaction force to your push? Why not?

30. ■ If a Mack truck and a motorcycle have a head-on collision, upon which vehicle is the impact force greater? Which vehicle undergoes the greater change in its motion? Explain your answers.

31. ● Two people of equal mass attempt a tug-of-war with a 12-m rope while standing on frictionless ice. When they pull on the rope, each person slides toward the other. How do their accelerations compare, and how far does each person slide before they meet?

32. ● Suppose that one person in the preceding exercise has twice the mass of the other. How far does each person slide before they meet?

33. ● Which team wins in a tug-of-war—the team that pulls harder on the rope or the team that pushes harder against the ground? Explain.

34. ● The photo shows Steve Hewitt and his daughter Gretchen. Is Gretchen touching her dad, or is he touching her? Explain.

35. ■ Why is it that a cat that falls from the top of a 50-story building will hit a safety net below no faster than if it fell from the 20th story?

36. ■ Free fall is motion in which gravity is the only force acting. (a) Explain why a skydiver who has reached terminal speed is not in free fall. (b) Explain why a satellite circling Earth above the atmosphere is in free fall.

37. ● How does the weight of a falling body compare with the air resistance it encounters just before it reaches terminal velocity? Just after?

38. ● You tell your friend that the acceleration of a skydiver decreases as falling progresses. Your friend then asks if this means that the skydiver is slowing down. What is your response?

39. ■ If and when Galileo dropped two balls from the top of the Leaning Tower of Pisa, air resistance was not really negligible. Assuming that both balls were the same size yet one was much heavier than the other, which ball struck the ground first? Why?

40. ■ If you simultaneously drop a pair of tennis balls from the top of a building, they strike the ground at the same time. If one of the tennis balls is filled with lead pellets, will it fall faster and hit the ground first? Which of the two will encounter more air resistance? Defend your answers.

PROBLEMS

● BEGINNER   ■ INTERMEDIATE   ◆ EXPERT

1. ● Calculate the weight of a 2000-kg elephant in newtons. What is its weight in pounds?

2. ● When two horizontal forces are exerted on a cart, 600 N forward and 400 N backward, the cart undergoes acceleration. What additional force is needed to produce nonaccelerated motion?

3. ● An astronaut of mass 100 kg recedes from her spacecraft by activating a small propulsion unit attached to her back. The force generated by a spurt is 25 N. Show that her acceleration is 0.25 m/s^2.

4. ● You push with a 20-N horizontal force on a 2-kg box of coffee resting on a horizontal surface. The force of friction on the box is 12 N. Show that the acceleration is 4 m/s^2.

5. ● Suppose that you push with a 40-N horizontal force on a 4-kg mass resting on a horizontal surface against a horizontal friction force of 24 N. Show that the acceleration is 4 m/s^2.

6. ● A rocket of mass 100,000 kg undergoes an acceleration of 2 m/s^2. Show that the force being developed by the rocket engines is 200,000 N.

7. ● During takeoff, a 747 jumbo jet of mass 330,000 kg experiences a 250,000-N thrust for each of four engines. Show that its acceleration is 3 m/s^2.

8. ● Calculate the horizontal force that must be applied to a 1-kg puck to make it accelerate on a horizontal friction-free air table with the same acceleration it would have if it were dropped and fell freely.

9. ● Irene exerts a net force of 10.0 N on a 6.70-kg shopping cart for 3.0 seconds. Show that the cart has an acceleration of 1.5 m/s^2.

10. ● Suppose that you are standing on a skateboard near a wall and that you push on the wall with a force of 30 N. How hard does the wall push on you? If your mass is 60 kg, show that your acceleration from the wall is 0.5 m/s^2.

11. ■ A firefighter of mass 80 kg slides down a vertical pole with an acceleration of 4 m/s^2. Show that the friction force that acts on the firefighter is 480 N.

12. ■ A boxer punches a sheet of paper in midair, bringing it from rest up to a speed of 25 m/s in 0.05 second. The mass of the paper is 0.003 kg. Show that the force of the punch on the paper is only 1.5 N.

13. ■ Suzie Skydiver with her parachute has a mass of 50 kg. (a) Before opening her chute, what force of air resistance will she encounter when she reaches terminal velocity? (b) What force of air resistance will she encounter when she reaches a lower terminal velocity after the chute is open? (c) Discuss why your answers are the same or different.

14. ◆ A falling 50-kg parachutist experiences an upward acceleration of 6.2 m/s^2 when she opens her parachute. Show that the drag force is 810 N when this occurs.

15. ◆ A force F acts in the forward direction on a cart of mass m. A friction force f opposes this motion. (a) Use Newton's second law and show that the acceleration of the cart is $\dfrac{F - f}{m}$.

(b) If the cart's mass is 4.0 kg, the applied force is 12.0 N, and the friction force is 6.0 N, show that the cart's acceleration is 1.5 m/s^2.

16. ◆ Phil and his rocket-powered sled have a combined mass M and are accelerating at a rate a when the sled runs into Mala, mass m, who tumbles aboard. (a) Show that the sled now accelerates at a rate equal to $\dfrac{M}{M + m}a$.

(b) If Phil and his sled have a combined mass of 70 kg, Mala's mass is 45 kg, and the initial acceleration of the sled was 3.6 m/s^2, show that when Mala joins Phil the acceleration of the sled is 2.2 m/s^2.

ACTIVE EXPLORATIONS

1. Write a letter to Grandma, similar to the one described in Activity 1 in Chapter 1. Tell her that Galileo introduced the concepts of acceleration and inertia and was familiar with forces, but didn't see the connection between these three concepts. Tell her how Isaac Newton did, and how the connection explains why heavy and light objects in free fall gain the same speed in the same time. In this letter, it's okay to use an equation or two, as long as you make it clear to Grandma that an equation is a shorthand notation of ideas you've explained.

2. The net force acting on an object and the resulting acceleration are always in the same direction. You can demonstrate this with a spool. If the spool is pulled horizontally to the right, in which direction will it roll?

3. Hold your hand with the palm down like a flat wing outside the window of a moving automobile. Then slightly tilt the front edge of your hand upward and notice the lifting effect as air is deflected downward from the bottom of your hand. Can you see Newton's laws at work here?

READINESS ASSURANCE TEST (RAT)

If you have a good handle on this chapter, if you really do, then you should be able to score at least 7 out of 10 on this RAT. If you score less than 7, you need to study further before moving on.

Choose the BEST answer to each of the following.

1. If gravity between the Sun and Earth suddenly vanished, Earth would continue moving in
   (a) a curved path.
   (b) a straight-line path.
   (c) an outward spiral path.
   (d) an inward spiral path.

2. If an object moves along a curved path, then it must be
   (a) accelerating.
   (b) acted on by a force.
   (c) both of these
   (d) none of these

3. A ball rolls down a curved ramp as shown. As its speed increases, its rate of gaining speed
   (a) increases.
   (b) decreases.
   (c) remains unchanged.

4. A heavy rock and a light rock in free fall have the same acceleration. The heavy rock does not have more acceleration because
   (a) the force of gravity on each is the same.
   (b) there is no air resistance.
   (c) the inertia of both rocks is the same.
   (d) all of these
   (e) none of these

5. When a 10-kg falling object encounters 10 N of air resistance, its acceleration is
   (a) less than g.
   (b) g.
   (c) more than g.
   (d) There is not enough information.

6. The amount of air resistance on a 0.8-N flying squirrel for terminal speed
   (a) is less than 0.8 N.
   (b) is 0.8 N.
   (c) is more than 0.8 N.
   (d) depends on the orientation of its body.

7. You drop a soccer ball off the edge of your school building. While falling,
   (a) its speed and acceleration both increase.
   (b) its speed increases and its acceleration decreases.
   (c) its speed and acceleration both decrease.
   (d) its speed decreases and its acceleration increases.

8. A karate chop delivers a force of 3000 N to a board that breaks. The force that acts on the hand during this event is
   (a) less than 3000 N.
   (b) 3000 N.
   (c) more than 3000 N.
   (d) There is not enough information.

9. A soccer ball is kicked to a 30-m/s speed. During the kick, the amount of force the player's foot exerts on the ball is
   (a) less than the amount of force on the foot.
   (b) the same as the amount of force on the foot.
   (c) more than the amount of force on the foot.
   (d) none of these

10. The force that propels a rocket is provided by
    (a) gravity.
    (b) Newton's laws of motion.
    (c) its exhaust gases.
    (d) the atmosphere against which the rocket pushes.

Answers to RAT
1b, 2c, 3b, 4c, 5a, 6b, 7b, 8b, 9b, 10c.

CHAPTER 2 ONLINE RESOURCES

The Physical Science Place

Interactive Figures
- 2.6, 2.9, 2.12, 2.21, 2.24, 2.25, 2.26, 2.27

Tutorials
- Parachuting and Newton's Second Law
- Newton's Third Law

Videos
- Newton's Second Law
- Force Causes Acceleration
- Friction

- Falling and Air Resistance
- Forces and Interactions
- Action and Reaction on Different Masses
- Action and Reaction on Rifle and Bullet

Quiz

Flashcards

Links

momentum
and energy

3.1 MOMENTUM AND IMPULSE

3.2 IMPULSE CHANGES MOMENTUM

3.3 CONSERVATION OF MOMENTUM

3.4 ENERGY AND WORK

3.5 WORK–ENERGY THEOREM

3.6 CONSERVATION OF ENERGY

3.7 POWER

3.8 MACHINES

3.9 EFFICIENCY

3.10 SOURCES OF ENERGY

■ We've learned that Galileo's concept of inertia is incorporated into Newton's first law of motion. We discussed inertia in terms of objects at rest and objects in motion. In this chapter, we will consider the inertia of moving objects. When we combine the ideas of inertia and motion, we are dealing with momentum. *Momentum* is a property of moving things. All things have energy, and when moving, they have energy of motion—*kinetic energy*. Things at rest have another kind of energy—*potential energy*. And all objects, whether at rest or moving, have an energy of being—$E = mc^2$. This chapter is about two of the most central concepts in mechanics—momentum and energy.

3.1 Momentum and Impulse

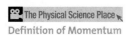
The Physical Science Place
Definition of Momentum

We know that it's harder to stop a large truck than a small car when both are moving at the same speed. We say the truck has more momentum than the car. By **momentum**, we mean *inertia in motion*, or more specifically, the mass of an object multiplied by its velocity.

$$\text{Momentum} = \text{mass} \times \text{velocity}$$

Or, in shorthand notation,

$$\text{Momentum} = mv$$

When direction is not an important factor, we can say

$$\text{Momentum} = \text{mass} \times \text{speed}$$

which we still abbreviate mv.*

We can see from the definition that a moving object can have a large momentum if it has a large mass, a high speed, or both. A moving truck has more momentum than a car moving at the same speed because the truck has more mass. But a fast car can have more momentum than a slow truck. And a truck at rest has no momentum at all.

If the momentum of an object changes, then either the mass or the velocity or both change. If the mass remains unchanged, as is most often the case, then the velocity changes and acceleration occurs. What produces acceleration? We know the answer is *force*. The greater the force acting on an object, the greater its change in velocity and, hence, the greater its change in momentum.

But something else is important in changing momentum: time—how long a time the force acts. If you apply a brief force to a stalled automobile, you produce a change in its momentum. Apply the same force over an extended period of time, and you produce a greater change in the automobile's momentum. A force sustained for a long time produces more change in momentum than does the same force applied briefly. So, both force and time interval are important in changing momentum.

The quantity *force × time interval* is called **impulse**. In shorthand notation,

$$\text{Impulse} = Ft$$

FIGURE 3.1

The boulder, unfortunately, has more momentum than the runner.

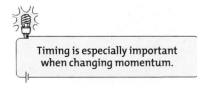

Timing is especially important when changing momentum.

FIGURE 3.2

When you push with the same force for twice the time, you impart twice the impulse and produce twice the change in momentum.

CHECK POINT

1. Compare the momentum of a 1-kg cart moving at 10 m/s with that of a 2-kg cart moving at 5 m/s.
2. Does a moving object have impulse?
3. Does a moving object have momentum?
4. For the same force, which cannon imparts a greater impulse to a cannonball—a long cannon or a short one?

Were these your answers?

1. Both have the same momentum (1 kg × 10 m/s = 2 kg × 5 m/s).
2. No, impulse is not something an object *has*, like momentum. Impulse is what an object can *provide* or what it can *experience* when it interacts with some other object. An object cannot possess impulse, just as it cannot possess force.
3. Yes, but, like velocity, in a relative sense—that is, with respect to a frame of reference, usually Earth's surface. The momentum possessed by a moving

---

* The symbol for momentum is p. In most physics textbooks, $p = mv$.

object with respect to a stationary point on Earth may be quite different from the momentum it possesses with respect to another moving object.

4. The long cannon imparts a greater impulse because the force acts over a longer time. (A greater impulse produces a greater change in momentum, so a long cannon imparts more speed to a cannonball than a short cannon does.)

■ 3.2 Impulse Changes Momentum

The greater the impulse exerted on something, the greater the change in momentum. The exact relationship is

$$\text{Impulse} = \text{change in momentum}$$

or*

$$Ft = \Delta(mv)$$

where Δ is the symbol for "change in."

The impulse–momentum relationship helps us analyze a variety of situations in which momentum changes. Here we will consider some ordinary examples in which impulse is related to increasing and decreasing momentum.

CASE 1: INCREASING MOMENTUM

To increase the momentum of an object, it makes sense to apply the greatest force possible for as long as possible. A golfer teeing off and a baseball player trying for a home run do both of these things when they swing as hard as possible and follow through with their swings. Following through extends the time of contact.

The forces involved in impulses usually vary from instant to instant. For example, a golf club that strikes a ball exerts zero force on the ball until it comes in contact; then the force increases rapidly as the ball is distorted (Figure 3.3). The force then diminishes as the ball comes up to speed and returns to its original shape. So when we speak of such forces in this chapter, we mean the *average* force.

CASE 2: DECREASING MOMENTUM OVER A LONG TIME

If you were in a car that was out of control and you had to choose between hitting a concrete wall or a haystack, you wouldn't have to call on your knowledge of physics to make up your mind. Common sense tells you to choose the haystack. But knowing the physics helps you understand *why* hitting a soft object is entirely different from hitting a hard one. In the case of hitting either the wall or the haystack and coming to a stop, it takes the *same* impulse to decrease your momentum to zero. The same impulse does not mean the same amount of force or the same amount of time; rather it means the same *product* of force and time. By hitting the haystack instead of the wall, you extend the *time during which your momentum is brought to zero*. A longer time interval reduces the force and decreases the resulting deceleration. For example, if the time interval is increased by a factor of 100, the force is reduced to a hundredth. Whenever we wish the force to be small, we extend the time of contact. Hence the reason for padded dashboards and airbags in motor vehicles.

When you jump from an elevated position down to the ground, what happens if you keep your legs straight and stiff? Ouch! Instead, you bend your knees

The Physical Science Place
Changing Momentum
Decreasing Momentum
Over a Short Time

The Physical Science Place
Momentum and
Collisions

FIGURE 3.3

The force of impact on a golf ball varies throughout the duration of impact.

---

* This relationship is derived by rearranging Newton's second law to make the time factor more evident. If we equate the formula for acceleration, $a = F/m$, with what acceleration actually is, $a = \Delta v/\Delta t$, we get $F/m = \Delta v/\Delta t$. From this we derive $F\Delta t = \Delta(mv)$. Calling Δt simply t, the time interval, $Ft = \Delta(mv)$.

If the change in momentum occurs over a long time, then the hitting force is small.

If the change in momentum occurs over a short time, then the hitting force is large.

when your feet make contact with the ground. By doing so you extend the time during which your momentum decreases to 10 to 20 times that of a stiff-legged, abrupt landing. The resulting force on your bones is reduced by a factor of 10 to 20. A wrestler thrown to the floor tries to extend his time of impact with the mat by relaxing his muscles and spreading the impact into a series of smaller ones as his foot, knee, hip, ribs, and shoulder successively hit the mat. Of course, falling on a mat is preferable to falling on a solid floor because the mat also increases the time during which the force acts.

The safety net used by circus acrobats is a good example of how to achieve the impulse needed for a safe landing. The safety net reduces the force experienced by a fallen acrobat by substantially increasing the time interval during which the force acts.

If you're about to catch a fast baseball with your bare hand, you extend your hand forward so you'll have plenty of room to let your hand move backward after you make contact with the ball. You extend the time of impact and thereby reduce the force of impact. Similarly, a boxer rides or rolls with the punch to reduce the force of impact (Figure 3.6).

CASE 3: DECREASING MOMENTUM OVER A SHORT TIME

When boxing, if you move into a punch instead of away, you're in trouble. It's the same as if you catch a high-speed baseball while your hand moves toward the ball instead of away upon contact. Or, when your car is out of control, if you drive it into a concrete wall instead of a haystack, you're really in trouble. In these cases of short impact times, the impact forces are large. Remember that for an object brought to rest, the impulse is the same no matter how it is stopped. But if the time is short, the force is large.

The idea of short time of contact explains how a karate expert can split a stack of bricks with the blow of her bare hand (Figure 3.7). She brings her arm and hand swiftly against the bricks with considerable momentum. This momentum is quickly reduced when she delivers an impulse to the bricks. The impulse is the force of her hand against the bricks multiplied by the time during which her hand

In both cases, the impulse provided by the boxer's jaw reduces the momentum of the punch. (a) When the boxer moves away (rides with the punch), he extends the time and diminishes the force. (b) If the boxer moves into the glove, the time is reduced and he must withstand a greater force.

makes contact with the bricks. By swift execution, she makes the time of contact very brief and correspondingly makes the force of impact huge. If her hand is made to bounce upon impact, as we will soon see, the force is even greater.

FIGURE 3.7

Cassy imparts a large impulse to the bricks in a short time and produces a considerable force.

CHECK POINT

1. If the boxer in Figure 3.6 increases the duration of impact to three times as long by riding with the punch, by how much is the force of impact reduced?
2. If the boxer instead moves *into* the punch to decrease the duration of impact by half, by how much is the force of impact increased?
3. A boxer being hit with a punch contrives to extend time for best results, whereas a karate expert delivers a force in a short time for best results. Isn't there a contradiction here?

Were these your answers?

1. The force of impact is only a third of what it would have been if he hadn't pulled back.
2. The force of impact is twice what it would have been if he had held his head still. Impacts of this kind account for many knockouts.
3. There is no contradiction because the best results for each are quite different. The best result for the boxer is reduced force, accomplished by maximizing time, and the best result for the karate expert is increased force delivered in minimum time.

BOUNCING

If a flowerpot falls from a shelf onto your head, you may be in trouble. If it bounces from your head, you may be in more serious trouble. Why? Because

---

FIGURING PHYSICAL SCIENCE

▪ Problem Solving

SAMPLE PROBLEM 1

An 8-kg bowling ball rolling at 2 m/s bumps into a padded guardrail and stops.
(a) What is the momentum of the ball just before hitting the guardrail?
(b) How much impulse acts on the ball?
(c) How much impulse acts on the guardrail?

SOLUTION:

(a) The momentum of the ball is $mv = (8\,\text{kg})(2\,\text{m/s}) = 16\,\text{kg} \cdot \text{m/s}$.
(b) In accord with the impulse–momentum relationship, the impulse on the ball is equal to its change in momentum. The momentum changes from 16 kg · m/s to zero. So $Ft = \Delta mv = (16\,\text{kg} \cdot \text{m/s}) - 0 = 16\,\text{kg} \cdot \text{m/s} = 16\,\text{N} \cdot \text{s}$. (Note that the units kg · m/s and N · s are equivalent.)
(c) In accord with Newton's third law, the force of the ball on the padded guardrail is equal and oppositely directed to the force of the guardrail on the ball. Because the time of the interaction is the same for both the ball and the guardrail, the impulses are also equal and opposite. So the amount of impulse on the ball is 16 N · s.

SAMPLE PROBLEM 2

An ostrich egg of mass m is thrown at a speed v into a sagging bedsheet and is brought to rest in time t.
(a) Show that the average force of egg impact is $\dfrac{mv}{t}$.
(b) If the mass of the egg is 1.0 kg, its speed when it hits the sheet is 2.0 m/s, and it is brought to rest in 0.2 s, show that the average force that acts is 10 N.
(c) Why is breakage less likely with a sagging sheet than with a taut one?

SOLUTION:

(a) From the impulse–momentum equation, $Ft = \Delta mv$, where in this case the egg ends up at rest, $\Delta mv = mv$, and simple algebraic rearrangement gives $F = \dfrac{mv}{t}$.

(b) $F = \dfrac{mv}{t} = \dfrac{(1.0\,\text{kg})\left(2.0\frac{\text{m}}{\text{s}}\right)}{(0.2\,\text{s})}$

$= 10\,\text{kg} \cdot \dfrac{\text{m}}{\text{s}^2} = 10\,\text{N}$.

(c) The time during which the tossed egg's momentum goes to zero is extended when it hits a sagging sheet. Extended time means less force in the impulse that brings the egg to a halt. Less force means less chance of breakage.

FIGURE 3.8

Howie Brand shows that the block topples when the swinging dart bounces from it. When he removes the rubber head of the dart so it doesn't bounce when it hits the block, no tipping occurs.

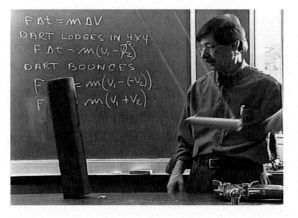

impulses are greater when an object bounces. The impulse required to bring an object to a stop and then to "throw it back again" is greater than the impulse required merely to bring the object to a stop. Suppose, for example, that you catch the falling pot with your hands. You provide an impulse to reduce its momentum to zero. If you throw the pot upward again, you have to provide additional impulse. This increased amount of impulse is the same that your head supplies if the flowerpot bounces from it.

The fact that impulses are greater when bouncing occurs was used with great success during the California gold rush. The waterwheels used in gold-mining operations were not very effective. A man named Lester A. Pelton recognized a problem with the flat paddles on the waterwheels. He designed a curved paddle that caused the incoming water to make a U-turn upon impact with the paddle. Because the water "bounced," the impulse exerted on the waterwheel was increased. Pelton patented his idea, and he probably made more money from his invention, the Pelton wheel, than any of the gold miners earned. Physics can indeed enrich your life in more ways than one.

FIGURE 3.9

The Pelton wheel. The curved blades cause water to bounce and make a U-turn, which produces a greater impulse to turn the wheel.

Impulse

CHECK POINT

1. In Figure 3.7, how does the force that Cassy exerts on the bricks compare with the force exerted on her hand?
2. How does the impulse resulting from the impact differ if her hand bounces back upon striking the bricks?

Were these your answers?

1. In accordance with Newton's third law, the forces are equal. Only the resilience of the human hand and the training she has undergone to toughen her hand allow this feat to be performed without broken bones.
2. The impulse is greater if her hand bounces back from the bricks upon impact. If the time of impact is not correspondingly increased, a greater force is then exerted on the bricks (and her hand!).

■ 3.3   Conservation of Momentum

Only an impulse external to a system can change the momentum of a system. Internal forces and impulses won't work. For example, consider the cannon being fired in Figure 3.10. The force on the cannonball inside the cannon barrel is equal and opposite to the force causing the cannon to recoil. Because these forces act for the same amount of time, the impulses are also equal and opposite. Recall Newton's third law about action and reaction forces. It applies to impulses, too. These impulses are internal to the system comprising the cannon and the cannonball, so they don't change the momentum of the cannon–cannonball system. Before the firing, the system is at rest and the momentum is zero. After the firing, the net momentum, or total momentum, is *still* zero. Net momentum is neither gained nor lost.

Momentum, like the quantities velocity and force, has both direction and magnitude. It is a *vector quantity*. Like velocity and force, momentum can be canceled. So although the cannonball in the preceding example gains momentum when fired and the recoiling cannon gains momentum in the opposite direction, there is no gain in the cannon–cannonball *system*. The momenta (plural form of *momentum*) of the cannonball and the cannon are equal in magnitude and opposite in direction.* They cancel to zero for the system as a whole. *If no net force or net impulse acts on a system, the momentum of that system cannot change.*

When momentum, or any quantity in physics, does not change, we say it is *conserved*. The idea that momentum is conserved when no external force acts is elevated to a central law of mechanics, called the **law of conservation of momentum**, which states:

> **In the absence of an external force, the momentum of a system remains unchanged.**

For any system in which all forces are internal—as, for example, cars colliding, atomic nuclei undergoing radioactive decay, or stars exploding—the net momentum of the system before and after the event is the same.

The Physical Science Place
Bowling Ball and Conservation of Momentum
Conservation of Momentum: Numerical Example

FIGURE 3.10

INTERACTIVE FIGURE

The net momentum before firing is zero. After firing, the net momentum is still zero, because the momentum of the cannon is equal and opposite to the momentum of the cannonball.

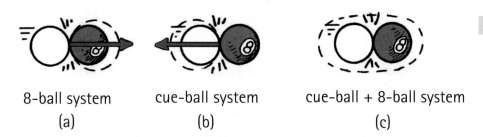

8-ball system
(a)

cue-ball system
(b)

cue-ball + 8-ball system
(c)

FIGURE 3.11

A cue ball hits an eight ball head-on. Consider this event in three systems: (a) An external force acts on the eight-ball system, and its momentum increases. (b) An external force acts on the cue-ball system, and its momentum decreases. (c) No external force acts on the cue-ball + eight-ball system, and momentum is conserved (simply transferred from one part of the system to the other).

---

* Here we neglect the momentum of ejected gases from the exploding gunpowder, which can be considerable. Firing a gun with blanks at close range is a definite no-no because of the considerable momentum of ejecting gases. More than one person has been killed by close-range firing of blanks. In 1998, a minister in Jacksonville, Florida, dramatizing his sermon before several hundred parishioners, including his family, shot himself in the head with a blank round from a .357-caliber Magnum. Although no slug emerged from the gun, exhaust gases did—enough to be lethal. So, strictly speaking, the momentum of the bullet (if any) + the momentum of the exhaust gases is equal to the opposite momentum of the recoiling gun.

CHECK POINT

1. Newton's second law states that if no net force is exerted on a system, no acceleration occurs. Does it follow that no change in momentum occurs?
2. Newton's third law states that the force a cannon exerts on a cannonball is equal and opposite to the force the cannonball exerts on the cannon. Does it follow that the *impulse* the cannon exerts on the cannonball is equal and opposite to the *impulse* the cannonball exerts on the cannon?

Were these your answers?

1. Yes, because no acceleration means that no change occurs in velocity or in momentum (mass × velocity). Another line of reasoning is simply that no net force means there is no net impulse and thus no change in momentum.
2. Yes, because the interaction between both occurs during the same *time* interval. Because time is equal and the forces are equal and opposite, the impulses, *Ft*, are also equal and opposite. Impulse is a vector quantity and can be canceled.

COLLISIONS

The collision of objects clearly illustrates the conservation of momentum. Whenever objects collide in the absence of external forces, the net momentum of both objects before the collision equals the net momentum of both objects after the collision.

$$\text{net momentum}_{\text{before collision}} = \text{net momentum}_{\text{after collision}}$$

This is true no matter how the objects might be moving before they collide.

When a moving billiard ball makes a head-on collision with another billiard ball at rest, the moving ball comes to rest and the other ball moves with the speed of the colliding ball. We call this an **elastic collision**; ideally, the colliding objects rebound without lasting deformation or the generation of heat (Figure 3.12). But momentum is conserved even when the colliding objects become entangled during the collision. This is an **inelastic collision**, characterized by deformation, or the generation of heat, or both. In a perfectly inelastic collision, the objects stick together. Consider, for example, the case of a freight car moving along a track and colliding with another freight car at rest (Figure 3.13). If the freight cars are of equal mass and are coupled by the collision, can we predict the velocity of the coupled cars after impact?

FIGURE 3.12

INTERACTIVE FIGURE

Elastic collisions of equally massive balls. (a) A green ball strikes a yellow ball at rest. (b) A head-on collision. (c) A collision of balls moving in the same direction. In each case, momentum is transferred from one ball to the other.

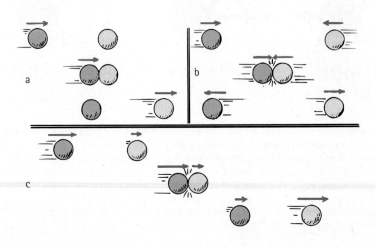

Conservation Laws

A conservation law specifies that certain quantities in a system remain precisely constant, regardless of what changes may occur within the system. It is a law of constancy during change. In this chapter, we see that momentum is unchanged during collisions. We say that momentum is conserved. We'll soon learn that energy is conserved as it transforms—the amount of energy in light, for example, transforms completely to thermal energy when the light is absorbed. In Appendix B we'll see that angular momentum is conserved—whatever the rotational motion of a planetary system, its angular momentum remains unchanged so long as it is free of outside influences. In Chapter 8, we'll learn that electric charge is conserved, which means that it can be nei-ther created nor destroyed. When we study nuclear physics, we'll see that these and other conservation laws rule in the submicroscopic world. Conservation laws are a source of deep insights into the simple regularity of nature and are often considered the most fundamental of physical laws. Can you think of things in your own life that remain constant as other things change?

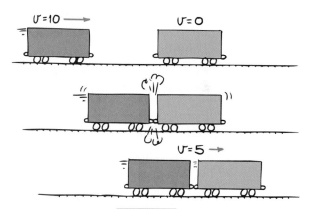

Suppose the single car is moving at 10 m/s, and we consider the mass of each car to be m. Then, from the conservation of momentum,

$$(\text{net } mv)_{\text{before}} = (\text{net } mv)_{\text{after}}$$

$$(m \times 10 \text{ m/s})_{\text{before}} = (2m \times V)_{\text{after}}$$

By simple algebra, $V = 5$ m/s. This makes sense: because twice as much mass is moving after the collision, the velocity must be half as much as the velocity before collision. Both sides of the equation are then equal.

FIGURE 3.13

INTERACTIVE FIGURE

Inelastic collision. The momentum of the freight car on the left is shared with the same-mass freight car on the right after collision.

> Momentum is conserved for all collisions, elastic and inelastic (whenever external forces don't interfere).

FIGURE 3.14

Will Maynez demonstrates his air track. Blasts of air from tiny holes provide a friction-free surface for the carts to glide on.

CHECK POINT
.

Consider the air track in Figure 3.14. Suppose a gliding cart with a mass of 0.5 kg bumps into, and sticks to, a stationary cart that has a mass of 1.5 kg. If the speed of the gliding cart before impact is v_{before}, how fast will the coupled carts glide after collision?

Was this your answer?

According to momentum conservation, the momentum of the 0.5-kg cart before the collision = momentum of both carts stuck together afterward.

$$(0.5 \text{ kg}) \, v_{before} = (0.5 \text{ kg} + 1.5 \text{ kg}) \, v_{after}$$

$$v_{after} = \frac{0.5 \text{ kg } v_{before}}{(0.5 \text{ kg} + 1.5 \text{ kg})} = \frac{0.5 \text{ kg } v_{before}}{2} = \frac{v_{before}}{4}$$

This makes sense, because four times as much mass will be moving after the collision, so the coupled carts will glide more slowly. The same momentum means that four times the mass glides $\frac{1}{4}$ as fast.

So we see that changes in an object's motion depend both on force and on how long the force acts. When "how long" means time, we refer to the quantity *force × time* as impulse. But "how long" can mean distance also. When we consider the quantity *force × distance*, we are talking about something entirely different—the concept of *energy*.

■ 3.4 Energy and Work

Link To Section 17.3

Perhaps the concept most central to all of science is energy. The combination of energy and matter makes up the universe: matter is substance, and energy is the mover of substance. The idea of matter is easy to grasp. Matter is stuff that we can see, smell, and feel. Matter has mass and occupies space. Energy, on the other hand, is abstract. We cannot see, smell, or feel most forms of energy. Surprisingly, the idea of energy was unknown to Isaac Newton, and its existence was still being debated in the 1850s. Although energy is familiar to us, it is difficult to define, because it is not only a "thing" but also both a thing and a process—similar to being both a noun and a verb. Persons, places, and things have energy, but we usually observe energy only when it is being transferred or being transformed. It appears in the form of electromagnetic waves from the Sun, and we feel it as thermal energy; it is captured by plants and binds molecules of matter together; it is in the foods we eat, and we receive it by digestion. Even matter itself is condensed, bottled-up energy, as set forth in Einstein's famous formula, $E = mc^2$, which we'll return to in the last part of this book. In general, **energy** is the property of a system that enables it to do *work*.

The Physical Science Place
Energy

When you push a crate across a floor you're doing work. By definition, *force × distance* equals the concept we call **work**.

When we lift a load against Earth's gravity, work is done. The heavier the load or the higher we lift the load, the more work is being done. Two things enter the picture whenever work is done: (1) application of a force and (2) the movement of something by that force. For the simplest case, in which the force is constant

The word *work*, in common usage, means physical or mental exertion. Don't confuse the physics definition of work with the everyday notion of work.

and the motion is in a straight line in the direction of the force,* we define the work done on an object by an applied force as the product of the force and the distance through which the object is moved. In shorter form:

$$\text{Work} = \text{force} \times \text{distance}$$

$$W = Fd$$

If we lift two loads one story up, we do twice as much work as we do in lifting one load the same distance, because the *force* needed to lift twice the weight is twice as much. Similarly, if we lift a load two stories instead of one story, we do twice as much work because the *distance* is twice as great.

We see that the definition of work involves both a force and a distance. A weightlifter who holds a barbell weighing 1000 N overhead does no work on the barbell. He may get really tired holding the barbell, but if it is not moved by the force he exerts, he does no work *on the barbell*. Work may be done on the muscles by stretching and contracting, which is force times distance on a biological scale, but this work is not done on the barbell. Lifting the barbell, however, is a different story. When the weightlifter raises the barbell from the floor, he does work on it.

The unit of measurement for work combines a unit of force (N) with a unit of distance (m); the unit of work is the newton-meter (N · m), also called the *joule* (J), which rhymes with *cool*. One joule of work is done when a force of 1 N is exerted over a distance of 1 m, as in lifting an apple over your head. For larger values, we speak of kilojoules (kJ, thousands of joules), or megajoules (MJ, millions of joules). The weightlifter in Figure 3.16 does work in kilojoules. To stop a loaded truck moving at 100 km/h requires megajoules of work.

FIGURE 3.15

He may expend energy when he pushes on the wall, but if the wall doesn't move, no work is done on the wall. Energy expended becomes thermal energy.

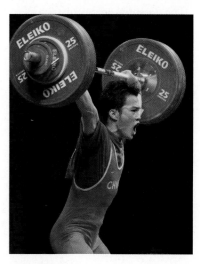

FIGURE 3.16

Work is done in lifting the barbell.

CHECK POINT

Assuming you have average strength, can you lift a 160-kg object with your bare hands? Can you do 1600 J of work on it?

Were these your answers?

An object with a mass of 160 kg weighs 1600 N, or 352 lb (the weight of a large refrigerator). So no, you cannot lift it without the use of some type of machine. If you can't move it, you can't do work on it. You'd do 1600 J of work on it if you could lift it a vertical distance of 1 m.

POTENTIAL ENERGY

An object may store energy by virtue of its position. The energy that is stored and held in readiness is called **potential energy** (PE) because in the stored state it has the potential for doing work. A stretched or compressed spring, for example, has the potential for doing work. When a bow is drawn, energy is stored in the bow. The bow can do work on the arrow. A stretched rubber band has

---

* More generally, work is the product of only the component of force that acts in the direction of motion and the distance moved. For example, if a force acts at an angle to the motion, the component of force parallel to the motion is multiplied by the distance moved. When a force acts at right angles to the direction of motion, with no force component in the direction of motion, no work is done. A common example is a satellite in a circular orbit; the force of gravity is at right angles to its circular path and no work is done on the satellite. Hence, it orbits with no change in speed.

The potential energy of Tenny's drawn bow equals the work (average force × distance) that she did in drawing the bow into position. When the arrow is released, most of the potential energy of the drawn bow will become the kinetic energy of the arrow.

An average apple weighs 1 N. When it is held 1 m above ground, then relative to the ground it has a PE of 1 J.

Gravitational potential energy always involves *two* interacting objects—one relative to the other. The ram of a pile driver, for example, interacts via gravitational force with Earth.

FIGURE 3.18

The potential energy of the 10-N ball is the same (30 J) in all three cases because the work done in elevating it 3 m is the same whether it is (a) lifted with 10 N of force, (b) pushed with 6 N of force up the 5-m incline, or (c) lifted with 10 N up each 1-m stair. No work is done in moving it horizontally (neglecting friction).

potential energy because of the relative position of its parts. If the rubber band is part of a slingshot, it is capable of doing work.

The chemical energy in fuels is also potential energy. It is actually energy of position at the submicroscopic level. This energy is available when the positions of electric charges within and between molecules are altered—that is, when a chemical change occurs. Any substance that can do work through chemical action possesses potential energy. Potential energy is found in fossil fuels, electric batteries, and the foods we consume.

Work is required to elevate objects against Earth's gravity. The potential energy due to elevated positions is called *gravitational potential energy*. Water in an elevated reservoir and the raised ram of a pile driver both have gravitational potential energy. Whenever work is done, energy is exchanged.

The amount of gravitational potential energy possessed by an elevated object is equal to the work done against gravity in lifting it. The work done equals the force required to move it upward multiplied by the vertical distance it is moved (remember $W = Fd$). The upward force required while moving at constant velocity is equal to the weight, mg, of the object, so the work done in lifting it through a height h is the product mgh.

$$\text{Gravitational potential energy} = \text{weight} \times \text{height}$$

$$PE = mgh$$

Note that the height is the distance above some chosen reference level, such as the ground or the floor of a building. The gravitational potential energy, mgh, is relative to that level and depends only on mg and h. We can see, in Figure 3.18, that the potential energy of the elevated ball does not depend on the path taken to get it there.

KINETIC ENERGY

If you push on an object, you can set it in motion. If an object is moving, then it is capable of doing work. It has energy of motion. We say it has *kinetic energy* (KE). The **kinetic energy** of an object depends on the mass of the object as well as its speed. It is equal to the mass multiplied by the square of the speed, multiplied by the constant $\frac{1}{2}$.

$$\text{Kinetic energy} = \frac{1}{2}\text{mass} \times \text{speed}^2$$

$$KE = \frac{1}{2}mv^2$$

(a)　　　　(b)　　　　(c)

PE = PE

FIGURE 3.19

He raises a block of ice by lifting it vertically. She pushes an identical block of ice up the ramp. Can you see that they do equal amounts of work? And can you see that when both blocks are raised to the same vertical height, they possess the same potential energy?

When you throw a ball, you do work on it to give it speed as it leaves your hand. The moving ball can then hit something and push it, doing work on what it hits. The kinetic energy of a moving object is equal to the work required to bring it from rest to that speed, or the work the object can do while being brought to rest:

Net force × distance = kinetic energy

or, in equation notation,

$$Fd = \frac{1}{2}mv^2$$

FIGURE 3.20

The potential energy of the elevated ram of the pile driver is converted to kinetic energy during its fall.

Note that the speed is squared, so if the speed of an object is doubled, its kinetic energy is quadrupled ($2^2 = 4$). Consequently, four times the work is required to double the speed. Whenever work is done, energy changes.

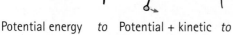

Potential energy *to* Potential + kinetic *to* Kinetic energy *to* Potential energy
And so on

FIGURE 3.21

Energy transitions in a pendulum. PE is relative to the lowest point of the pendulum, when it is vertical.

3.5 Work–Energy Theorem

When a car speeds up, its gain in kinetic energy comes from the work done on it. Or, when a moving car slows, work is done to reduce its kinetic energy. We can say*

Work = ΔKE

Work equals *change* in kinetic energy. This is the **work–energy theorem**.

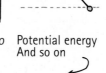

FIGURE 3.22

INTERACTIVE FIGURE

The pendulum bob will swing to its original height whether or not the peg is present.

FIGURE 3.23

The downhill "fall" of the roller coaster results in its roaring speed in the dip, and this kinetic energy sends it up the steep track to the next summit.

Link To Section 6.4

The work–energy theorem emphasizes the role of change. If there is no change in an object's energy, then we know that no net work was done on it. This theorem applies to changes in potential energy also. Recall our previous example of the weightlifter raising the barbell. When work was being done on the barbell, its potential energy was being changed. But when it was held stationary, no further work was being done on the barbell, as evidenced by no further change in its energy.

Similarly, if you push against a box on a floor and it doesn't slide, then you are not doing work on the box. There is no change in kinetic energy. But if you push harder and it slides, then you're doing work on it. When the amount of work done to overcome friction is small, the amount of work done on the box is practically matched by its gain in kinetic energy.

The work–energy theorem applies to decreasing speed as well. Energy is required to reduce the speed of a moving object or to bring it to a halt. When we apply the brakes to slow a moving car, we do work on it. This work is the friction force supplied by the brakes multiplied by the distance over which the friction force acts. The more kinetic energy something has, the more work is required to stop it.

Interestingly, the friction supplied by the brakes is the same whether the car moves slowly or quickly. Friction between solid surfaces doesn't depend on speed. The variable that makes a difference is the braking distance. A car moving at twice the speed of another takes four times ($2^2 = 4$) as much work to stop. Therefore, it takes four times as much distance to stop. Accident investigators are well aware that an automobile going 100 km/h has four times the kinetic energy it would have at 50 km/h. So a car going 100 km/h skids four times as far when its brakes are applied as it does when going 50 km/h. Kinetic energy depends on speed *squared*.

FIGURE 3.24

Because of friction, energy is transferred both into the floor and into the tire when the bicycle skids to a stop. An infrared camera reveals the heated tire track (the red streak on the floor, left) and the warmth of the tire (right). (Courtesy of Michael Vollmer.)

---

* This can be derived as follows: If we multiply both sides of $F = ma$ (Newton's second law) by d, we get $Fd = mad$. Recall from Chapter 2 that, for constant acceleration, $d = \frac{1}{2}at^2$, so we can say $Fd = ma(\frac{1}{2}at^2) = \frac{1}{2}maat^2 = \frac{1}{2}m(at)^2$; and substituting $v = at$, we get $Fd = \frac{1}{2}mv^2$. That is, work = KE, or more specifically, $W = \Delta KE$.

Automobile brakes convert kinetic energy to heat. Professional drivers are familiar with another way to slow a vehicle—shift to low gear to allow the engine to do the braking. Today's hybrid cars do the same and divert braking energy to electrical storage batteries, where it is used to complement the energy produced by gasoline combustion (Chapter 9 treats how they accomplish this).

Kinetic energy and potential energy are two of the many forms of energy, and they underlie other forms of energy, such as chemical energy, nuclear energy, sound, and light. Kinetic energy of random molecular motion is related to temperature; potential energies of electric charges account for voltage; and kinetic and potential energies of vibrating air define sound intensity. Even light energy originates from the motion of electrons within atoms. Every form of energy can be transformed into every other form.

CHECK POINT

1. When you are driving at 90 km/h, how much more distance do you need to stop than if you were driving at 30 km/h?
2. For the same force, why does a longer cannon impart more speed to a cannonball?

Were these your answers?

1. Nine times as much distance. The car has nine times as much kinetic energy when it travels three times as fast: $\frac{1}{2} m(3v)^2 = \frac{1}{2}m9v^2 = 9\left(\frac{1}{2}mv^2\right)$. The friction force is ordinarily the same in either case; therefore, nine times as much work requires nine times as much distance.
2. As learned earlier, a longer barrel imparts more impulse because of the longer *time* during which the force acts. The work–energy theorem similarly tells us that the longer the *distance* over which the force acts, the greater the change in kinetic energy. So we see two reasons for cannons with long barrels producing greater cannonball speeds.

KINETIC ENERGY AND MOMENTUM COMPARED

Momentum and kinetic energy are properties of moving things, but they differ from each other. Like velocity, momentum is a vector quantity and is therefore directional and capable of being canceled entirely. But kinetic energy is a non-vector (scalar) quantity, like mass, and can never be canceled. The momenta of two firecrackers approaching each other may cancel, but when they explode, there is no way their energies can cancel. Energies transform to other forms; momenta do not. Another difference is the velocity dependence of the two. Whereas momentum depends on velocity (mv), kinetic energy depends on the square of velocity $\left(\frac{1}{2}mv^2\right)$. An object that moves with twice the velocity of another object of the same mass has twice the momentum but four times the kinetic energy. So when a car traveling twice as fast crashes, it crashes with four times the energy.

If the distinction between momentum and kinetic energy isn't really clear to you, you're in good company. Failure to make this distinction resulted in disagreements and arguments between the best British and French physicists for two centuries.

FIGURE 3.25

The author puts kinetic energy and momentum into the hammer, which strikes the block that rests on physics buddy Paul Robinson, who is bravely sandwiched between beds of nails. Paul is not harmed. Why? Except for the flying cement fragments, every bit of the momentum of the hammer at impact is imparted to Paul, and subsequently to the table and Earth, which support him. But the momentum only provides the wallop; the energy does the damage. Most of the kinetic energy never gets to him, for it goes into smashing the block apart and into thermal energy. What energy remains is distributed over more than 200 nails that make contact with his body. The driving force per nail is not enough to puncture the skin.

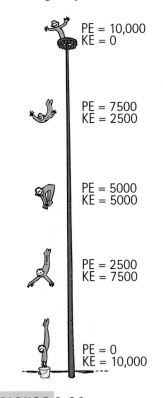

FIGURE 3.26

INTERACTIVE FIGURE

PE = 10,000
KE = 0

PE = 7500
KE = 2500

PE = 5000
KE = 5000

PE = 2500
KE = 7500

PE = 0
KE = 10,000

A circus diver at the top of a pole has a potential energy of 10,000 J. As he dives, his potential energy converts to kinetic energy. Note that, at successive positions one-fourth, one-half, three-fourths, and all the way down, the total energy is constant.

3.6　Conservation of Energy

Whenever energy is transformed or transferred, none is lost and none is gained. In the absence of work input or output or other energy exchanges, the total energy of a system before some process or event is equal to the total energy after.

Consider the changes in energy in the operation of the pile driver back in Figure 3.20. Work done to raise the ram, giving it potential energy, becomes kinetic energy when the ram is released. This energy transfers to the piling below. The distance the piling penetrates into the ground multiplied by the average force of impact is almost equal to the initial potential energy of the ram. We say *almost* because some energy goes into heating the ground and ram during penetration. Taking heat energy into account, we find that energy transforms without net loss or net gain. Quite remarkable!

The study of various forms of energy and their transformations has led to one of the greatest generalizations in physics—the **law of conservation of energy**:

Energy cannot be created or destroyed; it may be transformed from one form into another, but the total amount of energy never changes.

When we consider any system in its entirety, whether it be as simple as a swinging pendulum or as complex as an exploding supernova, one quantity isn't created or destroyed: energy. It may change form or it may simply be transferred from one place to another, but conventional wisdom tells us that the total energy score stays the same. This energy score takes into account the fact that the atoms that make up matter are themselves concentrated bundles of energy. When the nuclei (cores) of atoms rearrange themselves, enormous amounts of energy can be released. The Sun shines because some of this nuclear energy is transformed into radiant energy.

Enormous compression due to gravity and extremely high temperatures in the deep interior of the Sun fuse the nuclei of hydrogen atoms together to form helium nuclei. This is *thermonuclear fusion*, a process that releases radiant energy, a small part of which reaches Earth. Part of the energy reaching Earth falls on plants (and on other photosynthetic organisms), and part of this, in turn, is later stored in the form of coal. Another part supports life in the food chain that begins with plants (and other photosynthesizers), and part of this energy later is stored in oil. Part of the energy from the Sun goes into the evaporation of water from the ocean, and part of this returns to Earth in rain that may be trapped

---

FIGURING PHYSICAL SCIENCE

▪ Problem Solving

SAMPLE PROBLEM

Acrobat Art of mass *m* stands on the left end of a seesaw. Acrobat Bart of mass *M* jumps from a height *h* onto the right end of the seesaw, thus propelling Art into the air.
(a) Neglecting inefficiencies, how does the PE of Art at the top of his trajectory compare with the PE of Bart just before Bart jumps?
(b) Show that ideally Art reaches a height $\dfrac{M}{m}h$.

(c) If Art's mass is 40 kg, Bart's mass is 70 kg, and the height of the initial jump was 4 m, show that Art rises a vertical distance of 7 m.

SOLUTION:

(a) Neglecting inefficiencies, the entire initial PE of acrobat Bart before he drops goes into the PE of acrobat Art rising to his peak—that is, at Art's moment of zero KE.
(b) From $PE_{Bart} = PE_{Art} \Rightarrow Mgh_{Bart}$
$= mgh_{Art} \Rightarrow h_{Art} = \dfrac{M}{m}h.$

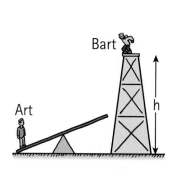

(c) $h_{Art} = \dfrac{M}{m}h = \left(\dfrac{70\text{ kg}}{40\text{ kg}}\right)4m = 7\text{ m}.$

---

behind a dam. By virtue of its elevated position, the water behind a dam has energy that may be used to power a generating plant below, where it is transformed to electric energy. The energy travels through wires to homes, where it is used for lighting, heating, cooking, and operating electrical gadgets. How wonderful that energy transforms from one form to another!

↻ Link To Section 17.5

▪ 3.7 Power

The definition of work says nothing about how long it takes to do the work. The same amount of work is done when carrying a bag of groceries up a flight of stairs, whether we walk up or run up. So why are we more out of breath after running upstairs in a few seconds than after walking upstairs in a few minutes? To understand this difference, we need to talk about a measure of how fast the work is done—*power*. **Power** is equal to the amount of work done per time it takes to do it:

$$\text{Power} = \frac{\text{work done}}{\text{time interval}}$$

The work done in climbing stairs requires more power when the worker is running up rapidly than it does when the worker is climbing slowly. A high-power automobile engine does work rapidly. An engine that delivers twice the power of another, however, does not necessarily move a car twice as fast or twice as far. Twice the power means that the engine can do twice the work in the same amount of time—or it can do the same amount of work in half the time. A powerful engine can produce greater acceleration.

Power is also the rate at which energy is changed from one form to another. The unit of power is the joule per second, called the *watt*. This unit was named in honor of James Watt, the 18th-century developer of the steam engine. One watt (W) of power is used when 1 J of work is done in 1 s. One kilowatt (kW) equals 1000 W. One megawatt (MW) equals 1 million watts.

fyi

▪ Your heart uses slightly more than 1 W of power in pumping blood through your body.

FIGURE 3.27

The three main engines of a space shuttle can develop 33,000 MW of power when fuel is burned at the enormous rate of 3400 kg/s. This is like emptying an average-size swimming pool in 20 s.

fyi

- Building a perpetual-motion machine (a device that can do work without energy input) is a no-no. But perpetual motion itself is a yes-yes. Atoms and their electrons, and stars and their planets, for example, are in a state of perpetual motion. Perpetual motion is the natural order of things.

The Physical Science Place
Machines: Pulleys

FIGURE 3.28

The lever.

$$\mathsf{{}^Fd = F^d}$$
50 N × 25 cm = 5000 N × 0.25 cm

FIGURE 3.29

Applied force × applied distance = output force × output distance.

fyi

- The principle of the lever was understood by Archimedes, a famous Greek scientist in the third century BC. He said, "Give me a place to stand, and I will move the Earth."

■ 3.8 Machines

A machine is a device for multiplying forces or simply changing the direction of forces. The principle underlying every machine is conservation of energy. Consider one of the simplest machines, the **lever** (Figure 3.28). At the same time that we do work on one end of the lever, the other end does work on the load. We see that the direction of force is changed: if we push down, the load is lifted up. If the little work done by friction forces is small enough to neglect, the work input equals the work output.

$$\text{Work input} = \text{work output}$$

Because work equals force times distance, **conservation of energy for machines** tells us that *input force × input distance = output force × output distance.*

$$(\text{force} \times \text{distance})_{\text{input}} = (\text{force} \times \text{distance})_{\text{output}}$$

The point of support on which a lever rotates is called the *fulcrum*. When the fulcrum of a lever is relatively close to the load, a small input force produces a large output force. This is because the input force is exerted through a large distance and the load is moved through a correspondingly short distance. So a lever can be a force multiplier. But no machine can multiply work or multiply energy. That's a conservation-of-energy no-no!

Today, a child can use the principle of the lever to jack up the front end of an automobile. By exerting a small force through a large distance, she can provide a large force that acts through a small distance. Consider the ideal example illustrated in Figure 3.29. Every time she pushes the jack handle down 25 cm, the car rises only a hundredth as far but with 100 times the force.

Another simple machine is a pulley. Can you see that it is a lever "in disguise"? When used as in Figure 3.30, it changes only the direction of the force; but, when used as in Figure 3.31, the output force is doubled. Force is increased and distance is decreased. As with any machine, forces can change while work input and work output are unchanged.

A block and tackle is a system of pulleys that multiplies force more than a single pulley can. With the ideal pulley system shown in Figure 3.32, the man pulls

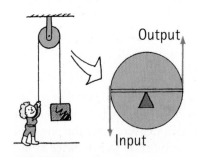

FIGURE 3.30

This pulley acts like a lever with equal arms. It changes only the direction of the input force.

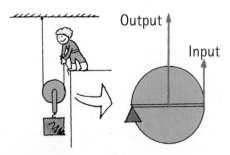

FIGURE 3.31

In this arrangement, a load can be lifted with half the input force. Note that the "fulcrum" is at the left end rather than in the center (as is the case in Figure 3.30).

7 m of rope with a force of 50 N and lifts a load of 500 N through a vertical distance of 0.7 m. The energy the man expends in pulling the rope is numerically equal to the increased potential energy of the 500-N block. Energy is transferred from the man to the load.

Any machine that multiplies force does so at the expense of distance. Likewise, any machine that multiplies distance, such as your forearm and elbow, does so at the expense of force. No machine or device can put out more energy than is put into it. No machine can create energy; it can only transfer energy or transform it from one form to another.

3.9 Efficiency

The three previous examples were of *ideal machines*; 100% of the work input appeared as work output. An ideal machine would operate at 100% efficiency. In practice, this doesn't happen, and we can never expect it to happen. In any transformation, some energy is dissipated to molecular kinetic energy—thermal energy. This makes the machine and its surroundings warmer.

Efficiency can be expressed by the ratio

$$\text{Efficiency} = \frac{\text{useful energy output}}{\text{total energy input}}$$

Even a lever converts a small fraction of input energy into heat when it rotates about its fulcrum. We may do 100 J of work but get out only 98 J. The lever is then 98% efficient, and we waste 2 J of work input as heat. In a pulley system, a larger fraction of input energy goes into heat. If we do 100 J of work, the forces of friction acting through the distances through which the pulleys turn and rub about their axles may dissipate 60 J of energy as heat. So the work output is only 40 J, and the pulley system has an efficiency of 40%. The lower the efficiency of a machine, the greater the amount of energy wasted as heat.*

A machine can multiply force, but never energy—no way!

FIGURE 3.32

Applied force × applied distance = output force × output distance.

Energy is nature's way of keeping score.

fyi

- Comparing transportation efficiencies, the most efficient is the human on a bicycle—far more efficient than train and car travel, and even that of fish and animals. Hooray for bicycles and cyclists who use them!

FIGURE 3.33

Energy transitions. The graveyard of mechanical energy is thermal energy.

* When you study thermodynamics in Chapter 6, you'll learn that an internal combustion engine *must* transform some of its fuel energy into thermal energy. A fuel cell, on the other hand, doesn't have this limitation. Watch for fuel cell–powered vehicles in the future!

An alternate definition of energy is anything that can be turned into heat.

CHECK POINT

Consider an imaginary miracle car that has a 100% efficient internal combustion engine and burns fuel that has an energy content of 40 megajoules per liter (MJ/L). If the air resistance and overall frictional forces on the car traveling at highway speed is 500 N, show that the distance the car could travel per liter at this speed is 80 km/L.

Was this your answer?

From the definition that *work = force × distance*, simple rearrangement gives *distance = work/force*. If all 40 million J of energy in 1 L were used to do the work of overcoming the air resistance and frictional forces, the distance would be

$$\text{Distance} = \frac{\text{work}}{\text{force}} = \frac{40{,}000{,}000 \text{ J/L}}{500 \text{ N}} = 80{,}000 \text{ m/L} = 80 \text{ km/L}$$

(This is about 190 miles per gallon [mpg].) The important point here is that, even with a hypothetically perfect engine, there is an upper limit of fuel economy dictated by the conservation of energy.

3.10 Sources of Energy

Except for nuclear power, the source of practically all our energy is the Sun. Even the energy we obtain from petroleum, coal, natural gas, and wood originally came from the Sun. That's because these fuels are created by photosynthesis—the process by which plants trap solar energy and store it as plant tissue.

Sunlight evaporates water, which later falls as rain; rainwater flows into rivers and into dams where it is directed to generator turbines. Then it returns to the sea, where the cycle continues. Even the wind, caused by unequal warming of Earth's surface, is a form of solar power. The energy of wind can be used to turn generator turbines within specially equipped windmills. Because wind power can't be turned on and off at will, it is presently a supplement to fossil and nuclear fuels for large-scale power production. Harnessing the wind is most practical when the energy it produces is stored for future use, such as in the form of hydrogen.

Hydrogen is the least polluting of all fuels. Most hydrogen in America is produced from natural gas, in a process that uses high temperatures and pressures to separate hydrogen from hydrocarbon molecules. The same is done with fossil fuels. A downside to separating hydrogen from carbon compounds is the unavoidable production of carbon dioxide, a greenhouse gas. A simpler and cleaner method that doesn't produce greenhouse gases is *electrolysis*—electrically splitting water into its constituent parts. Figure 3.34 shows how you can perform this in the lab or at home: Place two wires that are connected to the terminals of an ordinary battery into a glass of salted water. Be sure the wires don't touch each other. Bubbles of hydrogen form on one wire, and bubbles of oxygen form on the other. A fuel cell is similar, but runs backward. Hydrogen and oxygen gas are compressed at electrodes and electric current is produced, along with water. The space shuttle uses fuel cells to meet its electrical needs while producing drinking water for the astronauts. Here on Earth fuel-cell researchers are developing fuel cells for buses, automobiles, and trains.

fyi

- The power available in sunlight is about 1 kW/m². If all of the solar energy falling on a square meter could be harvested for power production, that energy would generate 1000 W. Some solar cells can convert 40% of the power, or about 400 W/m². Solar power via low-cost thin solar films used in building materials, including roofing and glass, are changing the way we produce and distribute energy.

Link To Sections 18.6, 18.7, 24.6

FIGURE 3.34

When electric current passes through conducting water, bubbles of hydrogen form at one wire and bubbles of oxygen form at the other. This is *electrolysis*. A fuel cell does the opposite—hydrogen and oxygen enter the fuel cell and are combined to produce electricity and water.

A hydrogen economy may likely start with railroad trains powered by fuel cells rather than with automobiles. Hydrogen can be obtained via solar cells, many along train tracks and on the rail ties themselves (Figure 3.35). Photovoltaic cells transform sunlight to electricity. They are familiar in solar-powered calculators, iPods, and flexible solar-powered shingles on rooftops. Solar cells can also supply the energy needed to produce hydrogen. It is important to know that hydrogen is not a *source* of energy. Energy is required to make hydrogen (to extract it from water and carbon compounds). As with electricity, the production of hydrogen needs an energy source; the hydrogen thus produced provides a way of storing and transporting that energy. Again, for emphasis, hydrogen is *not* an energy source.

The most concentrated source of usable energy is that stored in nuclear fuels—uranium and plutonium. For the same weight of fuel, nuclear reactions release about 1 million times more energy than do chemical or food reactions. Watch for renewed interest in this form of power that doesn't pollute the atmosphere. Interestingly, Earth's interior is kept hot because of nuclear power, which has been with us since time zero.

A byproduct of nuclear power in Earth's interior is geothermal energy. Geothermal energy is held in underground reservoirs of hot water. Geothermal energy is predominantly limited to areas of volcanic activity, such as Iceland, New Zealand, Japan, and Hawaii. In these locations, heated water near Earth's surface is tapped to provide steam for driving turbogenerators.

In locations where heat from volcanic activity is near the ground surface and groundwater is absent, another method holds promise for producing electricity: dry-rock geothermal power (Figure 3.36). With this method, water is put into cavities in deep, dry, hot rock. When the water turns to steam, it is piped to a turbine at the surface. After turning the turbine, it is returned to the cavity for reuse. In this way, electricity is produced inexpensively and cleanly.

As the world population increases, so does our need for energy, especially because per-capita demand is also growing. With the rules of physics to guide them, technologists are presently researching newer and cleaner ways to develop energy sources. But they race to keep ahead of a growing world population and

FIGURE 3.35

The power harvested by photovoltaic cells can be used to extract hydrogen for fuel-cell transportation. Plans for trains that run on solar power collected on railroad-track ties are presently at the drawing-board stage (see http://www.SuntrainUSA.com).

fyi

- Another source of energy is tidal power, by which the surging of tides turn turbines to produce power. Interestingly, this form of energy is neither nuclear nor from the Sun. It comes from the rotational energy of our planet.

FIGURE 3.36

Dry-rock geothermal power. (a) A hole is sunk several kilometers into dry granite. (b) Water is pumped into the hole at high pressure and fractures the surrounding rock to form a cavity with increased surface area. (c) A second hole is sunk to intercept the cavity. (d) Water is circulated down one hole and through the cavity, where it is superheated before rising through the second hole. After driving a turbine, it is recirculated into the hot cavity again, making a closed cycle.

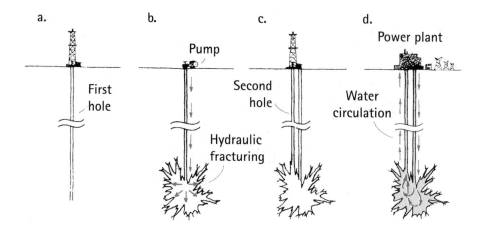

a.
First hole

b.
Pump
Hydraulic fracturing

c.
Second hole

d.
Power plant
Water circulation

Junk Science

Scientists have to be open to new ideas. That's how science grows. But a body of established knowledge exists that can't be easily overthrown. That includes energy conservation, which is woven into every branch of science and supported by countless experiments from the atomic to the cosmic scale. Yet no concept has inspired more "junk science" than energy. Wouldn't it be wonderful if we could get energy for nothing, to possess a machine that gives out more energy than is put into it? That's what many practitioners of junk science offer. Gullible investors put their money into some of these schemes. But none of them pass the test of being real science. Perhaps someday a flaw in the law of energy conservation will be discovered. If it ever is, scientists will rejoice at the breakthrough. But so far, energy conservation is as solid as any knowledge we have. Don't bet against it.

Inventors take heed: When introducing a new idea, first be sure it is in context with what is presently known. For example, it should be consistent with the conservation of energy.

greater demand in the developing world. Unfortunately, as long as controlling population is politically and religiously incorrect, human misery becomes the check to unrestrained population growth. H. G. Wells once wrote (in *The Outline of History*), "Human history becomes more and more a race between education and catastrophe."

SUMMARY OF TERMS

Momentum The product of the mass of an object and its velocity.

Impulse The product of the force acting on an object and the time during which it acts.

Law of conservation of momentum In the absence of an external force, the momentum of a system remains unchanged. Hence, the momentum before an event involving only internal forces is equal to the momentum after the event:

$$mv_{\text{(before event)}} = mv_{\text{(after event)}}$$

Elastic collision A collision in which colliding objects rebound without lasting deformation or the generation of heat.

Inelastic collision A collision in which the colliding objects become distorted, generate heat, and possibly stick together.

Energy The property of a system that enables it to do work.

Work The product of the force and the distance moved by the force:

$$W = Fd$$

(More generally, work is the component of force in the direction of motion times the distance moved.)

Potential energy The energy that matter possesses because of its position.

$$\text{Gravitational PE} = mgh$$

Kinetic energy Energy of motion, quantified by the relationship

$$\text{Kinetic energy} = \frac{1}{2}mv^2$$

Work–energy theorem The work done on an object equals the change in kinetic energy of the object.

$$\text{Work} = \Delta\text{KE}$$

(Work can also transfer other forms of energy to a system.)

Law of conservation of energy Energy cannot be created or destroyed; it may be transformed from one form into another, but the total amount of energy never changes.

Power The rate of doing work:

$$\text{Power} = \frac{\text{work}}{\text{time}}$$

(More generally, power is the rate at which energy is expended.)

Machine A device, such as a lever or pulley, that increases (or decreases) a force or simply changes the direction of a force.

Lever A simple machine consisting of a rigid rod pivoted at a fixed point called the fulcrum.

Conservation of energy for machines The work output of any machine cannot exceed the work input. In an ideal machine, where no energy is transformed into thermal energy, $\text{work}_{\text{input}} = \text{work}_{\text{output}}$; $(Fd)_{\text{input}} = (Fd)_{\text{output}}$.

Efficiency The percentage of the work put into a machine that is converted into useful work output. (More generally, useful energy output divided by total energy input.)

$$\text{Efficiency} = \frac{\text{useful energy output}}{\text{total energy input}}$$

REVIEW QUESTIONS

3.1 Momentum and Impulse

1. Which has a greater momentum, an automobile at rest or a moving skateboard?
2. When a ball is hit with a given force, why does contact over a long time impart more speed to the ball?

3.2 Impulse Changes Momentum

3. Why is it a good idea to have your hand extended forward when you are getting ready to catch a fast-moving baseball with your bare hand?
4. Why would it be a poor idea to have the back of your hand up against the outfield wall when you catch a long fly ball?
5. In karate, why is a force that is applied for a short time more advantageous?
6. In boxing, why is it advantageous to roll with the punch?
7. Which undergoes the greatest change in momentum: (1) a baseball that is caught, (2) a baseball that is thrown, or (3) a baseball that is caught and then thrown back, if all of the baseballs have the same speed just before being caught and just after being thrown?
8. In the preceding question, in which case is the greatest impulse required?

3.3 Conservation of Momentum

9. What does it mean to say that momentum (or any quantity) is *conserved*?
10. When a cannonball is fired, momentum is conserved for the *system* of cannon + cannonball. Would momentum be conserved for the system if momentum were not a vector quantity? Explain.
11. Railroad car A rolls at a certain speed and makes a perfectly elastic collision with car B of the same mass. After the collision, car A is observed to be at rest. How does the speed of car B compare with the initial speed of car A?
12. If the equally massive cars of the previous question stick together after colliding inelastically, how does their speed after the collision compare with the initial speed of car A?

3.4 Energy and Work

13. When is energy most evident?
14. Cite an example in which a force is exerted on an object without doing work on the object.
15. Which, if either, requires more work—lifting a 50-kg sack a vertical distance of 2 m or lifting a 25-kg sack a vertical distance of 4 m?
16. A car is raised a certain distance in a service-station lift and therefore has potential energy relative to the floor. If it were raised twice as high, how much potential energy would it have?
17. Two cars are raised to the same elevation on service-station lifts. If one car is twice as massive as the other, how do their potential energies compare?

18. A moving car has kinetic energy. If it speeds up until it is going four times as fast, how much kinetic energy does it have in comparison?

3.5 Work–Energy Theorem

19. Compared with some original speed, how much work must the brakes of a car supply to stop a car that is moving four times as fast? How does the stopping distance compare?
20. If you push a crate horizontally with 100 N across a 10-m factory floor, and friction between the crate and the floor is a steady 70 N, how much kinetic energy is gained by the crate?

3.6 Conservation of Energy

21. What will be the kinetic energy of a pile driver ram when it undergoes a 10 kJ decrease in potential energy?
22. An apple hanging from a limb has potential energy because of its height. If the apple falls, what becomes of this energy just before it hits the ground? When it hits the ground?

3.7 Power

23. If two equal-mass sacks are lifted equal distances in the same time, how does the power required for each compare? How does the power required compare for the case in which the lighter sack is moved its distance in half the time?

3.8 Machines

24. Can a machine multiply input force? Input distance? Input energy? (If your three answers are the same, seek help, for the last question is especially important.)
25. If a machine multiplies force by a factor of four, what other quantity is diminished, and by how much?

3.9 Efficiency

26. What is the efficiency of a machine that miraculously converts all the input energy to useful output energy?
27. What happens to the percentage of useful energy as it is transformed from one form to another?

3.10 Sources of Energy

28. What is the ultimate source of energies for the burning of fossil fuels, dams, and windmills?
29. What is the ultimate source of geothermal energy?
30. Can we correctly say that a new source of energy is hydrogen? Why or why not?

EXERCISES

● BEGINNER ■ INTERMEDIATE ◆ EXPERT

1. ● To bring a supertanker to a stop, its engines are typically cut off about 25 km from port. Why is it so difficult to stop or turn a supertanker?

2. ● In terms of impulse and momentum, why do airbags in cars reduce the chances of injury in accidents?

3. ● In terms of impulse and momentum, why are nylon ropes, which stretch considerably under tension, favored by mountain climbers?

4. ● Automobiles were previously manufactured to be as rigid as possible, whereas today's autos are designed to crumple upon impact. Why?

5. ● A lunar vehicle is tested on Earth at a speed of 10 km/h. When it travels as fast on the Moon, is its momentum more, less, or the same?

6. ● If you throw a raw egg against a wall, you'll break it; but, if you throw it with the same speed into a sagging sheet, it won't break. Explain, using concepts from this chapter.

7. ■ A boxer can punch a heavy bag for more than an hour without tiring, but tires quickly when boxing with an opponent for a few minutes. Why? (Hint: When the boxer's fist is aimed at the bag, what supplies the impulse to stop the punches? When the boxer's fist is aimed at the opponent, what or who supplies the impulse to stop the punches that don't connect?)

8. ■ Railroad cars are loosely coupled so that there is a noticeable time delay from the time the first car is moved until the last cars are moved from rest by the locomotive. Discuss the advisability of this loose coupling and slack between cars from the point of view of impulse and momentum.

9. ● If you throw a ball horizontally while standing on roller skates, you roll backward with a momentum that matches that of the ball. Will you roll backward if you go through the motions of throwing the ball, but instead hold on to it? Explain in terms of momentum conservation.

10. ● You are at the front of a floating canoe near a dock. You jump, expecting to land on the dock easily. Instead you land in the water. Explain in terms of momentum conservation.

11. ● A fully dressed person is at rest in the middle of a pond on perfectly frictionless ice and must get to shore. How can this be accomplished? Explain in terms of momentum conservation.

12. ■ The examples of the three previous exercises can be explained in terms of momentum conservation. Now also answer them in terms of Newton's third law.

13. ■ In the previous chapter, rocket propulsion was explained in terms of Newton's third law. That is, the force that propels a rocket is from the exhaust gases pushing against the rocket, the reaction to the force the rocket exerts on the exhaust gases. Explain rocket propulsion in terms of momentum conservation.

14. ■ Your friend says that the law of momentum conservation is violated when a ball rolls down a hill and gains momentum. What do you say?

15. ■ The momentum of an apple falling to the ground is not conserved because the external force of gravity acts on it. But momentum is conserved in a larger system. Explain.

16. ■ Drop a stone from the top of a high cliff. Identify the system wherein the net momentum is zero as the stone falls.

17. ■ Bronco dives from a hovering helicopter and finds his momentum increasing. Does this violate the conservation of momentum? Explain.

18. ■ An ice sailcraft is stalled on a frozen lake on a windless day. The skipper sets up a fan as shown. If all the wind bounces backward from the sail, will the craft be set in motion? If so, in what direction?

19. ■ Will your answer to the preceding exercise be different if the air is brought to a halt by the sail without bouncing?

20. ■ Discuss the advisability of simply removing the sail in the preceding exercises.

21. ■ To throw a ball, do you exert an impulse on it? Do you exert an impulse to catch it at the same speed? About how much impulse do you exert, in comparison, if you catch it and immediately throw it back again? (Imagine yourself on a skateboard.)

22. ◆ When vertically falling sand lands in a horizontally moving cart, the cart slows. Ignore any friction between the cart and the tracks. Give two reasons for this, one in terms of a horizontal force acting on the cart and one in terms of momentum conservation.

23. ◆ In a movie, the hero jumps straight down from a bridge onto a small boat that continues to move with no change in velocity. What physics is being violated here?

24. ◆ Suppose that three astronauts outside a spaceship decide to play catch. All the astronauts weigh the same on Earth and are equally strong. The first astronaut throws the second astronaut toward the third one and the game begins. Describe the motion of the astronauts as the game proceeds. How long will the game last?

25. ■ If your friend pushes a lawn mower four times as far as you do while exerting only half the force, which one of you does more work? How much more?

26. ● Which requires more work: stretching a strong spring a certain distance or stretching a weak spring the same distance? Defend your answer.

27. ● Two people who weigh the same climb a flight of stairs. The first person climbs the stairs in 30 s, while the second person climbs them in 40 s. Which person does more work? Which uses more power?

28. ■ When a rifle with a longer barrel is fired, the force of expanding gases acts on the bullet for a longer distance. What effect does this have on the velocity of the emerging bullet? (Do you see why long-range cannons have such long barrels?)

29. ■ A baseball and a golf ball have the same momentum. Which has the greater kinetic energy?

30. ● At what point in its motion is the KE of a pendulum bob at a maximum? At what point is its PE at a maximum? When its KE is at half its maximum value, how much PE does it possess?

31. ● A physics instructor demonstrates energy conservation by releasing a heavy pendulum bob, as shown in the sketch, allowing it to swing to and fro. What would happen if, in his exuberance, he gave the bob a slight shove as it left his nose? Explain.

32. ● Why does the force of gravity do work on a car that rolls down a hill, but no work when it rolls along a level part of the road?

33. ● On a playground slide, a child has potential energy that decreases by 1000 J while her kinetic energy increases by 900 J. What other form of energy is involved, and how much?

34. ■ Consider the identical balls released from rest on tracks A and B, as shown. When they reach the right ends of the tracks, which will have the greater speed? Why is this question easier to answer than the similar one (Exercise 33) in Chapter 1?

35. ■ If a golf ball and a Ping-Pong ball both move with the same kinetic energy, can you say which has the greater speed? Explain in terms of the definition of KE. Similarly, in a gaseous mixture of massive molecules and light molecules with the same average KE, can you say which have the greater speed?

36. ■ Does a car burn more gasoline when its lights are turned on? Does the overall consumption of gasoline depend on whether the engine is running while the lights are on? Defend your answer.

37. ■ This may seem like an easy question for a physics type to answer: With what force does a rock that weighs 10 N strike the ground if dropped from a rest position 10 m high? In fact, the question cannot be answered unless you have more information. What information, and why?

38. ■ In the absence of air resistance, a ball thrown vertically upward with a certain initial KE returns to its original level with the same KE. When air resistance is a factor affecting the ball, does it return to its original level with the same, less, or more KE? Does your answer contradict the law of energy conservation?

39. ■ You're on a rooftop and you throw one ball downward to the ground below and another upward. The second ball, after rising, falls and also strikes the ground below. If air resistance can be neglected, and if your downward and upward initial speeds are the same, how do the speeds of the balls compare upon striking the ground? (Use the idea of energy conservation to arrive at your answer.)

40. ■ When a driver applies brakes to keep a car going downhill at constant speed and constant kinetic energy, the potential energy of the car decreases. Where does this energy go? Where does most of it appear in a hybrid vehicle?

41. ■ Can something have energy without having momentum? Explain. Can something have momentum without having energy? Explain.

42. ● When the mass of a moving object is doubled with no change in speed, by what factor is its momentum changed? By what factor is its kinetic energy changed?

43. ● When the velocity of an object is doubled, by what factor is its momentum changed? By what factor is its kinetic energy changed?

44. ● Which, if either, has greater momentum: a 1-kg ball moving at 2 m/s or a 2-kg ball moving at 1 m/s? Which has greater kinetic energy?

45. ■ Two lumps of clay with equal and opposite momenta have a head-on collision and come to rest. Is momentum conserved? Is kinetic energy conserved? Why are your answers the same or different?

46. ■ If an automobile had a 100% efficient engine, transferring all of the fuel's energy to work, would the engine be warm to your touch? Would its exhaust heat the surrounding air? Would it make any noise? Would it vibrate? Would any of its fuel go unused?

47. ■ To combat wasteful habits, we often speak of "conserving energy," by which we mean turning off lights and hot water when they are not being used, and keeping thermostats at a moderate level. In this chapter, we also speak of "energy conservation." Distinguish between these two usages.

48. ■ Your friend says that one way to improve air quality in a city is to have traffic lights synchronized so that motorists can travel long distances at constant speed. What physics principle supports this claim?

49. ■ The energy we require to live comes from the chemically stored potential energy in food, which is transformed into other energy forms during the metabolism process. What happens to a person whose combined work and heat output is less than the energy consumed? What happens when the person's work and heat output is greater than the energy consumed? Can an undernourished person perform extra work without extra food? Defend your answers.

50. ◆ Consider the swinging-balls apparatus. If two balls are lifted and released, momentum is conserved as two balls pop out the other side with the same speed as the released balls at impact. But momentum would also be conserved if one ball popped out at twice the speed. Explain why this never happens.

PROBLEMS

● BEGINNER   ■ INTERMEDIATE   ◆ EXPERT

1. ● In Chapter 1 we learned that acceleration $a = \dfrac{\Delta v}{\Delta t}$, and in Chapter 2 we learned that the cause of acceleration involves net force, where $a = \dfrac{F}{m}$. Equate these two equations for acceleration and show that, for constant mass, $F\Delta t = \Delta(mv)$.

2. ● A 5-kg bag of groceries is tossed onto a table at 4 m/s and slides to a stop in 3 s. Begin with the equation you derived in Problem 1 and show that the force of friction is 6.7 N.

3. ● An 8-kg ball rolling at 2 m/s bumps into a pillow and stops in 0.5 s. (a) Show that the force exerted by the pillow is 32 N. (b) How much force does the ball exert on the pillow?

4. ● A car crashes into a wall at 25 m/s and is brought to rest in 0.1 s. Show that the average force exerted on a 75-kg test dummy by the seat belt is more than 18,000 N.

5. ■ At a ball game, consider a baseball of mass $m = 0.15$ kg that is moving at a speed $v = 40$ m/s as it is grabbed by a fan. (a) Show that the impulse supplied to bring the ball to rest is 6.0 N · s. (b) If the ball is stopped in 0.03 s, show that the average force of the ball on the catcher's hand is 200 N.

6. ■ Judy (mass 40.0 kg), standing on slippery ice, catches her leaping dog, Atti (mass 15 kg), moving horizontally at 3.0 m/s. Use the conservation of momentum to show that the speed of Judy and her dog after the catch is 0.8 m/s.

7. ■ A railroad diesel engine weighs four times as much as a freight car. The diesel engine coasts at 5 km/h into a freight car that is initially at rest. Use the conservation of momentum to show that after they couple together, the two coast at 4 km/h.

8. ■ A 5-kg fish swimming at 1 m/s swallows an absent-minded 1-kg fish swimming toward it at a velocity that brings both fish to a halt. Show that the speed of the smaller fish before lunch was 5 m/s.

9. ■ Comic-strip hero Superman meets an asteroid in outer space and hurls it at 800 m/s, as fast as a bullet. The asteroid is a thousand times as massive as Superman. In the strip, Superman is seen at rest after the throw. Taking physics into account, show that his recoil velocity would be 800,000 m/s.

10. ● Belly-Flop Bernie dives from atop a tall flagpole into a swimming pool below. His potential energy at the top is 10,000 J. Show that when his potential energy reduces to 2000 J, his kinetic energy is 8000 J.

11. ● A lever is used to lift a heavy load. When a 50-N force pushes one end of the lever down 1.2 m, the load rises 0.2 m. Show that the weight of the load is 300 N.

12. ● In raising a 5000-N piano with a pulley system, the movers note that, for every 2 m of rope pulled down, the piano rises 0.2 m. Ideally, show that the force required to lift the piano is 500 N.

13. ● How much power does a weightlifter expend when lifting a 50-kg barbell a vertical distance of 1.2 m in a time interval of 1.5 s?

14. ◆ A braking force is needed to bring a car of mass m moving at speed v to rest in time t.
    (a) Begin with the impulse–momentum relationship and show that the braking force is mv/t.
    (b) The mass of the car is 1200 kg and its initial speed is 25 m/s. Show that the braking force needed to stop it in 12 s is 1500 N.

15. ◆ A shopping cart full of groceries with total mass m moves down the aisle at a speed v. The cart is then stopped by a constant force F.
    (a) Use the impulse–momentum relationship and show that the time required to stop the cart is mv/F.
    (b) If the total mass of the cart is 20.0 kg, its initial speed is 3.0 m/s, and the stopping force is 15.0 N, show that the stopping time is 4.0 s.
    (c) Show that the initial kinetic energy of the loaded shopping cart was 90 J.

16. ◆ When an average force F is exerted over a certain distance on a shopping cart of mass m, its kinetic energy increases by $\frac{1}{2}mv^2$.
    (a) Use the work–energy theorem to show that the distance over which the force acts is $\dfrac{mv^2}{2F}$.
    (b) If twice the force is exerted over twice the distance, how does the resulting increase in kinetic energy compare with the original increase in kinetic energy?

ACTIVE EXPLORATIONS

1. When you get a bit ahead in your studies, cut classes some afternoon and visit your local pool or billiards parlor and bone up on momentum conservation. Note that, no matter how complicated the collision of balls, the momentum along the line of action of the cue ball before impact is the same as the combined momentum of all the balls along this direction after impact and that the components of momenta perpendicular to this line of action cancel to zero after impact, the same value as before impact in this direction. You'll see both the vector nature of momentum and its conservation more clearly when rotational skidding, "English," is not imparted to the cue ball. When English is imparted by striking the cue ball off center, rotational momentum, which is also conserved, somewhat complicates analysis. But, regardless of how the cue ball is struck, in the absence of external forces, both linear and rotational momentum are always conserved. Both pool and billiards offer a first-rate exhibition of momentum conservation in action.

2. Place a small rubber ball on top of a basketball or soccer ball and then drop them together. If vertical alignment nicely remains as they fall to the floor, you'll see that the small ball bounces unusually high. Can you reconcile this with energy conservation?

READINESS ASSURANCE TEST (RAT)

If you have a good handle on this chapter, if you really do, then you should be able to score at least 7 out of 10 on this RAT. If you score less than 7, you need to study further before moving on.

Choose the BEST answer to each of the following.

1. If the mass of a cart full of groceries decreases to half and its speed doubles, the momentum of the cart
   (a) remains unchanged.
   (b) is doubled.
   (c) is quadrupled.
   (d) decreases.
2. Whereas impulse involves force and time, work involves force and
   (a) energy.
   (b) acceleration.
   (c) distance.
   (d) power.
3. If the running speed of Fast Freddy doubles, what else doubles?
   (a) His momentum.
   (b) His kinetic energy.
   (c) both of these
   (d) neither of these
4. A 1-kg ball has the same speed as a 10-kg ball. Compared with the 1-kg ball, the 10-kg ball has
   (a) less momentum.
   (b) the same momentum.
   (c) 10 times as much momentum.
   (d) 100 times as much momentum.
5. Which of the following equations best illustrates the usefulness of automobile airbags?
   (a) $F = ma$.
   (b) $Ft = \Delta mv$.
   (c) $KE = \frac{1}{2}mv^2$.
   (d) $Fd = \Delta\frac{1}{2}mv^2$.
6. Which of the following equations is most useful for solving a problem that asks for the distance a fast-moving box slides across a post office floor and comes to a stop?
   (a) $F = ma$.
   (b) $Ft = \Delta mv$.
   (c) $KE = \frac{1}{2}mv^2$.
   (d) $Fd = \Delta\frac{1}{2}mv^2$.
7. A model airplane moves three times as fast as another identical model airplane. Compared to the kinetic energy of the slower airplane, the kinetic energy of the faster airplane is
   (a) the same for level flight.
   (b) twice as much.
   (c) four times as much.
   (d) more than four times as much.
8. Raising an auto in a service station requires work. Raising it twice as high requires
   (a) the same amount of work, but twice the power.
   (b) twice the work.
   (c) twice the power.
   (d) all of these
9. If a charging elephant has kinetic energy, it must also have
   (a) potential energy.
   (b) momentum.
   (c) work.
   (d) all of these
10. A machine cannot multiply
    (a) forces.
    (b) distances.
    (c) energy.
    (d) all of these

Answers to RAT

1a, 2c, 3a, 4c, 5b, 6d, 7d, 8b, 9b, 10c

MORE TO EXPLORE

Bodanis, David. $E = mc^2$: *A Biography of the World's Most Famous Equation*. New York: Berkley Publishing Group, 2002.

This is a truly delightful and engaging history of our understanding of energy. The book's charm is its emphasis on the people who contributed to energy, the equal sign, mass, the speed of light, and exponential notation. A treat!

CHAPTER 3 ONLINE RESOURCES

The Physical Science Place

Interactive Figures

- 3.10, 3.12, 3.13, 3.22, 3.26

Tutorials

- Momentum and Collisions
- Energy

Videos

- Definition of Momentum
- Changing Momentum

- Decreasing Momentum Over a Short Time
- Bowling Ball and Conservation of Momentum
- Conservation of Momentum: Numerical Example
- Machines: Pulleys

Quiz

Flashcards

Links

gravity, projectiles, and satellites

4.1 THE UNIVERSAL LAW OF GRAVITY

4.2 GRAVITY AND DISTANCE: THE INVERSE-
        SQUARE LAW

4.3 WEIGHT AND WEIGHTLESSNESS

4.4 UNIVERSAL GRAVITATION

4.5 PROJECTILE MOTION

4.6 FAST-MOVING PROJECTILES—SATELLITES

4.7 CIRCULAR SATELLITE ORBITS

4.8 ELLIPTICAL ORBITS

4.9 ESCAPE SPEED

■ It would be a mistake to say that Isaac Newton discovered gravity. That discovery dates back thousands of years earlier, when Earth dwellers experienced the consequences of tripping and falling. What Newton discovered was that gravity is universal—that it is not unique to Earth, as others of his time assumed.

Centuries before Newton's discovery, the circular motion of heavenly bodies was regarded as natural. Aristotle and his followers believed that the stars, the planets, and the Moon move in divine circles, free from any impelling forces. They assumed that this circular motion required no explanation. Newton, however, recognized that a force of some kind must act on the planets; otherwise, their paths would be straight lines. Others of his time, influenced by Aristotle, supposed that any force on a planet would be directed along its path. Newton, however, reasoned that the force on each planet would be directed toward a fixed central point—toward the Sun. This force of gravity was the same force that pulls an apple off a tree.

Newton's stroke of intuition, that the force between Earth and an apple is the same as the force that acts between moons and planets and everything else in our universe, was a revolutionary break with the prevailing notion that there were two sets of natural laws: one for earthly events and an altogether different set for motion in the heavens. This union of terrestrial laws and cosmic laws is called the *Newtonian synthesis*.

Link To Sections 26.1, 27.4

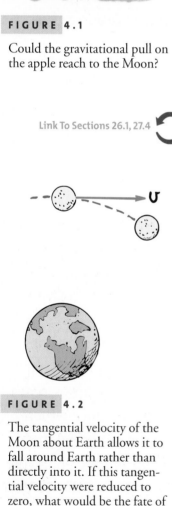

FIGURE 4.1

Could the gravitational pull on the apple reach to the Moon?

FIGURE 4.2

The tangential velocity of the Moon about Earth allows it to fall around Earth rather than directly into it. If this tangential velocity were reduced to zero, what would be the fate of the Moon?

■ 4.1   The Universal Law of Gravity

According to popular legend, Newton was sitting under an apple tree when the idea struck him that gravity extends beyond Earth. Perhaps he looked up through tree branches toward the origin of the falling apple and noticed the Moon. Perhaps the apple hit him in the head, as popular stories tell us. In any event, Newton had the insight to see that the force between Earth and a falling apple is the same force that pulls the Moon in an orbital path around Earth, a path similar to a planet's path around the Sun.

To test this hypothesis, Newton compared the fall of an apple with the "fall" of the Moon. He realized that the Moon falls in the sense that *it falls away from the straight line it would follow if there were no forces acting on it*. Because of its tangential velocity, it "falls around" the round Earth (as we shall investigate later in this chapter). By simple geometry, the Moon's distance of fall per second could be compared with the distance that an apple or anything that far away would fall in one second. Newton's calculations didn't check. Disappointed, but recognizing that brute fact must always win over a beautiful hypothesis, he placed his papers in a drawer, where they remained for nearly 20 years. During this period, he founded and developed the field of geometric optics, for which he first became famous.

Newton's interest in mechanics was rekindled with the advent of a spectacular comet in 1680 and another two years later. He returned to the Moon problem at the prodding of his astronomer friend, Edmund Halley, for whom the second comet was later named. He made corrections in the experimental data used in his earlier method and obtained excellent results. Only then did he publish what is one of the most far-reaching generalizations of the human mind: the **law of universal gravitation.***

Everything pulls on everything else in a beautifully simple way that involves only mass and distance. According to Newton, any body attracts any other body with a force that is directly proportional to the product of their masses and inversely proportional to the square of the distance separating them.

This statement can be expressed as

$$\text{Force} \sim \frac{\text{mass}_1 \times \text{mass}_2}{\text{distance}^2}$$

or symbolically as

$$F \sim \frac{m_1 m_2}{d^2}$$

* This is a dramatic example of the painstaking effort and cross-checking that go into the formulation of a scientific theory. Contrast Newton's approach with the failure to "do one's homework," the hasty judgments, and the absence of cross-checking that so often characterize the pronouncements of people advocating less-than-scientific theories.

where m_1 and m_2 are the masses of the bodies and d is the distance between their centers. Thus, the greater the masses m_1 and m_2, the greater the force of attraction between them, in direct proportion to the masses.* The greater the distance of separation d, the weaker the force of attraction, in inverse proportion to the square of the distance between their centers of mass.

CHECK POINT

1. In Figure 4.2, we see that the Moon falls around Earth rather than straight into it. If the Moon's tangential velocity were zero, how would it move?
2. According to the equation for gravitational force, what happens to the force between two bodies if the mass of one of the bodies is doubled? If both masses are doubled?
3. Gravitational force acts on all bodies in proportion to their masses. Why, then, doesn't a heavy body fall faster than a light body?

FIGURE 4.3

As the rocket gets farther from Earth, gravitational strength between the rocket and Earth decreases.

Were these your answers?

1. If the Moon's tangential velocity were zero, it would fall straight down and crash into Earth!
2. When one mass is doubled, the force between it and the other one doubles. If both masses double, the force is four times as much.
3. The answer goes back to Chapter 2. Recall Figure 2.9, in which heavy and light bricks fall with the same acceleration because both have the same ratio of weight to mass. Newton's second law ($a = F/m$) reminds us that greater force acting on greater mass does not result in greater acceleration.

Just as sheet music guides a musician playing music, equations guide a physics student to understand how concepts are connected.

THE UNIVERSAL GRAVITATIONAL CONSTANT, *G*

The proportionality form of the universal law of gravitation can be expressed as an exact equation when the constant of proportionality G is introduced. G is called the *universal gravitational constant*. Then the equation is

The Physical Science Place
Motion and Gravity
Orbits and Kepler's Laws

$$F = G\frac{m_1 m_2}{d^2}$$

In words, the force of gravity between two objects is found by multiplying their masses, dividing by the square of the distance between their centers, and then multiplying this result by the constant G. The magnitude of G is identical to the magnitude of the force between a pair of 1-kg masses that are 1 m apart: 0.0000000000667 N. This small magnitude indicates an extremely weak force. In standard units and in scientific notation,**

Link To Section 24.3

$$G = 6.67 \times 10^{-11} \, \text{N} \cdot \text{m}^2/\text{kg}^2$$

Interestingly, Newton could calculate the product of G and Earth's mass, but not either one alone. Calculating G alone was first done by the English physicist Henry Cavendish in the 18th century, a century after Newton's time.

Just as π relates circumference and diameter for circles, G relates gravitational force with mass and distance.

---

* Note the different role of mass here. Thus far, we have treated mass as a measure of inertia, which is called *inertial mass*. Now we see mass as a measure of gravitational force, which in this context is called *gravitational mass*. It is experimentally established that the two are equal, and, as a matter of principle, the equivalence of inertial and gravitational mass is the foundation of Einstein's general theory of relativity.

** The numerical value of G depends entirely on the units of measurement we choose for mass, distance, and time. The international system of choice uses the following units: for mass, the kilogram; for distance, the meter; and for time, the second. Scientific notation is discussed in Appendix A at the end of this book.

FIGURE 4.4

von Jolly's method of measuring G. Balls of mass m_1 and m_2 attract each other with a force F equal to the weights needed to restore balance.

You can never change only one thing! Every equation reminds us of this—you can't change a term on one side without affecting the other side.

Link To Sections 28.3, 28.5

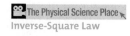

Saying that F is inversely proportional to the *square* of d means, for example, that if d gets bigger by a factor of 3, F gets *smaller* by a factor of 9.

Cavendish found G by measuring the tiny force between lead masses with an extremely sensitive torsion balance. A simpler method was later developed by Philipp von Jolly, who attached a spherical flask of mercury to one arm of a sensitive balance (Figure 4.4). After the balance was put in equilibrium, a 6-ton lead sphere was rolled beneath the mercury flask. The gravitational force between the two masses was measured by the weight needed on the opposite end of the balance to restore equilibrium. All the quantities—m_1, m_2, F, and d—were known, from which the constant G was calculated:

$$G = \frac{F}{\left(\frac{m_1 m_2}{d^2}\right)} = 6.67 \times 10^{-11} \text{ N/kg}^2/\text{m}^2 = 6.67 \times 10^{-11} \text{ N·m}^2/\text{kg}^2$$

The force of gravity is the weakest of the four known fundamental forces. (The other three are the electromagnetic force and two kinds of nuclear forces.) We sense gravitation only when masses like that of Earth are involved. If you stand on a large ship, the force of attraction between you and the ship is too weak for ordinary measurement. The force of attraction between you and Earth, however, can be measured. It is your weight.

Your weight depends not only on your mass but also on your distance from the center of Earth. At the top of a mountain, your mass is the same as it is anywhere else, but your weight is slightly less than it is at ground level. That's because your distance from Earth's center is greater.

Once the value of G was known, the mass of Earth was easily calculated. The force that Earth exerts on a mass of 1 kg at its surface is 9.8 N. The distance between the 1-kg mass and the center of Earth is Earth's radius, 6.4×10^6 m. Therefore, from $F = G(m_1 m_2/d^2)$, where m_1 is the mass of Earth,

$$9.8 \text{ N} = 6.67 \times 10^{-11} \text{ N·m}^2/\text{kg}^2 \frac{1 \text{ kg} \times m_1}{(6.4 \times 10^6 \text{ m})^2}$$

from which the mass of Earth is calculated to be $m_1 = 6 \times 10^{24}$ kg.

In the 18th century, when G was first measured, people all over the world were excited about it. Newspapers everywhere announced the discovery as one that measured the mass of the planet Earth. How exciting that Newton's formula gives the mass of the entire planet, with all its oceans, mountains, and inner parts yet to be discovered. G and the mass of Earth were measured when a great portion of Earth's surface was still undiscovered.

4.2 Gravity and Distance: The Inverse-Square Law

We can better understand how gravity is diluted with distance by considering how paint from a paint gun spreads with increasing distance (Figure 4.5). Suppose we position a paint gun at the center of a sphere with a radius of 1 m, and a burst of paint spray travels 1 m to produce a square patch of paint that is 1 mm thick. How thick would the patch be if the experiment were done in a sphere with twice the radius? If the same amount of paint travels in straight lines for 2 m, it spreads to a patch twice as tall and twice as wide. The paint is then spread over an area four times as big, and its thickness would be only $\frac{1}{4}$ mm.

Can you see from Figure 4.5 that for a sphere of radius 3 m, the thickness of the paint patch would be only $\frac{1}{9}$ mm? Can you see that the thickness of the paint

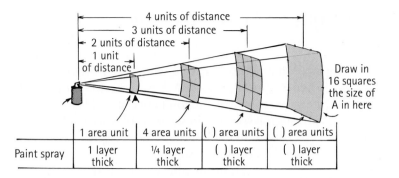

FIGURE 4.5

The inverse-square law. Paint spray travels radially away from the nozzle of the can in straight lines. Like gravity, the "strength" of the spray obeys the inverse-square law.

decreases as the square of the distance increases? This is known as the **inverse-square law**. The inverse-square law holds for gravity and for all phenomena in which the effect from a localized source spreads uniformly throughout the surrounding space: the electric field about an isolated electron, light from a match, radiation from a piece of uranium, and sound from a cricket.

Newton's law of gravity as written applies to particle and spherical bodies, as well as to nonspherical bodies sufficiently far apart. The distance term d in Newton's equation is the distance between the centers of masses of the objects. Note in Figure 4.6 that the apple that normally weighs 1 N at Earth's surface weighs only $\frac{1}{4}$ as much when it is twice the distance from Earth's center. The greater an object's distance from Earth's center, the less the object weighs. A child who weighs 300 N at sea level weighs only 299 N atop Mt. Everest. For greater distances, force is less. For very great distances, Earth's gravitational force approaches zero. The force *approaches* zero, but it never gets there. Even if you were transported to the far reaches of the universe, the gravitational influence of home would still be with you. It may be overwhelmed by the gravitational influences of nearer and/or more massive bodies, but it is there. The gravitational influence of every material object, however small or however far, is exerted through all of space.

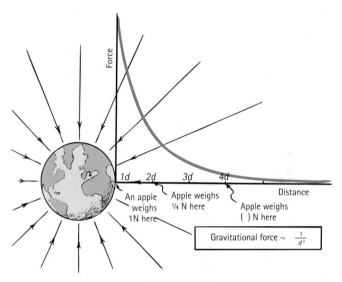

Gravitational force $\sim \dfrac{1}{d^2}$

FIGURE 4.6

INTERACTIVE FIGURE

The weight of an apple depends on its distance from Earth's center.

CHECK POINT

1. By how much does the gravitational force between two objects decrease when the distance between their centers is doubled? Tripled? Increased tenfold?

2. Consider an apple at the top of a tree that is pulled by Earth's gravity with a force of 1 N. If the tree were twice as tall, would the force of gravity be only $\frac{1}{4}$ as strong? Defend your answer.

Were these your answers?

1. It decreases to one-fourth, one-ninth, and one-hundredth the original value.

2. No, because an apple at the top of the twice-as-tall apple tree is not twice as far from Earth's center. The taller tree would need a height equal to the radius of Earth (6,370 km) for the apple's weight at its top to reduce to $\frac{1}{4}$ N. Before its weight decreases by 1%, an apple or any object must be raised 32 km—nearly four times the height of Mt. Everest. So, as a practical matter, we disregard the effects of everyday changes in elevation.

FIGURE 4.7

The person's weight (not her mass) decreases as she increases her distance from Earth's center.

■ 4.3 Weight and Weightlessness

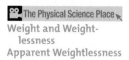
When you step on a bathroom scale, you effectively compress a spring inside. When the pointer stops, the elastic force of the deformed spring balances the gravitational attraction between you and Earth—nothing moves, because you and the scale are in static equilibrium. The pointer is cal-

ibrated to show your **weight**. If you stand on a bathroom scale in a moving elevator, you'll find variations in your weight. If the elevator accelerates upward, the springs inside the bathroom scale are more compressed and your weight reading is greater. If the elevator accelerates downward, the springs inside the scale are less compressed and your weight reading is less. If the elevator cable breaks and the elevator falls freely, the reading on the scale goes to zero. According to the scale's reading, you would be **weightless**. Would you really be weight-less? We can answer this question only if we agree on what we mean by *weight*.

In Chapter 1 we treated the weight of an object as the force due to gravity upon it. When in equilibrium on a firm surface, weight is evidenced by a support force, or, when in suspension, by a supporting rope tension. In either case, with no acceleration, weight equals *mg*. In future rotating habitats in space, where rotating environments act as a giant centrifuges, support force can occur without regard to gravity. So a broader definition of the weight of something is the force it exerts against a supporting floor or a weighing scale. According to this definition, you are as heavy as you feel; in an elevator that accelerates downward, the supporting force of the floor is less and you weigh less. If the elevator is in free fall, your weight is zero (Figure 4.10). Even in this weightless condition, however, a gravitational force is still acting on you, causing your downward acceleration. But gravity now is not felt as weight because there is no support force.

Astronauts in orbit are without a support force and are in a sustained state of weightlessness. They sometimes experience "space sickness" until they become accustomed to a state of sustained weightlessness. Astronauts in orbit are in a state of continual free fall.

The International Space Station (ISS), shown in Figure 4.11, provides a weightless environment. The station facility and astronauts all accelerate equally toward Earth, at somewhat less than 1 *g* because of their altitude. This accelera-tion is not sensed at all. With respect to the station, the astronauts experience zero *g*. Over extended periods of time, this causes loss of muscle strength and other detrimental changes in the body. Future space travelers, how-ever, need not be subjected to weight-lessness. Habitats that lazily rotate as giant wheels or pods at the end of a tether will

FIGURE 4.8

When you step on a weighing scale, two forces act on it; a downward force of gravity (your ordinary weight, *mg*, if there is no acceleration) and an upward support force. These equal and opposite forces squeeze a springlike device inside the scale that is cali-brated to show weight.

FIGURE 4.9

Both are weightless.

FIGURE 4.10

Your weight equals the force with which you press against the supporting floor. If the floor accelerates up or down, your weight varies (even though the gravitational force *mg* that acts on you remains the same).

FIGURE 4.11

The inhabitants in this laboratory and docking facility continually experience weightlessness. They are in free fall around Earth. Does a force of gravity act on them?

Inhabitants in a rotating space station won't get dizzy from rotation because their surroundings co-rotate (unless they look out windows).

likely replace today's nonrotating space habitats. Rotation effectively supplies a support force and nicely provides weight.

CHECK POINT

In what sense is drifting in space far away from all celestial bodies like stepping down off a stepladder?

Was this your answer?

In both cases, you'd experience weightlessness. Drifting in deep space, you would remain weightless because no discernable force acts on you. Stepping from a stepladder, you would be only momentarily weightless because of a momentary lapse of support force.

Astronauts inside an orbiting space vehicle have no weight, even though the force of gravity between them and Earth is only slightly less than at ground level.

4.4 Universal Gravitation

We all know that Earth is round. But why is Earth round? It is round because of gravitation. Everything attracts everything else, and so Earth has attracted itself together as far as it can! Any "corners" of Earth have been pulled in; as a result, every part of the surface is equidistant from the center of gravity. This makes it a sphere. Therefore, we see, from the law of gravitation, that the Sun, the Moon, and Earth are spherical because they have to be (although rotational effects make them slightly ellipsoidal).

If everything pulls on everything else, then the planets must pull on each other. The force that controls Jupiter, for example, is not just the force from the Sun; there are also pulls from the other planets. Their effect is small in comparison with the pull of the much more massive Sun, but it still shows. When Saturn is near Jupiter, its pull disturbs the otherwise smooth path traced by Jupiter. Both planets "wobble" about their expected orbits. The interplanetary forces causing this wobbling are called *perturbations*. By the 1840s, studies of the most recently discovered planet at the time, Uranus, showed that the deviations of its orbit could not be explained by perturbations from all other known planets. Either the law of gravitation was failing at this great distance from the Sun or an unknown eighth planet was perturbing the orbit of Uranus. An Englishman and a Frenchman, J. C. Adams and Urbain Leverrier, respectively, each assumed Newton's law to be valid, and they independently calculated where an eighth planet should be.

The Physical Science Place
Discovery of Neptune

fyi

- The space surrounding all objects with mass is energized with a *gravitational field*. Similarly, the space around a magnet is energized with a *magnetic field*, and the space about an electrically charged object is energized with an *electric field*.

fyi

- It's widely assumed that when Earth was no longer thought to be the center of the universe, both it and humankind were demoted in importance and were no longer considered special. On the contrary, writings of the time suggest most Europeans viewed humans as filthy and sinful because of Earth's lowly position—farthest from heaven, with hell at its center. Human elevation didn't occur until the Sun, viewed positively, took a center position. We became special by showing we're not so special.

Adams sent a letter to the Greenwich Observatory in England; at about the same time, Leverrier sent a letter to the Berlin Observatory in Germany. They both suggested that a certain area of the sky be searched for a new planet. Adams' request was delayed by misunderstandings at Greenwich, but Leverrier's request was heeded immediately. The planet Neptune was discovered that very night!

Subsequent tracking of the orbits of both Uranus and Neptune led to the discovery of Pluto in 1930 at the Lowell Observatory in Arizona. Whatever you may have learned in your early schooling, Pluto is no longer a planet. In 2006 Pluto was officially classified as a *dwarf planet*. Other objects of Pluto's size continue to be discovered beyond Neptune.* Pluto takes 248 years to make a single revolution about the Sun, so no one will see it in its discovered position again until 2178.

Recent evidence suggests that the universe is expanding and accelerating outward, pushed by an antigravity *dark energy* that makes up some 73% of the universe. Another 23% is composed of the yet-to-be-discovered particles of exotic *dark matter*. Ordinary matter, the stuff of stars, cabbages, and kings, makes up only about 4%. The concepts of dark energy and dark matter are late-20th- and 21st-century confirmations. The present view of the universe has progressed appreciably beyond what Newton and those of his time perceived.

Yet few theories have affected science and civilization as much as Newton's theory of gravity. The successes of Newton's ideas ushered in the Enlightenment. Newton had demonstrated that, by observation and reason, people could uncover the workings of the physical universe. How profound that all the moons and planets and stars and galaxies have such a beautifully simple rule to govern them, namely,

$$F = G\frac{m_1 m_2}{d^2}$$

The formulation of this simple rule is one of the major reasons for the success in science that followed, for it provided hope that other phenomena of the world might also be described by equally simple and universal laws.

This hope nurtured the thinking of many scientists, artists, writers, and philosophers of the 1700s. One of these was the English philosopher John Locke, who argued that observation and reason, as demonstrated by Newton, should be our best judge and guide in all things. Locke urged that all of nature and even society should be searched to discover any "natural laws" that might exist. Using Newtonian physics as a model of reason, Locke and his followers modeled a system of government that found adherents in the thirteen British colonies across the Atlantic. These ideas culminated in the Declaration of Independence and the Constitution of the United States of America.

■ 4.5 Projectile Motion

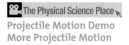
The Physical Science Place
Projectile Motion

The Physical Science Place
Projectile Motion Demo
More Projectile Motion

Without gravity, a rock tossed at an angle skyward would follow a straight-line path. Because of gravity, however, the path curves. A tossed rock, a cannonball, or any object that is projected by some means and continues in motion by its own inertia is called a **projectile**. To the cannoneers of earlier centuries, the curved paths of projectiles seemed very complex. Today these paths are surprisingly simple when we look at the horizontal and vertical components of velocity separately.

---

* Quaoar has a moon; Eris is 30% wider than Pluto and also has a moon. Object 2003 EL61 has two moons. Objects nicknamed Sedna and Buffy, discovered in 2005, are nearly the size of Pluto.

The horizontal component of velocity for a projectile is no more complicated than the horizontal velocity of a bowling ball rolling freely on the lane of a bowling alley. If the retarding effect of friction can be ignored, no horizontal force acts on the ball and its velocity is constant. It rolls of its own inertia and covers equal distances in equal intervals of time (Figure 4.12, top). The horizontal component of a projectile's motion is just like the bowling ball's motion along the lane.

The vertical component of motion for a projectile following a curved path is just like the motion described in Chapter 1 for a freely falling object. The vertical component is exactly the same as for an object falling freely straight down, as shown at the left in Figure 4.12. The faster the object falls, the greater the distance covered in each successive second. Or, if the object is projected upward, the vertical distances of travel decrease with time on the way up.

The curved path of a projectile is a combination of horizontal and vertical motion. Velocity is a vector quantity, and a velocity vector at an angle has horizontal and vertical components, as seen in Figure 4.13. When air resistance is small enough to ignore, the horizontal and vertical components of a projectile's velocity are completely independent of one another. Their combined effect produces the trajectories of projectiles.

FIGURE 4.12

(*left*) Drop it, and it accelerates downward and covers a greater vertical distance each second. (*right*) Roll a ball along a level surface, and its velocity is constant because no component of gravitational force acts horizontally.

PROJECTILES LAUNCHED HORIZONTALLY

Projectile motion is nicely analyzed in Figure 4.14, which shows a simulated multiple flash exposure of a ball rolling off the edge of a table. Investigate it carefully, for there's a lot of good physics there. On the left we notice equally timed sequential positions of the ball without the effect of gravity. Only the effect of the ball's horizontal component of motion is shown. Next we see vertical motion without a horizontal component. The curved path in the third view is best analyzed by considering the horizontal and vertical components of motion separately. There are two important things to notice. The first is that the ball's horizontal component of velocity doesn't change as the falling ball moves forward. The ball travels the same horizontal distance in equal times between each flash. That's because there is no component of gravitational force acting horizontally. Gravity acts only *downward*, so the only acceleration of the ball is *downward*. The second thing to notice is that the vertical positions become farther apart with time. The vertical distances traveled are the same as if the ball were simply dropped. Note that the curvature of the ball's path is the combination of horizontal motion, which remains constant, and vertical motion, which undergoes acceleration due to gravity.

The trajectory of a projectile that accelerates only in the vertical direction while moving at a constant horizontal velocity is a **parabola**. When air resistance

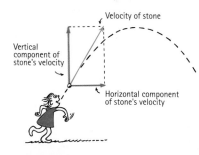

FIGURE 4.13

Vertical and horizontal components of a stone's velocity.

Horizontal motion with *no* gravity   Vertical motion only with gravity   Combined horizontal and vertical motion   Superposition of the preceding cases

FIGURE 4.14

INTERACTIVE FIGURE

Simulated photographs of a moving ball illuminated with a strobe light.

FIGURE 4.15

INTERACTIVE FIGURE

A strobe-light photograph of two golf balls released simultaneously from a mechanism that allows one ball to drop freely while the other is projected horizontally.

is small enough to neglect, as it is for a heavy object without great speed, the trajectory is parabolic.

CHECK POINT

At the instant a cannon fires a cannonball horizontally over a level range, another cannonball held at the side of the cannon is released and drops to the ground. Which ball, the one fired downrange or the one dropped from rest, strikes the ground first?

Was this your answer?

Both cannonballs hit the ground at the same time, because both fall *the same vertical distance*. Note that the physics is the same as the physics of Figures 4.14 through 4.16. We can reason this another way by asking which one would hit the ground first if the cannon were pointed at an *upward* angle. Then the dropped cannonball would hit first, while the fired ball is still airborne. Now consider the cannon pointing *downward*. In this case, the fired ball hits first. So projected upward, the dropped one hits first; downward, the fired one hits first. Is there some angle at which there is a dead heat, where both hit at the same time? Can you see that this occurs when the cannon is horizontal?

PROJECTILES LAUNCHED AT AN ANGLE

In Figure 4.17, we see the paths of stones thrown at an angle upward (left) and downward (right). The dashed straight lines at the top show the ideal trajectories of the stones if there were no gravity. Notice that the vertical distance that each

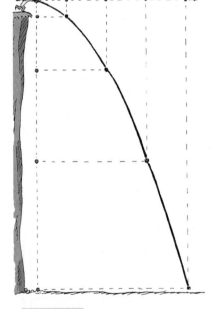

FIGURE 4.16

The vertical dashed line at left is the path of a stone dropped from rest. The horizontal dashed line at the top would be its path if there were no gravity. The curved solid line shows the resulting trajectory that combines horizontal and vertical motion.

FIGURE 4.17

Whether launched at an angle upward or downward, the vertical distance of fall beneath the idealized straight-line path is the same for equal times.

FIGURING PHYSICAL SCIENCE

■ Problem Solving

SAMPLE PROBLEM 1

A ball of mass 1.0 kg rolls off of a 1.25-m-high lab table and hits the floor 3.0 m from the base of the table.

1.25 m

3.0 m

(a) Show that the ball takes 0.5 s to hit the floor.
(b) Show that the ball leaves the table at 6.0 m/s.

SOLUTION:

(a) We want the time of the ball in the air. First, some physics. The time t it takes for any ball to hit the floor would be the same as if it were dropped from rest a vertical distance y. We say from rest because initially it moves horizontally off the desk, with zero velocity in the vertical direction.

From $y = \frac{1}{2}gt^2 \Rightarrow t^2 = \frac{2y}{g}$

Then $t = \sqrt{\frac{2y}{g}} = \sqrt{\frac{2(1.25\text{ m})}{10\frac{m}{s^2}}} = 0.5\text{ s}$

(b) The horizontal speed of the ball as it leaves the table, using time 0.5 s, is

$v_x = \frac{d}{t} = \frac{x}{t} = \frac{3.0\text{ m}}{0.5\text{ s}} = 6.0\frac{m}{s}$

Notice how the terms of the equations guide the solution. Notice also that the mass of the ball, not showing in the equations, is extraneous information (as would be the color of the ball).

SAMPLE PROBLEM 2

A horizontally moving tennis ball barely clears the net, a distance y above the surface of the court. To land within the tennis court the ball must not be moving too fast.

v = ?

d

(a) To remain within the court's border, a horizontal distance d from the bottom of the net, ignoring air resistance and any spin effects of the ball, show that the ball's maximum speed over the net is

$v = \frac{d}{\sqrt{\frac{2y}{g}}}$

(b) Suppose the height of the net is 1.00 m, and the court's border is 12.0 m from the bottom of the net. Use $g = 10\text{ m/s}^2$ and show that the maximum speed of the horizontally moving ball clearing the net is about 27 m/s (about 60 mi/h).
(c) Does the mass of the ball make a difference? Defend your answer.

SOLUTION:

(a) As with Sample Problem 1, the physics concept here involves projectile motion in the absence of air resistance, where horizontal and vertical components of velocity are independ-ent. We're asked for horizontal speed, so we write

$v_x = \frac{d}{t}$

where d is horizontal distance traveled in time t. As with Sample Problem 1, the time t of the ball in flight is the same as if we had just dropped it from rest a vertical distance y from the top of the net. As the ball clears the net, its highest point in its path, its vertical component of velocity is zero.

From $y = \frac{1}{2}gt^2 \Rightarrow t^2 = \frac{2y}{g} \Rightarrow t = \sqrt{\frac{2y}{g}}$

So $v = \frac{d}{t} = \frac{d}{\sqrt{\frac{2y}{g}}}$.

Can you see that solving in terms of symbols better shows that these two problems are one and the same? All the physics occurs in steps (a) and (b) in Sample Problem 1. These steps are combined in step (a) of Sample Problem 2.

(b) $v = \frac{d}{\sqrt{\frac{2y}{g}}} = \frac{12.0\text{ m}}{\sqrt{\frac{2(1.00\text{ m})}{10\frac{m}{s^2}}}}$

$= 26.8\frac{m}{s} \approx 27\frac{m}{s}$.

(c) We can see that the mass of the ball (in both problems) doesn't show up in the equations for motion, which tells us that mass is irrelevant. Recall from Chapter 2 that mass has no effect on a freely falling object—and the tennis ball is a freely falling object (as is every projectile when air resist-ance can be neglected).

---

stone falls beneath the idealized straight-line path is the same for equal times. This vertical distance is independent of what's happening horizontally.

Figure 4.18 shows specific vertical distances for a cannonball shot at an upward angle. If there were no gravity the cannonball would follow the straight-line path shown by the dashed line. But there is gravity, so this doesn't occur. What happens is that the cannonball continuously falls beneath the imaginary line until it finally strikes the ground. Note that the vertical distance it falls beneath any point on the dashed line is the same vertical distance it would have fallen if it had been dropped from rest and had been falling for the same amount of time. This distance, as introduced in Chapter 1, is given by $d = \frac{1}{2}gt^2$, where t is the elapsed time. For $g = 10\text{ m/s}^2$, this becomes $d = 5t^2$.

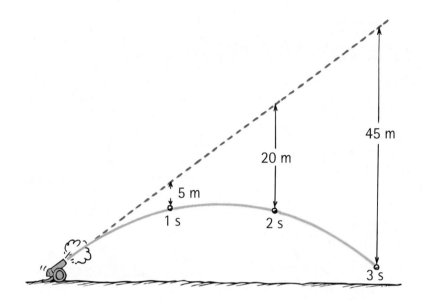

FIGURE 4.18

With no gravity, the projectile would follow a straight-line path (dashed line). But because of gravity, the projectile falls beneath this line the same vertical distance it would fall if it were released from rest. Compare the distances fallen with those given in Table 1.2 in Chapter 1. (With $g = 9.8$ m/s², these distances are more precisely 4.9 m, 19.6 m, and 44.1 m.)

We can put it another way: Shoot a projectile skyward at some angle and pretend there is no gravity. After so many seconds t, it should be at a certain point along a straight-line path. But because of gravity, it isn't. Where is it? The answer is that it's directly below this point. How far below? The answer in meters is $5t^2$ (or, more precisely, $4.9t^2$). How about that!

CHECK POINT

1. Suppose the cannonball in Figure 4.18 were fired faster. How many meters below the dashed line would it be at the end of the 5 s?
2. If the horizontal component of the cannonball's velocity is 20 m/s, how far downrange will the cannonball be in 5 s?

Were these your answers?

1. The vertical distance beneath the dashed line at the end of 5 s is 125 m [looking at magnitudes only: $d = 5t^2 = 5(5)^2 = 5(25) = 125$ m]. Interestingly enough, this distance doesn't depend on the angle of the cannon. If air resistance is neglected, any projectile will fall $5t^2$ meters below where it would have reached if there were no gravity.
2. With no air resistance, the cannonball will travel a horizontal distance of 100 m [$d = v_xt = (20 \text{ m/s})(5 \text{ s}) = 100$ m]. Note that because gravity acts only vertically and there is no acceleration in the horizontal direction, the cannonball travels equal horizontal distances in equal times. This distance is simply its horizontal component of velocity multiplied by the time (and not $5t^2$, which applies only to vertical motion under the acceleration of gravity).

Doing Physical Science

▪ Hands-On Dangling Beads

Make your own model of projectile paths. Divide a ruler or a stick into five equal spaces. At position 1, hang a bead from a string that is 1 cm long, as shown. At position 2, hang a bead from a string that is 4 cm long. At position 3, do the same with a 9-cm length of string. At position 4, use 16 cm of string, and for position 5, use 25 cm of string. If you hold the stick horizontally, you will have a version of Figure 4.16.

Hold it at a slight upward angle to show a version of Figure 4.17 (left). Hold it at a downward angle to show a version of Figure 4.17 (right).

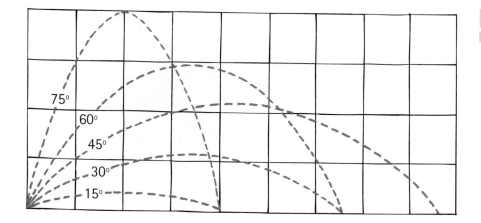

FIGURE 4.19

INTERACTIVE FIGURE

Ranges of a projectile shot at the same speed at different projection angles.

Figure 4.19 shows the paths of several projectiles, all with the same initial speed but different launching angles. The figure neglects the effects of air resistance, so the trajectories are all parabolas. Notice that these projectiles reach different *altitudes*, or heights above the ground. They also have different *horizontal ranges*, or distances traveled horizontally. The remarkable thing to note from Figure 4.19 is that the same range is obtained from two different launching angles when the angles add up to 90°! An object thrown into the air at an angle of 60°, for example, has the same range as if it were thrown at the same speed at an angle of 30°. For the smaller angle, of course, the object remains in the air for a shorter time. The greatest range occurs when the launching angle is 45°—and when air resistance is negligible.

Without the effects of air, the maximum range for a baseball would occur when it is batted 45° above the horizontal. Because of air resistance and lift due to spinning of the ball (Chapter 5), the best range occurs at batting angles noticeably less than 45°. Air resistance and spin are more significant for golf balls, where angles less than 38° or so result in maximum range. For heavy projectiles such as javelins and the shot, air has less effect on range. A javelin, being heavy and presenting a very small cross section to the air, follows an almost perfect parabola when thrown. So does a shot. For such projectiles, maximum range for equal launch speeds occurs for a launch angle of about 45° (slightly less because the launching height is above ground level). Aha, but launching speeds are *not* equal for such a projectile thrown at different angles. In throwing a javelin or putting a shot, a significant part of the launching *force* goes into combating gravity—the steeper the angle, the less

FIGURE 4.20

Maximum range is attained when a ball is batted at an angle of nearly 45°.

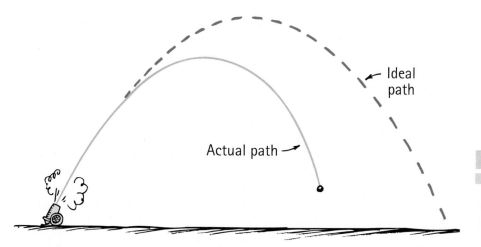

Ideal path

Actual path

FIGURE 4.21

INTERACTIVE FIGURE

In the presence of air resistance, the trajectory of a high-speed projectile falls short of the idealized parabolic path.

Hang Time Revisited

In Chapter 1, we stated that airborne time during a jump is independent of horizontal speed. Now we see why this is so—horizontal and vertical components of motion are independent of each other. The rules of projectile motion apply to jumping. Once one's feet are off the ground, only the force of gravity acts on the jumper (neglecting air resistance). Hang time depends only on the vertical component of liftoff velocity. However, the action of running can make a difference. When the jumper is running, the liftoff force during jumping can be somewhat increased by the pounding of the feet against the ground (and the ground pounding against the feet in action–reaction fashion), so hang time for a running jump can often exceed hang time for a standing jump. But once the runner's feet are off the ground, only the vertical component of liftoff velocity determines hang time.

speed the object has when leaving the thrower's hand. So gravity plays a role before and after launching. You can test this yourself: Throw a heavy boulder horizontally, then vertically—you'll find the horizontal throw to be considerably faster than the vertical throw. So maximum range for heavy projectiles thrown by humans is attained for angles of less than 45°—and not because of air resistance.

CHECK POINT

1. A baseball is batted at an angle into the air. Once the ball is airborne, and neglecting air resistance, what is the ball's acceleration vertically? Horizontally?
2. At what part of its trajectory does the baseball have minimum speed?
3. Consider a batted baseball following a parabolic path on a day when the Sun is directly overhead. How does the speed of the ball's shadow across the field compare with the ball's horizontal component of velocity?

Were these your answers?
1. Vertical acceleration is g because the force of gravity is vertical. Horizontal acceleration is zero because no horizontal force acts on the ball.
2. A ball's minimum speed occurs at the top of its trajectory. If it is launched vertically, its speed at the top is zero. If launched at an angle, the vertical component of velocity is zero at the top, leaving only the horizontal component. So the speed at the top is equal to the horizontal component of the ball's velocity at any point. Doesn't this make sense?
3. They are the same!

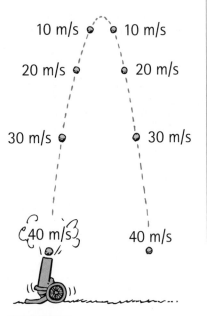

10 m/s   10 m/s

20 m/s   20 m/s

30 m/s   30 m/s

40 m/s   40 m/s

FIGURE 4.22

Without air resistance, speed lost while going up equals speed gained while coming down: Time going up equals time coming down.

When air resistance is small enough to be negligible, the time that a projectile takes to rise to its maximum height is the same as the time it takes to fall back to its initial level (Figure 4.22). This is because its deceleration by gravity while going up is the same as its acceleration by gravity while coming down. The speed it loses while going up is therefore the same as the speed gained while coming down. So the projectile arrives at its initial level with the same speed it had when it was initially projected.

Baseball games normally take place on level ground. For the short-range projectile motion on the playing field, Earth can be considered flat because the flight of the baseball is not affected by Earth's curvature. For very long-range projectiles, however, the curvature of Earth's surface must be taken into account. We'll now see that, if an object is projected fast enough, it falls all the way around Earth and becomes an Earth satellite.

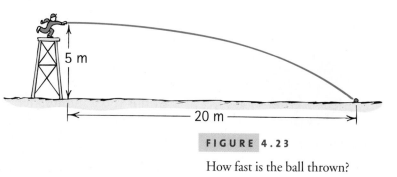

FIGURE 4.23

How fast is the ball thrown?

The boy on the tower in Figure 4.23 throws a ball 20 m downrange. What is his pitching speed?

Was this your answer?
The ball is thrown horizontally, so the pitching speed is horizontal distance divided by time. A horizontal distance of 20 m is given, but the time is not stated. However, knowing the vertical drop is 5 m, you remember that a 5-m drop takes 1 s! From the equation for constant speed (which applies to horizontal motion), $v = d/t = (20\,\text{m})/(1\,\text{s}) = 20$ m/s. It is interesting to note that the equation for constant speed, $v = d/t$, guides our thinking about the crucial factor in this problem—the *time*.

> Earth's curvature, dropping 5 m for each 8-km tangent, means that if you were floating in a calm ocean, you'd be able to see only the top of a 5-m mast on a ship 8 km away.

4.6 Fast-Moving Projectiles—Satellites

Consider the baseball pitcher on the cliff in Figure 4.24. If gravity did not act on the ball, the ball would follow a straight-line path shown by the dashed line. But gravity does act, so the ball falls below this straight-line path. In fact, as just discussed, 1 s after the ball leaves the pitcher's hand it has fallen a vertical distance of 5 m below the dashed line—whatever the pitching speed. It is important to understand this, for it is the crux of satellite motion.

An Earth **satellite** is simply a projectile that falls *around* Earth rather than *into* it. The speed of the satellite must be great enough to ensure that its falling distance matches Earth's curvature. A geometrical fact about the curvature of Earth is that its surface drops a vertical distance of 5 m for every 8000 m tangent

> A space shuttle is a projectile in a constant state of free fall. Because of its tangential velocity, it falls around Earth rather than vertically into it.

FIGURE 4.24

If you throw a stone at any speed, 1 s later it will have fallen 5 m below where it would have been without gravity.

FIGURE 4.25

Earth's curvature (not to scale).

FIGURE 4.26

If the speed of the stone and the curvature of its trajectory are great enough, the stone may become a satellite.

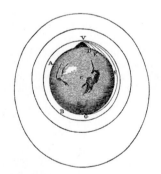

FIGURE 4.27

"The greater the velocity . . . with which (a stone) is projected, the farther it goes before it falls to the Earth. We may therefore suppose the velocity to be so increased, that it would describe an arc of 1, 2, 5, 10, 100, 1000 miles before it arrived at the Earth, till at last, exceeding the limits of the Earth, it should pass into space without touching." —Isaac Newton, *System of the World*

to the surface (Figure 4.24). If a baseball could be thrown fast enough to travel a horizontal distance of 8 km during the 1 s it takes to fall 5 m, then it would follow the curvature of Earth. This is a speed of 8 km/s. If this doesn't seem fast, convert it to kilometers per hour and you get an impressive 29,000 km/h (or 18,000 mi/h)!

At this speed, atmospheric friction would burn the baseball—or even a piece of iron—to a crisp. This is the fate of bits of rock and other meteorites that enter Earth's atmosphere and burn up, appearing as "falling stars." That is why satellites, such as the space shuttles, are launched to altitudes of 150 kilometers or more—to be above almost all of the atmosphere and to be nearly free of air resistance. A common misconception is that satellites orbiting at high altitudes are free from gravity. Nothing could be further from the truth. The force of gravity on a satellite 200 kilometers above Earth's surface is nearly as strong as it is at the surface. Otherwise the satellite would go in a straight line and leave Earth. The high altitude positions the satellite not beyond Earth's gravity, but beyond Earth's atmosphere, where air resistance is almost totally absent.

Satellite motion was understood by Isaac Newton, who reasoned that the Moon was simply a projectile circling Earth under the attraction of gravity. This concept is illustrated in a drawing by Newton (Figure 4.27). He compared the motion of the Moon to that of a cannonball fired from the top of a high mountain. He imagined that the mountaintop was above Earth's atmosphere, so that air resistance would not impede the motion of the cannonball. If fired with a low horizontal speed, a cannonball would follow a curved path and soon hit Earth below. If it were fired faster, its path would be less curved and it would hit Earth farther away. If the cannonball were fired fast enough, Newton reasoned, the curved path would become a circle and the cannonball would circle Earth indefinitely. It would be in orbit.

Both the cannonball and the Moon have tangential velocity (parallel to Earth's surface) sufficient to ensure motion *around* Earth rather than *into* it. Without resistance to reduce its speed, the Moon or any Earth satellite "falls" around Earth indefinitely. Similarly, the planets continuously fall around the Sun in closed paths. Why don't the planets crash into the Sun? They don't because of sufficient tangential velocities. What would happen if their tangential velocities were reduced to zero? The answer is simple enough: Their falls would be straight toward the Sun, and they would indeed crash into it. Any objects in the solar system without sufficient tangential velocities have long ago crashed into the Sun. What remains is the harmony we observe.

CHECK POINT

One of the beauties of physics is that there are usually different ways to view and explain a given phenomenon. Is the following explanation valid? "Satellites remain in orbit instead of falling to Earth because they are beyond the main pull of Earth's gravity."

Was this your answer?

No, no, a thousand times no! If any moving object were beyond the pull of gravity, it would move in a straight line and would not curve around Earth. Satellites remain in orbit because they *are* being pulled by gravity, not because they are beyond it. For the altitudes of most Earth satellites, Earth's gravitational force on a satellite is only a few percent weaker than it is at Earth's surface.

▪ 4.7 Circular Satellite Orbits

The Physical Science Place

Circular Orbits

An 8-km/s cannonball fired horizontally from Newton's mountain would follow Earth's curvature and glide in a circular path around Earth again and again (provided the cannoneer and the cannon got out of the way). Fired at a slower speed, the cannonball would strike Earth's surface; fired at a faster speed, it would overshoot a circular orbit, as we will discuss shortly. Newton calculated the speed for circular orbit, and because such a cannon-muzzle velocity was clearly impossible, he did not foresee the possibility of humans launching satellites (and he likely didn't consider multistage rockets).

Note that in circular orbit, the speed of a satellite is not changed by gravity; only the direction changes. We can understand this by comparing a satellite in circular orbit with a bowling ball rolling along a bowling lane. Why doesn't the gravity that acts on the bowling ball change its speed? The answer is that gravity pulls straight downward with no component of force acting forward or backward.

Consider a bowling lane that completely surrounds Earth, elevated high enough to be above the atmosphere and air resistance. The bowling ball rolls at constant speed along the lane. If a part of the lane were cut away, the ball would roll off its edge and would hit the ground below. A faster ball encountering the gap would hit the ground farther along the gap. Is there a speed at which the ball will clear the gap (like a motorcyclist who drives off a ramp and clears a gap to meet a ramp on the other side)? The answer is yes: 8 km/s will be enough to clear that gap—and any gap: even a 360° gap. The ball would be in circular orbit.

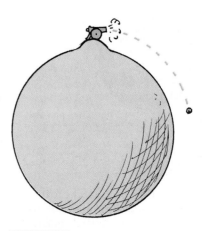

FIGURE 4.28

INTERACTIVE FIGURE

Fired fast enough, the cannonball goes into orbit.

FIGURE 4.29

(a) The force of gravity on the bowling ball is at 90° to its direction of motion, so it has no component of force to pull it forward or backward, and the ball rolls at constant speed. (b) The same is true even if the bowling alley is larger and remains "level" with the curvature of Earth.

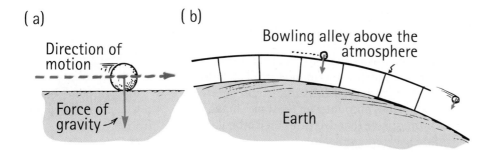

(a)
Direction of motion

Force of gravity

(b)
Bowling alley above the atmosphere

Earth

Note that a satellite in circular orbit is always moving in a direction perpendicular to the force of gravity that acts upon it. No component of force is acting in the direction of satellite motion to change its speed. Only a change in direction occurs. So we see why a satellite in circular orbit moves parallel to the surface of Earth at constant speed—a very special form of free fall.

For a satellite close to Earth, the period (the time for a complete orbit about Earth) is about 90 min. For higher altitudes, the orbital speed is less, the distance is more, and the period is longer. For example, communication satellites located in orbit 5.5 Earth radii above the surface of Earth have a period of 24 h. This period matches the period of daily Earth rotation. For an orbit around the equator, these satellites remain above the same point on the ground. The Moon is even farther away and has a period of 27.3 days. The higher the orbit of a satellite, the less its speed, the longer its path, and the longer its period.*

FIGURE 4.30

What speed will allow the ball to clear the gap?

---

* The speed of a satellite in circular orbit is given by $v = \sqrt{GM/d}$, and the period of satellite motion is given by $T = 2\pi\sqrt{d^3/GM}$, where G is the universal gravitational constant, M is the mass of Earth (or whatever body the satellite orbits), and d is the distance of the satellite from the center of Earth or other parent body.

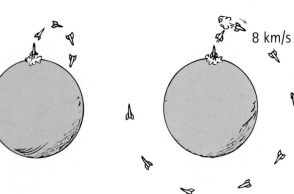

Putting a payload into Earth orbit requires control over the speed and direction of the rocket that carries it above the atmosphere. A rocket initially fired vertically is intentionally tipped from the vertical course. Then, once above the drag of the atmosphere, it is aimed horizontally, whereupon the payload is given a final thrust to orbital speed. We see this in Figure 4.31, where, for the sake of simplicity, the payload is the entire single-stage rocket. With the proper tangential velocity, it falls around Earth, rather than into it, and becomes an Earth satellite.

FIGURE 4.31

The initial thrust of the rocket lifts it vertically. Another thrust tips it from its vertical course. When it is moving horizontally, it is boosted to the required speed for orbit.

CHECK POINT

1. True or false: The space shuttle orbits at altitudes in excess of 150 km to be above both gravity and the atmosphere of Earth.
2. Satellites in close circular orbit fall about 5 m during each second of orbit. Why doesn't this distance accumulate and send satellites crashing into Earth's surface?

Were these your answers?

1. False. Satellites are above the atmosphere and air resistance—*not* gravity! It's important to note that Earth's gravity extends throughout the universe in accord with the inverse-square law.
2. In each second, the satellite falls about 5 m below the straight-line tangent it would have followed if there were no gravity. Earth's surface also curves 5 m beneath a straight-line 8-km tangent. The process of falling with the curvature of Earth continues from tangent line to tangent line, so the curved path of the satellite and the curve of Earth's surface "match" all the way around Earth. Satellites do, in fact, crash to Earth's surface from time to time when they encounter air resistance in the upper atmosphere that decreases their orbital speed.

> The initial vertical climb gets a rocket quickly through the denser part of the atmosphere. Eventually, the rocket must acquire enough tangential speed to remain in orbit without thrust, so it must tilt until its path is parallel to Earth's surface.

Link To Section 26.1

4.8 Elliptical Orbits

If a projectile just above the drag of the atmosphere is given a horizontal speed somewhat greater than 8 km/s, it overshoots a circular path and traces an oval path called an **ellipse**.

An ellipse is a specific curve: the closed path taken by a point that moves in such a way that the sum of its distances from two fixed points (called *foci*) is constant. For a satellite orbiting a planet, one focus is at the center of the planet; the other focus could be internal or external to the planet. An ellipse can be easily constructed by using a pair of tacks (one at each focus), a loop of string, and a pencil (Figure 4.32). The closer the foci are to each other, the closer the ellipse is to a circle. When both foci are together, the ellipse *is* a circle. So we can see that a circle is a special case of an ellipse.

Whereas the speed of a satellite is constant in a circular orbit, its speed varies in an elliptical orbit. For an initial speed greater than 8 km/s, the satellite

FIGURE 4.32

INTERACTIVE FIGURE

A simple method for constructing an ellipse.

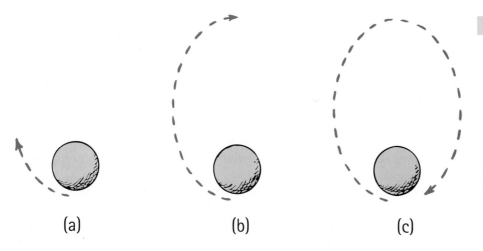

FIGURE 4.33

Elliptical orbit. When the speed of the satellite exceeds 8 km/s, (a) it overshoots a circular path and travels away from Earth against gravity. (b) At its maximum altitude it starts to come back toward Earth. (c) The speed it lost in going away is gained in returning, and the cycle repeats itself.

(a) (b) (c)

overshoots a circular path and moves away from Earth, against the force of gravity. It therefore loses speed. The speed it loses in receding is regained as it falls back toward Earth, and it finally rejoins its original path with the same speed it had initially (Figure 4.33). The procedure repeats over and over, and an ellipse is traced during each cycle.

Interestingly enough, the parabolic path of a projectile, such as a tossed baseball or a cannonball, is actually a tiny segment of a skinny ellipse that extends within and just beyond the center of Earth (Figure 4.34a). In Figure 4.34b, we see several paths of cannonballs fired from Newton's mountain. All these ellipses have the center of Earth as one focus. As muzzle velocity is increased, the ellipses are less *eccentric* (more nearly circular); and, when muzzle velocity reaches 8 km/s, the ellipse rounds into a circle and does not intercept Earth's surface. The cannonball coasts in circular orbit. At greater muzzle velocities, orbiting cannonballs trace the familiar external ellipses.

fyi

- When a spacecraft enters the atmosphere at too steep an angle, more than about 6°, it can burn up. If it comes in too shallow, it could bounce back into space like a pebble skipped across water.

FIGURE 4.34

(a) The parabolic path of the cannonball is part of an ellipse that extends within Earth. Earth's center is the far focus. (b) All paths of the cannonball are ellipses. For less than orbital speeds, the center of Earth is the far focus; for a circular orbit, both foci are Earth's center; for greater speeds, the near focus is Earth's center.

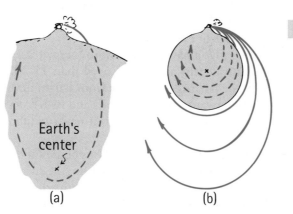

Earth's center

(a) (b)

CHECK POINT

The orbital path of a satellite is shown in the sketch. At which of the marked positions A through D does the satellite have the greatest speed? The lowest speed?

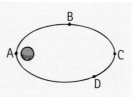

Were these your answers?

The satellite has its greatest speed as it whips around A and has its lowest speed at position C. After passing C, it gains speed as it falls back to A to repeat its cycle.

FIGURE 4.35

FIGURE 4.35

INTERACTIVE FIGURE

If Superman tosses a ball
8 km/s horizontally from the
top of a mountain high enough
to be just above air resistance
(a), then about 90 min later he
can turn around and catch it
(neglecting Earth's rotation).
Tossed slightly faster (b), it
takes an elliptical orbit and
returns in a slightly longer
time. Tossed at more than
11.2 km/s (c), it escapes Earth.
Tossed at more than 42.5 km/s
(d), it escapes the solar system.

4.9 Escape Speed

We know that a cannonball fired horizontally at 8 km/s from Newton's mountain would find itself in orbit. But what would happen if the cannonball were instead fired at the same speed *vertically*? It would rise to some maximum height, reverse direction, and then fall back to Earth. Then the old saying "What goes up must come down" would hold true, just as surely as a stone tossed skyward is returned by gravity (unless, as we shall see, its speed is great enough).

In today's spacefaring age, it is more accurate to say, "What goes up *may* come down," for a critical starting speed exists that permits a projectile to escape Earth. This critical speed is called the **escape speed** or, if direction is involved, the *escape velocity*. From the surface of Earth, escape speed is 11.2 km/s. If you launch a projectile at any speed greater than that, it leaves Earth, traveling slower and slower, never stopping due to Earth's gravity.* We can understand the magnitude of this speed from an energy point of view.

How much work would be required to lift a payload against the force of Earth's gravity to a distance extremely far ("infinitely far") away? We might think that the change of PE would be infinite because the distance is infinite. But gravity diminishes with distance by the inverse-square law. The force of gravity on the payload would be strong only near Earth. Most of the work done in launching a rocket occurs within 10,000 km or so of Earth. It turns out that the change of PE of a 1-km body moved from the surface of Earth to an infinite distance is 62 million J (62 MJ). So to put a payload infinitely far from Earth's surface requires at least 62 million joules of energy per kilogram of load. We won't go through the calculation here, but 62 MJ/kg corresponds to a speed of 11.2 km/s, whatever the total mass involved. This is the escape speed from the surface of Earth.**

If we give a payload any more energy than 62 MJ/kg at the surface of Earth or, equivalently, any more speed than 11.2 km/s, then, neglecting air resistance, the payload will escape from Earth, never to return. As the payload continues outward, its PE increases and its KE decreases. Earth's gravitational pull continuously slows it down but never reduces its speed to zero. The payload escapes.

The escape speeds from various bodies in the solar system are shown in Table 4.1. Note that the escape speed from the surface of the Sun is 620 km/s. Even at 150,000,000 km from the Sun (Earth's distance), the escape speed to break free of the Sun's influence is 42.5 km/s—considerably more than the escape speed from Earth. An object projected from Earth at a speed greater than 11.2 km/s but less than 42.5 km/s will escape Earth but not the Sun. Rather than recede forever, it will take up an orbit around the Sun.

The first probe to escape the solar system, *Pioneer 10*, was launched from Earth in 1972 with a speed of only 15 km/s. The escape was accomplished by

---

* Escape speed from any planet or any body is given by $v = \sqrt{2GM/d}$, where G is the universal gravitational constant, M is the mass of the attracting body, and d is the distance from its center. (At the surface of the body, d would simply be the radius of the body.) For a bit more mathematical insight, compare this formula with the one for orbital speed in the footnote on page 103.

** Interestingly enough, this might well be called the *maximum falling speed*. Any object, however far from Earth, released from rest and allowed to fall to Earth only under the influence of Earth's gravity would not exceed 11.2 km/s. (With air friction, it would be less.)

TABLE 4.1 ESCAPE SPEEDS AT THE SURFACE OF BODIES
IN THE SOLAR SYSTEM

Astronomical Body	Mass (Earth masses)	Radius (Earth radii)	Escape Speed (km/s)
Sun	333,000	109	620
Sun (at a distance of Earth's orbit)		23,500	42.2
Jupiter	318	11	60.2
Saturn	95.2	9.2	36.0
Neptune	17.3	3.47	24.9
Uranus	14.5	3.7	22.3
Earth	1.00	1.00	11.2
Venus	0.82	0.95	10.4
Mars	0.11	0.53	5.0
Mercury	0.055	0.38	4.3
Moon	0.0123	0.27	2.4

directing the probe into the path of oncoming Jupiter. It was whipped about by Jupiter's great gravitational field, picking up speed in the process—similar to the increase in the speed of a baseball encountering an oncoming bat. Its speed of departure from Jupiter was increased enough to exceed the escape speed from the Sun at the distance of Jupiter. *Pioneer 10* passed the orbit of Pluto in 1984. Unless it collides with another body, it will wander indefinitely through interstellar space. Like a note inside a bottle cast into the sea, *Pioneer 10* contains information about Earth that might be of interest to extraterrestrials, in hopes that it will one day "wash up" and be found on some distant "seashore."

It is important to stress that the escape speed of a body is the initial speed given by a brief thrust, after which there is no force to assist motion. One could escape Earth at *any* sustained speed more than zero, given enough time. For example, suppose a rocket is launched to a destination such as the Moon. If the rocket engines burn out when still close to Earth, the rocket needs a minimum speed of 11.2 km/s. But if the rocket engines can be sustained for long periods of time, the rocket could reach the Moon without ever attaining 11.2 km/s.

fyi

■ You won't fully appreciate the frontiers of physical science unless you're familiar with its foothills.

FIGURE 4.36

Pioneer 10, launched from Earth in 1972, passed the outermost planet in 1984 and is now wandering in our galaxy.

FIGURE 4.37

The European–U.S. spacecraft *Cassini* beams close-up images of Saturn and its giant moon Titan to Earth. It also measures surface temperatures, magnetic fields, and the size, speed, and trajectories of tiny surrounding space particles.

> Just as planets fall around the Sun, stars fall around the centers of galaxies. Those with insufficient tangential speeds are pulled into, and are gobbled up by, the galactic nucleus—usually a black hole.

> The mind that encompasses the universe is as marvelous as the universe that encompasses the mind.

It is interesting to note that the accuracy with which an unoccupied rocket reaches its destination is not accomplished by staying on a planned path or by getting back on that path if the rocket strays off course. No attempt is made to return the rocket to its original path. Instead, the control center in effect asks, "Where is it now and what is its velocity? What is the best way to reach its destination, given its present situation?" With the aid of high-speed computers, the answers to these questions are used to find a new path. Corrective thrusters direct the rocket to this new path. This process is repeated all the way to the goal.*

SUMMARY OF TERMS

Law of universal gravitation Every body in the universe attracts every other body with a force that, for two bodies, is directly proportional to the product of their masses and inversely proportional to the square of the distance separating them:

$$F = G\frac{m_1 m_2}{d^2}$$

Inverse-square law The intensity of an effect from a localized source spreads uniformly throughout the surrounding space and weakens with the inverse square of the distance:

$$\text{Intensity} = \frac{1}{\text{distance}^2}$$

Gravity follows an inverse-square law, as do the effects of electric, light, sound, and radiation phenomena.

Weight The force that an object exerts on a supporting surface (or, if suspended, on a supporting string), which is often, but not always, due to the force of gravity.

Weightless Being without a support force, as in free fall.

Projectile Any object that moves through the air or through space under the influence of gravity.

Parabola The curved path followed by a projectile under the influence of constant gravity only.

Satellite A projectile or small celestial body that orbits a larger celestial body.

Ellipse The oval path followed by a satellite. The sum of the distances from any point on the path to two points called foci is a constant. When the foci are together at one point, the ellipse is a circle. As the foci get farther apart, the ellipse becomes more *eccentric*.

Escape speed The speed that a projectile, space probe, or similar object must reach to escape the gravitational influence of Earth or of another celestial body to which it is attracted.

* Is there a lesson to be learned here? Suppose you find that you are off course. You may, like the rocket, find it more fruitful to follow a course that leads to your goal as best plotted from your present position and circumstances, rather than try to get back on the course you plotted from a previous position, perhaps under different circumstances.

REVIEW QUESTIONS

1. What did Newton discover about gravity?
2. What is the Newtonian synthesis?

4.1 The Universal Law of Gravity

3. In what sense does the Moon "fall"?
4. State Newton's law of universal gravitation in words. Then do the same with one equation.
5. What is the magnitude of gravitational force between two 1-kg bodies that are 1 m apart?
6. What is the magnitude of the gravitational force between Earth and a 1-kg body?

4.2 Gravity and Distance: The Inverse-Square Law

7. How does the force of gravity between two bodies change when the distance between them is quadrupled?
8. Where do you weigh more—at sea level or atop one of the peaks of the Rocky Mountains? Defend your answer.

4.3 Weight and Weightlessness

9. Would the springs inside a bathroom scale be more compressed or less compressed if you weighed yourself in an elevator that accelerated upward? Downward?
10. Would the springs inside a bathroom scale be more compressed or less compressed if you weighed yourself in an elevator that moved upward at *constant velocity*? Downward at *constant velocity*?
11. Explain why occupants of the International Space Station have no weight, yet are firmly in the grips of Earth's gravity.
12. When is your weight equal to *mg*?

4.4 Universal Gravitation

13. What was the cause of perturbations discovered in the orbit of the planet Uranus? What greater discovery did this lead to?
14. Why was the status of Pluto recently demoted to that of a dwarf planet?
15. What percentages of the universe are presently speculated to be composed of dark matter and of dark energy?

4.5 Projectile Motion

16. Why does the vertical component of velocity for a projectile change with time, whereas the horizontal component of velocity doesn't?

17. A stone is thrown upward at an angle. What happens to the horizontal component of its velocity as it rises? As it falls?
18. A stone is thrown upward at an angle. What happens to the vertical component of its velocity as it rises? As it falls?
19. A projectile is launched upward at an angle of 75° from the horizontal and strikes the ground a certain distance downrange. For what other angle of launch at the same speed would this projectile land just as far away?
20. A projectile is launched vertically at 100 m/s. If air resistance can be neglected, at what speed does it return to its initial level?

4.6 Fast-Moving Projectiles—Satellites

21. Why does a projectile that moves horizontally at 8 km/s follow a curve that matches the curvature of Earth?
22. Why is it important that the projectile in the previous question be above Earth's atmosphere?
23. When a satellite is above Earth's atmosphere, is it also beyond the pull of Earth's gravity? Defend your answer.

4.7 Circular Satellite Orbits

24. Why doesn't the force of gravity change the speed of a bowling ball as it rolls along a bowling lane?
25. Why doesn't the force of gravity change the speed of a satellite in circular orbit?
26. For orbits of greater altitude, is the period longer or shorter?

4.8 Elliptical Orbits

27. Why does the force of gravity change the speed of a satellite in an elliptical orbit?
28. At what part of an elliptical orbit does a satellite have the greatest speed? The least speed?

4.9 Escape Speed

29. What happens to a satellite close to Earth's surface if it is given a speed exceeding 11.2 km/s?
30. Although a space vehicle can outrun Earth's gravity, can it get entirely beyond Earth's gravity?

EXERCISES

● BEGINNER   ■ INTERMEDIATE   ◆ EXPERT

1. ● Comment on whether the following label on a consumer product should be cause for concern. *CAUTION: The mass of this product pulls on every other mass in the universe, with an attracting force that is proportional to the product of the masses and inversely proportional to the square of the distance between them.*

2. ● Gravitational force acts on all bodies in proportion to their masses. Why, then, doesn't a heavy body fall faster than a light body?

3. ● Is the force of gravity stronger on a piece of iron than on a piece of wood if both have the same mass? Defend your answer.

4. ● Is the force of gravity stronger on a crumpled piece of paper than on an identical piece of paper that has not been crumpled? Defend your answer.

5. ● A friend says that astronauts in orbit are weightless because they're beyond the pull of Earth's gravity. Correct your friend's ignorance.

6. ● Somewhere between Earth and the Moon, gravity from these two bodies on a space pod would cancel. Is this location nearer Earth or the Moon?

7. ● Is the acceleration due to gravity more or less atop Mt. Everest than at sea level? Defend your answer.

8. ● An astronaut lands on a planet that has the same mass as Earth but twice the diameter. How does the astronaut's weight differ from that on Earth?

9. ● An astronaut lands on a planet that has twice the mass of Earth and twice the diameter. How does the astronaut's weight differ from that on Earth?

10. ● If Earth somehow expanded to a larger radius, with no change in mass, how would your weight be affected? How would it be affected if Earth instead shrunk? (Hint: Let the equation for gravitational force guide your thinking.)

11. ● A small light source located 1 m in front of a 1-m^2 opening illuminates a wall behind. If the wall is 1 m behind the opening (2 m from the light source), the illuminated area covers 4 m^2. How many square meters are illuminated if the wall is 3 m from the light source? 5 m? 10 m?

12. ● The intensity of light from a central source varies inversely as the square of the distance. If you lived on a planet only half as far from the Sun as our Earth, how would the light intensity compare with that on Earth? How about a planet 10 times as far away as Earth?

13. ■ The planet Jupiter is more than 300 times as massive as Earth, so it might seem that a body on the surface of Jupiter would weigh 300 times as much as it would weigh on Earth. But it so happens that a body would scarcely weigh three times as much on the surface of Jupiter as it would on the surface of Earth. Can you think of an explanation for why this is so? (Hint: Let the terms in the equation for gravitational force guide your thinking.)

14. ● Why do the passengers in high-altitude jet planes feel the sensation of weight while passengers in an orbiting space vehicle, such as a space shuttle, do not?

15. ● If you were in a car that drove off the edge of a cliff, why would you be momentarily weightless? Would gravity still be acting on you?

16. ● What two forces act on you while you are in a moving elevator? When are these forces of equal magnitude and when are they not?

17. ● If you were in a freely falling elevator and you dropped a pencil, it would hover in front of you. Is a force of gravity acting on the pencil? Defend your answer.

18. ● Your friend says the primary reason astronauts in orbit feel weightless is because they are being pulled by other planets and stars. Why do you agree or disagree?

19. ■ Explain why the following reasoning is wrong. "The Sun attracts all bodies on Earth. At midnight, when the Sun is directly below, it pulls on you in the same direction as Earth pulls on you; at noon, when the Sun is directly overhead, it pulls on you in a direction opposite to Earth's pull on you. Therefore, you should be somewhat heavier at midnight and somewhat lighter at noon."

20. ■ Which requires more fuel—a rocket going from Earth to the Moon or a rocket returning from the Moon to Earth? Why?

21. ● Some people dismiss the validity of scientific theories by saying they are "only" theories. The law of universal gravitation is a theory. Does this mean that scientists still doubt its validity? Explain.

22. ● Suppose you roll a ball off a tabletop. Does the time to hit the floor depend on the speed of the ball? (Does a faster ball take a longer time to hit the floor?) Defend your answer.

23. ● A heavy crate accidentally falls from a high-flying airplane just as it flies directly above a shiny red Porsche smartly parked in a car lot. Relative to the Porsche, where does the crate crash?

24. ● In the absence of air resistance, why does the horizontal component of a projectile's motion not change, while the vertical component does?

25. ● At what point in its trajectory does a batted baseball have its minimum speed? If air resistance can be neglected, how does this compare with the horizontal component of its velocity at other points?

26. ● A friend claims that bullets fired by some high-powered rifles travel for many meters in a straight-line path before they start to fall. Another friend disputes this claim and states that all bullets from any rifle drop beneath a straight-line path a vertical distance given by $\frac{1}{2}gt^2$ and that the curved path is apparent for low velocities and less apparent for high velocities. Now it's your turn: Do all bullets drop the same vertical distance in equal times? Explain.

27. ● Two golfers each hit a ball at the same speed, but one at 60° with the horizontal and the other at 30°. Which ball goes farther? Which hits the ground first? (Ignore air resistance.)

28. ● A park ranger shoots a monkey hanging from a branch of a tree with a tranquilizing dart. The ranger aims directly at the monkey, not realizing that the dart will follow a parabolic path and thus will fall below the monkey. The monkey, however, sees the dart leave the gun and lets go of the branch to avoid being hit. Will the monkey be hit anyway? Does the velocity of the dart affect your answer, assuming that it is great enough to travel the horizontal distance to the tree before hitting the ground? Defend your answer.

29. ■ A projectile is fired straight upward at 141 m/s. How fast is it moving at the instant it reaches the top of its trajectory? Suppose instead that it were fired upward at 45°. What would be its speed at the top of its trajectory?

30. ● When you jump upward, your hang time is the time your feet are off the ground. Does hang time depend on your vertical component of velocity when you jump, your horizontal component of velocity, or both? Defend your answer.

31. ● The hang time of a basketball player who jumps a vertical distance of 2 ft (0.6 m) is about $\frac{2}{3}$ s. What is the hang time if the player reaches the same height while jumping 4 ft (1.2 m) horizontally?

32. ● Because the Moon is gravitationally attracted to Earth, why doesn't it simply crash into Earth?

33. ● Does the speed of a falling object depend on its mass? Does the speed of a satellite in orbit depend on its mass? Defend your answers.

34. ● If you have ever watched the launching of an Earth satellite, you may have noticed that the rocket starts vertically upward, then departs from a vertical course and continues its climb at an angle. Why does it start vertically? Why does it not continue vertically?

35. ● If a cannonball is fired from a tall mountain, gravity changes its speed all along its trajectory. But if it is fired fast enough to go into circular orbit, gravity does not change its speed at all. Explain.

36. ● A satellite can orbit at 5 km above the Moon, but not at 5 km above Earth. Why?

37. ● Would the speed of a satellite in close circular orbit about Jupiter be greater than, equal to, or less than 8 km/s?

38. ■ Why are satellites normally sent into orbit by firing them in an easterly direction, the direction in which Earth spins?

39. ■ Of all the United States, why is Hawaii the most efficient launching site for nonpolar satellites? (Hint: Look at the spinning Earth from above either pole and compare it to a spinning turntable.)

40. ■ Earth is closer to the Sun in December than in June. In which of these two months is Earth moving faster around the Sun?

41. ● What is the shape of the orbit when the velocity of the satellite is everywhere perpendicular to the force of gravity?

42. ■ A communications satellite with a 24-h period hovers over a fixed point on Earth. Why is it placed in orbit only in the plane of Earth's equator? (Hint: Think of the satellite's orbit as a ring around Earth.)

43. ■ If a flight mechanic drops a wrench from a high-flying jumbo jet, it crashes to Earth. If an astronaut on the orbiting space shuttle drops a wrench, does it crash to Earth also? Defend your answer.

44. ● How could an astronaut in a space shuttle "drop" an object vertically to Earth?

45. ● If you stopped an Earth satellite dead in its tracks, it would simply crash into Earth. Why, then, don't the communications satellites that "hover motionless" above the same spot on Earth crash into Earth?

46. ● The orbital velocity of Earth about the Sun is 30 km/s. If Earth were suddenly stopped in its tracks, it would simply fall radially into the Sun. Devise a plan whereby a rocket loaded with radioactive wastes could be fired into the Sun for permanent disposal. How fast and in what direction with respect to Earth's orbit should the rocket be fired?

47. ■ In an accidental explosion, a satellite breaks in half while in circular orbit about Earth. One half is brought momentarily to rest. What is the fate of the half brought to rest? What happens to the other half? (Hint: Think momentum conservation.)

48. ■ If Pluto were somehow stopped short in its orbit, it would fall into, rather than around, the Sun. How fast would it be moving when it hit the Sun?

49. ■ At which of the indicated positions does the satellite in elliptical orbit experience the greatest gravitational force? The greatest speed? The greatest velocity? The greatest momentum? The greatest kinetic energy? The greatest gravitational potential energy? The greatest total energy? The greatest acceleration?

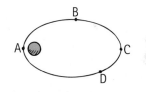

50. ◆ A rocket coasts in an elliptical orbit around Earth. To attain the greatest amount of KE for escape using a given amount of fuel, should it fire its engines at apogee (the point at which it is farthest from Earth) or at perigee (the point at which it is closest to Earth)? (Hint: Let the formula $Fd = \Delta KE$ be your guide to thinking. Suppose the thrust F is brief and of the same duration in either case. Then consider the distance d the rocket would travel during this brief burst at the apogee and at the perigee.)

PROBLEMS

● BEGINNER   ■ INTERMEDIATE   ◆ EXPERT

1. ● Consider a pair of planets that find that the distance between them is decreased by a factor of 5. Show that the force between them becomes 25 times as strong.

2. ● Many people mistakenly believe that the astronauts who orbit Earth are "above gravity." Earth's mass is 6×10^{24} kg, and its radius is 6.38×10^6 m (6380 km). Use the inverse-square law to show that in space-shuttle territory, 200 km above Earth's surface, the force of gravity on a shuttle is about 94% that at Earth's surface.

3. ■ The mass of a certain neutron star is 3.0×10^{30} kg (1.5 solar masses) and its radius is 8,000 m (8 km). Show that the force of gravity at the surface of this condensed, burned-out star is about 300 billion times that of Earth.

4. ■ A ball is thrown horizontally from a cliff at a speed of 10 m/s. Show that its speed one second later is 14.1 m/s.

5. ■ An airplane is flying horizontally with speed 1000 km/h (280 m/s) when an engine falls off. Neglecting air resistance, assume it takes 30 s for the engine to hit the ground.
(a) Show that the altitude of the airplane is 4500 m.
(b) Show that the horizontal distance that the aircraft engine falls is 8400 m.
(c) If the airplane somehow continues to fly as if nothing has happened, where is the engine relative to the airplane at the moment the engine hits the ground?

6. ■ A cannonball shot with an initial velocity of 141 m/s at an angle of 45° follows a parabolic path and hits a balloon at the top of its trajectory. Neglecting air resistance, show that the cannonball hits the balloon at a speed of 100 m/s.

7. ■ A certain satellite has a kinetic energy of 8 billion joules at perigee (the point at which it is closest to Earth) and 5 billion joules at apogee (the point at which it is farthest from Earth). As the satellite travels from apogee to perigee, how much work does the gravitational force do on it? Does its potential energy increase or decrease during this time, and by how much?

8. ■ Centripetal force is given by the equation $F_c = \dfrac{mv^2}{d}$,

where m is the mass of an object moving in a circular path at speed v and distance d from the center of the circular path. For the Moon circling planet Earth, gravity supplies the centripetal force. Equate centripetal force to gravitational force and show that the speed of the Moon in its orbit about Earth is $v = \sqrt{\dfrac{GM}{d}}$, where M is the mass of Earth and d is the distance between the centers of the Moon and Earth.

9. ■ Calculate the speed in meters per second at which Earth revolves about the Sun. You may assume the orbit is nearly circular.

10. ■ The Moon is about 3.8×10^5 km from Earth. Show that its average orbital speed about Earth is 1026 m/s.

11. ◆ The force of gravity on you by Earth is GmM/d^2, where G is the universal gravitational constant, m is your mass, M is the mass of Earth, and d is your distance from Earth's center.
(a) Use Newton's second law to show that your gravitational acceleration toward Earth at distance d from its center is $a = GM/d^2$.
(b) How does this equation support the finding that the acceleration due to gravity doesn't depend on the mass of an object in free fall?

12. ◆ The gravitational field about a massive object is defined to be the gravitational force per mass on an object in the vicinity of the massive object. The symbol for the gravitational field is boldfaced g (with magnitude the same as the magnitude of gravitational acceleration at that point, g).
(a) Show that the gravitational field a distance r from Earth's center is GM/r^2, where G is the universal gravitational constant and M is the mass of Earth.
(b) The value of g at Earth's surface is about 9.8 N/kg. Show that the value of g at a distance from Earth's center that is four times Earth's radius would be 0.6 m/s².

13. ◆ A rock thrown horizontally from a bridge hits the water below at a horizontal distance x directly below the throwing point. The rock travels a smooth parabolic path in time t.
(a) Show that the height of the bridge is $\frac{1}{2}gt^2$.
(b) What is the height of the bridge if the time the rock is airborne is 2 s?
(c) What information is here in Chapter 4 that wasn't in Chapter 2 for the solution of this problem?

14. ◆ A baseball is tossed at a steep angle into the air and makes a smooth parabolic path. Its time in the air is t and it reaches a maximum height h. Assume that air resistance is negligible.
(a) Show that the height reached by the ball is $gt^2/8$.
(b) If the ball is in the air for 4 s, show that the ball reaches a height of 19.6 m.
(c) If the ball reached the same height at when tossed at some other angle, would the time of flight be the same?

15. ◆ A penny moving at speed v slides off the horizontal surface of a coffee table a vertical distance y from the floor.
(a) Show that the penny lands a distance $v\sqrt{\dfrac{2y}{g}}$ from the base of the coffee table.
(b) If the speed is 3.5 m/s and the coffee table is 0.4 m tall, show that the distance the coin lands from the base of the table is 1.0 m.

16. ◆ Students in a lab measure the speed of a steel ball launched horizontally from a tabletop to be v. The tabletop is distance y above the floor. They place a tall tin coffee can of height $0.1y$ on the floor to catch the ball.
(a) Show that the can should be placed a horizontal distance from the base of the table of $v\sqrt{\dfrac{2(0.9)y}{g}}$.
(b) If the ball leaves the tabletop at a speed of 4.0 m/s, the tabletop is 1.5 m above the floor, and the can is 0.15 m tall, show that the center of the can should be placed a horizontal distance of 0.52 m from the base of the table.

ACTIVE EXPLORATIONS

1. Hold your hands outstretched in front of you, one twice as far from your eyes as the other, and make a casual judgment as to which hand looks bigger. Most people see them to be about the same size, while many see the nearer hand as slightly bigger. Almost no one, upon casual inspection, sees the nearer hand as four times as big; but, by the inverse-square law, the nearer hand should appear to be twice as tall and twice as wide and therefore seem to occupy four times as much of your visual field as the farther hand. Your belief that your hands are the same size is so strong that you likely overrule

this information. Now, if you overlap your hands slightly and view them with one eye closed, you'll see the nearer hand as clearly bigger. This raises an interesting question: What other illusions do you have that are not so easily checked?

2. Repeat the eyeballing experiment, only this time use two one-dollar bills—one regular, and the other folded along its middle lengthwise, and again widthwise, so it has $\frac{1}{4}$ the area. Now hold the two bills in front of your eyes. Where do you hold the folded one so that it looks the same size as the unfolded one? Nice enough activity to share with your friends?

3. With stick and strings, make a "trajectory stick" as shown on page 98.

READINESS ASSURANCE TEST (RAT)

If you have a good handle on this chapter, if you really do, then you should be able to score at least 7 out of 10 on this RAT. If you score less than 7, you need to study further before moving on.

Choose the BEST answer to each of the following.

1. The force of gravity between two planets depends on their
   (a) masses and distance apart.
   (b) planetary atmospheres.
   (c) rotational motions.
   (d) all of these

2. When the distance between two stars is reduced by $\frac{1}{5}$, the force between them
   (a) decreases by $\frac{1}{5}$.
   (b) decreases by $\frac{1}{25}$.
   (c) increases by a factor of 5.
   (d) increases by a factor of 25.

3. If the Sun were twice as massive, its pull on Mars would be
   (a) unchanged.
   (b) twice as much.
   (c) half as much.
   (d) four times as much.

4. When an astronaut in orbit is weightless, he or she is
   (a) beyond the pull of Earth's gravity.
   (b) still in the grips of Earth's gravity.
   (c) in the grips of interstellar gravity.
   (d) none of these

5. When no air resistance acts on a fast-moving baseball, its acceleration is
   (a) downward, *g*.
   (b) due to a combination of constant horizontal motion and accelerated downward motion.
   (c) opposite to the force of gravity.
   (d) zero.

6. A ball tossed at an angle of 30° with the horizontal goes as far downrange as one tossed at the same speed at an angle of
   (a) 45°.
   (b) 60°.
   (c) 75°.
   (d) none of these

7. When you toss a projectile sideways, it curves as it falls. It will become an Earth satellite if the curve it makes
   (a) matches the curve of Earth's surface.
   (b) results in a straight line.
   (c) spirals out indefinitely.
   (d) none of these

8. In circular orbit, the gravitational force on a satellite is
   (a) constant in magnitude.
   (b) at right angles to satellite motion.
   (c) in the direction of its acceleration.
   (d) all of these
   (e) none of these

9. The speed of a satellite in an elliptical orbit
   (a) remains constant.
   (b) acts at right angles to its motion.
   (c) varies.
   (d) all of these
   (e) none of these

10. A satellite in Earth orbit is above Earth's
    (a) atmosphere.
    (b) gravitational field.
    (c) both of these
    (d) neither of these

Answers to RAT
1a, 2d, 3b, 4b, 5a, 6b, 7a, 8d, 9c, 10a

MORE TO EXPLORE

Cole, K. C. *The Hole in the Universe: How Scientists Peered over the Edge of Emptiness and Found Everything*. New York: Harcourt, 2001. Enjoyable reading for the nonscientist.

Einstein, A., and L. Infeld. *The Evolution of Physics*. New York: Simon & Schuster, 1938. If you thought Einstein's writing was difficult to read, this book will change your mind.

For information on space-faring projections, visit the Web site of the National Space Society (NSS) at http://www.nss.org.

CHAPTER 4 ONLINE RESOURCES

The Physical Science Place

Interactive Figures
- 4.6, 4.14, 4.15, 4.19, 4.21, 4.28, 4.32, 4.35

Tutorials
- Motion and Gravity
- Orbits and Kepler's Laws
- Projectile Motion

Videos
- von Jolly's Method of Measuring the Attraction Between Two Masses
- Inverse-Square Law

- Weight and Weightlessness
- Apparent Weightlessness
- Discovery of Neptune
- Projectile Motion Demo
- More Projectile Motion
- Circular Orbits

Quiz

Flashcards

Links

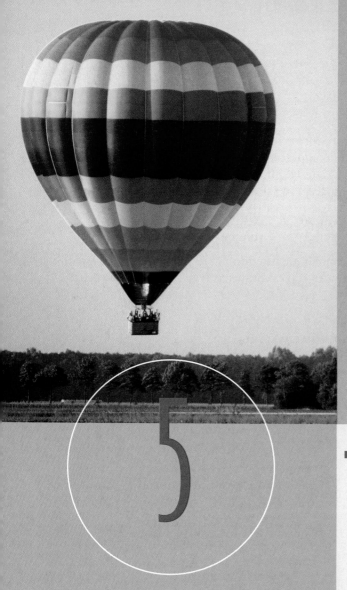

fluid mechanics

5.1 DENSITY

5.2 PRESSURE

5.3 BUOYANCY IN A LIQUID

5.4 ARCHIMEDES' PRINCIPLE

5.5 PRESSURE IN A GAS

5.6 ATMOSPHERIC PRESSURE

5.7 PASCAL'S PRINCIPLE

5.8 BUOYANCY IN A GAS

5.9 BERNOULLI'S PRINCIPLE

■ Liquids and gases have the ability to flow; hence, they are called *fluids*. Because they are both fluids we find that they obey similar mechanical laws. How is it that iron boats don't sink in water or that helium balloons don't sink from the sky? What determines whether an object will float or sink in both water and air? Why is gas compressible while liquid is not? Why is it impossible to breathe through a snorkel when you're under more than a meter of water? Why do your ears pop when riding an elevator? How do hydrofoils and airplanes attain lift? To discuss fluids, it is important to introduce two concepts—*density* and *pressure*.

FIGURE 5.1

When the volume of the bread is reduced, its density increases.

Link To Section 20.3

5.1 Density

An important property of a material, whether in the solid, liquid, or gaseous phase, is the measure of compactness: **density**. We think of density as the "lightness" or "heaviness" of materials of the same size. It is a measure of how much mass occupies a given space; it is the amount of matter per unit volume:

$$\text{Density} = \frac{\text{mass}}{\text{volume}}$$

The densities of a few materials are listed in Table 5.1. Mass is measured in grams or kilograms, and volume in cubic centimeters (cm^3) or cubic meters (m^3).*

TABLE 5.1 DENSITIES OF SOME MATERIALS

Material	Grams per Cubic Centimeter (g/cm^3)	Kilograms per Cubic Meter (kg/m^3)
Liquids		
Mercury	13.6	13,600
Glycerin	1.26	1,260
Seawater	1.03	1,025
Water at 4°C	1.00	1,000
Benzene	0.90	899
Ethyl alcohol	0.81	806
Solids		
Iridium	22.6	22,650
Osmium	22.6	22,610
Platinum	21.1	21,090
Gold	19.3	19,300
Uranium	19.0	19,050
Lead	11.3	11,340
Silver	10.5	10,490
Copper	8.9	8,920
Brass	8.6	8,600
Iron	7.8	7,874
Tin	7.3	7,310
Aluminum	2.7	2,700
Ice	0.92	919
Gases (atmospheric pressure at sea level)		
Dry air		
0°C	1.29	
10°C	1.25	
20°C	1.21	
30°C	1.16	
Helium	0.178	
Hydrogen	0.090	
Oxygen	1.43	

* A cubic meter is a sizable volume and contains a million cubic centimeters, so there are a million grams of water in a cubic meter (or, equivalently, a thousand kilograms of water in a cubic meter). Hence, 1 g/cm^3 = 1000 kg/m^3.

A gram of any material has the same mass as 1 cm^3 of water at a temperature of 4°C. So water has a density of 1 g/cm^3. Mercury's density is 13.6 g/cm^3, which means that it has 13.6 times as much mass as an equal volume of water. Iridium, a hard, brittle, silvery-white metal in the platinum family, is the densest substance on Earth.

A quantity known as weight density, commonly used when discussing liquid pressure, is expressed by the amount of weight per unit volume:*

$$\text{Weight density} = \frac{\text{weight}}{\text{volume}}$$

CHECK POINT

1. Which has the greater density—1 kg of water or 10 kg of water?
2. Which has the greater density—5 kg of lead or 10 kg of aluminum?
3. Which has the greater density—an entire candy bar or half a candy bar?

Were these your answers?

1. The density of any amount of water is the same: 1 g/cm^3 or, equivalently, 1000 kg/m^3, which means that the mass of water that would exactly fill a thimble of volume 1 cm^3 would be 1 g; or the mass of water that would fill a 1-m^3 tank would be 1000 kg. One kilogram of water would fill a tank only a thousandth as large, 1 L, whereas 10 kg would fill a 10-liter tank. Nevertheless, the important concept is that the ratio of mass/volume is the same for *any* amount of water.

2. Density is a *ratio* of weight or mass per volume, and this ratio is greater for any amount of lead than for any amount of aluminum—see Table 5.1.

3. Both the half and the entire candy bar have the same density.

■ 5.2 Pressure

Place a book on a bathroom scale; whether you place it on its back, on its side, or balanced on a corner, it still exerts the same force. The weight reading is the same. Now balance the book on the palm of your hand and you sense a difference—the *pressure* of the book depends on the area over which the force is distributed (Figure 5.2). There is a difference between force and pressure. **Pressure** is defined as the force exerted over a unit of area, such as a square meter or square foot:**

$$\text{Pressure} = \frac{\text{force}}{\text{area}}$$

FIGURE 5.2

Although the weight of both books is the same, the upright book exerts greater pressure against the table.

The Physical Science Place
Dam and Water

---

* Weight density is common to the United States Customary System (USCS) units, in which 1 ft^3 of fresh water (nearly 7.5 gallons) weighs 62.4 lb. So fresh water has a weight density of 62.4 lb/ft^3. Salt water is slightly denser at 64 lb/ft^3.

** Pressure may be measured in any unit of force divided by any unit of area. The standard international (SI) unit of pressure, the newton per square meter, is called the *pascal* (Pa), after the 17th-century theologian and scientist Blaise Pascal. A pressure of 1 Pa is very small and approximately equals the pressure exerted by a dollar bill resting flat on a table. Science types prefer kilopascals (1 kPa = 1000 Pa).

FIGURE 5.3

This water tower does more than store water. The height of the water above ground level ensures substantial and reliable water pressure to the many homes it serves.

FIGURE 5.4

Liquid pressure is the same for any given depth below the surface, regardless of the shape of the containing vessel.

> When measuring blood pressure, notice that you measure it in your upper arm—level with your heart.

CHECK POINT

Does a bathroom scale measure weight, pressure, or both?

Was this your answer?
A bathroom scale measures weight, the force that compresses an internal spring or equivalent. The weight reading is the same whether you stand on one or both feet (although the pressure on the scale is twice as much when standing on one foot).

PRESSURE IN A LIQUID

When you swim under water, you can feel the water pressure acting against your eardrums. The deeper you swim, the greater the pressure. What causes this pressure? It is simply the weight of the fluids directly above you—water plus air—pushing against you. As you swim deeper, more water is above you. Therefore, there's more pressure. If you swim twice as deep, twice the weight of water is above you, so the water's contribution to the pressure you feel is doubled. Added to the water pressure is the pressure of the atmosphere, which is equivalent to an extra 10.3-m depth of water. Because atmospheric pressure at Earth's surface is nearly constant, the pressure differences you feel under water depend only on changes in depth.

The pressure due to a liquid is precisely equal to the product of weight density and depth:*

$$\text{Liquid pressure} = \text{weight density} \times \text{depth}$$

Note that pressure does not depend on the volume of liquid. You feel the same pressure a meter deep in a small pool as you do a meter deep in the middle of the ocean. This is illustrated by the connecting vases shown in Figure 5.4. If the pressure at the bottom of a large vase were greater than the pressure at the bottom of a neighboring narrower vase, the greater pressure would force water sideways and then up the narrower vase to a higher level. We find, however, that this doesn't happen. Pressure depends on depth, not volume.

Water seeks its own level. This can be demonstrated by filling a garden hose with water and holding the two ends upright. The water levels are equal whether the ends are held close together or far apart. Pressure is depth dependent, not volume dependent. So we see there is an explanation for why water seeks its own level.

In addition to being depth dependent, liquid pressure is exerted equally in all directions. For example, if we are submerged in water, it makes no difference which way we tilt our heads—our ears feel the same amount of water pressure. Because a liquid can flow,

---

* This is derived from the definitions of pressure and density. Consider an area at the bottom of a vessel that contains liquid. The weight of the column of liquid directly above this area produces pressure. From the definition *weight density* = *weight/volume*, we can express this weight of liquid as *weight* = *weight density* × *volume*, where the volume of the column is simply the area multiplied by the depth. Then we get

$$\text{Pressure} = \frac{\text{force}}{\text{area}} = \frac{\text{weight}}{\text{area}} = \frac{\text{weight density} \times \text{volume}}{\text{area}} = \frac{\text{weight density} \times (\text{area} \times \text{depth})}{\text{area}}$$

$$= \text{weight density} \times \text{depth}$$

For the total pressure we should add to this equation the pressure due to the atmosphere on the surface of the liquid.

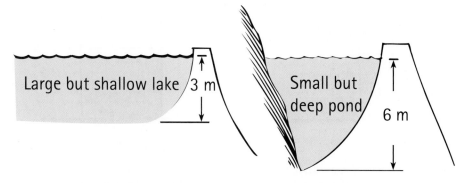

FIGURE 5.5

The average water pressure acting against the dam depends on the average depth of the water and not on the volume of water held back. The large shallow lake exerts only half the average pressure that the small deep pond exerts.

Link To Section 23.2

the pressure isn't only downward. We know pressure acts upward when we try to push a beach ball beneath the water's surface. The bottom of a boat is certainly pushed upward by water pressure. And we know water pressure acts sideways when we see water spurting sideways from a leak in an upright can. Pressure in a liquid at any point is exerted in equal amounts in all directions.

When liquid presses against a surface, a net force is directed perpendicular to the surface (Figure 5.6). If there is a hole in the surface, the liquid spurts at right angles to the surface before curving downward because of gravity (Figure 5.7). At greater depths the pressure is greater and the speed of the exiting liquid is greater.*

FIGURE 5.6

The forces due to liquid pressure against a surface combine to produce a net force that is perpendicular to the surface.

FIGURE 5.7

The force vectors act in a direction perpendicular to the inner container surface and increase with increasing depth.

5.3 Buoyancy in a Liquid

Anyone who has ever lifted a submerged object out of water is familiar with buoyancy, the apparent loss of weight of submerged objects. For example, lifting a large boulder off the bottom of a riverbed is a relatively easy task as long as the boulder is below the surface. When it is lifted above the surface, however, the force required to lift it is considerably more. This is because when the boulder is submerged, the water exerts an upward force on it—opposite in direction to gravity. This upward force is called the **buoyant force** and is a consequence of greater pressure at greater depth. Figure 5.8 shows why the buoyant force acts upward. Pressure is exerted everywhere against the object in a direction perpendicular to its surface. The arrows represent the magnitude and direction of forces at different places. Forces that produce pressures against the sides due to equal depths cancel

The Physical Science Place
Buoyancy

FIGURE 5.8

The greater pressure against the bottom of a submerged object produces an upward buoyant force.

* The speed of liquid exiting the hole is $\sqrt{2gh}$, where h is the depth below the free surface. Interestingly, this is the same speed that water or anything else would have if freely falling the same distance h.

FIGURE 5.9

When a stone is submerged, it displaces a volume of water equal to the volume of the stone.

Water displaced

FIGURE 5.10

The raised level due to placing a stone in the container is the same as if a volume of water equal to the volume of the stone were poured in.

FIGURE 5.11

A liter of water occupies a volume of 1000 cm³, has a mass of 1 kg, and weighs 9.8 N. Its density may therefore be expressed as 1 kg/L and its weight density as 9.8 N/L. (Seawater is slightly denser, about 10 N/L).

one another. Pressure is greatest against the bottom of the boulder simply because the bottom of the boulder is deeper. Because the upward forces against the bottom are greater than the downward forces against the top, the forces do not cancel, and there is a net force upward. This net force is the buoyant force.

If the weight of the submerged object is greater than the buoyant force, the object sinks. If the weight is equal to the buoyant force acting upward on the submerged object, it remains at any level, like a fish. If the buoyant force is greater than the weight of the completely submerged object, it rises to the surface and floats.

Understanding buoyancy requires understanding the meaning of the expression "volume of water displaced." If a stone is placed in a container that is already up to its brim with water, some water overflows (Figure 5.9). Water is *displaced* by the stone. A little thought tells us that the *volume of the stone*—that is, the amount of space it occupies or its number of cubic centimeters—is equal to the *volume of water displaced*. Place any object in a container partially filled with water, and the level of the surface rises (Figure 5.10). How high? That would be to exactly the level that would be reached by pouring in a volume of water equal to the volume of the submerged object. This is a good method for determining the volume of irregularly shaped objects: *A completely submerged object always displaces a volume of liquid equal to its own volume.*

5.4 Archimedes' Principle

Link To Section 22.2

The relationship between buoyancy and displaced liquid was first discovered in the third century BC by the Greek scientist Archimedes. It is stated as follows:

An immersed body is buoyed up by a force equal to the weight of the fluid it displaces.

The Physical Science Place
Archimedes' Principle

This relationship is called **Archimedes' principle**. It applies to liquids and gases, which are both fluids. If an immersed body displaces 1 kg of fluid, the buoyant force acting on it is equal to the weight of 1 kg.* By *immersed*, we mean either *completely* or *partially submerged*. If we immerse a sealed 1-L container halfway into the water, it displaces half a liter of water and is buoyed up by the weight of half a liter of water. If

---

* A kilogram is not a unit of force but a unit of mass. So, strictly speaking, the buoyant force is not 1 kg, but the *weight* of 1 kg, which is 9.8 N. We could as well say that the buoyant force is 1 *kilogram weight*, not simply 1 kg.

FIGURE 5.12

A 3-kg block weighs more in air than it does in water. When the block is submerged in water, its loss in weight is the buoyant force, which equals the weight of water displaced.

we immerse it completely (submerge it), it is buoyed up by the weight of a full liter (or 1 kg) of water. Unless the completely submerged container is compressed, the buoyant force equals the weight of 1 kg at *any* depth. This is because, at any depth, it can displace no greater volume of water than its own volume. And the weight of this volume of water (not the weight of the submerged object!) is equal to the buoyant force.

If a 25-kg object displaces 20 kg of fluid upon immersion, its apparent weight equals the weight of 5 kg. Notice in Figure 5.12 that the 3-kg block has an apparent weight equal to the weight of 1 kg when submerged. The apparent weight of a submerged object is its weight out of water minus the buoyant force.

CHECK POINT

1. Does Archimedes' principle tell us that if an immersed block displaces 10 N of fluid, the buoyant force on the block is 10 N?
2. A 1-L container completely filled with lead has a mass of 11.3 kg and is submerged in water. What is the buoyant force acting on it?
3. A boulder is thrown into a deep lake. As it sinks deeper and deeper into the water, does the buoyant force on it increase? Decrease?

Were these your answers?

1. Yes. Looking at it in a Newton's-third-law way, when the immersed block pushes 10 N of fluid aside, the fluid reacts by pushing back on the block with 10 N.
2. The buoyant force is equal to the weight of 1 kg (9.8 N) because the volume of water displaced is 1 L, which has a mass of 1 kg and a weight of 9.8 N. The 11.3 kg of the lead is irrelevant; 1 L of anything submerged in water displaces 1 L and is buoyed upward with a force 9.8 N, the weight of 1 kg. (Get this straight before going further!)
3. Buoyant force remains the same. It doesn't change as the boulder sinks because the boulder displaces the same volume of water at any depth. Because water is practically incompressible, its density is very nearly the same at all depths; hence, the weight of water displaced, or the buoyant force, is practically the same at all depths.

Perhaps your instructor will summarize Archimedes' principle by way of a numerical example to show that the difference between the upward-acting and the downward-acting forces on a submerged cube (due to differences of pressure) is numerically identical to the weight of fluid displaced. It makes no difference how deep the cube is placed, because, although the pressures are greater with increasing depths, the *difference* between the pressure up against the bottom of the cube and the pressure exerted downward against the top of the cube is the same at any depth (Figure 5.13). Whatever the shape of the submerged body, the buoyant force is equal to the weight of fluid displaced.

FIGURE 5.13

The difference in the upward and downward forces acting on the submerged block is the same at any depth.

FIGURE 5.14

An iron block sinks, while the same quantity of iron shaped like a bowl floats.

Buoyant force →

Weight of iron

Only in the special case of floating does the buoyant force acting on an object equal the object's weight.

FLOTATION

The Physical Science Place
Flotation

Iron is much denser than water and therefore sinks, but an iron ship floats. Why is this so? Consider a solid 1-ton block of iron. Iron is nearly eight times as dense as water, so when it is submerged it displaces only $\frac{1}{8}$ ton of water, which is certainly not enough to prevent it from sinking. Suppose we reshape the same iron block into a bowl, as shown in Figure 5.14. It still weighs 1 ton. When we place it in the water, it settles into the water, displacing a greater volume of water than before. The deeper it is immersed, the more water it displaces and the greater the buoyant force acting on it. When the buoyant force equals 1 ton, the iron sinks no further.

When the iron boat displaces a weight of water equal to its own weight, it floats. This is called the **principle of flotation**:

A floating object displaces a weight of fluid equal to its own weight.

FIGURE 5.15

The weight of a floating object equals the weight of the water displaced by the submerged part.

Every ship, submarine, or dirigible airship must be designed to displace a weight of fluid equal to its own weight. Thus, a 10,000-ton ship must be built wide enough to displace 10,000 tons of water before it immerses too deep in the water. The same applies to vessels in air. A dirigible or huge balloon that weighs

Physics in History

▪ Archimedes and the Gold Crown

According to legend, Archimedes (287–212 BC) had been given the task of determining whether a crown made for King Hiero II of Syracuse was of pure gold or contained some less expensive metals such as silver. Archimedes' problem was to determine the density of the crown without destroying it. He could weigh the crown, but determining its volume was a problem. The story tells us that Archimedes came to the solution when he noted the rise in water level while immersing his body in the public baths of Syracuse. Legend reports that he excitedly rushed naked through the streets shouting "Eureka! Eureka!" ("I have found it! I have found it!").

What Archimedes discovered was a simple and accurate way of finding the volume of an irregular object— the displacement method of determining volumes. Once he knew both the weight and volume, he could calculate the density. Then the density of the crown could be compared with the density of gold. Archimedes' insight preceded Newton's law of motion, from which Archimedes' principle can be derived, by almost 2000 years.

Earth Science

■Floating Mountains

Mountains float on Earth's semi-liquid mantle just as icebergs float in water. Both the mountains and icebergs are less dense than the material they float upon. Just as most of an iceberg is below the water surface (90%), most of a mountain (about 85%) extends into the dense semiliquid mantle. If you could shave off the top

of an iceberg, the iceberg would be lighter and be buoyed up to nearly its original height before its top was shaved. Similarly, when mountains erode they are lighter, and are pushed up from below to float to nearly their original heights. So when a kilometer of mountain erodes away, some 85% of a kilometer of mountain returns. That's why it takes so long for mountains to weather away. Mountains, like icebergs, are bigger than they appear

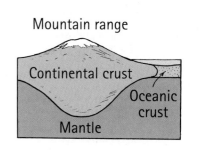

Mountain range

to be. The concept of floating mountains is isostacy—Archimedes' principle for rocks.

100 tons displaces at least 100 tons of air. If it displaces more, it rises; if it displaces less, it descends. If it displaces exactly its weight, it hovers at constant altitude.

Because the buoyant force upon a body equals the weight of the fluid it displaces, denser fluids exert more buoyant force upon a body than less-dense fluids of the same volume. A ship therefore floats higher in salt water than in fresh water because salt water is slightly denser than fresh water. In the same way, a solid chunk of iron floats in mercury even though it sinks in water.

CHECK POINT

Fill in the blanks for these statements:
1. The volume of a submerged body is equal to the _____ of the fluid displaced.
2. The weight of a floating body is equal to the _____ of the fluid displaced.
3. Why is it easier to float in salt water than in fresh water?

Were these your answers?
1. volume
2. weight
3. When you're floating, the weight of water you displace equals your weight. Salt water is denser, so you don't "sink" as far to displace your weight. You'd float even higher in mercury (density 13.6 g/cm^3), and you'd sink completely in alcohol (density 0.8 g/cm^3).

Notice in our discussion of liquids that Archimedes' principle and the law of flotation were stated in terms of *fluids*, not liquids. That's because although

FIGURE 5.16

A floating object displaces a weight of fluid equal to its own weight.

People who can't float are, 9 times out of 10, males. Most males are more muscular and slightly denser than females. Also, cans of diet soda float whereas cans of regular soda sink in water. What does this tell you about their relative densities?

FIGURE 5.17

The same ship empty and loaded. How does the weight of its load compare to the weight of additional water displaced?

Liquids and gases are both fluids. A gas takes the shape of its container. A liquid does so only below its surface.

Link To Section 24.6

FIGURE 5.18

When the density of gas in the tire is increased, pressure is increased.

FIGURE 5.19

When the volume of gas is decreased, density and therefore pressure are increased.

liquids and gases are different phases of matter, they are both fluids, with much the same mechanical principles. Let's turn our attention to the mechanics of gases in particular.

5.5 Pressure in a Gas

The primary difference between a gas and a liquid is the distance between molecules. In a gas, the molecules are far apart and free from the cohesive forces that dominate their motions in the liquid and solid phases. Molecular motions in a gas are less restricted. A gas expands, fills all space available to it, and exerts a pressure against its container. Only when the quantity of gas is very large, such as Earth's atmosphere or a star, do the gravitational forces limit the size or determine the shape of the mass of gas.

BOYLE'S LAW

The air pressure inside the inflated tires of an automobile is considerably greater than the atmospheric pressure outside. The density of air inside is also greater than that of the air outside. To understand the relation between pressure and density, think of the molecules of air (primarily nitrogen and oxygen) inside the tire. The air molecules behave like tiny billiard balls, randomly moving and banging against the inner walls, producing a jittery force that appears to our coarse senses as a steady push. This pushing force, averaged over the wall area, provides the pressure of the enclosed air.

Suppose there are twice as many molecules in the same volume (Figure 5.18). Then the air density is doubled. If the molecules move at the same average speed—or, equivalently, if they have the same temperature—then the number of collisions is doubled. This means that the pressure is doubled. So pressure is proportional to density.

We double the density of air in the tire by doubling the amount of air. We can also double the density of a *fixed* amount of air by compressing it to half its volume. Consider the cylinder with the movable piston in Figure 5.19. If the piston is pushed downward so that the volume is half the original volume, the density of molecules is doubled, and the pressure is correspondingly doubled. Decrease the volume to a third of its original value, and the pressure is increased by three, and so forth (provided the temperature remains the same).

Notice in these examples with the piston that the product of pressure and volume remains the same. For example, a doubled pressure multiplied by a halved volume gives the same value as a tripled pressure multiplied by a one-third volume. In general, we can state that the product of pressure and volume for a given mass of gas is a constant *as long as the temperature does not change. Pressure × volume* for a sample of gas at some initial time is equal to any *different pressure × different volume* of the same sample of gas at some later time. In shorthand notation,

$$P_1 V_1 = P_2 V_2$$

where P_1 and V_1 represent the original pressure and volume, respectively, and P_2 and V_2 the second pressure and volume. This relationship is called **Boyle's law**, after Robert Boyle, the 17th-century physicist who is credited with its discovery.*

Boyle's law applies to ideal gases. An ideal gas is one in which the disturbing effects of the forces between molecules and the finite size of the individual

* A general law that takes temperature changes into account is $P_1 V_1 / T_1 = P_2 V_2 / T_2$, where T_1 and T_2 represent the initial and final *absolute* temperatures, measured in SI units called kelvins (Chapter 6).

molecules can be neglected. Air and other gases under normal pressures and temperatures approach ideal gas conditions.

Link To Section 25.2

CHECK POINT

1. A piston in an airtight pump is withdrawn so that the volume of the air chamber is tripled. What is the change in pressure?
2. A scuba diver breathes compressed air beneath the surface of water. If she holds her breath while returning to the surface, what happens to the volume of her lungs?

Were these your answers?

1. The pressure in the piston chamber is reduced to one-third. This is the principle that underlies a mechanical vacuum pump.
2. When she rises toward the surface, the surrounding water pressure on her body decreases, allowing the volume of air in her lungs to increase—ouch! A first lesson in scuba diving is to not hold your breath when ascending. To do so can be fatal.

5.6 Atmospheric Pressure

We live at the bottom of an ocean of air. The atmosphere, much like the water in a lake, exerts a pressure. One of the most celebrated experiments demonstrating the pressure of the atmosphere was conducted in 1654 by Otto von Guericke, burgermeister of Magdeburg and inventor of the vacuum pump. Von Guericke placed together two copper hemispheres about 0.5 m in diameter to form a sphere, as shown in Figure 5.20. He set a gasket made of a ring of leather soaked in oil and wax between them to make an airtight joint. When he evacuated the sphere with his vacuum pump, two teams of eight horses each were unable to pull the hemispheres apart.

The Physical Science Place
Air Has Weight
Air Is Matter
Air Has Pressure

Link To Sections 24.1, 24.4

Interestingly, von Guericke's demonstration preceded knowledge of Newton's third law. The forces on the hemispheres would have been the same if he had used only one team of horses and tied the other end of the rope to a tree!

FIGURE 5.20

The famous "Magdeburg hemispheres" experiment of 1654, demonstrating atmospheric pressure. Two teams of horses couldn't pull the evacuated hemispheres apart. Were the hemispheres sucked together or pushed together? By what?

FIGURE 5.21

Is the piston pulled up or pushed up?

FIGURE 5.22

You don't notice the weight of a bag of water while you're submerged in water. Similarly, you don't notice that the air around you has weight.

FIGURE 5.23

The mass of air that would occupy a bamboo pole that extends to the "top" of the atmosphere is about 1 kg. This air has a weight of about 10 N.

When the air pressure inside a cylinder like that shown in Figure 5.21 is reduced, an upward force is exerted on the piston. This force is large enough to lift a heavy weight. If the inside diameter of the cylinder is 12 cm or greater, a person can be lifted by this force.

What do the experiments of Figures 5.20 and 5.21 demonstrate? Do they show that air exerts pressure or that there is a "force of suction"? If we say there is a force of suction, then we assume that a vacuum can exert a force. But what is a vacuum? It is an absence of matter; it is a condition of nothingness. How can nothing exert a force? The hemispheres are not sucked together, nor is the piston holding the weight sucked upward. The pressure of the atmosphere is pushing against the hemispheres and the piston.

Just as water pressure is caused by the weight of water, **atmospheric pressure** is caused by the weight of air. We have adapted so completely to the invisible air that we sometimes forget it has weight. Perhaps a fish "forgets" about the weight of water in the same way. The reason we don't feel this weight crushing against our bodies is that the pressure inside our bodies equals that of the surrounding air. There is no net force for us to sense.

At sea level, 1 m^3 of air at 20°C has a mass of about 1.2 kg. To estimate the mass of air in your room, estimate the number of cubic meters there, multiply by 1.2 kg/m^3, and you'll have the mass. Don't be surprised if it's heavier than your kid sister. If your kid sister doesn't believe air has weight, maybe it's because she's always surrounded by air. Hand her a plastic bag of water and she'll tell you it has weight. But hand her the same bag of water while she's submerged in a swimming pool, and she won't feel the weight. We don't notice that air has weight because we're submerged in air.

Whereas water in a lake has the same density at any level (assuming constant temperature), the density of air in the atmosphere decreases with altitude. Although 1 m^3 of air at sea level has a mass of about 1.2 kg, at 10 km, the same volume of air has a mass of about 0.4 kg. To compensate for this, airplanes are pressurized; the additional air needed to fully pressurize a 747 jumbo jet, for example, is more than 1000 kg. Air is heavy, if you have enough of it.

Consider the mass of air in an upright 30-km-tall hollow bamboo pole that has an inside cross-sectional area of 1 cm^2. If the density of air inside the pole matches the density of air outside, the enclosed mass of air would be about 1 kg. The weight of this much air is about 10 N. So the air pressure at the bottom of the bamboo pole would be about 10 N/cm^2. Of course, the same is true without the bamboo pole. There are 10,000 cm^2 in 1 m^2, so a column of air 1 m^2 in cross section that extends up through the atmosphere has a mass of about 10,000 kg. The weight of this air is about 100,000 N. This weight produces a pressure of 100,000 N/m^2—or equivalently, 100,000 pascals (Pa), or 100 kilopascals (kPa). To be more precise, the average atmospheric pressure at sea level is 101.3 kPa.*

The pressure of the atmosphere is not uniform. Besides altitude variations, there are variations in atmospheric pressure at any one locality due to moving fronts and storms. Measurement of changing air pressure is important to meteorologists in predicting weather.

---

* The pascal is the SI unit of measurement. The average pressure at sea level (101.3 kPa) is often called 1 atmosphere (atm). In British units, the average atmospheric pressure at sea level is 14.7 lb/in^2 (pounds per square inch, or psi).

CHECK
■ POINT

1. Estimate the mass of air in kilograms in a classroom that has a 200-m² floor area and a 4-m-high ceiling. (Assume a chilly 10°C temperature).
2. Why doesn't the pressure of the atmosphere break windows?

Were these your answers?

1. The mass of air is 1000 kg. The volume of air is 200 m² × 4 m = 800 m³; each cubic meter of air has a mass of about 1.25 kg, so 800 m³ × 1.25 kg/m³ = 1000 kg (about a ton).
2. Atmospheric pressure is exerted on *both* sides of a window, so no net force is exerted on the window. If for some reason the pressure is reduced or increased on one side only, as in a strong wind, then watch out!

BAROMETERS

An instrument used for measuring the pressure of the atmosphere is called a **barometer**. A simple mercury barometer is illustrated in Figure 5.25. A glass tube, longer than 76 cm and closed at one end, is filled with mercury and tipped upside down in a dish of mercury. The mercury in the tube flows out of the submerged open bottom until the difference in the mercury levels in the tube and the dish is 76 cm. The empty space trapped above, except for some mercury vapor, is a pure vacuum.

The explanation for the operation of such a barometer is similar to that of children balancing on a seesaw. The barometer "balances" when the weight of liquid in the tube exerts the same pressure as the atmosphere outside. Whatever the width of the tube, a 76-cm column of mercury weighs the same as the air that would fill a vertical 30-km tube of the same width. If the atmospheric pressure increases, then the atmosphere pushes down harder on the mercury in the dish and pushes the mercury higher in the tube. Then the increased height of the mercury column exerts an equal balancing pressure.

760 mm

Water could instead be used to make a barometer, but the glass tube would have to be much longer—13.6 times as long, to be exact. The density of mercury is 13.6 times the density of water. That's why a tube of water 13.6 times longer than one of mercury (of the same cross section) is needed to provide the same weight as mercury in the tube. A water barometer would have to be 13.6 × 0.76 m, or 10.3 m high—too tall to be practical.

What happens in a barometer is similar to what happens when you drink through a straw. By sucking, you reduce the air pressure in the straw when it is placed in a drink. Atmospheric pressure on the drink then pushes the liquid up into the reduced-pressure region. Strictly speaking, the liquid is not sucked up; it is pushed up the straw by the pressure of the atmosphere. If the atmosphere is prevented from pushing on the surface of the drink, as in the party-trick bottle with the straw through an airtight cork stopper, one can suck and suck and get no drink.

If you understand these ideas, you can understand why there is a 10.3-m limit on the height to which water can be lifted with vacuum pumps. The old-fashioned farm-type pump shown in Figure 5.27 operates by producing a partial vacuum in a pipe that extends down into the water below.

FIGURE 5.24

The weight of air that presses down on a 1-m² surface at sea level is about 100,000 N. So atmospheric pressure is about 10^5 N/m², or about 100 kPa.

FIGURE 5.25

A simple mercury barometer. Mercury is pushed up into the tube by atmospheric pressure.

Workers in underwater construction work in an environment of compressed air. The air pressure in their underwater chambers is at least as much as the combined pressure of water and atmosphere outside.

FIGURE 5.26

Strictly speaking, they do not suck the soda up the straws. They instead reduce pressure in the straws, which allows the weight of the atmosphere to press the liquid up into the straws. Could they drink a soda this way on the Moon?

FIGURE 5.27

The atmosphere pushes water from below up into a pipe that is evacuated of air by the pumping action.

When the pump handle is raised, air in the pipe is "thinned" as it expands to fill a larger volume. Atmospheric pressure on the well surface pushes water up into the pipe, causing water to overflow at the spout.

FIGURE 5.28

The aneroid barometer.

Atmospheric pressure on the surface of the water simply pushes the water up into the region of reduced pressure inside the pipe. Can you see that, even with a perfect vacuum, the maximum height to which water can be lifted in this way is 10.3 m?

A small portable instrument that measures atmospheric pressure is the *aneroid barometer* (Figure 5.28). A metal box partially exhausted of air with a slightly flexible lid bends in or out with changes in atmospheric pressure. Motion of the lid is indicated on a scale by a mechanical spring-and-lever system. Atmospheric pressure decreases with increasing altitude, so a barometer can be used to determine elevation. An aneroid barometer calibrated for altitude is called an *altimeter* (altitude meter). Some of these instruments are sensitive enough to indicate a change in elevation as you walk up a flight of stairs.*

Reduced air pressures are produced by pumps, which work by virtue of a gas tending to fill its container. If a space with less pressure is provided, gas flows from the region of higher pressure to the one of lower pressure. A vacuum pump simply provides a region of lower pressure into which the normally fast-moving gas molecules randomly move. The air pressure is repeatedly lowered by piston and valve action (Figure 5.29).

FIGURE 5.29

A mechanical vacuum pump. When the piston is lifted, the intake valve opens and air moves in to fill the empty space. When the piston is moved downward, the outlet valve opens and the air is pushed out. What changes would you make to convert this pump into an air compressor?

Piston

Piston

Intake   Outlet   Intake   Outlet

* Evidence of a noticeable pressure difference over a 1-m or less difference in elevation is any small helium-filled balloon that rises in air. The atmosphere really does push with more force against the lower bottom than against the higher top!

5.7 Pascal's Principle

One of the most important facts about fluid pressure is that a change in pressure at one part of the fluid is transmitted undiminished to other parts. For example, if the pressure of city water is increased at the pumping station by 10 units of pressure, the pressure everywhere in the pipes of the connected system is increased by 10 units of pressure (providing the water is at rest). This rule is called **Pascal's principle**:

A change in pressure at any point in an enclosed fluid at rest is transmitted undiminished to all points in the fluid.

Pascal's principle was discovered in the 17th century by theologian and scientist Blaise Pascal, for whom the SI unit of pressure, the pascal (1 Pa = 1 N/m²), is named.

Fill a U-tube with water and place pistons at each end, as shown in Figure 5.30. Pressure exerted against the left piston is transmitted throughout the liquid and against the bottom of the right piston. (The pistons are simply "plugs" that can slide freely but snugly inside the tube.) The pressure that the left piston exerts against the water is exactly equal to the pressure the water exerts against the right piston. This is nothing to write home about. But suppose you make the tube on the right side wider and use a piston of larger area; then the result is impressive. In Figure 5.31 the piston on the right has 50 times the area of the piston on the left (say the left has 100 cm² and the right 5000 cm²). Suppose a 10-kg load is placed on the left piston. Then an additional pressure due to the weight of the load is transmitted throughout the liquid and up against the larger piston. Here is where the difference between force and pressure comes in. The additional pressure is exerted against every square centimeter of the larger piston. Because there is 50 times the area, 50 times as much force is exerted on the larger piston. Thus, the larger piston supports a 500-kg load—50 times the load on the smaller piston!

This *is* something to write home about, for we can multiply forces using such a device. One newton of input produces 50 N of output. By further increasing the area of the larger piston (or reducing the area of the smaller piston), we can multiply force, in principle, by any amount. Pascal's principle underlies the operation of the hydraulic press.

The hydraulic press does not violate energy conservation, because a decrease in the distance moved compensates for the increase in force. When the small piston in Figure 5.31 is moved downward 10 cm, the large piston is raised only one-fiftieth of this, or 0.2 cm. The input force multiplied by the distance moved by the smaller piston is equal to the output force multiplied by the distance moved by the larger piston; this is one more example of a simple machine operating on the same principle as a mechanical lever.

Pascal's principle applies to all fluids, whether gases or liquids. A typical application of Pascal's principle for gases and liquids is the automobile lift seen in many service stations (Figure 5.32). Increased air pressure produced by an air compressor is transmitted through the air to the surface of oil in an underground reservoir. The oil in turn transmits the pressure to a piston, which lifts the automobile. The relatively

FIGURE 5.30

The force exerted on the left piston increases the pressure in the liquid and is transmitted to the right piston.

FIGURE 5.31

A 10-kg load on the left piston supports 500 kg on the right piston.

FIGURE 5.32

Pascal's principle in a service station.

FIGURE 5.33

Pascal's principle at work in the hydraulic devices on this common but incredible machine. We can only wonder whether Pascal envisioned the extent to which his principle would allow huge loads to be so easily lifted.

fyi

- Pascal was an invalid at age 18 and remained so until his death at age 39. He is remembered scientifically for hydraulics, which changed the technological landscape more than he imagined. He is remembered theologically for his many assertions, one of which relates to centuries of human landscape: "Men never do evil so cheerfully and completely as when they do so from religious conviction."

The Physical Science Place
Buoyancy of Air

FIGURE 5.34

All bodies are buoyed up by a force equal to the weight of air they displace. Why, then, don't all objects float like this balloon?

low pressure that exerts the lifting force against the piston is about the same as the air pressure in automobile tires.

Hydraulics is employed by modern devices ranging from very small to enormous. Note the hydraulic pistons in almost all construction machines where heavy loads are involved (Figure 5.33).

CHECK POINT

1. As the automobile in Figure 5.32 is being lifted, how does the change in oil level in the reservoir compare to the distance the automobile moves?
2. If a friend commented that a hydraulic device is a common way of multiplying energy, what would you say?

Were these your answers?

1. The car moves up a greater distance than the oil level drops, because the area of the piston is smaller than the surface area of the oil in the reservoir.
2. No, no, no! Although a hydraulic device, like a mechanical lever, can multiply *force*, it always does so at the expense of distance. Energy is the product of force and distance. Increase one, decrease the other. *No device has ever been found that can multiply energy!*

5.8 Buoyancy in a Gas

A crab lives at the bottom of its ocean floor and looks upward at jellyfish and other lighter-than-water marine life drifting above it. Similarly, we live at the bottom of our ocean of air and look upward at balloons and other lighter-than-air objects drifting above us. A balloon is suspended in air and a jellyfish is suspended in water for the same reason: each is buoyed upward by a displaced weight of fluid equal to its own weight. Objects in water are buoyed upward because the pressure acting up against the bottom of the object exceeds the pressure acting down against the top. Likewise, air pressure acting upward against an object immersed in air is greater than the pressure above pushing down. The buoyancy in both cases is numerically equal to the weight of fluid displaced. Archimedes' principle applies to air just as it does for water:

An object surrounded by air is buoyed up by a force equal to the weight of the air displaced.

We know that a cubic meter of air at ordinary atmospheric pressure and room temperature has a mass of about 1.2 kg, so its weight is about 12 N. Therefore, any 1-m^3 object in air is buoyed up with a force of 12 N. If the mass of the 1-m^3 object is greater than 1.2 kg (so that its weight is greater than 12 N), it falls to the ground when released. If an object of this size has a mass of less than 1.2 kg, buoyant force is greater than weight and it rises in the air. Any object that has a mass that is less than the mass of an equal volume of air rises in the air. Stated another way, any object less dense than air rises in air. Gas-filled balloons that rise in air are less dense than air.

No gas at all in a balloon would mean no weight (except for the weight of the balloon's material), but such a balloon would be crushed by atmospheric pressure. The gas used in balloons prevents the atmosphere from collapsing them. Hydrogen is the lightest gas, but it is seldom used because it is highly flammable. In sport balloons, the gas is simply heated air. In balloons intended to reach very high altitudes or to remain aloft for a long time, helium is commonly used.

Its density is small enough that the combined weight of the helium, the balloon, and the cargo is less than the weight of air they displace. Low-density gas is used in a balloon for the same reason that cork is used in life preservers. The cork possesses no strange tendency to be drawn toward the water's surface, and the gas possesses no strange tendency to rise. Cork and gases are buoyed upward like anything else. They are simply light enough for the buoyancy to be significant.

Unlike water, the "top" of the atmosphere has no sharply defined surface. Furthermore, unlike water, the atmosphere becomes less dense with altitude. Whereas cork floats to the surface of water, a released helium-filled balloon does not rise to any atmospheric surface. Will a lighter-than-air balloon rise indefinitely? How high will a balloon rise? We can state the answer in several ways. A gas-filled balloon rises only so long as it displaces a weight of air greater than its own weight. Because air becomes less dense with altitude, a lesser weight of air is displaced per given volume as the balloon rises. When the weight of displaced air equals the total weight of the balloon, upward motion of the balloon ceases. We can also say that when the buoyant force on the balloon equals its weight, the balloon ceases rising. Equivalently, when the density of the balloon (including its load) equals the density of the surrounding air, the balloon ceases rising. Helium-filled toy rubber balloons usually break some time after being released into the air when the expansion of the helium they contain stretches the rubber until it ruptures.

CHECK POINT

Is a buoyant force acting on you? If so, why are you not buoyed up by this force?

Was this your answer?
A buoyant force *is* acting on you, and you *are* buoyed upward by it. You aren't aware of it only because your weight is so much greater.

Large helium-filled dirigible airships are designed so that when they are loaded, they slowly rise in air; that is, their total weight is a little less than the weight of air displaced. When in motion, the ship may be raised or lowered by means of horizontal "elevators."

Thus far we have treated pressure only as it applies to stationary fluids. Motion produces an additional influence.

■ 5.9 Bernoulli's Principle

Consider a continuous flow of liquid or gas through a pipe: the volume of fluid that flows past any cross section of the pipe in a given time is the same as that flowing past any other section of the pipe—even if the pipe widens or narrows. For continuous flow, a fluid speeds up when it goes from a wide to a narrow part of the pipe. This is evident for a broad, slow-moving river that flows more swiftly as it enters a narrow gorge. It is also evident when water flowing from a garden hose speeds up when you squeeze the end of the hose to make the stream narrower.

The motion of a fluid in steady flow follows imaginary *streamlines*, represented by thin lines in Figure 5.35 and in other figures that follow. Streamlines are the smooth paths of bits of fluid. The lines are closer together in narrower regions,

FIGURE 5.35

Because the flow is continuous, water speeds up when it flows through the narrow and/or shallow part of the brook.

This is stated in the equation of continuity:

$$A_1 v_1 = A_2 v_2$$

The product $A_1 v_1$ at point 1 equals the product $A_2 v_2$ at point 2.

FIGURE 5.36

Water speeds up when it flows into the narrower pipe. The close-together streamlines indicate increased speed and decreased internal pressure.

fyi

- The friction of both liquids and gases sliding over one another is called *viscosity* and is a property of all fluids.

FIGURE 5.37

Internal pressure is greater in slower-moving water in the wide part of the pipe, as evidenced by the more-squeezed air bubbles. The bubbles are bigger in the narrow part because internal pressure there is less.

Recall from Chapter 3 that a large change in momentum is associated with a large impulse. So when water from a firefighter's hose hits you, the impulse can knock you off your feet. Interestingly, the pressure *within* that water is relatively small!

FIGURE 5.38

The paper rises when Tim blows air across its top surface.

where the flow speed is greater. (Streamlines are visible when smoke or other visible fluids are passed through evenly spaced openings, as in a wind tunnel.)

Daniel Bernoulli, an 18th-century Swiss scientist, studied fluid flow in pipes. His discovery, now called **Bernoulli's principle**, can be stated as follows:

> **Where the speed of a fluid increases, internal pressure in the fluid decreases.**

Where streamlines of a fluid are closer together, flow speed is greater and pressure within the fluid is less. Changes in internal pressure are evident for water containing air bubbles. The volume of an air bubble depends on the surrounding water pressure. Where water gains speed, pressure is lowered and bubbles become bigger. In water that slows, pressure is greater and bubbles are squeezed to a smaller size.

Bernoulli's principle is a consequence of the conservation of energy, although, surprisingly, he developed it long before the concept of energy was formalized.* The full energy picture for a fluid in motion is quite complicated. Simply stated, more speed and kinetic energy mean less pressure, and more pressure means less speed and kinetic energy.

Bernoulli's principle applies to a smooth, steady flow (called *laminar* flow) of constant-density fluid. At speeds above some critical point, however, the flow may become chaotic (called *turbulent* flow) and follow changing, curling paths called *eddies*. This exerts friction on the fluid and dissipates some of its energy. Then Bernoulli's equation doesn't apply well.

The decrease of fluid pressure with increasing speed may at first seem surprising, particularly if you fail to distinguish between the pressure *within* the fluid, internal pressure, and the pressure *by* the fluid on something that interferes with its flow. Internal pressure within flowing water and the external pressure it can exert on whatever it encounters are two different pressures. When the momentum of moving water or anything else is suddenly reduced, the impulse it exerts is relatively huge. A dramatic example is the use of high-speed jets of water to cut steel in modern machine shops. The water has very little internal pressure, but the pressure the stream exerts on the steel interrupting its flow is enormous.

APPLICATIONS OF BERNOULLI'S PRINCIPLE

Hold a sheet of paper in front of your mouth, as shown in Figure 5.38. When you blow across the top surface, the paper rises. That's because the internal pressure of moving air against the top of the paper is less than the atmospheric pressure beneath it.

Anyone who has ridden in a convertible car with the canvas top up has noticed that the roof puffs upward as the car moves. This is Bernoulli's principle again. The pressure outside—on top of the fabric, where air is moving—is less than the static atmospheric pressure on the inside.

Consider wind blowing across a peaked roof. The wind gains speed as it flows over the roof, as the crowding of streamlines in Figure 5.39 indicates. Pressure

---

* In mathematical form: $\frac{1}{2}mv^2 + mgy + pV$ = constant (along a streamline), where m is the mass of some small volume V, v its speed, g the acceleration due to gravity, y its elevation, and p its internal pressure. If mass m is expressed in terms of density ρ, where $\rho = m/V$, and each term is divided by V, Bernoulli's equation reads: $\frac{1}{2}\rho v^2 + \rho gy + p$ = constant. Then all three terms have units of pressure. If y does not change, an increase in v means a decrease in p, and vice versa. Note that when v is zero, Bernoulli's equation reduces to $\Delta p = \rho g \Delta y$ (weight density × depth).

along the streamlines is reduced where they are closer together. The greater pressure inside the roof can lift it off the house. During a severe storm, the difference in outside and inside pressure doesn't need to be very much. A small pressure difference over a large area produces a force that can be formidable.

FIGURE 5.39

Air pressure above the roof is less than air pressure beneath the roof.

If we think of the blown-off roof as an airplane wing, we can better understand the lifting force that supports a heavy aircraft. In both cases, a greater pressure below pushes the roof or the wing into a region of lesser pressure above. Wings come in a variety of designs. What they all have in common is that air is made to flow faster over the wing's top surface than under its lower surface. This is mainly accomplished by a tilt in the wing, called its *angle of attack*. Then air flows faster over the top surface for much the same reason that air flows faster in a narrowed pipe or in any other constricted region. Most often, but not always, different speeds of airflow over and beneath a wing are enhanced by a difference in the curvature (*camber*) of the upper and lower surfaces of the wing. The result is more-crowded streamlines along the top

FIGURE 5.40

The vertical vector represents the net upward force (lift) that results from more air pressure below the wing than above the wing. The horizontal vector represents air drag.

Link To Section 23.4

wing surface than along the bottom. When the average pressure difference over the wing is multiplied by the surface area of the wing, we have a net upward force—lift. Lift is greater when there is a large wing area and when the plane is traveling fast. A glider has a very large wing area relative to its weight, so it does not have to be going very fast for sufficient lift. At the other extreme, a fighter plane designed for high-speed flight has a small wing area relative to its weight. Consequently, it must take off and land at high speeds.

We all know that a baseball pitcher can throw a ball in such a way that it curves to one side as it approaches home plate. This is accomplished by imparting a large spin to the ball. Similarly, a tennis player can hit a ball so it curves. A thin layer of air is dragged around the spinning ball by friction, which is enhanced by the baseball's threads or the tennis ball's fuzz. The moving layer of air produces a crowding of streamlines on one side. Note in Figure 5.41b that the streamlines are more crowded at B than at A for the direction of spin shown. Air pressure is greater at A, and the ball curves as shown.

Recent findings show that many insects increase lift by employing motions similar to those of a curving baseball. Interestingly, most insects do not flap their wings up and down. They flap them forward and backward, with a tilt that provides an angle of attack. Between flaps, their wings make semicircular motions to create lift.

a   b

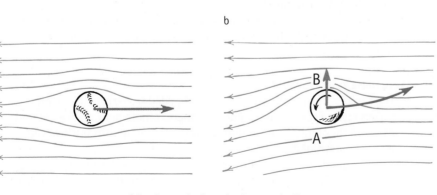

Motion of air relative to ball

FIGURE 5.41

(a) The streamlines are the same on either side of a non-spinning baseball. (b) A spinning ball produces a crowding of streamlines. The resulting "lift" (red arrow) causes the ball to curve (blue arrow).

FIGURE 5.42

Why does the liquid in the reservoir go up the tube?

A familiar sprayer, such as a perfume atomizer, uses Bernoulli's principle. When you squeeze the bulb, air rushes across the open end of a tube inserted into the perfume. This reduces the pressure in the tube, whereupon atmospheric pressure on the liquid below pushes it up into the tube, where it is carried away by the stream of air.

Bernoulli's principle explains why trucks passing closely on the highway are drawn to each other, and why passing ships run the risk of a sideways collision. Water flowing between the ships travels faster than water flowing past the outer sides. Streamlines are closer together between the ships than outside, so water pressure acting against the hulls is reduced between the ships. Unless the ships are steered to compensate for this, the greater pressure against the outer sides of the ships forces them together. Figure 5.43 shows how to demonstrate this in your kitchen sink or bathtub.

FIGURE 5.43

Try this in your sink. Loosely moor a pair of toy boats side by side. Then direct a stream of water between them. The boats draw together and collide. Why?

FIGURE 5.44

The curved shape of an umbrella can be disadvantageous on a windy day.

Bernoulli's principle plays a small role when your bathroom shower curtain swings toward you in the shower when the water is on full blast. The pressure in the shower stall is reduced with fluid in motion, and the relatively greater pressure outside the curtain pushes it inward. Like so much in the complex real world, this is but one physics principle that applies. More important is the convection of air in the shower. In any case, the next time you're taking a shower and the curtain swings in against your legs, think of Daniel Bernoulli.

CHECK POINT

1. On a windy day, waves in a lake or the ocean are higher than their average height. How does Bernoulli's principle contribute to the increased height?

2. Blimps, airplanes, and rockets operate under three very different principles. Which operates by way of buoyancy? Bernoulli's principle? Newton's third law?

Were these your answers?

1. The troughs of the waves are partially shielded from the wind, so air travels faster over the crests. Pressure there is more reduced than down below in the troughs. The greater pressure in the troughs pushes water into the even higher crests.
2. Blimps operate by way of buoyancy, airplanes by the Bernoulli principle, and rockets by way of Newton's third law. Interesting, Newton's third law also plays a significant role in airplane flight—wing pushes air downward; air pushes wing upward.

SUMMARY OF TERMS

Density The amount of matter per unit volume.

$$\text{Density} = \frac{\text{mass}}{\text{volume}}$$

Weight density is expressed as weight per unit volume.

Pressure The ratio of force to the area over which that force is distributed:

$$\text{Pressure} = \frac{\text{force}}{\text{area}}$$

Liquid pressure = weight density × depth

Buoyant force The net upward force that a fluid exerts on an immersed object.

Archimedes' principle An immersed body is buoyed up by a force equal to the weight of the fluid it displaces (for both liquids and gases).

Principle of flotation A floating object displaces a weight of fluid equal to its own weight.

Boyle's law The product of pressure and volume is a constant for a given mass of confined gas regardless of changes in either pressure or volume individually, so long as temperature remains unchanged:

$$P_1V_1 = P_2V_2$$

Atmospheric pressure The pressure exerted against bodies immersed in the atmosphere resulting from the weight of air pressing down from above. At sea level, atmospheric pressure is about 101 kPa.

Barometer Any device that measures atmospheric pressure.

Pascal's principle A change in pressure at any point in an enclosed fluid at rest is transmitted undiminished to all points in the fluid.

Bernoulli's principle The pressure in a fluid moving steadily without friction or outside energy input decreases when the fluid velocity increases.

REVIEW QUESTIONS

1. Give two examples of a fluid.

5.1 Density

2. What happens to the volume of a loaf of bread that is squeezed? What happens to the mass? What happens to the density?
3. Distinguish between mass density and weight density. What are the mass density and the weight density of water?

5.2 Pressure

4. Distinguish between force and pressure.
5. How does the pressure exerted by a liquid change with depth in the liquid? How does the pressure exerted by a liquid change as the density of the liquid changes?
6. Discounting the pressure of the atmosphere, if you swim twice as deep in water, how much more water pressure is exerted on your ears? If you swim in salt water, is the pressure greater than in fresh water at the same depth? Why or why not?
7. How does water pressure 1 m below the surface of a small pond compare to water pressure 1 m below the surface of a huge lake?
8. If you punch a hole in the side of a container filled with water, in what direction does the water initially flow outward from the container?

5.3 Buoyancy in a Liquid

9. Why does buoyant force act upward on an object submerged in water?
10. How does the volume of a completely submerged object compare with the volume of water displaced?

5.4 Archimedes' Principle

11. State Archimedes' principle.
12. What is the difference between being immersed and being submerged?
13. How does the buoyant force on a fully submerged object compare with the weight of water displaced?
14. What is the mass in kilograms of 1 L of water? What is its weight in newtons?
15. If a 1-L container is immersed halfway in water, what is the volume of water displaced? What is the buoyant force on the container?
16. Does the buoyant force on a floating object depend on the weight of the object or on the weight of the fluid displaced by the object? Or are these two weights the same for the special case of floating? Defend your answer.
17. What weight of water is displaced by a 100-ton floating ship? What is the buoyant force that acts on this ship?

5.5 Pressure in a Gas

18. By how much does the density of air increase when it is compressed to half its volume?
19. What happens to the air pressure inside a balloon when the balloon is squeezed to half its volume at constant temperature?

5.6 Atmospheric Pressure

20. What is the approximate mass in kilograms of a column of air that has a cross-sectional area of 1 cm² and extends from sea level to the upper atmosphere? What is the weight in newtons of this amount of air?
21. How does the downward pressure of the 76-cm column of mercury in a barometer compare with the air pressure at the bottom of the atmosphere?

22. How does the weight of mercury in a barometer tube compare with the weight of an equal cross section of air from sea level to the top of the atmosphere?
23. Why would a water barometer have to be 13.6 times as tall as a mercury barometer?
24. When you drink liquid through a straw, is it more accurate to say that the liquid is pushed up the straw rather than sucked up? What exactly does the pushing? Defend your answer.

5.7 Pascal's Principle

25. What happens to the pressure in all parts of a confined fluid when the pressure in one part is increased?
26. Does Pascal's principle provide a way to get more energy from a machine than is put into it? Defend your answer.

5.8 Buoyancy in a Gas

27. A balloon that weighs 1 N is suspended in air, drifting neither up nor down. How much buoyant force acts upon it? What happens if the buoyant force decreases? Increases?

5.9 Bernoulli's Principle

28. What are streamlines? Is pressure greater or less in regions of crowded streamlines?
29. Does Bernoulli's principle refer to internal pressure changes in a fluid or to pressures that a fluid can exert on objects it encounters?
30. What do peaked roofs, convertible tops, and airplane wings have in common when air moves faster across their top surfaces?

EXERCISES

● BEGINNER ■ INTERMEDIATE ◆ EXPERT

1. ● Stand on a bathroom scale and read your weight. When you lift one foot up so you're standing on one foot, does the reading change? Does a scale read force or pressure?
2. ● The photo shows physics teacher Marshall Ellenstein walking barefoot on broken glass bottles in his class. What physics concept is Marshall demonstrating, and why is he careful that the broken pieces are small and numerous? (The Band-Aids on his feet are for humor!)

3. ● In a deep dive, a whale is appreciably compressed by the pressure of the surrounding water. What happens to the whale's density?
4. ● The density of a rock doesn't change when it is submerged in water. Does your density change when you are submerged in water? Defend your answer.
5. ● Why are people who are confined to bed less likely to develop bedsores on their bodies if they use a waterbed rather than an ordinary mattress?
6. ● If water faucets upstairs and downstairs are turned fully on, does more water per second flow out the downstairs faucet? Or is the volume of water flowing from the faucets the same?
7. ■ Which do you suppose exerts more pressure on the ground—an elephant or a woman standing on spike heels? (Which is more likely to make dents in a linoleum floor?) Can you approximate a rough calculation for each?

8. ■ Suppose you wish to lay a level foundation for a home on hilly and bushy terrain. How can you use a garden hose filled with water to determine equal elevations for distant points?
9. ● When you are bathing on a stony beach, why do the stones hurt your feet less when you get in deep water?
10. ■ If liquid pressure were the same at all depths, would there be a buoyant force on an object submerged in the liquid? Explain.
11. ● How much force is needed to push a nearly weightless but rigid 1-L carton beneath a surface of water?
12. ● Why is it inaccurate to say that heavy objects sink and that light objects float? Give exaggerated examples to support your answer.
13. ■ Compared to an empty ship, would a ship loaded with a cargo of Styrofoam sink deeper into water or rise in water? Defend your answer.
14. ■ A barge filled with scrap iron is in a canal lock. If the iron is thrown overboard, does the water level at the side of the lock rise, fall, or remain unchanged? Explain.
15. ■ Would the water level in a canal lock go up or down if a battleship in the lock were to sink?
16. ■ A balloon is weighted so that it is barely able to float in water. If it is pushed beneath the surface, does it rise back to the surface, stay at the depth to which it is pushed, or sink? Explain. (Hint: Does the balloon's density change?)

17. ■ A ship sailing from the ocean into a freshwater harbor sinks slightly deeper into the water. Does the buoyant force on it change? If so, does it increase or decrease?
18. ■ Suppose you are given the choice between two life preservers that are identical in size, the first a light one filled with Styrofoam and the second a very heavy one filled with lead pellets. If you submerge these life preservers in the water, upon which is the buoyant force greater? Upon which is the buoyant force ineffective? Why are your answers different?

19. ■ The relative densities of water, ice, and alcohol are 1.0, 0.9, and 0.8, respectively. Do ice cubes float higher or lower in a mixed alcoholic drink? What can you say about a cocktail in which the ice cubes lie submerged at the bottom of the glass?

20. ◆ When an ice cube in a glass of water melts, does the water level in the glass rise, fall, or remain unchanged? Does your answer change if the ice cube contains many air bubbles? Does your answer change if the ice cube contains many grains of heavy sand?

21. ■ A half-filled bucket of water is on a spring scale. Does the reading of the scale increase or remain the same if a fish is placed in the bucket? (Is your answer different if the bucket is initially filled to the brim?)

22. ■ Count the tires on a large tractor trailer that is unloading food at your local supermarket, and you may be surprised to count 18 tires. Why so many tires? (Hint: See Active Exploration 5.)

23. ● How does the density of air in a deep mine compare with the air density at Earth's surface?

24. ■ Two teams of eight horses each were unable to pull the Magdeburg hemispheres apart (Figure 5.20). Why? Suppose two teams of nine horses each could pull them apart. Then would one team of nine horses succeed if the other team were replaced with a strong tree? Defend your answer.

25. ● Before boarding an airplane, you buy a bag of chips (or any item sealed in an airtight foil package) and, while in flight, you notice that the bag is puffed up. Explain why this occurs.

26. ● Why do you suppose that airplane windows are smaller than bus windows?

27. ● We can understand how pressure in water depends on depth by considering a stack of bricks. The pressure below the bottom brick is determined by the weight of the entire stack. Halfway up the stack, the pressure is half because the weight of the bricks above is half. To explain atmospheric pressure, we should consider compressible bricks, like foam rubber. Why is this so?

28. ● The "pump" in a vacuum cleaner is merely a high-speed fan. Would a vacuum cleaner pick up dust from a rug on the Moon? Explain.

29. ● If you could somehow replace the mercury in a mercury barometer with a denser liquid, would the height of the liquid column be greater or less than with mercury? Why?

30. ● Would it be slightly more difficult to draw soda through a straw at sea level or on top of a very high mountain? Explain.

31. ● Your friend says that the buoyant force of the atmosphere on an elephant is significantly greater than the buoyant force of the atmosphere on a small helium-filled balloon. What do you say?

32. ■ Why is it so difficult to breathe when snorkeling at a depth of 1 m, and practically impossible at a 2-m depth? Why can't a diver simply breathe through a hose that extends to the surface?

33. ■ When you replace helium in a balloon with hydrogen, which is less dense, does the buoyant force on the balloon change if the balloon remains the same size? Explain.

34. ■ A steel tank filled with helium gas doesn't rise in air, but a balloon containing the same helium easily does. Why?

35. ■ Two identical balloons of the same volume are pumped up with air to more than atmospheric pressure and suspended on the ends of a stick that is horizontally balanced. One of the balloons is then punctured. Is there a change in the stick's balance? If so, which way does it tip?

36. ● The force of the atmosphere at sea level against the outside of a 10-m² store window is about 1 million N. Why does this not shatter the window? Why might the window shatter in a strong wind blowing past the window?

37. ■ In the hydraulic arrangement shown, the larger piston has an area that is 50 times that of the smaller piston. The strongman hopes to exert enough force on the large piston to raise the 10 kg that rest on the small piston. Do you think he will be successful? Defend your answer.

38. ● When a steadily flowing gas flows from a larger-diameter pipe to a smaller-diameter pipe, what happens to (a) its speed, (b) its pressure, and (c) the spacing between its streamlines?

39. ■ What physics principle underlies the following three observations? When passing an oncoming truck on the highway, your car tends to sway toward the truck. The canvas roof of a convertible automobile bulges upward when the car is traveling at high speeds. The windows of older passenger trains sometimes break when a high-speed train passes by on the next track.

40. ◆ How is an airplane able to fly upside down?

PROBLEMS

● BEGINNER   ■ INTERMEDIATE   ◆ EXPERT

1. ● Suppose that you balance a 5-kg ball on the tip of your finger, which has an area of 1 cm². Show that the pressure on your finger is 49 N/cm², which is 490 kPa.

2. ● A 6-kg piece of metal displaces 1 L of water when submerged. Show that its density is 6000 kg/m³. How does this compare with the density of water?

3. ● The depth of water behind the Hoover Dam in Nevada is 220 m. Ignore the pressure due to the atmosphere. Show that the water pressure at the base of the dam is 2160 kPa.

4. ■ A rectangular barge, 5 m long and 2 m wide, floats in fresh water. Suppose that a 400-kg crate of auto parts is loaded onto the barge. Show that the barge floats 4 cm deeper.

5. ■ Suppose that the barge in the preceding problem can be pushed only 15 cm deeper into the water before the water overflows to sink it. Show that it could carry three, but not four, 400-kg crates.

6. ● A merchant in Kathmandu sells you a solid gold 1-kg statue for a very reasonable price. When you get home, you wonder whether you got a bargain, so you lower the statue into a container of water and measure the volume of displaced water. Show that for 1 kg of pure gold, the volume of water displaced is 51.8 cm^3.

7. ■ A vacationer floats lazily in the ocean with 90% of his body below the surface. The density of the ocean water is 1,025 kg/m^3. Show that the vacationer's average density is 923 kg/m^3.

8. ● On a perfect fall day, you are hovering at low altitude in a hot-air balloon, accelerated neither upward nor downward. The total weight of the balloon, including its load and the hot air in it, is 20,000 N. Show that the volume of the displaced air is 1,700 m^3.

9. ■ In the hydraulic pistons shown in the sketch, the small piston has a diameter of 2 cm. The large piston has a diameter of 6 cm. How much more force can the larger piston exert compared with the force applied to the smaller piston?

10. ■ The wings of a certain airplane have a total surface area of 100 m^2. At a particular speed, the difference in air pressure below and above the wings is 4% of atmospheric pressure. Show that the lift on the airplane is 4×10^5 N.

ACTIVE EXPLORATIONS

1. Try to float an egg in water. Then dissolve salt in the water until the egg floats. How does the density of an egg compare to that of tap water? To salt water?

2. Punch a couple of holes in the bottom of a water-filled container, and water spurts out because of water pressure. Now drop the container, and, as it freely falls, note that the water no longer spurts out. If your friends don't understand this, could you figure it out and then explain it to them?

3. Place a wet Ping-Pong ball in a can of water held high above your head. Then drop the can on a rigid floor. Because of surface tension, the ball is pulled beneath the surface as the can falls. What happens when the can comes to an abrupt stop is worth watching!

4. Try this in the bathtub or when you're washing dishes: lower a drinking glass, mouth downward, over a small floating object. What do you observe? How deep must the glass be pushed in order to compress the enclosed air to half its volume? (You won't be able to do this in your bathtub unless it's 10.3 m deep!)

5. You can find the pressure exerted by the tires of your car on the road and compare it with the air pressure in the tires. For this project, you need to get the weight of your car from the manual or a dealer, and then divide by 4 to get the approximate weight held up by one tire. You can closely approximate the area of contact of a tire with the road by tracing the edges of tire contact on a sheet of paper marked with 1-inch-×-1-inch squares beneath the tire. After you get the pressure of the tire on the road, compare it with the air pressure in the tire. Are they nearly equal? Which one is greater?

6. You ordinarily pour water from a full glass into an empty glass simply by placing the full glass above the empty glass and tipping. Have you ever poured air from one glass to another? The procedure is similar. Lower two glasses in water, mouths downward. Let one fill with water by tilting its mouth upward. Then hold the mouth of the water-filled glass downward above the air-filled glass. Slowly tilt the lower glass and let the air escape, filling the upper glass. You are pouring air from one glass into another!

7. Raise a filled glass of water above the waterline, but with its mouth beneath the surface. Why does the water not run out? How tall would a glass have to be before water began to run out? (You won't be able to do this indoors unless you have a ceiling that is at least 10.3 m higher than the waterline.)

8. Place a card over the open top of a glass filled to the brim with water, and then invert it. Why does the card stay in place? Try it sideways.

9. Invert a water-filled soft-drink bottle or small-necked jar. Notice that the water doesn't simply fall out, but gurgles out of the container instead. Air pressure doesn't allow the water out until some air has pushed its way up inside the bottle to occupy the space above the liquid. How would an inverted, water-filled bottle empty if you tried this on the Moon?

10. Heat a small amount of water to boiling in an aluminum soft-drink can and invert it quickly into a dish of cold water. What happens is surprisingly dramatic!

11. Make a small hole near the bottom of an open tin can. Fill the can with water, which then proceeds to spurt from the hole. If you cover the top of the can firmly with the palm of your hand, the flow stops. Explain.

12. Lower a narrow glass tube or drinking straw into water and place your finger over the top of the tube. Lift the tube from the water and then lift your finger from the top of the tube. What happens? (You'll do this often in chemistry experiments.)

13. Blow across the top of a sheet of paper as Tim does in Figure 5.38. Try this with those of your friends who are not taking a physical science course. Then explain it to them!

14. Push a pin through a small card and place it over the hole of a thread spool. Try to blow the card from the spool by blowing through the hole. Try it in all directions.

15. Hold a spoon in a stream of water as shown and feel the effect of the differences in pressure.

READINESS ASSURANCE TEST (RAT)

If you have a handle on this chapter, if you really do, then you should be able to score at least 7 out of 10 on this RAT. If you score less than 7, you need to study further before moving on.

Choose the BEST answer to each of the following.

1. Pumice is a volcanic rock that floats in water. The density of pumice is
   (a) less than the density of water.
   (b) equal to the density of water.
   (c) more than the density of water.
   (d) irrelevant; because it's a rock, it sinks!

2. The pressure at the bottom of a pond does NOT depend on
   (a) the acceleration due to gravity.
   (b) water density.
   (c) the depth of the pond.
   (d) the surface area of the pond.
   (e) any of these

3. A completely submerged object always displaces its own
   (a) weight of fluid.
   (b) volume of fluid.
   (c) density of fluid.
   (d) all of these
   (e) none of these

4. A rock suspended by a weighing scale weighs 5 N out of water and 3 N when submerged in water. What is the buoyant force on the rock?
   (a) 3 N
   (b) 5 N
   (c) 8 N
   (d) none of these

5. A block of wood with a flat rock tied to its top floats in a bucket of water. If the wood and rock are turned over so the rock is submerged beneath the wood,
   (a) the water level at the side of the bucket rises.
   (b) the water level at the side of the bucket falls.
   (c) the water level at the side of the bucket remains the same.
   (d) both rock and wood sink.

6. In a vacuum, an object has
   (a) no buoyant force.
   (b) no mass.
   (c) no weight.
   (d) none of these

7. Consider two mercury barometers, one having a cross-sectional area of 1 cm^2 and the other 2 cm^2. Mercury in the smaller tube rises
   (a) to the same height as the other.
   (b) twice as high as the other.
   (c) four times as high as the other.
   (d) none of these

8. In a hydraulic press operation, it is impossible for the
   (a) output piston to move farther than the input piston.
   (b) force output to exceed the force input.
   (c) output piston's speed to exceed the input piston's speed.
   (d) energy output to exceed energy input.
   (e) none of these

9. The flight of a blimp best illustrates
   (a) Archimedes' principle.
   (b) Pascal's principle.
   (c) Bernoulli's principle.
   (d) Boyle's law.

10. Wind speeding up as it blows over the top of a hill
    (a) increases atmospheric pressure there.
    (b) decreases atmospheric pressure there.
    (c) doesn't affect atmospheric pressure there.

Answers to RAT

1a, 2d, 3b, 4d, 5c, 6a, 7a, 8d, 9a, 10b

CHAPTER 5 ONLINE RESOURCES

The Physical Science Place

Videos

- Dam and Water
- Buoyancy
- Archimedes' Principle
- Flotation
- Air Has Weight
- Air Is Matter
- Air Has Pressure
- Buoyancy of Air

Quiz

Flashcards

Links

6

thermal energy and thermodynamics

6.1 TEMPERATURE

6.2 ABSOLUTE ZERO

6.3 HEAT

6.4 QUANTITY OF HEAT

6.5 THE LAWS OF THERMODYNAMICS

6.6 ENTROPY

6.7 SPECIFIC HEAT CAPACITY

6.8 THERMAL EXPANSION

6.9 EXPANSION OF WATER

■ What's the difference between a cup of hot tea and a cup of cool tea? The answer involves molecular motion. In the hot cup the molecules that constitute the tea are moving faster than those in the cooler cup. Matter in all forms is made up of constantly jiggling particles, namely atoms and/or molecules. When they jiggle at a very slow rate, they form solids. When they jiggle faster, they slide over one another and we have a liquid. When the same particles move so fast that they disconnect and fly loose, we have a gas. When they move still faster, atoms dissociate to form a plasma. So whether a substance is a solid, a liquid, a gas, or a plasma depends on the motion of its particles. In this and the following chapter we will investigate the effects of particle motions. We call the energy that a body has by virtue of its energetic jostling of atoms and molecules **thermal energy**.

FIGURE 6.1

Can we trust our sense of hot
and cold? Do both fingers feel the
same temperature when they are
dipped in the warm water? Try
this and see (feel) for yourself.

The particles that make up matter
are atoms and/or molecules. A
molecule is two or more atoms
bonded in a specific way, as
discussed in Chapter 2.

FIGURE 6.2

A testament to Fahrenheit outside his home (now in Gdansk,
Poland).

FIGURE 6.3

Fahrenheit and Celsius scales
on a thermometer.

6.1 Temperature

When you touch a hot stove, thermal energy enters your hand because the stove is warmer than your hand. When you touch a piece of ice, however, thermal energy passes out of your hand and into the colder ice. The quantity that indicates how warm or cold an object is relative to some standard is called **temperature**. We express the temperature of matter by a number that corresponds to the degree of hotness on some chosen scale. A common thermometer measures temperature by means of the expansion and contraction of a liquid, usually mercury or colored alcohol.

The most common temperature scale used worldwide is the Celsius scale, named in honor of the Swedish astronomer Anders Celsius (1701–1744), who first suggested the scale of 100 equal parts (*degrees*) between the freezing point and boiling point of water. The number 0 is assigned to the temperature at which water freezes, and the number 100 to the temperature at which water boils (at standard atmospheric pressure).

The most common temperature scale used in the United States is the Fahrenheit scale, named after its originator, the German physicist D. G. Fahrenheit

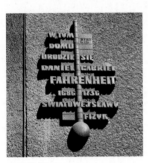

(1686–1736). On this scale the number 32 is assigned to the temperature at which water freezes, and the number 212 is assigned to the temperature at which water boils. The Fahrenheit scale will become obsolete if and when the United States changes to the metric system.

Arithmetic formulas used for converting from one temperature scale to the other are common in classroom exams. Because such arithmetic exercises are not really physics, we won't be concerned with these conversions (perhaps important in a math class, but not here). Besides, the conversion between Celsius and Fahrenheit temperatures is closely approximated in the side-by-side scales of Figure 6.3.*

Temperature is proportional to the average translational kinetic energy per particle that makes up a substance. By *translational* we mean to-and-fro linear motion. For a gas, we refer to how fast the gas particles are bouncing back and forth; for a liquid, we refer to how fast they slide and jiggle past each other; and for a solid, we refer to how fast the particles move as they vibrate and jiggle in place. Note that temperature does *not* depend on how much of the substance you have. If you have a cup of hot water and then pour half of the water onto the floor, the water remaining in the cup hasn't changed its temperature. The water remaining in the cup contains half the thermal energy that the full cup of water contained, because there are only half as many water molecules in the cup as before. Temperature is a *per-particle property*; *thermal energy* is related to the sum total kinetic energy of all of the particles in your sample.** Twice as much hot water has twice the thermal energy, even though its temperature (the average KE per particle) is the same.

When we measure the temperature of something with a conventional thermometer, thermal energy flows between the thermometer and the object whose temperature we are measuring. When the object and the thermometer have the

---

* Okay, if you really want to know, the formulas for temperature conversion are $C = \frac{5}{9}F - 32$ and $F = \frac{9}{5}C + 32$, where C is the Celsius temperature and F is the Fahrenheit temperature.
** Rather than the term *thermal energy*, physicists prefer the term *internal energy*, to emphasize that the energy is internal to a body.

Volume = 1 + $\frac{100}{273}$ 100°C

Volume = 1   0°C

Volume = 1 − $\frac{100}{273}$ −100°C

Volume = 1 − $\frac{273}{273}$ = 0   −273°C

same average kinetic energy per particle, we say that they are in *thermal equilibrium*. When we measure something's temperature, we are really reading the temperature of the thermometer when it and the object have reached thermal equilibrium.

6.2 Absolute Zero

As thermal motion increases, a solid object first melts and becomes a liquid. With more thermal motion it then vaporizes. As the temperature further increases, molecules dissociate into atoms, and atoms lose some or all of their electrons, thereby forming a cloud of electrically charged particles—a *plasma*. Plasmas exist in stars, where the temperature is millions of degrees Celsius. Temperature has no upper limit.

In contrast, a definite limit exists at the lower end of the temperature scale. Gases expand when heated, and they contract when cooled. Nineteenth-century experiments found something quite amazing. They found that if one starts out with a gas, any gas, at 0°C and changes its temperature, while pressure is held constant, the volume changes by $\frac{1}{273}$ for each degree Celsius change in temperature. When a gas was cooled from 0°C to −10°C, its volume decreased by $\frac{10}{273}$ and it contracted to $\frac{263}{273}$ of its original volume. If a gas at 0°C could be cooled down by 273°C, it would contract $\frac{273}{273}$ volumes and be reduced to zero volume. Clearly, we cannot have a substance with zero volume.

Experimenters got similar results for pressure. Starting at 0°C, the pressure of a gas held in a container of fixed volume decreased by $\frac{1}{273}$ for each Celsius degree its temperature was lowered. If it were cooled to 273°C below zero, it would have no pressure at all. In practice, every gas converts to a liquid before becoming this cold. Nevertheless, these decreases by $\frac{1}{273}$ increments suggested the idea of a lowest temperature: −273°C. That's the lower limit of temperature, **absolute zero**. At this temperature, molecules have lost all available kinetic energy.* No more energy can be removed from a substance at absolute zero. It can't get any colder.

The absolute temperature scale is called the *Kelvin scale*, named after the famous British mathematician and physicist William Thomson, First Baron Kelvin. Absolute zero is 0 K (short for "0 kelvins"; note that the word *degrees* is not used with Kelvin temperatures).** There are no negative numbers on the Kelvin scale. Its temperature divisions are identical to the divisions on the Celsius scale. Thus, the melting point of ice is 273 K, and the boiling point of water is 373 K.

---

* Even at absolute zero, molecules still possess a small amount of kinetic energy, called the *zero-point energy*. Helium, for example, has enough motion at absolute zero to prevent it from freezing. The explanation for this involves quantum theory.

** When Thomson became a baron he took his title from the Kelvin River, which ran through his estate. In 1968 the term *degrees Kelvin* (°K) was officially changed to simply *kelvins* (lowercase k), which is abbreviated K (capital K). The precise value of absolute zero (0 K) is −273.15°C.

fyi

- Thermal contact is not required with infrared thermometers that show digital temperature readings by measuring the infrared radiation emitted by all bodies.

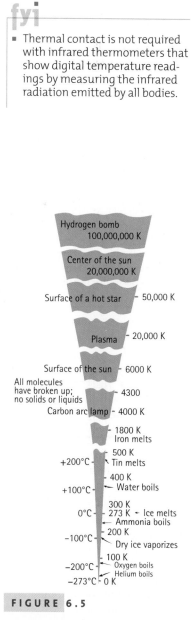

Hydrogen bomb
100,000,000 K

Center of the sun
20,000,000 K

Surface of a hot star — 50,000 K

Plasma — 20,000 K

Surface of the sun — 6000 K

All molecules have broken up; no solids or liquids — 4300

Carbon arc lamp — 4000 K

— 1800 K Iron melts

+200°C — 500 K Tin melts

+100°C — 400 K Water boils

— 300 K

0°C — 273 K ← Ice melts
— Ammonia boils

−100°C — 200 K Dry ice vaporizes

— 100 K

−200°C — Oxygen boils
Helium boils

−273°C — 0 K

FIGURE 6.5

Some absolute temperatures.

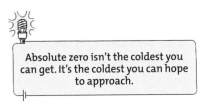

Absolute zero isn't the coldest you can get. It's the coldest you can hope to approach.

FIGURE 6.6

The temperature of the sparks is very high, about 2000°C. That's a lot of energy per molecule of spark. But because there are only a few molecules per spark, the total amount of thermal energy in the sparks is safely small. Temperature is one thing; transfer of thermal energy is another.

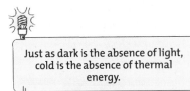

Just as dark is the absence of light, cold is the absence of thermal energy.

Hot stove

FIGURE 6.7

The pot on the left contains 1 L of water. The pot on the right contains 3 L of water. Although both pots absorb the same quantity of heat, the temperature increases three times as much in the pot with the smaller amount of water.

CHECK POINT

1. Which is larger, a Celsius degree or a kelvin?
2. A sample of hydrogen gas has a temperature of 0°C. If the gas is heated until its hydrogen molecules have doubled their kinetic energy, what is its temperature?

Were these your answers?

1. Neither. They are equal.
2. The 0°C gas has an absolute temperature of 273 K. Twice as much kinetic energy means that it has twice the absolute temperature, or two times 273 K. This would be 546 K, or 273°C.

6.3 Heat

When you place a warm object and a cool object in close proximity, thermal energy transfers in a direction from the warmer object to the cooler object. A physicist defines **heat** as the thermal energy transferred from one thing to another due to a temperature difference.

According to this definition, matter contains *thermal energy*—not heat. Once thermal energy has been transferred to an object or substance, it ceases to be heat. Again, for emphasis: a substance does not contain heat—it contains thermal energy. Heat is thermal energy in transit.

For substances in thermal contact, thermal energy flows from the higher-temperature substance into the lower-temperature substance until thermal equilibrium is reached. This does not mean that thermal energy necessarily flows from a substance with more thermal energy into one with less thermal energy. For example, a bowl of warm water contains more thermal energy than does a red-hot thumbtack. If the tack is placed into the water, thermal energy doesn't flow from the warm water to the tack. Instead, it flows from the hot tack to the cooler water. Thermal energy never flows unassisted from a low-temperature substance into a higher-temperature one.

CHECK POINT

1. You apply a flame to 1 L of water for a certain time and its temperature rises by 2°C. If you apply the same flame for the same time to 2 L of water, by how much does its temperature rise?
2. If a fast marble hits a random scatter of slow marbles, does the fast marble usually speed up or slow down? Which lose(s) kinetic energy and which gain(s) kinetic energy, the initially fast-moving marble or the initially slow ones? How do these questions relate to the direction of heat flow?

Were these your answers?

1. Its temperature rises by only 1°C, because 2 L of water contains twice as many molecules, and each molecule receives only half as much energy on the average. So the average kinetic energy, and thus the temperature, increases by half as much.
2. A fast-moving marble slows when it hits slower-moving marbles. It gives up some of its kinetic energy to the slower ones. Likewise with heat. Molecules with more kinetic energy that make contact with slower molecules give some of their excess kinetic energy to the slower ones. The direction of energy transfer is from hot to cold. For both the marbles and the molecules, however, the total energy of the system before and after contact is the same.

▪ 6.4 Quantity of Heat

Heat is a form of energy and is measured in joules. In the U.S. heat has traditionally been measured in calories, another measure of thermal energy. In science courses, the joule is usually preferred. It takes 4.18 J (or equivalently, 1 calorie) of heat to change the temperature of 1 g of water by 1°C.*

The energy ratings of foods and fuels are determined from the energy released when they are burned. (Metabolism is really "burning" at a slow rate). A common heat unit for labeling food is the kilocalorie (kcal), which is 1000 calories (cal), the heat needed to change the temperature of 1 kg of water by 1°C. To differentiate this unit and the smaller calorie, the food unit is usually called a *Calorie*, with a capital C. So 1 Calorie is really 1000 calories.

What we've learned thus far about heat and thermal energy is summed up in the *laws of thermodynamics*. The word **thermodynamics** stems from Greek words meaning "movement of heat."

▪ CHECK POINT

Which raises the temperature of water more, adding 4.18 J or 1 calorie?

Was this your answer?
They have the same effect. This is like asking which is longer, a 1.6-km-long track or a 1-mi-long track. They're the same length, just expressed in different units.

▪ 6.5 The Laws of Thermodynamics

When thermal energy transfers as heat, the energy lost in one place is gained in another in accord with conservation of energy. When the law of energy conservation is applied to thermal systems, we call it the **first law of thermodynamics**. We state it generally in the following form:

When heat flows to or from a system, the system gains or loses an amount of heat equal to the amount of heat transferred.

When we add heat energy to a system, whether it is a steam engine, Earth's atmosphere, or the body of a living creature, this added energy increases the thermal energy of the system if it remains in the system and/or does external work if it leaves the system. More specifically, the first law of thermodynamics states the following:

**Heat added = increase in thermal energy
+ external work done by the system**

Suppose that you put an air-filled, rigid, airtight can on a hot plate and add a certain amount of thermal energy to the can. **Warning:** *Do not actually do this.* Because the can has a fixed volume, the walls of the can don't move, so no work is done. All of the heat going into the can increases the thermal energy of the enclosed air, so its temperature rises. Now suppose instead that the can is a flexible container that can expand. The heated air does work as the sides of the can

---

Temperature is measured in degrees. Heat is measured in joules (or calories). In the U.S. we speak of low-calorie foods and drinks. Most of the world speaks of low-joule foods and drinks.

Link To Section 3.5

FIGURE 6.8

To the weight watcher, the peanut contains 10 Calories; to the scientist, it releases 10,000 calories (41,800 J) of energy when burned or digested.

Link To Section 17.5

fyi

▪ The only weight-loss plan endorsed by the first law of thermodynamics: Burn more calories than you consume and you will lose weight—guaranteed.

Link To Sections 24.6, 25.2

---

* So 1 calorie = 4.18 J. Another common unit of heat is the British thermal unit (Btu). The Btu is defined as the amount of heat required to change the temperature of 1 lb of water by 1°F. One Btu is equal to 1054 J.

FIGURE 6.9

When you push down on the piston, you do work on the air inside. What happens to its temperature?

Link To Section 17.6

The laws of thermodynamics can be stated this way: You can't win (because you can't get any more energy out of a system than you put into it), you can't break even (because you can't get as much useful energy out as you put in), and you can't get out of the game (entropy in the universe is always increasing).

expand, exerting a force for some distance on the surrounding atmosphere. Because some of the added heat goes into doing work, less of the added heat goes into increasing the thermal energy of the enclosed air. Can you see that the temperature of the enclosed air is lower when it does work than when it doesn't do work? The first law of thermodynamics makes good sense.

The **second law of thermodynamics** restates what we've learned about the direction of heat flow:

Heat never spontaneously flows from a cold substance to a hot substance.

When heat flow is spontaneous—that is, without the assistance of external work—the direction of flow is always from hot to cold. In winter, heat flows from inside a warm home to the cold air outside. In summer, heat flows from the hot air outside into the home's cooler interior. Heat can be made to flow the other way *only* when work is done on the system or by adding energy from another source. This occurs with heat pumps that move heat from cooler outside air into a home's warmer interior, or with air conditioners that remove heat from a home's cool interior to the warmer air outside. Without external effort, the direction of heat flow is always from hot to cold. The second law, like the first, makes logical sense.*

The **third law of thermodynamics** restates what we've learned about the lowest limit of temperature:

No system can reach absolute zero.

As investigators attempt to reach this lowest temperature, it becomes more difficult to get closer to it. Physicists have been able to record temperatures that are less than a millionth of 1 K—but never as low as 0 K.

6.6 Entropy

The first law of thermodynamics states that energy can be neither created nor destroyed. It speaks of the *quantity* of energy. The second law speaks of the *quality* of energy, as energy becomes more diffuse and ultimately degenerates into waste.

With this broader perspective, the second law can be stated another way:

In natural processes, high-quality energy tends to transform into lower-quality energy—order tends to disorder.

Processes in which disorder returns to order without external help don't occur in nature. Interestingly, time is given a direction via this thermodynamic rule. Time's arrow always points from order to disorder.**

---

* The laws of thermodynamics were the rage back in the 1800s. At that time, horses and buggies were yielding to steam-driven locomotives. There is the story of the engineer who explained the operation of a steam engine to a peasant. The engineer cited in detail the operation of the steam cycle, how expanding steam drives a piston that in turn rotates the wheels. After some thought, the peasant asked, "Yes, I understand all that. But where's the horse?" This story illustrates how difficult it is to abandon our way of thinking about the world when a newer method comes along to replace established ways. Are we different today?

** In the previous century when movies were new, audiences were amazed to see a train come to a stop inches away from a heroine tied to the tracks. This was filmed by starting with the train at rest, inches away from the heroine, and then moving *backward*, gaining speed. When the film was reversed, the train was seen to move *toward* the heroine. (Next time, watch closely for the telltale smoke that *enters* the smokestack.)

The idea of ordered energy tending to disordered energy is embodied in the concept of *entropy*.* **Entropy** is the measure of how energy spreads to disorder in a system. When disorder increases, entropy increases. The molecules of an automobile's exhaust, for example, cannot spontaneously recombine to form more highly organized gasoline molecules. Warm air that spreads throughout a room when the oven door is open cannot spontaneously return to the oven. Whenever a physical system is allowed to spread its energy freely, it always does so in a manner such that entropy increases, while the energy of the system available for doing work decreases.**

However, when work is input to a system, as in living organisms, the entropy of the system can decrease. All living things, from bacteria to trees to human beings, extract energy from their surroundings and use this energy to increase their own organization. The process of extracting energy (for instance, breaking down a highly organized food molecule into smaller molecules) increases entropy elsewhere, so life forms plus their waste products have a net increase in entropy. Energy must be transformed within the living system to support life. When it is not, the organism soon dies and tends toward disorder.

FIGURE 6.10

Entropy.

■ 6.7 Specific Heat Capacity

While eating, you've likely noticed that some foods remain hotter much longer than others. Whereas the filling of hot apple pie can burn your tongue, the crust does not, even when the pie has just been removed from the oven. Or a piece of toast may be comfortably eaten a few seconds after coming from the hot toaster, whereas you must wait several minutes before eating soup that has the same high temperature.

Different substances have different thermal capacities for storing energy. If we heat a pot of water on a stove, we might find that it requires 15 minutes to rise from room temperature to its boiling temperature. But an equal mass of iron on the same stove would rise through the same temperature range in only about 2 minutes. For silver, the time would be less than a minute. Equal masses of different materials require different quantities of heat to change their temperatures by a specified number of degrees.†

As mentioned earlier, a gram of water requires 1 calorie of energy to raise the temperature 1°C. It takes only about one-eighth as much energy to raise the temperature of a gram of iron by the same amount. Water absorbs more heat than iron for the same change in temperature. We say water has a higher **specific heat capacity** (sometimes simply called *specific heat*):

The specific heat capacity of any substance is defined as the quantity of heat required to change the temperature of a unit mass of the substance by 1°C.

Water is useful in the cooling systems of automobiles and other engines because it absorbs a great quantity of heat for small increases in temperature. Water also takes longer to cool.

FIGURE 6.11

The filling of hot apple pie may be too hot to eat, even though the crust is not.

---

* Entropy can be expressed mathematically. The increase in entropy ΔS of a thermodynamic system is equal to the amount of heat added to the system ΔQ divided by the temperature T at which the heat is added: $\Delta S = \Delta Q/T$.

** Interestingly enough, the American writer Ralph Waldo Emerson, who lived during the time when the second law of thermodynamics was the new science topic of the day, philosophically speculated that not everything becomes more disordered with time and cited the example of human thought. Ideas about the nature of things grow increasingly refined and better organized as they pass through the minds of succeeding generations. Human thought is evolving toward more order.

† In the case of silver and iron, silver atoms are about twice as massive as iron atoms. A given mass of silver contains only about half as many atoms as an equal mass of iron, so only about half the heat is needed to raise the temperature of the silver. Hence, the specific heat of silver is about half that of iron's.

▪ Problem Solving

If the specific heat capacity c is known for a substance, then the heat transferred = specific heat capacity × mass × change in temperature. This can be expressed by the formula

$$Q = cm\Delta T$$

where Q is the quantity of heat, c is the specific heat of the substance, m is the mass, and ΔT is the corresponding change in temperature of the substance. When mass m is in grams, using the specific heat capacity of water as 1.0 cal/g·°C gives Q in calories.

SAMPLE PROBLEM 1

What would be the final temperature of a mixture of 50 g of 20°C water and 50 g of 40°C water?

SOLUTION:

The heat gained by the cooler water equals the heat lost by the warmer water. Because the masses of water are the same, the final temperature is midway, 30°C. So we'll end up with 100 g of 30°C water.

SAMPLE PROBLEM 2

Consider mixing 100 g of 25°C water with 75 g of 40°C water. Show that the final temperature of the mixture is 31.4°C.

SOLUTION:

Here we have different masses of water that are mixed together. We equate the heat gained by the cool water to the heat lost by the warm water. We can express this equation formally, then let the expressed terms lead to a solution:

Heat gained by cool water = heat lost by warm water

$$cm_1\Delta T_1 = cm_2\Delta T_2$$

ΔT_1 doesn't equal ΔT_2 as in Sample Problem 1 because of different masses of water. Some thinking shows that ΔT_1 is the final temperature T minus 25°C, because T will be greater than 25°C. ΔT_2 is 40°C minus T, because T will be less than 40°C. Then,

$$c(100\text{ g})(T - 25) = c(75\text{ g})(40 - T)$$

$$100T - 2500 = 3000 - 75T$$

$$T = 31.4°C$$

SAMPLE PROBLEM 3

Radioactive decay in Earth's interior provides enough energy to keep the interior hot, generate magma, and provide warmth to natural hot springs. This is due to the average release of about 0.03 J/kg each year. Show that the time it takes for a chunk of thermally insulated rock to increase 500°C in temperature (assuming that the specific heat of the rock sample is 800 J/kg·°C) is 13.3 million years.

SOLUTION:

Here we switch to rock, but the same concept applies. And we switch to specific heat expressed in joules per kilogram per degree Celsius. No particular mass is specified, so we'll work with quantity of heat/mass (for our answer should be the same for a small chunk of rock or a huge chunk).

From $Q = cm\Delta T$ we divide by m and get $Q/m = c\Delta T = (800\text{ J/kg}·°C)(500°C) = 400{,}000$ J/kg. The time required is $(400{,}000\text{ J/kg}) \div (0.03\text{ J/kg}·\text{yr}) = 13.3$ million years. Small wonder it remains hot down there!

FIGURE 6.12

Because water has a high specific heat capacity and is transparent, it takes more energy to warm the water than to warm the land. Solar energy striking the land is concentrated at the surface, but energy striking the water extends beneath the surface and so is "diluted."

We can think of specific heat capacity as thermal inertia. Recall that inertia is a term used in mechanics to signify the resistance of an object to a change in its state of motion. Specific heat capacity is like thermal inertia because it signifies the resistance of a substance to a change in temperature.

THE HIGH SPECIFIC HEAT CAPACITY OF WATER

Water has a much higher capacity for storing thermal energy than almost any other substance. The reason for water's high specific heat involves the various ways that energy can be absorbed. Energy absorbed by any substance increases the jiggling motion of molecules, which raises the temperature. Or absorbed energy may increase the amount of internal vibration or rotation within the molecules, which becomes potential energy, which does not raise the temperature. Usually absorption of energy involves a combination of both. When we compare water molecules with atoms in a metal, we find many more ways for water molecules to absorb energy without increasing translational kinetic energy. So water has a much higher specific heat capacity than metals—and most other common materials.

CHECK
  POINT

1. Which has a higher specific heat capacity, water or sand? In other words, which takes longer to warm in sunlight (or longer to cool at night)?
2. Why does a piece of watermelon stay cool for a longer time than sandwiches do when both are removed from a picnic cooler on a hot day?

Were these your answers?

1. Water has the higher specific heat capacity. In the same sunlight, the temperature of water increases more slowly than the temperature of sand. And water cools more slowly at night. (Walking or running barefoot across scorching sand in daytime is a different experience from doing the same in the evening!) The low specific heat capacity of sand and soil, as evidenced by how quickly they warm in the morning Sun and how quickly they cool at night, affects local climates.
2. Water in the melon has more "thermal inertia" than sandwich ingredients, and it resists changes in temperature much more. This thermal inertia is specific heat capacity.

Water's high specific heat capacity affects the world's climate. Look at a world globe and notice the high latitude of Europe. Water's high specific heat helps keep Europe's climate appreciably milder than regions of the same latitude in northeastern regions of Canada. Both Europe and Canada receive about the same amount of sunlight per square kilometer. Fortunately for Europeans, the Atlantic Ocean current known as the Gulf Stream carries warm water northeast from the Caribbean Sea, retaining much of its thermal energy long enough to reach the North Atlantic Ocean off the coast of Europe. There the water releases 4.18 J of energy for each gram of water that cools by 1°C. The released energy is carried by westerly winds over the European continent.*

 Link To Section 24.2

A similar effect occurs in the United States. The winds in North America are mostly westerly. On the West Coast, air moves from the Pacific Ocean to the land. In winter months, the ocean water is warmer than the air. Air blows over the warm water and then moves over the coastal regions. This produces a warm climate. In summer, the opposite occurs. Air blowing over the water carries cooler air to the coastal regions. The East Coast does not benefit from the

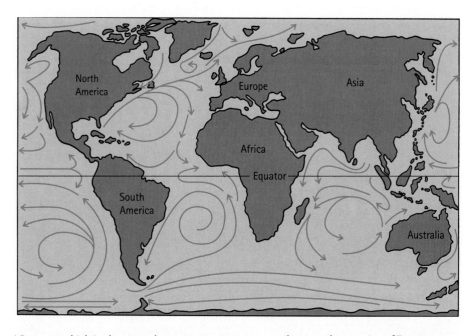

FIGURE 6.13

Many ocean currents, shown in blue, distribute heat from the warmer equatorial regions to the colder polar regions.

* Jet streams high in the atmosphere are an even greater contributor to the warming of Europe.

moderating effects of water because the direction of air is from the land to the Atlantic Ocean. Land, with a lower specific heat capacity, gets hot in the summer but cools rapidly in the winter.

Islands and peninsulas do not have the temperature extremes that are common in interior regions of a continent. The high summer and low winter temperatures common in Manitoba and the Dakotas, for example, are largely due to the absence of large bodies of water. Europeans, islanders, and people living near ocean air currents should be glad that water has such a high specific heat capacity. San Franciscans certainly are!

CHECK POINT
■
Bermuda is close to North Carolina, but, unlike North Carolina, it has a tropical climate year-round. Why?

Was this your answer?
Bermuda is an island. The surrounding water warms it when it might otherwise be too cold, and cools it when it might otherwise be too warm.

6.8 Thermal Expansion

The Physical Science Place
How a Thermostat Works

As the temperature of a substance increases, its molecules jiggle faster and move farther apart. The result is *thermal expansion*. Most substances expand when heated and contract when cooled. Sometimes the changes aren't noticeable, and sometimes they are. Telephone wires are longer and sag more on a hot summer day than in winter. Railroad tracks that were laid on cold winter days expand and buckle in the hot summer (Figure 6.14). Metal lids on glass fruit jars can often be loosened by heating them under hot water. If one part of a piece of glass is heated or cooled more rapidly than adjacent parts, the resulting expansion or contraction may break the glass. This is especially true of thick glass. Pyrex glass is an exception because it is specially formulated to expand very little with increasing temperature.

Thermal expansion accounts for the creaky noises often heard in the attics of old houses on cold nights.

FIGURE 6.14

Thermal expansion. Extreme heat on a July day caused the buckling of these railroad tracks.

FIGURE 6.15

One end of the bridge rides on rockers to allow for thermal expansion. The other end (not shown) is anchored.

FIGURE 6.16

This gap in the roadway of a bridge is called an expansion joint; it allows the bridge to expand and contract. (Was this photo taken on a warm day or a cold day?)

Thermal expansion must be taken into account in structures and devices of all kinds. A civil engineer uses reinforcing steel with the same expansion rate as concrete. A long steel bridge usually has one end anchored while the other rests on rockers (Figure 6.15). Notice also that many bridges have tongue-and-groove gaps called *expansion joints* (Figure 6.16). Similarly, concrete roadways and sidewalks are intersected by gaps, which are sometimes filled with tar, so that the concrete can expand freely in summer and contract in winter.

The fact that different substances expand at different rates is nicely illustrated with a bimetallic strip (Figure 6.17). This device is made of two strips of different metals welded together, one of brass and the other of iron. When heated, the greater expansion of the brass bends the strip. This bending may be used to turn a pointer, regulate a valve, or close a switch.

FIGURE 6.17

A bimetallic strip. Brass expands more when heated than iron does, and it contracts more when cooled. Because of this behavior, the strip bends as shown.

A practical application of a bimetallic strip wrapped into a coil is the thermostat (Figure 6.18). When a room becomes too cold, the coil bends toward the brass side and activates an electrical switch that turns on the heater. When the room gets too warm, the coil bends toward the iron side, which breaks the electrical circuit and turns off the heater. Bimetallic strips are used in oven thermometers, refrigerators, electric toasters, and various other devices.

With increases in temperature, liquids expand more than solids. We notice this when gasoline overflows from a car's tank on a hot day. If the tank and its contents expanded at the same rate, no overflow would occur. This is why a gas tank being filled shouldn't be "topped off," especially on a hot day.

6.9 Expansion of Water

Water, like most other substances, expands when heated. But interestingly, it *doesn't* expand in the temperature range between 0°C and 4°C. Something quite fascinating happens in this range.

Ice has a crystalline structure, with open-structured crystals. Water molecules in this open structure have more space between them than they do in the liquid phase (Figure 6.19). This means that ice is less dense than water. When ice melts, not all the open-structured crystals collapse. Some remain in the ice-water mixture, making up a microscopic slush that slightly "bloats" the water—

FIGURE 6.18

A thermostat. When the bimetallic coil expands, the drop of liquid mercury rolls away from the electrical contacts and breaks the electrical circuit. When the coil contracts, the mercury rolls against the contacts and completes the circuit.

Water molecules in a liquid are denser than water molecules frozen in ice, in which they have an open crystalline structure.

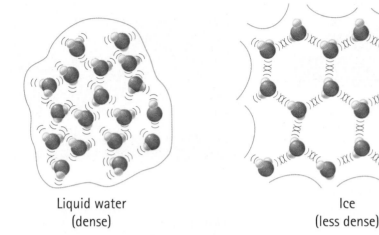

Liquid water
(dense)

Ice
(less dense)

increases its volume slightly (Figure 6.21). This results in ice water being less dense than slightly warmer water. As the temperature of water at 0°C is increased, more of the remaining ice crystals collapse. The melting of these ice crystals further decreases the volume of the water. Two opposite processes occur for the water at the same time—contraction and expansion. Volume decreases as ice crystals collapse, while volume increases due to greater molecular motion. The collapsing effect dominates until the temperature reaches 4°C. After that, expansion overrides contraction because most of the ice crystals have melted (Figure 6.22).

When ice water freezes to become solid ice, its volume increases tremendously. As solid ice cools further, like most substances, it contracts. The density of ice at any temperature is much lower than the density of water, which is why ice floats on water. This behavior of water is very important in nature. If water were most dense at 0°C it would settle to the bottom of a pond or lake instead of forming at the surface.

A pond freezes from the surface downward. In a cold winter the ice is thicker than in a mild winter. Water at the bottom of an ice-covered pond is 4°C, which is relatively warm for organisms that live there. Interestingly, very deep bodies of water are not ice-covered even in the coldest of winters. This is because all the water must be cooled to 4°C before lower temperatures can be reached. For deep water, the winter is not long enough to reduce an entire pond to 4°C. Any 4°C water lies at the bottom. Because of water's high specific heat and poor ability to conduct heat, the bottom of deep bodies of water in cold regions remains at a constant 4°C year round. Fish should be glad that this is so.

The six-sided structure of a snowflake is a result of the six-sided ice crystals that make it up. The crystals are made mostly from water vapor, not liquid water. Most snowflakes are not as symmetrical as this one.

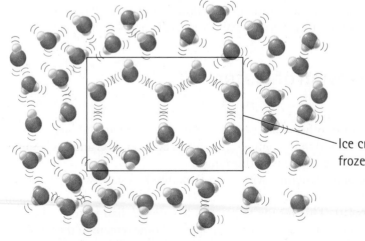

Ice crystals in nearly frozen liquid water

Close to 0°C, liquid water contains crystals of ice. The open structure of these crystals increases the volume of the water slightly.

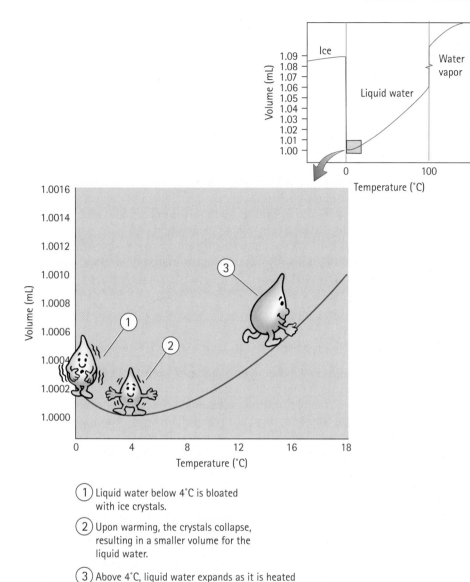

FIGURE 6.22

Between 0°C and 4°C, the volume of liquid water decreases as temperature increases. Above 4°C, thermal expansion exceeds contraction and volume increases as temperature increases.

Because water is most dense at 4°, colder water rises and freezes on the surface. This means that fish remain in relative warmth!

① Liquid water below 4°C is bloated with ice crystals.

② Upon warming, the crystals collapse, resulting in a smaller volume for the liquid water.

③ Above 4°C, liquid water expands as it is heated because of greater molecular motion.

FIGURE 6.23

As water cools, it sinks until the entire pond is at 4°C. Then, as water at the surface cools further, it floats on top and can freeze. Once ice is formed, temperatures lower than 4°C can extend down into the pond.

1. What was the precise temperature at the bottom of Lake Michigan on New Year's Eve in 1901?
2. What's inside the open spaces of the ice crystals shown in Figure 6.19? Is it air, water vapor, or nothing?

Were these your answers?

1. The temperature at the bottom of any body of water that has 4°C water in it is 4°C at the bottom, for the same reason that rocks are at the bottom. Both 4°C water and rocks are more dense than water at any other temperature. Water is a poor heat conductor, so if the body of water is deep and in a region of long winters and short summers, the water at the bottom is likely to remain a constant 4°C year round.
2. There's nothing at all in the open spaces. It's empty space—a void. If there were air or vapor in the open spaces, the illustration should show molecules there—oxygen and nitrogen for air and H_2O for water vapor.

Life at the Extremes

Some deserts, such as those on the plains of Spain, the Sahara in Africa, and the Gobi in central Asia, reach surface temperatures of 60°C (140°F). Too hot for life? Not for certain species of ants of the genus *Cataglyphis*, which thrive at this searing temperature. At this extremely high temperature, the desert ants can forage for food without the presence of lizards, which would otherwise prey upon them. Resilient to heat, these ants can withstand higher temperatures than any other creatures in the desert. How they are able to do this is currently being researched. They scavenge the desert surface for the corpses of creatures that did not find cover in time, touching the hot sand as little as possible while often sprinting on four legs with two held high in the air. Although their foraging paths zigzag over the desert floor, their return paths are almost straight lines to their nest holes. They attain speeds of 100 body lengths per second. During an average six-day life, most of these ants retrieve 15 to 20 times their weight in food.

From deserts to glaciers, a variety of creatures have invented ways to survive the harshest corners of the world. A species of worm thrives in the glacial ice in the Arctic. Insects in the Antarctic ice pump their bodies full of antifreeze to ward off becoming frozen solid. Some fish that live beneath the ice are able to do the same. Some bacteria thrive in boiling hot springs as a result of having heat-resistant proteins.

An understanding of how creatures survive at the extremes of temperature can provide clues for practical solutions to the physical challenges faced by humans. Astronauts who venture from Earth, for example, will need all the techniques available for coping with unfamiliar environments.

SUMMARY OF TERMS

Temperature A measure of the hotness or coldness of substances, related to the average kinetic energy per molecule in a substance, measured in degrees Celsius, degrees Fahrenheit, or kelvins.

Absolute zero The theoretical temperature at which a substance possesses no kinetic energy.

Thermal energy The total energy (kinetic plus potential) of the submicroscopic particles that make up a substance.

Heat The thermal energy that flows from a substance of higher temperature to a substance of lower temperature, commonly measured in calories or joules.

Thermodynamics The study of heat and its transformation to different forms of energy.

First law of thermodynamics A restatement of the law of energy conservation, usually as it applies to systems involving changes in temperature: Whenever heat flows into or out of a system, the gain or loss of thermal energy equals the amount of heat transferred.

Second law of thermodynamics Heat never spontaneously flows from a cold substance to a hot substance. Also, in natural processes, high-quality energy tends to transform into lower-quality energy—order tends to disorder.

Third law of thermodynamics No system can reach absolute zero.

Entropy The measure of energy dispersal of a system. Whenever energy freely transforms from one form to another, the direction of transformation is toward a state of greater disorder and, therefore, toward one of greater entropy.

Specific heat capacity The quantity of heat required to raise the temperature per unit mass of a substance by 1°C.

REVIEW QUESTIONS

6.1 Temperature

1. What are the temperatures for freezing water on the Celsius and Fahrenheit scales? For boiling water?
2. Is the temperature of an object a measure of the total kinetic energy of molecules in the object or a measure of the average kinetic energy per molecule in the object?
3. What is meant by the statement "a thermometer measures its own temperature"?

6.2 Absolute Zero

4. By how much does the pressure of a gas in a rigid vessel decrease when the temperature is decreased by 1°C?
5. What pressure would you expect in a rigid container of 0°C gas if you cooled it by 273°C?
6. What are the temperatures for freezing water and boiling water on the Kelvin temperature scale?
7. How much energy can be taken from a system at 0 K?

6.3 Heat

8. When you touch a cold surface, does cold travel from the surface to your finger or does thermal energy travel from your finger to the cold surface? Explain.
9. Distinguish between temperature and heat.
10. Distinguish between heat and thermal energy.
11. What determines the direction of heat flow?
12. Is cold the opposite of thermal energy or the lack of it?

6.4 Quantity of Heat

13. How is the energy value of foods determined?
14. Distinguish between a calorie and a Calorie.
15. Distinguish between a calorie and a joule.

6.5 The Laws of Thermodynamics

16. State the first law of thermodynamics.

17. How does the law of the conservation of energy relate to the first law of thermodynamics?
18. State the second law of thermodynamics.
19. How does the second law of thermodynamics relate to the direction of heat flow?
20. State the third law of thermodynamics.

6.6 Entropy

21. Under what condition can the entropy of a system be decreased?

6.7 Specific Heat Capacity

22. Which warms up faster when heat is applied—iron or silver?
23. Does a substance that heats up quickly have a high or a low specific heat capacity?
24. How does the specific heat capacity of water compare with the specific heat capacities of other common materials?

6.8 Thermal Expansion

25. Why does a bimetallic strip bend with changes in temperature?
26. Which generally expands more for an equal increase in temperature—solids or liquids?

6.9 Expansion of Water

27. When the temperature of ice-cold water is increased slightly, does it undergo a net expansion or a net contraction?
28. What is the reason for ice being less dense than water?
29. At what temperature do the combined effects of contraction and expansion produce the smallest volume for water?
30. Why does ice form at the surface of a pond instead of at the bottom?

EXERCISES

● BEGINNER   ■ INTERMEDIATE   ◆ EXPERT

1. ● In your room are things such as tables, chairs, and other people. Which of these things has a temperature (a) lower than, (b) greater than, and (c) equal to the temperature of the air?
2. ● Why can't you establish whether you are running a high temperature by touching your own forehead?
3. ● Why wouldn't you expect all the molecules in a gas to have the same speed?
4. ● Which is greater, an increase in temperature of 1°C or an increase of 1°F?
5. ● Which has the greater amount of thermal energy, an iceberg or a cup of hot coffee? Explain.
6. ● On which temperature scale does the average kinetic energy of molecules double when the temperature doubles?

7. ■ The temperature of the Sun's interior is about 10^7 degrees. Does it matter whether this is degrees Celsius or kelvins? Defend your answer.
8. ■ When air is rapidly compressed, why does its temperature increase?
9. ● Which of the laws of thermodynamics has exceptions?
10. ● What happens to the gas pressure within a sealed gallon can when it is heated? Cooled? Why?
11. ● After a car is driven for some distance, why does the air pressure in the tires increase?
12. ● If you drop a hot rock into a pail of water, the temperature of the rock and the water change until both are equal. The rock cools and the water warms. Does this hold true if the hot rock is dropped into the Atlantic Ocean? Defend your answer.

13. ● In the old days, on a cold winter night, it was common to bring a hot object to bed with you. Which would be better to keep you warm through the cold night—a 10-kg iron brick or a 10-kg jug of hot water at the same temperature? Explain.

14. ● Desert sand is very hot in the day and very cool at night. What does this tell you about its specific heat?

15. ● Why does adding the same amount of heat to two different objects not necessarily produce the same increase in temperature?

16. ● What role does specific heat capacity play in a watermelon staying cool after removal from a cooler on a hot day?

17. ● Why does the presence of large bodies of water tend to moderate the climate of nearby land—make it warmer in cold weather and cooler in hot weather?

18. ● If the winds at the latitude of San Francisco and Washington, D.C., were from the east rather than from the west, why might San Francisco be able to grow only cherry trees and Washington, D.C., only palm trees?

19. ● State an exception to the claim that all substances expand when heated.

20. ● Would a bimetallic strip function if the two different metals happened to have the same rates of expansion? Is it important that they expand at different rates? Defend your answer.

21. ■ Steel plates are commonly attached to each other with rivets, which are slipped into holes in the plates and rounded over with hammers. The hotness of the rivets makes them easier to round over, but their hotness has another important advantage in providing a tight fit. What is it?

22. ■ A method for breaking boulders used to be putting them in a hot fire, then dousing them with cold water. Why would this fracture the boulders?

23. ■ An old remedy for a pair of nested drinking glasses that stick together is to run water at different temperatures into the inner glass and over the surface of the outer glass. Which water should be hot, and which cold?

24. ■ A metal ball is just able to pass through a metal ring. When Ellyn Daugherty heats the ball, it does not pass

through the ring. What happens if she instead heats the ring (as shown)—does the size of the hole increase, stay the same, or decrease?

25. ■ After a machinist very quickly slips a hot, snugly fitting iron ring over a very cold brass cylinder, the two cannot be separated intact. Can you explain why this is so?

26. ■ How does the combined volume of the billions and billions of hexagonal open spaces in the structures of ice crystals in a piece of ice compare with the portion of ice that floats above the water line?

27. ■ Suppose you cut a small gap in a metal ring. If you heat the ring, does the gap become wider or narrower?

28. ■ State whether water at the following temperatures expands or contracts when warmed a little: 0°C; 4°C; 6°C.

29. ■ Suppose that water is used in a thermometer instead of mercury. If the temperature is at 4°C and then changes, why can't the thermometer indicate whether the temperature is rising or falling?

30. ■ If cooling occurred at the bottom of a pond instead of at the surface, would a lake freeze from the bottom up? Explain.

PROBLEMS

● BEGINNER ■ INTERMEDIATE ◆ EXPERT

The quantity of heat Q released or absorbed from a substance of specific heat c and mass m undergoing a change in temperature ΔT is

$$Q = cm\Delta T$$

1. ● Will Maynez burns a 0.6-g peanut beneath 50 g of water, which increases in temperature from 22°C to 50°C. (The specific heat capacity of water is 1.0 cal/g · °C.)
   (a) Assuming 40% efficiency, show that the peanut's food value is 3500 calories.
   (b) Then show how the food value in calories per gram is 5.8 kcal/g (or 5.8 Calories per gram).

2. ■ Pounding a nail into wood makes the nail warmer. Consider a 5-g steel nail 6 cm long and a hammer that exerts an average force of 500 N on the nail when it is being driven into a piece of wood. The nail becomes hotter. Show that the increase in the nail's temperature is 13.3°C. (Assume that the specific heat capacity of steel is 450 J/kg·°C.)

3. ■ If you wish to warm 100 kg of water by 20°C for your bath, show that the amount of heat is 2,000 kcal (2,000 Calories). Then show that this is equivalent to 8370 kJ.

4. ■ The specific heat capacity of copper is 0.092 cal/g·°C. Show that the amount of heat needed to raise the temperature of a 10-g piece of copper from 0°C to 100°C is 92 cal. How does this compare with the heat needed to raise the temperature of the same mass of water through the same temperature difference?

5. ● In lab you submerge 100 g of 40°C nails in 100 g of 20°C water. (The specific heat of iron is 0.12cal/g·°C.) Equate the heat gained by the water to the heat lost by the nails and show that the final temperature of the water becomes 22.1°C.

To solve the problems below, you will need to know about the average coefficient of linear expansion, α, which differs for different materials. We define α to be the change in length (L) *per unit length—or the fractional change in length—for a temperature change of 1°C. That is, ΔL/L per °C. For aluminum, α = 24 × 10²⁶/°C, and for steel, α = 11 × 10²⁶/°C. The change in length ΔL of a material is given by ΔL = L α ΔT.*

6. ● Consider a bar 1 m long that expands 0.5 cm when heated. Show that when similarly heated, a 100-m bar of the same material becomes 100.5 m long.

7. ● Suppose that the 1.3-km main span of steel for the Golden Gate Bridge had no expansion joints. Show that for an increase in temperature of 15°C the bridge would be 0.21 m longer.

8. ■ Imagine a 40,000-km steel pipe that forms a ring to fit snugly all around the circumference of Earth. Suppose that people along its length breathe on it so as to raise its temperature by 1°C. The pipe gets longer. It is also no longer snug. How high does it stand above ground level? Show that the answer is an astounding 70 m higher! (To simplify, consider only the expansion of its radial distance from the center of Earth, and apply the geometry formula that relates circumference C and radius r, $C = 2\pi r$.)

ACTIVE EXPLORATION

How much energy is in a nut? Burn it and find out. The heat of the flame is energy released upon the formation of chemical bonds (carbon dioxide, CO_2, and water, H_2O). Pierce a nut (pecan or walnut halves work best) with a bent paper clip that holds the nut above the table surface. Above this, secure a can of water so that you can measure its temperature change when the nut burns. Use about 10 cm³ (10 mL) of water and a Celsius thermometer. As soon as you ignite the nut with a match, place the can of water above it and record the increase in water temperature as soon as the flame burns out. The number of calories released by the burning nut can be calculated by the formula $Q = cm\Delta T$, where c is its specific heat (1cal/g°·C), m is the mass of water, and ΔT is the change in temperature. The energy in food is expressed in terms of the Calorie, which is 1000 of the calories you'll measure. So to find the number of Calories, divide your result by 1000.

READINESS ASSURANCE TEST (RAT)

If you have a good handle on this chapter, if you really do, then you should be able to score at least 7 out of 10 on this RAT. If you score less than 7, you need to study further before moving on.

Choose the BEST answer to each of the following.

1. Temperature is generally proportional to a substance's
   (a) thermal energy.
   (b) vibrational kinetic energy.
   (c) average translational kinetic energy.
   (d) rotational kinetic energy.

2. When three-quarters of a container of hot water is poured into a second empty container, the second container then has

   (a) $\frac{3}{4}$ the thermal energy.
   (b) $\frac{3}{4}$ the original volume of water.
   (c) the same temperature.
   (d) all of these

3. Water freezes at a temperature of
   (a) 0°C.
   (b) 273 K.
   (c) both of these
   (d) none of these

4. Heat is simply another word for
   (a) temperature.
   (b) thermal energy.
   (c) thermal energy that flows from hot to cold.
   (d) all of these

5. The law of thermodynamics for which exceptions can occur is the
(a) first law.
(b) second law.
(c) third law.
(d) all of these

6. Your closet gets messier day by day. In this case entropy is
(a) decreasing.
(b) increasing.
(c) hanging steady.
(d) none of these

7. To say that water has a high specific heat capacity is to say that water
(a) requires a lot of energy in order to increase in temperature.
(b) gives off a lot of energy in cooling.
(c) absorbs a lot of energy for a small increase in temperature.
(d) all of these

8. A bimetallic strip used in thermostats relies on the fact that different metals have different
(a) specific heat capacities.
(b) thermal energies at different temperatures.
(c) rates of thermal expansion.
(d) all of these

9. The density of water at 4°C increases slightly when it is
(a) cooled.
(b) warmed.
(c) both of these
(d) neither of these

10. Water at 4°C sinks to the bottom of a pond because
(a) of the presence of microscopic ice crystals.
(b) marine life in cold climates need it.
(c) ice forms first at the water surface.
(d) like a rock, 4°C water is denser than surrounding water.

Answers to RAT
1c, 2d, 3c, 4c, 5b, 6b, 7d, 8c, 9d, 10d

CHAPTER 6 ONLINE RESOURCES

The Physical Science Place

Videos
- Low Temperatures with Liquid Nitrogen
- How a Thermostat Works

Quiz

Flashcards

Links

heat transfer and change of phase

7.1 CONDUCTION

7.2 CONVECTION

7.3 RADIATION

7.4 NEWTON'S LAW OF COOLING

7.5 GLOBAL WARMING AND THE GREENHOUSE EFFECT

7.6 HEAT TRANSFER AND CHANGE OF PHASE

7.7 BOILING

7.8 MELTING AND FREEZING

7.9 ENERGY AND CHANGE OF PHASE

■ Whenever a substance changes phase, a transfer of energy accompanies the phase change. When water changes to ice, for example, energy is extracted from the water. Or when water is changed to steam, energy must be given to the water. Phase changes are always accompanied by a transfer of thermal energy.

Thermal energy can also transfer when substances don't change phase—when heat transfers from warmer to cooler substances. We begin this chapter with this simpler case of heat transfer. When substances with different temperatures come into contact, those that are warm become cooler and those that are cool become warmer. They tend to reach a common temperature. This process occurs in three ways: by *conduction*, by *convection*, and by *radiation*.

7.1 Conduction

If you hold one end of an iron nail in a flame, the nail quickly becomes too hot to hold. If you hold one end of a short glass rod in a flame, the rod takes much longer before it becomes too hot to hold. In both cases, heat at the hot end travels along the entire length. This method of heat transfer is called **conduction**. Thermal conduction occurs by collisions between particles and their immediate neighbors. Because the heat travels quickly through the nail we say that it is a good *conductor* of heat. Materials that are poor conductors are called *insulators*.

Solids (such as metals) whose atoms or molecules have loosely held electrons are good conductors of heat. These mobile electrons move quickly and transfer energy to other electrons, which migrate quickly throughout the solid. Poor conductors (such as glass, wool, wood, paper, cork, and plastic foam) are made up of molecules that hold tightly to their electrons. In these materials, molecules vibrate in place, and transfer energy only through interactions with their immediate neighbors. Because the electrons are not mobile, energy is transferred much more slowly in insulators.

Wood is a good insulator, and it is often used for cookware handles. Even when a pot is hot, you can briefly grasp the wooden handle with your bare hand without harm. An iron handle of the same temperature would surely burn your hand. Wood is a good insulator even when it's red hot. This explains how fire-walking *Conceptual Chemistry* author John Suchocki can walk barefoot on red-hot wood coals without burning his feet (as shown in the chapter-opener photo). (**CAUTION**: Don't try this on your own; even experienced fire walkers sometimes receive bad burns when conditions aren't just right.) The main factor here is the poor conductivity of wood—even red-hot wood. Although its temperature is high, very little internal energy is conducted to the feet. A fire walker must be careful that no iron nails or other good conductors are among the hot coals. Ouch!

Air is a very poor conductor. Hence, you can briefly put your hand in a hot pizza oven without harm. But don't touch the metal in the hot oven. Ouch again! The good insulating properties of such things as wool, fur, and feathers are largely due to the air spaces they contain. Be glad that air is a poor conductor; if it weren't, you'd feel quite chilly on a 20°C (68°F) day!

Snow is a poor conductor because its flakes are formed of crystals that trap air and provide insulation. That's why a blanket of snow keeps the ground warm in winter. Animals in the forest find shelter from the cold in snow banks and in holes in the snow. The snow doesn't provide them with energy—it simply slows down the loss of body heat that the animals generate. The same principle explains why igloos, arctic dwellings built from compacted snow, can shield their inhabitants from the cold.

Interestingly, insulation doesn't prevent the flow of internal energy. Insulation simply slows down the *rate* at which internal energy flows. Even a warm, well-insulated house gradually cools. Insulation such as rock wool or fiberglass placed

FIGURE 7.1

The tile floor feels colder than the wooden floor, even though both are at the same temperature. Tile is a better heat conductor than wood, and it more quickly conducts internal energy from your feet.

What can be both good and poor at the same time? Answer: Any good insulator is a poor conductor. Or any good conductor is a poor insulator.

FIGURE 7.2

When you stick a nail into ice, does cold flow from the ice to your hand, or does energy flow from your hand to the ice?

FIGURE 7.3

Conduction of heat from Lil's hand to the wine is minimized by the long stem of the wine glass.

in the walls and ceiling of a house slows down the transfer of internal energy from a warm house to the cooler outside (in winter) and from the warmer outside to the cool house (in summer).

CHECK POINT

1. In desert regions that are hot in the day and cold at night, the walls of houses are often made of mud. Why is it important that the mud walls be thick?
2. Wood is a better insulator than glass. Yet fiberglass is commonly used to insulate buildings. Why?

Were these your answers?

1. A wall of appropriate thickness retains the warmth of the house at night by slowing the flow of internal energy from inside to outside, and it keeps the house cool in the daytime by slowing the flow of internal energy from outside to inside. Such a wall has *thermal inertia.*
2. Fiberglass is a good insulator, many times better than glass, because of the air that is trapped among its fibers.

FIGURE 7.4

Snow patterns on the roof of a house show areas of conduction and insulation. Bare parts show where heat from the inside has conducted through the roof and melted the snow.

7.2 Convection

On a hot day you can see ripples in the air as hot air rises from an asphalt road. Likewise, if you put an ice cube into a clear glass of hot water you can see ripples as the cold water from the melting ice cube descends in the glass. Transfer of heat by the motion of fluid as it rises or sinks is called **convection.** Unlike conduction, convection occurs only in fluids (liquids and gases). Convection involves bulk motion of a fluid (currents) rather than interactions at the molecular level.

We can see why warm air rises. When warmed, it expands, becomes less dense, and is buoyed upward in the cooler surrounding air like a balloon buoyed upward. When the rising air reaches an altitude at which the air density is the same, it no longer rises. We see this occurring when smoke from a fire rises and then settles off as it cools and its density matches that of the surrounding air.

To see for yourself that expanding air cools, do the experiment shown in Figure 7.7. Expanding air really does cool.*

A dramatic example of cooling by expansion occurs with steam expanding through the nozzle of a pressure cooker (Figure 7.8). The combined cooling effects of expansion and rapid mixing

a

b

Thermal conduction is the process by which energy is transferred by heat through a material between two points at different temperatures.

FIGURE 7.5

Convection currents in a gas (air) and a liquid.

fyi

- Convection ovens are simply ovens with a fan inside. Cooking is speeded up by the circulation of heated air.

FIGURE 7.6

The tip of a heater element submerged in water produces convection currents, which are revealed as shadows (caused by deflections of light in water of different temperatures).

---

* Where does the energy go in this case? It goes into work done on the surrounding air as the expanding air pushes outward.

Link To Sections 22.2, 24.6, 25.2

FIGURE 7.7

Blow warm air onto your hand from your wide-open mouth. Now reduce the opening between your lips so the air expands as you blow. Try it now. Do you notice a difference in the temperature of exhaled air? Does air cool as it expands?

FIGURE 7.8

The hot steam expands as it leaves the pressure cooker and is cool to Millie's touch.

Link To Sections 24.4, 25.3

with cooler air allow you to hold your hand comfortably in the jet of condensed vapor. (**CAUTION**: If you try this, be sure to place your hand high above the nozzle at first and then lower it slowly to a comfortable distance above the nozzle. If you put your hand directly at the nozzle where no steam is visible, watch out! Steam is invisible and is clear of the nozzle before it expands and cools. The cloud of "steam" you see is actually condensed water vapor, which is much cooler than live steam.)

Cooling by expansion is the opposite of what occurs when air is compressed. If you've ever compressed air with a tire pump, you probably noticed that both the air and the pump became quite hot. Compressing air warms it.

Convection currents stir the atmosphere and produce winds. Some parts of Earth's surface absorb energy from the Sun more readily than others. This results in uneven heating of the air near the ground. We see this effect at the seashore, as Figure 7.9 shows. In the daytime, the ground warms up more than the water. Then warmed air close to the ground rises and is replaced by cooler air that moves in from above the water. The result is a sea breeze. At night, the process reverses because the shore cools off more quickly than the water, and then the warmer air is over the sea. If you build a fire on the beach, you'll see that the smoke sweeps inland during the day and then seaward at night.

CHECK POINT

Explain why you can hold your fingers beside the candle flame without harm, but not above the flame.

Was this your answer?

Hot air travels upward by air convection. Because air is a poor conductor, very little energy travels sideways to your fingers.

FIGURE 7.9

Convection currents produced by unequal heating of land and water. During the day, warm air above the land rises, and cooler air over the water moves in to replace it. At night, the direction of air flow is reversed, because now the water is warmer than the land.

▪ 7.3 Radiation

Energy travels from the Sun through space and then through Earth's atmosphere and warms Earth's surface. This transfer of energy cannot involve conduction or convection, for there is no medium between the Sun and Earth. Energy must be transmitted some other way—by **radiation**.* The transferred energy is called *radiant energy*.

Radiant energy exists in the form of *electromagnetic waves*, ranging from the longest wavelengths to the shortest: radio waves, microwaves, infrared waves (invisible waves below red in the visible spectrum), visible waves, ultraviolet waves, X-rays, and gamma rays. We'll treat waves further in Chapters 11 and 12.

The wavelength of radiation is related to the frequency of vibration. Frequency is the rate of vibration of a wave source. Nellie Newton in Figure 7.11 shakes a rope at a low frequency (left) and at a higher frequency (right). Note that shaking at a low frequency produces a long, lazy wave, and shaking at a higher frequency produces a wave of shorter wavelength. We shall see in later chapters that vibrating electrons emit electromagnetic waves. Low-frequency vibrations produce long-wavelength waves, and high-frequency vibrations produce waves with shorter wavelengths.

EMISSION OF RADIANT ENERGY

Every object at any temperature above absolute zero emits radiant energy. The peak frequency $\overline{f}$ of radiant energy is directly proportional to the Kelvin temperature T of the emitter:

$$\overline{f} \sim T$$

If an object is hot enough, some of the radiant energy it emits is in the range of visible light. At a temperature of about 500°C, an object begins to emit the longest waves we can see, red light. Higher temperatures produce a yellowish light. At about 1500°C, all the different waves to which the eye is sensitive are emitted and we see an object as "white hot." A blue-hot star is hotter than a white-hot star, and a red-hot star is less hot. Because a blue-hot star has twice the light frequency of a red-hot star, it has twice the surface temperature of a red-hot star.**

Because the surface of the Sun has a high temperature (by earthly standards), it emits radiant energy at a high frequency—much of it in the visible portion of the

FIGURE 7.10

Types of radiant energy (electromagnetic waves).

Link To Section 24.5

FIGURE 7.11

A wave of long wavelength is produced when the rope is shaken gently (at a low frequency). When shaken more vigorously (at a high frequency), a wave of shorter wavelength is produced.

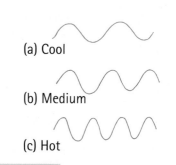

(a) Cool

(b) Medium

(c) Hot

FIGURE 7.12

(a) A low-temperature (cool) source emits primarily low-frequency, long-wavelength waves. (b) A medium-temperature source emits primarily medium-frequency, medium-wavelength waves. (c) A high-temperature (hot) source emits primarily high-frequency, short-wavelength waves.

* The radiation we are talking about here is electromagnetic radiation, including visible light. Don't confuse this with *radioactivity*, a process of the atomic nucleus that we'll discuss in Chapter 16.
** The *amount* of radiant energy Q emitted by an object is proportional to the fourth power of the Kelvin temperature T:

$$Q \sim T^4$$

So whereas a blue-hot star with twice the radiation frequency of a red-hot star has twice the Kelvin temperature, a blue-hot star with twice the temperature emits 16 times as much energy as a same-size red-hot star.

The amount of radiation emitted also depends on surface characteristics, and is referred to as the *emissivity* of the object—ranging from close to 0 for very shiny surfaces and close to 1 for very black ones. A perfectly black surface emits what is called *blackbody radiation* and has an emissivity of 1.

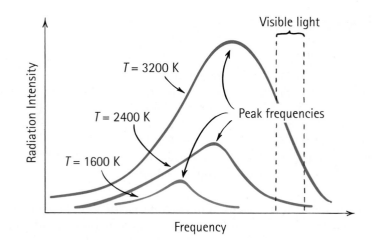

FIGURE 7.13

INTERACTIVE FIGURE

Radiation curves for different temperatures. The average frequency of radiant energy is directly proportional to the absolute temperature of the emitter.

Radiation emitted by Earth is *terrestrial radiation*. Radiation emitted by the Sun is *solar radiation*. Both are regions in the electromagnetic spectrum. (What do you call radiation from that special someone?)

Link To Section 25.4

FIGURE 7.14

Both the Sun and Earth emit the same kind of radiant energy. The Sun's glow is visible to the eye; Earth's glow consists of longer waves and isn't visible to the eye.

fyi

- Everything around you both radiates and absorbs energy continuously!

electromagnetic spectrum. The surface of Earth, by comparison, is relatively cool, and so the radiant energy it emits has a frequency lower than that of visible light. The radiation emitted by Earth is in the form of *infrared waves*—below our threshold of sight. Radiant energy emitted by Earth is called **terrestrial radiation**.

The Sun's radiant energy stems from nuclear reactions in its deep interior. Likewise, nuclear reactions in Earth's interior warm Earth (visit the depths of any mine and you'll find that it's warm down there year-round). Much of this internal energy conducts to the surface to become terrestrial radiation.

All objects—you, your instructor, and everything in your surroundings—continually emit radiant energy over a range of frequencies. Objects with everyday temperatures emit mostly low-frequency infrared waves. When the higher-frequency infrared waves are absorbed by your skin, as when you stand beside a hot stove, you feel the sensation of heat. So it is common to refer to infrared radiation as *heat radiation.* Common infrared sources that give the sensation of heat are the Sun, a lamp filament, and burning embers in a fireplace.

Heat radiation underlies infrared thermometers. You simply point the thermometer at something whose temperature you want, press a button, and a digital temperature reading appears. The radiation emitted by the object in question provides the reading. Typical classroom infrared thermometers operate in the range of about −30°C to 200°C.

CHECK POINT

Which of these do not emit radiant energy? (a) the Sun; (b) lava from a volcano; (c) red-hot coals; (d) this textbook.

Was this your answer?

All the above emit radiant energy—even your textbook, which, like the other substances listed, has a temperature. According to the rule $\bar{f} \sim T$, the book therefore emits radiation whose peak frequency $\bar{f}$ is quite low compared with the radiation frequencies emitted by the other substances. Everything with any temperature above absolute zero emits radiant energy. That's right—*everything!*

ABSORPTION OF RADIANT ENERGY

If everything is radiating energy, why doesn't everything finally run out of it? The answer is that everything is also *absorbing* energy. Good emitters of radiant energy are also good absorbers; poor emitters are poor absorbers. For example, a radio dish antenna constructed to be a good emitter of radio waves is also, by design, a good receiver (absorber) of them. A poorly designed transmitting antenna is also a poor receiver.

The surface of any material, hot or cold, both absorbs and emits radiant energy. If the surface absorbs more energy than it emits, it is a net absorber and its temperature rises. If it emits more than it absorbs, it is a net emitter and its temperature drops. Whether a surface plays the role of net emitter or net absorber depends on whether its temperature is above or below that of its surroundings. In short, if it's hotter than its surroundings, the surface is a net emitter and cools; if it's colder than its surroundings, it is a net absorber and becomes warmer.

FIGURE 7.15

When the black rough-surfaced container and the shiny polished one are filled with hot (or cold) water, the blackened one cools (or warms) faster.

CHECK POINT

1. If a good absorber of radiant energy were a poor emitter, how would its temperature compare with the temperature of its surroundings?
2. A farmer turns on the propane burner in his barn on a cold morning and heats the air to 20°C (68°F). Why does he still feel cold?

Were these your answers?

1. If a good absorber were not also a good emitter, there would be a net absorption of radiant energy and the temperature of the absorber would remain higher than the temperature of the surroundings. Things around us approach a common temperature only because good absorbers are, by their nature, also good emitters.
2. The walls of the barn are still cold. He radiates more energy to the walls than the walls radiate back at him, and he feels chilly. (On a winter day, you are comfortable inside your home or classroom only if the walls are warm—not just the air.)

Link To Section 27.2

A hot pizza put outside on a winter day is a net emitter. The same pizza placed in a hotter oven is a net absorber.

REFLECTION OF RADIANT ENERGY

Absorption and reflection are opposite processes. A good absorber of radiant energy reflects very little of it, including visible light. Hence, a surface that reflects very little or no radiant energy looks dark. So a good absorber appears dark, and a perfect absorber reflects no radiant energy and appears completely black. The pupil of the eye, for example, allows light to enter with no reflection, which is why it appears black. (An exception occurs in flash photography when pupils appear pink, which occurs when very bright light is reflected off the eye's pink inner surface and back through the pupil.)

Look at the open ends of pipes in a stack; the holes appear black. Look at open doorways or windows of distant houses in the daytime, and they, too, look black. Openings appear black because the light that enters them is reflected back and forth on the inside walls many times and is partly absorbed at each reflection. As a result, very little or none of the light remains to come back out of the opening and travel to your eyes (Figure 7.16).

Good reflectors, on the other hand, are poor absorbers. Clean snow is a good reflector and therefore does not melt rapidly in sunlight. If the snow is dirty, it absorbs radiant energy from the Sun and melts faster. Dropping black soot from an aircraft onto snow-covered mountains is a technique sometimes used in flood control to accomplish controlled melting at favorable times, rather than a sudden runoff of melted snow.

FIGURE 7.16

Radiation that enters the opening has little chance of leaving because most of it is absorbed. For this reason, the opening to any cavity appears black to us.

Link To Section 11.5

fyi

- Emission and absorption in the visible part of the spectrum are affected by color—but not so in the infrared part of the spectrum, where surface texture has more effect. In the infrared, a dull finish emits/absorbs better than a polished one, whatever the color.

FIGURE 7.17

The hole looks perfectly black and indicates a black interior, when in fact the interior has been painted a bright white.

CHECK POINT

Which would be more effective in heating the air in a room, a heating radiator painted black or silver?

Was this your answer?

Interestingly, the color of paint is a small factor, so either color can be used. That's because radiators do very little heating by radiation. Their hot surfaces warm surrounding air by conduction, the warmed air rises, and warmed convection currents heat the room. (A better name for this type of heater would be a *convector*.) Now if you're interested in *optimum* efficiency, a silver-painted radiator radiates less, becomes and remains hotter, and does a better job of heating the air.

7.4 Newton's Law of Cooling

Left to themselves, objects hotter than their surroundings eventually cool to match the surrounding temperature. The rate of cooling depends on how much hotter the object is than its surroundings. A hot apple pie cools more each minute if it is put in a cold freezer than if it is left on the kitchen table. That's because in the freezer, the temperature difference between the pie and its surroundings is greater. Similarly, the rate at which a warm house leaks internal energy to the cold outdoors depends on the difference between the inside and outside temperatures.

The rate of cooling of an object—whether by conduction, convection, or radiation—is approximately proportional to the temperature difference ΔT between the object and its surroundings.

$$\text{Rate of cooling} \sim \Delta T$$

This is known as **Newton's law of cooling**. (Guess who is credited with discovering this?)

The law applies also to warming. If an object is cooler than its surroundings, its rate of warming up is also proportional to ΔT.* Frozen food warms up faster in a warm room than in a cold room.

---

* A warm object that contains a source of energy may remain warmer than its surroundings indefinitely. The internal energy it emits doesn't necessarily cool it, and Newton's law of cooling doesn't apply. Thus an automobile engine that is running remains warmer than the automobile's body and the surrounding air. But after the engine is turned off, it cools in accordance with Newton's law of cooling and gradually approaches the same temperature as its surroundings. Likewise the Sun will remain hotter than its surroundings as long as its nuclear furnace is functioning—another 5 billion years or so.

The Thermos Bottle

A common Thermos bottle, a double-walled glass container with a vacuum between its silvered walls, nicely summarizes heat transfer. When a hot or cold liquid is poured into such a bottle, it remains at very nearly the same temperature for many hours. This is because the transfer of internal energy by conduction, convection, and radiation is severely inhibited.

1. Heat transfer by conduction through the vacuum is impossible. Some internal energy escapes by conduction through the glass and stopper, but this is a slow process, because glass, plastic, and cork are poor conductors.
2. The vacuum also prevents heat loss through the walls by convection, because there is no air between the walls.
3. Heat loss by radiation is prevented by the silvered surfaces of the walls, which reflect radiant energy back into the bottle.

CHECK POINT

Because a hot cup of tea loses internal energy more rapidly than a lukewarm cup of tea, would it be correct to say that a hot cup of tea will cool to room temperature before a lukewarm cup of tea will?

Was this your answer?
No! Although the *rate* of cooling is greater for the hotter cup, it has further to cool to reach thermal equilibrium. The extra time is equal to the time it takes to cool to the initial temperature of the lukewarm cup of tea. Cooling rate and cooling time are not the same thing.

Link To Section 24.5

7.5 Global Warming and the Greenhouse Effect

Link To Section 28.2

An automobile parked in the street in the bright Sun on a hot day with closed windows can get very hot inside—appreciably hotter than the outside air. This is an example of the *greenhouse effect*, so named for the same temperature-raising effect in florists' glass greenhouses. Understanding the greenhouse effect requires knowing about two concepts.

The first concept has been previously stated—that all things radiate, and the wavelength of radiation depends on the temperature of the object emitting the radiation. High-temperature objects radiate short waves; low-temperature objects radiate long waves. The second concept we need to know is that the transparency of things such as air and glass depends on the wavelength of radiation. Air is transparent to both infrared (long) waves and visible (short) waves, unless the air contains excess water vapor and carbon dioxide, in which case it is opaque to infrared. Glass is transparent to visible light waves, but opaque to infrared waves. (The physics of transparency and opacity is discussed in Chapter 11.)

Now to why that car gets so hot in bright sunlight: Compared with the car, the Sun's temperature is very high. This means the waves the Sun radiates are very short. These short waves easily pass through both Earth's atmosphere and the glass windows of the car. So energy from the Sun gets into the car interior, where, except for reflection, it is absorbed. The interior of the car warms up. The car interior radiates its own waves, but because it is not as hot as the Sun, the waves are longer. The reradiated long waves encounter glass that isn't transparent to them. So the reradiated energy remains in the car, which makes the car's

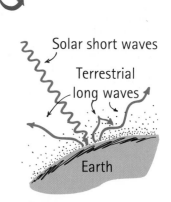

FIGURE 7.18

The hot Sun emits short waves, and the cool Earth emits long waves. Water vapor, carbon dioxide, and other *greenhouse gases* in the atmosphere retain heat that would otherwise be radiated from Earth to space.

Short-wavelength radiation from the Sun is transmitted through the glass.

Long-wavelength reradiated energy is not transmitted out through the glass and is trapped inside.

FIGURE 7.19

Glass is transparent to short-wavelength radiation but opaque to long-wavelength radiation. Reradiated energy from the plant is of long wavelength because the plant has a relatively low temperature.

Link To Section 25.1

A significant role of glass in a florist greenhouse is to prevent convection of cooler outside air with warmer inside air. So the greenhouse effect actually plays a bigger role in global warming than it does in the warming of florist greenhouses.

Link To Sections 16.7, 17.5

interior even warmer (which is why leaving your pet in a car on a hot sunny day is a no-no).

The same effect occurs in Earth's atmosphere, which is transparent to solar radiation. The surface of Earth absorbs this energy, and reradiates part of this as longer-wavelength terrestrial radiation. Atmospheric gases (mainly water vapor and carbon dioxide) absorb and re-emit much of this long-wavelength terrestrial radiation back to Earth. Terrestrial radiation that cannot escape Earth's atmosphere warms Earth. This global warming process is very nice, for Earth would be a frigid $-18°C$ otherwise. Over the last 500,000 years the average temperature of Earth has fluctuated between 19°C and 27°C and is presently at the high point, 27°C—and climbing. Our present environmental concern is that increased levels of carbon dioxide and other atmospheric gases in the atmosphere may further increase the temperature and produce a new thermal balance unfavorable to the biosphere.

An important credo is, "You can never change only one thing." Change one thing, and you change another. A slightly higher Earth temperature means slightly warmer oceans, which means changes in weather and storm patterns. Warmer oceans also mean slightly increased evaporation, which means slightly increased snowfall in polar regions. The fraction of Earth presently beneath ice and snow is greater than the total area used for farmlands—and is shrinking at a historically unprecedented rate. These white areas reflect more solar radiation, which potentially could lead to a significant drop in global temperature. So overheating Earth today conceivably could cool it tomorrow and trigger the next ice age! Or it might not. We don't know.

What we do know is that energy consumption is related to population size. We are seriously questioning the idea of continued growth. (Please take the time to read Appendix D, "Exponential Growth and Doubling Time"—very important material.)

CHECK POINT

What does it mean to say that the greenhouse effect is like a one-way valve?

Was this your answer?

Both the atmosphere of Earth and glass in a florist greenhouse are transparent only to incoming short-wavelength light and block outgoing long waves. Because of the blockage, radiation travels only in one direction.

7.6 Heat Transfer and Change of Phase

Matter exists in four common phases (states). Ice, for example, is the *solid* phase of water. When internal energy is added, the increased molecular motion breaks down the frozen structure and it becomes the *liquid* phase, water. When more energy is added, the liquid changes to the *gaseous* phase. Add still more energy, and the molecules break into ions and electrons, giving the *plasma* phase. Plasma (not to be confused with blood plasma) is the illuminating gas found in TV screens and fluorescent and other vapor lamps. The Sun, stars, and much of the space between them is in the plasma phase. Whenever matter changes phase, a transfer of internal energy is involved.

EVAPORATION

Water changes to the gaseous phase by the process of **evaporation**. In a liquid, molecules move randomly at a wide variety of speeds. Think of the water molecules as tiny billiard balls, moving helter-skelter, continually bumping into one another. During their bumping, some molecules gain kinetic energy while others lose kinetic energy. Molecules at the surface that gain kinetic energy by being bumped from below are the ones to break free from the liquid. They leave the surface and escape into the space above the liquid. In this way, they become gas.

When fast-moving molecules leave the water, the molecules left behind are the slower-moving ones. What happens to the overall kinetic energy in a liquid when the high-energy molecules leave? The answer: the average kinetic energy of molecules left in the liquid decreases. The temperature (which measures the average kinetic energy of the molecules) decreases and the water is cooled.

When our bodies begin to overheat, our sweat glands produce perspiration. This is part of nature's thermostat; the evaporation of sweat cools us and helps maintain a stable body temperature. Many animals do not have sweat glands and must cool themselves by other means (Figures 7.21 and 7.22).

CHECK POINT

Would evaporation be a cooling process if there were no transfer of molecular kinetic energy from water to the air above?

Was this your answer?
No. A liquid cools only when kinetic energy is carried away by evaporating molecules. This is similar to billiard balls that gain speed at the expense of others that lose speed. Those that leave (evaporate) are gainers, while losers remain behind and lower the temperature of the water.

In solid carbon dioxide (dry ice), molecules jump directly from the solid to the gaseous phase—that's why it's called *dry ice*. This form of evaporation is called **sublimation**. Mothballs are well known for their sublimation. Even frozen water undergoes sublimation. Because water molecules are so tightly held in a solid, frozen water sublimes much more slowly than liquid water evaporates. Sublimation accounts for the loss of much snow and ice, especially on high, sunny mountain tops. Sublimation also explains why ice cubes left in the freezer for a long time get smaller.

FIGURE 7.20

When wet, the cloth covering on the canteen promotes cooling. As the faster-moving water molecules evaporate from the wet cloth, the temperature of the cloth decreases and cools the metal. The metal, in turn, cools the water within. The water in the canteen can become a lot cooler than the outside air.

 Link To Section 19.4

 fyi

- Water evaporating from your body takes energy with it, which is why you feel cool when emerging from water on a warm and windy day.

 Link To Section 23.6

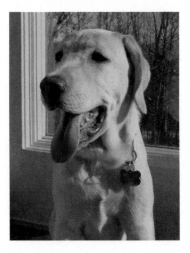

FIGURE 7.21

Sam, like other dogs, has no sweat glands (except between his toes). He cools by panting. In this way, evaporation occurs in the mouth and within the bronchial tract.

FIGURE 7.22

Pigs have no sweat glands and therefore cannot cool by the evaporation of perspiration. Instead, they wallow in the mud to cool themselves.

FIGURE 7.23

The exchange of molecules at the interface between liquid and gaseous water.

Liquid water   Water vapor

The Physical Science Place
Condensation Is a
Warming Process

FIGURE 7.24

Internal energy is released by steam when it condenses inside the "radiator."

FIGURE 7.25

If you're chilly outside the shower stall, step back inside and be warmed by the condensation of the excess water vapor there.

FIGURE 7.26

The toy drinking bird operates by the evaporation of ether inside its body and by the evaporation of water from the outer surface of its head. The lower body contains liquid ether, which evaporates rapidly at room temperature. As it (a) vaporizes, it (b) creates pressure (inside arrows), which pushes ether up the tube. Ether in the upper part does not vaporize because the head is cooled by the evaporation of water from the outer felt-covered beak and head. When the weight of ether in the head is sufficient, the bird (c) pivots forward, permitting the ether to run back to the body. Each pivot wets the felt surface of the beak and head, and the cycle is repeated.

CONDENSATION

The opposite of evaporation is **condensation**—the changing of a gas to a liquid. When gas molecules near the surface of a liquid are attracted to the liquid, they strike the surface with increased kinetic energy and become part of the liquid. This kinetic energy is absorbed by the liquid. The result is increased temperature. So whereas the liquid left behind is cooled with evaporation, with condensation the object upon which the vapor condenses is warmed. Condensation is a warming process.

A dramatic example of warming by condensation is the energy released by steam when it condenses. The steam gives up a lot of energy when it condenses to a liquid and moistens the skin. That's why a burn from 100°C steam is much more damaging than a burn from 100°C boiling water. This energy release by condensation is used in steam-heating systems.

When taking a shower, you may have noticed that you feel warmer in the moist shower region than outside the shower. You can quickly sense this difference when you step outside. Away from the moisture, the rate of evaporation is much higher than the rate of condensation, and you feel chilly. When you remain in the moist shower stall, the rate of condensation is higher and you feel warmer. So now you know why you can dry yourself with a towel much more comfortably if you remain in the shower stall. If you're in a hurry and don't mind the chill, dry yourself off in the hallway.

On a July afternoon in dry Phoenix or Santa Fe, you'll feel a lot cooler than in New York City or New Orleans, even when the temperatures are the same. In the drier cities, the rate of evaporation from your skin is much greater than the rate of condensation of water molecules from the air onto your skin. In humid locations, the rate of condensation is greater than the rate of evaporation. You feel the warming effect as vapor in the air condenses on your skin. You are literally being bombarded by the impact of H₂O molecules in the air slamming into you. (We will explore condensation in the atmosphere when we study weather and climate in Chapter 25.)

a   b   c   d

Condensation Crunch

Put a small amount of water in an aluminum soft-drink can and heat it on a stove until steam issues from the opening. When this occurs, air has been driven out and replaced by steam. Then, with a pair of tongs, quickly invert the can into a pan of water. Crunch! The can is crushed by atmospheric pressure! Why? When the molecules of steam inside the can hit the inner wall, they bounce—the metal certainly doesn't absorb them. But when steam molecules encounter water in the pan, they stick to the water surface. Condensation occurs, leaving a very low pressure in the can, whereupon the surrounding atmospheric pressure crunches the can. Here we see, dramatically, how pressure is reduced by condensation. (This demonstration nicely underlies the condensation cycle of a steam engine—perhaps something for future study.)

CHECK POINT

Place a dish of water anywhere in your room. If the water level in the dish remains unchanged from one day to the next, can you conclude that no evaporation or condensation is occurring?

Was this your answer?
Not at all, for significant evaporation and condensation occur continuously at the molecular level. The fact that the water level remains constant indicates equal rates of evaporation and condensation.

■ 7.7   Boiling

Evaporation occurs at the surface of a liquid. A change of phase from liquid to gas can also occur beneath the surface under proper conditions. The gas that forms beneath the surface of a liquid produces bubbles. The bubbles are buoyed upward to the surface, where they escape into the surrounding air. This change of phase is called **boiling**.

The pressure of the vapor within the bubbles in a boiling liquid must be great enough to resist the pressure of the surrounding liquid. Unless the vapor pressure is great enough, the surrounding pressures collapse any bubbles that tend

Pressure of atmosphere plus water

FIGURE 7.27

The motion of vapor molecules in the bubble of steam (much enlarged) creates a gas pressure (called the vapor pressure) that counteracts the atmospheric and water pressure against the bubble.

Link To Section 24.4

The Physical Science Place
Boiling Is a Cooling Process

FIGURE 7.28

Heating warms the water from below, and boiling cools it from above.

When we say that we boil water, it is common to mean we are heating it. Actually, the boiling process cools the water.

FIGURE 7.29

The tight lid of a pressure cooker holds pressurized vapor above the water surface, and this inhibits boiling. In this way, the boiling temperature of the water is increased to more than 100°C.

FIGURE 7.30

Apparatus to demonstrate that water freezes and boils at the same time in a vacuum. A gram or two of water is placed in a dish that is insulated from the base by a polystyrene cup.

to form. At temperatures below the boiling point, the vapor pressure is not great enough. Bubbles do not form until the boiling point is reached.

Boiling, like evaporation, is a cooling process. At first thought, this may seem surprising—perhaps because we usually associate boiling with heating. However, heating water is one thing; boiling it is another. When 100°C water at atmospheric pressure is boiling, it is in thermal equilibrium. The water in the pot is being cooled by boiling as fast as it is being heated by energy from the heat source (Figure 7.28). If cooling did not occur, continued application of heat to a pot of boiling water would raise its temperature.

When pressure on the surface of a liquid increases, boiling is hampered. Then the temperature needed for boiling rises. The boiling point of a liquid depends on the pressure on the liquid—which is most evident with a pressure cooker (Figure 7.29). In such a device, vapor pressure builds up inside and prevents boiling. This results in a water temperature that is higher than the normal boiling point. Note that the high temperature of the water cooks the food, not the boiling process itself.

Lower atmospheric pressure (as at high altitudes) decreases the boiling temperature. For example, in Denver, Colorado, the "mile-high city," water boils at 95°C instead of at 100°C. If you try to cook food in boiling water that is cooler than 100°C, you must wait a longer time for proper cooking. A three-minute boiled egg in Denver is yucky. If the temperature of the boiling water is very low, food does not cook at all.

CHECK POINT

1. Because boiling is a cooling process, would it be a good idea to cool your hot, sticky hands by dipping them into boiling water?
2. Rapidly boiling water has the same temperature as simmering water, both 100°C. Why, then, do the directions for cooking spaghetti often call for rapidly boiling water?

Were these your answers?

1. No, no, no! When we say boiling is a cooling process, we mean that the water left behind in the pot (and not your hands!) is being cooled relative to the higher temperature it would attain otherwise. Because of the cooling effect of the boiling, the water remains at 100°C instead of getting hotter. A dip in 100°C water would be extremely uncomfortable for your hands!
2. Good cooks know that the reason for the rapidly boiling water is not higher temperature, but simply a way to keep the spaghetti strands from sticking together.

A dramatic demonstration of the cooling effect of evaporation and boiling is shown in Figure 7.30. Here we see a shallow dish of room-temperature water in a vacuum jar. When the pressure in the jar is slowly reduced by a vacuum pump, the water begins to boil. As in all evaporation, the highest-energy molecules escape from the water, and the water left behind is cooled. As the pressure is further reduced, more and more of the faster-moving molecules boil away until the remaining liquid water reaches approximately 0°C. Continued cooling by boiling causes ice to form over the surface of the bubbling water. Boiling and freezing occur at the same time! Frozen bubbles of boiling water are a remarkable sight.

If you spray some drops of coffee into a vacuum chamber, they boil until they freeze. Even after they are frozen, the water molecules continue to evaporate into

FIGURE 7.31

Ron Hipschman at the Exploratorium removes a freshly frozen piece of ice from the "Water Freezer" exhibit, a vacuum chamber as depicted in Figure 7.30.

The Physical Science Place
Pressure Cooker: Boiling and Freezing at the Same Time

the vacuum, until little crystals of coffee solids remain. This is how freeze-dried coffee is produced. The low temperature of this process tends to keep the chemical structure of the coffee solids from changing. When hot water is added, much of the original flavor of the coffee is preserved.

fyi

- Mountaineering pioneers in the 19th century, without altimeters, used the boiling point of water to determine their altitudes.

7.8 Melting and Freezing

Melting occurs when a substance changes phase from a solid to a liquid. To visualize what happens, imagine a group of people holding hands and jumping around. The more violent the jumping, the more difficult it is to keep holding hands. If the jumping is violent enough, continuing to hold hands might become impossible. A similar thing happens to the molecules of a solid when it is heated. As heat is absorbed by the solid, its molecules vibrate more and more violently. If enough heat is absorbed, the attractive forces between the molecules no longer hold them together. The solid melts.

Freezing occurs when a liquid changes to a solid phase—the opposite of melting. As energy is removed from a liquid, molecular motion slows until molecules move so slowly that attractive forces between them bind them together. The liquid freezes when its molecules vibrate about fixed positions and form a solid.

Link To Sections 20.7, 24.7

FIGURE 7.32

(a) In a mixture of ice and water at 0°C, ice crystals gain and lose water molecules at the same time. The ice and water are in thermal equilibrium. (b) When salt is added to the water, fewer water molecules enter the ice because fewer of them are at the interface.

= water molecule entering ice crystal

= water molecule leaving ice crystal

(a)   Ice   Liquid water   (b)   Ice   Aqueous solution of NaCl

fyi

■ Why is rock salt spread on icy roads in winter? A short answer is that salt makes ice melt. Salt in water separates into sodium and chlorine ions. When these ions join water molecules, heat is given off, which melts microscopic parts of an icy surface. The melting process is enhanced by the pressure of automobiles rolling along the salt-covered icy surface, which forces the salt into the ice. The only difference between the rock salt applied to roads in winter and the substance you sprinkle on popcorn is the size of the crystals.

At atmospheric pressure, ice forms at 0°C. With impurities in the water, the freezing point is lowered. "Foreign" molecules get in the way and interfere with crystal formation. In general, adding anything to water lowers its freezing temperature. Antifreeze is a practical application of this process.

■ 7.9 Energy and Change of Phase

If you heat a solid sufficiently, it melts and becomes a liquid. If you heat the liquid, it vaporizes and becomes a gas. Energy must be put into a substance to change its phase in the direction from solid to liquid to gas. Conversely, energy must be extracted from a substance to change its phase in the direction from gas to liquid to solid (Figure 7.33).

The cooling cycle of a refrigerator nicely illustrates these concepts. A motor pumps a special fluid through the system, where it undergoes the cyclic process of vaporization and condensation. When the fluid vaporizes, internal energy is drawn from objects stored inside the refrigerator. The gas that forms, with its added energy, is directed to and condenses to a liquid in outside coils in the back—appropriately called *condensation coils*. The next time you're near a refrigerator, place your hand near the condensation coils in the back and you'll feel the heat that has been extracted from the inside.

An air conditioner uses the same principle and simply pumps heat energy from one part of the unit to another. If the roles of vaporization and condensation are reversed, the air conditioner becomes a heater.

Energy is absorbed when change of phase is in this direction

Solid ⇌ Liquid ⇌ Gas

Energy is released when change of phase is in this direction

FIGURE 7.33

Energy changes with change of phase.

Link To Section 23.6

The amount of energy needed to change any substance from solid to liquid (and vice versa) is called the **heat of fusion** for the substance. For water, this is 334 J/g. The amount of energy required to change any substance from liquid to gas (and vice versa) is called the **heat of vaporization** for the substance. For water, this is a whopping 2256 J/g.

In premodern times, farmers in cold climates prevented jars of food from freezing by taking advantage of water's high heat of fusion. They simply kept large tubs of water in their cellars. The outside temperature could drop to well below freezing, but not in the cellars, where water was releasing internal energy while undergoing freezing. Canned food requires subzero temperatures to freeze because of its salt or sugar content. So farmers had only to replace frozen tubs of water with unfrozen ones, and the cellar temperatures wouldn't fall below 0°C.

Heat of vaporization is either the energy required to separate molecules from the liquid phase or the energy released when gases condense to the liquid phase.

FIGURE 7.34

The energy of sunlight simply and nicely harnessed.

CHECK POINT

In the process of water vapor condensing in the air, the slower-moving molecules are the ones that condense. Does condensation warm or cool the surrounding air?

> Heat of fusion is either the energy needed to separate molecules from the solid phase or the energy released when bonds form in a liquid that change it to the solid phase.

Was this your answer?

As slower-moving molecules are removed from the air, there is an increase in the average kinetic energy of molecules that remain in the air. Therefore, the air is warmed. The change of phase is from gas to liquid, which releases energy (Figure 7.33).

FIGURE 7.35

Paul Ryan tests the hotness of molten lead by dragging his wetted finger through it.

Water's high heat of vaporization allows you to briefly touch your wetted finger to a hot skillet on a hot stove without harm. You can even touch it a few times in succession as long as your finger remains wet. Energy that ordinarily would flow into and burn your finger goes instead into changing the phase of the moisture on your finger. Similarly, you are able to judge the hotness of a hot clothes iron.

Paul Ryan, former supervisor in the Department of Public Works in Malden, Massachusetts, has for years used molten lead to seal pipes in certain plumbing operations. He startles onlookers by dragging his finger through molten lead to judge its hotness (Figure 7.35). He is sure that the lead is very hot and his finger is thoroughly wet before he does this. (Do not try this on your own: if the lead is not hot enough, it will stick to your finger—ouch!)

In Chapter 25 we'll discuss the role of thermal energy in global warming.

SUMMARY OF TERMS

Conduction The transfer of internal energy by molecular and electronic collisions within a substance (especially a solid).

Convection The transfer of internal energy in a gas or liquid by means of currents in the heated fluid. The fluid flows, carrying energy with it.

Radiation The transfer of energy by means of electromagnetic waves.

Terrestrial radiation The radiant energy emitted by Earth.

Newton's law of cooling The rate of loss of internal energy from an object is proportional to the temperature difference between the object and its surroundings.

$$\text{Rate of cooling} \sim \Delta T$$

Evaporation The change of phase at the surface of a liquid as it passes to the gaseous phase.

Sublimation The change of phase directly from solid to gas, bypassing the liquid phase.

Condensation The change of phase from gas to liquid; the opposite of evaporation. Warming of the liquid results.

Boiling A rapid state of evaporation that takes place within the liquid as well as at its surface. As with evaporation, cooling of the liquid results.

Melting The process of changing phase from solid to liquid, as from ice to water.

Freezing The process of changing phase from liquid to solid, as from water to ice.

Heat of fusion The amount of energy needed to change any substance from solid to liquid (and vice versa). For water, this is 334 J/g (or 80 cal/g).

Heat of vaporization The amount of energy required to change any substance from liquid to gas (and vice versa). For water, this is 2256 J/g (or 540 cal/g).

REVIEW QUESTIONS

1. What are the three common ways in which heat is transferred?

7.1 Conduction

2. What is the role of "loose" electrons in heat conductors?
3. What is the explanation for a barefoot fire walker being able to walk safely on red-hot wooden coals?
4. Does a good insulator prevent heat from getting through it, or does it simply slow its passage?

7.2 Convection

5. By what means is heat transferred by convection?
6. What happens to the temperature of air when it expands?
7. Why is Millie's hand not burned when she holds it above the escape valve of the pressure cooker (Figure 7.8)?
8. Why does the direction of coastal winds change from day to night?

7.3 Radiation

9. How does the frequency of radiant energy relate to the absolute temperature of the radiating source?
10. What is terrestrial radiation? How does it differ from solar radiation?
11. Because all objects emit energy to their surroundings, why don't the temperatures of all objects continuously decrease?
12. Why does the pupil of the eye appear black?

7.4 Newton's Law of Cooling

13. Which undergoes a greater rate of cooling, a red-hot poker in a warm oven or a red-hot poker in a cold room? (Or do both cool at the same rate?)
14. Does Newton's law of cooling apply to warming as well as to cooling?

7.5 Global Warming and the Greenhouse Effect

15. What would be the consequence to Earth's temperature if the greenhouse effect were completely eliminated?

16. What is meant by the expression, "You can never change only one thing?"

7.6 Heat Transfer and Change of Phase

17. What are the four common phases of matter?
18. Do all the molecules in a liquid have about the same speed, or do they have a wide variety of speeds?
19. What is evaporation, and why is it a cooling process? What is it that cools?
20. What is sublimation?
21. What is condensation, and why is it a warming process? What is it that warms?
22. Why is a steam burn more damaging than a burn from boiling water of the same temperature?

7.7 Boiling

23. Distinguish between evaporation and boiling.
24. Why does water not boil at 100°C when it is under greater-than-normal atmospheric pressure?
25. Is it the boiling of the water or the higher temperature of the water that cooks food faster in a pressure cooker?

7.8 Melting and Freezing

26. Why does increasing the temperature of a solid make it melt?
27. Why does decreasing the temperature of a liquid make it freeze?
28. Why doesn't water freeze at 0°C when foreign ions are present?

7.9 Energy and Change of Phase

29. Does a liquid give off energy or absorb energy when it turns into a gas? When it turns into a solid?
30. Does a gas give off energy or absorb energy when it turns into a liquid? How about a solid when it turns to a liquid?

EXERCISES

● BEGINNER ■ INTERMEDIATE ◆ EXPERT

1. ● Wrap a fur coat around a thermometer. Does the temperature rise?
2. ● What is the explanation for a feather quilt being so warm on a cold winter night?
3. ● What is the purpose of the layer of copper or aluminum on the bottom of a piece of stainless-steel cookware?
4. ● In terms of physics, why do restaurants serve baked potatoes wrapped in aluminum foil?

5. ● Many tongues have been injured by licking a piece of metal on a very cold day. Why would no harm result if a piece of wood were licked on the same day?
6. ● Wood is a better insulator than glass. Yet fiberglass is commonly used as an insulator in wooden buildings. Explain.
7. ■ Visit a snow-covered cemetery and note that the snow does not slope upward against the gravestones but,

instead, forms depressions around them, as shown. What is your explanation for this?

8. ■ Wood has a very low conductivity. Does it still have a low conductivity if it is very hot—that is, in the stage of smoldering red-hot coals? Could you safely walk across a bed of red-hot wooden coals with bare feet? Although the coals are hot, does much heat conduct from them to your feet if you step quickly? Could you do the same on pieces of red-hot iron? Explain. (**CAUTION**: Coals can stick to your feet, so OUCH—don't try it!)

9. ● A friend says that in a mixture of gases in thermal equilibrium, the molecules have the same average kinetic energy. Do you agree or disagree? Defend your answer.

10. ● A friend says that in a mixture of gases in thermal equilibrium, the molecules have the same average speed. Do you agree or disagree? Defend your answer.

11. ● What does the high specific heat of water have to do with the convection currents in the air at the seashore?

12. ■ How do the average kinetic energies per molecule compare in a mixture of hydrogen and oxygen gases at the same temperature?

13. ■ In a mixture of hydrogen and oxygen gases at the same temperature, which molecules move faster? Why?

14. ■ Which atoms have the greater average speed in a mixture, U-238 or U-235? How would this affect diffusion through a porous membrane of otherwise identical gases made from these isotopes?

15. ■ Machines used for making snow at ski areas blow a mixture of compressed air and water through a nozzle. The temperature of the mixture may initially be well above the freezing temperature of water, yet crystals of snow are formed as the mixture is ejected from the nozzle. Explain how this happens.

16. ● Turn an incandescent lamp on and off quickly while you are standing near it. You feel its heat, but you find when you touch the bulb that it is not hot. Explain why you felt heat from the lamp.

17. ● A number of bodies at different temperatures placed in a closed room share radiant energy and ultimately come to the same temperature. Would this thermal equilibrium be possible if good absorbers were poor emitters and poor absorbers were good emitters? Defend your answer.

18. ■ From the rules that a good absorber of radiation is a good radiator and a good reflector is a poor absorber, state a rule relating the reflecting and radiating properties of a surface.

19. ● The heat of volcanoes and natural hot springs comes from trace amounts of radioactive minerals in common rock in Earth's interior. Why isn't the same kind of rock at Earth's surface warm to the touch?

20. ◆ Suppose that, at a restaurant, you are served coffee before you are ready to drink it. In order that it be hottest when you are ready for it, would you be wiser to add cream to it right away or just before you are ready to drink it?

21. ■ Is it important to convert temperatures to the Kelvin scale when we use Newton's law of cooling? Why or why not?

22. ◆ If you wish to save fuel and you're going to leave your warm house for a half hour or so on a very cold day, should you turn your thermostat down a few degrees, turn it off altogether, or let it remain at the room temperature you desire?

23. ◆ If you wish to save fuel and you're going to leave your cool house for a half hour or so on a very hot day, should you turn your air conditioning thermostat up a bit, turn it off altogether, or let it remain at the room temperature you desire?

24. ● Why is whitewash sometimes applied to the glass of florists' greenhouses? Would you expect this practice to be more prevalent in winter or summer months?

25. ● If the composition of the upper atmosphere were changed so that it permitted a greater amount of terrestrial radiation to escape, what effect would this have on Earth's climate?

26. ● You can determine wind direction by wetting your finger and holding it up in the air. Explain.

27. ■ If all the molecules in a liquid had the same speed, and some were able to evaporate, would the remaining liquid be cooled? Explain.

28. ■ Where does the energy come from that keeps the dunking bird in Figure 7.26 operating?

29. ■ Why does wrapping a bottle in a wet cloth at a picnic often produce a cooler bottle than placing the bottle in a bucket of cold water?

30. ■ Why does the boiling temperature of water decrease when the water is under reduced pressure, such as at a higher altitude?

31. ■ Place a jar of water on a small stand within a saucepan of water so that the bottom of the jar is held above the bottom of the pan. When the pan is put on a stove, the water in the pan boils, but the water in the jar does not. Why?

32. ● Room-temperature water boils spontaneously in a vacuum—on the Moon, for example. Could you cook an egg in this boiling water? Explain.

33. ● Your inventor friend proposes a design for cookware that allows boiling to take place at a temperature of less than 100°C so that food can be cooked with the consumption of less energy. Comment on this idea.

34. ● When you boil potatoes, is your cooking time reduced if the water is vigorously boiling instead of gently boiling?

35. ■ Why does putting a lid over a pot of water on a stove shorten the time it takes for the water to come to a boil, whereas, after the water is boiling, the use of the lid only slightly shortens the cooking time?

36. ■ In the power plant of a nuclear submarine, the temperature of the water in the reactor is above 100°C. How is this possible?

37. ■ A piece of metal and an equal mass of wood are both removed from a hot oven at equal temperatures and dropped onto blocks of ice. The metal has a lower specific heat capacity than the wood. Which melts more ice before cooling to 0°C?

38. ■ Why is it that in cold winters, a tub of water placed in a farmer's canning cellar helps prevent canned food from freezing?

39. ■ Why does spraying fruit trees with water before a frost help protect the fruit from freezing?

40. ● Why does a hot dog pant?

PROBLEMS

● BEGINNER ■ INTERMEDIATE ◆ EXPERT

1. ● The specific heat capacity of ice is about 0.5 cal/g·°C. Suppose it remains at that value all the way to absolute zero. Show that the heat required to change a 1-g ice cube at absolute zero (−273°C) to 1 g of boiling water is 320 cal.
2. ● A small block of ice at 0°C is subjected to 10 g of 100°C steam and melts completely. Show that the mass of the block of ice can be no more than 80 g.
3. ■ A 10-kg iron ball is dropped onto a pavement from a height of 100 m. Suppose half of the heat generated goes into warming the ball. Show that the temperature increase of the ball is 1.1°C. (In SI units, the specific heat capacity of iron is 450 J/kg·°C.) Why is the answer the same for an iron ball of any mass?
4. ■ A block of ice at 0°C is dropped from a height that causes it to completely melt upon impact. Assume that there is no air resistance and that all the energy goes into melting the ice. Show that the height necessary for this to occur is at least 34 km. [Hint: Equate the joules of gravitational potential energy to the product of the mass of ice and its heat of fusion (in SI units, 335,000 J/kg). Do you see why the answer doesn't depend on mass?]
5. ■ Fifty grams of hot water at 80°C is poured into a cavity in a very large block of ice at 0°C. The final temperature of the water in the cavity is then 0°C. Show that the mass of ice that melts is 50 g.
6. ■ A 50-g chunk of 80°C iron is dropped into a cavity in a very large block of ice at 0°C. Show that the mass of ice that melts is 5.5 g. (The specific heat capacity of iron is 0.11 cal/g·°C.)
7. ■ The heat of vaporization of ethyl alcohol is about 200 cal/g. Show that if 2 kg of this refrigerant were allowed to vaporize in a refrigerator, it could freeze 5 kg of 0°C water to ice.

ACTIVE EXPLORATIONS

1. If you live where there is snow, do as Benjamin Franklin did about 200 years ago: Lay samples of light and dark cloth on the snow and note the differences in the rate of melting beneath the samples of cloth.
2. Hold the bottom end of a test tube full of cold water in your hand. Heat the top part in a flame until the water boils. The fact that you can still hold the bottom shows that water is a poor conductor of heat. This is even more dramatic when you use steel wool to wedge chunks of ice at the bottom; then the water above can be brought to a boil without melting the ice. Try it and see.

3. Wrap a piece of paper around a thick iron bar and place it in a flame. Note that the paper does not catch fire. Can you figure out why? (Paper generally does not ignite until its temperature reaches 233°C.)

4. Place a Pyrex funnel mouth down in a saucepan full of water so that the straight tube of the funnel sticks above the water. Rest a part of the funnel on a nail or coin so that water can get under it. Place the pan on a stove, and watch the water as it begins to boil. Where do the bubbles form first? Why? As the bubbles rise, they expand rapidly and push water ahead of them. The funnel confines the water, which is forced up the tube and driven out at the top. Now do you know how a geyser and a coffee percolator operate?

5. Watch the spout of a teakettle of boiling water. Notice that you cannot see the steam that issues from the spout. The cloud that you see farther away from the spout is not steam, but condensed water droplets. Now hold the flame of a candle in the cloud of condensed steam. Can you explain your observations?
6. You can make rain in your kitchen. Put a cup of water in a Pyrex saucepan or a Silex coffeemaker and heat it slowly over a low flame. When the water is warm, place a saucer filled with ice cubes on top of the container. As the water below is heated, droplets form at the bottom of the cold saucer and combine until they are large enough to fall, producing a steady "rainfall" as the water below is gently heated. How does this resemble, and how does it differ from, the way in which natural rain is formed?

7. Measure the temperature of boiling water and the temperature of a boiling solution of salt and water. How do the temperatures compare?
8. If you suspend an open-topped container of water in a pan of boiling water, with its top above the surface of the boiling water, water in the inner container reaches 100°C but does not boil. Can you explain why this is so?

READINESS ASSURANCE TEST (RAT)

If you have a good handle on this chapter, if you really do, then you should be able to score at least 7 out of 10 on this RAT. If you score less than 7, you need to study further before moving on.

Choose the BEST answer to the following.

1. A fire walker walking barefoot across hot wooden coals depends on wood's
   (a) good conduction.
   (b) poor conduction.
   (c) low specific heat capacity.
   (d) low radiation.
2. Thermal convection is linked mostly to
   (a) radiant energy.
   (b) fluids.
   (c) insulators.
   (d) all of these
3. When air rapidly expands, its temperature normally
   (a) increases.
   (b) decreases.
   (c) remains unchanged.
   (d) is unaffected, but not always.
4. A high-temperature source radiates relatively
   (a) short wavelengths.
   (b) long wavelengths.
   (c) low frequencies of radiation.
   (d) none of these
5. Which star has the higher temperature?
   (a) A red-hot star.
   (b) A white-hot star.
   (c) A blue-hot star.
   (d) There is not enough information.
6. Compared with radiation from the Sun, terrestrial radiation has a
   (a) longer wavelength.
   (b) lower frequency.
   (c) both of these
   (d) neither of these
7. The origin of much of the thermal energy in Earth's interior is
   (a) high pressures.
   (b) low thermal conductivity of rock.
   (c) trapped radiant energy.
   (d) radioactive decay.
8. A hot pizza placed in the snow is a net
   (a) absorber.
   (b) emitter.
   (c) both of these
   (d) neither of these
9. When a liquid changes phase to a gas it
   (a) absorbs energy.
   (b) emits energy.
   (c) neither absorbs nor emits energy.
   (d) becomes more conducting.
10. When a liquid is brought to a boil, the boiling process tends to
    (a) resist a further change of phase.
    (b) heat the liquid.
    (c) cool the liquid.
    (d) radiate energy from the system.

Answers to RAT
1b, 2b, 3b, 4a, 5c, 6c, 7d, 8b, 9a, 10c

CHAPTER 7 ONLINE RESOURCES

The Physical Science Place

Interactive Figure

- 7.13

Videos

- The Secret to Walking on Hot Coals
- Air Is a Poor Conductor
- Condensation Is a Warming Process

- Boiling Is a Cooling Process
- Pressure Cooker: Boiling and Freezing at the Same Time

Quiz

Flashcards

Links

static and current electricity

8

■ Electricity is everywhere, including the lightning in the sky and the batteries that power your iPod. A study and understanding of electricity requires a step-by-step approach, because one concept is the building block for the next. This has been the case in our study of physics thus far, but more so with what now follows. So please give extra care to the study of this material. It can be difficult, confusing, and frustrating if you're hasty, but with careful effort, it can be comprehensible and rewarding. We start with static electricity, electricity at rest, and complete the chapter with current electricity. Let's begin.

8.1 ELECTRIC CHARGE

8.2 COULOMB'S LAW

8.3 ELECTRIC FIELD

8.4 ELECTRIC POTENTIAL

8.5 VOLTAGE SOURCES

8.6 ELECTRIC CURRENT

8.7 ELECTRIC RESISTANCE

8.8 OHM'S LAW

8.9 ELECTRIC CIRCUITS

8.10 ELECTRIC POWER

8.1 Electric Charge

Try this: Tie a thread around the middle of a plastic drinking straw and then hang the straw by the thread. Rub half of the hanging straw with a piece of wool. If you rub another straw with wool and bring the rubbed ends of the straws near each other, the two straws *repel*.

If instead you rub a glass test tube with silk and bring the rubbed glass near the hanging straw, the two rubbed ends *attract*. And if you replace the hanging straw with a glass test tube and rub it and another test tube with silk, the two rubbed test tubes *repel*.

The ability of rubbed straws and test tubes to exert forces through space is due to a property we call electric *charge*. Although it may seem like magic, it is no more (or less!) magical than the ability of masses to exert gravitational forces on each other through space.

More than two centuries ago, America's first great scientist, Benjamin Franklin, did similar experiments. He formed the following hypotheses:

1. Every neutral (uncharged) substance has its own appropriate level of electric fluid.

2. Rubbing two materials together transfers "electric fluid" from one material to the other.

3. If an object gains electric fluid, it becomes *positively charged* with electric fluid. Likewise, if an object loses electric fluid, it becomes *negatively charged* with electric fluid.

Franklin couldn't see which fluid was transferred when the glass was rubbed with silk. He decided to call the glass *positively charged*, which means that the silk ended up with a negative charge because it lost electric fluid to the glass. Likewise in our example, the wool ends up positively charged and the plastic ends up negatively charged.

Once these charges have been assigned, we can see the most fundamental rule of electrical behavior:

Like charges repel; opposite charges attract.

Electrical forces arise from particles in atoms. In the simple model of the atom proposed in the early 1900s by Ernest Rutherford and Niels Bohr, a positively charged nucleus is surrounded by negatively charged electrons (Figure 8.2). The nucleus attracts the electrons and holds them in orbit, similar to the way the Sun holds the planets in orbit. But with a difference—electrons repel other electrons (whereas gravitational forces only attract).

The following are some important facts about atoms:

1. Every atom has a positively charged nucleus surrounded by negatively charged electrons.

2. All electrons are identical; that is, each has the same mass and the same quantity of negative charge as every other electron.

3. The nucleus is composed of protons and neutrons. (The common form of hydrogen, which has no neutrons, is the only exception.) All protons are positively charged and identical; similarly, all neutrons are identical. A proton has nearly 2000 times the mass of an electron, but its positive charge is equal in magnitude to the negative charge of the electron. A neutron has slightly greater mass than a proton and has no charge.

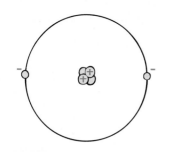

FIGURE 8.1

A plastic straw rubbed with wool is suspended by a thread. When another straw that has also been rubbed with wool is brought nearby, the two straws repel each other.

Charge is like a baton in a relay race. It can be passed from one object to another but it cannot be lost.

FIGURE 8.2

INTERACTIVE FIGURE

Model of a helium atom. The atomic nucleus is made up of two protons and two neutrons. The positively charged protons attract two negatively charged electrons. What is the net charge of this atom?

4. Atoms usually have as many electrons as protons, so the atom has zero *net* charge.

Just *why* electrons repel electrons and are attracted to protons is beyond the scope of this book. At our level of understanding we simply say that this is nature as we find it—that this electric behavior is fundamental, or basic.

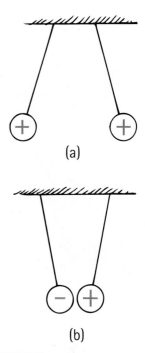

(a)

(b)

FIGURE 8.3

INTERACTIVE FIGURE

(a) Like charges repel.
(b) Unlike charges attract.

CHECK POINT

1. Beneath the complexities of electrical phenomena lies a fundamental rule from which nearly all other electrical effects stem. What is this fundamental rule?
2. How does the charge of an electron differ from the charge of a proton?

Were these your answers?
1. Like charges repel; opposite charges attract.
2. The charges of the two particles are equal in magnitude, but opposite in sign.

CONSERVATION OF CHARGE

Electrons and protons have electric charge. A neutral atom has as many electrons as protons, so it has no net charge. The total positive charge balances the total negative charge exactly. If an electron is removed from an atom, the atom is no longer neutral. The atom then has one more positive charge (proton) than negative charge (electron) and is positively charged.

A charged atom is called an *ion*. A *positive* ion has a net positive charge because it has lost one or more electrons. A *negative* ion has a net negative charge because it has gained one or more extra electrons.

Matter is made of atoms, and atoms are made of electrons and protons (and neutrons as well). An object that has equal numbers of electrons and protons has no net electric charge. But if the numbers do not balance, the object is then electrically charged. An imbalance comes about by adding or removing electrons.

Although the innermost electrons in an atom are held very tightly to the oppositely charged atomic nucleus, the outermost electrons of many atoms are held very loosely and can be easily dislodged. How much energy is required to tear an electron away from an atom varies for different substances. The electrons are held more firmly in rubber or plastic than in wool or fur, for example. Hence, when a plastic straw is rubbed with a piece of wool, electrons transfer from the wool to the plastic straw. The plastic then has an excess of electrons and is negatively charged. The wool, in turn, has a deficiency of electrons and is positively charged. If you rub a glass or plastic rod with silk, you'll find that the rod becomes positively charged. The silk hangs on to electrons more tightly than the glass or plastic rod does. Electrons are rubbed off the rod and onto the silk. In summary:

> An object that has unequal numbers of electrons and protons is electrically charged. If it has more electrons than protons, the object is negatively charged. If it has fewer electrons than protons, then it is positively charged.

Franklin didn't explain charge transfer in terms of transfer of electrons because electrons were unknown in his day. Later it was found that electrons are neither created nor destroyed but are simply transferred from one material to another. Charge is conserved. In every event, whether large-scale or at the atomic and nuclear level, the principle of *conservation of charge* applies. No case

FIGURE 8.4

When a rubber rod is rubbed with fur, electrons transfer from the fur to the rod. The rod is then negatively charged. Is the fur charged? By how much, compared with the rod? Positively or negatively?

Conservation of charge is another of the conservation principles. Recall from previous chapters the conservation of momentum and the conservation of energy.

FIGURE 8.5

Why do you get a slight shock from the doorknob after scuffing across the carpet?

of the creation or destruction of net electric charge has ever been found. The conservation of charge is a cornerstone in physics, ranking with the conservation of energy and momentum.

Any object that is electrically charged has an excess or deficiency of some whole number of electrons—electrons cannot be divided into fractions of electrons. This means that the charge of the object is a whole-number multiple of the charge of an electron. It cannot have a charge equal to the charge of 1.5 or 1000.5 electrons, for example. In all measurements to date, objects have a charge that is a whole-number multiple of the charge of a single electron.

CHECK POINT

If you scuff electrons onto your shoes while walking across a rug, are you negatively or positively charged?

Was this your answer?

When your rubber- or plastic-soled shoes drag across the rug, they pick up electrons from the rug in the same way you charge a rubber rod by rubbing it with cloth. You have more electrons after you scuff your shoes, so you are negatively charged (and the rug is positively charged).

Electronics Technology and Sparks

Electric charge can be dangerous. Two hundred years ago, young boys called *powder monkeys* ran barefooted below the decks of warships to bring sacks of black gunpowder to the cannons above. It was ship law that this task be done barefoot. Why? Because it was important that no static charge build up on the powder on their bodies as they ran to and fro. Bare feet scuffed the decks much less than shoes and ensured no charge accumulation that might produce an igniting spark and an explosion.

Static charge is a danger in many industries today—not because of explosions, but because delicate electronic circuits may be destroyed by static charges. Some circuit components are sensitive enough to be "fried" by sparks of static electricity. Electronics technicians frequently wear clothing of special fabrics with ground wires between their sleeves and their socks. Some wear special wristbands that are connected to a grounded surface so that static charges do not build up—when moving a chair, for example. The smaller the electronic circuit, the more hazardous are sparks that may short-circuit the circuit elements.

fyi

- Static electricity is a problem at gasoline pumps. Even the tiniest of sparks ignite vapors coming from the gasoline and cause fires—frequently lethal. A good rule is to touch metal and discharge static charge from your body before you fuel. Also, don't use a cell phone when fueling.

An ordinary coin contains about 10^{24} electrons, all repelling one another. Why don't these electrons fly off the coin?

8.2 Coulomb's Law

Electrical force, like gravitational force, decreases inversely as the square of the distance between charges. This relationship, which was discovered by Charles Coulomb in the 18th century, is called **Coulomb's law**. It states that for two charged objects that are much smaller than the distance between them, the force between them varies directly as the product of their charges and inversely as the square of the separation distance. The force acts along a straight line from one charge to the other. Coulomb's law can be expressed as

$$F = k\frac{q_1 q_2}{d^2}$$

where d is the distance between the charged particles, q_1 represents the quantity of charge of one particle, q_2 represents the quantity of charge of the second particle, and k is the proportionality constant.

Ionized Bracelets: Science or Pseudoscience?

Surveys indicate that most Americans believe that ionized bracelets can reduce joint or muscle pain. Manufacturers claim that ionized bracelets relieve such pain. Are they correct? In 2002, the claim was tested by researchers at the Mayo Clinic in Jacksonville, Florida, who randomly assigned 305 participants to wear an ionized bracelet for 28 days and another 305 participants to wear a placebo bracelet for the same duration. The study volunteers were men and women 18 and older who had self-reported musculoskeletal pain at the beginning of the study.

Neither the researchers nor the participants knew which volunteers wore an ionized bracelet and which wore a placebo bracelet. Both types of bracelets were identical, were supplied by the manufacturer, and were worn according to the manufacturer's recommendations. Interestingly, both groups reported significant relief from pain. No difference was found in the amount of self-reported pain relief between the group wearing the ionized bracelets and the group wearing the placebo bracelets. Apparently, just believing that the bracelet relieves pain does the trick!

Interestingly, the brain initiates the creation of endorphins (which bind to opiate receptor sites) when the person expects to get relief from pain. The placebo effect is very real and measurable via blood titrations. So there's some merit in the old adage that wishing hard for something makes it come true. But this has nothing to do with the physics, chemistry, or biological interaction with the bracelet. Hence, ionized bracelets join the ranks of pseudoscientific devices.

In any society that thrives more on capturing attention than on informing, pseudoscience is big business.

The unit of charge is called the **coulomb**, abbreviated C. It turns out that a charge of 1 C is the charge associated with 6.25 billion billion electrons. This might seem like a great number of electrons, but it only represents the amount of charge that flows through a common 100-W lightbulb in a little more than a second.

The proportionality constant k in Coulomb's law is similar to G in Newton's law of gravity. Instead of being a very small number, like G, k is a very large number, approximately

$$k = 9,000,000,000 \text{ N} \cdot \text{m}^2/\text{C}^2.$$

In scientific notation, $k = 9.0 \times 10^9 \text{ N} \cdot \text{m}^2/\text{C}^2$. The unit $\text{N} \cdot \text{m}^2/\text{C}^2$ is not central to our interest here; it simply converts the right-hand side of the equation to the unit of force, the newton (N). What is important is the large magnitude of k. If, for example, a pair of like charges of 1 C each were 1 m apart, the force of repulsion between the two would be 9 billion N.* That would be about 10 times the weight of a battleship! Obviously, such quantities of net charge do not usually exist in our everyday environment.

So Newton's law of gravitation for masses is similar to Coulomb's law for electrically charged bodies. The most important difference between gravitational and electrical forces is that electrical forces may be either attractive or repulsive, whereas gravitational forces are only attractive. Coulomb's law underlies the bonding forces between molecules that are essential in the field of chemistry.

Link To Sections 12.3, 15.6, 15.8

(a)   (b)

FIGURE 8.6

(a) A gravitational force holds the satellite in orbit about the planet, and (b) an electrical force holds the electron in orbit about the proton. In both cases, there is no contact between the bodies. We say that the orbiting bodies interact with the force fields of the planet and proton and are everywhere in contact with these fields. Thus, the force that one electric charge exerts on another can be described as the interaction between one charge and the field set up by the other.

---

* Contrast this to the gravitational force of attraction between two 1-kg masses 1 m apart: 6.67×10^{-11} N. This is an extremely small force. For the force to be 1 N, the masses at 1 m apart would have to be nearly 123,000 kg each! Gravitational forces between ordinary objects are exceedingly small, and differences in electrical forces between ordinary objects can be exceedingly huge. We don't sense them because the positives and negatives normally balance out, and, even for highly charged objects, the imbalance of electrons to protons is normally less than one part in a trillion trillion.

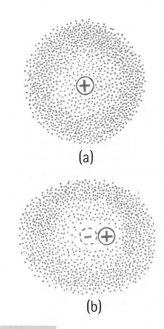

(a)

(b)

FIGURE 8.7

(a) The center of the negative "cloud" of electrons coincides with the center of the positive nucleus in an atom. (b) When an external negative charge is brought nearby to the right, as on a charged balloon, the electron cloud is distorted so that the centers of negative and positive charge no longer coincide. The atom is electrically polarized.

FIGURE 8.8

The negatively charged balloon polarizes molecules in the wooden wall and creates a positively charged surface, so the balloon sticks to the wall.

CHECK POINT

1. The proton is the nucleus of the hydrogen atom, and it attracts the electron that orbits it. Relative to this force, does the electron attract the proton with less force, more force, or the same amount of force?
2. If a proton at a particular distance from a charged particle is repelled with a given force, by how much does the force decrease when the proton is three times as distant from the particle? Five times as distant?
3. What is the sign of charge of the particle in this case?

Were these your answers?

1. The same amount of force, in accord with Newton's third law—basic mechanics! Recall that a force is an interaction between two things—in this case, between the proton and the electron. They pull on each other equally.
2. In accord with the inverse-square law, at three times the distance, the force decreases to $\frac{1}{9}$ its original value. At five times the distance, the force decreases to $\frac{1}{25}$ of its original value.
3. Positive

CHARGE POLARIZATION

If you charge an inflated balloon by rubbing it on your hair and then place the balloon against a wall, it sticks. This is because the charge on the balloon alters the charge distribution in the atoms or molecules in the wall, effectively inducing an opposite charge on the wall. The molecules cannot move from their relatively stationary positions, but their "centers of charge" are moved. The positive part of the atom or molecule is attracted toward the balloon while the negative part is repelled. This has the effect of distorting the atom or molecule (Figure 8.7). The atom or molecule is said to be **electrically polarized**. We will see in Part 2 how polarization plays an important role in chemistry.

CHECK POINT

You know that a balloon rubbed on your hair sticks to a wall. In a humorous vein, does it follow that your oppositely charged head would also stick to the wall?

Was this your answer?

No, unless you're an airhead (having a head mass about the same as that of an air-filled balloon). The force that holds a balloon to the wall cannot support your heavier head.

8.3 Electric Field

Electrical forces, like gravitational forces, can act between things that are not in contact with each other. For both electricity and gravity, a force field exists that influences distant charges and masses, respectively. The properties of space surrounding any mass are altered such that another mass introduced to this region experiences a force. This "alteration in space" is called its *gravitational field*. We can think of any other mass as interacting with the field and not directly with the mass that produces it. For example, when an apple falls from a tree, we say it is interacting with the mass of Earth, but we can also think of the apple as interacting with the gravitational field of Earth. It is common to

Microwave Oven

Imagine an enclosure filled with Ping-Pong balls among a few batons, all at rest. Now imagine that the batons suddenly rotate backward and forward, striking neighboring Ping-Pong balls. Almost immediately, most of the Ping-Pong balls are energized, vibrating in all directions. A microwave oven works similarly. The batons are water molecules made to rotate to and fro in rhythm with microwaves in the enclosure. The Ping-Pong balls are the other molecules that make up the bulk of material being cooked.

H_2O molecules are electrically polarized, with opposite charges on opposite sides. When an electric field is imposed on them, they align with the field as a compass needle aligns with a magnetic field. When the field is made to oscillate, the H_2O molecules oscillate also—and quite energetically when the frequency of the waves matches the natural rotational frequency of the H_2O. So, food is cooked by converting H_2O molecules into flip-flopping energy sources that impart thermal motion to surrounding food molecules. Without polar molecules in the food, a microwave oven wouldn't work.

That's why microwaves pass through foam, paper, or ceramic plates and reflect from metals with no effect. They do energize, however, water molecules.

A note of caution is due when boiling water in a microwave oven. Water can sometimes heat faster than bubbles can form, and the water then heats beyond its boiling point—it becomes superheated. If the water is bumped or jarred just enough to cause the bubbles to form rapidly, they'll violently expel the hot water from its container. More than one person has had boiling water blast into his or her face.

---

think of distant rockets and the like as interacting with gravitational fields rather than bodies responsible for the fields. The field plays an intermediate role in the force between bodies. More important, the field stores energy. So similar to a gravitational field, the space around every electric charge is energized with an **electric field**—an energetic aura that extends through space.*

If you place a charged particle in an electric field, it experiences a force. The direction of the force on a positive charge is the same direction as the field. The electric field about a proton extends radially from the proton. About an electron, the field is in the opposite direction (Figure 8.9). As with electric force, the electric field about a particle obeys the inverse-square law. Some electric field configurations are shown in Figure 8.10, and photographs of field patterns are shown in Figure 8.11. In the next chapter, we'll see how bits of iron similarly align with magnetic fields.

Perhaps your instructor will demonstrate the effects of the electric field that surrounds the charged dome of a Van de Graaff generator (Figure 8.12). Charged objects in the field of the dome are either attracted or repelled, depending on their sign of charge.

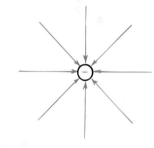

FIGURE 8.9

INTERACTIVE FIGURE

Electric field representations about a negative charge.

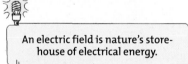

An electric field is nature's storehouse of electrical energy.

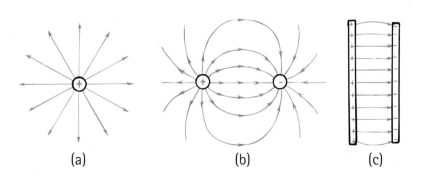

(a)   (b)   (c)

FIGURE 8.10

INTERACTIVE FIGURE

Some electric field configurations. (a) Lines of force about a single positive charge. (b) Lines of force for a pair of equal but opposite charges. Note that the lines emanate from the positive charge and terminate on the negative charge. (c) Uniform lines of force between two oppositely charged parallel plates.

---

* An electric field is a vector quantity, having both magnitude and direction. The magnitude of the field at any point is simply the force per unit of charge. If a charge q experiences a force F at some point in space, then the electric field E at that point is $E = F/q$.

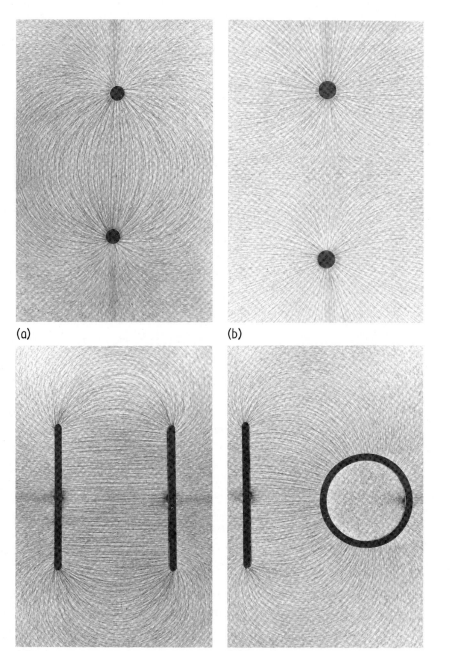

FIGURE 8.11

Bits of thread suspended in an oil bath line up end-to-end along the direction of the field.
(a) Equal and opposite charges.
(b) Equal like charges.
(c) Oppositely charged plates.
(d) Oppositely charged cylinder and plate.

Static charge on the surface of any electrically conducting surface arranges itself such that the electric field inside the conductor cancels to zero. Note the randomness of threads inside the cylinder of Figure 8.11d, where no field exists.

fyi

- Whatever the intensity of the electric field about a charged Van de Graaff generator, the electric field inside the dome cancels to zero. This is true for the interiors of all metals that carry static charge.

(a)   (b)

(c)   (d)

FIGURE 8.12

Both Lori and the spherical dome of the Van de Graaff generator are electrically charged.

CHECK
POINT

Both Lori and the dome of the Van de Graaff generator in Figure 8.12 are charged. Why does Lori's hair stand out?

Was this your answer?

She and her hair are charged. Each hair is repelled by others around it—evidence that *like charges repel*. Even a small charge produces an electrical force greater than the weight of strands of hair. Fortunately, the electrical force is not great enough to make her arms stand out!

■ 8.4 Electric Potential

In our study of energy in Chapter 3, we learned that an object has gravitational potential energy because of its location in a gravitational field. Similarly, a charged object has potential energy by virtue of its location in an electric field. Just as work is required to lift a massive object against the gravitational field of the Earth, work is required to push a charged particle against the electric field of a charged body. This work changes the electric potential energy of the charged particle.* Similarly, work done in compressing a spring increases the potential energy of the spring (Figure 8.14a). Likewise, the work done in pushing a charged parti-

(a)   (b)

FIGURE 8.13

(a) The PE (gravitational potential energy) of a mass held in a gravitational field. (b) The PE of a charged particle held in an electric field. When the mass and particle are released, how does the KE (kinetic energy) acquired by each compare with the decrease in PE?

cle closer to the charged sphere in Figure 8.14b increases the potential energy of the charged particle. We call the energy possessed by the charged particle that is due to its location **electric potential energy.** If the particle is released, it accelerates in a direction away from the sphere, and its electric potential energy changes to kinetic energy.

If we push a particle with twice the charge, we do twice as much work. Twice the charge in the same location has twice the electric potential energy; with three times the charge, there is three times as much potential energy; and so on. When working with electricity, rather than dealing with the total potential energy of a charged body, it is convenient to consider the electric potential energy *per charge*. We simply divide the amount of energy in any case by the amount of charge. The concept of potential energy per charge is called **electric potential**; that is,

$$\text{Electric potential} = \frac{\text{electric potential energy}}{\text{amount of charge}}$$

> In a nutshell: *Electric potential* and *potential* mean the same thing—electrical potential energy per unit charge—in units of volts. On the other hand, *potential difference* is the same as *voltage*—the *difference* in electrical potential between two points—also in units of volts.

The unit of measurement for electric potential is the volt, so electric potential is often called *voltage*. A potential of 1 volt (V) equals 1 joule (J) of energy per 1 coulomb (C) of charge.

$$1 \text{ volt} = \frac{1 \text{ joule}}{1 \text{ coulomb}}$$

Thus, a 1.5-V battery gives 1.5 J of energy to every 1 C of charge flowing through the battery. *Electric potential* and *voltage* are the same thing, and they are commonly used interchangeably.

The significance of voltage is that a definite value for it can be assigned to a location. We can speak about the voltages at different locations in an electric field whether or not charges occupy those locations. The same is true of voltages at various locations in an electric circuit. Later in this chapter, we will see that the location of the positive terminal of a 12-V battery is maintained at a voltage 12 V higher than the location of the negative terminal. When a conducting medium connects this voltage difference, any charges in the medium move between these locations.

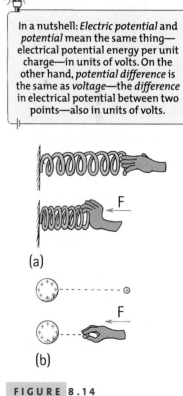

(a)

(b)

FIGURE 8.14

(a) The spring has more elastic PE when compressed. (b) The small charge similarly has more PE when pushed closer to the charged sphere. In both cases, the increased PE is the result of work input.

---

* This work is positive if it increases the electric potential energy of the charged particle and negative if it decreases it.

FIGURE 8.15

The larger test charge has more PE in the field of the charged dome, but the electric potential of any amount of charge at the same location is the same.

📽 The Physical Science Place

Caution on Handling
    Electric Wires
Birds and High-Voltage
    Wires

FIGURE 8.16

Although the voltage of the charged balloon is high, the electric potential energy is low because of the small amount of charge.

High voltage at low energy is similar to the harmless high-temperature sparks emitted by a fireworks sparkler. Recall that temperature is average kinetic energy per molecule, which means total energy is a lot only for lots of molecules. Similarly, high voltage means a lot of energy only for lots of charge.

FIGURE 8.17

Although the Wimshurst machine can generate thousands of volts, it puts out no more energy than the work that Jim Stith puts into it by cranking the handle.

CHECK POINT

1. If there were twice as many coulombs in the test charge near the charged sphere in Figure 8.15, would the *electric potential energy* of the test charge relative to the charged sphere be the same, or would it be twice as great? Would the *electric potential* of the test charge be the same, or would it be twice as great?
2. What does it mean to say that the battery in your car is rated at 12 V?

Were these your answers?

1. The result of twice as many coulombs is twice as much *electric potential energy* because it takes twice as much work to put the charge there. But the *electric potential* would be the same. Twice the energy divided by twice the charge gives the same potential as one unit of energy divided by one unit of charge. Electric potential is not the same thing as electric potential energy. Be sure you understand this before you study further.
2. It means that one of the battery terminals is 12 V higher in potential than the other one. We'll soon learn that when a circuit is connected between these terminals, each coulomb of charge in the resulting current is given 12 J of energy as it passes through the battery (and 12 J of energy "spent" in the circuit).

Rub a balloon on your hair, and the balloon becomes negatively charged—perhaps to several thousand volts! That would be several thousand joules of energy, if the charge were 1 C. However, 1 C is a fairly respectable amount of charge. The charge on a balloon rubbed on hair is typically much less than a millionth of a coulomb. Therefore, the amount of energy associated with the charged balloon is very, very small. A high voltage means a lot of energy only if a lot of charge is involved. Electrical potential energy differs from electric potential (or voltage).

8.5 Voltage Sources

When the ends of a heat conductor are at different temperatures, heat energy flows from the higher temperature to the lower temperature. The flow ceases when both ends reach the same temperature. Any material having free charged particles that easily flow through it when an electric force acts on them is called an electric **conductor**. Both heat and electric conductors are characterized by electric charges that are free to move. Similar to heat flow, when the ends of an electrical conductor are at different electric potentials—when there is a **potential difference**—charges in the conductor flow from the higher potential to the lower potential. The flow of charge persists until both ends reach the same potential. Without a potential difference, no flow of charge occurs.

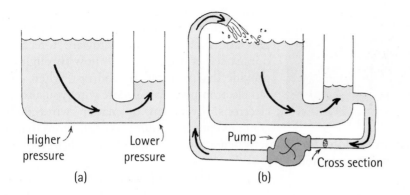

FIGURE 8.18

(a) Water flows from the reservoir of higher pressure to the reservoir of lower pressure. The flow ceases when the difference in pressure ceases. (b) Water continues to flow because a difference in pressure is maintained with the pump.

To attain a sustained flow of charge in a conductor, some arrangement must be provided to maintain a difference in potential while charge flows from one end to the other. The situation is analogous to the flow of water from a higher reservoir to a lower one (Figure 8.18a). Water flows in a pipe that connects the reservoirs only as long as a difference in water level exists. The flow of water in the pipe, like the flow of charge in a wire, ceases when the pressures at each end are equal. (We imply this phenomenon when we say that water seeks its own level.) A continuous flow is possible if the difference in water levels—hence the difference in water pressures—is maintained with the use of a suitable pump (Figure 8.18b).

A sustained electric current requires a suitable pumping device to maintain a difference in electric potential—to maintain a voltage. Chemical batteries or generators are "electrical pumps" that can maintain a steady flow of charge. These devices do work to pull negative charges apart from positive ones. In chemical batteries, this work is done by the chemical disintegration of zinc or lead in acid, and the energy stored in the chemical bonds is converted to electric potential energy.

Generators separate charge by electromagnetic induction, a process we will describe in the next chapter. The work that is done (by whatever means) in separating the opposite charges is available at the terminals of the battery or generator. This energy per charge provides the difference in potential (voltage) that provides the "electrical pressure" to move electrons through a circuit joined to those terminals.

fyi

- Chemical batteries don't respond well to sudden surges of charge. An alternative that does respond well to spurts of energy input is a spinning flywheel. Unlike the ones used by potters for spinning clay, modern flywheels are made of lightweight composite materials that are strong and can be spun at high speeds without coming apart. Rotational kinetic energy is then converted to other forms of energy. Watch for flywheels as energy-storing devices.

Link To Section 18.6

FIGURE 8.19

An unusual source of voltage. The electric potential between the head and tail of the electric eel (*Electrophorus electricus*) can be up to 650 V.

A battery doesn't supply electrons to a circuit; it instead supplies energy to electrons that already exist in the circuit.

fyi

- When a common automobile battery provides an electrical pressure of 12 V to a circuit connected across its terminals, 12 J of energy are supplied to each coulomb of charge that is made to flow in the circuit.

8.6 Electric Current

Just as a water current is a flow of H_2O molecules, **electric current** is a flow of charged particles. In circuits of metal wires, electrons make up the flow of charge. One or more electrons from each metal atom are free to move throughout the atomic lattice. These charge carriers are called *conduction electrons*. Protons, on the other hand, do not move in a solid because they are bound within the nuclei of atoms that are more or less locked in fixed positions. In fluids, however, positive ions as well as electrons may constitute the flow of an electric charge.

FIGURE 8.20

Each coulomb of charge that is made to flow in a circuit that connects the ends of this 1.5-V flashlight cell is energized with 1.5 J.

An important difference between water flow and electron flow has to do with their conductors. If you purchase a water pipe at a hardware store, the clerk doesn't sell you the water to flow through it. You provide that yourself. By contrast, when you buy "an electron pipe," an electric wire, you also get the electrons. Every bit of matter, wires included, contains enormous numbers of electrons that swarm about in random directions. When a source of voltage sets them moving, we have an electric current.

(a)

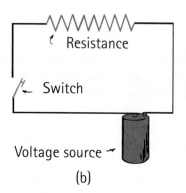

(b)

FIGURE 8.21

Analogy between (a) a simple hydraulic circuit and (b) an electrical circuit. Much effort is expended in building particle accelerators that accelerate electrons to speeds approaching the speed of light. If electrons in a common circuit were to travel that fast, one would only have to bend a wire at a sharp angle to cause those high-momentum electrons to fail to make the turn and to fly off into the air. There'd be no need for accelerators! In fact, electrons in circuits move fairly slowly.

The *rate* of electrical flow is measured in *amperes*. An **ampere** is the rate of flow of 1 coulomb of charge per second. (That's a flow of 6.25 billion billion electrons per second.) In a wire that carries 4 amperes to a car headlight bulb, for example, 4 C of charge flows past any cross section in the wire each second. In a wire that carries 8 amperes, twice as many coulombs flow past any cross section each second.

The speed of electrons as they drift through a wire is surprisingly slow. This is because electrons continually bump into atoms in the wire. The net speed, or *drift speed,* of electrons in a typical circuit is much less than 1 cm/s. The electric signal, however, travels at nearly the speed of light. That's the speed at which the electric *field* in the wire is established.

Also interesting is that a current-carrying wire is not electrically charged. Under ordinary conditions, there are as many conduction electrons swarming through the atomic lattice as there are positively charged atomic nuclei. The numbers of electrons and protons balance, so whether a wire carries a current or not, the net charge of the wire is normally zero at every moment.

There is often some confusion between charge flowing *through* a circuit and voltage placed, or impressed, *across* a circuit. We can distinguish between these ideas by considering a long pipe filled with water. Water flows through the pipe if there is a difference in pressure across (or between) its ends. Water flows from the high-pressure end to the low-pressure end. Only the water flows, not the pressure. Similarly, electric charge flows because of the differences in electrical pressure (voltage). You say that *charges* flow through a circuit because of an applied voltage across the circuit. You don't say that *voltage* flows through a

FIGURE 8.22

The electric field lines between the terminals of a battery are directed through a conductor, which joins the terminals. A thick metal wire is shown here, but the path from one terminal to the other is usually an electric circuit. (If you touch this conducting wire, you won't be shocked, but the wire will heat quickly and may burn your hand!)

History of 110 Volts

In the early days of electrical lighting, high voltages burned out electric light filaments, so low voltages were more practical. The hundreds of power plants built in the United States prior to 1900 adopted 110 V (or 115 or 120 V) as their standard. The tradition of 110 V was decided upon because it made the bulbs of the day glow as brightly as a gas lamp. By the time electrical lighting became popular in Europe, engineers had figured out how to make lightbulbs that would not burn out so fast at higher voltages. Power transmission is more efficient at higher voltages, so Europe adopted 220 V as its standard. The U.S. remained with 110 V (today, it is officially 120 V) because of the initial huge expense in the installation of 110-V equipment. Interestingly, in ac circuits 120 V is the *root-mean-square* average of the voltage. The actual voltage in a 120-V ac circuit varies between +170 V and −170 volts, delivering the same power to an iron or a toaster as a 120-V dc circuit.

circuit. Voltage doesn't go anywhere, for it is the charges that move. Voltage produces current (if there is a complete circuit).

DIRECT CURRENT AND ALTERNATING CURRENT

Electric current may be direct or alternating. **Direct current (dc)** refers to charges flowing in one direction. A battery produces direct current in a circuit because the terminals of the battery always have the same sign. Electrons move from the repelling negative terminal toward the attracting positive terminal, and they always move through the circuit in the same direction.

Alternating current (ac) acts as the name implies. Electrons in the circuit are moved first in one direction and then in the opposite direction, alternating to and fro about relatively fixed positions. This is accomplished in a generator or alternator by periodically switching the sign at the terminals. Nearly all commercial ac circuits involve currents that alternate back and forth at a frequency of 60 cycles per second. This is 60-hertz current [one cycle per second is called a *hertz* (Hz)]. In some countries, 25-Hz, 30-Hz, or 50-Hz current is used. Throughout the world, most residential and commercial circuits are ac because electric energy in the form of ac can easily be stepped up to high voltage to be transmitted great distances with small heat losses, then stepped down to convenient voltages where the energy is consumed. Why this occurs is quite fascinating, and it will be touched on in the next chapter. The rules of electricity in this chapter apply to both dc and ac.

FIGURE 8.23

Time graphs of dc and ac.

■ 8.7 Electric Resistance

How much current is in a circuit depends not only on voltage but also on the **electrical resistance** of the circuit. Just as narrow pipes resist water flow more than wide pipes, thin wires resist electrical current more than thicker wires. And length contributes to resistance also. Just as long pipes have more resistance than short ones, long wires offer more electrical resistance. And most important is the material from which the wires were made. Copper has a low electrical resistance, while a strip of rubber has an enormous resistance. Temperature also affects electrical resistance. The greater the jostling of atoms within a conductor (the higher the temperature), the greater its resistance. The resistance

fyi

- The danger from car batteries is not so much electrocution as it is explosion. If you touch both terminals with a metal wrench, for instance, you can create a spark that can ignite hydrogen gas in the battery and send pieces of battery and acid flying!

The Physical Science Place
Alternating Current

fyi

- Conversion from ac to dc is accomplished with an electronic device that allows electron flow in one direction only—a *diode*. A more familiar type diode is the *light-emitting diode* (LED). Photons are emitted when electrons cross a "band gap" in the device. The photon energy most often corresponds to the frequency of red light, which is why most LEDs emit red light. You see these on instrument panels of many kinds, including VCRs and DVD players. Interestingly, when electrical input and light output are reversed, the resulting device is a solar cell!

- Some materials, such as germanium or silicon, can be made to alternate between being conductors and insulators. These are *semiconductors*. Between pairs of them the transfer of an electron through their junction can cause emission of light, as in a light-emitting diode (LED). Conversely, the absorption of light can lead to an electric current, as in a solar cell.

FIGURE 8.24

The conduction electrons that surge to and fro in the filament of the lamp do not come from the voltage source. They are within the filament to begin with. The voltage source simply provides them with surges of energy. When switched on, the resistance of the very thin tungsten filament heats up to 3000°C and roughly doubles its resistance.

Current is a flow of charge, pressured into motion by voltage, and hampered by resistance.

The Physical Science Place
Ohm's Law

The unit of electrical resistance is the ohm, Ω. Like the song of old, "Ω, Ω on the Range."

of some materials reaches zero at very low temperatures. These materials are referred to as *superconductors*.

Electrical resistance is measured in units called *ohms*. The Greek letter *omega*, Ω, is commonly used as the symbol for the ohm. This unit was named after Georg Simon Ohm, a German physicist who, in 1826, discovered a simple and very important relationship among voltage, current, and resistance.

SUPERCONDUCTORS

In common household wiring, flowing electrons collide with atomic nuclei in the wire and convert their kinetic energy to thermal energy in the wire. Early-20th-century investigators discovered that certain metals in a bath of liquid helium at 4 K lost all electrical resistance. The electrons in these conductors traveled pathways that avoided atomic collisions, permitting them to flow indefinitely. These materials are called **superconductors**, having zero electrical resistance to the flow of charge. No current is lost and no heat is generated in superconductivity. For decades, it was generally thought that zero electrical resistance could occur only in certain metals near absolute zero. Then, in 1986, superconductivity was achieved at 30 K, which spurred hopes of finding superconductivity above 77 K, the point at which nitrogen liquefies. Nitrogen is easier to handle than liquid helium, which is needed for creating colder conditions. The historic leap came in the following year with a nonmetallic compound that lost its resistance at 90 K.

Various ceramic oxides have since been found to be superconducting at temperatures above 100 K. These ceramic materials are "high-temperature" superconductors. High-temperature superconductor (HTS) cables, already in use, carry more current at a lower voltage, which means large power transformers can be located farther away from urban centers—allowing the development of green space. Watch for additional growth of HTS cables in delivering electric power.

8.8 Ohm's Law

The relationship between voltage, current, and resistance is summarized by a statement called **Ohm's law**. Ohm discovered that the amount of current in a circuit is directly proportional to the voltage established across the circuit and is inversely proportional to the resistance of the circuit:

$$\text{Current} = \frac{\text{voltage}}{\text{resistance}}$$

Or, in units form,

$$\text{Amperes} = \frac{\text{volts}}{\text{ohms}}$$

So, for a given circuit of constant resistance, current and voltage are proportional to each other.* This means we'll get twice the current for twice the voltage. The greater the voltage, the greater the current. But if the resistance is doubled for a circuit, the current is half what it would have been otherwise. The greater the resistance, the smaller the current. Ohm's law makes good sense.

---

* Many texts use V as the symbol for voltage, I for current, and R for resistance, and express Ohm's law as $V = IR$. It then follows that $I = V/R$, or $R = V/I$, so that, if any two variables are known, the third can be found. (The names of the units are often abbreviated: V for volts, A for amperes, and Ω (the capital Greek letter omega) for ohms.

FIGURING PHYSICAL SCIENCE

▪ Problem Solving

SAMPLE PROBLEM 1

How much current flows through a lamp with a resistance of 60 Ω when the voltage across the lamp is 12 V?

SOLUTION:

From Ohm's law:

$$\text{Current} = \frac{\text{voltage}}{\text{resistance}} = \frac{12\ \text{V}}{60\ \Omega} = 0.2\ \text{A}$$

SAMPLE PROBLEM 2

What is the resistance of a toaster that draws a current of 12 A when connected to a 120-V circuit?

SOLUTION:

Rearranging Ohm's law:

$$\text{Resistance} = \frac{\text{voltage}}{\text{current}} = \frac{120\ \text{V}}{12\ \text{A}} = 10\ \Omega$$

SAMPLE PROBLEM 3

At 100,000 Ω, how much current flows through your body if you touch the terminals of a 12-V battery?

SOLUTION:

$$\text{Current} = \frac{\text{voltage}}{\text{resistance}} = \frac{12\ \text{V}}{100,000\ \Omega}$$

$$= 0.00012\ \text{A}$$

SAMPLE PROBLEM 4

If your skin is very moist, so that your resistance is only 1000 Ω, and you touch the terminals of a 12-V battery, how much current do you receive?

SOLUTION:

$$\text{Current} = \frac{\text{voltage}}{\text{resistance}} = \frac{12\ \text{V}}{1000\ \Omega}$$

$$= 0.012\ \text{A}$$

Ouch!

The resistance of a typical lamp cord is much less than 1 Ω, and a typical lightbulb has a resistance of more than 100 Ω. An iron or electric toaster has a resistance of 15 to 20 Ω. The current inside these and all other electrical devices is regulated by circuit elements called resistors (Figure 8.25), whose resistance may be a few ohms or millions of ohms. Resistors heat up when current flows through them, but for small currents the heating is slight.

FIGURE 8.25

Resistors. The symbol of resistance in an electric circuit is ‑‑⋁⋁⋁‑‑.

ELECTRIC SHOCK

The damaging effects of shock are the result of current passing through the human body. What causes electric shock in the body—current or voltage? From Ohm's law, we can see that this current depends on the voltage that is applied and also on the electrical resistance of the human body. The resistance of one's body depends on its condition, and it ranges from about 100 Ω, if it is soaked with salt water, to about 500,000 Ω, if the skin is very dry. If we touch the two electrodes of a battery with dry fingers, completing the circuit from one hand to the other, we offer a resistance of about 100,000 Ω. We usually cannot feel 12 V, and 24 V just barely tingles. If our skin is moist, 24 V can be quite uncomfortable. Table 8.1 describes the effects of different amounts of current on the human body.

TABLE 8.1   EFFECT OF ELECTRIC CURRENTS ON THE BODY

Current	Effect
0.001 A	Can be felt
0.005 A	Is painful
0.010 A	Causes involuntary muscle contractions (spasms)
0.015 A	Causes loss of muscle control
0.070 A	Goes through the heart; causes serious disruption; probably fatal if current lasts for more than 1 s

FIGURE 8.26

The bird can stand harmlessly on one wire of high potential, but it had better not reach over and touch a neighboring wire! Why not?

FIGURE 8.27

The third prong connects the body of the appliance directly to ground. Any charge that builds up on an appliance is therefore conducted to the ground.

FIGURE 8.28

This table lamp has an insulating body and doesn't need the third (ground) wire.

In order for you to receive a shock, there must be a *difference* in electric potential between one part of your body and another part. Most of the current passes along the path of least electrical resistance connecting these two points. Suppose you fall from a bridge and manage to grab a high-voltage power line, halting your fall. So long as you touch nothing else of different potential, you receive no shock at all. Even if the wire is a few thousand volts above ground potential and you hang by it with two hands, no appreciable charge flows from one hand to the other. This is because there is no appreciable difference in electric potential between your hands. If, however, you reach over with one hand and grab a wire of different potential . . . *zap*! We have all seen birds perched on high-voltage wires. Every part of their bodies is at the same high potential as the wire, so they feel no ill effects.

Interestingly, the source of electrons in the current that shocks you is your own body. As in all conductors, the electrons are already there. It is the energy given to the electrons that you should be wary of. They are energized when a voltage difference exists across different parts of your body.

Most electric plugs and sockets today are wired with three, instead of two, connections. The principal two flat prongs on an electrical plug are for the current-carrying double wire, one part "live" and the other neutral, while the third round prong is grounded—connected directly to the ground (Figure 8.27). Appliances such as irons, stoves, washing machines, and dryers are connected with these three wires. If the live wire accidentally comes into contact with the metal surface of the appliance, and you touch the appliance, you could receive a dangerous shock. This won't occur when the appliance casing is grounded via the ground wire, which ensures that the appliance casing is at zero ground potential.

Injury by Electric Shock

Many people are killed each year by current from common 120-V electric circuits. If your hand touches a faulty 120-V light fixture while your feet are on the ground, there's likely a 120-V "electrical pressure" between your hand and the ground. Resistance to current is usually greatest between your feet and the ground, and so the current is usually not enough to do serious harm. But if your feet and the ground are wet, there is a low-resistance electrical path between you and the ground. The 120 V across this lowered resistance may produce a harmful current in your body.

Pure water is not a good conductor. But the ions that are normally found in water make it a fair conductor. Dissolved materials in water, especially small quantities of salt, lower the resistance even more. There is usually a layer of salt remaining on your skin from perspiration, which, when wet, lowers your skin resistance to a few hundred ohms or less. Handling electrical devices while taking a bath is a definite no-no.

Injury by electric shock occurs in three forms: (1) burning of tissues by heating, (2) contraction of muscles, and (3) disruption of cardiac rhythm. These conditions are caused by the delivery of excessive power for too long a time in critical regions of the body.

Electric shock can upset the nerve center that controls breathing. In rescuing shock victims, the first thing to do is remove them from the source of the electricity. Use a dry wooden stick or some other nonconductor so that you don't get electrocuted yourself. Then apply artificial respiration. It is important to continue artificial respiration. There have been cases of victims of lightning who did not breathe without assistance for several hours, but who were eventually revived and who completely regained good health.

CHECK POINT

What causes electric shock—current or voltage?

Was this your answer?
Electric shock *occurs* when current is produced in the body, but the current is *caused* by an impressed voltage.

Myth: Lightning never strikes the same place twice.
Fact: Lightning does favor certain spots, mainly high locations. The Empire State Building is struck by lightning about 25 times every year.

8.9 Electric Circuits

Any path along which electrons can flow is a *circuit*. For a continuous flow of electrons, there must be a complete circuit with no gaps. A gap is usually provided by an electric switch that can be opened or closed to either cut off energy or allow energy to flow. Most circuits have more than one device that receives electric energy. These devices are commonly connected in a circuit in one of two ways: in *series* or in *parallel*. When connected in series, they form a single pathway for electron flow between the terminals of the battery, generator, or wall outlet (which is simply an extension of these terminals). When connected in parallel, they form branches, each of which is a separate path for the flow of electrons. Both series and parallel connections have their own distinctive characteristics. In the following sections, we shall briefly discuss circuits using these two types of connections.

We often think of current flowing through a circuit, but don't say this around somebody who is picky about grammar, for the expression "current flows" is redundant. More properly, charge flows—which *is* current.

The Physical Science Place
Electricity and Circuits

The Physical Science Place
Electric Circuits

SERIES CIRCUITS

A simple **series circuit** is shown in Figure 8.29. Three lamps are connected in series with a battery. The same current exists almost immediately in all three lamps when the switch is closed. The current does not "pile up" or accumulate in any lamp but flows *through* each lamp. Electrons that make up this current leave the negative terminal of the battery, pass through each of the resistive filaments in the lamps in turn, and then return to the positive terminal of the battery. (The same amount of current passes through the battery.) This is the only path of the electrons through the circuit. A break anywhere in the path results in an open circuit, and the flow of electrons ceases. Such a break occurs when the switch is opened, when the wire is accidentally cut, or when one of the lamp filaments burns out. The circuit shown in Figure 8.29 illustrates the following characteristics of series connections:

FIGURE 8.29

INTERACTIVE FIGURE

A simple series circuit. The 6-V battery provides 2 V across each lamp.

1. Electric current has a single pathway through the circuit. This means that the current passing through the resistance of each electrical device along the pathway is the same.

2. This current is resisted by the resistance of the first device, the resistance of the second, and that of the third, so the total resistance to current in the circuit is the sum of the individual resistances along the circuit path.

3. The current in the circuit is numerically equal to the voltage supplied by the source divided by the total resistance of the circuit. This is in accord with Ohm's law.

4. The total voltage impressed across a series circuit divides among the individual electrical devices in the circuit so that the sum of the "voltage drops"

All batteries degrade. The lithium-ion cells popular in notebook computers, cameras, and cell phones erode faster when highly charged and warm. So keep yours at about half charge in a cool or cold environment to extend battery life.

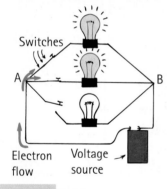

FIGURE 8.30

INTERACTIVE FIGURE

A simple parallel circuit. A 6-V battery provides 6 V across the top two lamps.

across the resistance of each individual device is equal to the total voltage supplied by the source. This characteristic follows from the fact that the amount of energy given to the total current is equal to the sum of energies given to each device.

5. The voltage drop across each device is proportional to its resistance. This follows from the fact that more energy is dissipated when a current passes through a large resistance than when the same current passes through a small resistance.

CHECK POINT

1. What happens to the current in the other lamps if one lamp in a series circuit burns out?
2. What happens to the brightness of each lamp in a series circuit when more lamps are added to the circuit?

Were these your answers?

1. If one of the lamp filaments burns out, the path connecting the terminals of the voltage source breaks and current ceases. All lamps go out.
2. Adding more lamps in a series circuit produces a greater circuit resistance. This decreases the current in the circuit and therefore in each lamp, which causes dimming of the lamps. Energy is divided among more lamps, so the voltage drop across each lamp is less.

The rules above hold for ac or dc circuits. It is easy to see the main disadvantage of a series circuit: if one device fails, current in the entire circuit ceases. Some cheap Christmas tree lights are connected in series. When one bulb burns out, it's fun and games (or frustration) trying to locate which one to replace.

Most circuits are wired so that it is possible to operate several electrical devices at once, each independently of the other. In your home, for example, a lamp can be turned on or off without affecting the operation of other lamps or electrical devices. This is because these devices are connected not in series but in parallel with one another.

PARALLEL CIRCUITS

A simple **parallel circuit** is shown in Figure 8.30. Three lamps are connected to the same two points, A and B. Electrical devices connected to the same two points of an electrical circuit are said to be *connected in parallel*. Electrons leaving the negative terminal of the battery need travel through only one lamp filament before returning to the positive terminal of the battery. In this case, current branches into three separate pathways from A to B. A break in any one path does not interrupt the flow of charge in the other paths. Each device operates independently of the other devices (whether the circuit is ac or dc).

The circuit shown in Figure 8.30 illustrates the following major characteristics of parallel connections:

1. Each device connects the same two points, A and B, of the circuit. The voltage is therefore the same across each device.

2. The total current in the circuit divides among the parallel branches. Because the voltage across each branch is the same, the amount of current in each branch is inversely proportional to the resistance of the branch.

3. The total current in the circuit equals the sum of the currents in its parallel branches.

4. As the number of parallel branches is increased, the overall resistance of the circuit is *decreased*. Overall resistance is lowered with each added path between any two points of the circuit. This means the overall resistance of the circuit is less than the resistance of any one of the branches.

CHECK POINT

1. What happens to the current in the other lamps if one of the lamps in a parallel circuit burns out?
2. What happens to the brightness of each lamp in a parallel circuit when more lamps are added in parallel to the circuit?

Were these your answers?

1. If one lamp burns out, the other lamps are unaffected. The current in each branch, according to Ohm's law, is equal to voltage/resistance, and because neither voltage nor resistance is affected in the other branches, the current in those branches is unaffected. The total current in the overall circuit (the current through the battery), however, is decreased by an amount equal to the current drawn by the lamp in question before it burned out. But the current in any other single branch is unchanged.

2. The brightness of each lamp is unchanged as other lamps are introduced (or removed). Only the total resistance and total current in the total circuit changes, which is to say that the current in the battery changes. (There is resistance in a battery also, which we assume is negligible here.) As lamps are introduced, more paths are available between the battery terminals, which effectively decreases total circuit resistance. This decreased resistance is accompanied by an increased current, the same increase that feeds energy to the lamps as they are introduced. Although changes of resistance and current occur for the circuit as a whole, no changes occur in any individual branch in the circuit.

PARALLEL CIRCUITS AND OVERLOADING

Electricity is usually fed into a home by way of two wires called *lines*. These lines are very low in resistance and are connected to wall outlets in each room—sometimes through two or more separate circuits. An electric potential of about 110 to 120 V ac is applied across these lines by a transformer in the neighborhood. (A transformer, as we shall see in the next chapter, is a device that steps down the higher voltage supplied by the power utility.) As more devices are connected to a circuit, more pathways for current result. This lowers the combined resistance of the circuit. Therefore, more current exists in the circuit, which is sometimes a problem. Circuits that carry more than a safe amount of current are said to be *overloaded*.

We can see how overloading occurs in Figure 8.32. The supply line is connected to a toaster that draws 8 amperes, a heater that draws 10 amperes, and a lamp that draws 2 amperes. When only the toaster is operating and drawing 8 amperes, the total line current is 8 amperes. When the heater is also operating, the total line current increases to 18 amperes (8 amperes to the toaster plus 10 amperes to the heater). If you turn on the lamp, the line current increases to 20 amperes. Connecting additional devices increases the current still more. Connecting too many devices into the same circuit results in overheating the wires, which can cause a fire.

FIGURE 8.31

New Zealand physics instructor David Housden constructs a parallel circuit by fastening lamps to extended terminals of a common battery. He asks his class to predict the relative brightnesses of two identical lamps in one wire about to be connected in parallel.

In a parallel circuit, *most* current travels in the path of least resistance—but not all. *Some* current travels in each path.

FIGURE 8.32

Circuit diagram for appliances connected to a household circuit.

Electrical Energy and Technology

Try to imagine everyday home life before the advent of electrical energy. Think of homes without electric lights, refrigerators, heating and cooling systems, telephones, and radio and TV. We may romanticize a better life without these, but only if we overlook the many hours of daily toil devoted to laundry, cooking, and heating homes. We'd also have to overlook how difficult it was to reach a doctor in times of emergency before the advent of the telephone—when all the doctor had in his bag were laxatives, aspirins, and sugar pills—and when infant death rates were staggering.

We have become so accustomed to the benefits of technology that we are only faintly aware of our dependency on dams, power plants, mass transportation, electrification, modern medicine, and modern agricultural science for our very existence. When we enjoy a good meal, we give little thought to the technology that went into growing, harvesting, and delivering the food on our table. When we turn on a light, we give little thought to the centrally controlled power grid that links the widely separated power stations by long-distance transmission lines. These lines serve as the productive life force of industry, transportation, and the electrification of civilization. Anyone who thinks of science and technology as "inhuman" fails to grasp the ways in which they make our lives more human.

FIGURE 8.33

A safety fuse.

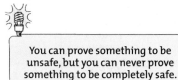

You can prove something to be unsafe, but you can never prove something to be completely safe.

FIGURE 8.34

Electrician Dave Hewitt with a safety fuse and a circuit breaker. He favors the old fuses, which he's found more reliable.

SAFETY FUSES

To prevent overloading in circuits, fuses are connected in series along the supply line. In this way the entire line current must pass through the fuse. The fuse shown in Figure 8.33 is constructed with a wire ribbon that heats up and melts at a given current. If the fuse is rated at 20 amperes, it passes 20 amperes, but no more. A current above 20 amperes melts the fuse, which "blows out" and breaks the circuit. Before a blown fuse is replaced, the cause of overloading should be determined and remedied. Sometimes insulation that separates the wires in a circuit wears away and allows the wires to touch. This greatly reduces the resistance in the circuit and is called a *short circuit*.

In modern buildings, fuses have been largely replaced by circuit breakers, which use magnets or bimetallic strips to open a switch when the current is excessive. Utility companies use circuit breakers to protect their lines all the way back to the generators.

■ 8.10 Electric Power

The moving charges in an electric current do work. This work, for example, can heat a circuit or turn a motor. The rate at which work is done—that is, the rate at which electric energy is converted into another form, such as mechanical energy, heat, or light—is called **electric power**. Electric power is equal to the product of current and voltage.*

$$\text{Power} = \text{current} \times \text{voltage}$$

If the voltage is expressed in volts and the current in amperes, then the power is expressed in watts. So, in units form,

$$\text{Watts} = \text{amperes} \times \text{volts}$$

If a lamp rated at 120 W operates on a 120-V line, you can figure that it draws a current of 1 ampere (120 W = 1 ampere × 120 V). A 60-W lamp

---

* Recall from Chapter 4 that *power = work/time*; 1 W = 1 J/s. Note that the units for mechanical power and electrical power agree (work and energy are both measured in joules):

$$\text{Power} = \frac{\cancel{\text{charge}}}{\text{time}} \times \frac{\text{energy}}{\cancel{\text{charge}}} = \frac{\text{energy}}{\text{time}}$$

FIGURING PHYSICAL SCIENCE

▪ Problem Solving

SAMPLE PROBLEM 1

If a 120-V line to a socket is limited to 15 A by a safety fuse, will it operate a 1200-W hair dryer?

SOLUTION:

Yes. From the expression *watts = amperes × volts,* we can see that current = 1200 W/120 V = 10 A, so the hair dryer will operate when connected to the circuit. But two hair dryers on the same circuit will blow the fuse.

SAMPLE PROBLEM 2

At 30¢/kWh, what does it cost to operate the 1200-W hair dryer for 1 h?

SOLUTION:

1200 W = 1.2 kW; 1.2 kW × 1 h × 30¢/1 kWh = 36¢.

---

draws 0.5 ampere on a 120-V line. This relationship becomes a practical matter when you wish to know the cost of electrical energy, which is usually a small fraction of a dollar per kilowatt-hour, depending on the locality. A kilowatt is 1000 W, and a kilowatt-hour represents the amount of energy consumed in 1 h at the rate of 1 kW.* Therefore, in a locality where electric energy costs 25¢/kWh, a 100-W electric lightbulb can operate for 10 hours at a cost of 25 cents, or a half nickel for each hour. A toaster or iron, which draws much more current and therefore much more energy, costs about 10 times as much to operate.

FIGURE 8.35

The power and voltage on the lightbulb read "100 W 120 V." Does it *have* 100 W, or does it use 100 W when lit? How many amperes flow through it when it is lit?

fyi

▪ The brightness of a bulb depends on how much power it uses, that is, on how much electricity is converted into heat each second. A tungsten lightbulb that uses 100 W is brighter than one that uses 60 W. Because of this, many people mistakenly believe that a watt is a unit of brightness, but it isn't. A 13-W fluorescent lightbulb is as bright as a 60-W conventional (incandescent) bulb. Does that mean that a conventional bulb wastes electricity? Yes. The extra electric power used just heats the bulb. That's why tungsten bulbs are much hotter to the touch than equally bright, fluorescent bulbs.

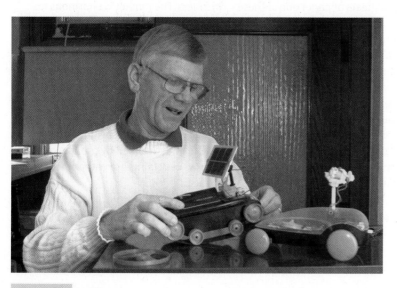

FIGURE 8.36

Roy Unruh harnesses solar energy to produce electricity, which in turn powers demonstration vehicles.

fyi

▪ Your heart uses slightly more than 1 W of power in pumping blood through your body.

---

*Because *power = energy/time*, simple rearrangement gives energy = *power × time*; thus, energy can be expressed in the unit *kilowatt-hours* (kWh).

Magnetic Therapy*

Back in the eighteenth century, a celebrated "magnetizer" from Vienna, Franz Mesmer, brought his magnets to Paris and established himself as a healer in Parisian society. He healed patients by waving magnetic wands above their heads.

At that time, Benjamin Franklin, the world's leading authority on electricity, was visiting Paris as a U.S. representative. He suspected that Mesmer's patients did benefit from his ritual, but only because it kept them away from the bloodletting practices of other physicians. At the urging of the medical establishment, King Louis XVI appointed a royal commission to investigate Mesmer's claims. The commission included Franklin and Antoine Lavoisier, the founder of modern chemistry. The commissioners designed a series of tests in which some subjects thought they were receiving Mesmer's treatment when they weren't, while others received the treatment but were led to believe they had not. The results of these blind experiments established beyond any doubt that Mesmer's success was due solely to the power of suggestion. To this day, the report is a model of clarity and reason. Mesmer's reputation was destroyed, and he retired to Austria.

Now some two hundred years later, with increased knowledge of magnetism and physiology, hucksters of magnetism are attracting even larger followings. But there is no government commission of Franklins and Lavoisiers to challenge their claims. Instead, magnetic therapy is another of the untested and unregulated "alternative therapies" given official recognition by Congress in 1992.

Although testimonials about the benefits of magnets are many, there is no scientific evidence whatever for magnets boosting body energy or combating aches and pains. None. Yet millions of therapeutic magnets are sold in stores and catalogs. Consumers are buying magnetic bracelets, insoles, wrist and knee bands, back and neck braces, pillows, mattresses, lipstick, and even water. They are told that magnets have powerful effects on the body, mainly increasing blood flow to injured areas. The idea that blood is attracted by a magnet is bunk, for the type of iron that occurs in blood doesn't respond to a magnet. Furthermore, most therapeutic magnets are of the refrigerator type, with a very limited range. To get an idea of how quickly the field of these magnets drops off, see how many sheets of paper one of these magnets will hold on a refrigerator or any iron surface. The magnet will fall off after a few sheets of paper separate it from the iron surface. The field doesn't extend much more than one millimeter, and it wouldn't penetrate the skin, let alone into muscles. And even if it did, there is no scientific evidence that magnetism has any beneficial effects on the body at all. But, again, testimonials are another story.

Sometimes an outrageous claim has some truth to it. For example, the practice of bloodletting in previous centuries was, in fact, beneficial to a small percentage of men. These men suffered the genetic disease (*hemochromatosis*, excess iron in the blood—women being less afflicted partly due to menstruation). Although the number of men who benefited from bloodletting was small, testimonials of its success prompted the widespread practice that killed many.

No claim is so outrageous that testimonials can't be found to support it. Claims that the Earth is flat or claims for the existence of flying saucers are quite harmless and may amuse us. Magnetic therapy may likewise be harmless for many ailments, but not when it is used to treat a serious disorder in place of modern medicine. Pseudoscience may be promoted to intentionally deceive or it may be the result of flawed and wishful thinking. In either case, pseudoscience is very big business. The market is enormous for therapeutic magnets and other such fruits of unreason.

Scientists must keep open minds, must be prepared to accept new findings, and must be ready to be challenged by new evidence. But scientists also have a responsibility to inform the public when they are being deceived and, in effect, robbed by pseudoscientists whose claims are without substance.

*Adapted from Voodoo Science: The Road from Foolishness to Fraud, by Robert L. Park; Oxford University Press, 2000.

SUMMARY OF TERMS

Coulomb's law The relationship among electrical force, charge, and distance: If the charges are alike in sign, the force is repelling; if the charges are unlike, the force is attractive.

Coulomb The SI unit of electrical charge. One coulomb (symbol C) is equal in magnitude to the total charge of 6.25×10^{18} electrons.

Electrically polarized Term applied to an atom or molecule in which the charges are aligned so that one side has a slight excess of positive charge and the other side a slight excess of negative charge.

Electric field Defined as force per unit charge, it can be considered an energetic "aura" surrounding charged objects. About a charged point, the field decreases with distance according to the inverse-square law, like a gravitational field. Between oppositely charged parallel plates, the electric field is uniform.

Electric potential energy The energy a charge possesses by virtue of its location in an electric field.

Electric potential The electric potential energy per amount of charge, measured in volts and often called *voltage*.

Conductor Any material having free charged particles that easily flow through it when an electric force acts on them.

Potential difference The difference in potential between two points, measured in volts and often called *voltage difference*.

Electric current The flow of electric charge that transports energy from one place to another.

Ampere The unit of electric current; the rate of flow of 1 C of charge per second.

Direct current (dc) An electric current flowing in one direction only.

Alternating current (ac) Electric current that repeatedly reverses its direction; the electric charges vibrate about relatively fixed points. In the United States, the vibrational rate is 60 Hz.

Electrical resistance The property of a material that resists the flow of an electric current through it. It is measured in ohms (Ω).

Superconductor Any material with zero electrical resistance, wherein electrons flow without losing energy and without generating heat.

Ohm's law The statement that the current in a circuit varies in direct proportion to the potential difference or voltage and inversely with the resistance:

$$\text{Current} = \frac{\text{voltage}}{\text{resistance}}$$

A current of 1 A is produced by a potential difference of 1 V across a resistance of 1 Ω.

Series circuit An electric circuit with devices connected in such a way that the same electric current flows through each of them.

Parallel circuit An electric circuit with two or more devices connected in such a way that the same voltage acts across each one, and any single one completes the circuit independently of all the others.

Electric power The rate of energy transfer, or the rate of doing work; the amount of energy per unit time, which can be measured by the product of current and voltage:

$$\text{Power} = \text{current} \times \text{voltage}$$

It is measured in watts (or kilowatts), where $1 \text{ A} \times 1 \text{ V} = 1 \text{ W}$.

REVIEW QUESTIONS

8.1 Electric Charge

1. Which part of an atom is positively charged, and which part is negatively charged?
2. How does the charge of one electron compare with that of another electron?
3. How do the masses of electrons compare with the masses of protons?
4. How does the number of protons in the atomic nucleus normally compare with the number of electrons that orbit the nucleus?
5. What kind of charge does an object acquire when electrons are stripped from it?
6. What is meant by saying that charge is conserved?

8.2 Coulomb's Law

7. How is Coulomb's law similar to Newton's law of gravitation? How is it different?

8. How does a coulomb of charge compare with the charge of a single electron?
9. How does the magnitude of electrical force between a pair of charged particles change when the particles are moved twice as far apart? Three times as far apart?
10. How does an electrically polarized object differ from an electrically charged object?

8.3 Electric Field

11. Give two examples of common force fields.
12. How is the direction of an electric field defined?

8.4 Electric Potential

13. In terms of the units that measure them, distinguish between electric potential energy and electric potential.
14. A balloon may easily be charged to several thousand volts. Does that mean it has several thousand joules of energy? Explain.

8.5 Voltage Sources

15. What condition is necessary for a sustained flow of electric charge through a conducting medium?
16. How much energy is given to each coulomb of charge passing through a 6-V battery?

8.6 Electric Current

17. Does electric charge flow *across* a circuit or *through* a circuit? Does voltage flow *across* a circuit or is it *impressed across* a circuit? Explain.
18. Distinguish between dc and ac.
19. Does a battery produce dc or ac? Does the generator at a power station produce dc or ac?

8.7 Electric Resistance

20. Which has the greater resistance, a thick wire or a thin wire of the same length?
21. What is the unit of electrical resistance?

8.8 Ohm's Law

22. What is the effect on current through a circuit of steady resistance when the voltage is doubled? What if both voltage and resistance are doubled?

23. Which has the greater electrical resistance, wet skin or dry skin?
24. What is the function of the third prong on the plug of an electric appliance?
25. What is the source of electrons that makes a shock when you touch a charged conductor?

8.9 Electric Circuits

26. In a circuit consisting of two lamps connected in series, if the current in one lamp is 1 A, what is the current in the other lamp?
27. If 6 V were impressed across the circuit in question 26, and the voltage across the first lamp were 2 V, what would be the voltage across the second lamp?
28. How does the total current through the branches of a parallel circuit compare with the current through the voltage source?
29. As more lines are opened at a fast-food restaurant, the resistance to the motion of people trying to get served is reduced. How is this similar to what happens when more branches are added to a parallel circuit?

8.10 Electric Power

30. What is the relationship among electric power, current, and voltage?

EXERCISES

● BEGINNER   ■ INTERMEDIATE   ◆ EXPERT

1. ● When you comb your hair, you scuff electrons from your hair onto the comb. Is your hair then positively or negatively charged? How about the comb?
2. ● When one material is rubbed against another, electrons jump readily from one to the other, but protons do not. Why is this? (Think in atomic terms.)
3. ● If electrons were positive and protons were negative, would Coulomb's law be written the same or differently?
4. ● The five thousand billion billion freely moving electrons in a penny repel one another. Why don't they fly out of the penny?
5. ● Two equal charges exert equal forces on each other. What if one charge has twice the magnitude of the other? How do the forces they exert on each other compare?
6. ● How does the magnitude of electric force compare with the charge between a pair of charged particles when they are brought to half their original distance of separation? To one-quarter their original distance? To four times their original distance? (What law guides your answers?)
7. ● Suppose that the strength of the electric field about an isolated point charge has a certain value at a distance of 1 m. How does the electric field strength compare at a distance of 2 m from the point charge? What law guides your answer?
8. ● Why is a good conductor of electricity also a good conductor of heat?
9. ● If you put in 10 J of work to push 1 C of charge against an electric field, what is its voltage with respect to its starting position? When you release it, what is its kinetic energy if it flies past its starting position?

10. ● What is the voltage at the location of a 0.0001-C charge that has an electric potential energy of 0.5 J (both voltage and potential relative to the same reference point)?
11. ● What happens to the brightness of light emitted by a lightbulb when the current in it increases?
12. ● Your tutor tells you that an ampere and a volt really measure the same thing, and the different terms only make a simple concept seem confusing. Why should you consider getting a different tutor?
13. ● In which of the circuits below does a current exist to light the bulb?

14. ● Does more current flow out of a battery than into it? Does more current flow into a lightbulb than out of it? Explain.
15. ● Sometimes you hear someone say that a particular appliance "uses up" electricity. What is it that the appliance actually consumes, and what becomes of it?
16. ● A simple lie detector consists of an electric circuit, one part of which is part of your body, such as a circuit that connects one of your fingers to another of your fingers. A sensitive meter shows the current that flows when a small voltage is applied. How does this technique indicate that

a person is lying? (And when does this technique not indicate when someone is lying?)

17. ● Only a small percentage of the electric energy fed into a common lightbulb is transformed into light. What happens to the rest?

18. ● Does a lamp with a thick filament draw more current or less current than a lamp with a thin filament?

19. ● Is the current in a lightbulb connected to a 220-V source greater or less than that in the same bulb when it is connected to a 110-V source?

20. ● Which does less damage—plugging a 110-V appliance into a 220-V circuit or plugging a 220-V appliance into a 110-V circuit? Explain.

21. ● If a current of one- or two-tenths of an ampere were to flow into one of your hands and out the other, you would probably be electrocuted. But if the same current were to flow into your hand and out the elbow above the same hand, you could survive, even though the current might be large enough to burn your flesh. Explain.

22. ● Would you expect to find dc or ac in the filament of a lightbulb in your home? How about in the headlight of an automobile?

23. ● Are automobile headlights wired in parallel or in series? What is your evidence?

24. ● What unit is represented by (a) joule per coulomb, (b) coulomb per second, and (c) watt-second?

25. ● To connect a pair of resistors so that their equivalent resistance is greater than the resistance of either one, should you connect them in series or in parallel?

26. ● To connect a pair of resistors so that their equivalent resistance is less than the resistance of either one, should you connect them in series or in parallel?

27. ● A friend says that a battery provides not a source of constant current, but a source of constant voltage. Do you agree or disagree, and why?

28. ● A friend says that adding bulbs in series to a circuit provides more obstacles to the flow of charge, so there is less current with more bulbs, but adding bulbs in parallel provides more paths so more current can flow. Do you agree or disagree, and why?

29. ● Estimate the number of electrons that a power company delivers annually to the homes of a typical city of 50,000 people.

30. ● If electrons flow very slowly through a circuit, why does it not take a noticeably long time for a lamp to glow when you turn on a distant switch?

31. ● Consider a pair of flashlight bulbs connected to a battery. Do they glow brighter if they are connected in series or in parallel? Does the battery run down faster if they are connected in series or in parallel?

32. ■ An electroscope is a simple device consisting of a metal ball that is attached by a conductor to two thin leaves of metal foil protected from air disturbances in a jar, as shown. When the ball is touched by a charged body, the leaves that normally hang straight down spread apart. Why? (Electroscopes are useful not only as charge detectors but also for measuring the quantity of charge: the greater the charge transferred to the ball, the more the leaves diverge.)

33. ■ The leaves of a charged electroscope collapse in time. At higher altitudes, they collapse more rapidly. Why is this true? (Hint: The existence of cosmic rays was first indicated by this observation.)

34. ■ Strictly speaking, is a penny slightly more massive if it has a negative charge or a positive charge? Explain.

35. ■ When a car is moved into a painting chamber, a mist of paint is sprayed around it. When the body of the car is given a sudden electric charge and the mist of paint is attracted to it, presto—the car is quickly and uniformly painted. What does the phenomenon of polarization have to do with this?

36. ■ If you place a free electron and a free proton in the same electric field, how do the forces acting on them compare? How do their accelerations compare? Their directions of travel?

37. ■ One example of a water system is a garden hose that waters a garden. Another is the cooling system of an automobile. Which of these exhibits behavior more analogous to that of an electric circuit? Why?

38. ■ Is it correct to say that the energy from a car battery ultimately comes from fuel in the gas tank? Defend your answer.

39. ■ A 1-mi-long copper wire has a resistance of 10 Ω. What is its new resistance when it is shortened by (a) cutting it in half or by (b) doubling it over and using it as if it were one wire of half the length but twice the cross-sectional area?

40. ■ A car's headlights dissipate 40 W on low beam and 50 W on high beam. Is there more or less resistance in the high-beam filament?

41. ■ Why might the wingspans of birds be a consideration in determining the spacing between parallel wires on power poles?

42. ■ If several bulbs are connected in series to a battery, they may feel warm to the touch even though they are not visibly glowing. What is your explanation?

43. ■ In the circuit shown, how do the brightnesses of the identical lightbulbs compare? Which lightbulb draws the most current? What happens if bulb A is unscrewed? If bulb C is unscrewed?

44. ■ As more and more bulbs are connected in series to a flashlight battery, what happens to the brightness of each bulb? Assuming that the heating inside the battery is negligible, what happens to the brightness of each bulb when more and more bulbs are connected in parallel?

45. ■ Are these circuits equivalent to one another? Why or why not?

46. ■ A battery has internal resistance, so if the current it supplies goes up, the voltage it supplies goes down. If too many bulbs are connected in parallel across a battery, does their brightness diminish? Explain.

47. ■ Your friend says that electric current takes the path of least resistance. Why is it more accurate in the case of a parallel circuit to say that *most* current travels in the path of least resistance?

48. ◆ If a 60-W bulb and a 100-W bulb are connected in series in a circuit, across which bulb is the greater voltage drop? How about if they are connected in parallel?

PROBLEMS

● BEGINNER   ■ INTERMEDIATE   ◆ EXPERT

1. ● Two pellets, each with a charge of 1 microcoulomb $(10^{-6}$ C), are located 3 cm (0.03 m) apart. Show that the electric force between them is 10 N.

2. ● Two point charges are separated by 6 cm. The attractive force between them is 20 N. Show that when they are separated by 12 cm, the force between them is 5 N. (Why can you solve this problem without knowing the magnitudes of the charges?)

3. ● If the charges attracting each other in the preceding problem have equal magnitudes, show that the magnitude of each charge is 2.8 microcoulombs.

4. ● A droplet of ink in an industrial ink-jet printer carries a charge of 1.6×10^{-10} C and is deflected onto paper by a force of 3.2×10^{-4} N. Show that the strength of the electric field required to produce this force is $2 \times 10^{6} \frac{N}{C}$.

5. ● When an electric field does 12 J of work on a charge of 0.0001 C, (a) show that the change in voltage is 120,000 V. (b) When the same electric field does 24 J of work on a charge of 0.0002 C, show that the voltage change is the same.

6. ■ The current driven by voltage V in a circuit of resistance R is given by Ohm's law, $I = V/R$. Show that the resistance of a circuit carrying current I and driven by voltage V is given by the equation $R = V/I$.

7. ■ The same voltage V is impressed on each of the branches of a parallel circuit. The voltage source provides a total current I_{total} to the circuit, and "sees" a total equivalent resistance of R_{eq} in the circuit. That is, $V = I_{total}R_{eq}$. The total current is equal to the sum of the currents through each branch of the parallel circuit. In a circuit with n branches, $I_{total} = I_1 + I_2 + I_3 + \cdots + I_n$. Use Ohm's law ($I = V/R$) and show that the equivalent resistance of a parallel circuit with n branches is given by

$$\frac{1}{R_{eq}} = \frac{1}{R_1} + \frac{1}{R_2} + \frac{1}{R_3} + \cdots + \frac{1}{R_n}.$$

8. ● The wattage marked on a lightbulb is not an inherent property of the bulb; rather, it depends on the voltage to which it is connected, usually 110 or 120 V. Show that the current in a 60-W bulb connected in a 120-V circuit is 0.5 A.

9. ● Rearrange the equation *current = voltage/resistance* to express resistance in terms of current and voltage. Then consider the following: A certain device in a 120-V circuit has a current rating of 20 A. Show that the resistance of the device is 6 Ω.

10. ● Using the formula *power = current × voltage*, show that the current drawn by a 1200-W hair dryer connected to 120 V is 10 A. Then using your same method for the solution to the previous problem, show that the resistance of the hair dryer is 12 Ω.

11. ■ The power of in an electric circuit is given by the equation $P = IV$. Use Ohm's law to express V and show that power can be expressed by the equation $P = I^2R$.

12. ■ The total charge that an automobile battery can supply without being recharged is given in terms of ampere-hours. A typical 12-V battery has a rating of 60 ampere-hours (60 A for 1 h, 30 A for 2 h, and so on). Suppose that you forget to turn off the headlights in your parked automobile. If each of the two headlights draws 3 A, show that your battery will go dead in about 10 h.

13. ■ Suppose you operate a 100-W lamp continuously for 1 week when the power utility rate is 20¢/kWh. Show that this costs you $3.36.

14. ■ An electric iron connected to a 110-V source draws 9 A of current. Show that the amount of heat generated in 1 min is almost 60 kJ.

15. ■ For the electric iron of the previous problem, show that the number of coulombs that flow through it in 1 min is 540 C.

16. ◆ A certain lightbulb with a resistance of 95 Ω is labeled "150 W." Was this bulb designed for use in a 120-V circuit or a 220-V circuit?

17. ◆ In periods of peak demand, power companies lower their voltage. This saves them power (and saves you money)! To see the effect, consider a 1200-W toaster that draws 10 A when connected to 120 V. Suppose the voltage is lowered by 10 percent to 108 V. By how much does the current decrease? By how much does the power decrease? (**CAUTION:** The 1200-W label is valid only when 120 V is applied. When the voltage is lowered, it is the resistance of the toaster, not its power, which remains constant.)

ACTIVE EXPLORATIONS

1. Write a letter to your favorite uncle and bring him up to speed on your progress with physics. Relate most of the terms in this chapter and how learning to distinguish among them contributes to your understanding. Select four of the terms and discuss them. Relate the terms to practical examples.
2. Demonstrate charging by friction and discharging from points with a friend who stands at the far end of a carpeted room. With leather shoes, scuff your way across the rug until your noses are close together. This can be a delightfully tingling experience, depending on how dry the air is and how pointed your noses are.
3. Briskly rub a comb against your hair or a woolen garment and bring it near a small but smooth stream of running water. Is the stream of water charged? (Before you say yes, note the behavior of the stream when an opposite charge is brought nearby.)

4. An electric cell is made by placing two plates of different materials that have different affinities for electrons in a conducting solution. You can make a simple 1.5-V cell by placing a strip of copper and a strip of zinc in a tumbler of salt water. The voltage of a cell depends on the materials used and the solution they are placed in, not the size of the plates. A battery is actually a series of cells.

    An easy cell to construct is the citrus cell. Stick a paper clip and a piece of copper wire into a lemon. Hold the ends of the wire close together, but not touching, and place the ends on your tongue. The slight tingle you feel and the metallic taste you experience result from a slight current of electricity pushed by the citrus cell through the wires when your moist tongue closes the circuit.

Paper clip

Lemon

Copper wire

READINESS ASSURANCE TEST (RAT)

If you have a good handle on this chapter, if you really do, then you should be able to score at least 7 out of 10 on this RAT. If you score less than 7, you need to study further before moving on.

Choose the BEST answer to each of the following.

1. When you brush your hair and scrape electrons from your hair, the charge of your hair is
   (a) positive.
   (b) negative.
   (c) both of these
   (d) neither of these
2. According to Coulomb's law, a pair of particles that are placed twice as close to each other experience forces
   (a) twice as strong.
   (b) four times as strong.
   (c) half as strong.
   (d) one-quarter as strong.
3. An electric field surrounds all
   (a) electric charge.
   (b) electrons.
   (c) protons.
   (d) all of these
4. When you increase the potential energy of a charged particle, you increase its ability to
   (a) do work.
   (b) charge other particles.

   (c) conduct.
   (d) transform to heat.
5. A 10-Ω resistor carries 10 amperes. The voltage across the resistor is
   (a) zero.
   (b) more than zero but less than 10 V.
   (c) 10 V.
   (d) more than 10 V.
6. You can touch and discharge a 10,000-V Van de Graaff generator with little harm because, although the voltage is high, there is relatively little
   (a) resistance.
   (b) energy.
   (c) grounding.
   (d) all of these
   (e) none of these
7. Compared with the amount of current in the filament of a lamp, the amount of current in the connecting wire is
   (a) definitely less.
   (b) often less.
   (c) more.
   (d) the same.
   (e) incredibly, all of these
8. As more lamps are connected to a series circuit, the overall current in the power source
   (a) increases.
   (b) decreases.
   (c) remains the same.
   (d) none of these

9. As more lamps are connected to a parallel circuit, the overall current in the power source
   (a) increases.
   (b) decreases.
   (c) remains the same.

10. What is the power rating of a lamp connected to a 12-V source when it carries 1.5 A?
   (a) 8 W
   (b) 12 W
   (c) 18 W
   (d) none of these

Answers to RAT
1a, 2b, 3d, 4a, 5d, 6b, 7d, 8b, 9a, 10c.

CHAPTER 8 ONLINE RESOURCES

The Physical Science Place

Interactive Figures
- 8.2, 8.3, 8.9, 8.10, 8.29, 8.30

Tutorials
- Electrostatics
- Electricity and Circuits

Videos
- Electric Potential
- Van de Graaff Generator

- Caution on Handling Electric Wires
- Birds and High-Voltage Wires
- Alternating Current
- Ohm's Law
- Electric Circuits

Quiz

Flashcards

Links

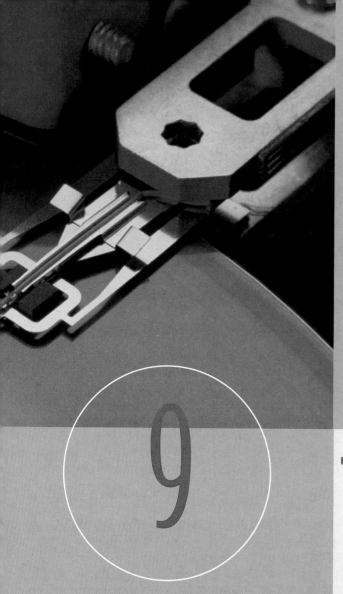

magnetism and electromagnet induction

9.1 MAGNETIC POLES

9.2 MAGNETIC FIELDS

9.3 MAGNETIC DOMAINS

9.4 ELECTRIC CURRENTS AND MAGNETIC FIELDS

9.5 MAGNETIC FORCES ON MOVING CHARGES

9.6 ELECTROMAGNETIC INDUCTION

9.7 GENERATORS AND ALTERNATING CURRENT

9.8 POWER PRODUCTION

9.9 THE TRANSFORMER—BOOSTING OR LOWERING VOLTAGE

9.10 FIELD INDUCTION

■ The term *magnetism* comes from Magnesia, the name of an ancient city in Asia Minor, where the Greeks found certain very unusual stones more than 2000 years ago. These stones, called *lodestones*, possess the unusual property of attracting pieces of iron. Such magnets were first fashioned into compasses and used for navigation by the Chinese in the 12th century AD.

In the 16th century, William Gilbert, Queen Elizabeth's physician, made artificial magnets by rubbing pieces of iron against lodestones. He suggested that a compass always points north and south because Earth itself has magnetic properties. Later, in 1750, John Michell in England found that magnetic poles obey the inverse-square law, and his results were confirmed by Charles Coulomb. The subjects of magnetism and electricity developed almost independently until 1820, when a Danish physicist named Hans

The Physical Science Place
Oersted's Discovery

In days gone by, Dick Tracy comic strips, in addition to predicting the advent of cell phones, featured the heading, "He who controls magnetism controls the universe."

Christian Oersted discovered, in a classroom demonstration, that an electric current affects a magnetic compass.* In his demonstration he saw that magnetism was related to electricity. Shortly thereafter, the French physicist André Marie Ampère proposed that electric currents are the source of all magnetic phenomena.

◼ 9.1 Magnetic Poles

Anyone who has played around with magnets knows that magnets exert forces on one another. A **magnetic force** is similar to an electrical force in that a magnet can both attract and repel without touching (depending on which end of the magnet is held near another) and the strength of its interaction depends on the distance between magnets. Whereas electric charges produce electrical forces, regions called *magnetic poles* give rise to magnetic forces.

If you suspend a bar magnet at its center by a piece of string, you've got a compass. One end, called the *north-seeking pole*, points northward. The opposite end, called the *south-seeking pole*, points southward. More simply, these are called the *north* and *south* poles. All magnets have both a north and a south pole (some have more than one of each). Refrigerator magnets have narrow strips of alternating north and south poles. These magnets are strong enough to hold sheets of paper against a refrigerator door, but they have a very short range because the north and south poles cancel a short distance from the magnet. In a simple bar magnet, the magnetic poles are located at the two ends. A common horseshoe magnet is a bar magnet bent into a U shape. Its poles are also located at its two ends.

If the north pole of one magnet is brought near the north pole of another magnet, they repel. The same is true of a south pole near a south pole. If opposite poles are brought together, however, attraction occurs.**

Like poles repel; opposite poles attract.

This rule is similar to the rule for the forces between electric charges, in which like charges repel one another and unlike charges attract. But there is a very important difference between magnetic poles and electric charges. Whereas electric charges can be isolated, magnetic poles cannot. Electrons and protons are entities by themselves. A cluster of electrons need not be accompanied by a cluster of protons, and vice versa. But a north magnetic pole never exists without the presence of a south pole, and vice versa. The north and south poles of a magnet are like the head and tail of the same coin.

FIGURE 9.1

A horseshoe magnet.

fyi

◼ Interestingly, the north pole of a magnet points north because it's attracted to Earth's magnetic *south* pole! Earth's magnetic north pole is in Antarctica. Magnetic and geographic poles don't match.

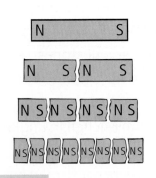

FIGURE 9.2

If you break a magnet in half, you have two magnets. Break these in half, and you have four magnets, each with a north and south pole. Continue breaking the pieces further and further and you find that you always get the same results. Magnetic poles exist in pairs.

---

* We can only speculate about how often such relationships become evident when they "aren't supposed to" and are dismissed as "something wrong with the apparatus." Oersted, however, was keen enough to see that nature was revealing another of its secrets.

** The force of interaction between magnetic poles is given by $F: \dfrac{p_1 p_2}{d^2}$, where p_1 and p_2 represent magnetic pole strengths and d represents the separation distance between the poles. Note the similarity of this relationship to Coulomb's law and Newton's law of universal gravitation.

If you break a bar magnet in half, each half still behaves as a complete magnet. Break the pieces in half again, and you have four complete magnets. You can continue breaking the pieces in half and never isolate a single pole. Even if your pieces were one atom thick, there would still be two poles on each piece, which suggests that the atoms themselves are magnets.

CHECK POINT

Does every magnet necessarily have a north and a south pole?

Was this your answer?
Yes, just as every coin has two sides, a "head" and a "tail." (Some "trick" magnets have more than two poles, but none has only one.)

9.2 Magnetic Fields

If you sprinkle some iron filings on a sheet of paper placed on a magnet, you'll see that the filings trace out an orderly pattern of lines that surround the magnet. The space around the magnet is energized by a **magnetic field**. The shape of the field is revealed by magnetic field lines that spread out from one pole and return to the other pole. It is interesting to compare the field patterns in Figures 9.3 and 9.5 with the electric field patterns in Figures 8.10 and 8.11 in the previous chapter.

The direction of the field outside the magnet is, by convention, from the north pole to the south pole. Where the lines are closer together, the field is stronger. We can see that the magnetic field strength is greater at the poles. If we place another magnet or a small compass anywhere in the field, its poles tend to align with the magnetic field.

A magnetic field is produced by the motion of electric charge.* Where, then, is this motion in a common bar magnet? The answer is, in the electrons of the atoms that make up the magnet. These electrons are in constant motion. Two kinds of electron motion produce magnetism: electron spin and electron revolution. A common science model views electrons as spinning about their own axes like tops, while they revolve about the nuclei of their atoms like planets revolving around the Sun. In most common magnets, electron spin is the main contributor to magnetism.

FIGURE 9.3

INTERACTIVE FIGURE

Top view of iron filings sprinkled on a sheet of paper on top of a magnet. The filings trace out a pattern of magnetic field lines in the surrounding space.

Torque   No torque

FIGURE 9.4

When the compass needle is not aligned with the magnetic field, the oppositely directed forces produce a pair of torques (called a couple) that twist the needle into alignment.

(a)   (b)

FIGURE 9.5

The magnetic field patterns for a pair of magnets. (a) Opposite poles are nearest to each other. (b) Like poles are nearest to each other.

---

* Interestingly, because motion is relative, the magnetic field is relative. For example, when an electron moves by you, a definite magnetic field is associated with the moving electron. But if you move along with the electron, so that there is no motion relative to you, you find no magnetic field associated with the electron. Magnetism is relativistic, as first explained by Albert Einstein when he published his first paper on special relativity, "On the Electrodynamics of Moving Bodies."

Link To Section 26.4

FIGURE 9.6

Fred Myers shows that the magnetic field of a ceramic magnet penetrates flesh and the plastic coating on a paper clip.

fyi

- Both the spinning motion and the orbital motion of every electron in an atom produce magnetic fields. These fields combine constructively or destructively to produce the magnetic field of the atom. The resulting field is greatest for iron atoms. (Electrons don't actually spin like a rotating planet, but behave as if it they were—the concept of spin is a quantum effect.)

The Physical Science Place
Magnetic Fields

FIGURE 9.7

A microscopic view of magnetic domains in a crystal of iron. Each domain consists of billions of aligned iron atoms. In this view, orientation of the domains is random.

Every spinning electron is a tiny magnet. A pair of electrons spinning in the same direction creates a stronger magnet. A pair of electrons spinning in opposite directions, however, work against each other. The magnetic fields cancel. This is why most substances are not magnets. In most atoms, the various fields cancel one another because the electrons spin in opposite directions. In such materials as iron, nickel, and cobalt, however, the fields do not cancel each other entirely. Each iron atom has four electrons whose spin magnetism is uncanceled. Each iron atom, then, is a tiny magnet. The same is true, to a lesser extent, of nickel and cobalt atoms. Most common magnets are therefore made from alloys containing iron, nickel, cobalt, and aluminum in various proportions.

Most of the iron objects around you are magnetized to some degree. A filing cabinet, a refrigerator, and even cans of food on your pantry shelf have north and south poles induced by Earth's magnetic field. If you pass a compass from their bottoms to their tops, you can easily identify their poles. (See Active Exploration 2 at the end of this chapter, where you are asked to turn cans upside down and note how many days go by for the poles to reverse themselves.)

9.3 Magnetic Domains

The magnetic field of an individual iron atom is so strong that interactions among adjacent atoms cause large clusters of them to line up with one another. These clusters of aligned atoms are called **magnetic domains**. Each domain is perfectly magnetized and is made up of billions of aligned atoms. The domains are microscopic (Figure 9.7), and there are many of them in a crystal of iron.

Not every piece of iron is a magnet, because the domains in ordinary iron are not aligned. In a common iron nail, for example, the domains are randomly oriented. But when you bring a magnet nearby, they can be induced into alignment. (It is interesting to listen, with an amplified stethoscope, to the clickety-clack of domains aligning in a piece of iron when a strong magnet approaches.) The domains align themselves much as electrical charges in a piece of paper align themselves (become polarized) in the presence of a charged rod. When you

Unmagnetized iron

Slightly magnetized iron

Strongly magnetized iron

When a magnet is broken into two pieces, each piece is an equally strong magnet

FIGURE 9.8

INTERACTIVE FIGURE

Pieces of iron in successive stages of magnetism. The arrows represent domains; the head is a north pole and the tail is a south pole. Poles of neighboring domains neutralize each other's effects, except at the ends.

Link To Section 22.4

remove the nail from the magnet, ordinary thermal motion causes most or all of the domains in the nail to return to a random arrangement.

Permanent magnets can be made by placing pieces of iron or similar magnetic materials in a strong magnetic field. Alloys of iron differ; soft iron is easier to magnetize than steel. It helps to tap the material to nudge any stubborn domains into alignment. Another way is to stroke the material with a magnet. The stroking motion aligns the domains. If a permanent magnet is dropped or heated outside the strong magnetic field from which it was made, some of the domains are jostled out of alignment and the magnet becomes weaker.

fyi

■ A magnetic stripe on a credit card contains millions of tiny magnetic domains held together by a resin binder. Data is encoded in binary code, with zeros and ones distinguished by the frequency of domain reversals.

CHECK POINT

■
1. Why doesn't a magnet pick up a penny or a piece of wood?
2. How can a magnet attract a piece of iron that is not magnetized?

Were these your answers?
1. A penny and a piece of wood have no magnetic domains that can be induced into alignment.
2. Like the compass needle in Figure 9.4, domains in the unmagnetized piece of iron are induced into alignment by the magnetic field of the magnet. One domain pole is attracted to the magnet and the other domain pole is repelled. Does this mean the net force is zero? No, because the force is slightly greater on the domain pole closest to the magnet than it is on the farther pole. That's why there is a net attraction. In this way, a magnet attracts unmagnetized pieces of iron (Figure 9.9).

FIGURE 9.9

Wai Tsan Lee shows iron nails becoming induced magnets.

■ 9.4 Electric Currents and Magnetic Fields

A moving charge produces a magnetic field. A current of charges, then, also produces a magnetic field. The magnetic field that surrounds a current-carrying wire can be demonstrated by arranging an assortment of compasses around the wire (Figure 9.10). The magnetic field about the current-carrying

Link To Section 22.3

Magnetic
compasses

Electric
current

FIGURE 9.10

The compasses show the circular shape of the magnetic field surrounding the current-carrying wire.

FIGURE 9.11

Magnetic field lines about a current-carrying wire become bunched up when the wire is bent into a loop.

wire makes up a pattern of concentric circles. When the current reverses direction, the compass needles turn around, showing that the direction of the magnetic field changes also.*

If the wire is bent into a loop, the magnetic field lines become bunched up inside the loop (Figure 9.11). If the wire is bent into another loop that overlaps the first, the concentration of magnetic field lines inside the loops is doubled. It follows that the magnetic field intensity in this region is increased as the number of loops is increased. The magnetic field intensity is appreciable for a current-carrying coil that has many loops.

ELECTROMAGNETS

If a piece of iron is placed in a current-carrying coil of wire, the alignment of magnetic domains in the iron produces a particularly strong magnet known as an **electromagnet**. The strength of an electromagnet can be increased simply by increasing the current through the coil. Strong electromagnets are used to control charged-particle beams in high-energy accelerators. They also levitate and propel prototypes of high-speed trains (Figure 9.13).

Electromagnets powerful enough to lift automobiles are a common sight in junkyards. The strength of these electromagnets is limited mainly by overheating

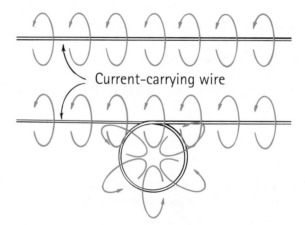

Current-carrying wire

FIGURE 9.12

Iron filings sprinkled on paper reveal the magnetic field configurations about (a) a current-carrying wire, (b) a current-carrying loop, and (c) a coil of loops.

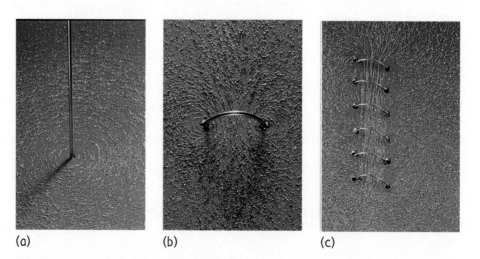

(a)   (b)   (c)

* Earth's magnetism is generally accepted as being the result of electric currents that accompany thermal convection in the molten parts of Earth's interior. Earth scientists have found evidence that Earth's poles periodically reverse places—more than 20 reversals have occurred in the past 5 million years. This is perhaps the result of changes in the direction of electric currents within Earth.

of the current-carrying coils. The most powerful electromagnets omit the iron core and employ superconducting coils through which large electrical currents flow with ease.

SUPERCONDUCTING ELECTROMAGNETS

Ceramic superconductors (Chapter 8) have the interesting property of expelling magnetic fields. Because magnetic fields cannot penetrate the surface of a superconductor, magnets levitate above them. The reasons for this behavior, which are beyond the scope of this book, involve quantum mechanics. One of the more exciting applications of super-conducting electromagnets is the levitation of high-speed trains for transportation. Prototype trains have already been demonstrated in the United States, Japan, and Germany. Watch for the growth of this relatively new technology.

■ 9.5 Magnetic Forces on Moving Charges

charged particle at rest does not interact with a static magnetic field. However, if the charged particle moves in a magnetic field, the magnetic character of a charge in motion becomes evident: The charged particle experiences a deflecting force.* The force is greatest when the particle moves in a direction perpendicular to the magnetic field lines. At other angles, the force is less, and it becomes zero when the particle moves parallel to the field lines. In any case, the direction of the force is always perpendicular to the magnetic field lines and the velocity of the charged particle (Figure 9.15). So a moving charge is deflected when it crosses through a magnetic field, but when it travels parallel to the field, no deflection occurs.

This deflecting force is very different from the forces that occur in other interactions, such as the gravitational forces between masses, the electric forces between charges, and the magnetic forces between magnetic poles. The force that acts on a moving charged particle, such as an electron in an electron beam, does not act along the line that joins the sources of interaction. Instead, it acts perpendicularly both to the magnetic field and to the electron beam.

We are fortunate that charged particles are deflected by magnetic fields. This fact was employed in guiding electrons onto the inner surface of early television tubes to produce pictures. Also, charged particles from outer space are deflected by Earth's magnetic field. Otherwise the harmful cosmic rays bombarding Earth's surface would be much more intense.

FIGURE 9.13

A magnetically levitated train—a *magplane*. Whereas conventional trains vibrate as they ride on rails at high speeds, magplanes can travel vibration-free at high speeds because they make no physical contact with the guideway they float above.

■■ The Physical Science Place ↘

Magnetic Forces on
    Current-Carrying
    Wires

FIGURE 9.14

A permanent magnet levitates above a superconductor because its magnetic field cannot penetrate the supercon-ducting material.

FIGURE 9.15

A beam of electrons is deflected by a magnetic field.

* When particles of electric charge q and velocity v move perpendicularly into a magnetic field of strength B, the force F on each particle is simply the product of the three variables: $F = qvB$. For non-perpendicular angles, v in this relationship must be the component of velocity perpendicular to B.

FIGURE 9.16

Earth's magnetic field deflects the many charged particles that make up cosmic radiation.

Link To Section 24.4

MAGNETIC FORCE ON CURRENT-CARRYING WIRES

Simple logic tells you that if a charged particle moving through a magnetic field experiences a deflecting force, then a current of charged particles moving through a magnetic field also experiences a deflecting force. If the particles are deflected while moving inside a wire, the wire is also deflected (Figure 9.17).

If we reverse the direction of current, the deflecting force acts in the opposite direction. The force is strongest when the current is perpendicular to the magnetic field lines. The direction of force is not along the magnetic field lines nor along the direction of current. The force is perpendicular to both field lines and current. It is a sideways force—perpendicular to the wire.

We see that, just as a current-carrying wire deflects a magnet such as a compass needle (as discovered by Oersted in a physics classroom in 1820), a magnet deflects a current-carrying wire. When discovered, these complementary links between electricity and magnetism created much excitement. Almost immediately, people began harnessing the electromagnetic force for useful purposes—with great sensitivity in electric meters and with great force in electric motors.

FIGURE 9.17

INTERACTIVE FIGURE

A current-carrying wire experiences a force in a magnetic field. (Can you see that this is a simple extension of Figure 9.15?)

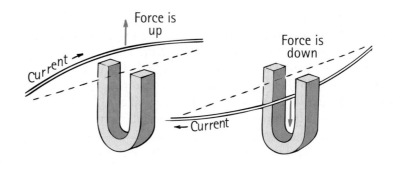

fyi

- In an advanced course, you'll learn the "simple" right-hand rule.

CHECK POINT

What law of physics tells you that if a current-carrying wire produces a force on a magnet, a magnet must produce a force on a current-carrying wire?

Was this your answer?
Newton's third law, which applies to all forces in nature.

ELECTRIC METERS

The simplest meter to detect electric current is a magnetic compass. The next simplest meter is a compass in a coil of wires (Figure 9.18). When an electric current passes through the coil, each loop produces its own effect on the needle, so even a very small current can be detected. Such a current-indicating instrument is called a *galvanometer*.

FIGURE 9.18

A very simple galvanometer.

FIGURE 9.19

Both the ammeter and the voltmeter are basically galvanometers. (The electrical resistance of the instrument is designed to be very low for the ammeter and very high for the voltmeter.)

A more common design is shown in Figure 9.20. It employs more loops of wire and is therefore more sensitive. The coil is mounted for movement, and the magnet is held stationary. The coil turns against a spring, so the greater the current in its windings, the greater its deflection. A galvanometer may be calibrated to measure current (amperes), in which case it is called an *ammeter*. Or it may be calibrated to measure electric potential (volts), in which case it is called a *voltmeter*.*

ELECTRIC MOTORS

If we change the design of the galvanometer slightly so that deflection makes a complete turn rather than a partial rotation, we have an *electric motor*. The principal difference is that the current in a motor is made to change direction each time the coil makes a half rotation. This happens in a cyclic fashion to produce continuous rotation, which has been used to run clocks, operate gadgets, and lift heavy loads.

In Figure 9.21 we see the principle of the electric motor in bare outline. A permanent magnet produces a magnetic field in a region where a rectangular loop of wire is mounted to turn about the axis shown by the dashed line. When a current passes through the loop, it flows in opposite directions in the upper and lower sides of the loop. (It must do this because if charge flows into one end of the loop, it must flow out the other end.) If the upper portion of the loop is forced to the left, then the lower portion is forced to the right, as if it were a galvanometer. But, unlike a galvanometer, the current is reversed during each half revolution by means of stationary contacts on the shaft. The parts of the wire that brush against these contacts are called *brushes*. In this way, the current in the loop alternates so that the forces in the upper and lower regions do not change directions as the loop rotates. The rotation is continuous as long as current is supplied.

We have described here only a very simple dc motor. Larger motors, dc or ac, are usually manufactured by replacing the permanent magnet by an electromagnet that is energized by the power source. Of course, more than a single loop is used. Many loops of wire are wound about an iron cylinder, called an *armature*, which then rotates when the wire carries current.

The advent of electric motors brought to an end much human and animal toil in many parts of the world. Electric motors have greatly changed the way people live.

FIGURE 9.20

A common galvanometer design.

fyi

- The galvanometer is named after Luigi Galvani (1737–1798), who, while dissecting a frog's leg, discovered that dissimilar metals touching the leg caused it to twitch. This chance discovery led to the invention of the chemical cell and the battery. The next time you pick up a galvanized pail, think of Luigi Galvani in his anatomy laboratory.

FIGURE 9.21

INTERACTIVE FIGURE

A simplified motor.

---

* To some degree, measuring instruments change what is being measured—ammeters and voltmeters included. Because an ammeter is connected in series with the circuit it measures, its resistance is made very low. That way, it doesn't appreciably lower the current it measures. Because a voltmeter is connected in parallel, its resistance is made very high, so that it draws very little current for its operation. In the lab part of your course you'll likely learn how to connect these instruments in simple circuits.

MRI: Magnetic Resonance Imaging

agnetic resonance imaging scanners provide high-resolution pictures of the tissues inside a body. Superconducting coils produce a strong magnetic field (up to 60,000 times as strong as the intensity of Earth's magnetic field) that is used to align the protons of hydrogen atoms in the body of the patient.

Like electrons, protons have a "spin" property, so they align with a magnetic field. Unlike a compass needle that aligns with Earth's magnetic field, the proton's axis wobbles about the applied magnetic field. Wobbling protons are slammed with a burst of radio waves tuned to push the proton's spin axis sideways, perpendicular to the applied magnetic field. When the radio waves pass and the protons quickly return to their wobbling pattern, they emit faint electromagnetic signals whose frequencies depend slightly on the chemical environment in which the proton resides. The signals, which are detected by sensors, are then analyzed by a computer to reveal varying densities of hydrogen atoms in the body and their interactions with surrounding tissue. The images clearly distinguish between fluid and bone, for example.

MRI was formerly called NMRI (nuclear magnetic resonance imaging), because hydrogen nuclei resonate with the applied fields. Because of public phobia about anything "nuclear," this diagnostic technique is now called MRI. (Tell your friends that every atom in their bodies contains a nucleus!)

CHECK POINT

What is the major similarity between a galvanometer and a simple electric motor? What is the major difference?

Was this your answer?

A galvanometer and a motor are similar in that they both use coils positioned in a magnetic field. When a current passes through the coils, forces on the wires rotate the coils. The major difference is that the maximum coil rotation in a galvanometer is half a turn, whereas the coil in a motor (which is wrapped on an armature) rotates through many complete turns. This is accomplished by alternating the direction of the current with each half turn of the armature.

Applications of Induction

Note that a magnetic field does not induce voltage: a *change* in the field over some *time interval* does. If the field changes in a closed loop, and the loop is an electrical conductor, then both voltage and current are induced.

■ 9.6 Electromagnetic Induction

In the early 1800s, the only current-producing devices were voltaic cells, which produced small currents by dissolving metals in acids. These were the forerunners of modern batteries. The question arose as to whether electricity could be produced from magnetism. The answer was provided in 1831 by two physicists, Michael Faraday in England and Joseph Henry in the United States—each working without knowledge of the other. Their discovery changed the world by making electricity commonplace—powering industries by day and lighting up cities at night.

FIGURE 9.22

When the magnet is plunged into the coil, charges in the coil are set in motion, and voltage is induced in the coil.

Faraday and Henry both discovered **electromagnetic induction**—that electric current could be produced in a wire simply by moving a magnet into or out of a coil of wire (Figure 9.22). No battery or other voltage source was needed—only the motion of a magnet in a wire loop. They discovered that voltage is caused, or induced, by the relative motion between a wire and a magnetic field. Whether the magnetic field moves near a stationary conductor or vice versa, voltage is induced either way (Figure 9.23).

FIGURE 9.23

Voltage is induced in the wire loop whether the magnetic field moves past the wire or the wire moves through the magnetic field.

FIGURE 9.24

INTERACTIVE FIGURE

When a magnet is plunged into a coil with twice as many loops as another, twice as much voltage is induced. If the magnet is plunged into a coil with three times as many loops, three times as much voltage is induced.

The greater the number of loops of wire moving in a magnetic field, the greater the induced voltage (Figure 9.24). Pushing a magnet into a coil with twice as many loops induces twice as much voltage; pushing into a coil with 10 times as many loops induces 10 times as much voltage; and so on. It may seem that we get something (energy) for nothing simply by increasing the number of loops in a coil of wire, but we don't. We find that it is more difficult to push the magnet into a coil made up of more loops. This is because the induced voltage produces a current, which makes an electromagnet, which repels the magnet in our hand. So we must do more work against this "back force" to induce more voltage (Figure 9.25).

The amount of voltage induced depends on how fast the magnetic field lines are entering or leaving the coil. Very slow motion produces hardly any voltage at all. Rapid motion induces a greater voltage.

fyi

- A long helically wound coil of insulated wire is called a *solenoid*.

FARADAY'S LAW

Electromagnetic induction is summarized by **Faraday's law:**

> **The induced voltage in a coil is proportional to the number of loops, multiplied by the rate at which the magnetic field changes within those loops.**

The amount of *current* produced by electromagnetic induction depends on the resistance of the coil and the circuit that it connects, as well as the induced voltage.* For example, we can plunge a magnet into and out of a closed rubber loop and into and out of a closed loop of copper. The voltage induced in each is the same, providing the loops are the same size and the magnet moves with the same speed. But the current in each is quite different. The electrons in the rubber sense the same voltage as those in the copper, but their bonding to the fixed atoms prevents the movement of charge that so freely occurs in the copper.

The Physical Science Place
Faraday's Law

FIGURE 9.25

It is more difficult to push the magnet into a coil with many loops because the magnetic field of each current loop resists the motion of the magnet.

* Current also depends on the *inductance* of the coil. Inductance measures the tendency of a coil to resist a change in current because the magnetism produced by one part of the coil opposes the change of current in other parts of the coil. In ac circuits it is comparable to resistance in dc circuits. To reduce "information overload" we will not treat inductance in this book.

FIGURE 9.26

Guitar pickups are tiny coils with magnets inside them. The magnets magnetize the steel strings. When the strings vibrate, voltage is induced in the coils and boosted by an amplifier, and sound is produced by a speaker.

FIGURE 9.27

When Jean Curtis powers the large coil with ac, an alternating magnetic field is established in the iron bar and thence through the metal ring. Current is therefore induced in the ring, which then establishes its own magnetic field, which always acts in a direction to oppose the field producing it. The result is mutual repulsion—levitation.

CHECK POINT

If you push a magnet into a coil, as shown in Figure 9.25, you'll feel a resistance to your push. Why is this resistance greater in a coil with more loops?

Was this your answer?

Simply put, more work is required to provide more energy. You can also look at it this way: When you push a magnet into a coil, you induce electric current and cause the coil to become an electromagnet. The more loops in the coil, the stronger the electromagnet that you produce and the stronger it pushes back against you. (If the electromagnetic coil attracted your magnet instead of repelling it, energy would have been created from nothing and the law of energy conservation would have been violated. So the coil must repel the magnet.)

We have mentioned two ways in which voltage can be induced in a loop of wire: by moving the loop near a magnet, or by moving a magnet near the loop. There is a third way—by changing a current in a nearby loop. All three of these cases possess the same essential ingredient—a changing magnetic field in the loop.

We see electromagnetic induction all around us. On the road, we see it operate when a car drives over buried coils of wire to activate a nearby traffic light. When iron parts of a car move over the buried coils, the effect of Earth's magnetic field on the coils is changed, inducing a voltage to trigger the changing of the traffic lights. Similarly, when you walk through the upright coils in the security system at an airport, any metal you carry slightly alters the magnetic field in the coils. This change induces voltage, which sounds an alarm. When the mag-

netic strip on the back of a credit card is scanned, induced voltage pulses identify the card. Something similar occurs in the recording head of a tape recorder: magnetic domains in the tape are sensed as the tape moves past a current-carrying coil. Electromagnetic induction is at work in computer hard drives, iPods, and devices galore. As we soon see, it underlies the electromagnetic waves that we call light.

9.7 Generators and Alternating Current

When a magnet is repeatedly plunged into and back out of a coil of wire, the direction of the induced voltage alternates. As the magnetic field strength inside the coil is increased (as the magnet enters), the induced voltage in the coil is directed one way. When the magnetic field strength diminishes (as the magnet leaves), the voltage is induced in the opposite direction. The frequency of the alternating voltage that is induced is equal to the frequency of the changing magnetic field within the loop.

Rather than moving the magnet, it is more practical to move the coil. This is best accomplished by rotating the coil in a stationary magnetic field (Figure 9.28). This arrangement is called a **generator**. It is essentially the opposite of a motor. Whereas a motor converts electrical energy into mechanical energy, a generator converts mechanical energy into electrical energy.

FIGURE 9.28

INTERACTIVE FIGURE

A simple generator. Voltage is induced in the loop when it is rotated in the magnetic field.

Because the voltage induced by the generator alternates, the current produced is ac, an alternating current.* The alternating current in our homes is produced by generators standardized so that the current goes through 60 full cycles of change in magnitude and direction each second—60 hertz.

fyi

- A motor and a generator are actually the same device, with input and output reversed.

FIGURE 9.29

As the loop rotates, the magnitude and direction of the induced voltage (and current) changes. One complete rotation of the loop produces one complete cycle in voltage (and current).

■ 9.8 Power Production

Fifty years after Faraday and Henry discovered electromagnetic induction, Nikola Tesla and George Westinghouse put those findings to practical use and showed the world that electricity could be generated reliably and in sufficient quantities to light entire cities.

Tesla built generators that were much like those still in use, but quite a bit more complicated than the simple model we have discussed. Tesla's generators had armatures consisting of bundles of copper wires that were made to spin within strong magnetic fields by means of a turbine, which, in turn, was spun by the energy of steam or falling water. The rotating loops of wire in the armature cut through the magnetic field of the surrounding electromagnets, thereby inducing alternating voltage and current.

We can look at this process from an atomic point of view. When the wires in the spinning armature cut through the magnetic field, oppositely directed electromagnetic forces act on the negative and positive charges. Electrons respond to this force by

Two hundred years ago, people got light from whale oil. Whales should be glad that humans discovered electricity!

FIGURE 9.30

Steam drives the turbine, which is connected to the armature of the generator.

---

* By means such as appropriately designed *brushes* (contacts that brush against the rotating *armature*), the ac in the loop(s) can be converted to dc to make a dc generator.

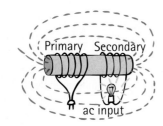

FIGURE 9.31

A simple transformer.

momentarily swarming relatively freely in one direction throughout the crystalline copper lattice; the copper atoms, which are actually positive ions, are forced in the opposite direction. But the ions are anchored in the lattice, so they barely move at all. Only the electrons move significantly, sloshing back and forth in alternating fashion with each rotation of the armature. The energy produced by this electronic sloshing is tapped at the electrode terminals of the generator.

It's important to know that generators don't produce energy—they simply convert energy from some other form to electric energy. As we discussed in Chapter 3, energy from a source, whether fossil or nuclear fuel or wind or water, is converted to mechanical energy to drive the turbine. The attached generator converts most of this mechanical energy to electrical energy. Some people think that electricity is a primary source of energy. It is not. It is a carrier of energy that requires a source.

FIGURE 9.32

A practical transformer. Both primary and secondary coils are wrapped on the inner part of the iron core (yellow), which guides alternating magnetic field lines (green) produced by ac in the primary. The alternating field induces ac voltage in the secondary. Thus power at one voltage from the primary is transferred to the secondary at a different voltage.

■ 9.9 The Transformer—Boosting or Lowering Voltage

FIGURE 9.33

This common transformer lowers 120 V to 6 V or 9 V. It also converts ac to dc by means of a *diode* inside—a tiny electronic device that acts as a one-way valve.

When changes in the magnetic field of a current-carrying coil of wire are intercepted by a second coil of wire, voltage is induced in the second coil. This is the principle of the **transformer**—a simple electromagnetic-induction device consisting of an input coil of wire (the primary) and an output coil of wire (the secondary). The coils need not physically touch each other, but they are normally wound on a common iron core so that the magnetic field of the primary passes through the secondary. The primary is powered by an ac voltage source, and the secondary is connected to some external circuit. Changes in the primary current produce changes in its magnetic field. These changes extend to the secondary, and, by electromagnetic induction, voltage is induced in the secondary. If the number of turns of wire in both coils is the same, voltage input and voltage output are the same. Nothing is gained. But if the secondary has more turns than the primary, then greater voltage is induced in the secondary. This is a *step-up transformer*. If the secondary has fewer turns than the primary, the ac voltage induced in the secondary is lower than that in the primary. This is a *step-down transformer*.

The relationship between primary and secondary voltages relative to the number of turns is as follows:

$$\frac{\text{Primary voltage}}{\text{Number of primary turns}} = \frac{\text{Secondary voltage}}{\text{Number of secondary turns}}$$

It might seem that we get something for nothing with a transformer that steps up the voltage, but we don't. When voltage is stepped up, current in the secondary is less than in the primary. The transformer actually transfers energy from one coil to the other. The rate of transferring energy is *power*. The power used in the secondary is supplied by the primary. The primary gives no more than the secondary uses, in accord with the law of energy conservation. If any slight power losses due to heating of the core can be neglected, then

$$\text{Power into primary} = \text{power out of secondary}$$

Electric power is equal to the product of voltage and current, so we can say that

$$(\text{Voltage} \times \text{current})_{\text{primary}} = (\text{voltage} \times \text{current})_{\text{secondary}}$$

The ease with which voltages can be stepped up or down with a transformer is the principal reason that most electric power is ac rather than dc.

FIGURE 9.34

A common neighborhood transformer, which typically steps 2400 V down to 240 V for houses and small businesses. Inside the home or business, the 240 V can divide to a safer 120 V.

9.10 Field Induction

Electromagnetic induction explains the induction of voltages and currents. Actually, the more basic concept of *fields* is at the root of both voltages and currents. The modern view of electromagnetic induction states that electric and magnetic fields are induced. These, in turn, produce the voltages we have considered. So induction occurs whether or not a conducting wire or any material medium is present. In this more general sense, Faraday's law states:

An electric field is induced in any region of space in which a magnetic field is changing with time.

There is a second effect, an extension of Faraday's law. It is the same except that the roles of electric and magnetic fields are interchanged. It is one of nature's many symmetries. This effect, which was advanced by the British physicist James Clerk Maxwell in about 1860, is known as **Maxwell's counterpart to Faraday's law:**

A magnetic field is induced in any region of space in which an electric field is changing with time.

In each case, the strength of the induced field is proportional to the rates of change of the inducing field. The induced electric and magnetic fields are at right angles to each other.

fyi

■ Enormous intergalactic magnetic fields that spread far beyond the galaxies have been recently detected. These giant magnetic fields make up an important part of the cosmic energy store and play a significant role in shaping the evolution of galaxies and large-scale grouping of galaxies.

FIGURE 9.35

Voltage generated in power stations is stepped up with transformers before being transferred across country by overhead cables. Then other transformers reduce the voltage before supplying it to homes, offices, and factories.

Nanotechnology

The age of microtechnology was ushered in some 60 years ago with the invention of the solid state transistor. Engineers were quick to grasp the idea of integrating many transistors together to create logic boards that could perform calculations and run programs. The more transistors they could squeeze into a circuit, the more powerful the logic board. The race thus began to squeeze more and more transistors together into tinier and tinier circuits. The scales achieved were in the realm of the micron (10^{-6} meters), thus the term *micro*technology. Few at the time of the transistor's invention realized the impact that microtechnology would have on society—from personal computers, to cell phones, to the Internet.

Today, we are at the beginings of a similar revolution. Technological advances have recently brought us past the realm of microns to the realm of the nanometer (10^{-9} meters), which is the realm of individual atoms and molecules—a realm where we have reached the basic building blocks of matter. Technology that works on this scale is called *nano*technology. No one knows exactly how nanotechnology will impact society, but people are quickly coming to realize its vast potential, which is likely much greater than that of microtechnology.

Nanotechnology generally concerns the manipulations of objects from 1 to 100 nanometers in scale. For perspective, a DNA molecule is about 2.0 nm wide, while a water molecule is only about 0.2 nm. Like microtechnology, nanotechnology is interdisciplinary, requiring the cooperative efforts of chemists, engineers, physicists, molecular biologists, and many others. Interestingly, there are already many products on the market that contain components developed through nanotechnology. These include sun-

screens, mirrors that don't fog, dental bonding agents, automotive catalytic converters, stain-free clothing, water filtration systems, the heads to computer hard drives, and much more. Nanotechnology, however, is still in its infancy, and it will likely be decades before its potential is fully realized. Recall, for example, that personal computers didn't blossom until the 1990's, some 40 years after the first solid state transistor. Most experts agree that the first big benefits will arise in computer science and medicine.

Computer circuits are currently made by projecting the image of a circuit design onto a photosensitive material which captures the image of the circuit much like photographic paper captures a photo image. Upon chemical treatments, the captured image of the circuit can be developed into an actual circuit. The circuit structures can be shrunk to about 500 nm via size-reducing lenses. Smaller structures are not practical because the scales are smaller than the wavelength of the light itself. Even the light waves are too big!

The alternative is nanotechnology, where circuits can be built atom by atom. Pioneering tools that allow this to happen are the scanning probe microscopes discussed in Chapter 12, which are not only able to produce images of individual atoms, but allow the operator to move individual atoms into desired positions.

A whole new approach involves the design of logic boards in which molecules (not electric circuits) read, process, and write information. The most promising molecule for such *molecular computation* is DNA, the one that holds our genetic code. Molecular computing can run a massive number of calculations in parallel (at the same time) and may one day outshine even the fastest of integrated circuits. Molecular computing, in

An artist's depiction of the ultrafine needle of the scanning probe microscope detecting the arrangement of individual atoms.

turn, may then soon be eclipsed by other novel approaches, such as quantum or photon computing, also made possible by nanotechnology.

The ultimate expert on nanotechnology is nature. Living organisms, for example, are complex systems of interacting biomolecules all functioning within the realm of nanometers. In this sense, the living organism is nature's nanomachine. We need look no further than our own bodies to find evidence of the feasibility and power of nanotechnology. With nature as our teacher, we have much to learn. With such knowledge, we would be well equipped to understand exact causes of nearly any disease or disorder (aging included) and empowered to develop innovative cures.

Watch for the advent of some of the fruits of nanotechnology: wall paint that can change color or be used to display video. Solar cells that capture sunlight so efficiently that they render fossil fuels obsolete. Robots with so much processing power that we confuse them with conscious life forms. Nanomachines that can "photocopy" three-dimensional objects, including living organisms. Medicines that more than double the average human life span. Nanotechnology truly opens a new frontier.

FIGURE 9.36

In turning the crank of the generator, Sheron Snyder does work, which is transformed to voltage and current, which, in turn, is transformed into light.

Maxwell saw the link between electromagnetic waves and light. If electric charges are set into vibration in the range of frequencies that match those of light, waves are produced that are light! Maxwell discovered that light is simply electromagnetic waves in the range of frequencies to which the eye is sensitive.

On the eve of his discovery, Maxwell had a date with the young woman he was later to marry. Story has it that while they were walking in a garden, she remarked about the beauty and wonder of the stars. Maxwell asked her how she would feel if she knew that she was walking with the only person in the world who knew what starlight really was. In fact, at that time, James Clerk Maxwell was the only person in the entire world to know that light of any kind is energy carried in waves of electric and magnetic fields that continually regenerate each other.

The laws of electromagnetic induction were discovered at about the time the American Civil War was being fought. From a long view of human history, there can be little doubt that events such as the American Civil War will pale into provincial insignificance in comparison with the more significant event of the 19th century: the discovery of the electromagnetic laws.

> Each of us needs a knowledge filter to tell us the difference between what is true and what only pretends to be true. The best knowledge filter ever invented is science.

SUMMARY OF TERMS

Magnetic force (1) Between magnets, it is the attraction of unlike magnetic poles for each other and the repulsion between like magnetic poles. (2) Between a magnetic field and a moving charge, it is a deflecting force due to the motion of the charge: the deflecting force is perpendicular to the velocity of the charge and perpendicular to the magnetic field lines. This force is greatest when the charge moves perpendicular to the field lines and is smallest (zero) when it moves parallel to the field lines.

Magnetic field The region of magnetic influence around a magnetic pole or a moving charged particle.

Magnetic domains Clustered regions of aligned magnetic atoms. When these regions themselves are aligned with one another, the substance containing them is a magnet.

Electromagnet A magnet whose field is produced by an electric current. It is usually in the form of a wire coil with a piece of iron inside the coil.

Electromagnetic induction The induction of voltage when a magnetic field changes with time. If the magnetic field within a closed loop changes in any way, a voltage is induced in the loop:

$$\text{Voltage induced} \sim \text{number of loops} \times \frac{\text{magnetic field change}}{\text{time}}$$

Faraday's law The law of electromagnetic induction, in which the induced voltage in a coil is proportional to the number of loops multiplied by the rate at which the magnetic field changes within those loops. (The induction of voltage is actually the result of a more fundamental phenomenon: the induction of an electric field.)

Generator An electromagnetic induction device that produces electric current by rotating a coil within a stationary magnetic field.

Transformer A device for transferring electric power from one coil of wire to another by means of electromagnetic induction.

Maxwell's counterpart to Faraday's law A magnetic field is induced in any region of space in which an electric field is changing with time. Correspondingly, an electric field is induced in any region of space in which a magnetic field is changing with time.

REVIEW QUESTIONS

1. By whom, and in what setting, was the relationship between electricity and magnetism discovered?

9.1 Magnetic Poles

2. In what way is the rule for the interaction between magnetic poles similar to the rule for the interaction between electric charges?
3. In what way are magnetic poles very different from electric charges?

9.2 Magnetic Fields

4. What produces a magnetic field?
5. What two kinds of motion are exhibited by electrons in an atom?

9.3 Magnetic Domains

6. What is a magnetic domain?
7. Why is iron magnetic and wood not magnetic?

9.4 Electric Currents and Magnetic Fields

8. What is the shape of a magnetic field about a current-carrying wire?
9. What happens to the direction of the magnetic field about an electric current when the direction of the current is reversed?
10. Why is the magnetic field strength inside a current-carrying loop of wire greater than the field strength about a straight section of wire?
11. How is the strength of a magnetic field in a coil affected when a piece of iron is placed inside? Defend your answer.

9.5 Magnetic Forces on Moving Charges

12. In what direction relative to a magnetic field does a charged particle move in order to experience maximum deflecting force? Minimum deflecting force?
13. What effect does Earth's magnetic field have on the intensity of cosmic rays striking Earth's surface?
14. What relative direction between a magnetic field and a current-carrying wire results in the greatest force on the wire? In the smallest force?

15. What happens to the direction of the magnetic force on a wire in a magnetic field when the current in the wire is reversed?
16. What is a galvanometer called when it is calibrated to read current? To read voltage?
17. Is it correct to say that an electric motor is a simple extension of the physics that underlies a galvanometer?

9.6 Electromagnetic Induction

18. What important discovery did physicists Michael Faraday and Joseph Henry make?
19. State Faraday's law.
20. What are the three ways in which voltage can be induced in a wire?

9.7 Generators and Alternating Current

21. How does the frequency of induced voltage compare with how frequently a magnet is plunged into and out of a coil of wire?
22. What are the basic differences and similarities between a generator and an electric motor?
23. Why does the voltage induced in a generator alternate?

9.8 Power Production

24. What commonly supplies the energy input to a turbine?
25. Is it correct to say that a generator produces electric energy? Defend your answer.

9.9 The Transformer—Boosting or Lowering Voltage

26. Is it correct to say that a transformer boosts electric energy? Defend your answer.
27. Which of these does a transformer change: voltage, current, energy, power?

9.10 Field Induction

28. What is induced by the rapid alternation of a magnetic field?
29. What is induced by the rapid alternation of an electric field?
30. What important connection did Maxwell discover about electric and magnetic fields?

EXERCISES

● BEGINNER ■ INTERMEDIATE ◆ EXPERT

1. ● If every iron atom is a tiny magnet, why aren't all iron materials themselves magnets?
2. ● If you place a chunk of iron near the north pole of a magnet, attraction occurs. Why does attraction also occur if you place the iron near the south pole of the magnet?
3. ● What is different about the magnetic poles of common refrigerator magnets compared with those of common bar magnets?

4. ● What surrounds a stationary electric charge? A moving electric charge?
5. ■ "An electron always experiences a force in an electric field, but not always in a magnetic field." Defend this statement.
6. ● Why does a magnet attract an ordinary nail or paper clip but not a wooden pencil?

7. ⬤ A friend tells you that a refrigerator door, beneath its layer of white-painted plastic, is made of aluminum. How could you check to see if this is true (without any scraping)?

8. ◼ One way to make a compass is to stick a magnetized needle into a piece of cork and float it in a glass bowl full of water, as shown. The needle aligns itself with the horizontal component of Earth's magnetic field. As the north pole of this compass is attracted northward, does the needle float toward the north side of the bowl? Defend your answer.

9. ⬤ What is the net magnetic force on a compass needle? By what mechanism does a compass needle line up with a magnetic field?

10. ⬤ Cans of food in your kitchen pantry are likely magnetized. Why?

11. ⬤ We know that a compass points northward because Earth is a giant magnet. Does the northward-pointing needle point northward when the compass is brought to the Southern Hemisphere?

12. ⬤ When a current-carrying wire is placed in a strong magnetic field, no force acts on the wire. What orientation of the wire is likely?

13. ⬤ Magnet A has twice the magnetic field strength of magnet B, and at a certain distance it pulls on magnet B with a force of 50 N. With how much force, then, does magnet B pull on magnet A?

14. ⬤ In Figure 9.17, we see a magnet exerting a force on a current-carrying wire. Does a current-carrying wire exert a force on a magnet? Why or why not?

15. ⬤ A strong magnet attracts a paper clip to itself with a certain force. Does the paper clip exert a force on the strong magnet? If not, why not? If so, does it exert as much force on the magnet as the magnet exerts on it? Defend your answers.

16. ◼ When steel naval ships are built, the location of the shipyard and the orientation of the ship while in the shipyard are recorded on a brass plaque permanently fixed to the ship. Why?

17. ⬤ Can an electron at rest in a magnetic field be set into motion by the magnetic field? What if it were at rest in an electric field?

18. ⬤ A cyclotron is a device for accelerating charged particles to high speeds as they follow an expanding spiral path. The charged particles are subjected to both an electric field and a magnetic field. One of these fields increases the speed of the charged particles, and the other field causes them to follow a curved path. Which field performs which function?

19. ⬤ A beam of high-energy protons emerges from a cyclotron. Do you suppose a magnetic field is associated with these particles? Why or why not?

20. ◼ A magnetic field can deflect a beam of electrons, but it cannot do work on the electrons to change their speed. Why?

21. ◼ Two charged particles are projected into a magnetic field that is perpendicular to their velocities. If the charges are deflected in opposite directions, what does this tell you about the particles?

22. ⬤ Residents of northern Canada are bombarded by more intense cosmic radiation than are residents of Mexico. Why is this so?

23. ⬤ What changes in cosmic-ray intensity at Earth's surface would you expect during periods in which Earth's magnetic field is passing through a zero phase while undergoing pole reversals?

24. ◼ In a mass spectrometer, ions are directed into a magnetic field, where they curve around in the field and strike a detector. If a variety of singly ionized atoms travel at the same speed through the magnetic field, would you expect them all to be deflected by the same amount? Or would you expect different ions to be bent by different amounts?

25. ⬤ Historically, replacing dirt roads with paved roads reduced friction between vehicles and the surface of the road. Replacing paved roads with steel rails reduced friction further. What will be the next step in reducing friction between vehicles and the surfaces over which they move? What friction will remain after surface friction has been eliminated?

26. ◼ Do a pair of parallel current-carrying wires exert forces on each other?

27. ⬤ When Tim pushes the wire between the poles of the magnet, the galvanometer registers a pulse. When he lifts the wire, another pulse is registered. How do the pulses differ?

28. ◼ Why is a generator armature harder to rotate when it is connected to a circuit and supplying electric current?

29. ◼ Does a cyclist coast farther if the lamp connected to his wheel-activated generator is turned off? Explain.

30. ◼ If your metal car moves over a wide, closed loop of wire embedded in a road surface, is Earth's magnetic field within the loop altered? Does this produce a current

pulse? Can you think of a practical application for this at a traffic intersection?

31. ● At the security area of an airport, you walk through a weak ac magnetic field inside a large coil of wire. What is the result of a small piece of metal on your person that slightly alters the magnetic field in the coil?

32. ● A piece of plastic tape coated with iron oxide is magnetized more in some parts than in others. When the tape is moved past a small coil of wire, what happens in the coil? What has been a practical application of this?

33. ● How do the input and output parts of a generator and a motor compare?

34. ■ Your friend says that if you crank the shaft of a dc motor manually, the motor becomes a dc generator. Do you agree or disagree? Defend your position.

35. ■ If you place a metal ring in a region in where a magnetic field is rapidly alternating, the ring may become hot to your touch. Why?

36. ■ A magician places an aluminum ring on a table, underneath which is hidden an electromagnet. When the magician says "abracadabra" (and pushes a switch that starts current flowing through the coil under the table), the ring jumps into the air. Explain his "trick."

37. ■ How could a lightbulb near, yet not touching, an electromagnet be lit? Is ac or dc required? Defend your answer.

38. ● Two separate but similar coils of wire are mounted close to each other, as shown below. The first coil is connected to a battery and has a direct current flowing through it. The second coil is connected to a galvanometer. How does the galvanometer respond when the switch in the first circuit is closed? After being closed when the current is steady? When the switch is opened?

Primary   Secondary

39. ● Why will more voltage be induced with the apparatus shown above if an iron core is inserted in the coils?

40. ● Why does a transformer require alternating voltage?

41. ● How does the current in the secondary of a transformer compare with the current in the primary when the secondary voltage is twice the primary voltage?

42. ■ In what sense can a transformer be thought of as an electrical lever? What does it multiply? What does it not multiply?

43. ■ In the circuit shown, how many volts are impressed across, and how many amperes flow through, the lightbulb?

1 A
120 V ac
100 turns   10 turns   100 turns

44. ■ In the circuit shown, how many volts are impressed across, and how many amperes flow through, the meter?

12 V dc
1 A
100 turns

45. ■ How would you answer the previous question if the input were 12 V ac?

46. ● Can an efficient transformer step up energy? Defend your answer.

47. ● A friend says that changing electric and magnetic fields generate one another, and this gives rise to visible light when the frequency of change matches the frequencies of light. Do you agree? Explain.

48. ■ Would electromagnetic waves exist if changing magnetic fields could produce electric fields but changing electric fields could not in turn produce magnetic fields? Explain.

49. ◆ When a bar magnet is dropped through a vertical length of copper pipe, it falls noticeably more slowly than it does when it is dropped through a vertical length of plastic pipe. If the copper pipe is long enough, the dropped magnet reaches a terminal falling speed. Propose an explanation.

50. ◆ What is wrong with this scheme? To generate electricity without fuel, arrange a motor to run a generator that produces electricity that is stepped up with transformers so that the generator can run the motor and simultaneously furnish electricity for other uses.

PROBLEMS

● BEGINNER   ■ INTERMEDIATE   ◆ EXPERT

1. ● A portable CD player requires 12 V to operate correctly. A transformer nicely allows the device to be powered from a 120-V outlet. If the primary has 500 turns, show that the secondary should have 50 turns.

2. ● A model electric train requires 6 V to operate. When it is connected to a 120-V household circuit, a transformer is needed. If the primary coil of the transformer has 240 windings, show that the secondary coil should have 12 turns.

3. ● A transformer for a laptop computer converts a 120-V input to a 24-V output. Show that the primary coil has five times as many turns as the secondary coil has.

4. ● If the output current for the transformer in the previous problem is 1.8 A, show that the input current is 0.36 A.

5. ● A transformer has an input of 9 V and an output of 36 V. If the input is changed to 12 V, show that the output would be 48 V.

6. ● An ideal transformer has 50 turns in its primary and 250 turns in its secondary. 12-V ac is connected to the primary. Show that (a) 60 V ac is available at the secondary; (b) 6 A of current is in a 10-Ω device connected to the secondary; and (c) the power supplied to the primary is 360 W.

7. ● Neon signs require about 12,000 V for their operation. Consider a neon-sign transformer that operates off 120-V lines. Show that there should be 100 times as many turns in the secondary as in the primary.

8. ■ 100 kW (10^5 W) of power is delivered to the other side of a city by a pair of power lines, between which the voltage is 12,000 V.

(a) Show that the current in the lines is 8.3 A.
(b) If each of the two lines has a resistance of 10 Ω, show that there is a 83-V change of voltage *along* each line. (Think carefully. This voltage change is along each line, *not between* the lines.)
(c) Show that the power expended as heat in both lines together is 1.38 kW (distinct from power delivered to customers).
(d) How do your calculations support the importance of stepping voltages up with transformers for long-distance transmission?

ACTIVE EXPLORATIONS

1. An iron bar can be magnetized easily by aligning it with the magnetic field lines of Earth and striking it lightly a few times with a hammer. This works best if the bar is tilted down to match the dip of Earth's magnetic field. The hammering jostles the domains so that they can better fall into alignment with Earth's field. The bar can be demagnetized by striking it when it is in an east–west direction.

2. Earth's magnetic field induces some degree of magnetism in most of the iron objects around you. With a compass you can see that cans of food on your pantry shelf have north and south poles. When you pass the compass from their bottoms to their tops, you can easily identify their poles. Mark the poles, either N or S. Then turn the cans upside down and note how many days it takes for the poles to reverse themselves. Explain to your friends why the poles reverse.

READINESS ASSURANCE TEST (RAT)

If you have a good handle on this chapter, if you really do, then you should be able to score at least 7 out of 10 on this RAT. If you score less than 7, you need to study further before moving on.

Choose the BEST answer to each of the following.

1. Moving electric charged particles can interact with
   (a) an electric field.
   (b) a magnetic field.
   (c) both of these
   (d) neither of these

2. The magnetic field lines about a current-carrying wire form
   (a) circles.
   (b) radial lines.
   (c) eddy currents.
   (d) energy loops.

3. A magnetic force can act on an electron even when it
   (a) is at rest.
   (b) moves parallel to magnetic field lines.
   (c) both of these
   (d) neither of these

4. A magnetic force acting on a beam of electrons can change its
   (a) direction.
   (b) energy.
   (c) both of these
   (d) neither of these

5. A motor and a generator are
   (a) similar devices.
   (b) very different devices with different applications.
   (c) forms of transformers.
   (d) energy sources.

6. If you change the magnetic field in a closed loop of wire, you induce in the loop
   (a) a current.
   (b) a voltage.
   (c) an electric field.
   (d) all of these
   (e) none of these

7. A voltage is induced in a wire loop when the magnetic field within that loop
   (a) changes.
   (b) aligns with the electric field.
   (c) is at right angles to the electric field.
   (d) converts to magnetic energy.

8. An efficient transformer in an ac electric circuit can change
   (a) current.
   (b) energy.
   (c) power.
   (d) all of these
   (e) none of these
9. An electric field is induced in any region of space in which
   (a) a magnetic field's orientation is at right angles to the electric field.
   (b) the accompanying electric field undergoes changes in time.
   (c) a magnetic field changes with time.
   (d) all of these
10. Electricity and magnetism connect to form
    (a) mass.
    (b) energy.
    (c) ultrahigh-frequency sound
    (d) light.

Answers to RAT

1c, 2a, 3d, 4a, 5a, 6d, 7a, 8a, 9c, 10d

CHAPTER 9 ONLINE RESOURCES

The Physical Science Place

Interactive Figures
- 9.3, 9.8, 9.17, 9.21, 9.24, 9.28

Tutorial
- Magnetic Fields

Videos
- Oersted's Discovery
- Magnetic Forces on Current-Carrying Wires

- Applications of Induction
- Faraday's Law

Quiz

Flashcards

Links

waves and sound

10.1 VIBRATIONS AND WAVES

10.2 WAVE MOTION

10.3 TRANSVERSE AND LONGITUDINAL WAVES

10.4 SOUND WAVES

10.5 REFLECTION AND REFRACTION OF SOUND

10.6 FORCED VIBRATIONS AND RESONANCE

10.7 INTERFERENCE

10.8 DOPPLER EFFECT

10.9 BOW WAVES AND THE SONIC BOOM

10.10 MUSICAL SOUNDS

■ Many things in the world about us wiggle and jiggle—the surface of a bell, a string on a guitar, the reed in a clarinet, lips on the mouthpiece of a trumpet, and the vocal cords of your larynx when you speak or sing. All these things *vibrate*. When they vibrate in air, they make the air molecules they touch wiggle and jiggle too, in exactly the same way, and these vibrations spread out in all directions, getting weaker, losing energy as heat, until they die out completely. But if these vibrations were to reach your ear instead, they would be transmitted to a part of your brain, and you would hear sound.

10.1 Vibrations and Waves

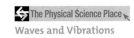

In a general sense, anything that moves back and forth, to and fro, from side to side, in and out, or up and down is vibrating. A **vibration** is a wiggle in time. A wiggle in space and time is a **wave**. A wave extends from one location to another. Light and sound are both vibrations that propagate throughout space as waves, but as waves of two very different kinds. Sound is the propagation of vibrations through a material medium—a solid, a liquid, or a gas. If no medium exists to vibrate, then no sound is possible. Sound cannot travel in a vacuum. But light can, because (as we discuss in Chapter 11) light is a vibration of nonmaterial electric and magnetic fields—a vibration of pure energy. Although light can pass through many materials, it needs none. This is evident when it propagates through the vacuum between the Sun and Earth.

The relationship between a vibration and a wave is shown in Figure 10.1. A marking pen on a bob attached to a vertical spring vibrates up and down and traces a waveform on a sheet of paper that is moved horizontally at constant speed. The waveform is actually a *sine curve*, a pictorial representation of a wave. As for a water wave, the high points are called *crests*, and the low points are the *troughs*. The straight dashed line represents the "home" position, or midpoint, of the vibration. The term **amplitude** refers to the distance from the midpoint to the crest (or to the trough) of the wave. So the amplitude equals the maximum displacement from equilibrium.

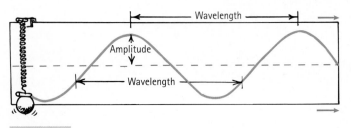

FIGURE 10.1

INTERACTIVE FIGURE

When the bob vibrates up and down, a marking pen traces out a sine curve on the paper, which is moved horizontally at constant speed.

The **wavelength** of a wave is the distance from the top of one crest to the top of the next one, or, equivalently, the distance between successive identical parts of the wave. The wavelengths of waves at the beach are measured in meters, the wavelengths of ripples in a pond in centimeters, and the wavelengths of light in billionths of a meter (nanometers). All waves have a vibrating source.

How frequently a vibration occurs is described by its *frequency*. The **frequency** of a vibrating pendulum, or of an object on a spring, specifies the number of to-and-fro vibrations it makes in a given time (usually in 1 s). A complete to-and-fro oscillation is one vibration. If it occurs in 1 s, the frequency is one vibration per second. If two vibrations occur in 1 s, the frequency is two vibrations per second.

The unit of frequency is called the **hertz** (Hz), after Heinrich Hertz, who demonstrated the existence of radio waves in 1886. One vibration per second is 1 Hz; two vibrations per second is 2 Hz, and so on. Higher frequencies are measured in kilohertz (kHz), and still higher frequencies in megahertz (MHz). AM radio waves are usually measured in kilohertz, while FM radio waves are measured in megahertz. A station at 960 kHz on the AM radio dial, for example, broadcasts radio waves that have a frequency of 960,000 vibrations per second. A station at 101.7 MHz on the FM dial broadcasts radio waves with a frequency of 101,700,000 hertz. These radio-wave frequencies are the frequencies at which electrons are forced to vibrate in the antenna of a radio station's transmitting tower. Still higher frequencies are measured in gigahertz (GHz), 1 billion vibrations per second. Cell phones operate in the GHz range, which means electrons inside are jiggling in unison billions of times per second! The frequency of the vibrating electrons and the frequency of the wave produced are the same.

The **period** of a wave or vibration is the time it takes for a complete vibration—for a complete cycle. Period can be calculated from frequency, and vice versa. Suppose, for example, that a pendulum makes two vibrations in 1 s. Its

FIGURE 10.2

The source of any wave is something that vibrates. Electrons in the transmitting antenna vibrate 940,000 times each second and produce 940-kHz radio waves. Radio waves can't be seen or heard, but they send a pattern that tells a radio or a TV set what sounds or pictures to make.

frequency is 2 Hz. The time needed to complete one vibration—that is, the period of vibration—is $\frac{1}{2}$ s. Or if the vibration frequency is 3 Hz, then the period is $\frac{1}{3}$ s. The frequency and period are the inverse of each other:

$$\text{Frequency} = \frac{1}{\text{period}}$$

Or, vice versa,

$$\text{Period} = \frac{1}{\text{Frequency}}$$

The frequency of a wave matches the frequency of its vibrating source. This is true not only of sound waves, but, as we'll see in the next chapter, of light waves also. The waves we're learning about, strictly speaking, are *periodic waves*—having distinct periods.

CHECK POINT

1. An electric razor completes 60 cycles every second. What is (a) its frequency? (b) its period?
2. If the difference in height between the crest and trough of a wave is 60 cm, what is the amplitude of the wave?

Were these your answers?
1. (a) 60 cycles per second or 60 Hz; (b) $\frac{1}{60}$ second.
2. The amplitude is 30 cm, half of the crest-to-trough height distance.

10.2 Wave Motion

If you drop a stone into a calm pond, waves travel outward in expanding circles. Energy is carried by the wave, traveling from one place to another. The water itself goes nowhere. This can be seen by waves encountering a floating leaf. The leaf bobs up and down, but it doesn't travel with the waves. The waves move along, not the water. The same is true for waves of wind over a field of tall grass on a gusty day. Waves travel across the grass, while the individual grass plants remain in place; they swing to and fro between definite limits, but they go nowhere. When you speak, molecules in air propagate the disturbance through the air at about 340 m/s. The disturbance, not the air itself, travels across the room at this speed. In these examples, when the wave motion ceases, the water, the grass, and the air return to their initial positions. A characteristic of wave motion is that the medium transporting the wave returns to its initial condition after the disturbance has passed.

FIGURE 10.3

Diane Riendeau uses a classroom wave machine to demonstrate how a vibration produces a wave.

WAVE SPEED

The speed of periodic wave motion is related to the frequency and wavelength of the waves. Consider the simple case of water waves (Figures 10.4 and 10.5). Imagine that we fix our eyes on a stationary point on the water's surface and observe the waves passing by that point. We can measure how much time passes between the arrival of one crest and the arrival of the next one (the period), and we can also observe the distance between crests (the wavelength). We know that speed is defined as distance divided by time. In this case, the distance is one wavelength and the time is one period, so the speed of a wave = wavelength/period.

FIGURE 10.4

A top view of water waves.

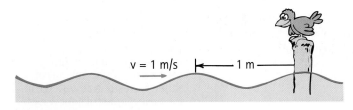

FIGURE 10.5

INTERACTIVE FIGURE

If the wavelength is 1 m, and one wavelength per second passes the pole, then the speed of the wave is 1 m/s.

Link To Sections 22.6, 24.3

fyi

■ It is customary to express the speed of a wave by the equation $v = f\lambda$, where v is wave speed, f is wave frequency, and λ (the Greek letter lambda) is wavelength.

Be clear about the distinction between *frequency* and *speed*. How frequently a wave vibrates is altogether different from how fast it moves from one location to another.

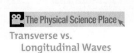
The Physical Science Place

Transverse vs. Longitudinal Waves

Link To Section 22.6

FIGURE 10.6

INTERACTIVE FIGURE

Both waves transfer energy from left to right. (a) When the end of the Slinky is pushed and pulled rapidly along its length, a longitudinal wave is produced. (b) When its end is shaken up and down (or side to side), a transverse wave is produced.

For example, if the wavelength is 10 m and the time between crests at a point on the surface is 0.5 s, the wave is traveling 10 m in 0.5 s and its speed is 10 m divided by 0.5 s, or 20 m/s.

Because period is equal to the inverse of frequency, the formula **wave speed** = *wavelength/period* can also be written as

$$\text{wave speed} = \text{wavelength} \times \text{frequency}$$

This relationship applies to all kinds of waves, whether they are water waves, sound waves, or light waves.

CHECK POINT

1. If a train of freight cars, each 10 m long, rolls by you at the rate of three cars each second, what is the speed of the train?
2. If a water wave oscillates up and down three times each second and the distance between wave crests is 2 m, (a) what is its frequency? (b) what is its wavelength? (c) what is its wave speed?

Were these your answers?

1. 30 m/s. We can see this in two ways. According to the definition of speed in Chapter 2, $v = \dfrac{d}{t} = \dfrac{3 \times 10\,\text{m}}{1\,\text{s}} = 30$ m/s, because 30 m of train passes you in 1 s. If we compare our train to wave motion, where wavelength corresponds to 10 m and frequency is 3 Hz, then

$$\text{Speed} = \text{frequency} \times \text{wavelength} = 3\,\text{Hz} \times 10\,\text{m} = 30\,\text{m/s}$$

2. (a) 3 Hz; (b) 2 m; (c) Wave speed = frequency × wavelength = 3/s × 2 m = 6 m/s.

10.3 Transverse and Longitudinal Waves

Fasten one end of a Slinky to a wall and hold the free end in your hand. If you shake the free end up and down, you produce vibrations that are at right angles to the direction of wave travel. The right-angled, or sideways, motion is called *transverse motion*. This type of wave is called a **transverse wave**. Waves in the stretched strings of musical instruments and on the surfaces of liquids are transverse waves. We will see later that electromagnetic waves, some of which are radio waves and light waves, are also transverse waves.

A **longitudinal wave** is one in which the direction of wave travel is along the direction in which the source vibrates. You produce a longitudinal wave with your Slinky when you shake it back and forth along the Slinky's axis

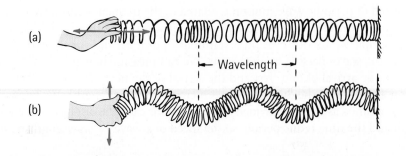

(Figure 10.6a). The vibrations are then parallel to the direction of energy transfer. Part of the Slinky is compressed, and a wave of **compression** travels along it. Between successive compressions is a stretched region called a **rarefaction**. Both compressions and rarefactions travel parallel to the Slinky. Together they make up the longitudinal wave. Figure 10.6b shows the generation of a transverse wave.

If you study earthquakes, you'll learn about two types of waves that travel in the ground. One type is longitudinal (P waves), and the other type is transverse (S waves). These travel at different speeds, which provides investigators with a means of determining the source of the waves. Furthermore, the transverse waves cannot travel through liquid matter, while the longitudinal waves can, which provides a means of determining whether matter below ground is molten or solid.

10.4 Sound Waves

Think of the air molecules in a room as tiny randomly moving Ping-Pong balls. If you vibrate a Ping-Pong paddle in the midst of the balls, you set them vibrating to and fro. The balls vibrate in rhythm with your vibrating paddle. In some regions they are momentarily bunched up (compressions), and in other regions in between they are momentarily spread out (rarefactions). The vibrating prongs of a tuning fork do the same to air molecules. Vibrations made up of compressions and rarefactions spread from the tuning fork throughout the air, and a *sound wave* is produced.

Sound requires a medium. It can't travel in a vacuum because there's nothing to compress and stretch.

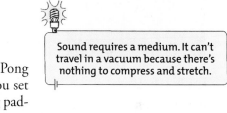

The wavelength of a sound wave is the distance between successive compressions or, equivalently, the distance between successive rarefactions. Each molecule in the air vibrates to and fro about some equilibrium position as the waves move by.

Our subjective impression about the frequency of sound is described as **pitch**. A high-pitched sound, such as that from a tiny bell, has a high vibration frequency. Sound from a large bell has a low pitch because its vibrations are of a low frequency. Pitch is the high or low we perceive a sound to be, depending on the frequency of the sound wave.

FIGURE 10.8

Compressions and rarefactions travel (both at the same speed and in the same direction) from the tuning fork through the air in the tube. The wavelength is the distance between successive compressions (or rarefactions).

The human ear can normally hear pitches from sound ranging from about 20 Hz to about 20,000 Hz. As we age, this range shrinks. So by the time you can afford to trade in your old stereo for an expensive sound system, you may not be able to tell the difference. Sound waves of frequencies lower than 20 Hz are called *infrasonic waves*, and those of frequencies higher than 20,000 Hz are called *ultrasonic waves*. We cannot hear infrasonic or ultrasonic sound waves.* But dogs and some other animals can.

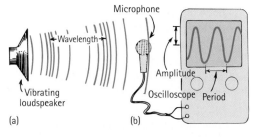

FIGURE 10.9

(a) The radio loudspeaker is a paper cone that vibrates in rhythm with an electric signal. The sound that is produced sets up similar vibrations in the microphone. The vibrations are displayed on an oscilloscope. (b) The waveform on the oscilloscope screen is a graph of pressure against time, showing how air pressure near the microphone rises and falls as sound waves pass. When the loudness increases, the amplitude of the waveform increases.

---

* In hospitals, concentrated beams of ultrasound are used to break up kidney stones and gallstones, eliminating the need for surgery.

■ Elephants communicate with one another with infrasonic waves. Their large ears help them detect these low-frequency sound waves.

FIGURE 10.10

Waves of compressed and rarefied air, generated by the vibrating cone of the loudspeaker, reproduce the sound of music.

■ A sound wave traveling through the ear canal vibrates the eardrum, which vibrates three tiny bones, which in turn vibrate the fluid-filled cochlea. Inside the cochlea, tiny hair cells convert the pulse into an electrical signal to the brain.

Most sound is transmitted through air, but any elastic substance—solid, liquid, or gas—can transmit sound.* Air is a poor conductor of sound compared with solids and liquids. You can hear the sound of a distant train clearly by placing your ear against the rail. When swimming, have a friend at a distance click two rocks together beneath the surface of water while you are submerged. Observe how well water conducts the sound. Sound cannot travel in a vacuum because there is nothing to compress and expand. The transmission of sound requires a medium.

Pause to reflect on the physics of sound while you are quietly listening to your radio sometime. The radio loudspeaker is a paper cone that vibrates in rhythm with an electrical signal. Air molecules next to the vibrating cone of the speaker are themselves set into vibration. These, in turn, vibrate against neighboring molecules, which, in turn, do the same, and so on. As a result, rhythmic patterns of compressed and rarefied air emanate from the loudspeaker, showering the entire room with undulating motions. The resulting vibrating air sets your eardrum into vibration, which, in turn, sends cascades of rhythmic electrical impulses along nerves in the cochlea of your inner ear and into the brain. And thus you listen to the sound of music.

SPEED OF SOUND

If, from a distance, we watch a person chopping wood or hammering, we can easily see that the blow occurs a noticeable time before its sound reaches our ears. Thunder is often heard seconds after a flash of lightning is seen. These common experiences show that sound requires time to travel from one place to another. The speed of sound depends on wind conditions, temperature, and humidity. It does not depend on the loudness or the frequency of the sound; all sounds travel at the same speed in a given medium. The speed of sound in dry air at 0°C is about 330 m/s, which is nearly 1200 km/s. Water vapor in the air increases this speed slightly. Sound travels faster through warm air than through cold air. This is to be expected, because the faster-moving molecules in warm air bump into each other more frequently and, therefore, can transmit a pulse in less time.** For each 1-degree increase in temperature above 0°C, the speed of sound in air increases by 0.6 m/s. Thus, in air at a normal room temperature of about 20°C, sound travels at about 340 m/s. In water, the speed of sound is about 4 times its speed in air; in steel, about 15 times its speed in air.

Loudspeakers

The loudspeakers of your radio and other sound-producing systems change electrical signals into sound waves. The electrical signals pass through a coil wound around the neck of a paper cone. This coil, which acts as an electromagnet, is located near a permanent magnet. When current flows one way, magnetic force pushes the electromagnet toward the permanent magnet, pulling the cone inward. When current flows in the opposite direction, the cone is pushed outward. Vibrations in the electric signal cause the cone to vibrate. Vibrations of the cone then produce sound waves in the air.

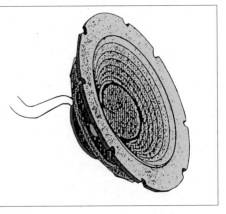

---

* An elastic substance is "springy," has resilience, and can transmit energy with little loss. Steel, for example, is elastic, but lead and putty are not.
** The speed of sound in a gas is about $\frac{3}{4}$ the average speed of its molecules.

1. Do compressions and rarefactions in a sound wave travel in the same direction or in opposite directions from one another?
2. What is the approximate distance of a thunderstorm when you note a 3-s delay between the flash of lightning and the sound of thunder?

Were these your answers?
1. They travel in the same direction.
2. Assuming the speed of sound in air is about 340 m/s, in 3 s sound travels 340 m/s × 3 s = 1020 m. There is no appreciable time delay for the flash of light, so the storm is slightly more than 1 km away.

■ 10.5 Reflection and Refraction of Sound

Like light, when sound encounters a surface, it can either be returned by the surface, or continue through it. When it is returned, the process is **reflection**. We call the reflection of sound an *echo*. The fraction of sound energy reflected from a surface is large if the surface is rigid and smooth, but it is less if the surface is soft and irregular. The sound energy that is not reflected is transmitted or absorbed.

Sound reflects from a smooth surface in the same way that light does—the angle of incidence (the angle between the direction of the sound and the reflecting surface) is equal to the angle of reflection (Figure 10.11). Sometimes, when sound reflects from the walls, ceiling, and floor of a room, the surfaces are too reflective and the sound becomes garbled. Sound due to multiple reflections is called a **reverberation**. On the other hand, if the reflective surfaces are too absorbent, the sound level is low and the room may sound dull and lifeless. Reflected sound in a room makes it sound lively and full, as you have probably experienced while singing in the shower. The designer of an auditorium or concert hall must find a balance between reverberation and absorption. The study of sound properties is called *acoustics*.

It is often advantageous to position highly reflective surfaces behind the stage to direct sound out to the audience. In some concert halls, reflecting surfaces are suspended above the stage. Ones such as those in Davies Hall in San Francisco are large shiny plastic surfaces that also reflect light (Figure 10.12). A listener can look up at these reflectors and see the reflected images of the members of the orchestra (the plastic reflectors are somewhat curved, which increases the field of view). Both sound and light obey the same law of reflection. Thus, if a reflector is oriented so that you can see a particular musical instrument, rest assured that you can also hear it. Sound from the instrument follows the line of sight to the reflector and then to you. In some halls, absorbers rather than reflectors are used to improve the acoustics.

Refraction occurs when sound continues through a medium and bends. Sound waves bend when parts of the wave fronts travel at different speeds. This may happen when sound waves are affected by uneven winds, or when sound travels through air of uneven temperatures. On a warm day, the air near the ground may be appreciably warmer than the air above, so the speed of sound near the ground increases. Sound waves therefore tend to bend away from the ground, resulting in sound that does not seem to transmit well (Figure 10.13).

The refraction of sound occurs under water, where the speed of sound varies with temperature. This poses a problem for surface vessels that bounce ultrasonic waves off the bottom of the ocean to chart its features, but it's a blessing to

FIGURE 10.11

The angle of incident sound is equal to the angle of reflected sound.

FIGURE 10.12

The plastic plates above the orchestra reflect both light and sound. Adjusting them is quite simple: what you see is what you hear.

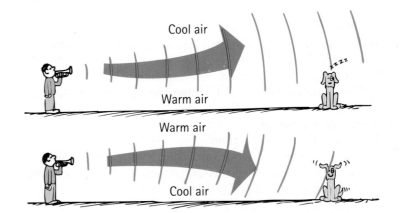

FIGURE 10.13

Sound waves are bent in air of uneven temperatures.

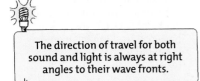

The direction of travel for both sound and light is always at right angles to their wave fronts.

FIGURE 10.14

The 14-week-old fetus that became Megan Hewitt Abrams, who is more recently seen on page 503.

Link To Sections 24.3, 24.7

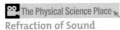

The Physical Science Place
Refraction of Sound

submarines that wish to escape detection. Because the ocean has layers of water that are at different temperatures, the refraction of sound leaves gaps or "blind spots" in the water. This is where submarines hide. If not for refraction, submarines would be much easier to detect.

Physicians use the multiple reflections and refractions of ultrasonic waves to "see" the interior of the body without the use of X-rays. High-frequency sound (ultrasound) that enters the body is reflected more strongly from the organs' exteriors than from their interiors, producing an outline of the organs (Figure 10.14). This ultrasound echo technique is nothing new to bats and dolphins, who can emit ultrasonic squeaks and locate objects by their echoes.

FIGURE 10.15

A dolphin emits ultrahigh-frequency sound to locate and identify objects in its environment. It senses distance by the time delay between sending sound and receiving the echo, and it senses direction by differences in time for the echo to reach the dolphin's two ears. A dolphin's main diet is fish. Because fish hear mainly low frequencies, they are not alerted to the fact that they are being hunted.

Dolphins and Acoustical Imaging

The dominant sense of the dolphin is hearing, because vision is not a very useful sense in the often murky and dark depths of the ocean. Whereas sound is a passive sense for us, it is an active sense for the dolphin, which sends out sounds and then perceives its surroundings by means of the echoes that return. The ultrasonic waves emitted by a dolphin enable it to "see" through the bodies of other animals and people. Skin, muscle, and fat are almost transparent to dolphins, so they "see" a thin outline of the body—but the bones, teeth, and gas-filled cavities are clearly apparent. Dolphins can "see" physical evidence of cancers, tumors, and heart attacks—which humans have only recently been able to detect with ultrasound.

What's more fascinating, the dolphin can reproduce the sonic signals that paint the mental image of its surroundings; thus, it is probably able to communicate its experiences to other dolphins by communicating the full acoustic image of what it has "seen," placing the image directly in the minds of other dolphins. It needs no word or symbol for "fish," for example, but can communicate an image of the real thing—perhaps with emphasis highlighted by selective filtering, as we similarly communicate a musical concert to others via various means of sound reproduction. Small wonder that the language of the dolphin is very unlike our own!

FIGURING PHYSICAL SCIENCE

▪Problem Solving

SAMPLE PROBLEM 1

An oceanic depth-sounding vessel surveys the ocean floor with ultrasonic sound that travels 1530 m/s in seawater. How deep is the water if the time delay of the echo from the ocean floor is 2 s?

SOLUTION:

The round trip is 2 s, meaning 1 s down and 1 s up. Then,

$$d = vt = 1530 \text{ m/s} \times 1 \text{ s} = 1530 \text{ m}$$

(Radar works similarly; microwaves rather than sound waves are transmitted.)

SAMPLE PROBLEM 2

While sitting on the dock of the bay, Otis notices incoming waves with distance d between crests. The incoming crests lap against the pier pilings at a rate of one every 2 s.

(a) Find the frequency of the waves.
(b) Show that the speed of the waves is given by fd.
(c) Suppose the distance d between wave crests is 1.8 m. Show that the speed of the waves is slightly less than 1.0 m/s.

SOLUTION:

(a) The frequency of the waves is given: one per 2 s, or $f = 0.5$ Hz
(b) $v = f\lambda = fd$.
(c) $v = f\lambda = fd = 0.5 \text{ Hz} (1.8 \text{ m})$
$= 0.5\left(\frac{1}{s}\right)(1.8 \text{ m}) = 0.9 \frac{m}{s}.$

▪ 10.6 Forced Vibrations and Resonance

If you strike an unmounted tuning fork, its sound is rather faint. Repeat with the handle of the fork held against a table after striking it, and the sound is louder. This is because the table is forced to vibrate, and its larger surface sets more air in motion. The table is forced into vibration by a fork of any frequency. This is an example of **forced vibration**. The vibration of a factory floor caused by the running of heavy machinery is another example of forced vibration. A more pleasing example is given by the sounding boards of stringed instruments.

If you drop a wrench and a baseball bat on a concrete floor, you easily notice the difference in their sounds. This is because each vibrates differently when striking the floor. They are not forced to vibrate at a particular frequency; instead, each vibrates at its own characteristic frequency. Any object composed of an elastic material, when disturbed, vibrates at its own special set of frequencies, which together form its characteristic sound. We speak of an object's **natural frequency**, which depends on such factors as the elasticity and shape of the object. Bells and tuning forks, of course, vibrate at their own characteristic frequencies. Interestingly, most things, from atoms to planets and almost everything else in between, have springiness to them, and they vibrate at one or more natural frequencies.

When the frequency of forced vibrations on an object matches the object's natural frequency, a dramatic increase in amplitude occurs. This phenomenon is called **resonance**. Literally, *resonance* means "resounding" or "sounding again." Putty doesn't resonate, because it isn't elastic, and a dropped handkerchief is too limp to resonate. In order for something to resonate, it needs both a force to pull it back to its starting position and enough energy to maintain its vibration.

A common experience illustrating resonance occurs when you are on a swing. When pumping a swing, you pump in rhythm with the natural frequency of the swing. More important than the force with which you pump is the timing. Even

fyi

▪ Owls have extremely sensitive ears. Hunting at night, owls tune in to the soft rustles and squeaks of rodents and other small mammals. Like humans, owls locate sound sources by using the fact that sound waves often reach one ear milliseconds before the other. An owl moves its head as it glides toward its prey; when sounds from the target reach both ears at once, the meal is dead ahead. In some owls, one ear is also higher than the other, further sharpening their prey-locating ability.

 The Physical Science Place
Resonance
Resonance and Bridges

fyi

▪ Parrots, like humans, use their tongues to craft and shape sound. Tiny changes in tongue position produce big differences in the sound first produced in the syrinx, a voice box organ nestled between the trachea and lungs.

FIGURE 10.16

Stages of resonance. (a) The first compression meets the fork and gives it a tiny and momentary push; (b) the fork bends and then (c) returns to its initial position just at the time a rarefaction arrives and (d) overshoots in the opposite direction. Just when it returns to its initial position, (e) the next compression arrives to repeat the cycle. Now it bends farther because it is moving.

FIGURE 10.17

Ryan demonstrates resonance with a pair of tuning forks with matched frequencies.

FIGURE 10.18

In 1940, four months after being completed, the Tacoma Narrows Bridge in the state of Washington was destroyed by wind-generated resonance. The mild gale produced a fluctuating force in resonance with the natural frequency of the bridge, steadily increasing the amplitude until the bridge collapsed.

Why does Hollywood persist in playing engine noises whenever a spacecraft in outer space passes by? Wouldn't seeing them float by silently be far more dramatic?

small pumps, or small pushes from someone else, if delivered in rhythm with the frequency of the swinging motion, produce large amplitudes.

A common classroom demonstration of resonance is illustrated with a pair of tuning forks adjusted to the same frequency and spaced a meter or so apart (Figure 10.17). When one of the forks is struck, it sets the other fork into vibration. This is a small-scale version of pushing a friend on a swing—it's the timing that's important. When a series of sound waves impinge on the fork, each compression gives the prong of the fork a tiny push. Because the frequency of these pushes corresponds to the natural frequency of the fork, the pushes successively increase the amplitude of its vibration. This is because the pushes occur at the right time and repeatedly occur in the same direction as the instantaneous motion of the fork. The motion of the second fork is called a *sympathetic vibration*.

If the forks are not adjusted for matched frequencies, the timing of pushes is off, and resonance doesn't occur. When you tune your radio, you are similarly adjusting the natural frequency of the electronics in the device to match one of the many surrounding signals. The device then resonates to one station at a time, instead of playing all stations at once.

Resonance is not restricted to wave motion. It occurs whenever successive impulses are applied to a vibrating object in rhythm with its natural frequency. Cavalry troops marching across a footbridge near Manchester, England, in 1831 inadvertently caused the bridge to collapse when they marched in rhythm with the bridge's natural frequency. Since then, it is customary to order troops to "break step" when crossing bridges. A more recent bridge disaster was caused by wind-generated resonance (Figure 10.18).

10.7 Interference

An intriguing property of all waves is **interference**. Consider transverse waves. When the crest of one wave overlaps the crest of another, their individual effects add together. The result is a wave of increased amplitude. This is *constructive interference* (Figure 10.19). When the crest of one wave overlaps the trough of another, their individual effects are reduced. The high part of one

+ = Reinforcement   + = Cancellation

FIGURE 10.19

Constructive and destructive interference in a transverse wave.

wave simply fills in the low part of another. This is *destructive interference*.

Wave interference is easiest to observe in water. In Figure 10.20, we see the interference pattern produced when two vibrating objects touch the surface of water. We can see the regions in which the crest of one wave overlaps the trough of another to produce a region of zero amplitude. At points along such regions, the waves arrive out of step. We say they are *out of phase* with one another.

Interference is a property of all wave motion, whether the waves are water waves, sound waves, or light waves. We see a comparison of interference for transverse waves and for longitudinal waves in Figure 10.22. For the transverse waves of light we see constructive interference where crests and

FIGURE 10.20

Two sets of overlapping water waves produce an interference pattern.

The superposition of two identical transverse waves in phase produces a wave of increased amplitude.

The superposition of two identical longitudinal waves in phase produces a wave of increased intensity.

Two identical transverse waves that are out of phase destroy each otherwhen they are superimposed.

Two identical longitudinal waves that are out of phase destroy each other when they are superimposed.

FIGURE 10.21

New Zealander Jennie McKelvie showing interference with a classroom ripple tank.

FIGURE 10.22

Constructive (top two panels) and destructive (bottom two panels) wave interference in transverse and longitudinal waves.

When a mirror image of a sound signal combines with the sound itself, the sound is canceled.

troughs of one wave superimpose on another. Such waves hitting a screen show bright light. Dark light appears where destructive interference occurs—where crests overlap troughs. Similar effects occur for the interference of longitudinal sound waves, shown by the regions of compressions and rarefactions.

Destructive sound interference is at the heart of *antinoise technology*. Some noisy devices such as jackhammers are now equipped with microphones that send the sound of the device to electronic microchips, which create mirror-image wave patterns of the sound signals. This mirror-image sound signal is fed to earphones worn by the operator. In this way, sound compressions (or rarefactions) from the hammer are canceled by mirror-image rarefactions (or compressions) in the earphones. The combination of signals cancels the jackhammer noise. Antinoise devices are also common in some aircraft, which are much quieter inside than before this technology was introduced. Are automobiles next, perhaps eliminating the need for mufflers?

Sound interference is dramatically illustrated when monaural sound is played by stereo speakers that are out of phase. Speakers are out of phase when the input wires to one speaker are interchanged (positive and negative wire inputs reversed). For a monaural signal, this means that when one speaker is sending a

When the positive and negative wire inputs to one of the stereo speakers have been interchanged, the speakers are then out of phase. When the speakers are far apart, monaural (not stereo) sound is not as loud as it is from properly phased speakers. When they are brought face to face, very little sound is heard. Interference is nearly complete, as the compressions of one speaker fill in the rarefactions of the other.

compression of sound, the other is sending a rarefaction. The sound produced is not as full and not as loud as from speakers properly connected in phase. The longer waves are canceled by interference. Shorter waves are canceled as the speakers are brought closer together, and when the two speakers are brought face to face against each other, very little sound is heard! Only the highest frequencies survive cancellation. You must try this experiment to appreciate it.

The interference of light is evident in the bright colors seen in reflections from thin films of gasoline on water. Reflections from the gasoline and water surfaces interfere, canceling colors and producing their complementary colors (discussed in the next chapter).

BEATS

When two tones of slightly different frequencies are sounded together, a fluctuation in the loudness of the combined sounds is heard; the sound is loud, then faint, then loud, then faint, and so on. This periodic variation in the loudness of sound is called **beats**, and it is due to interference. If you strike two slightly mismatched tuning forks, one fork vibrates at a different frequency from the other, and the vibrations of the forks are momentarily in step, then out of step, then in again, and so on. When the combined waves reach our ears in step—say, when a compression from one fork overlaps a compression from the other—the sound is at a maximum. A moment later, when the forks are out of step, a compression from one fork meets a rarefaction from the other, resulting in a minimum. The sound that reaches our ears throbs between maximum and minimum loudness and produces a tremolo effect.

Beats can occur with any kind of wave, and they can provide a practical way to compare frequencies. To tune a piano, for example, a piano tuner listens for beats produced between a standard tuning fork and those of a particular string

Ken Ford tows gliders in quiet comfort when he wears his noise-canceling earphones. In larger aircraft, sound from the engines is processed and emitted as antinoise from loudspeakers inside the cabin to provide passengers with a quieter ride.

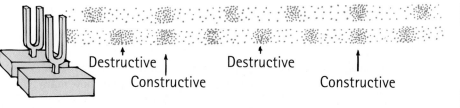

FIGURE 10.26

The interference of two sound sources of slightly different frequencies produces beats.

📽 The Physical Science Place

Interference and Beats

on the piano. When the frequencies are identical, the beats disappear. The members of an orchestra tune up their instruments by listening for beats between their instruments and a standard tone produced by a piano or some other instrument.

STANDING WAVES

Another fascinating effect of interference is *standing waves*. Tie a rope to a wall and shake the free end up and down. The wall is too rigid to shake, so the waves are reflected back along the rope. By shaking the rope just right, you can cause the incident and reflected waves to interfere and form a **standing wave**, in which parts of the rope, called the *nodes*, are stationary. You can hold your fingers on either side of the rope at a node, and the rope doesn't touch them. Other parts of the rope, however, would make contact with your fingers. The positions on a standing wave with the largest displacements are known as *antinodes*. Antinodes occur halfway between nodes.

Standing waves are produced when two sets of waves of equal amplitude and wavelength pass through each other in opposite directions. Then the waves are steadily in and out of phase with each other and produce stable regions of constructive and destructive interference (Figure 10.27).

Standing waves are set up in the strings of musical instruments when plucked, bowed, or struck. They are produced in the air in an organ pipe, a flute, or a clarinet—and in the air of a soft-drink bottle when air is blown over the top. Standing waves appear in a tub of water or a cup of coffee when sloshed back and forth at the appropriate frequency. Standing waves can be produced with either transverse or longitudinal vibrations.

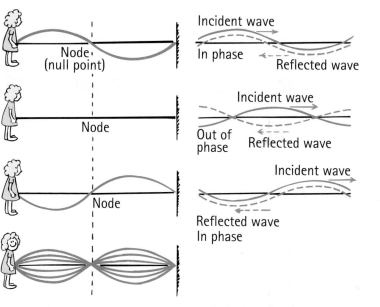

FIGURE 10.27

INTERACTIVE FIGURE

The incident and reflected waves interfere to produce a standing wave.

fyi

- See the production of standing waves at http://www2.biglobe. ne.jp/~norimari/science/JavaEd/ e-wave4.html.

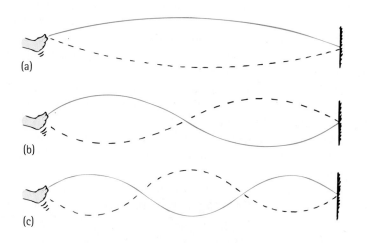

FIGURE 10.28

INTERACTIVE FIGURE

(a) Shake the rope until you set up a standing wave of one loop ($\frac{1}{2}$ wavelength).
(b) Shake with twice the frequency and produce a wave with two loops (1 wavelength).
(c) Shake with three times the frequency and produce three loops ($\frac{3}{2}$ wavelengths).

CHECK POINT

1. Is it possible for one wave to cancel another wave so that no amplitude remains?
2. Suppose you set up a standing wave of three segments, as shown in Figure 10.28c. If you shake with a frequency twice as great, how many wave segments occur in your new standing wave? How many wave-lengths?

Were these your answers?

1. Yes. This is called destructive interference. When a standing wave is set up in a rope, for example, parts of the rope have no amplitude—the nodes.
2. If you impart twice the frequency to the rope, you produce a standing wave with twice as many segments (six). Because a full wavelength has two segments, you have three complete wavelengths in your standing wave.

The Physical Science Place
Doppler Effect

The Physical Science Place
Doppler Effect

Link To Section 27.6

FIGURE 10.29

Top view of water waves made by a stationary bug jiggling in still water.

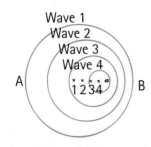

FIGURE 10.30

INTERACTIVE FIGURE

Water waves made by a bug swimming in still water toward point B.

■ 10.8 Doppler Effect

Consider a bug in the middle of a quiet puddle. A pattern of water waves is produced when it jiggles its legs and bobs up and down (Figure 10.29). The bug is not traveling anywhere but merely treads water in a stationary position. The waves it creates are concentric circles because wave speed is the same in all directions. If the bug bobs in the water at a constant frequency, the distance between wave crests (the wavelength) is the same in all directions. Waves encounter point A as frequently as they encounter point B. Therefore, the frequency of wave motion is the same at points A and B, or anywhere in the vicinity of the bug. This wave frequency remains the same as the bobbing frequency of the bug.

Suppose the jiggling bug moves across the water at a speed less than the wave speed. In effect, the bug chases part of the waves it has produced. The wave pattern is distorted and is no longer composed of concentric circles (Figure 10.30). The center of the outer wave originated when the bug was at the center of that circle. The center of the next smaller wave originated when the bug was at the center of that circle, and so forth. The centers of the circular waves move in the direction of the swimming bug. Although the bug maintains the same bobbing frequency as before, an observer at B would see the waves coming more often. The observer would measure a higher frequency. This is because each successive wave has a shorter distance to travel and therefore arrives at B sooner than if the bug weren't moving toward B. An observer at A, on the other hand, measures a lower frequency because of the longer time between wave-crest arrivals. This occurs because each successive wave travels farther to get to A as a result of the bug's motion. This change in frequency due to the motion of the source (or due to the motion of the receiver) is called the **Doppler effect** (after the Austrian physicist and mathematician Christian Johann Doppler, who lived from 1803 to 1853).

Water waves spread over the flat surface of the water. Sound and light waves, on the other hand, travel in three-dimensional space in all directions like an expanding balloon. Just as circular waves are closer together in front of the swimming bug, spherical sound or light waves ahead of a moving source are closer together and reach an observer more frequently. The Doppler effect holds for all types of waves.

The Doppler effect is evident when you hear the changing pitch of an ambulance or fire-engine siren. When the siren is approaching you, the crests of the sound waves encounter your ear more frequently, and the pitch is higher than normal. And when the siren passes you and moves away, the crests of the waves encounter your ear less frequently, and you hear a drop in pitch.

The Doppler effect also occurs for light. When a light source approaches, its measured frequency increases; when it recedes, its frequency decreases. An increase in light frequency is called a *blueshift*, because the increase is toward a higher frequency, or toward the blue end of the color spectrum. A decrease in frequency is called a *redshift*, referring to a shift toward a lower frequency, or toward the red end of the color spectrum. Galaxies, for example, show a redshift in the light they emit as they move away from us in the expanding universe. Measuring this shift allows us to calculate their speed. A rapidly spinning star shows a redshift on the side turning away from us and a blueshift on the side turning toward us. This enables us to calculate the star's spin rate.

FIGURE 10.31

INTERACTIVE FIGURE

The pitch of sound increases when the source moves toward you, and it decreases when the source moves away.

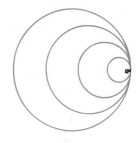

FIGURE 10.32

The wave pattern made by a bug swimming at wave speed.

CHECK POINT

When a light or sound source moves toward you, is there an increase or a decrease in the wave speed?

Was this your answer?
Neither! The frequency of a wave undergoes a change when the source is moving, not the wave speed.

10.9 Bow Waves and the Sonic Boom

When a source of waves travels as fast as the waves it produces, a *wave barrier* is produced. Consider the bug in our previous example. If it swims as fast as the waves it makes, the bug keeps up with the waves it produces. Instead of moving ahead of the bug, the waves superimpose on one another directly, forming a hump in front of the bug (Figure 10.32). Thus, the bug encounters a wave barrier. The bug must expend much effort to swim over the hump before it can swim faster than wave speed.

The same thing happens when an aircraft travels at the speed of sound. The waves overlap to produce a barrier of compressed air on the leading edges of the wings and on other parts of the aircraft. The aircraft requires considerable thrust to push through this barrier (Figure 10.33). Once through, the aircraft can fly faster than the speed of sound without similar opposition. The aircraft is *supersonic*. It is like the bug, which, once it has passed its wave barrier, finds the medium ahead relatively smooth and undisturbed.

When the bug swims faster than wave speed, it produces a pattern of overlapping waves, ideally shown in Figure 10.34. The bug overtakes and outruns the waves it produces. The overlapping waves form a V shape, called a **bow wave**, which appears to be dragging behind the bug. Overlapping waves produce the familiar bow wave generated by a speedboat knifing through the water.

Some wave patterns created by sources moving at various speeds are shown in Figure 10.35. Note that after the speed of the source exceeds wave speed, increased speed produces a narrower V shape.*

FIGURE 10.33

Condensation of water vapor by rapid expansion of air can be seen in the rarefied region behind the wall of compressed air.

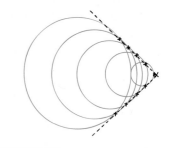

FIGURE 10.34

Idealized wave pattern made by a bug swimming faster than wave speed.

---

* Bow waves generated by boats in water are more complex than is indicated here. Our idealized treatment serves as an analogy for the production of the less complex shock waves in air.

FIGURE 10.35

Idealized patterns made by a bug swimming at successively greater speeds. Overlapping at the edges occurs only when the bug swims faster than wave speed.

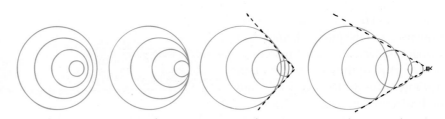

v less than v_w v equals v_w v exceeds v_w v greatly exceeds v_w

FIGURE 10.36

The shock wave of a bullet piercing a sheet of Plexiglas. Light is deflected as it passes through the compressed air that makes up the shock wave, making it visible. Look carefully and see the second shock wave originating at the tail of the bullet.

FIGURE 10.37

The shock wave actually consists of two cones—a high-pressure cone with its apex at the bow and a low-pressure cone with its apex at the tail. A graph of the air pressure at ground level between the cones takes the shape of the letter N.

Whereas a speedboat knifing through the water generates a two-dimensional bow wave at the surface of the water, a supersonic aircraft similarly generates a three-dimensional *shock wave*. Just as a bow wave is produced by overlapping circles that form a V, a **shock wave** is produced by overlapping spheres that form a cone. And just as the bow wave of a speedboat spreads until it reaches the shore of a lake, the conical wake generated by a supersonic aircraft spreads until it reaches the ground.

The bow wave of a speedboat that passes by can splash and douse you if you are at the water's edge. You could say that, in a sense, you are hit by a "water boom." In the same way, when the conical shell of compressed air that sweeps behind a supersonic aircraft reaches listeners on the ground below, the sharp crack they hear is described as a **sonic boom**.

We don't hear a sonic boom from slower-than-sound (subsonic) aircraft because the sound waves reach our ears gradually and are perceived as a continuous tone. Only when the craft moves faster than sound do the waves overlap to reach the listener in a single burst. The sudden increase in pressure is much the same in effect as the sudden expansion of air produced by an explosion. Both processes direct a burst of high-pressure air to the listener. The ear is hardpressed to distinguish between the high pressure caused by an explosion and that produced by many overlapping waves.

A water skier is familiar with the fact that next to the high hump of the V-shaped bow wave is a V-shaped depression. The same is true of a shock wave, which consists of two cones: a high-pressure cone generated at the bow of the supersonic aircraft and a low-pressure cone that follows toward (or at) the tail of the aircraft. The edges of these cones are visible in the photograph of the supersonic bullet in Figure 10.36. Between these two cones, the air pressure rises sharply to above atmospheric pressure, then falls below atmospheric pressure before sharply returning to normal beyond the inner tail cone (Figure 10.37). This overpressure, suddenly followed by underpressure, intensifies the sonic boom.

FIGURE 10.38

A shock wave.

A common misconception is that sonic booms are produced when an aircraft breaks through the sound barrier—that is, just when the aircraft exceeds the speed of sound. This is essentially the same as saying that a boat produces a bow wave when it overtakes its own waves. This is not true. A shock wave and its resulting sonic boom are swept continuously behind an aircraft that is traveling faster than sound, just as a bow wave is swept continuously behind a speedboat. In Figure 10.39, listener B is in the process of hearing a sonic boom. Listener C has already heard it, and listener A will hear it shortly. The aircraft that generated this shock wave may have broken through the sound barrier hours ago!

The moving source need not be "noisy" to produce a shock wave. Once an object is moving faster than the speed of sound, it makes sound. A supersonic bullet passing overhead produces a crack, which is a small sonic boom. If the bullet were larger and disturbed more air in its path, the crack would be more boomlike. When a lion tamer cracks a circus whip, the cracking sound is actually a sonic boom produced because the tip of the whip is traveling faster than the speed of sound. Both the bullet and the whip are not in themselves sound sources, but when they travel at supersonic speeds, they produce their own sound as they generate shock waves.

▪ 10.10 Musical Sounds

Most of the sounds we hear are noises. The impact of a falling object, the slamming of a door, the roaring of a motorcycle, and most of the sounds from traffic in city streets are noises. Noise corresponds to an irregular vibration of the eardrum produced by an irregularly vibrating source. Graphs that indicate the varying pressure of the air on the eardrum are shown in Figures 10.41a and 10.41b. In part (a), we see the erratic pattern of noise. In part (b), the sound of music has shapes that repeat themselves periodically. These are periodic tones, or musical "notes." (But musical instruments can make noise as well!) Such graphs can be displayed on the screen of an oscilloscope when the electrical signal from a microphone is fed into the input terminal of this useful device.

FIGURE 10.39
The shock wave has not yet reached listener A, but it is now reaching listener B, and it has already reached listener C.

Don't confuse *supersonic* with *ultrasonic*. *Supersonic* has to do with speed—faster than sound. *Ultrasonic* has to do with frequency—higher than we can hear.

FIGURE 10.40
Physics chanteuse Lynda Williams, physics instructor at Santa Rosa Junior College, puts herself fully into the physics of music.

FIGURE 10.41
Graphical representations of noise and music.

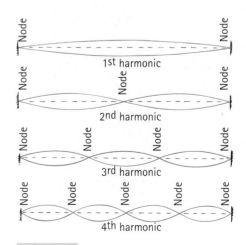
FIGURE 10.42
Modes of vibration of a guitar string.

FIGURE 10.43

A composite vibration of the fundamental mode and the third harmonic.

Who better appreciates music—one who is knowledgeable about it, or the casual listener?

Link To Section 28.1

Piano C

Clarinet C

FIGURE 10.45

Sounds from the piano and clarinet differ in quality.

fyi

- Scratched CD? Gently wipe a bit of toothpaste on it. The abrasives that polish teeth can also buff scratches from a disc.

FIGURE 10.46

Does each listener hear the same music?

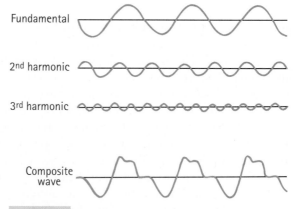

Fundamental

2nd harmonic

3rd harmonic

Composite wave

FIGURE 10.44

Sine waves combine to produce a composite wave.

We have no trouble distinguishing between the tone from a piano and the tone from a clarinet of the same musical pitch (frequency). Each of these tones has a characteristic sound that differs in **quality**, or timbre, a mixture of harmonics of different intensities. Most musical sounds are composed of a superposition of many frequencies called **partial tones**, or simply *partials*. The lowest frequency, called the **fundamental frequency**, determines the pitch of the note. Partial tones that are whole multiples of the fundamental frequency are called **harmonics**. A tone that has twice the frequency of the fundamental is the second harmonic, a tone with three times the fundamental frequency is the third harmonic, and so on (Figure 10.42).* The variety of partial tones gives a musical note its characteristic quality.

Thus, if we strike middle C on the piano, we produce a fundamental tone with a pitch of about 262 Hz and also a blending of partial tones of two, three, four, five, and so on times the frequency of middle C. The number and relative loudness of the partial tones determine the quality of sound associated with the piano. Sound from practically every musical instrument consists of a fundamental and partials. Pure tones, those having only one frequency, can be produced electronically. Electronic synthesizers, for example, produce pure tones and mixtures of these to produce a vast variety of musical sounds.

The quality of a tone is determined by the presence and relative intensity of the various partials. The ear recognizes the different partials and can therefore differentiate the different sounds produced by a piano and a clarinet. A pair of tones of the same pitch with different qualities has either different partials or a difference in the relative intensity of the partials.

Amazingly, when listening to music we can discern what instruments are being played, what notes are playing, and what their relative loudness is. Whether the music is live or electronic, our ears break the overall sound signal into its component parts automatically. How this incredible feat is accomplished has to do with Fourier Analysis, which concludes our study of sound.

* Not all partial tones present in a complex tone are integer multiples of the fundamental. Unlike the harmonics of woodwinds and brasses, stringed instruments, such as the piano, produce "stretched" partial tones that are nearly, but not quite, harmonics.

Fourier Analysis

The French mathematician Joseph Fourier made one of the most interesting discoveries about music in 1822. He discovered that wave motion could be reduced to simple sine waves. A sine wave is the simplest of waves, having a single frequency, as shown in Figure 10.44. All periodic waves, however complicated, can be broken down into constituent sine waves of different amplitudes and frequencies. The mathematical operation for doing this is called **Fourier analysis.** We will not explain the mathematics here, but we will simply point out that, by such analysis, one can find the pure sine tones that constitute the tone of, say, a violin. When these pure tones are sounded together, as by striking a number of tuning forks or by selecting the proper keys on an electric organ,

they combine to produce the tone of the violin. The lowest-frequency sine wave is the fundamental, and it determines the pitch of the note. The higher-frequency sine waves are the partials, which give the characteristic quality. Thus, the waveform of any musical sound is no more than a sum of simple sine waves.

Because the waveform of music is a multitude of various sine waves, to duplicate sound accurately by radio, tape recorder, or CD player, we should be able to process as large a range of frequencies as possible. The notes of a piano keyboard range from 27 Hz to 4200 Hz, but to duplicate the music of a piano composition accurately, the sound system must have a range of frequencies up to 20,000 Hz. The greater the range of the frequencies of an electrical sound system, the closer

the musical output approximates the original sound, hence the wide range of frequencies that can be produced in a high-fidelity sound system.

Our ear performs a sort of Fourier analysis automatically. It sorts out the complex jumble of air pulsations that reach it, and it transforms them into pure tones. And we recombine various groupings of these pure tones when we listen. What combinations of tones we have learned to focus our attention on determines what we hear when we listen to a concert. We can direct our attention to the sounds of the various instruments and discern the faintest tones from the loudest; we can delight in the intricate interplay of instruments and still detect the extraneous noises of others around us. This is a most incredible feat.

SUMMARY OF TERMS

Vibration A wiggle in time.

Wave A wiggle in both space and time.

Amplitude For a wave or vibration, the maximum displacement on either side of the equilibrium (midpoint) position.

Wavelength The distance between successive crests, troughs, or identical parts of a wave.

Frequency For a vibrating body or medium, the number of vibrations per unit time. For a wave, the number of crests that pass a particular point per unit time.

Hertz The SI unit of frequency. One hertz (symbol Hz) equals one vibration per second.

Period The time required for a vibration or a wave to make a complete cycle; equal to 1/frequency.

Wave speed The speed with which waves pass a particular point:

$$\text{wave speed} = \text{frequency} \times \text{wavelength}$$

Transverse wave A wave in which the medium vibrates in a direction perpendicular (transverse) to the direction in which the wave travels. Light consists of transverse waves.

Longitudinal wave A wave in which the medium vibrates in a direction parallel (longitudinal) with the direction in which the wave travels. Sound consists of longitudinal waves.

Compression A condensed region of the medium through which a longitudinal wave travels.

Rarefaction A rarefied region, or region of lessened pressure, of the medium through which a longitudinal wave travels.

Pitch The subjective impression of the frequency of sound.

Reflection The return of a sound wave; an echo.

Reverberation Reechoed sound.

Refraction The bending of a wave, either through a nonuniform medium or from one medium to another, caused by differences in wave speed.

Forced vibration The setting up of vibrations in an object by a vibrating force.

Natural frequency A frequency at which an elastic object naturally tends to vibrate, so that minimum energy is required to produce a forced vibration or to continue vibration at that frequency.

Resonance The response of a body when a forcing frequency matches its natural frequency.

Interference A property of all types of waves; a result of superposing different waves, often of the same wavelength. *Constructive interference* results from crest-to-crest reinforcement; *destructive interference* results from crest-to-trough cancellation.

Beats A series of alternate reinforcements and cancellations produced by the interference of two waves of slightly different frequency, heard as a throbbing effect in sound waves.

Standing wave A stationary wave pattern formed in a medium when two sets of identical waves pass through the medium in opposite directions.

Doppler effect The change in frequency of wave motion resulting from motion of the sender or the receiver.

Bow wave The V-shaped wave made by an object moving across a liquid surface at a speed greater than the wave speed.

Shock wave The cone-shaped wave made by an object moving at supersonic speed through a fluid.

Sonic boom The loud sound resulting from a shock wave.

Quality The characteristic timbre of a musical sound, which is governed by the number and relative intensities of partial tones.

Partial tone One of the frequencies present in a complex tone. When a partial tone is an integer multiple of the lowest frequency, it is a harmonic.

Fundamental frequency The lowest frequency of vibration, or the first harmonic. In a string, the vibration makes a single segment.

Harmonic A partial tone that is an integer multiple of the fundamental frequency. The vibration that begins with the fundamental vibrating frequency is the first harmonic, twice the fundamental is the second harmonic, and so on in sequence.

REVIEW QUESTIONS

10.1 Vibrations and Waves

1. What is the source of all waves?
2. Distinguish between these different parts of a wave: period, amplitude, wavelength, and frequency.
3. How do frequency and period relate to each other?

10.2 Wave Motion

4. In one word, what is it that moves from source to receiver in wave motion?
5. Does the medium in which a wave travels move with the wave? Give examples to support your answer.
6. What is the relationship among frequency, wavelength, and wave speed?

10.3 Transverse and Longitudinal Waves

7. In what direction are the vibrations in a transverse wave, relative to the direction of wave travel? In a longitudinal wave?
8. Distinguish between a compression and a rarefaction.

10.4 Sound Waves

9. Does sound travel faster in warm air or in cold air? Defend your answer.
10. How does the speed of sound in water compare with the speed of sound in air? How does the speed of sound in steel compare with the speed of sound in air?

10.5 Reflection and Refraction of Sound

11. What is the law of reflection for sound?
12. What is a reverberation?
13. How does wave speed relate to the phenomenon of refraction?
14. Does sound tend to bend upward or downward when its speed near the ground is lower than its speed higher up?
15. There is a difference between the way in which we passively see our surroundings in daylight and the way in which we actively probe our surroundings with a searchlight in the darkness. Which of these ways of perceiving

our surroundings is more like the way in which a dolphin perceives its environment?

10.6 Forced Vibrations and Resonance

16. Why does a struck tuning fork sound louder when its handle is held against a table?
17. Distinguish between forced vibrations and resonance.
18. When you listen to a radio, why do you hear only one station at a time instead of all stations at once?
19. Why do troops "break step" when crossing a bridge?

10.7 Interference

20. What kinds of waves exhibit interference?
21. Distinguish between constructive interference and destructive interference.
22. What does it mean to say that one wave is out of phase with another?
23. What physical phenomenon underlies beats?
24. What is a node? What is an antinode?

10.8 Doppler Effect

25. In the Doppler effect, does frequency change? Does wavelength change? Does wave speed change?
26. Can the Doppler effect be observed with longitudinal waves, with transverse waves, or with both?

10.9 Bow Waves and the Sonic Boom

27. How do the speed of a wave source and the speed of the waves themselves compare when a wave barrier is being produced? How do they compare when a bow wave is being produced?
28. How does the V shape of a bow wave depend on the speed of the wave source?
29. True or false: A sonic boom occurs only when an aircraft is breaking through the sound barrier. Defend your answer.

10.10 Musical Sounds

30. Distinguish between a musical sound and noise.

EXERCISES

● BEGINNER   ■ INTERMEDIATE   ◆ EXPERT

1. ● What is the source of wave motion?
2. ● If we double the frequency of a vibrating object, what happens to its period?
3. ● You dip your finger repeatedly into a puddle of water and make waves. What happens to the wavelength if you dip your finger more frequently?
4. ● How does the frequency of vibration of a small object floating in water compare to the number of waves passing it each second?
5. ● What kind of motion should you impart to the nozzle of a garden hose so that the resulting stream of water approximates a sine curve?
6. ● What kind of motion should you impart to a stretched coiled spring (or to a Slinky) to produce a transverse wave? A longitudinal wave?
7. ● If a gas tap is turned on for a few seconds, someone a couple of meters away hears the gas escaping long before he or she smells it. What does this indicate about the speed of sound and the motion of molecules in the sound-carrying medium?
8. ● A cat can hear sound frequencies up to 70,000 Hz. Bats send and receive ultrahigh-frequency squeaks up to 120,000 Hz. Which animals hear sound of shorter wavelengths, cats or bats?
9. ● What does it mean to say that a radio station is "at 101.1 on your FM dial"?
10. ● Sound from source A has twice the frequency of sound from source B. Compare the wavelengths of sound from the two sources.
11. ● Suppose a sound wave and an electromagnetic wave have the same frequency. Which has the longer wavelength?
12. ● At the stands of a racetrack, you notice smoke from the starter's gun before you hear it fire. Explain.
13. ● In an Olympic competition, a microphone picks up the sound of the starter's gun and sends it electrically to speakers at every runner's starting block. Why?
14. ● At the instant that a high-pressure region is created just outside the prongs of a vibrating tuning fork, what is being created inside between the prongs?
15. ● Why is it so quiet after a snowfall?
16. ● If a bell is ringing inside a bell jar, we can no longer hear it when the air is pumped out, but we can still see it. What differences in the properties of sound and light does this indicate?
17. ● Why is the Moon described as a "silent planet"?
18. ● As you pour water into a glass, you repeatedly tap the glass with a spoon. As the tapped glass is being filled, does the pitch of the sound increase or decrease? (What should you do to answer this question?)
19. ● If the speed of sound depended on its frequency, would you enjoy a concert sitting in the second balcony?
20. ● If the frequency of sound is doubled, what change occurs in its speed? What change occurs in its wavelength? Defend your answer.
21. ● Why does sound travel faster in warm air?
22. ● Why does sound travel faster in moist air? (Hint: At the same temperature, water-vapor molecules have the same average kinetic energy as the heavier nitrogen and oxygen molecules in the air. How, then, do the average speeds of H_2O molecules compare with those of N_2 and O_2 molecules?)
23. ● Why is an echo weaker than the original sound?
24. ● What two physics mistakes occur in a science-fiction movie that shows a distant explosion in outer space that you see and hear at the same time?
25. ● A rule of thumb for estimating the distance in kilometers between an observer and a lightning stroke is to divide the number of seconds in the interval between the flash and the sound by 3. Is this rule correct?
26. ■ If a single disturbance some unknown distance away sends out both transverse and longitudinal waves that travel with distinctly different speeds in the medium, such as in the ground during an earthquake, how could the distance to the disturbance be determined?
27. ● Why are marchers at the end of a long parade following a band out of step with marchers near the front?
28. ● What is the danger posed by people in the balcony of an auditorium stamping their feet in a steady rhythm?
29. ● Why is the sound of a harp soft in comparison with the sound of a piano?
30. ● If the handle of a tuning fork is held solidly against a tabletop, the sound from the tuning fork becomes louder. Why? How does this affect the length of time the fork keeps vibrating? Explain.
31. ● What physics principle does Manuel use when he pumps in rhythm with the natural frequency of the swing?

32. ● The sitar, an Indian musical instrument, has a set of strings that vibrate and produce music, even though the player never plucks them. These "sympathetic strings" are identical to the plucked strings and are mounted below them. What is your explanation?
33. ● A special device can transmit sound that is out of phase with the sound of a noisy jackhammer to the jackhammer operator by means of earphones. Over the noise of the jackhammer, the operator can easily hear your voice while you are unable to hear his. Explain.
34. ■ Two sound waves of the same frequency can interfere with each other, but two sound waves must have different frequencies in order to make beats. Why?
35. ■ Walking beside you, your friend takes 50 strides per minute while you take 48 strides per minute. If you start in step, when will you be in step again?
36. ■ Suppose a piano tuner hears three beats per second when listening to the combined sound from his tuning

fork and the piano note being tuned. After slightly tightening the string, he hears five beats per second. Should the string be loosened or tightened?

37. ■ A railroad locomotive is at rest with its whistle shrieking, and then it starts moving toward you.
(a) Does the frequency that you hear increase, decrease, or stay the same?
(b) How about the wavelength reaching your ear?
(c) How about the speed of sound in the air between you and the locomotive?

38. ■ When you blow your horn while driving toward a stationary listener, he hears an increase in the frequency of the horn. Would the listener hear an increase in the frequency of the horn if he were also in a car traveling at the same speed in the same direction as you are? Explain.

39. ● How does the Doppler effect aid police in detecting speeding motorists?

40. ● Astronomers find that light emitted by a particular element at one edge of the Sun has a slightly higher frequency than light from that element at the opposite edge. What do these measurements tell us about the Sun's motion?

41. ● Would it be correct to say that the Doppler effect is the apparent change in the speed of a wave due to motion of the source? (Why is this question a test of reading comprehension as well as a test of physics knowledge?)

42. ● Does the conical angle of a shock wave open wider, narrow down, or remain constant as a supersonic aircraft increases its speed?

43. ● If the sound of an airplane does not originate in the part of the sky where the plane is seen, does this imply that the airplane is traveling faster than the speed of sound? Explain.

44. ● Does a sonic boom occur at the moment when an aircraft exceeds the speed of sound? Explain.

45. ■ Why is it that a subsonic aircraft, no matter how loud it may be, cannot produce a sonic boom?

PROBLEMS

● BEGINNER   ■ INTERMEDIATE   ◆ EXPERT

1. ● A nurse counts 72 heartbeats in 1 min. Show that the period and frequency of the heartbeats are 0.83 s and 1.2 Hz, respectively.

2. ● We know that speed $v =$ distance/time. Show that when the distance traveled is one wavelength λ and the time of travel is the period T (which equals 1/frequency) you get $v = f\lambda$.

3. ● Microwave ovens typically cook food using microwaves with frequency 2.45 GHz (gigahertz, 10^9 Hz). Show that the wavelength of these microwaves is 12.2 cm.

4. ● For years, marine scientists were mystified by sound waves detected by underwater microphones in the Pacific Ocean. These so-called T waves were among the purest sounds in nature. Eventually the researchers traced the source to underwater volcanoes whose rising columns of bubbles resonated like organ pipes. A typical T wave has a frequency of 7 Hz. Knowing that the speed of sound in seawater is 1530 m/s, show that the wavelength of a T wave is 219 m.

5. ● An oceanic depth-sounding vessel surveys the ocean bottom with ultrasonic waves that travel 1530 m/s in seawater. Show that when the time delay of an echo to the ocean floor below is 6 s, the depth of the water is 4590 m.

6. ● A bat flying in a cave emits a sound and receives its echo 0.1 s later. Show that the distance to the wall of the cave is 17 m.

7. ● Susie hammers on a block of wood when she is 85 m from a large brick wall. Each time she hits the block, she hears an echo 0.5 s later. With this information, show that the speed of sound is 230 m/s.

8. ● Imagine an old hermit type who lives in the mountains. Just before going to sleep, he yells "WAKE UP," and the sound echoes off the nearest mountain and returns 8 hours later. Show that the mountain is almost 5000 km distant.

9. ● On a keyboard, you strike a note having a frequency of 256 Hz. (a) Show that the period of one vibration of this tone is 0.00391 s. (b) As the sound leaves the instrument at a speed of 340 m/s, show that its wavelength in air is 1.33 m.

10. ■ (a) If you were so foolish as to play your keyboard instrument under water, where the speed of sound is 1,500 m/s, show that the wavelength of the middle-C tone in water would be 5.86 m. (b) Explain why middle C (or any other tone) has a longer wavelength in water than in air.

11. ● What beat frequencies are possible with tuning forks of frequencies 256, 259, and 261 Hz?

12. ■ As shown in the drawing, the half-angle of the shock-wave cone generated by a supersonic aircraft is 45°. What is the speed of the plane relative to the speed of sound?

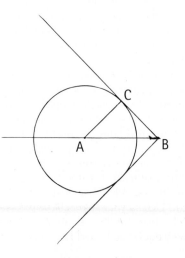

ACTIVE EXPLORATIONS

1. Tie a rubber tube, a spring, or a rope to a fixed support and shake it to produce standing waves. See how many nodes you can produce.

2. Test to see which of your ears has better hearing by covering one ear and finding how far away your open ear can hear the ticking of a clock; repeat for the other ear. Notice also how the sensitivity of your hearing improves when you cup your hands behind your ears.

3. Do the activity suggested in Figure 10.24 with a stereo sound system. Simply reverse the wire inputs to one of the speakers so the two are out of phase. When monaural sound is played and the speakers are brought face to face, the lowering of volume is truly amazing! If the speakers are well insulated, you hear almost no sound at all.

4. For this activity, you'll need an isolated loudspeaker (bare of its casing) and a sheet of plywood or cardboard—the bigger the better. Cut a hole in the middle part of the sheet that is about the size of the speaker. Listen to music from the isolated speaker, and then hear the difference when it's placed against the hole. The sheet diminishes the amount of sound from the back of the speaker that interferes with sound coming from the front side, producing a much fuller sound. Now you know why speakers are mounted in enclosures.

5. Wet your finger and rub it slowly around the rim of a thin-rimmed, stemmed glass while you hold the base of the glass firmly to a tabletop with your other hand. The friction of your finger excites standing waves in the glass, much like the wave made on the strings of a violin by the friction from a violin bow. Try it with a metal bowl.

6. Swing a buzzer of any kind over your head in a circle. You won't hear the Doppler shift, but your friends off to the side will. The pitch will increase as it approaches them, and decrease when it recedes. Then switch places with a friend so you can hear it too.

7. Make the lowest-pitched vocal sound you are capable of; then keep doubling the pitch to see how many octaves your voice can span.

8. Blow over the top of two identical empty bottles and see if the tones produced are of the same pitch. Then put one in a freezer and try the procedure again. Sound travels more slowly in the cold denser air of the cold bottle and the note is lower. Try it and see.

READINESS ASSURANCE TEST (RAT)

If you have a good handle on this chapter, if you really do, then you should be able to score at least 7 out of 10 on this RAT. If you score less than 7, you need to study further before moving on.

Choose the BEST answer to each of the following.

1. When we consider the distance a pendulum swings to and fro, we're talking about its
   (a) frequency.
   (b) period.
   (c) wavelength.
   (d) amplitude.

2. If the frequency of a particular wave is 30 Hz, its period is
   (a) 1/30 second.
   (b) 30 seconds.
   (c) more than 30 seconds.
   (d) none of these.

3. In Europe, alternating electric current vibrates to and fro 50 cycles in 1 s. The frequency of these vibrations is
   (a) 50 Hz with a period of 1/50 s.
   (b) 1/50 Hz with a period of 50 s.
   (c) 50 Hz with a period of 50 s.
   (d) 1/50 Hz with a period of 1/50 s.

4. If you dip your finger repeatedly onto the surface of still water, you produce waves. The more frequently you dip your finger,
   (a) the lower the wave frequency and the longer the wavelengths.
   (b) the higher the wave frequency and the shorter the wavelengths.
   (c) strangely, both of these
   (d) neither of these

5. The vibrations along a longitudinal wave move in a direction
   (a) parallel to the wave direction.
   (b) perpendicular to the wave direction.
   (c) both of these
   (d) neither of these

6. A common example of a transverse wave is
   (a) sound.
   (b) light.
   (c) both of these
   (d) neither of these

7. When your radio set is tuned to an incoming radio signal, what is occurring?
   (a) refraction
   (b) forced vibration
   (c) resonance
   (d) diffraction

8. When sound or light undergoes interference, it can sometimes
   (a) build up to an amplitude greater than the sum of amplitudes.

(b) cancel completely.
(c) both of these
(d) neither of these

9. What does NOT occur with the Doppler effect are changes in
   (a) frequency due to motion.
   (b) the speed of sound due to motion.
   (c) both of these
   (d) neither of these

10. A sonic boom is the result of wave
    (a) interference.
    (b) resonance.
    (c) superposition.
    (d) reflection and refraction.

Answers to RAT

1d, 2a, 3a, 4b, 5a, 6b, 7c, 8b, 9b, 10c.

MORE TO EXPLORE

Chiaverina, Chris, and Tom Rossing. *Light Science: Physics for the Visual Arts*. New York: Springer, 1999. Enjoyable reading by two fun physicists!

For more on the acoustics of concert halls, see http://www.concerthalls.org.

CHAPTER 10 ONLINE RESOURCES

The Physical Science Place

Interactive Figures

- 10.1, 10.5, 10.6, 10.27, 10.28, 10.30, 10.31

Tutorials

- Waves and Vibrations
- Doppler Effect

Videos

- Transverse vs. Longitudinal Waves
- Refraction of Sound

- Resonance
- Resonance and Bridges
- Interference and Beats
- Doppler Effect

Quiz

Flashcards

Links

light

11.1 ELECTROMAGNETIC SPECTRUM

11.2 TRANSPARENT AND OPAQUE MATERIALS

11.3 REFLECTION

11.4 REFRACTION

11.5 COLOR

11.6 DISPERSION

11.7 POLARIZATION

■ Light is the only thing we can really see. But what is light? We know that during the day, the primary source of light is the Sun, and a secondary source is the brightness of the sky. Other common sources are white-hot filaments in lightbulbs, glowing gases in glass tubes, and flames. We find that light originates from the accelerated motion of electrons. Light is an electromagnetic phenomenon, and it is only a tiny part of a larger whole—a wide range of electromagnetic waves called the *electromagnetic spectrum*. We begin our study of light by investigating its electromagnetic properties, how it interacts with materials, and how it reflects. We'll see its transverse wave nature in how it refracts and how we see its colors, quite spectacularly as rainbows. We'll conclude this exciting chapter with the phenomenon of polarization.

The Physical Science Place
Light and Spectroscopy

■ 11.1 Electromagnetic Spectrum

If you shake the end of a stick back and forth in still water, you create waves on the water's surface. If you similarly shake an electrically charged rod to and fro in empty space, you create electromagnetic waves in space. We learned in Chapter 9 why this is so: The shaking stick creates an electric current around which is generated a magnetic field, and the changing magnetic field induces an electric field—electromagnetic induction. The changing electric field in turn induces a changing magnetic field. The vibrating electric and magnetic fields regenerate each other to make up an **electromagnetic wave**.

FIGURE 11.1

If you shake an electrically charged object to and fro, you produce an electromagnetic wave.

FIGURE 11.2

INTERACTIVE FIGURE

The electric and magnetic fields of an electromagnetic wave in free space are perpendicular to each other and to the direction of motion of the wave.

Link To Sections 28.1, 28.2

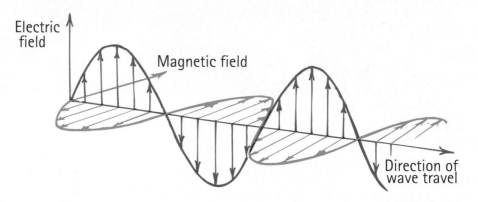

Light is energy carried in an electromagnetic wave emitted by vibrating electrons in atoms.

In a vacuum, all electromagnetic waves move at the same speed, differing only in frequency. The classification of electromagnetic waves according to frequency, from radio waves to gamma rays, is the **electromagnetic spectrum** (Figure 11.3). Electromagnetic waves have been measured from 0.01 Hz to radio frequencies up to 108 MHz. Then come ultrahigh frequencies (UHF), followed by microwaves, beyond which are infrared waves, often called *heat waves*. Further still is visible light, which makes up less than a millionth of 1 percent of the measured electromagnetic spectrum. The higher the frequency of the wave, the shorter its wavelength.*

FIGURE 11.3

INTERACTIVE FIGURE

The electromagnetic spectrum is a continuous range of waves extending from radio waves to gamma rays. The descriptive names of the sections are merely a historical classification, for all waves are the same in their basic nature, differing principally in frequency and wavelength; all of the waves have the same speed.

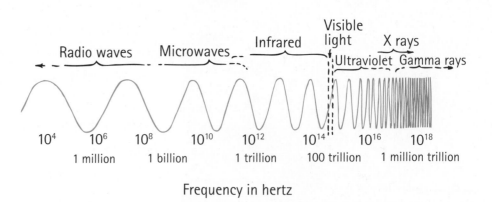

* The relationship is $c = f\lambda$, where c is the speed of light (constant), f is the frequency, and λ is the wavelength. It is common to describe sound and radio by frequency and light by wavelength. In this book, however, we'll favor the single concept of frequency in describing light.

CHECK
 POINT

Is it correct to say that a radio wave is a low-frequency light wave? Is a radio wave also a sound wave?

Were these your answers?

Yes and no. Both radio waves and light waves are electromagnetic waves that originate in the vibrations of electrons. Radio waves have lower frequencies than light waves, so a radio wave might be considered a low-frequency light wave (and a light wave might be considered a high-frequency radio wave). But a radio wave is definitely not a sound wave, which we learned in the previous chapter is a mechanical vibration of matter. (Don't confuse a radio wave with the sound that a loudspeaker emits.)

11.2 Transparent and Opaque Materials

Vibrating electrons emit most electromagnetic waves. When light is incident on matter, some of the electrons in the matter are forced into vibration. Electrons vibrations are then transmitted to the vibrations of other electrons in the material. This is similar to the way that sound is transmitted (Figure 11.4).

The Physical Science Place

Light and Transparent Materials

FIGURE 11.4

Just as a sound wave can force a sound receiver into vibration, a light wave can force the electrons in materials into vibration.

Materials such as glass and water allow light to pass through without absorption, usually in straight lines. These materials are **transparent** to light. To understand how light penetrates a transparent material, visualize the electrons in an atom as if they were connected to the atomic nucleus by springs (Figure 11.5).* An incident light wave sets the electrons into vibration.

The vibration of electrons in a material is similar to the vibrations of ringing bells and tuning forks. Bells ring at a particular frequency, and tuning forks vibrate at a particular frequency—and so do the electrons of atoms and molecules. Different atoms and molecules have different "spring strengths." Electrons in glass have a natural vibration frequency in the ultraviolet range. When ultraviolet rays in sunlight shine on glass, resonance occurs as the wave builds and maintains a large amplitude of electron vibration, just as pushing someone at the resonant frequency on a swing builds a large amplitude. Resonating atoms in the glass can hold on to the energy of the ultraviolet light for quite a long time (about 100 millionths of a second). During this time, the atom undergoes about 1 million vibrations, collides with neighboring atoms, and transfers

In air, light travels a million times faster than sound.

FIGURE 11.5

The electrons of atoms have certain natural frequencies of vibration, which can be modeled as particles connected to the atomic nucleus by springs. As a result, atoms and molecules behave somewhat like optical tuning forks.

---

* Electrons, of course, are not really connected by springs. Here we present a visual "spring model" of the atom to help us understand the interaction of light with matter. The worth of a model lies not in whether it is "true" but in whether it is useful—in explaining observations and predicting new ones. The simplified model that we present here—of an atom whose electrons vibrate as if on springs, with a time interval between absorbing energy and re-emitting it—is quite useful for understanding how light passes through a transparent material.

FIGURE 11.6

A light wave incident on a pane of glass sets up vibrations in the molecules that produce a chain of absorptions and re-emissions, which pass the light energy through the material and out the other side. Because of the time delay between absorptions and re-emissions, the light travels through the glass more slowly than through empty space.

3 of many atoms

Glass

fyi

■ Materials such as glass are transparent only for creatures that see in the "visible" part of the spectrum. Other creatures that are tuned to different frequency ranges see glass as opaque and other materials as transparent.

absorbed energy as thermal energy. Thus, glass is not transparent to ultraviolet. Glass absorbs ultraviolet.

At lower wave frequencies, such as those of visible light, electrons in the glass are forced into vibration at a lower amplitude. The atoms or molecules in the glass hold the energy for less time, with less chance of collision with neighboring atoms and molecules, and less of the energy is transformed to heat. Instead, the energy of vibrating electrons is re-emitted as light. Glass is transparent to all the frequencies of visible light. The frequency of re-emitted light passed from molecule to molecule is identical to the frequency of the original light that produced the vibration. However, there is a slight time delay between absorption and reemission.

This time delay lowers the average speed of light through the material (Figure 11.6). Light of different frequencies travels at different average speeds through different materials. We say *average speeds*, for the speed of light in a vacuum is a

FIGURE 11.7

When the raised ball is released and hits the others, the ball that emerges from the opposite side is not the same ball that initiated the transfer of energy. Likewise, each photon that emerges from a pane of glass is not the same photon that was incident on the glass. Both the emerging ball and emerging photon are different from, though identical to, the incident ones.

constant 300,000 kilometers per second. We call this speed of light c.* The speed of light in the atmosphere is slightly less than it is in a vacuum, but is usually rounded off as c. In water, light travels at 75 percent of its speed in a vacuum, or $0.75c$. In glass, light travels about $0.67c$, depending on the type of glass. In a diamond, light travels at less than half its speed in a vacuum, only $0.41c$. Light travels even slower in a silicon carbide crystal called *carborundum*. When light emerges from these materials into the air, it travels at its original speed.

Link To Sections 20.3, 24.5

Infrared waves, which have frequencies lower than those of visible light, vibrate not only the electrons but the entire molecules in the structure of the glass and in many other materials. This molecular vibration increases the thermal energy and temperature of the material, which is why infrared waves are often called *heat waves*. Glass is transparent to visible light, but not to ultraviolet and infrared light.

FIGURE 11.8

Clear glass blocks both infrared and ultraviolet, but it is transparent to all the frequencies of visible light.

Ultraviolet

Visible

Infrared

Glass

---

* The presently accepted value is 299,792 km/s, which is often rounded to 300,000 km/s. (This corresponds to 186,000 mi/s.)

CHECK POINT

1. Why is glass transparent to visible light but opaque to ultraviolet and infrared?
2. Pretend that while you are at a social gathering, you make several momentary stops across the room to greet people who are "on your wavelength." How is this analogous to light traveling through glass?

Were these your answers?

1. The natural frequency of vibration for electrons in glass is the same as the frequency of ultraviolet light, so resonance in glass occurs when ultraviolet waves shine on glass. The absorbed energy is transferred to other atoms as heat, not re-emitted as light, so the glass is opaque at ultraviolet frequencies. In the range of visible light, forced vibration of electrons occurs at smaller amplitudes—vibrations are more subtle. So re-emission of light (rather than the generation of heat) occurs, and the glass is transparent. Lower-frequency infrared light causes whole molecules, rather than electrons, to resonate; again, heat is generated and the glass is opaque.
2. Your average speed across the room would be less because of the time delays associated with your momentary stops. Likewise, the speed of light in glass is less because of the time delays in interactions with atoms along its path.

Most things around us are **opaque**—they absorb light without re-emission. Books, desks, chairs, and people are opaque. Energetic vibrations produced by incident light on the atoms of these materials is turned into random kinetic energy—into thermal energy. The materials become slightly warmer.

Metals are opaque to visible light. The outer electrons of atoms in metals are not bound to any particular atom. They are loose and free to wander, with very little restraint, throughout the material (which is why metal conducts electricity and heat so well). When light shines on metal and sets these free electrons into vibration, their energy does not "spring" from atom to atom in the material. It is reflected instead. That's why metals are shiny.

Earth's atmosphere is transparent to some ultraviolet light, to all visible light, and to some infrared light. But the atmosphere is opaque to high-frequency ultraviolet light. The small amount of ultraviolet light that does penetrate causes sunburns. If all ultraviolet light penetrated the atmosphere, we would be fried to a crisp. Clouds are semitransparent to ultraviolet light, which is why you can get a sunburn on a cloudy day. Ultraviolet light is not only harmful to your skin, it is also damaging to tar roofs. Now you know why tarred roofs are often covered with gravel.

Have you noticed that things look darker when they are wet than when they are dry? Light incident on a dry surface, such as sand, bounces directly to your eye. But light incident on a wet surface bounces around inside the transparent wet region before it reaches your eye. What happens with each bounce? Absorption! So sand and other things look darker when wet.

CHECK POINT

What are two common fates for light shining on a material that isn't absorbed?

Was this your answer?

Transmission and/or reflection. Most light incident on a pane of glass, for example, is transmitted through the pane. But some reflects from its surface. How much transmits and how much reflects varies with different conditions.

FIGURE 11.9

Metals are shiny because light that shines on them forces free electrons into vibration, which then emit their "own" light waves as reflection.

Link To Section 25.3

Lateral Inhibition

The human eye can do what no camera film can do: it can perceive degrees of brightness that range from about 500 million to 1. The difference in brightness between the Sun and Moon, for example, is about 1 million to 1. But because of an effect called lateral inhibition, we don't perceive the actual differences in brightness. The brightest places in our visual field are prevented from outshining the rest, because whenever a receptor cell on our retina sends a strong brightness signal to our brain, it also signals neighboring cells to dim their responses. In this way, we even out our visual field, which allows us to discern detail in very bright areas and in dark areas as well.

Lateral inhibition exaggerates the difference in brightness at the edges of places in our visual field. Edges, by definition, separate one thing from another. So we accentuate differences rather than similarities. This is illustrated in the pair of shaded rectangles to the right. They appear to be different shades of brightness because of the edge that separates them. But cover the edge with your pencil or your finger, and they look equally bright (try it now)! That's because both rectangles are equally bright; each rectangle is shaded from lighter to darker, moving from left to right. Our eye concentrates on the boundary where the dark edge of the left rectangle joins the light edge of the right rectangle, and our eye–brain system assumes that the rest of the rectangle is the same. We pay attention to the boundary and ignore the rest.

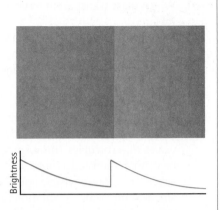

Questions to ponder: Is the way the eye picks out edges and makes assumptions about what lies beyond similar to the way in which we sometimes make judgments about other cultures and other people? Don't we, in the same way, tend to exaggerate the differences on the surface while ignoring the similarities and subtle differences within?

The Physical Science Place
Image Formation in a Mirror

Link To Section 24.1

11.3 Reflection

When this page is illuminated by sunlight or lamplight, electrons in the atoms of the paper are set into vibration. The energized electrons re-emit the light by which we see the page. Light undergoes **reflection** (properly called *specula reflection*). When the page is illuminated by white light, it appears white because the electrons re-emit all the visible frequencies. They reflect all of the light. Very little absorption occurs. The ink on the page is a different story. Except for a bit of reflection, the ink absorbs all the visible frequencies and therefore appears black.

LAW OF REFLECTION

Anyone who has played pool or billiards knows that when a ball bounces from a surface, the angle of incidence is equal to the angle of rebound. The same is true of light. This is the **law of reflection**, which holds for all angles:

The angle of reflection equals the angle of incidence.

The law of reflection is illustrated with arrows representing light rays in Figure 11.10. Instead of measuring the angles of incident and reflected rays from the reflecting surface, it is customary to measure them from a line perpendicular to the plane of the reflecting surface. This imaginary line is called the *normal*. The incident ray, the normal, and the reflected ray all lie in the same plane.

If you place a candle in front of a mirror, rays of light radiate from the flame in all directions. Figure 11.11 shows only four of the infinite number of rays leaving one of the infinite number

FIGURE 11.10

INTERACTIVE FIGURE

The law of reflection.

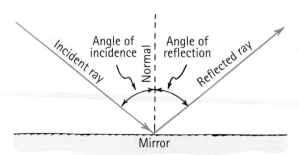

of points on the candle. When these rays meet the mirror, they reflect at angles equal to their angles of incidence. The rays diverge from the flame. Note that they also diverge when reflecting from the mirror. These divergent rays appear to emanate from behind the mirror (dashed lines). You see an image of the candle at this point. The light rays do not actually come from this point, so the image is called a *virtual image*. The image is as far behind the mirror as the object is in front of the mirror, and image and object have the same size—as long as the mirror is flat. A flat mirror is called a *plane mirror*.

When the mirror is curved, the sizes and distances of object and image are no longer equal. We will not study curved mirrors in this text, except to say that a curved mirror behaves as a succession of flat mirrors, each at a slightly different angular orientation from the one next to it. At each point, the angle of incidence is equal to the angle of reflection (Figure 11.13). Note that in a curved mirror, unlike in a plane mirror, the normals (shown by the dashed black lines) at different points on the surface are not parallel to one another.

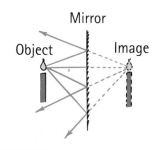

FIGURE 11.11

A virtual image is formed behind the mirror and is located at the position where the extended reflected rays (dashed lines) converge.

FIGURE 11.12

Marjorie's image is as far behind the mirror as she is in front of it. Note that she and her image have the same color of clothing—evidence that light doesn't change frequency upon reflection. Interestingly, her left-and-right axis is no more reversed than her up-and-down axis. The axis that is reversed, as shown to the right, is her front-and-back axis. That's why it appears that her left hand faces the right hand of her image.

Your image behind a plane mirror is as if your twin stood behind a pane of clear glass at a distance as far behind the glass as you are in front of it.

Whether the mirror is plane or curved, the eye–brain system cannot ordinarily distinguish between an object and its reflected image. So the illusion that an object exists behind a mirror (or, in some cases, in front of a concave mirror) is merely due to the fact that the light from the object enters the eye in exactly the same manner, physically, as it would have entered if the object really were at the image location.

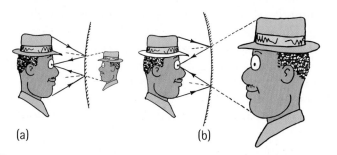

(a)   (b)

FIGURE 11.13

(a) The virtual image formed by a convex mirror (a mirror that curves outward) is smaller and closer to the mirror than the object. (b) When the object is close to a concave mirror (a mirror that curves inward like a "cave"), the virtual image is larger and farther away than the object. In either case, the law of reflection holds for each ray.

CHECK POINT

If you wish to take a picture of your image while standing 5 m in front of a plane mirror, for what distance should you set your camera to provide the sharpest focus?

Was this your answer?
Set the distance for 10 m, the distance between the camera and your image.

FIGURE 11.14

Diffuse reflection. Although reflection of each single ray obeys the law of reflection, the many different surface angles that light rays encounter in striking a rough surface produce reflection in many directions.

FIGURE 11.15

The open-mesh parabolic dish is a diffuse reflector for short-wavelength light but a polished reflector for long-wavelength radio waves.

FIGURE 11.16

A magnified view of the surface of ordinary paper.

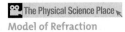
The Physical Science Place
Model of Refraction

Only part of the light that strikes a surface is reflected. For example, on a surface of clear glass and for normal incidence (light perpendicular to the surface), only about 4 percent is reflected from each surface. On a clean and polished aluminum or silver surface, however, about 90 percent of the incident light is reflected.

DIFFUSE REFLECTION

In contrast to specula reflection is **diffuse reflection**, which occurs when light is incident on a rough surface and reflected in many directions (Figure 11.14). If the surface is so smooth that the distances between successive elevations on the surface are less than about one-eighth the wavelength of the light, there is very little diffuse reflection, and the surface is said to be *polished*. A surface therefore may be polished for radiation of long wavelengths but rough for light of short wavelengths. The wire-mesh "dish" shown in Figure 11.15 is very rough for light waves and is hardly mirrorlike. But for long-wavelength radio waves, it is "polished" and is an excellent reflector.

Light reflecting from this page is diffuse. The page may be smooth to a radio wave, but to a light wave it is rough. Smoothness is relative to the wavelength of the illuminating waves. Rays of light striking this page encounter millions of tiny flat surfaces facing in all directions. The incident light, therefore, is reflected in all directions. This is desirable, for it enables us to see this page and other objects from any direction or position. You can see the road ahead of your car at night, for instance, because of diffuse reflection by the rough road surface. When the road is wet, however, it is smoother with less diffuse reflection, and therefore more difficult to see. Most of our environment is seen by diffuse reflection.

■ 11.4 Refraction

As we learned in Section 11.2, light slows down when it enters glass, and it travels at different speeds in different materials.* It travels at 300,000 km/s in a vacuum, at a slightly lower speed in air, and at about three-fourths that speed in water. Unless the light is perpendicular to the surface of penetration, bending occurs. This is the phenomenon of **refraction**.

To gain a better understanding of the bending of light in refraction, look at the pair of toy cart wheels in Figure 11.17. The wheels roll from a smooth sidewalk onto a grass lawn. If the wheels meet the grass at an angle, as the figure shows, they are deflected from their straight-line course. Note that the left

---

* Just how much the speed of light differs from its speed in a vacuum is given by the index of refraction, n, of the material:

$$n = \frac{\text{speed of light in vacuum}}{\text{speed of light in material}}$$

For example, the speed of light in a diamond is 124,000 km/s, and so the index of refraction for diamond is

$$n = \frac{300{,}000 \text{ km/s}}{124{,}000 \text{ km/s}} = 2.42$$

For a vacuum, $n = 1$.

FIGURE 11.17

The direction of the rolling wheels changes when one wheel slows down before the other does.

FIGURE 11.18

The direction of the light waves changes when one part of the wave slows down before the other part.

FIGURE 11.19

INTERACTIVE FIGURE

Refraction. The angles of incidence and refraction are in accord with Snell's law (see footnote).

wheel slows first when it interacts with the grass on the lawn. The right wheel maintains its higher speed while on the sidewalk. It pivots about the slower-moving left wheel because it travels farther in the same time. So the direction of the rolling wheels is bent toward the "normal," the black dashed line perpendicular to the grass-sidewalk border in Figure 11.17.

Figure 11.18 shows how a light wave bends in a similar way. Note the direction of light, indicated by the blue arrow (the light ray). Also note the *wave fronts* drawn at right angles to the ray. (If the light source were close, the wave fronts would appear circular; but if the distant Sun is the source, the wave fronts are practically straight lines.) The wave fronts are everywhere at right angles to the light rays. The bending of the wave (sound or light) is caused by a change of speed.*

Figure 11.20 shows a beam of light entering water at the left and exiting at the right. The path would be the same if the light entered from the right and exited at the left. The light paths are reversible for both reflection and refraction. If you see someone's eyes by way of a reflective or refractive device, such as a mirror or a prism, then that person can see you by way of the device also (unless the device is optically coated to produce a one-way effect).

Refraction causes many illusions. One of them is the apparent bending of a stick that is partially submerged in water. The submerged part appears closer to the surface than it actually is. The same is true when you look at a fish in water. The fish appears nearer to the surface and closer than it really is (Figure 11.21). If we look straight down into water, an object submerged 4 meters beneath the

A light ray is always at right angles to its wave front.

FIGURE 11.20

When light slows down in going from one medium to another, as it does in going from air to water, it bends toward the normal. When it speeds up in traveling from one medium to another, as it does in going from water to air, it bends away from the normal.

FIGURE 11.21

Because of refraction, a submerged object appears to be nearer to the surface than it actually is.

Although wave speed and wavelength change when undergoing refraction, frequency remains unchanged. Refraction doesn't change the color of light.

* The quantitative law of refraction, called Snell's law, is credited to Willebrord Snell, a 17th-century Dutch astronomer and mathematician: $n_1 \sin \theta_1 = n_2 \sin \theta_2$, where n_1 and n_2 are the indices of refraction of the media on either side of the surface, and θ_1 and θ_2 are the respective angles of incidence and refraction. If three of these values are known, the fourth can be calculated from this relationship.

For a wave explanation of refraction (and diffraction), read about Huygens' principle, pages 558–560, *Conceptual Physics—10th Edition*.

FIGURE 11.22

Because of atmospheric refraction, when the Sun is near the horizon it appears to be higher in the sky.

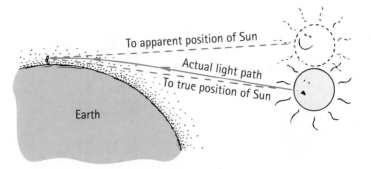

surface appears to be only 3 meters deep. Because of refraction, submerged objects appear to be magnified.

Refraction occurs in Earth's atmosphere. Whenever we watch a sunset, we see the Sun for several minutes after it has sunk below the horizon (Figure 11.22). Earth's atmosphere is thin at the top and dense at the bottom. Because light travels faster in thin air than in dense air, parts of the wave fronts of sunlight at high altitude travel faster than parts closer to the ground. Light rays bend. The density of the atmosphere changes gradually, so light rays bend gradually and follow a curved path. So we gain additional minutes of daylight each day. Furthermore, when the Sun (or Moon) is near the horizon, the rays from the lower edge are bent more than the rays from the upper edge. This shortens the vertical diameter, causing the Sun to appear elliptical (Figure 11.23).

A mirage occurs when refracted light appears as if it were reflected light. Mirages are a common sight on a desert when the sky appears to be reflected from water on the distant sand. But when you approach what seems to be water, you find dry sand. Why is this so? The air is very hot close to the sand surface and cooler above the sand. Light travels faster through the thinner hot air near the surface than through the denser cool air above. So wave fronts near the ground travel faster than they do above. The result is upward bending (Figure 11.24). So we see an upside-down view that looks as if reflection were occurring from a water surface. We see a mirage, which is formed by real light and can be photographed (Figure 11.25). A mirage is not, as many people think, a trick of the mind.

When we look at an object over a hot stove or over a hot pavement, we see a wavy, shimmering effect. This is due to varying densities of air caused by changes in temperature. The twinkling of stars results from similar variations in the sky, where light passes through unstable layers in the atmosphere.

FIGURE 11.23

The Sun is distorted by differential refraction.

One of the many beauties of physics is the redness of a fully eclipsed Moon—resulting from the refraction of sunsets and sunrises that completely circle the world. This refracted light shines on an otherwise dark Moon.

FIGURE 11.24

Light from the top of the tree gains speed in the warm and less dense air near the ground. When the light grazes the surface and bends upward, the observer sees a mirage.

FIGURE 11.25

A mirage. The apparent wetness of the road is not a reflection of the sky by water but a refraction of skylight through the warmer and less-dense air near the road surface.

Your Eye

With all of today's technology, the most remarkable optical instrument known is your eye. Light enters through your cornea, which does about 70% of the necessary bending of the light before it passes through your pupil (the aperture, or opening, in the iris). Light then passes through your lens, which provides the extra bending power needed to focus images of nearby objects on your extremely sensitive retina. (Only recently have artificial detectors been made with greater sensitivity to light than the human eye.) An image of the visual field outside your eye is spread over the retina. The retina is not uniform. A spot in the center of the retina, called the fovea, is the region of most acute vision. You see greater detail here than at any other part of your retina. There is also a spot on your retina where the nerves carrying all the information exit the eye on their way to the brain. This is your blind spot.

You can demonstrate that you have a blind spot in each eye. Simply hold this book at arm's length, close your left eye, and look at the round dot and the X to its right with your right eye only. You can see both the dot and the X at this distance. Now move the book slowly toward your face, with your right eye fixed on the dot, and you'll reach a position about 20–25 cm from your eye where the X disappears. When both eyes are open, one eye "fills in" the part to which your other eye is blind. Now repeat with only the left eye open, looking this time at the X, and the dot will disappear. But note that your brain fills in the two intersecting lines. Amazingly, your brain fills in the "expected" view even with one eye closed. Instead of seeing nothing, your brain graciously fills in the appropriate background. Repeat this for small objects on various backgrounds. You not only see what's there—you see what's not there!

The light receptors in your retina do not connect directly to your optic nerve but are instead interconnected with many other cells. Through these interconnections, a certain amount of information is combined and "digested" in your retina. In this way, the light signal is "thought about" before it goes to the optic nerve and then to the main body of your brain. So some brain functioning occurs in your eye. Amazingly, your eye does some of your "thinking."

CHECK POINT

If the speed of light were the same in air of various temperatures and densities, would there still be slightly longer daytimes, twinkling stars at night, mirages, and slightly squashed suns at sunset?

Was this your answer?

No.

FIGURE 11.26

Sunlight passing through a prism separates into a color spectrum. The colors of things depend on the colors of the light that illuminates them.

Link To Section 20.3

The Physical Science Place
Color

■ 11.5 Color

Roses are red and violets are blue; colors intrigue artists and physical science types too. To the scientist, the colors of objects are not in the substances of the objects themselves or even in the light they emit or reflect. Color is a physiological experience and is in the eye of the beholder. So when we say that light from a rose petal is red, in a stricter sense we mean that it appears red. Many organisms, including people with defective color vision, do not see the rose as red at all.

Different frequencies of light are perceived as different colors; the lowest frequency we see appears, to most people, as the color red, and the highest appears as violet. Between them range the infinite number of hues that make up the

FIGURE 11.27

The square on the top reflects all the colors illuminating it. In sunlight, it is white. When illuminated with blue light, it is blue. The square on the bottom absorbs all the colors illuminating it. In sunlight, it is warmer than the white square.

The Physical Science Place

Colored Shadows
Yellow-Green Peak of
    Sunlight
Why the Sky Is Blue and
    the Sunset Is Red

fyi

■ Carbon is ordinarily black in color, but not when chemically bonded with water in foods such as bread and potatoes. Water is removed when you overheat your toast, which is why burnt toast is black.

FIGURE 11.28

The bunny's dark fur absorbs all the radiant energy in incident sunlight and therefore appears black. Light fur on other parts of the body reflects light of all frequencies and therefore appears white.

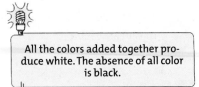

All the colors added together produce white. The absence of all color is black.

color spectrum of the rainbow. By convention, these hues are grouped into seven colors: red, orange, yellow, green, blue, indigo, and violet. These colors together appear white. The white light from the Sun is a composite of all the visible frequencies.

Except for such light sources as lamps, lasers, and gas discharge tubes, most of the objects around us reflect rather than emit light. They reflect only part of the light that is incident upon them, the part that provides their color.

SELECTIVE REFLECTION

A rose, for example, doesn't emit light; it reflects light. If we pass sunlight through a prism and then place the petal of a deep-red rose in various parts of the spectrum, the petal appears brown or black in all regions of the spectrum except in the red region. In the red part of the spectrum, the petal also appears red, but the green stem and leaves appear black. This shows that the petal has the ability to reflect red light, but it cannot reflect other colors; the green leaves have the ability to reflect green light and, likewise, cannot reflect other colors. When the rose is held in white light, the petals appear red and the leaves appear green, because the petals reflect the red part of the white light and the leaves reflect the green part of the white light. To understand why objects reflect specific colors of light, we turn our attention to the atom.

Light is reflected from objects in a manner similar to the way sound is "reflected" from a tuning fork when another tuning fork nearby sets it into vibration. A tuning fork can be made to vibrate even when the frequencies are not matched, although at significantly reduced amplitudes. The same is true of atoms and molecules. Electrons can be forced into vibration by the vibrating electric fields of electromagnetic waves. Once vibrating, these electrons emit their own electromagnetic waves, just as vibrating acoustical tuning forks emit sound waves.

Interestingly, the petals of most yellow flowers, such as daffodils, reflect red and green as well as yellow. Yellow daffodils reflect a broad band of frequencies. The reflected colors of most objects are not pure single-frequency colors but are a mixture of frequencies.

An object can reflect only frequencies present in the illuminating light. An incandescent lamp emits light of lower average frequencies than sunlight, enhancing any reds viewed in this light. In a fabric having only a little bit of red in it, the red is more apparent under an incandescent lamp than it is under a fluorescent lamp. Fluorescent lamps are richer in the higher frequencies, and so blues are enhanced in their light. How a color appears depends on the light source (Figure 11.30).

SELECTIVE TRANSMISSION

The color of a transparent object depends on the color of the light it transmits. A red piece of glass appears red because it absorbs colors of white light except red, so red light is transmitted. Similarly, a blue piece of glass appears blue because it transmits primarily blue and absorbs the other colors. These pieces of glass contain dyes or *pigments*—fine particles that selectively absorb light of particular frequencies and selectively transmit others. Light of some of the frequencies is absorbed by the pigments. The rest is re-emitted from atom to atom in the glass. The energy of the absorbed light increases the kinetic energy of the atoms, and the glass is warmed. Ordinary window glass doesn't have a color because it transmits light of all visible frequencies equally well.

FIGURE 11.29

Only energy having the
frequency of blue light is
transmitted; energy of the
other frequencies, or of
the complementary color
yellow, is absorbed and
warms the glass.

Red
Orange
Yellow
Green
Blue
Violet

Blue glass

CHECK POINT

1. Why do the leaves of a red rose become warmer than the petals when illuminated with red light?
2. When illuminated with green light, why do the petals of a red rose appear black?

Were these your answers?
1. The leaves absorb rather than reflect red light, so the leaves become warmer.
2. The petals absorb rather than reflect the green light. Because green is the only color illuminating the rose, and green contains no red to be reflected, the rose reflects no color at all and appears black.

MIXING COLORED LIGHTS

White light is dispersed by a prism into a rainbow-colored spectrum. The distribution of sunlight (Figure 11.31) is uneven, and the light is most intense in the yellow-green part of the spectrum. How fascinating it is that our eyes have evolved to have maximum sensitivity in this range. That's why fire engines and tennis balls are yellow-green for better visibility.

All the colors combined produce white. Interestingly, we see white also from the combination of only red, green, and blue light. We can understand this by dividing the solar radiation curve into three regions, as in Figure 11.32. Three types of cone-shaped receptors in our eyes perceive color. Each is stimulated only by certain frequencies of light. Light of lower visible frequencies stimulates the cones that are sensitive to low frequencies and appears red. Light of middle frequencies stimulates the cones that are sensitive to middle frequencies and appears green. Light of higher frequencies stimulates the cones that are sensitive to higher frequencies and appears blue. When all three types of cones are stimulated equally, we see white.

FIGURE 11.30

Color depends on the light source.

Link To Sections 7.3, 27.2

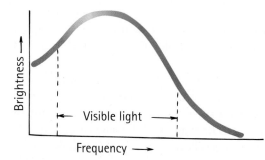

FIGURE 11.31

The radiation curve of sunlight is a graph of brightness versus frequency. Sunlight is brightest in the yellow-green region, which is in the middle of the visible range.

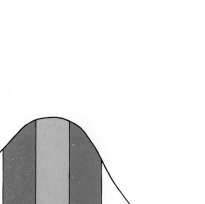

Red Green Blue

FIGURE 11.32

The radiation curve of sunlight divided into three regions—red, green, and blue. These are the additive primary colors.

It's interesting to note that the "black" you see on the darkest scenes on a TV tube is simply the color of the tube face itself, which is more a light gray than black. Because our eyes are sensitive to the contrast with the illuminated parts of the screen, we see this gray as black.

FIGURE 11.33
INTERACTIVE FIGURE

Color addition by the mixing of colored lights. When three projectors shine red, green, and blue light on a white screen, the overlapping parts produce different colors. White is produced where all three overlap.

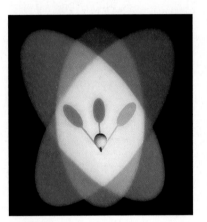

FIGURE 11.34
INTERACTIVE FIGURE

The white golf ball appears white when it is illuminated with red, green, and blue lights of equal intensities. Why are the shadows cast by the ball cyan, magenta, and yellow?

Project red, green, and blue lights on a screen and where they all overlap, white is produced. If two of the three colors overlap, or are added, then another color sensation is produced (Figure 11.33). By adding various amounts of red, green, and blue, the colors to which each of our three types of cones are sensitive, we can produce any color in the spectrum. For this reason, red, green, and blue are called the **additive primary colors**. A close examination of the picture on most color television tubes reveals that the picture is an assemblage of tiny spots, each less than a millimeter across. When the screen is lit, some of the spots are red, some are green, and some are blue; the mixtures of these primary colors at a distance provide a complete range of colors, plus white.

COMPLEMENTARY COLORS

Here's what happens when two of the three additive primary colors are combined:

$$Red + blue = magenta$$
$$Red + green = yellow$$
$$Blue + green = cyan$$

We say that magenta is the opposite of green, cyan is the opposite of red, and yellow is the opposite of blue. The addition of any color to its opposite color results in white.

$$Magenta + green = white (= red + blue + green)$$
$$Cyan + red = white (= blue + green + red)$$
$$Yellow + blue = white (= red + green + blue)$$

When two colors are added together to produce white, they are called **complementary colors**. Every hue has some complementary color that makes white when added to it.

The fact that a color and its complement combine to produce white light is pleasantly used in lighting stage performances. Blue and yellow lights shining on performers, for example, produce the effect of white light—except where one of the two colors is absent, as in the shadows. The shadow of the blue lamp is illuminated by the yellow lamp, and thus it appears yellow. Similarly, the shadow cast by the yellow lamp appears blue. This is a most intriguing effect.

We can see this effect in Figure 11.34, where red, green, and blue lights shine on the golf ball. Note the shadows cast by the ball. The middle shadow is cast by the green spotlight and is not dark because it is illuminated by the red and blue lights, which produces magenta. The shadow cast by the blue light appears yellow because it is illuminated by red and green light. Can you see why the shadow cast by the red light appears cyan?

FIGURE 11.35

Paul Robinson displays a variety of colors when he is illuminated by only red, green, and blue lamps. Can you account for the other resulting colors that appear?

FIGURING PHYSICAL SCIENCE

■ Problem Solving

SAMPLE PROBLEMS

1. From Figure 11.33 or 11.34, find the complements of cyan, yellow, and red.
2. Red + cyan = _____
3. White − cyan = _____
4. White − red = _____

SOLUTIONS:

1. red, blue, cyan
2. white
3. red
4. cyan; interestingly enough, the cyan color of the sea is the result of the removal of red light from white sunlight. The natural frequency of water molecules coincides with the frequency of infrared light, so infrared is strongly absorbed by water. To a lesser extent, red light is also absorbed by water—enough so that it appears a greenish-blue or cyan color.

MIXING COLORED PIGMENTS

Every artist knows that if you mix red, green, and blue paint, the result is not white but a muddy dark brown. Mixing red and green paint certainly does not produce yellow, so the rule for adding colored lights doesn't apply here. The mixing of pigments in paints and dyes is entirely different from mixing lights. Pigments are tiny particles that absorb specific colors. For example, pigments that produce the color red absorb the complementary color cyan. So something painted red absorbs cyan, which is why it reflects red. In effect, cyan has been subtracted from white light. Something painted blue absorbs yellow, so it reflects all the colors except yellow. Remove yellow from white and you've got blue. The colors magenta, cyan, and yellow are the **subtractive primary colors**. The variety of colors that you see in the colored photographs in this or any book are the result of magenta, cyan, and yellow dots. Light illuminates the book, and light of some frequencies is subtracted from the light reflected. The rules of color subtraction differ from the rules of light addition.

FIGURE 11.36

Seen through a magnifying glass, the color green on a printed page consists of cyan and yellow dots.

WHY THE SKY IS BLUE

Not all colors are the result of the addition or subtraction of light. Some colors, like the blue of the sky, are the result of selective scattering.* Consider the analogous case of sound: If a beam of a particular frequency of sound is directed to a tuning fork of a similar frequency,

FIGURE 11.37

The vivid colors of Sneezlee represent many frequencies of light. The photo, however, is a mixture of only yellow, magenta, cyan, and black.

(a)   (b)   (c)

(d)   (e)   (f)

FIGURE 11.38

Only three colors of ink (plus black) are used to print color photographs—(a) magenta, (b) yellow, and (c) cyan, which when combined produce the colors shown in (d). The addition of black (e) produces the finished result (f).

* This type of scattering, called *Rayleigh scattering*, occurs whenever the scattering particles are much smaller than the wavelength of incident light and have resonances at frequencies higher than those of the scattered light.

FIGURE 11.39

A beam of light falls on an atom and increases the vibrational motion of electrons in the atom. The vibrating electrons, in turn, re-emit light in various directions. Light is scattered.

FIGURE 11.40

In clean air, the scattering of high-frequency light gives us a blue sky. When the air is full of particles larger than oxygen and nitrogen molecules, light of lower frequencies is also scattered, which adds to the high-frequency scattered light to give us a whitish sky.

Atmospheric soot heats Earth's atmosphere by absorbing light while cooling local regions by blocking sunlight from reaching the ground. Soot particles in the air can trigger severe rains in one region and droughts and dust storms in another.

Link To Section 24.4

Colors in distant landscapes are duller, and color contrasts tend to diminish. That's why a color photograph normally conveys more depth than a black-and-white photograph of the same scene.

the tuning fork is set into vibration and redirects the beam in multiple directions. The tuning fork *scatters* the sound. A similar process occurs with the scattering of light from atoms and particles that are far apart from one another. This is what happens in the atmosphere.

We know that atoms behave like tiny optical tuning forks and re-emit light waves that shine on them. Very tiny particles act in a similar way. The tinier the particle, the higher the frequency of light it will re-emit. This is similar to the way in which small bells ring with higher notes than larger bells. The nitrogen and oxygen molecules that make up most of the atmosphere are like tiny bells that "ring" with high frequencies when they are energized by sunlight. Like sound from the bells, the re-emitted light is sent in all directions. The re-emitted light is said to be *scattered* in all directions.

Of the visible frequencies of sunlight, violet is scattered the most by nitrogen and oxygen in the atmosphere. Then the other colors are scattered in order: blue, green, yellow, orange, and red. Red is scattered only a tenth as much as violet. Although violet light is scattered more than blue, our eyes are not very sensitive to violet light. Therefore, the blue scattered light is what predominates in our vision, so we see a blue sky!

On clear, dry days, the sky is a much deeper blue than it is on clear, humid days. Places where the upper air is exceptionally dry, such as Italy and Greece, have beautiful blue skies that have inspired painters for centuries. Where the atmosphere contains a lot of particles of dust and other particles larger than oxygen and nitrogen molecules, light of the lower frequencies also undergoes significant scattering. This causes the sky to appear less blue, with a whitish appearance. After a heavy rainstorm, when the airborne particles have been washed away, the sky becomes a deeper blue.

The grayish haze in the skies over large cities is the result of particles emitted by automobile and truck engines and by factories. Even when idling, a typical automobile engine emits more than 100 billion particles per second. Most are invisible, but they act as tiny centers to which other particles adhere. These are the primary scatterers of lower-frequency light. With the largest of these particles, absorption rather than scattering occurs, and a brownish haze is produced. Yuck!

WHY SUNSETS ARE RED

Light that isn't scattered is light that is transmitted. Because red, orange, and yellow light are the least scattered by the atmosphere, light of these low frequencies is better transmitted through the air. Red is scattered the least, and it passes through more atmosphere than any other color. So the thicker the atmosphere through which a beam of sunlight travels, the more time there is to scatter all the higher-frequency parts of the light. So red light travels through the atmosphere best. As Figure 11.41 shows, sunlight travels through more atmosphere at sunset, which is why sunsets are red.

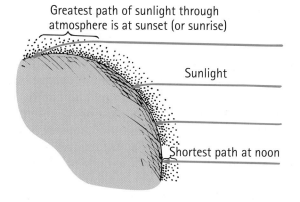

Greatest path of sunlight through atmosphere is at sunset (or sunrise)

Sunlight

Shortest path at noon

FIGURE 11.41

INTERACTIVE FIGURE

A sunbeam must travel through more kilometers of atmosphere at sunset than at noon. As a result, more blue is scattered from the beam at sunset than at noon. By the time a beam of initially white light reaches the ground, only light of the lower frequencies survives to produce a red sunset.

At noon, sunlight travels through the least amount of atmosphere to reach Earth's surface. Only a small amount of blue is scattered, which makes the Sun appear yellowish. As the day progresses and the Sun descends lower in the sky, as Figure 11.41 indicates, the path through the atmosphere is longer, and more violet and blue are scattered from the sunlight. The Sun becomes progressively redder, going from yellow to orange and finally to a red-orange at sunset. Sunsets and sunrises are unusually colorful following volcanic eruptions because particles larger than atmospheric molecules are more abundant in the air.

WHY CLOUDS ARE WHITE

Clouds are made up of clusters of water droplets in a variety of sizes. These clusters of different sizes result in a variety of scattered colors. The tiniest clusters tend to produce blue clouds; slightly larger clusters, green clouds; and still larger clusters, red clouds. The overall result is a white cloud. Electrons close to one another in a cluster vibrate in phase. This results in a greater intensity of scattered light than there would be if the same number of electrons were vibrating separately. Hence, clouds are bright!

Larger clusters of droplets absorb much of the light incident upon them, and so the scattered intensity is less. Therefore, clouds composed of larger clusters darken to a deep gray. Further increase in the size of the clusters causes them to fall as raindrops, and we have rain.

The next time you find yourself admiring a crisp blue sky, or delighting in the shapes of bright clouds, or watching a beautiful sunset, think about all those ultratiny optical tuning forks vibrating away. You'll appreciate these daily wonders of nature even more!

CHECK POINT

1. If molecules in the sky were to scatter low-frequency light more than high-frequency light, what color would the sky be? What color would sunsets be?
2. Distant dark mountains are bluish in color. What is the source of this blueness? (Hint: Exactly what is between us and the mountains we see?)

Were these your answers?
1. If light of low frequencies were scattered, the noontime sky would appear reddish orange. At sunset, more reds would be scattered by the longer distance traveled by the sunlight, and the sunlight would be predominantly blue and violet. So sunsets would appear blue!
2. If we look at distant dark mountains, very little light from them reaches us, and the blueness of the atmosphere between us and the mountains predominates. The blueness is of the low-altitude "sky" between us and the mountains. That's why distant mountains appear blue.

FIGURE 11.42

A cloud is composed of water droplets of various sizes. The tiniest droplets scatter blue light, slightly larger ones scatter green light, and still larger ones scatter red light. The overall result is a white cloud.

Link To Section 25.3

FIGURE 11.43

The wave appears cyan because seawater absorbs red light. The spray at the crest of the wave appears white because, like clouds, it is composed of a variety of tiny water droplets that scatter all the visible frequencies.

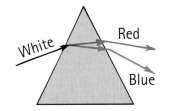

FIGURE 11.44

Dispersion by a prism makes the components of white light visible.

11.6 Dispersion

We have seen that light is absorbed when it resonates with electrons of atoms and molecules in a material. Such a material is opaque to light. Also recall that transparency occurs for light of frequencies near (but not at) the resonant frequencies of the material. Light is slowed because of the absorption/re-emission sequence, and the closer to the resonant frequencies, the slower the light. This was shown in Figure 11.6. The grand result is that high-frequency light in a transparent medium travels slower than low-frequency light. Violet light travels about 1% slower in ordinary glass than red light. Light of colors between red and violet travel at their own respective speeds in glass.

Because light of various frequencies travels at different speeds in transparent materials, different colors of light refract by different amounts. When white light is refracted twice, as in a prism, the separation of light by colors is quite noticeable. This separation of light into colors arranged by frequency is called **dispersion** (Figure 11.44). Because of dispersion, there are rainbows!

RAINBOWS

For you to see a rainbow, the Sun must shine on drops of water in a cloud or in falling rain. The drops act as prisms that disperse light. When you face a rainbow, the Sun is behind you, in the opposite part of the sky. Seen from an airplane near midday, the bow forms a complete circle. As we will see, all rainbows would be completely round if the ground were not in the way.

You can see how a raindrop disperses light in Figure 11.45. Follow the ray of sunlight as it enters the drop near its top surface. Some of the light here is reflected (not shown), and the remainder is refracted into the water. At this first refraction, the light is dispersed into its spectrum colors, red being deviated the least and violet the most. When the light reaches the opposite side of the drop, each color is partly refracted out into the air (not shown) and partly reflected back into the water. Arriving at the lower surface of the drop, each color is again partly reflected (not shown) and partly refracted back into the air. This refraction at the second surface, like that in a prism, increases the dispersion already produced at the first surface.*

Although each drop disperses a full spectrum of colors, an observer is in a position to see only a single color from any one drop (Figure 11.46). If violet

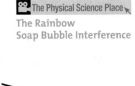

The Physical Science Place
The Rainbow
Soap Bubble Interference

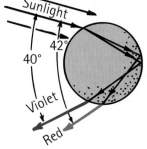

FIGURE 11.45

Dispersion of sunlight by a single raindrop.

FIGURE 11.46

Sunlight incident on two raindrops, as shown, emerges from them as dispersed light. The observer sees the red light from the upper drop and the violet light from the lower drop. Millions of drops produce the entire spectrum of visible light.

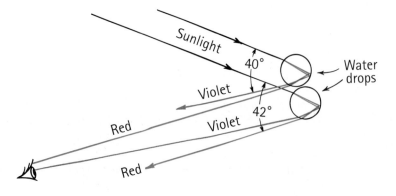

---

* We're simplifying when we indicate that the red ray disperses at 42°. Actually, the angle between the incoming and outgoing rays can be anywhere between zero and about 42° (zero degrees corresponding to a full 180-degree reversal of the light). The strongest concentration of light intensity for red, however, is near the maximum angle of 42°, as shown in Figures 11.45 and 11.46.

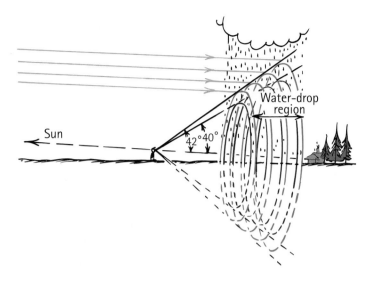

FIGURE 11.47

When your eye is located between the Sun (not shown, off to the left) and the water-drop region, the rainbow you see is the edge of a three-dimensional cone that extends through the water-drop region. Violet is dispersed by drops that form a 40° conical surface; red is seen from drops along a 42° conical surface, with other colors in between. (Innumerable layers of drops form innumerable two-dimensional arcs, like the four sets suggested here.)

light from a single drop reaches an observer's eye, red light from the same drop is incident elsewhere toward the feet. To see red light, one must look to a drop higher in the sky. The color red is seen where the angle between a beam of sunlight and the dispersed light is 42°. The color violet is seen where the angle between the sunbeams and dispersed light is 40°.

Why does the light dispersed by the raindrops form a bow? The answer involves a bit of geometry. First of all, a rainbow is not the flat two-dimensional arc it appears to be. The rainbow you see is actually a three-dimensional cone of dispersed light. The apex of this cone is at your eye. To understand this, consider a glass cone, the shape of those paper cones you sometimes see at drinking fountains. If you held the tip of such a glass cone against your eye, what would you see? You'd see the glass as a circle. Likewise with a rainbow. All the drops that disperse the rainbow's light toward you lie in the shape of a cone—a cone of different layers with drops that deflect red to your eye on the outside, orange beneath the red, yellow beneath the orange, and so on, all the way to violet on the inner conical surface (Figure 11.47). The thicker the region containing water drops, the thicker the conical edge you look through and the more vivid the rainbow.

FIGURE 11.48

Only the raindrops along the dashed line disperse red light to the observer at a 42° angle; hence, the light forms a bow.

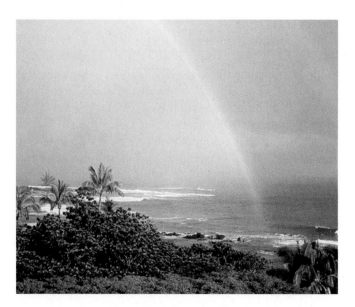

FIGURE 11.49

Two refractions and a reflection in water droplets produce light at all angles up to about 42°, with the intensity concentrated where we see the rainbow at 40° to 42°. Light doesn't exit the water droplet at angles greater than 42° unless it undergoes two or more reflections inside the drop. Thus the sky is brighter inside the rainbow than outside it. Notice the weak secondary rainbow.

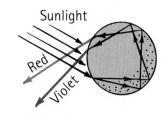

FIGURE 11.50

Double reflection in a drop produces a secondary bow.

Your cone of vision intersects the cloud of drops and creates your rainbow. It is ever so slightly different from the rainbow seen by a person nearby. So, when a friend says, "Look at the pretty rainbow," you can reply, "Okay, move aside so I can see it, too." Everybody sees his or her own personal rainbow.

Another fact about rainbows: A rainbow always faces you squarely. When you move, your rainbow appears to move with you. So you can never approach the side of a rainbow or see it end-on as in the exaggerated view of Figure 11.47. You *can't* reach its end. Thus the saying "looking for the pot of gold at the end of the rainbow" means pursuing something you can never reach.

Often a larger, secondary bow with its colors reversed can be seen arching at a greater angle around the primary bow. We won't treat this secondary bow except to say that it is formed by similar circumstances and is a result of double reflection within the raindrops (Figure 11.50). Because of this extra reflection (and extra refraction loss), the secondary bow is much dimmer and reversed.

CHECK POINT

1. Suppose you point to a wall with your arm extended. Then you sweep your arm around, making an angle of about 42° to the wall. If you rotate your arm in a full circle while keeping the same angle, what shape does your arm describe? What shape does your finger sweep out on the wall?
2. If light traveled at the same speed in raindrops as it does in air, would we have rainbows?

Were these your answers?
1. Your arm describes a cone, and your finger sweeps out a circle. Likewise with rainbows.
2. No.

The Physical Science Place
Polarization and 3D Viewing

FIGURE 11.51

A vertically plane-polarized plane wave and a horizontally plane-polarized plane wave.

11.7 Polarization

As we learned in Chapter 10, waves can be either longitudinal or transverse. Sound waves are longitudinal, which means the vibratory motion of the medium is along the direction of wave travel. The fact that light waves exhibit **polarization** demonstrates that light waves are transverse.

If you shake a rope either up and down or from side to side as shown in Figure 11.51, you produce a transverse wave along the rope. The plane of vibration is the same as the plane of the wave. If you shake it up and down, the wave vibrates in a vertical plane. If you shake it back and forth, the wave vibrates in a horizontal plane. We say that such a wave is *plane-polarized*—that the waves traveling along the rope are confined to a single plane. Polarization is a property of transverse waves. (Polarization does not occur among longitudinal waves—there is no such thing as polarized sound.)

A single vibrating electron can emit an electromagnetic wave that is plane-polarized. The plane of polarization matches the vibrational direction of the

(a)   (b)

FIGURE 11.52

(a) A vertically plane-polarized wave from a charge vibrating vertically. (b) A horizontally plane-polarized wave from a charge vibrating horizontally.

electron. That means that a vertically accelerating electron emits light that is vertically polarized. A horizontally accelerating electron emits light that is horizontally polarized (Figure 11.52).*

A common light source, such as an incandescent lamp, a fluorescent lamp, or a candle flame, emits light that is unpolarized. This is because the electrons that emit the light are vibrating in many random directions. There are as many planes of vibration as the vibrating electrons producing them. A few planes are represented in Figure 11.53a. We can represent all these planes by radial lines, shown in Figure 11.53b. (Or, more simply, the planes can be represented by vectors in two mutually perpendicular directions, as shown in Figure 11.53c.) The vertical vector represents all the components of vibration in the vertical direction. The horizontal vector represents all the components of vibration horizontally. The simple model of Figure 11.53c represents unpolarized light. Polarized light would be represented by a single vector.

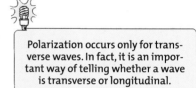

Polarization occurs only for transverse waves. In fact, it is an important way of telling whether a wave is transverse or longitudinal.

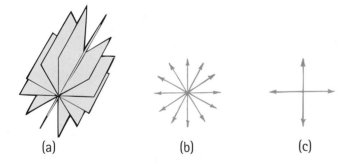

(a)   (b)   (c)

FIGURE 11.53

Representations of plane-polarized waves.

All transparent crystals having a noncubic natural shape have the property of polarizing light. These crystals divide unpolarized light into two internal beams polarized at right angles to each other. Some crystals strongly absorb one beam while transmitting the other (Figure 11.54). This makes them excellent polarizers. Herapathite is such a crystal. Microscopic herapathite crystals are aligned and embedded between cellulose sheets. They make up Polaroid filters, popular in sunglasses. Other Polaroid sheets consist of certain aligned molecules rather than tiny crystals.

If you look at unpolarized light through a Polaroid filter, you can rotate the filter in any direction and the light appears unchanged. But if the light is polarized, rotating the filter allows you to block out more and more of the light until it is completely blocked out. An ideal Polaroid filter transmits 50% of incident unpolarized light. That 50% is polarized. When two Polaroid filters are arranged so that their polarization axes are aligned, light can pass through both, as shown in the rope analogy (Figure 11.55a). If their axes are at right angles to each other (in this case, we say the filters are crossed), almost no light penetrates the pair (Figure 11.55b). (A small amount of shorter wavelengths do get through.) When Polaroid filters are used in pairs like this, the first one is called the *polarizer* and the second one is called the *analyzer*.

fyi

■ Cosmic microwave background (CMB) fills all of space and approaches us from every direction. It is an echo of the Big Bang that got our universe started some 14 billion years ago. Recent findings show this radiation to be polarized. Polarization observations are unaffected by gravity and provide a clear and detailed look at the early cosmos.

FIGURE 11.54

One component of the incident unpolarized light is absorbed, resulting in emerging polarized light.

* Light may also be circularly polarized and elliptically polarized, which are also transverse polarizations. But we will not study these cases.

FIGURE 11.55

A rope analogy illustrates the effect of crossed Polaroids.

Nonpolarized light vibrates in all directions
Horizontal and vertical components
Vertical component passes through first polarizer...
...And the second

Vertical component does not pass through this second polarizer

(a)　　　　(b)

FIGURE 11.56

Polaroid sunglasses block out horizontally vibrating light. When the lenses overlap at right angles, light doesn't get through.

Much of the light reflected from nonmetallic surfaces is polarized. The glare from glass or water is a good example. Except for light that hits vertically, the reflected ray has more vibrations parallel to the reflecting surface. The part of the ray that penetrates the surface has more vibrations at right angles to the surface (Figure 11.56). Skipping flat rocks off the surface of a pond provides an appropriate analogy. When the rocks hit parallel to the surface, they are easily reflected by the surface. But when they hit with their faces at right angles to the surface, they "refract" into the water. The glare from reflecting surfaces can be dimmed a lot with the use of Polaroid sunglasses. The polarization axes of the lenses are vertical because most of the glare reflects from horizontal surfaces.

CHECK POINT

Which pair of glasses is best suited for automobile drivers? (The polarization axes are shown by the straight lines.)

(a)　　　　(b)　　　　(c)

Was this your answers?

Glasses A are best suited because the vertical axis blocks horizontally polarized light, which makes up much of the glare from horizontal surfaces. Glasses C are suited for viewing 3-D movies.

FIGURE 11.57

Most glare from nonmetallic surfaces is polarized. Here we see that the components of incident light parallel to the surface are reflected, and the components perpendicular to the surface pass through the surface into the medium. Because most of the glare we encounter is from horizontal surfaces, the polarization axes of Polaroid sunglasses are vertical.

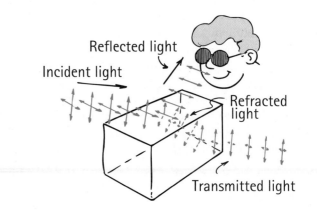

Reflected light
Incident light
Refracted light
Transmitted light

FIGURE 11.58

INTERACTIVE FIGURE

Light is transmitted when the axes of the Polaroids are aligned (a), but absorbed when Ludmila rotates one so that the axes are at right angles to each other (b). When she inserts a third Polaroid at an angle between the crossed Polaroids, light is again transmitted (c). Why? (For the answer, after you have given this some thought see Appendix C, "More About Vectors.")

SUMMARY OF TERMS

Electromagnetic wave An energy-carrying wave emitted by vibrating electrical charges (often electrons) and composed of oscillating electric and magnetic fields that regenerate one another.

Electromagnetic spectrum The range of electromagnetic waves that extends in frequency from radio waves to gamma rays.

Transparent The term applied to materials through which light can pass without absorption, usually in straight lines.

Opaque The property of absorbing light without re-emission (opposite of transparent).

Reflection The return of light rays from a surface in such a way that the angle at which a given ray is returned is equal to the angle at which it strikes the surface (also called *specula reflection*).

Law of reflection The angle of incidence equals the angle of reflection. The incident and reflected rays lie in a plane that is normal to the reflecting surface.

Diffuse reflection Reflection in irregular directions from an irregular surface.

Refraction The bending of an oblique ray of light when it passes from one transparent medium to another. This is caused by a difference in the speed of light in the transparent media. When the change in medium is abrupt (say, from air to water), the bending is abrupt; when the change in medium is gradual (say, from cool air to warm air), the bending is gradual, which accounts for mirages.

Additive primary colors The three colors—red, green, and blue—that, when added in certain proportions, can produce any color in the spectrum.

Complementary colors Any two colors that, when added, produce white light.

Subtractive primary colors The three colors of absorbing pigments—magenta, yellow, and cyan—that, when mixed in certain proportions, can reflect any color in the spectrum.

Dispersion The separation of light into colors arranged by frequency.

Polarization The alignment of the transverse electric vectors that make up electromagnetic radiation. Such waves of aligned vibrations are said to be *polarized*.

REVIEW QUESTIONS

11.1 Electromagnetic Spectrum

1. What is the principal difference between a radio wave and light? Between light and an X-ray?
2. How does the frequency of an electromagnetic wave compare with the frequency of the vibrating electrons that produce it?

11.2 Transparent and Opaque Materials

3. In what region of the electromagnetic spectrum is the resonant frequency of electrons in glass?
4. What is the fate of the energy in ultraviolet light incident on glass?
5. What is the fate of the energy in infrared light incident on glass? Of visible light?

6. How does the average speed of light in glass compare with its speed in a vacuum?
7. How does the speed of light that emerges from a pane of glass compare with the speed of light incident on the glass?

11.3 Reflection

8. What is the law of reflection?
9. Relative to the distance of an object in front of a plane mirror, how far behind the mirror is the image?
10. Does the law of reflection hold for curved mirrors? Explain.
11. In what sense does the law of reflection hold for a diffuse reflector?

11.4 Refraction

12. What is the angle between a light ray and its wave front?
13. What is the relationship to refraction and the speed of light?
14. Does light travel faster in thin air or in dense air? What does this difference in speed have to do with the length of daylight?
15. What is a mirage?

11.5 Color

16. Which has the higher frequency, red light or blue light?
17. What is the color of the peak frequency of solar radiation? To what color of light are our eyes most sensitive?
18. What are the three primary colors? The three subtractive primary colors?

19. Why are red and cyan called complementary colors?
20. Why does the sky sometimes appear whitish?
21. Why does the Sun look reddish at sunrise and sunset but not at noon?
22. What is the evidence for a cloud being composed of particles having a variety of sizes?
23. What is absorbed by water to give it a cyan color?

11.6 Dispersion

24. Which travels more slowly in glass, red light or violet light?
25. What prevents rainbows from being seen as complete circles?
26. Why is a secondary rainbow dimmer than a primary bow?

11.7 Polarization

27. Is polarization a property of transverse waves, longitudinal waves, or both?
28. How does the direction of polarization of light compare with the direction of vibration of the electrons that produced it?
29. Why does light pass through a pair of Polaroid filters when the axes are aligned but not when the axes are at right angles to each other?
30. How much unpolarized light does an ideal Polaroid filter transmit?

EXERCISES

● **BEGINNER** ■ **INTERMEDIATE** ◆ **EXPERT**

1. ● What is the fundamental source of electromagnetic radiation?
2. ● Which have the longest wavelength: light waves, X-rays, or radio waves?
3. ● Which has the shorter wavelength, ultraviolet or infrared? Which has the higher frequency?
4. ● We hear people talk of "ultraviolet light" and "infrared light." Why are these terms misleading? Why are we less likely to hear people talk of "radio light" and "X-ray light"?
5. ● Which requires a physical medium in which to travel, light or sound? Or do both require a physical medium? Explain.
6. ● Do radio waves travel at the speed of sound, at the speed of light, or at some speed in between?
7. ● What do radio waves and light have in common? What is different about them?
8. ● What evidence can you cite to support the idea that light can travel in a vacuum?
9. ● Short wavelengths of visible light interact more frequently with the atoms in glass than do longer wavelengths. Does this interaction time tend to speed up or slow down the average speed of light in glass?
10. ● What determines whether a material is transparent or opaque?
11. ● You can get a sunburn on a cloudy day, but you can't get a sunburn even on a sunny day if you are behind glass. Explain.

12. ■ Peter Hopkinson stands astride a large mirror and boosts class interest with this zany demonstration. How does he accomplish his apparent levitation in midair?

13. ● The person's eye at point P looks into the mirror. Which of the numbered cards can she see reflected in the mirror?

14. ● Cowboy Joe wishes to shoot his assailant by ricocheting a bullet off a mirrored metal plate. To do so, should he simply aim at the mirrored image of his assailant? Explain.

15. ● Trucks often have signs on the back that say, "If you can't see my mirrors, I can't see you." Explain the physics here.

16. ● Why is the lettering on the front of some vehicles, for example ambulances, "backward"?

AMBULANCE

17. ■ We see the bird and its reflection. Why do we not see the bird's feet in the reflection?

18. ■ What must be the minimum length of a plane mirror in order for you to see a full view of yourself?

19. ■ What effect does your distance from the plane mirror have on your answer to the preceding question? (Try it and see!)

20. ■ Hold a pocket mirror almost at arm's length from your face and note the amount of your face you can see. To see more of your face, should you hold the mirror closer or farther, or would you have to have a larger mirror? (Try it and see!)

21. ■ From a steamy mirror, wipe away just enough steam to allow you to see your full face. How tall is the wiped area compared with the vertical dimension of your face?

22. ■ A pair of toy cart wheels are rolled obliquely from a smooth surface onto two plots of grass, a rectangular plot and a triangular plot, as shown. The ground is on a slight incline, so that, after slowing down in the grass, the wheels speed up again when emerging onto the smooth surface. Finish the sketches by showing some positions of the wheels inside each plot and on the other side of each plot, thereby indicating their direction of travel.

23. ■ A pulse of red light and a pulse of blue light enter a glass block normal to its surface at the same time. Strictly speaking, after passing through the block, which pulse exits first?

24. ■ During a lunar eclipse, the Moon is not completely dark, but it is often a deep red in color. Explain this in terms of the refraction of all the sunsets and sunrises around the world.

25. ● Suppose that sunlight falls on both a pair of reading glasses and a pair of dark sunglasses. Which pair of glasses would you expect to become warmer? Defend your answer.

26. ● In a dress shop with only fluorescent lighting, a customer insists on taking dresses into the daylight at the doorway to check their color. Is she being reasonable? Explain.

27. ■ The radiation curve of the Sun (Figure 11.32) shows that the brightest light from the Sun is yellow-green. Why then do we see the Sun as whitish instead of yellow-green? (Hint: Take into account the wideness of the solar radiation curve.)

28. ■ A spotlight is coated so that it won't transmit yellow light from its white-hot filament. What color is the emerging beam of light?

29. ■ How could you use the spotlights at a play to make the yellow clothes of the performers suddenly change to black?

30. ● What colors of ink do color ink-jet printers use to produce a full range of colors? Do the colors form by color addition or by color subtraction?

31. ■ Below is a photo of science author Suzanne Lyons with her son Tristan wearing red and her daughter Simone wearing green. Below that is the negative of the photo, which shows these colors differently. What is your explanation?

32. ● Check Figure 11.33 to see if the first three statements that follow are accurate. Then provide the missing word in the last statement. (All colors are combined by the addition of light.)

Red + green + blue = white
Red + green = yellow = white − blue
Red + blue = magenta = white − green
Green + blue = cyan = white − _____

33. ● In which of these cases does a ripe banana appear black? When it is illuminated with (a) red light; (b) yellow light; (c) green light; (d) blue light.

34. ■ When white light is shined on red ink that has dried on a clear glass plate, the color that is transmitted is red. But the color that is reflected is not red. What is it?

35. ● Stare intently for at least 30 s at an American flag. Then turn your gaze to a white wall. What colors do you see in the image of the flag that appears on the wall?

36. ■ Why can't we see stars in the daytime?

37. ■ Why is the sky a darker blue when you are at high altitudes? (Hint: What color is the "sky" on the Moon?)

38. ■ Why does smoke from a campfire look bluish against trees near the ground but yellowish against the sky?

39. ■ Tiny particles, like tiny bells, scatter high-frequency waves more than low-frequency waves. Large particles, like large bells, mostly scatter low frequencies. Intermediate-size particles and bells mostly scatter intermediate frequencies. What does this have to do with the whiteness of clouds?

40. ● Very big particles, like droplets of water, absorb more radiation than they scatter. What does this have to do with the darkness of rain clouds?

41. ■ The atmosphere of Jupiter is more than 1000 km thick. From the surface of this planet, would you expect to see a white Sun?

42. ■ You're explaining to a youngster at the seashore why the water is cyan colored. The youngster points to the white-caps of overturning waves and asks why they are white. What is your answer?

43. ■ When you stand with your back to the Sun, you see a rainbow as a circular arc. Could you move off to one side and then see the rainbow as the segment of an ellipse rather than the segment of a circle (such as Figure 11.47 suggests)? Defend your answer.

44. ● Two observers standing apart from one another do not see the "same" rainbow. Explain.

45. ■ A rainbow viewed from an airplane may form a complete circle. Where does the shadow of the airplane appear? Explain.

46. ● What percentage of light is transmitted by two ideal Polaroid filters, one on top of the other with their polarization axes aligned? With their polarization axes at right angles to each other?

47. ■ How can a single Polaroid filter be used to show that the sky is partially polarized? (Interestingly enough, unlike humans, bees and many insects can discern polarized light, and they use this ability for navigation.)

48. ◆ Light does not pass through a pair of Polaroid filters when they are aligned perpendicularly. But if a third Polaroid filter is sandwiched between the other two with its alignment halfway between the alignments of the others (that is, with its axis making a 45° angle with each of the other two alignment axes), some light does get through. Why?

PROBLEMS

● BEGINNER ■ INTERMEDIATE ◆ EXPERT

1. ● Electrons on a radio broadcasting tower are forced to oscillate up and down the tower 535,000 times each second. What is the wavelength of the radio waves that are produced?

2. ● Consider a pulse of laser light aimed at the Moon that bounces back to Earth. The distance between Earth and the Moon is 3.8×10^8 m. Show that the round-trip time for the light is 2.5 s.

3. ● The nearest star beyond the Sun is Alpha Centauri, which is 4.2×10^{16} m away. If we were to receive a radio message from this star today, show that it would have been sent 4.4 years ago.

4. ● Blue-green light has a frequency of about 6×10^{14} Hz. Using the relationship $c = f\lambda$, show that its wavelength in air is 5×10^{-7} m. How much larger is this wavelength compared to the size of an atom, which is about 10^{-10} m?

5. ■ A certain radar installation that is used to track airplanes transmits electromagnetic radiation with a

wavelength of 3 cm. (a) Show that the frequency of this radiation is 10 GHz. (b) Show that the time required for a pulse of radar waves to reach an airplane 5 km away and return would be 3.3×10^{-5} s.

6. ● A spider hangs by a strand of silk at an eye level 20 cm in front of a plane mirror. You are behind the spider, 50 cm from the mirror. Show that the distance between your eye and the image of the spider in the mirror is 70 cm.

7. ■ When light strikes glass perpendicularly, about 4% of the light is reflected at each surface. Show that the amount of light transmitted through a pane of window glass is approximately 92%.

8. ■ The average speed of light slows to $0.75c$ when it refracts through a particular piece of plastic. (a) What change is there in the light's frequency in the plastic? (b) In its wavelength?

ACTIVE EXPLORATIONS

1. Which eye do you use more? To test which you favor, hold a finger up at arm's length. With both eyes open, look past it at a distant object. Now close your right eye. If your finger appears to jump to the right, then you use your right eye more.

2. Stare at a piece of colored paper for 45 s or so. Then look at a plain white surface. The cones in your retina receptive to the color of the paper become fatigued, so you see an afterimage of the complementary color when you look at a white area. This is because the fatigued cones send a weaker signal to the brain. All the colors produce white, but all the colors minus one produce the color that is complementary to the missing color. Try it and see!

3. Simulate your own sunset: add a few drops of milk to a glass of water and look at a lightbulb through the glass. The bulb appears to be red or pale orange, while light scattered to the side appears blue. Try it and see.

4. Write a letter to your grandparents and tell them the reasons for the blueness of the sky, the redness of sunrises and sunsets, and why the clouds are normally white. Explain how knowing the reasons adds to, not subtracts from, your appreciation of nature.

5. Set up two pocket mirrors at right angles and place a coin between them. You'll see four coins. Change the angle of the mirrors and see how many images of the coin you can see. With the mirrors at right angles, look at your face.

Then wink. What do you see? You now see yourself as others see you. Hold a printed page up to the double mirrors and compare its appearance with the reflection of a single mirror.

Left eye   Right eye

6. Rotate a pair of mirrors, keeping them at right angles to each other. Does your image rotate also? Then place the mirrors 60° apart so that you can see your face. Again rotate the mirrors, and see if your image rotates also. Amazing?

90°   60°

7. Make a pinhole camera, as illustrated below. Cut out one end of a small cardboard box and cover the end with tissue or wax paper. Make a clean-cut pinhole at the other end. (If the cardboard is thick, make it through a piece of aluminum foil placed over an opening in the cardboard.) Aim the camera at a bright object in a darkened room, and you see an upside-down image on the tissue paper. If, in a dark room, you replace the tissue paper with unexposed photographic film, cover the back so it is light-tight, and cover the pinhole with a removable flap, you are ready to take a picture. Exposure times differ, depending principally on the kind of film and the amount of light. Try different exposure times, starting with about 3 s. Also try boxes of various lengths. You'll find everything in focus in your photographs, but the pictures will not have clear-cut, sharp outlines. The lens on a commercial camera is much bigger than the pinhole and therefore admits more light in less time—hence the term *snapshot*.

READINESS ASSURANCE TEST (RAT)

If you have a good handle on this chapter, if you really do, then you should be able to score at least 7 out of 10 on this RAT. If you score less than 7, you need to study further before moving on.

Choose the BEST answer to each of the following.

1. The electromagnetic spectrum spans waves ranging from lowest to highest frequencies. The smallest portion of the electromagnetic spectrum is that of
   (a) radio waves.
   (b) microwaves.
   (c) visible light.
   (d) gamma rays.
2. Strictly speaking, the photons of light that shine on glass are
   (a) also the ones that travel through and exit the other side.
   (b) not the ones that travel through and exit the other side.
   (c) absorbed and transformed to thermal energy.
   (d) diffracted.
3. The law of reflection applies to
   (a) light.
   (b) sound.
   (c) both of these
   (d) neither of these
4. When a light ray passes at an angle from air to water, the ray inside the water bends
   (a) toward the normal.
   (b) away from the normal.
   (c) either away or toward the normal.
   (d) parallel to the normal.
5. A red rose does not appear red when illuminated only with
   (a) red light.
   (b) orange light.
   (c) white light.
   (d) cyan light.
6. Red, green, and blue light overlap to form
   (a) red light.
   (b) green light.
   (c) blue light.
   (d) white light.
7. When light travels from one medium to another, and changes speed in doing so, we call the process
   (a) reflection.
   (b) interference.
   (c) dispersion.
   (d) none of these
8. When light incident on a prism separates into a spectrum, we call the process
   (a) polarization.
   (b) interference.
   (c) dispersion.
   (d) none of these
9. A rainbow illustrates the properties of
   (a) reflection.
   (b) refraction.
   (c) dispersion.
   (d) all of these
10. Polarization occurs for waves that are
    (a) translational.
    (b) longitudinal.
    (c) both of these
    (d) neither of these

Answers to RAT
1c, 2b, 3c, 4a, 5d, 6d, 7d, 8c, 9c, 10a

CHAPTER 11 ONLINE RESOURCES

The Physical Science Place

Interactive Figures

- 11.2, 11.3, 11.10, 11.19, 11.33, 11.34, 11.41, 11.58

Tutorials

- Light and Spectroscopy
- Color

Videos

- Light and Transparent Materials
- Image Formation in a Mirror
- Model of Refraction

- Colored Shadows
- Yellow-Green Peak of Sunlight
- Why the Sky Is Blue and the Sunset Is Red
- The Rainbow
- Soap Bubble Interference
- Polarization and 3D Viewing

Quiz

Flashcards

Links

part two
chemistry

Like everyone, I'm made of atoms, which are so small and numerous that I inhale billions of trillions with each breath of air. I exhale some of them right away, but other atoms stay for awhile and become part of me, which I may exhale later. Some of my atoms are in each breath you take, and stay to become part of you (and likewise, yours become part of me). There are way more atoms in a breath of air than the total number of humans since time zero, so in each breath you inhale, you recycle atoms that once were a part of every person who lived. Hey, in this sense, we're all one!

atoms and the periodic table

12

12.1 ATOMS ARE ANCIENT AND EMPTY

12.2 THE ELEMENTS

12.3 PROTONS AND NEUTRONS

12.4 THE PERIODIC TABLE

12.5 PHYSICAL AND CONCEPTUAL MODELS

12.6 IDENTIFYING ATOMS USING THE SPECTROSCOPE

12.7 THE QUANTUM HYPOTHESIS

12.8 ELECTRON WAVES

12.9 THE SHELL MODEL

■ We humans have long tinkered with the materials around us and used them to our advantage. Once we learned how to control fire, we were able to create many new substances. Moldable wet clay, for example, was found to harden to ceramic when heated by fire. By 5000 BC, pottery fire pits gave way to furnaces hot enough to convert copper ores to metallic copper. By 1200 BC, even hotter furnaces were converting iron ores to iron. This technology allowed for the mass production of metal tools and weapons and made possible the many achievements of ancient Chinese, Egyptian, and Greek civilizations.

Fast-forward to the 21st century, and we've since learned that all the materials around us are made of remarkably small particles called atoms. We have learned how to manipulate these atoms to produce a vast array of new and useful modern materials, including pharmaceuticals that help prolong our lives. We have even learned how to move atoms, one by one, into desired positions.

The opening photograph for this chapter, for example, shows a group of titanium atoms that scientists pushed into a circle. This is in the forefront of our present technology, which allows us to build nanodevices and new materials atom by atom. In this chapter, we will explore both the nature of atoms and the amazing chart that tells their story—the periodic table.

12.1 Atoms Are Ancient and Empty

The Physical Science Place
Atomic Structure

The Physical Science Place
Evidence for Atoms
Atoms Are Recyclable

Link To Sections 14.2, 27.4

If a typical atom were expanded to a diameter of 3 km, about as big as a medium-sized airport, the nucleus would be about the size of a basketball. Atoms are mostly empty space.

The origin of most atoms goes back to the birth of the universe. Hydrogen, H, the lightest atom, was the original atom, and hydrogen atoms make up more than 90% of the atoms in the known universe. Heavier atoms are produced in stars, which are massive collections of hydrogen atoms pulled together by gravitational forces. The great pressures deep in a star's interior cause hydrogen atoms to fuse into heavier atoms. With the exception of hydrogen, therefore, all the atoms that occur naturally on Earth—including those in your body—are the products of stars. A tiny fraction of these atoms came from our own star, the Sun, but most are from stars that ran their course long before our solar system came into being. You are made of stardust, as is everything that surrounds you.

So most atoms are ancient. They have existed through imponderable ages, recycling through the universe in innumerable forms, both nonliving and living. In this sense, you don't "own" the atoms that make up your body—you are simply their present caretaker. Many more caretakers will follow.

Atoms are so small that each breath you exhale contains more than 10 billion trillion of them. This is more than the number of breaths in Earth's atmosphere. Within a few years, the atoms of your breath are uniformly mixed throughout the atmosphere. What this means is that anyone anywhere on Earth inhaling a breath of air takes in numerous atoms that were once part of you. And, of course, the reverse is true: you inhale atoms that were once part of everyone who has ever lived. We are literally breathing one another.

Atoms are so small that they can't be seen with visible light. That's because they are even smaller than the wavelengths of visible light. We could stack microscope on top of microscope and never "see" an atom. Photographs of atoms, such as in Figure 12.1, are obtained with a scanning probe microscope. Discussed further in Section 12.5, this is a nonlight imaging device that bypasses light and optics altogether.

Today we know the atom is made of smaller, subatomic particles—*electrons*, *protons*, and *neutrons*. We also know that atoms differ from one another only in the number of subatomic particles they contain. Protons and neutrons are bound together at the atom's center to form a larger particle—the **atomic nucleus**. The nucleus is a relatively heavy particle that makes up most of an atom's mass. Surrounding the nucleus are the tiny **electrons**, as shown in Figure 12.2.

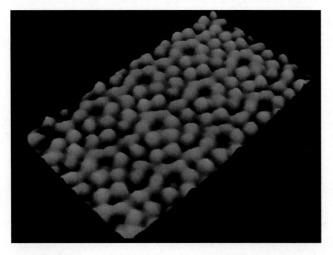

FIGURE 12.1

An image of carbon atoms obtained with a scanning probe microscope.

A friend claims there are atoms in his brain that were once in the brain of Albert Einstein. Is your friend's claim likely correct or nonsense?

Was this your answer?
Your friend is correct! In addition, there are atoms in your friend's and everyone else's body that were once part of Mother Teresa and everybody else, too! The arrangements of these atoms, however, are now quite different. What's more, the atoms of which you and your friend are composed will be found in the bodies of all the people on Earth who are yet to be.

We and all materials around us are mostly empty space. How can this be? Electrons move about the nucleus in an atom defining the volume of space that the atom occupies. But because electrons are very small, and because they are widely separated from each other and from the nucleus, atoms are indeed mostly empty space.

So why don't atoms simply pass through one another? How is it that we are supported by the floor despite the empty nature of its atoms? Although subatomic particles are much smaller than the volume of the atom, the range of their electric field is several times larger than that volume. In the outer regions of any atom are electrons, which repel the electrons of neighboring atoms. Two atoms therefore can get only so close to each other before they start repelling (provided they don't join in a chemical bond, as is discussed in Chapter 15).

When the atoms of your hand push against the atoms of a wall, electrical repulsions between electrons in your hand and electrons in the wall prevent your hand from passing through the wall. These same electrical repulsions prevent us from falling through the solid floor. They also allow us the sense of touch. Interestingly, when you touch someone, your atoms and those of the other person do not meet. Instead, atoms from the two of you get close enough so that you sense an electrical repulsion. A tiny, though imperceptible, gap still exists between the two of you (Figure 12.3).

12.2 The Elements

You know that atoms make up the matter around you, from stars to steel to chocolate ice cream. Given all these different types of material, you might think that there must be many different kinds of atoms. But the number of different kinds of atoms is surprisingly small. The great variety of substances results from the many ways a few kinds of atoms can be combined. Just as the three colors red, green, and blue can be combined to form any color on a television screen or the 26 letters of the alphabet make up all the words in a dictionary, only a few kinds of atoms combine in different ways to produce all substances. To date, we know of slightly more than 100 distinct atoms. Of these, about 90 are found in nature. The remaining atoms have been created in the laboratory.

Any material made of only one type of atom is classified as an **element**. A few examples are shown in Figure 12.4. Pure gold, for example, is an element—it contains only gold atoms. Nitrogen gas is an element because it contains only nitrogen atoms. Likewise, the graphite in your pencil is an element—carbon.

FIGURE 12.2

Electrons whiz around the atomic nucleus, forming what can be best described as a cloud that is more dense where the electrons tend to spend most of their time. Electrons, however, are invisible to us. Hence, such a cloud can only be imagined. Furthermore, if this illustration were drawn to scale, the atomic nucleus would be too small to be seen. In short, atoms are not well suited to graphical depictions.

FIGURE 12.3

As close as Tracy and Ian are in this photograph, none of their atoms meet. The closeness between us is in our hearts.

Link To Section 14.5

Most materials are made from more than one kind of atom. Water, H_2O, for example, is made from the combination of hydrogen and oxygen atoms. These materials are called *compounds*, which we discuss further in Chapter 14.

Any element consists of only one kind of atom. Gold consists of only gold atoms, a flask of gaseous nitrogen consists of only nitrogen atoms, and the carbon of a graphite pencil consists of only carbon atoms.

Graphite is made up solely of carbon atoms. All of the elements are listed in a chart called the **periodic table**, shown in Figure 12.5.

As you can see from the periodic table, each element is designated by its **atomic symbol**, which comes from the letters of the element's name. For example, the atomic symbol for carbon is C, and that for chlorine is Cl. In many cases, the atomic symbol is derived from the element's Latin name. Gold has the atomic symbol Au after its Latin name, *aurum*. Lead has the atomic symbol Pb after its Latin name, *plumbum* (Figure 12.6). Elements with symbols derived from Latin names are usually those that were discovered earliest.

Note that only the first letter of an atomic symbol is capitalized. The symbol for the element cobalt, for instance, is Co, but CO is a combination of two elements: carbon, C, and oxygen, O.

■ 12.3 Protons and Neutrons

Let us take a closer look at the atom and investigate the particles found in the atomic nucleus. A **proton** carries a positive charge and is relatively heavy—nearly 2000 times as massive as an electron. The proton and electron have the same quantity of charge, but the opposite sign. The number of protons in the nucleus of any atom is equal to the number of electrons whirling about the nucleus. So the opposite charges of protons and electrons balance each other, producing a zero net charge. For example, an oxygen atom has a total of eight electrons and eight protons and is thus electrically neutral.

Scientists have agreed to identify elements by **atomic number,** which is the number of protons each atom of a given element contains. The modern periodic table lists the elements in order of increasing atomic number. Hydrogen, with one proton per atom, has atomic number 1; helium, with two protons per atom, has atomic number 2; and so on.

The periodic table lists all the known elements.

1 H																	2 He
3 Li	4 Be											5 B	6 C	7 N	8 O	9 F	10 Ne
11 Na	12 Mg											13 Al	14 Si	15 P	16 S	17 Cl	18 Ar
19 K	20 Ca	21 Sc	22 Ti	23 V	24 Cr	25 Mn	26 Fe	27 Co	28 Ni	29 Cu	30 Zn	31 Ga	32 Ge	33 As	34 Se	35 Br	36 Kr
37 Rb	38 Sr	39 Y	40 Zr	41 Nb	42 Mo	43 Tc	44 Ru	45 Rh	46 Pd	47 Ag	48 Cd	49 In	50 Sn	51 Sb	52 Te	53 I	54 Xe
55 Cs	56 Ba	57 La	72 Hf	73 Ta	74 W	75 Re	76 Os	77 Ir	78 Pt	79 Au	80 Hg	81 Tl	82 Pb	83 Bi	84 Po	85 At	86 Rn
87 Fr	88 Ra	89 Ac	104 Rf	105 Db	106 Sg	107 Bh	108 Hs	109 Mt	110 Uun	111 Uuu	112 Uub						

58 Ce	59 Pr	60 Nd	61 Pm	62 Sm	63 Eu	64 Gd	65 Tb	66 Dy	67 Ho	68 Er	69 Tm	70 Yb	71 Lu
90 Th	91 Pa	92 U	93 Np	94 Pu	95 Am	96 Cm	97 Bk	98 Cf	99 Es	100 Fm	101 Md	102 No	103 Lr

FIGURE 12.6

A plumb bob, a heavy weight attached to a string and used by carpenters and surveyors to establish a straight vertical line, gets it name from the lead (*plumbum*, Pb) that is still sometimes used as the weight. Plumbers got their name because they once worked with lead pipes.

CHECK POINT

How many protons are there in an iron atom, Fe (atomic number 26)?

Was this your answer?
The atomic number of an atom and its number of protons are the same. Thus, there are 26 protons in an iron atom. Another way to put this is that all atoms that contain 26 protons are, by definition, iron atoms.

If we compare the electric charges and masses of different atoms, we see that the atomic nucleus must be made up of more than just protons. Helium, for example, has twice the electric charge of hydrogen but four times the mass. The added mass is due to another subatomic particle found in the nucleus, the neutron. The **neutron** has about the same mass as the proton, but it has no electric charge. Any object that has no net electric charge is said to be electrically neutral, and that is where the neutron got its name. We discuss the important role that neutrons play in holding the atomic nucleus together in Chapter 13.

Link To Sections 8.2, 13.3

Both protons and neutrons are called **nucleons**, a term that denotes their location in the atomic nucleus. Table 12.1 summarizes the basic facts about electrons, protons, and neutrons.

TABLE 12.1 SUBATOMIC PARTICLES

	Particle	Charge	Mass Compared to Electron	Actual Mass* (kg)
	Electron	−1	1	9.11×10^{-31}**
Nucleons	Proton	+1	1836	1.673×10^{-27}
	Neutron	0	1841	1.675×10^{-27}

* Not measured directly but calculated from experimental data.
** 9.11×10^{-31} kg = 0.00000000000000000000000000000911 kg.

ISOTOPES AND ATOMIC MASS

For any element, no set number of neutrons are in the nucleus. For example, most hydrogen atoms (atomic number 1) have no neutrons. A small percentage, however, have one neutron, and a smaller percentage have two neutrons. Similarly, most iron atoms (atomic number 26) have 30 neutrons, but a small percentage have 29 neutrons. Atoms of the same element that contain different numbers of neutrons are **isotopes** of one another.

We identify isotopes by their mass number, which is the total number of protons and neutrons (in other words, the number of nucleons) in the nucleus. As Figure 12.7 shows, a hydrogen isotope with only one proton is called hydrogen-1, where 1 is the mass number. A hydrogen isotope with one proton and one neutron is therefore hydrogen-2, and a hydrogen isotope with one proton and two neutrons is hydrogen-3. Similarly, an iron isotope with 26 protons and 30 neutrons is called iron-56, and one with only 29 neutrons is iron-55.

Hydrogen-1	Hydrogen-2	Hydrogen-3	Iron-56	Iron-55
1 proton	1 proton	1 proton	26 protons	26 protons
0 neutron	1 neutron	2 neutrons	30 neutrons	29 neutrons
(protium)	(deuterium)	(tritium)		

Hydrogen isotopes   Iron isotopes

FIGURE 12.7

Isotopes of an element have the same number of protons but different numbers of neutrons and hence different mass numbers. The three hydrogen isotopes have special names: protium for hydrogen-1, deuterium for hydrogen-2, and tritium for hydrogen-3. Of these three isotopes, hydrogen-1 is most common. For most elements, such as iron, the isotopes have no special names and are indicated merely by mass number.

An alternative method of indicating isotopes is to write the mass number as a superscript and the atomic number as a subscript to the left of the atomic symbol. For example, an iron isotope with a mass number of 56 and an atomic number of 26 is written:

Mass number

$$^{56}_{26}\text{Fe}$$ —Atomic symbol

Atomic number

The total number of neutrons in an isotope can be calculated by subtracting its atomic number from its mass number:

$$\begin{array}{r} \text{mass number} \\ - \text{ atomic number} \\ \hline \text{number of neutrons} \end{array}$$

For example, uranium-238 has 238 nucleons. The atomic number of uranium is 92, which tells us that 92 of these 238 nucleons are protons. The remaining 146 nucleons must be neutrons:

$$\begin{array}{r} 238 \text{ protons and neutrons} \\ - 92 \text{ protons} \\ \hline 146 \text{ neutrons} \end{array}$$

Atoms interact with one another electrically. Therefore the way any atom behaves in the presence of other atoms is determined largely by the charged particles it contains, especially its electrons. Isotopes of an element differ only by mass, not by electric charge. For this reason, isotopes of an element share many characteristics—in fact, as chemicals they cannot be distinguished from one another. For example, a sugar molecule containing seven neutrons per carbon nucleus is digested no differently from a sugar molecule containing six neutrons per carbon nucleus. Interestingly, about 1% of the carbon we eat is the carbon-13 isotope containing seven neutrons per nucleus. The remaining 99% of the carbon in our diet is the more common carbon-12 isotope containing six neutrons per nucleus.

Most water molecules, H_2O, consist of hydrogen atoms with no neutrons. The few that do, however, are heavier, and because of this difference they can be isolated. Such water is appropriately called "heavy water."

FIGURING PHYSICAL SCIENCE

▪ Calculating Atomic Mass

About 99% of all carbon atoms are the isotope carbon-12, and most of the remaining 1% are the heavier isotope carbon-13. This small amount of carbon-13 raises the average mass of carbon from 12.000 amu to the slightly greater value of 12.011 amu.

To arrive at the atomic mass presented in the periodic table, first multiply the mass of each naturally occurring isotope of an element by the fraction of its abundance and then add up all the fractions.

SAMPLE PROBLEM 1

Carbon-12 has a mass of 12.0000 amu and makes up 98.89% of naturally occurring carbon. Carbon-13 has a mass of 13.0034 amu and makes up 1.11% of naturally occurring carbon. Use this information to show that the atomic mass of carbon shown in the periodic table, 12.011 amu, is correct.

SOLUTION:

Recognize that 98.89% and 1.11% expressed as decimals are 0.9889 and 0.0111, respectively.

	Contributing Mass of ^{12}C	Contributing Mass of ^{13}C	
Fraction of Abundance	0.9889	0.0111	
Mass (amu)	× 12.000	× 13.0034	step 1
	11.867	0.144	

automic mass = 11.867 + 0.144 = 12.011   step 2

YOUR TURN

Chlorine-35 has a mass of 34.97 amu, and chlorine-37 has a mass of 36.95 amu. Determine the atomic mass of chlorine, Cl (atomic number 17), if 75.53% of all chlorine atoms are the chlorine-35 isotope and 24.47% are the chlorine-37 isotope.

The total mass of an atom is called its **atomic mass**. This is the sum of the masses of all the atom's components (electrons, protons, and neutrons). Because electrons are so much less massive than protons and neutrons, their contribution to atomic mass is negligible. A special unit has been developed for atomic masses. This is the *atomic mass unit*, amu, where 1 atomic mass unit is equal to 1.661×10^{-24} gram, which is slightly less than the mass of a single proton. As shown in Figure 12.8, the atomic masses listed in the periodic table are in atomic mass units. The atomic mass of an element as presented in the periodic table is actually the average atomic mass of its various isotopes.

CHECK POINT

Distinguish between mass number and atomic mass.

Was this your answer?
Both terms include the word *mass* and so are easily confused. Focus your attention on the second word of each term, however, and you'll get it right every time. Mass number is a count of the *number* of nucleons in an isotope. An atom's mass number requires no units because it is simply a count. Atomic mass is a measure of the total *mass* of an atom, which is given in atomic mass units.

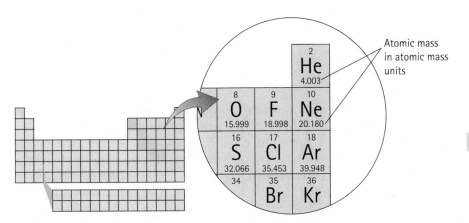

FIGURE 12.8

Helium, He, has an atomic mass of 4.003 amu, and neon, Ne, has an atomic mass of 20.180 amu.

About 50,000 pounds of synthetic diamonds are produced from **carbon** each year.

Alloys of **titanium** are relatively strong and resistant to corrosion, which makes them useful for hip implants.

Helium is formed underground as a by product of radioactive decay.

If this **silver** mug were filled with boiling water, the handle would quickly become too hot to handle because silver is one of the best conductors of heat.

Cylinders of 99.9999% pure **silicon** are sliced into wafers for the manufacture of integrated circuits.

Zinc has a low melting point and is commonly used in making coins.

Mercury freezes at −40°C and is a liquid at room temperature.

Bromine is a dark orange liquid that readily vaporizes at room temperature.

☐ Metal   ☐ Metalloid   ☐ Nonmetal

FIGURE 12.9

The periodic table is color-coded to show metals, nonmetals, and metalloids.

The Physical Science Place
Atoms and Isotopes

Please put to rest any fear you may have about needing to memorize the periodic table, or even parts of it—better to focus on the many great concepts behind its organization.

■ 12.4 The Periodic Table

So the periodic table is a listing of all the known elements with their atomic masses, atomic numbers, and atomic symbols. But the periodic table contains much more information. The way the table is organized, for example, tells us a lot about the elements' properties. Let's look at how the elements are grouped as metals, nonmetals, and metalloids.

As shown in Figure 12.9, most of the known elements are metals, which are defined as elements that are shiny, opaque, and good conductors of electricity and heat. Metals are *malleable*, which means they can be hammered into different shapes or bent without breaking. They are also *ductile*, which means they can be drawn into wires. All but a few metals are solid at room temperature. The exceptions are mercury, Hg; gallium, Ga; cesium, Cs; and francium, Fr, which are all liquids at a warm room temperature of 30°C (86°F). Another interesting exception is hydrogen, H, which takes on the properties of a liquid metal only at very high pressures (Figure 12.10). Under normal conditions, hydrogen behaves as a nonmetallic gas.

The nonmetallic elements, with the exception of hydrogen, are on the right side of the periodic table. Nonmetals are very poor conductors of electricity and heat, and may also be transparent. Solid nonmetals are neither malleable nor ductile. Rather, they are brittle and shatter when hammered. At 30°C (86°F), some nonmetals are solid (carbon, C), others are liquid (bromine, Br), and still others are gaseous (helium, He).

Six elements are classified as metalloids: boron, B; silicon, Si; germanium, Ge; arsenic, As; antimony, Sb; and tellurium, Te. Situated between the metals and the nonmetals in the periodic table, the metalloids have both metallic and nonmetallic characteristics. For example, these elements are weak conductors of electricity, which makes them useful as semiconductors in the integrated circuits of computers. Note from the periodic table how germanium, Ge (atomic number 32), is closer to the metals than to the nonmetals. Because of this positioning, we can deduce that germanium has more metallic properties than silicon, Si (atomic number 14), and is a slightly better conductor of electricity. So we find that integrated circuits fabricated with germanium operate faster than those fabricated with silicon. Because silicon is much more abundant and less expensive to obtain, however, silicon computer chips remain the industry standard.

PERIODS AND GROUPS

Two other important ways in which the elements are organized in the periodic table are by horizontal rows and vertical columns. Each horizontal row is called a **period**, and each vertical column is called a **group** (or sometimes a *family*). As shown in Figure 12.11, there are 7 periods and 18 groups.

Across any period, the properties of elements gradually change. This gradual change is called a periodic trend. As is shown in Figure 12.12, one periodic trend is that atomic size tends to decrease as you move from left to right across any period. Note that the trend repeats from one horizontal row to the next. This

FIGURE 12.10

Geoplanetary models suggest that hydrogen exists as a liquid metal deep beneath the surfaces of Jupiter (shown here) and Saturn. These planets are composed mostly of hydrogen. Inside them, the pressure exceeds 3 million times Earth's atmospheric pressure. At this tremendously high pressure, hydrogen is pressed to a liquid-metal phase.

FIGURE 12.11

The 7 periods (horizontal rows) and 18 groups (vertical columns) of the periodic table. Note that not all periods contain the same number of elements. Also note that, for reasons explained later, the sixth and seventh periods each include a subset of elements, which are listed apart from the main body.

FIGURE 12.12

The size of atoms gradually decreases in moving from left to right across any period. Atomic size is a periodic (repeating) property.

fyi

- The *carat* is the common unit used to describe the mass of a gem. A 1-carat diamond, for example, has a mass of 0.20 g. The *karat* is the common unit used to describe the purity of a precious metal, such as gold. A 24-karat gold ring is as pure as can be. A gold ring that is 50% pure is 12 karat.

Link To Section 20.4

fyi

- As the tungsten filament inside a lightbulb is heated, minute particles of tungsten evaporate. Over time, these particles are deposited on the inner surface of the bulb, causing the bulb to blacken. As it loses its tungsten, the filament eventually breaks and the bulb "burns out." A remedy is to replace the air inside the bulb with a halogen gas, such as iodine or bromine. In such a *halogen* bulb, the evaporated tungsten combines with the halogen rather than depositing on the bulb, which remains clear. Furthermore, the tungsten becomes unstable and splits from the halogen when it touches the hot filament. The halogen returns as a gas while the tungsten is deposited onto the filament, thereby restoring the filament. This is why halogen lamps have such long lifetimes.

phenomenon of repeating trends is called periodicity, a term used to indicate that the trends recur in cycles. Each horizontal row is called a period because it corresponds to one full cycle of a trend.

CHECK POINT

Which are larger: atoms of cesium, Cs (atomic number 55), or atoms of radon, Rn (atomic number 86)?

Was this your answer?
Perhaps you tried looking at Figure 12.12 to answer this question and quickly became frustrated because the sixth-period elements are not shown. Well, relax. Look at the trends and you'll see that, in any one period, all atoms to the left are larger than those to the right. Accordingly, cesium is positioned at the far left of period 6, and so you can reasonably predict that its atoms are larger than those of radon, which is positioned at the far right of period 6. The periodic table is a road map to understanding the elements.

Down any group (vertical column), the properties of elements tend to be remarkably similar, which is why these elements are said to be "grouped" or "in a family." As Figure 12.13 shows, several groups have traditional names that describe the properties of their elements. Early in human history, people discovered that ashes mixed with water produce a slippery solution useful for removing grease. By the Middle Ages, such mixtures were described as being alkaline, a term derived from the Arabic word for ashes, *al-qali*. Alkaline mixtures found many uses, particularly in the preparation of soaps (Figure 12.14). We now know that alkaline ashes contain compounds of group 1 elements, most notably potassium carbonate, also known as potash. Because of this history, group 1 elements, which are metals, are called the *alkali metals*.

Elements of group 2 also form alkaline solutions when mixed with water. Furthermore, medieval alchemists noted that certain minerals (which we now know are made up of group 2 elements) do not melt or change when put in fire. These fire-resistant substances were known to the alchemists as "earth." As a holdover from these ancient times, group 2 elements are known as the *alkaline-earth metals*.

Over toward the right side of the periodic table, elements of group 16 are known as the *chalcogens* ("ore-forming" in Greek) because the top two elements of this group, oxygen and sulfur, are so commonly found in ores. Elements of group 17 are known as the *halogens* ("salt-forming" in Greek) because of their tendency to form various salts. Group 18 elements are all unreactive gases that tend not to combine with other elements. For this reason, they are called the *noble gases*,

FIGURE 12.13

The common names for various groups of elements.

presumably because the nobility of earlier times were above interacting with common folk.

The elements of groups 3 through 12 are all metals that do not form alkaline solutions with water. These metals tend to be harder than the alkali metals and less reactive with water; hence they are used for structural purposes. Collectively they are known as the *transition metals*, a name that denotes their central position in the periodic table. The transition metals include some of the most familiar and important elements—iron, Fe; copper, Cu; nickel, Ni; chromium, Cr; silver, Ag; and gold, Au. They also include many lesser-known elements that are nonetheless important in modern technology. People with hip implants appreciate the transition metals titanium, Ti; molybdenum, Mo; and manganese, Mn, because these noncorrosive metals are used in implant devices.

FIGURE 12.14

Ashes and water make a slippery alkaline solution once used to clean hands.

CHECK POINT

The elements copper, Cu; silver, Ag; and gold, Au, are three of the few metals that can be found naturally in their elemental state. These three metals have found great use as currency and jewelry for a number of reasons, including their resistance to corrosion and their remarkable colors. How is the fact that these metals have similar properties reflected in the periodic table?

Was this your answer?

Copper (atomic number 29), silver (atomic number 47), and gold (atomic number 79) are all in the same group in the periodic table (group 11), which suggests they should have similar—though not identical—properties.

> A uranium atom is 40 times as heavy as a lithium atom, but only slightly larger in size because its more highly charged nucleus pulls harder on its electrons. But it has more electrons to pull, a balancing act that barely changes the atom's size.

Within the sixth period is a subset of 14 metallic elements (atomic numbers 58 to 71) that are quite unlike any of the other transition metals. A similar subset (atomic numbers 90 to 103) is found within the seventh period. These two subsets are the *inner transition metals*. Inserting the inner transition metals into the main body of the periodic table as in Figure 12.15 results in a long and cumbersome table. So that the table can fit nicely on a standard paper size, these elements are commonly placed below the main body of the table, as shown in Figure 12.16.

The sixth-period inner transition metals are called the *lanthanides* because they fall after lanthanum, La. Because of their similar physical and chemical properties, they tend to occur mixed together in the same locations on Earth. Also because of their similarities, lanthanides are unusually difficult to purify. Recently, the commercial use of lanthanides has increased. Several lanthanide elements, for example, are used in the fabrication of the light-emitting diodes (LEDs) of computer monitors and flat-screen televisions.

fyi

- From 1943 to 1986, the Hanford nuclear facility in central Washington state produced 72 tons of plutonium, nearly two-thirds the nation's supply. Creating this much plutonium generated an estimated 450 billion gallons of radioactive and hazardous liquids, which were discharged into the local environment. Today, some 53 million gallons of high-level radioactive and chemical wastes are stored in 177 underground tanks, many of them leaking into the ground water.

FIGURE 12.15

Inserting the inner transition metals between atomic groups 3 and 4 results in a periodic table that is not easy to fit on a standard sheet of paper.

The typical display of the inner transition metals. The count of elements in the sixth period goes from lanthanum (La, 57) to cerium (Ce, 58) on through to lutetium (Lu, 71) and then back to hafnium (Hf, 72). A similar jump is made in the seventh period.

The seventh-period inner transition metals are called the *actinides* because they fall after actinium, Ac. They, too, all have similar properties and hence are not easily purified. The nuclear power industry faces this obstacle because it requires purified samples of two of the most publicized actinides: uranium, U, and plutonium, Pu. Actinides heavier than uranium are not found in nature but are synthesized in the laboratory.

■ 12.5 Physical and Conceptual Models

Atoms are so small that the number of them in a baseball is roughly equal to the number of Ping-Pong balls that could fit inside a hollow sphere as big as Earth, as Figure 12.17 illustrates. This number is incredibly large—beyond our intuitive grasp. Atoms are so incredibly small that we can never see them in the usual

If Earth were filled with nothing but Ping-Pong balls, the number of balls would be roughly equal to the number of atoms in a baseball. Put differently, if a baseball were the size of Earth, one of its atoms would be the size of a Ping-Pong ball.

sense. This is because light travels in waves, and atoms are smaller than the wavelengths of visible light, which is the light that allows the human eye to see things. As illustrated in Figure 12.18, the diameter of an object visible under the highest magnification must be larger than the wavelengths of visible light.

Microscopic objects can be seen through a microscope that works with visible light, but submicroscopic particles cannot. (a) A bacterium is visible because it is larger than the wavelengths of visible light. We can see the bacterium through the microscope because the bacterium reflects visible light. (b) An atom is invisible because it is smaller than the wavelengths of visible light and so does not reflect the light toward our eyes.

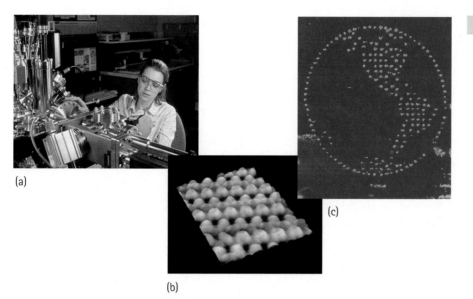

(a)

(b)

(c)

FIGURE 12.19

(a) Scanning probe microscopes are relatively simple devices used to create submicroscopic imagery. (b) An image of gallium and arsenic atoms obtained with a scanning probe microscope. (c) Each dot in the world's tiniest map consists of a few thousand gold atoms, each atom moved into its proper place by a scanning probe microscope.

Although we cannot see atoms *directly*, we can generate images of them *indirectly*. In the mid-1980s, researchers developed the *scanning probe microscope*, which produces images by dragging an ultrathin needle back and forth over the surface of a sample. Bumps the size of atoms on the surface cause the needle to move up and down. This vertical motion is detected and translated by a computer into a topographical image that corresponds to the positions of atoms on the surface (Figure 12.19). A scanning probe microscope can also be used to push individual atoms into desired positions. This ability opened the field of nanotechnology, in which incredibly small electronic circuits and motors are built atom by atom.

CHECK POINT

■ Why are atoms invisible?

Was this your answer?
An individual atom is smaller than the wavelengths of visible light and so is unable to reflect that light. Atoms are invisible, therefore, because visible light passes right by them. The atomic images generated by scanning probe microscopes are not photographs taken by a camera. Rather, they are computer renditions generated from the movements of an ultrathin needle.

A very small or very large visible object can be represented with a **physical model**, which is a model that replicates the object at a more convenient scale. Figure 12.20a, for instance, shows a large-scale physical model of a microorganism that a biology student uses to study the microorganism's internal structure. Because atoms are invisible, however, we cannot use a physical model to represent them. In other words, we cannot simply scale up the atom to a larger size, as we might with a microorganism. (A scanning probe microscope merely shows the *positions* of atoms and not actual images of atoms, which do not have the solid surfaces implied in the scanning probe image of Figure 12.19.) So, rather than describing the atom with a physical model, chemists use what is known as a **conceptual model**, which describes a system. The more accurate a conceptual model, the more accurately it predicts the behavior of the system. The weather is best described using a conceptual model like the one shown in Figure 12.20b. Such a model shows how the various components of the system—humidity, atmospheric pressure, temperature, electric charge, the motion of large masses of

Link To Section 25.6

(a)

(b)

FIGURE 12.20

(a) This large-scale model of a microorganism is a physical model. (b) Weather forecasters rely on conceptual models such as this one to predict the behavior of weather systems.

air—interact with one another. Other systems that can be described by conceptual models are the economy, population growth, the spread of diseases, and team sports.

CHECK POINT

A basketball coach describes a playing strategy to her team by way of sketches on a game board. Do the illustrations represent a physical model or a conceptual model?

Was this your answer?
The sketches are a conceptual model the coach uses to describe a system (the players on the court), with the hope of predicting an outcome (winning the game).

Like the weather, the atom is a complex system of interacting components, and it is best described with a conceptual model. You should therefore be careful not to interpret any visual representation of an atomic conceptual model as a re-creation of an actual atom. In Section 12.7, for example, you will be introduced to the planetary model of the atom, wherein electrons are shown orbiting the atomic nucleus much as planets orbit the Sun. This planetary model is limited, however, in that it fails to explain many properties of atoms. Thus newer and more accurate (and more complicated) conceptual models of the atom have since been introduced. In these models, electrons appear as a cloud hovering around the atomic nucleus, but even these models have their limitations. Ultimately, the best models of the atom are purely mathematical.

In this textbook, our focus is on conceptual atomic models that are easily represented by visual images, including the planetary model and a model in which electrons are grouped in units called shells. Despite their limitations, such images are excellent guides to learning about the behavior of atoms, especially for the beginning student. As we discuss in the following sections, scientists developed these models to help explain how atoms emit light.

We can't "see" an atom because they're too small. We can't see the farthest star either. There's much that we can't see. But that doesn't prevent us from thinking about such things or even collecting indirect evidence.

■ 12.6 Identifying Atoms Using the Spectroscope

Recall from Chapter 11 that we see white light when all frequencies of visible light reach our eye at the same time. By passing white light through a prism or through a diffraction grating, the color components of the light can be separated,

Link To Section 27.2

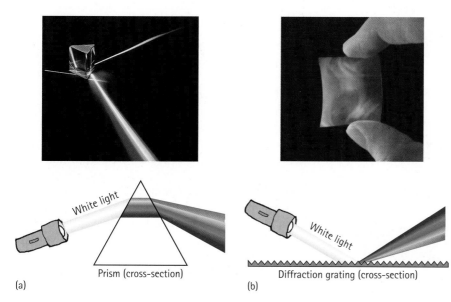

FIGURE 12.21

White light is separated into its color components by (a) a prism and (b) a diffraction grating.

(a) Prism (cross-section)

(b) Diffraction grating (cross-section)

as shown in Figure 12.21. (Remember—each color of visible light corresponds to a different frequency.) A **spectroscope**, shown in Figure 12.22, is an instrument used to observe the color components of any light source. The spectroscope allows us to analyze the light emitted by elements as they are made to glow.

Light is given off by atoms subjected to various forms of energy, such as heat or electricity. The atoms of a given element emit only certain frequencies of light, however. As a consequence, each element emits a distinctive glow when energized. Sodium atoms emit bright yellow light, which makes them useful as the light source in streetlamps because our eyes are very sensitive to yellow light. To name just one more example, neon atoms emit a brilliant red-orange light, which makes them useful as the light source in neon signs.

When we view the light from glowing atoms through a spectroscope, we see that the light consists of a number of discrete (separate from one another) frequencies rather than a continuous spectrum like the one shown in Figure 12.22. The pattern of frequencies formed by a given element—some of which are shown in Figure 12.23—is referred to as that element's **atomic spectrum**. The atomic spectrum is an element's fingerprint. You can identify the elements in a light source by analyzing the light through a spectroscope and looking for characteristic patterns. If you don't have the opportunity to work with a spectroscope in your laboratory, check out the activities at the end of this chapter.

fyi

- A star's age is revealed by its elemental makeup. The first and oldest stars were composed of hydrogen and helium because those were the only elements available at that time. Heavier elements were produced after many of these early stars exploded in supernovae. Later stars incorporated these heavier elements in their formation. In general, the younger a star, the greater amounts of these heavier elements it contains.

Link To Section 27.2

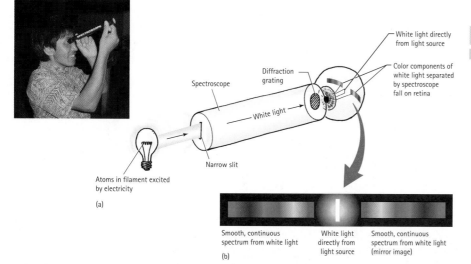

FIGURE 12.22
INTERACTIVE FIGURE

(a) In a spectroscope, light emitted by atoms passes through a narrow slit before being separated into particular frequencies by a prism or (as shown here) a diffraction grating. (b) This is what the eye sees when the slit of a diffraction-grating spectroscope is pointed toward a white-light source. Spectra of colors appear to the left and right of the slit.

FIGURE 12.23

Elements heated by a flame glow their characteristic color. This is commonly called a flame test and is used to test for the presence of an element in a sample. When viewed through a spectroscope, the color of each element is revealed to consist of a pattern of distinct frequencies known as an atomic spectrum.

Link To Section 28.2

FIGURE 12.24

Light is quantized, which means it consists of a stream of energy packets. Each packet is called a quantum, also known as a photon.

How might you deduce the elemental composition of a star?

Was this your answer?
Aim a well-built spectroscope at the star and study its spectral patterns. In the late 1800s, this was done with our own star, the Sun. Spectral patterns of hydrogen and some other known elements were observed, in addition to one pattern that could not be identified. Scientists concluded that this unidentified pattern belonged to an element not yet discovered on Earth. They named this element helium after the Greek word for "sun," *helios*.

■ 12.7 The Quantum Hypothesis

An important step toward our present understanding of atoms and their spectra was taken by the German physicist Max Planck (1858–1947). In 1900, Planck hypothesized that light energy is *quantized* in much the same way matter is. The mass of a gold brick, for example, equals some whole-number multiple of the mass of a single gold atom. Similarly, an electric charge is always some whole-number multiple of the charge on a single electron. Mass and electric charge are therefore said to be *quantized* in that they consist of some number of fundamental units.

What Planck did with his **quantum hypothesis** was to recognize that a beam of light energy is not the continuous (nonquantized) stream of energy we think it is. Instead, the beam consists of zillions of small, discrete packets of energy, each packet called a **quantum**, as represented in Figure 12.24. A few years later, Einstein recognized that these quanta of light behave much like tiny particles of matter. To emphasize their particulate nature, each quantum of light was called a *photon*, a name coined because of its similarity to the words *electron*, *proton*, and *neutron*.

Using Planck's quantum hypothesis, the Danish scientist Niels Bohr (1885–1962) explained the formation of atomic spectra as follows. First, an electron has more potential energy when it is farther from the nucleus. This is analogous to the greater potential energy an object has when it is held higher above the ground. Second, Bohr recognized that when an atom absorbs a photon of light, it is absorbing energy. This energy is acquired by one of the electrons. Because this electron has gained energy, it must move away from the nucleus.

Light source   Light beam

One quantum (photon) of light

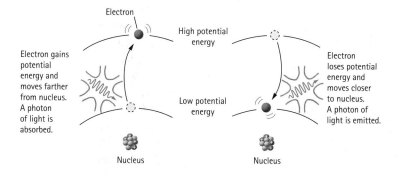

FIGURE **12.25**

An electron is lifted away from the nucleus as the atom it is in absorbs a photon of light and drops closer to the nucleus as the atom releases a photon of light.

Bohr also realized that the opposite is true: when a high-potential-energy electron in an atom loses some of its energy, the electron moves closer to the nucleus and the energy lost from the electron is emitted from the atom as a photon of light. Both absorption and emission are illustrated in Figure 12.25.

CHECK POINT

Which has more energy: a photon of red light or a photon of infrared light?

Was this your answer?
Red light has a higher frequency than infrared light, which means a photon of red light has more energy than a photon of infrared light. Recall that a photon is a single discrete packet (a quantum) of radiant energy.

Bohr reasoned that because light energy is quantized, the energy of an electron in an atom must also be quantized. In other words, an electron cannot have just any amount of potential energy. Rather, within the atom there must be a number of distinct energy levels, analogous to steps on a staircase. Where you are on a staircase is restricted to where the steps are—you cannot stand at a height that is, say, halfway between any two adjacent steps. Similarly, an atom has only a limited number of permitted energy levels, and an electron can never have an amount of energy between these permitted energy levels. Bohr gave each energy level a **principal quantum number n**, where n is always some integer. The lowest energy level has a principal quantum number $n = 1$. An electron for which $n = 1$ is as close to the nucleus as possible, and an electron for which $n = 1$, $n = 3$, and so forth is farther away from the nucleus.

Just as I can't stand between two adjacent steps, an electron can't exist between two energy levels.

Using these ideas, Bohr developed a conceptual model in which an electron moving around the nucleus is restricted to certain distances from the nucleus, with these distances determined by the amount of energy the electron has. Bohr saw this as similar to how the planets are held in orbit around the Sun at given distances from the Sun. The allowed energy levels for any atom, therefore, could be graphically represented as orbits around the nucleus, as shown in Figure 12.26. Bohr's quantized model of the atom thus became known as the *planetary model*.

Bohr used his planetary model to explain why atomic spectra contain only a limited number of light frequencies. According to the model, photons are emitted by atoms as electrons move from higher-energy outer orbits to lower-energy inner orbits. The energy of an emitted photon is equal to the difference in energy between the two orbits. Because an electron is restricted to discrete orbits, only particular light frequencies are emitted, as atomic spectra show.

Interestingly, any transition between two orbits is always instantaneous. In other words, the electron doesn't "jump" from a higher to a lower orbit the way a squirrel jumps from a higher branch in a tree to a lower one. Rather, an electron takes no time to move between two orbits. Bohr was serious when he stated that electrons could never exist between permitted energy levels!

FIGURE **12.26**

Bohr's planetary model of the atom, in which electrons orbit the nucleus much as planets orbit the Sun, is a graphical representation that helps us understand how electrons can possess only certain quantities of energy.

Recall from Chapter 11 that a photon behaves as a particle when it is being emitted by an atom or being absorbed by photographic film or other detectors, but it behaves as a wave in traveling from a source to the place where it is detected.

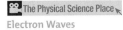
The Physical Science Place
Electron Waves

FIGURE 12.27

(a) An electron microscope makes practical use of the wave nature of electrons. The wavelengths of electron beams are typically shorter than the wavelengths of visible light by a factor of a thousand, and so the electron microscope can distinguish detail not visible with optical microscopes. (b) Detail of a female mosquito head as seen with an electron microscope at a "low" magnification of 200×. Note the remarkable resolution.

CHECK POINT

Is the Bohr model of the atom a physical model or a conceptual model?

Was this your answer?
The Bohr model is a conceptual model. It is not a scaled-up version of an atom, but instead is a representation that accounts for the atom's behavior.

Bohr's planetary atomic model proved to be a tremendous success. By utilizing Planck's quantum hypothesis, Bohr's model solved the mystery of atomic spectra. Despite its successes, though, Bohr's model was limited because it did not explain why energy levels in an atom are quantized. Bohr was quick to point out that his model was to be interpreted only as a crude beginning, and the picture of electrons whirling about the nucleus like planets about the Sun was not to be taken literally (a warning to which popularizers of science paid no heed).

12.8 Electron Waves

If light has both wave properties and particle properties, why can't a material particle, such as an electron, also have both? This question was posed by the French physicist Louis de Broglie (1892–1987) while he was still a graduate student in 1924. His revolutionary answer was that every particle of matter is somehow endowed with a wave to guide it as it travels. The more slowly an electron moves, the more its behavior is that of a particle with mass. The more quickly it moves, however, the more its behavior is that of a wave of energy. This duality is an extension of Einstein's famous equation $E = mc^2$, which tells us that matter and energy are interconvertible. We talk more about this relationship in the next chapter.

A practical application of the wave properties of fast-moving electrons is the electron microscope, which focuses not visible-light waves, but rather electron waves. Because electron waves are much shorter than visible-light waves, electron microscopes can show far greater detail than optical microscopes, as Figure 12.27 shows.

(a)

(b)

Mechanical   Wire
vibrator   loop

(a)

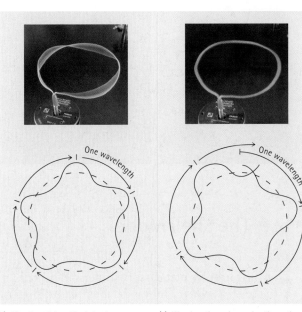

(b) Wavelength is self-reinforcing.   (c) Wavelength produces chaotic motion.

FIGURE 12.28

For the fixed circumference of a wire loop, only some wavelengths are self-reinforcing. (a) The loop affixed to the post of a mechanical vibrator at rest. Waves are sent through the wire when the post vibrates. (b) Waves created by vibration at particular rates are self-reinforcing. (c) Waves created by vibration at other rates are not self-reinforcing.

In an atom, an electron moves at very high speeds—on the order of 2 million m/s—and therefore exhibits many of the properties of a wave. An electron's wave nature can be used to explain why electrons in an atom are restricted to particular energy levels. Permitted energy levels are a natural consequence of electron waves closing in on themselves in a synchronized manner.

As an analogy, consider the wire loop shown in Figure 12.28. This loop is affixed to a mechanical vibrator that can be adjusted to create waves of different wavelengths in the wire. Waves passing through the wire that meet up with themselves, as shown in Figure 12.28b, form a stationary wave pattern called a standing wave. This pattern results because the peaks and valleys of successive waves are perfectly matched, which makes the waves reinforce one another. With other wavelengths, as shown in Figure 12.28c, successive waves are not synchronized. As a result, the waves do not build to great amplitude.

The only waves that an electron exhibits while confined to an atom are those that are self-reinforcing. These resemble a standing wave centered on the atomic nucleus. Each standing wave corresponds to one of the permitted energy levels. Only the frequencies of light that match the difference between any two of these permitted energy levels can be absorbed or emitted by an atom.

The wave nature of electrons also explains why they do not spiral closer and closer to the positive nucleus that attracts them. By viewing each electron orbit as a self-reinforcing wave, we see that the circumference of the smallest orbit can be no smaller than a single wavelength.

fyi

- Electron waves are three-dimensional, which makes them difficult to visualize, but scientists have come up with ways of visualizing them. This includes *probability clouds* and *atomic orbitals*, which you would learn about in a follow-up course on chemistry.

Link To Section 10.1

CHECK POINT

What must an electron be doing in order to have wave properties?

Was this your answer?

According to de Broglie, particles of matter behave like waves by virtue of their motion. An electron must therefore be moving in order to have wave properties. In atoms, electrons move at speeds of about 2 million m/s, and so their wave nature is most pronounced.

The quality of a song depends on the arrangement of musical notes. In a similar fashion, the properties of an element depend on the arrangements of electrons in its atoms.

FIGURE 12.29

(a) A cutaway view of the seven shells, with the number of electrons each shell can hold indicated. (b) A two-dimensional, cross-sectional view of the shells. (c) An easy-to-draw cross-sectional view that resembles Bohr's planetary model.

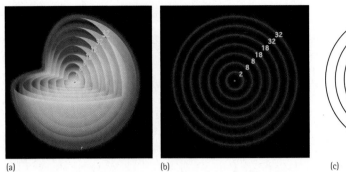

(a)   (b)   (c)

■ 12.9 The Shell Model

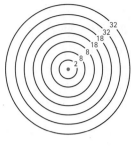

The Physical Science Place

Bohr's Shell Model

Link To Section 15.2

F or the purposes of a cursory understanding of how atoms behave, we turn to the *shell model*. This model is similar to Bohr's planetary model in that it is highly simplified. We can use it, however, to help explain the organization of the periodic table.

According to the shell model, electrons behave as though they are arranged in a series of concentric shells. A **shell** is defined as a region of space about the atomic nucleus within which electrons may reside. An important aspect of this model is that there are at least seven shells and each shell can hold only a limited number of electrons. As shown in Figure 12.29, the innermost shell can hold 2 electrons; the second and third shells, 8 electrons each; the fourth and fifth shells, 18 electrons each; and the sixth and seventh shells, 32 electrons each.

A series of seven such concentric shells accounts for the seven periods of the periodic table. Furthermore, the number of elements in each period is equal to the shell's capacity for electrons. The first shell, for example, has a capacity for only two electrons. That's why we find only two elements, hydrogen and helium, in the first period (Figure 12.30). Hydrogen is the element whose

FIGURE 12.30

The first three periods of the periodic table according to the shell model. Elements in the same period have electrons in the same shells. Elements in the same period differ from one another by the number of electrons in the outermost shell.

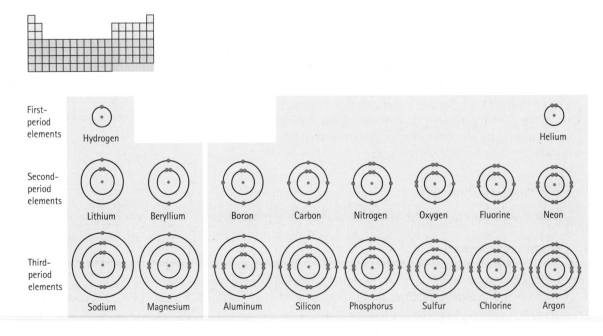

atoms have only one electron. This electron resides within the first shell, which is the shell closest to the nucleus. Each helium atom has two electrons, both of which are also within the first shell, which is thus filled to its maximum capacity. Similarly, the second and third shells each have a capacity for eight electrons, and so eight elements are found in both the second and third periods.

Two-time Nobel laureate Linus Pauling (1901–1994) was an early proponent of teaching beginning chemistry students a shell model from which the organization of the periodic table could be described. In this model, as is described here, orbitals are grouped according to energy level. However, this shell model differs from that found in advanced physics and chemistry textbooks, which identify a shell as a group of orbitals that all have the same principal quantum number.

Linus Pauling

The electrons of the outermost occupied shell in any atom are directly exposed to the external environment and are the first to interact with other atoms. Most notably, they are the ones that participate in chemical bonding, as we shall be discussing in Chapter 15. The electrons in the outermost shell, therefore, are quite important. They are called **valence electrons**. The term *valence* is derived from the Latin *valentia*, "strength," and it refers to the "combining power" of an atom.

Look carefully at Figure 12.30. Can you see that the valence electrons of atoms above and below one another (within the same group) are similarly organized? For example, atoms of the first group—hydrogen, lithium, and sodium—each have a single valence electron. The atoms of the second group—beryllium and magnesium—each have two valence electrons. Similarly, atoms of the last group—helium, neon, and argon—each have their outermost shells filled to capacity with valence electrons—two for helium, and eight for both neon and argon. In general, the valence electrons of atoms in the same group of the periodic table are similarly organized. This explains why elements of the same group have similar properties—a concept first presented in Section 12.5.

What do poets and scientists have in common? They both use metaphors to help us understand abstract concepts and relationships. The "shell," for example, is a metaphor that helps us visualize an invisible reality. Scientific models are essentially equivalent to the metaphorical language used in poetry.

CHECK POINT

Do atoms really consist of shells that look like those depicted in Figure 12.29?

Link To Section 15.1

Was this your answer?

The shell model is NOT a depiction of the "appearance of an atom." Rather, it is a conceptual model that allows us to account for observed behavior. An atom, therefore, does not actually contain a series of concentric shells; it merely behaves as though it does.

Remember that the shell model is not to be interpreted as an actual representation of the atom's physical structure. Rather, it serves as a tool to help us understand and predict how atoms behave. In Chapter 15 we will use a simplified version of this model to show how atoms join together to form molecules, which are tightly held groups of atoms. In the next chapter, however, we will explore in greater detail the nature of the atomic nucleus, which is a potential source of enormous amounts of energy.

fyi

- According to Einstein's theory of special relativity, at 60% of the speed of light, gold's innermost electrons experience only 52 seconds for each one of our minutes. A diamond may be forever, but the innermost electrons of gold are 8 s/min slow!

SUMMARY OF TERMS

Atomic nucleus The dense, positively charged center of every atom.

Electron An extremely small, negatively charged subatomic particle found outside the atomic nucleus.

Element Any material that is made up of only one type of atom.

Periodic table A chart in which all known elements are listed in order of atomic number.

Atomic symbol An abbreviation for an element or atom.

Proton A positively charged subatomic particle of the atomic nucleus.

Atomic number A count of the number of protons in the atomic nucleus.

Neutron An electrically neutral subatomic particle of the atomic nucleus.

Nucleon Any subatomic particle found in the atomic nucleus. Another name for either a proton or a neutron.

Isotopes Any member of a set of atoms of the same element whose nuclei contain the same number of protons but different numbers of neutrons.

Atomic mass The mass of an element's atoms listed in the periodic table as an average value based on the relative abundance of the element's isotopes.

Period A horizontal row in the periodic table.

Group A vertical column in the periodic table, also known as a family of elements.

Physical model A representation of an object on some convenient scale.

Conceptual model A representation of a system that helps us predict how the system behaves.

Spectroscope A device that uses a prism or diffraction grating to separate light into its color components.

Atomic spectrum The pattern of frequencies of electromagnetic radiation emitted by the atoms of an element, considered to be an element's "fingerprint."

Quantum hypothesis The idea that light energy is contained in discrete packets called quanta.

Quantum A small, discrete packet of light energy.

Principal quantum number n An integer that specifies the quantized energy level of an atomic orbital.

Shell A set of overlapping atomic orbitals of similar energy levels. In other words, a region of space in which electrons of similar energy levels in an atom have a 90% chance of being located.

Valence electron An electron that is located in the outermost occupied shell in an atom and can participate in chemical bonding.

REVIEW QUESTIONS

12.1 Atoms Are Ancient and Empty

1. Which is the oldest element?
2. Is it possible to see an atom using visible light?
3. What is at the center of every atom?

12.2 The Elements

4. How many types of atoms can you expect to find in a pure sample of any element?
5. Distinguish between an atom and an element.
6. What is the atomic symbol for the element cobalt?

12.3 Protons and Neutrons

7. What role does atomic number play in the periodic table?
8. Distinguish between atomic number and mass number.
9. Distinguish between mass number and atomic mass.

12.4 The Periodic Table

10. Are most elements metallic or nonmetallic?
11. How many periods are there in the periodic table? How many groups?
12. What happens to the properties of elements across any period of the periodic table?

12.5 Physical and Conceptual Models

13. If a baseball were the size of Earth, about how large would its atoms be?
14. When we use a scanning probe microscope, do we see atoms directly or only indirectly?
15. What is the difference between a physical model and a conceptual model?

12.6 Identifying Atoms Using the Spectroscope

16. What does a spectroscope do to the light coming from an atom?
17. What causes an atom to emit light?
18. Why do we say atomic spectra are like fingerprints of the elements?

12.7 The Quantum Hypothesis

19. What was Planck's quantum hypothesis?
20. Which has more potential energy: an electron close to an atomic nucleus or one far from an atomic nucleus?
21. Did Bohr think of his planetary model as an accurate representation of what an atom looks like?

12.8 Electron Waves

22. Who first proposed that electrons exhibit the properties of a wave?
23. About how fast does an electron travel around the atomic nucleus?
24. How does the speed of an electron change its fundamental nature?

12.9 The Shell Model

25. Does the periodic table explain the shell model, or does the shell model explain the periodic table?
26. Which electrons are most responsible for the properties of an atom?
27. What is the relationship between the maximum number of electrons each shell can hold and the number of elements in each period of the periodic table?

EXERCISES

● BEGINNER   ■ INTERMEDIATE   ◆ EXPERT

1. ● A cat strolls across your backyard. An hour later, a dog with its nose to the ground follows the trail of the cat. Explain what is going on from a molecular point of view.
2. ● If all the molecules of a body remained part of that body, would the body have any odor?
3. ■ Where did the atoms that make up a newborn baby originate?
4. ● Where did the carbon atoms in Leslie's hair originate? (Shown below is a photo of coauthor Leslie at 16.)

5. ■ In what sense can you truthfully say that you are a part of every person around you?
6. ● Considering how small atoms are, what are the chances that at least one of the atoms exhaled in your first breath will be in your last breath?
7. ● Is the head of a politician really made of 99.99999999% empty space?
8. ● If two protons and two neutrons are removed from the nucleus of an oxygen-16 atom, a nucleus of which element remains?
9. ● If an atom has 43 electrons, 56 neutrons, and 43 protons, what is its approximate atomic mass? What is the name of this element?
10. ● The nucleus of an electrically neutral iron atom contains 26 protons. How many electrons does this iron atom have?
11. ● Evidence for the existence of neutrons did not come until many years after the discoveries of the electron and the proton. Give a possible explanation.
12. ■ Which has more atoms: a 1-g sample of carbon-12 or a 1-g sample of carbon-13? Explain.
13. ■ Why are the atomic masses listed in the periodic table not whole numbers?
14. ● Which contributes more to an atom's mass: electrons or protons? Which contributes more to an atom's size?

15. ● What is the approximate mass of an oxygen atom in atomic mass units? What is the approximate mass of two oxygen atoms? How about an oxygen molecule?
16. ● What is the approximate mass of a carbon atom in atomic mass units? How about a carbon dioxide molecule?
17. ● What is the approximate mass of a hydrogen atom in atomic mass units? How about a water molecule?
18. ◆ When we breathe we inhale oxygen, O_2, and exhale carbon dioxide, CO_2, plus water vapor, H_2O. Which likely has more mass, the air that we inhale or the same volume of air we exhale? Does breathing cause you to lose or gain weight?
19. ■ As a tree respires it takes in carbon dioxide, CO_2, and water vapor, H_2O, from the air while also releasing oxygen, O_2. Does the tree lose or gain weight as it respires? Explain.
20. ● Does it make sense to say that this textbook is about 99.9% empty space?
21. ● The atoms that compose your body are mostly empty space, and structures such as the chair you're sitting on are composed of atoms that are also mostly empty space. So why don't you fall through the chair?
22. ■ Which of the following diagrams best represents the size of the atomic nucleus relative to the size of the atom:

23. ● A beam of protons and a beam of neutrons of the same energy are both harmful to living tissue. The beam of neutrons, however, is less harmful. Suggest why.
24. ■ As depicted in Figure 12.19, are gallium atoms really red and arsenic atoms green?
25. ■ With scanning probe microscopy technology, we see not actual atoms, but rather images of them. Explain.
26. ■ Why is it not possible for a scanning probe microscope to make images of the inside of an atom?
27. ● What do the components of a conceptual model have in common?
28. ● Would you use a physical model or a conceptual model to describe the following: a gold coin, a dollar bill, a car engine, air pollution, a virus, the spread of sexually transmitted disease?

29. ● What is the function of an atomic model?
30. ● What is the relationship between the light emitted by an atom and the energies of the electrons in the atom?
31. ● How might you distinguish a sodium-vapor streetlight from a mercury-vapor streetlight?
32. ■ How can a hydrogen atom, which has only one electron, create so many spectral lines?
33. ■ Which color of light comes from a greater energy transition, red or blue?
34. ◆ How does the wave model of electrons orbiting the nucleus account for the fact that the electrons can have only discrete energy values?
35. ■ What might the spectrum of an atom look like if the atom's electrons were not restricted to particular energy levels?
36. ◆ Some older cars vibrate loudly when driving at particular speeds. For example, at 65 mi/h the car may be most quiet, but at 60 mi/h the car rattles uncomfortably. How is this analogous to the quantized energy levels of an electron in an atom?
37. ◆ Does a shell have to contain electrons in order to exist?

38. ■ Place the proper number of electrons in each shell:

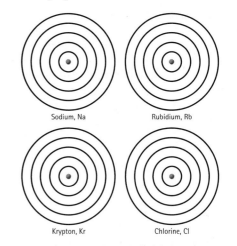

Sodium, Na   Rubidium, Rb

Krypton, Kr   Chlorine, Cl

39. ■ Use the shell model to explain why a potassium atom, K, is larger than a sodium atom, Na.
40. ■ Use the shell model to explain why a lithium atom, Li, is larger than a beryllium atom, Be.

PROBLEMS

● BEGINNER   ■ INTERMEDIATE   ◆ EXPERT

1. ■ The isotope lithium-7 has a mass of 7.0160 amu, and the isotope lithium-6 has a mass of 6.0151 amu. Given the information that 92.58% of all lithium atoms found in nature are lithium-7 and 7.42% are lithium-6, show that the atomic mass of lithium is 6.941 amu.

2. ■ The element bromine, Br (atomic number 35), has two major isotopes of similar abundance, both around 50%. The atomic mass of bromine is reported in the periodic table as 79.904 amu. Choose the most likely set of mass numbers for these two bromine isotopes: (a) ^{80}Br, ^{81}Br; (b) ^{79}Br, ^{8}Br; (c) ^{79}Br, ^{81}Br.

ACTIVE EXPLORATIONS

1. Purchase some "rainbow" glasses from a nature, toy, or hobby store. The lenses of these glasses are diffraction gratings. Looking through them, you will see light separated into its color components. Certain light sources, such as the Moon or a car's headlights, are separated into a continuous spectrum—in other words, all the colors of the rainbow appear in a continuous sequence from red to violet. Other light sources, however, emit a distinct number of discontinuous colors. Examples include streetlights, neon signs, sparklers, and fireworks. The spectral patterns you see from these light sources are the atomic spectra of elements heated in the light sources. You'll be able to see the patterns best when you are at least 50 m from the light source. This distance makes the spectrum appear as a series of dots similar to the series of lines shown in Figure 12.23.

2. Stretch a rubber band between your two thumbs and pluck one length of it. Note that no matter where along the length you pluck, the area of greatest oscillation is always at the midpoint. This is a self-reinforcing wave

that occurs as overlapping waves bounce back and forth from thumb to thumb.

Under regular light, it is difficult to see the waves traveling back and forth. For a better view, pluck the rubber band in front of a computer monitor or a television screen that uses a cathode ray tube (the older nonflat kind). The light from these devices, which acts like a strobe light, makes the waves appear to slow down.

Vary the tension in the rubber band to see different effects.

3. You can "quantize" your whistle by whistling down a long tube, such as the tube from a roll of wrapping paper. First, without the tube, whistle from a high pitch to a low pitch. Do it in a single breath and as loud as you can. Next, try the same thing while holding the tube to your lips. A-ha! Note that some frequencies simply cannot be whistled, no matter how hard you try. These frequencies are forbidden because their wavelengths are not a multiple of the length of the tube.

Try experimenting with tubes of different lengths. To hear yourself more clearly, use a flexible plastic tube and twist the outer end toward your ear.

When your whistle is confined to the tube, the consequence is a quantization of its frequencies. When an electron wave is confined to an atom, the consequence is a quantization of the electron's energy.

People watching you perform this activity may not believe that the audible "steps" of your whistling down the tube are not intentional. Explain quantization to them before allowing them to attempt this activity for themselves. Try to count the number of steps in your tubular whistle, understanding that each step is analogous to an energy level in an atom. Does a longer tube create fewer or more steps than a shorter tube? Why is it so difficult to whistle down a garden hose?

If you punch a few holes along the tube, you alter the frequencies of the standing waves that can form in the tube, with the result that different pitches are produced. This is the underlying principle in such musical instruments as flutes and saxophones.

READINESS ASSURANCE TEST (RAT)

If you have a good handle on this chapter, if you really do, then you should be able to score 7 out of 10 on this RAT. If you score less than 7, you need to study further before moving on.

Choose the BEST answer to each of the following.

1. Which are older, the atoms in the body of an elderly person or those in the body of a baby?
   (a) The baby's atoms are older, because this is surely a trick question.
   (b) The elderly person's atoms are older, because the person has been around much longer.
   (c) The atoms are of the same age, which is appreciably older than the solar system.
   (d) It depends on their diet.

2. You could swallow a capsule of germanium, Ge (atomic number 32), without significant ill effects. If a proton were added to each germanium nucleus, however, you would not want to swallow the capsule because the germanium would
   (a) become arsenic.
   (b) become radioactive.
   (c) expand and likely lodge in your throat.
   (d) change in flavor.

3. If an atom were the size of a baseball, its nucleus would be about the size of
   (a) a walnut.
   (b) a raisin.
   (c) a flea.
   (d) an atom.

4. Why are the atomic masses listed in the periodic table not whole numbers?
   (a) Scientists have yet to make the precise measurements.
   (b) That would be too much of a coincidence.
   (c) The atomic masses are average atomic masses.
   (d) Today's instruments can measure the atomic masses to many decimal places.

5. An element found in another galaxy exists as two isotopes. If 80.0% of the atoms have an atomic mass of 80.00 amu and the other 20.0% have an atomic mass of 82.00 amu, what is the approximate *atomic mass* of the element?
   (a) 80.4 amu
   (b) 81.0 amu
   (c) 81.6 amu
   (d) 64.0 amu
   (e) 16.4 amu

6. List the following atoms in order of increasing atomic size: thallium, Tl; germanium, Ge; tin, Sn; phosphorus, P.
   (a) Ge < P < Sn < Tl
   (b) Tl < Sn < P < Ge
   (c) Tl < Sn < Ge < P
   (d) P < Ge < Sn < Tl

7. Would you use a physical model or a conceptual model to describe each of the following: the brain; the mind; the solar system; the beginning of the universe?
   (a) conceptual; physical; conceptual; physical
   (b) conceptual; conceptual; conceptual; conceptual
   (c) physical; conceptual; physical; conceptual
   (d) physical; physical; physical; physical

8. How does the wave model of electrons orbiting the nucleus account for the fact that the electrons can have only discrete energy values?
   (a) Electrons can vibrate only at particular frequencies.
   (b) When an electron wave is confined, it is reinforced only at particular frequencies.
   (c) The energy values of an electron occur only where its wave properties have a maximum amplitude.
   (d) The wave model accounts for the shells an electron may occupy, not its energy levels.

9. How can a hydrogen atom, which has only one electron, have so many spectral lines?
   (a) The electron can move at various speeds.
   (b) The protons in the nucleus are also giving off various light frequencies.
   (c) One electron can be boosted to many different energy levels.
   (d) The atomic radius of the hydrogen atom is relatively large.

10. How many electrons are there in the third shell of sodium, Na (atomic number 11)?
    (a) none
    (b) one
    (c) two
    (d) three

Answers to RAT

1c, 2a, 3c, 4c, 5a, 6d, 7c, 8b, 9c, 10b

EXPLORING FURTHER

http://particleadventure.org
An award-winning interactive tour of quarks, neutrinos, antimatter, extra dimensions, dark matter, accelerators, and much more.

http://www.astro.uiuc.edu/~kaler/sow/spectra.html
A detailed Web page for learning about the spectral patterns of stars and how they are used to study the universe.

http://www.superstringtheory.com
If you think the wave nature of the electron is bizarre, explore this site for information on and references to the potentially revolutionary theory that particles, forces, space, and time are merely manifestations of incredibly tiny strings that exist in 11 dimensions.

http://www.nano.org.uk
Home page for the Institute of Nanotechnology, whose mission is to organize international scientific events, conferences, and educational courses designed to encourage academic, corporate, and governmental interests in nanotechnology.

http://www.nano.gov
Home page for the U.S. government sponsored National Nanotechnology Initiative (NNI).

http://www.foresight.org
Home page for the Foresight Institute, a nonprofit organization dedicated to helping prepare society for anticipated advanced nanotechnologies.

CHAPTER 12 ONLINE RESOURCES

The Physical Science Place

Interactive Figure
- 12.22

Tutorials
- Atomic Structure
- Atoms and Isotopes
- Bohr's Shell Model

Videos
- Evidence for Atoms

- Atoms Are Recyclable
- Electron Waves

Quiz

Flashcards

Links

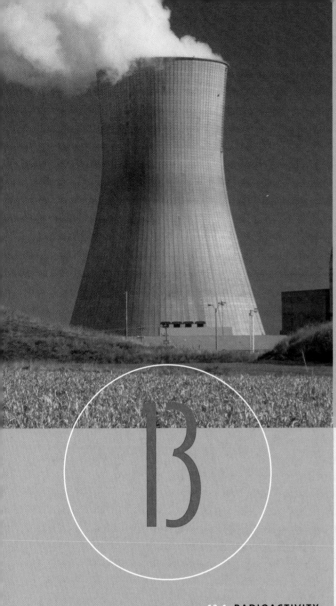

the atomic nucleus
and radioactivity

13.1 RADIOACTIVITY

13.2 THE ATOMIC NUCLEUS AND THE STRONG NUCLEAR FORCE

13.3 HALF-LIFE AND TRANSMUTATION

13.4 RADIOMETRIC DATING

13.5 NUCLEAR FISSION

13.6 MASS–ENERGY EQUIVALENCE: $E = mc^2$

13.7 NUCLEAR FUSION

■ The atomic nucleus and its processes are one of the most misunderstood and controversial areas of science. Distrust of anything *nuclear*, or anything *radioactive*, is much like the fears of electricity more than a century ago. The distrust of electricity in households stemmed from ignorance. Indeed, electricity can be dangerous, and even lethal, when improperly handled. But with safeguards and well-informed consumers, society has determined that the benefits of electricity outweigh its risks. Today we are making similar decisions about nuclear technology's risks and benefits. These decisions should be made with an adequate understanding of the atomic nucleus and its inner processes.

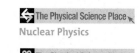

The Physical Science Place
Nuclear Physics

The Physical Science Place
Radioactive Decay

Radioactivity has been around since Earth's beginning.

Natural background (cosmic rays, Earth minerals) 81%

Medicine and diagnostics 15%

Consumer products (television sets, smoke detectors) 4%

FIGURE 13.1

Origins of radiation exposure for an average individual in the United States.

fyi

- Once alpha and beta particles slow down, they combine to form harmless helium. This happens primarily deep underground. As the newly formed helium seeps toward the surface it becomes concentrated within natural gas deposits. Some natural gas deposits, such as those in Texas, contain as much as 7% helium. This helium is purified and sold for various applications, such as blimps and helium balloons. Interestingly, gas fields within the United States contain about two-thirds of the world's supply of helium.

FIGURE 13.2

INTERACTIVE FIGURE

In a magnetic field, alpha rays bend one way, beta rays bend the other way, and gamma rays don't bend at all. Note that the alpha rays bend less than do the beta rays. This occurs because alpha particles have more inertia (mass) than beta particles.

13.1 Radioactivity

Elements with unstable nuclei are said to be *radioactive*. They eventually break down and eject energetic particles and emit high-frequency electromagnetic radiation. This process is **radioactivity**, which, because it involves the decay of the atomic nucleus, is often called *radioactive decay*.

A common misconception is that radioactivity is new in the environment, but it has been around far longer than the human race. Interestingly, the deeper you go below Earth's surface, the hotter it gets. At a mere depth of 30 km the temperature is more than 500°C. At greater depths it is so hot that rock melts into magma, which can rise to Earth's surface to escape as lava. Superheated subterranean water can escape violently to form geysers or more gently to form a soothing natural hot spring. The main reason it gets hotter down below is because Earth contains an abundance of radioactive isotopes and is heated as it absorbs radiation from these isotopes. So volcanoes, geysers, and hot springs are all powered by radioactivity. Even the drifting of continents (see Chapter 22) is a consequence of Earth's internal radioactivity. Radioactivity is as natural as sunshine and rain.

ALPHA, BETA, AND GAMMA RAYS

All elements with an atomic number greater than 82 (lead) are radioactive. These elements, and others, emit three distinct types of radiation, named by the first three letters of the Greek alphabet, α, β, γ— *alpha*, *beta*, and *gamma*. Alpha rays carry a positive electrical charge, beta rays carry a negative charge, and gamma rays carry no charge. The three rays can be separated by placing a magnetic field across their paths (Figure 13.2).

An **alpha particle** is the combination of two protons and two neutrons (in other words, it is the nucleus of the helium atom, atomic number 2). Alpha particles are relatively easy to shield because of their relatively large size and their double positive charge ($+2$). For example, they do not normally penetrate through light materials such as paper or clothing. Because of their great kinetic energies, however, alpha particles can cause significant damage to the surface of a material, especially living tissue. When traveling through only a few centimeters of air, alpha particles pick up electrons and become nothing more than harmless helium. As a matter of fact, that's where the helium in a child's balloon comes from—practically all of Earth's helium atoms were at one time energetic alpha particles.

Alpha particle = helium nucleus (+2 electric charge)

Gamma ray = ultrahigh-energy nonvisible light (no electric charge)

Magnet

Beta particle = electron (−1 electric charge)

Radium sample   Lead block

A **beta particle** is an electron ejected from a nucleus. Once ejected, it is indistinguishable from an electron in a cathode ray or electrical circuit, or one orbiting the atomic nucleus. The difference is that a beta particle originates inside the nucleus—from a neutron. As we shall soon see, the neutron becomes a proton once it loses the electron that is a beta particle. A beta particle is normally faster than an alpha particle and carries only a single negative charge (-1). Beta particles are not as easy to stop as alpha particles are, and they can penetrate light materials such as paper or clothing. They can penetrate fairly deeply into skin, where they have the potential for harming or killing living cells. But they are not able to penetrate deeply into denser materials such as aluminum. Beta particles, once stopped, simply become part of the material they are in, like any other electron.

Gamma rays are the high-frequency electromagnetic radiation emitted by radioactive elements. Like visible light, a gamma ray is pure energy. The amount of energy in a gamma ray, however, is much greater than in visible light, ultraviolet light, or even X-rays. Because they have no mass or electric charge and because of their high energies, gamma rays can penetrate through most materials. However, they cannot penetrate unusually dense materials such as lead, which absorbs them. Delicate molecules inside cells throughout our bodies that are zapped by gamma rays suffer structural damage. Hence, gamma rays are generally more harmful to us than alpha or beta particles (unless the alphas or betas are ingested).

CHECK POINT

■ Pretend you are given three radioactive rocks—one an alpha emitter, one a beta emitter, and one a gamma emitter. You can throw away one, but of the remaining two, you must hold one in your hand and place the other in your pocket. What can you do to minimize your exposure to radiation?

Was this your answer?
Hold the alpha emitter in your hand because the skin on your hand shields you. Put the beta emitter in your pocket because beta particles are likely stopped by the combined thickness of your clothing and skin. Throw away the gamma emitter because gamma rays penetrate your body from any of these locations. Ideally, of course, you should distance yourself as much as possible from all of the rocks.

Common rocks and minerals in our environment contain significant quantities of radioactive isotopes because most of them contain trace amounts of uranium. People who live in brick, concrete, or stone buildings are exposed to greater amounts of radiation than people who live in wooden buildings.

FIGURE 13.3

INTERACTIVE FIGURE

A gamma ray is simply electromagnetic radiation, much higher in frequency and energy than light and X-rays.

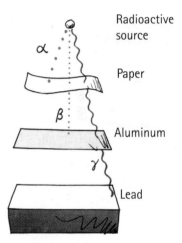

FIGURE 13.4

INTERACTIVE FIGURE

Alpha particles are the least penetrating and can be stopped by a few sheets of paper. Beta particles readily pass through paper, but not through a sheet of aluminum. Gamma rays penetrate several centimeters into solid lead.

FIGURE 13.5

The shelf life of fresh strawberries and other perishables is markedly increased when the food is subjected to gamma rays from a radioactive source. The strawberries on the right were treated with gamma radiation, which kills the microorganisms that normally lead to spoilage. The food is only a receiver of radiation and is not transformed into an emitter of radiation, as can be confirmed with a radiation detector.

Link To Section 20.7

FIGURE 13.6

A commercially available radon test kit for the home. The canister is unsealed in the area to be sampled. Radon seeping into the canister is adsorbed by activated carbon within the canister. After several days, the canister is resealed and sent to a laboratory that determines the radon level by measuring the amount of radiation emitted by the adsorbed radon.

The leading source of naturally occurring radiation is radon-222, an inert gas arising from uranium deposits. Radon is a heavy gas that tends to accumulate in basements after it seeps up through cracks in the floor. Levels of radon vary from region to region, depending on local geology. You can check the radon level in your home with a radon detector kit (Figure 13.6). If levels are abnormally high, corrective measures such as sealing the basement floor and walls and maintaining adequate ventilation are recommended.

About one-fifth of our annual exposure to radiation comes from nonnatural sources, primarily medical procedures. Television sets, fallout from nuclear testing, and the coal and nuclear power industries are also contributors. The coal industry far outranks the nuclear power industry as a source of radiation. The global combustion of coal annually releases about 13,000 tons of radioactive thorium and uranium into the atmosphere. Both these minerals are found naturally in coal deposits, so their release is a natural consequence of burning coal. Worldwide, the nuclear power industries generate about 10,000 tons of radioactive waste each year. Most of this waste, however, is contained and *not* released into the environment.

RADIATION DOSAGE

Radiation dosage is commonly measured in *rads* (*r*adiation *a*bsorbed *d*ose), a unit of absorbed energy. One **rad** is equal to 0.01 J of radiant energy absorbed per kilogram of tissue.

The capacity for nuclear radiation to cause damage is not just a function of its level of energy, however. Some forms of radiation are more harmful than others. For example, suppose you have two arrows, one with a pointed tip and one with a suction cup at its tip. Shoot the two of them at an apple at the same speed and both have the same kinetic energy. The one with the pointed tip, however, invariably does more damage to the apple than the one with the suction cup. Similarly, some forms of radiation cause greater harm than other forms, even when we receive the same number of rads from both forms.

The unit of measure for radiation dosage based on potential damage is the **rem** (*r*oentgen *e*quivalent *m*an).* In calculating the dosage in rems, we multiply the number of rads by a factor that corresponds to different health effects of different types of radiation determined by clinical studies. For example, 1 rad of alpha particles has the same biological effect as 10 rads of beta particles.** We call both of these dosages 10 rems:

Particle	Radiation Dosage		Factor		Health Effect
alpha	1 rad	×	10	=	10 rems
beta	10 rad	×	1	=	10 rems

CHECK POINT

Would you rather be exposed to 1 rad of alpha particles or 1 rad of beta particles?

Was this your answer?

Multiply these quantities of radiation by the appropriate factor to get the dosages in rems. Alpha: 1 rad × 10 = 10 rems; beta: 1 rad × 1 = 1 rem. The factors show us that, physiologically speaking, alpha particles are 10 times as damaging as beta particles.

---

* This unit is named for the discoverer of X-rays, Wilhelm Roentgen.
** This is true even though beta particles have more penetrating power, as discussed earlier.

Lethal doses of radiation begin at 500 rems. A person has about a 50% chance of surviving a dose of this magnitude received over a short period of time. During radiation therapy, a patient may receive localized doses in excess of 200 rems each day for a period of weeks (Figure 13.7).

All the radiation we receive from natural sources and from medical procedures is only a fraction of 1 rem. For convenience, the smaller unit *millirem* is used, where 1 millirem (mrem) is 1/1000 of a rem.

The average person in the United States is exposed to about 360 mrem a year, as Table 13.1 indicates. About 80% of this radiation comes from natural sources, such as cosmic rays and Earth itself. A typical chest X-ray exposes a person to 5–30 mrem (0.005–0.030 rem), less than 1/10,000 of the lethal dose. Interestingly, the human body is a significant source of natural radiation, primarily from the potassium we ingest. Our bodies contain about 200 g of potassium. Of this quantity, about 20 mg is the radioactive isotope potassium-40, which is a gamma ray emitter. Between every heartbeat about 5,000 potassium-40 isotopes in the average human body undergo spontaneous radioactive decay. Radiation is indeed everywhere.

When radiation encounters the intricately structured molecules in the watery, ion-rich brine that makes up our cells, the radiation can create chaos on the atomic scale. Some molecules are broken, and this change alters other molecules, which can be harmful to life processes.

Cells can repair most kinds of molecular damage caused by radiation if the radiation is not too severe. A cell can survive an otherwise lethal dose of radiation if the dose is spread over a long period of time to allow intervals for healing. When radiation is sufficient to kill cells, the dead cells can be replaced by new ones. Sometimes a radiated cell survives with a damaged DNA molecule. New cells arising from the damaged cell retain the altered genetic information, producing a *mutation*. Usually the effects of a mutation are insignificant, but occasionally the mutation results in cells that do not function as well as unaffected ones, sometimes leading to a cancer. If the damaged DNA is in an individual's reproductive cells, the genetic code of the individual's offspring may retain the mutation.

FIGURE 13.7

Nuclear radiation is focused on harmful tissue, such as a cancerous tumor, to selectively kill or shrink the tissue in a technique known as *radiation therapy*. This application of nuclear radiation has saved millions of lives—a clear-cut example of the benefits of nuclear technology. The inset shows the internationally used symbol indicating an area where radioactive material is being handled or produced.

TABLE 13.1   ANNUAL RADIATION EXPOSURE

Source	Typical Dose (mrem) Received Annually
Natural Origin	
Cosmic radiation	26
Ground	33
Air (radon-222)	198
Human tissues (K-40; Ra-226)	35
Human Origin	
Medical procedures	
Diagnostic X-rays	40
Nuclear medicine	15
TV tubes, other consumer products	11
Weapons-test fallout	1
Commercial fossil-fuel power plants	<1
Commercial nuclear power plants	≪1

FIGURE 13.8

The film badges worn by Tammy and Larry contain audible alerts for both radiation surge and accumulated exposure. Information from the individualized badges is periodically downloaded to a database for analysis and storage.

FIGURE 13.9

Tracking fertilizer uptake with a radioactive isotope.

Without the nuclear strong force—strong interaction—there would be no atoms beyond hydrogen.

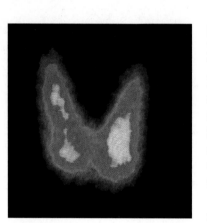

FIGURE 13.10

The thyroid gland, located in the neck, absorbs much of the iodine that enters the body through food and drink. Images of the thyroid gland, such as the one shown here, can be obtained by giving a patient the radioactive isotope iodine-131. These images are useful in diagnosing metabolic disorders.

RADIOACTIVE TRACERS

In scientific laboratories radioactive samples of all the elements have been made. This is accomplished by bombardment with neutrons or other particles. Radioactive materials are extremely useful in scientific research and industry. To check the action of a fertilizer, for example, researchers combine a small amount of radioactive material with the fertilizer and then apply the combination to a few plants. The amount of radioactive fertilizer taken up by the plants can be easily measured with radiation detectors. From such measurements, scientists can inform farmers of the proper amount of fertilizer to use. Radioactive isotopes used to trace such pathways are called *tracers*.

In a technique known as medical imaging, tracers are used to diagnose internal disorders. This technique works because the path the tracer takes is influenced only by its physical and chemical properties, not by its radioactivity. The tracer may be introduced alone or along with some other chemical that helps target the tracer to a particular type of tissue in the body.

13.2 The Atomic Nucleus and the Strong Nuclear Force

As described in Chapter 12, the atomic nucleus occupies only a few quadrillionths of the volume of the atom, leaving most of the atom as empty space. The nucleus is composed of *nucleons*, which, as discussed in Chapter 12, is the collective name for protons and neutrons.

We know that electrical charges of like sign repel one another. So how do positively charged protons in the nucleus stay clumped together? This question led to the discovery of an attraction called the **strong nuclear force**, which acts between all nucleons. This force is very strong but only over extremely short distances (about 10^{-15} m, the diameter of a typical atomic nucleus). Repulsive electrical interactions, on the other hand, are relatively long-ranged. Figure 13.11 suggests a comparison of the strengths of these two forces over distance. For

FIGURE 13.11

INTERACTIVE FIGURE

(a) Two protons near each other experience both an attractive strong nuclear force and a repulsive electric force. At this tiny separation distance, the strong nuclear force overcomes the electric force, and the protons stay together. (b) When the two protons are relatively far from each other, the electric force is more significant and the protons repel each other. This proton–proton repulsion in large atomic nuclei reduces nuclear stability.

(a)   (b)

protons that are close together, as in small nuclei, the attractive strong nuclear
force easily overcomes the repulsive electrical force. But for protons that are far
apart, such as those on opposite edges of a large nucleus, the attractive strong
nuclear force may be weaker than the repulsive electrical force.

A large nucleus is not as stable as a small one. In a helium nucleus, which has
two protons, each proton feels the repulsive effect of only one other proton. In a
uranium nucleus, however, each of the 92 protons feels the repulsive effects of
the other 91 protons! The nucleus is unstable. We see that there is a limit to the
size of the atomic nucleus. For this reason, all nuclei with more than 83 protons
are radioactive.

(a) Nucleons close together

CHECK POINT

> Two protons in the atomic nucleus repel each other, but they are also
> attracted to each other. Why?

Was this your answer?
> Although two protons repel each other by the electric force, they also
> attract each other by the strong nuclear force. Both of these forces act
> simultaneously. As long as the attractive strong nuclear force is stronger
> than the repulsive electric force, the protons remain together. When the
> electric force overcomes the strong nuclear force, however, the protons fly
> apart from each other.

(b) Nucleons far apart

FIGURE 13.12

(a) All nucleons in a small
atomic nucleus are close to one
another; hence, they experience an attractive strong
nuclear force. (b) Nucleons on
opposite sides of a larger
nucleus are not as close to one
another, and so the attractive
strong nuclear forces holding
them together are much
weaker. The result is that the
large nucleus is less stable.

Neutrons serve as a "nuclear cement" holding the atomic nucleus together.
Protons attract both protons and neutrons by the strong nuclear force. Protons
also repel other protons by the electric force. Neutrons, on the other hand, have
no electric charge and so only attract other protons and neutrons by the strong
nuclear force. The presence of neutrons therefore adds to the attraction among
nucleons and helps hold the nucleus together (Figure 13.13).

The more protons there are in a nucleus, the more neutrons are needed to
help balance the repulsive electric forces. For light elements, it is sufficient to
have about as many neutrons as protons. The most common isotope of
carbon, C-12, for instance, has equal numbers of each—six protons and six
neutrons. For large nuclei, more neutrons than protons are needed. Because
the strong nuclear force diminishes rapidly over dis-
tance, nucleons must be practically touching in order
for the strong nuclear force to be effective. Nucleons on
opposite sides of a large atomic nucleus are not as
attracted to one another. The electric force, however,
does not diminish by much across the diameter of a
large nucleus and so begins to win out over the strong
nuclear force. To compensate for the weakening of the
strong nuclear force across the diameter of the nucleus,
large nuclei have more neutrons than protons. Lead, for
example, has about one and a half times as many neutrons as protons.

All nucleons, both protons and
neutrons, attract one another by
the strong nuclear force.

Only protons repel one another
by the electric force.

So we see that neutrons are stabilizing and large nuclei require an abundance
of them. But neutrons are not always successful in keeping a nucleus intact.
Interestingly, neutrons are not stable when they are by themselves. A lone neutron is radioactive, and spontaneously transforms to a proton and an electron
(Figure 13.14a). A neutron seems to need protons around to keep this from
happening. After the size of a nucleus reaches a certain point, the neutrons so
outnumber the protons that there are not enough protons in the mix to prevent

FIGURE 13.13

The presence of neutrons helps
hold the nucleus together by
increasing the effect of the
strong nuclear force, represented by the single-headed
arrows.

FIGURE 13.14

(a) A neutron near a proton is stable, but a neutron by itself is unstable and decays to a proton by emitting an electron.
(b) Destabilized by an increase in the number of protons, the nucleus begins to shed fragments, such as alpha particles.

The radioactive half-life of a material is also the time for its decay rate to reduce to half.

the neutrons from turning into protons. As neutrons in a nucleus change into protons, the stability of the nucleus decreases because the repulsive electric force becomes increasingly significant. The result is that pieces of the nucleus fragment away in the form of radiation, as indicated in Figure 13.14b.

CHECK POINT

What role do neutrons serve in the atomic nucleus? What is the fate of a neutron when alone or distant from one or more protons?

Was this your answer?
Neutrons serve as a nuclear cement in nuclei and add to nuclear stability. But when alone or away from protons, a neutron becomes radioactive and spontaneously transforms to a proton and an electron.

13.3 Half-Life and Transmutation

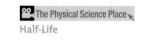
Half-Life

The rate of decay for a radioactive isotope is measured in terms of a characteristic time, the **half-life**. This is the time it takes for half of an original quantity of an element to decay. For example, radium-226 has a half-life of 1620 years, which means that half of a radium-226 sample will be converted to other elements by the end of 1620 years. In the next 1620 years, half of the remaining radium will decay, leaving only one-fourth the original amount of radium. (After 20 half-lives, the initial quantity of radium-226 will be diminished by a factor of about one million.)

Half-lives are remarkably constant and not affected by external conditions. Some radioactive isotopes have half-lives that are less than a millionth of a second, while others have half-lives of more than a billion years. Uranium-238 has a half-life of 4.5 billion years. All uranium eventually decays in a series of steps to lead. In 4.5 billion years, half the uranium presently in Earth today will be lead.

It is not necessary to wait through the duration of a half-life in order to measure it. The half-life of an element can be calculated at any given moment by measuring the rate of decay of a known quantity. This is easily done using a radiation detector (Figure 13.16). In general, the shorter the half-life of a substance, the faster it disintegrates, and the more radioactivity per amount is detected.

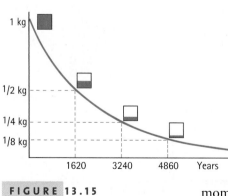

FIGURE 13.15

INTERACTIVE FIGURE

Every 1620 years the amount of radium decreases by half.

(a)

(b)

FIGURE 13.16

Some radiation detectors.
(a) A Geiger counter detects incoming radiation by its ionizing effect on enclosed gas in the tube. (b) A scintillation counter detects incoming radiation by flashes of light that are produced when charged particles or gamma rays pass through it.

CHECK POINT

1. If a radioactive isotope has a half-life of 1 day, how much of an original sample is left at the end of the second day? The third day?
2. Which gives a higher counting rate on a radiation detector: a radioactive material with a short half-life or a radioactive material with a long half-life?

Were these your answers?

1. One-fourth of the original sample is left at the end of the second day—the three-fourths that underwent decay is then a different element altogether. At the end of 3 days, 1/8 of the original sample remains.
2. The material with the shorter half-life is more active and show a higher counting rate on a radiation detector.

When a radioactive nucleus emits an alpha or a beta particle, there is a change in atomic number, which means that a different element is formed. (Recall from Chapter 12 that an element is defined by its atomic number, which is the number of protons in the nucleus.) The changing of one chemical element to another is called **transmutation**. Transmutation occurs in natural events, and is also initiated artificially in the laboratory.

Link To Section 12.3

NATURAL TRANSMUTATION

Consider uranium-238, the nucleus of which contains 92 protons and 146 neutrons. When an alpha particle is ejected, the nucleus loses two protons and two neutrons. Because an element is defined by the number of protons in its nucleus, the 90 protons and 144 neutrons left behind are no longer identified as being uranium. Instead we have the nucleus of a different element—thorium. This transmutation can be written as a nuclear equation:

$$^{238}_{92}\text{U} \rightarrow\ ^{234}_{90}\text{Th} +\ ^{4}_{2}\text{He}$$

We see that $^{238}_{92}\text{U}$ transmutes to the two elements written to the right of the arrow. When this transmutation occurs, energy is released, partly in the form of kinetic energy of the alpha particle ($^{4}_{2}\text{He}$), partly in the kinetic energy of the thorium atom, and partly in the form of gamma radiation. In this and all such equations, the mass numbers at the top balance ($238 = 234 + 4$) and the atomic numbers at the bottom also balance ($92 = 90 + 2$).

Thorium-234, the product of this reaction, is also radioactive. When it decays, it emits a beta particle. Because a beta particle is an electron, the atomic number of the resulting nucleus is *increased* by 1. So after beta emission by thorium with 90 protons, the resulting element has 91 protons. It is no longer thorium, but the element protactinium. Although the atomic number has increased by 1 in this process, the mass number (protons + neutrons) remains the same. The nuclear equation is

$$^{234}_{90}\text{Th} \rightarrow \,^{234}_{91}\text{Pa} + \,^{0}_{-1}\text{e}$$

We write an electron as $^{0}_{-1}e$. The superscript 0 indicates that the electron's mass is insignificant relative to that of protons and neutrons. The subscript -1 is the electric charge of the electron.

So we see that when an element ejects an alpha particle from its nucleus, the mass number of the resulting atom is decreased by 4, and its atomic number is decreased by 2. The resulting atom is an element two spaces back in the periodic table of the elements. When an element ejects a beta particle from its nucleus, the mass of the atom is practically unaffected, meaning there is no change in mass number, but its atomic number increases by 1. The resulting atom belongs to an element one place forward in the periodic table. Gamma radiation results in no change in either the mass number or the atomic number. So we see that radioactive elements can decay backward or forward in the periodic table.

The successions of radioactive decays of $^{238}_{92}\text{U}$ to $^{206}_{82}\text{Pb}$, an isotope of lead, is shown in Figure 13.17. Each gray arrow shows an alpha decay, and each red arrow shows a beta decay. Notice that some of the nuclei in the series can decay in both ways. This is one of several similar radioactive series that occur in nature.

FIGURE 13.17

U-238 decays to Pb-206 through a series of alpha and beta decays.

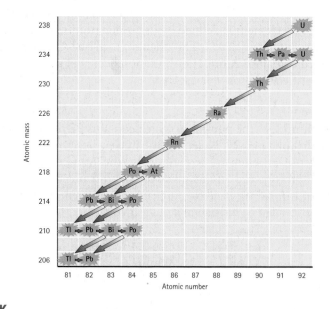

1. Complete the following nuclear reactions.
   a. $^{226}_{88}\text{Ra} \rightarrow \,^{?}_{?}? + \,^{0}_{-1}\text{e}$
   b. $^{209}_{84}\text{Po} \rightarrow \,^{205}_{82}\text{Pb} + \,^{?}_{?}?$
2. What finally becomes of all the uranium that undergoes radioactive decay?

Were these your answers?
1. a. $^{226}_{88}\text{Ra} \rightarrow \,^{226}_{89}\text{Ac} + \,^{0}_{-1}\text{e}$
   b. $^{209}_{84}\text{Po} \rightarrow \,^{205}_{82}\text{Pb} + \,^{4}_{2}\text{He}$
2. All uranium ultimately becomes lead. On the way to becoming lead, it exists as a series of elements, as indicated in Figure 13.17.

ARTIFICIAL TRANSMUTATION

Ernest Rutherford, in 1919, was the first of many investigators to succeed in transmuting a chemical element. He bombarded nitrogen gas with alpha particles from a piece of radioactive ore. The impact of an alpha particle on a nitrogen nucleus transmutes nitrogen into oxygen:

$$_2^4He + \,_7^{14}N \rightarrow \,_8^{17}O + \,_1^1H$$

Rutherford used a device called a *cloud chamber* to record this event (Figure 13.18). In a cloud chamber, moving charged particles show a trail of ions along their path in a way similar to the ice crystals that show the trail of jet planes high in the sky. From a quarter of a million cloud-chamber tracks photographed on movie film, Rutherford showed seven examples of atomic transmutation. Analysis of tracks bent by a strong external magnetic field showed that when an alpha particle collided with a nitrogen atom, a proton bounced out and the heavy atom recoiled a short distance. The alpha particle disappeared. The alpha particle was absorbed in the process, transforming nitrogen to oxygen.

Since Rutherford's announcement in 1919, experimenters have carried out many other nuclear reactions, first with natural bombarding projectiles from radioactive ores and then with still more energetic projectiles—protons and electrons hurled by huge particle accelerators. Artificial transmutation produces the hitherto unknown synthetic elements from atomic number 93 to 118. All of these artificially made elements have short half-lives. If they ever existed naturally when Earth was formed, they have long since decayed.

FIGURE 13.18

A cloud chamber. Charged particles moving through supersaturated vapor leave trails. When the chamber is in a strong electric or magnetic field, bending of the tracks provides information about the charge, mass, and momentum of the particles.

FIGURE 13.19

Tracks of elementary particles in a bubble chamber, a similar yet more complicated device than a cloud chamber. Two particles have been destroyed at the points where the spirals emanate, and four others created in the collision.

■ 13.4 Radiometric Dating

Earth's atmosphere is continuously bombarded by cosmic rays, and this bombardment causes many atoms in the upper atmosphere to transmute. These transmutations result in many protons and neutrons being "sprayed out" into the environment. Most of the protons are stopped as they collide with the atoms of the upper atmosphere, stripping electrons from these atoms to become hydrogen atoms. The neutrons, however, keep going for longer distances because they have no electrical charge and therefore do not interact electrically with matter. Eventually, many of them collide with the nuclei in the denser lower atmosphere. A nitrogen nucleus that captures a neutron, for instance, becomes an isotope of carbon by emitting a proton:

$$_0^1n + \,_7^{14}N \rightarrow \,_6^{14}C + \,_1^1H$$

The alchemists of old tried in vain to cause the transmutation of one element to another. Despite their fervent efforts and rituals, they never came close to succeeding. Ironically, natural transmutations were going on all around them.

This carbon-14 isotope, which makes up less than one-millionth of 1% of the carbon in the atmosphere, is radioactive and has eight neutrons. (The most common isotope, carbon-12, has six neutrons and is not radioactive.) Because

Link To Section 21.2

both carbon-12 and carbon-14 are forms of carbon, they have the same chemical properties. Both these isotopes can chemically react with oxygen to form carbon dioxide, which is taken in by plants. This means that all plants contain a tiny bit of radioactive carbon-14. All animals eat plants (or at least plant-eating animals), and therefore have a little carbon-14 in them. In short, all living things on Earth contain some carbon-14.

Carbon-14 is a beta emitter and decays back to nitrogen by the following reaction:

$$^{14}_{6}C \rightarrow {}^{14}_{7}N + {}^{0}_{-1}e$$

fyi

- A 1-g sample of carbon from recently living matter contains about 50 trillion billion (5×10^{22}) carbon atoms. Of these carbon atoms, about 65 billion (6.5×10^{10}) of them are the radioactive C-14 isotope. This gives the carbon a beta disintegration rate of about 13.5 decays per minute.

Because plants continue to take in carbon dioxide as long as they live, any carbon-14 lost by decay is immediately replenished with fresh carbon-14 from the atmosphere. In this way, a radioactive equilibrium is reached at which there is a constant ratio of about one carbon-14 atom to every 100 billion carbon-12 atoms. When a plant dies, replenishment of carbon-14 stops. Then the percentage of carbon-14 decreases at a constant rate given by its half-life. The longer a plant or other organism is dead, therefore, the less carbon-14 it contains relative to the constant amount of carbon-12.

The half-life of carbon-14 is about 5760 years. This means that half of the carbon-14 atoms that are now present in a plant or animal that dies today will decay in the next 5760 years. Half of the remaining carbon-14 atoms will then decay in the following 5760 years, and so forth.

With this knowledge, scientists can calculate the age of carbon-containing artifacts, such as wooden tools or skeletons, by measuring their current level of radioactivity. This process, known as **carbon-14 dating**, enables us to probe as much as 50,000 years into the past. Beyond this time span, too little carbon-14 remains to permit accurate analysis.

22,920 years ago   17,190 years ago   11,460 years ago   5730 years ago   Present

FIGURE 13.20

The amount of radioactive carbon-14 in the skeleton diminishes by half every 5730 years, with the result that today the skeleton contains only a fraction of the carbon-14 it originally had. The red arrows symbolize relative amounts of carbon-14.

Carbon-14 dating would be an extremely simple and accurate dating method if the amount of radioactive carbon in the atmosphere had been constant over the ages. But it hasn't been. Fluctuations in the Sun's magnetic field as well as changes in the strength of Earth's magnetic field affect cosmic-ray intensities in Earth's atmosphere, which in turn produce fluctuations in the production of C-14. In addition, changes in Earth's climate affect the amount of carbon dioxide in the atmosphere. The oceans are great reservoirs of carbon dioxide. When the oceans are warm, they release more carbon dioxide into the atmosphere than when they are cold. We'll return to the oceans and their important interplay with carbon dioxide in Chapters 18 and 24.

One ton of ordinary granite contains about 9 g of uranium and 20 g of thorium. Basalt rocks contain 3.5 g and 7.7 g of the same elements, respectively.

CHECK POINT

Suppose an archaeologist extracts a gram of carbon from an ancient ax handle and finds it one-fourth as radioactive as a gram of carbon extracted from a freshly cut tree branch. About how old is the ax handle?

Was this your answer?

Assuming the ratio of C-14 to C-12 was the same when the ax was made, the ax handle is as old as two half-lives of C-14, or about 11,460 years old.

The dating of older, but nonliving, things is accomplished with radioactive minerals, such as uranium. The naturally occurring isotopes U-238 and U-235 decay very slowly and ultimately become isotopes of lead—but not the common lead isotope Pb-208. For example, U-238 decays through several stages to finally become Pb-206, whereas U-235 finally becomes the isotope Pb-207. Lead isotopes 206 and 207 that now exist were at one time uranium. The older the uranium-bearing rock, the higher the percentage of these remnant isotopes.

From the half-lives of uranium isotopes and the percentage of lead isotopes in uranium-bearing rock, it is possible to calculate the date at which the rock was formed. We'll return to isotopic dating when we investigate Earth's dynamic interior in Chapter 22.

◾ 13.5 Nuclear Fission

In 1938, two German scientists, Otto Hahn and Fritz Strassmann, made an accidental discovery that was to change the world. While bombarding a sample of uranium with neutrons in the hope of creating new, heavier elements, they were astonished to find chemical evidence for the production of barium, an element with about half the mass of uranium. Hahn wrote of this news to his former colleague Lise Meitner, who had fled from Nazi Germany to Sweden because of her Jewish ancestry. From Hahn's evidence, Meitner concluded that the uranium nucleus, activated by neutron bombardment, had split in half. Soon thereafter, Meitner, working with her nephew, Otto Frisch, also a physicist, published a paper in which the term *nuclear fission* was first coined.

In the nucleus of every atom is a delicate balance between attractive nuclear forces and repulsive electric forces between protons. In all known nuclei, the nuclear forces dominate. In uranium, however, this domination is tenuous. If a uranium nucleus stretches into an elongated shape (Figure 13.21), the electrical forces may push it into an even more elongated shape. If the elongation passes a certain point, electrical forces overwhelm strong nuclear forces, and the nucleus splits. This is **nuclear fission**.

The energy released by the fission of one U-235 nucleus is relatively enormous—about 7 million times the energy released by the explosion of one TNT molecule. This energy is mainly in the form of kinetic energy of the fission fragments that fly apart from one another, with some energy given to ejected neutrons and the rest to gamma radiation.

A typical uranium fission reaction is

fyi

■ Otto Hahn, rather than Lise Meitner, received the Nobel Prize for the work on nuclear fission. Notoriously, Hahn didn't even acknowledge Meitner's role, pushing her to obscurity. See more about this in the readable book $E = mc^2$, by David Bodanis.

The Physical Science Place
Plutonium

(1) The greater force is the strong nuclear force.

(2) Critical deformation occurs.

(3) The greater force is the electric force, which results in a splitting of the nucleus.

FIGURE 13.21

Nuclear deformation may result in repulsive electrical forces overcoming attractive nuclear forces, in which case fission occurs.

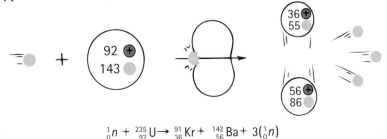

$$\,^{1}_{0}n + \,^{235}_{92}U \rightarrow \,^{91}_{36}Kr + \,^{142}_{56}Ba + 3(\,^{1}_{0}n)$$

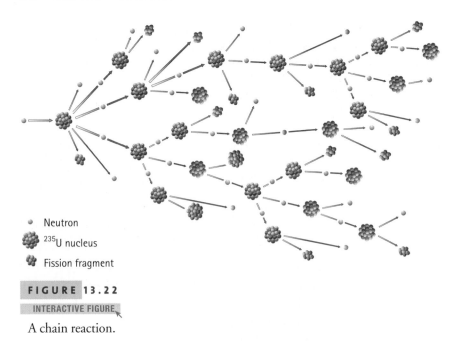

- Neutron
- ^{235}U nucleus
- Fission fragment

FIGURE 13.22

INTERACTIVE FIGURE

A chain reaction.

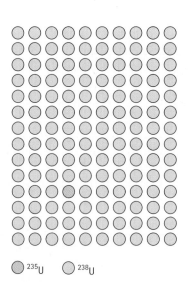

^{235}U ^{238}U

FIGURE 13.23

Only 1 part in 140 of naturally occurring uranium is U-235.

Neutrons escape surface

Neutrons trigger more reactions

FIGURE 13.24

The exaggerated view shows that a chain reaction in a small piece of pure U-235 runs its course before it can cause a large explosion because neutrons leak from the surface too soon. The surface area of the small piece is large relative to the mass. In a larger piece, more uranium and less surface is presented to the neutrons.

Note in this reaction that 1 neutron starts the fission of a uranium nucleus and that the fission produces 3 neutrons. (A fission reaction may produce fewer or more than 3 neutrons.) These product neutrons can cause the fissioning of 3 other uranium atoms, releasing 9 more neutrons. If each of these 9 neutrons succeeds in splitting a uranium atom, the next step in the reaction produces 27 neutrons, and so on. Such a sequence, illustrated in Figure 13.22, is called a **chain reaction**—a self-sustaining reaction in which the products of one reaction event stimulate further reaction events.

Why do chain reactions not occur in naturally occurring uranium ore deposits? They would if all uranium atoms fissioned so easily. Fission occurs mainly for the rare isotope U-235, which makes up only 0.7% of the uranium in pure uranium metal. When the more abundant isotope U-238 absorbs neutrons created by fission of U-235, the U-238 typically does not undergo fission. So any chain reaction is snuffed out by the neutron-absorbing U-238, as well as by the rock in which the ore is imbedded.

If a chain reaction occurred in a baseball-size chunk of pure U-235, an enormous explosion would result. If the chain reaction were started in a smaller chunk of pure U-235, however, no explosion would occur. This is because of geometry: the ratio of surface area to mass is larger in a small piece than in a large one (just as there is more skin on six small potatoes with a combined mass of 1 kg than there is on a single 1-kg potato). So there is more surface area on a bunch of small pieces of uranium than on a large piece. In a small piece of U-235, neutrons leak through the surface before an explosion can occur. In a bigger piece, the chain reaction builds up to enormous energies before the neutrons get to the surface and escape (Figure 13.24). For masses greater than a certain amount, called the **critical mass**, an explosion of enormous magnitude may take place.

Consider a large quantity of U-235 divided into two pieces, each with a mass less than critical. The units are *subcritical*. Neutrons in either piece readily reach a surface and escape before a sizable chain reaction builds up. But if the pieces are suddenly driven together, the total surface area decreases. If the timing is right and the combined mass is greater than critical, a violent explosion takes place. This is what happens in a nuclear fission bomb (Figure 13.25).

Constructing a fission bomb is a formidable task. The difficulty is separating enough U-235 from the more abundant U-238. Scientists took more than two years to extract enough U-235 from uranium ore to make the bomb that was detonated at Hiroshima in 1945. To this day uranium isotope separation remains a difficult process.

FIGURE 13.25

Simplified diagram of a uranium fission bomb.

CHECKPOINT

A 1-kg ball of U-235 is at critical mass, but the same ball broken up into small chunks is not. Explain.

Was this your answer?

The small chunks have more combined surface area than the ball from which they came (just as the combined surface area of gravel is greater than the surface area of a boulder of the same mass). Neutrons escape via the surface before a sustained chain reaction can build up.

NUCLEAR FISSION REACTORS

The awesome energy of nuclear fission was introduced to the world in the form of nuclear bombs, and this violent image still colors our thinking about nuclear power, making it difficult for many people to recognize its potential usefulness. Currently, about 20% of electric energy in the United States is generated by *nuclear fission reactors* (whereas most electric power is nuclear in some other countries—about 75% in France). These reactors are simply nuclear furnaces. They, like fossil fuel furnaces, do nothing more elegant than boil water to produce steam for a turbine (Figure 13.26). The greatest practical difference is the amount of fuel involved: A mere kilogram of uranium fuel, less than the size of a baseball, yields more energy than 30 freight-car loads of coal.

A fission reactor contains four components: nuclear fuel, control rods, moderator (to slow neutrons, which is required for fission), and liquid (usually water) to transfer heat from the reactor to the turbine and generator. The nuclear fuel is primarily U-238 plus about 3% U-235. Because the U-235 isotopes are so highly diluted with U-238, an explosion like that of a nuclear bomb is not possible. The reaction rate, which depends on the number of neutrons that initiate the fission of other U-235 nuclei, is controlled by rods inserted into the reactor. The control rods are made of a neutron-absorbing material, usually the metal cadmium or boron.

■ With the rise of the German Nazis in the 1930s, many scientists, especially those of Jewish ancestry, fled mainland Europe to America. This included dozens of brilliant theoretical physicists who eventually played key roles in the development of nuclear fission. Of these physicists, Leo Szilard (1898–1964) first envisioned the idea of a chain nuclear reaction. With Albert Einstein's consent, Szilard drafted a letter that was signed by Einstein and delivered to President Roosevelt in 1939. This letter outlined the possibility of the chain reaction and its implications for a nuclear bomb. Within six years the first test nuclear bomb was exploded in the desert in New Mexico. In 1945, Szilard generated a petition in which 68 of the scientists involved in the nuclear program asked President Truman not to drop the atomic bomb on a populous Japanese city, such as Nagasaki. Held back by the military, this petition never reached the president.

FIGURE 13.26

Diagram of a nuclear fission power plant. Note that the water in contact with the fuel rods is completely contained, and radioactive materials are not involved directly in the generation of electricity.

FIGURE 13.27

The nuclear reactor is housed within a dome-shaped containment building that is designed to prevent the release of radioactive isotopes in the event of an accident.

Nuclear power plants release practically no atmospheric pollutants. This includes the global warming greenhouse gas carbon dioxide.

fyi

- In 1972 it was discovered that Earth also once contained nuclear reactors that operated much like today's nuclear power plants. These natural reactors occurring within uranium deposits went extinct about 1.7 billion years ago, but they confirm that power from atomic nuclei is as old as Earth itself.

An average ton of coal contains 1.3 parts per million (ppm) of uranium and 3.2 ppm of thorium. That's why the average coal-burning power plant is a far greater source of airborne radioactive material than a nuclear power plant.

Heated water around the nuclear fuel is kept under high pressure to keep it at a high temperature without boiling. It transfers heat to a second lower-pressure water system, which operates the turbine and electric generator in a conventional fashion. In this design, two separate water systems are used so that no radioactivity reaches the turbine or the outside environment.

A significant disadvantage of fission power is the generation of radioactive waste products. Light atomic nuclei are most stable when composed of equal numbers of protons and neutrons, as discussed earlier, and heavy nuclei need more neutrons than protons for stability. For example, U-235 has 143 neutrons but only 92 protons. When uranium fissions into two medium-weight elements, the extra neutrons in their nuclei make them unstable. They are radioactive, most with very short half-lives, but some with half-lives of thousands of years. Safely disposing of these waste products as well as materials made radioactive in the production of nuclear fuels requires special storage casks and procedures. Although fission has been successfully producing electricity for a half century, disposing of radioactive wastes in the United States remains problematic.

American policy has been to deeply bury radioactive wastes, but many nuclear scientists disagree with deep burial as a desirable solution. A shift in thinking is occurring. Spent nuclear waste can be recycled to fuel an *Integral Fast Reactor* (IFR), which feeds on the wastes of present nuclear reactors. Other devices are being researched that convert long-life isotopes to ones of shorter half-life. Rather than burying nuclear wastes, for many years the French have been tending and monitoring them in underground storage facilities. Just as the tailings of gold mines and other mines were considered worthless a century ago but are today being reworked for their commercial value, so it may well be for today's radioactive wastes. These wastes need not plague future generations indefinitely, as has been commonly thought.

The benefits of fission power include plentiful electricity and the conservation of many billions of tons of fossil fuels. Every year these fuels are turned to heat, smoke, and megatons of poisonous gases such as sulfur oxides. Notably, fossil fuels are far more precious as sources of organic molecules, which, as we discuss in Chapter 19, can be used to create medicines, clothing, automobiles, and much more.

CHECK POINT

Coal contains tiny quantities of radioactive materials, enough that more environmental radiation surrounds a typical coal-fired power plant than a fission power plant. What does this indicate about the shielding typically surrounding the two types of power plants?

Was this your answer?

Coal-fired power plants are as American as apple pie, with no required (and expensive) shielding to restrict the emissions of radioactive particles. Nukes, on the other hand, are required to have shielding to ensure strictly low levels of radioactive emissions.

THE BREEDER REACTOR

One of the fascinating features of fission power is the breeding of fission fuel from nonfissionable U-238. This breeding occurs when small amounts of fissionable isotopes are mixed with U-238 in a reactor. Fission liberates neutrons that convert the relatively abundant nonfissionable U-238 to U-239, which beta decays to Np-239, which in turn beta decays to fissionable plutonium—Pu-239 (Figure 13.28). So in addition to the abundant energy produced, fission fuel is bred from the relatively abundant U-238 in the process.

FIGURE 13.28

INTERACTIVE FIGURE

Pu-239, like U-235, undergoes fission when it captures a neutron.

Breeding occurs to some extent in all fission reactors, but a reactor specifically designed to breed more fissionable fuel than is put into it is called a *breeder reactor.* Using a breeder reactor is like filling your car's gas tank with water, adding some gasoline, then driving the car and having more gasoline after the trip than at the beginning! The basic principle of the breeder reactor is very attractive, for after a few years of operation a breeder-reactor power plant can produce vast amounts of power while breeding twice as much fuel as its original fuel.

The downside is the enormous complexity of successful and safe operation. The United States gave up on breeders about two decades ago, and only Russia, France, Japan, and India are still investing in them. Officials in these countries point out that the supplies of naturally occurring U-235 are limited. At present rates of consumption, all natural sources of U-235 may be depleted within a century. If countries then decide to turn to breeder reactors, they may well find themselves digging up the radioactive wastes they once buried.

13.6 Mass–Energy Equivalence: $E = mc^2$

In the early 1900s, Albert Einstein discovered that mass is actually "congealed" energy. Mass and energy are two sides of the same coin, as stated in his celebrated equation $E = mc^2$. In this equation E stands for the energy that any mass has at rest, m stands for mass, and c is the speed of light. The quantity c^2 is the proportionality constant of energy and mass. This relationship between energy and mass is the key to understanding why and how energy is released in nuclear reactions.

The more energy associated with a particle, the greater the mass of the particle. Is the mass of a nucleon inside a nucleus the same as that of the same nucleon outside a nucleus? This question can be answered by considering the work that would be required to separate nucleons from a nucleus. From physics we know that work, which is expended energy, equals *force × distance*. Think of the amount of force required to pull a nucleon out of the nucleus through a sufficient distance to overcome the attractive strong nuclear force, comically indicated in Figure 13.29. Enormous work would be required. This work is energy added to the nucleon that is pulled out.

According to Einstein's equation, this newly acquired energy reveals itself as an increase in the nucleon's mass. The mass of a nucleon outside a nucleus is greater than the mass of the same nucleon locked inside a nucleus. As discussed in Chapter 14, a carbon-12 atom—the nucleus of which is made up of six protons and six neutrons—has a mass of exactly 12.00000 atomic mass units (amu). Therefore on average, each nucleon contributes a mass of 1 amu. However, outside the nucleus, a proton has a mass of 1.00728 amu and a neutron has a mass of 1.00867 amu. Thus we see that the combined mass of six free protons and six free neutrons—$(6 \times 1.00728) + (6 \times 1.00867) = 12.09570$—is

FIGURE 13.29

Work is required to pull a nucleon from an atomic nucleus. This work increases the energy and hence the mass of the nucleon outside the nucleus.

$E = mc^2$ says that mass is congealed energy. Mass and energy are two sides of the same coin.

FIGURE 13.30

The mass spectrometer. Electrically charged isotopes are directed into the semicircular "drum," where they are forced into semicircular paths by a strong magnetic field. Lighter isotopes have less inertia (mass), and so they easily change direction and are pulled into curves of smaller radii. Heavier isotopes have greater inertia (mass), and so they are pulled into curves of large radii. The mass of an isotope, therefore, is directly proportional to how far away from the slit it lands.

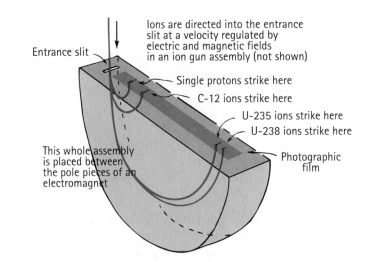

Ions are directed into the entrance slit at a velocity regulated by electric and magnetic fields in an ion gun assembly (not shown)

Entrance slit

Single protons strike here
C-12 ions strike here
U-235 ions strike here
U-238 ions strike here

This whole assembly is placed between the pole pieces of an electromagnet

Photographic film

fyi

■ Mass spectrometers are ultra-sensitive and provide quick results, which makes them ideal for detecting molecules associated with explosives at airport security stations. The security agent swabs luggage with a soft cloth, which is then placed in the spectrometer. Molecules on the swab vaporize into a chamber where they are ionized and then identified according to their characteristic masses.

greater than the mass of one carbon-12 nucleus. The greater mass reflects the energy required to pull the nucleons apart from one another. Thus, what mass a nucleon has depends on where the nucleon is.

The masses of the isotopes of various elements can be very accurately measured with a mass spectrometer (Figure 13.30). This important device uses a magnetic field to deflect ions of these isotopes into circular arcs. The greater the inertia (mass) of the ion, the more it resists deflection, and the greater the radius of its curved path. The magnetic force sweeps lighter ions into shorter arcs and heavier ions into larger arcs.

A graph of the nuclear masses for the elements from hydrogen through uranium is shown in Figure 13.31. The graph slopes upward with increasing atomic number as expected: elements are more massive as atomic number increases. The slope curves because there are proportionally more neutrons in the more massive atoms.

A more important graph results from the plot of nuclear mass *per nucleon* from hydrogen through uranium (Figure 13.32). This is perhaps the most important graph in this book, for it is the key to understanding the energy associated with nuclear processes. To obtain the average mass per nucleon, you divide the total mass of a nucleus by the number of nucleons in the nucleus. (Similarly, if you divide the total mass of a roomful of people by the number of people in the room, you get the average mass per person.)

FIGURE 13.31

The plot shows how nuclear mass increases with increasing atomic number.

FIGURE 13.32

This graph shows that the average mass of a nucleon depends on which nucleus it is in. Individual nucleons have the most mass in the lightest nuclei, the least mass in iron, and intermediate mass in the heaviest nuclei.

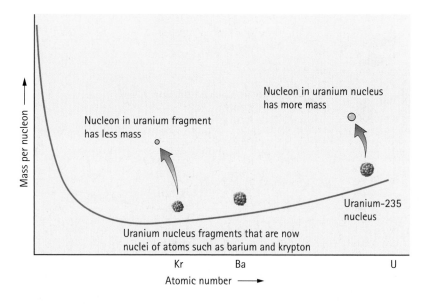

FIGURE 13.33

The mass of each nucleon in a uranium nucleus is greater than the mass of each nucleon in any one of its nuclear fission fragments. This lost mass is mass that has been transformed into energy, which is why nuclear fission is an energy-releasing process.

Note that the masses of the nucleons are different when combined in different nuclei. The greatest mass per nucleon occurs for the proton alone, hydrogen, because it has no binding energy to pull its mass down. Progressing beyond hydrogen, the mass per nucleon is smaller, and is least for one in the nucleus of the iron atom. Beyond iron, the process reverses itself as nucleons have progressively more and more mass in atoms of increasing atomic number. This continues all the way to uranium and elements heavier than uranium.

From Figure 13.32 we can see how energy is released when a uranium nucleus splits into two nuclei of lower atomic number. Uranium, being toward the right-hand side of the graph, is shown to have a relatively large amount of mass per nucleon. When the uranium nucleus splits in half, however, smaller nuclei of lower atomic numbers are formed. As shown in Figure 13.33, these nuclei are lower on the graph than uranium, which means that they have a smaller amount of mass per nucleon. Thus, nucleons lose mass in their transition from being in a uranium nucleus to being in one of its fragments. When this decrease in mass is multiplied by the speed of light squared (c^2 in Einstein's equation), the product is equal to the energy yielded by each uranium nucleus as it undergoes fission.

The graph of Figure 13.32 (and Figures 13.33 and 13.34) reveals the energy of the atomic nucleus, a primary source of energy in the universe—which is why it can be considered the most important graph in this book.

Link To Sections 27.4, 28.3, 28.6

CHECK POINT

Correct the following incorrect statement: When a heavy element such as uranium undergoes fission, there are fewer nucleons after the reaction than before.

Was this your answer?

When a heavy element such as uranium undergoes fission, there aren't fewer nucleons after the reaction. Instead, there's *less mass* in the same number of nucleons.

We can think of the mass-per-nucleon graph as an energy valley that starts at hydrogen (the highest point) and slopes steeply to the lowest point (iron), then slopes gradually up to uranium. Iron is at the bottom of the energy valley and is the most stable nucleus. It is also the most tightly bound nucleus; more energy per nucleon is required to separate nucleons from its nucleus than from any other nucleus.

All nuclear power today is by way of nuclear fission. A more promising long-range source of energy is found on the left side of the energy valley.

The mass of each nucleon in a hydrogen-2 nucleus is greater than the mass of each nucleon in a helium-4, which results from the fusion of two hydrogen-2 nuclei. This lost mass has been converted to energy, which is why nuclear fusion is an energy-releasing process.

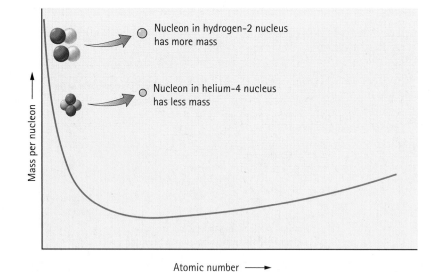

Nucleon in hydrogen-2 nucleus has more mass

Nucleon in helium-4 nucleus has less mass

Mass per nucleon →

Atomic number →

fyi

- A common reaction is the fusion of H-2 and H-3 nuclei to become He-4 plus a neutron. Most of the energy released is in the kinetic energy of the ejected neutron, with the rest of the energy in the kinetic energy of the recoiling He-4 nucleus. Interestingly, without the neutron energy carrier, a fusion reaction won't occur. The intensity of fusion reactions is measured by the accompanying neutron flux.

Link To Sections 26.1, 27.4

13.7 Nuclear Fusion

Notice in the graphs of Figures 13.32 and 13.33 that the steepest part of the energy valley goes from hydrogen to iron. Energy is released as light nuclei combine. This combining of nuclei is **nuclear fusion**—the opposite of nuclear fission. We see from Figure 13.34 that, as we move along the list of elements from hydrogen to iron, the average mass per nucleon decreases. Thus when two small nuclei fuse, say two hydrogen isotopes, the mass of the resulting helium-4 nucleus is less than the mass of the two small nuclei before fusion. Energy is released as smaller nuclei fuse.

For a fusion reaction to occur, the nuclei must collide at a very high speed in order to overcome their mutual electric repulsion. The required speeds correspond to the extremely high temperatures found in the Sun and other stars. Fusion brought about by high temperatures is called **thermonuclear fusion**. In the high temperatures of the Sun, approximately 657 million tons of hydrogen are converted into 653 million tons of helium *each second*. The missing 4 million tons of mass are discharged as radiant energy.

Such reactions are, quite literally, nuclear burning. Thermonuclear fusion is analogous to ordinary chemical combustion. In both chemical and nuclear burning, a high temperature starts the reaction; the release of energy by the reaction maintains a high enough temperature to spread the fire. The net result of the chemical reaction is a combination of atoms into more tightly bound molecules. In nuclear fusion reactions, the net result is more tightly bound nuclei. In both cases mass decreases as energy is given off.

The mass of a nucleus is not equal to the sum of the mass of its parts. (a) The fission fragments of a heavy nucleus such as uranium are less massive than the uranium nucleus. (b) Two protons and two neutrons are more massive in their free states than when combined to form a helium nucleus.

(a)

(b)

CHECK POINT

1. Fission and fusion are opposite processes, yet each releases energy. Isn't this contradictory?
2. To get nuclear energy release from the element iron, should iron be fissioned or fused?
3. Predict whether the temperature of the core of a star increases or decreases when iron and elements of higher atomic number than iron in the core are fused.

Were these your answers?

1. No, no, no! This is contradictory only if the same element is said to release energy by both the processes of fission and fusion. Only the fusion of light elements and the fission of heavy elements result in a decrease in nucleon mass and a release of energy.
2. Neither, because iron is at the very bottom of the "energy valley." Fusing a pair of iron nuclei produces an element to the right of iron on the curve, where mass per nucleon is higher. If you split an iron nucleus, the products lie to the left of iron on the curve—also a higher mass per nucleon. So no energy is released. For energy release, "Decrease Mass" is the name of the game—any game, chemical or nuclear.
3. In the fusion of iron and any nuclei beyond, energy is absorbed and the star core cools at this late stage of its evolution. This, however, leads to the star's collapse, which then greatly increases it temperature. Interestingly, elements beyond iron are not manufactured in normal fusion cycles in stellar sources, but are manufactured when stars violently explode—supernovae.

Before the development of the atomic bomb, the temperatures required to initiate nuclear fusion on Earth were unattainable. When researchers found that the temperature inside an exploding atomic bomb is four to five times the temperature at the center of the Sun, the thermonuclear bomb was but a step away. This first thermonuclear bomb, a hydrogen bomb, was detonated in 1952. Whereas the critical mass of fissionable material limits the size of a fission bomb (atomic bomb), no such limit is imposed on a fusion bomb (thermonuclear or hydrogen bomb). Just as there is no limit to the size of an oil-storage depot, there is no theoretical limit to the size of a fusion bomb. Like the oil in the storage depot, any amount of fusion fuel can be stored safely until ignited. Although a mere match can ignite an oil depot, nothing less energetic than an atomic bomb can ignite a thermonuclear bomb. We can see that there is no such thing as a "baby" hydrogen bomb. A typical thermonuclear bomb stockpiled by the United States today, for example, is about 1000 times as destructive as the atomic bomb detonated over Hiroshima at the end of World War II.

The hydrogen bomb is another example of a discovery used for destructive rather than constructive purposes. The potential constructive possibility is the controlled release of vast amounts of clean energy.

CONTROLLING FUSION

Carrying out fusion reactions under controlled conditions requires temperatures of millions of degrees. A variety of techniques exist for attaining high temperatures. No matter how the temperature is produced, a problem is that all materials melt and vaporize at the temperatures required for fusion. One solution is to confine the reaction in a nonmaterial container.

A nonmaterial container is a magnetic field, which can exist at any temperature and can exert powerful forces on charged particles in motion. "Magnetic walls" of sufficient strength provide a kind of magnetic straitjacket for hot gases

$${}_1^2\text{H} + {}_1^2\text{H} \rightarrow {}_2^3\text{He} + {}_0^1\text{n} + 3.26 \text{ MeV}$$

$${}_1^2\text{H} + {}_1^3\text{H} \rightarrow {}_2^4\text{He} + {}_0^1\text{n} + 17.6 \text{ MeV}$$

FIGURE 13.36

Fusion reactions of hydrogen isotopes. Most of the energy released is carried by the neutrons, which are ejected at high speeds.

FIGURE 13.37

A cross-sectional view of the ITER (rhymes with "fitter") planned to be built and operating in Cadarache, France, before 2020.

called plasmas. Magnetic compression further heats the plasma to fusion temperatures. At this writing, fusion by magnetic confinement has been only partially successful—a sustained and controlled reaction has so far been out of reach.

Although no nuclear fusion power plants are currently operating, an international project now exists whose goal is to prove the feasibility of nuclear fusion power in the near future. This fusion power project is the International Thermonuclear Experimental Reactor (ITER). After construction at the chosen site in Cadarache, France, the first sustainable fusion reaction may begin as early as 2015 (Figure 13.37). The reactor will house electrically charged hydrogen gas (plasma) heated to more than 100 million °C, which is hotter than the center of the Sun. In addition to producing about 500 MW of power, the reactor could be the energy source for the creation of hydrogen, H_2, which could be used to power fuel cells, such as those incorporated into automobiles.

If people are one day to dart about the universe in the same way we jet about Earth today, their supply of fuel is ensured. The fuel for fusion—hydrogen—is found in every part of the universe, not only in the stars but also in the space between them. About 91% of the atoms in the universe are estimated to be hydrogen. For people of the future, the supply of raw materials is also ensured because all the elements known to exist result from the fusing of more and more hydrogen nuclei. Future humans might synthesize their own elements and produce energy in the process, just as the stars have always done.

SUMMARY OF TERMS

Radioactivity The process in which unstable atomic nuclei break down and emit radiation.

Alpha particle The nucleus of a helium atom, which consists of two neutrons and two protons, ejected by certain radioactive elements.

Beta particle An electron (or positron) emitted during the radioactive decay of certain nuclei.

Gamma ray High-frequency electromagnetic radiation emitted by the nuclei of radioactive atoms.

Rad A quantity of radiant energy equal to 0.01 J absorbed per kilogram of tissue.

Rem A unit for measuring the ability of radiation to harm living tissue.

Strong nuclear force The force of interaction between all nucleons, effective only at extremely close distances.

Half-life The time required for half the atoms in a sample of a radioactive isotope to decay.

Transmutation The conversion of an atomic nucleus of one element into an atomic nucleus of another element through a loss or gain in the number of protons.

Carbon-14 dating The process of estimating the age of once-living material by measuring the amount of a radioactive isotope of carbon present in the material.

Nuclear fission The splitting of the nucleus of a heavy atom, such as uranium-235, into two main parts, accompanied by the release of much energy.

Chain reaction A self-sustaining reaction in which the products of one reaction event stimulate further reaction events.

Critical mass The minimum mass of fissionable material in a reactor or nuclear bomb that will sustain a chain reaction.

Nuclear fusion The combination of the nuclei of light atoms to form heavier nuclei, with the release of much energy.

Thermonuclear fusion Nuclear fusion produced by high temperature.

REVIEW QUESTIONS

13.1 Radioactivity

1. Which has the greatest penetrating power—alpha particles, beta particles, or gamma rays?
2. Is the human body naturally radioactive?
3. Is radioactivity on Earth relatively new? Defend your answer.
4. Distinguish between a *rad* and a *rem*.
5. How are radioactive isotopes used in medical imaging?

13.2 The Atomic Nucleus and the Strong Nuclear Force

6. Why doesn't the repulsive electric force of protons in the atomic nucleus cause the protons to fly apart?
7. Which have more neutrons than protons, large nuclei or small nuclei?
8. How are the strong nuclear force and the electric force different from each other?
9. What role do neutrons play in the atomic nucleus?

13.3 Half-Life and Transmutation

10. What is meant by the half-life of a radioactive sample?
11. What is the half-life of radium-226?
12. What change in atomic number occurs when a nucleus emits an alpha particle? A beta particle?
13. What is the long-range fate of all the uranium that exists in the world today?

13.4 Radiometric Dating

14. Which is radioactive, carbon-12 or carbon-14?
15. Why is there more carbon-14 in living bones than in once-living ancient bones of the same mass?
16. Why is carbon-14 dating useless for dating old coins but not old pieces of cloth?
17. Why is lead found in all deposits of uranium ores?
18. What does the proportion of lead and uranium in rock tell us about the age of the rock?

13.5 Nuclear Fission

19. Why does a chain reaction not occur in uranium mines?
20. Is a chain reaction more likely to occur in two separate pieces of uranium-235 or in the same pieces stuck together?
21. How is a nuclear reactor similar to the furnace in a fossil-fuel power plant? How is it different?
22. How does a breeder reactor breed nuclear fuel?

13.6 Mass–Energy Equivalence: $E = mc^2$

23. Is work required to pull a nucleon out of an atomic nucleus? Does the nucleon, once outside, then have more energy than it did when it was inside the nucleus? In what form is this energy?
24. Which ions are least deflected in a mass spectrometer?
25. In which atomic nucleus do nucleons have the least mass?
26. How does the mass per nucleon in uranium compare with the mass per nucleon in the fission fragments of uranium?

13.7 Nuclear Fusion

27. If the graph in Figure 13.34 is seen as an energy valley, what can be said of nuclear transformations that progress toward iron?
28. When a pair of hydrogen isotopes is fused, is the mass of the product nucleus more or less than the sum of the masses of the two hydrogen nuclei?
29. What kind of containers are used to contain plasmas at temperatures of millions of degrees?
30. From where does the Sun gets its energy?

EXERCISES

● BEGINNER ■ INTERMEDIATE ◆ EXPERT

1. ■ Why is a sample of radioactive material always a little warmer than its surroundings?
2. ■ Is it possible for a hydrogen nucleus to emit an alpha particle? Defend your answer.
3. ■ Just after an alpha particle leaves the nucleus, would you expect it to speed up? Defend your answer.
4. ● How do the electric charges of alpha particles, beta particles, and gamma rays differ from one another?
5. ■ Why are alpha particles and beta particles deflected in opposite directions in a magnetic field? Why are gamma rays undeflected?
6. ● The alpha particle has twice the electric charge of the beta particle but deflects less in a magnetic field. Why?
7. ● Which type of radiation—alpha, beta, or gamma—results in the greatest change in mass number? The greatest change in atomic number?
8. ● Which type of radiation—alpha, beta, or gamma—results in the least change in mass number? The least change in atomic number?
9. ● A pair of protons in an atomic nucleus repel each other, but they are also attracted to each other. Explain.
10. ● If an atom has 104 electrons, 157 neutrons, and 104 protons, what is its approximate atomic mass? What is the name of this element?

11. ● Which type of radiation—alpha, beta, or gamma—predominates on the inside of a high-flying commercial airplane? Why?
12. ● When the isotope bismuth-213 emits an alpha particle, what new element results? What new element results if it instead emits a beta particle?
13. ● When $^{226}_{88}$Ra decays by emitting an alpha particle, what is the atomic number of the resulting nucleus? What is the resulting atomic mass?
14. ● What are the atomic number and atomic mass of the element formed when $^{218}_{84}$Po emits a beta particle? What are they if the polonium emits an alpha particle?
15. ■ Elements above uranium in the periodic table do not exist in any appreciable amounts in nature because they have short half-lives. Yet several elements below uranium in the table have equally short half-lives but exist in appreciable amounts in nature. How can you account for this?
16. ● Your friend says that the helium used to inflate balloons is a product of radioactive decay. Another friend says no way. With whom do you agree?
17. ● Another friend, fretful about living near a fission power plant, wishes to get away from radiation by traveling to the high mountains and sleeping out at night on granite outcroppings. What comment do you have about this?

18. ● Still another friend has journeyed to the mountain foothills to escape the effects of radioactivity altogether. While bathing in the warmth of a natural hot spring she wonders aloud how the spring gets its heat. What do you tell her?

19. ■ Coal contains only minute quantities of radioactive materials, and yet more environmental radiation surrounds a coal-fired power plant than a fission power plant. What does this indicate about the shielding that typically surrounds these two types of plants?

20. ● A friend checks the local background radiation with a Geiger counter, which ticks audibly. Another friend, who normally fears most that which is understood least, tries to keep away from the region of the Geiger counter and looks to you for advice. What do you say?

21. ◆ When food is irradiated with gamma rays from a cobalt-60 source, does the food become radioactive? Defend your answer.

22. ● How is carbon-14 produced in the atmosphere?

23. ■ Radium-226 is a common isotope on Earth, but has a half-life of about 1600 years. Given that Earth is some 5 billion years old, why is there any radium at all?

24. ■ Is carbon dating advisable for measuring the age of materials a few years old? How about a few thousand years old? A few million years old?

25. ■ Why is carbon-14 dating not accurate for estimating the age of materials more than 50,000 years old?

26. ■ The age of the Dead Sea Scrolls was determined by carbon-14 dating. Could this technique have worked if they had been carved on stone tablets? Explain.

27. ■ If you make an account of 1000 people born in the year 2000 and find that half of them are still living in 2060, does this mean that one-quarter of them will be alive in 2120 and one-eighth of them alive in 2180? What is different about the death rates of people and the "death rates" of radioactive atoms?

28. ■ Why doesn't uranium ore spontaneously undergo a chain reaction?

29. ● Why will nuclear fission probably never be used directly for powering automobiles? How could it be used indirectly?

30. ◆ Does the average distance a neutron travels through fissionable material before escaping increase or decrease when two pieces of fissionable material are assembled into one piece? Does this assembly increase or decrease the probability of an explosion?

31. ◆ Which shape is likely to need more material for a critical mass, a cube or a sphere? Explain.

32. ■ Why does a neutron make a better nuclear bullet than a proton or an electron?

33. ■ Why does plutonium not occur in appreciable amounts in natural ore deposits?

34. ● What is the function of control rods in a nuclear reactor?

35. ◆ Why is carbon better than lead as a moderator in nuclear reactors?

36. ◆ Uranium-235 releases an average of 2.5 neutrons per fission, while plutonium-239 releases an average of 2.7 neutrons per fission. Which of these elements might you therefore expect to have the smaller critical mass?

37. ● If a nucleus of $^{232}_{90}$Th absorbs a neutron and the resulting nucleus undergoes two successive beta decays, which nucleus results?

38. ● Which process would release energy from gold, fission or fusion? From carbon? From iron?

39. ■ If a uranium nucleus were to fission into three fragments of approximately equal size instead of two, would more energy or less energy be released? Defend your answer using Figures 13.32 and 13.33.

40. ■ Is the mass of an atomic nucleus greater or less than the sum of the masses of the nucleons composing it? Why don't the nucleon masses add up to the total nuclear mass?

41. ◆ The original reactor built in 1942 was just "barely" critical because the natural uranium that was used contained less than 1% of the fissionable isotope U-235 (half-life 713 million years). What if Earth had been 9 billion years old instead of 4.5 billion years old? Would this reactor have reached critical stage with natural uranium?

42. ● Heavy nuclei can be made to fuse—for instance by firing one gold nucleus at another one. Does such a process yield energy or cost energy? Explain.

43. ● Light nuclei can be split. For example, a deuteron, which is a proton–neutron combination, can split into a separate proton and separate neutron. Does such a process yield energy or cost energy? Explain.

44. ● Is work required to pull a nucleon out of an atomic nucleus? Does the nucleon, once outside the nucleus, have more mass than it had inside the nucleus?

45. ■ Which produces more energy, the fissioning of a single uranium nucleus or the fusing of a pair of deuterium nuclei? The fissioning of a gram of uranium or the fusing of a gram of deuterium? (Why do your answers differ?)

46. ● Sustained nuclear fusion has yet to be achieved and remains a hope for abundant future energy. Yet the energy that has always sustained us has been the energy of nuclear fusion. Explain.

47. ◆ If a fusion reaction produces no appreciable radioactive isotopes, why does a hydrogen bomb produce significant radioactive fallout?

48. ■ Explain how radioactive decay has always warmed Earth from the inside and how nuclear fusion has always warmed Earth from the outside.

49. ■ Ordinary hydrogen is sometimes called a perfect fuel, because of its almost unlimited supply on Earth, and when it burns, harmless water is the product of the combustion. So why don't we abandon fission energy and fusion energy, not to mention fossil-fuel energy, and just use hydrogen?

50. ● Speculate about some worldwide changes likely to follow the advent of successful fusion reactors.

PROBLEMS

● **BEGINNER**   ■ **INTERMEDIATE**   ◆ **EXPERT**

1. ■ The isotope cesium-137, which has a half-life of 30 years, is a product of nuclear power plants. How long will it take for this isotope to decay to about $\frac{1}{16}$ its original amount?
2. ■ A certain radioactive element has a half-life of 1 h. If you start with a 1-g sample of the element at noon, how much is left at 3:00 PM? At 6:00 PM? At 10:00 PM?
3. ■ A sample of a particular radioisotope is placed near a Geiger counter, which is observed to register 160 counts per minute. Eight hours later the detector counts at a rate of 10 counts per minute. What is the half-life of the material?

4. ◆ Suppose that you measure the intensity of radiation from carbon-14 in an ancient piece of wood to be 6% of what it would be in a freshly cut piece of wood. How old is this artifact?
5. ◆ Suppose that you want to find out how much gasoline is in an underground storage tank. You pour in one gallon of gasoline that contains some radioactive material with a long half-life that gives off 5000 counts per minute. The next day, you remove a gallon from the underground tank and measure its radioactivity to be 10 counts per minute. How much gasoline is in the tank?

ACTIVE EXPLORATIONS

1. Write a letter to Grandma to dispel any notion she or her friends might have about radioactivity being something new in the world. Tie this to the idea that many people have the strongest views on that which they least understand.

2. Write a letter to Grandpa discussing nuclear power. Cite both the ups and downs of it, and explain how the comparison affects your personal view of nuclear power. Also explain how nuclear fission and nuclear fusion differ.

READINESS ASSURANCE TEST (RAT)

If you have a good handle on this chapter, if you really do, then you should be able to score 7 out of 10 on this RAT. If you score less than 7, you need to study further before moving on.

Choose the BEST answer to each of the following.

1. Is it at all possible for a hydrogen nucleus to emit an alpha particle?
   (a) Yes, because alpha particles are the simplest form of radiation.
   (b) No, because it would require the nuclear fission of hydrogen, which is impossible.
   (c) Yes, but it does not occur very frequently.
   (d) No, because it does not contain enough nucleons.
2. In bombarding atomic nuclei with proton "bullets," the protons must be accelerated to high energies to make contact with the target nuclei
   (a) because the target nuclei are so small.
   (b) because the target nuclei are negatively charged.
   (c) in order to penetrate through the electrons that surround each target nucleus.
   (d) because the target nuclei are positively charged.
3. Why would you expect alpha particles to be less able to penetrate materials than beta particles of the same kinetic energy?
   (a) Alpha particles pick up electrons to become harmless helium atoms.
   (b) Alpha particles carry twice the electric charge.
   (c) Alpha particles move much slower for a given kinetic energy.
   (d) two of these

4. What evidence supports the contention that the strong nuclear force is stronger than the electrical interaction at short internuclear distances?
   (a) Protons can exist side-by-side within an atomic nucleus.
   (b) Neutrons spontaneously decay into protons and electrons.
   (c) Uranium deposits are always slightly warmer than their immediate surroundings.
   (d) Radio interference arises adjacent to any radioactive source.
5. How is it possible for an element to decay "forward in the periodic table"—that is, to decay to an element of higher atomic number?
   (a) The decay is instigated by a collision with a proton.
   (b) As a beta particle is released, a neutron transforms into a proton.
   (c) This occurs only during nuclear fusion.
   (d) It is not possible. Radioactive decay always results in an isotope with the same or lower atomic number.
6. The isotope cesium-137, which has a half-life of 30 years, is a product of nuclear power plants. How long will it take for this isotope to decay to about half its original amount?
   (a) 0 years
   (b) 15 years
   (c) 30 years
   (d) 60 years
   (e) 90 years

7. Why, after a uranium fuel rod reaches the end of its fuel cycle (typically three years) does most of its energy come from the fissioning of plutonium?
(a) Because fissionable Pu-239 is formed as the U-238 absorbs neutrons from the fissioning U-235.
(b) Because fissionable Pu-239 is formed as the U-235 absorbs neutrons from the fissioning U-238.
(c) Because fissionable Pu-239 is formed as the U-238 absorbs alpha particles from the fissioning U-235.
(d) Because fissionable Pu-239 is formed as the U-235 absorbs alpha particles from the fissioning U-238.

8. Uranium-235 releases an average of 2.5 neutrons per fission, while plutonium-239 releases an average of 2.7 neutrons per fission. Which of these elements might you therefore expect to have the smaller critical mass?
(a) plutonium-239
(b) uranium-235
(c) The number of neutrons per fission will not affect the size of critical mass.
(d) The critical mass of these elements cannot be determined from only the number of neutrons per fission.

9. If uranium were to split into 90 pieces of equal size instead of 2, would more energy or less energy be released?
(a) Less energy would be released because of less mass per nucleon.
(b) Less energy would be released because of more mass per nucleon.
(c) More energy would be released because of less mass per nucleon.
(d) More energy would be released because of more mass per nucleon.

10. To predict the approximate energy release of either a fission or a fusion reaction, explain how a physicist uses a table of nuclear masses and the equation $E = mc^2$.
(a) Find the change in mass and divided by the speed of light squared.
(b) Find the change in mass and multiply by the speed of light squared.
(c) Find the change in mass, square it, and multiply by the speed of light.
(d) Take the square root of the change in mass and multiply by the speed of light.

Answers to RAT

1d, 2d, 3d, 4a, 5b, 6c, 7a, 8a, 9b, 10b

EXPLORING FURTHER

http://www.iaea.or.at
The Web site for the International Atomic Energy Agency, which monitors almost all issues related to nuclear technology. A good starting point for exploring applications of many of the concepts discussed in this chapter.

http://www.iter.org
The Web site for the International Thermonuclear Experimental Reactor project. Explore this site for the latest on the science and politics of this important project.

http://www.ocrwm.doe.gov
The Office of Civilian Radioactive Waste Management (OCRWM) is a program of the U.S. Department of Energy assigned to develop and manage a federal system for disposing of spent nuclear fuel from commercial nuclear reactors and high-level radioactive waste from national defense activities. Here you'll find the official position of the U.S. government regarding Yucca Mountain, Nevada, as a potential nuclear waste repository.

http://www.ne.doe.gov
The U.S. Department of Energy, Office of Nuclear Energy, Science and Technology.

http://www.state.nv.us/nucwaste
The state of Nevada is home to a number of nuclear weapons tests sites as well as Yucca Mountain, a potential national long-term storage facility for nuclear wastes. This is the Web site of the State of Nevada, Office of the Governor, Agency for Nuclear Projects, Nuclear Waste Project Office.

CHAPTER 13 ONLINE RESOURCES

The Physical Science Place

Interactive Figures

- 13.2, 13.3, 13.4, 13.11, 13.15, 13.22, 13.28, 13.34, 13.35

Tutorial

- Nuclear Physics

Videos

- Radioactive Decay

- Half-Life
- Plutonium

Quiz

Flashcards

Links

elements of chemistry

14.1 CHEMISTRY: THE CENTRAL SCIENCE

14.2 THE SUBMICROSCOPIC WORLD

14.3 PHYSICAL AND CHEMICAL PROPERTIES

14.4 DETERMINING PHYSICAL AND CHEMICAL CHANGES

14.5 ELEMENTS TO COMPOUNDS

14.6 NAMING COMPOUNDS

■ As you progress through this physical science course, you will note an accumulating list of key terms in boldface type. Why an increase of new terms? Why can't physical science be described in everyday English without the addition of new vocabulary? Consider this: in the laboratory, scientists perform experiments, make many observations, and then draw conclusions. Over time, the result is a growing body of new knowledge that inevitably exceeds the capacity of everyday language. For example, in the language of chemistry, we say that there are more than 100 kinds of *atoms*, and that any material consisting of a single kind of atom is an *element*. (The elements gold and carbon are shown in this chapter's opening photograph.) Atoms can link together to form a *molecule*, and a molecule consisting of atoms from different elements is a *compound*. And on and on, one term building on another, as we attempt to describe the nature of matter beyond its casual appearance.

Practice articulating and paraphrasing the concepts represented by the boldface terms. Do this aloud to yourself (or to a friend), minimizing looking at the book. When you can express these concepts in your own words—in your own "plain English"—you'll have the insight to do well in this course and beyond.

Rather than memorizing key terms, you will serve yourself far better by focusing on the underlying concept each term represents. Bear in mind, a term is only a label. It is possible to know the term without understanding the concepts behind it—just as it is possible to understand a concept without knowing the term that labels that concept. So although this new vocabulary is useful for communication, it does not guarantee conceptual understanding. If you focus first on the concepts, the vocabulary represented will come to you much more naturally.

◼ 14.1 Chemistry: The Central Science

When you wonder what the land, sky, or ocean is made of, you are thinking about chemistry. When you wonder how a rain puddle dries up, how a car acquires energy from gasoline, or how your body extracts energy from the food you eat, you are again thinking about chemistry. By definition, chemistry is the study of matter and the transformations it can undergo. Matter is anything that occupies space. It is the stuff that makes up all material things; anything you can touch, taste, smell, see, or hear is matter. The scope of chemistry, therefore, is very broad.

Chemistry is often described as a central science because it touches all the other sciences. It springs from the principles of physics, and it serves as the foundation for the most complex science of all—biology. Indeed, many of the great advances in the life sciences today, such as genetic engineering, are applications of some very exotic chemistry. Chemistry sets the foundation for the major Earth sciences—geology, oceanography, meteorology. It is also an important component of space science, as described in Figure 14.1. Just as we learned about the origin of the Moon from the chemical analysis of moon rocks in the early 1970s, we are now learning about the history of Mars and other planets from the chemical information gathered by space probes.

Progress in science is made as scientists conduct research. Research is any activity aimed at the systematic discovery and interpretation of new knowledge. Many scientists focus on **basic research**, which leads us to a greater understanding of how the natural world operates. The foundation of knowledge laid down

The Physical Science Place

What Is Chemistry?

Link To Section 26.3

FIGURE 14.1

Special materials of chemistry, such as rocket fuels, metals for spaceships, and fabrics for the space suits, were required to allow astronauts to reach and explore the surface of the Moon.

Transparent matrix of processed silicon dioxide

Chemically disinfected drinking water

Caffeine solution

Thermoset polymer

Prescription medicines stored in refrigerator

Chlorofluorocarbon-free refrigerating fluids

Electrical energy from a fossil fuel or nuclear power plant

Metal alloy

Roasting carbohydrates, fats, proteins, and vitamins

Natural gas laced with odoriferous sulfur compounds

Fertilizer-grown vegetables

FIGURE 14.2

Most of the material items in any modern house are shaped by some human-devised chemical process.

by basic research frequently leads to useful applications. Research that focuses on developing these applications is known as **applied research**. Most chemists choose applied research as their major focus. Applied research in chemistry has provided us with medicine, food, water, shelter, and many of the material goods that characterize modern life. Just a few examples are shown in Figure 14.2.

Over the course of the past century, we excelled at manipulating atoms and molecules to create materials to suit our needs. At the same time, however, we made mistakes in caring for the environment. Waste products were dumped into rivers, buried in the ground, or vented into the air without regard for possible long-term consequences. Many people believed that Earth was so large that its resources were virtually unlimited and that it could absorb wastes without being significantly harmed.

Most nations now recognize this as a dangerous attitude. As a result, government agencies, industries, and concerned citizens are involved in extensive efforts to clean up toxic-waste sites. Such regulations as the international ban on ozone-destroying chlorofluorocarbons have been enacted to protect the environment. Members of the American Chemistry Council, who produce 90% of the chemicals manufactured in the United States, have adopted a program called Responsible Care, in which they have pledged to manufacture without causing environmental damage. The Responsible Care program emblem is shown in Figure 14.3. By using chemistry wisely, most waste products can be minimized, recycled, engineered into salable commodities, or rendered environmentally benign.

Chemistry has influenced our lives in profound ways, and it will continue to do so in the future. For this reason, it is in everyone's interest to become acquainted with the basic concepts of chemistry.

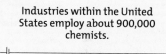

Industries within the United States employ about 900,000 chemists.

FIGURE 14.3

The Responsible Care symbol of the American Chemistry Council.

Many major advances were made in both the physical sciences and the life sciences over the course of the 20th century. Advances made in the physical sciences, such as our understanding of the chemistry of life, however, will likely propel the life sciences to even more fantastic advances in the 21st century.

CHECK POINT

Chemists have learned how to produce aspirin using petroleum as a starting material. Is this an example of basic or applied research?

Was this your answer?

This is an example of applied research, because the primary goal was to develop a useful commodity. However, the ability to produce aspirin from petroleum depended on an understanding of atoms and molecules developed from many years of basic research.

There are far more atoms in a glass of water than there are grains of sand within this towering sand dune.

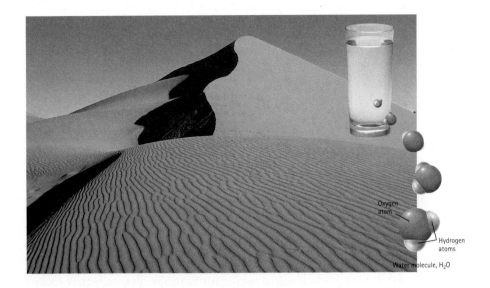

Link To Sections 12.1, 20.7

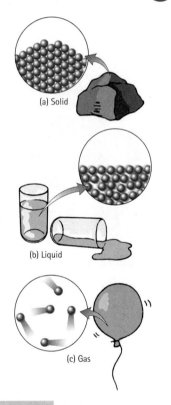

(a) Solid

(b) Liquid

(c) Gas

The familiar bulk properties of a solid, a liquid, and a gas. (a) The submicroscopic particles of the solid phase vibrate about fixed positions. (b) The submicroscopic particles of the liquid phase slip past one another. (c) The fast-moving submicroscopic particles of the gaseous phase are separated by large average distances.

14.2 The Submicroscopic World

From afar, a sand dune appears to be a smooth, continuous material. Up close, however, the dune reveals itself to be made of tiny particles of sand. In a similar fashion, as discussed in Chapter 12, everything around us—no matter how smooth it may appear—is made of the basic units you know as *atoms*. Atoms are so small, however, that a single grain of sand contains on the order of 125 million trillion of them. There are roughly 250,000 times more atoms in a single grain of sand than there are grains of sand in the dunes shown in Figure 14.4.

As small as atoms are, there is much we have learned about them. We know, for example, that there are more than 100 different types of atoms, and they are listed in the widely recognized periodic table. Some atoms link together to form larger but still incredibly small basic units of matter called **molecules**. As shown in Figure 14.4, for example, two hydrogen atoms and one oxygen atom link together to form a single molecule of water, which you know as H_2O. Water molecules are so small that an 8-oz glass of water contains about a trillion trillion of them.

Our world can be studied at different levels of magnification. At the *macroscopic* level, matter is large enough to be seen, measured, and handled. A handful of sand and a glass of water are macroscopic samples of matter. At the *microscopic* level, physical structure is so fine that it can be seen only with a microscope. A biological cell is microscopic, as is the detail on a dragonfly's wing. Beyond the microscopic level is the **submicroscopic**—the realm of atoms and molecules and an important focus of chemistry.

Recall from Chapter 7 that matter exists in *phases*. At the submicroscopic level, solid, liquid, and gaseous phases are distinguished by how the submicroscopic particles hold together. This is illustrated in Figure 14.5. In solid matter, such as rock, the attractions between particles are strong enough to hold all the particles together in some fixed three-dimensional arrangement. The particles can vibrate about fixed positions, but they cannot move past one another.

The addition of heat causes these vibrations to increase until, at a certain temperature, the vibrations are rapid enough to disrupt the fixed arrangements. Rock melts into magma (a topic of much discussion in Part 3). Likewise, ice melts into water. The particles can then slip past one another and tumble around much like a bunch of marbles in a bag. This is the liquid phase of matter, and the mobility of the submicroscopic particles gives rise to the liquid's fluid character—its ability to flow and to assume the shape of its container.

Further heating causes the submicroscopic particles in a liquid to move so fast that the attractions they have for one another are unable to hold them together. They then separate from one another, forming a gas. For magma, this doesn't easily happen, because the particles are strongly attracted to one another. Water molecules separate into a gas at 100°C. For a substance like helium, the submicroscopic particles are already in the gaseous phase at room temperature.

Moving at an average speed of 500 m/s (1100 mi/h), the particles of a gas are widely separated from one another. Matter in the gaseous phase therefore occupies much more volume than it does in the solid or liquid phase. Applying pressure to a gas squeezes the gas particles closer together, which decreases the volume. The amount of air an underwater diver needs to breathe for many minutes, for example, can be squeezed (compressed) into a tank small enough to be carried on the diver's back.

fyi

- Coffee and tea are decaffeinated using carbon dioxide in a fourth phase of matter known as a *supercritical fluid*. This phase behaves like a gaseous liquid, which is attained by adding lots of pressure and heat. Supercritical carbon dioxide is relatively easy to produce. To get water to form a supercritical fluid, however, requires pressures in excess of 217 atm and a temperature of 374°C. Supercritical water is very corrosive. Also, so much oxygen can dissolve in supercritical water that flames can burn within this medium, which is ideal for the destruction of toxic wastes.

■ 14.3 Physical and Chemical Properties

Properties that describe the look or feel of a substance, such as color, hardness, density, texture, and phase, are called **physical properties**. Every substance has its own set of characteristic physical properties that we can use to identify that substance (Figure 14.6).

The physical properties of a substance can change when conditions change, but that does not mean that a different substance is created. Cooling liquid water to below 0°C causes the water to transform to solid ice, but the substance is still water, no matter what the phase. The only difference is the relative orientation of the H_2O molecules to one another. In the liquid phase, the water molecules tumble around one another, whereas in the ice phase, they vibrate about fixed positions. The freezing of water is an example of what chemists call a physical change. During a **physical change**, a substance changes its phase or some other physical property, but not its chemical composition, as Figure 14.7 shows.

Gold
Opacity: opaque
Color: yellowish
Phase at 25°C: solid
Density: 19.3 g/mL

Diamond
Opacity: transparent
Color: colorless
Phase at 25°C: solid
Density: 3.5 g/mL

Water
Opacity: transparent
Color: colorless
Phase at 25°C: liquid
Density: 1.0 g/mL

FIGURE 14.6

Gold, diamond, and water can be identified by their physical properties. If a substance has all the physical properties listed under gold, for example, it must be gold.

CHECK POINT

The melting of gold is a physical change. Why?

Was this your answer?
During a physical change, a substance changes only one or more of its physical properties; its chemical identity does not change. Because melted gold is still gold but in a different form, its melting represents only a physical change.

Chemical properties characterize the ability of a substance to react with other substances or to transform from one substance to another. Figure 14.8 shows three examples. The methane of natural gas has the chemical property of reacting with oxygen to produce carbon dioxide and water, along with appreciable heat energy. Similarly, baking soda has the chemical property of reacting with vinegar to produce carbon dioxide and water while absorbing a small

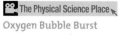
The Physical Science Place

Oxygen Bubble Burst

Link To Section 17.1

Two physical changes. (a) Liquid water and ice may appear to be different substances, but a submicroscopic view shows that both consist of water molecules. (b) At 25°C, the atoms in a sample of mercury are a certain distance apart, yielding a density of 13.53 g/mL. At 100°C, the atoms are farther apart, meaning that each milliliter now contains fewer atoms than at 25°C, and the density is now 13.35 g/mL. The physical property we call density has changed with temperature, but the identity of the substance remains unchanged: mercury is mercury.

Water molecules (H_2O) of liquid water

Water molecules (H_2O) of solid water (ice)

(a)

Atoms of liquid mercury (Hg) at 25°C

Atoms of liquid mercury (Hg) at 100°C (expanded)

(b)

The Physical Science Place
Fire Water

A chemical property of a substance is its tendency to change into another substance. For example, it is a chemical property of iron to transform into rust.

amount of heat energy. Copper has the chemical property of reacting with carbon dioxide and water to form a greenish-blue solid known as *patina*. Copper statues exposed to the carbon dioxide and water in the air become coated with patina. The patina is not copper, it is not carbon dioxide, and it is not water. It is a new substance formed by the reaction of these chemicals with one another.

All three of these transformations involve a change in the way the atoms in the molecules are *chemically bonded* to one another. A **chemical bond** is the force of attraction between two atoms that holds them together. A methane molecule, for example, is made of a single carbon atom bonded to four hydrogen atoms, and an oxygen molecule is made of two oxygen atoms bonded to each other. Figure 14.9 shows the chemical change in which the atoms in a methane molecule and those in two oxygen molecules first pull apart and then form new bonds with different partners, resulting in the formation of molecules of carbon dioxide and water.

Any change in a substance that involves a rearrangement of the way atoms are bonded is called a **chemical change**. Thus the transformation of methane to carbon dioxide and water is a chemical change, as are the other two transformations shown in Figure 14.8.

The chemical properties of substances allow them to transform to new substances. Natural gas and baking soda transform to carbon dioxide, water, and heat. Copper transforms to patina.

Methane
Reacts with oxygen to form carbon dioxide and water, giving off lots of heat during the reaction.

Baking soda
Reacts with vinegar to form carbon dioxide and water, absorbing heat during the reaction.

Copper
Reacts with carbon dioxide and water to form the greenish-blue substance called patina.

Oxygen

Methane

Water

Carbon dioxide

The chemical change shown in Figure 14.10 occurs when an electric current is passed through water. The energy of the current causes the bonds holding atoms together to break apart. Loose atoms then form new bonds with different atoms, which results in the formation of new molecules. Thus, water molecules are changed to molecules of hydrogen and oxygen, two substances that are very different from water. The hydrogen and oxygen are both gases at room temperature, and they can be seen as bubbles rising to the surface.

In the language of chemistry, materials undergoing a chemical change are said to be *reacting*. Methane reacts with oxygen to form carbon dioxide and water. Water reacts when exposed to electricity to form hydrogen gas and oxygen gas. Thus, the term *chemical change* means the same thing as *chemical reaction*. During a **chemical reaction**, new materials are formed by a change in the way atoms are bonded together. We shall explore chemical bonds and the reactions in which they are formed and broken in Chapter 15, 17, and 18.

CHECK POINT

Each sphere in the following diagrams represents an atom. Joined spheres represent molecules. One set of diagrams shows a physical change, and the other shows a chemical change. Which is which?

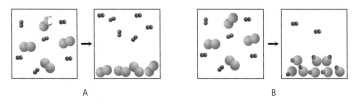

A   B

Was this your answer?

Remember that a chemical change (also known as a chemical reaction) involves molecules breaking apart so that the atoms are free to form new bonds with new partners. Be careful to distinguish this breaking apart from a mere change in the relative positions of a group of molecules. In set A, the molecules before and after the change are the same. They differ only in their positions relative to one another. Set A, therefore, represents only a physical change. In set B, new molecules, consisting of bonded red and blue spheres, appear after the change. These molecules represent a new material, and so set B represents a chemical change.

■ 14.4 Determining Physical and Chemical Changes

How can you determine whether an observed change is physical or chemical? This can be tricky because in both cases changes in physical appearance occur. Water, for example, looks quite different after it freezes, just as a car

Liquid water, H₂O

Gaseous oxygen, O₂   CHEMICAL CHANGE

Gaseous hydrogen, H₂

FIGURE 14.10

Water can be transformed to hydrogen gas and oxygen gas by applying the energy of an electric current. This is a chemical change, because new materials (the two gases) are formed as the atoms originally found in the water molecules are rearranged.

FIGURE 14.11

The transformation of water to ice and the transformation of iron to rust both involve changes in physical appearance. The formation of ice is a physical change, whereas the formation of rust is a chemical change.

looks quite different after it rusts (Figure 14.11). The freezing of water is a physical change because liquid water and frozen water are both forms of water—only the orientation of the water molecules to one another changes. The rusting of a car, by contrast, is the result of the transformation of iron to rust. This is a chemical change because iron and rust are two different materials, each consisting of a different arrangement of atoms. As we shall see in the next two sections, iron is an element, and rust is a compound consisting of iron and oxygen atoms.

Two powerful guidelines can help you assess physical and chemical changes. First, in a physical change, a change in appearance is the result of a new set of conditions imposed on the same material. Restoring the original conditions restores the original appearance: frozen water melts upon warming. Second, in a chemical change, a change in appearance is the result of the formation of a new material that has its own unique set of physical properties. The more evidence you have suggesting that a different material has been formed, the greater the likelihood that the change is a chemical change. Iron is a material that can be used to build cars. Rust is not. This suggests that the rusting of iron is a chemical change.

CHECK POINT

Evan, shown to the left, has grown an inch in height over the past year. Is this best described as a physical or a chemical change?

Was this your answer?
Are new materials being formed as Evan grows? Absolutely—created out of the food he eats. His body is very different from, say, the peanut butter sandwich he ate yesterday. Yet, through some very advanced chemistry, his body is able to absorb the atoms of that peanut butter sandwich and rearrange them into new materials. Biological growth, therefore, is best described as a chemical change.

Figure 14.12 shows potassium chromate, a material whose color depends on its temperature. At room temperature, potassium chromate is a bright canary yellow. At higher temperatures, it is a deep reddish orange. Upon cooling, the canary color

FIGURE 14.12

Potassium chromate changes color as its temperature changes. This change in color is a physical change. A return to the original temperature restores the original bright yellow color.

COOL   HOT   COOL

COOL ——————————— HOT ——————————— COOL

FIGURE 14.13

When heated, orange ammonium dichromate undergoes a chemical change to ammonia, water vapor, and chromium(III) oxide. A return to the original temperature does not restore the orange color, because the ammonium dichromate is no longer there.

returns, suggesting that the change is physical. With a chemical change, reverting to the original conditions does not restore the original appearance. Ammonium dichromate, shown in Figure 14.13, is an orange material that, when heated, explodes into ammonia, water vapor, and green chromium(III) oxide. When the test tube is returned to the original temperature, there is no trace of orange ammonium dichromate. In its place are new substances having completely different physical properties.

14.5 Elements to Compounds

As briefly described in Chapter 12, the terms *element* and *atom* are often used in a similar context. You might hear, for example, that gold is an element made of gold atoms. Generally, *element* is used in reference to an entire macroscopic or microscopic sample, and *atom* is used when speaking of the submicroscopic particles in the sample. The important distinction is that elements are made of atoms and not the other way around.

The fundamental unit of an element is indicated by its **elemental formula**. For elements in which the fundamental units are individual atoms, the elemental formula is simply the chemical symbol: Au is the elemental formula for gold, and Li is the elemental formula for lithium, to name just two examples. For elements in which the fundamental units are two or more atoms bonded into molecules, the elemental formula is the chemical symbol followed by a subscript indicating the number of atoms in each molecule. For example, elemental nitrogen, shown in Figure 14.1, commonly consists of molecules containing two nitrogen atoms per molecule. Thus, N_2 is the usual elemental formula given for nitrogen. Similarly, O_2 is the elemental formula for the oxygen we breathe, and S_8 is the elemental formula for sulfur.

Physical change? Chemical change? It's not always easy to distinguish between the two. Because of many subtleties that are recognized only after years of study and laboratory experience, you'll not soon achieve a firm handle on how to categorize many observed changes. It's okay to learn a little now, and to entrust a lot that remains for some future time or perhaps to others who chose to specialize within this field.

Link To Section 12.2

fyi

- Carbon is the only element that can form bonds with itself indefinitely. Sulfur's practical limit is S_8 and nitrogen's limit is around N_{12}. The elemental formula for a 1-carat diamond, however, is about $C_{10,000,000,000,000,000,000,000}$.

CHECK POINT

The oxygen we breathe, O_2, is converted to ozone, O_3, in the presence of an electric spark. Is this a physical or chemical change?

Was this your answer?

When atoms regroup, the result is an entirely new substance, and that is what happens here. The oxygen we breathe, O_2, is odorless and life-giving. Ozone, O_3, can be toxic, and it has a pungent smell commonly associated with electric motors. The conversion of O_2 to O_3 is therefore a chemical change. However, both O_2 and O_3 are elemental forms of oxygen.

FIGURE 14.14

The compounds sodium chloride and ammonia are represented by their chemical formulas, NaCl and NH₃. A chemical formula shows the ratio of atoms that constitute the compound.

Sodium atom
Chlorine atom

Sodium chloride, NaCl

Hydrogen atom

Nitrogen atom

Ammonia, NH₃

When atoms of different elements bond to one another, they make a **compound**. Sodium atoms and chlorine atoms, for example, bond to make the compound sodium chloride, commonly known as table salt. Nitrogen atoms and hydrogen atoms join to make the compound ammonia, which is a common household cleaner.

A compound is represented by its **chemical formula**, in which the symbols for the elements are written together. The chemical formula for sodium chloride is NaCl, and the formula for ammonia is NH₃. Numerical subscripts indicate the ratio in which the atoms combine. By convention, the subscript 1 is understood and omitted. So the chemical formula NaCl tells us that the compound sodium chloride has one sodium atom for every chlorine atom; the chemical formula NH₃ tells us that the compound ammonia has one nitrogen atom for every three hydrogen atoms, as Figure 14.14 shows.

Compounds have physical and chemical properties that are completely different from the properties of their elemental components. The sodium chloride, NaCl, shown in Figure 14.15 is very different from the elemental sodium and the elemental chlorine used in its formation. Elemental sodium, Na, consists of nothing but sodium atoms, which form a soft, silvery metal that can be cut easily with a knife. Its melting point is 97.5°C, and it reacts violently with water. Elemental chlorine, Cl₂, consists of chlorine molecules. This material, a yellow-green gas at room temperature, is very toxic, and it was used as a chemical warfare agent during World War I. Its boiling point is −34°C. The compound sodium chloride, NaCl, is a translucent, brittle, colorless crystal with a melting point of 800°C. Sodium chloride does not react chemically with water the way sodium does; not only is it not toxic to humans, which chlorine is, but the very opposite is true—it is an essential component of all living organisms. Sodium chloride is not sodium, nor is it chlorine; it is uniquely sodium chloride, a tasty chemical when sprinkled lightly over popcorn.

FIGURE 14.15

Sodium metal and chlorine gas react together to form sodium chloride. Although the compound sodium chloride is composed of sodium and chlorine, the physical and chemical properties of sodium chloride are very different from the physical and chemical properties of either sodium metal or chlorine gas.

Sodium metal and chlorine gas react to form sodium chloride

Hydrogen sulfide, H_2S, is one of the smelliest compounds. Rotten eggs get their characteristic bad smell from the hydrogen sulfide they release. Can you infer from this information that elemental sulfur, S_8, is just as smelly?

Was this your answer?
   No, you cannot. In fact, the odor of elemental sulfur is negligible compared with that of hydrogen sulfide. Compounds are truly different from the elements from which they are formed. Hydrogen sulfide, H_2S, is as different from elemental sulfur, S_8, as water, H_2S, is from elemental oxygen, O_2.

> A compound is uniquely different from the elements from which it is made. For example, water is a liquid, while the elements that compose it, hydrogen and oxygen, are gases. The harmless compound known as table salt is composed of two very dangerous chemicals: metallic sodium and chlorine gas.

▪ 14.6 Naming Compounds

A system for naming the countless number of possible compounds has been developed by the International Union for Pure and Applied Chemistry (IUPAC). This system is designed so that a compound's name reflects the elements it contains and how those elements are joined. Anyone familiar with the system, therefore, can deduce the chemical identity of a compound from its systematic name.

As you might imagine, this system is very intricate. There is no need for you to learn all its rules. Instead, learning some guidelines will prove most helpful. These guidelines alone will not enable you to name every compound. However, they will acquaint you with how the system works for many simple compounds consisting of only two elements.

GUIDELINE 1   The name of the element farther to the left in the periodic table is followed by the name of the element farther to the right, with the suffix *-ide* added to the name of the latter:

$NaCl$	Sodium chloride	HCl	Hydrogen chloride
Li_2O	Lithium oxide	MgO	Magnesium oxide
CaF_2	Calcium fluoride	Sr_3P_2	Strontium phosphide

GUIDELINE 2   When two or more compounds have different numbers of the same elements, prefixes are added to remove the ambiguity. The first four prefixes are *mono-* (one), *di-* (two), *tri-* (three), and *tetra-* (four). The prefix *mono-*, however, is commonly omitted from the beginning of the first word of the name:

Carbon and oxygen
CO	Carbon monoxide
CO_2	Carbon dioxide

Nitrogen and oxygen
NO_2	Nitrogen dioxide
N_2O_4	Dinitrogen tetroxide

Sulfur and oxygen
SO_2	Sulfur dioxide
SO_3	Sulfur trioxide

fyi

- Hydrogen is touted as the fuel of the future. It burns clean, producing only energy and water vapor. Much has to happen, however, before we can convert from fossil fuels to hydrogen. For example, we need an efficient method for generating hydrogen. Ideally, a system will be developed that produces hydrogen using the energy of direct sunlight. Also, the infrastructure for distributing hydrogen needs to be built. These are no small tasks, but current economic and environmental pressures appear to be pushing us in this direction.

GUIDELINE 3   Many compounds are not usually referred to by their systematic names. Instead, they are assigned common names that are more convenient or have been used traditionally for many years. Some common names are water for H_2O, ammonia for NH_3, and methane for CH_4.

CHECK POINT

What is the systematic name for NaF?

Was this your answer?
This compound is a cavity-fighting substance added to some tooth-pastes—sodium fluoride.

SUMMARY OF TERMS

Basic research A branch of scientific research that focuses on a greater understanding of how the natural world operates.

Applied research A branch of scientific research that focuses on developing applications built on the principles discovered through basic research.

Molecule A submicroscopic particle consisting of a group of atoms.

Submicroscopic On the scale of atoms and molecules, which are so small that we cannot observe them directly with optical microscopes.

Physical property Any physical attribute of a substance, such as color, density, or hardness.

Physical change A change in which a substance changes one or more of its physical properties without transforming into a new substance.

Chemical property A property that characterizes the ability of a substance to undergo a change that transforms it into a different substance.

Chemical bond The force of attraction between two atoms that holds them together. As discussed in Chapter 19, the nature of this force is electrical.

Chemical change A change in which the atoms of one or more substances are rearranged into one or more new substances.

Chemical reaction Synonymous with chemical change.

Elemental formula A notation that uses the atomic symbol and (sometimes) a numerical subscript to denote how atoms of the element are bonded together.

Compound A material in which atoms of different elements are bonded to one another.

Chemical formula A notation used to indicate the composition of a compound, consisting of the atomic symbols for the different elements of the compound and numerical subscripts indicating the ratio in which the atoms combine.

REVIEW QUESTIONS

14.1 Chemistry: The Central Science

1. What is the difference between basic research and applied research?
2. Why is chemistry often called the central science?
3. What do members of the American Chemistry Council pledge in the Responsible Care program?

14.2 The Submicroscopic World

4. Are atoms made of molecules or are molecules made of atoms?
5. How are the particles in a solid arranged differently from those in a liquid?
6. How does the arrangement of particles in a gas differ from the arrangements in liquids and solids?
7. Which occupies the greatest volume: 1 g of ice, 1 g of liquid water, or 1 g of water vapor?

14.3 Physical and Chemical Properties

8. What is a physical property?
9. What doesn't change during a physical change?
10. What is a chemical property?
11. What is a chemical bond?
12. What changes during a chemical reaction?

14.4 Determining Physical and Chemical Changes

13. Why is the freezing of water considered a physical change?
14. Why is it sometimes difficult to decide whether an observed change is physical or chemical?
15. Why is the rusting of iron considered a chemical change?
16. What are some clues that help us determine whether an observed change is physical or chemical?

14.5 Elements to Compounds

17. Distinguish between an atom and an element.
18. How many atoms are in a sulfur molecule that has the elemental formula S_8?
19. What is the difference between an element and a compound?
20. How many atoms are there in one molecule of H_3PO_4? How many atoms of each element are there in one molecule of H_3PO_4?
21. Are the physical and chemical properties of a compound necessarily similar to those of the elements from which it was composed?

14.6 Naming Compounds

22. Which element within a compound is shown first within the compound's name?
23. What is the IUPAC systematic name for the compound KF?
24. What is the chemical formula for the compound titanium dioxide?
25. Why are common names often used for chemical compounds instead of systematic names?

EXERCISES

● BEGINNER   ■ INTERMEDIATE   ◆ EXPERT

1. ● While you are visiting a foreign country, a non-English-speaking citizen tries to give you verbal directions to a local museum in that country's language. After multiple attempts he is unsuccessful. An onlooker sees your frustration and concludes that you are not smart enough to understand simple directions. Another onlooker sympathizes with you because he knows how difficult it is to navigate through an unfamiliar city. Which onlooker is correct?

2. ● If someone can explain an idea to you using small familiar words, what does this say about how well that person understands the idea?

3. ● Person A can explain an idea to a group of college students. Person B can explain the same idea, to the same depth, to a group of elementary school students. Who has demonstrated a greater command of the idea?

4. ● What is the best way to really prove to yourself that you understand an idea?

5. ● In what sense is a color computer monitor or television screen similar to our view of matter? Place a drop (and only a drop) of water on your computer monitor or television screen for a closer look.

6. ■ Of physics, chemistry, and biology, which science is the most complex?

7. ■ Is chemistry the study of the submicroscopic, the microscopic, the macroscopic, or all three? Defend your answer.

8. ● A cotton ball is dipped in alcohol and wiped across a tabletop. Explain what happens to the alcohol molecules deposited on the tabletop. Is this a physical or chemical change?

9. ■ A skillet is lined with a thin layer of cooking oil followed by a layer of unpopped popcorn kernels. Upon heating, the kernels all pop, thereby escaping the skillet. Identify any physical or chemical changes.

10. ● A cotton ball dipped in alcohol is wiped across a tabletop. Would the resulting smell of the alcohol be more or less noticeable if the tabletop were much warmer? Explain.

11. ■ Use Exercise 9 as an analogy to describe what occurs in Exercise 10. Does it make sense to think that the alcohol is made of very tiny particles (molecules) rather than being an infinitely continuous material?

12. ● Alcohol wiped across a tabletop rapidly disappears. What happens to the temperature of the tabletop? Why?

13. ● Red Kool-Aid crystals are added to a still glass of water. The crystals sink to the bottom. Twenty-four hours later the entire solution is red, even though no one stirred the water. Explain.

14. ● Red Kool-Aid crystals are added to a still glass of hot water. The same amount of crystals is added to a second still glass filled with the same amount of cold water. With no stirring, which would you expect to become uniform in color first: the hot water or the cold water? Why?

15. ● With no one looking, you add 5 mL of a cinnamon solution to a blue balloon, which you tie shut. You also add 5 mL of fresh water to a red balloon, which you also tie shut. You heat the two balloons in a microwave until they each inflate to about the size of a grapefruit. Your brother then comes along, examines the inflated balloons, and tells you that the blue balloon is the one that contains cinnamon. How did he know?

16. ■ Which has stronger attractions among its submicroscopic particles: a solid at 25°C or a gas at 25°C? Explain.

17. ◆ The leftmost diagram shows the moving particles of a gas within a rigid container. Which of the three boxes on the right—(a), (b), or (c)—best represents this material after the addition of heat?

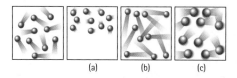

(a)   (b)   (c)

18. ◆ The leftmost diagram shows two phases of a single substance. In the middle box, draw what these particles would look like if heat were taken away. In the box on the right, show what they would look like if heat were added. If each particle represents a water molecule, what is the temperature of the box on the left?

19. ● Which occupies the greatest volume: 1 g of ice, 1 g of liquid water, or 1 g of water vapor?

20. ■ Gas particles travel at speeds of up to 500 m/s. Why, then, does it take so long for gas molecules to travel the length of a room?

21. ■ Humidity is a measure of the amount of water vapor in the atmosphere. Why is humidity always very low inside your kitchen freezer?

22. ■ You combine 50 mL of small BBs with 50 mL of large BBs and get a total of 90 mL of BBs of mixed size. Explain.

23. ■ You combine 50 mL of water with 50 mL of purified alcohol and get a total of 98 mL of mixture. Explain.

24. ■ In the winter, Vermonters make a tasty treat called "sugar on snow" in which they pour boiled-down maple syrup onto a scoop of clean fresh snow. As the syrup hits the snow it forms a delicious taffy. Identify the physical changes involved in the making of sugar on snow. Identify any chemical changes.

25. ■ Oxygen, O_2, has a boiling point of 90 K($-183°C$), and nitrogen, N_2, has a boiling point of 77 K($-196°C$). Which is a liquid and which is a gas at 80 K($-193°C$)?

26. ■ State whether each of the following is an example of a physical or chemical property of matter.
(a) Graphite conducts electricity.
(b) Bismuth, Bi, loses its iridescence upon melting.
(c) A copper penny is smushed into an embossed souvenir.

27. ■ State whether each of the following is an example of a physical or chemical property of matter.
(a) Carbon dioxide escapes upon the opening of a soda can.
(b) A bronze statue turns green.
(c) A silver spoon tarnishes.

28. ■ Classify the following changes as physical or chemical. Even if you are incorrect in your assessment, you should be able to defend why you chose as you did.
(a) Grape juice turns to wine. _____
(b) Wood burns to ashes. _____
(c) Water begins to boil. _____
(d) A broken leg mends itself. _____
(e) Grass grows. _____
(f) An infant gains 10 pounds. _____
(g) A rock is crushed to powder. _____

29. ● Each sphere in the diagrams represents an atom. Joined spheres represent molecules. Which box contains a liquid phase? Why can you not assume that box B represents a lower temperature?

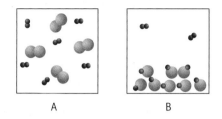

A   B

30. ● What physical and chemical changes occur when a wax candle burns?

31. ♦ Octane is a component of gasoline. It reacts with oxygen, O_2, to form carbon dioxide and water. Is octane an element or a compound? How can you tell?

32. ● Which elements are some of the oldest known? What is your evidence?

33. ■ Name 10 elements you have access to macroscopic samples of as a consumer here on Earth.

34. ● Oxygen atoms are used to make water molecules. Does this mean that oxygen, O_2, and water, H_2O, have similar properties? Why do we drown when we breathe in water, despite all of the oxygen atoms present in this material?

35. ■ If you eat metallic sodium or inhale chlorine gas, you stand a strong chance of dying. Let these two elements react with each other, however, and you can safely sprinkle the compound on your popcorn for better taste. What is going on?

36. ● Common names of chemical compounds are generally much shorter than the corresponding systematic names. The systematic names for water, ammonia, and methane, for example, are dihydrogen monoxide, H_2O; trihydrogen nitride, NH_3; and tetrahydrogen carbide, CH_4. For these compounds, which would you rather use: common names or systematic names? Which do you find more descriptive?

37. ● What is the chemical formula for the compound dihydrogen sulfide?

38. ● What is the chemical name for a compound with the formula Ba_3N_2?

39. ● What is the common name for dioxygen oxide?

40. ● What is the common name for oxygen oxide?

ACTIVE EXPLORATIONS

Fire Water

Place a large pot of cool water on top of a gas stove (or Bunsen burner, if performed in the laboratory) and set the flame on high. What product from the combustion of the natural gas do you see condensing on the outside of the pot? Where did it originate? Would more or less of this product form if the pot contained ice water? Where does this product go as the pot gets warmer? What physical and chemical changes can you identify?

Oxygen Bubble Bursts

Compounds can be broken down to their component elements. For example, when you pour a solution of the compound hydrogen peroxide, H_2O_2, over a cut, an enzyme in your blood decomposes it to produce oxygen gas, O_2, as evidenced by the bubbling that occurs. This oxygen at high concentrations at the site of the injury kills off microorganisms. A similar enzyme is found in baker's yeast.

What You Need

Packet of baker's yeast; 3% hydrogen peroxide solution; short, wide drinking glass; tweezers; matches

Safety Note

Wear safety glasses and remove all combustibles, such as paper towels, from the area. Keep your fingers well away from the flame because it will glow more brightly as it is exposed to the oxygen.

Procedure

1. Pour the yeast into the glass. Add a couple of capfuls of the hydrogen peroxide and watch the oxygen bubbles form.
2. Test for the presence of oxygen by holding a lighted match with the tweezers and placing the flame near the bubbles. Look for the flame to glow more brightly as the escaping oxygen passes over it. Describe oxygen's physical and chemical properties.

READINESS ASSURANCE TEST (RAT)

If you have a good handle on this chapter, if you really do, then you should be able to score 7 out of 10 on this RAT. If you score less than 7, you need to study further before moving on.

Choose the BEST answer to each of the following.

1. Chemistry is the study of
   (a) matter.
   (b) transformations of matter.
   (c) only microscopic phenomena.
   (d) only macroscopic phenomena.
   (e) both (a) and (b)
2. Imagine that you can see individual molecules. You watch a small collection of molecules that are moving around slowly while vibrating and bumping against each other. The slower-moving molecules then start to line up, but as they do so their vibrations increase. Soon all the molecules are aligned and vibrating about fixed positions. What is happening?
   (a) The sample is being cooled and the material is freezing.
   (b) The sample is being heated and the material is melting.
   (c) The sample is being cooled and the material is condensing.
   (d) The sample is being heated and the material is boiling.
   (e) The sample is unchanged.
3. The phase in which atoms and molecules no longer move is the
   (a) solid phase.
   (b) liquid phase.
   (c) gas phase.
   (d) none of these
4. Oxygen, O_2, has a boiling point of 90 K($-183°$C), and nitrogen, N_2, has a boiling point of 77 K($-196°$C). Which is a liquid and which is a gas at 50 K($-223°$C)?
   (a) Oxygen is a liquid and nitrogen is a gas.
   (b) Nitrogen is a liquid and oxygen is a gas.
   (c) They are both liquids.
   (d) They are both gases.
5. Based on the information given in the diagrams, which substance has the lower boiling point: one made from molecule A, or one made from molecule B?

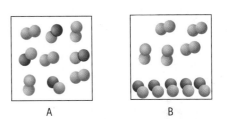

A   B

   (a) molecule A, which is the first to transform into a liquid
   (b) molecule B, which is the first to transform into a liquid
   (c) molecule A, which remains in the gaseous phase
   (d) molecule B, which remains in the gaseous phase
6. Does the following transformation represent a physical change or a chemical change?

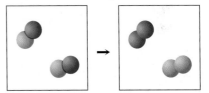

   (a) chemical, because of the formation of elements
   (b) physical, because a new material has been formed
   (c) chemical, because the atoms are connected differently
   (d) physical, because of a change in phase
7. Which of the following is an example of a chemical change?
   (a) water freezing into ice crystals
   (b) aftershave or perfume on your skin generating a smell
   (c) a piece of metal expanding when heated, but returning to its original size when cooled
   (d) breaking a glass window
   (e) gasoline being used in the engine of a car producing exhaust
8. If you burn 50 g of wood and produce 10 g of ash, what is the total mass of all the products produced from the burning of this wood?
   (a) more than 50 g
   (b) 10 g
   (c) less than 10 g

(d) 50 g

(e) none of these

9. If you have one molecule of TiO_2, how many molecules of O_2 does it contain?

(a) One; TiO_2 is a mixture of Ti and O_2.

(b) None; O_2 is a different molecule from TiO_2.

(c) Two; TiO_2 is a mixture of Ti and 2 O.

(d) Three; TiO_2 contains three molecules.

(e) none of these

10. What is the name of the compound $CaCl_2$?

(a) carbon chloride

(b) dichlorocalcium

(c) calc two

(d) dicalcium chloride

(e) calcium chloride

Answers to RAT

1e, 2a, 3d, 4d 5c, 6c, 7c, 8a, 9c, 10e

EXPLORING FURTHER

http://www.chemsoc.org

Chemistry news updates, an online chemistry magazine, and much more at this site run by the Royal Society of Chemistry.

http://www.chemsoc.org/viselements/pages/periodic _table.html

The Visual Elements project of the Royal Society of Chemistry, providing animations of almost all the elements. A high-speed Internet connection is required.

http://www.gsi.de

Web site of the heavy-ion research facility in Darmstadt, Germany, where many of the heaviest but shortest-lived elements are being created.

http://www.newton.dep.anl.gov

The Division of Educational Programs of the Argonne National Laboratory presents the Newton Bulletin Board Service, which features "Ask a Scientist." Explore the

chemistry archives for the answers to more than 1500 student questions compiled since 1991.

http://www.newscientist.com

Web site of the British weekly science and technology newsmagazine. Current events and many "hot issues" in science are presented.

http://www.csicop.org

Home page of the Committee for Skeptical Inquiry. This organization of Nobel laureates and other respected scientists takes on the claims of pseudoscience with all the rigor required of any scientific claim.

http://www.randi.org

This is the Web site of James Randi, a widely known magician who is active in critically examining claims for the paranormal.

CHAPTER 14 ONLINE RESOURCES

The Physical Science Place

Interactive Figure

■ 14.9

Tutorial

■ What Is Chemistry?

Videos

■ Oxygen Bubble Burst

■ Fire Water

Quiz

Flashcards

Links

how atoms bond and molecules attract

15.1 ELECTRON-DOT STRUCTURES

15.2 THE FORMATION OF IONS

15.3 IONIC BONDS

15.4 METALLIC BONDS

15.5 COVALENT BONDS

15.6 POLAR COVALENT BONDS

15.7 MOLECULAR POLARITY

15.8 MOLECULAR ATTRACTIONS

■ Millions of years ago, the Great Plains region of what is now the United States was an ocean. As sea levels fell while the North American continent rose, many isolated pockets of seawater, called saline lakes, remained. Over time, these lakes evaporated, leaving behind the solids that had been dissolved in the seawater. Most abundant was sodium chloride, which formed the cubic crystals referred to by mineralogists as the mineral *halite*. When conditions were right, halite crystals like the ones in this chapter's opening photograph grew to be several centimeters across.

Why do halite crystals have such a distinct shape? As we will see in this chapter, the macroscopic properties of any substance can be traced to how its submicroscopic parts are held together. The sodium and chloride ions in a halite crystal, for example, hold together in a cubic orientation, and, as a result, the macroscopic object we know as a halite crystal is also cubic.

Similarly, the macroscopic properties of substances made of molecules are a result of how the atoms in the molecules hold together. For example, many of water's interesting properties result from the angle between the hydrogen and oxygen atoms in the water molecule. Because of this angled orientation, one side of the molecule has a slight negative charge and the opposite side has a slight positive charge. The water molecule is electrically polarized (Chapter 8). The polarization of water molecules gives rise to such phenomena as the inability of water and oil to mix and water's high boiling temperature.

The force of attraction that holds ions or atoms together is the electric force between oppositely charged particles. Chemists refer to this ion-binding or atom-binding force as a chemical bond. In this chapter, we will explore three types of chemical bonds: the *ionic bond*, which holds ions together in a crystal; the *covalent bond*, which holds atoms together in a molecule; and the *metallic bond*, which holds atoms together in a piece of metal. Then later in this chapter, we will explore how the behavior of ions and molecules gives rise to macroscopic phenomena, such as the mixing of salt and water.

15.1 Electron-Dot Structures

An atomic model is needed to help us understand how atoms bond. We begin this chapter with a brief overview of the shell model presented in Section 12.9. Recall how electrons are arranged around an atomic nucleus. Rather than moving in neat orbits like planets around the Sun, electrons are wavelike entities that swarm in various volumes of space called *shells*.

As was shown in Figure 12.29, seven shells are available to the electrons in an atom, and the electrons fill these shells in order, from innermost to outermost. Furthermore, the maximum number of electrons allowed in the first shell is 2, and for the second and third shells it is 8. The fourth and fifth shells can each hold 18 electrons, and the sixth and seventh shells can each hold 32 electrons.* These numbers match the number of elements in each period (horizontal row) of the periodic table. Figure 15.1 shows how this model applies to the first four elements of group 18.

Electrons in the outermost occupied shell of any atom may play a significant role in that atom's chemical properties, including its ability to form chemical bonds. To indicate their importance, we call these electrons *valence electrons* (as described in Section 15.5), and we call the shell they occupy the **valence shell**. Valence electrons can be conveniently represented as a series of dots surrounding an atomic symbol. This notation is called an **electron-dot structure**, or

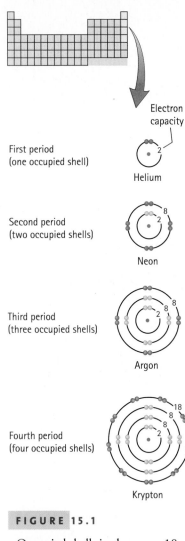

First period
(one occupied shell)

Helium

Second period
(two occupied shells)

Neon

Third period
(three occupied shells)

Argon

Fourth period
(four occupied shells)

Krypton

FIGURE 15.1

Occupied shells in the group 18 elements helium through krypton. Each of these elements has a filled outermost occupied shell, and the number of electrons in each corresponds to the number of elements in the period to which a particular group 18 element belongs.

---

* As a point of reference for physicists reading this text, these shells of orbitals are grouped by similar energy levels rather than by principal quantum number. They are the "argonian" shells developed by Linus Pauling to explain chemical bonding and the organization of the periodic table.

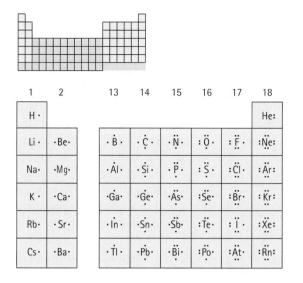

FIGURE 15.2

The valence electrons of an atom are shown in its electron-dot structure. Note that the first three periods here parallel Figure 12.30. Also note that for larger atoms, not all the electrons in the valence shell are valence electrons. Krypton, Kr, for example, has 18 electrons in its valence shell, as shown in Figure 15.1, but only 8 of these are classified as valence electrons.

Link To Section 12.9

sometimes a *Lewis dot symbol* (in honor of the American chemist G. N. Lewis, who first proposed the concepts of shells and valence electrons).

Figure 15.2 shows the electron-dot structures for the atoms important in our discussions of ionic and covalent bonds. For our discussion of metallic bonds at the end of this chapter, we'll focus only on the valence electrons of metal atoms and not on their electron-dot structures.

When you look at the electron-dot structure of an atom, you immediately know two important things about that element. You know how many valence electrons it has and how many of these electrons are *paired*. Chlorine, for example, has three sets of paired electrons and one unpaired electron, and carbon has four unpaired electrons:

Too much detail to learn? What would the scientists of 200 years ago give for the information that today is so readily available to you?

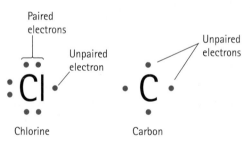

Paired valence electrons are relatively stable. In other words, they usually do not form chemical bonds with other atoms. For this reason, electron pairs in an electron-dot structure are called **nonbonding pairs**. (Do not take this term literally, however, because in Chapter 18 you'll see that under the right conditions, even "nonbonding" pairs can form a chemical bond.)

Valence electrons that are *unpaired*, by contrast, have a strong tendency to participate in chemical bonding. By doing so, they become paired with an electron from another atom. The ionic and covalent bonds discussed in this chapter all result from either a transfer or a sharing of unpaired valence electrons.

FIGURE 15.3

Gilbert Newton Lewis (1875–1946) revolutionized chemistry with his theory of chemical bonding, which he published in 1916. He worked most of his life in the chemistry department of the University of California, Berkeley, where he was not only a productive researcher but also an exceptional teacher. Among his teaching innovations was the idea of providing students with problem sets as a follow-up to lectures and readings.

CHECK POINT

Where are valence electrons located, and why are they important?

Was this your answer?
Valence electrons are located in the outermost occupied shell of an atom. They are important because they play a leading role in determining the chemical properties of the atom.

■ 15.2 The Formation of Ions

When the number of protons in the nucleus of an atom equals the number of electrons in the atom, the charges balance and the atom is electrically neutral. If one or more electrons are lost or gained, as illustrated in Figure 15.4 and 15.5, the balance is upset and the atom takes on a net electric charge. Any atom with a net electric charge is an **ion**. When electrons are lost, protons outnumber electrons and the ion has a positive net charge. When electrons are gained, electrons outnumber protons and the ion has a negative net charge.

Chemists use a superscript to the right of the atomic symbol to indicate the magnitude and sign of an ion's charge. Thus, as shown in Figures 15.4 and 15.5, the positive ion formed from the sodium atom is written Na^{1+} and the negative ion formed from the fluorine atom is written F^{1-}. Usually the numeral 1 is omitted when indicating either a 1^+ or 1^- charge. Hence, these two ions are most frequently written Na^+ and F^-.

To give two more examples, a calcium atom that loses two electrons is written Ca^{2+}, and an oxygen atom that gains two electrons is written O^{2-}. (Note that the convention is to write the numeral before the sign, not after it: $2+$, not $+2$.)

We can use the shell model to deduce the type of ion an atom tends to form. According to this model, *atoms tend to lose or gain electrons that result in an outermost occupied shell filled to capacity*. Let's take a moment to consider this point, looking to Figures 15.4 and 15.5 as visual guides.

If an atom has only one or a few electrons in its valence shell, it tends to give up (lose) these electrons so that the next shell inward, which is already filled, becomes the outermost occupied shell. The sodium atom of Figure 15.4, for example, has one electron in its valence shell, which is the third shell. In forming an ion, the sodium atom loses this electron, thereby making the second shell, which is already filled to capacity, the outermost occupied shell. Because the sodium atom has only one valence electron to lose, it tends to form the $1+$ ion.

If the valence shell of an atom is almost filled, that atom attracts electrons from another atom and so forms a negative ion. The fluorine atom of Figure 15.5, for example, has one space available in its valence shell for an additional electron. After this additional electron is gained, the fluorine atom achieves a filled valence shell. Fluorine therefore tends to form the $1-$ ion.

You can use the periodic table as a quick reference when determining the type of ion an atom tends to form. As Figure 15.6 shows, each atom of any group 1 element, for example, has only one valence electron and so tends to form the $1+$ ion. Each atom of any group 17 element has room for one additional electron in its valence shell and therefore tends to form the $1-$ ion. Atoms of the noble-gas elements tend not to form ions of any type because their valence shells are already filled to capacity.

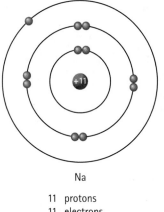

Na

11 protons
11 electrons
0 net charge

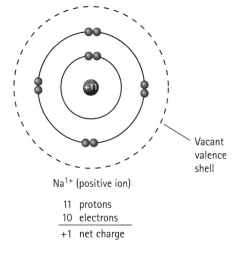

Vacant valence shell

Na^{1+} (positive ion)

11 protons
10 electrons
+1 net charge

FIGURE 15.4

An electrically neutral sodium atom contains 11 negatively charged electrons surrounding the 11 positively charged protons of the nucleus. When this atom loses an electron, the result is a positive ion.

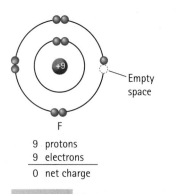

Empty space

F

9 protons
9 electrons
0 net charge

Filled valence shell

F^{1-} (negative ion)

9 protons
10 electrons
−1 net charge

FIGURE 15.5

An electrically neutral fluorine atom contains 9 protons and 9 electrons. When this atom gains an electron, the result is a negative ion.

= Weak nuclear attraction for valence electrons; tendency to form positive ions

= Strong nuclear attraction for valence electrons; tendency to form negative ions

= Strong nuclear attraction for valence electrons but valence shell is already filled; no tendency to form ions of either type

FIGURE 15.6

The periodic table is your guide to the types of ions that atoms tend to form.

CHECK POINT

What type of ion does the magnesium atom, Mg, tend to form?

Was this your answer?
The magnesium atom (atomic number 12) is found in group 2 and has two valence electrons to lose (see Figure 15.2). Therefore, it tends to form the 2+ ion.

The Physical Science Place

Intermolecular Forces

As is indicated in Figure 15.6, the attraction between an atom's nucleus and its valence electrons is weakest for elements on the left in the periodic table and strongest for elements on the right. From sodium's position in the table, we can see that a sodium atom's single valence electron is not held very strongly, which explains why it is so easily lost. The attraction the sodium nucleus has for its second-shell electrons, however, is much stronger, which is why the sodium atom rarely loses more than one electron.

At the other side of the periodic table, the nucleus of a fluorine atom holds strongly onto its valence electrons, which explains why the fluorine atom tends not to lose any electrons to form a positive ion. Instead, fluorine's nuclear pull on the valence electrons is strong enough to accommodate even an additional electron "imported" from some other atom.

The nucleus of a noble-gas atom pulls so strongly on its valence electrons that they are very difficult to remove. Because no space is available in the valence shell of a noble-gas atom, no additional electrons are gained. Thus, a noble-gas atom tends not to form ions of any sort.

Electrons are negatively charged. So gaining an electron results in a negative ion, and losing an electron results in a positive ion.

Link To Section 12.9

CHECK POINT

Why does the magnesium atom tend to form the 2+ ion?

Was this your answer?
Magnesium is on the left in the periodic table, and so atoms of this element do not hold on to the two valence electrons very strongly. Because these electrons are not held very tightly, they are easily lost, which is why the magnesium atom tends to form the 2+ ion.

Using our shell model to explain the formation of ions works well for groups 1, 2, and 13 through 18. This model is too simplified to work well for the transition metals of groups 3 through 12, however, or for the inner transition metals. In general, these metal atoms tend to form positive ions, but the number of electrons lost varies. For example, depending on conditions, an iron atom may lose two electrons to form the Fe^{2+} ion, or it may lose three electrons to form the Fe^{3+} ion.

MOLECULES CAN FORM IONS

So we see that atoms form ions by losing or gaining electrons. Interestingly, molecules can also become ions. In most cases, this occurs whenever a molecule loses or gains a proton—equivalent to the hydrogen ion, H^+. (Recall that a hydrogen atom is a proton together with an electron. The hydrogen ion, H^+, therefore, is simply a proton.) For example, a water molecule, H_2O, can gain a hydrogen ion, H^+ (a proton), to form the hydronium ion, H_3O^+:

Water   Hydrogen ion   Hydronium ion
                (proton)

Similarly, the carbonic acid molecule, H_2CO_3, can lose two protons to form the carbonate ion, CO_3^{2-}:

Carbonic acid   Carbonate ion   Hydrogen ions
                                          (protons)

Link To Section 18.1

How these reactions occur will be explored in later chapters. For now, you should understand that the hydronium and carbonate ions are examples of **polyatomic ions**, which are molecules that carry a net electric charge. Table 15.1 lists some commonly encountered polyatomic ions.

TABLE 15.1   COMMON POLYATOMIC IONS

Name	Formula
Hydronium ion	H_3O^+
Ammonium ion	NH_4^+
Bicarbonate ion	HCO_3^-
Acetate ion	$CH_3CO_2^-$
Nitrate ion	NO_3^-
Cyanide ion	CN^-
Hydroxide ion	OH^-
Carbonate ion	CO_3^{2-}
Sulfate ion	SO_4^{2-}
Phosphate ion	PO_4^{3-}

Electron transfer	Ions formed	Ionic bond
Sodium and chlorine atoms	Sodium and chloride ions	Sodium chloride, NaCl
①	②	③

FIGURE 15.7

(1) An electrically neutral sodium atom loses its valence electron to an electrically neutral chlorine atom. (2) This electron transfer results in two oppositely charged ions. (3) The ions are then held together by an ionic bond. The spheres drawn around these and subsequent illustrations of electron-dot structures indicate the relative sizes of the atoms and ions. Note that the sodium ion is smaller than the sodium atom because the lone electron in the third shell has gone once the ion forms, leaving the ion with only two occupied shells. The chloride ion is larger than the chlorine atom because the addition of that one electron to the third shell makes the shell expand due to the repulsions among the electrons.

15.3 Ionic Bonds

When an atom that tends to lose electrons is placed in contact with an atom that tends to gain them, the result is an electron transfer and the formation of two oppositely charged ions. This occurs when sodium and chlorine are combined. As shown in Figure 15.7, the sodium atom loses one of its electrons to the chlorine atom, resulting in the formation of a positive sodium ion and a negative chloride ion. The two oppositely charged ions are attracted to each other by the electric force, which holds them close together. This electric force of attraction between two oppositely charged ions is called an **ionic bond**.

A sodium ion and a chloride ion together make the chemical compound sodium chloride, commonly known as table salt. This and all other chemical compounds containing ions are referred to as **ionic compounds**. All ionic compounds are completely different from the elements from which they are made. As discussed in Section 14.5, sodium chloride is not sodium, nor is it chlorine. Rather, it is a collection of sodium and chloride ions that form a unique material with its own physical and chemical properties.

The Physical Science Place
Ionic Bonds

CHECK POINT

Is the transfer of an electron from a sodium atom to a chlorine atom a physical change or a chemical change?

Was this your answer?
Recall from Chapter 14 that only a chemical change involves the formation of new material. Thus, this or any other electron transfer, because it results in the formation of a new substance, is a chemical change.

The ionic bond is merely the electrical force of attraction that holds ions of opposite charge together, in accord with Coulomb's law (Chapter 8).

As Figure 15.8 shows, ionic compounds typically consist of elements that are found on opposite sides of the periodic table. Also, because of how the metals

Potassium iodide, KI

(a)

Sodium fluoride, NaF

(b)

FIGURE 15.8

(a) The ionic compound potassium iodide, KI, is added in minute quantities to commercial salt because the iodide ion, I^-, it contains is an essential dietary mineral. (b) The ionic compound sodium fluoride, NaF, is often added to municipal water supplies and toothpastes because it is a good source of the tooth-strengthening fluoride ion, F^-.

FIGURE 15.9

A calcium atom loses two electrons to form a calcium ion, Ca^{2+}. These two electrons may be picked up by two fluorine atoms, transforming the atoms to two fluoride ions. Calcium ions and fluoride ions then join to form the ionic compound calcium fluoride, CaF_2, which occurs naturally as the mineral fluorite.

Bonds forming

Fluorine atom   Calcium atom   Fluorine atom

Ionic bonds formed

Calcium fluoride, CaF_2

Fluorite

Link To Section 20.2

and nonmetals are organized in the periodic table, positive ions are generally derived from metallic elements and negative ions are generally derived from nonmetallic elements.

For all ionic compounds, positive and negative charges must balance. In sodium chloride, for example, there is one sodium $1+$ ion for every chloride $1-$ ion. Charges must also balance in compounds containing ions that carry multiple charges. The calcium ion, for example, carries a charge of $2+$, but the fluoride ion carries a charge of only $1-$. Because two fluoride ions are needed to balance each calcium ion, the formula for calcium fluoride is CaF_2, as Figure 15.9 illustrates. Calcium fluoride occurs naturally in the drinking water of some communities, where it is a good source of the tooth-strengthening fluoride ion, F^-.

An aluminum ion carries a $3+$ charge, and an oxide ion carries a $2-$ charge. Together, these ions make the ionic compound aluminum oxide, Al_2O_3, the main component of such gemstones as rubies and sapphires. Figure 15.10 illustrates the formation of aluminum oxide. The three oxide ions in Al_2O_3 carry a total charge of $6-$, which balances the total $6+$ charge of the two aluminum ions. As mentioned earlier, rubies and sapphires differ in color because of the impurities they contain. Rubies are red because of minor amounts of chromium ions, and sapphires are blue because of minor amounts of iron and titanium ions.

CHECK POINT

What is the chemical formula for the ionic compound magnesium oxide?

Was this your answer?
Because magnesium is a group 2 element, you know a magnesium atom must lose two electrons to form a Mg^{2+} ion. Because oxygen is a group 16 element, an oxygen atom gains two electrons to form an O^{2-} ion. These charges balance in a one-to-one ratio, and so the formula for magnesium oxide is MgO.

FIGURE 15.10

Two aluminum atoms lose a total of six electrons to form two aluminum ions, Al^{3+}. These six electrons may be picked up by three oxygen atoms, transforming the atoms to three oxide ions, O^{2-}. The aluminum and oxide ions then join to form the ionic compound aluminum oxide, Al_2O_3.

Bonds forming

Aluminum atom   Oxygen atom

Ionic bonds formed

Aluminum oxide, Al_2O_3

Ruby

Sapphire

An ionic compound typically contains a multitude of ions grouped together in a highly ordered three-dimensional array. In sodium chloride, for example, each sodium ion is surrounded by six chloride ions, and each chloride ion is surrounded by six sodium ions (Figure 15.11). Overall, there is one sodium ion for each chloride ion, but there are no identifiable sodium–chloride pairs. Such an orderly array of ions is known as an ionic crystal. As mentioned at the beginning of this chapter, on the atomic level, the crystalline structure of sodium chloride is cubic, which is why macroscopic crystals of table salt are also cubic. Smash a large cubic sodium chloride crystal with a hammer, and what do you get? Smaller cubic sodium chloride crystals!

Similarly, the crystalline structures of other ionic compounds, such as calcium fluoride and aluminum oxide, are a consequence of how the ions pack together. We go into more detail about the crystalline structures of minerals in Chapter 20.

● Sodium ion, Na$^+$

○ Chloride ion, Cl$^-$

(a)

(b)

FIGURE 15.11

(a) Sodium chloride, as well as other ionic compounds, forms ionic crystals in which every internal ion is surrounded by ions of the opposite charge. (For simplicity, only a small portion of the ion array is shown here. A typical NaCl crystal involves millions of ions.) (b) A view of crystals of table salt through a microscope shows their cubic structure. The cubic shape is a consequence of the cubic arrangement of sodium and chloride ions.

15.4 Metallic Bonds

In Section 12.4 you learned about the properties of metals. They conduct electricity and heat, are opaque to light, and deform—rather than fracture—under pressure. Because of these properties, metals are used to build homes, appliances, cars, bridges, airplanes, and skyscrapers. Metal wires across the landscape transmit communication signals and electric power. We wear metal jewelry, exchange metal currency, and drink from metal cans. Yet what gives a metal its metallic properties? We can answer this question by looking at the behavior of its atoms.

The outer electrons of most metal atoms tend to be weakly held to the atomic nucleus. Consequently, these electrons are easily dislodged, leaving behind positively charged metal ions. The many electrons dislodged from a large group of metal atoms flow freely through the resulting metal ions, as is depicted in Figure 15.12. This "fluid" of electrons holds the positively charged metal ions together in the type of chemical bond known as a **metallic bond**.

The mobility of electrons in a metal accounts for the metal's significant ability to conduct electricity and heat. Also, metals are opaque and shiny because the free electrons easily vibrate to the oscillations of any light falling on them, reflecting most of it. Furthermore, the metal ions are not rigidly held to fixed positions, as ions are in an ionic crystal. Rather, because the metal ions are held together by a "fluid" of electrons, these ions can move into various orientations relative to one another, which occurs when a metal is pounded, pulled, or molded into a different shape.

Two or more metals can be bonded to each other by metallic bonds. This occurs, for example, when molten gold and molten palladium are blended to form the homogeneous solution known

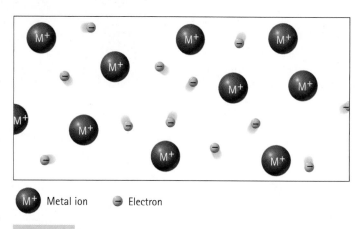

Ⓜ Metal ion   ⊖ Electron

FIGURE 15.12

Metal ions are held together by freely flowing electrons. These loose electrons form a kind of "electronic fluid," which flows through the lattice of positively charged ions.

FIGURE 15.13

The gold color of the Sacagawea U.S. dollar coin is achieved by an outer surface made of an alloy of 77% copper, 12% zinc, 7% manganese, and 4% nickel. The interior of the coin is pure copper.

fyi

- Metal ores are ionic compounds in which the metal atoms have lost electrons to become positive ions. To convert the ores to metals requires that electrons be given back to the metal ions. This is done by heating the ore with electron-releasing materials, such as carbon, in hot furnaces that reach about 1500°C. The metal emerges in a molten state that can be cast into a variety of useful shapes.

Link To Section 20.4 ⟳

as white gold. The quality of the white gold can be modified simply by changing the proportions of gold and palladium. White gold is an example of an **alloy**, which is any mixture composed of two or more metallic elements. By playing around with proportions, metal workers can readily modify the properties of an alloy. For example, in designing the Sacagawea dollar coin, shown in Figure 15.13, the U.S. Mint needed a metal with a gold color—so that it would be easy to recognize—and also have the same electrical characteristics as the Susan B. Anthony dollar coin—so that the new coin could substitute for the Anthony coin in vending machines.

Only a few metals—gold and platinum are two examples—appear in nature in metallic form. Deposits of these natural metals, also known as *native metals*, are quite rare. For the most part, metals found in nature are chemical compounds. Iron, for example, is most frequently found as iron oxide, Fe_2O_3, and copper is found as chalcopyrite, $CuFeS_2$. Geologic deposits containing relatively high concentrations of metal-containing compounds are called *ores*. The metals industry mines these ores from the ground, as shown in Figure 15.14, and then processes them into metals. Although metal-containing compounds occur just about everywhere, only ores are concentrated enough to make the extraction of the metal economical.

FIGURE 15.14

The world's biggest open-pit mine is the copper mine at Bingham Canyon, Utah.

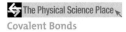
The Physical Science Place
Covalent Bonds

The Physical Science Place
Covalent Bonds

15.5 Covalent Bonds

Imagine two children playing together and sharing their toys. Perhaps a force that keeps the children together is their mutual attraction to the toys they share. In a similar fashion, two atoms can be held together by their mutual attraction for electrons they share. A fluorine atom, for example, has a strong attraction for one additional electron to fill its outermost occupied shell. As shown in Figure 15.15, a fluorine atom can obtain an additional electron by holding on to the unpaired valence electron of another fluorine atom. This

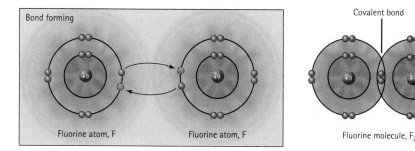

FIGURE 15.15

The effect of the positive nuclear charge (represented by red shading) of a fluorine atom extends beyond the atom's outermost occupied shell. This positive charge can cause the fluorine atom to become attracted to the unpaired valence electron of a neighboring fluorine atom. Then the two atoms are held together in a fluorine molecule by the attraction they both have for the two shared electrons. Each fluorine atom achieves a filled valence shell.

results in a situation in which the two fluorine atoms are mutually attracted to the same two electrons. This type of electrical attraction in which atoms are held together by their mutual attraction for shared electrons is called a **covalent bond**, in which *co-* signifies sharing and *-valent* indicates that valence electrons are being shared.

A substance composed of atoms held together by covalent bonds is a **covalent compound**. The fundamental unit of most covalent compounds is a **molecule**, which we can now formally define as any group of atoms held together by covalent bonds. Figure 15.16 uses the element fluorine to illustrate this principle.

When writing electron-dot structures for covalent compounds, chemists often use a straight line to represent the two electrons involved in a covalent bond. In some representations, the nonbonding electron pairs are ignored. This occurs in instances where these electrons play no significant role in the process being illustrated. Here are two frequently used ways of showing the electron-dot structure for a fluorine molecule without using spheres to represent the atoms:

$$:\ddot{F}\!\!-\!\!\ddot{F}: \qquad F\!\!-\!\!F$$

Gaseous elemental fluorine

Fluorine molecule, F_2

FIGURE 15.16

Molecules are the fundamental units of the gaseous covalent compound fluorine, F_2. Notice that in this model of a fluorine molecule, the spheres overlap, whereas the spheres shown earlier for ionic compounds do not. Now you know that this difference in representation is because of the difference in bond types.

Remember—the straight line in both versions represents two electrons, one from each atom. Thus, we now have two types of electron pairs to keep track of. The term *nonbonding pair* refers to any pair that exists in the electron-dot structure of an individual atom, and the term *bonding pair* refers to any pair that results from formation of a covalent bond. In a nonbonding pair, both electrons originate in the same atom; in a bonding pair, one electron comes from one of the atoms participating in the covalent bond, and the other electron comes from the other atom participating in the bond.

Recall from Section 15.3 that an ionic bond is formed when an atom that tends to lose electrons makes contact with an atom that tends to gain them. A covalent bond, by contrast, is formed when two atoms that tend to gain electrons are brought into contact with each other. Atoms that tend to form covalent bonds are therefore primarily atoms of the nonmetallic elements in the upper right corner of the periodic table (with the exception of the noble-gas elements, which are very stable and tend not to form bonds).

Hydrogen tends to form covalent bonds because, unlike the other group 1 elements, it has a fairly strong attraction for an additional electron. Two hydrogen atoms, for example, covalently bond to form a hydrogen molecule, H_2, as shown in Figure 15.17.

Before bonding

H· Hydrogen atom   ·H Hydrogen atom

Covalent bond formed

Hydrogen molecule, H_2

FIGURE 15.17

Two hydrogen atoms form a covalent bond as they share their unpaired electrons.

FIGURE 15.18

INTERACTIVE FIGURE

The two unpaired valence electrons of oxygen pair with the unpaired valence electrons of two hydrogen atoms to form the covalent compound water.

Water molecule, H_2O

fyi

- Spectroscopic studies of interstellar dust within our galaxy have revealed the presence of more than 120 kinds of molecules, such as hydrogen chloride, HCl; water, H_2O; acetylene, H_2C_2; formic acid, HCO_2H; methanol, CH_3OH; methyl amine, NH_2CH_3; acetic acid, CH_3CO_2H; and even the amino acid glycine, $NH_2CH_2CO_2H$. Notably, about half of these interstellar molecules are carbon-based organic molecules. As discussed in Chapter 13, the atoms originated from the nuclear fusion of ancient stars. How interesting that these atoms then join together to form molecules even in the deep vacuum of outer space. Chemistry is truly everywhere.

The number of covalent bonds an atom can form is equal to the number of additional electrons it can attract, which is the number needed to fill its valence shell. Hydrogen attracts only one additional electron, and so it forms only one covalent bond. Oxygen, which attracts two additional electrons, finds them when it encounters two hydrogen atoms and reacts with them to form water, H_2O, as Figure 15.18 shows. In water, not only does the oxygen atom have access to two additional electrons by covalently bonding to two hydrogen atoms, but each hydrogen atom has access to an additional electron by bonding to the oxygen atom. Each atom thus achieves a filled valence shell.

Nitrogen attracts three additional electrons and thus can form three covalent bonds, as occurs in ammonia, NH_3, shown in Figure 15.19. Likewise, a carbon atom can attract four additional electrons and is thus able to form four covalent bonds, as occurs in methane, CH_4. Note that the number of covalent bonds formed by these and other nonmetallic elements parallels the type of negative ions they tend to form (see Figure 15.6). This makes sense because covalent-bond formation and negative-ion formation are both applications of the same concept: nonmetallic atoms tend to gain electrons until their valence shells are filled.

Diamond is a very unusual covalent compound consisting of carbon atoms covalently bonded to one another in four directions. The result is a covalent crystal, which, as shown in Figure 15.20, is a highly ordered, three-dimensional network of covalently bonded atoms. This network of carbon atoms forms a very strong and rigid structure, which is why diamonds are so hard. Also, because a diamond is a group of atoms held together only by covalent bonds, it

FIGURE 15.19

(a) A nitrogen atom attracts the three electrons in three hydrogen atoms to form ammonia, NH_3, a gas that can dissolve in water to make an effective cleanser. (b) A carbon atom attracts the four electrons in four hydrogen atoms to form methane, CH_4, the primary component of natural gas. In these and most other cases of covalent-bond formation, the result is a filled valence shell for all the atoms involved.

(a)

Ammonia molecule, NH_3

(b)

Methane molecule, CH_4

FIGURE 15.20

The crystalline structure of diamond is nicely illustrated with sticks to represent the covalent bonds. The molecular nature of a diamond is responsible for its extreme hardness.

can be characterized as a single molecule! Unlike most other molecules, a diamond molecule is large enough to be visible to the naked eye, and so it is more appropriately referred to as a *macromolecule*.

CHECK POINT

How many electrons make up a covalent bond?

Was this your answer?
Two—one from each participating atom.

It is possible to have more than two electrons shared between two atoms, and Figure 15.21 shows a few examples. Molecular oxygen, O_2, consists of two oxygen atoms connected by four shared electrons. This arrangement is called a *double covalent bond* or, for short, a *double bond*. As another example, the covalent compound carbon dioxide, CO_2, consists of two double bonds connecting two oxygen atoms to a central carbon atom.

Some atoms can form triple covalent bonds, in which six electrons—three from each atom—are shared. One example is molecular nitrogen, N_2. Any double or triple bond is often referred to as a *multiple covalent bond*. Multiple bonds higher than these, such as the quadruple covalent bond, are not commonly observed.

fyi

- Astronomers have recently discovered an expired star that has a solid core made of diamond. This star-sized diamond is about 4000 km wide, which amounts to about 10 billion trillion trillion carats. It has been named "Lucy" after the Beatles song "Lucy in the Sky with Diamonds." In about 7 billion years, our own star, the Sun, is also likely to crystallize into a huge diamond ball.

Link To Section 20.3

$$:\ddot{O}=\ddot{O}:$$
Oxygen, O_2

$$:\ddot{O}=C=\ddot{O}:$$
Carbon dioxide, CO_2

$$:N\equiv N:$$
Nitrogen, N_2

FIGURE 15.21

Double covalent bonds in molecules of oxygen, O_2, and carbon dioxide, CO_2, and a triple covalent bond in a molecule of nitrogen, N_2.

15.6 Polar Covalent Bonds

If the two atoms in a covalent bond are identical, their nuclei have the same positive charge, and therefore the electrons are shared evenly. We can represent these electrons as being centrally located by using an electron-dot structure with the electrons situated exactly halfway between the two atomic symbols. Alternatively, we can draw a cloud in which the positions of the two bonding

There are always two electrons per covalent bond. A double bond, therefore, consists of four electrons, while a triple bond consists of six electrons.

electrons over time are shown as a series of dots. Where the dots are most concentrated is where the electrons have the greatest probability of being located:

> One side of the hydrogen–fluorine bond has a greater density of electrons and is slightly negative, while the opposite side is slightly positive. This makes up a *dipole*, an extension of electric-charge polarization, which was discussed in Chapter 8.

In a covalent bond between nonidentical atoms, the nuclear charges are different, and consequently the bonding electrons may be shared unevenly. This occurs in a hydrogen–fluorine bond, where electrons are more attracted to fluorine's greater nuclear charge:

> A dipole is a vector quantity possessing both magnitude and direction. When two dipoles are equal and opposite, they effectively cancel each other out.

The bonding electrons spend more time around the fluorine atom. For this reason, the fluorine side of the bond is slightly negative and, because the bonding electrons have been drawn away from the hydrogen atom, the hydrogen side of the bond is slightly positive. This separation of charge is called a **dipole** (pronounced *die*-pole) and is represented either by the characters $\delta-$ and $\delta+$ (read "slightly negative" and "slightly positive," respectively) or by a crossed arrow pointing to the negative side of the bond:

$$\overset{\delta+\quad\delta-}{\text{H}-\text{F}} \qquad \overset{\longmapsto}{\text{H}-\text{F}}$$

Link To Section 8.2

So atoms forming a chemical bond engage in a tug-of-war for electrons. How strongly an atom is able to tug on bonding electrons has been measured experimentally and quantified as the atom's **electronegativity**. The range of electronegativities runs from 0.7 to 3.98, as Figure 15.22 shows. The greater an atom's electronegativity, the greater its ability to pull electrons toward itself when bonded. Thus, in hydrogen fluoride, fluorine has a greater electronegativity, or pulling power, than hydrogen.

Electronegativity is greatest for elements at the upper right of the periodic table and lowest for elements at the lower left. Noble gases are not considered in electronegativity discussions because, as previously mentioned, they rarely participate in chemical bonding.

When the two atoms in a covalent bond have the same electronegativity, no dipole is formed (as is the case with H_2) and the bond is classified as a **nonpolar** bond. When the electronegativities of the atoms differ, a dipole may form (as with HF) and the bond is classified as a **polar** bond. Just how polar a bond is depends on the difference between the electronegativity values of the two atoms—the greater the difference, the more polar the bond.

H 2.2																	He –
Li 0.98	Be 1.57											B 2.04	C 2.55	N 3.04	O 3.44	F 3.98	Ne –
Na 0.93	Mg 1.31											Al 1.61	Si 1.9	P 2.19	S 2.58	Cl 3.16	Ar –
K 0.82	Ca 1.0	Sc 1.36	Ti 1.54	V 1.63	Cr 1.66	Mn 1.55	Fe 1.83	Co 1.88	Ni 1.91	Cu 1.90	Zn 1.65	Ga 1.81	Ge 2.01	As 2.18	Se 2.55	Br 2.96	Kr
Rb 0.82	Sr 0.95	Y 1.22	Zr 1.33	Nb 1.6	Mo 2.16	Tc 1.9	Ru 2.2	Rh 2.28	Pd 2.20	Ag 1.93	Cd 1.69	In 1.78	Sn 1.96	Sb 2.05	Te 2.1	I 2.66	Xe –
Cs 0.79	Ba 0.89	La 1.10	Hf 1.3	Ta 1.5	W 2.36	Re 1.9	Os 2.2	Ir 2.8	Pt 2.2	Au 2.54	Hg 2.00	Tl 2.04	Pb 2.33	Bi 2.02	Po 2.0	At 2.2	Rn –
Fr 0.7	Ra 0.9	Ac 1.1	Rf –	Db –	Sg –	Bh –	Hs –	Mt –	Uun –	Uuu –	Uub –						

FIGURE 15.22

The experimentally measured electronegativities of elements.

As can be seen in Figure 15.22, the greater the distance between two atoms in the periodic table, the greater the difference in their electronegativities, and hence the greater the polarity of the bond between them. So a chemist can predict which bonds are more polar than others without reading the electronegativities. Bond polarity can be inferred by looking at the relative positions of the atoms in the periodic table—the farther apart they are, especially when one is at the lower left and one is at the upper right, the greater the polarity of the bond between them.

CHECK POINT

List these bonds in order of increasing polarity: P—F, S—F, Ga—F, Ge—F (F, fluorine, atomic number 9; P, phosphorus, atomic number 15; S, sulfur, atomic number 16; Ga, gallium, atomic number 31; Ge, germanium, atomic number 32):

(least polar) _____ , _____ , _____ , _____ (most polar)

Was this your answer?

If you answered the question, or attempted to, before reading this answer, hooray for you! You're doing more than reading the text—you're learning physical science. The greater the difference in electronegativities between two bonded atoms, the greater the polarity of the bond, and so the order of increasing polarity is S—F < P—F < Ge—F < Ga—F.

Note that this answer can be obtained by looking only at the relative positions of these elements in the periodic table rather than by calculating the differences in their electronegativities.

The magnitude of bond polarity is sometimes indicated by the size of the crossed arrow or the $\delta-$ and $\delta+$ symbols used to depict a dipole, as shown in Figure 15.23.

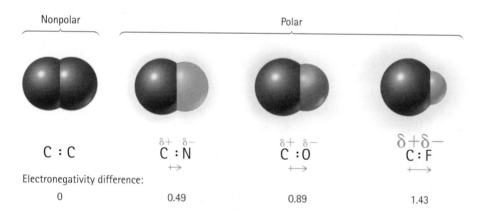

Nonpolar   Polar

C : C

$\overset{\delta+ \;\; \delta-}{C : N}$

$\overset{\delta+ \;\; \delta-}{C : O}$

$\overset{\delta+\delta-}{C : F}$

Electronegativity difference:

0   0.49   0.89   1.43

FIGURE 15.23

These bonds are in order of increasing polarity from left to right, a trend indicated by the larger and larger crossed arrows and $\delta-/\delta+$ symbols. Which of these pairs of elements are farthest apart in the periodic table?

Note that the electronegativity difference between atoms in an ionic bond can also be calculated. For example, the bond in NaCl has an electronegativity difference of 2.23, far greater than the difference of 1.43 shown for the C—F bond in Figure 15.23.

What is important to understand here is that there is no black-and-white distinction between ionic and covalent bonds. Rather, there is a gradual change from one to the other as the atoms that bond are located farther apart in the periodic table. This continuum is illustrated in Figure 15.24. Atoms on opposite sides of the periodic table have great differences in electronegativity, and hence the bonds between them are highly polar—in other words, ionic. Nonmetallic atoms of the same type have the same electronegativities, and so their bonds are nonpolar covalent. The polar covalent bond with its uneven sharing of electrons and slightly charged atoms is between these two extremes.

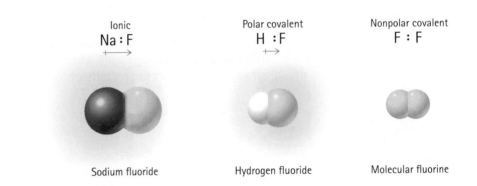

Ionic
Na : F

Polar covalent
H : F

Nonpolar covalent
F : F

Sodium fluoride

Hydrogen fluoride

Molecular fluorine

FIGURE 15.24

The ionic bond and the non-polar covalent bond represent the two extremes of chemical bonding. The ionic bond involves a transfer of one or more electrons, and the non-polar covalent bond involves the equitable sharing of electrons. The character of a polar covalent bond falls between these two extremes.

15.7 Molecular Polarity

When all the bonds in a molecule are nonpolar, the molecule as a whole is also nonpolar—as is the case with H_2, O_2, and N_2. When a molecule consists of only two atoms and the bond between them is polar, the polarity of the molecule is the same as the polarity of the bond—as with HF, HCl, and ClF.

Complexities arise when assessing the polarity of a molecule containing more than two atoms. Consider carbon dioxide, CO_2, shown in Figure 15.25. The cause of the dipole in either one of the carbon–oxygen bonds is oxygen's greater pull on the bonding electrons (because oxygen is more electronegative than carbon). At the same time, however, the oxygen atom on the opposite side of the carbon pulls those electrons back to the carbon. The net result is an even distribution of bonding electrons around the entire molecule. So dipoles that are of equal strength but pull in opposite directions in a molecule effectively cancel each other, with the result that the molecule as a whole is nonpolar.

Figure 15.26 illustrates a similar situation for boron trifluoride, BF_3, in which three fluorine atoms are oriented 120° from one another around a central boron atom. Because the angles are all the same, and because each fluorine atom pulls on the electrons of its boron–fluorine bond with the same force, the resulting polarity of this molecule is zero.

Nonpolar molecules have only relatively weak attractions to other nonpolar molecules. The covalent bonds in a carbon dioxide molecule, for example, are

The Physical Science Place
Bonds and Bond Polarity

O=C=O

FIGURE 15.25

There is no net dipole in a carbon dioxide molecule, and so the molecule is nonpolar. This is analogous to two people in a tug-of-war. As long as they pull with equal forces but in opposite directions, the rope remains stationary.

FIGURE 15.26

The three dipoles of a boron trifluoride molecule oppose one another at 120° angles, which makes the overall molecule nonpolar. This is analogous to three people pulling with equal force on ropes attached to a central ring. As long as they all pull with equal force and all maintain the 120° angles, the ring remains stationary.

Gaseous N₂

Nonpolar molecule

Relatively weak attraction

Nitrogen at −196°C

Liquid N₂

FIGURE 15.27

Nitrogen is a liquid at temperatures below its chilly boiling point of −196°C. Nitrogen molecules are not very attracted to one another because they are nonpolar. As a result, the small amount of heat energy available at −196°C is enough to separate them and allow them to enter the gaseous phase.

many times stronger than any forces of attraction that might occur between two adjacent carbon dioxide molecules. This lack of attraction between nonpolar molecules explains the low boiling points of many nonpolar substances. Recall from Section 14.2 that boiling is a process wherein the molecules of a liquid separate from one another as they go into the gaseous phase. When only weak attractions exist between the molecules of a liquid, less heat energy is required to liberate the molecules from one another and allow them to enter the gaseous phase. This translates into a relatively low boiling point for the liquid, as, for example, in molecular nitrogen, N_2, shown in Figure 15.27. The boiling points of hydrogen (H_2), oxygen (O_2), carbon dioxide (CO_2), and boron trifluoride, (BF_3), are also quite low for the same reason.

There are many instances in which the dipoles of different bonds in a molecule do not cancel each other. Reconsider the rope analogy of Figure 15.26. As long as everyone pulls equally, the ring stays put. Imagine, however, that one person begins to ease off on the rope. Now the pulls are no longer balanced, and the ring begins to move away from the person who is slacking off, as Figure 15.28 shows. Likewise, if one person began to pull harder, the ring would move away from the other two people.

A similar situation occurs in molecules in which polar covalent bonds are not equal and opposite. Perhaps the most relevant example is water, H_2O. Each

Link To Section 1.4

FIGURE 15.28

If one person eases off in a three-way tug-of-war but the other two continue to pull, the ring moves in the direction of the purple arrow.

(a)   (b)

FIGURE 15.29

(a) The individual dipoles in a water molecule add together to give a large overall dipole for the whole molecule, shown in purple. (b) The region around the oxygen atom is therefore slightly negative, and the region around the two hydrogens is slightly positive.

A water molecule is a natural dipole—a bit positive on one end and negative on the other. What's the net charge of a dipole?

hydrogen–oxygen covalent bond has a relatively large dipole because of the great electronegativity difference. Because of the bent shape of the molecule, however, the two dipoles, shown in blue in Figure 15.29, do not cancel each other the way the C—O dipoles in Figure 15.25 do. Instead, the dipoles in the water molecule work together to give an overall dipole, shown in purple, for the molecule.

CHECK POINT

Which of these molecules is polar and which is nonpolar?

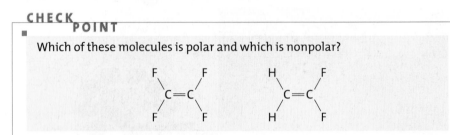

Was this your answer?

Symmetry is often the greatest clue for determining polarity. Because the molecule on the left is symmetrical, the dipoles on the two sides cancel each other. This molecule is therefore nonpolar:

Because the molecule on the right is less symmetrical (more "lopsided"), it is the polar molecule. Because carbon is more electronegative than hydrogen, the dipoles of the two hydrogen–carbon bonds point toward the carbon. Because fluorine is more electronegative than carbon, the dipoles of the carbon–fluorine bonds point toward the fluorines. Because the general direction of all dipole arrows is toward the fluorines, so is the average distribution of the bonding electrons. The fluorine side of the molecule is therefore slightly negative, and the hydrogen side is slightly positive.

Figure 15.30 illustrates how polar molecules electrically attract one another and, as a result, are relatively difficult to separate. In other words, polar molecules can be thought of as being "sticky," which is why it takes more energy to

FIGURE 15.30

Water molecules attract one another because each contains a slightly positive side and a slightly negative side. The molecules position themselves such that the positive side of one faces the negative side of a neighbor.

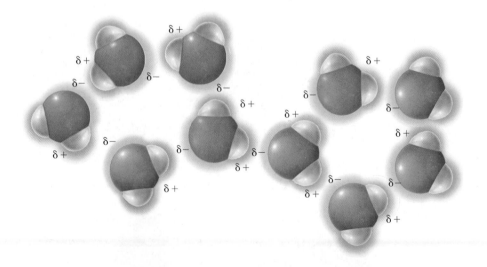

TABLE 15.2 BOILING POINTS OF SOME POLAR AND NONPOLAR SUBSTANCES

Substance	Boiling point (°C)
Polar	
Hydrogen fluoride, HF	20
Water, H_2O	100
Ammonia, NH_3	−33
Nonpolar	
Hydrogen, H_2	−253
Oxygen, O_2	−183
Nitrogen, N_3	−196
Boron trifluoride, BF_3	−100
Carbon dioxide, CO_2	−79

separate them—to change phase. For this reason, substances composed of polar molecules typically have higher boiling points than substances composed of nonpolar molecules, as Table 15.2 shows.

Water boils at 100°C, whereas carbon dioxide boils at −79°C. This 179°C difference is quite dramatic when you consider that a carbon dioxide molecule is more than twice as massive as a water molecule.

Because molecular "stickiness" can play a lead role in determining a substance's macroscopic properties, molecular polarity is a central concept of chemistry. Figure 15.31 describes an interesting example.

FIGURE 15.31

Oil and water are difficult to mix, as is evident from this oil spill off the coast of Spain in 2002. It's not, however, that oil and water repel each other. Rather, water molecules are so attracted to themselves because of their polarity that they pull themselves together. The nonpolar oil molecules are thus excluded and left to themselves. Being less dense than water, oil floats on the surface, where it poses great danger to birds and other wildlife.

■ 15.8 Molecular Attractions

So far you have learned that the atoms of a molecule are held together by covalent bonds. Furthermore, the molecule, behaving as a fundamental unit, may have electrical attractions with neighboring molecules. As discussed

The Physical Science Place
Polar Attraction

In the first part of this chapter, we talked about how molecules form. Now we see how molecules mix together.

in the previous section, the greater the polarity of the molecule, the greater its attraction to neighboring molecules. This explains how water has such a high boiling point—the water molecules, being quite polar, are so attracted to one another that a lot of energy is required to separate them from one another into the gaseous phase. In this section, we explore further how the physical properties of a material, such as boiling point, can be deduced from the polarity of its molecules. In addition to discussing the attractions among the molecules within a single substance, we'll explore attractions that occur between the fundamental units of different substances, such as water and salt.

As shown in Table 5.3, there are four types of electrical attractions involving molecules. The strength of even the strongest of these attractions is much weaker than any chemical bond. The attraction between two adjacent water molecules, for example, is only about $\frac{1}{20}$ as strong as the chemical bonds holding the hydrogen and oxygen atoms together in the water molecule. Although molecule-to-molecule attractions are relatively weak, their effects on the physical properties of substances are most significant.

TABLE 15.3 ELECTRICAL ATTRACTIONS BETWEEN A MOLECULE AND ITS NEIGHBOR

Attraction	Relative Strength
Ion–dipole	Strongest
Dipole–dipole	
Dipole–induced dipole	
Induced dipole–induced dipole	Weakest

Link To Section 8.2

IONS AND DIPOLES

Recall from Section 15.7 that a *polar* molecule is one in which the bonding electrons are unevenly distributed. One side of the molecule carries a slight negative charge, and the opposite side carries a slight positive charge. This separation of charge makes up a *dipole*.

So what happens to polar molecules, such as water molecules, when they are near an ionic compound, such as sodium chloride? The opposite charges electrically attract one another. A positive sodium ion attracts the negative side of a water molecule, and a negative chloride ion attracts the positive side of a water molecule. This phenomenon is illustrated in Figure 15.32. Such an attraction between an ion and the dipole of a polar molecule is called an *ion–dipole* attraction.

Ion–dipole attractions are much weaker than ionic bonds. However, a large number of ion–dipole attractions can act collectively to disrupt ionic bonds. This is what happens to sodium chloride in water. Attractions exerted by the water molecules break the ionic bonds and pull the ions away from one another. The result, represented in Figure 15.33, is a solution of sodium chloride in water. (A solution in water is called an *aqueous solution*.)

An attraction between two polar molecules is called a *dipole–dipole* attraction. An unusually strong dipole–dipole attraction is the **hydrogen bond**. This attraction occurs between molecules that have a hydrogen atom covalently bonded to a

FIGURE 15.32

Electrical attractions are shown as a series of overlapping arcs. The blue arcs indicate negative charge, and the red arcs indicate positive charge.

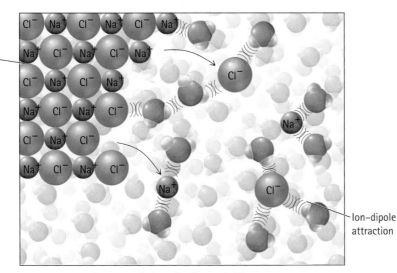

Sodium chloride crystal

Ion–dipole attraction

Aqueous solution of sodium chloride

FIGURE 15.33

Sodium and chloride ions tightly bound in a crystal lattice are separated from one another by the collective attraction exerted by many water molecules to form an aqueous solution of sodium chloride.

highly electronegative atom, usually nitrogen, oxygen, or fluorine. Recall from Section 15.6 that the electronegativity of an atom describes how well that atom is able to pull bonding electrons toward itself. The greater the atom's electronegativity, the better it is able to gain electrons and thus the more negative is its charge.

Look at Figure 15.34 to see how hydrogen bonding works. The hydrogen side of a polar molecule (water, in this example) has a positive charge because the more electronegative oxygen atom pulls more strongly on the electrons of the covalent bond. The hydrogen is therefore electrically attracted to a pair of nonbonding electrons on the negatively charged atom of another molecule (in this case, another water molecule). This mutual attraction between hydrogen and the negatively charged atom of another molecule is a hydrogen bond.

Even though the hydrogen bond is much weaker than any covalent or ionic bond, the effects of hydrogen bonding can be very pronounced. For example, water owes many of its properties to hydrogen bonds. The hydrogen bond is also of great importance in the chemistry of the large molecules, such as DNA and proteins, that are found in living organisms.

Dipole–dipole attraction (hydrogen bond)

FIGURE 15.34

The dipole–dipole attraction between two water molecules is a hydrogen bond because it involves hydrogen atoms bonded to highly electronegative oxygen atoms.

INDUCED DIPOLES

In many molecules, the electrons are distributed evenly, and so there is no dipole. The oxygen molecule, O_2, is an example. Such a nonpolar molecule can be induced to become a temporary dipole, however, when it is brought close to a water molecule (or to any other polar molecule), as Figure 15.35 illustrates. The slightly negative side of the water molecule pushes the electrons in the oxygen molecule away. Thus, the oxygen molecule's electrons are pushed to the side that is farthest from the water molecule. The result is a temporarily uneven distribution of electrons called an **induced dipole**. The resulting attraction between the permanent dipole (water) and the induced dipole (oxygen) is a *dipole–induced dipole* attraction.

Dipole – induced dipole attraction

Isolated oxygen molecule (nonpolar)

(a)

Induced dipole (oxygen molecule)

(b)

Permanent dipole (water molecule)

FIGURE 15.35

INTERACTIVE FIGURE

(a) An isolated oxygen molecule has no dipole; its electrons are distributed evenly. (b) An adjacent water molecule induces a redistribution of electrons in the oxygen molecule. (The slightly negative side of the oxygen molecule is shown larger than the slightly positive side because the slightly negative side contains more electrons.)

FIGURE 15.36

The electrical attraction between water and oxygen molecules is relatively weak, which explains why not much oxygen is able to dissolve in water. For example, water fully aerated at room temperature contains only about 1 oxygen molecule for every 200,000 water molecules. The gills of a fish, therefore, must be highly efficient at extracting molecular oxygen from water.

FIGURE 15.37

Temporary dipoles induced in the normally nonpolar molecules in plastic wrap make it stick to glass.

Nonpolar argon | Temporary dipole in argon

FIGURE 15.38

The electron distribution in an atom is normally even. At any given moment, however, the electron distribution may be somewhat uneven, resulting in a temporary dipole.

CHECK POINT

How does the electron distribution in an oxygen molecule change when the hydrogen side of a water molecule is nearby?

Was this your answer?
Because the hydrogen side of the water molecule is slightly positive, the electrons in the oxygen molecule are pulled toward the water molecule, inducing in the oxygen molecule a temporary dipole in which the larger side is nearest the water molecule (rather than as far away as possible, as it was in Figure 15.35).

Remember, induced dipoles are only temporary. If the water molecule in Figure 15.35b were removed, the oxygen molecule would return to its normal, nonpolar state. As a consequence, dipole–induced dipole attractions are weaker than dipole–dipole attractions. But dipole–induced dipole attractions are strong enough to hold relatively small quantities of oxygen dissolved in water, as depicted in Figure 15.36. This attraction between water and molecular oxygen is vital for fish and other forms of aquatic life that rely on molecular oxygen dissolved in water.

Dipole–induced dipole attractions are also responsible for holding plastic wrap to glass, as shown in Figure 15.37. These wraps are made of very long nonpolar molecules that are induced to have dipoles when placed in contact with glass, which is highly polar. As we discuss next, the molecules of a nonpolar material, such as plastic wrap, can also induce dipoles among themselves. This explains why plastic wrap sticks not only to polar materials such as glass but also to itself.

CHECK POINT

Distinguish between a dipole–dipole attraction and a dipole–induced dipole attraction.

Was this your answer?
The dipole–dipole attraction is stronger and involves two permanent dipoles. The dipole–induced dipole attraction is weaker and involves a permanent dipole and a temporary one.

Individual atoms and nonpolar molecules, on average, have a fairly even distribution of electrons. Because of the randomness of electron motion, however, at any given moment the electrons in an atom or a nonpolar molecule may be bunched to one side. The result is a temporary dipole, as shown in Figure 15.38.

Just as the permanent dipole of a polar molecule can induce a dipole in a nonpolar molecule, a temporary dipole can do the same thing. This gives rise to the relatively weak *induced dipole–induced dipole* attraction, illustrated in Figure 15.39.

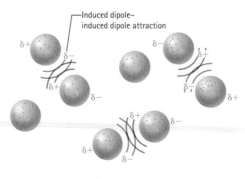

FIGURE 15.39

Because the normally even distribution of electrons in atoms can momentarily become uneven, atoms can be attracted to one another through induced dipole–induced dipole attractions.

Induced dipole–induced dipole attraction

Iodine molecule

$\delta+$ $\delta-$

$\delta+$

$\delta+$ $\delta-$

$\delta-$

$\delta+$

Electrons can bunch to one side without experiencing strong electrical repulsions among themselves.

(a) Iodine, I_2, a solid at room temperature

Fluorine molecules

Negligible intermolecular attractions

Electrons are confined to a small space; therefore they repel one another and remain fairly evenly distributed.

(b) Fluorine, F_2, a gas at room temperature

FIGURE 15.40

(a) Temporary dipoles form more readily in larger atoms, such as those in an iodine molecule, because in larger atoms, electrons bunched to one side are still relatively far apart from one another and not so repelled by the electric force. (b) In smaller atoms, such as those in a fluorine molecule, electrons cannot bunch to one side as well because the repulsive electric force increases as the electrons bunch closer.

Electrons repel electrons, which means they resist bunching together to one side of the atom. In a large atom, however, the electrons find it fairly easy to do just that. By analogy, consider a cruise ship with only 10 passengers. Because the ship is so large, these 10 passengers could easily congregate to one side. On a much smaller life raft, however, the same 10 passengers would find it necessary to space themselves as evenly apart from each other as possible, lest they tip over. In a similar fashion, larger atoms can form temporary dipoles much more easily than smaller atoms, as is illustrated in Figure 15.40. So larger atoms—and molecules made of larger atoms—have the strongest induced dipole–induced dipole attractions. In other words, they are more "sticky." Iodine, I_2, for example, is stickier than fluorine, F_2, which explains why iodine is a solid at room temperature while fluorine is a gas, even though they are both nonpolar materials.

Fluorine is one of the smallest atoms, and nonpolar molecules made with fluorine atoms exhibit only very weak induced dipole–induced dipole attractions. This is the principle behind the Teflon nonstick surface. The Teflon molecule, part of which is shown in Figure 15.41, is a long chain of carbon atoms chemically bonded to fluorine atoms, and the fluorine atoms exert essentially no attractions on any material in contact with the Teflon surface—scrambled eggs in a frying pan, for instance.

F   F   F   F   F   F
| | | | | |
··· —C — C — C — C — C — C— ···
| | | | | |
F   F   F   F   F   F

FIGURE 15.41

Few things stick to Teflon because of the high proportion of fluorine atoms that it contains. The structure depicted here is only a portion of the full length of the molecule.

FIGURE 15.42

(a) Two nonpolar methane molecules are attracted to each other by induced dipole–induced dipole attractions, but there is only one attraction per molecule. (b) Two nonpolar octane molecules are similar to methane, but they are longer. The number of induced dipole–induced dipole attractions between these two molecules is therefore greater.

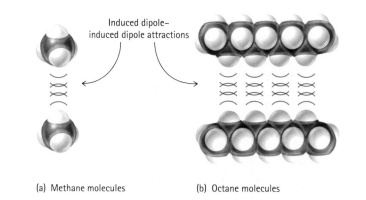

Induced dipole–induced dipole attractions

(a) Methane molecules   (b) Octane molecules

CHECK POINT

What is the distinction between a dipole–induced dipole attraction and an induced dipole–induced dipole attraction?

Was this your answer?
The dipole–induced dipole attraction is stronger and involves a permanent dipole and a temporary one. The induced dipole–induced dipole attraction is weaker and involves two temporary dipoles.

Induced dipole–induced dipole attractions help explain why natural gas is a gas at room temperature but gasoline is a liquid. The major component of natural gas is methane, CH_4, and one of the major components of gasoline is octane, C_8H_{18}. We can see in Figure 15.42 that the number of induced dipole–induced dipole attractions between two methane molecules is appreciably less than the number between two octane molecules. You know that two small pieces of Velcro are easier to pull apart than two long pieces. Like short pieces of Velcro, methane molecules can be pulled apart with little effort. That's why methane has a low boiling point, −161°C, and is a gas at room temperature. Octane molecules, like long strips of Velcro, are relatively difficult to pull apart because of the larger number of induced dipole–induced dipole attractions. The boiling point of octane, 125°C, is therefore much higher than that of methane, and octane is a liquid at room temperature. (The greater mass of octane also plays a role in making its boiling point higher.)

Induced dipole–induced dipole attractions, also known as *dispersion forces*, also explain how the gecko can race up a glass wall and support its entire body weight with only a single toe. A gecko's feet are covered with billions of microscopic hairs called *spatulae*, each of which is about $\frac{1}{300}$ as thick as a human hair. The force of attraction between these hairs and the wall is the weak induced dipole–induced dipole attraction. But because there are so many hairs, the surface area of contact is relatively great, and hence the total force of attraction is enough to prevent the gecko from falling (Figure 15.43). Research is currently underway to develop a synthetic dry glue based on gecko adhesion. Velcro, watch out!

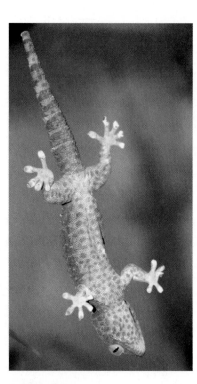

FIGURE 15.43

If the gecko's foot is so sticky, how does the gecko keep its feet clean? Answer: the gecko's foot is extremely nonpolar. Dirt may stick to it briefly, but after a few steps, the dirt sticks better to the surface on which the gecko walks. Of course, there is at least one surface a gecko finds very difficult to climb—Teflon.

CHECK POINT

Methanol, CH_3OH, which can be used as a fuel, is not much larger than methane, CH_4, but it is a liquid at room temperature. Suggest why.

Was this your answer?
The polar oxygen–hydrogen covalent bond in each methanol molecule leads to hydrogen bonding between molecules. These relatively strong interparticle attractions hold methanol molecules together as a liquid at room temperature.

SUMMARY OF TERMS

Valence shell The outermost occupied shell of an atom.

Electron-dot structure A shorthand notation of the shell model of the atom, in which valence electrons are shown around an atomic symbol.

Nonbonding pairs Two paired valence electrons that tend not to participate in a chemical bond.

Ion An electrically charged particle created when an atom either loses or gains one or more electrons.

Polyatomic ion An ionically charged molecule.

Ionic bond A chemical bond in which an attractive electric force holds ions of opposite charge together.

Ionic compound Any chemical compound containing ions.

Metallic bond A chemical bond in which positively charged metal ions are held together within a "fluid."

Alloy A mixture of two or more metallic elements.

Covalent bond A chemical bond in which atoms are held together by their mutual attraction for two or more electrons they share.

Covalent compound An element or chemical compound in which atoms are held together by covalent bonds.

Molecule A group of atoms held tightly together by covalent bonds.

Dipole A separation of charge that occurs in a chemical bond because of differences in the electronegativities of the bonded atoms.

Electronegativity The ability of an atom to attract a bonding pair of electrons to itself when bonded to another atom.

Nonpolar The property of a chemical bond that has no dipole.

Polar The property of a chemical bond that has a dipole.

Hydrogen bond A strong dipole–dipole attraction between a slightly positive hydrogen atom on one molecule and a pair of nonbonding electrons on another molecule.

Induced dipole A dipole temporarily created in an otherwise nonpolar molecule, induced by a neighboring charge.

REVIEW QUESTIONS

15.1 Electron-Dot Structures

1. How many electrons can occupy the first shell? How many can occupy the second shell?
2. Which electrons are represented by an electron-dot structure?
3. How do the electron-dot structures of elements in the same group in the periodic table compare with one another?
4. How many nonbonding pairs are in the valence shell of an oxygen atom? How many unpaired valence electrons?

15.2 The Formation of Ions

5. How does an ion differ from an atom?
6. To become a negative ion, does an atom lose or gain electrons?
7. Why does the fluorine atom tend to gain only one electron?
8. What do molecules lose or gain to become polyatomic ions?

15.3 Ionic Bonds

9. Which elements tend to form ionic bonds?
10. What is the electric charge on the calcium ion in calcium chloride, $CaCl_2$?
11. Suppose an oxygen atom gains two electrons to become an oxygen ion. What is its electric charge?
12. What is an ionic crystal?

15.4 Metallic Bonds

13. Do metals more readily gain or lose electrons?
14. What is an alloy?
15. What is a native metal?

15.5 Covalent Bonds

16. Which elements tend to form covalent bonds?
17. How many electrons are shared in a double covalent bond?
18. How many covalent bonds can an oxygen atom form?

15.6 Polar Covalent Bonds

19. What is a dipole?
20. Which element in the periodic table has the greatest electronegativity? Which has the least electronegativity?
21. Which is more polar: a carbon–oxygen bond or a carbon–nitrogen bond?

15.7 Molecular Polarity

22. How can a molecule be nonpolar when it consists of atoms that have different electronegativities?
23. Why do nonpolar substances boil at relatively low temperatures?
24. Which has a greater degree of symmetry: a polar molecule or a nonpolar molecule?
25. Why don't oil and water mix?
26. Which would you describe as "stickier": a polar molecule or a nonpolar one?

15.8 Molecular Attractions

27. What is the primary difference between a chemical bond and an attraction between two molecules?
28. Which is stronger, the ion–dipole attraction or the induced dipole–induced dipole attraction?
29. What is a hydrogen bond?
30. Are induced dipoles permanent?

EXERCISES

1. ● How is the number of unpaired valence electrons in an atom related to the number of bonds that the atom can form?

2. ● Why does the fluorine atom tend to gain only one electron?

3. ■ An atom loses an electron to another atom. Is this an example of a physical or chemical change?

4. ● Magnesium ions carry a 2+ charge, and chloride ions carry a 1− charge. What is the chemical formula for the ionic compound magnesium chloride?

5. ● Barium ions carry a 2+ charge, and nitrogen ions carry a 3− charge. What is the chemical formula for the ionic compound barium nitride?

6. ● Take money away from your bank account and the bank shows a negative credit. Take an electron away from an atom, however, and the atom shows up positive. Explain.

7. ● Sulfuric acid, H_2SO_4, loses two protons to form what polyatomic ion?

8. ● What molecule loses a proton to form the hydroxide ion, OH^-?

9. ■ Which should be larger, the potassium atom, K, or the potassium ion, K^+?

10. ◆ Which should have a higher melting point, sodium chloride, NaCl, or aluminum oxide, Al_2O_3?

11. ◆ Two fluorine atoms join together to form a covalent bond. Why don't two potassium atoms do the same thing?

12. ■ What element is well suited to forming either ionic or covalent bonds?

13. ■ What happens when hydrogen's electron gets close to the valence shell of a fluorine atom?

14. ■ Is there an abrupt or gradual change between ionic and covalent bonds? Explain.

15. ■ Classify the following bonds as ionic, covalent, or neither (O, atomic number 8; F, atomic number 9; Na, atomic number 11; Cl, atomic number 17; U, atomic number 92).
    (a) O with F _____
    (b) Ca with Cl _____
    (c) Na with Na _____
    (d) U with Cl _____

16. ■ Atoms of nonmetallic elements form covalent bonds, but they can also form ionic bonds. How is this possible?

17. ■ Atoms of metallic elements can form ionic bonds, but they are not very good at forming covalent bonds. Why?

18. ■ Phosphine is a covalent compound of phosphorus, P, and hydrogen, H. What is its chemical formula?

19. ■ What is the source of an atom's electronegativity?

20. ● Which bond is most polar: H—N, N—C, C—O, C—C, O—H, or C—H?

21. ■ Which molecule is most polar: S=C=S, O=C=O, or O=C=S?

22. ● In each molecule, which atom carries the greater positive charge:
    (a) H—Cl
    (b) Br—F

23. (c) C≡O
    (d) Br—Br

23. ■ List these bonds in order of increasing polarity: N—N, N—F, N—O, H—F.
    _____ < _____ < _____ < _____
    (least polar)   (most polar)

24. ◆ True or False? The greater the nuclear charge of an atom, the greater the electronegativity. Explain.

25. ◆ True or False? The more shells in an atom, the lower its electronegativity. Explain.

26. ● Why don't oil and water mix?

27. ■ Water, H_2O, and methane, CH_4, have about the same mass and differ by only one type of atom. Why is the boiling point of water so much higher than that of methane?

28. ● An individual carbon–oxygen bond is polar. Yet carbon dioxide, CO_2, which has two carbon–oxygen bonds, is nonpolar. Why?

29. ■ Three kids sitting equally apart around a table are sharing jelly beans. One of the kids, however, tends only to take jelly beans and only rarely gives one away. If each jelly bean represents an electron, who ends up being slightly negative? Who ends up being slightly positive? Is the negative kid just as negative as one of the positive kids is positive? Would you describe this as a polar or nonpolar situation? How about if all three kids were equally greedy?

30. ■ Which is stronger: the covalent bond that holds atoms together within a molecule or the electrical attraction between two neighboring molecules?

31. ■ Why is a water molecule more attracted to a calcium ion than a sodium ion?

32. ■ The charges in sodium chloride are all balanced—for every positive sodium ion there is a corresponding negative chloride ion. Since its charges are balanced, how can sodium chloride be attracted to water, and vice versa?

33. ● Is a water molecule more attracted to another water molecule or to a sodium ion? Explain.

34. ◆ Why are there no strong hydrogen bonds in liquid hydrogen sulfide, H_2S?

35. ■ Why are ion–dipole attractions stronger than dipole–dipole attractions?

36. ■ Chlorine, Cl_2, is a gas at room temperature, but bromine, Br_2, is a liquid. Why?

37. ● How are oxygen molecules attracted to water molecules?

38. ■ List the following compounds in order of increasing boiling point: CI_4, CBr_4, CCl_4, CF_4.

39. ■ Which should have a greater solubility in water: NH_3 or NCl_3? Why?

40. ● Why is the surface area of a gecko's foot so extensive?

ACTIVE EXPLORATIONS

Up Close with Crystals

View crystals of table salt with a magnifying glass or, better yet, a microscope if one is available. If you have a microscope, crush the crystals with a spoon and examine the resulting powder. Purchase some sodium-free salt, which is potassium chloride, KCl, and examine these ionic crystals, both intact and crushed. Sodium chloride and potassium chloride both form cubic crystals, but there are significant differences. What are they?

Gumdrop Molecules

Use toothpicks and gumdrops or jelly beans of different colors to build models of the molecules shown in Figures 15.17

through 15.19 and Figure 15.21, letting the different colors represent different elements.

Once you have become proficient at building these models, test your expertise by building models for difluoromethane, CH_2F_2; ethane, C_2H_6; hydrogen peroxide, H_2O_2; and acetylene, C_2H_2. Keep in mind that each carbon atom must have four covalent bonds, each oxygen must have two, and each fluorine and hydrogen must have only one. Hint: one of these molecules has a triple bond.

READINESS ASSURANCE TEST (RAT)

If you have a good handle on this chapter, if you really do, then you should be able to score 7 out of 10 on this RAT. If you score less than 7, you need to study further before moving on.

Choose the BEST answer to each of the following.

1. An atom loses an electron to another atom. Is this an example of a physical or chemical change?
   (a) a chemical change involving the formation of ions
   (b) a physical change involving the formation of ions
   (c) a chemical change involving the formation of covalent bonds
   (d) a physical change involving the formation of covalent bonds

2. Aluminum ions carry a 3+ charge, and chloride ions carry a 1− charge. What would be the chemical formula for the ionic compound aluminum chloride?
   (a) Al_3Cl
   (b) $AlCl_3$
   (c) Al_3Cl_3
   (d) $AlCl$

3. Which would you expect to have a higher melting point: sodium chloride, NaCl, or aluminum oxide, Al_2O_3?
   (a) The aluminum oxide has a higher melting point because it is a larger molecule and has a greater number of molecular interactions.
   (b) NaCl has a higher melting point because it is a solid at room temperature.
   (c) The aluminum oxide has a higher melting point because of the greater charges of the ions, and hence the greater force of attractions between them.
   (d) The aluminum oxide has a higher melting point because of the covalent bonds within the molecule.

4. Atoms of metallic elements can form ionic bonds, but they are not very good at forming covalent bonds. Why?
   (a) These atoms are too large to be able to come in close contact with other atoms.
   (b) They have a great tendency to lose electrons.
   (c) Their valence shells are already filled with electrons.
   (d) They are on the wrong side of the periodic table.

5. In terms of the periodic table, is there an abrupt or gradual change between ionic and covalent bonds?
   (a) An abrupt change occurs across the metalloids.
   (b) Any element of the periodic table can form a covalent bond.
   (c) There is a gradual change: the farther apart, the more ionic.
   (d) Whether an element forms one or the other depends on nuclear charge and not the relative positions in the periodic table.

6. A hydrogen atom does not form more than one covalent bond because it
   (a) has only one shell of electrons.
   (b) has only one electron to share.
   (c) loses its valence electron so readily.
   (d) has such a strong electronegativity.

7. When nitrogen and fluorine combine to form a molecule, the most likely chemical formula is
   (a) N_3F.
   (b) N_2F.
   (c) NF_4.
   (d) NF.
   (e) NF_3.

8. A substance consisting of which molecule shown below should have a higher boiling point?

$$S{=}C{=}O \qquad O{=}C{=}O$$

(a) the molecule on the left, SCO, because it comes later in the periodic table

(b) the molecule on the left, SCO, because it has less symmetry

(c) the molecule on the right, OCO, because it has more symmetry

(d) the molecule on the right, OCO, because it has more mass

9. Why are ion–dipole attractions stronger than dipole–dipole attractions?

(a) The chemical bond in an ion–dipole molecule is similar to a covalent bond.

(b) Like charge (dipole) does not attract like charge (another dipole).

(c) Dipole areas are subject to changing from positive to negative regions on the molecule.

(d) The magnitude of the electric charge associated with an ion is much greater.

10. Chlorine, Cl_2, is a gas at room temperature, but bromine, Br_2, is a liquid. Explain.

(a) Chlorine atoms are larger and this makes the formation of induced dipole–induced dipole attractions more favorable.

(b) Bromine atoms are larger and this makes the formation of induced dipole–induced dipole attractions more favorable.

(c) The smaller chlorine molecules are able to pack together in a tighter physical orientation.

(d) The bromine ions are held together by ionic bonds.

Answers to RAT

1a, 2b, 3c, 4b, 5c, 6b, 7e, 8b, 9d, 10c

EXPLORING FURTHER

http://www.ada.org/public/topics/fluoride/index.asp

This is the fluoride page of the American Dental Association, with many links to information regarding fluorides and fluoridation of drinking water and toothpastes. The ADA was one of the original water fluoridation supporters starting in the 1940s.

http://www.fluoridealert.org/50-reasons.htm

St. Lawrence University chemistry professor Paul Connett, Ph.D., provides 50 reasons to oppose fluoridation of public drinking water, complete with hyperlinks to peer-reviewed professional journal articles.

http://www.saltinstitute.org/idd.htm

Numerous reports in the literature demonstrate the effectiveness of iodized salt in controlling the medical condition called goiter. Check out this site for historical case studies that first pointed to this conclusion. An interesting project would be to compare the research efforts behind iodized salt and fluoridated drinking water.

http://www.soils.wisc.edu/virtual_museum

Home page of the Virtual Museum of Minerals and Molecules, curated by Phillip Barak of the University of Minnesota and Ed Nater of the University of Wisconsin. On this site, you will find molecular models that you can manipulate in three dimensions. To do so, your browser must be equipped with the Chime plug-in, which you may download for free after registering at http://www.mdl.com/my_account/register1.jsp.

CHAPTER 15 ONLINE RESOURCES

The Physical Science Place

Interactive Figures

- 15.18, 15.35

Tutorials

- Intermolecular Forces
- Covalent Bonds
- Bonds and Bond Polarity
- Polar Attraction

Videos

- Ionic Bonds
- Covalent Bonds

Quiz

Flashcards

Links

16

mixtures

16.1 MOST MATERIALS ARE MIXTURES

16.2 THE CHEMIST'S CLASSIFICATION OF MATTER

16.3 SOLUTIONS

16.4 SOLUBILITY

16.5 SOAPS AND DETERGENTS

16.6 SOFTENING HARD WATER

16.7 PURIFYING THE WATER WE DRINK

16.8 WASTEWATER TREATMENT

■ How can fresh water be prepared from seawater? Is it true that a fish can drown in water? When you stir sugar into water, the sugar crystals disappear, but where do they go? What are clouds made of, and what do they have in common with the blood that runs through our veins? When tap water is left boiling on the stove too long, it evaporates completely but leaves a chalky residue in the pot. What is this residue and where did it originate? How is municipal water treated so that it's safe enough to pipe into our homes for consumption? How does a wastewater treatment facility treat wastewater? The answers to these questions involve an understanding of mixtures.

FIGURE 16.1

Earth's atmosphere is a mixture of gaseous elements and compounds. Some of them are shown here.

The Physical Science Place

Mixtures Can Be Separated by Physical Means

FIGURE 16.2

Tap water provides us with H₂O as well as many other compounds, many of which are flavorful and help us grow, as Graham demonstrates at ages 7 and 21. Bottoms up!

■ 16.1 Most Materials Are Mixtures

A **mixture** is a combination of two or more substances in which each substance retains its own properties. Most materials we encounter are mixtures: mixtures of elements, mixtures of compounds, or mixtures of elements and compounds. Stainless steel, for example, is a mixture of the elements iron, chromium, nickel, and carbon. Seltzer water is a mixture of a liquid compound, water, and a gaseous compound, carbon dioxide. Our atmosphere, as Figure 16.1 illustrates, is a mixture of the elements nitrogen, oxygen, and argon, plus small amounts of such compounds as carbon dioxide and water vapor.

Tap water is a mixture containing mostly water but also many other compounds. Depending on your location, your water may contain compounds of calcium, magnesium, chlorine, fluorine, iron, and potassium; trace amounts of compounds of lead, mercury, and cadmium; organic compounds; and dissolved oxygen, nitrogen, and carbon dioxide. Although minimizing any toxic components in your drinking water is important, removing all other substances from it is unnecessary, undesirable, and impossible. Some of the dissolved solids and gases give water its characteristic taste, and many of them promote human health: fluoride compounds protect teeth, chlorine destroys harmful bacteria, and as much as 10% of our daily requirement of iron, potassium, calcium, and magnesium is obtained from drinking water (Figures 16.2 and 16.3).

CHECK POINT

So far, you have learned about three kinds of matter: elements, compounds, and mixtures. Which box below contains only an element? Which contains only a compound? Which contains a mixture?

Was this your answer?

The molecules in box A each contain two different types of atoms and so represent a compound. The molecules in box B each consist of the same atoms and so represent an element. Box C is a mixture of the compound and the element.

Note how the molecules of the compound and those of the element remain intact in the mixture. That is, when the mixture forms, atoms are not exchanged between the components.

There is a difference between the way substances—either elements or compounds—combine to form mixtures and the way elements combine to form compounds. Each substance in a mixture retains its chemical identity. The sugar molecules in the teaspoon of sugar in Figure 16.4, for example, are identical to the sugar molecules already in the tea. The only difference is that the sugar molecules in the tea are mixed with other substances, mostly water. The formation of a mixture, therefore, is a physical change. In contrast, as discussed in Section 14.5, there is a change in chemical identity when elements join to form compounds. Recall that sodium chloride is not a mixture of sodium and chlorine atoms but is instead a compound, which means it is entirely different from the elements that compose it. The formation of a compound is therefore a chemical change.

MIXTURES CAN BE SEPARATED BY PHYSICAL MEANS

The components of mixtures can be separated from one another by taking advantage of differences in the components' physical properties. A mixture of solids and liquids, for example, can be separated using filter paper through which the liquids pass but the solids do not. This is how coffee is often made: the caffeine and flavor molecules in the hot water pass through the filter and into the coffeepot, while the solid coffee grounds remain behind. This method of separating a solid-liquid mixture is called *filtration*, and it is a common technique used by chemists.

Mixtures can also be separated by taking advantage of a difference in boiling or melting points. Seawater is a mixture of water and a variety of compounds, mostly sodium chloride. Whereas water boils at 100°C, sodium chloride doesn't even melt until 800°C. One way to separate water from the mixture we call seawater, therefore, is to heat the seawater to about 100°C. At this temperature, the liquid water readily transforms to water vapor, but the sodium chloride stays behind dissolved in the remaining water. As the water vapor rises, it can be channeled into a cooler container, where it condenses into a liquid without the dissolved solids. This process of collecting a vaporized substance, called **distillation**,

- The toxicity of any substance is in the dose. Fresh water, for example, can be lethal if you drink too much of it. Why? Because too much would flush out dissolved ions that are absolutely essential for your health. Similarly, although small amounts of fluoride protect against tooth decay, larger amounts can cause your teeth to mottle. Worse still, excessive fluoride binds with calcium in your blood, forming lethal calcium fluoride crystals. Again for emphasis: the toxicity of any substance is in the dose!

FIGURE 16.3

Most of the oxygen in the air bubbles produced by an aquarium aerator escapes into the atmosphere. Some of the oxygen, however, mixes with the water. Fish depend on this oxygen to survive. Without this dissolved oxygen, which fish extract from the water with their gills, the fish would promptly drown. So fish don't "breathe" water. They breathe the O_2 that is dissolved in the water.

> Chemists have devised many ingenious ways of separating the components of a mixture. Most of these techniques employ the simple principle of separating the components by differences in their physical properties.

FIGURE 16.4

Table sugar is a compound consisting only of sucrose molecules. Once these molecules are mixed into hot tea, they become interspersed among the water and tea molecules and form a sugar–tea–water mixture. No new compounds are formed, and so this is an example of a physical change.

Symbol for sugar molecule, which is sucrose, $C_{12}H_{22}O_{11}$ Sugar   Sugar in water

FIGURE 16.5

(a) A simple distillation setup used to separate one component from a mixture. The mixture is boiled in the flask on the left. The rising vapors contain only the volatile components of the mixture. The vapors are channeled into a downward-slanting tube kept cool by cold water flowing across its outer surface. The vapors inside the cool tube condense and collect in the flask on the right. (b) A whiskey still functions by the same principle. A mixture containing alcohol is heated to the point where the alcohol, some flavoring molecules, and some water are vaporized. These vapors travel through the copper pipes, where they then condense to a liquid.

(a)   (b)

is illustrated in Figure 16.5. Distillation is a very effective, though costly, way of isolating fresh water from seawater. We explore this in further detail in Section 16.7. After all the water has been distilled from seawater, dry solids remain. These solids, also a mixture of compounds, contain a variety of commercially valuable compounds, such as sodium chloride and potassium bromide (Figure 16.6).

FIGURE 16.6

At the southern end of San Francisco Bay are areas where the seawater has been partitioned off by earthen dikes. These are evaporation ponds, where the water is allowed to evaporate, leaving behind the solids that were dissolved in the seawater. These solids are further refined for commercial sale. The remarkable color of the ponds results from suspended particles of iron oxide and other minerals, which are easily removed during refining.

■ 16.2 The Chemist's Classification of Matter

If a material is **pure**, it consists of only a single element or a single compound. Pure gold, for example, contains nothing but the element gold. Pure table salt contains nothing but the compound sodium chloride. If a material is **impure**, it is a mixture and contains two or more elements or compounds. This classification scheme is shown in Figure 16.7.

Because atoms and molecules are so small, it is impractical to prepare a sample that is truly pure—that is, truly 100% of a single material. For example, if just one atom or molecule out of a trillion trillion were different, then the 100% pure

FIGURE 16.7

A chemical classification of matter.

Orange juice may be 100% natural, but it is never 100% pure.

status would be lost. Samples, however, can be "purified" by various methods such as distillation. *Pure* is understood to be a relative term. When we compare the purity of two samples, the purer one contains fewer impurities. A sample of water that is 99.9% pure has a greater proportion of impurities than does a purer sample of water that is 99.9999% pure.

Sometimes naturally occurring mixtures are labeled as being pure, as in "pure orange juice." Such a statement merely means that nothing artificial has been added. According to a chemist's definition, however, orange juice is anything but pure, as it contains a wide variety of materials, including water, pulp, flavorings, vitamins, and sugars.

Mixtures may be heterogeneous or homogeneous. In a **heterogeneous mixture**, the different components can be seen as individual substances, such as pulp in orange juice, sand in water, or oil globules dispersed in vinegar. The different components are visible. **Homogeneous mixtures** have the same composition throughout. Any one region of the mixture has the same ratio of substances as any other region, and the components cannot be seen as individual identifiable entities. The distinction is shown in Figure 16.8.

A homogeneous mixture may be either a solution or a suspension. In a **solution**, all components are in the same phase. The atmosphere we breathe is a gaseous solution consisting of the gaseous elements nitrogen and oxygen as well as minor amounts of other gaseous materials. Salt water is a liquid solution because both the water and the dissolved sodium chloride are found in a single liquid phase. An example of a solid solution is white gold, which is a homogeneous mixture of the elements gold and palladium. We shall be discussing solutions in more detail in the next section.

FIGURE 16.8

(a) In heterogeneous mixtures, the different components can be seen with the naked eye.
(b) In homogeneous mixtures, the different components are mixed at a much finer level and so are not readily distinguished.

Granite

"Snow" in snow globe

Pizza

(a) Heterogeneous mixtures

Air

Clear seawater

White gold

(b) Homogeneous mixtures

The path of light becomes visible when the light passes through a suspension.

A **suspension** is a homogeneous mixture in which the different components are in different phases, such as solids in liquids or liquids in gases. In a suspension, the mixing is so thorough that the different phases cannot be readily distinguished. Milk is a suspension because it is a homogeneous mixture of proteins and fats finely dispersed in water. Blood is a suspension composed of finely dispersed blood cells in water. Another example of a suspension is clouds, which are homogeneous mixtures of tiny water droplets suspended in air. Shining a light through a suspension, as in Figure 16.9, results in a visible cone as the light is reflected by the suspended components.

The easiest way to distinguish a suspension from a solution in the laboratory is to spin a sample in a centrifuge. This device, spinning at thousands of revolutions per minute, separates the components of suspensions but not those of solutions, as Figure 16.10 shows.

CHECK POINT

Impure water can be purified by which of these?

(a) removing the impure water molecules
(b) removing everything that is not water
(c) breaking down the water to its simplest components
(d) adding some disinfectant such as chlorine

Was this your answer?

The answer is (b): impure water can be purified by removing everything that isn't water. H_2O is a compound made of the elements hydrogen and oxygen in a 2-to-1 ratio. Every H_2O molecule is exactly the same as every other, and there's no such thing as an impure H_2O molecule. Just about anything, including you, beach balls, rubber ducks, dust particles, and bacteria, can be found in water. When something other than water is found in water, we say that the water is impure. It is important to see that the impurities are in the water and not part of the water, which means that it is possible to remove them by a variety of physical means, such as filtration or distillation.

Blood, because it is a suspension, can be centrifuged into its components, which include the blood plasma (a yellowish solution) and white and red blood cells. The components of the plasma, however, cannot be separated from one another here because a centrifuge has no effect on solutions.

Blood (a suspension)   Centrifuge

Blood plasma (a solution)
White blood cells
Red blood cells

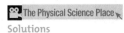
Solutions

16.3 Solutions

What happens when table sugar, known chemically as sucrose, is stirred into water? Is the sucrose destroyed? We know it isn't, because it sweetens the water. Does the sucrose disappear because it somehow ceases to occupy space or because it fits within the nooks and crannies of the water? Not so, because adding sucrose changes the volume. This may not be noticeable at first, but if you continue adding sucrose to a glass of water, you'll see that the water level rises, just as it would if you were adding sand.

Sucrose stirred into water loses its crystalline form. Each sucrose crystal consists of billions of sucrose molecules packed neatly together. When the crystal is exposed to water (as was first shown in Figure 16.4 and is shown again here in Figure 16.11), an even greater number of water molecules pull on the sucrose molecules via hydrogen bonds formed between the sucrose molecules and the water molecules. With a little stirring, the sucrose molecules soon mix throughout the water. In place of sucrose crystals and water, we have a homogeneous mixture of sucrose molecules in water. As discussed ear-

lier, *homogeneous* means that a sample taken from any part of a mixture is the same as a sample taken from any other part. In our sucrose example, this means that the sweetness of the first sip of the solution is the same as the sweetness of the last sip.

Recall that a homogeneous mixture consisting of a single phase is called a *solution*. Sugar in water is a solution in the liquid phase. Solutions aren't always liquids, however. They can also be solid or gaseous, as Figure 16.12 shows. Gemstones are solid solutions. A ruby, for example, is a solid solution of trace quantities of red chromium compounds in transparent aluminum oxide. A blue sapphire is a solid solution of trace quantities of light green iron compounds and blue titanium compounds in aluminum oxide. Another important example of solid solutions is metal alloys, which are mixtures of different metallic elements. The alloy known as brass is a solid solution of copper and zinc, for instance, and the alloy stainless steel is a solid solution of iron, chromium, nickel, and carbon.

An example of a gaseous solution is the air we breathe. By volume, this solution is 78% nitrogen gas, 21% oxygen gas, and 1% other gaseous materials, including water vapor and carbon dioxide. The air we exhale is a gaseous solution of 75% nitrogen, 14% oxygen, 5% carbon dioxide, and around 6% water vapor. So we see that the air we breathe undergoes a chemical change before being exhaled.

In describing solutions, the component present in the largest amount is the **solvent**, and any other components are **solutes**. For example, when a teaspoon of table sugar is mixed with 1 L of water, we identify the sugar as the solute and the water as the solvent.

FIGURE 16.11

Water molecules pull the sucrose molecules in a sucrose crystal away from one another. This pulling away does not, however, affect the covalent bonds within each sucrose molecule, which is why each dissolved sucrose molecule remains intact as a single molecule.

Link To Section 20.3

To most people, solutions mean finding the answers. To chemists, however, solutions are things that are still all mixed up.

(a)

(b)

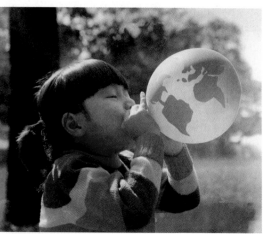
(c)

FIGURE 16.12

Solutions may occur in (a) the solid phase, (b) the liquid phase, or (c) the gaseous phase.

The process of a solute mixing with a solvent is called **dissolving**. To make a solution, a solute must dissolve in a solvent; that is, the solute and solvent must form a homogeneous mixture. Whether one material dissolves in another is a function of their electrical attractions for each other.

CHECK POINT

What is the solvent in the gaseous solution we call air?

Was this your answer?
Nitrogen is the solvent, because it is the component that is present in the greatest quantity.

There is a limit to how much of a given solute can be dissolved in a given solvent, as Figure 16.13 illustrates. We know that when you add table sugar to a glass of water, for example, the sugar rapidly dissolves. As you continue to add sugar, however, there comes a point when it no longer dissolves. Instead, it collects at the bottom of the glass, even after stirring. At this point, the water is saturated with sugar, meaning that the water cannot accept any more sugar. When this happens, we have a **saturated solution**, defined as one in which no more solute can be dissolved. A solution that has not reached the limit of solute that will dissolve is called an **unsaturated solution**.

FIGURE 16.13

A maximum of 200 g of sucrose dissolves in 100 mL of water at 20°C. (a) Mixing 150 g of sucrose in 100 mL of water at 20°C produces an unsaturated solution. (b) Mixing 200 g of sucrose in 100 mL of water at 20°C produces a saturated solution. (c) If 250 g of sucrose is mixed with 100 mL of water at 20°C, 50 g of sucrose remains undissolved. (As we discuss later, the concentration of a saturated solution varies with temperature.)

(a) 150 g sucrose in 100 mL water at 20°C

(b) 200 g sucrose in 100 mL water at 20°C

(c) 250 g sucrose in 100 mL water at 20°C — 50 g sucrose undissolved

The quantity of solute dissolved in a solution is described in mathematical terms by the solution's **concentration**, which is the amount of solute dissolved per amount of solution:

$$\text{Concentration} = \text{amount of solute/amount of solution}$$

For example, a sucrose–water solution may have a concentration of 1 g of sucrose for every liter of solution. This can be compared with concentrations of other solutions. A sucrose–water solution containing 2 g of sucrose per liter of solution, for example, is more concentrated, and one containing only 0.5 g of sucrose per liter of solution is less concentrated, or more dilute.

Chemists are often more interested in the number of solute particles in a solution than in the number of grams of solute. Submicroscopic particles, however, are so small that the number of them in any observable sample is incredibly large. To avoid awkwardly large numbers, scientists use a unit

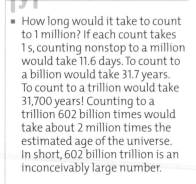

fyi

- How long would it take to count to 1 million? If each count takes 1 s, counting nonstop to a million would take 11.6 days. To count to a billion would take 31.7 years. To count to a trillion would take 31,700 years! Counting to a trillion 602 billion times would take about 2 million times the estimated age of the universe. In short, 602 billion trillion is an inconceivably large number.

called the mole. One **mole** of any type of particle is, by definition, 6.02×10^{23} particles. (This superlarge number is about 602 billion trillion, or 602,000,000,000,000,000,000,000 particles. Interestingly, the term *mole* is derived from the Latin word *moles*, meaning heap, mass, or pile.) One mole of gold atoms, for example, is 6.02×10^{23} gold atoms, and 1 mole of sucrose molecules is 6.02×10^{23} sucrose molecules.

Even if you've never heard the term *mole* in your life before now, you are already familiar with the basic idea. Saying "one mole" is just a shorthand way of saying "six point oh two times ten to the twenty-third particles." Just as "a couple of" means 2 of something and "a dozen of" means 12 of something, "a mole of" means 6.02×10^{23} of some elementary unit, such as atoms, molecules, or ions. It's as simple as that:

- a couple of coconuts = 2 coconuts

- a dozen donuts = 12 donuts

- a mole of molecules = 6.02×10^{23} molecules

One mole of gold atoms, for example, is 6.02×10^{23} gold atoms, and 1 mole of sucrose molecules is 6.02×10^{23} sucrose molecules. A stack containing "1 mole" of pennies would reach a height of about 860 quadrillion km, which is roughly equal to the diameter of our galaxy, the Milky Way. And "1 mole" of marbles would be enough to cover the entire land area of the 50 United States to a depth greater than 1.1 km.

But sucrose molecules are so small that there are 6.02×10^{23} of them in only 342 g of sucrose, which is about a cupful. Thus, because 342 g of sucrose contains 6.02×10^{23} molecules of sucrose, we can use our shorthand wording and say that 342 g of sucrose contains 1 mole of sucrose. As Figure 16.14 shows, therefore, an aqueous solution that has a concentration of 342 g of sucrose per liter of solution also has a concentration of 6.02×10^{23} sucrose molecules per liter of solution or, by definition, a concentration of 1 mole of sucrose per liter of solution. The number of grams tells you the mass of solute in a given solution, and the number of moles indicates the actual number of molecules.

A common unit of concentration used by chemists is **molarity**, which is the solution's concentration expressed in moles of solute per liter of solution:

$$\text{Molarity} = \text{number of moles of solute/liters of solution}$$

A solution that contains 1 mole of solute per liter of solution is a 1-molar solution, which is often abbreviated 1 M. A 2-molar (2 M) solution contains 2 moles of solute per liter of solution.

The difference between referring to the number of molecules of solute and referring to the number of grams of solute can be illustrated by the following question. A saturated aqueous solution of sucrose contains 200 g of sucrose and 100 g of water. Which is the solvent: sucrose or water?

As shown in Figure 16.15, there are 3.5×10^{23} molecules of sucrose in 200 g of sucrose, but there are almost 10 times as many molecules of water in 100 g of water—3.3×10^{24} molecules. As defined earlier, the solvent is the component present in the largest amount, but what do we mean by *amount*? If *amount* means number of molecules, then water is the solvent. If *amount* means mass, then sucrose is the solvent. So the answer depends on how you look at it. From a chemist's point of view, *amount* typically means the number of molecules, and so water is the solvent in this case.

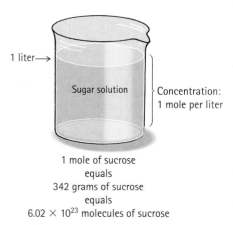

1 mole of sucrose
equals
342 grams of sucrose
equals
6.02×10^{23} molecules of sucrose

FIGURE 16.14

An aqueous solution of sucrose that has a concentration of 1 mole of sucrose per liter of solution contains 6.02×10^{23} sucrose molecules (342 g) in every liter of solution.

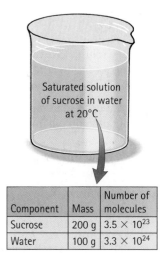

Component	Mass	Number of molecules
Sucrose	200 g	3.5×10^{23}
Water	100 g	3.3×10^{24}

FIGURE 16.15

Although 200 g of sucrose is twice as massive as 100 g of water, there are about 10 times as many water molecules in 100 g of water as there are sucrose molecules in 200 g of sucrose. How can this be? Each water molecule is about $\frac{1}{20}$ as massive as (and much smaller than) each sucrose molecule, which means that about 10 times as many water molecules can fit within half the mass.

- Would a "mole" of stacked pennies stretch across our galaxy? Estimate the answer yourself. First measure the number of stacked pennies in 1 cm. To find the length of one "mole" of stacked pennies, take the number of particles in one mole (6.02×10^{23}) and divide it by the number of pennies in 1 cm. Your answer will be in centimeters. To convert to kilometers, divide by the number of centimeters in a kilometer, which is 100,000. Tall stack!

CHECK POINT

1. How many moles of sucrose are in 0.5 L of a 2 M solution? How many molecules of sucrose is this?
2. Does 1 L of a 1 M solution of sucrose in water contain 1 L of water, less than 1 L of water, or more than 1 L of water?

Were these your answers?

1. First you need to understand that 2 M means 2 moles of sucrose per liter of solution. To obtain the amount of solute, you should multiply solution concentration by amount of solution:

$$(2 \text{ moles/L})(0.5 \text{ L}) = 1 \text{ mole}$$

which is the same as 6.02×10^{23} molecules.

2. The definition of molarity refers to the number of liters of solution, not to the number of liters of solvent. When sucrose is added to a given volume of water, the volume of the solution increases. So if 1 mole of sucrose is added to 1 L of water, the result is more than 1 L of solution. Therefore, 1 L of a 1 M solution requires less than 1 L of water.

FIGURING PHYSICAL SCIENCE

■Calculating for Solutions

From the formula for the concentration of a solution, we can derive equations for the amount of solute and the amount of solution:

Concentration of solution = amount of solute/volume of solution

Amount of solute = concentration of solution × volume of solution

Volume of solution = amount of solute/ concentration of solution

In solving for any of these values, the units must always match. If concentration is given in grams per liter of solution, for example, the amount of solute must be in grams and the amount of solution must be in liters.

Note that these equations are set up for calculating the volume of solution rather than volume of solvent. The volume of solution is greater than the volume of solvent because, in addition to containing the solvent, the solution also contains the solute. As discussed at the beginning of Section 16.3, for example, the volume of an aqueous

solution of sucrose depends not only on the volume of water but also on the volume of dissolved sucrose.

SAMPLE PROBLEM 1

How many grams of sucrose are in 3 L of an aqueous solution that has a concentration of 2 g of sucrose per liter of solution?

SOLUTION:

This question asks for amount of solute, and so you should use the second of the three formulas given above:

Amount of solute = 2 g/1 L × 3 L = 6 g

SAMPLE PROBLEM 2

A solution you are using in an experiment has a concentration of 10 g of solute per liter of solution. If you pour enough of this solution into an empty laboratory flask to make the flask contain 5 g of the solute, how many liters of the solution have you poured into the flask?

SOLUTION:

This question asks for amount of solution, and you should use the third formula given above:

Volume of solution = 5 g/10 g/L = 0.5 L

SAMPLE PROBLEM 3

At 20°C, a saturated solution of sodium chloride in water has a concentration of about 380 g of sodium chloride per liter of solution. How many grams of sodium chloride are required to make 3 L of a saturated solution?

SOLUTION:

Multiply the solution concentration by the final volume of the solution. This provides the amount of solute required: (380 g/L)(3 L) = 1140 g.

SAMPLE PROBLEM 4

A student is told to use 20 g of sodium chloride to make an aqueous solution that has a concentration of 10 g of sodium chloride per liter of solution. How many liters of solution does she end up with?

SOLUTION:

Divide the amount of solute by the solution concentration to obtain the amount of solution prepared:

20 g/10 g/L = 2L.

■ 16.4 Solubility

The **solubility** of a solute is its *ability* to dissolve in a solvent. As can be expected, this ability mainly depends on the submicroscopic attractions between solute particles and solvent particles. If a solute has any appreciable solubility in a solvent, then that solute is said to be **soluble** in that solvent.

Solubility also depends on attractions of solute particles for one another and attractions of solvent particles for one another. As shown in Figure 16.16, for example, a sucrose molecule has many polar hydrogen–oxygen bonds. Sucrose molecules, therefore, can form multiple hydrogen bonds with one another. These hydrogen bonds are strong enough to make sucrose a solid at room temperature and to give it the relatively high melting point of 185°C. In order for sucrose to dissolve in water, the water molecules must first pull sucrose molecules away from one another. This puts a limit on the amount of sucrose that can dissolve in water—eventually, a point is reached at which there are not enough water molecules to separate the sucrose molecules from one another. As we discussed in Section 16.3, this is the point of saturation, and any additional sucrose added to the solution does not dissolve.

When the molecule-to-molecule attractions among solute molecules are comparable to the molecule-to-molecule attractions among solvent molecules, there is no practical point of saturation. As shown in Figure 16.17, for example, the hydrogen bonds among water molecules are about as strong as those between ethanol molecules. These two liquids therefore mix together quite well and in just about any proportion. We can even add ethanol to water until the ethanol, rather than the water, can be considered the solvent.

A solute that has no practical point of saturation in a given solvent is said to be *infinitely soluble* in that solvent. Ethanol, for example, is infinitely soluble in water. Also, all gases are generally infinitely soluble in other gases because they can be mixed together in just about any proportion.

Let's now look at the other extreme of solubility, in which a solute has very little solubility in a given solvent. An example is oxygen, O_2, in water. In contrast to sucrose, which has a solubility of 200 g per 100 mL of water, only 0.004 g of oxygen can dissolve in 100 mL of water. We can account for oxygen's low solubility in water by noting that the only electrical attractions that occur

Sucrose

FIGURE 16.16

A sucrose molecule contains many hydrogen–oxygen covalent bonds, in which the hydrogen atoms are slightly positive and the oxygen atoms are slightly negative. These dipoles in any given sucrose molecule result in the formation of hydrogen bonds with neighboring sucrose molecules.

The Physical Science Place
Solubility

The Physical Science Place
Solubility Changes with Temperature and Pressure

Ethanol

Ethanol and water

Water

FIGURE 16.17

Ethanol and water molecules are about the same size, and they both form hydrogen bonds. As a result, ethanol and water readily mix with each other.

FIGURE 16.18

Glass is frosted by dissolving its outer surface in hydrofluoric acid.

between oxygen molecules and water molecules are relatively weak dipole–induced dipole attractions. More important, however, is the fact that the stronger attraction of water molecules for one another—through the hydrogen bonds that the water molecules form with one another—effectively excludes oxygen molecules from intermingling.

A material that does not dissolve in a solvent to any appreciable extent is said to be **insoluble** in that solvent. We consider many substances insoluble in water, including sand and glass. Just because a material is not soluble in one solvent, however, does not mean it won't dissolve in another. Sand and glass, for example, are soluble in hydrofluoric acid, HF, which is used to give glass the decorative frosted look shown in Figure 16.18. Also, although Styrofoam is insoluble in water, it is soluble in acetone, a solvent used in fingernail polish remover. Pour a little acetone into a Styrofoam cup, and the acetone soon causes the Styrofoam to deform, as demonstrated in Figure 16.19.

CHECK POINT

■ Why isn't sucrose infinitely soluble in water?

Was this your answer?

The attraction between two sucrose molecules is much stronger than the attraction between a sucrose molecule and a water molecule. Because of this, sucrose dissolves in water only as long as the number of water molecules far exceeds the number of sucrose molecules. When there are too few water molecules to dissolve any additional sucrose, the solution is saturated.

FIGURE 16.19

Is this cup melting or dissolving?

FIGURE 16.20

The solubility of many water-soluble solids increases with temperature, while the solubility of others is only very slightly affected by temperature.

SOLUBILITY CHANGES WITH TEMPERATURE

You probably know from experience that water-soluble solids usually dissolve better in hot water than in cold water. A highly concentrated solution of sucrose in water, for example, can be made by heating the solution almost to the boiling point. This is how syrups and hard candy are made.

Solubility increases with increasing temperature because hot water molecules have greater kinetic energy and therefore can collide with the solid solute more vigorously. The vigorous collisions help disrupt electrical particle-to-particle attractions in the solid.

Although the solubilities of many solid solutes—sucrose, to name just one example—are greatly increased by temperature increases, the solubilities of other solid solutes, such as sodium chloride, are only mildly affected, as Figure 16.20 shows. This difference involves a number of factors, including the strength of the chemical bonds in the solute molecules and the way those molecules are packed together. Some chemicals, such as calcium carbonate, $CaCO_3$, actually become *less* soluble as the water temperature increases. This explains why the inner surface of tea kettles are often coated with calcium carbonate residues.

When a sugar solution saturated at a high temperature is allowed to cool, some of the sugar usually comes out of solution and forms a **precipitate**. When this occurs, the solute, sugar in this case, is said to have *precipitated* from the solution. Let's put on our quantitative thinking caps and consider another example. At 100°C, the solubility of sodium nitrate, $NaNO_3$, in water is 165 g per 100 mL of water. As we cool this solution, the solubility of $NaNO_3$ decreases, as shown in Figure 16.21, and this change in solubility causes some of the dissolved $NaNO_3$ to precipitate (come out of solution). At 20°C, the solubility of $NaNO_3$ is only 87 g per 100 mL of water. So if we cool the 100°C solution to 20°C, 78 g (165 g − 87 g) precipitates, as shown in Figure 16.21.

Grease is soluble in paint thinner, which is why paint thinner can be used to clean one's hands of grease. But body oils are also soluble in paint thinner, which is why hands cleaned with paint thinner feel dry and chapped.

100°C   20°C

Saturated solution   Saturated solution   — 78 g of $NaNO_3$ precipitates out of solution

165 g $NaNO_3$ in 100 mL water   87 g $NaNO_3$ in 100 mL water

FIGURE 16.21

The solubility of sodium nitrate is 165 g per 100 mL of water at 100°C but only 87 g per 100 mL at 20°C. Cooling a 100°C saturated solution of $NaNO_3$ to 20°C causes 78 g of the solute to precipitate.

SOLUBILITY OF GASES

In contrast to the solubilities of most solids, the solubilities of gases in liquids *decrease* with increasing temperature, as Table 16.1 shows. This effect occurs because with an increase in temperature, the solvent molecules have more kinetic energy. This makes it more difficult for a gaseous solute to remain in solution because the solute molecules are ejected by the high-energy solvent molecules.

Link To Sections 20.5, 20.8

TABLE 16.1   TEMPERATURE-DEPENDENT SOLUBILITY OF OXYGEN GAS IN WATER AT A PRESSURE OF 1 ATM

Temperature (°C)	O_2 Solubility (g O_2/L H_2O)
0	0.0141
10	0.0109
20	0.0092
25	0.0083
30	0.0077
35	0.0070
40	0.0065

Perhaps you have noticed that warm carbonated beverages go flat faster than cold ones. The higher temperature causes the molecules of carbon dioxide gas to leave the liquid solvent at a higher rate.

The solubility of a gas in a liquid also depends on the pressure of the gas immediately above the liquid. In general, a higher gas pressure above the liquid means more of the gas dissolves. A gas at a high pressure has many gas particles crammed into a given volume. The "empty" space in an unopened soft drink bottle, for example, is crammed with carbon dioxide molecules in the gaseous phase. With nowhere else to go, many of these molecules dissolve in the liquid,

Air is a gaseous solution, and one of its minor components is water vapor. The process of this water coming "out of solution" in the form of rain or snow is called "precipitation." The rain or snow is the "precipitate."

FIGURE 16.22

(a) The carbon dioxide gas above the liquid in an unopened soft drink bottle consists of many tightly packed carbon dioxide molecules that are forced by pressure into solution. (b) When the bottle is opened, the pressure is released, and carbon dioxide molecules originally dissolved in the liquid can escape into the air.

as shown in Figure 16.22. Alternatively, we might say that the great pressure forces the carbon dioxide molecules into solution. When the bottle is opened, the "head" of highly pressurized carbon dioxide gas escapes. Now the gas pressure above the liquid is lower than before. As a result, the solubility of the carbon dioxide drops, and the carbon dioxide molecules that were once squeezed into the solution begin to escape into the air above the liquid.

The rate at which carbon dioxide molecules leave an opened soft drink is relatively slow. You can increase the rate by pouring in granulated sugar, salt, or sand. The microscopic nooks and crannies on the surfaces of the grains serve as *nucleation sites* where carbon dioxide bubbles can form rapidly and then escape by buoyant forces. Shaking the beverage also increases the surface area of the liquid-to-gas interface, making it easier for the carbon dioxide to escape from the solution. Once the solution is shaken, the rate at which carbon dioxide escapes becomes so great that the beverage froths over. You also increase the rate at which carbon dioxide escapes when you pour the beverage into your mouth, which abounds in nucleation sites. You can feel the resulting tingly sensation.

CHECK POINT

You open two cans of soft drink, one from a warm kitchen shelf and the other from the coldest depths of your refrigerator. Which fizzes more in your mouth?

Was this your answer?

The solubility of carbon dioxide in water decreases with increasing temperature. The warm drink, therefore, fizzes in your mouth more than the cold one does.

16.5 Soaps and Detergents

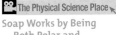
fyi

■ It is not just dipole–induced dipole attractions that keep carbon dioxide dissolved within water. As we'll discuss in Chapter 18, carbon dioxide reacts with water to form carbonic acid, which is much more soluble in water. When a can of carbonated soda is opened, much of this carbonic acid quickly transforms back into water and carbon dioxide, which quickly bubbles out of solution because of its low solubility.

irt and grease together make *grime*. Because grime contains many nonpolar components, it is difficult to remove from hands or clothing with water alone. To remove most grime, we can use a nonpolar solvent, such as turpentine, which dissolves the grime because of strong induced dipole–induced dipole attractions. Turpentine is good for removing the grime left on hands after such activities as changing a car's motor oil. Rather than washing our dirty hands and clothes with nonpolar solvents, however, we have a more pleasant alternative—soap and water. Soap works because soap molecules have both nonpolar and polar properties. A typical soap molecule has two parts: a long, nonpolar tail of carbon and hydrogen atoms and a polar head containing at least one ionic bond.

Because most of a soap molecule is nonpolar, it attracts nonpolar grime molecules via induced dipole–induced dipole attractions, as Figure 16.23 illustrates. In fact, grime quickly finds itself surrounded in three dimensions by the nonpolar tails of soap molecules. This attraction is usually sufficient to lift the grime away from the surface being cleaned. With the nonpolar tails facing inward toward the grime, the polar heads are all directed outward, where they are attracted to water molecules by relatively strong ion–dipole attractions. If the water is flowing, the whole conglomeration of grime and soap molecules flows with it, away from your hands or clothes and then down the drain.

For the past several centuries, soaps have been prepared by treating animal fats with sodium hydroxide, NaoH, also known as caustic lye. In this reaction, which is still used today, each fat molecule is broken down into three *fatty acid* soap molecules and one glycerol molecule:

FIGURE 16.23

Nonpolar grime attracts and is surrounded by the nonpolar tails of soap molecules. The polar heads of the soap molecules are attracted by ion–dipole attractions to water molecules, which then carry the soap–grime combination away.

Fat molecule   Three fatty acid soap molecules   Glycerol molecule

In the 1940s, chemists began developing a class of synthetic soaplike compounds, known as *detergents*, that offer several advantages over true soaps, such as stronger grease penetration and lower price.

The chemical structure of detergent molecules is similar to that of soap molecules in that both possess a polar head attached to a nonpolar tail. The polar head in a detergent molecule, however, typically consists of either a sulfate group, $-OSO_3^-$, or a sulfonate group, $-SO_3^-$, and the nonpolar tail can have an assortment of structures.

One of the most common sulfate detergents is sodium lauryl sulfate, a main ingredient of many toothpastes. A common sulfonate detergent is sodium dodecyl benzenesulfonate, also known as a linear alkylsulfonate, or LAS, often found in dishwashing liquids. Both of these detergents are biodegradable, which means that microorganisms can break down the molecules once they are released into the environment.

$CH_3CH_2CH_2CH_2CH_2CH_2CH_2CH_2CH_2CH_2CH_2CH_2$

Sodium lauryl sulfate

$CH_3CH_2CH_2CH_2CH_2CH_2CH_2CH_2CH_2CH_2CH_2CH_2$

Sodium dodecyl benzenesulfonate

fyi

- "Dry cleaning" is the process of washing clothes without water. The most common dry-cleaning solvent is perchloroethylene, C_2Cl_4, which is gentle on the clothing and can clean a full load in under 10 min. After a washing cycle, the solvent is centrifuged out of the machine, filtered, distilled, and recycled for the next load. Clothes come out of the machine already dry and ready for folding. Perchloroethylene, also known as perc, is relatively safe, but it is mildly carcinogenic and can cause dizziness in those who work with it. An up-and-coming alternative to perc is supercritical carbon dioxide, CO_2.

What type of attractions hold soap or detergent molecules to grime?

Was this your answer?
If you haven't yet formulated an answer, why not back up and reread the question? You've got only four choices: ion-dipole, dipole–dipole, dipole-induced dipole, and induced dipole-induced dipole. The answer is induced dipole-induced dipole attractions, because the interaction is between two nonpolar entities—the grime and the nonpolar tail of a soap or detergent molecule.

16.6 Softening Hard Water

Water containing large amounts of calcium and magnesium ions is said to be **hard water**, and it has many undesirable qualities. For example, when hard water is heated, its calcium and magnesium ions tend to bind with negatively charged ions also found in the water to form solid compounds, like those shown in Figure 16.24. These can clog water heaters and boilers. You'll also find coatings of these calcium and magnesium compounds on the inside surfaces of a well-used teakettle (because the solubility of these compounds decreases with increasing temperature, as discussed earlier).

Hard water also inhibits the cleansing actions of soaps and, to a lesser extent, detergents. The sodium ions of soap and detergent molecules carry a 1+ charge, and calcium and magnesium ions carry a 2+ charge (note their positions in the periodic table). The negatively charged portion of the polar head of a soap or detergent molecule is more attracted to the double positive charge of calcium and magnesium ions than to the single positive charge of sodium ions. Soap or detergent molecules, therefore, give up their sodium ions to bind selectively with calcium or magnesium ions:

Soap or detergent molecules bound to calcium or magnesium ions tend to be insoluble in water. As they come out of solution, they form a scum, which can appear as a ring around the inside of your bathtub. Because the soap or detergent molecules are tied up with calcium and magnesium ions, more soap or detergent must be added to maintain cleaning effectiveness.

Many detergents today contain sodium carbonate, Na_2CO_3, commonly known as washing soda. The calcium and magnesium ions in hard water are more attracted to the carbonate ion with its two negative charges than they are to a soap or detergent molecule with its single negative charge. With the calcium and magnesium ions bound to the carbonate ion, as shown in Figure 16.25, the soap or detergent is free to do its job. Because it removes the ions that make water hard, sodium carbonate is known as a water-softening agent.

In some homes, the water is so hard that it must be passed through a *water-softening unit*. In a typical unit, illustrated in Figure 16.26, hard water is passed through a large tank filled with tiny beads of a water-insoluble resin known as an *ion-exchange resin*. The surface of the resin contains many negatively charged ions bound to positively charged sodium ions. As calcium and magnesium ions pass over the resin, the ions displace the sodium ions and thereby become bound to the

FIGURE 16.24

Hard water causes calcium and magnesium compounds to build up on the inner surfaces of water pipes, especially those used to carry hot water.

Carbonate ion, CO$_3^{2-}$

or

Sodium carbonate, Na$_2$CO$_3$

(a)

(b)

Calcium carbonate

Magnesium carbonate

FIGURE **16.25**

(a) Sodium carbonate is added to many detergents as a water-softening agent. (b) The doubly positive calcium and magnesium ions of hard water preferentially bind with the doubly negative carbonate ion, freeing the detergent molecules to do their job.

resin. The calcium and magnesium ions can do this because their positive charge (2+) is greater than that of the sodium ions (1+). The calcium and magnesium ions therefore have a greater attraction for the negative sites on the resin. The net result is that for every calcium or magnesium ion that binds, two sodium ions are set free. In this way, the resin *exchanges* ions. The water that exits the unit is now free of calcium and magnesium ions, but contains sodium ions in their place.

Eventually, all the sites for calcium and magnesium on the resin are filled, and then the resin needs to be either discarded or recharged. It is recharged by flushing it with a concentrated solution of sodium chloride, NaCl. The abundant sodium ions displace the calcium and magnesium ions (ions are *exchanged* once again), freeing up the binding sites on the resin.

Most modern water softeners are equipped with meters that let you know the rate at which you consume water. This is a great way to keep tabs on your water-conservation efforts.

■ 16.7 Purifying the Water We Drink

As was discussed earlier, it is impossible to obtain 100% pure water. However, we can purify water to meet our needs. We do this by taking advantage of the differences in physical properties of water and the solutes or particulates

FIGURE **16.26**

(1) Negatively charged sites on the unused ion-exchange resin are occupied by sodium ions. (2) As hard water passes over the resin, sodium ions are displaced by calcium and magnesium ions. (3) After the resin becomes saturated with calcium and magnesium ions, it is no longer effective at softening water.

Hard water

Resin beads

Ion exchange

Softened water

$$3\,Ca(OH)_2 \;+\; Al_2(SO_4)_3 \longrightarrow 2\,Al(OH)_3 \;+\; 3\,CaSO_4$$

Slaked lime   Aluminum   Aluminum
               sulfate   hydroxide
                                 (gelatinous)

Slaked lime
Aluminum
sulfate

Aluminum
hydroxide

Impurities in water

① Slaked lime and aluminum sulfate added to water react to form gelatinous aluminum hydroxide.

② Impurities captured by aluminum hydroxide as it settles.

③ Gelatinous aluminum hydroxide and impurities collect at bottom of basin.

FIGURE 16.27

Slaked lime, $Ca(OH)_2$, and aluminum sulfate, $Al_2(SO_4)_3$, react to form aluminum hydroxide, $Al(OH)_3$, and calcium sulfate, $CaSO_4$, which together form a gelatinous material.

Lack of clean drinking water is one of the world's leading causes of death, especially among children of developing nations.

FIGURE 16.28

Volatile impurities are removed from drinking water by cascading it through the columns of air within each of these stacks.

it contains. For the remainder of this chapter, we turn our attention to some of the details involved in the production of drinkable water and the treatment of wastewater.

Water that is safe for drinking is said to be *potable*. In the United States, potable water is currently used for everything from cooking to flushing our toilets. The first step most public utilities take to produce potable water from natural sources is to remove any dirt particles or pathogens, such as bacteria. This is done by mixing the water with certain minerals, such as slaked lime and aluminum sulfate, which coagulates into a gelatinous material, aluminum hydroxide, that intersperses throughout the water (Figure 16.27). This is done in a large settling basin. Slow stirring causes the gelatinous material to clump together and settle to the bottom of the basin. As these clumps form and settle, they carry with them many of the dirt particles and bacteria. The water is then filtered through sand and gravel.

To improve the odor and flavor of the water, many treatment facilities also *aerate* the water by cascading it through a column of air, as shown in Figure 16.28. Aeration removes many unpleasant-smelling volatile chemicals, such as sulfur compounds. At the same time, air dissolves into the water giving it a better taste—without dissolved air, the water tastes flat. As a final step, the water is treated with a disinfectant, usually chlorine gas, Cl_2, but sometimes ozone, O_3, and then stored in a holding tank that feeds into the city mains.

Developed countries have the technology and infrastructure to produce vast quantities of water suitable for drinking; as a result, many citizens take their drinking water for granted. The number of public water treatment facilities in developing nations, however, is relatively small. In these locations, many people drink their water in the form of a hot beverage, such as tea, which is disinfected through boiling. Alternatively, disinfecting iodine tablets can be used.

Fuel for boiling and tablets for disinfecting, however, are not always available. As a result, more than 400 people in the world (mostly children) die every hour from preventable diseases or infections such as cholera, typhoid fever, dysentery, and hepatitis, which they contract by drinking contaminated water. In response, several American manufacturers have developed tabletop systems that bathe water with pathogen-killing ultraviolet light. One prototype model, shown in Figure 16.29, disinfects 15 gal/min, weighs about 15 lb, and is powered by photovoltaic solar cells, which permit it to run unsupervised in remote locations.

FIGURE 16.29

Small-scale water-disinfecting units, such as the one shown here, hold great value in regions of the world where potable water is scarce.

Aside from pathogens, untreated water from wells or rivers may contain toxic metals that seep into the water supply from natural geologic formations. Many wells in Bangladesh, for example, are made very deep so as to avoid the pathogens that run rampant in the surface waters of this region. The water obtained from these deep wells, however, is highly contaminated with arsenic—a naturally occurring element in Earth's crust. The arsenic is in the underlying rock, which formed from river sediments carried down from the Himalayas. Because this region is so densely populated, as many as 70 million people may be subject to some level of arsenic poisoning, which manifests itself as skin lesions and a higher susceptibility to cancer. Low-cost methods for removing arsenic from well water are greatly needed, as are worldwide recognition of this problem and the political, economic, and social support to overcome it.

> In the early 1990s, an anti-water-chlorination campaign in Peru led the government to stop chlorinating the drinking water. Within months there were 1.3 million new cases of cholera resulting in 13,000 deaths.

CHECK POINT

At a water-treatment facility, how does adding slaked lime and aluminum sulfate to water purify the water?

Was this your answer?
The water entering a water treatment plant is usually a heterogeneous mixture containing suspended solids. The added slaked lime and aluminum sulfate capture these suspended solids, which then sink to the bottom, where they are easily removed.

DESALINATION

With the depletion of sources of natural fresh water in many regions, there has been growing interest in techniques for generating fresh water from Earth's far larger reserves of seawater or from *brackish* (moderately salty) groundwater. Worldwide, *desalination* plants operate in about 120 countries with a combined capacity to produce about 16 billion liters daily. In many areas of the Caribbean, North Africa, and the Middle East, desalinized water is the main source of municipal supply (Figure 16.30). In the United States, more than a thousand desalination plants produce a combined capacity of more than 400 million liters daily. Most of the treated water in the United States is used for industrial purposes, and it comes from brackish sources or from water that is high in dissolved minerals.

The two primary methods of removing salts from seawater or brackish water are *distillation* and *reverse osmosis*. These techniques are also highly effective in

fyi

- In 1908, Jersey City, New Jersey, became the first American city to begin chlorinating its drinking water. By 1910, as disinfecting drinking water with chlorine became more widespread, the death rate from typhoid fever dropped to 20 per 100,000. In 1935, the death rate fell to 3 per 100,000. By 1960, fewer than 20 people in the entire United States died from typhoid fever annually.

FIGURE 16.30

Saudi Arabia is the world's leading producer of desalinized water. Its desalination plants, such as the one shown here, have a combined generating capacity of about 4 billion liters per day.

Link To Section 7.6

removing a host of other contaminants, such as pathogens, fertilizers, and pesticides. Distillation and reverse osmosis, therefore, are also used to purify naturally occurring fresh water. Many popular brands of bottled water, for example, contain fresh water that has been treated either by distillation or by reverse osmosis.

Distillation involves vaporizing water with heat and then condensing the vapors into purified liquid water (Section 16.1). More than 60% of Earth's desalinized water is produced using this technique. Because water has such a high heat of vaporization, however, this technique is energy intensive. Today, most distilling plants heat the water by burning large quantities of fossil fuels, which, unfortunately, generates excessive levels of pollution relative to the volume of fresh water produced. Solar distillers avoid the burning of fuels, but they require about 1 m^2 of surface area to produce 4 L of fresh water per day, as shown in Figure 16.31. For a single home or a small village, this surface area requirement may be easily accommodated. For larger urban areas, where open land is scarce, solar distillation is less practical, especially when the maintenance costs of vast fields of solar distillers are taken into account.

For many regions, *reverse osmosis* is a preferable method of water desalination. In order to understand reverse osmosis, you must first understand osmosis. Osmosis involves a semipermeable membrane. A **semipermeable membrane** contains submicroscopic pores that allow the passage of water

FIGURE 16.31

These solar distillers are popular in the remote communities along the Texas–Mexico border, where the waters from the Rio Grande basin are saline and tainted by the runoff of agricultural chemicals from upstream irrigation.

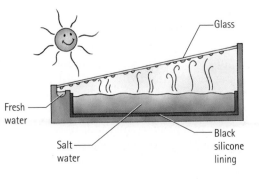

Glass

Fresh water

Salt water

Black silicone lining

molecules but not of larger solute ions or solute molecules. When a body of fresh water is partitioned from a body of salt water by a semipermeable membrane, water molecules pass from the fresh water into the salt water at a higher rate than they pass from the salt water into the fresh water. The reason for this is the presence of more water molecules along the freshwater face of the membrane than along the saltwater face. The result is a net movement of fresh water into the body of salt water, as illustrated in Figure 16.32. This net flow of water across a semipermeable membrane into a more concentrated solution is called **osmosis**.

The result of osmosis is a buildup in volume of the salt water and a decrease in volume of the fresh water. These changes in volume, in turn, allow for a buildup in pressure, called *osmotic pressure*. For the system in Figure 16.33a, osmotic pressure is the consequence of the salt water's greater height. As osmotic pressure builds, the rate at which water molecules are able to pass from the salt water into the fresh water increases. The water molecules in the salt water are being squeezed back across the membrane by the osmotic pressure. Eventually, the rates of water molecules passing in both directions across the membrane are the same and the system reaches equilibrium, as shown in Figure 16.33b. If external pressure is applied to the salt water, even more water molecules are squeezed across the membrane from the salt water into the fresh water, as shown in Figure 16.33c. Water forced across a semipermeable membrane into a less concentrated solution is **reverse osmosis**. So we see that reverse osmosis is a mechanism for generating fresh water from salt water.

The osmotic pressure for seawater, however, is an astounding 24.8 atm (365 psi). Generating pressures greater than this has its share of technical difficulties and is an energy-intensive process. Nonetheless, engineers have succeeded in building durable reverse osmosis units, shown in Figure 16.34, that can be networked together to generate fresh water from seawater at rates of millions of gallons per day. Reverse osmosis desalination facilities treating brackish waters, which require much lower external pressures, are proportionately more economical.

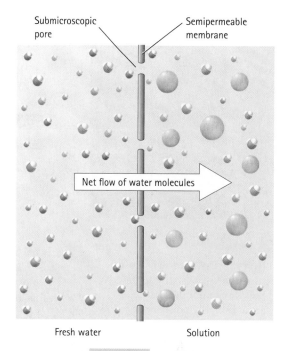

FIGURE 16.32

INTERACTIVE FIGURE

Osmosis. The submicroscopic pores of a semipermeable membrane allow only water molecules to pass. Because more water molecules exist along the freshwater face of the membrane than along the solution face, more water molecules are available to migrate into the solution than are available to migrate into the fresh water.

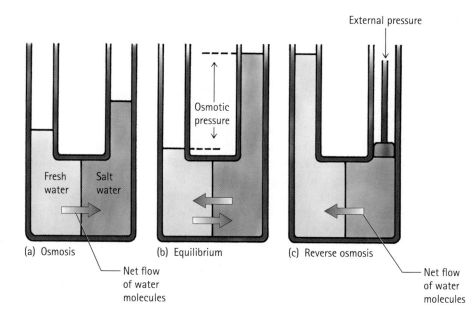

(a) Osmosis   (b) Equilibrium   (c) Reverse osmosis

FIGURE 16.33

(a) Osmosis results in a greater volume of salt water, which causes the pressure to increase on the salt side of the membrane. (b) When the pressure on the salt side becomes high enough, equal numbers of water molecules pass in both directions. (c) The application of external pressure forces water molecules to pass from the salt water to the fresh water, so that now the salt-to-fresh rate exceeds the fresh-to-salt rate.

Semipermeable
membrane

Pressurized
salt water

Concentrated
salt solution

Desalinated
water

FIGURE 16.34

An industrial reverse osmosis
unit consists of many semipermeable membranes packed
around highly pressurized salt
water. As desalinated water is
pushed out one side, the
remaining salt water, which is
now even more concentrated,
exits on the other side. A network of reverse osmosis units
operating parallel to one
another can produce enormous
volumes of fresh water from
salt water.

CHECK POINT

Biological membranes, including cucumber membranes, are semipermeable. A cucumber shrivels to a smaller size when it is left in a solution of salt
water. Is this an example of osmosis or of reverse osmosis?

Was this your answer?

No external pressure is involved, which rules out reverse osmosis. Instead,
the shriveling of the cucumber tells us that the cucumber's cells are losing water to the more concentrated salt water. This is osmosis, whereby
water molecules migrate across a semipermeable membrane into
regions of higher salt concentrations. If you added a few other ingredients to the solution, such as spices and the right kinds of microorganisms, you would have a pickle.

FIGURE 16.35

Carbonating your own water is
not only fun, it's cheaper and
more ecologically sound than pur-
chasing your soda from a store.

BOTTLED WATER

Desalinated seawater and brackish water are important new sources of fresh
water. Although this fresh water is more costly than fresh water from natural
sources, one could argue that the higher cost reflects fresh water's true value. In
the United States, natural sources of fresh water are relatively plentiful, allowing
companies to sell fresh water at rates of a fraction of a penny per liter. Nonetheless, consumers are still willing to purchase bottled water at up to $2 per liter!
Each year, Americans spend about $400 million on bottled water, and the mar-
ket continues to grow rapidly. It is easy to project a growing reliance on distillation and reverse osmosis.

In 2005, about 100 billion liters of bottled water were sold worldwide, but
mostly in developed nations where tap water is potable. It is difficult to estimate
the amount of energy consumed in shipping this bottled water from its source
to the customer, but because water is so dense, the amount is likely huge.
Assuming it takes about 100 mL of gasoline to ship each liter of bottled water,
worldwide this would translate to about 25 million tons of carbon dioxide, not
to mention other pollutants. Studies show that only about 20% of drinking
water bottles are recycled. In California alone, more than 1 billion water bottles
wind up in the trash each year. The plastic from these bottles could be used to
make 74 million square feet of carpet. People who drink bottled water should
know of its hefty ecological price tag.

In an effort to overcome the multiple ecological negatives of bottled water
and a price up to 1000 times that of tap water, many bottled water marketers are
now focusing on supposed peripheral benefits of their product. This includes
the addition of dissolved oxygen, which as shown in Table 16.1, can be no

more than 0.0083 g/L at room temperature. For comparison, a single breath of air contains about 100 times the amount of molecular oxygen found in a half-liter of "oxygenated" water. Furthermore, most of the gases accumulated in your gut simply pass out the opposite end of your mouth, assuming you haven't burped. Worse still are claims of bottled water that contains "functional" water in which the structure of water has been modified using "subtle energy" to make it more nutritious. Subtle energy, of course, cannot be detected by current science, but the effects on your health, they claim, can be most dramatic. Buyer beware!

Most bottled water sold today is simply municipal water that has been purified via reverse osmosis. Many homeowners have discovered that it is less expensive and more ecologically sound to install a small reverse osmosis unit within their own home. For fun, carbonators can also be installed so that you can have your very own soda fountain, as shown in Figure 16.35.

fyi

- Naturally occurring water is alive with bacteria, which break down organic matter. *Aerobic bacteria* decompose organic matter only in the presence of O_2, transforming organic matter into odorless carbon dioxide, water, nitrates, and sulfates. *Anaerobic bacteria* decompose organic matter in the absence of oxygen resulting in methane, which is flammable, and foul-smelling nitrogen- and sulfur-containing compounds. Cesspools owe their wretched stink to a lack of dissolved oxygen and the resulting anaerobic decomposition.

16.8 Wastewater Treatment

The contents of the sewer systems that underlie most municipalities must be treated before being released into a body of water. The level of treatment depends in great part on whether the treated water is to be released into a river or into the ocean. Wastewater destined for a river requires the highest level of treatment for the benefit of communities downstream. However, in a facility located in a region surrounded by very deep ocean water, as is the facility shown in Figure 16.36, treatment requirements are less stringent.

Human waste loses its form by the time it reaches the wastewater facility, and the wastewater appears as a murky stream. In this stream, however, are many insoluble products—including small plastic items, such as tampon applicators, and gritty material, such as coffee grounds and sand. Hardened balls of grease from discarded cooking fats are also found. The initial step in all wastewater treatments, therefore, is to screen out these insolubles. (You should know that wastewater treatment experts point out that these insolubles—even cooking grease—should be disposed of as solid waste and not washed down the drain or flushed down the toilet.)

After screening, the next level of municipal wastewater treatment is *primary* treatment. In primary treatment, screened wastewater enters a large settling basin, where suspended solids settle out as sludge (Figure 16.37). After a period of time, the sludge is removed from the bottom of the settling basin and is often sent directly to a landfill as solid wastes. Some facilities, however, are equipped with large furnaces in which dried sludge is burned, sometimes along with other municipal wastes, such as paper products. The resulting ash is more compact and takes up less space in a landfill.

Wastewater effluent from primary treatment, as well as higher levels of treatment, is commonly disinfected with either chlorine gas or ozone before its release into the environment. A great advantage of using chlorine gas is that it remains in the water for an extended time after leaving the facility. This provides residual protection against diseases. The chlorine, however, reacts with organic compounds within the effluent to form chlorinated hydrocarbons, many of which are known carcinogens (cancer-causing agents). Also, chlorine kills only bacteria, leaving viruses unharmed. Ozone is more advantageous in that it kills both bacteria and viruses. Also, no carcinogenic by-products result from treating wastewater effluent with ozone. A disadvantage

FIGURE 16.36

In Honolulu, about 280 million liters of wastewater pass through the largest of several wastewater facilities each day. This water can be piped to depths of hundreds of meters below sea level, whence it continues to flow toward the bottom of the ocean. Water treatment requirements are therefore much less stringent than those at mainland facilities, where the effluent is not so easily discarded.

FIGURE 16.37

A schematic for primary-level wastewater treatment. The rotating skimmer on the settling basin removes buoyant materials and artifacts not captured by the screening process.

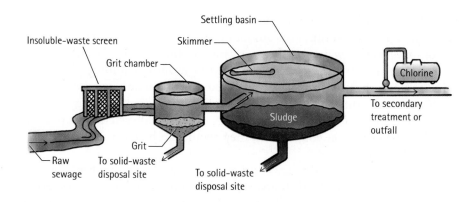

fyi

- In Hong Kong, about 80% of all toilets flush using seawater. Developed since the 1960s, this system now saves the equivalent of about 25% of freshwater consumption. Also, effluent from freshwater activities, such as personal hygiene and dishwashing, is treated and reused for watering city trees and other nonpotable uses.

For dwellings in remote locations, such as summer cabins, many people choose composting toilets, which use no water. Rather, they allow human waste to decompose aerobically (with oxygen) as air is vented over the waste, which is buried in peat moss. Dried, odor-free compost, which is removed every few months, is useful as a garden fertilizer.

of ozone, however, is that it provides no residual protection for the effluent once it is released. Most facilities in the United States use chlorine for disinfecting, whereas European facilities tend to favor ozone. In a few locations, chlorine and ozone gases have been replaced by strong ultraviolet lamps, which, like ozone, kill both bacteria and viruses but provide no long-term residual protection.

The potential for pathogens to grow in primary effluent is extremely high and, by virtue of the Clean Water Act of 1972, the release of primary effluent is not permitted in most places. A frequently used *secondary* level of treatment, shown in Figure 16.38, involves passing the primary effluent first through an aeration tank. This supplies the oxygen necessary for continued decomposition of organic matter by oxygen-dependent bacteria, known as *aerobic bacteria*. The effluent is then sent into a tank where any fine particles not removed in primary treatment can settle. Because sludge from this settling step is high in aerobic bacteria, some of it is recycled back to the aeration tank to increase efficiency. The remainder of the sludge is hauled off to the landfill or an incinerator.

Many municipalities also require a third level, a *tertiary* level, of wastewater treatment. A number of tertiary processes are used, and most involve filtrations of some sort. A common method is to pass secondary-level effluent through a bed of finely powdered carbon, which captures most particulate matter and many of the organic molecules not removed in earlier stages. The advantage of tertiary-level treatment is greater protection of our water resources. Unfortunately, tertiary treatment is costly and is normally used only when the need is vital. Primary and secondary levels of treatment are also not without great cost.

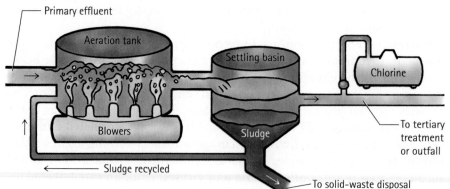

FIGURE 16.38

A schematic for secondary treatment of wastewater from a municipal system.

CHECK
POINT

Distinguish between the main functions of primary, secondary, and tertiary wastewater treatment.

Was this your answer?

Primary wastewater treatment removes the bulk of solid waste and sludge from the sewage effluent using screening devices and large settling basins. Secondary treatment decreases the biochemical oxygen demand of the effluent by aeration. Tertiary treatment removes pathogens and wastes not removed by earlier treatments by filtering the effluent through beds of powdered carbon or other fine particles.

SUMMARY OF TERMS

Mixture A combination of two or more substances in which each substance retains its properties.

Distillation A purifying process in which a vaporized substance is collected by exposing it to cooler temperatures over a receiving flask, which collects the condensed purified liquid.

Pure Having a uniform composition, or being without impurities. In chemistry, the term refers to a material that consists of a single element or compound.

Impure In chemistry, this term refers to a material that is a mixture of more than one element or compound.

Heterogeneous mixture A mixture in which the various components can be seen as individual substances.

Homogeneous mixture A mixture in which the components are so finely mixed that the composition is the same throughout.

Solution A homogeneous mixture in which all components are in the same phase.

Suspension A homogeneous mixture in which the various components are in different phases.

Solvent The component in a solution that is present in the largest amount.

Solute Any component in a solution that is not the solvent.

Dissolving The process of mixing a solute in a solvent to produce a homogeneous mixture.

Saturated solution A solution containing the maximum amount of solute that will dissolve in its solvent.

Unsaturated solution A solution that is capable of dissolving additional solute.

Concentration A quantitative measure of the amount of solute in a solution.

Mole The amount of any pure substance that contains as many atoms, molecules, ions, or other elementary units as the number of atoms in 12 g of carbon-12. This is equal to 6.02×10^{23} particles.

Molarity A unit of concentration equal to the number of moles of a solute per liter of solution.

Solubility The ability of a solute to dissolve in a given solvent.

Soluble Capable of dissolving to an appreciable extent in a given solvent.

Insoluble Not capable of dissolving to any appreciable extent in a given solvent.

Precipitate A solute that has come out of solution.

Hard water Water containing large amounts of calcium and magnesium ions.

Semipermeable membrane A membrane that allows only the passage of molecules small enough to fit through its submicroscopic pores.

Osmosis The diffusion of water or some other fluid through a semipermeable membrane from a solution with a low concentration of solutes to a solution with a higher concentration of solutes.

Reverse osmosis A technique for purifying water by forcing it through a semipermeable membrane.

REVIEW QUESTIONS

16.1 Most Materials Are Mixtures

1. What defines a material as being a mixture?
2. How can the components of a mixture be separated from one another?
3. How does distillation separate the components of a mixture?

16.2 The Chemist's Classification of Matter

4. Why is it not practical to have a macroscopic sample that is 100% pure?
5. How is a solution different from a suspension?
6. How can a solution be distinguished from a suspension?

16.3 Solutions

7. What happens to the volume of a sugar solution as more sugar is dissolved in it?
8. Why is a ruby considered a solution?
9. Distinguish between a solute and a solvent.
10. What does it mean to say that a solution is concentrated?
11. Is 1 mole of particles a very large number of particles or a very small number?

16.4 Solubility

12. By what means are ethanol and water molecules attracted to each other?
13. What does it mean to say that two materials are infinitely soluble in each other?
14. What type of electrical attraction is responsible for oxygen's ability to dissolve in water?
15. Why does the solubility of a gas solute in a liquid solvent decrease with increasing temperature?

16.5 Soaps and Detergents

16. Which portion of a soap molecule is nonpolar?
17. Water and soap are attracted to each other by what type of electrical attraction?
18. Grime molecules are likely attracted to each other by what type of electrical attraction?
19. What is the difference between a soap and a detergent?

16.6 Softening Hard Water

20. What component of hard water makes it hard?
21. Why are soap molecules so attracted to calcium and magnesium ions?
22. Calcium and magnesium ions are more attracted to sodium carbonate than to soap. Why?

16.7 Purifying the Water We Drink

23. Why is treated water sprayed into the air before being piped to users?
24. What are two ways in which people disinfect water in areas where municipal treatment facilities are not available?
25. What naturally occurring element has been contaminating the water supply of Bangladesh?
26. What are the two main techniques for desalinizing water?
27. Name one disadvantage of solar distillation.

16.8 Wastewater Treatment

28. Why can wastewater treatment requirements in Hawaii be less stringent than those in most locations on the U.S. mainland?
29. What is the first step in treating raw sewage?
30. What is the source of the energy used to aerate wastewater in an advanced integrated pond system?

EXERCISES

● BEGINNER ■ INTERMEDIATE ◆ EXPERT

1. ● Classify each of the following as (a) a homogeneous mixture, (b) a heterogeneous mixture, (c) an element, or (d) a compound: milk ____ steel ____ ocean water ____ blood ____ sodium ____ planet Earth ____
2. ■ Which box contains an element? A compound? A mixture? How many different types of molecules are shown altogether in all three boxes?

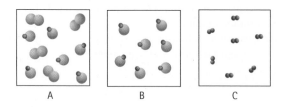

A   B   C

3. ■ How might you separate a mixture of sand and salt? How about a mixture of iron and sand?
4. ● Mixtures can be separated into their components by taking advantage of differences in the chemical properties of the components. Why might this separation method be less convenient than taking advantage of differences in the physical properties of the components?
5. ■ Why can't the elements of a compound be separated from one another by physical means?

6. ● How can a solution be distinguished from a suspension?

A   B   C

7. ■ Which of the above best represents a suspension?
8. ■ Which of the above best represents a solution?
9. ■ Which of the above best represents a compound?
10. ● Many dry cereals are fortified with iron, which is added to the cereal in the form of small iron particles. How might these particles be separated from the cereal?
11. ■ Explain why for these three substances, the solubility in 20°C water goes down as the molecules get larger but the boiling point goes up:

Substance	Boiling point/ Solubility
CH₃—O—H	65°C infinite
CH₃CH₂CH₂CH₂—O—H	117°C 8 g/100 mL
CH₃CH₂CH₂CH₂CH₂—O—H	138°C 2.3 g/100 mL

12. ◆ The boiling point of 1,4-butanediol is 230°C. Would you expect this compound to be soluble or insoluble in room-temperature water? Explain.

$$H\diagdown O—CH_2CH_2CH_2CH_2—O\diagup^H$$

1,4-Butanediol

13. ● What do tap water and distilled water contain besides water?

14. ● Why does oxygen have such a low solubility in water?

15. ● Why does the solubility of a gas solute in a liquid solvent decrease with increasing temperature?

16. ● Based on atomic size, which would you expect to be more soluble in water: helium, He, or nitrogen, N_2?

17. ■ If nitrogen, N_2, were pumped into your lungs at high pressure, what would happen to its solubility in your blood?

18. ■ The air a scuba diver breathes is pressurized to counteract the pressure exerted by the water surrounding the diver's body. Breathing the high-pressure air causes excessive amounts of nitrogen to dissolve in body fluids, especially the blood. If a diver ascends to the surface too rapidly, the nitrogen bubbles out of the body fluids (much as carbon dioxide bubbles out of a soda immediately after the container is opened). This results in a painful and potentially lethal medical condition known as the *bends*. Why does breathing a mixture of helium and oxygen rather than air help divers avoid getting the bends?

19. ● Why are noble gases infinitely soluble in other noble gases?

20. ■ Why does oxygen have some solubility in water?

21. ● Distinguish between a saturated solution and an unsaturated solution.

22. ■ At 10°C, which is more concentrated: a saturated solution of sodium nitrate, $NaNO_3$, or a saturated solution of sodium chloride, $NaCl$? (See Figure 16.20.)

23. ■ The volume of many liquid solvents expands with increasing temperature. What happens to the concentration of a solution made with such a solvent as the temperature of the solution is increased?

24. ● What is the relationship between a precipitate and a solute?

25. ■ Hydrogen chloride, HCl, is a gas at room temperature. Would you expect this material to be very soluble or not very soluble in water?

26. ■ Would you expect to find more dissolved oxygen in ocean water around the North Pole or in ocean water close to the equator? Why?

27. ■ How necessary is soap for removing salt from your hands? Why?

28. ■ When you set a pot of tap water on the stove to boil, you often see bubbles start to form well before boiling temperature is reached. Explain this observation.

29. ■ Why might softened water not be good for people trying to reduce their dietary sodium-ion intake?

30. ● A scum forms on the surface of boiling hard water. What is this scum? Why does it form?

31. ● Why do hot water heaters lose their efficiency quicker in households with hard water?

32. ● Phosphate ions, PO_4^{3-}, were once added to detergents to assist in cleaning. What function did they serve?

33. ● Phosphate ions, PO_4^{3-}, are no longer added to detergents because they cause excessive growth of algae in aquatic habitats receiving the wastewater. What chemical has replaced them?

34. ■ Might reverse osmosis be used to obtain fresh water from a sugar–water solution? Explain.

35. ◆ Cells at the top of a tree have a higher concentration of sugars than cells at the bottom. How might this fact help a tree move water upward from its roots?

36. ● Why is flushing a toilet with clean water from a municipal supply about as wasteful as flushing it with bottled water?

37. ■ What reverses with reverse osmosis?

38. ■ How does a semipermeable membrane allow the passage of the solvent but not the solute?

39. ■ Why is it significantly less costly to purify fresh water through reverse osmosis than to purify salt water through reverse osmosis?

40. ■ Some people fear drinking distilled water because they have heard it leaches minerals from the body. Using your knowledge of chemistry, explain how these fears have no basis and how distilled water is in fact very good for drinking.

41. ● Many homeowners get their drinking water piped up from wells dug into their property. Sometime this well water smells bad because of trace quantities of the gaseous compound hydrogen sulfide, H_2S. How might this odor be removed from water already taken from the tap?

42. ● Is the decomposition of food by bacteria in our digestive systems aerobic or anaerobic? What evidence supports your answer?

43. ● What do most water treatment plants use to filter water?

44. ● Where does most of the solid mass of raw sewage end up after being collected at a treatment facility?

45. ● How are the disinfecting properties of ultraviolet light and ozone similar to each other?

PROBLEMS

● BEGINNER   ■ INTERMEDIATE   ◆ EXPERT

1. ■ Assume the total number of molecules in a sample of liquid is about 3 million trillion. One million trillion of these are molecules of some poison, while 2 million trillion of these are water molecules. Show that the purity of the water is about 67%.

2. ■ Assume the total number of molecules in a glass of liquid is about 1,000,000 million trillion. One million trillion of these are molecules of some poison, while 999,999 million trillion of these are water molecules. What is the purity of the water? In other words, what percentage of all the molecules in the glass are water?

3. ■ You drink a small glass of water that is 99.9999% pure water and 0.0001% some poison. Assume the glass contains about a million million trillion molecules, which is

about 30 mL. Show that you just drank 1 million trillion poison molecules. Should you be concerned?

4. ■ Show that 45 g of sodium chloride is needed to make 15 L of a solution that has a concentration of 3.0 g of sodium chloride per liter of solution.

5. ■ If water is added to 1 mole of sodium chloride in a flask until the volume of the solution is 1 L, show that the molarity of the water is 1 molar (1 M). What is the molarity when water is added to 2 moles of sodium chloride to make 0.5 L of solution?

6. ◆ A student is told to use 20.0 g of sodium chloride to make an aqueous solution that has a concentration of 10.0 g of sodium chloride per liter of solution. Assuming that 20.0 g of sodium chloride has a volume of 7.5 mL, about how much water does she use in making this solution?

ACTIVE EXPLORATIONS

Bottoms Up and Bubbles Out

What's in a glass of water? Separate the components of your tap water to find out. *Safety note:* Wear safety glasses for step 1 because some splattering may occur.

Procedure

1. Put on your safety glasses and add tap water to a cooking pot. Boil the water to dryness. (Turn off the burner before the water is completely gone. The heat from the pot will complete the evaporation.)

2. Examine the resulting residue by scraping it with a knife. These are the solids you ingest with every glass of water you drink.

3. To see the gases dissolved in your water, fill a clean cooking pot with water and let it stand at room temperature for several hours. Note the bubbles that adhere to the inner sides of the pot. Where do the bubbles originate? What do you suppose they contain?

    For further experimentation, perform step 3 in two pots side by side. In one pot, use warm water from the kitchen faucet. In the second pot, use boiled water that has cooled to the same temperature. You'll find that boiling *deaerates* the water—that is, it removes the atmospheric gases. Chemists sometime need to use deaerated water, which is made by allowing boiled water to cool in a sealed container. Why don't fish live very long in deaerated water?

Circular Rainbows

Black ink contains pigments of many different colors. Acting together, these pigments absorb all the frequencies of visible light. Because no light is reflected, the ink appears black. We can use molecular attractions to separate the components of black ink through a technique called *paper chromatography*.

What You Need

Black felt-tip pen or black water-soluble marker; piece of porous paper, such as a paper towel, table napkin, or coffee filter; solvent, such as water, acetone (fingernail polish remover), rubbing alcohol, or white vinegar

Procedure

1. Place a concentrated dot of ink at the center of the piece of porous paper.

2. Carefully place one drop of solvent on top of the dot, and watch the ink spread radially with the solvent. Because the different components of the ink have different affinities for the solvent (based on the attractions between component molecules and solvent molecules), they travel with the solvent at different rates.

3. Just after the drop of solvent is completely absorbed, add a second drop at the same location as the first one, then a third, and so on until the ink components have separated to your satisfaction.

    How the components separate depends on several factors, including your choice of solvent and your technique. It's also interesting to watch the leading edge of the moving ink under a strong magnifying glass or microscope.

    Paper chromatography was originally developed to separate plant pigments from one another. The separated pigments had different colors, which is how this technique got its name—chroma is Latin for "color." Mixtures need not be colored, however, to be separable by chromatography. All that's required is that the components have distinguishable affinities for the moving solvent and the stationary medium, such as paper, through which the solvent will pass.

Overflowing Sweetness

Just because a solid dissolves in a liquid doesn't imply the solid no longer occupies space.

What You Need

Tall glass, warm water, a container larger than the tall glass, 4 Tbsp table sugar

Procedure

1. Fill the glass to its brim with the warm water, and then carefully pour all the water into the larger container.

2. Add the sugar to the empty glass.

3. Return half of the warm water to the glass and stir to dissolve all the sugar.

4. Return the remaining water, and as you get close to the top, ask a friend to predict whether the water level will be less than before, about the same as before, or more than before so that the water spills over the edge of the glass.

If your friend doesn't understand the result, ask him or her what would happen if you had added the sugar to the glass when the glass was full of water.

Micro Water Purifier

You can build a relatively inefficient but fun-to-watch distiller at home. *Safety note:* Wear safety glasses and avoid the steam produced in this experiment—steam burns can be particularly harmful.

What You Need

A deep cooking pot; water; food coloring; salt; heavy ceramic coffee mug; plastic wrap; rubber band; scissors; ice cube

Procedure

1. Fill the cooking pot to a depth of about 1 cm. Add several drops of food coloring or some salt (or both) to the water and stir.
2. Place the coffee mug in the center of the pot. The height of the mug should be at least an inch below the height of the pot.
3. Lay plastic wrap loosely across the top of the pot and secure it with a rubber band. The seal should not be airtight. Instead, leave two of the edges open to prevent a buildup of pressure within the pot as the water is brought to a boil. Use scissors to trim away the excess plastic wrap along the perimeter. Place an ice cube at the center of the plastic wrap, which should then sag above the mug.
4. Put on your safety glasses and turn the burner on low to bring the water in your distiller to a low boil. Look for signs of cloud formation below the ice cube. Once boiling begins, the mug may jostle in the pot. Turn the heat down or off if the jostling becomes too pronounced.
5. Boil the water for only as long as ice is still present on the plastic wrap. The melted ice can be removed with a sponge. Distillation continues well after the stove has been turned off. Examine the water in the mug. Why isn't the food coloring or the salt carried over into the mug? How much distilled water are you able to collect per ice cube? How might you modify your distiller so that it works well using only sunlight to drive the distillation?

READINESS ASSURANCE TEST (RAT)

If you have a good handle on this chapter, if you really do, then you should be able to score 7 out of 10 on this RAT. If you score less than 7, you need to study further before moving on.

Choose the BEST answer to each of the following.

1. Someone argues that he or she doesn't drink tap water because it contains thousands of molecules of some impurity in each glass. How would you respond in defense of the water's purity, if it indeed does contain thousands of molecules of some impurity per glass?
   (a) Impurities aren't necessarily bad; in fact, they may be good for you.
   (b) The water contains water molecules and each water molecule is pure.
   (c) There's no defense. If the water contains impurities it should not be drunk.
   (d) Compared to the billions of water molecules, a thousand molecules of something else is practically nothing.
2. What is the difference between a compound and a mixture?
   (a) They both consist of atoms from different elements.
   (b) Their atoms are bonded together differently.
   (c) One is a solid and the other is a liquid.
   (d) The components of a mixture are not chemically bonded together.
3. Is the air in your house a homogeneous or heterogeneous mixture?
   (a) homogeneous, because it is mixed very well
   (b) heterogeneous, because of the dust particles it contains
   (c) homogeneous, because it is all at the same temperature
   (d) heterogeneous, because it consists of different types of molecules
4. Why is half-frozen fruit punch always sweeter than the same fruit punch completely melted?
   (a) because the sugar sinks to the bottom
   (b) because crystallization is a purifying process
   (c) because the half-frozen fruit punch is warmer
   (d) because sugar molecules are less soluble in a half-frozen solution
5. Suggest why sodium chloride, NaCl, is insoluble in gasoline. Consider the electrical attractions.
   (a) Because the NaCl molecule is so small, the gasoline doesn't have much opportunity to interact with it through any electrical attractions.

(b) Because gasoline is a very polar molecule, the NaCl can form only dipole–induced dipole bonds, which are very weak, giving it a low solubility in gasoline.

(c) Because gasoline is so strongly attracted to itself, NaCl is excluded.

(d) NaCl is composed of ions that are too attracted to themselves. Gasoline is nonpolar, so NaCl and gasoline do not interact very well.

6. Fish don't live very long in water that has just been boiled and brought back to room temperature. Suggest why.

(a) There is now a higher concentration of dissolved CO_2 in the water.

(b) The nutrients in the water have been destroyed.

(c) Because some of the water evaporated while boiling, the salts in the water are now more concentrated. This has a negative effect on the fish.

(d) The boiling process removes the air that was dissolved in the water. Upon cooling, the water is void of its usual air content, and hence the fish drown.

7. How many grams of sugar (sucrose) are in 5 L of sugar water that has a concentration of 0.5 g per liter of solution?

(a) 50 g
(b) 25 g
(c) 2.5 g
(d) 1.5 g

8. What is an advantage of using chlorine gas to disinfect drinking water supplies?

(a) It provides residual protection against pathogens.
(b) It gives the water a fresh taste.
(c) Residual chlorine in water helps whiten teeth.
(d) Excess chlorine is absorbed in our bodies as a mineral supplement.
(e) all of these

9. Why do red blood cells, which contain an aqueous solution of dissolved ions and minerals, burst when placed in fresh water?

(a) The dissolved ions provide a pressure that eventually bursts open the cell.
(b) More water molecules enter the cell than leave the cell.
(c) The fresh water dissolves the blood cell wall.
(d) all of these

10. A stagnant pond smells worse than a babbling brook because

(a) odors are not transported downstream.
(b) of the type of aquatic life it attracts.
(c) it lacks sufficient dissolved oxygen.
(d) all of these

Answers to RAT

1d, 2d, 3b, 4b, 5d, 6d, 7c, 8a, 9c, 10c

EXPLORING FURTHER

http://www.sugar.org

According to this site, the Sugar Association is proud to provide reliable, science-based information about pure, natural sugar. Reliance on sound science has positioned the association as a leader in communicating accurate information about the nutritional and functional uses of sugar to consumers, professionals, and the media.

http://www.idadesal.org

Home page of the International Desalination Association, whose goals are the development and promotion

of the appropriate use of desalination technology worldwide.

http://www.bicn.com/acic

To learn more about the arsenic crisis in Bangladesh, visit the home page of the Arsenic Crisis Information Centre. A detailed description of the chemistry involved in the removal of arsenic can be found at http://www.eng2.uconn.edu/~nikos/asrt-brochure.html.

CHAPTER 16 ONLINE RESOURCES

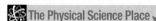 The Physical Science Place

Interactive Figure

- 16.32

Tutorial

- Solubility

Videos

- Mixtures Can Be Separated by Physical Means
- Solutions

- Solubility Changes with Temperature and Pressure
- Soap Works by Being Both Polar and Nonpolar

Quiz

Flashcards

Links

how chemicals react

17.1 CHEMICAL EQUATIONS

17.2 COUNTING ATOMS AND MOLECULES BY MASS

17.3 REACTION RATES

17.4 CATALYSTS

17.5 ENERGY AND CHEMICAL REACTIONS

17.6 CHEMICAL REACTIONS ARE DRIVEN BY ENTROPY

■ The heat of a lightning bolt causes multiple chemical reactions in the atmosphere, including one in which nitrogen and oxygen react to form nitrogen monoxide, NO. The nitrogen monoxide formed in this manner then reacts with atmospheric oxygen and water vapor to form nitric acid, HNO_3 and nitrous acid, HNO_2. These acids are carried by rain into the ground, where they form ions that are absorbed by growing plants—a process that involves further chemical reactions.

Scientists have learned how to control chemical reactions to produce many useful materials—nitrates and other nitrogen-based fertilizers from atmospheric nitrogen, metals from rocks, and plastics and pharmaceuticals from petroleum. These materials and the thousands of others produced by chemical reactions, as well as the abundant energy released when fossil fuels take part in the chemical reaction called combustion, have dramatically improved our living conditions.

The goal of this chapter is to provide you with a stronger handle on the basics of chemical reactions, which were introduced in Chapter 14. Then, in the following chapter, we'll look at two specific classes of chemical reactions: acid–base reactions and oxidation–reduction reactions.

■ 17.1 Chemical Equations

As was discussed in Chapter 14, during a chemical reaction, atoms rearrange to create one or more new compounds. This activity is neatly summed up in written form as a **chemical equation**. A chemical equation shows the reacting substances, called **reactants**, to the left of an arrow that points to the newly formed substances, called **products**:

$$\text{reactants} \longrightarrow \text{products}$$

Link To Section 14.3

The Physical Science Place

Chemical Reactions and Equations

Typically, reactants and products are represented by their elemental or chemical formulas. Sometimes molecular models or, simply, names may be used instead. Phases are also often shown: (s) for solid, (l) for liquid, and (g) for gas. Compounds dissolved in water are designated (aq) for aqueous solution. Lastly, numbers are placed in front of the reactants or products to show the ratio in which they either combine or form. These numbers are called *coefficients*, and they represent numbers of individual atoms and molecules. For instance, to represent the chemical reaction in which coal burns in the presence of oxygen to form gaseous carbon dioxide, we write the chemical equation using coefficients of 1:

$$1\ C(s) + 1\ O_2(g) \longrightarrow 1\ CO_2(g) \quad \text{(balanced)}$$

Reactants   Products

One of the most important principles of chemistry is the **law of mass conservation**. The law of mass conservation states that matter is neither created nor destroyed during a chemical reaction.* The atoms present at the beginning of a reaction merely rearrange to form new molecules. This means that no atoms are lost or gained during any reaction. The chemical equation must therefore be *balanced*. In a balanced equation, each atom must appear on both sides of the arrow the same number of times. The equation for the formation of carbon dioxide is balanced because each side shows one carbon atom and two oxygen atoms. You can count the number of atoms in the models to see this for yourself.

---

* For all practical purposes this law holds true. Technically, however, any energy released or absorbed by a chemical reaction arises from the transformation of matter into energy, or vice versa. The amount of matter lost or gained in a chemical reaction, however, is so small that, for all practical purposes, we can ignore this detail. Not so for the nuclear reactions discussed in Chapter 13. For these reactions, matter/energy conversions are much more pronounced.

In another chemical reaction, two hydrogen gas molecules, H_2, react with one oxygen gas molecule, O_2, to produce two molecules of water, H_2O, in the gaseous phase:

$$2\,H_2(g) + 1\,O_2(g) \longrightarrow 2\,H_2O(g) \qquad \text{(balanced)}$$

This equation for the formation of water is also balanced—there are four hydrogen and two oxygen atoms before and after the arrow.

A coefficient in front of a chemical formula tells us the number of times that element or compound must be counted. For example, $2\,H_2O$ indicates two water molecules, which contain a total of four hydrogen atoms and two oxygen atoms.

By convention, the coefficient 1 is omitted so that the above chemical equations are typically written

$$C(s) + O_2(g) \longrightarrow CO_2(g) \qquad \text{(balanced)}$$

$$2\,H_2(g) + O_2(g) \longrightarrow 2\,H_2O(g) \qquad \text{(balanced)}$$

CHECK POINT
■

How many oxygen atoms are indicated by the following balanced equation?

$$3\,O_2(g) \longrightarrow 2\,O_3(g)$$

Was this your answer?

Before the reaction, these six oxygen atoms are found in three O_2 molecules. After the reaction, these same six atoms are found in two O_3 molecules.

An unbalanced chemical equation shows the reactants and products without the correct coefficients. For example, the equation:

$$NO(g) \longrightarrow N_2O(g) + NO_2(g) \qquad \text{(not balanced)}$$

is not balanced because there is one nitrogen atom and one oxygen atom before the arrow, but three nitrogen atoms and three oxygen atoms after the arrow.

You can balance unbalanced equations by adding or changing coefficients to produce correct ratios. (It's important not to change subscripts, however, because to do so changes the compound's identity—H_2O is water, but H_2O_2 is hydrogen peroxide!) For example, to balance the equation above, add a 3 before the NO:

$$3\,NO(g) \longrightarrow N_2O(g) + NO_2(g) \qquad \text{(balanced)}$$

Now there are three nitrogen atoms and three oxygen atoms on each side of the arrow, and the law of mass conservation is not violated.

Practicing chemists develop a skill for balancing equations. This skill involves creative energy and, like other skills, improves with experience. More important than being an expert at balancing equations is knowing why they need to be balanced. And the reason is the law of mass conservation, which tells us that atoms are neither created nor destroyed in a chemical reaction—they are simply rearranged. So every atom present before the reaction must be present after the reaction, even though the groupings of atoms are different.

fyi

■ Chemical explosions typically involve the transformation of an unstable solid or liquid chemical into more stable gases that occupy much more volume. Upon detonation, one mole of nitroglycerin, $C_3H_5N_3O_9$, produces 7.25 moles of gases including carbon dioxide, CO_2; nitrogen, N_2; oxygen, O_2; and water vapor, H_2O. The volume change is dramatic—from less than 0.3 L to about 170 L, which is an increase of about 600%. For nitroglycerin and similar high explosives, these gases expand at supersonic speeds, creating a powerful and destructive shock wave.

CHECK POINT

Write a balanced equation for the reaction showing hydrogen gas, H_2, and nitrogen gas, N_2, forming ammonia gas, NH_3:

$$\_\_\_H_2(g) + \_\_\_N_2(g) \longrightarrow \_\_\_NH_3(g)$$

Was this your answer?

Initially, we see two hydrogen atoms before the reaction arrow and three on the right. This can be remedied by placing a coefficient of 3 by the hydrogen, H_2, and a coefficient of 2 by the ammonia, NH_3. This makes for six hydrogen atoms both before and after the reaction arrow. Meanwhile, the coefficient of 2 by the ammonia also makes for two nitrogen atoms after the arrow, which balances out the two nitrogen atoms appearing before the arrow. The full balanced equation, therefore, is:

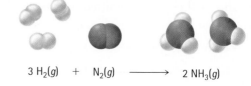

$$3\ H_2(g) \quad + \quad N_2(g) \quad \longrightarrow \quad 2\ NH_3(g)$$

Chemists use many methods to balance equations. Look to the *Conceptual Physical Science—4th Edition Practice Book* supplement for an example. Your instructor may also share with you his or her favorite methods. For more practice balancing equations, see the questions at the end of this chapter.

17.2 Counting Atoms and Molecules by Mass

In any chemical reaction, a specific number of reactants react to form a specific number of products. For example, when carbon and oxygen combine to form carbon dioxide, they always combine in the ratio of one carbon atom to one oxygen molecule. A chemist who wants to carry out this reaction in the laboratory would be wasting chemicals and money if she were to combine, say, four carbon atoms for every one oxygen molecule. The excess carbon atoms would have no oxygen molecules to react with and would remain unchanged.

How is it possible to measure out a specific number of atoms or molecules? Rather than counting these particles individually, chemists can use a scale that measures the mass of bulk quantities. Because different atoms and molecules have different masses, however, a chemist can't simply measure out equal masses of each. Say, for example, he needs the same number of carbon atoms as oxygen molecules. Measuring equal masses of the two materials would not provide equal numbers.

You know that 1 kg of Ping-Pong balls contains more balls than 1 kg of golf balls, as Figure 17.1 illustrates. Likewise, because different atoms and molecules have different masses, there are different numbers of them in a 1-g sample of each. Because carbon atoms are less

FIGURE 17.1

The number of balls in a given mass of Ping-Pong balls is very different from the number of balls in the same mass of golf balls.

Equal masses

massive than oxygen molecules, there are more carbon atoms in 1 g of carbon than
there are oxygen molecules in 1 g of oxygen. So, clearly, equal masses of these two
particles do not yield equal numbers of carbon atoms and oxygen molecules.

If we know the *relative masses* of different materials, we can measure equal
numbers. Golf balls, for example, are about 20 times as massive as Ping-Pong
balls, which is to say the relative mass of golf balls to Ping-Pong balls is 20 to 1.
Measuring out 20 times as much mass of golf balls as Ping-Pong balls, therefore,
gives equal numbers of each, as is shown in Figure 17.2.

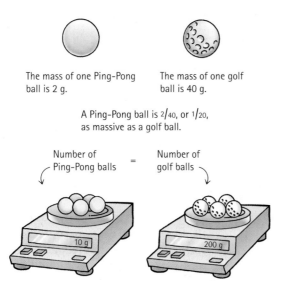

FIGURE 17.2

The number of golf balls in
200 g of golf balls equals the
number of Ping-Pong balls in
10 g of Ping-Pong balls.

CHECK POINT

A customer wants to buy a 1:1 mixture of blue and red jelly beans.
Each blue bean is twice as massive as each red bean. If the clerk
measures out 5 lb of red beans, how many pounds of blue beans
must she measure out?

Was this your answer?
Because each blue jelly bean has twice the mass of each red one, the
clerk needs to measure out twice as much mass of blues in order to
have the same count, which means 10 lb of blues. If the clerk did not
know that the blue beans were twice as massive as the red ones, she
would not know what mass of blues was needed for the 1:1 ratio.
Likewise, a chemist would be at a loss in setting up a chemical reaction if she did not know the relative masses of the reactants.

The masses of elements shown in the periodic table are relative
masses. Using these masses we can measure out equal numbers of atoms
or molecules. For example, as illustrated in Figure 17.3, the mass of carbon is 12.011 amu. (As discussed in Section 12.3, one *atomic mass unit*
[amu] equals 1.661×10^{-24} gram.) The **formula mass** of a substance
is the sum of the atomic masses of the elements in its chemical formula.
Therefore, the formula mass of an oxygen molecule, O_2, is about 32
amu (15.999 amu + 15.999 amu). A carbon atom, therefore, is about $\frac{12}{32}$ as
massive as an oxygen molecule. To measure out equal numbers of carbon atoms
and oxygen molecules we could measure out 12 g of carbon and 32 g of molecular
oxygen. Any proportion equal to $\frac{12}{32}$, such as $\frac{6}{16}$ or $\frac{3}{8}$, would do. For example, 3 g
of carbon would have the same number of particles as 8 g of molecular oxygen.

FIGURE 17.3

To have equal numbers of carbon atoms and oxygen mole-
cules requires measuring out $\frac{3}{8}$
as much carbon as oxygen.

1. Reacting 3 g of carbon, C, with 8 g of molecular oxygen, O_2, results in 11 g of carbon dioxide, CO_2. Does it follow that 1.5 g of carbon will react with 4 g of oxygen to form 5.5 g of carbon dioxide?
2. Would reacting 5 g of carbon with 8 g of oxygen also result in 11 g of carbon dioxide?

Were these your answers?
1. The quantities are only half as much, but their ratio is the same as when 11 g of carbon dioxide is formed: $1.5 : 4 : 5.5 = 3 : 8 : 11$.
2. Many students make the common error of thinking that no reaction will occur if the proper ratios of reactants are not provided. You should understand, however, that in a 5-g sample of carbon, 3 g of carbon is available for reacting. This 3 g will react with the 8 g of oxygen to form 11 g of carbon dioxide. There will be 2 g of carbon unreacted after the reaction. Reacting this remaining 2 g of carbon would require more oxygen.

CONVERTING BETWEEN GRAMS AND MOLES

Atoms and molecules react in specific ratios. In the laboratory, however, chemists work with bulk quantities of materials, which are measured by mass. Chemists therefore need to know the relationship between the mass of a given sample and the number of atoms or molecules contained in that mass. The key to this relationship is the *mole*. Recall from Section 16.3 that the mole is a unit equal to 6.02×10^{23}. This number is known as **Avogadro's number**, in honor of the 18th-century scientist Amedeo Avogadro.

As Figure 17.4 illustrates, if you express the numeric value of the atomic mass of any element in *grams*, the number of atoms in a sample of the element having this mass is always 6.02×10^{23}, which is 1 mole. For example, a 22.990-g sample of sodium metal, Na (atomic mass $= 22.990$ amu), contains 6.02×10^{23} sodium atoms, and a 207.2-g sample of lead, Pb (atomic mass $= 207.2$ amu), contains 6.02×10^{23} lead atoms.

The same concept holds for compounds. Express the numeric value of the formula mass of any compound in grams, and a sample having that mass contains 6.02×10^{23} molecules of that compound. For example, there are 6.02×10^{23} O_2 molecules in 31.998 g of molecular oxygen, O_2 (formula mass $= 31.998$ amu), and 6.02×10^{23} CO_2 molecules in 44.009 g of carbon dioxide, CO_2 (formula mass $= 44.009$ amu).

fyi

■ An Avogadro's number of grains of sand would fill the United States to a depth of about 2 m. There are about 6.4 billion people on Earth. You would need about 94 trillion Earth-size populations to have an Avogadro's number of people. If you collected 1 million hydrogen atoms every second, it would take you about 19 billion years to come up with a whole gram of hydrogen—the universe itself is only about 13 billion years old. A stack of an Avogadro's number of pennies would be about 800,000 trillion km, which is about the diameter of our galaxy. Placed side by side, these pennies would reach to the Andromeda galaxy, which is about a million light-years away.

FIGURE 17.4

Express the numeric value of the atomic mass of any element in grams, and that many grams contains 6.02×10^{23} atoms.

CHECK
  ■   POINT

1. How many atoms are there in a 6.941-g sample of lithium, Li (atomic mass = 6.941 amu)?
2. How many molecules are there in an 18.015-g sample of water, H_2O (formula mass = 18.015 amu)?

Were these your answers?

1. Because this number of grams of lithium is numerically equal to the atomic mass, the sample contains 6.02×10^{23} lithium atoms, which is 1 mole.
2. Because this number of grams of water is numerically equal to the formula mass, the sample contains 6.02×10^{23} water molecules, which is 1 mole.

The **molar mass** of any substance, be it element or compound, is defined as the mass of 1 mole of the substance. Thus the units of molar mass are grams per mole. For instance, the atomic mass of carbon is 12.011 amu, which means that 1 mole of carbon has a mass of 12.011 g, and we say that the molar mass of carbon is 12.011 g/mole. The molar mass of molecular oxygen, O_2, (formula mass 31.998 amu) is 31.998 g/mole. For convenience, values such as these are often rounded off to the nearest whole number. The molar mass of carbon, therefore, might also be presented as 12 g/mole, and that of molecular oxygen as 32 g/mole.

CHECK
  ■   POINT

What is the molar mass of water (formula mass 18 amu)?

Was this your answer?

From the formula mass, you know that 1 mole of water has a mass of 18 g. Therefore the molar mass is 18 g/mole.

Because 1 mole of any substance always contains 6.02×10^{23} particles, the mole is an ideal unit for chemical reactions. For example, 1 mole of carbon (12 g) reacts with 1 mole of molecular oxygen (32 g) to give 1 mole of carbon dioxide (44 g).

In many instances, the ratio in which chemicals react is not 1 : 1. As shown in Figure 17.5, for example, 2 moles (4 g) of molecular hydrogen react with 1 mole (32 g) of molecular oxygen to give 2 moles (36 g) of water. Note how the coefficients of the balanced chemical equation can be conveniently interpreted as the number of moles of reactants or products. A chemist therefore need only convert these numbers of moles to grams in order to know how much mass of each reactant he or she should measure out to have the proper proportions.

$2\,H_2$	$+$	$1\,O_2$	$\longrightarrow$	$2\,H_2O$
2 moles		1 mole		2 moles
which is		which is		which is
4 g		32 g		36 g
which is		which is		which is
12.04×10^{23} molecules		6.02×10^{23} molecules		12.04×10^{23} molecules

FIGURE 17.5

Two moles of H_2 react with 1 mole of O_2 to give 2 moles of H_2O. This is the same as saying 4 g of H_2 reacts with 32 g of O_2 to give 36 g of H_2O or, equivalently, that 12.04×10^{23} H_2 molecules react with 6.02×10^{23} O_2 molecules to give 12.04×10^{23} H_2O molecules.

FIGURING PHYSICAL SCIENCE

▪ Masses of Reactants and Products

The coefficients of a chemical equation tell us the ratio by which reactants react and products form. The following equation, for example, tells us that every 1 mole of methane, CH_4, reacts to produce 2 moles of water, H_2O.

$$CH_4 + 2O_2 \longrightarrow CO_2 + 2H_2O$$

So if you were given 16 g of methane, CH_4, how many grams of water, H_2O, would form? From the text you should know that 16 g of methane, CH_4, is 1 mole (formula mass 16 amu). So 16 g of methane would yield 2 moles of water. But how many grams is 2 moles of water? Well, if 1 mole of water, H_2O, equals 18 g (formula mass 18 amu), then 2 moles equals 36 g. Thus, 16 g of methane, CH_4, reacting with oxygen, O_2, would yield 36 g of water, H_2O. Let's look at this process from a step-by-step mathematical point of view.

SAMPLE PROBLEM 1

What mass of water is produced when 16 g of methane, CH_4 (formula mass 16 amu), reacts with oxygen, O_2, in the reaction below?

$$CH_4 + 2O_2 \longrightarrow CO_2 + 2H_2O$$

SOLUTION:

Step 1. Convert the given mass to moles:

Conversion factor

$$(16 \text{ g } CH_4)\left(\frac{1 \text{ mole } CH_4}{16 \text{ g } CH_4}\right) = 1 \text{ mole } CH_4$$

Step 2. Use the coefficients of the balanced equation to find out how many moles of H_2O are produced from this many moles of CH_4:

Conversion factor

$$(1 \text{ mole } CH_4)\left(\frac{2 \text{ moles } H_2O}{1 \text{ mole } CH_4}\right)$$
$$= 2 \text{ moles } H_2O$$

Step 3. Now that you know how many moles of H_2O are produced, convert this value to grams of H_2O:

Conversion factor

$$(2 \text{ moles } H_2O)\left(\frac{18 \text{ g } H_2O}{1 \text{ mole } H_2O}\right) = 36 \text{ g } H_2O$$

The method of converting grams of a substance to moles (step 1), then from moles of this substance to moles of that substance (step 2), followed by moles of that substance to grams (step 3) is called *stoichiometry*. Using stoichiometry, a scientist can calculate the amounts of reactants or products in any chemical reaction. The methods of stoichiometry are developed much further in general chemistry courses. For this course, all you need to do is be familiar with what stoichiometry is all about, which is keeping tabs on atoms and molecules as they react to form products. Nonetheless, for a special assignment, you might try your analytical-thinking skills on the following problems. First try to deduce the answer based on what you know about the law of mass conservation, and then follow the steps given here to check your answers.

SAMPLE PROBLEM 2

Show that 44 g of carbon dioxide, CO_2, is produced when 16 g of methane, CH_4, reacts with oxygen, O_2. How many grams of oxygen, O_2, are needed for this reaction?

SOLUTION:

The 16 g of methane, CH_4, is 1 mole, which reacts with oxygen to produce

---

Cooking and chemistry are similar in that both require measuring ingredients. Just as a cook looks to a recipe to find the necessary quantities measured by the cup or the tablespoon, a chemist looks to the periodic table to find the necessary quantities measured by the number of grams per mole for each element or compound.

▪ 17.3 Reaction Rates

Link To Section 21.4

A balanced chemical equation helps determine the amount of products that can be formed from given amounts of reactants. But the equation tells us little about what occurs on the submicroscopic level during the reaction. In this and the following section, we explore that submicroscopic level to show how the *rate* of a reaction can be changed, either by changing the concentration or the temperature of the reactants or by adding what is known as a *catalyst*.

Some chemical reactions, such as the rusting of iron, are slow, while others, such as the burning of gasoline, are fast. The speed of any reaction is indicated by its reaction rate, which is an indicator of how quickly the reactants transform to products. As shown in Figure 17.6, initially a flask may contain only reactant molecules. Over time, these reactants form product molecules, and, as a result,

1 mole of carbon dioxide, CO_2. One mole of carbon dioxide (formula mass 44 amu) is 44 g. So the 16 g of methane reacts with oxygen to produce 44 g of carbon dioxide plus 36 g of water. The mass of the reactants (44 g + 36 g = 80 g) must be equal to the mass of the products (16 g + ? = 80 g). So we can calculate that the methane reacted with 64 g of oxygen, which, interestingly enough, is 2 moles, as shown in the equation.

SAMPLE PROBLEM 3

How many grams of ozone (O_3, 48 amu) can be produced from 64 g of oxygen (O_2, 32 amu) in the reaction below?

$$3\,O_2 \longrightarrow 2\,O_3$$

SOLUTION:

According to the law of mass conservation, the amount of mass in the products must equal the amount of mass in the reactants. Given that this reaction involves only one reactant and one product, you should not be surprised to learn that 64 g of reactant produces 64 g of product. Here are the stepwise calculations:

Step 1. Convert grams of O_2 to moles of O_2:

$$(64\,\text{g}\,O_2)\left(\frac{1\text{ mole }O_2}{32\,\text{g}\,O_2}\right) = 2\text{ moles }O_2$$

Step 2. Convert moles of O_2 to moles of O_3:

$$(2\text{ moles }O_2)\left(\frac{2\text{ moles }O_3}{3\text{ moles }O_2}\right)$$
$$= 1.33\text{ moles }O_3$$

Step 3. Convert moles of O_3 to grams of O_3:

$$(1.33\text{ moles }O_3)\left(\frac{48\text{ g }O_3}{1\text{ mole }O_3}\right) = 64\text{ g }O_3$$

SAMPLE PROBLEM 4

What mass of nitrogen monoxide (NO, 30 amu) is formed when 28 g of nitrogen (N_2, 28 amu) reacts with 32 g of oxygen (O_2, 32 amu) in the reaction below?

$$N_2 + O_2 \longrightarrow 2\,NO$$

How about when 28 g of nitrogen, (N_2), is combined with 40 g of oxygen, O_2?

SOLUTION:

There are several ways to answer this problem. One way would be to recognize that 28 g of N_2 is 1 mole of N_2 and 32 g of O_2 is 1 mole of O_2. According to the balanced equation, combining 1 mole of N_2 with 1 mole of O_2 yields 2 moles of NO. The mass of 2 moles of NO is

$$(2\text{ moles NO})\left(\frac{30\text{ g NO}}{1\text{ mole NO}}\right) = 60\text{ g NO}$$

which is the sum of the masses of the reactants, as it must be because of the law of mass conservation.

Combining 40 g of oxygen with 28 g of nitrogen would be 8 g too much oxygen. Only 32 g of this oxygen would react with the nitrogen, producing 60 g of NO, leaving 8 g of oxygen left over unreacted.

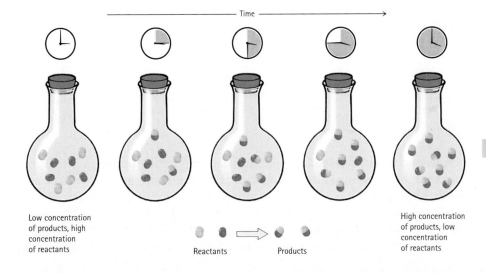

Time

Low concentration of products, high concentration of reactants

Reactants   Products

High concentration of products, low concentration of reactants

FIGURE 17.6

Over time, the reactants in this reaction flask may transform to products. If this happens quickly, the reaction rate is high. If this happens slowly, the reaction rate is low.

During a reaction, reactant molecules collide with one another.

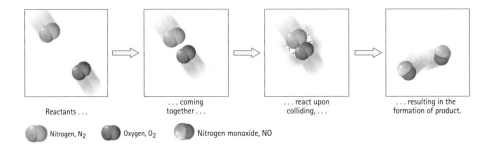

Reactants . . . | . . . coming together . . . | . . . react upon colliding, . . . | . . . resulting in the formation of product.

Nitrogen, N$_2$ Oxygen, O$_2$ Nitrogen monoxide, NO

the concentration of product molecules increases. The **reaction rate**, therefore, can be defined either as how quickly the concentration of products increases or as how quickly the concentration of reactants decreases.

What determines the rate of a chemical reaction? The answer is complex, but one important factor is that reactant molecules must physically come together. Because molecules move rapidly, this physical contact is appropriately described as a collision. We can illustrate the relationship between molecular collisions and reaction rate by considering the reaction of gaseous nitrogen and gaseous oxygen to form gaseous nitrogen monoxide, as shown in Figure 17.7.

Because reactant molecules must collide in order for a reaction to occur, the rate of a reaction can be increased by increasing the number of collisions. An effective way to increase the number of collisions is to increase the concentration of the reactants. Figure 17.8 shows that with higher concentrations, more molecules are in a given volume, which makes collisions between molecules more probable. As an analogy, consider a group of people on a dance floor—as the number of people increases, so does the rate at which they bump into one another. An increase in the concentration of nitrogen and oxygen molecules, therefore, leads to a greater number of collisions between these molecules; hence, a greater number of nitrogen monoxide molecules form in a given period of time.

Not all collisions between reactant molecules lead to products, however, because the molecules must collide in a certain orientation in order to react. Nitrogen and oxygen, for example, are much more likely to form nitrogen monoxide when the molecules collide in the parallel orientation shown in Figure 17.7. When they collide in the perpendicular orientation shown in Figure 17.9, nitrogen monoxide does not form. For larger molecules, which can have numerous orientations, this orientation requirement is even more restrictive.

Less concentrated

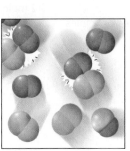

More concentrated

The more concentrated a sample of nitrogen and oxygen, the greater the probability that N$_2$ and O$_2$ molecules will collide and form nitrogen monoxide.

The orientation of reactant molecules in a collision can determine whether a reaction occurs. A perpendicular collision between N$_2$ and O$_2$ tends not to result in formation of a product molecule.

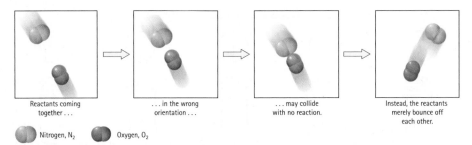

Reactants coming together . . . | . . . in the wrong orientation . . . | . . . may collide with no reaction. | Instead, the reactants merely bounce off each other.

Nitrogen, N$_2$ Oxygen, O$_2$

A second reason that not all collisions lead to product formation is that the reactant molecules must also collide with enough kinetic energy to break their bonds. Only then is it possible for the atoms in the reactant molecules to change bonding partners and form product molecules. The bonds in N$_2$ and O$_2$ molecules, for example, are quite strong. In order for these bonds to be broken, collisions between the molecules must contain enough energy to break the bonds. As

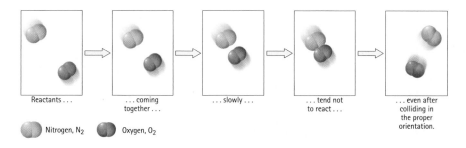

Reactants . . .   . . . coming
                    together . . .   . . . slowly . . .   . . . tend not
                                                             to react . . .   . . . even after
                                                                               colliding in
                                                                               the proper
                                                                               orientation.

Nitrogen, N_2 Oxygen, O_2

FIGURE 17.10

INTERACTIVE FIGURE

Slow-moving molecules may collide with insufficient force to break their bonds. As a result, they cannot react to form product molecules.

a result, collisions between slow-moving N_2 and O_2 molecules, even those that collide in the proper orientation, may not form NO, as is shown in Figure 17.10.

The higher the temperature of a material, the faster its molecules move and the more forceful the collisions between them. Higher temperatures, therefore, increase reaction rates. The nitrogen and oxygen molecules that make up our atmosphere, for example, are continually colliding with one another. At the ambient temperatures of our atmosphere, however, these molecules do not generally have sufficient kinetic energy for the formation of nitrogen monoxide. The heat of a lightning bolt, however, dramatically increases the kinetic energy of these molecules to the point that a large portion of the collisions in the vicinity of the bolt result in the formation of nitrogen monoxide. As discussed in the opening of this chapter, the nitrogen monoxide formed in this manner undergoes further atmospheric reactions to form chemicals known as nitrates that plants depend on for survival. This is an example of *nitrogen fixation*, which you may have explored already in a course on the life sciences.

The life sciences involve fantastic applications of chemistry, nitrogen fixation being just one example. Others include photosynthesis, cellular respiration, and molecular genetics. So there are distinct advantages to learning about chemistry and other physical sciences *before* advancing to the life sciences.

Link To Section 3.4

CHECK POINT

An internal-combustion engine works by drawing a mixture of air and gasoline vapors into a chamber. The action of a piston then compresses these gases into a smaller volume before ignition by the spark of a spark plug. What is the advantage of squeezing the vapors to a smaller volume?

Spark plug

Piston

Fuel/air intake   Compression   Power stroke

Was this your answer?

Squeezing the vapors to a smaller volume effectively increases their concentration and, hence, the number of collisions between molecules. This, in turn, promotes the chemical reaction. As discussed in Section 7.2, compression also increases the temperature, which further favors the chemical reaction.

The energy required to break bonds can also come from the absorption of electromagnetic radiation. As the radiation is absorbed by reactant molecules, the atoms in the molecules may start to vibrate so rapidly that the bonds between them are easily broken. In many instances, direct absorption of electromagnetic radiation is sufficient to break chemical bonds and to initiate a chemical reaction. The common atmospheric pollutant nitrogen dioxide, NO_2, for example, may transform to nitrogen monoxide and atomic oxygen merely on exposure to sunlight:

$$NO_2 + sunlight \longrightarrow NO + O$$

FIGURE 17.11

Reactant molecules must gain a minimum amount of energy, called the activation energy, E_a, before they can transform to product molecules.

The Physical Science Place
Equilibrium

Kinetic energies not sufficient to overcome energy barrier

Kinetic energies sufficient to overcome energy barrier

FIGURE 17.12

Because fast-moving reactant molecules possess sufficient energy to pass over the energy barrier, they are the first ones to transform to product molecules.

In order for two chemicals to be able to react, they must first collide in the proper orientation. Second, they must have sufficient kinetic energy to initiate the breaking of chemical bonds so that new bonds can form. These are all aspects of a broad theory known as the *molecular-kinetic theory.*

Whether they result from collisions, absorption of electromagnetic radiation, or both, broken bonds are a necessary first step in most chemical reactions. The energy required for this initial breaking of bonds can be viewed as an *energy barrier.* The minimum energy required to overcome this energy barrier is known as the **activation energy** (E_a).

In the reaction between nitrogen and oxygen to form nitrogen monoxide, the activation energy is so high (because the bonds in N_2 and O_2 are strong) that only the fastest-moving nitrogen and oxygen molecules possess sufficient energy to react. Figure 17.11 shows the activation energy in this chemical reaction as a vertical hump.

The activation energy of a chemical reaction is analogous to the energy a car needs to drive over the top of a hill. Without sufficient energy to climb to the top of the hill, the car cannot get to the other side. Likewise, reactant molecules can transform to product molecules only if the reactant molecules possess an amount of energy equal to or greater than the activation energy.

At any given temperature, there is a wide distribution of kinetic energies in reactant molecules. Some are moving slowly, and others are moving quickly. As we discussed in Section 6.1, the temperature of a material is related to the average of all these kinetic energies. The few fast-moving reactant molecules in Figure 17.12 have enough energy to pass over the energy barrier and are the first to transform to product molecules.

When the temperature of the reactants is increased, the number of reactant molecules possessing sufficient energy to pass over the barrier also increases, which is why reactions are generally faster at higher temperatures. Conversely, at lower temperatures, fewer molecules have sufficient energy to pass over the barrier. Hence, reactions are generally slower at lower temperatures.

CHECK POINT

What kitchen device is used to lower the rate at which microorganisms grow on food?

Was this your answer?

The refrigerator! Microorganisms, such as bread mold, are everywhere and difficult to avoid. By lowering the temperature of microorganism-contaminated food, the refrigerator decreases the rate of the chemical reactions that these microorganisms depend on for growth, thereby increasing the food's shelf life.

FIGURE 17.13

This alligator became immobilized on the pavement after being caught in the cold night air. By midmorning, shown here, the temperature had warmed sufficiently to allow the alligator to get up and walk away.

Most chemical reactions are influenced by temperature in this manner, including reactions that occur in living bodies. The body temperature of animals that regulate their internal temperature, such as humans, is fairly constant. However, the body temperature of some animals, such as the alligator shown in Figure 17.13, rises and falls with the temperature of the environment. On a warm day, the chemical reactions occurring in an alligator are "up to speed," and the animal is more active. On a chilly day, however, the chemical reactions proceed at a lower rate, and, as a consequence, the alligator's movements are unavoidably sluggish.

■ 17.4 Catalysts

As discussed in the previous section, increasing the concentration or the temperature of the reactants can cause a chemical reaction to go faster. A third way to increase the rate of a reaction is to add a **catalyst**, which is any substance that increases the rate of a chemical reaction by lowering its activation energy. The catalyst may participate as a reactant, but it is then regenerated as a product and is thus available to catalyze subsequent reactions.

The conversion of ozone, O_3, to oxygen, O_2, is normally sluggish because the reaction has a relatively high activation energy, as shown in Figure 17.14a. However, when chlorine atoms act as a catalyst, the energy barrier is lowered, as shown in Figure 17.14b, and the reaction can proceed faster.

FIGURE 17.14

(a) The relatively high activation energy (energy barrier) indicates that only the most energetic ozone molecules can react to form oxygen molecules. (b) The presence of chlorine atoms lowers the activation energy, which means more reactant molecules have sufficient energy to form product. The chlorine allows the reaction to proceed in two steps, and the two smaller activation energies correspond to these steps. (Note that the convention is to write the catalyst above the reaction arrow.)

(a) Without catalyst

(b) With chlorine catalyst

fyi

- Before the fall of the Soviet Union, numerous oil-drilling sites in Siberia were allowed to vent natural gas freely into the atmosphere, presumably because the natural gas had no commercial value. After the fall of the Soviet Union, the wells were capped to prevent this venting. Within weeks, instruments at the Mauna Loa weather observatory on the other side of the planet noted a significant drop in atmospheric levels of methane and its byproduct, carbon dioxide. The effect that we humans have on global atmospheric conditions is very measurable.

Link To Section 20.9

Before it reaches the catalytic converter, the exhaust contains such pollutants as NO, CO, and hydrocarbons.

After it has passed through the catalytic converter, the exhaust contains water vapor, N_2, O_2, and CO_2.

FIGURE 17.15

A catalytic converter reduces the pollution caused by automobile exhaust by converting such harmful combustion products as NO, CO, and hydrocarbons to harmless N_2, O_2, and CO_2. The catalyst is typically platinum, Pt; palladium, Pd; or rhodium, Rd.

Atomic chlorine lowers the energy barrier of this reaction by providing an alternate pathway involving intermediate reactions, each having a lower activation energy than the uncatalyzed reaction. This alternate pathway involves two steps. Initially, the chlorine reacts with the ozone to form chlorine monoxide and oxygen:

$$Cl \;+\; O_3 \longrightarrow ClO \;+\; O_2$$

Chlorine   Ozone   Chlorine monoxide   Oxygen

The chlorine monoxide then reacts with another ozone molecule to re-form the chlorine atom as well as to produce two additional oxygen molecules:

$$ClO \;+\; O_3 \longrightarrow Cl \;+\; 2\,O_2$$

Chlorine monoxide   Ozone   Chlorine   Oxygen

Although chlorine is depleted in the first reaction, it is regenerated in the second reaction. As a result, there is no net consumption of chlorine. At the same time, however, two ozone molecules are rapidly converted to three oxygen molecules. The chlorine is therefore a catalyst for the conversion of ozone to oxygen because the chlorine increases the speed of the reaction while not being consumed by the reaction.

Chlorine atoms in the stratosphere catalyze the destruction of Earth's ozone layer. Evidence indicates that chlorine atoms are generated in the stratosphere as a byproduct of human-made chlorofluorocarbons (CFCs), once widely produced as the cooling fluid of refrigerators and air conditioners. Destruction of the ozone layer is a serious concern because of its role in protecting us from the Sun's harmful ultraviolet rays. One chlorine atom in the ozone layer is estimated to catalyze the transformation of 100,000 ozone molecules to oxygen molecules in the one or two years before the chlorine atom is removed by natural processes.

Chemists have been able to harness the power of catalysts for numerous beneficial purposes. The exhaust that comes from an automobile engine, for example, contains a wide assortment of pollutants, such as nitrogen monoxide, carbon monoxide, and uncombusted fuel vapors (hydrocarbons). To reduce the amount of these pollutants entering the atmosphere, most automobiles are equipped with *catalytic converters*, as shown in Figure 17.15. Metal catalysts in a converter speed up reactions that convert exhaust pollutants to less toxic substances. Nitrogen monoxide is transformed to nitrogen and oxygen, carbon monoxide is transformed to carbon dioxide, and unburned fuel is converted to carbon dioxide and water vapor. Because catalysts are not consumed by the reactions they facilitate, a single catalytic converter may continue to operate effectively for the lifetime of the car.

Catalytic converters, along with microchip-controlled fuel–air ratios, have led to a significant drop in the per-vehicle emission of pollutants. A typical car in 1960 emitted about 11 g of uncombusted fuel, 4 g of nitrogen oxide, and 84 g of carbon monoxide per mile traveled. An improved vehicle in 2000 emitted less than 0.5 g of uncombusted fuel, less than 0.5 g of nitrogen oxide, and only about 3 g of carbon monoxide per mile traveled. This improvement, however, has been offset by an increase in the number of cars being driven, as exemplified by the traffic jam shown in Figure 17.16. It is also offset by the growing popularity of SUVs (sport-utility vehicles), which bypass pollution requirements.

The chemical industry depends on catalysts because they lower manufacturing costs by lowering required temperatures and by providing greater product yields without being consumed. Indeed, more than 90% of all manufactured goods are produced with the assistance of catalysts. Without catalysts, the price of gasoline would be much higher, as would be the price of such consumer goods as rubber, plastics, pharmaceuticals, automobile parts, clothing, and food grown with chemical fertilizers. Living organisms rely on special types of catalysts known as *enzymes*, which allow exceedingly complex biochemical reactions to occur with ease. You may learn more about the nature and behavior of enzymes in a life science course.

CHECK POINT

How does a catalyst lower the activation energy of a chemical reaction?

Was this your answer?
The catalyst provides an alternate and easier-to-achieve pathway along which the chemical reaction can proceed.

FIGURE 17.16

The exhaust from automobiles today is much cleaner than before the advent of the catalytic converter, but many more cars are on the road. In 1960, there were about 74 million registered motor vehicles in the United States. In 2003, there were about 243 million.

17.5 Energy and Chemical Reactions

As we have discussed in the preceding two sections, reactants must have a certain amount of energy in order to overcome the activation energy so that a chemical reaction can proceed. Once a reaction is complete, however, there may be either a net release or a net absorption of energy. Reactions in which there is a net release of energy are called **exothermic**. Rocket ships lift off into space and campfires glow red hot as a result of exothermic reactions. Reactions in which there is a net absorption of energy are called **endothermic**. Photosynthesis, for example, involves a series of endothermic reactions that are driven by the energy of sunlight. Both exothermic and endothermic reactions, illustrated in Figure 17.17, can be understood through the concept of bond energy.

 Link To Section 6.5

During a chemical reaction, chemical bonds are broken and atoms rearrange to form new chemical bonds. Such breaking and forming of chemical bonds involves changes in energy. As an analogy, consider a pair of magnets. To separate them requires an input of "muscle energy." Conversely, when the two separated magnets collide, they become slightly warmer than they were, and this warmth is evidence of energy released. The magnets must absorb energy if they are to break apart, and release energy as they come together. The same principle applies to atoms. To pull bonded atoms apart requires an energy input. When atoms combine, there is an energy output, usually in the form of faster-moving atoms and molecules, electromagnetic radiation, or both.

The amount of energy required to pull two bonded atoms apart is the same as the amount released when they are brought together. This energy, whether it is

FIGURE 17.17

Chemical reactions that occur when wood is burning have a net release of energy. Chemical reactions that occur in a photosynthetic plant have a net absorption of energy.

the energy that is absorbed as a bond breaks or the energy that is released as a bond forms, is called **bond energy**. Each chemical bond has its own characteristic bond energy. The hydrogen–hydrogen bond energy, for example, is 436 kJ/mole. This means that 436 kJ of energy is absorbed as 1 mole of hydrogen–hydrogen bonds break apart, and 436 kJ of energy is released upon the formation of 1 mole of hydrogen–hydrogen bonds. Different bonds involving different elements have different bond energies, as Table 17.1 shows. You can refer to the table as you study this section, but please do not memorize these bond energies. Instead, focus on understanding what they mean.

TABLE 17.1   SELECTED BOND ENERGIES

Bond	Bond Energy (kJ/mole)	Bond	Bond Energy (kJ/mole)
H—H	436	N—N	159
H—C	414	O—O	138
H—N	389	Cl—Cl	243
H—O	464	C=O	803
H—F	569	N=O	631
H—S	339	O=O	498
H—Cl	431	C≡C	837
C—C	347	N≡N	946

Remember, in a chemical reaction, the bonds being formed are different from the bonds that were broken. The bond energies of the bonds being formed, therefore, are also different from those of the bonds that were broken.

By convention, a positive bond energy represents the amount of energy absorbed as a bond breaks, and a negative bond energy represents the amount of energy released as a bond forms. Thus, when you are calculating the net energy released or absorbed during a reaction, you'll need to be careful about plus and minus signs. It is standard practice when doing such calculations to assign a plus sign to energy absorbed and a minus sign to energy released. For instance, when dealing with a reaction in which 1 mole of H—H bonds are broken, you'll write +436 kJ to indicate energy absorbed, and when dealing with the formation of 1 mole of H—H bonds, you'll write −436 kJ to indicate energy released. We'll do some sample calculations in a moment.

CHECK POINT

Do all covalent single bonds have the same bond energy?

Was this your answer?

No. Bond energy depends on the types of atoms bonding. The H–H single bond, for example, has a bond energy of 436 kJ/mole, but the H–O single bond has a bond energy of 464 kJ/mole. All covalent single bonds do not have the same bond energy.

AN EXOTHERMIC REACTION INVOLVES A NET RELEASE OF ENERGY

For any chemical reaction, the total amount of energy absorbed in breaking bonds in reactants is always different from the total amount of the energy released as bonds form in the products. Consider the reaction in which hydrogen and oxygen react to form water:

$$\text{H—H} + \text{H—H} + \text{O}{=}\text{O} \longrightarrow \text{H—O}\diagdown_{\text{H}} + \text{H}\diagdown_{\text{O}}\diagup^{\text{H}}$$

In the reactants, hydrogen atoms are bonded to hydrogen atoms, and oxygen atoms are double-bonded to oxygen atoms. The total amount of energy absorbed as these bonds break is $+1370$ kJ.

Link To Section 3.6

Type of bond	Number of moles	Bond energy	Total energy
H—H	2	+436 kJ/mole	+872 kJ
O=O	1	+498 kJ/mole	+498 kJ
		Total energy absorbed:	+1370 kJ

In the products there are four hydrogen–oxygen bonds. The total amount of energy released as these bonds form is -1856 kJ.

Type of bond	Number of moles	Bond energy	Total energy
H—O	4	−464 kJ/mole	−1856 kJ
		Total energy released:	−1856 kJ

The amount of energy released in this reaction exceeds the amount of energy absorbed. The net energy of the reaction is found by adding the two quantities:

$$\text{Net energy of reaction} = \text{energy absorbed} + \text{energy released}$$
$$= +1370 \text{ kJ} + (-1856 \text{ kJ})$$
$$= -486 \text{ kJ}$$

The negative sign on the net energy indicates that there is a net release of energy, and so the reaction is exothermic. For any exothermic reaction, energy can be considered a product and is thus sometimes included after the arrow of the chemical equation:

$$2 H_2 + O_2 \longrightarrow 2 H_2O + \text{energy}$$

In an exothermic reaction, the potential energy of atoms in the product molecules is lower than their potential energy in the reactant molecules. This is illustrated in the reaction profile shown in Figure 17.18. The lowered potential energy of the atoms in the product molecules is due to their being more tightly held together. This is analogous to two attracting magnets, whose potential

fyi

- NASA scientists routinely test various materials for their durability against atomic oxygen, O, which is abundant in the low orbit of the space shuttle. They discovered that atomic oxygen effectively transforms surface organic materials into gaseous carbon dioxide. The scientists realized atomic oxygen's usefulness for restoring paintings damaged by smoke or other organic contaminants. Together with art conservationists they used atomic oxygen to restore certain damaged paintings, and it worked spectacularly.

FIGURE 17.18

In an exothermic reaction, the product molecules are at a lower potential energy than the reactant molecules. The net amount of energy released by the reaction is equal to the difference in potential energies of the reactants and products.

FIGURE 17.19

A space shuttle uses exothermic chemical reactions to lift off from Earth's surface.

energy decreases as they come closer together. The loss of potential energy is balanced by a gain in kinetic energy. Like two free-floating magnets coming together and accelerating to higher speeds, the potential energy of the reactants is converted to faster-moving atoms and molecules, electromagnetic radiation, or both. This kinetic energy released by the reaction is equal to the difference between the potential energy of the reactants and the potential energy of the products, as is indicated in Figure 17.18.

It is important to understand that the energy released by an exothermic reaction is not created by the reaction. This is in accord with the *law of conservation of energy*, which tells us that energy is neither created nor destroyed in a chemical reaction (or any process). Instead, energy is merely converted from one form to another. During an exothermic reaction, energy that was once in the form of the potential energy of chemical bonds is released as the kinetic energy of fast-moving molecules and/or as electromagnetic radiation.

The amount of energy released in an exothermic reaction depends on the amounts of the reactants. The reaction of large amounts of hydrogen and oxygen, for example, provides the energy to lift the space shuttle shown in Figure 17.19 into orbit. There are two compartments in the large central tank, to which the orbiter is attached—one filled with liquid hydrogen and the other filled with liquid oxygen. Upon ignition, these two liquids mix and react chemically to form water vapor, which produces the needed thrust as it is expelled out the rocket cones. Additional thrust is obtained by a pair of solid-fuel rocket boosters containing a mixture of ammonium perchlorate, NH_4ClO_4, and powdered aluminum. On ignition, these chemicals react to form products that are expelled at the rear of the rocket. The balanced equation representing this reaction is

$$3\,NH_4ClO_4 + 3\,Al \longrightarrow Al2O_3 + AlCl_3 + 3\,NO + 6\,H_2O + energy$$

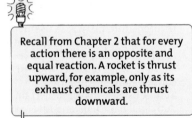

Recall from Chapter 2 that for every action there is an opposite and equal reaction. A rocket is thrust upward, for example, only as its exhaust chemicals are thrust downward.

Link To Section 7.6

CHECK POINT

Where does the net energy released in an exothermic reaction go?

Was this your answer?
This energy goes into increasing the speeds of reactant atoms and molecules and often into electromagnetic radiation.

AN ENDOTHERMIC REACTION INVOLVES A NET ABSORPTION OF ENERGY

When the amount of energy released in product formation is *less* than the amount of energy absorbed when reactant bonds break, the reaction is endothermic. An example is the reaction of atmospheric nitrogen and oxygen to form nitrogen monoxide, which is the same reaction used for many of the discussions earlier in this chapter:

$$N\equiv N + O=O \longrightarrow N=O + N=O$$

The amount of energy absorbed as the chemical bonds in the reactants break is

Type of bond	Number of moles	Bond energy	Total energy
N≡N	+1	+946 KJ/mole	+946 KJ
O=O	+1	+498 kJ/mole	+498 kJ
		Total energy absorbed:	+1444 kJ

The amount of energy released upon the formation of bonds in the products is

Type of bond	Number of moles	Bond energy	Total energy
N=O	2	−631 kJ/mole	−1262 KJ
		Total energy released:	−1262 KJ

As before, the net energy of the reaction is found by adding the two quantities:

$$\text{Net energy of reaction} = \text{energy absorbed} + \text{energy released}$$
$$= +1444\,\text{kJ} + (-1262\,\text{kJ})$$
$$= +182\,\text{kJ}$$

The positive sign indicates a net *absorption* of energy, meaning the reaction is endothermic. For any endothermic reaction, energy can be considered a reactant and is thus sometimes included before the arrow of the chemical equation:

$$\text{Energy} + N_2 + O_2 \longrightarrow 2\,NO$$

In an endothermic reaction, the potential energy of atoms in the product molecules is higher than their potential energy in the reactant molecules. This is illustrated in the reaction profile shown in Figure 17.20. Raising the potential energy of the atoms in the product molecules requires a net input of energy, which must come from some external source, such as electromagnetic radiation, electricity, or heat. Thus, nitrogen and oxygen react to form nitrogen monoxide only with the application of much heat, as occurs adjacent to a lightning bolt or in an internal-combustion engine.

FIGURE 17.20

In an endothermic reaction, the product molecules are at a higher potential energy than the reactant molecules. The net amount of energy absorbed by the reaction is equal to the difference in potential energies of the reactants and products.

■ 17.6 Chemical Reactions Are Driven by Entropy

As was discussed in Section 6.6, energy tends to disperse. It flows from where it is concentrated to where it is spread out. The energy of a hot pan, for example, does not stay concentrated in the pan once the pan is taken off the stove. Instead, the energy disperses away from the pan and into the cooler surroundings. Similarly, the concentrated chemical energy found in gasoline disperses into the heat of many smaller lower-energy molecules upon combustion. Some of this heat is used by the engine to get the car moving. The rest spreads into the engine block, into the radiator fluid, or out the exhaust pipe.

Link To Section 6.6

Processes that result in the dispersion of energy, such as exothermic reactions, tend to occur on their own. Endothermic reactions, by contrast, require that energy be concentrated around the reactants. Endothermic reactions, therefore, do not occur on their own.

A quick way to determine whether a reaction might be favorable is to assess whether the reaction leads to an overall dispersal of energy, which is the same thing as an increase in entropy.

Because energy naturally tends to disperse, a reaction that leads to an increase in entropy will likely occur, while a reaction that leads to a decrease in entropy will _not_ likely occur.

Scientists consider this tendency of energy to disperse one of the central reasons for both physical and chemical processes. In other words, processes that result in the dispersion of energy tend to occur on their own—they are favored. This includes the cooling of a hot pan or the burning of gasoline. In both cases, energy is dispersed to the environment.

The opposite holds true, too. Processes that result in the concentration of energy tend _not_ to occur—they are not favored. Heat from the room, for example, does not spontaneously move back into the pan to heat it. Likewise, low-energy exhaust molecules coming out of your car's tailpipe do not spontaneously come back together to form higher-energy gasoline molecules. The natural flow of energy is always a one-way trip from where it is concentrated to where it is less concentrated or "spread out."

Entropy is the term we use to describe this natural spreading of energy. The concept of entropy was described in Section 6.6. Applied to chemistry, entropy helps us answer a fundamental question: If you put two materials together, will they react to form new materials? If the reaction results in an overall increase in entropy (a dispersal of energy), then the answer is yes. Conversely, if the reaction results in an overall decrease in entropy (a concentration of energy), then the reaction will _not_ occur by itself.

Using this concept of entropy, you are now in a position to understand why exothermic reactions are self-sustaining, but most endothermic reactions need a continual prodding. Exothermic reactions spread energy out to the surroundings, much like a cooling hot pan. This is an increase in entropy; hence, exothermic reactions are favored to occur. An endothermic reaction, by contrast, requires that the reactants absorb energy from the surroundings. This is a concentration of energy, which is counter to energy's natural tendency to disperse. Endothermic reactions, therefore, can be sustained only with continual input of some external source of energy.* For photosynthesis, the source of this energy is the Sun, which is a hothouse of entropy-producing exothermic nuclear reactions.

FIGURE 17.21

Some of the Sun's dispersed energy is used to drive endothermic reactions that allow for the functioning of living organisms.

CHECK POINT

Sugar crystals form naturally within a supersaturated solution of sugar water. Does the formation of these crystals, in which the sugar molecules are aligned in an orderly fashion, result in an increase or decrease in entropy?

Was this your answer?
The formation of these sugar crystals results in an _increase_ in entropy. Your clue to an increase in entropy here is that the crystals form "on their own" without the input of an external source of energy. Interestingly, energy is released when molecules come together to form a solid. For example, when water freezes, the heat released is called the _heat of fusion_ (see Section 7.9). This release of heat is the dispersal of energy, which is, by definition, an increase in entropy.

---

* There are examples of endothermic reactions that proceed spontaneously absorbing heat from the environment while increasing entropy. A classic example is the mixing of a salt in water (see the Active Explorations at the end of this chapter). In such cases, entropy increases not by a release of energy but by the dispersion of energy-containing atoms and molecules into solution.

SUMMARY OF TERMS

Chemical equation A representation of a chemical reaction in which reactants are drawn before an arrow that points to the products.

Reactants The reacting substances in a chemical reaction.

Products The new materials formed in a chemical reaction.

Law of mass conservation Matter is neither created nor destroyed during a chemical reaction—atoms merely rearrange, without any apparent loss or gain of mass, to form new molecules.

Formula mass The sum of the atomic masses of the atoms in a chemical compound or element.

Avogadro's number The number of particles—6.02×10^{23}—contained in 1 mole of anything.

Molar mass The mass of 1 mole of a substance.

Reaction rate A measure of how quickly the concentration of products in a chemical reaction increases or the concentration of reactants decreases.

Activation energy The minimum energy required for a chemical reaction to proceed.

Catalyst Any substance that increases the rate of a chemical reaction without itself being consumed by the reaction.

Exothermic A term that describes a chemical reaction in which there is a net release of energy.

Endothermic A term that describes a chemical reaction in which there is a net absorption of energy.

Bond energy The amount of energy that is either absorbed as a chemical bond breaks or is released as a chemical bond forms.

REVIEW QUESTIONS

17.1 Chemical Equations

1. What is the purpose of coefficients in a chemical equation?
2. How many chromium atoms and how many oxygen atoms are indicated on the right side of this balanced chemical equation?

$$4\,Cr(s) + 3\,O_2(g) \longrightarrow 2\,Cr_2O_3(g)$$

3. What do the letters (s), (l), (g), and (aq) stand for in a chemical equation?
4. Why is it important that a chemical equation be balanced?
5. Why is it important never to change a subscript in a chemical formula when balancing a chemical equation?

17.2 Counting Atoms and Molecules by Mass

6. Why don't equal masses of golf balls and Ping-Pong balls contain the same number of balls?
7. Why don't equal masses of carbon atoms and oxygen molecules contain the same number of particles?
8. What is the mass of a single sodium atom in atomic mass units?
9. What is the formula mass of nitrogen monoxide, NO, in atomic mass units?
10. If you had 1 mole of marbles, how many marbles would you have?
11. If you had 2 moles of pennies, how many pennies would you have?
12. How many moles of water are there in 18 g of water?
13. How many molecules of water are there in 18 g of water?
14. Why is saying you have 1 mole of water molecules the same as saying you have 6.02×10^{23} water molecules?

17.3 Reaction Rates

15. Why don't all collisions between reactant molecules lead to product formation?
16. What generally happens to the rate of a chemical reaction with increasing temperature?
17. Which reactant molecules are the first to pass over the energy barrier?
18. What term is used to describe the minimum amount of energy required for a reaction to proceed?

17.4 Catalysts

19. What catalyst is effective in the destruction of atmospheric ozone, O_3?
20. What is the purpose of a catalytic converter?
21. What does a catalyst do to the energy barrier of a reaction?
22. What net effect does a chemical reaction have on a catalyst?
23. Why are catalysts so important to our economy?

17.5 Energy and Chemical Reactions

24. If it takes 436 kJ to break a bond, how many kilojoules are released when the same bond is formed?
25. Is energy consumed at any time during an exothermic reaction?
26. What is released by an exothermic reaction?
27. What is absorbed by an endothermic reaction?

17.6 Chemical Reactions Are Driven by Entropy

28. As energy disperses, where does it go?
29. What is always increasing?
30. Why are exothermic reactions self-sustaining?

EXERCISES

● BEGINNER ■ INTERMEDIATE ◆ EXPERT

1. ● Balance these equations:
   (a) ___ Fe(s) + ___ O$_2$(g) → ___ Fe$_2$O$_3$(s)
   (b) ___ H$_2$(g) + ___ N$_2$(g) → ___ NH$_3$(g)
   (c) ___ Cl$_2$(g) + ___ KBr(aq) → ___ Br$_2$(l) + ___ KCl(aq)
   (d) ___ CH$_4$(g) + ___ O$_2$(g) → ___ CO$_2$(g) + ___ H$_2$O(l)

2. ● Balance these equations:
   (a) ___ Fe(s) + ___ S(s) → ___ Fe$_2$S$_3$(s)
   (b) ___ P$_4$(s) + ___ H$_2$(g) → ___ PH$_3$(g)
   (c) ___ NO(g) + ___ Cl$_2$(g) → ___ NOCl(g)
   (d) ___ SiCl$_4$(l) + ___ Mg(s) → ___ Si(s) + ___ MgCl$_2$(s)

3. ● Is the following chemical equation balanced?

   $$2\,C_4H_{10}(g) + 13\,O_2(g) \longrightarrow 8\,CO_2(g) + 10\,H_2O(l)$$

4. ■ Is the following chemical equation balanced?

   $$4\,C_6H_7N_5O_{16}(s) + 19\,O_2(g) \longrightarrow 24\,CO_2(g)$$
   $$+\ 20\,NO_2(g) + 14\,H_2O(g)$$

5. ● Which equations are balanced?
   (a) Mg(s) + 2 HCl(aq) → MgCl$_2$(aq) + H$_2$(g)
   (b) 3 Al(s) + 3 Br$_2$(l) → Al$_2$Br$_3$(s)
   (c) 2 HgO(s) → 2 Hg(l) + O$_2$(g)

Use the following illustrations to answer Exercises 6–9.

6. ■ Is this reaction balanced?
7. ◆ Assume the illustrations above are two frames of a movie—one from before the reaction and the other from after the reaction. How many diatomic molecules are represented in this movie?
8. ● There is an excess of at least one of the reactant molecules, shown below. Which one?

A   B   C   D   E

9. ■ Which equation best describes this reaction?
   (a) 2 AB$_2$ + 2 DCB$_3$ + B$_2$ → 2 DBA$_4$ + 2 CA$_2$
   (b) 2 AB$_2$ + 2 CDA$_3$ + B$_2$ → 2 C$_2$A$_4$ + 2 DBA
   (c) 2 AB$_2$ + 2 CDA$_3$ + A$_2$ → 2 DBA$_4$ + 2 CA$_2$
   (d) 2 BA$_2$ + 2 DCA$_3$ + A$_2$ → 2 DBA$_4$ + 2 CA$_2$

10. ● The reactants shown schematically on the left in the illustration below represent methane, CH$_4$, and water, H$_2$O. Write out the full balanced chemical equation that is depicted.

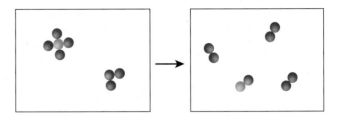

11. ● The reactants shown schematically on the left in the illustration below represent iron oxide, Fe$_2$O$_3$, and carbon monoxide, CO. Write out the full balanced chemical equation that is depicted.

12. ■ What are the formula masses of water, H$_2$O; propene, C$_3$H$_6$; and 2-propanol, C$_3$H$_8$O?

13. ◆ Which has more atoms: 17.031 g of ammonia, NH$_3$, or 72.922 g of hydrogen chloride, HCl?

14. ◆ Which has the greatest number of molecules: 28 g of nitrogen, N$_2$; 32 g of oxygen, O$_2$; 32 g of methane, CH$_4$; or 38 g of fluorine, F$_2$?

15. ● Two atomic mass units equal how many grams?

16. ● What is the mass of an oxygen atom in atomic mass units?

17. ● What is the mass of a water molecule in atomic mass units?

18. ■ What is the mass of an oxygen atom in grams?

19. ■ What is the mass of a water molecule in grams?

20. ■ Is it possible to have a sample of oxygen that has a mass of 14 amu? Explain.

21. ● Which is greater: 1.01 amu of hydrogen or 1.01 g of hydrogen?

22. ■ Which has the greater mass, 1.204×10^{24} molecules of molecular hydrogen or 1.204×10^{24} molecules of water?

23. ◆ How many grams of gallium are in a 145-g sample of gallium arsenide, GaAs?

24. ◆ How many atoms of arsenic are in a 145-g sample of gallium arsenide, GaAs?

25. ● How does formula mass differ from atomic mass?

26. ■ How is it possible for a jet airplane carrying 110 tons of jet fuel to emit 340 tons of carbon dioxide?

27. ● What two aspects of a collision between two reactant molecules determine whether the collision results in the formation of product molecules?

28. ● Can a catalyst react with a reactant?

29. ● Why does refrigerated food take longer to spoil?

30. ● Does a refrigerator prevent or delay food spoilage? Explain.

31. ■ Why does a glowing splint of wood burn only slowly in air but burst into flames when placed in pure oxygen?

32. ● Why is heat often added to chemical reactions performed in the laboratory?

33. ■ An Alka-Seltzer antacid tablet bubbles vigorously in room-temperature water but only slowly in a 50 : 50 mix of alcohol and water also at room temperature. Propose an explanation involving the relationship between reaction speed and frequency of molecular collisions.

34. ■ In the following reaction sequence for the catalytic formation of ozone from molecular oxygen, which chemical compound is the catalyst: nitrogen monoxide or nitrogen dioxide?

$$O_2 + 2\,NO \longrightarrow 2\,NO_2$$

$$2\,NO_2 \longrightarrow 2\,NO + 2\,O$$

$$2\,O + 2\,O_2 \longrightarrow 2\,O_3$$

35. ● Many people hear about atmospheric ozone depletion and wonder why we don't simply replace what has been destroyed. Knowing about chlorofluorocarbons and knowing how catalysts work, explain how this would not be a lasting solution.

36. ● In an endothermic reaction, which has greater potential energy: the reactants or the products?

37. ● Are the chemical reactions that take place in a disposable battery exothermic or endothermic? What evidence supports your answer? Is the reaction going on in a rechargeable battery while it is recharging exothermic or endothermic?

38. ● What role does entropy play in chemical reactions?

39. ● Why do exothermic reactions typically favor the formation of products?

40. ● Under what conditions does a hot pie not lose heat to its surroundings?

41. ● As the Sun shines on a snow-capped mountain, much of the snow sublimes instead of melts. How is this favored by entropy?

42. ◆ Estimate whether entropy increases or deceases with the following reaction. Use data from Table 17.1 to confirm your estimate.

$$2\,C(s) + 3\,H_2(g) \longrightarrow C_2H_6(g)$$

43. ■ Exothermic reactions are favored because they release heat to the environment. Would an exothermic reaction be more favored or less favored if it were carried out within a superheated chamber?

44. ◆ In the laboratory, endothermic reactions are usually performed at elevated temperatures, while exothermic reactions are usually performed at lower temperatures. What are some possible reasons for this?

45. ■ Wild plants readily grow "all by themselves," yet the molecules of the growing plant have *less* entropy than the materials used to make the plant. How can this *decrease* in entropy exist for a process that occurs all by itself?

PROBLEMS

● BEGINNER   ■ INTERMEDIATE   ◆ EXPERT

1. ■ Show that there are 1.0×10^{22} carbon atoms in a 1-carat pure diamond, which has a mass of 0.20 g.

2. ■ How many gold atoms are there in a 5.00-g sample of pure gold, Au (197 amu)?

3. ● Show that one mole of $KClO_3$ contains 122.55 g.

4. ■ Small samples of oxygen gas needed in the laboratory can be generated by any number of simple chemical reactions, such as

$$2\,KClO_3(s) \longrightarrow 2\,KCl(s) + 3\,O_2(g)$$

According to this balanced chemical equation, how many moles of oxygen gas are produced from the reaction of 2 moles of $KClO_3$ solid?

5. ◆ Small samples of oxygen gas needed in the laboratory can be generated by any number of simple chemical reactions, such as

$$2\,KClO_3(s) \longrightarrow 2\,KCl(s) + 3\,O_2(g)$$

What mass of oxygen (in grams) is produced when 122.55 g of $KClO_3$ (formula mass 122.55 amu) takes part in this reaction?

6. ● Show that the formula mass of 2-propanol, C_3H_8O, is 60 amu, and that the formula mass of propene, C_3H_6, is 42 amu, and also that the formula mass of water, H_2O, is 18 amu.

7. ■ How many grams of water, H_2O, and propene, C_3H_6, can be formed from the reaction of 6.0 g of 2-propanol, C_3H_8O?

$$C_3H_8O \longrightarrow C_3H_6 + H_2O$$

8. ■ How many moles of water, H_2O, can be produced from the reaction of 16 g of methane, CH_4, with an unlimited supply of oxygen, O_2? How many grams of water is this? The reaction is

$$CH_4 + 2\,O_2 \longrightarrow CO_2 + 2\,H_2O$$

9. ■ Use the bond energies in Table 17.1 and the accounting format shown in Section 17.5 to determine whether these reactions are exothermic or endothermic:

$$H_2 + Cl_2 \longrightarrow 2\,HCl$$

$$2\,HC \equiv CH + 5\,O_2 \longrightarrow 4\,CO_2 + 2\,H_2O$$

10. ■ Use the bond energies in Table 17.1 and the accounting format shown in Section 17.5 to determine whether these reactions are exothermic or endothermic:

$$N_2H_4 \longrightarrow 2\,H_2 + N_2$$

$$2\,H_2O_2 \longrightarrow O_2 + 2\,H_2O$$

ACTIVE EXPLORATIONS

Warming and Cooling Water Mixtures

Recall from Section 15.8 that chemical bonds and intermolecular attractions are both consequences of the electric force, the difference being that chemical bonds are generally many times stronger than molecule-to-molecule attractions. So just as the formation and breaking of chemical bonds involves energy, so does the formation and breaking of molecular attractions. For molecule-to-molecule attractions, the amount of energy absorbed or released per gram of material is relatively small. Physical changes involving the formation or breaking of molecule-to-molecule attractions, therefore, are much safer to perform, which makes them more suitable for an out-of-laboratory activity. Experience the exothermic and endothermic nature of physical changes for yourself by performing the following two activities:

1. Hold some room-temperature water in the cupped palm of your hand over a sink. Pour an equal amount of room-temperature rubbing alcohol into the water. Is this mixing an exothermic or endothermic process? What's going on at the molecular level?
2. Add lukewarm water to two plastic cups. (Do *not* use insulating Styrofoam cups.) Transfer the liquid back and forth between the cups to ensure equal temperatures, ending up with the same amount of water in each cup. Add several tablespoons of table salt to one cup and stir. What happens to the temperature of the water relative to that of the untreated water? (Hold the cups up to your cheeks to tell.) Is this an exothermic or endothermic process? What's going on at the molecular level?

READINESS ASSURANCE TEST (RAT)

If you have a good handle on this chapter, if you really do, then you should be able to score 7 out of 10 on this RAT. If you score less than 7, you need to study further before moving on.

Choose the BEST answer to each of the following.

1. What coefficients balance the following equation:

   $$\_\_\_ P_4(s) + \_\_\_ H_2(g) \longrightarrow \_\_\_ PH_3(g)$$

   (a) 4, 2, 3
   (b) 1, 6, 4
   (c) 1, 4, 4
   (d) 2, 10, 8
2. What is the formula mass of sulfur dioxide, SO_2?
   (a) about 16 amu
   (b) about 32 amu
   (c) about 60 amu
   (d) about 64 amu
3. Which has the greatest number of atoms?
   (a) 28 g of nitrogen, N_2
   (b) 32 g of oxygen, O_2
   (c) 16 g of methane, CH_4
   (d) 38 g of fluorine, F_2
4. How many molecules of aspirin (formula mass 180 amu) are in a 0.250-g sample?
   (a) 6.02×10^{23}
   (b) 8.38×10^{20}
   (c) 1.51×10^{23}
   (d) There is not enough information.
5. The yeast in bread dough feeds on sugar to produce carbon dioxide. Why does the dough rise faster in a warmer area?
   (a) There is a greater number of effective collisions among reacting molecules.
   (b) Atmospheric pressure decreases with increasing temperature.

   (c) The yeast tends to "wake up" with warmer temperatures, which is why baker's yeast is best stored in the refrigerator.
   (d) The rate of evaporation increases with increasing temperature.
6. What can you deduce about the activation energy of a reaction that takes billions of years to go to completion? How about a reaction that takes only fractions of a second?
   (a) The activation energy of both these reactions must be very low.
   (b) The activation energy of both these reactions must be very high.
   (c) The slow reaction must have a high activation energy and the fast reaction must have a low activation energy.
   (d) The slow reaction must have a low activation energy and the fast reaction must have a high activation energy.
7. What role do CFCs play in the catalytic destruction of ozone?
   (a) Ozone is destroyed on binding to a CFC molecule that has been energized by ultraviolet light.
   (b) There is no strong scientific evidence that CFCs play a significant role in the catalytic destruction of ozone.
   (c) CFC molecules activate chlorine atoms into their catalytic action.
   (d) CFC molecules migrate to the upper stratosphere, where they generate chlorine atoms on being destroyed by ultraviolet light.
8. Is the synthesis of ozone, O_3, from oxygen, O_2, an example of an exothermic or endothermic reaction?
   (a) exothermic, because ultraviolet light is emitted during its formation
   (b) endothermic, because ultraviolet light is emitted during its formation

(c) exothermic, because ultraviolet light is absorbed during its formation

(d) endothermic, because ultraviolet light is absorbed during its formation

9. How much energy, in kilojoules, is released or absorbed from the reaction of 1 mole of nitrogen, N_2, with 3 moles of molecular hydrogen, H_2, to form 2 moles of ammonia, NH_3? Consult Table 17.1 for bond energies.
   (a) $+899$ kJ/mole
   (b) -993 kJ/mole
   (c) $+80$ kJ/mole
   (d) -80 kJ/mole

10. How is it possible to cause an endothermic reaction to proceed when the reaction causes energy to become less dispersed?
    (a) The reaction should be placed in a vacuum.
    (b) The reaction should be cooled down.
    (c) The concentration of the reactants should be increased.
    (d) The reaction should be heated.

Answers to RAT

1b, 2d, 3c, 4b, 5a, 6c, 7d, 8d, 9c, 10d

EXPLORING FURTHER

http://www.thecatalyst.org/wwwchem.html
This site has been developed as a resource for high school chemistry teachers, but anyone studying chemistry should find the links helpful. You might follow the link to the history of chemistry, for example, to learn more about Amedeo Avogadro and that huge number named after him.

http://dbhs.wvusd.k12.ca.us/webdocs/ChemTeamIndex.html
An excellent collection of tutorials developed by John Park of the ChemTeam of Diamond Bar High School, California. Tutorials applicable to this chapter include Chemical Reactions & Reaction Types, Kinetic-Molecular Theory & Gas Laws, The Mole, Kinetics, Stoichiometry, and Thermochemistry.

http://www.wxumac.demon.co.uk
Nitrogen monoxide, also known as nitric oxide, NO, is a precursor to nitrate fertilizers and a common atmospheric pollutant, but it also plays a multitude of vital roles in our human biology. Use *nitric oxide* as a keyword in your Internet search engine to find a plethora of Web sites such as this one devoted to the many roles this small but important molecule plays in our physiology and in various diseases, such as Alzheimer's, Parkinson's, asthma, heart disease, and infections.

http://www.secondlaw.com
http://www.entropysimple.com
These sites emphasize the "big picture" of how the second law of thermodynamics applies to our everyday experiences, including our sense of time. Many practical and down-to-earth applications are provided. A great follow-up to Section 17.6, these sites will help you see this law as one of the simplest yet most profound laws of nature.

CHAPTER 17 ONLINE RESOURCES

The Physical Science Place

Interactive Figures

■ 17.7, 17.8, 17.9, 17.10

Tutorials

■ Chemical Reactions and Equations
■ Equilibrium

Quiz

Flashcards

Links

two classes of chemical reactions

18

18.1 ACIDS DONATE PROTONS; BASES ACCEPT THEM

18.2 RELATIVE STRENGTHS OF ACIDS AND BASES

18.3 ACIDIC, BASIC, AND NEUTRAL SOLUTIONS

18.4 ACIDIC RAIN AND BASIC OCEANS

18.5 LOSING AND GAINING ELECTRONS

18.6 HARNESSING THE ENERGY OF FLOWING ELECTRONS

18.7 ELECTROLYSIS

18.8 CORROSION AND COMBUSTION

■ During a chemical reaction, the atoms of the reactants change partners to form new materials we call products. As wood burns, for example, the atoms of cellulose molecules break away from each other in order to combine with the atoms of oxygen molecules to form carbon dioxide, water vapor, plus lots of heat.

In this chapter, we explore two main classes of chemical reactions: acid–base reactions and oxidation–reduction reactions. Acid–base reactions involve the transfer of *protons* from one reactant to another. These sorts of reactions within your stomach help you digest your food. They play a key role in global warming. Most consumer goods can trace their origins to acid–base chemical reactions. Oxidation–reduction reactions involve the transfer of one or more *electrons* from one reactant to another. The burning of wood is an oxidation–reduction reaction, as are the reactions your body uses to transform the food you eat into biochemical energy. Oxidation–reduction reactions are responsible for the rusting of a car. They are also the source of a battery's electrical energy.

FIGURE 18.1

Examples of acids. (a) Citrus fruits contain many types of acids, including ascorbic acid, $C_6H_8O_6$, which is vitamin C. (b) Vinegar contains acetic acid, $C_2H_4O_2$, and can be used to preserve foods. (c) Many toilet bowl cleaners are formulated with hydrochloric acid, HCl. (d) All carbonated beverages contain carbonic acid, H_2CO_3; many also contain phosphoric acid, H_3PO_4.

(a)   (b)   (c)   (d)

The Physical Science Place

Nature of Acids and Bases

Link To Section 15.2

18.1 Acids Donate Protons; Bases Accept Them

The term *acid* comes from the Latin *acidus*, which means "sour." The sour taste of vinegar and citrus fruits is due to the presence of acids. Acids are essential in the chemical industry. For example, more than 85 billion pounds of sulfuric acid are produced annually in the United States, making this the number-one manufactured chemical. Sulfuric acid is used to make fertilizers, detergents, paint dyes, plastics, pharmaceuticals, and storage batteries, as well as to produce iron and steel. It is so important in the manufacturing of goods that its production is considered a standard measure of a nation's industrial strength. Figure 18.1 shows only a few of the acids we commonly encounter.

Bases are characterized by their bitter taste and slippery feel. Interestingly, bases themselves are not slippery. Rather, they cause skin oils to transform into slippery solutions of soap. Most commercial preparations for unclogging drains contain sodium hydroxide, NaOH (also known as lye), which is extremely basic and hazardous when concentrated. Bases are also heavily used in industry. Each year in the United States, about 25 billion pounds of sodium hydroxide are manufactured for use in the production of various chemicals and in the pulp and paper industry. Solutions containing bases are often called *alkaline*, a term derived from the Arabic *al-qali* ("the ashes"). Ashes are slippery when wet because of the presence of the base potassium carbonate, K_2CO_3. Figure 18.2 shows some familiar bases.

Acids and bases may be defined in several ways. For our purposes, an appropriate definition is the one suggested in 1923 by the Danish chemist Johannes Brønsted (1879–1947) and the English chemist Thomas Lowry (1874–1936). In the Brønsted–Lowry definition, an **acid** is any chemical that donates a hydrogen

FIGURE 18.2

Examples of bases. (a) Reactions involving sodium bicarbonate, $NaHCO_3$, cause baked goods to rise. (b) Ashes contain potassium carbonate, K_2CO_3. (c) Soap is made by reacting bases with animal or vegetable oils. The soap itself, then, is slightly alkaline. (d) Powerful bases, such as sodium hydroxide, NaOH, are used in drain cleaners.

(a)   (b)   (c)   (d)

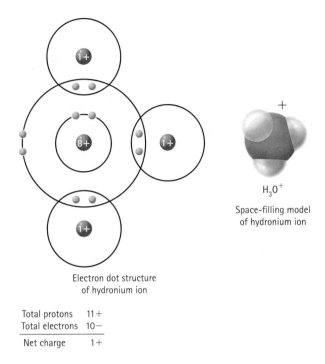

Electron dot structure
of hydronium ion

Space-filling model
of hydronium ion

H_3O^+

Total protons	11+
Total electrons	10−
Net charge	1+

FIGURE 18.3

The hydronium ion's positive charge is a consequence of the extra proton this molecule has acquired. Hydronium ions, which play a role in many acid–base reactions, are poly-atomic ions, which, as mentioned in Section 15.2, are molecules that carry a net electric charge.

ion, H^+, and a **base** is any chemical that accepts a hydrogen ion. Recall that a hydrogen atom consists of one electron surrounding a one-proton nucleus. A hydrogen ion, H^+, formed from the loss of an electron, therefore, is nothing more than a lone proton. Thus, it is also sometimes said that an acid is a chemical that donates a proton and a base is a chemical that accepts a proton.

Consider what happens when hydrogen chloride is mixed into water:

$$HCl + H_2O \longrightarrow Cl^- + H_3O^+$$

H^+ donor H^+ acceptor
(acid)   (base)

Hydrogen chloride donates a hydrogen ion to one of the nonbonding electron pairs on a water molecule, resulting in a third hydrogen bonded to the oxygen. In this case, hydrogen chloride behaves as an acid (proton donor) and water behaves as a base (proton acceptor). The products of this reaction are a chloride ion and a **hydronium ion**, H_3O^+, which, as Figure 18.3 shows, is a water molecule with an extra proton.

When added to water, ammonia behaves as a base as its nonbonding electrons (see Section 15.1) accept a hydrogen ion from water, which, in this case, behaves as an acid:

$$H_2O + NH_3 \longrightarrow OH^- + NH_4^+$$

H^+ donor H^+ acceptor
(acid)   (base)

Here's a BAAD acronym for remembering how acids and bases handle protons: Bases Accept, Acids Donate.

Acid	→	Base
Proton donor		Proton acceptor

The hydrogen ion, H^+ does not readily exist in water because any hydrogen ion formed is quickly picked up by a water molecule and transformed to the hydronium ion, H_3O^+.

Recall that a hydrogen ion with a positive charge is simply a lone proton.

Hydrogen atom   Positive hydrogen ion (lone proton)

Hydroxide ions have a net negative charge, which is a consequence of having lost a proton. Like hydronium ions, they play a part in many acid–base reactions.

Electron dot structure
of hydroxide ion

Space-filling model
of hydroxide ion

Total protons	9+	
Total electrons	10−	
Net charge	1−	

How we behave depends on who
we're with. Likewise for chemicals.

Link To Section 24.2

This reaction results in the formation of an ammonium ion and a **hydroxide ion**, which, as shown in Figure 18.4, is a water molecule without the nucleus of one of the hydrogen atoms.

An important aspect of the Brønsted–Lowry definition is that it uses a *behavior* to define a substance as an acid or a base. We say, for example, that hydrogen chloride *behaves* as an acid when mixed with water, which *behaves* as a base. Similarly, ammonia *behaves* as a base when mixed with water, which under this circumstance *behaves* as an acid. Because acid–base is seen as a behavior, there is really no contradiction when a chemical like water behaves as a base in one instance but as an acid in another instance. By analogy, consider yourself. You are who you are, but your behavior changes depending on whom you are with. Likewise, it is a chemical property of water to behave as a base (to accept H^+) when mixed with hydrogen chloride and as an acid (to donate H^+) when mixed with ammonia.

The products of an acid–base reaction can also behave as acids or as bases. An ammonium ion, for example, may donate a hydrogen ion back to a hydroxide ion to re-form ammonia and water:

$$H_2O \; + \; NH_3 \; \longleftarrow \; OH^- + \; NH_4^+$$

H^+ acceptor (base)	H^+ donor (acid)

Forward and reverse acid–base reactions proceed simultaneously and can therefore be represented as occurring at the same time by using two oppositely facing arrows:

$$H_2O \; + \; NH_3 \; \rightleftharpoons \; OH^- + \; NH_4^+$$

H^+ donor (acid)	H^+ acceptor (base)	H^+ acceptor (base)	H^+ donor (acid)

When the equation is viewed from left to right, the ammonia behaves as a base because it accepts a hydrogen ion from the water, which therefore acts as an acid. Viewed in the reverse direction, the equation shows that the ammonium ion behaves as an acid because it donates a hydrogen ion to the hydroxide ion, which therefore behaves as a base.

CHECK
POINT

■

Identify the acid or base behavior of each participant in the reaction

$$H_2PO_4^- + H_3O^+ \rightleftharpoons H_3PO_4 + H_2O$$

Was this your answer?
In the forward reaction (left to right), $H_2PO_4^-$ gains a hydrogen ion to become H_3PO_4. In accepting the hydrogen ion, $H_2PO_4^-$ is behaving as a base. It gets the hydrogen ion from the H_3O^+, which is behaving as an acid. In the reverse direction, H_3PO_4 loses a hydrogen ion to become $H_2PO_4^-$ and is thus behaving as an acid. The recipient of the hydrogen ion is H_2O, which is behaving as a base as it transforms to H_3O^+.

A SALT IS THE IONIC PRODUCT OF AN ACID–BASE REACTION

In everyday language, the word *salt* implies sodium chloride, NaCl, table salt. In the language of chemistry, however, **salt** is a general term meaning any ionic compound formed from the reaction between an acid and a base. Hydrogen chloride and sodium hydroxide, for example, react to produce the salt sodium chloride and water:

$$HCl \;+\; NaOH \longrightarrow NaCl \;+\; H_2O$$

| Hydrogen chloride (acid) | Sodium hydroxide (base) | Sodium chloride (salt) | Water |

Similarly, the reaction between hydrogen chloride and potassium hydroxide yields the salt potassium chloride and water:

$$HCl \;+\; KOH \longrightarrow KCl \;+\; H_2O$$

| Hydrogen chloride (acid) | Potassium hydroxide (base) | Potassium chloride (salt) | Water |

Potassium chloride is the main ingredient in "salt-free" table salt, as noted in Figure 18.5.

Salts are generally far less corrosive than the acids and bases from which they are formed. A corrosive chemical has the power to disintegrate a material or wear away its surface. Hydrogen chloride is a remarkably corrosive acid, which makes it useful for cleaning toilet bowls and etching metal surfaces. Sodium hydroxide is a very corrosive base used for unclogging drains. Mixing hydrogen chloride and sodium hydroxide together in equal portions, however, produces an aqueous solution of sodium chloride—salt water, which is not nearly as destructive as either starting material.

There are as many salts as there are acids and bases. Sodium cyanide, NaCN, is a deadly poison. "Saltpeter," which is potassium nitrate, KNO_3, is useful as a fertilizer and in the formulation of gunpowder. Calcium chloride, $CaCl_2$, is commonly used to de-ice walkways, and sodium fluoride, NaF, helps prevent tooth decay. The acid–base reactions forming these salts are shown in Table 18.1.

The reaction between an acid and a base is called a **neutralization** reaction. As can be seen in the color-coding of the neutralization reactions in Table 18.1, the positive ion of a salt comes from the base and the negative ion comes from the acid. The remaining hydrogen and hydroxide ions join to form water.

Not all neutralization reactions result in the formation of water. In the presence of hydrogen chloride, for example, the drug pseudoephedrine behaves as a base by accepting H^+ from a hydrogen chloride. The negative Cl^- then joins the pseudoephedrine–H^+ ion to form the salt pseudoephedrine hydrochloride,

FIGURE 18.5

"Salt-free" table salt substitutes contain potassium chloride in place of sodium chloride. Caution is advised in using these products, however, because excessive quantities of potassium salts can lead to serious illness. Furthermore, sodium ions are a vital component of our diet and should never be totally excluded. For a good balance of these two important ions, you might inquire about commercially available half-and-half mixtures of sodium chloride and potassium chloride, such as the one shown here.

- What makes one acid strong and another weak? Briefly, it involves the stability of the negative ion that remains after the proton has been donated. Hydrogen chloride is a strong acid because the chloride ion can accommodate the negative charge rather well. Acetic acid, however, is a weaker acid because the resulting oxygen ion is less able to accommodate the negative charge.

TABLE 18.1   ACID-BASE REACTIONS AND THE SALTS FORMED

Acid		Base		Salt		Water
HCN Hydrogen cyanide	+	NaOH Sodium hydroxide	→	NaCH Sodium cyanide	+	H_2O
HNO_3 Nitric acid	+	KOH Potassium hydroxide	→	KNO_3 Potassium nitrate	+	H_2O
2 HCl Hydrogen chloride	+	$Ca(OH)_2$ Calcium hydroxide	→	$CaCl_2$ Calcium chloride	+	$2H_2O$
HF Hydrogen fluoride	+	NaOH Sodium hydroxide	→	NaF Sodium fluoride	+	H_2O

which is a common nasal decongestant, shown in Figure 18.6. This salt is soluble in water and can be absorbed through the digestive system.

CHECK POINT

Is a neutralization reaction best described as a physical change or a chemical change?

Was this your answer?
New chemicals are formed during a neutralization reaction, meaning the reaction is a chemical change.

FIGURE 18.6

Hydrogen chloride and pseudoephedrine react to form the salt *pseudoephedrine hydrochloride*, which, because of its solubility in water, is readily absorbed into the body.

Pseudoephedrine
(base)

Pseudoephedrine hydrochloride
(salt)

18.2 Relative Strengths of Acids and Bases

In general, the stronger an acid, the more readily it donates hydrogen ions. Likewise, the stronger a base, the more readily it accepts hydrogen ions. An example of a strong acid is hydrogen chloride, HCl, and an example of a strong base is sodium hydroxide, NaOH. The corrosiveness of these materials is a result of their strength.

One way to assess the strength of an acid or base is to measure how much of it remains after it has been added to water. If little remains, the acid or base is strong.

If a lot remains, the acid or base is weak. To illustrate this concept, consider what happens when the strong acid hydrogen chloride is added to water and what happens when the weak acid acetic acid, $C_2H_4O_2$ (the active ingredient of vinegar), is added to water.

Being an acid, hydrogen chloride donates hydrogen ions to water, forming chloride ions and hydronium ions. Because HCl is such a strong acid, nearly all of it is converted to these ions, as is shown in Figure 18.7.

Because acetic acid is a weak acid, it has much less tendency to donate hydrogen ions to water. When this acid is dissolved in water, only a small portion of the acetic acid molecules are converted to ions, a process that occurs as the polar O — H bonds are broken (the C — H bonds of acetic acid are unaffected by the water because of their nonpolarity). The majority of acetic acid molecules remain intact in their original non-ionized form, as shown in Figure 18.8.

Figures 18.7 and 18.8 show the submicroscopic behavior of strong and weak acids in water. As molecules and ions are too small to see, how then does a chemist measure the strength of an acid? One way is by measuring a solution's ability to conduct an electric current, as Figure 18.9 illustrates. Pure water contains practically no ions to conduct electricity. When a strong acid is dissolved in water, many ions are generated, as indicated in Figure 18.7. The presence of these

Some Acids and Bases Are Stronger Than Others

Link To Section 8.6

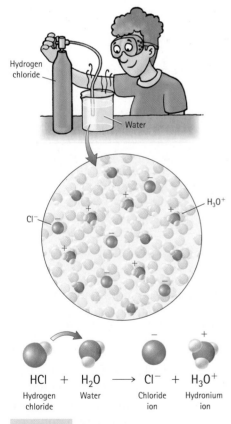

FIGURE 18.7

Immediately after gaseous hydrogen chloride is added to water, it reacts with the water to form hydronium ions and chloride ions. That very little HCl remains (none shown here) lets us know that HCl acts as a strong acid.

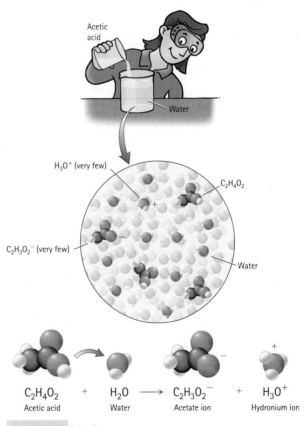

FIGURE 18.8

When liquid acetic acid is added to water, only a few acetic acid molecules react with water to form ions. The majority of the acetic acid molecules remain in their non-ionized form, which implies that acetic acid is a weak acid.

(a)   (b)   (c)

FIGURE 18.9

(a) The pure water in this circuit cannot conduct electricity because it contains practically no ions. The lightbulb in the circuit therefore remains unlit. (b) Because HCl is a strong acid, nearly all of its molecules break apart in water, giving a high concentration of ions, which can conduct an electric current that lights the bulb. (c) Acetic acid, $C_2H_4O_2$, is a weak acid; in water, only a small portion of its molecules break up into ions. Because fewer ions are generated, only a weak current exists, and the bulb is therefore dimmer.

ions allows for the flow of a large electric current. A weak acid dissolved in water generates only a few ions, as indicated in Figure 18.8. The presence of fewer ions means there can be only a small electric current.

This same trend is seen with strong and weak bases. Strong bases, for example, tend to accept hydrogen ions more readily than weak bases do. In solution, a strong base allows the flow of a large electric current, and a weak base allows the flow of a small electric current.

CHECK POINT

According to the aqueous solutions illustrated here, which is the stronger base, NH_3 or NaOH?

Aqueous solution of NH_3 Aqueous solution of NaOH

Was this your answer?
The solution on the right contains the greater number of ions, meaning that sodium hydroxide, NaOH, is the stronger base. Ammonia, NH_3, is the weaker base, indicated by the relatively few ions in the solution on the left.

Just because an acid or base is strong doesn't mean a solution of that acid or base is corrosive. The corrosive action of an acidic solution is caused by the hydronium ions rather than by the acid that generated those hydronium ions. Similarly, the corrosive action of a basic solution results from the hydroxide ions it contains, regardless of the base that generated those hydroxide ions. A *very* dilute solution of a strong acid or a strong base may have little corrosive action because in such solutions there are only a few hydronium or hydroxide ions. (Almost all the molecules of the strong acid or base break up into ions, but because the solution is dilute, only a few acid or base molecules are present to begin with. As a result, there are only a few hydronium or hydroxide ions.) You shouldn't be too alarmed, therefore, when you discover that some toothpastes are formulated with small amounts of sodium hydroxide, one of the strongest bases known.

On the other hand, a concentrated solution of a weak acid, such as the acetic acid in vinegar, may be just as corrosive as or even more corrosive than a dilute solution of a strong acid, such as hydrogen chloride. The relative strengths of two acids in solution or two bases in solution, therefore, can be compared only when the two solutions have the same concentration.

18.3 Acidic, Basic, and Neutral Solutions

A substance whose ability to behave as an acid is about the same as its ability to behave as a base is said to be **amphoteric**. Water is a good example. Because it is amphoteric, water can react with itself. In behaving as an acid, a water molecule donates a hydrogen ion to a neighboring water molecule, which, in accepting the hydrogen ion, is behaving as a base. This reaction produces a hydroxide ion and a hydronium ion, which react together to re-form the water molecule:

$$H_2O \;+\; H_2O \;\rightleftharpoons\; OH^+ \;+\; H_3O$$

Water   Water   Hydroxide   Hydronium
                            ion   ion

When a water molecule gains a hydrogen ion, a second water molecule must lose a hydrogen ion. So for every hydronium ion formed, a hydroxide ion also forms. In pure water, therefore, the total number of hydronium ions must be the same as the total number of hydroxide ions. Experiments reveal that the concentration of hydronium and hydroxide ions in pure water is extremely low—about 0.0000001 M for each, where M stands for molarity or moles per liter (Section 16.3). Water by itself, therefore, is a very weak acid as well as a very weak base, as evidenced by the unlit lightbulb in Figure 18.9a.

CHECK POINT

Do water molecules react with one another?

Was this your answer?
Yes, but not to any large extent. When they do react, they form hydronium and hydroxide ions. (Note: Make sure you understand this point because it serves as a basis for most of the rest of the chapter.)

Further experiments reveal an interesting rule pertaining to the concentrations of hydronium and hydroxide ions in any solution that contains water. The concentration of hydronium ions in any aqueous solution multiplied by the concentration of the hydroxide ions in the solution always equals the constant K_w, which is a very, very small number:

Concentration H_3O^+ × concentration OH^- = K_w = 0.00000000000001

Concentration is usually given as molarity, which is indicated by abbreviating this equation using brackets:

$$[H_3O^+] \times [OH^-] = K_w = 0.00000000000001$$

The brackets mean this equation is read "the molarity of H_3O^+ times the molarity of OH^- equals K_w." Writing in scientific notation, we have

$$[H_3O^+][OH^-] = K_w = 1.0 \times 10^{-14}$$

For pure water, the value of K_w is the concentration of hydronium ions, 0.0000001 M, multiplied by the concentration of hydroxide ions, 0.0000001 M, which can be written in scientific notation as

$$[1.0 \times 10^{-7}][1.0 \times 10^{-7}] = K_w = 1.0 \times 10^{-14}$$

The constant value of K_w is quite significant because it means that *no matter what is dissolved in the water*, the product of the hydronium-ion and hydroxide-ion concentrations always equals 1.0×10^{-14}. So if the concentration of H_3O^+ goes up, the concentration of OH^- must go down, and the product of the two remains 1.0×10^{-14}.

CHECK POINT

1. In pure water, the hydroxide-ion concentration is 1.0×10^{-7} M. What is the hydronium-ion concentration?
2. What is the concentration of hydronium ions in a solution if the concentration of hydroxide ions is 1.0×10^{-3} M?

Was this your answer?
1. 1.0×10^{-7} M, because in pure water $[H_3O^+] = [OH^-]$.
2. 1.0×10^{-11} M, because $[H_3O^+][OH^-]$ must equal $1.0 \times 10^{-14} = K_w$.

In an **acidic** solution, $[H_3O^+] > [OH^-]$.

In a **basic** solution, $[H_3O^+] < [OH^-]$.

In a **neutral** solution, $[H_3O^+] = [OH^-]$.

FIGURE 18.10

The relative concentrations of hydronium and hydroxide ions determine whether a solution is acidic, basic, or neutral.

An aqueous solution can be described as acidic, basic, or neutral, as Figure 18.10 summarizes. An **acidic solution** is one in which the hydronium-ion concentration is higher than the hydroxide-ion concentration. An acidic solution is made by adding an acid to water. The effect of this addition is to increase the concentration of hydronium ions, which necessarily decreases the concentration of hydroxide ions. A **basic solution** is one in which the hydroxide-ion concentration is higher than the hydronium-ion concentration. A basic solution is made by adding a base to water. This addition increases the concentration of hydroxide ions, which necessarily decreases the concentration of hydronium ions. A **neutral solution** is one in which the hydronium-ion concentration equals the hydroxide-ion concentration. Pure water is an example of a neutral solution—not because it contains so few hydronium and hydroxide ions but because it contains equal numbers of these ions. A neutral solution is also obtained when equal quantities of acid and base are combined, which explains why acids and bases are said to *neutralize* each other.

CHECK POINT

How does adding ammonia, NH_3, to water make a basic solution when there are no hydroxide ions in the formula for ammonia?

Was this your answer?
Ammonia indirectly increases the hydroxide-ion concentration by reacting with water:

$$NH_3 + H_2O \longrightarrow NH_4^+ + OH^-$$

This reaction raises the hydroxide-ion concentration, which has the effect of lowering the hydronium-ion concentration. With the hydroxide-ion concentration now higher than the hydronium-ion concentration, the solution is basic.

THE pH SCALE IS USED TO DESCRIBE ACIDITY

The *pH scale* is a numeric scale used to express the acidity of a solution. Mathematically, **pH** is equal to the negative logarithm of the hydronium-ion concentration:

$$pH = -\log[H_3O^+]$$

Note again that brackets are used to represent molar concentrations, meaning $[H_3O^+]$ is read "the molar concentration of hydronium ions." For understanding the logarithm function, see the Figuring Physical Science box on page 448.

Consider a neutral solution that has a hydronium-ion concentration of 1.0×10^{-7} M. To find the pH of this solution, we first take the logarithm of this value, which is -7 (see the Figuring Physical Science box on logarithms, page 448). The pH is by definition the negative of this value, which means $-(-7) = 7$. Hence, in a neutral solution, in which the hydronium-ion concentration equals 1.0×10^{-7}M, the pH is 7.

Acidic solutions have pH values less than 7. For an acidic solution in which the hydronium-ion concentration is 1.0×10^{-4}M, for example, pH = $-\log(1.0 \times 10^{-4}) = 4$. The more acidic a solution is, the greater its hydronium-ion concentration and the lower its pH.

Basic solutions have pH values greater than 7. For a basic solution in which the hydronium-ion concentration is 1.0×10^{-8}M, for example, pH = $-\log(1.0 \times 10^{-8}) = 8$. The more basic a solution is, the smaller its hydronium-ion concentration and the higher its pH.

Figure 18.11 shows typical pH values of some familiar solutions, and Figure 18.12 shows two common ways of determining pH values.

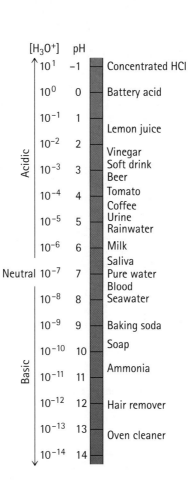

$[H_3O^+]$	pH	
10^1	−1	Concentrated HCl
10^0	0	Battery acid
10^{-1}	1	
		Lemon juice
10^{-2}	2	Vinegar
10^{-3}	3	Soft drink / Beer
10^{-4}	4	Tomato
10^{-5}	5	Coffee / Urine
		Rainwater
10^{-6}	6	Milk
		Saliva
10^{-7}	7	Pure water
		Blood
10^{-8}	8	Seawater
10^{-9}	9	Baking soda
10^{-10}	10	Soap
10^{-11}	11	Ammonia
10^{-12}	12	Hair remover
10^{-13}	13	Oven cleaner
10^{-14}	14	

Acidic / Neutral / Basic

FIGURE 18.11

The pH values of some common solutions.

The Physical Science Place
The pH Scale

fyi

- The outer surface of hair is made of microscopic scale-like structures called cuticles that, like window shutters, can open and close. Alkaline solutions cause the cuticles to open up, which makes the hair "porous." Acidic solutions cause the cuticles to close down, which makes the hair "resistant." A beautician can control how long hair retains artificial coloring by modifying the pH of the hair-coloring solution. With an acidic solution, the cuticles close shut so that the dye binds only to the outside of each shaft of hair. This results in a temporary hair coloring, which may come off with the next hair washing. With an alkaline solution, the dye can penetrate through the cuticles into the hair for a more permanent effect.

(a)

(b)

FIGURE 18.12

(a) The pH of a solution can be measured electronically using a pH meter. (b) A rough estimate of the pH of a solution can be obtained with litmus paper, which is coated with a dye that changes color with pH.

FIGURING PHYSICAL SCIENCE

■ Logarithms and pH

The logarithm of a number can be found on any scientific calculator by keying in the number and pressing the [log] button. The calculator finds the power to which 10 is raised to give the number. The logarithm of 10^2, for example, is 2 because that is the power to which 10 is raised to give the number 10^2. If you know that 10^2 is equal to 100, then you'll understand that the logarithm of 100 also is 2. Check this out on your calculator. Similarly, the logarithm of 1000 is 3 because 10 raised to the third power, 10^3, equals 1000. (Note: we speak here of the base-10 logarithm, not the natural logarithm of base e.)

Any positive number, including a very small one, has a logarithm. The logarithm of 0.0001, which equals 10^{-4}, for example, is −4 (the power to which 10 is raised to equal this number).

SAMPLE PROBLEM 1

What is the logarithm of 0.01?

SOLUTION:

The number 0.01 is 10^{-2}, the logarithm of which is −2 (the power to which 10 is raised).

The concentration of hydronium ions in most solutions is typically much less than 1 M. Recall, for example, that in neutral water the hydronium-ion concentration is 0.0000001 M (10^{-7} M). The logarithm of any number smaller than 1 (but greater than zero) is a negative number. The definition of pH includes the minus sign so as to transform the logarithm of the hydronium-ion concentration to a positive number.

When a solution has a hydronium-ion concentration of 1 M, the pH is 0 because 1 M $= 10^0$ M. A 10 M solution has a pH of −1 because 10 M $= 10^1$ M.

SAMPLE PROBLEM 2

What is the pH of a solution that has a hydronium-ion concentration of 0.001 M?

SOLUTION:

The number 0.001 is 10^{-3}, so

$$pH = -\log[H_3O^+]$$
$$= -\log 10^{-3}$$
$$= -(-3) = 3$$

SAMPLE PROBLEM 3

What is the logarithm of 10^5?

SOLUTION:

"What is the logarithm of 10^5?" can be rephrased as "To what power is 10 raised to give the number 10^5?" The answer is 5.

SAMPLE PROBLEM 4

What is the logarithm of 100,000?

SOLUTION:

You should know that 100,000 is the same as 10^5. Thus the logarithm of 100,000 is 5.

SAMPLE PROBLEM 5

What is the pH of a solution with a hydronium-ion concentration of 10^{-9} M? Is this solution acidic, basic, or neutral?

SOLUTION:

The pH is 9, which means this is a basic solution:

$$pH = -\log[H_3O^+]$$
$$= -\log 10^{-9}$$
$$= -(-9)$$
$$= 9$$

■ 18.4 Acidic Rain and Basic Oceans

As previously mentioned, rainwater is naturally acidic. One source of this acidity is carbon dioxide, the same gas that gives fizz to soda drinks. The atmosphere contains 810 billion tons of CO_2, most of it from such natural sources as volcanoes and decaying organic matter but a growing amount (about 135 billion tons) from human activities.

Water in the atmosphere reacts with carbon dioxide to form *carbonic acid*:

$$CO_2(g) + H_2O(\ell) \longrightarrow H_2CO_3(aq)$$

Carbon   Water   Carbonic
dioxide   acid

Carbonic acid, as its name implies, behaves as an acid and lowers the pH of water. The CO_2 in the atmosphere brings the pH of rainwater to about 5.6—noticeably below the neutral pH value of 7. Because of local fluctuations, the

fyi

■ Above temperatures of 374°C and pressures of 218 atm, water transforms into a state of matter known as a supercritical fluid, which resembles both a liquid and a gas. For a neutral solution of supercritical water, the pH equals about 2, which means that it is highly corrosive. Research is underway to learn how supercritical water might be used to destroy toxic chemicals, such as chemical warfare agents.

normal pH of rainwater varies between 5 and 7. This natural acidity of rainwater may accelerate the erosion of land, and under certain circumstances it can lead to the formation of underground caves.

By convention, *acid rain* is a term used for rain with a pH lower than 5. Acid rain is created when airborne pollutants, such as sulfur dioxide, are absorbed by atmospheric moisture. Sulfur dioxide is readily converted to sulfur trioxide, which reacts with water to form *sulfuric acid*:

$$2\,SO_2(g) \;+\; O_2(g) \;\longrightarrow\; SO_3(g)$$

Sulfur   Oxygen   Sulfur
dioxide   trioxide

$$SO_3(g) \;+\; H_2O(\ell) \;\longrightarrow\; H_2SO_4(aq)$$

Sulfur   Water   Sulfuric
trioxide   acid

Each year about 20 million tons of SO_2 are released into the atmosphere by the combustion of sulfur-containing coal and oil. Sulfuric acid is much stronger than carbonic acid, and, as a result, rain laced with sulfuric acid eventually corrodes metal, paint, and other exposed substances. Each year, the damage costs billions of dollars. The cost to the environment is also high (Figure 18.13). Many rivers and lakes receiving acid rain become less capable of sustaining life. Much vegetation that receives acid rain doesn't survive. This is particularly evident in heavily industrialized regions.

CHECK POINT

When sulfuric acid, H_2SO_4, is added to water, what makes the resulting aqueous solution corrosive?

Was this your answer?
Because H_2SO_4 is a strong acid, it readily forms hydronium ions when dissolved in water. Hydronium ions are responsible for the corrosive action.

The Physical Science Place
Rainwater Is Acidic and
Ocean Water Is Basic
Beach Sand Composition

Link To Section 24.5

fyi

- Acid rain remains a serious problem in many regions of the world. Significant progress, however, has been made toward fixing the problem. In the United States, for example, sulfur dioxide and nitrogen oxide emissions have been reduced by nearly half since 1980. Also, in 2005 the EPA implemented the Clean Air Interstate Rule (CAIR), which is designed to reduce levels of these pollutants even further especially for areas downwind of heavily industrialized regions.

Link To Section 23.3

(a)

(b)

FIGURE 18.13

(a) These two photographs show the same obelisk in New York City's Central Park before and after the effects of acid rain. (b) Many forests downwind from heavily industrialized areas, such as in the northeastern United States and in Europe, have been noticeably hard-hit by acid rain.

FIGURE 18.14

(a) The damaging effects of acid rain do not appear in bodies of fresh water lined with calcium carbonate, which neutralizes any acidity. (b) Lakes and rivers lined with inert materials are not protected.

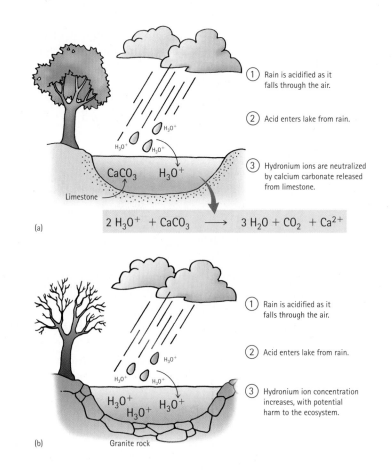

(a)

① Rain is acidified as it falls through the air.

② Acid enters lake from rain.

③ Hydronium ions are neutralized by calcium carbonate released from limestone.

Limestone

$CaCO_3$ H_3O^+

H_3O^+ H_3O^+ H_3O^+

$$2\ H_3O^+ + CaCO_3 \longrightarrow 3\ H_2O + CO_2 + Ca^{2+}$$

(b)

① Rain is acidified as it falls through the air.

② Acid enters lake from rain.

③ Hydronium ion concentration increases, with potential harm to the ecosystem.

H_3O^+ H_3O^+

H_3O^+ H_3O^+

Granite rock

FIGURE 18.15

Most chalks are made from calcium carbonate, which is the same chemical found in limestone. The addition of even a weak acid, such as the acetic acid of vinegar, produces hydronium ions that react with the calcium carbonate to form several products, the most notable being carbon dioxide, which rapidly bubbles out of solution. Try this for yourself! If the bubbling is not as vigorous as shown here, then the chalk is made of other mineral components.

The environmental impact of acid rain depends on local geology, as Figure 18.14 illustrates. In certain regions, such as the midwestern United States, the ground contains significant quantities of the alkaline compound calcium carbonate (limestone), deposited when these lands were submerged under oceans, as has occurred several times over the past 500 million years. Acid rain pouring into these regions is often neutralized by the calcium carbonate before any damage is done. (Figure 18.15 shows calcium carbonate neutralizing an acid.) In the northeastern United States and many other regions, however, the ground contains very little calcium carbonate and is composed primarily of chemically less reactive materials, such as granite. In these regions, the effect of acid rain on lakes and rivers accumulates.

One demonstrated solution to this problem is to raise the pH of acidified lakes and rivers by adding calcium carbonate—a process known as *liming*. The cost of transporting the calcium carbonate, coupled with the need to monitor treated water systems closely, limits liming to only a small fraction of the vast number of water systems already affected. Furthermore, as acid rain continues to pour into these regions, the need to lime also continues.

A longer-term solution to acid rain is to prevent most of the generated sulfur dioxide and other pollutants from entering the atmosphere in the first place. Toward this end, smokestacks have been designed or retrofitted to minimize the quantities of pollutants released. Though this process is costly, the positive effects of these adjustments have been demonstrated. An ultimate long-term solution, however, would be a shift from fossil fuels to cleaner energy sources, such as nuclear and solar energy.

CHECK POINT

What kind of lakes are protected against the negative effects of acid rain?

Was this your answer?
Lakes that have a floor consisting of basic minerals, such as limestone, are more resistant to acid rain because the chemicals of the limestone (mostly calcium carbonate, $CaCO_3$) neutralize any incoming acid.

It should come as no surprise that the amount of carbon dioxide put into the atmosphere by human activities is growing. What is surprising, however, is that studies indicate that the atmospheric concentration of CO_2 is not increasing proportionately. A likely explanation has to do with the oceans (Figure 18.16). When atmospheric CO_2 dissolves in any body of water—a raindrop, a lake, or the ocean—it forms carbonic acid. In fresh water, this carbonic acid transforms back to water and carbon dioxide, which is released back into the atmosphere. Carbonic acid in the ocean, however, is quickly neutralized by dissolved alkaline substances such as calcium carbonate (the ocean is alkaline, pH ≈ 8.2). The products of this neutralization eventually end up on the ocean floor as insoluble solids. Thus, carbonic acid neutralization in the ocean prevents CO_2 from being released back into the atmosphere. The ocean, therefore, is a carbon dioxide *sink*—most of the CO_2 that goes in doesn't come out. So pushing more CO_2 into our atmosphere means pushing more of it into our vast oceans. This is another of the many ways in which the oceans regulate our global environment.

Nevertheless, as Figure 18.17 shows, the concentration of atmospheric CO_2 is increasing. Carbon dioxide is being produced faster than the ocean can absorb it, and this may alter Earth's environment. Carbon dioxide is a *greenhouse gas*, which means it helps keep Earth's surface warm by preventing infrared radiation from escaping into outer space. Without greenhouse gases in the atmosphere, Earth's surface would average a frigid −18°C. However, with increasing concentration of CO_2 in the atmosphere, we may experience higher average temperatures. Higher temperatures may significantly alter global weather patterns as well as raise the average sea level, as the polar ice caps melt and the volume of seawater increases because of thermal expansion. Global warming is explored in greater detail in Chapter 25.

So the pH of rain depends in great part on the concentration of atmospheric CO_2, which depends on the pH of the oceans. These systems are interconnected with global temperatures, which naturally connect to the countless living systems

The pollution humans release knows no political boundaries. Iron smelters operating in China, for example, release pollutants that are readily detected in Seattle, Washington.

Link To Section 25.1

Link To Section 26.3

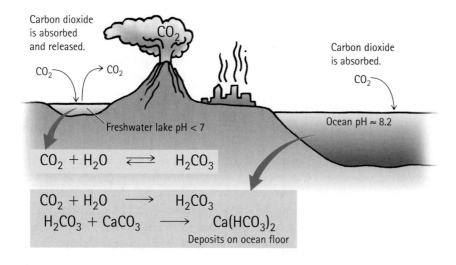

Freshwater lake pH < 7

Ocean pH ≈ 8.2

$$CO_2 + H_2O \rightleftharpoons H_2CO_3$$

$$CO_2 + H_2O \longrightarrow H_2CO_3$$
$$H_2CO_3 + CaCO_3 \longrightarrow Ca(HCO_3)_2$$
Deposits on ocean floor

FIGURE 18.16

Carbon dioxide forms carbonic acid on entering any body of water. In fresh water, this reaction is reversible, and the carbon dioxide is released back into the atmosphere. In the alkaline ocean, the carbonic acid is neutralized to such compounds as calcium bicarbonate, $Ca(HCO_3)_2$, which precipitate to the ocean floor. As a result, most of the atmospheric carbon dioxide that enters our oceans remains there.

FIGURE 18.17

Researchers at the Mauna Loa weather observatory in Hawaii have recorded increasing concentrations of atmospheric carbon dioxide since they began collecting data in the 1950s. This famous graph is known as the Keeling curve, after the scientist Charles Keeling, who initiated this project and first noted the trends. Interestingly, the oscillations within the Keeling curve reflect seasonal changes in CO_2 levels.

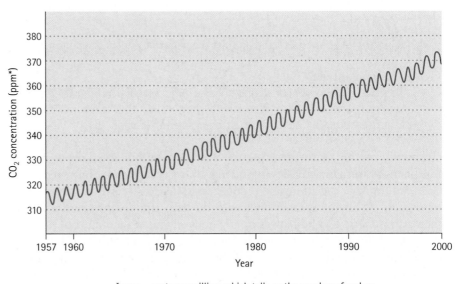

* ppm = parts per million, which tells us the number of carbon dioxide molecules for every million molecules of air.

fyi

- Scientists have experimented with ways of enhancing the ocean's ability to absorb atmospheric carbon dioxide. Adding powdered iron to a small plot of the ocean, they found, has the effect of fostering the growth of microorganisms that enhance the rate at which carbon dioxide is absorbed. Might this be a solution to the problem of global warming? Might adding too much iron initiate another ice age or alter the ocean's ecology? We don't know.

FIGURE 18.18

In the exothermic formation of sodium chloride, sodium metal is oxidized by chlorine gas, and chlorine gas is reduced by sodium metal.

on Earth. How true it is: All the parts are intricately connected, down to the level of atoms and molecules!

■ 18.5 Losing and Gaining Electrons

Oxidation is the process whereby a reactant loses one or more electrons. **Reduction** is the opposite process, whereby a reactant gains one or more electrons. Oxidation and reduction are complementary processes that occur at the same time. They always occur together; you cannot have one without the other. The electrons lost by one chemical in an oxidation reaction don't simply disappear; they are gained by another chemical in a reduction reaction.

An oxidation–reduction reaction occurs when sodium and chlorine react to form sodium chloride, as shown in Figure 18.18. The equation for this reaction is

$$2\,Na(s)\ +\ Cl_2(g)\ \longrightarrow\ 2\,NaCl(s)$$

To see how electrons are transferred in this reaction, we can look at each reactant individually. Each electrically neutral sodium atom changes to a positively charged ion. At the same time, we can say that each atom loses an electron and is therefore oxidized:

$$2\,Na(s)\ \longrightarrow\ 2\,Na^+ + 2e^- \quad \text{Oxidation}$$

Each electrically neutral chlorine molecule changes to two negatively charged ions. Each of these atoms gains an electron and is therefore reduced:

$$Cl_2\ +\ 2e^-\ \longrightarrow\ 2Cl^- \quad \text{Reduction}$$

The net result is that the two electrons lost by the sodium atoms are transferred to the chlorine atoms. Therefore, each of the two equations shown above actually represents one half of an entire process, which is why they are each called a **half reaction**. In other words, an electron won't be lost from a sodium atom without the presence of a chlorine atom available to pick up that electron. Both half reactions are required to represent the *whole* oxidation–reduction process. Half reactions are useful for showing which reactant loses electrons and which reactant gains them, which is why half reactions are used throughout this chapter.

Because the sodium causes reduction of the chlorine, the sodium is acting as a *reducing agent*. A reducing agent is any reactant that causes another reactant to be reduced. Note that sodium is oxidized when it behaves as a reducing agent—it loses electrons. Conversely, the chlorine causes oxidation of the sodium and so is acting as an *oxidizing agent*. Because it gains electrons in the process, an oxidizing agent is reduced. Just remember that **l**oss of **e**lectrons is **o**xidation, and **g**ain of **e**lectrons is **r**eduction. Here is a helpful mnemonic adapted from a once-popular children's story: **Leo** the lion went "**ger**."

Different elements have different oxidation and reduction tendencies—some lose electrons more readily, while others gain electrons more readily, as Figure 18.19 illustrates.

Little tendency to lose or gain electrons

Tendency to gain electrons

Tendency to lose electrons

■ More likely to behave as oxidizing agent (be reduced)

■ More likely to behave as reducing agent (be oxidized)

FIGURE 18.19

The ability of an atom to gain or lose electrons is indicated by its position in the periodic table. Those at the upper right tend to gain electrons, and those at the lower left tend to lose them.

CHECK POINT

True or false?
1. Reducing agents are oxidized in oxidation–reduction reactions.
2. Oxidizing agents are reduced in oxidation–reduction reactions.

Was this your answer?
Both statements are true.

Whether a reaction classifies as an oxidation–reduction reaction is not always immediately apparent. The chemical equation, however, can provide some important clues. First, look for changes in the ionic states of elements. Sodium metal, for example, consists of neutral sodium atoms. In the formation of sodium chloride, these atoms transform into positively charged sodium ions, which occurs as sodium atoms lose electrons (oxidation). A second way to identify a reaction as an oxidation–reduction reaction is to look to see whether an element is gaining or losing oxygen atoms. As the element gains the oxygen, it is losing electrons to that oxygen because of the oxygen's high electronegativity. The gain of oxygen, therefore, is oxidation (loss of electrons), while the loss of oxygen is reduction (gain of electrons). For example, hydrogen, H_2, reacts with oxygen, O_2, to form water, H_2O, as follows:

$$H—H + H—H + O=O \longrightarrow H—O—H + H—O—H$$

Note that the element hydrogen becomes attached to an oxygen atom through this reaction. The hydrogen, therefore, is oxidized.

A third way to identify a reaction as an oxidation–reduction reaction is to see whether an element is gaining or losing hydrogen atoms. The gain of hydrogen is reduction, while the loss of hydrogen is oxidation. For the formation of water shown above, we see that the element oxygen is gaining hydrogen atoms, which means that the oxygen is being reduced—that is, the oxygen is gaining electrons from the hydrogen, which is why the oxygen atom within water is slightly negative as discussed in Section 15.7. The three ways of identifying a reaction as an oxidation–reduction type of reaction are summarized in Figure 18.20.

When we say a substance was oxidized, we're saying that it lost electrons. When we say a substance was reduced, we're saying that it gained electrons.

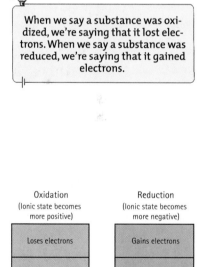

Oxidation (Ionic state becomes more positive)	Reduction (Ionic state becomes more negative)
Loses electrons	Gains electrons
Gains oxygen	Loses oxygen
Loses hydrogen	Gains hydrogen

FIGURE 18.20

Oxidation results in a greater positive charge, which can be achieved by losing electrons, gaining oxygen atoms, or losing hydrogen atoms. Reduction results in a greater negative charge, which can be achieved by gaining electrons, losing oxygen atoms, or gaining hydrogen atoms.

CHECK POINT

In the following equation, is carbon oxidized or reduced?

$$CH_4 + 2O_2 \longrightarrow CO_2 + 2H_2O$$

Was this your answer?
As the carbon of methane, CH_4, forms carbon dioxide, CO_2, it is losing hydrogen and gaining oxygen, which tells us that the carbon is being oxidized.

Oxidation $Fe \longrightarrow Fe^{2+} + 2\ e^-$

Reduction $Cu^{2+} + 2\ e^- \longrightarrow Cu$

FIGURE 18.21

A nail made of iron placed in a solution of Cu^{2+} ions oxidizes to Fe^{2+} ions, which dissolve in the water. At the same time, copper ions are reduced to metallic copper, which coats the nail. (Negatively charged ions, such as chloride ions, Cl^-, must also be present to balance these positively charged ions in solution.)

Link To Section 3.10

This side immediately builds up a positive charge that attracts electrons, preventing them from migrating.

This side immediately builds up a negative charge that repels electrons, preventing them from entering.

FIGURE 18.22

An iron nail is placed in water and connected by a conducting wire to a solution of copper ions. Nothing happens, because this arrangement results in a buildup of charge that prevents the further flow of electrons.

■ 18.6 Harnessing the Energy of Flowing Electrons

Electrochemistry is the study of the relationship between electrical energy and chemical change. It involves either the use of an oxidation–reduction reaction to produce an electric current or the use of an electric current to produce an oxidation–reduction reaction.

To understand how an oxidation–reduction reaction can generate an electric current, consider what happens when a reducing agent is placed in direct contact with an oxidizing agent: electrons flow from the reducing agent to the oxidizing agent. This flow of electrons is an electric current, which is a form of kinetic energy that can be harnessed for useful purposes.

Iron atoms, Fe, for example, are better reducing agents than copper ions, Cu^{2+}. So when a piece of iron metal and a solution containing copper ions are placed in contact with each other, electrons flow from the iron to the copper ions, as Figure 18.21 illustrates. The result is the oxidation of iron atoms and the reduction of copper ions.

The elemental iron and copper ions need not be in physical contact for electrons to flow between them. If they are in separate containers but bridged by a conducting wire, the electrons can flow from the iron through the wire to the copper ions. The resulting electric current in the wire can be attached to some useful device, such as a lightbulb. But alas, an electric current is not sustained by this arrangement.

The reason the electric current is not sustained is shown in Figure 18.22. An initial flow of electrons through the wire immediately results in a buildup of electric charge in both containers. The container on the left builds up positive charge as it accumulates Fe^{2+} ions from the nail. The container on the right builds up negative charge as electrons accumulate on this side. This situation prevents any further migration of electrons through the wire. Recall that electrons are negative, and so they are repelled by the negative charge in the right container and attracted to the positive charge in the left container. The net result is that the electrons do not flow through the wire, and the bulb remains unlit.

Oxidation $Fe \longrightarrow Fe^{2+} + 2\,e^-$ Reduction $Cu^{2+} + 2\,e^- \longrightarrow Cu$

FIGURE 18.23

The salt bridge completes the electric circuit. Electrons freed as the iron is oxidized pass through the wire to the container on the right. Nitrate ions, NO_3^-, from the salt bridge flow into the left container to balance the positive charges of the Fe^{2+} ions that form, thereby preventing any buildup of positive charge. Meanwhile, Na^+ ions from the salt bridge enter the right container to balance the Cl^- ions "abandoned" by the Cu^{2+} ions as the Cu^{2+} ions pick up electrons to become metallic copper.

The solution to this problem is to allow ions to migrate into either container so that neither builds up any positive or negative charge. This is accomplished with a *salt bridge*, which may be a U-shaped tube filled with a salt, such as sodium nitrate, $NaNO_3$, and closed with semiporous plugs. Figure 18.23 shows how a salt bridge allows the ions it holds to enter either container, permitting the flow of electrons through the conducting wire and creating a complete electric circuit.

> A salt bridge may be as simple as a paper towel soaked in salt water.

BATTERIES

So we can see that, with the proper setup, it is possible to harness electrical energy from an oxidation–reduction reaction. The apparatus shown in Figure 18.23 is one example. Such devices are called *voltaic cells*. Instead of two containers, a voltaic cell can be an all-in-one, self-contained unit, in which case it is called a *battery*. Batteries are either disposable or rechargeable, and here we explore some examples of each. Although the two types differ in design and composition, they function by the same principle: two materials that oxidize and reduce each other are connected by a medium through which ions travel to balance an external flow of electrons.

Let's look at disposable batteries first. The common *dry-cell battery*, which was invented in the 1860s, is still used today, and it is probably the cheapest disposable energy source for flashlights, toys, and the like. The basic design consists of a zinc cup filled with a thick paste of ammonium chloride, NH_4Cl; zinc chloride, $ZnCl_2$; and manganese dioxide, MnO_2. Immersed in this paste is a porous stick of graphite that projects to the top of the battery, as shown in Figure 18.24.

Graphite is a good conductor of electricity. Chemicals in the paste receive electrons at the graphite stick and so are reduced. The reaction for the ammonium ions is

$$2NH_4^+(aq) + 2e^- \longrightarrow 2NH_3(g) + H_2(g) \quad \text{Reduction}$$

An **electrode** is any material that conducts electrons into or out of a medium in which electrochemical reactions are occurring. The electrode where chemicals

Reduction $2\,NH_4^+ + 2e^- \longrightarrow 2\,NH_3 + H_2$

- Graphite rod (cathode)
- Zinc cup (anode)
- Paste (NH_4Cl, $ZnCl_2$, MnO_2)
- Membrane

Oxidation $Zn \longrightarrow Zn^{2+} + 2e^-$

FIGURE 18.24

A common dry-cell battery with a graphite rod immersed in a paste of ammonium chloride, manganese dioxide, and zinc chloride.

If you store your extra flashlight batteries in the refrigerator, they'll last longer.

FIGURE 18.25

Alkaline batteries last a lot longer than dry-cell batteries and give a steadier voltage, but they are more expensive.

Link To Section 8.5

are reduced is called a **cathode**. For any battery, such as the one shown in Figure 18.24, the cathode is always positive (+), which indicates that electrons are naturally attracted to this location. The electrons gained by chemicals at the cathode originate at the **anode**, which is the electrode where chemicals are oxidized. For any battery, the anode is always negative (−), which indicates that electrons are streaming away from this location. The anode in Figure 18.24 is the zinc cup, where zinc atoms lose electrons to form zinc ions:

$$Zn(s) \longrightarrow Zn^{2+}(aq) + 2e^- \quad \text{Oxidation}$$

The reduction of ammonium ions in a dry-cell battery produces two gases—ammonia, NH_3, and hydrogen, H_2—that need to be removed to avoid a pressure buildup and a potential explosion. Removal is accomplished by having the ammonia and hydrogen react with the zinc chloride and manganese dioxide:

$$ZnCl_2(aq) + 2 NH_3(g) \longrightarrow Zn(NH_3)_2Cl_2(s)$$

$$2 MnO_2(s) + H_2(g) \longrightarrow Mn_2O_3(s) + H_2O(l)$$

The life of a dry-cell battery is relatively short. Oxidation causes the zinc cup to deteriorate, and eventually the contents leak out. Even while the battery is not operating, the zinc corrodes as it reacts with ammonium ions. This zinc corrosion can be inhibited by storing the battery in a refrigerator. As discussed in Section 17.3, chemical reactions slow down with decreasing temperature. Chilling a battery, therefore, slows down the rate at which the zinc corrodes, which increases the life of the battery.

Another type of disposable battery, the more expensive *alkaline battery*, shown in Figure 18.25, avoids many of the problems of dry-cell batteries by operating in a strongly alkaline paste. In the presence of hydroxide ions, the zinc oxidizes to insoluble zinc oxide:

$$Zn(s) + 2 OH^-(aq) \longrightarrow ZnO(s) + H_2O(l) + 2e^- \quad \text{Oxidation}$$

At the same time, manganese dioxide is reduced:

$$2 MnO_2(s) + H_2O(l) + 2e^- \longrightarrow Mn_2O_3(s) + 2 H_2O(aq) \quad \text{Reduction}$$

Note how these two reactions avoid the use of the zinc-corroding ammonium ion (which means alkaline batteries last a lot longer than dry-cell batteries) and also prevent formation of any gaseous products. Furthermore, these reactions are better suited to maintaining a given voltage during longer periods of operation.

The small mercury and lithium disposable batteries used for calculators and cameras are variations of the alkaline battery. In the mercury battery, mercuric oxide, HgO, is reduced rather than manganese dioxide. Manufacturers are phasing out these batteries because of the environmental hazard posed by mercury, which is poisonous. In the lithium battery, lithium metal is used as the source of electrons rather than zinc. Not only can lithium maintain a higher voltage than zinc, but it also is about $\frac{1}{13}$ as dense, which allows for a lighter battery.

Disposable batteries have relatively short lives because electron-producing chemicals are consumed. The main feature of *rechargeable* batteries is the reversibility of the oxidation and reduction reactions. A noteworthy example is the nickel metal hydride, NiMH, battery. Charging this battery causes the nickel metal to extract hydrogen from water to form the negatively charged hydride ion, shown below as H: where the two dots represent two electrons.*

$$H_2O + \quad Ni \quad + \quad 2e^- \quad \longrightarrow \quad H\!:\!Ni \quad + \quad HO^-$$

Water \quad Nickel metal \quad\quad Nickel hydride \quad Hydroxide ion

* The nickel in this reaction is actually an intermetallic compound of nickel and various rare-earth elements, such as lanthanum, La.

The role of the nickel is to stabilize the two electrons on the hydrogen, which, because it contains an added electron, is called the *hydride* ion, much as chlorine with an extra electron is called the *chloride* ion, using the *-ide* suffix. A fully charged battery thus contains an abundance of nickel hydride. As the battery provides electricity, the hydride ion releases electrons, which allows it to join with the hydroxide ion to reform water:

$$\text{H} : \text{Ni} \quad + \quad \text{HO}^- \quad \longrightarrow \quad \text{H}_2\text{O} \quad + \quad \text{Ni} \quad + \quad 2e^-$$

Nickel hydride   Hydroxide ion   Water   Nickel metal

> Aside from the initial charge of a brand-new battery, the energy in a car battery ultimately comes from fuel in the gas tank through the process of recharging.

So recharging a rechargeable battery simply means regenerating the chemicals, such as nickel hydride, that can release electrons on demand. For the NiMH battery, this chemical is nickel hydride, H : Ni. For a traditional car battery, this chemical is simply lead, Pb, which transforms into lead sulfate, $PbSO_4$, as it releases electrons. As the car battery is recharged, the $PbSO_4$ is transformed back into lead, Pb.

Rechargeable lithium-ion batteries have found a wide range of applications, such as powering laptop computers and cell phones. Safer lithium phosphate iron batteries are used for hybrid cars, such as the popular Toyota Prius shown in Figure 18.26. Hybrids have improved gas mileage because as the car slows down, its kinetic energy is transformed into the electric potential energy of the battery rather than being wasted as heat from the car's brake pads. The captured electrical energy of the battery is subsequently used to assist the gas-powered engine to get the car moving. Also, the hybrid's battery system allows the engine to shut off when the car is merely idling or moving slowly, as in heavy traffic.

Continued improvements in battery technology are permitting next-generation hybrids, known as *plug-in hybrids*, which have much larger batteries and smaller fuel tanks. These hybrids can be plugged into electrical outlets, charged overnight, and driven the next day for up to 60 mi without using any gasoline. This is significant because the typical driver in the United States drives less than 40 mi a day. Furthermore, utility companies can sell electricity at cheaper rates at night because their massive generators are underutilized at that time. Alternatively, the plug-in hybrid can be charged via residential photovoltaic panels or a small wind turbine. Cars that remain plugged in during the day would be available to contribute energy back into the grid during peak demand—and owners of such cars could receive rebates for this energy. Furthermore, during an electricity blackout, a family's plug-in hybrid would store enough energy to serve as emergency backup for household needs. Plug-in hybrids with their large and highly efficient batteries offer much in the way of moving individuals and the nation as a whole toward energy conservation and independence.

FIGURE 18.26

As of 2006, about 500,000 Prius hybrids have been sold worldwide, about 266,000 of them in the United States. In coming years, look for hybrid vehicles that can be plugged into your home electrical outlet, charged at night, and driven the next day using no gasoline for up to 60 mi.

CHECK POINT

What chemicals are produced as a nickel metal hydride battery recharges?

Was this your answer?

Nickel hydride, H : Ni, and hydroxide ions, HO⁻.

FUEL CELLS

A *fuel cell* is a device that converts the chemical energy of a fuel to electrical energy. Fuel cells are by far the most efficient means of generating electricity. A hydrogen–oxygen fuel cell is shown in Figure 18.27. It has two compartments, one for entering hydrogen fuel and the other for entering oxygen fuel,

fyi

- As a battery provides electricity, electrons move from the negative anode to the positive cathode. When a rechargeable battery is being recharged, however, what was once the negative anode now becomes the negative cathode, which is where electrons are needed for reduction. Because electrons won't travel to a negative cathode on their own, they must be forced to do so. Hence, recharging is an energy-consuming process.

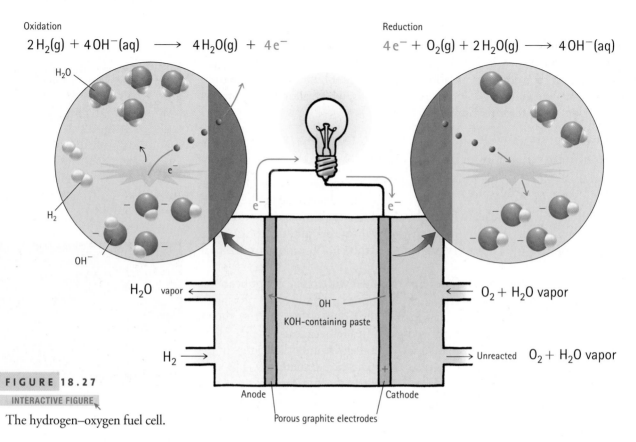

Oxidation

$$2H_2(g) + 4OH^-(aq) \longrightarrow 4H_2O(g) + 4e^-$$

Reduction

$$4e^- + O_2(g) + 2H_2O(g) \longrightarrow 4OH^-(aq)$$

FIGURE 18.27

INTERACTIVE FIGURE

The hydrogen–oxygen fuel cell.

Because fuel cells can do so much, there is little doubt that they will be spread throughout the marketplace in the not-too-distant future.

FIGURE 18.28

Because this bus is powered by a fuel cell, its tailpipe emits mostly water vapor.

separated by a set of porous electrodes. Hydrogen is oxidized on contact with hydroxide ions at the hydrogen-facing electrode (the anode). The electrons from this oxidation flow through an external circuit and provide electric power before meeting up with oxygen at the oxygen-facing electrode (the cathode). The oxygen readily picks up the electrons (in other words, the oxygen is reduced) and reacts with water to form hydroxide ions. To complete the circuit, these hydroxide ions migrate across the porous electrodes and through an ionic paste of potassium hydroxide, KOH, to join with hydrogen at the hydrogen-facing electrode.

As the oxidation equation shown at the top of Figure 18.27 demonstrates, the hydrogen and hydroxide ions react to produce energetic water molecules that arise in the form of steam. This steam may be used for heating or for generating electricity in a steam turbine. Furthermore, the water that condenses from the steam is pure water, suitable for drinking!

Although fuel cells are similar to dry-cell batteries, they don't run down as long as fuel is supplied. The space shuttle uses hydrogen–oxygen fuel cells to meet its electrical needs. The cells also produce more than 100 gal of drinking water for the astronauts during a typical week-long mission. Back on Earth, researchers are developing fuel cells for buses and automobiles. As shown in Figure 18.28, experimental fuel-cell buses are already operating in several cities, such as Vancouver, British Columbia, and Chicago, Illinois. These vehicles produce very few pollutants and can run much more efficiently than vehicles that burn fossil fuels.

In the future, commercial buildings as well as individual homes may be outfitted with fuel cells as an alternative to receiving electricity (and heat) from regional power stations. Researchers are also working on miniature fuel cells that could replace the batteries used for portable electronic devices, such as cell phones and laptop computers. Such devices could operate for extended periods of time on a single "ampoule" of fuel available at your local supermarket.

Amazingly, a car powered by a hydrogen–oxygen fuel cell requires only about 3 kg of hydrogen to travel 500 km. However, this quantity of hydrogen gas at room temperature and atmospheric pressure would occupy a volume of about 36,000 L, the volume of about four midsize cars! Thus, the major hurdle to the development of fuel-cell technology lies not with the cell but with the fuel. This volume of gas could be compressed to a much smaller volume, as it is in the experimental buses in Vancouver.

Compressing a gas requires energy, however—and, as a consequence, the inherent efficiency of the fuel cell is lost. Chilling hydrogen to its liquid phase, which occupies much less volume, poses similar problems. Instead, researchers are looking for novel ways of providing fuel cells with hydrogen. In one design, hydrogen is generated within the fuel cell from chemical reactions involving liquid fuels, such as methanol, CH_3OH. Alternatively, certain porous materials, including the recently developed carbon nanofibers shown in Figure 18.29, can hold large volumes of hydrogen on their surfaces, behaving in effect like hydrogen "sponges." The hydrogen is "squeezed" out of these materials on demand by controlling the temperature—the warmer the material, the more hydrogen that is released.

FIGURE 18.29

Carbon nanofibers consist of near-submicroscopic tubes of carbon atoms. They outclass almost all other known materials in their ability to absorb hydrogen molecules. With carbon nanofibers, a volume of 36,000 L of hydrogen can be reduced to a mere 35 L. Carbon nanofibers are a recent discovery, however, and much research is still required to confirm their applicability to hydrogen storage and to develop the technology.

CHECK POINT

As long as fuel is available to it, a given fuel cell can supply electrical energy indefinitely. Why can't batteries do the same?

Was this your answer?
Batteries generate electricity as the chemical reactants they contain are reduced and oxidized. Once these reactants are consumed, the battery can no longer generate electricity. A rechargeable battery can be made to operate again, but only after the energy flow is interrupted so that the reactants can be replenished.

18.7 Electrolysis

E lectrolysis is the use of electrical energy to produce chemical change. The recharging of a car battery is an example of electrolysis. Another, shown in Figure 18.30, is passing an electric current through water, a process that breaks the water down into its elemental components:

$$\text{electrical energy} + 2\,H_2O(l) \longrightarrow 2\,H_2(g) + O_2(g)$$

Electrolysis is used to purify metals from metal ores. An example is aluminum, the third most abundant element in Earth's crust. Aluminum occurs naturally bonded to oxygen in an ore called bauxite. Aluminum metal wasn't known until about 1827, when it was prepared by reacting bauxite with hydrochloric acid.

FIGURE 18.30

The electrolysis of water produces hydrogen gas and oxygen gas in a 2 : 1 ratio by volume, in accord with the chemical formula for water: H_2O. In order for this process to work, ions must be dissolved in the water so that electric charge can be conducted between the electrodes.

Link To Section 3.10

fyi

- Our bodies require lots of energy for living. We get this energy from special high-energy molecules, such as ATP, which the body produces by oxidizing food molecules with oxygen. If you were to stop breathing, say by choking, your cells would be deprived of oxygen and no longer able to produce these high-energy molecules. The result is a prompt death. But instead of dying, why doesn't the body simply turn off until oxygen becomes available again? Lethal damage occurs because many cellular mechanisms continue to operate even at very low oxygen levels. With some parts working and others not working, the cell is thrown so far off balance that it dies. The trick is to make sure that all cellular processes shut down together. This explains how people who fall into frozen waters can sometimes be resuscitated even though they haven't been breathing for more than an hour—their cells were shut down uniformly because of the rapid onslaught of the extreme cold.

FIGURE 18.31

The melting point of aluminum oxide (2030°C) is too high for efficiently electrolyzing to aluminum metal. When the oxide is mixed with the mineral cryolite, the melting point of the oxide drops to a more reasonable 980°C. A strong electric current passed through the molten aluminum oxide–cryolite mixture generates aluminum metal at the cathode, where aluminum ions pick up electrons and are thus reduced to elemental aluminum.

This reaction gave the aluminum ion, Al^{3+}, which was reduced to aluminum metal, with sodium metal acting as the reducing agent:

$$Al^{3+} + 3\,Na \longrightarrow Al + 3\,Na^+$$

This chemical process was expensive. The price of aluminum at that time was about \$100,000 per pound, and it was considered a rare and precious metal. In 1855, aluminum dinnerware and other items were exhibited in Paris with the crown jewels of France. Then, in 1886, two men working independently, Charles Hall (1863–1914) in the United States and Paul Heroult (1863–1914) in France, almost simultaneously discovered a process whereby aluminum could be produced from aluminum oxide, Al_2O_3, a main component of bauxite. In what is now known as the Hall–Heroult process, shown in Figure 18.31, a strong electric current is passed through a molten mixture of aluminum oxide and cryolite, Na_3AlF_6, a naturally occurring mineral. The fluoride ions of the cryolite react with the aluminum oxide to form various aluminum fluoride ions, such as $AlOF_3^{2-}$, which are then oxidized to the aluminum hexafluoride ion, AlF_6^{3-}. The Al^{3+} in this ion is then reduced to elemental aluminum, which collects at the bottom of the reaction chamber. This process, which is still in use by manufacturers today, greatly facilitated mass production of aluminum metal, and, by 1890, the price of aluminum had dropped to about \$2 per pound.

Today, worldwide production of aluminum is about 16 million tons annually. For each ton produced from ore, about 16,000 kWh of electrical energy is required, as much as a typical American household consumes in 18 months. Processing recycled aluminum, on the other hand, consumes only about 700 kWh for every ton. Thus, recycling aluminum not only reduces litter but also helps reduce the load on power companies, which in turn reduces air pollution. Furthermore, reserves of high-quality aluminum oxide ores are already depleted in the United States. Recycling aluminum, therefore, also helps minimize the need for developing new bauxite mines in foreign countries.

Oxidation $2\,AlOF_3^{2-} + 6\,F^- + C \longrightarrow 2\,AlF_6^{3-} + CO_2 + 4\,e^-$

Reduction $AlF_6^{3-} + 3\,e^- \longrightarrow Al + 6\,F^-$

CHECK POINT

Is the exothermic reaction in a hydrogen–oxygen fuel cell an example of electrolysis?

Was this your answer?

No. During electrolysis, electrical energy is used to produce chemical change. In the hydrogen-oxygen fuel cell, chemical change is used to produce electrical energy.

■ 18.8 Corrosion and Combustion

If you look to the upper right of the periodic table, you will find one of the most common oxidizing agents—oxygen. In fact, if you haven't guessed already, the term *oxidation* is derived from the name of this element. Oxygen can pluck electrons from many other elements, especially those that lie at the lower left of the periodic table. Two common oxidation–reduction reactions involving oxygen as the oxidizing agent are *corrosion* and *combustion*.

CHECK POINT

Oxygen is a good oxidizing agent, but so is chlorine. What does this indicate about their relative positions in the periodic table?

Was this your answer?

Chlorine and oxygen must lie in the same area of the periodic table. Both have strong effective nuclear charges and are strong oxidizing agents.

Corrosion is the process whereby a metal deteriorates. Corrosion caused by atmospheric oxygen is a widespread and costly problem. About one-quarter of the steel produced in the United States, for example, goes into replacing corroded iron at a cost of billions of dollars annually. Iron corrodes when it reacts with atmospheric oxygen and water to form iron oxide trihydrate, which is the naturally occurring reddish-brown substance you know as rust, shown in Figure 18.32:

$$4\,\text{Fe} + 3\,\text{O}_2 + 3\,\text{H}_2\text{O} \longrightarrow 2\,\text{Fe}_2\text{O}_3 \cdot 3\,\text{H}_2\text{O}$$

Iron   Oxygen   Water   Rust

Another common metal oxidized by oxygen is aluminum. The product of aluminum oxidation is aluminum oxide, Al_2O_3, which is not water soluble. Because of its insolubility, aluminum oxide forms a protective coat that shields the metal from further oxidation. This coat is so thin that it's transparent, which is why aluminum maintains its metallic shine.

A protective, water-insoluble oxidized coat is the principle underlying a process called *galvanization*. Zinc has a slightly greater tendency to oxidize than does iron. For this reason, many iron objects, such as the nails pictured in Figure 18.33, are *galvanized* by coating them with a thin layer of zinc. The zinc oxidizes to zinc oxide, an inert, insoluble substance that protects the iron underneath it from rusting.

FIGURE 18.32

Rust itself does not harm the iron structures on which it forms. The loss of metallic iron ruins the structural integrity of these objects.

FIGURE 18.33

The galvanized nail (*bottom*) is protected from rusting by the sacrificial oxidation of zinc.

FIGURE 18.34

Zinc strips help protect the iron hull of an oil tanker from oxidizing. The zinc strip shown here is attached to the hull's interior surface.

The metals used for cathodic protection are "sacrificing" themselves to be anodes (to lose electrons) so that the desired metal, such as the copper pipe, is spared from oxidation. These sacrificing metals, therefore, are sometimes called *sacrificial anodes*.

FIGURE 18.35

As electrons flow into the hubcap and give it a negative charge, positively charged chromium ions move from the solution to the hubcap and are reduced to chromium metal, which deposits as a coating on the hubcap. The solution is supplied with ions as chromium atoms in the cathode are oxidized to Cr^{2+} ions.

In a technique called *cathodic protection*, iron structures can be protected from oxidation by placing them in contact with certain metals, such as zinc or magnesium, that have a greater tendency to oxidize. This forces the iron to accept electrons, which means that it is behaving as a cathode. (Rusting occurs only where iron behaves as an anode). Ocean tankers, for example, are protected from corrosion by strips of zinc affixed to their hulls, as shown in Figure 18.34. Similarly, outdoor steel pipes are protected by being connected to magnesium rods inserted into the ground.

Yet another way to protect iron and other metals from oxidation is to coat them with a corrosion-resistant metal, such as chromium, platinum, or gold. *Electroplating* is the operation of coating one metal with another by electrolysis, and it is illustrated in Figure 18.35. The object to be electroplated is connected to a negative battery terminal and then submerged in a solution containing ions of the metal to be used as the coating. The positive terminal of the battery is connected to an electrode made of the coating metal. The circuit is completed when this electrode is submerged in the solution. Dissolved metal ions are attracted to the negatively charged object, where they pick up electrons and are deposited as metal atoms. The ions in solution are replenished by the forced oxidation of the coating metal at the positive electrode.

Combustion is an oxidation–reduction reaction between a nonmetallic material and molecular oxygen. Combustion reactions are characteristically exothermic (energy-releasing). A violent combustion reaction is the formation of water from hydrogen and oxygen. As discussed in Section 17.5, the energy from this reaction is used to power rockets into space. More common examples of combustion include the burning of wood and fossil fuels. The combustion of these and other carbon-based chemicals forms carbon dioxide and water. Consider, for example, the combustion of methane, the major component of natural gas:

$$CH_4 \ + \ 2\,O_2 \ \longrightarrow \ CO_2 \ + \ 2\,H_2O \ + \ \text{energy}$$

Methane   Oxygen   Carbon   Water
           dioxide

In combustion, electrons are transferred as polar covalent bonds are formed in place of nonpolar covalent bonds, or vice versa. (This is in contrast with the other examples of oxidation–reduction reactions presented in this chapter, which involve the formation of ions from atoms or, conversely, atoms from ions.) This

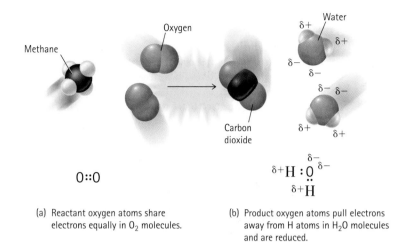

(a) Reactant oxygen atoms share electrons equally in O_2 molecules.

(b) Product oxygen atoms pull electrons away from H atoms in H_2O molecules and are reduced.

FIGURE 18.36

(a) Neither atom in an oxygen molecule can preferentially attract the bonding electrons. (b) The oxygen atom of a water molecule pulls the bonding electrons away from the hydrogen atoms on the water molecule, making the oxygen slightly negative and the two hydrogens slightly positive.

fyi

- There are two kinds of matches: the "strike anywhere" type usually having a "bulls-eye" looking tip, and the "safety match," which requires you to strike the match on a strip on the packaging. Both involve the burning of sulfur within the tip of the match. Getting the sulfur to burn using only the oxygen in the air, however, is difficult, which is why the sulfur is blended with an oxidizing agent, such as potassium chlorate, $KClO_3$. For the "strike anywhere" match a third ingredient, red phosphorus, P_4, is included. The heat of friction causes the red phosphorus to convert into white phosphorus—an alternate form of phosphorus that burns rapidly in air. This initiates the reduction-oxidation reaction between the sulfur and the potassium chlorate, which in turn ignites the burning of the matchstick. Safety matches work the same way, except that the red phosphorus is embedded within the striking strip, which is the only place where the match can be lit.

concept is illustrated in Figure 18.36, which compares the electronic structures of the combustion starting material, molecular oxygen, and the combustion product, water. Molecular oxygen is a nonpolar covalent compound. Although each oxygen atom in the molecule has a fairly strong electronegativity, the four bonding electrons are pulled equally by both atoms and thus cannot congregate on one side or the other. After combustion, however, the electrons are shared between the oxygen and hydrogen atoms in a water molecule and are pulled to the oxygen. This gives the oxygen a slight negative charge, which is another way of saying it has gained electrons and has thus been reduced. At the same time, the hydrogen atoms in the water molecule develop a slight positive charge, which is another way of saying they have lost electrons and have thus been oxidized. This gain of electrons by oxygen and loss of electrons by hydrogen is an energy-releasing process. Typically, the energy is released either as molecular kinetic energy (heat) or as light (the flame).

Interestingly, combustion oxidation–reduction reactions occur throughout your body. You can visualize a simplified model of your metabolism by reviewing Figure 18.36 and substituting a food molecule for the methane. Food molecules relinquish their electrons to the oxygen molecules you inhale. The products are carbon dioxide, water vapor, and energy. You exhale the carbon dioxide and water vapor, but much of the energy from the reaction is used to keep your body warm and to drive the many other biochemical reactions necessary for life.

SUMMARY OF TERMS

Acid A substance that donates hydrogen ions.

Base A substance that accepts hydrogen ions.

Hydronium ion A water molecule after accepting a hydrogen ion.

Hydroxide ion A water molecule after losing a hydrogen ion.

Salt An ionic compound formed from the reaction between an acid and a base.

Neutralization A reaction in which an acid and base combine to form a salt.

Amphoteric Describes a substance that can behave either as an acid or as a base.

Acidic solution A solution in which the hydronium-ion concentration is higher than the hydroxide-ion concentration.

Basic solution A solution in which the hydroxide-ion concentration is higher than the hydronium-ion concentration.

Neutral solution A solution in which the hydronium-ion concentration is equal to the hydroxide-ion concentration.

pH A measure of the acidity of a solution, equal to the negative of the base-10 logarithm of the hydronium-ion concentration.

Oxidation The process whereby a reactant loses one or more electrons.

Reduction The process whereby a reactant gains one or more electrons.

Half reaction One portion of an oxidation–reduction reaction, represented by an equation showing electrons as either reactants or products.

Electrochemistry The branch of chemistry concerned with the relationship between electrical energy and chemical change.

Electrode Any material that conducts electrons into or out of a medium in which electrochemical reactions are occurring.

Cathode The electrode where reduction occurs.

Anode The electrode where oxidation occurs.

Electrolysis The use of electrical energy to produce chemical change.

Corrosion The deterioration of a metal, typically caused by atmospheric oxygen.

Combustion An exothermic oxidation–reduction reaction between a nonmetallic material and molecular oxygen.

REVIEW QUESTIONS

18.1 Acids Donate Protons; Bases Accept Them

1. What are the Brønsted–Lowry definitions of *acid* and *base*?
2. When an acid is dissolved in water, what ion does the water form?
3. When a chemical loses a hydrogen ion, is it behaving as an acid or a base?

18.2 Relative Strengths of Acids and Bases

4. What does it mean to say that an acid is strong in aqueous solution?
5. Why does a solution of a strong acid conduct electricity better than a solution of a weak acid with the same concentration?
6. When can a solution of a weak base be more corrosive than a solution of a strong base?

18.3 Acidic, Basic, and Neutral Solutions

7. Is water a strong acid or a weak acid?
8. What is true about the relative concentrations of hydronium and hydroxide ions in an acidic solution? How about a neutral solution? A basic solution?
9. What does the pH of a solution indicate?
10. As the hydronium-ion concentration of a solution increases, does the pH of the solution increase or decrease?

18.4 Acidic Rain and Basic Oceans

11. What is the product of the reaction between carbon dioxide and water?
12. What does sulfur dioxide have to do with acid rain?
13. How do humans generate the air pollutant sulfur dioxide?
14. Why aren't atmospheric levels of carbon dioxide rising as rapidly as might be expected based on the increased output of carbon dioxide resulting from human activities?

18.5 Losing and Gaining Electrons

15. Which elements have the greatest tendency to behave as oxidizing agents?
16. Write an equation for the half reaction in which a potassium atom, K, is oxidized.
17. What is the difference between an oxidizing agent and a reducing agent?
18. What elements have the greatest tendency to behave as reducing agents?

18.6 Harnessing the Energy of Flowing Electrons

19. What is electrochemistry?
20. What is the purpose of the salt bridge in a voltaic cell?
21. What type of reaction occurs at the cathode?
22. What type of reaction occurs at the anode?
23. What is the prime difference between a battery and a fuel cell?

18.7 Electrolysis

24. What is electrolysis, and how does it differ from what goes on inside a battery?
25. What is an example of a metal produced primarily by electrolysis?

18.8 Corrosion and Combustion

26. What do the oxidation of zinc and the oxidation of aluminum have in common?
27. What metal coats a galvanized nail?
28. What are some differences between corrosion and combustion?
29. What is iron forced to accept during cathodic protection?
30. What happens to the polarity of oxygen atoms as they transform from molecular oxygen, O_2, into water molecules, H_2O?

EXERCISES

● BEGINNER   ■ INTERMEDIATE   ◆ EXPERT

1. ■ An acid and a base react to form a salt, which consists of positive and negative ions. Which forms the positive ions: the acid or the base? Which forms the negative ions?

2. ● Water is formed from the reaction between an acid and a base. Why is water not classified as a salt?

3. ■ Identify the acid or base behavior of each substance in these reactions:

   (a) H_3O^+ + Cl^- $\rightleftharpoons$ H_2O + HCl

   _____   _____   _____   _____

   (b) H_2PO_4 + H_2O $\rightleftharpoons$ H_3O^+ + HPO_4^-

   _____   _____   _____   _____

4. ● What happens to the corrosive properties of an acid and a base after they neutralize each other? Why?

5. ● Why do we use the pH scale to indicate the acidity of a solution rather than simply stating the concentration of hydronium ions?

6. ◆ The amphoteric reaction between two water molecules is endothermic, which means the reaction requires the input of heat energy in order to proceed:

   $$Energy + H_2O + H_2O \longrightarrow H_3O^+ + OH^-$$

   The warmer the water, the more heat energy is available for this reaction, and the more hydronium and hydroxide ions are formed.

   (a) Which has a lower pH: pure water that is hot or pure water that is cold?

   (b) Is it possible for water to be neutral but have a pH less than or greater than 7.0?

7. ◆ What is the concentration of hydronium ions in a solution that has a pH of −3? Why is such a solution impossible to prepare?

8. ■ What happens to the pH of an acidic solution as pure water is added?

9. ■ Can an acidic solution be made less acidic by adding an acidic solution?

10. ● Many of the smelly molecules of cooked fish are alkaline compounds. How might these smelly molecules be conveniently transformed into less smelly salts just before eating the fish?

11. ◆ How readily an acid donates a hydrogen ion is a function of how well the acid is able to accommodate the resulting negative charge it gains after donating. Which should be the stronger acid: water or hypochlorous acid? Why?

Water   Hypochlorous acid

12. ■ Pour vinegar onto beach sand from the Caribbean and the result is a lot of froth and bubbles. Pour vinegar onto beach sand from California, however, and nothing happens. Why?

13. ◆ Why is the H—F bond so much stronger than the H—I bond? (Hint: Think atomic size.)

14. ■ Which bond is easier to break: H—F or H—I?

15. ◆ Which is the stronger acid: H—F or H—I?

16. ◆ The main component of bleach is sodium hypochlorite, NaOCl, which consists of sodium ions, Na^+, and hypochlorite ions, ^-OCl. What products are formed when this compound is reacted with the hydrochloric acid, HCl, of toilet bowl cleaner?

17. ■ What happens to the pH of soda water as it loses its carbonation?

18. ● How might you tell whether or not your toothpaste contained calcium carbonate, $CaCO_3$, or perhaps baking soda, $NaHCO_3$, without looking at the ingredients label?

19. ● Why do lakes lying in granite basins tend to become acidified by acid rain more readily than lakes lying in limestone basins?

20. ● How might warmer oceans accelerate global warming?

21. ● What happens to the pH of water as you blow bubbles into it through a drinking straw?

22. ◆ What happens to the pH of a 1 M solution of hydrochloric acid, HCl, as carbon dioxide gas is bubbled into it?

23. ● What element is oxidized in the following equation and what element is reduced?

   $$I_2 + 2 Br^- \longrightarrow 2 I^- + Br_2$$

24. ● What element is behaves as the oxidizing agent in the following equation and what element behaves as the reducing agent?

   $$Sn^{2+} + 2 Ag \longrightarrow Sn + 2 Ag^+$$

25. ● Hydrogen sulfide, H_2S, burns in the presence of oxygen, O_2, to produce water, H_2O, and sulfur dioxide, SO_2. Through this reaction, is sulfur oxidized or reduced?

   $$2 H_2S + 3 O_2 \longrightarrow 2 H_2O + 2 SO_2$$

26. ● Unsaturated fatty acids, such as $C_{12}H_{22}O_2$, react with hydrogen gas, H_2, to form saturated fatty acids, such as $C_{12}H_{24}O_2$. Are the unsaturated fatty acids being oxidized or reduced through this process?

27. ■ Which atom is oxidized, the red one or the blue one?

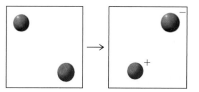

28. ■ In the previous exercise, which atom behaves as the oxidizing agent, the red one or the blue one?

29. ● What correlation might you expect between an element's electronegativity (Section 15.6) and its ability to behave as an oxidizing agent? How about its ability to behave as a reducing agent?

30. ■ Iron atoms, Fe, are better reducing agents than copper ions, Cu^{2+}. In which direction do electrons flow when an iron nail is submerged in a solution of Cu^{2+} ions?

31. ◆ A major source of chlorine gas, Cl_2, is from the electrolysis of brine, which is concentrated salt water, $NaCl(aq)$. What other two products result from this electrolysis reaction? Write the balanced chemical equation.

32. ■ Pennies manufactured after 1982 are made of zinc metal, Zn, within a coat of copper metal, Cu. Zinc is more easily oxidized than copper. Why, then, don't these pennies quickly corrode?

33. ■ The general chemical equation for photosynthesis is shown below. Through this reaction, is the carbon oxidized or reduced?

$$6\,CO_2 + 6\,H_2O \longrightarrow C_6H_{12}O_6 + 6\,O_2$$

34. ● A chemical equation for the combustion of propane, C_3H_8, is shown below. Through this reaction, is the carbon oxidized or reduced?

$$C_3H_8 + 5\,O_2 \longrightarrow 3\,CO_2 + 4\,H_2O$$

35. ■ Chemical equations need to be balanced not only in terms of the number of atoms, but also by the charge. In other words, just as the same number of atoms should appear before and after the arrow of an equation, so should the same charge. Take this into account to balance the following chemical equation:

$$Sn^{2+} + Ag \longrightarrow Sn + Ag^+$$

36. ■ Perform Exercise 35 before attempting to balance both the atoms and charges of the following chemical equation:

$$Fe^{3+} + I^- \longrightarrow Fe^{2+} + I_2$$

37. ● The type of iron that the human body needs for good health is the Fe^{2+} ion. Cereals fortified with iron, however, usually contain small grains of elemental iron, Fe. What must the body do to this elemental iron to make good of it? Oxidation or reduction?

38. ■ Why are combustion reactions generally exothermic?

39. ■ Water is 88.88% oxygen by mass. Oxygen is exactly what a fire needs to grow brighter and stronger. So why doesn't a fire grow brighter and stronger when water is added to it?

40. ■ Iron atoms have a greater tendency to oxidize than copper atoms do. Is this good news or bad news for a home in which much of the plumbing consists of iron and copper pipes connected together? Explain.

41. ■ Copper atoms have a greater tendency to be reduced than iron atoms do. Was this good news or bad news for the Statue of Liberty, whose copper exterior was originally held together by steel rivets?

42. ■ When lightning strikes, nitrogen molecules, N_2, and oxygen molecules, O_2, in the air react to form nitrates, NO_3^-, which come down in the rain to help fertilize the soil. Is this an example of oxidation or reduction?

43. ● Why is the air over an open flame always moist?

44. ■ How might electrolysis be used to raise the hull of a sunken ship?

45. ■ As we digest and subsequently metabolize food, is the food gradually oxidized or reduced? What evidence do you have?

PROBLEMS

● BEGINNER   ■ INTERMEDIATE   ◆ EXPERT

1. ● Show that the hydroxide-ion concentration in an aqueous solution is 1×10^{-4} M when the hydronium-ion concentration is 1×10^{-10} M.

2. ● When the hydronium-ion concentration of a solution is 1×10^{-10} M, what is the pH of the solution? Is the solution acidic or basic?

3. ● When the hydronium-ion concentration of a solution is 1×10^{-4} M, what is the pH of the solution? Is the solution acidic or basic?

4. ● Show that an aqueous solution with a pH of 5 has a hydroxide-ion concentration of 1×10^{-9} M.

5. ■ When the pH of a solution is 1, the concentration of hydronium ions is 10^{-1} M = 0.1 M. Assume that the volume of this solution is 500 mL. What is the pH after 500 mL of pure water is added? You will need a calculator with a logarithm function to answer this question.

6. ■ Show that the pH of a solution is -0.301 when its hydronium-ion concentration equals 2 M. Is the solution acidic or basic?

ACTIVE EXPLORATIONS

Rainbow Cabbage

Many pH indicators are found in plants; the pigment of red cabbage is a good example. This pigment is red at low pH values (1 to 5), light purple around neutral pH values (6 to 7), light green at moderately alkaline pH values (8 to 11), and dark green at very alkaline pH values (12 to 14). *Safety note:* Wear safety glasses. Do not use bleach products.

What You Need

1. A head of red cabbage; a small pot; water; four colorless plastic cups; toilet bowl cleaner; vinegar; baking soda; ammonia cleanser

Procedure

1. Boil a cup of shredded cabbage in 2 cups of water (5 min). Strain and collect the broth.
2. Pour one-fourth of the broth into each cup and allow to cool.
3. Add a small amount of toilet bowl cleaner to the first cup, a small amount of vinegar to the second cup, baking soda to the third, and ammonia solution to the fourth.
4. Use the different colors to estimate the pHs.
5. Mix some of the acidic and basic solutions together and note the rapid change in pH (indicated by the change in color).

The change in color of a pH indicator is not permanent. To demonstrate this, add a teaspoon of baking soda to the cup to which you originally added the vinegar. (Why does this addition of baking soda also result in bubbling?) Add vinegar again to bring the color back to red.

Silver Lining

Tarnish on silverware is a coating of silver sulfide, Ag_2S, an ionic compound consisting of two silver ions, Ag^+, and one sulfide ion, S^{2-}. Tarnishing begins when silver atoms in the silverware come into contact with airborne hydrogen sulfide, H_2S, a smelly gas produced by the digestion of food in mammals and other organisms. The half reaction for the silver and hydrogen sulfide is

$$4\,Ag + 2\,H_2S \longrightarrow 4\,Ag^+ + 4\,H^+ + 2\,S^{2-} + 4e^- \quad \text{Oxidation}$$

The silver ions and sulfide ions combine to form blackish silver sulfide, while, at the same time, the hydrogen ions and electrons combine with atmospheric oxygen to form water:

$$4\,H^+ + 4e^- + O_2 \longrightarrow 2\,H_2O \quad \text{Reduction}$$

The balanced chemical equation for the tarnishing of silver is the combination of these two half reactions:

$$4\,Ag + 2\,H_2S + O_2 \longrightarrow 2\,Ag_2S + 2\,H_2O$$

From these equations, we can see that the hydrogen sulfide causes the silver to lose electrons to oxygen. To restore the silver to its shiny elemental state, we need to return the electrons it lost. The oxygen won't relinquish electrons back to silver, but, with the proper connection, aluminum atoms will.

What You Need

A very clean aluminum pot (or a non-aluminum pot and aluminum foil); water; baking soda; a piece of tarnished silver

Procedure

1. Place about a liter of water and several heaping tablespoons of baking soda in the aluminum pot (or in the non-aluminum pot containing a piece of aluminum foil).
2. Bring the water to boiling, and then remove the pot from the heat source.
3. Slowly immerse the tarnished silver; you'll see an immediate effect as the silver and aluminum make contact. (Add more baking soda if you don't.) Also, as the silver ions accept electrons from the aluminum and are thereby reduced to shiny silver atoms, the sulfide ions are free to re-form hydrogen sulfide gas, which is released back into the air. You may smell it!

The baking soda serves as a conductive ionic solution that permits electrons to move from the aluminum atoms to the silver ions. What is the advantage of this approach over polishing the silver with an abrasive paste?

Polishing with an abrasive paste removes both the thin layer of tarnish and some silver atoms. Silver-plated pieces are therefore susceptible to losing their thin coating of silver. The aluminum method, by contrast, restores the silver lost to the tarnishing. For pieces too large to fit in the pot, try rubbing lightly with a paste of baking soda and water, using aluminum foil as your rubbing cloth.

READINESS ASSURANCE TEST (RAT)

If you have a good handle on this chapter, if you really do, then you should be able to score 7 out of 10 on this RAT. If you score less than 7, you need to study further before moving on.

Choose the BEST answer to each of the following.

1. What is the relationship between the hydroxide ion and a water molecule?
   (a) A hydroxide ion is a water molecule plus a proton.
   (b) A hydroxide ion and a water molecule are the same thing.
   (c) A hydroxide ion is a water molecule minus a hydrogen nucleus.
   (d) A hydroxide ion is a water molecule plus two extra electrons.

2. Sodium hydroxide, NaOH, is a strong base, which means that it readily accepts hydrogen ions. What products are formed when sodium hydroxide accepts a hydrogen ion from a water molecule?
   (a) water and sodium hydroxide
   (b) sodium hydroxide and hydronium ions
   (c) sodium ions and hydronium ions
   (d) sodium ions and water

3. When the hydronium-ion concentration equals 1 M, what is the pH of the solution? Is the solution acidic, basic, or neutral?
   (a) pH = 0; acidic
   (b) pH = 1; acidic
   (c) pH = 10; basic
   (d) pH = 7; neutral

4. A weak acid is added to a concentrated solution of hydrochloric acid. Does the solution become more acidic, become less acidic, or stay the same?
   (a) more acidic, because more hydronium ions are being added to the solution
   (b) less acidic, because the solution becomes more dilute with a less concentrated solution of hydronium ions being added to the solution
   (c) no change in acidity, because the concentration of the hydrochloric acid is too high to be changed by the weak solution
   (d) less acidic, because the concentration of hydroxide ions increases

5. Why might a small piece of chalk be useful for alleviating acid indigestion?
   (a) Calcium carbonate is a base that reacts to neutralize any excess acids.
   (b) Calcium carbonate can absorb excess acid and carry it out of the system.
   (c) Calcium carbonate can add an extra lining to the stomach to protect it from the excess acid.
   (d) Calcium carbonate can turn the acid into a gas, which can relieve the buildup of excess acid.

6. What element is oxidized in the following equation and what element is reduced?

$$Sn^{2+} + 2\,Ag \longrightarrow Sn + 2\,Ag^+$$

   (a) The tin ion, Sn^{2+}, is oxidized; the silver, Ag, is reduced.
   (b) The tin ion, Sn^{2+}, is reduced; the silver, Ag, is oxidized.
   (c) Both the tin ion, Sn^{2+}, and the silver, Ag, are reduced.
   (d) Both the tin ion, Sn^{2+}, and the silver, Ag, are oxidized.

7. How does an atom's electronegativity relate to its ability to become oxidized?
   (a) The greater the electronegativity of an atom, the greater its ability to become oxidized.
   (b) The lower the electronegativity of an atom, the lower its ability to become oxidized.
   (c) The greater the electronegativity of an atom, the lower its ability to become oxidized.
   (d) Electronegativity does not effect the atom's ability to become oxidized.

8. Sodium metal is
   (a) oxidized in the production of aluminum.
   (b) reduced in the production of aluminum.
   (c) both oxidized and reduced in the production of aluminum.
   (d) neither oxidized nor reduced in the production of aluminum.

9. Why does a battery that has thick zinc walls last longer than one that has thin zinc walls?
   (a) Thick zinc walls prevent the battery from overheating.
   (b) Thicker zinc walls prevent electrons from being lost into the surrounding environment.
   (c) Thicker zinc walls last longer at holding in the battery acid.
   (d) The zinc walls are transformed into zinc ions as the battery provides electricity.

10. The general chemical equation for photosynthesis is shown below. Through this reaction, are the oxygens of the water molecules, H_2O, oxidized or reduced?

$$6\,CO_2 + 6\,H_2O \longrightarrow C_6H_{12}O_6 + 6\,O_2$$

   (a) The oxygens of the water molecules are oxidized.
   (b) The oxygens of the water molecules are reduced.
   (c) The oxygens of some of these water molecules are oxidized while others are reduced.
   (d) The oxygens of the water molecules are neither oxidized nor reduced.

Answers to RAT

1c, 2a, 3a, 4b, 5a, 6b, 7c, 8a, 9d, 10a

EXPLORING FURTHER

http://www.epa.gov
Go to the home page of the Environmental Protection Agency and use *acid rain* as a keyword in the agency's search engine to find numerous articles on this subject.

http://mlso.hao.ucar.edu/cgi-bin/mlso_homepage.cgi
This site itemizes the atmospheric projects of the Climate Monitoring and Diagnostic Laboratory of the Mauna Loa Solar Observatory. Links to the Network for the Detection of Stratospheric Changes are included.

http://householdproducts.nlm.nih.gov/products.htm
Use the Household Products Database of the National Library of Medicine to learn more about the chemical nature of various household products such as personal care, arts and crafts, pet care, pesticides, yard care, and auto products.

http://www.aluminum.org
At the Web site of the Aluminum Association, Inc., you will find basic facts about the aluminum industry, recycling efforts, and the impact of our aluminum use on the environment.

http://www.calcars.org/vehicles.html
Learn more about plug-in hybrid vehicles (PHEVs) on this site for the California Cars Initiative, which supports the boosting of fuel efficiencies to more than 100 mi/gal.

http://www.fuelcellworld.org
Use *fuel cells* as a search keyword to find a number of private companies and organizations that are dedicated to improving the efficiency of fuel cells and publicizing their use. Fuel cells are certainly a wave of the future.

CHAPTER 18 ONLINE RESOURCES

The Physical Science Place

Interactive Figure

- 18.27

Tutorials

- Nature of Acids and Bases
- Strong and Weak Acids and Bases
- The pH Scale

Videos

- Some Acids and Bases Are Stronger Than Others
- Rainwater Is Acidic and Ocean Water Is Basic
- Beach Sand Composition

Quiz

Flashcards

Links

Vanillin

Tetramethylpyrazine

organic compounds

19.1 **HYDROCARBONS**

19.2 **UNSATURATED HYDROCARBONS**

19.3 **FUNCTIONAL GROUPS**

19.4 **ALCOHOLS, PHENOLS, AND ETHERS**

19.5 **AMINES AND ALKALOIDS**

19.6 **CARBONYL COMPOUNDS**

19.7 **POLYMERS**

■ Carbon atoms have the unique ability to link together and thereby form molecules made up of many carbon atoms. Add to this the fact that carbon atoms can also bond with atoms of other elements, and you see the possibility of an endless number of different carbon-based molecules. Each molecule has its own unique set of physical, chemical, and biological properties. The flavor of vanilla, for example, is perceived when the compound *vanillin* is absorbed by the sensory organs in the nose. This compound consists of a ring of carbon atoms with oxygen atoms attached in a particular fashion. Vanillin is the essential ingredient in anything that has the flavor of vanilla—without vanillin, there is no vanilla flavor. The flavor of chocolate, on the other hand, is generated when not just one but a wide assortment of carbon-based molecules are absorbed in the nose. One of the more significant of these molecules is *tetramethylpyrazine*, which has a ring of nitrogen and carbon atoms attached in a particular fashion.

Life is based on carbon's ability to bond with other carbon atoms to form diverse structures. Reflecting this fact, the branch of chemistry that is the study of carbon-containing compounds has come to be known as **organic chemistry**. The term *organic* is derived from *organism* and is not necessarily related to the environmental-friendly form of farming. Today, more than 13 million organic compounds are known, and about 100,000 new ones are added to the list each year. This includes those discovered in nature and those synthesized in the laboratory. (By contrast, there are only 200,000 to 300,000 known *inorganic* compounds, those based on elements other than carbon.)

Because organic compounds are so closely tied to living organisms and because they have many applications—flavorings, fuels, polymers, medicines, agriculture, and more—it is important to have a basic understanding of them. We begin with the simplest organic compounds—those consisting of only carbon and hydrogen.

■ 19.1 Hydrocarbons

The Physical Science Place

Introduction to Organic Molecules
Organic Molecules and Isomers

Organic compounds that contain only carbon and hydrogen are called **hydrocarbons**, which differ from one another by the number of carbon and hydrogen atoms they contain. The simplest hydrocarbon is methane, CH_4, with only one carbon per molecule. Methane is the main component of natural gas. The hydrocarbon octane, C_8H_{18}, has eight carbons per molecule and is a component of gasoline. The hydrocarbon polyethylene contains hundreds of carbon and hydrogen atoms per molecule. Polyethylene is a plastic used to make many familiar items, such as milk containers and plastic bags.

Methane, CH_4 Octane, C_8H_{18} Polyethylene

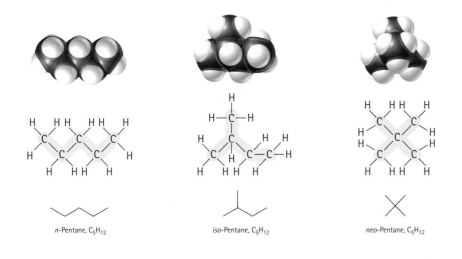

n-Pentane, C_5H_{12} iso-Pentane, C_5H_{12} neo-Pentane, C_5H_{12}

FIGURE 19.1

These three hydrocarbons all have the same molecular formula. We can see their different structural features by highlighting the carbon framework in two dimensions. Easy-to-draw stick structures that use lines for all carbon–carbon covalent bonds are also useful.

Hydrocarbons also differ in the way the carbon atoms connect to one another. Figure 19.1 shows the three hydrocarbons *n*-pentane, *iso*-pentane, and *neo*-pentane. These hydrocarbons have the same molecular formula, C_5H_{12}, but they are structurally different from one another. The carbon framework of *n*-pentane is a chain of five carbon atoms. In *iso*-pentane, the carbon chain branches so that the framework is a *four*-carbon chain branched at the second carbon. In *neo*-pentane, a central carbon atom is bonded to four surrounding carbon atoms.

We can see the different structural features of *n*-pentane, *iso*-pentane, and *neo*-pentane more clearly by drawing the molecules in two dimensions, as shown in the middle row of Figure 19.1. Alternatively, we can represent them by the *stick structures* shown in the bottom row. A stick structure is a commonly used shorthand notation for representing an organic molecule. Each line (stick) represents a covalent bond, and carbon atoms are understood to exist at the end of any line or wherever two or more straight lines meet (unless another type of atom is drawn at the end of the line). Any hydrogen atoms bonded to the carbons are also typically not shown. Instead, their presence is only implied, so that the focus can remain on the skeletal structure that is formed by the carbon atoms.

When every carbon atom in a hydrocarbon except the two terminal ones is bonded to only two other carbon atoms, the molecule is called a *straight-chain hydrocarbon*. (Do not take this name literally, for, as the *n*-pentane structures in Figure 19.1 show, this is a straight-chain hydrocarbon despite the zigzag nature of the drawings representing it.) When at least one carbon atom in a hydrocarbon is bonded to either three or four carbon atoms, the molecule is a *branched hydrocarbon*. Therefore, both *iso*-pentane and *neo*-pentane are branched hydrocarbons.

Molecules such as *n*-pentane, *iso*-pentane, and *neo*-pentane have the same molecular formula, which means they have the same number of the same kinds of atoms. The way these atoms are put together, however, is different. We say that they each have their own **configuration**, which refers to how the atoms are connected. Different configurations result in different chemical structures. Molecules with the same molecular formula but different configurations (and hence different structures) are known as **structural isomers**. Structural isomers differ from each other and have different physical and chemical properties. For example, *n*-pentane has a boiling point of 36°C, *iso*-pentane's boiling point is 30°C, and *neo*-pentane's boiling point is 10°C.

The number of possible structural isomers for a chemical formula increases rapidly as the number of carbon atoms increases. There are three structural

In looking at the stick structures, remember that each corner or end represents a carbon atom and that each carbon atom must be bonded four times. Because hydrogen atoms are assumed, they're not usually depicted.

FIGURE 19.2

Three conformations for a molecule of *n*-pentane. The molecule looks different in each conformation, but the five-carbon framework is the same in all three conformations. In a sample of liquid *n*-pentane, the molecules are found in all conformations— not unlike a bucket of worms.

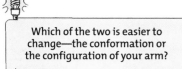

Which of the two is easier to change—the conformation or the configuration of your arm?

isomers for compounds that have the formula C_5H_{12}, 18 for C_8H_{18}, 75 for $C_{10}H_{22}$, and a whopping 366,319 for $C_{20}H_{42}$!

A carbon-based molecule can have different spatial orientations called **conformations**. Flex your wrist, elbow, and shoulder joints, and you'll find your arm passing through a range of conformations. Likewise, organic molecules can twist and turn about their carbon–carbon single bonds and thus have a range of conformations. The structures in Figure 19.2, for example, are different conformations of *n*-pentane. In the language of organic chemistry, we say that the *configuration* of a molecule, such as *n*-pentane, has a broad range of *conformations*. Change the configuration of *n*-pentane, however, and you no longer have *n*-pentane. Rather, you have a different structural isomer, such as *iso*-pentane, which has its own range of different conformations.

fyi

- As discussed in Section 18.4, we are placing unusually large amounts of carbon dioxide into the atmosphere by burning fossil fuels. The atmosphere, however, is not the only possible repository for the carbon dioxide we produce. The smokestacks of power plants, for example, can be modified to capture CO_2, which is then lique-fied and pumped kilometers deep into the ground. This idea is not so far-fetched. Underground storage of carbon dioxide is already being employed at the Salah natural gas refinery in Algeria. Such a system, however, would have its costs. The price of electricity from a CO_2-capturing coal-fired power plant would rise by about 20%. The long-term costs of not implementing such systems, however, may be even greater.

CHECK POINT

Which carbon–carbon bond was rotated to go from the "before" conformation of *iso*-pentane to the "after" conformation?

Was this your answer?

The best way to answer any question about the conformation of a molecule is to play around with molecular models that you can hold in your hand. In this case, bond (c) rotates in such a way that the carbon at the right end of bond (d) comes up out of the plane of the page, momentarily points straight at you, and then plops back into the plane of the page below bond (c). This rotation is similar to that of the arm of an arm wrestler who, with the arm just above the table while on the brink of losing, suddenly gets a surge of strength and swings the opponent's arm (and his or her own) through a half-circle arc and wins.

Hydrocarbons are obtained primarily from coal and petroleum. Most of the coal and petroleum that exists today was formed between 280 million and 395 million years ago when plant and animal matter decayed in the absence of oxygen. At that time, Earth was covered with extensive swamps that, because they were close to sea level, periodically became submerged. The organic matter of the swamps was buried beneath layers of marine sediments and was eventually transformed to either coal or petroleum.

Coal is a solid material that contains many large, complex hydrocarbon molecules. Most of the coal mined today is used to produce steel and to generate electricity at coal-burning power plants.

Link To Section 21.4

FIGURE 19.3

A schematic for the fractional distillation of petroleum into its useful hydrocarbon components.

Petroleum, also called crude oil, is a liquid readily separated into its hydrocarbon components through a process known as *fractional distillation*, shown in Figure 19.3. The crude oil is heated in a pipe still to a temperature high enough to vaporize most of the components. The hot vapor flows into the bottom of a fractionating tower, which is warmer at the bottom than at the top. As the vapor rises in the tower and cools, the various components begin to condense. Hydrocarbons that have high boiling points, such as tar and lubricating stocks, condense first at warmer temperatures. Hydrocarbons that have low boiling points, such as gasoline, travel to the cooler regions at the top of the tower before condensing. Pipes drain the various liquid hydrocarbon fractions from the tower. Natural gas, which is primarily methane, does not condense. It remains a gas and is collected at the top of the tower.

> The lower the boiling point of the hydrocarbon, the higher it travels up the fractionation tower.

FIGURE 19.4

(a) A straight-chain hydrocarbon, such as *n*-hexane, can be ignited from the heat generated as gasoline is compressed by the piston—before the spark plug fires. This upsets the timing of the engine cycle, giving rise to a knocking sound. (b) Branched hydrocarbons, such as *iso*-octane, burn less readily and are ignited not by compression alone but only when the spark plug fires.

FIGURE 19.5

Octane ratings are posted on gasoline pumps. Interestingly, the engines of modern cars are designed to run best on 87 octane grade fuel. With higher octane fuels, not only do you lose performance, but you lose your money as well.

Differences in the strength of molecular attractions explain why different hydrocarbons condense at different temperatures. As discussed in Section 15.8, in our comparison of induced dipole–induced dipole attractions in methane and octane, larger hydrocarbons experience many more of these attractions than smaller hydrocarbons do. For this reason, the larger hydrocarbons condense readily at high temperatures and so are found at the bottom of the tower. Smaller molecules, because they experience fewer attractions to their neighbors, condense only at the cooler temperatures found at the top of the tower.

The gasoline obtained from the fractional distillation of petroleum consists of a wide variety of hydrocarbons having similar boiling points. Some of these components burn more efficiently than others in a car engine. The straight-chain hydrocarbons, such as *n*-hexane, burn too quickly, causing what is called *engine knock*, as illustrated in Figure 19.4. Gasoline hydrocarbons that have more branching, such as *iso*-octane, burn slowly and result in the engine running more smoothly. These two compounds, *n*-hexane and *iso*-octane, are used as standards in assigning *octane ratings* to gasoline. An octane rating of 100 is arbitrarily assigned to *iso*-octane, and *n*-hexane is assigned an octane rating of 0. The antiknock performance of a particular gasoline is compared with that of various mixtures of *iso*-octane and *n*-hexane, and an octane rating is assigned. Figure 19.5 shows the octane information that appears on a typical gasoline pump.

CHECK POINT

Which structural isomer shown in Figure 19.1 should have the highest octane rating?

Was this your answer?

The structural isomer with the greatest amount of branching in the carbon framework likely has the highest octane rating, making *neo*-pentane the clear winner. For your information, the ratings are as follows:

Compound	Octane Rating
n-Pentane	61.7
iso-Pentane	92.3
neo-Pentane	116

19.2 Unsaturated Hydrocarbons

Recall from Section 15.1 that carbon has four unpaired valence electrons. As shown in Figure 19.6, each of these electrons is available for pairing with an electron from another atom, such as hydrogen, to form a covalent bond.

In all the hydrocarbons discussed so far, including the methane shown in Figure 19.6, each carbon atom is bonded to four neighboring atoms by four

fyi

- Saturated fats in our diet have long carbon chains that contain single bonds. These chains tend to be straight, so they pack well together (like wooden matches in a box). Hence, saturated fats, such as lard, are solid at room temperature. The carbon chains of unsaturated fats have double bonds and take on a bent shape, so they don't pack well together. Unsaturated fats, such as vegetable oils, are liquid at room temperature.

FIGURE 19.6

Carbon has four valence electrons. Each electron pairs with an electron from a hydrogen atom in the four covalent bonds of methane.

Methane

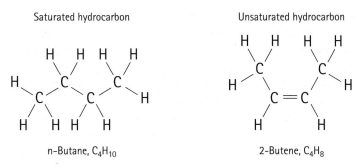

Saturated hydrocarbon

n-Butane, C₄H₁₀

Unsaturated hydrocarbon

2-Butene, C₄H₈

FIGURE 19.7

The carbons of the hydrocarbon *n*-butane are *saturated*, each being bonded to four other atoms. Because of the double bond, two of the carbons of the unsaturated hydrocarbon 2-butene are bonded to only three other atoms, which makes the molecule an unsaturated hydrocarbon.

single covalent bonds. Such hydrocarbons are known as **saturated hydrocarbons**. The term *saturated* means that each carbon has as many atoms bonded to it as possible. We now explore cases in which one or more carbon atoms in a hydrocarbon are bonded to fewer than four neighboring atoms. This occurs when at least one of the bonds between a carbon and a neighboring atom is a multiple bond. (See Section 15.5 for a review of multiple bonds.)

A hydrocarbon containing a multiple bond—either double or triple—is known as an **unsaturated hydrocarbon**. Because of the multiple bond, two of the carbons are bonded to fewer than four other atoms. These carbons are thus said to be *unsaturated*.

Figure 19.7 compares the saturated hydrocarbon *n*-butane with the unsaturated hydrocarbon 2-butene. The number of atoms that are bonded to each of the two middle carbons of *n*-butane is four, whereas each of the two middle carbons of 2-butene is bonded to only three other atoms—a hydrogen and two carbons.

An important unsaturated hydrocarbon is benzene, C_6H_6, which may be drawn as three double bonds contained within a flat hexagonal ring, as shown in Figure 19.8a. Unlike the double-bond electrons in most other unsaturated hydrocarbons, the electrons of the double bonds in benzene are not fixed between any two carbon atoms. Instead, these electrons can move freely around the ring. This is commonly repre-

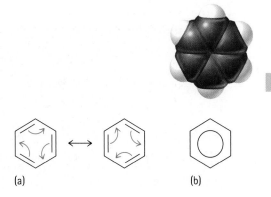

(a)   (b)

FIGURE 19.8

(a) The double bonds of benzene, C_6H_6, can migrate around the ring. (b) For this reason, they are often represented by a circle within the ring.

sented by drawing a circle within the ring, as shown in Figure 19.8b, rather than by individual double bonds.

Many organic compounds contain one or more benzene rings in their structure. Because many of these compounds are fragrant, any organic molecule that contains a benzene ring is classified as an **aromatic compound** (even if it is not particularly fragrant). Figure 19.9 shows a few examples. Toluene, a common solvent used as a paint thinner, is toxic and gives airplane glue its distinctive odor. Some aromatic compounds, such as naphthalene, contain two or more benzene rings fused together. At one time, mothballs were made of naphthalene. Most mothballs sold today, however, are made of the less toxic 1,4-dichlorobenzene.

An example of an unsaturated hydrocarbon containing a triple bond is acetylene, C_2H_2. A confined flame of acetylene burning in oxygen is hot enough to melt iron, which makes acetylene a choice fuel for welding (Figure 19.10).

FIGURE 19.9

The structures for three odoriferous organic compounds that contain one or more benzene rings: toluene, naphthalene, and 1,4-dichlorobenzene.

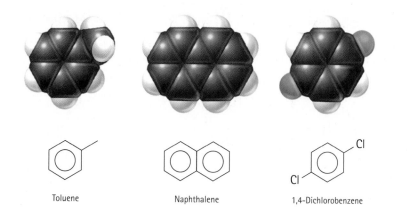

Toluene   Naphthalene   1,4-Dichlorobenzene

CHECK
POINT

Prolonged exposure to benzene increases the risk of developing certain cancers. The structure of aspirin contains a benzene ring. Does this indicate that prolonged exposure to aspirin increases a person's risk of developing cancer?

Benzene ring

Aspirin

Was this your answer?

No. Although benzene and aspirin both contain a benzene ring, these two molecules have different overall structures and quite different chemical properties. Each carbon-containing organic compound has its own set of unique physical, chemical, and biological properties. Although benzene may cause cancer, aspirin works as a safe remedy for headaches.

FIGURE 19.10

The unsaturated hydrocarbon acetylene, C_2H_2, when burned in this torch, produces a flame that is hot enough to melt iron.

$H-C\equiv C-H$

Acetylene

19.3 Functional Groups

The Physical Science Place
Functional Groups

Carbon atoms can bond to one another and to hydrogen atoms in many ways, which results in an incredibly large number of hydrocarbons. But carbon atoms can bond to atoms of other elements as well, further increasing the number of possible organic molecules. In organic chemistry, any atom other than carbon or hydrogen in an organic molecule is called a **heteroatom**, where *hetero-* indicates that the atom is different from either carbon or hydrogen.

A hydrocarbon structure can serve as a framework for the attachment of various heteroatoms. This is analogous to a Christmas tree serving as the scaffolding on which ornaments are hung. Just as the ornaments give character to the tree, so do heteroatoms give character to an organic molecule. Heteroatoms have profound effects on the properties of an organic molecule.

Consider ethane, C_2H_6, and ethanol, C_2H_6O, which differ from each other by only a single oxygen atom. Ethane has a boiling point of $-88°C$, making it a gas at room temperature, and it does not dissolve in water very well. Ethanol, by contrast, has a boiling point of $+78°C$, making it a liquid at room temperature. It is infinitely soluble in water, and it is the active ingredient of alcoholic beverages. Consider further ethylamine, C_2H_7N, which has a nitrogen atom on the same basic two-carbon framework. This compound is a corrosive, pungent, highly toxic gas—very different from either ethane or ethanol.

Organic molecules are classified according to the functional groups they contain. A **functional group** is defined as a combination of atoms that behave as a unit. Most functional groups are distinguished by the heteroatoms they contain; some common groups are listed in Table 19.1.

The remainder of this section introduces the classes of organic molecules shown in Table 19.1. The role heteroatoms play in determining the properties

> The chemistry of hydrocarbons is surely interesting, but start adding heteroatoms to these organic molecules and the chemistry becomes extraordinarily interesting. The organic chemicals of living organisms, for example, all contain heteroatoms.

TABLE 19.1 FUNCTIONAL GROUPS IN ORGANIC MOLECULES

General Structure	Name	Class
Hydroxyl group	Hydroxyl group	Alcohols
Phenolic group	Phenolic group	Phenols
Ether group	Ether group	Ethers
Amine group	Amine group	Amines
Ketone group	Ketone group	Ketones
Aldehyde group	Aldehyde group	Aldehydes
Amide group	Amide group	Amides
Carboxyl group	Carboxyl group	Carboxylic acids
Ester group	Ester group	Esters

of each class is the underlying theme. As you study this material, focus on understanding the chemical and physical properties of the various classes of compounds; doing so will give you a greater appreciation of the remarkable diversity of organic molecules and their many applications.

CHECK POINT

What is the significance of heteroatoms in an organic molecule?

Was this your answer?
Heteroatoms largely determine an organic molecule's "personality."

19.4 Alcohols, Phenols, and Ethers

Hydroxyl group

Alcohols are organic molecules in which a *hydroxyl group* is bonded to a saturated carbon. The hydroxyl group consists of an oxygen bonded to a hydrogen. Because of the polarity of the oxygen–hydrogen bond, low-mass alcohols are often soluble in water, which is itself very polar. Some common alcohols and their melting and boiling points are listed in Table 19.2.

More than 11 billion pounds of methanol, CH_3OH, are produced annually in the United States. Most of it is used for making formaldehyde and acetic acid, important starting materials in the production of plastics. In addition, methanol is used as a solvent, an octane booster, and an anti-icing agent in gasoline. Sometimes called *wood alcohol* because it can be obtained from wood, methanol should never be ingested because, in the body, it is metabolized to formaldehyde and formic acid. Formaldehyde is harmful to the eyes, can lead to blindness, and was once used to preserve dead biological specimens. Formic acid, the active ingredient in an ant bite, can lower the pH of the blood to dangerous levels. Methanol has its own inherent toxicities. Ingesting only about 15 mL (about 3 Tbsp) of methanol may lead to blindness, and about 30 mL can cause death.

TABLE 19.2   SOME SIMPLE ALCOHOLS

Structure	Scientific Name	Common Name	Melting Point (°C)	Boiling Point (°C)
	Methanol	Methyl alcohol	−97	65
	Ethanol	Ethyl alcohol	−115	78
	2-Propanol	Isopropyl alcohol	−126	97

Ethanol, C_2H_5OH, on the other hand, is the "alcohol" of alcoholic beverages, and it is one of the oldest chemicals manufactured by humans. Ethanol is prepared by feeding the sugars of various plants to certain yeasts, which produce ethanol through a biological process known as *fermentation*. Ethanol is also widely used as an industrial solvent. For many years, ethanol intended for this purpose was made by fermentation, but today industrial-grade ethanol is more cheaply manufactured from petroleum byproducts, such as ethene, as Figure 19.11 illustrates.

Ethene   Water   Ethanol

FIGURE 19.11

Ethanol can be synthesized from the unsaturated hydrocarbon ethene, with phosphoric acid as a catalyst.

The liquid produced by fermentation has an ethanol concentration no greater than about 12% because at this concentration the yeast cells begin to die. This is why most wines have an alcohol content of about 12%—they are produced solely by fermentation. To attain the higher ethanol concentrations found in "hard" alcoholic beverages such as gin and vodka, the fermented liquid must be distilled. In the United States, the ethanol content of distilled alcoholic beverages is measured as *proof*, which is twice the percentage of ethanol. An 86-proof whiskey, for example, is 43% ethanol by volume. The term *proof* evolved from a crude method once employed to test alcohol content. Gunpowder was wetted with a beverage of suspect alcohol content. If the beverage was primarily water, the powder did not ignite. If the beverage contained a significant amount of ethanol, the powder burned, thus providing "proof" of the beverage's worth.

A third well-known alcohol is isopropyl alcohol, also called 2-propanol. This is the rubbing alcohol you buy at the drugstore. Although 2-propanol has a relatively high boiling point, it evaporates readily, leading to a pronounced cooling effect when it is applied to skin—an effect once used to reduce fevers. (Isopropyl alcohol is very toxic if ingested. See the activities at the end of this chapter to understand why. In place of isopropyl alcohol, washcloths wetted with cold water are nearly as effective in reducing fever, and they are far safer.) You are probably most familiar with the use of isopropyl alcohol as a topical disinfectant.

Phenols contain a phenolic group, which consists of a hydroxyl group attached to a benzene ring. Because of the presence of the benzene ring, the hydrogen of the hydroxyl group is readily lost in an acid–base reaction, which makes the phenolic group mildly acidic.

The reason for this acidity is illustrated in Figure 19.12. How readily an acid donates a hydrogen ion is a function of how well the acid can accommodate the resulting negative charge it gains after donating the hydrogen ion. After phenol donates the hydrogen ion, it becomes a negatively charged phenoxide ion. The negative charge of the phenoxide ion, however, is not restricted to the oxygen atom. Recall that the electrons of the benzene ring can migrate around the ring. In a similar manner, the electrons responsible for the negative charge of the phenoxide ion can migrate around the ring, as shown in Figure 19.12. Just as several people can easily hold a hot potato by quickly passing it around, the phenoxide ion can easily hold the negative charge because the charge gets passed around. Because the negative charge of the ion is so nicely accommodated, the phenolic group is more acidic than it would be otherwise.

fyi

- Just as the body metabolizes methanol into formaldehyde, HCOH, it metabolizes ethanol into acetaldehyde, CH_3COH. Acetaldehyde won't cause you to go blind, but it does provide for some painful side effects, which people who drink too much experience as part of their "hangover."

Link To Section 7.6

Phenolic group

FIGURE 19.12

The negative charge of the phenoxide ion can migrate to select positions on the benzene ring. This mobility helps accommodate the negative charge, which is why the phenolic group readily donates a hydrogen ion.

Phenol (acidic) → Phenoxide ion + Hydrogen ion

The simplest phenol, shown in Figure 19.13, is called phenol. In 1867, Joseph Lister (1827–1912) discovered the antiseptic value of phenol, which, when applied to surgical instruments and incisions, greatly increased surgery survival rates. Phenol was the first purposefully used antibacterial solution, or *antiseptic*. Phenol damages healthy tissue, however, and so a number of milder phenols have since been introduced. The phenol 4-*n*-hexylresorcinol, for example, is commonly used in throat lozenges and mouthwashes. This compound has even greater antiseptic properties than phenol, and yet it does not damage tissue. Listerine brand mouthwash (named after Joseph Lister) contains the antiseptic phenols thymol and methyl salicylate.

We're classifying organic molecules based on the functional groups they contain. As you will see shortly, however, organic molecules may contain more than one type of functional group. A single organic molecule, therefore, might be classified as both a phenol and an ether.

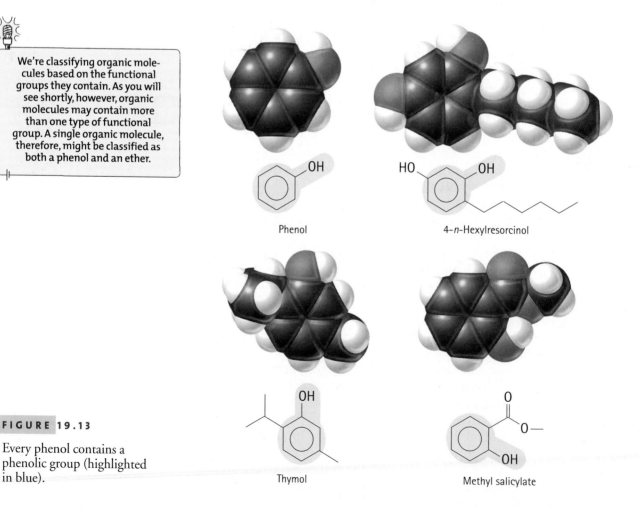

Phenol

4-*n*-Hexylresorcinol

Thymol

Methyl salicylate

FIGURE 19.13

Every phenol contains a phenolic group (highlighted in blue).

CHECK POINT

Why are alcohols less acidic than phenols?

Was this your answer?
An alcohol does not contain a benzene ring adjacent to the hydroxyl group. If the alcohol were to donate the hydroxyl hydrogen, the result would be a negative charge on the oxygen. Without an adjacent benzene ring, this negative charge has nowhere to go. As a result, an alcohol behaves only as a very weak acid, much the way water does.

Ethanol: Soluble in water, boiling point 78°C

Dimethyl ether: Insoluble in water, boiling point −25°C

FIGURE 19.14

The oxygen in an alcohol, such as ethanol, is bonded to one carbon atom and one hydrogen atom. The oxygen in an ether, such as dimethyl ether, is bonded to two carbon atoms. Because of this difference, alcohols and ethers of similar molecular mass have vastly different physical properties.

Ethers are organic compounds structurally related to alcohols. The oxygen atom in an ether group, however, is bonded not to a carbon and a hydrogen but rather to two carbons. As we see in Figure 19.14, ethanol and dimethyl ether have the same chemical formula, C_2H_6O, but their physical properties are vastly different. Whereas ethanol is a liquid at room temperature (boiling point 78°C) and mixes quite well with water, dimethyl ether is a gas at room temperature (boiling point −25°C) and is much less soluble in water.

Ethers are not very soluble in water because without the hydroxyl group they are unable to form strong hydrogen bonds with water (Section 15.8). Furthermore, without the polar hydroxyl group, the molecular attractions among ether molecules are relatively weak. As a result, little energy is required to separate ether molecules from one another. This is why low-formula-mass ethers have relatively low boiling points and evaporate so readily.

Diethyl ether, shown in Figure 19.15, was one of the first anesthetics. The anesthetic properties of this compound were discovered in the early 1800s, and its use revolutionized the practice of surgery. Because of its high volatility at room temperature, inhaled diethyl ether rapidly enters the bloodstream. Because this ether has low solubility in water and high volatility, it quickly leaves the bloodstream. Because of these physical properties, a surgical patient can be brought in and out of anesthesia (a state of unconsciousness) simply by regulating the gases breathed. Modern gaseous anesthetics have fewer side effects than diethyl ether, but they operate on the same principle.

Ether group

Diethyl ether, boiling point 35°C

FIGURE 19.15

Diethyl ether is the systematic name for the "ether" historically used as an anesthetic.

■ 19.5 Amines and Alkaloids

Amine group

Amines are organic compounds that contain the amine group—a nitrogen atom bonded to one, two, or three saturated carbons. Amines are typically less soluble in water than are alcohols because the nitrogen–hydrogen bond is not quite as polar as the oxygen–hydrogen bond. The lower polarity of amines also means their boiling points are typically somewhat lower than those of alcohols of similar formula mass. Table 19.3 lists three simple amines.

TABLE 19.3 THREE SIMPLE AMINES

Structure	Name	Melting Point (°C)	Boiling Point (°C)
	Ethylamine	−18	17
	Diethylamine	−50	55
	Triethylamine	−7	89

One of the most notable physical properties of many low-formula-mass amines is their offensive odor. Figure 19.16 shows two appropriately named amines, putrescine and cadaverine, which are partly responsible for the odor of decaying flesh.

Amines are typically alkaline because the nitrogen atom readily accepts a hydrogen ion from water, as Figure 19.17 illustrates.

FIGURE 19.16

Low-formula-mass amines such as these tend to have offensive odors.

H_2N NH_2
Putrescine
(1,4-butanediamine)

H_2N NH_2
Cadaverine
(1,5-pentanediamine)

FIGURE 19.17

Ethylamine acts as a base and accepts a hydrogen ion from water to become the ethylammonium ion. This reaction generates a hydroxide ion, which increases the pH of the solution.

A group of naturally occurring complex molecules that are alkaline because they contain nitrogen atoms are often called *alkaloids*. Because many alkaloids have medicinal value, there is great interest in isolating these compounds from plants or marine organisms that contain them. As shown in Figure 19.18, an alkaloid reacts with an acid to form a salt that is usually quite soluble in water. This is in contrast to the non-ionized form of the alkaloid, known as a *free base*, which is typically insoluble in water.

FIGURE 19.18

All alkaloids are bases that react with acids to form salts. An example is the alkaloid caffeine, shown here reacting with phosphoric acid.

Most pharmaceuticals that can be administered orally contain nitrogen heteroatoms in the water-soluble salt form.

Most alkaloids exist in nature not in their free-base form but rather as the salts of naturally occurring acids known as *tannins*, a group of phenol-based organic acids that have complex structures. The alkaloid salts of these acids are usually much more soluble in hot water than in cold water. The caffeine in coffee and tea exists in the form of the tannin salt, which is why coffee and tea are more effectively brewed in hot water. As Figure 19.19 relates, tannins are also responsible for the stains caused by these beverages.

CHECK POINT

Why do most caffeinated soft drinks also contain phosphoric acid?

Was this your answer?
Phosphoric acid, as shown in Figure 19.18, reacts with caffeine to form the caffeine–phosphoric acid salt, which is much more soluble in cold water than the naturally occurring tannin salt.

■ 19.6 Carbonyl Compounds

A **carbonyl group** consists of a carbon atom double-bonded to an oxygen atom. It occurs in the organic compounds known as ketones, aldehydes, amides, carboxylic acids, and esters.

FIGURE 19.19

Tannins are responsible for the brown stains in coffee mugs or on a coffee drinker's teeth. Because tannins are acidic, they can be readily removed with an alkaline cleanser. Use a little laundry bleach on the mug, and brush your teeth with baking soda.

Acetone
(a)

Propionaldehyde
(b)

FIGURE 19.20

(a) When the carbon of a carbonyl group is bonded to two carbon atoms, the result is a ketone. An example is acetone. (b) When the carbon of a carbonyl group is bonded to at least one hydrogen atom, the result is an aldehyde. An example is propionaldehyde.

Ketone group

Aldehyde group

A **ketone** is a carbonyl-containing organic molecule in which the carbonyl carbon is bonded to two carbon atoms. A familiar example of a ketone is *acetone*, which is often used in fingernail-polish remover and is shown in Figure 19.20a. In an **aldehyde**, the carbonyl carbon is bonded either to one carbon atom and one hydrogen atom, as in Figure 19.20b, or, in the special case of formaldehyde, to two hydrogen atoms.

Many aldehydes are particularly fragrant. A number of flowers, for example, owe their pleasant odor to the presence of simple aldehydes. The smells of lemons, cinnamon, and almonds are due to the aldehydes citral, cinnamaldehyde, and benzaldehyde, respectively. The structures of these three aldehydes are shown in Figure 19.21. Another aldehyde, vanillin, which was introduced at the beginning of this chapter, is the key flavoring molecule derived from seed pods of the vanilla orchid. You may have noticed that vanilla seed pods and vanilla extract are fairly expensive. Imitation vanilla flavoring is less expensive because it is merely a solution of the compound vanillin, which is economically synthesized from waste chemicals from the wood-pulp industry. Imitation vanilla does not taste the same as natural vanilla extract, however, because, in addition to vanillin, many other flavorful molecules contribute to the complex taste of

FIGURE 19.21

Aldehydes are responsible for many familiar fragrances.

Citral

Cinnamonaldehyde

Benzaldehyde

Vanillin

natural vanilla. Many books manufactured in the days before "acid-free" paper smell of vanilla because of the vanillin formed and released as the paper ages, a process that is accelerated by the acids the paper contains.

An **amide** is a carbonyl-containing organic molecule in which the carbonyl carbon is bonded to a nitrogen atom. The active ingredient of most mosquito repellents is an amide whose chemical name is *N,N*-diethyl-*m*-toluamide but is commercially known as DEET, shown in Figure 19.22. This compound is actually not an insecticide. Rather, it causes certain insects, especially mosquitoes, to lose their sense of direction, which effectively protects DEET wearers from being bitten.

A **carboxylic acid** is a carbonyl-containing organic molecule in which the carbonyl carbon is bonded to a hydroxyl group. As its name implies, this functional group can donate hydrogen ions. Organic molecules that contain it are therefore acidic. An example is acetic acid, $C_2H_4O_2$, which, after water, is the main ingredient of vinegar. You may recall that this organic compound was used as an example of a weak acid back in Chapter 18.

As with phenols, the acidity of a carboxylic acid results in part from the ability of the functional group to accommodate the negative charge of the ion that forms after the hydrogen ion has been donated. As shown in Figure 19.23, a carboxylic acid transforms to a carboxylate ion as it loses the hydrogen ion. The negative charge of the carboxylate ion can then pass back and forth between the two oxygens. This spreading out helps accommodate the negative charge.

An interesting example of an organic compound that contains both a carboxylic acid and a phenol is salicylic acid, found in the bark of willow trees and illustrated in Figure 19.24a. At one time brewed for its antipyretic (fever-reducing) effect, salicylic acid is an important analgesic (painkiller), but it causes nausea and stomach upset because of its relatively high acidity, a result of the presence of two acidic functional groups. In 1899, Friedrich Bayer and Company, in Germany, introduced a chemically modified version of this compound in which the acidic phenolic group was transformed into an ester functional group. The result was the less acidic and more tolerable acetylsalicylic acid, the chemical name for aspirin, shown in Figure 19.24b.

An **ester** is an organic molecule similar to a carboxylic acid except that in the ester, the hydroxyl hydrogen is replaced by a carbon. Unlike carboxylic acids, esters are not acidic because they lack the hydrogen of the hydroxyl group. Like aldehydes, many simple esters have notable fragrances and are often used as flavorings. Some familiar ones are listed in Table 19.4.

Amide group

Carboxyl group

Ester group

N,N-Diethyl-*m*-toluamide
(DEET)

FIGURE 19.22

N,N-diethyl-*m*-toluamide is an example of an amide. Amides contain the amide group, shown highlighted in blue.

Carboxyl group in acetic acid

Carboxylate ion in acetate ion

Hydrogen ion

FIGURE 19.23

The negative charge of the carboxylate ion can pass back and forth between the two oxygen atoms of the carboxyl group.

FIGURE 19.24

(a) Salicylic acid, which is found in the bark of willow trees, is an example of a molecule that contains both a carboxyl group and a phenolic group. (b) Aspirin, acetylsalicylic acid, is less acidic than salicylic acid because it no longer contains the acidic phenolic group, which has been converted to an ester.

(a) Salicylic acid

(b) Aspirin
(acetylsalicylic acid)

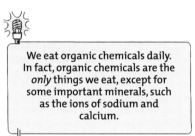

We eat organic chemicals daily. In fact, organic chemicals are the *only* things we eat, except for some important minerals, such as the ions of sodium and calcium.

CHECK POINT

Identify all the functional groups in these four molecules (ignore the sulfur group in penicillin G):

Acetaldehyde

Penicillin G

Testosterone

Morphine

Were these your answers?
Acetaldehyde: aldehyde. Penicillin G: amide (two amide groups), carboxylic acid. Testosterone: alcohol and ketone. Morphine: alcohol, phenol, ether, and amine.

TABLE 19.4   SOME ESTERS AND THEIR FLAVORS AND ODORS

Structure	Name	Flavor/Odor
	Ethyl formate	Rum
	Isopentyl acetate	Banana
	Octyl acetate	Orange
	Ethyl butyrate	Pineapple
	Methyl butyrate	Apple
	Isobutyl formate	Raspberry
	Methyl salicylate	Wintergreen

fyi

- In the 1800s most salicylic acid used by people was produced not from willow bark but from coal tar. Tar residues within the salicylic acid had a nasty taste. This, combined with salicylic acid's stomach irritation, led many to view the salicylic acid cure to be worse than the disease. Felix Hoffman, a chemist working at Bayer, added the acetyl group to the phenol group of salicylic acid in 1897. According to Bayer, Hoffman was inspired by his father, who had been complaining about salicylic acid's side effects. To market the new drug, Bayer invented the name *aspirin*, in which "a" is for acetyl; "spir" is for the spirea flower, another natural source of salicylic acid; and "in" is used as a common suffix for medications. After World War I, Bayer, a German company, lost the rights to use the name *aspirin*. Bayer didn't regain these rights until 1994 for a steep price of $1 billion.

19.7 Polymers

Polymers are exceedingly long molecules that consist of repeating molecular units called **monomers**, as Figure 19.25 illustrates. Monomers have relatively simple structures consisting of anywhere from 4 to 100 atoms per molecule. When monomers are chained together, they can form polymers consisting of hundreds of thousands of atoms per molecule. These large molecules are still too small to be seen with the unaided eye. They are, however, giants in the submicroscopic world—if a typical polymer molecule were as thick as a kite string, it would be 1 km long.

The Physical Science Place
Polymers
Polymers from Monomers

FIGURE 19.25

A polymer is a long molecule consisting of many smaller monomer molecules linked together.

490 PART TWO CHEMISTRY

TABLE 19.5   ADDITION AND CONDENSATION POLYMERS

Addition Polymers	Repeating Unit	Common Uses	Recycling Code
Polyethylene (PE)	$\cdots\overset{\displaystyle H}{\underset{\displaystyle H}{C}}-\overset{\displaystyle H}{\underset{\displaystyle CH_3}{C}}\cdots$	Plastic bags, bottles	♺2 HDPE ♺4 LDPE
Polypropylene (PP)	$\cdots\overset{\displaystyle H}{\underset{\displaystyle H}{C}}-\overset{\displaystyle H}{\underset{\displaystyle CH_3}{C}}\cdots$	Indoor–outdoor carpets	♺5 PP
Polystyrene (PS)	$\cdots\overset{\displaystyle H}{\underset{\displaystyle H}{C}}-\overset{\displaystyle H}{\underset{\displaystyle \bigcirc}{C}}\cdots$	Plastic utensils, insulation	♺6 PS
Polyvinyl chloride (PVC)	$\cdots\overset{\displaystyle H}{\underset{\displaystyle H}{C}}-\overset{\displaystyle H}{\underset{\displaystyle Cl}{C}}\cdots$	Shower curtains, tubing	♺3 V
Polyvinylidene chloride (Saran)	$\cdots\overset{\displaystyle H}{\underset{\displaystyle H}{C}}-\overset{\displaystyle Cl}{\underset{\displaystyle Cl}{C}}\cdots$	Plastic wrap	—
Polytetrafluoroethylene (Teflon)	$\cdots\overset{\displaystyle F}{\underset{\displaystyle F}{C}}-\overset{\displaystyle F}{\underset{\displaystyle F}{C}}\cdots$	Nonstick coating	—
Polyacrylonitrile (Orlon)	$\cdots\overset{\displaystyle H}{\underset{\displaystyle H}{C}}-\overset{\displaystyle H}{\underset{\displaystyle C\equiv N}{C}}\cdots$	Yarn, paints	—
Polymethyl methacrylate (Lucite, Plexiglas)	$\cdots\overset{\displaystyle H}{\underset{\displaystyle H}{C}}-\overset{\displaystyle CH_3}{\underset{\displaystyle \overset{O}{\underset{}{C}}-OCH_3}{C}}\cdots$	Windows, bowling balls	—
Polyvinyl acetate (PVA)	$\cdots\overset{\displaystyle H}{\underset{\displaystyle H}{C}}-\overset{\displaystyle H}{\underset{\displaystyle \underset{O}{O}-C-CH_3}{C}}\cdots$	Adhesives, chewing gum	—

TABLE 19.5   CONTINUED

Addition Polymers	Repeating Unit	Common Uses	Recycling Code
Nylon		Carpeting, clothing	—
Polyethylene terephthalate		Clothing, plastic bottles	♻ 1 PET
Melamine–formaldehyde resin (Melmac, Formica)		Dishes, countertops	—

Many of the molecules that constitute living organisms are polymers, including DNA, proteins, the cellulose of plants, and the complex carbohydrates of starchy foods. For now, we focus on the human-made polymers, also known as synthetic polymers, that make up the class of materials that are commonly known as plastics.

We will begin by exploring the two major types of synthetic polymers used today—*addition polymers* and *condensation polymers*.

As shown in Table 19.5, addition and condensation polymers have a wide variety of uses. Solely the product of human design, these polymers pervade

FIGURE 19.26

The addition polymer polyethylene is formed as electrons from the double bonds of ethylene monomer molecules split away and become unpaired valence electrons. Each unpaired electron then joins with an unpaired electron of a neighboring carbon atom to form a new covalent bond that links two monomer units.

FIGURE 19.27

(a) The polyethylene strands of HDPE can pack closely together, much like strands of uncooked spaghetti. (b) The polyethylene strands of LDPE are branched, which prevents the strands from packing well.

(a) Molecular strands of HDPE

(b) Molecular strands of LDPE

modern living. In the United States, for example, synthetic polymers have surpassed steel as the most widely used material.

ADDITION POLYMERS

Addition polymers form simply by the joining of monomer units. For this to happen, each monomer must contain at least one double bond. As shown in Figure 19.26, polymerization occurs when two of the electrons from each double bond split away from each other to form new covalent bonds with neighboring monomer molecules. During this process, no atoms are lost, so the total mass of the polymer is equal to the sum of the masses of all the monomers.

Nearly 12 million tons of polyethylene is produced annually in the United States; that's about 90 lb per U.S. citizen. The monomer from which it is synthesized, ethylene, is an unsaturated hydrocarbon produced in large quantities from petroleum.

Two principal forms of polyethylene are produced by using different catalysts and reaction conditions. High-density polyethylene (HDPE), shown schematically in Figure 19.27a, consists of long strands of straight-chain molecules packed closely together. The tight alignment of neighboring strands makes HDPE a relatively rigid, tough plastic useful for such things as bottles and milk jugs. Low-density polyethylene (LDPE), shown in Figure 19.27b, is made of strands of highly branched chains, an architecture that prevents the strands from packing closely together. This makes LDPE more bendable than HDPE and gives it a lower melting point. HDPE holds its shape in boiling water; LDPE deforms. LDPE is most useful for such items as plastic bags, photographic film, and electrical-wire insulation.

Other addition polymers are created by using different monomers. The only requirement is that the monomer must contain a double bond. The monomer propylene, for example, yields polypropylene, as shown in Figure 19.28. Polypropylene is a tough plastic material useful for pipes, hard-shell suitcases, and appliance parts. Fibers of polypropylene are used for upholstery, indoor–outdoor carpets, and even thermal underwear.

FIGURE 19.28

Propylene monomers polymerize to form polypropylene.

Styrene monomers

Polystyrene

FIGURE 19.29

Styrene monomers polymerize to form polystyrene. Another important addition polymer is polyvinyl chloride (PVC), which is tough and easily molded. Floor tiles, shower curtains, and pipes are most often made of PVC, shown in Figure 19.30.

Figure 19.29 shows that using styrene as the monomer yields polystyrene. Transparent plastic cups are made of polystyrene, as are thousands of other household items. Blowing gas into liquid polystyrene generates Styrofoam, which is widely used for coffee cups, packing material, and insulation.

The addition polymer polyvinylidene chloride (trade name Saran), shown in Figure 19.31, is used to make plastic wrap for food. The large chlorine atoms in this polymer help it stick to such surfaces as glass by dipole–induced dipole attractions, as discussed in Section 15.8.

The addition polymer polytetrafluoroethylene, shown in Figure 19.32, is what you know as Teflon. In contrast to the chlorine-containing Saran, fluorine-containing Teflon has a nonstick surface because the fluorine atoms tend not to experience any molecular attractions. In addition, because carbon–fluorine bonds are unusually strong, Teflon can be heated to high temperatures before decomposing. These properties make Teflon an ideal coating for cooking surfaces. It is also relatively inert, which is why many corrosive chemicals are shipped or stored in Teflon containers.

Polyvinyl chloride (PVC)

FIGURE 19.30

PVC is tough and easily molded, which is why it often is used to fabricate many household items.

FIGURE 19.31

The large chlorine atoms in polyvinylidene chloride make this addition polymer sticky.

Polyvinylidene chloride (Saran)

$$\cdots \ \underset{\underset{H}{|}}{\overset{\overset{H}{|}}{C}} - \underset{\underset{Cl}{|}}{\overset{\overset{Cl}{|}}{C}} - \underset{\underset{H}{|}}{\overset{\overset{H}{|}}{C}} - \underset{\underset{Cl}{|}}{\overset{\overset{Cl}{|}}{C}} - \underset{\underset{H}{|}}{\overset{\overset{H}{|}}{C}} - \underset{\underset{Cl}{|}}{\overset{\overset{Cl}{|}}{C}} - \underset{\underset{H}{|}}{\overset{\overset{H}{|}}{C}} - \underset{\underset{Cl}{|}}{\overset{\overset{Cl}{|}}{C}} - \underset{\underset{H}{|}}{\overset{\overset{H}{|}}{C}} - \underset{\underset{Cl}{|}}{\overset{\overset{Cl}{|}}{C}} \ \cdots$$

fyi

- So if nothing sticks to Teflon, how is Teflon made to adhere to a pan as a coating? That's a trade secret, but rumor has it that there are microscopic pits in the metal pan that help the Teflon adhere physically. Of course, we all know that the Teflon is fairly easy to scrape out of the pan, which is why manufacturers recommend that you stir-fry with a wooden utensil.

CHECK POINT

What do all monomers that are used to make addition polymers have in common?

Was this your answer?
A double covalent bond between two carbon atoms

CONDENSATION POLYMERS

A **condensation polymer** is formed when the joining of monomer units is accompanied by the loss of a small molecule, such as water or hydrochloric acid. Any monomer capable of becoming part of a condensation polymer must have a functional group on each end. When two such monomers come together to form a condensation polymer, one functional group of the first monomer links with one functional group of the other monomer. The result is a two-monomer unit that has two terminal functional groups, one from each of the two original monomers. Each of these terminal functional groups in the two-monomer unit is now free to link with one of the functional groups of a third monomer, and then a fourth, and so on. In this way a polymer chain is built.

FIGURE 19.32

The fluorine atoms in polytetrafluoroethylene tend not to experience molecular attractions, which is why this addition polymer is used as a non-stick coating and lubricant.

Polytetra-fluoroethylene (Teflon)

$$\cdots \ \underset{\underset{F}{|}}{\overset{\overset{F}{|}}{C}} - \underset{\underset{F}{|}}{\overset{\overset{F}{|}}{C}} - \underset{\underset{F}{|}}{\overset{\overset{F}{|}}{C}} - \underset{\underset{F}{|}}{\overset{\overset{F}{|}}{C}} - \underset{\underset{F}{|}}{\overset{\overset{F}{|}}{C}} - \underset{\underset{F}{|}}{\overset{\overset{F}{|}}{C}} - \underset{\underset{F}{|}}{\overset{\overset{F}{|}}{C}} - \underset{\underset{F}{|}}{\overset{\overset{F}{|}}{C}} - \underset{\underset{F}{|}}{\overset{\overset{F}{|}}{C}} - \underset{\underset{F}{|}}{\overset{\overset{F}{|}}{C}} \ \cdots$$

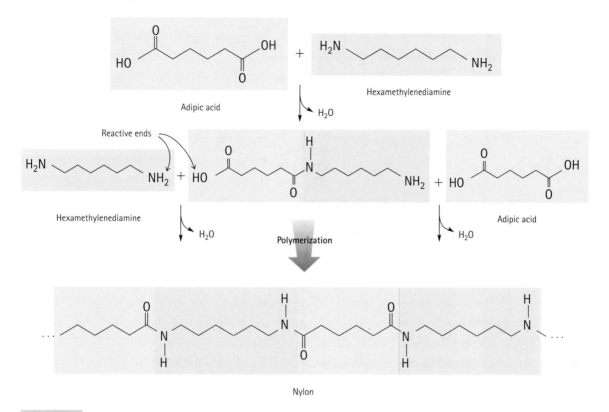

FIGURE 19.33

Adipic acid and hexamethylenediamine polymerize to form the condensation copolymer nylon.

Figure 19.33 shows this process for the condensation polymer called nylon, which was created in 1937 by DuPont chemist Wallace Carothers (1896–1937). This polymer is composed of two different monomers, which classifies it as a *copolymer*. One monomer is adipic acid, which contains two reactive end groups, both carboxyl groups. The second monomer is hexamethylenediamine, in which two amine groups are the reactive end groups. One end of an adipic acid molecule and one end of a hexamethylenediamine molecule can be made to react with each other, splitting off a water molecule in the process. After two monomers have joined, reactive ends still remain for further reactions, which leads to a growing polymer chain. Aside from its use in hosiery, nylon also finds important uses in the manufacture of ropes, parachutes, clothing, and carpets.

Another widely used condensation polymer is polyethylene terephthalate (PET), which is formed from the copolymerization of ethylene glycol and terephthalic acid, as shown in Figure 19.34. Plastic soda bottles are made from this polymer. Also, PET fibers are sold as Dacron polyester, a product used in clothing and stuffing for pillows and sleeping bags. Thin films of PET, which are called Mylar, can be coated with metal particles to make magnetic recording tape or those metallic-looking balloons you see for sale at most grocery store check-out counters.

Monomers that contain three reactive functional groups can also form polymer chains. These chains become interlocked in a rigid three-dimensional network that lends considerable strength and durability to the polymer. Once formed, these condensation polymers cannot be remelted or reshaped, which makes them hard-set, or *thermoset*, polymers. Hard plastic dishes (Melmac) and countertops (Formica) are made of this material. A similar polymer, Bakelite, made from formaldehyde and phenols that contain multiple oxygen atoms, is used to bind plywood and particle board. Bakelite was synthesized in the early 1900s, and it was the first widely used polymer.

Polyethylene terephthalate (PET)

FIGURE 19.34

Terephthalic acid and ethylene glycol polymerize to form the condensation copolymer polyethylene terephthalate.

fyi

■ Rigid polymers such as PVC can be made soft by incorporating small molecules called plasticizers. Pure PVC, for example, is a tough material great for making pipes. Mixed with a plasticizer, the PVC becomes soft and flexible and thus useful for making shower curtains, toys, and many other products now found in most households. One of the more commonly used plasticizers is the phthalates, some of which have been shown to disrupt the development of reproductive organs, especially in the fetus and in growing children. Governments and manufacturers are now working to phase out these plasticizers. But some phthalates, such as DINP, have been shown to be safe. For social and political simplicity, should all phthalates be banned or just the ones shown to be harmful? This question has yet to be resolved.

CHECK POINT

■ The structure of 6-aminohexanoic acid is the following:

Is this compound a suitable monomer for forming a condensation polymer? If so, what is the structure of the polymer formed, and what small molecule is split off during the condensation?

Was this your answer?
Yes, because the molecule has two reactive ends. You know both ends are reactive because they are the ends shown in Figure 19.33. The only difference here is that both types of reactive ends are on the same molecule. Monomers of 6-aminohexanoic acid combine by splitting off water molecules to form the polymer known as nylon-6:

The synthetic-polymers industry has grown remarkably over the past half century. Annual production of polymers in the United States alone has grown from 3 billion pounds in 1950 to more than 100 billion pounds in 2003. Today, it is a challenge to find any consumer item that does *not* contain a plastic of one sort or another. Try finding one yourself.

In the future, watch for new kinds of polymers with a wide range of remarkable properties. One interesting application is shown in Figure 19.35. We already have polymers that conduct electricity, others that emit light, others that replace body parts, and still others that are stronger but much lighter

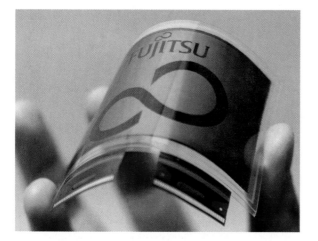

FIGURE 19.35

Flexible and flat video displays can now be fabricated from polymers.

than steel. Imagine synthetic polymers that mimic photosynthesis by transforming solar energy to chemical energy, or that efficiently separate fresh water from the oceans. These are not dreams. They are realities that chemists have already been demonstrating in the laboratory. Polymers hold a clear promise for the future.

SUMMARY OF TERMS

Organic chemistry The study of carbon-containing compounds.

Hydrocarbon A chemical compound that contains only carbon and hydrogen atoms.

Configuration A term used to describe how the atoms within a molecule are connected. For example, two structural isomers consist of the same number and same kinds of atoms, but in different configurations.

Structural isomers Molecules that have the same molecular formula but different chemical structures.

Conformation One of a wide range of possible spatial orientations of a particular configuration.

Saturated hydrocarbon A hydrocarbon that contains no multiple covalent bonds, with each carbon atom bonded to four other atoms.

Unsaturated hydrocarbon A hydrocarbon that contains at least one multiple covalent bond.

Aromatic compound Any organic molecule that contains a benzene ring.

Heteroatom Any atom other than carbon or hydrogen in an organic molecule.

Functional group A specific combination of atoms that behaves as a unit in an organic molecule.

Alcohol An organic molecule that contains a hydroxyl group bonded to a saturated carbon.

Phenol An organic molecule in which a hydroxyl group is bonded to a benzene ring.

Ether An organic molecule that contains an oxygen atom bonded to two carbon atoms.

Amine An organic molecule that contains a nitrogen atom bonded to one or more saturated carbon atoms.

Carbonyl group A carbon atom double-bonded to an oxygen atom; found in ketones, aldehydes, amides, carboxylic acids, and esters.

Ketone An organic molecule that contains a carbonyl group, the carbon of which is bonded to two carbon atoms.

Aldehyde An organic molecule that contains a carbonyl group, the carbon of which is bonded either to one carbon atom and one hydrogen atom or to two hydrogen atoms.

Amide An organic molecule that contains a carbonyl group, the carbon of which is bonded to a nitrogen atom.

Carboxylic acid An organic molecule that contains a carbonyl group, the carbon of which is bonded to a hydroxyl group.

Ester An organic molecule that contains a carbonyl group, the carbon of which is bonded to one carbon atom and one oxygen atom bonded to another carbon atom.

Polymer A long organic molecule made of many repeating units.

Monomers The small molecular units from which a polymer is formed.

Addition polymer A polymer formed by the joining of monomer units with no atoms being lost as the polymer forms.

Condensation polymer A polymer formed by the joining of monomer units accompanied by the loss of small molecules, such as water.

REVIEW QUESTIONS

19.1 Hydrocarbons

1. What are some examples of hydrocarbons?
2. What are some uses of hydrocarbons?
3. How do two structural isomers differ from each other?
4. How are two structural isomers similar to each other?
5. What physical property of hydrocarbons is used in fractional distillation?
6. What types of hydrocarbons are more abundant in higher-octane gasoline?
7. A saturated carbon atom is bonded to how many atoms?

19.2 Unsaturated Hydrocarbons

8. What is the difference between a saturated hydrocarbon and an unsaturated hydrocarbon?
9. How many multiple bonds must a hydrocarbon have in order to be classified as unsaturated?
10. Aromatic compounds contain what kind of ring?

19.3 Functional Groups

11. What is a heteroatom?
12. Why do heteroatoms make such a difference in the physical and chemical properties of an organic molecule?

19.4 Alcohols, Phenols, and Ethers

13. Which of the following molecules should have the higher boiling point, and why?

$$CH_3CH_2CH_2CH_3$$
$$CH_3CH_2CH_2CH_2 - OH$$

14. Why are low-formula-mass alcohols soluble in water?
15. What distinguishes an alcohol from a phenol?
16. What distinguishes an alcohol from an ether?

19.5 Amines and Alkaloids

17. Which heteroatom is characteristic of an amine?
18. Do amines tend to be acidic, neutral, or basic?
19. Are alkaloids found in nature?
20. What are some examples of alkaloids?

19.6 Carbonyl Compounds

21. Which elements make up the carbonyl group?
22. How are ketones and aldehydes related to each other? How are they different from each other?
23. What is one commercially useful property of aldehydes?
24. How are amides and carboxylic acids related to each other? How are they different from each other?
25. From what naturally occurring compound is aspirin prepared?
26. Identify each molecule as hydrocarbon, alcohol, or carboxylic acid:

$$CH_3CH_2CH_2CH_3$$
$$CH_3CH_2CH_2CH_2 - OH$$

19.7 Polymers

27. What happens to the double bond of a monomer participating in the formation of an addition polymer?
28. What is released in the formation of a condensation polymer?
29. Why is plastic food wrap a stickier plastic than polyethylene?
30. What is a copolymer?

EXERCISES

● BEGINNER ■ INTERMEDIATE ◆ EXPERT

1. ● What property of carbon allows for the formation of so many different organic molecules?
2. ■ Draw all the structural isomers for hydrocarbons that have the molecular formula C_6H_{14}.
3. ■ How many structural isomers are shown here?

4. ● According to Figure 19.3, which has a higher boiling point: gasoline or kerosene?

5. ■ The temperatures in a fractionating tower at an oil refinery are important, but so are the pressures. Where might the pressure in a fractionating tower be greatest, at the bottom or at the top? Defend your answer.
6. ● There are five atoms in the methane molecule, CH_4. One out of these five is a carbon atom, which is $\frac{1}{5} \times 100 = 20\%$ carbon. What is the percentage of carbon in ethane, C_2H_6? Propane, C_3H_8? Butane, C_4H_{10}?
7. ● Do heavier hydrocarbons tend to produce more or less carbon dioxide on combustion compared to lighter hydrocarbons? Why?

8. ● What is the chemical formula for the following structure?

9. ● What is the chemical formula for the following structure?

10. ● What is the chemical formula for the following structure?

11. ■ Shown below is the structure of 2-methyl-pentane. What is the structure of 3-methyl-pentane?

2-methyl-pentane

12. ■ List the following compounds in order of least oxidized to most oxidized.

(a)   (b)   (c)   (d)

13. ● Circle the longest chain of carbon atoms in the following structure. How many carbon atoms are in this chain?

14. ● Number the carbon atoms within the longest chain of carbon atoms for the structure in Exercise 13. How many branches extend off this chain?

15. ■ A single carbon branch within a hydrocarbon is referred to as a *methyl* group; a two-carbon branch is referred to as an *ethyl* group. What then is the formal name for the structure shown in Exercise 13?
(a) 3-methyl-4-ethyl-6-methyl-octane
(b) 3,6-dimethyl-4-ethyl-octane
(c) 2-ethyl-4-ethyl-5-methyl-heptane
(d) 2,4-diethyl-5-methyl-heptane

16. ■ What is the name of the following structure?

17. ◆ Carbon–carbon single bonds can rotate; carbon-carbon double bonds cannot rotate (see the activity titled "Twisting Jelly Beans" at the end of the chapter). How many different structures are shown below?

18. ■ Which of the structures shown in Exercise 17 are 3-methyl-2-pentene, where "2" indicates the placement of the double bond? Circle them.

19. ◆ How many different isomers of 3-methyl-2-pentene exist?

20. ● Why do ethers typically have lower boiling points than alcohols?

21. ■ What is the percent volume of water in 80-proof vodka?

22. ■ One of the skin-irritating components of poison oak is tetrahydrourushiol:

The long, nonpolar hydrocarbon tail embeds itself in a person's oily skin, where the molecule initiates an allergic response. Scratching the itch spreads tetrahydrourushiol molecules over a greater surface area, causing the zone of irritation to grow. Is this compound an alcohol or a phenol? Defend your answer.

23. ◆ Why is 2,4,5-trifluorophenol much more acidic than phenol?

2,4,5-trifluorophenol

24. ■ Cetyl alcohol, $C_{16}H_{33}OH$, is a common ingredient of soaps and shampoos. It was once commonly obtained from whale oil, which is where it gets its name (*cetyl* is derived from *cetacean*). Draw the chemical structure for this compound.

25. ■ A common ingredient in products such as sunscreen lotions and shampoo is triethyl amine, also known as TEA. What is the chemical structure for this compound?

26. ■ Another common ingredient in products such as sunscreen lotions and shampoo is triethanol amine. What is the chemical structure for this compound?

27. ■ The phosphoric acid salt of caffeine has the following structure:

Caffeine–phosphoric acid salt

This molecule behaves as an acid in that it can donate a hydrogen ion, created from the hydrogen atom bonded to the positively charged nitrogen atom. What are all the products formed when 1 mole of this salt reacts with 1 mole of sodium hydroxide, NaOH, a strong base?

28. ◆ Draw all the structural isomers for amines that have the molecular formula C_3H_9N.

29. ◆ In water, does the following molecule act as an acid, a base, neither, or both?

Lysergic acid diethylamide

30. ◆ If you saw the label phenylephrine·HCl on a decongestant, would you worry that consuming it would expose you to the strong acid hydrochloric acid? Explain.

Phenylephrine–hydrochloric acid salt

31. ■ Suggest an explanation for why aspirin has a sour taste.
32. ■ An amino acid is an organic molecule that contains both an amine group and a carboxyl group. At an acidic pH, which of the following structures is most likely? Explain your answer.

33. ■ Identify the following functional groups in this organic molecule—amide, ester, ketone, ether, alcohol, aldehyde, amine:

34. ● Benzaldehyde is a fragrant oil. If stored in an uncapped bottle this compound slowly transforms into benzoic acid along the surface. Is this an oxidation or a reduction?

Benzaldehyde   Benzoic acid

35. ● What products are formed upon the reaction of benzoic acid with sodium hydroxide, NaOH? One of these products is a common food preservative. Can you name it?
36. ● The amino acid lysine is shown below. What functional group must be removed in order to produce cadaverine, shown in Figure 19.16?

Lysine

37. ◆ Would you expect polypropylene to be more or less dense than low-density polyethylene? Why?
38. ■ Many polymers emit toxic fumes when burning. Which polymer in Table 19.5 produces hydrogen cyanide, HCN? Which two produce toxic hydrogen chloride gas, HCl?
39. ■ Hydrocarbons release a lot of energy when ignited. Where does this energy come from?
40. ■ The copolymer styrene-butadiene rubber (SBR), shown below, is used for making tires as well as bubble gum. Is it an addition polymer or a condensation polymer?

SBR

ACTIVE EXPLORATIONS

Twisting Jelly Beans

Two carbon atoms connected by a single bond can rotate relative to each other. As we discussed in Section 19.1, this ability to rotate can give rise to numerous conformations (spatial orientations) of an organic molecule. Is it also possible for two carbon atoms connected by a double bond to rotate relative to each other? Perform this quick activity to see for yourself.

What You Need

Jelly beans (or gumdrops); round toothpicks

Procedure

1. Attach one jelly bean to each end of a single toothpick. Hold one of the jelly beans firmly with one hand while rotating the second jelly bean with your other hand. Observe how there is no restriction on the different orientations of the two jelly beans relative to each other.
2. Hold two toothpicks side by side and attach one jelly bean to each end such that each jelly bean has both toothpicks poked into it. As before, hold one jelly bean while rotating the other. What kind of rotations are possible now? Relate what you observe to the carbon–carbon double bond. Which structure of Figure 19.7 do you suppose has more possible conformations: *n*-butane or 2-butene? What do you suppose is true about the ability of atoms connected by a carbon–carbon triple bond to twist relative to each other?

Rubbing Eggs

Isopropyl alcohol, also known as rubbing alcohol, is very toxic if ingested because it destroys digestive proteins and other important biomolecules in your stomach. Do this activity to see firsthand the destructive action of isopropyl alcohol on proteins. Crack open an egg and place the egg white and the yolk into two separate bowls. Pour a capful of isopropyl alcohol into the egg white and observe what happens. In the second bowl, stir the yolk with a fork. Add another capful of isopropyl alcohol to the stirred yolk and observe what happens. The same sort of destruction would occur to your own stomach proteins, as well as various tissues, upon ingesting the isopropyl alcohol. Not good! Our skin, however, is more impervious to the destructive powers of isopropyl alcohol, which therefore serves as a good topical antiseptic.

READINESS ASSURANCE TEST (RAT)

If you have a good handle on this chapter, if you really do, then you should be able to score 7 out of 10 on this RAT. If you score less than 7, you need to study further before moving on.

Choose the BEST answer to each of the following.

1. Why does the melting point of hydrocarbons increase as the number of carbon atoms per molecule increases?
(a) An increase in the number of carbon atoms per molecule also means an increase in the density of the hydrocarbon.
(b) Many induced dipole–induced dipole molecular attractions hold the molecules together.
(c) Larger hydrocarbon chains tend to be branched.
(d) The molecular mass also increases.
2. How many structural isomers are there for hydrocarbons that have the molecular formula C_4H_{10}?
(a) none   (b) one   (c) two   (d) three
3. Which contains more hydrogen atoms: a five-carbon saturated hydrocarbon molecule or a five-carbon unsaturated hydrocarbon molecule?
(a) The unsaturated hydrocarbon has more hydrogen atoms.
(b) The saturated hydrocarbon has more hydrogen atoms.
(c) They both have the same number of hydrogen atoms.
(d) It depends on whether the unsaturation is due to a double or triple bond.
4. Heteroatoms make a difference in the physical and chemical properties of an organic molecule because
(a) they add extra mass to the hydrocarbon structure.
(b) each heteroatom has its own characteristic chemistry.
(c) they can enhance the polarity of the organic molecule.
(d) all of these
5. Why might a high-formula-mass alcohol be insoluble in water?
(a) A high-formula-mass alcohol would be too attracted to itself to be soluble in water.
(b) The bulk of a high-formula-mass alcohol likely consists of nonpolar hydrocarbons.
(c) Such an alcohol would likely be in a solid phase.
(d) In order for two substance to be soluble in each other, their molecules need to be of comparable mass.
6. Alkaloid salts are not very soluble in the organic solvent diethyl ether. What might happen to the free-base form of caffeine (an alkaloid) dissolved in diethyl ether if gaseous hydrogen chloride, HCl, were bubbled into the solution?
(a) A second layer of water would form.
(b) Nothing would happen; the HCl gas would merely bubble out of solution.
(c) The diethyl ether insoluble caffeine salt would form as a white precipitate.
(d) The acid/base reaction would release heat, which would cause the diethyl ether to start evaporating.

7. Explain why caprylic acid, $CH_3(CH_2)_6COOH$, dissolves in a 5% aqueous solution of sodium hydroxide, but caprylaldehyde, $CH_3(CH_2)_6CHO$, does not.
(a) With two oxygens the caprylic acid is about twice as polar the caprylaldehyde.
(b) The caprylaldehyde is a gas at room temperature.
(c) The caprylaldehyde behaves as a reducing agent, which neutralizes the sodium hydroxide.
(d) The caprylic acid reacts to form the water-soluble salt.

8. How many oxygen atoms are bonded to the carbon of the carbonyl of an ester functional group?
(a) none   (b) one   (c) two   (d) three

9. One solution to the problem of our overflowing landfills is to burn plastic objects instead of burying them. What would be some of the advantages and disadvantages of this practice?
(a) Disadvantage: toxic air pollutants. Advantage: reduced landfill volume.

(b) Disadvantage: loss of vital petroleum-based resource. Advantage: generation of electricity.
(c) Disadvantage: discourages recycling. Advantage: provides new jobs.
(d) all of these

10. Which would you expect to be more viscous, a polymer made of long molecular strands or one made of short molecular strands? Why?
(a) long molecular strands, because they tend to tangle among themselves
(b) short molecular strands, because of a greater density
(c) long molecular strands, because of a greater molecular mass
(d) short molecular strands, because their ends are typically polar

Answers to RAT
1b, 2c, 3b, 4d, 5b, 6c, 7d, 8c, 9d, 10a

EXPLORING FURTHER

http://www.icco.org
Through the home page of the International Cocoa Organization, you can find answers to many questions regarding the chemistry of chocolate and its path from the cocoa tree to your mouth.

http://en.wikipedia.org/wiki/Oil_refinery
Wikipedia has a well-illustrated Web page describing all the essential components and processes of oil refineries. You will also find many hyperlinks to terms related to the topics of this chapter.

http://www.CO2captureproject.org
The CO_2 Capture Project is an international effort that addresses the issue of reducing emissions in a manner that

will contribute to an environmentally acceptable and competitively priced continuous energy supply for the world.

http://www.bayeraspirin.com
Visit the home page for Bayer Aspirin to learn everything you ever wanted to know about aspirin and its many wonders.

http://www.dinp-facts.com
Learn more about the risks and benefits of the PVC plasticizer diisononyl phthalate (DINP) from this Web site sponsored by the European Council for Plasticisers.

CHAPTER 19 ONLINE RESOURCES

The Physical Science Place

Tutorials

- Introduction to Organic Molecules
- Organic Molecules and Isomers
- Functional Groups
- Polymers
- Polymers from Monomers

Quiz

Flashcards

Links

part three

earth science

You're right, Megan, it is the perfect place. We are standing on the solid geosphere, but all the while, the hydrosphere and atmosphere are at work weathering the rock we stand on. The hydrosphere is where life on Earth began, and the atmosphere provides the oxygen animals need and the carbon dioxide plants need. Plus the atmosphere shields us from harmful UV rays. Our planet is unique in our solar system. It is our home and we need to learn more about it to be able to preserve it.

Hey Emily, the seashore is the perfect place to see interactions of the geosphere, hydrosphere, and atmosphere!

rocks and minerals

20.1 THE GEOSPHERE IS MADE UP OF ROCKS AND MINERALS

20.2 MINERALS

20.3 MINERAL PROPERTIES

20.4 CLASSIFICATION OF ROCK-FORMING MINERALS

20.5 THE FORMATION OF MINERALS AND ROCK

20.6 ROCK TYPES

20.7 IGNEOUS ROCKS

20.8 SEDIMENTARY ROCKS

20.9 METAMORPHIC ROCKS

20.10 THE ROCK CYCLE

■ Our Earth is an enormous, interconnected system. As a system, Earth science topics can be organized into "spheres"—the geosphere, hydrosphere, and atmosphere. Each sphere is separate, but each sphere touches and is interconnected to the other spheres. Earth processes do not occur in isolation; events in one sphere affect one or both of the other spheres. We focus first on the *geosphere*, which encompasses Earth beneath our feet—rocks and minerals, plate tectonics, earthquakes, volcanoes, and how the processes of erosion and deposition influence landforms. Next, we turn our attention to the *hydrosphere*, which includes Earth's fresh water—rivers and streams, glaciers, and groundwater—and saline water, Earth's oceans. We end our study of Earth by focusing on the *atmosphere*—weather and climate.

The elements that make up Earth are found in its rocks and minerals. These rocks and minerals form Earth's *crust*, *mantle*, and *core*. The crust rides atop a dozen or so **tectonic plates** that move in response to heat flow and convection in Earth's interior (Figure 20.1). These plates can split apart from, collide with, and slide past one another. As the plates move, Earth's surface changes. Heat escaping from the interior is sometimes seen as volcanic eruptions, which creates new rock. Internal movement is reflected in surface movement (*earthquakes*) that break apart and deform the land. As the rock that comprises Earth's crust is touched by the hydrosphere (water) and the atmosphere (air) it begins to erode and weather and is broken up into smaller rocks. The process continues as the three spheres interact to influence landforms at Earth's surface.

FIGURE 20.1

Earth's crustal surface is a mosaic of tectonic plates that move in response to heat flow and convection in Earth's interior. As the plates move, Earth's surface changes.

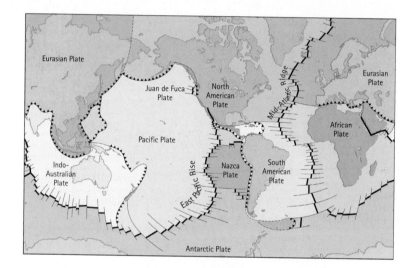

▪ 20.1 The Geosphere Is Made Up of Rocks and Minerals

We begin our study of Earth science by examining the ground beneath our feet—Earth's geosphere. The geosphere is made up of rocks, and rocks are made up of minerals. Knowing about rocks and minerals helps us understand the structure and makeup of our geosphere, much like knowing about concrete, steel, and glass helps us understand the design and architecture of a building. For example, the kinds of minerals found in volcanic rocks provide evidence that molten rock erupted from Earth's interior to the surface. And the size and type of minerals in a metamorphic rock can reveal the rate of crystallization and the conditions of formation (the temperature and pressure) that occurred deep below Earth's surface.

Minerals are the building blocks of rocks, and elements, in turn, are the building blocks of minerals. You may recall from our discussion of the periodic table in Chapter 12 that there are 112 known elements, and that many of these elements are rare. But out of this very large number, it may surprise you to learn that just eight elements make up 98% of Earth's entire mass (Figure 20.2)! All of the other elements combined make up the remaining 2%.

Link To Section 12.2

Earth's elements are not distributed evenly. For example, most of the iron on Earth is concealed deep in the planet's interior, where it forms the central core. Lighter elements, such as silicon and oxygen, are mostly distributed in the mid-to-outer portions of the planet. To explain this lopsided distribution we need to examine the very beginnings of Earth.

Our solar system formed about 4.5 billion years ago, when dust, gases, and rocky and metallic debris orbiting the newly forming Sun collided and coalesced into the planets, asteroids, and comets we know today. One such rocky mass became our Earth, which formed as chunks of all sizes accumulated. When first formed, the elements were distributed evenly throughout because that is the way they accumulated. But all that was about to change.

With each collision, heat was released because of the conversion of kinetic energy to heat energy—impact heating. As Earth grew, gravitational attraction toward the planet's center attracted even more debris. The attraction became strong enough that the young Earth actually squeezed itself into a smaller volume, which produced even more heat. A third source of heat came from the decay of naturally occurring, widely distributed radioactive elements. Although the amount of heat generated in any cubic meter of rock is small, when we think of the trillions of cubic meters of rock within Earth the amount of heat produced can be quite large. These three sources of heat—*impact heating*, *gravitational contraction heating*, and *radioactive decay heating*—acted together to bring young Earth to its melting point.

So, in a molten or nearly molten state and under the influence of gravity, dense, heavy iron-rich material sank to Earth's center and less-dense, silicon- and oxygen-rich material rose toward the surface (see Chapter 5). The same type of density segregation occurs in a mixture of oil and water. The heavier water sinks to form a layer at the bottom and the less-dense oil rises to form a layer at the top. In Earth, density segregation led to the formation of a dense, iron-rich *core*, a less-dense, rocky *mantle*, and an even less-dense, rocky *crust* (Figure 20.3).

Figure 20.4 shows the current composition of Earth's crust. Compare the composition of the crust to that of Earth as a whole and you will notice that practically the same few elements appear in both. The percentages, however, are quite different. As expected, the crust is composed of mostly lighter elements. In fact, almost half the mass of Earth's crust is the element oxygen (O) and about a fourth is the element silicon (Si)!

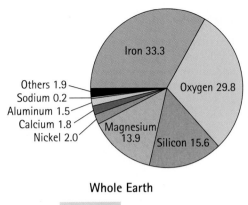

Whole Earth

FIGURE 20.2

Only eight of the chemical elements are found in abundance on Earth.

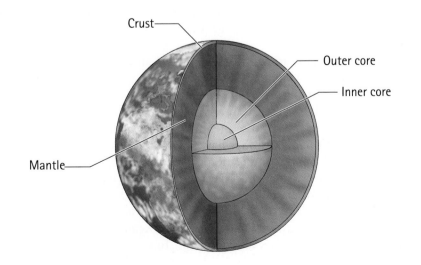

FIGURE 20.3

Earth has a layered internal structure. The layers—the crust, mantle, and core—differ in composition and density. As a whole, Earth's average density is 5.5 g/cm³. Because samples of Earth's crust average 2.7 g/cm³, material below the crust must have much higher densities. Although not directly measured, the mantle averages 4.5 g/cm³, and the core (solid inner and liquid outer) has an estimated average density of 13.5 g/cm³.

FIGURE 20.4

Percentage of elements in Earth's crust, by mass. Oxygen and silicon make up more than 75% of Earth's crust.

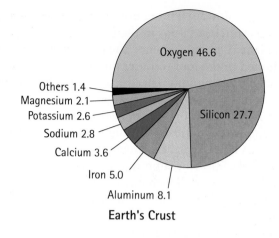

Earth's Crust

fyi

- The dominance of oxygen in Earth's crust becomes even more apparent when you consider its abundance, not by mass, but in terms of numbers of atoms: 63 out of every 100 atoms making up Earth's crust are oxygen atoms. Oxygen isn't just important as a constituent of air—in the solid state, it makes up most of Earth's crust.

CHECK POINT

1. Earth's interior is separated into layers, much as oil and vinegar in salad dressing separate into their respective layers. What caused this to happen?
2. How is the composition of the "whole Earth" different from "Earth's crust"?

Were these your answers?

1. Gravity. Heat from collision impacts, gravitational contraction, and radioactive decay softened our planet to a point at which its components could easily move around. In this state, dense elements sank toward Earth's center and lighter elements migrated upward to the surface.
2. Iron makes up one-third of Earth as a whole, but most of this iron is in Earth's interior. Oxygen and silicon dominate Earth's outer layers. And as we will soon see, this is why the most common group of rock-forming minerals—the silicate group—is dominated by these two elements.

20.2 Minerals

In everyday usage, minerals are a part of our diet ("vitamins and minerals"), and minerals provide the raw materials needed for industry (aluminum for cans, iron for steel, etc.). From these two simple examples, it is easy to see the importance of minerals in the geosphere and in our lives. But what exactly is a mineral? A **mineral** is a naturally formed, inorganic crystalline solid, composed of a determinable and ordered arrangement of atoms with a specific chemical composition.

The definition is fairly straightforward. To be naturally formed means that it is not manufactured in a laboratory—cubic zirconia and synthetic gems are not minerals. To be a crystalline solid means that the atoms that make up a mineral are always arranged in an orderly geometric pattern. For example, glass, although solid, has no crystal structure. It is *amorphous* and so is not a mineral. The same types of minerals always have the same geometric arrangement of atoms. A definite chemical composition means that for two samples to be considered the same mineral, they must have the same basic chemical composition. Minerals typically have a range of compositions, but that range has fixed limits.

The "inorganic" part of the definition requires some explaining. *Inorganic* means that it is not made of organic molecules. *Organic* means something different to a scientist than it does to a layperson. To a scientist, *organic* means that

fyi

- Minerals found in rocks and in dietary supplements are similar yet different. The minerals in rocks are naturally occurring, inorganic, crystalline solids, with a definite chemical composition. Minerals found in dietary supplements are human-made inorganic compounds that contain elements necessary for life functions. However, the elements used to make dietary supplements ultimately come from the naturally occurring minerals of Earth's crust.

the substance's composition is dominated by carbon, oxygen, and hydrogen. Organic substances are almost always made by living things. Inorganic substances are usually made by processes that do not involve living things, although they can be made by living things. Seashells and pearls, although made by animals, are not considered organic because they are made of calcium carbonate, which contains carbon and oxygen, but not hydrogen.

CHECK POINT

1. Are synthetic diamonds minerals?
2. Obsidian is a kind of glass formed in volcanoes. Is it a mineral?
3. What is a crystal?

Were these your answers?
1. No. To be a mineral, it must be naturally formed.
2. No. Obsidian, though it is naturally formed and has a specific range of chemical compositions, is a type of glass—it is amorphous and so does not have a crystalline structure. It does not meet all the criteria needed to be a mineral.
3. A crystal is a solid that has a crystalline structure—the atoms, ions, or molecules within it are arranged in a definite repeating pattern.

20.3 Mineral Properties

Minerals are classified by chemical composition (which elements are present) and crystal structure (how the elements are arranged). A mineral's observable physical properties depend on its inner microscopic properties. Microscopic properties such as composition, crystal structure, and the strength of chemical bonds determine a mineral's crystal form, hardness (resistance to scratching), fracture or cleavage (how a mineral breaks), color, and density. Most minerals can be identified by these easily observable physical properties. Other physical properties that can help identify minerals are luster (the way a mineral reflects light) and streak (the color of a mineral in its powdered form). In this section, we discuss the physical properties of minerals as expressions of their inner structures.

CRYSTAL FORM

Have you ever seen table salt (halite) under a magnifying glass? If so, you may have marveled at its perfect geometric form (Figure 20.5). Crystals are well known for the striking geometric shapes that they can exhibit. A crystal's shape, or its **crystal form**, is an expression of the orderly arrangement of its atoms. When you look at a fully formed crystal, what you see is the actual arrangement of atoms in its structure. Every mineral has a unique combination of composition and crystal

fyi

- The term *organic* is derived from *organism*. Some organisms produce minerals. For example, many marine organisms make calcium carbonate minerals (mostly aragonite) for their shells, and oysters can make pearls, which are also mostly aragonite. The minerals opal and fluorite and some phosphates can also be precipitated by organisms. For example, teeth and bone material contain the mineral apatite. These minerals may be produced by organisms, but they are not organic. As a rule, organic compounds—those containing carbon, oxygen, and hydrogen atoms—all together—are not minerals. So although the hydrocarbons of petroleum and coal are referred to as mineral fuels, they are not minerals. And despite being a naturally occurring crystalline solid with a very ordered internal arrangement of atoms, table sugar ($C_{12}H_{22}O_{11}$) is not a mineral.

Minerals differ from one another in their combination of elements and/or in the internal arrangement of their constituent atoms.

(a) Crystalline structure form of halite

Chlorine ion
Sodium ion
Basic structural form

(b) Grains of the mineral halite (table salt)

FIGURE 20.5

The basic structural form of the mineral halite (table salt) is cubic. This form is repeated over and over in three dimensions. The internal order of halite crystals is reflected in its macroscopic mineral grains.

(a) (b) (c) (d) (e) (f)

FIGURE 20.6

Many minerals are easily recognized by their crystal form. (a) Amethyst, the purple variety of quartz, has a hexagonal crystal form with pointed ends. (b) Pyrite, or "fool's gold," typically forms cubic crystals marked with parallel lines called striations. (c) Rosasite has radiating bluish-green crystals that group into balls. (d) Rhodochrosite (whose name means "rose-colored") has a rhombohedral crystal form. Some minerals have distinctive growth patterns. (e) The mineral hematite often grows in a grape-clustered form. (f) Asbestos minerals have a fibrous form.

form (Figure 20.6). Unfortunately, well-shaped crystals are rare in nature because of space constraints—most crystals grow in cramped spaces.

Just as buildings are made of different materials—some stone, some brick, some wood—minerals are made of different elements. Some minerals have the same combination of elements, but their atoms are arranged differently, which makes them different minerals. So, carrying our analogy with buildings further, different architecture using the same materials can result in very different minerals. Sometimes, two or more minerals contain the same elements in the same proportions, but their atoms are arranged differently. As a result, their crystalline structure and the properties they display are different. Such minerals are called **polymorphs** (*poly* = many, *morph* = form) of each other. Graphite and diamond are polymorphs because they both consist entirely of the same element, carbon, but the carbon atoms are arranged differently. As a result, graphite and diamond show vastly different properties (Figure 20.7). Because the formation of these similar-yet-different minerals depends on temperature and pressure, a polymorph is a good indicator of the geological conditions at the time and place of its formation.

CHECK POINT

Many minerals can be identified by their physical properties—crystal form, hardness, fracture, cleavage, luster, color, streak, and density. Why is identifying a mineral by its crystal form usually difficult?

Was this your answer?
Well-shaped crystals are rare in nature because minerals typically grow in cramped spaces.

HARDNESS

Hardness does not refer to how easily a mineral breaks, but rather a mineral's resistance to scratching. For example, a quartz crystal can scratch a feldspar crystal because quartz is harder than feldspar. The ability of one mineral to scratch another and the resistance of a mineral to being scratched are measures of hardness. We use

FIGURE 20.7

Both graphite and diamond are pure carbon. (a) Diamond, the hardest substance known, has a tightly packed symmetrical structure. (b) Graphite has an open, layered structure and is a very soft mineral. When you rub graphite between your fingers, individual graphite molecules glide over one another like cards in a deck, giving it a slippery feel. This slippery effect is why graphite is used as a dry lubricant. Graphite also glides easily when it is stroked onto paper, leaving a mark—hence its use in pencils. (Graphite is also preferable to lead in pencils because it is less toxic.)

(a) Diamond (b) Graphite

TABLE 20.1 MOHS SCALE OF HARDNESS

Mineral	Hardness	Object of Similar Hardness
Talc	1	
Gypsum	2	Fingernail (2.5)
Calcite	3	Copper wire or coin (3.5)
Fluorite	4	
Apatite	5	Steel knife blade, glass (5.5)
Feldspar	6	Unglazed porcelain tile (6.5)
Quartz	7	
Topaz	8	
Corundum	9	
Diamond	10	

the **Mohs scale of hardness** (Table 20.1) to compare the hardness of different minerals.

Why are some minerals harder than others? Hardness depends on the strength of a mineral's chemical bonds—the stronger its bonds, the harder the mineral. The factors that influence bond strength are ionic charge, atom or ion size, and packing (Chapter 15). Strong bonds are generally found between highly charged ions—the greater the attraction, the stronger the bond. Size affects bond strength as well, because small atoms and ions can generally pack closer together than large atoms and ions. Closely packed atoms and ions have a smaller distance between one another, and thus they form stronger bonds because they attract one another with more force. Gold, with its large atoms, is soft. Its atoms are rather loosely packed and loosely bonded. Diamond, with its small carbon atoms and tightly packed structure, is very hard—the hardest mineral known (Figure 20.7a).

> The stronger the bond, the harder the mineral.

Link To Sections 5.1, 11.2, 11.5, 15.5

CLEAVAGE AND FRACTURE

If you shatter a sample of the mineral calcite with a hammer, it tends to break along its *planes of weakness*—planes along which chemical bonds are weak or few in number. **Cleavage** is the tendency for a mineral to break along such planes of weakness. Planes of weakness are determined by crystal structure and chemical bond strength.

Some minerals show a greater tendency toward cleavage than others. In general, minerals that have strong bonds between flat (planar) crystal surfaces show poor cleavage, while those with weak bonds along planar surfaces show more developed cleavage. The minerals muscovite (mica) and calcite both have very distinct cleavage. For example, mica's crystal structure consists of atoms arranged in sheets. The atoms *within* the individual sheets are connected by strong bonds, but *between* the sheets, the bonds are weak. So muscovite cleaves where its bonding is weak—between its planar sheets (Figure 20.8). You can even peel muscovite off in thin layers. Shiny flakes of muscovite are used in glittering body paints, and they add shimmer to auto body paints as well.

Minerals that have no planar alignment of bonds, like quartz, cannot display cleavage and always **fracture**. A fracture that is smooth and curved, so that it resembles broken glass, is called *conchoidal*. The minerals quartz and olivine display smooth conchoidal fractures (Figure 20.9). But most minerals fracture irregularly. The degree and type of cleavage or fracture are useful guides for identifying minerals.

(a)   (b)

FIGURE 20.8

A mineral's cleavage is very useful in its identification. (a) Muscovite, a mineral of the mica group, has perfect cleavage in one direction. It breaks apart into sheets. (b) Calcite (calcium carbonate) has perfect cleavage in three directions (though *not* at right angles, like a cube). It breaks apart into smaller rhombohedral shapes.

FIGURE 20.9

This quartz specimen shows its crystal form and conchoidal fracture. When quartz breaks, it develops a curved, smooth surface that resembles broken glass—a conchoidal fracture.

The density of a mineral depends on a number of factors—the masses of the mineral's constituent atoms and the packing of these atoms, which, in turn, is a function of the atoms' sizes.

Ruby

Sapphire

FIGURE 20.10

The mineral corundum (Al_2O_3) comes in a variety of colors as a result of chemical impurities. The addition of small amounts of chromium in place of aluminum produces the red gemstone *ruby*. With the addition of small amounts of iron and titanium, the result is the blue gemstone *sapphire*.

CHECK POINT

1. When pieces of calcite and fluorite are scraped together, which scratches which?
2. The mineral muscovite displays very distinct cleavage, yet the mineral quartz fractures. How does this relate to each mineral's crystal structure?

Were these your answers?

1. Looking at Table 20.1, we see that fluorite is harder than calcite. So fluorite scratches calcite.
2. Muscovite forms as a layered sheetlike structure. The bonds between the different layers are weaker than the bonds within the individual layers. Micaceous minerals cleave between the layers. Quartz has a more complicated structure with no layering and no planes of weakness. Therefore, quartz fractures.

COLOR

Although color is an obvious feature of a mineral, it is not a reliable means of identification. Some minerals—copper and turquoise are two examples—have a distinctive color. But most minerals either occur in a variety of colors or can be colorless.

Chemical impurities in a mineral affect color. For example, the common mineral quartz, SiO_2, can be found in many colors, depending on slight impurities. It can be clear and colorless if it has no impurities, or it can be milky white from tiny fluid inclusions. Rose-colored quartz results from small amounts of titanium; purple quartz (amethyst) results from small amounts of iron. The color of the mineral corundum, Al_2O_3, is commonly white or grayish. But impurities in corundum give us rubies and sapphires (Figure 20.10).

DENSITY

Density is a property of all matter, minerals included. In practical terms, the density of a mineral tells us how heavy a mineral feels for its size. More specifically, a mineral's density is the ratio of its mass to its volume. The densities of some minerals are shown in Table 20.2.

Gold's particularly high density of 19.3 g/cm³ is nicely taken advantage of by miners panning for gold. Fine gold pieces hidden in a mixture of mud and sand settle to the bottom of the pan when the

TABLE 20.2 DENSITY OF VARIOUS MINERALS (g/cm³)

Borax	1.7	Pyrite	5.0
Quartz	2.65	Hematite	5.26
Talc	2.8	Copper	8.9
Mica	3.0	Silver	10.5
Chromite	4.6	Gold	19.3

mixture is swirled in water. Water and less-dense materials spill out when the mixture is swirled. After a succession of douses and swirls, only the substance with the highest density remains in the pan—gold!

The physical properties of a mineral all relate back to the mineral's chemistry.

■ 20.4 Classification of Rock-Forming Minerals

More than 4000 known minerals exist on Earth, and new ones are discovered every year. With so many minerals, how can they all be classified in a simple, systematic fashion? First of all, most minerals are actually quite rare. In fact, only about a dozen minerals make up most of the rocks exposed at Earth's surface. These are the *rock-forming minerals*.

We classify minerals based on their chemical composition. Doing so produces the two main categories: the **silicates** and the **nonsilicates** (Figure 20.11). Look back at Figure 20.4 and you will understand how Earth scientists arrived at this simple division. Oxygen is the most abundant element in Earth's crust, and silicon is the second most abundant. Minerals that contain both silicon (Si) and

fyi

■ All silicate minerals have the same fundamental structure of atoms, the silicon-oxygen tetrahedron—four oxygen atoms joined to one silicon atom $(SiO_4)^{4-}$. The powerful bond that unites the oxygen and silicon ions is akin to the cement that holds Earth's crust together.

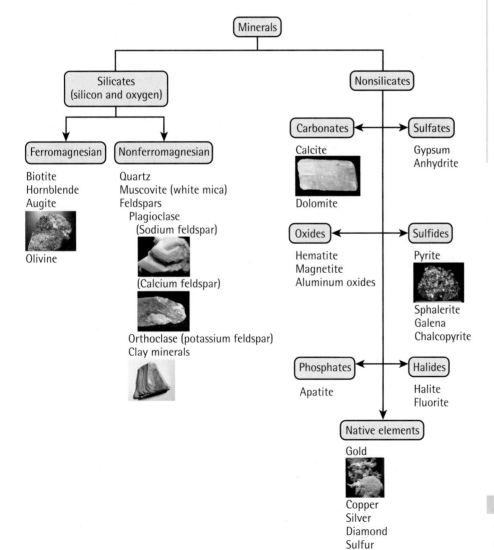

FIGURE 20.11

Classification of common rock-forming minerals.

A familiar example of crystallization is ice crystals that form in water when the temperature drops below 0°C. So just as there is water and ice, there is magma and rock.

Link To Section 12.4

oxygen (O) as part of their chemical composition are called silicates. Minerals that do not contain these two elements are called nonsilicates.

Silicon has a great affinity for oxygen. In fact, silicon has such a strong tendency to bond with oxygen that silicon is never found in nature as a pure element; it is *always* chemically combined with oxygen. For these reasons, the silicates are the most common mineral group, making up more than 90% of Earth's crust. Most silicates also contain the rest of the eight most common elements, which include Fe, Mg, Ca, and Al, but the basic building block of *all* silicates is Si and O.

The silicates are subdivided into two groups—*ferromagnesian silicates* and *nonferromagnesian silicates*. As the name indicates, ferromagnesian silicates contain iron (Fe), magnesium (Mg), or both, plus the standard oxygen and silicon atoms. Olivine, $(Mg,Fe)_2SiO_4$, is a ferromagnesian silicate. Because of the presence of iron and magnesium, ferromagnesian silicates tend to be dense and dark in color. Nonferromagnesian silicates do not contain significant amounts of iron or magnesium; therefore, they generally have relatively low densities and are light in color. The most abundant mineral in the crust is feldspar, a nonferromagnesian silicate that contains aluminum, sodium, potassium, and/or calcium, plus silicon and oxygen. Feldspar makes up more than 50% of the crust. Quartz (SiO_2), the second most common mineral in Earth's crust, is composed of only silicon and oxygen. If you have ever collected rocks and minerals, you probably have some quartz and feldspar specimens in your collection.

Nonsilicate minerals make up just 8% of Earth's crust by mass. The nonsilicates include carbonates, oxides, sulfides, native elements such as gold and silver, and a few others. The carbonates are the most abundant nonsilicate minerals. Two common carbonate minerals are calcite and dolomite. Calcite consists of the chemical compound calcium carbonate, $CaCO_3$. Dolomite is a mixture of calcium carbonate and magnesium carbonate, $CaMg(CO_3)_2$. Calcite and dolomite are the main minerals found in the group of rocks called *limestone*.

For oxide minerals, oxygen is combined with one or more metals, such as iron, chromium, manganese, tin, or uranium. The oxide mineral group is important economically because it contains many useful *ores*. An ore is a mineral deposit rich in valuable metals that can be extracted for a profit. Sulfide minerals are also important as ore deposits. The most common sulfide mineral is pyrite (fool's gold), FeS_2.

■ 20.5 The Formation of Minerals and Rock

So far we have explored the definition of a mineral and the different properties and classification of minerals. Now we turn our attention to how minerals form. Understanding how minerals form is a stepping-stone to understanding how rocks form. Rocks, after all, are made of minerals.

Minerals are formed by **crystallization**—the growth of a crystalline solid from a liquid or gas. Crystallization starts when atoms begin to bond with each other in a particular geometric pattern. As the numbers of atoms and bonds increase, a single crystal forms, with edges reflecting the shape of the underlying geometric pattern. As more and more atoms bond to the microscopic crystal, repeating the underlying pattern, the crystal grows.

Minerals commonly crystallize from two different sources: from **magma**—molten rock—and from water solutions. As we shall see, *igneous* rocks are formed from magma, and certain *sedimentary* rocks are formed from water solutions.

CRYSTALLIZATION IN MAGMA

Magma is composed primarily of the elements found in the silicate group of minerals—namely, silicon and oxygen, plus aluminum, potassium, sodium, calcium, iron, and magnesium. When magma starts to cool, atoms in the hot liquid lose kinetic energy. Attractive forces then pull the atoms into orderly crystalline structures. Minerals crystallize from cooling magma in a systematic fashion, based on their respective melting points.

The crystallization points of silicate minerals strongly depend on the amount of silica they contain. As magma cools, the first minerals to crystallize have the highest (hottest) melting points and the lowest amount of *silica* (Si combined with O); the last minerals to crystallize have the lowest (coldest) melting points but the highest silica content (Figure 20.12). Conversely, high-silica minerals melt at lower temperatures than do low-silica minerals. Consider the minerals quartz and feldspar. When a rock with both quartz and feldspar melts, the quartz melts before the feldspar because quartz contains more silica than feldspar (in fact, quartz is pure silica!). In a cooling magma, on the other hand, quartz crystallizes after, and at a lower temperature than, feldspar.

Magma consists of liquid—the molten portion—and newly formed crystals. In general, as minerals crystallize in a cooling magma, many settle out from the liquid portion of the magma. Also, some of the remaining liquid may migrate from the region where crystallization first took place. The result is that the molten liquid becomes separated from the newly formed crystals, which allows the composition of the remaining liquid to change continuously as crystallization proceeds.

The liquid composition changes because certain constituents are removed from the liquid and incorporated into the first-formed minerals. Constituents that did not become part of the new minerals are left behind in the liquid and are eventually incorporated into later-forming minerals. This happens because most minerals of different compositions do not crystallize simultaneously or randomly, but rather in a temperature-dependent sequence as described above. Later-forming minerals contain more silica and different suites of elements—in varying proportions—than early-formed minerals. The remaining liquid thus becomes depleted in the constituents of minerals that have already crystallized and enriched in the constituents of minerals that have yet to crystallize.

To simplify understanding the process of crystallization in a cooling magma, consider the following analogy. Suppose you have all the pieces for a game of checkers—12 red pieces and 12 black ones. All 24 pieces are mixed together to form a single group, which represents the magma before any minerals have formed. So the "liquid" consists of 50% red pieces and 50% black pieces. Now suppose 3 red and 2 black pieces are removed to form the first "mineral." Now 9 red and 10 black pieces remain in the "liquid," which is now represented by only 19 checkers. The "liquid" is now roughly 47% red and 53% black. Thus, the "liquid" has become depleted in red and enriched in black—its composition has changed. If the black checkers represent silica molecules, can you now see that the crystallization process enriches magma in silica? This crystallization process allows a single magma to generate a range of igneous minerals and rocks.

Silicon Content	Melting Temperatures	Order of Melting	Order of Crystallization
Low	High	Last	First
High	Low	First	Last

FIGURE 20.12

High-silica-content minerals have low melting points. They are the first minerals to melt and the last minerals to crystallize. Low-silica-content minerals have high melting points. They are the last minerals to melt and the first minerals to crystallize.

Just as ice melts at the same temperature at which water freezes, a mineral's melting point is the same temperature at which it begins to crystallize from hot magma.

1. Olivine and pyroxene are both ferromagnesian silicate minerals that crystallize from cooling magma. In the sequence of crystallization, olivine crystallizes first, pyroxene second. Which mineral contains more silica?
2. Having a high melting point can be thought of as requiring a higher (hotter) temperature to melt. A low melting point means that the temperature does not need to be as high for the mineral to melt. How does having a high melting point translate to crystallizing first?

Were these your answers?

1. The mineral pyroxene contains more silica than olivine. As magma cools, minerals with less silica crystallize before minerals with more silica.
2. When the temperature is high, the constituents of possible minerals are in the liquid state. Remember that water freezes at the same temperature at which ice melts. So think of the melting point as being equivalent to the freezing—that is, crystallization—point. Ice forms when the water temperature drops below the freezing/melting point. When magma begins to cool, minerals with high melting temperatures crystallize first, because the magma temperature drops below their respective freezing/melting points. The minerals with lower melting points do not form yet—they stay in the liquid state. They cannot begin to crystallize until lower temperatures are reached. So minerals with low melting points crystallize last. When all is said and done, the liquid magma freezes to become rock.

FIGURE 20.13

Calcium carbonate precipitating from dripping water in a cave forms icicle-shaped stalactites hanging down from the ceiling and cone-shaped stalagmites protruding upward from the ground.

CRYSTALLIZATION IN WATER SOLUTIONS

Minerals crystallize from water solutions in two main settings. The first is associated with the final stages of magma crystallization—magma generally contains from 1% to 6% water. When a body of magma is nearly solidified, this very hot water circulates through fractures in the new rock and often into the surrounding rock—*hydrothermal activity*. These water solutions usually contain many dissolved mineral constituents. The solutions become chemically saturated as the temperature decreases, causing various minerals to *precipitate*. These minerals are often deposited in cracks, and sometimes within the rock matrix itself. Many of the important ore deposits we find today were formed in this manner.*

Similar to hydrothermal minerals, **chemical sediments** are also formed by the precipitation of mineral constituents from water solutions. But in this second setting, chemical sediments form where temperatures are much cooler than Earth's interior, such as in a body of water on the surface of Earth. Chemical sediments fall into two categories: carbonates and evaporites.

Carbonates are minerals and rocks composed mostly of calcium carbonate, $CaCO_3$, which has the mineral name *calcite*. Dolomite, $CaMg(CO_3)_2$, is also a common carbonate mineral. Carbonates can form in two ways—by inorganic precipitation, and, as we will see later in this chapter, as a result of biologic activity. Many seashells are composed of calcium carbonate secreted by organisms. Cave dripstones, such as stalactites and stalagmites, provide a great example of calcium carbonate precipitating inorganically from dripping water (Figure 20.13). They form because groundwater (Chapter 23) picks up calcium and carbonate ions as it moves through limestone formations in a cave. The dripping water is saturated with dissolved calcite, so the dripstones form as evaporation removes very small amounts of water during the dripping process.

Evaporites are minerals and rocks precipitated when a restricted body of seawater, or the water of a salty lake, evaporates. Examples are gypsum, anhydrite,

---

* Hydrothermal activity is rich in rare metals. Such metals are chemically happier "hiding out" with the water until they are so concentrated that they can form their own minerals—such as gold ore.

and halite. These names apply both to individual minerals and to rocks made of a single type of evaporite mineral. Evaporites precipitate out of water solutions in a way that is very similar to the crystallization of minerals from magma. The difference, though, is that **solubility** rather than melting point determines which minerals crystallize first. As evaporation proceeds, the minerals with the lowest solubility—the most difficult to dissolve—such as gypsum precipitate first, followed by the minerals that dissolve more easily (those with higher solubilities), such as anhydrite and then halite. Although carbonates make up the bulk of chemical sediments, evaporites are a small but important group.

CHECK POINT

1. When water evaporates from a body of water, what type of mineral is formed?
2. A mineral with low solubility does not dissolve easily. A mineral with high solubility dissolves easily. How is this factor of "dissolvability" related to crystallization?

Were these your answers?

1. Evaporite minerals precipitate out of solution as water evaporates.
2. Minerals that dissolve easily stay dissolved longer in a solution than those that do not dissolve so easily. So as a restricted body of water dries up, the first minerals to crystallize are those that *do not* dissolve so easily—those with lower solubilities. Minerals that are easily dissolved (high solubility) remain in solution longer—they are the last to crystallize.

Now that we have seen the different ways that minerals form, we can begin to learn about the combinations of minerals called rock. We know that minerals formed from the crystallization of magma make up igneous rock, and minerals formed from the precipitation or evaporation of water make up some types of sedimentary rock. Now we will also see that igneous rock breaks down to form sedimentary rock, and that a third rock type—metamorphic rock—forms from rocks that already exist. Minerals, in their many forms, are the building blocks of the many different rocks on Earth.

20.6 Rock Types

A **rock** is defined as an aggregate of minerals. Some rocks are aggregates of fossil shell fragments, solid organic matter, or any combination of two or three of these components. Just as we think of minerals as chemical mixtures or compounds, rocks can be thought of as physical mixtures. In some rocks, the grains are "cemented" together; in others, the grains are tightly interlocked. In many rocks you can see mineral crystals. Granite, one of the most common rocks in Earth's continental crust, contains visible crystals of the minerals feldspar, quartz, hornblende, and others (Figure 20.14). On the other hand, in rocks such as basalt, shale, or slate, individual grains are difficult to distinguish—they are too small to be seen with the unaided eye.

Rocks are divided into the following three categories (Figure 20.15), based on how they were formed:

Igneous rocks are formed by the cooling and crystallization of hot, molten rock. The word *igneous* means "formed by fire." *Plutonic* igneous rocks are formed when molten rock below Earth's surface—magma—cools. Granite is a common plutonic rock. *Volcanic* igneous rocks are formed when molten rock at Earth's surface—lava—cools. Basalt is a common volcanic rock.

Link To Section 16.4

FIGURE 20.14

A rock is an aggregate of one or more minerals, and sometimes shell fragments and/or solid organic matter. This granite is an aggregate of minerals—the minerals feldspar, quartz, and hornblende.

Feldspar
(Mineral)
+

Quartz
(Mineral)
+

Hornblende
(Mineral)

Sedimentary rocks are formed at or near Earth's surface from the cementation or compaction of *sediment*—rock, mineral, shell, or solid organic fragments carried by water, wind, or ice and deposited in low-lying areas. Sedimentary rocks also form when minerals precipitate out of water solutions at or near Earth's surface. Sandstone, shale, and limestone are common sedimentary rocks.

Metamorphic rocks are formed from older, preexisting rocks (igneous, sedimentary, or metamorphic) that were transformed in Earth's interior by high temperature, high pressure, or both—without melting. The word *metamorphic* means "changed in form." For example, marble is metamorphosed limestone, and slate is metamorphosed shale.

▪ 20.7 Igneous Rocks

Link To Section 13.1

Earth's crust consists primarily of various rocks of igneous origin. On the continents, the most common igneous rocks are granite and andesite. On the ocean floor, basalt is the most common kind of rock. All igneous rock originated as magma.

GENERATION OF MAGMA

We've learned that many minerals form from cooling magma. But where does magma come from? Do we have magma because Earth's interior is molten? The answer is no. Earth's interior is mostly solid, not molten. And the one layer of Earth that is liquid is composed of molten iron, not magma, and is not encountered until a depth of almost 3,000 km.

(a)   Basalt   Granite

(b)   Sandstone   Limestone

(c)   Marble   Slate

FIGURE 20.15

The three main types of rock. (a) Basalt and granite are igneous rocks. (b) Sandstone and limestone are sedimentary rocks. (c) Marble and slate are metamorphic rocks.

Simply stated, magma is derived from rocks that have melted. Just as water cools to form ice, magma cools and solidifies to form the minerals that eventually become rock. But how does rock melt to become magma? Temperatures recorded in mines and drill holes indicate that Earth's temperature increases through most of the continental crust at an average of 30°C for each kilometer of depth (Figure 20.16). But increased temperature is not enough to cause rocks to melt, even though the temperature at sufficient depth is actually much hotter than that of magma.

If the temperature at depth is hotter than magma, why are the rocks at depth solid? The answer can be found by using water as an analogy. Recall (from Section 7.7) that a phase change occurs when water is heated to the boiling point. We know that water boils at 100°C at sea level. If we increase the pressure, the temperature needs to be higher for boiling to occur. So phase changes depend on pressure as well as temperature. In the case of hot rocks at depth, they are under enormous pressure from the weight of the rock above—enough pressure to prevent melting, even at hotter-than-magma temperatures.

But rocks sometimes melt to form magma, and three reasons explain why. The dominant reason, in terms of the amount of magma produced, is that hot rock rises upward from depth to levels where pressure is reduced enough to induce melting.*

Another mechanism for generating magma is the addition of water to rock, which lowers the rock's melting point. To understand fluid-induced melting, we

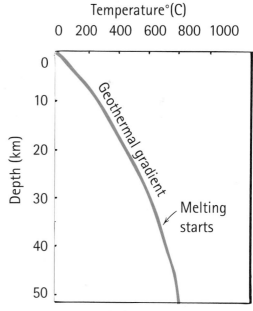

Temperature°(C)

FIGURE 20.16

The temperature inside Earth increases about 30°C for each kilometer of depth from the surface deep into the continental crust (the gradient is much lower in the deep crust and mantle). This increase of temperature with depth is known as the *geothermal gradient*.

---

* The layer below Earth's crust, the mantle, behaves like an elastic solid. Although movement is extremely slow, rock actually flows in convection currents that allow hot materials to rise and cool materials to sink.

can use the behavior of water as an analogy once again. A "foreign" substance in water lowers its freezing point (it must be colder to freeze). By the same token, the foreign substance also lowers ice's melting point. For example, salt placed on 0°C ice causes the ice to melt even though the temperature has not changed. The salt lowers the freezing point of water, so 0°C is no longer cold enough to keep the ice frozen. In the case of rock, water is the foreign substance that lowers the melting point of the rock—enough to cause melting.

These two magma-generating mechanisms produce magma that rises upward through preexisting rock. And this gives us the third mechanism—namely, that rock can melt when its temperature rises, often because of the presence of hotter materials that have risen from deeper areas. We will explore the first two mechanisms in more detail in Chapter 22, where we can link them to the plate tectonic model. For now, we will restrict our discussion to the general formation of magma from melting rock.

Link To Section 14.2

Keep in mind that rock is an aggregate of solids, so the melting of rock into magma occurs over a broad temperature range. As rock is heated, the first minerals to melt are those with the lowest melting points. This is similar to the process of crystallization from magma, but the changes of phase occur in reverse order. If all the minerals in a rock could melt simultaneously, the composition of the resulting magma would be the same as the composition of the original rock. But melting does not occur this way—**partial melting** is the rule.

Magma produced by the partial melting of rock is made only from the constituents of those minerals that have melted—the ones with the lowest melting points. Thus, the magma produced has a different composition from the partially melted rock that produced it. In this way, magmas with a variety of compositions are produced. And because magma cools to form igneous rock, there are many different types of igneous rocks. Another consequence of partial melting is that newly generated magma contains more silica than its parent rock, because high-silica minerals are the first to melt. Lower-silica minerals are left unmelted.

A great way to visualize partial melting is to consider a bowl that contains equal quantities of water, butter (in chunks), and cheese (also in chunks). First, we freeze the mixture. The frozen mixture is now a "rock" whose composition is 33% water. Intuitively, we know that the melting points of ice, butter, and cheese are different. When we add heat to our frozen mixture, the first substance to melt is the water. The substances that are harder to melt—the butter and cheese—remain unmelted in the "rock." We now have a liquid that is 100% water, so the percentage of water in the liquid is higher than the percentage of water that was in the "rock."

Link To Section 7.9

THREE TYPES OF MAGMA, THREE MAJOR IGNEOUS ROCKS

There are three major types of magma—*basaltic, andesitic,* and *granitic.* In general, the different magmas occur in different geologic settings. These three magma types will be tied to the plate tectonic model in Chapter 22.

The largest region of Earth's interior is the solid mantle, which is composed of low-silica-content igneous rocks. When these rocks undergo partial melting, basaltic magma is formed. Basaltic magma is still relatively low in silica—it is about 50% silica. When solidified at Earth's surface, basaltic lava forms the dark igneous rock known as *basalt,* which is the kind of rock that makes up the Hawaiian Islands and the oceanic crust. Andesitic magma is about 60% silica. The rock known as *andesite,* which is produced from andesitic lava, gets its name from the Andes Mountains in South America, where it is very common. Granitic magma, at about 70% silica, cooling slowly at depth forms *granite* and other similar granitic rocks. Of all the igneous rocks in the crust, oceanic and continental crust combined, approximately 80% was formed from basaltic magma, 10% from andesitic magma, and 10% from granitic magma.

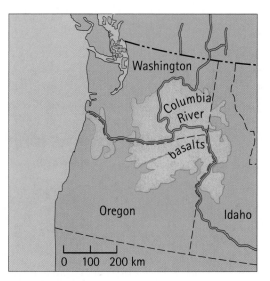

FIGURE 20.17

The flood basalts that produced the Columbia Plateau covered more than 200,000 km^2 of the preexisting land surface.

(a)

(b)

(c)

CHECK POINT

1. If 80% of all igneous rocks are formed from basaltic magma, why do we see so much granite?
2. If basaltic magma is produced from the partial melting of mantle rock, how is most granitic magma produced?

Were these your answers?

1. Basalt is the most common igneous rock on the ocean floor. Look at a globe to see that the oceans cover about 71% of Earth's surface. We see so much granite because it is the most common igneous rock on continental land.
2. When magma rises toward Earth's surface, it makes contact with the surrounding rock. Rock in the continental crust contains more silica than rock in the mantle. These silica-rich rocks are partially melted by the rising magma and incorporated into the melt, which increases the magma's silica content. Additionally, the process of crystallization enriches magma in silica. As magma crystallizes and the liquid becomes separated from the crystals, the liquid portion of the magma changes, becoming more silica enriched. Eventually granitic magma (70% silica) is produced, and when this cools, granitic-type rocks form.

IGNEOUS ROCKS AT EARTH'S SURFACE

Igneous rocks are divided into two general categories based on where they formed. Igneous rocks formed by the eruption of molten rock at Earth's surface are called **volcanic** (or *extrusive*) **rocks**. Molten magma that moves upward from inside Earth and flows onto the surface is called **lava**. The term *lava* refers both to the molten rock itself and to the solid rock that forms from it.

Lava may be extruded through fractures and fissures (long, planar cracks) in Earth's surface or through a central vent—a volcano. Eruptions from a volcano are more familiar to us because they are very dramatic to see, but the outpourings of lava from fissures are much more common. Most fissure eruptions occur underwater as basaltic lava erupts where the ocean floor is spreading apart. Fissure eruptions also occur on land. Episodic and unusually voluminous lava outpourings known as *flood basalts* have flooded large areas of Earth throughout its history, creating extensive lava plains. The Columbia Plateau in the Pacific Northwest is the result of extensive flood basalts (Figure 20.17), as is the Deccan Plateau in India.

Volcanoes are vents where magma rises to Earth's surface and erupts as lava. Three main types of volcanoes exist—shield, cinder cone, and composite (Figure 20.18).

FIGURE 20.18

The three types of volcanoes. (a) Shield volcanoes, such as Mauna Loa, have broad, gentle slopes that average between 1° and 10° (from the horizontal). (b) Cinder cones, such as Sunset Crater in Arizona, generally have smooth steep slopes of 25° to 40° and bowl-shaped summit craters. (c) Composite cones, such as picturesque Mt. Fuji, are also very steep. On average, the slope of a composite cone starts out at 30° at the summit and gradually flattens to 10° at the base.

FIGURE 20.19

(a) Mauna Loa, a shield volcano on the island of Hawaii,
is the largest volcano on Earth.
(b) When compared with
other large volcanoes, its
immense size and volume is
dramatic.

fyi

■ Volcanic rocks (such as basalt)
form at Earth's surface where
they cool quickly; they tend to
have microscopic crystals. Plutonic rocks (such as granite) form
below Earth's surface where they
cool slowly; they tend to have
much larger crystals easily seen
without a microscope.

Shield volcanoes are built by a steady supply of easily flowing basaltic lava that flows out in all directions to make a broad, gently sloping cone. Mauna Loa in Hawaii, the largest volcano on Earth, is a shield volcano standing 4145 m above sea level and more than 9750 m above the deep ocean floor (Figure 20.19).

Cinder cone volcanoes are very steep but small in comparison to shield volca-noes. Cinder cones are not restricted to a particular type of lava. They are formed from the piling up of ash, cinders, and rocks that have been explosively erupted from a single vent to form a symmetrical, steep-sided cone. Two well-known examples of cinder cones are Sunset Crater in Arizona and Parícutin in Mexico.

A *composite cone* is a volcano built up of alternating layers of lava, ash, and mud. Composite cones have a steep-sided summit and gently sloping lower flanks. Mt. St. Helens in Washington is an example of an active composite cone. In 1980, it erupted violently, disrupting the lives of thousands of people and transforming more than 200 mi^2 of lush forest into a burned, gray landscape.

Composite volcanoes tend to erupt explosively because their magmas and lavas usually do not flow easily. The viscous magma traps volcanic gases, which increases the pressure inside the volcano. We can compare the gases in a magma to gases in a bottle of carbonated soda. If we cover the top of the bottle and shake vigorously, the gases separate from the soda and form bubbles. When we remove the cover, pressure is released and gases and liquid explode from the bot-tle. The gases in magma behave in much the same way. In a volcanic blast, the pressure and temperature increase, and the whole mass of viscous magma and overlying rock explodes into dust and rubble. When combined with abundant volcanic ash, this mixture can expand and destroy everything in its path. Exam-ples of this kind of volcanic activity occurred at Mt. Vesuvius in AD 79, at Mt. Pelee in 1902, and at Mt. St. Helens in 1980.

Three principal types of volcanic rocks exist—basalt, andesite, and *rhyolite*. Rhyolite forms when granitic lava erupts at Earth's surface. Rhyolite comes in several different forms (Figure 20.20). It can be a typical fine-grained volcanic rock, or the lava can cool quickly to form volcanic glass—pumice if the rock is riddled with tiny holes from former gas bubbles, obsidian if it lacks such holes, or volcanic ash if it is erupted explosively.

(a)   (b)   (c)

IGNEOUS ROCKS BENEATH EARTH'S SURFACE

When magma cools beneath Earth's surface, the igneous rock that forms is called **plutonic** (or *intrusive*) **rock**. The word *plutonic* is derived from Pluto, the mythological god of the underworld. All intrusive igneous rock bodies are called *plutons*. Being intrusive, plutonic rocks can be studied only after they are exposed by uplift and erosion at Earth's surface. The most common plutonic rock is granite.

FIGURE 20.20

The volcanic rock rhyolite comes in several forms. (a) Fine-grained rhyolite. (b) Pumice. (c) Obsidian.

FIGURE 20.21

Intrusive igneous features in cross-sectional view.

Plutons occur in a great variety of shapes and sizes, ranging from small pipelike *dikes* to large, expansive *batholiths* (Figure 20.21). Batholiths, the largest plutons, are defined as having more than 100 km^2 of surface exposure. Batholiths are created by numerous intrusive events over millions of years—they form the cores of many major mountain systems around the world. Generally speaking, they are the crystallized magma chambers that fed long-since-eroded volcanoes. Many modern mountains are actually the exposed batholith cores of larger mountains that have long since eroded away. Some of the largest batholiths in North America include the Coast Range batholith and the Sierra Nevada batholith (Figure 20.22). It is interesting to note that the Sierra Nevada is gaining height with time—long after all the magma has crystallized—because its rate of uplift is greater than its rate of erosion.

■ 20.8 Sedimentary Rocks

Sedimentary rocks are the most common rocks in the uppermost part of the crust. They cover two-thirds of Earth's surface, forming a thin, extensive blanket over older igneous and metamorphic rocks. Because sedimentary rocks contain the remains of organisms and older rocks, they provide information about geological events that have occurred over time at Earth's surface.

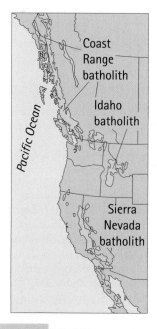

FIGURE 20.22

Some of the largest batholiths in North America include the Coast Range batholith and the Sierra Nevada batholith.

FIGURE 20.23

The rocks on this mountain peak in Wales have been split apart by mechanical weathering. In this cold environment, water in the cracks of the rocks has frozen and expanded, thereby causing the rocks to crack and break apart.

- When water freezes, the water molecules arrange themselves into six-sided crystalline structures that have much open space. So water expands upon freezing. Out in nature, the freezing process mechanically breaks rock apart. Ice wedging occurs as water seeps into small cracks, freezes, and expands, thereby widening the cracks. Eventually, with repeated freezing, cracks can widen to split the rock!

Forces of ice on rock →   ← Ice

THE FORMATION OF SEDIMENTARY ROCK

Sedimentary rock forms in a long process with four stages: *weathering, erosion, deposition,* and *sedimentation.* **Weathering** is the disintegration or decomposition of rock—in place—at or near Earth's surface. Agents such as water, wind, ice, and reactive chemicals weather the rock—breaking it into smaller pieces, cracking its surface, rounding and smoothing its edges and corners, and sometimes transforming its chemical composition. Two types of weathering occur— *mechanical* and *chemical.* Both produce sediment. *Mechanical weathering* physically breaks rocks into smaller and smaller pieces (Figure 20.23). For example, the freezing and thawing of ice can widen preexisting cracks in rock. In *chemical weathering,* reactions with water decompose rock, analogous to the way bacteria decomposes organic matter but with different chemical reactions (Figure 20.24). Because liquid water and water vapor are everywhere (excepting liquid water in polar regions), chemical weathering produces more sediment than mechanical weathering.

FIGURE 20.24

Granite rocks in the Mojave Desert of California have been worn into rounded shapes by chemical weathering. Rainwater chemically decays the rock's outer layers, making the rock easier to erode. Rainwater erodes the rock by physically washing away its weakened outer layers, leaving rounded boulders behind. The rounding occurs because chemical weathering is faster where surface area is greater, and corners and edges have the most surface area.

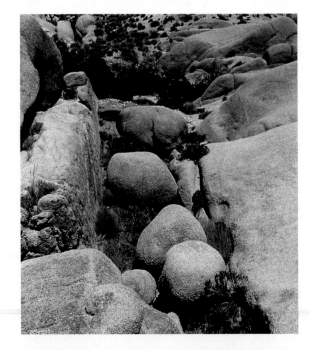

As rock weathers, it also erodes. **Erosion** is the process by which weathered rock particles are removed and transported away by water, wind, or ice. The principal difference between weathering and erosion is that erosion does not occur in place—it involves movement. Running water is the dominant agent of erosion. Pieces of mechanically weathered rock are normally quite angular and jagged when they are first produced. During transportation, especially by water, the various particles collide with one another and break. This decreases their size and rounds off their sharp edges. When transportation stops, deposition and sedimentation begin.

Deposition is the stage in which eroded particles come to rest. Sediments are deposited in horizontal layers, with each successive layer younger than the one beneath it. The larger a sediment particle, the stronger a current must be to carry it. As flowing water slows down, larger particles are the first to be deposited, while smaller particles remain with the flow. In this way, sediments are sorted according to size as they are deposited (Figure 20.25).

In the process of **sedimentation**, sediments are deposited horizontally one layer at a time. As the deposited sediment accumulates, it begins to change into sedimentary rock. We say that the sediments *lithify*—they undergo *lithification*, a term that means "conversion into rock" (*lith* = rock). Lithification occurs through the processes of *compaction* and, usually, *cementation*. Compaction is the first step. As the weight of overlying sediments presses down on deeper layers, sediment particles are squeezed and compacted together. This compaction squeezes much of the water out of the pores between the sediment particles.

The remaining "pore water" often contains dissolved compounds, such as silica, calcium carbonate, and iron oxide. These compounds can precipitate from solution and partially fill the pore spaces with mineral matter. The mineral matter glues the particles together and acts as a cementing agent. This is the process of cementation. Silica cement, the most durable, produces some of the hardest and most resistant sedimentary rocks. When iron oxide acts as a cementing agent, it produces the red or orange stain often seen in sedimentary rocks. The rock colors of Bryce Canyon National Park in Utah provide a beautiful example of iron oxide stain (Figure 20.26).

(a) Well-sorted sediments

(b) Poorly sorted sediments

FIGURE 20.25

A deposit that contains particles of similar shapes and sizes is called *well sorted*. A *poorly sorted* deposit contains particles of many different shapes and sizes. In general, poorly sorted sediments traveled a short distance before being deposited, and well-sorted, well-rounded sediments traveled a long distance before being deposited.

fyi

- Sediment shape and size give us clues to the method of sediment transport. Glacial deposits are poorly sorted and very angular because they are trapped in ice during transport. Wind-blown deposits tend to be very well sorted and small. Wind can generally only move small particles.

FIGURE 20.26

The red and orange colors in the sedimentary rocks at Bryce Canyon in Utah are caused by the presence of iron oxide.

fyi

- Living things, especially micro-organisms, are unimaginably abundant in rivers, lakes, and oceans. They "drink" the water and use the minerals dissolved in the water to make their shells, skeletons, and cell walls. A great deal of calcite was formed in this way. When these creatures die, they settle to the bottom and form layers of chemical sediment.

(a)

(b)

(c)

FIGURE 20.27

Clastic sedimentary rocks. (a) Shale, the most abundant sedimentary rock, which is composed of very fine mud-sized particles; (b) sandstone, which is composed of sand-sized particles; and (c) conglomerate, which is made up of a poorly sorted variety of rounded particles, mostly gravel sized, but also with sands and muds.

TABLE 20.3 CLASSIFICATION OF CLASTIC SEDIMENTARY ROCKS

Sediment		Particle Size	Rock
COARSE			
Gravel	Boulder	256 mm	Conglomerate
	Cobble	64 mm	
	Pebble	2 mm	
Sand			Sandstone
		0.062 mm	
Mud	Silt		Siltstone
		0.0039 mm	
	Clay		Mudstone Shale
FINE			

CHECK POINT

Which type of weathering produces the most sediment: mechanical or chemical?

Was this your answer?

Look to the agents of weathering to guide your thinking. Water is the primary agent of weathering. Liquid water and water vapor interact chemically with rock to break it down. Less commonly, freezing and thawing physically assault and disintegrate rock. In this way, water performs a double duty. Because liquid water and water vapor are present nearly everywhere, chemical weathering is the main producer of sediment.

CLASSIFYING SEDIMENTARY ROCKS

Sedimentary rocks are divided into two groups, based on the kinds of sediments they contain: *clastic* and *chemical*. Clastic sedimentary rocks are composed of small pieces of weathered rock, shell debris, and/or solid organic matter. Clastic rocks are classified by particle size (Table 20.3). The classification scheme is meaningful because it allows us to visualize the depositional environment in which the rock was formed. For example, larger particle sizes are deposited in energetic environments, such as a raging river, and small particle sizes settle out of the water in tranquil environments, such as the middle of a lake or the center of the ocean. Going from very small grain size to larger grain size, the three most abundant clastic sedimentary rocks are *shale* (which for our purposes includes mudstones), *sandstone*, and *conglomerate* (Figure 20.27).

Shale (or *mudstone*), formed by the compaction of superfine silt and clay-sized particles, is finely layered and has the ability to split into thin flakes parallel to the depositional layers. The extremely fine particle size indicates deposition in a low-energy environment characterized by quiet waters, such as deep ocean basins, flood plains, deltas, lakes, or lagoons. The color of shale provides clues to the environment of formation. Gray to black shale indicates the presence of organic matter, which can be preserved only in environments with little or no oxygen, such as swampy areas. If sufficient oxygen had been present in the depositional environment, bacteria would have decomposed the organic matter very quickly. Black shale is commercially important, because it is the main source rock from which crude oil formed. Red to brown shale contains ferric oxide (red) or ferric hydroxide (brown). These oxides indicate an oxygenated depositional environment, where deep water was well mixed with water from near the surface. Green shale simply does not contain organic matter, ferric oxide, or ferric hydroxide.

Sandstones are classified into three types, based on their mineral makeup. When quartz is the primary mineral, the rock is simply called *quartz sandstone*. Quartz sandstone is composed of well-sorted, well-rounded quartz particles. Sandstone that contains considerable amounts of the mineral feldspar is called *arkose*. The particles in arkose tend to be angular and not as well sorted as those in quartz sandstone. Sandstone made of a mixture of minerals and angular rock fragments is called *graywacke* (pronounced "gray-wack-ee"). Sandstones form in a variety of environments, including dunes, beaches, marine sand bars, river channels, canyons, and underwater canyons—all places where moderate-energy waters deposit similarly sized grains.

Conglomerate is composed of gravel-sized and smaller rounded rock and mineral fragments. Large rock fragments must have been transported by water currents strong enough to carry them, which indicates dynamic, high-energy environments, such as rapids and fast-moving streams. Because these strong currents round out the rock fragments, the roundness of their edges and corners is a good indication of the distances they have traveled. Conglomerates are often found in river channels and along rapidly eroding coastlines.

CHEMICAL SEDIMENTARY ROCKS

Recall that chemical sediments, and in turn chemical rocks, form by the precipitation of minerals from water solutions. Some carbonates, such as travertine, are formed by the inorganic precipitation of calcium carbonate (Figure 20.28). Evaporites, such as halite, are formed by chemical precipitation caused by the evaporation of salty waters. In general, chemical sedimentary rocks form where there are no clastic sediments.

Precipitation also occurs indirectly when water-dwelling organisms take up dissolved constituents and use them to build shells or hard body parts or to perform other life processes in which solids are created and then discharged into the surrounding water. Chemical sediments created by organisms are called *biochemical sediments.*

Limestone, made up of the mineral calcite ($CaCO_3$), is the most abundant carbonate rock. About 90% of all limestone rocks form as a result of biologic activity. Many marine organisms make their shells out of calcite. When such organisms die, their shells accumulate on the sea floor. The shells begin to dissolve, forming a noncrystalline *ooze* of calcium carbonate. This ooze eventually recrystallizes into calcite, which then forms limestone. Because of compaction and how easily calcium carbonate dissolves, the original textures and structures of the seashells are often obliterated. Sometimes, however, the shells and shell fragments can still be seen as in the fossiliferous limestone, coquina (Figure 20.29).

Warm climates favor carbonate deposition—carbonates are actually more soluble in cold water than in warm water. Evaporite deposits require a dry climate that causes the evaporation of lakes or seawater. As the water dries out, evaporite minerals precipitate and are left behind. Vast carbonate and evaporite deposits on the continents are evidence that expansive, shallow seas have periodically covered the land surfaces in the past.

FIGURE 20.28

These "jaws" of travertine at Pamukkale in Turkey were created by the precipitation of calcium carbonate, $CaCO_3$, from a hot mineral spring.

Link To Section 16.4

FIGURE 20.29

Coquina is just one example of a biochemical limestone. This coquina, composed of shells and shell fragments, is a fossiliferous limestone—full of fossils!

(a)

(b)

(c)

(d)

FIGURE 20.30

Some of the many processes of fossilization. (a) *Permineralization* occurs when porous remains of an organism become filled with water that is rich in dissolved minerals. Petrified wood is formed by this process. (b) An *impression* or *cast* is made by an organism (or part of an organism) that has been buried quickly before it can decompose. Although the organism does eventually decompose, it is only after the sediment has hardened around the organism's form that its shape can be preserved as a cast. (c) *Replacement* occurs when the remains of organisms are replaced by a mineral. Pyrite has replaced the original shell in this specimen. (d) *Carbonization* occurs when an organism is preserved as a thin film of carbon.

CHECK POINT

1. What makes coal exceptional among rocks?
2. Is coal a chemical sedimentary rock or a clastic sedimentary rock?

Were these your answers?
1. It is composed of organic matter rather than minerals.
2. It is a chemical sedimentary rock, or, more specifically, a biochemical rock.

FOSSILS: CLUES TO LIFE IN THE PAST

Because sedimentary rocks (clastic and chemical) form at Earth's surface, they often contain the remains of life forms—fossils. Fossils not only tell the story of life on Earth, they also give us important clues about Earth's geologic past. As we shall see in Chapter 21, fossils can indicate where and when sediments were deposited. They also help us match rocks from different places that are of similar geologic age. Some fossils are made of whole organisms, but most fossils are just parts of an organism. Other fossils are simply an impression, or print, made in the rock before it hardened. Plants commonly leave their impression as a thin film of carbon. Organisms have become fossilized in many ways (Figure 20.30).

20.9 Metamorphic Rocks

What happens when a mass of rock is brought to a location that has much higher temperature and pressure than the environment in which it formed? The changes in physical and chemical conditions to which the rock is exposed can transform the rock. New rock is made from old. The new rock is stable under the new conditions, although the preexisting rock was not.

The changes in rocks that happen as physical and chemical conditions change are called **metamorphism**. All rocks, whether igneous, sedimentary, or metamorphic, can undergo metamorphism. An everyday example of metamorphism is potter's clay. Potter's clay is soft at room temperature. But when heated, it becomes a hard ceramic. Similarly, limestone subjected to enough heat and pressure becomes marble. And shale is metamorphosed to slate. Rocks may also be drastically stretched or compressed. It is important to note that during metamorphism, minerals do not melt. Once minerals melt, metamorphism has ended and igneous activity has begun. In metamorphism, change occurs instead by *recrystallization* of preexisting minerals or by *mechanical deformation* of rock.

Recrystallization occurs when the minerals in a rock change because the rock was subjected to higher temperatures and pressures than the conditions under which it formed. The constituents of the metamorphosing minerals actually migrate and recombine to form new minerals. Recrystallization may occur with or without the exchange of fluid. For example, consider sedimentary rocks that contain such fluids as water or carbon dioxide. The fluids in the rock, which are enclosed in pore spaces, can act as catalysts to initiate or speed up metamorphic reactions. If temperature and pressure are high enough, the rock loses pore space as the fluid in the rock is squeezed out. The released fluid can then react chemically with the surrounding rock, contributing constituents to new minerals that are forming. These fluids can also initiate magma generation far away from the site of metamorphism as they migrate upward and drive fluid-induced melting (see Chapter 22). Metamorphic reactions can also occur

Swamp   pressure and heat   more pressure and heat   metamorphism

1. PEAT   2. LIGNITE   3. BITUMINOUS   4. ANTHRACITE

FIGURE 20.31

The formation of coal. (1) Peat forms from plant remains in a stagnant swamp. (2) Peat is buried by sediments. Over time, and subjected to increased heat and pressure, peat transforms into lignite—a soft brown coal. (3) More time, burial, heat, and pressure lead to the formation of bituminous coal—a soft, black, high-energy coal used for power production. (4) Metamorphism converts bituminous coal to anthracite, a hard, black, high-energy coal.

without the involvement of fluids. This happens in the case of low-temperature metamorphism.

Mechanical deformation occurs when a rock is subjected to physical stress. It may or may not involve elevated temperatures. For example, surface rocks that become deeply buried are subjected to increased pressure. Such stress may cause the rocks to flow like a plastic, bending them into intricate folds. Or the increased pressure may deform and flatten the rock, shear it, or break it and grind it into fragments. Such physical stress occurs deep in Earth's crust.

CHECK POINT

Is recrystallization, due to elevated temperature and pressure, the opposite of partial melting?

Was this your answer?

No. The process of crystallization from magma, not recrystallization, is the opposite of partial melting. Recrystallization occurs within a rock because of exposure to high temperatures and/or pressures, without melting. Rock may not even undergo any chemical changes during recrystallization.

> Plants and animals were energized by the Sun before they became fossil fuels. So the energy from fossil fuels is delayed solar power.

LINK TO Biology

▪ Fossil Fuels

When ancient plants and animals died, most of the organic matter they were made of was quickly decomposed by bacteria and converted to nutrients consumed by other organisms. Material that escaped bacterial decay was either preserved as sparsely distributed organic matter or converted to biochemical sediments to become coal, oil, or gas.

Coal, oil, and gas are all fossils, in the sense that they are the remains of past organisms. However, these remains have been so changed over time that the forms and even the composition of the accumulated organisms are beyond recognition.

Most coal deposits were formed about 300 million years ago, when steamy swamps covered much of Earth's surface. As plants and trees died, their remains sank to the bottoms of stagnant swamps. These oxygen-poor environments hindered decay, and plant matter accumulated layer upon layer to form a dense and soggy organic material called *peat*. Over time, sediments of sand and clay buried the peat. With more burial, the increase in heat and pressure caused the peat to lithify to form lignite and bituminous coals. (Figure 20.31). Although coal is composed of organic matter rather than minerals, it is considered a chemical sedimentary rock. When it is subjected to even more pressure and heat, coal can transform into the metamorphic

rock called *anthracite*, which is the highest grade of coal.

The source of oil and gas is fossilized microscopic organic matter found in buried marine sediments. When buried sediment with considerable amounts of this organic matter is subjected to low heat over a long enough period of time, chemical changes take place that create oil. Under the pressure of the overlying sediments, tiny oil droplets are squeezed out of the source rocks and into overlying porous rocks. The porous rocks—commonly sandstones—become oil reservoirs. Just as in the metamorphism of rocks, deeper burial results in higher temperatures. If the temperature gets high enough, natural gas is generated rather than oil.

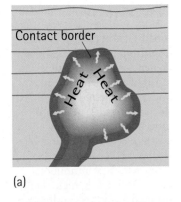

Contact border

Heat Heat

(a)

Granite pluton

Temperature decreasing

(b)   1 km

FIGURE 20.32

(a) Contact metamorphism is the result of rising molten magma that intrudes a rock body. (b) Surrounding the solidified intrusive rock is a zone of alteration. Alteration is greatest at the contact area, and it decreases farther away from the contact area.

Link To Section 6.3

TYPES OF METAMORPHISM: CONTACT AND REGIONAL

The most common types of metamorphism are *contact metamorphism* and *regional metamorphism*. Each type of metamorphism is characterized by differences in mechanical deformation and recrystallization.

Contact metamorphism occurs when a body of rock is intruded by magma (Figure 20.32). The high temperature of the magma produces a zone of alteration that surrounds the intrusion. The alteration is greatest at the contact, which is the interface between the intrusive rock and the surrounding rock. The width of the altered zone may range from a few centimeters to several hundred meters. Around a small intrusive body, such as a dike, the altered zone is very narrow and may resemble "baked" rock, with a texture and appearance like ordinary brick. But with a larger intrusive body, such as a batholith, the altered zone may be 100 m thick or more. One of the most common changes is an increase in crystal size due to recrystallization. Crystal size is greatest at the contact and decreases with increasing distance from that point. The water content of the rock also changes with distance from the contact. At the contact, where temperature is high, water content is low because it has boiled away. So we find dry, high-temperature minerals, such as garnet and pyroxene, at the contact. Farther away, we find water-rich, low-temperature minerals, such as muscovite and chlorite (another sheet silicate). Contact metamorphism is typically associated with high temperatures and high water content—lots of chemical activity and little or no mechanical deformation.

Regional metamorphism is the alteration of rock by both heat and pressure over an entire region rather than just near a contact between rock bodies. During the process of mountain building, Earth's crust is severely compressed into a mass of highly deformed rock. This deformation can be seen in the folded and fractured rock layers in many mountain ranges (Figure 20.33). Regionally metamorphosed rocks are found in all the major mountain belts of the world. Regional metamorphism combines recrystallization with mechanical deformation. Large sections of rock can be heated if the rock is buried deeply enough, simply because Earth is hotter at greater depths.

The effects of regional metamorphism are most pronounced in the cores of deformed mountains. Rocks develop a distinct "foliated" texture, which forms because of the great pressures generated by converging tectonic plates (Chapter 22). Zoned sequences of minerals are also characteristic of regional metamorphism. For example, one geographic area may have rocks with one set of minerals, while the adjacent area has rocks with a different set of minerals. Because of the large-scale nature of regional metamorphism, these zones tend to be broad

FIGURE 20.33

This satellite photo reveals regional-scale folding of metamorphic rocks in the Appalachian Mountains of central Pennsylvania.

and extensive. Areas of regional metamorphism are the hunting grounds of gem prospectors, because the heat and pressure that accompany these changes can produce beautiful minerals.

CLASSIFYING METAMORPHIC ROCKS

Metamorphic rocks are defined by their appearance and the minerals they contain. For classification and identification, metamorphic rocks can be divided into two groups: *foliated* and *nonfoliated*.

FOLIATED METAMORPHIC ROCKS When rock is subjected to increased pressure, some of its minerals realign into parallel planes as they recrystallize. The face of each of these parallel planes is perpendicular to the main direction of the compressive force. This leads to a layered appearance called *foliation*. Foliation is a prominent visual feature of regionally metamorphosed rocks, and it is very different from the layering seen in sedimentary rock. Deposition does not cause the foliated texture in metamorphic rocks. Rather, sheet-structured minerals, such as the micas, grow and orient themselves with their sheets perpendicular to the direction of maximum pressure (Figure 20.34). The new rock, which now has parallel flakes, or plates, of mica, is said to be foliated. The most common foliated metamorphic rocks—slate, schist, and gneiss—are derived from sedimentary rocks that have the appropriate chemical composition to favor mica formation (Figure 20.35).

Slate is the "lowest-grade" foliated metamorphic rock, which means that it was formed under relatively low temperature and pressure. Slate, which is metamorphosed shale, is a foliated rock composed of very small particles and tiny mica flakes. The most obvious characteristic of slate is its excellent rock cleavage, which allows it to be split into thin slabs. The best pool tables and chalkboards are made from slate quarried in metamorphic areas where slaty cleavage is well developed. Slate is also commonly used as roofing tile and floor tile.

Schist is one of the most easily recognizable metamorphic rocks because it is scaly looking with large micas that reflect light like tinted windows. Schist forms under higher temperature and pressure conditions than slate, which causes the mineral grains to grow large enough to be identified with the naked eye. Schists usually contain about 50% platy minerals—most commonly muscovite and biotite. The larger mica flakes give the rock a highly reflective surface that is quite striking. Schists are named according to the major minerals in the rock (biotite schist, staurolite-garnet schist, and so on).

Gneiss (pronounced "nice") is a foliated metamorphic rock that contains alternating layers of dark platy minerals and lighter granular minerals. The layers give this metamorphic rock its characteristic banded appearance. This appearance

FIGURE 20.34

As compressive forces squeeze platy and sheet-structured minerals, the grains align themselves perpendicular to the main direction of force. Arrows indicate the direction of compressive force.

Link To Section 17.4

FIGURE 20.35

Common foliated metamorphic rocks: (a) slate, (b) schist, and (c) gneiss.

(a)

(b)

(c)

(a)

(b)

FIGURE 20.36

Nonfoliated metamorphic rocks: (a) marble and (b) quartzite.

results from even greater temperature and pressure conditions than those that create schist. The most common granular minerals found in gneiss are quartz and feldspar. These are also the most common granular minerals in granite. In fact, some gneisses are actually metamorphosed granites.

NONFOLIATED METAMORPHIC ROCKS Nonfoliated metamorphic rocks can form because of increased temperature and pressure or because of increased temperature alone. Even under high pressure, foliation cannot develop if the rock lacks the chemical composition needed for micas (or other elongated crystals) to form. Similarly, even if the chemical composition contains the necessary constituents but the pressure is not high enough, such as in contact metamorphism, foliation cannot develop. Two common nonfoliated rocks that cannot become foliated are marble and quartzite.

Marble (Figure 20.36a) is a crystalline, metamorphosed limestone. Pure marble is white and is virtually 100% calcite, which is neither platy nor elongated. Because of its color and its relative softness (hardness 3), marble is a popular building stone. Often the limestone from which marble formed contained impurities that produce various colors in the marble. Thus, marble can vary in color from pink to gray, green, or even black.

Quartzite (Figure 20.36b) is metamorphosed quartz sandstone, and it is therefore very hard (hardness 7). Quartz is another mineral that is not platy or elongated. The recrystallization of quartzite can be so complete that when struck the rock splits across the original quartz particles, rather than between them. Although pure quartzite is white, it commonly contains impurities that can cause it to be a variety of colors, such as pink, green, or light gray.

CHECK POINT

1. Under conditions of extreme temperature, when can a rock no longer undergo metamorphism?
2. Why does recrystallization occur in metamorphic rock?

Were these your answers?

1. When it melts. Once a rock melts, it becomes magma. And when magma cools to form rock, the new rock, by definition, is igneous rock.
2. Recrystallization occurs because the rock is exposed to high temperatures or pressures.

20.10 The Rock Cycle

The Physical Science Place
The Rock Cycle Activity

The Physical Science Place
The Rock Cycle

Earth is a dynamic, ever-changing, and active planet. Made up of atoms and molecules, Earth's elements combine to make minerals, which are formed by the process of crystallization from either magma or water solutions. The minerals formed are determined by the elements present and the conditions that lead to their formation. These factors in turn determine the arrangement of atoms in each mineral and the strength of the bonds that hold the atoms together. More than 90% of Earth's minerals are silicates—composed predominantly of silicon and oxygen plus other elements such as aluminum, iron, calcium, sodium, potassium, and magnesium. Minerals combine to make rocks—the igneous, sedimentary, and metamorphic rocks that we see all around us.

Most minerals (and hence, rocks) are formed by the crystallization of magma. And magma forms when rock melts. The type of magma formed depends on the

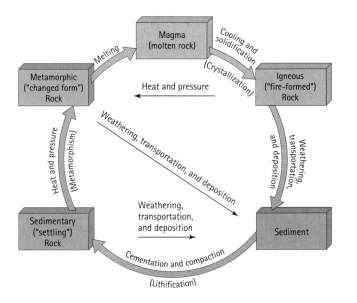

FIGURE 20.37

The rock cycle: Igneous rock, subjected to heat and pressure far below Earth's surface, may become metamorphic rock; metamorphic or sedimentary rocks at Earth's surface may decompose to become sediment that in turn becomes new sedimentary rock. Whatever the route, molten rock rises from the depths of Earth, cools, and solidifies to form a crust that, over eons, is reworked by shifting and erosion, only to return eventually to become magma in Earth's interior.

type of rocks and minerals that melt. A single magma can transform into different magmas (basaltic, andesitic, and granitic) during migration and crystallization, and then into different types of igneous rocks.

Although most of Earth's crust is composed of igneous and metamorphic rock, the rock we see at the surface is mainly sedimentary. Sedimentary rock forms from the remains of rock that has been weathered and eroded. Sedimentary rock provides a record of environmental and biological changes on Earth's surface. And when sedimentary rock is buried deep within Earth or involved in mountain building, great temperatures and pressures can transform it into metamorphic rock. Under the proper conditions, metamorphic rock can melt and become magma, which eventually solidifies as igneous rock to complete the **rock cycle** (Figure 20.37).

The focus of this chapter has been on the formation of minerals and rocks. In the next chapter, we turn our attention to how minerals and rocks can be used to unravel Earth's immense span of history. Formed more than 4.5 billion years ago, our Earth has experienced many changes: changes in the atmosphere, changes in sea level, changes in landforms, and changes in life and life forms. And many of these changes are revealed in the archives of the rock record. On to reading the rock record!

SUMMARY OF TERMS

Tectonic plates Sections into which Earth's crust is broken up; they move in response to heat flow and convection in Earth's interior.

Mineral A naturally formed inorganic crystalline solid, composed of an ordered arrangement of atoms with a specific chemical composition.

Crystal form The outward expression of the orderly internal arrangement of atoms in a crystal.

Polymorphs Two or more minerals that contain the same elements in the same proportions but have different crystal structures.

Mohs scale of hardness A ranking of a mineral's hardness, which is its resistance to scratching.

Cleavage The tendency of a mineral to break along planes of weakness.

Fracture A break that does not occur along a plane of weakness.

Density The ratio between the mass of a substance and its volume.

Silicate A mineral that contains both silicon and oxygen and (usually) other elements in its chemical composition; silicates are the largest and most common rock-forming mineral group.

Nonsilicate A mineral that does not contain silica (silicon + oxygen).

Crystallization The growth of a solid from liquid or gas, whose atoms come together in specific chemical proportions and crystalline arrangements.

Magma Molten rock in Earth's interior.

Chemical sediments Sediments that form by the precipitation of minerals from water on Earth's surface.

Solubility A measure of the ease with which a mineral can be dissolved. Low-solubility minerals are difficult to dissolve. High-solubility minerals are easier to dissolve.

Rock An aggregate of minerals. Some rocks are aggregates of fossil shell fragments, solid organic matter, or any combination of these components.

Igneous rocks Rocks formed by the cooling and crystallization of hot, molten rock material called magma (or lava).

Sedimentary rocks Rocks formed from the accumulation of weathered material (sediments) that has been eroded by water, wind, or ice.

Metamorphic rocks Rocks formed from preexisting rocks that have been changed or transformed by high temperature, high pressure, or both.

Partial melting The incomplete melting of rocks, resulting in magmas of various compositions.

Volcanic rocks Extrusive igneous rocks formed by the eruption of molten rock at Earth's surface. Basalt is a volcanic rock.

Lava Molten magma that moves upward from inside Earth and flows onto the surface. The term *lava* refers both to the molten rock itself and to the solid rocks that form from it.

Plutonic rock Intrusive igneous rock formed from magma that cools beneath Earth's surface. Granite is a plutonic rock.

Weathering Disintegration and/or decomposition of rock at or near Earth's surface.

Erosion The wearing away of rocks, and the processes by which rock particles are transported by water, wind, or ice.

Deposition The stage of sedimentary rock formation in which eroded particles come to rest.

Sedimentation The stage of sedimentary rock formation in which sediments accumulate and change (lithify) into sedimentary rock through the processes of compaction and, usually, cementation.

Metamorphism The changes in rock that happen as physical and chemical conditions change.

Recrystallization A process that occurs when rocks are subjected to high temperatures and pressures and go through a change in minerals; often accompanied by the loss of H_2O or CO_2.

Mechanical deformation Metamorphism caused by stress, such as increased pressure.

Rock cycle A sequence of events involving the formation, destruction, alteration, and reformation of rocks as a result of the generation and movement of magma; the weathering, erosion, transportation, and deposition of sediment; and the metamorphism of preexisting rocks.

REVIEW QUESTIONS

20.1 The Geosphere Is Made Up of Rocks and Minerals

1. How did density segregation contribute to Earth's internal layers?
2. What three sources of heat contributed to melting and density segregation of early Earth?
3. What is the most abundant element for Earth as a whole?
4. What is the most abundant element in Earth's crust? What is the second most abundant element?

20.2 Minerals

5. What is a mineral?
6. What does *inorganic* in the definition of a mineral imply?

20.3 Mineral Properties

7. What physical properties are used to identify minerals?
8. Minerals are characterized by their orderly internal arrangement of atoms—their crystal forms. Yet most mineral samples do not display their crystal forms. Why not?
9. What is a polymorph?

20.4 Classification of Rock-Forming Minerals

10. What is the difference between a silicate mineral and a nonsilicate mineral?
11. Silicate minerals can be subdivided into ferromagnesian silicates and nonferromagnesian silicates. What two factors contribute to this subdivision?
12. What is the most abundant mineral in Earth's crust? What is the second most abundant mineral?

20.5 The Formation of Minerals and Rock

13. What is the process of crystallization?
14. Briefly describe two sources from which minerals crystallize.
15. As minerals crystallize in cooling magma, which minerals are the first to crystallize—the minerals with lower amounts of silica or the minerals with higher amounts of silica?
16. When water evaporates from a body of water, what type of sediment is left behind?

20.6 Rock Types

17. Name the three major types of rocks and describe the conditions of their origin.

20.7 Igneous Rocks

18. What are the most common igneous rocks, and where do they generally occur?
19. What is meant by partial melting?
20. With respect to the silica content of the parent rock, partial melting will produce what type of magma?
21. What determines a rock's melting point?
22. In Earth's interior, does temperature increase or decrease with depth?

20.8 Sedimentary Rocks

23. How does weathering produce sediment? Distinguish between weathering and erosion.

24. What is a clastic sedimentary rock?
25. What are the three most common clastic sedimentary rocks?
26. What is the most abundant carbonate rock?
27. How are most carbonate rocks formed?

20.9 Metamorphic Rocks

28. What is metamorphism? What causes it?
29. Distinguish between foliated and nonfoliated metamorphic rocks.
30. In contact metamorphism, water-rich, low-temperature minerals are found far away from the contact zone. Give two examples of such minerals.

EXERCISES

● BEGINNER ■ INTERMEDIATE ◆ EXPERT

Again, please do not be intimidated by the large number of exercises and problems in this and other meatier chapters. If your course work is to cover many chapters, your instructor will likely assign only a few exercises and/or problems from each.

1. ● What do we call minerals that have the same combination of elements but a different arrangement of elements?
2. ■ What is the difference between the minerals that make up a rock and the minerals we find in common dietary supplements?
3. ■ Silicon is essential for the computer industry in making microchips. Can silicon be mined directly from Earth? Defend your answer.
4. ■ What two minerals make up most of the sand in the world?
5. ■ What two mineral groups provide most of the ore that society needs?
6. ■ If a rock contains mineral A (30% silica) and mineral B (25% silica), which would melt first as temperature rises?
7. ■ If a magma contains molten forms of mineral A (30% silica) and mineral B (25% silica), which would crystallize first as the magma cools?
8. ◆ The factors that influence bond strength influence mineral hardness. What are these factors?
9. ■ If a magma contains molten forms of mineral A (30% silica) and mineral B (25% silica), which would crystallize last as the magma cools?
10. ■ If a rock contains mineral A (30% silica) and mineral B (25% silica), which would melt last as temperature increases?
11. ◆ Is it possible for crystallization to enrich magma in more than just silica? Defend your answer.
12. ■ If high-silica minerals are the last to crystallize, why aren't high-silica minerals the last to melt?
13. ◆ Why is halite commonly the last mineral to precipitate from evaporating seawater?
14. ● Would you expect to find any fossils in limestone? Why or why not?
15. ● How do chemical sediments produce rock? Name two rock types that form by chemical sedimentation.
16. ● Why is color not always the best way to identify a mineral?

17. ■ While you are out hiking in the wilderness you find a shiny, glassy-looking mineral. What physical test could you use to determine if this mineral is a diamond?
18. ● What makes gold so soft (easily scratched) while quartz and diamond are so much harder?
19. ● Imagine that we have a liquid with a density of 3.5 g/cm^3. Knowing that objects of higher density will sink in the liquid, will a piece of quartz sink or float in the liquid? How about a piece of chromite?
20. ■ Is cleavage the same thing as crystal form? Why or why not?
21. ◆ If the volcanic glass obsidian is not considered a mineral, why is it considered a rock?
22. ■ Relate the shape and sorting of sand particles to the way in which they were most likely transported.
23. ● In which parts of Earth's crust (oceanic and/or continental crust) do we find the two common igneous rocks, basalt and granite?
24. ◆ Aside from silica content, what two factors can change a rock's melting point?
25. ■ Which of these is a true statement about silicate minerals? (a) Melting point decreases as silica percentage increases. (b) Melting point increases as silica percentage increases.
26. ● Are the Hawaiian Islands made up primarily of igneous, sedimentary, or metamorphic rock?
27. ■ Where does most magma originate?
28. ■ What patterns of alteration are characteristic of contact metamorphism?
29. ◆ Can metamorphic rocks exist on an island of purely volcanic origin? Defend your answer.
30. ■ Are high-silica minerals "easier" to melt than those with low silica?
31. ■ What mainly determines a rock's initial melting temperature?
32. ● What general rock feature does a geologist look for in a sedimentary rock to determine the distance the rock has traveled from its place of origin?
33. ◆ What feature of clastic sedimentary rock enables the flow of oil after it has been formed?
34. ◆ Which of these rocks—granite, sandstone, limestone, or halite—is the first to weather in a wet (humid) climate? Why?

35. ■ In what two ways does sediment turn into sedimentary rock?
36. ♦ In a conglomerate rock, why are pebbles of granite very common and pebbles of marble relatively uncommon?
37. ♦ Which type of rock is most sought by petroleum prospectors—igneous, sedimentary, or metamorphic? Why?
38. ♦ Cite two examples of sedimentary rocks that provide information about past geologic events at Earth's surface.
39. ● What is a fossil? How are fossils used in the study of geology?
40. ● What kind of weathering is imposed on a rock when it is smashed into small pieces? When it is dissolved in acid?
41. ● Which type(s) of rock is (are) made from previously existing rock? Which does not require high temperature and pressure for its formation?
42. ■ What properties of slate make it good roofing material?
43. ● Name two mica minerals that can give a metamorphic rock its foliation.
44. ■ How is foliation different from sedimentary layering?
45. ■ Why do we find folded and fractured rock layers in zones of regional metamorphism?
46. ● Where on Earth's surface are lava flows most common?
47. ● Each of the following statements describes one or more characteristics of a particular metamorphic rock.

For each statement, name the metamorphic rock being described.
(a) Foliated rock, sometimes derived from granite
(b) Hard, nonfoliated, single-mineral rock, formed under high to moderate pressure
(c) Foliated rock, possessing excellent rock cleavage; generally used in making blackboards
(d) Nonfoliated rock composed of carbonate minerals
(e) Foliated rock containing about 50% platy minerals; named according to the major minerals in the rock.
48. ■ What feature helps distinguish schist and gneiss from quartzite and marble?
49. ♦ Earth's temperature increases with depth. Why, then, don't all rocks melt in Earth's interior?
50. ♦ Is Earth's interior mostly magma?
51. ● How does gneiss differ from granite?
52. ■ Why is schist so easily recognized?
53. ♦ If a rock contains both quartz and pyroxene (a silicate mineral), which melts first as the rock is heated?
54. ♦ If a magma contains molten forms of quartz and olivine (a silicate mineral), which crystallizes first as the magma cools?
55. ● What are the two processes by which rock is changed during metamorphism?

PROBLEM

● BEGINNER ■ INTERMEDIATE ♦ EXPERT

1. ■ Gold has a density of 19.3g/cm^3. A 5-gal pail of water (density of water $= 1.0 \text{ g/cm}^3$) has a mass of about 18 kg. What is the mass of a 5-gal pail of gold?

READINESS ASSURANCE TEST (RAT)

If you have a good handle on this chapter, if you really do, then you should be able to score at least 7 out of 10 on this RAT. If you score less than 7, you need to study further before moving on.

Choose the BEST answer to each of the following.

1. The silicates are the largest mineral group because silicon and oxygen are
   (a) the hardest elements on Earth's surface.
   (b) the two most abundant elements in Earth's crust.
   (c) found in the common mineral quartz.
   (d) stable at Earth's surface.
2. Compaction and cementation of sediments leads to
   (a) magma generation.
   (b) lithification.
   (c) formation of pore water.
   (d) metamorphism.
3. Why are silicon and oxygen concentrated near Earth's surface while iron is concentrated at the core?
   (a) Earth's materials separated early in its history through the process of density segregation.

   (b) Silicon and oxygen are less dense than iron.
   (c) both of these
   (d) neither of these
4. Which minerals crystallize first from cooling magma?
   (a) minerals with the lowest melting point
   (b) minerals with the highest melting point
   (c) minerals with the highest solubility
   (d) minerals with high silica content
5. In Earth's interior, what two factors increase with depth?
   (a) pressure and water content
   (b) temperature and pressure
   (c) temperature and water content
   (d) silica content and water content
6. In a sedimentary rock, the degree of particle roundness can indicate
   (a) the duration and/or length of travel.
   (b) where the sediment particles originated.
   (c) where the particles were deposited.
   (d) how the particles were cemented and compacted.
7. The characteristics of regional metamorphism include
   (a) folded and faulted rock layers.

(b) distinctly foliated rocks.
(c) zoned sequences of minerals.
(d) all of these
8. Listed in order of increasing silica content, the three main types of magma are
(a) basaltic, andesitic, and granitic.
(b) granitic, andesitic, and basaltic.
(c) basaltic, granitic, and andesitic.
(d) andesitic, granitic, and basaltic.
9. Rocks in Earth's interior can melt into magma when
(a) they move upward to where pressure is reduced.
(b) water is added, which helps lower the melting point of the rocks.

(c) all of these
(d) none of these
10. Sedimentary rocks often contain the remains of ancient life forms—fossils. And fossils in these rocks help us understand
(a) Earth's geologic and biologic history.
(b) Earth's early formation.
(c) the zoned sequences of minerals.
(d) the differentiation of life forms.

Answers to RAT

1b, 2b, 3c, 4b, 5b, 6a, 7d, 8a, 9c, 10a

ACTIVE EXPLORATIONS

1. Look at some crystals of table salt under a microscope or a magnifying glass and observe their generally cubic shapes. There's no machine at the salt factory specifically designed to give salt crystals these cubic shapes, as opposed to round or triangular ones. The cubic shape occurs naturally and is a reflection of how the atoms of salt are organized—cubically. Smash a few of these salt cubes and then look at them again carefully. What you'll see are smaller salt cubes! Use the cleavage properties of crystals to explain these results.

2. A physical property of any material is its density—its mass per volume. Pennies made after 1982 contain both copper and zinc. Pennies made before this date are pure copper. Zinc is less dense than copper, so post-1982 pennies are less dense and have less mass than pre-1982 pennies. Dig into your penny collection and find 20 pre-1982 and 20 post-1982 pennies. Measure their masses on a sensitive scale, such as a home postage scale. Alternatively, hold the pennies in opposite hands to see if you can feel the difference in their masses. How few pennies can you hold and still feel the difference? Try holding single pennies on your left and right index fingers. Can you tell the difference with your eyes closed? Try this with a friend.

3. Rocks never fully melt; they partially melt. So what happens as a rock partially melts? To understand the partial melting of rocks, let us use the ice, butter, and cheese analogy from the text.
Step 1: Mix cubes of ice, butter, and cheese together with a little water and put this mixture in the freezer. Once this mixture has frozen, you will have made a "rock."
Step 2: Put this "rock" in a bowl in the refrigerator overnight, and then examine the contents of the bowl the following morning or afternoon.
1. What melts?
2. What does not melt?
3. How is the melt different from the original "rock?"

For the partial melting of real rocks:
1. What melts? What doesn't?
2. How is the melt different from the original rock?
3. After the melt cools and solidifies into rock, how is that new igneous rock different from the rock that originally melted?

EXPLORING FURTHER

http://minerals.usgs.gov
The Mineral Resources Program home page, sponsored by the U.S. Geological Survey, provides current, impartial information on the occurrence, quality, quantity, and availability of mineral resources. The site is text intensive, emphasizing research and statistics.

http://webmineral.com
Winner of the Rockhounds Informative Site Award, this Web site provides an extensive mineralogy database. It contains classification charts and alphabetical reference listings. Check out the beautiful mineral picture gallery, which includes mineral origin and composition descriptions.

http://sln.fi.edu/fellows/fellow1/oct98/create/index.html
Part of the Rock Hounds site sponsored by Loogootee Community Schools, this page details how rocks are

formed. Go to the main page for more rock facts, along with some suggested teaching resources.

http://wrgis.wr.usgs.gov/docs/parks/rxmin/content.html
A cooperative endeavor of the U.S. Geological Survey Western Earth Surface Processes Team and the National Park Service, this site provides extensive information about types of rocks. It includes pictures and descriptions, an easy classification chart, and some information about minerals.

http://hvo.wr.usgs.gov
Sponsored by the U.S. Geological Survey, this is the home page for the Hawaiian Volcano Observatory. Find out current information about Hawaii's active volcanoes, along with history of dormant volcanoes. The site describes the hazards of both volcanoes and earthquakes.

CHAPTER 20 ONLINE RESOURCES

The Physical Science Place

Tutorial
- The Rock Cycle Activity

Video
- The Rock Cycle

Quiz

Flashcards

Links

geologic time – reading the rock record

21.1 THE ROCK RECORD—RELATIVE DATING

21.2 RADIOMETRIC DATING

21.3 GEOLOGIC TIME

21.4 PRECAMBRIAN TIME (4500 TO 543 MILLION YEARS AGO)

21.5 THE PALEOZOIC ERA (543 TO 248 MILLION YEARS AGO)

21.6 THE MESOZOIC ERA (248 TO 65 MILLION YEARS AGO)

21.7 THE CENOZOIC ERA (65 MILLION YEARS AGO TO THE PRESENT)

21.8 EARTH HISTORY IN A CAPSULE

■ Earth is some 4.5 billion years old. This vast span of time, called *geologic time*, is difficult to comprehend. But we can try to imagine it with the following thought exercise. Let's imagine that we can compress 4.5 billion years, the age of Earth, into a single year. Then planet Earth would have begun forming from matter in the primordial solar system on January 1. The oldest known Earth rocks would appear at the end of February. Simple bacterial life would appear in the sea at the end of March, but complex plants and animals would not emerge until late October or early November. Dinosaurs would rule Earth in mid-December and would disappear by December 26. *Homo sapiens* (humans) would appear at 11:50 PM on the evening of December 31. All of recorded human history would take place in the last minute of New Year's Eve!

How can we unravel this vast amount of time to tell Earth's story? Earth's history is recorded in the rocks of its crust. Scientists use an assumption called *uniformitarianism* to relate what we know about present-day processes to past events.

"The present is the key to the past."

Simply put, uniformitarianism states that the natural laws (like the laws of physics) we know about today have been constant over the geologic past. The rock record is like a long and detailed diary, containing the history of Earth-shaping events. The book is incomplete, however. Many pages, especially in the early part, are missing, and many others are tattered, torn, and difficult to read. But enough pages have been preserved to provide an account of the remarkable events of Earth's 4.5 billion years of history.

21.1 The Rock Record—Relative Dating

Sedimentary rock layers provide good evidence of relative rock ages. This is because the rock layers were deposited one atop the other. The lower layers were formed before the upper layers, and so they are older than the upper layers. Perhaps the world's most spectacular display of the rock record is the Grand Canyon of the Colorado River in Arizona (Figure 21.1). The many layers of rock exposed in the canyon walls and the thickness of these layers are testimony to great geologic activity over millions of years. The conditions under which the sedimentary layers were deposited varied widely—changing from season to season, year to year, and millennium to millennium. Some layers reveal climatic cycles that span centuries, other layers indicate times when the land surface became submerged beneath shallow seas, and still other layers show periods of increased rainfall accompanied by gradual uplift of the entire area. Millions of years after the top layer was deposited, erosion from the Colorado River cut through all these accumulated layers of sedimentary rock like a knife cutting into a layer cake, forming the canyon we see today.

To fully understand Earth's history, we need to know the age of each rock formation. As we will soon discuss, the actual age of a rock is determined by *radiometric dating* (the measurement of radioactive isotopes relative to their decay products). But before the technology existed to determine the actual age of a rock, scientists relied solely upon relative dating. **Relative dating** is the ordering of rocks in sequence by their comparative ages. Thus, relative

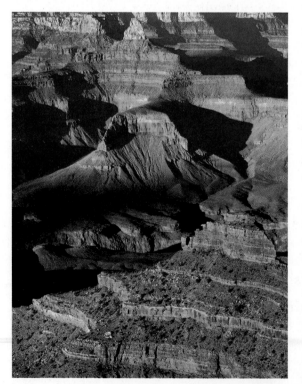

The lowermost layers of the Grand Canyon are older than the uppermost layers, which illustrates the principle of superposition.

FIGURE 21.2

Dikes cutting into a rock body are younger than the rock into which they cut. In the diagram, dike A cuts into dike B, and dike B cuts into dike C. From the principle of cross-cutting relationships, A is the youngest dike, B the next youngest, and C the oldest of the three. The horizontal layers, which are cut by all three dikes, are all older than C.

dating doesn't reveal the date when an event occurred, but rather its timing relative to other episodes in Earth's past. Relative dating, which is still widely used today, employs several key principles:

1. **Original horizontality:** Layers of sediment are originally deposited as an even, horizontal surface. Therefore, layers that we see today that are inclined at an angle—from slight to very steep—were moved into that position by disturbances, such as earthquakes and mountain building, after deposition.

2. **Superposition:** In an undeformed (horizontal) sequence of sedimentary rocks, each layer was deposited on top of older preexisting layers below. Like newspapers in a recycling bin, older papers are found below newer, more recent papers.

3. **Cross-cutting relationships:** An igneous intrusion or fault that cuts through preexisting rock is younger than the rock through which it cuts (Figure 21.2).

4. **Inclusions:** Inclusions are pieces of one rock contained within another. Any inclusion is older than the rock containing it, just as small pieces of rock incorporated in a slab of concrete were formed before the concrete was formed (Figure 21.3).

These four principles are fairly straightforward and can be used to determine the ages of rock formations relative to one another—which formation was formed first, second, and so on—for a particular area or rock outcrop. But to reconstruct the relative age of rocks for a larger area, additional information is needed. Once again, the Grand Canyon provides the ideal example. The rocks of the Grand Canyon not only chronicle a great span of time, but because the rock layers stretch continuously for hundreds of miles, they also show the range and extent of the depositional area. But not all rock layers stretch on like those found in the Grand Canyon. Over spans of time, rock layers can be broken by faulting and/or folding, or covered by younger sediments, so that all we see are geographically separated rock outcrops. In cases like this, we use the principle of lateral continuity.

5. **Lateral continuity:** Sedimentary layers are deposited in all directions over a large area unless some sort of obstruction or barrier limits their deposition. Faulting, folding, and erosion can separate originally continuous layers into isolated outcrops.

Lateral continuity can be used to match isolated rock outcrops to other rock outcrops over a large area (Figure 21.4). If the various rock layers in isolated outcrops have similar characteristics (such as color, mineralogy, grain size, and

FIGURE 21.3

The rocks locked in the sedimentary layer existed before the sedimentary layer formed, illustrating the principle of inclusions.

FIGURE 21.4

Lateral continuity can be used to determine the relative age of rocks in widely separated areas. Outcrops A and B share similar characteristics, but are separated by 500 km. Is it possible that these outcrops were, at one time, continuous? By looking at the sequence of layers, can you tell which rocks are older and which are younger?

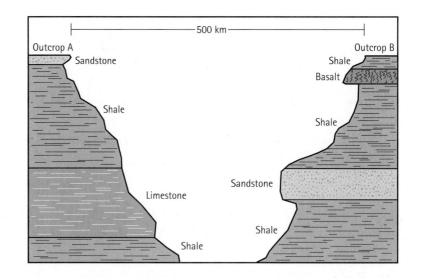

fossil content) and consistent vertical sequences, we can assume that the layers were once continuous. So when combined with other relative dating principles—namely superposition and original horizontality—lateral continuity allows relative age relationships to be applied over a much larger area.

As discussed in Chapter 20, fossils are the remains or impressions of ancient animals and/or plants preserved in rock. Today it is common knowledge that fossils record the evolution of life on Earth, but in the 1700s fossils were thought of as mere curiosities. Then along came William "Strata" Smith (1769–1839). Working as a surveyor for a canal project, Smith observed that certain rock layers contained different kinds of fossils, and that these fossil-bearing layers followed a consistent, predictable sequence. Smith noted the succession of rock types and the fossils within each layer and then used the fossils to correlate (match up) rock layers at various locations. He discovered that fossils could be used to chronologically order the vertical sequence of rock layers. With this knowledge, Smith established the principle of faunal succession.

6. **Faunal succession:** Fossil organisms follow one another in a definite, irreversible time sequence. Fossil assemblages change through time as some species become extinct and new ones appear. Such changes are reflected in the rock record. In this way, fossils provide a key tool for recognizing the relative age of rocks.

Smith's observation that fossils are found in rocks in a definite order not only helped refine relative dating, but also helped with the correlation of rock layers on a worldwide scale. Fossils—identified and categorized—could be correlated to specific phases in Earth's geologic history. Once scientists established a time period, the fossils in the rock could be used to identify other rocks of the same age in other regions of Earth.

In many ways, studying the rock record is like a detective studying a crime scene. Both studies involve looking for clues. Just as detectives have their methods for solving crimes, Earth scientists rely on the principles of relative dating—age relationships and the correlation of rock layers. They correlate the rocks, the sequences of rock layers, and the fossils within the rock layers. As with a crime scene, the case cannot be solved until sufficient evidence is found. And the Earth science detective often has little evidence on which to operate, and sometimes the evidence is completely lost. Earth's internal processes can fold and distort rock layers (Chapter 22) and external processes can weather

fyi

■ Fossils are great time indicators; they are also very important environmental indicators. For example, certain present-day corals are found in warm tropical waters. When a similar fossilized coral is found, we can safely assume that the area where it was found was once covered by a warm and tropical shallow sea. Fossils help unravel Earth's history.

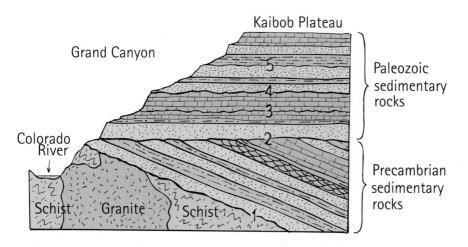

FIGURE 21.5

The age of the Grand Canyon can be deciphered by its sequence of rock layers. As in other places, the sequence is not continuous, and there are time gaps. (1) A nonconformity separating older metamorphic rocks from sedimentary layers. (2) An angular unconformity separating older tilted layers from horizontal layers. Time gaps are also represented between horizontal sedimentary layers. The disconformities (3)–(5) are difficult to identify, and they often require both a good eye and a knowledge of fossils.

and erode rock layers (Chapter 20 and Chapter 23). Whatever the cause, the result is lost evidence.

GAPS IN THE ROCK RECORD

Although most sedimentary rock layers were deposited without interruption, nowhere is there a continuous sequence from Earth's formation to the present time. Deposition is ongoing, but the processes of weathering and erosion, crustal uplifts, and other geologic processes remove rock layers or interrupt deposition. Such occurrences have created breaks or gaps in the rock record, as Figure 21.5 shows. We find these gaps, called **unconformities,** by observing the relationships of layers and fossils.

The most easily recognized of all unconformities is an **angular unconformity.** In an angular unconformity, tilted or folded sedimentary rocks are covered by younger, relatively horizontal rock layers. They are easy to recognize because rock layers below the unconformity are at an angle relative to the rock layers on top of the unconformity. An angular unconformity forms when older, previously horizontal rock layers are uplifted and tilted by movements within Earth (Figure 21.6). During and after the uplift, erosion wears down the tilted layers so that rocks at the surface are eroded to a more-or-less even plane, which creates a new, flat land surface. After the period of erosion is over, more sediment layers are deposited over the tilted ones, and these younger layers are horizontal. The angular unconformity is the "erosional surface" that separates the tilted layers from the horizontal layers, and it represents the long interval of time during which uplift and erosion took place. Can you see that angular unconformities usually represent ancient mountain-building events? The part of the rock record representing this long interval is now missing because of erosion, and the unconformity is the evidence that remains.

When overlying sedimentary rocks are found on an eroded surface of metamorphic or igneous plutonic rocks, the unconformity is called a *nonconformity*. The igneous or metamorphic rocks formed deep beneath Earth's surface but were present at the surface when the overlying sedimentary rocks were deposited on top of them. Therefore, a nonconformity shows that a great deal of uplift and erosion occurred before the sedimentary layers were deposited, with a large stretch of time "missing" from the rock record. A more subtle type of unconformity, a *disconformity*, occurs as a time gap between parallel layers of sedimentary rock. Because the rocks above and below the disconformity can be very similar, such time gaps are often quite difficult to identify.

The Physical Science Place
Formation of an Angular Unconformity

FIGURE 21.6

INTERACTIVE FIGURE

The sequence of events that create an angular unconformity.

(a) Sediments are deposited layer upon layer beneath the sea.

(b) During mountain building solidified sediment layers become folded and deformed. Erosion begins.

(c) As mountain building wanes, the exposed surface is eroded to a more or less even plain.

(d) As the land subsides below sea level, younger sediments are deposited on the former erosional surface.

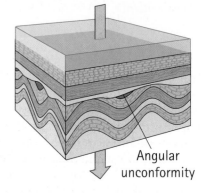

Angular unconformity

CHECK POINT

1. If a granitic intrusion, such as a dike, cuts into or across sedimentary layers, which is older: the granite or the sedimentary layers?
2. Look at outcrops A and B in Figure 21.4. The outcrops are separated by 500 km; is it possible that these rock layers were once continuous? If so, which rocks are older and which are younger?

Were these your answers?

1. The intrusion is new rock in the making. Therefore, the sedimentary layers are older than the intrusions that cut into them.
2. Yes, it is possible. If continuous, the sandstone capping outcrop A matches the sandstone sandwiched between the two shales in outcrop B. So all rock layers above sandstone B are younger. For outcrop A, original horizontality and superposition tell us that the bottom shale is oldest, followed by limestone, shale, then sandstone. We assume the same sequence in outcrop B even though we cannot see the limestone layer. If fossils were present, we could further support our conclusion.

21.2 Radiometric Dating

Relative dating tells us which parts of Earth's crust are older or younger, but it doesn't tell us the actual age of a rock—the amount of time that has passed since the rock solidified. The actual age of a rock can be determined by **radiometric dating,** a process that measures the ratio of radioactive isotopes to their decay products. The half-life of a radioactive isotope is the amount of time

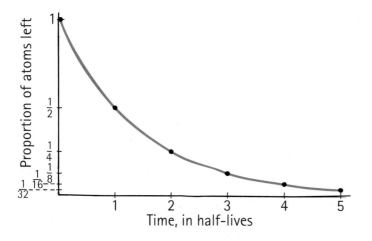

FIGURE 21.7

The amount of parent material versus the number of half-lives remaining as the radioactive parent decays.

it takes for half of the material to decay to its daughter product (Figure 21.7). Some common radioactive isotopes used for dating and estimates of geologic time are shown in Table 21.1.

To date geologically young objects, and especially for dating organic matter, carbon-14 is the isotope of choice. Carbon-14 has a relatively short half-life (5,760 years), so it is useful for dating geologically recent events, within the last 50,000 years or so (see Section 13.4). To date older materials, radioactive elements such as uranium are used. Many common rocks contain trace amounts of uranium, and only a small amount is needed to perform the laboratory analysis. Uranium-238 decays to its stable daughter isotope, lead-206, and uranium-235 decays to the stable isotope lead-207. No other natural sources exist for these two isotopes of lead; therefore, any lead-206 and lead-207 found in a rock today were at one time uranium. If, for example, a sample contains equal numbers of uranium-235 and lead-207 atoms, the age of the sample is one uranium-235 half-life—704 million years. If, on the other hand, a sample of uranium contains only a small amount of lead-207, the sample is younger than one half-life (a single half-life has not yet passed).

fyi

- Radiometric dating is based on the assumption that once a mineral has crystallized, any daughter product found within it originated only from the decay of the unstable parent—that is, that no daughter product was initially present. Another important assumption is that there is no "leakage" of parent or daughter products into or out of the mineral. For example, if a mineral is reheated by metamorphism, its "time clock" is reset, complicating the age estimation. Radiometric dating and other dating techniques are subject to some uncertainty. In fact, for relative and radiometric dating, cross-checking increases accuracy.

CHECK POINT

1. Could carbon-14 be used for dating 100-million-year-old rocks?
2. How can we determine the age of sedimentary rock layers?

Were these your answers?

1. No. Carbon-14 has a half-life of 5760 years and can be used to date only relatively younger rocks. Any carbon-14 (from calcite, for example) in rocks this old would have long since been reduced to undetectable amounts.
2. If we know the maximum age (meaning the rock can be no older than the age of the datable minerals within it) of an overlying and an underlying rock layer, we can bracket the age of the sedimentary layer in between by using the principle of superposition.

Link To Section 13.4

TABLE 21.1   ISOTOPES MOST COMMONLY USED FOR RADIOMETRIC DATING

Radioactive Parent	Stable Daughter Product	Half-Life Value
Uranium-238	Lead-206	4.5 billion years
Uranium-235	Lead-207	704 million years
Potassium-40	Argon-40	1.3 billion years
Carbon-14	Nitrogen-14	5760 years

Radioactivity is everywhere! In fact, all elements that have an atomic number greater than 82 (lead) are radioactive, but not necessarily dangerously so.

fyi

- Radiometric dating can give us the age of minerals and/or organic matter, but it cannot directly give us the age of sedimentary rocks. Remember, sedimentary rocks are made from the remains of preexisting rocks. Therefore, we can date only the minerals in the rock, not the sedimentary rock itself. The rock can be no older than the age of the datable minerals within it (principle of inclusions). So how do we date sedimentary rock layers? We use relative dating combined with absolute dates from radiometric dating to provide limits on the age of the rock. The more techniques, the better the date.

21.3 Geologic Time

The geologic time scale (Figure 21.8) was developed through the use of relative dating, and specific dates have been applied to it with radiometric dating. The geologic time scale subdivides Earth's 4.5-billion-year history into time units of different sizes. *Eons* are the largest units of geologic time. The eon we are living in began about 543 million years ago. It is called the Phanerozoic, which means "visible life." The Phanerozoic eon is subdivided into three *eras*: the Paleozoic era (time of ancient life), the Mesozoic era (time of middle life), and the Cenozoic era (time of recent life). Each of the three eras is further divided into *periods*, which are further divided into *epochs*.

Note that most of Earth's history occurred before the Paleozoic era. This vast span of time, called **Precambrian time,** accounts for about 4 billion years of Earth's history! The Precambrian is divided into three eons: the Hadean, the Archean, and the Proterozoic.

CHECK POINT

1. Describe the present time in Earth's history in terms of all units of the geologic time scale from eons to epochs.
2. The time units on the geologic scale are Ma. What does *Ma* stand for?

Were these your answers?
1. We are living in the Phanerozoic eon, in the Cenozoic era, in the Quaternary period, and in the Holocene epoch.
2. *Ma* stands for *Mega-annum*, which means "one million years (ago)."

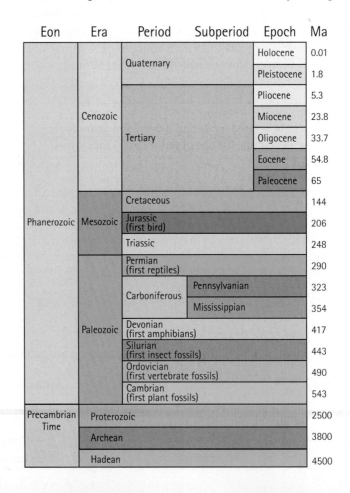

FIGURE 21.8

INTERACTIVE FIGURE

The geologic time scale divides Earth's history into time units of different size. Units of time on this scale are Ma, which stands for *Mega-annum*, or "one million years (ago)." For example, the Paleozoic era began about 543 Ma, or 543 million years ago.

Eon	Era	Period	Subperiod	Epoch	Ma
Phanerozoic	Cenozoic	Quaternary		Holocene	0.01
				Pleistocene	1.8
		Tertiary		Pliocene	5.3
				Miocene	23.8
				Oligocene	33.7
				Eocene	54.8
				Paleocene	65
	Mesozoic	Cretaceous			144
		Jurassic (first bird)			206
		Triassic			248
	Paleozoic	Permian (first reptiles)			290
		Carboniferous	Pennsylvanian		323
			Mississippian		354
		Devonian (first amphibians)			417
		Silurian (first insect fossils)			443
		Ordovician (first vertebrate fossils)			490
		Cambrian (first plant fossils)			543
Precambrian Time	Proterozoic				2500
	Archean				3800
	Hadean				4500

21.4 Precambrian Time (4500 to 543 Million Years Ago)

Precambrian time ranges from about 4.5 billion years ago, when Earth formed, to about 543 million years ago, when abundant macroscopic life appeared. The Precambrian—the time about which we know the least—comprises almost 90% of Earth's history! Most of the rocks that formed in this early part of Earth's history have been eroded away, metamorphosed, or recycled into Earth's interior. Organisms of that time did not have easily fossilized hard body parts, which evolved only later in the history of life. Thus, fossils from this huge span of time are scarce.

The beginning of the Precambrian—the Hadean eon—was likely a time of considerable volcanic activity and frequent meteorite impact.* Large and small chunks of interplanetary debris left over from the formation of the solar system continually smashed into Earth to scar its surface. Earth was an oceanless planet covered with volcanoes erupting gases and steam from its scorching interior. Intense convection caused by the tremendous amount of heat escaping from the interior left Earth's earliest crust in turmoil. The Hadean atmosphere consisted mostly of gases that had erupted from the many volcanoes. Carbon dioxide may have comprised 80% or more of this atmosphere. Water vapor made up most of the balance, with molecular nitrogen, ammonia, sulfur dioxide, and nitric oxide as minor constituents. There was no free oxygen.

In the mid-Precambrian—the Archean eon—Earth's surface began to cool appreciably (Figure 21.9). This set the stage for the formation of the oceans. With cooler temperatures, water vapor in the atmosphere condensed to form clouds. Out of the clouds poured torrents of rain, enough to cover Earth's surface with shallow seas. Earth's interior was still quite hot and active, and erupting volcanoes dotted Earth's surface to form small islands. These small volcanic islands, carried from below by heat-driven motion in Earth's interior, collided with other small islands to form larger islands. These larger islands were the precursors of early continents. Evidence of these early collisions can be found in certain folded and faulted rocks (for example, rock beds near Acasta, Canada) that now form the "cores" of present-day continents.

Life emerged during the Archean, and this event touched all other events that followed. The first simple organisms for which fossils have been found are dated at 3.5 billion years old. Stromatolites**—algae-like colonies of bacteria layered with precipitated,

- Very early in Earth's history (between 4.5 and 4.3 billion years ago), a Mars-sized object struck Earth. The impact created tremendous heat and produced orbiting debris that coalesced to form our Moon. The impact also knocked Earth off its axis of rotation to the present tilt of 23.5°. Without such an impact, Earth's axis of rotation would not be tilted relative to the Sun and we would have no seasons!

FIGURE 21.9

This artwork of the Archean eon of the Precambrian shows characteristic features of the time, including space debris, such as meteorites and comets, crashing into Earth's surface. The bright green material on the edge of the water is primitive bacteria (*archaea*), while cyanobacteria thrive in the darker green, round structures called stromatolites in the water.

---

* A meteorite is any solid rock object from interplanetary space that has fallen to Earth's surface without being vaporized during its passage through the atmosphere. We will learn more about these objects in Chapter 26, when we study the formation of the solar system.
** Stromatolites still exist today—they are valuable "living fossils." They can be found in hypersaline lakes and protected marine lagoons, such as Shark's Bay in Western Australia.

LINK TO Chemistry

▪ Banded Iron Formations

Earth's rocks reveal our planet's history. Even Earth's early atmosphere can be determined by studying minerals in the rock record. Four billion years ago, the atmosphere was about 80% carbon dioxide, 10% nitrogen, and 10% water vapor. Today the atmosphere is about 78% nitrogen and 21% oxygen. Quite a difference! What happened to all the carbon dioxide? Where did all the oxygen we have today come from? One process sums it up: photosynthesis.

Much of the early carbon dioxide was dissolved in seawater, where it reacted with dissolved calcium—and to some degree, magnesium—to form limestone. So the first limestones had their origins during the earliest stages of Earth's atmosphere. As photosynthesizing organisms died, their remains sank to the bottom of the ocean to be incorporated into the sediment—more limestone.

Oxygen also dissolves in water. When oxygen production began because of photosynthesis, much of it dissolved in seawater. But before the increased concentration of dissolved oxygen, the oxygen-poor environment allowed dissolved iron to accumulate in the oceans. So dissolved iron was abundant in the early oceans. Then when dissolved oxygen became plentiful, it reacted with the dissolved iron, causing iron in the water to precipitate. So when the air contained little or no oxygen, there was little oxygen in the seawater, but large amounts of dissolved iron. When oxygen concentrations rose to a certain threshold level, iron precipitated out of the seawater to form a layer of solid iron oxide minerals on the seafloor. When oxygen levels decreased, so did iron precipitation. This cyclical pattern of varying oxygen levels is depicted in the alternating layers of iron oxide minerals and iron-free sediment—the banded iron formation.

This process continued from about 2.6 to 1.9 billion years ago, a period of around 700 million years. That's how the iron-rich rocks formed, rocks that now support our industrial society. The world's iron and steel industry is based almost exclusively on iron ores associated with Precambrian banded iron formations.

Link To Section 17.3

fyi
▪ The rains reduced not only the amount of water vapor in the atmosphere, but also the amount of carbon dioxide, which dissolved into water droplets and newly formed oceans. In this way, the atmosphere started to become more nitrogen enriched.

carbonate sediments—thrived in shallow seas and are the oldest known fossils of microorganisms. The success of these early organisms depended on their ability to survive in the primitive, oxygen-poor environment. These microorganisms, cyanobacteria, had a simple version of photosynthesis. Photosynthetic organisms combine CO_2 and the Sun's energy to make simple sugars—a biologically usable energy source.

Oxygen is a waste product of the photosynthesis reaction and is expelled. The expelled oxygen, however, did not yet accumulate in the atmosphere; instead, it stayed dissolved in seawater and went into the production of certain rocks. The dissolved oxygen reacted with dissolved iron, which was abundant in the early oceans. The reaction produced solid iron oxides, which chemically precipitated to form *banded iron formations*. These ancient rock formations are a testimony to the great amount of oxygen that was expelled during this time.

The Proterozoic marks the end of Earth's longest time span—and this eon lasted another 2 billion years! Let us highlight a few changes that occurred during this expanse of time, starting with the configuration of land. All continental landmasses merged to form a single supercontinent—*Rodinia* (Russian for "homeland"). Rodinia stretched between the North and South Poles and then split to form other continental configurations. Climatically, Earth cooled to the point at which glaciers covered much of its surface. The reasons for such drastic cooling are unclear, but the fact that many of the continents were near either the North Pole or South Pole is certainly a factor.

Life in the Proterozoic flourished in new directions. Evolution produced single-celled organisms with a nucleus (eukaryotes). Fossils of such are found in rocks dated at approximately 1.5 billion years ago. Another important change was the debut of multicellular plants and animals (700 Ma). Proterozoic rocks in southern

Australia contain diverse fossils of soft-bodied animals and provide us with the first evidence of an animal community that lived in shallow marine water.

Probably the most important highlight is the accumulation of free oxygen in the atmosphere. The early Proterozoic atmosphere was still mostly nitrogen, with a little water vapor and carbon dioxide. But free oxygen released by photosynthesizing plants in the oceans began to collect in the air. Up until this time, expelled oxygen had been sequestered in the great iron deposits around the world. But as oxygen production outpaced its uptake by iron, free oxygen began to enrich the atmosphere. Then, with sufficient free oxygen (O_2) in the atmosphere, a primitive ozone layer (O_3) began to develop above Earth's surface. This ozone layer was very significant—it reduced the amount of harmful ultraviolet (UV) radiation reaching Earth's surface. The accumulation of free oxygen in Earth's atmosphere and the added protection from the newly formed ozone layer led to the emergence of new life.

CHECK POINT

1. In what two ways was the development of free oxygen essential to the development of new life forms?
2. Where did the carbon dioxide that characterized the early atmosphere go?

Were these your answers?
1. Free oxygen, in the form of O_3, provided protection from harmful UV rays; in the form of O_2, it provided oxygen for respiration.
2. Most of the carbon dioxide dissolved in the oceans, combined with dissolved calcium, and formed limestone ($CaCO_3$). Some was consumed by one-celled photosynthesizing organisms, which when dead, were also incorporated into rocks.

■ 21.5 The Paleozoic Era (543 to 248 Million Years Ago)

More is known about the **Paleozoic era** than the Precambrian, but the Paleozoic is very short in comparison. The Paleozoic era began about 543 million years ago, and it lasted about 295 million years. During this time, sea levels rose and fell worldwide several times. This allowed shallow seas to partially cover the continents and marine life to flourish. Varying sea levels greatly influenced the progression and diversification of life forms—from marine invertebrates to fishes, amphibians, and reptiles. An important event in the Paleozoic era was the evolution of shelled organisms. In fact, it is because of shelled organisms that we know so much more about the Paleozoic than Precambrian time. Shelled organisms have hard parts that are more likely to be preserved as fossils. The Paleozoic era is divided into six periods, each characterized by changes in life forms and changes in land configurations.

THE CAMBRIAN PERIOD (543 TO 490 MILLION YEARS AGO)

The Cambrian period marks the beginning of the Paleozoic era. Global temperatures were significantly warmer than during the preceding icy Proterozoic. In fact, fossil evidence

FIGURE 21.10

This "clock" shows the time when major groups of organisms appeared on Earth. The Paleozoic marked the development of a variety of animal and plant life.

FIGURE 21.11

Trilobites are the dominant fossils of the Cambrian period.

suggests that temperatures were much warmer than they are today. As for land configuration, there were no continental lands located at the poles and no significant ice formation. The ice sheets and glaciers from the Proterozoic melted, which resulted in rising sea levels. Low-elevation areas of the continents were flooded, and much of the land became covered by shallow seas. This flooding expanded the habitats for early Paleozoic marine organisms. These organisms thrived and evolved to produce a great diversity of life. Hence, this part of Earth's history has become known as the *Cambrian explosion*.

Almost all major groups of marine organisms came into existence during the Cambrian, as shown by abundant fossil evidence. The most important event of this time was that organisms evolved the ability to secrete calcium carbonate and calcium phosphate for the formation of outer skeletons, or shells. This helped organisms become less vulnerable to predators, and it provided protection against harmful UV rays. With outer skeletons, many organisms moved to shallower marine habitats.

So because of hard body parts, the fossil record of Cambrian life is well preserved and dominated by the skeletons of shallow marine invertebrates. A variety of organisms flourished, including the trilobite, the armored "cockroaches" of the Cambrian sea (Figure 21.11).

THE ORDOVICIAN PERIOD (490 TO 443 MILLION YEARS AGO)

During the Ordovician, the ancestral continental landmasses of South America, Africa, Australia, Antarctica, and India were merging together to form a new supercontinent called *Gondwanaland*. Throughout the Ordovician, Gondwanaland shifted southward, finally settling on top of the South Pole. With so much landmass over the South Pole, the latter part of the Ordovician was one of the coldest times in Earth history. Ice and massive glaciers covered much of Gondwanaland, draining the shallow seas. As sea level dropped, many shallow-water invertebrates were deprived of their habitat.

Fossil records show that the early to mid-Ordovician period was a time of great diversity and abundant marine life (Figure 21.12). The Ordovician also marks the earliest unquestionable arrival of vertebrates, including the jawless fishes. The end of the Ordovician period is marked by a surge in the rate of extinctions, probably the result of the widespread cooling and glaciation described above. The extinctions mainly affected shallow-water marine groups because of habitat loss, while deep-water organisms were relatively untouched.

fyi

- How do warming temperatures raise sea levels? Several ways: When ice caps and glaciers on land melt, the meltwater flows into the ocean, raising its level (the North Pole ice cap floats on open ocean; its melting would not raise sea level). Also, ocean water expands in volume as its temperature rises—this is an example of thermal expansion, which was discussed in Chapter 6.

FIGURE 21.12

The hagfish is one of the few remaining jawless fishes, a group that flourished in the Ordovician period.

THE SILURIAN PERIOD (443 TO 417 MILLION YEARS AGO)

During the Silurian period, the supercontinent of Gondwanaland remained close to the South Pole, and the ancestral continents of North America, Europe, and Siberia were located near the equator (Figure 21.13). The world's climate began to stabilize and warm, which resulted in the melting of many large glaciers and a general rise in sea level. Shallow seas moved in to cover many continental interiors. In North America, inland shallow seas were surrounded by coral reefs. These reefs blocked the circulation of water between the inland seas and the open oceans. As water in the inland seas evaporated, deposits of gypsum and other evaporite minerals were left behind. Evaporite beds formed in the Silurian are found today in Ohio, Michigan, and New York.

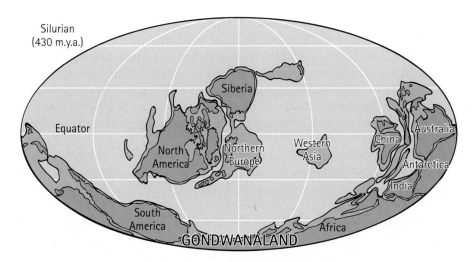

Silurian
(430 m.y.a.)

FIGURE 21.13

This map represents the ancestral continents as they may have been positioned during the Silurian period.

The Silurian period brought the emergence of terrestrial life—plants. The earliest known land plants with a well-developed "circulatory" system* appeared during the Silurian. These plants were closely connected to their water origins and inhabited only low wetlands. As plants moved ashore, so did other terrestrial organisms. Air-breathing scorpions and millipedes were common land animals during this period in Earth's history.

THE DEVONIAN PERIOD (417 TO 354 MILLION YEARS AGO)

During the Devonian period, the ancestral continents of Europe and North America merged to form another supercontinent known as *Laurasia* (ancestral Europe, North America, Siberia, and North China), which was positioned near the equator. The supercontinent of Gondwanaland remained in the Southern Hemisphere.

The Devonian climate was generally warm and moist. Plants spread over the land surface, and lowland forests of trees and ferns flourished. In the seas, fishes diversified into many new groups, which is why the Devonian is known as the "age of the fishes." Some groups, such as sharks and bony fishes, are still present today. Among the bony fishes, lobe-finned fishes are of particular interest because they gave rise to land-living, terrestrial vertebrates. Some lobe-finned fishes evolved internal nostrils, which enabled them to breathe air. In addition, the fins of these fishes were lobed and muscular, enabling the animals to support their bodies and "walk." Today, lungfishes and the coelacanth (pronounced "SEE-la-kanth") are the only lobe-finned fishes still in existence. The first truly terrestrial vertebrates, descendants of lobe-finned fishes, appeared during the late Devonian. These vertebrates share many features with the amphibians of today. For example, like today's amphibians, they laid unshelled eggs and could live only in wet environments (Figure 21.14).

THE CARBONIFEROUS PERIOD (354 TO 290 MILLION YEARS AGO)

The Carboniferous period includes both the Mississippian and Pennsylvanian subperiods. During this time, the Paleozoic ocean between Laurasia and Gondwanaland began to close. Throughout the Carboniferous and into the Permian, these two large landmasses began merging to become the supercontinent of **Pangaea** (Greek for "all lands"). The collision of these landmasses contributed to great mountain building—the Appalachian Mountains in North America, the Hercynian and Caledonian Mountains in Europe, and the Ural Mountains in Russia.

fyi

- The coelacanth was thought to have become extinct after the Mesozoic era. However, in 1938, the first living specimen was caught off the coast of East Africa. Since then, other specimens have been discovered in the Madagascar area. The coelacanth is now considered a "living fossil."

FIGURE 21.14

This artwork of a late Devonian forest shows an *Acanthostega tetrapod*, an amphibian, climbing over a rock while a dragonfly flies above. Tetrapods flourished in the Devonian, having evolved from fishes.

* The "circulatory" system that distributes water and other resources for a plant is called a *vascular system*. We use the common term *circulatory* because it is easy to visualize.

FIGURE **21.15**

Warm, moist climatic conditions contributed to the lush vegetation and swampy forests of the Carboniferous period. These forests produced most of the coal deposits around the world.

Pangaea stretched more or less continuously from pole to pole, with a large part of Gondwanaland at the South Pole and the northernmost tip of Laurasia at the North Pole. Such a stretch of land contributed to alternating periods of ice ages and warming throughout the Carboniferous. By the late Carboniferous, southern Gondwanaland was completely buried by large sheets of ice. The majority of the Laurasian landmass was still located along the equator and was characterized by warm, moist, tropical conditions that contributed to lush vegetation, forests, and swamps.

The name *Carboniferous*, or "carbon-bearing," refers to the swampy conditions that produced widespread coal deposits and characterized this part of Earth's history. In fact, dense swamps covered large portions of what are now North America, Europe, China, and Siberia (Figure 21.15). As plants and trees died, their remains settled to the bottoms of the stagnant swamps and decayed to produce coal (see the section titled "Link to Biology: Fossil Fuels" in Chapter 20). Most of the coal used today is derived from these Carboniferous coal swamps.

The Carboniferous period witnessed many life changes. Insects underwent rapid changes that led to diverse forms, including giant cockroaches and dragonflies with wingspans of 80 cm. The first amniotes, the vertebrate group that includes today's reptiles and mammals, also appeared at this time. Amniotes are characterized by a shelled, or *amniote*, egg. The amniote egg provides a completely self-contained environment for an embryo. The shell protects the embryo from drying out, allowing animals to complete the transition, begun by amphibians in the Devonian period, from aquatic environments to land. Thanks to the amniote egg, reptiles do not need to lay their eggs in water the way amphibians do.

THE PERMIAN PERIOD (290 TO 248 MILLION YEARS AGO)

The Permian period marks the end of the Paleozoic era. During this period, all the world's landmasses were still joined together as Pangaea (Figure 21.16). Like a chain reaction, the formation of Pangaea had many consequences. Continental collisions that started in the Carboniferous continued to bring about crustal disturbances that affected not only the edges of the continents but also their inner regions. The ancestral Rocky Mountains,* for example, owe their formation to the dramatic collision that produced Pangaea. In turn, mountain uplift blocked moisture-laden winds from the continental interiors, causing much of Pangaea's interior to become dry and arid. So as land configurations changed, so did the Permian climate.

FIGURE **21.16**

With the collision of continental landmasses, the supercontinent Pangaea was formed.

* The present-day Rocky Mountains formed much later, around 70 million years ago.

At the beginning of the Permian period, glaciers continued to cover much of the Southern Hemisphere, and tropical equatorial regions were covered by swampy forests. By the mid-Permian, however (after Pangaea was fully formed), the climate became warmer and milder—glaciers decreased and continental interiors became drier. All landmasses together also changed the level of the oceans—sea level was lowered. When sea level is lowered, more of the continental shelf (the area between land and the open ocean) is exposed. Because so much life lives in these shallow areas, these organisms began to lose their habitat. This is but one of the factors that may have led to the extinction of many life forms at the end of the period.

The late Permian is noted as being one of the greatest extinctions of animals in Earth's history. Marine invertebrates were affected more than terrestrial life. About 95% of all marine species and 70% of all land species became extinct. The cause of the extinction is not well understood. One likely explanation, of course, is the domino effect caused by the formation of Pangaea. The redistribution of land and water, the changes in landmass elevations, the change in the world's climate (temperature and precipitation), and the lowering of sea level all come into play.

The new land configuration radically decreased the amount of shallow marine environments. Overall, a large, single landmass has less coastline than that of the same landmass broken into several pieces. Each former piece had been encircled by coast. Put together, most of the former coasts of individual pieces became inland parts of Pangaea. All landmasses joined together also altered the oceanic and atmospheric circulation patterns that existed before the joining, which in turn altered the climate. And the long duration of lowered sea level, about 20–25 million years, undoubtedly placed more stress on already-stressed marine habitats. Whatever happened took a less drastic toll on terrestrial life. Terrestrial life, although affected, continued to evolve, and it expanded rapidly as new land habitats emerged, perhaps due in part to the lowered sea level.

The evolution of reptiles continued throughout the Permian. The reptiles must have been well suited to their environment, for they ruled Earth for 200 million years! (By comparison, modern humans have inhabited Earth for about 100,000 years.) Two major groups of reptiles appeared during the Permian: the diapsids and the synapsids. The synapsids, which include ancestors of the earliest mammals, dominated the Permian. The diapsids were less noticeable than the synapsids in the Permian, but it was the diapsids that eventually gave rise to the dinosaurs early in the Mesozoic era.

> Think of continental collisions like the crumpling of a flat cardboard cereal box. Initially the box is a "flat" surface that takes up only horizontal space. If you push the ends of the box together, the length decreases and the height increases. The box becomes "folded." That's what happens when continents collide; stretches of long, flat crust are shortened and pushed upward into mountains, becoming folded in the process. And mountains, like the crumpled box, occupy both horizontal and vertical space.

fyi

- In the next chapter, we will learn about plate tectonics. In short, Earth's outer layer is not a single shell. It is broken into a dozen or so "plates" that shift in response to heat flow in Earth's interior. Throughout Earth's history, the various plates have collided with one another, split apart from one another, and changed their configuration. For now, we will discuss how Earth's surface has changed through time and in Chapter 22, we will develop an understanding of the mechanisms that made these changes occur.

CHECK POINT

1. As continental landmasses merged together, coastlines and continental shelf areas were reduced by two different processes. Look at Figure 21.16 and explain the possible two processes.
2. Why would the consolidation of landmasses into the supercontinent Pangaea increase the seasonal variation they experienced?

Were these your answers?

1. Some of the former coastlines and continental shelves were no longer next to the ocean—they were now in the middle of Pangaea! With all landmasses crunched together, less water was displaced so sea level went down, meaning some former continental shelf areas became dry land.
2. The high specific heat capacity of water has a moderating effect on temperature (as explained in Chapter 6). A consolidated landmass has a large interior area far away from the moderating effect of water, thus increasing the seasonal temperature variation.

■ 21.6 The Mesozoic Era (248 to 65 Million Years Ago)

The **Mesozoic era** is separated from the Paleozoic era because of profound changes in fossils. Reptiles that survived the Permian extinction at the end of the Paleozoic evolved to become the rulers of the Mesozoic world. The Mesozoic era consists of three periods—the Triassic, Jurassic, and Cretaceous—and is informally known as "the age of reptiles." The most significant evolutionary event of this era was the rise of the dinosaurs. Mammals evolved from reptiles early in the Mesozoic, but they were relatively small and insignificant compared to the great dinosaurs.

Land plants greatly diversified during the Mesozoic. True pines and redwoods appeared, and they rapidly spread throughout the land. Flowering plants arose in the Cretaceous period, and they diversified so quickly that by the end of the period, they were the dominant plants. The emergence of the flowering plants also accelerated the evolution and specialization of insects.

The major geological event of the Mesozoic was the breakup of Pangaea. Just as Pangaea was formed in different phases throughout the Paleozoic, Pangaea broke apart phase by phase throughout the Mesozoic and into the Cenozoic (Figure 21.17). The first phase, during the Triassic/Jurassic (about 200 million years ago), occurred as Pangaea began to break apart into its former landmasses of Gondwanaland in the south and Laurasia in the north. The breakup was in response to the formation of a *rift zone* (a series of cracks related to plates spreading apart) that developed between what are today's North American and African continents. Out of the rift zone, basaltic lava erupted to form new crust that further separated the diverging continental landmasses.* Denser than granitic continental crust, this new oceanic crust formed at lower elevations and was the beginning of the central Atlantic Ocean basin. As the rift zone spread, Europe split from North Africa. In this manner, Laurasia became completely separated from Gondwanaland. Also during this time, rifting occurred within Gondwanaland. For example, Africa split from Antarctica and Madagascar as a volcanic rift zone formed in the western Indian Ocean basin.

Recall the mountain building that resulted from Pangaea's formation. As Pangaea broke apart, mountains on the western coast of North America began to form as the eastern side of the "Pacific" oceanic crust collided with the western side of the North American continental crust. The difference in rock density caused a different kind of collision from a continent-continent collision. Instead of a head-on collision like that of two continents, when oceanic crust collides with continental crust, the denser oceanic crust sinks below the less-dense continental crust. This *subduction* of oceanic crust produced widespread deformation, volcanism, and mountain building along the entire Western coast—from Alaska to Chile.

The second phase of the breakup occurred in the early Cretaceous, about 140 million years ago, as Gondwanaland continued to fragment. The rifting and separation of South America from Africa opened up the South Atlantic Ocean basin, and as ancestral India/Madagascar rifted away from Antarctica/Australia, the eastern Indian Ocean basin opened up. Of all the former continental unions that existed in Paleozoic time, only that of Europe and Asia has survived to the present time.

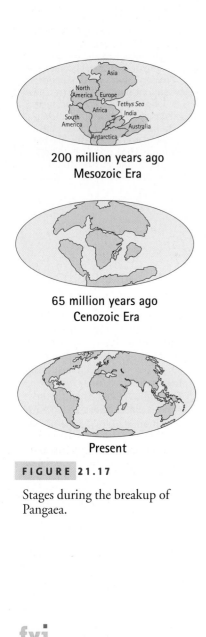

200 million years ago
Mesozoic Era

65 million years ago
Cenozoic Era

Present

FIGURE 21.17

Stages during the breakup of Pangaea.

■ The Andes Mountains and the Sierra Nevada have their beginnings in the great volcanic activity that rimmed the eastern Pacific basin during the Mesozoic.

---

* This process is called *seafloor spreading*. It is important to stress that Earth does not get larger when new crust is formed. Just as there are areas of crust formation, there are areas of crust destruction. So as new seafloor forms, old seafloor is destroyed at another location. The process of crust destruction is called *subduction*, and the area where it occurs is called a *subduction zone*. We will discuss these processes in detail in Chapter 22, but for now this short explanation may provide clarity.

The separation of landmasses had several worldwide consequences. Coastlines were once again more extensive, which increased shallow marine habitat. Land-based life—plants and animals—also changed. As formerly connected habitats became separated, members of the same species also became separated. Now in new, unconnected habitats, organisms that shared a common ancestor or were closely related began to diverge from one another—they became less and less alike. In this way, new species evolved according to their new habitats.

Also, with most landmasses no longer centered at the equator, oceanic circulation and atmospheric weather patterns changed once again. Warmer water from equatorial regions circulated northward, where it warmed northern landmasses. Climate varied from very hot and humid during the Triassic to cooler temperatures and pronounced seasons in the Cretaceous. Overall, however, the Mesozoic climate was still much warmer than conditions today. The rifting activity that initiated Pangaea's breakup resulted in a worldwide rise of sea level, and with shallow seas covering the continents, climate conditions became milder.

THE CRETACEOUS EXTINCTION

The end of the Cretaceous period, 65 million years ago, brought another mass extinction that killed more than 60% of Earth's species. Many dinosaurs, flying reptiles, and marine reptiles were wiped out, as were other organisms, both on land and in the seas.

The cause of this great extinction is still a source of some debate among scientists. The most popular explanation is a hypothesis by Luis and Walter Alvarez (1980). They propose that the extinction was caused by the impact of a very large meteorite. Support for their hypothesis comes from an abundance of the element iridium found in sediments that mark the boundary between the Cretaceous and Tertiary periods (the *K-T boundary*). The concentration of iridium in a meteorite is about the same as the concentration of iridium in Earth as a whole. But most of the iridium in Earth is found in the deeper interior, not in the crust. Thus, the iridium concentration in a meteorite is much higher than the concentration of iridium in Earth's crust. Yet all over the world, the concentration of iridium at the K-T boundary is much greater than it is in sediments above or below the boundary. This strongly suggests that iridium was probably spread worldwide by an impact event. The K-T boundary layer was deposited about 65 million years ago—the time of the great Cretaceous extinction.

The Alvarez team hypothesized that a large meteorite hit Earth with such force that a gigantic light-blocking cloud of dust developed. The dust cloud lasted for months, perhaps even longer. The huge cloud stopped photosynthesis, catastrophically reduced the food supply, and chilled Earth. Finally, as the dust settled, the layer of iridium-enriched sediment was deposited. Other killing mechanisms associated with a meteorite impact of this size would include acid rain, tsunamis, and wildfires. The Chicxulub crater, located off the Yucatan coast in Mexico, is the hypothesized site of impact (Figure 21.18). The crater is large enough (180 km in diameter) and dates back to the close of the Cretaceous period.

An alternative to the Alvarez hypothesis suggests that the iridium layer may have been generated from massive volcanic eruptions. The ash and debris from these eruptions also could have blocked out the Sun. A third possibility is that large-scale volcanic eruptions could have been caused by the impact of an extraterrestrial object. Whatever the cause, the Cretaceous extinction dramatically marked the close of the Mesozoic era.

FIGURE 21.18

Artwork of the Chicxulub crater on the Yucatan Peninsula of Mexico, soon after its formation. The crater is about 180 km in diameter. The impact that produced the crater may have caused the mass extinction that ended the reign of the dinosaurs, 65 million years ago at the end of the Cretaceous.

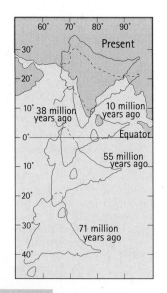

FIGURE 21.19

The formation of the Himalayas was a result of the collision of India with Asia. Because this was a continent-to-continent collision, the Himalayas have an unusually large thickness of continental crust.

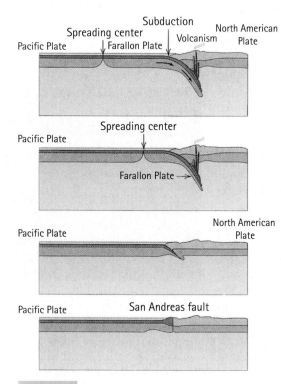

FIGURE 21.20

Subduction sequence of the Farallon Plate beneath the North American Plate. As the spreading center and the continental margin approached one another, the San Andreas Fault began to form between the Pacific Plate and the North American Plate.

CHECK POINT

1. Could the breakup of Pangaea account for the extinction at the end of the Cretaceous?
2. In terms of the number of species destroyed, which was the bigger extinction—the Permian or Cretaceous?

Were these your answers?

1. There is no doubt that the reconfiguring of landmasses when Pangaea broke up put added stress on many areas of life. Climate changed, shallow marine habitats changed, species became separated and diverged from one another, moving landmasses triggered widespread volcanic activity, and so on. But the breakup of Pangaea cannot explain the high concentration of iridium at the K-T boundary.
2. The Permian was the bigger mass extinction, having killed about 95% of all marine species and 70% of all land species. The Cretaceous extinction resulted in the demise of about 60% of Earth's species.

21.7 The Cenozoic Era (65 Million Years Ago to the Present)

The **Cenozoic era,** known as the "age of the mammals," is made up of two periods—the Tertiary and the Quaternary. From oldest to youngest, these two periods are broken into the Paleocene, Eocene, Oligocene, Miocene, and Pliocene epochs (for the Tertiary period) and the Pleistocene and Holocene epochs (for the Quaternary period). We are currently living in the Holocene epoch.

The third and final phase of Pangaea's breakup took place during the early Cenozoic. North America and Greenland split away from Europe, and Antarctica split from Australia. Earth's surface was very active and landmass collisions were numerous. Considerable mountain-building activity occurred when Africa-Arabia collided with Europe to produce the Alps, and India collided with Asia to produce the Himalayas. As shown in Figure 21.19, India moved toward Asia at a rate of 15–20 cm/yr—a tectonic speed record! As India partially wedged itself below Asia, it generated an unusually thick accumulation of continental land. This continental land atop continental land provided additional height to the Himalayas.

The late-Mesozoic collision that began the formation of the Sierra Nevada and Andes Mountains continued (and continues) into the Cenozoic. Off the western coastline of North America, a zone of seafloor spreading—a midocean ridge—formed in the Pacific Ocean. This ridge separated two different oceanic tectonic plates—the Pacific on the west and the Farallon on the east (Figure 21.20). As the Farallon Plate moved eastward (to descend below the continental landmass of North America), the spreading ridge began to approach the North American continental margin. This not only reactivated mountain building throughout the west but also contributed to widespread deformation. The collision of the spreading ridge, formally called the Pacific Ridge System, with the westward moving North American continent occurred about 30 million years ago and gave birth to the San Andreas Fault (Figure 21.21). In time, as the fault grew, Baja California was torn

FIGURE 21.21

INTERACTIVE FIGURE

The San Andreas Fault is the result of an encounter between the North American Plate and the Pacific Ridge System. As the fault grew longer, the area of Baja California was torn from the Mexican mainland.

away from the Mexican mainland, and the Gulf of California was created. Because these plates are still moving, western California and Baja California will eventually become completely detached from the mainland or will find themselves joined to western Canada.

The Hawaiian Island/Emperor Seamount chain (Figure 21.22) provides additional evidence of Cenozoic deformation: the change in direction of the Pacific Plate. The bend in the chain of islands occurred between 30 and 40 million years ago (mid-Tertiary) when the direction of plate motion changed from nearly due north to northwesterly. The change in direction occurred at about the same time in Earth history as the collision of northern Mexico (the North American Plate) with the Pacific Ridge System.

Climates cooled during much of the Cenozoic era, resulting in the widespread glaciation that characterized the Pleistocene. Although this ice age continues today, there have been many alternations between glacial and interglacial conditions. During the glacial peak, as much as one-third of Earth's surface was covered by ice. On land, great thicknesses of ice formed continental glaciers. The great weight of ice depressed the land and altered the courses of many streams and rivers. The glaciers eroded and scratched the land in some places and deposited huge *moraines,** leaving behind abundant evidence of the extent of their former existence.

---

* A glacial moraine is a sediment deposit created by glaciers. Moraines mark the former boundaries of a glacier after it has melted. This topic will be discussed further in Chapter 23.

FIGURE **21.22**

The Hawaiian Island/Emperor Seamount chain. The bend in the chain shows the change in direction of the Pacific Plate, which is likely a result of the collision of northern Mexico with the Pacific Ridge. The red numbers indicate the age (in millions of years) of the individual islands and seamounts.

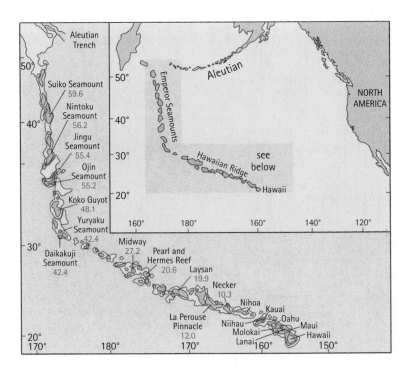

CENOZOIC LIFE

After the mass extinctions at the end of the Mesozoic era, many environmental niches—habitats—were left vacant. These openings allowed the relatively rapid evolution of mammals. Bats, some large land mammals, and such marine animals as whales and dolphins evolved to occupy niches left vacant by the extinction of many of the Mesozoic reptiles. Later in the Cenozoic, global temperatures cooled and the Ice Age ensued. The cool temperatures had a profound impact on life. Mammoths, rhinos, bison, reindeer, and musk oxen all evolved warm, woolly coats for protection from the frigid cold.

Humans evolved in the Quaternary period during the Pleistocene epoch. Extensive glaciation caused sea level to drop as water became bound in glaciers and ice caps. Even though the distribution of landmasses was essentially the same as it is today, the lowered sea level resulted in "land bridge" connections between landmasses that are now separated by water. One of these land bridges existed across the present-day Bering Strait, and it provided the route for the human migration from Asia to North America.

The expansion of humans, not only into North America but also throughout the world, coincided with a period of extinctions that occurred during the Pleistocene. The extinctions primarily involved large terrestrial mammals, while marine animals were for the most part unaffected. In North America, many large mammals became extinct after humans arrived, and in Africa, mammalian extinctions can be related to the appearance of the Stone Age hunters.

The cause of the Pleistocene extinction is a much-debated issue. The extreme climatic variation that existed at the time could have been partly responsible. However, though large-scale glaciation was occurring in some regions, the climate in many areas was relatively mild. This leads scientists to believe that harsh climate likely played only a small role in the Pleistocene extinctions, and that humans essentially hunted and ate the large mammals to extinction. However, the debate is not completely settled. Some skeptics state that it is unclear whether *Homo sapiens* were present in sufficient numbers or had sufficient technological know-how to kill off entire species.

Following the Pleistocene is the Holocene epoch. To observe the Holocene environment, just look around. It is the most recent 10,000 years or so of Earth's

It would not have been possible for large mammals to evolve if dinosaurs had not gone extinct. Such mammals would have been tasty treats for large meat-eating dinosaurs!

LINK TO   Global Thermodynamics

▪ Is It Cold Outside?

Yes it is, relatively speaking. For 90% of Earth's history, there were no glaciers of continental scale anywhere. In fact, because such glaciers exist today, mainly in the polar ice caps and in Greenland, we are technically now in an *ice age*. Because these continental-scale glaciers are restricted to polar regions, we are in what is known as an *interglacial* period of an ice age.

Ice ages have occurred five times over the course of Earth's history. The first one for which we have evidence occurred more than 2 billion years ago. Another began about 840 million years ago and lasted about 200 million years! There were two ice ages during the Paleozoic era, but none in the Mesozoic. For the first 50 million years or so of the Cenozoic era, there were also no ice ages. The present ice age actually began 8–10 million years ago, but the extensive glaciation that characterized the Pleistocene epoch began about 1 million years ago.

So what causes ice ages? There is, most likely, no single explanation, but most scientists agree that global-scale cooling leading to ice ages is caused by the right combination of three things: (1) the arrangement of continents around the globe, (2) the amount of sunlight being reflected back into space, and (3) the geometry of Earth's rotation on its axis and revolution around the Sun.

The arrangement of continents greatly influences ocean and atmospheric currents, which are the main mechanisms for redistributing thermal energy around the globe. Continents grouped in one location are easier to warm, as equatorial waters flow with less obstruction toward the poles. For continents that are spread out around the globe, as they are today, circulation "cells" are smaller and heat redistribution is more local and less efficient.

When sea level is lower, for whatever reason, more land area is exposed. The increased amount of land area tends to increase the amount of sunlight reflected back into space. This phenomenon results in cooler temperatures globally. Cloud cover and/or dust in the atmosphere also causes sunlight to be reflected back into space, reducing the absorption of solar radiation.

The *Milankovitch effect* refers to a combination of factors that affect the distribution of solar radiation over Earth's surface during the year: (1) variations in the angle at which Earth's rotational axis is tilted (currently about 23.5°), (2) the wobbling of Earth's rotational axis, and (3) variations in the eccentricity ("ovalness") of Earth's orbit around the Sun. Certain combinations of these factors, which recur periodically, lead to reduced solar radiation at high northern latitudes during the summer. If the reduction of that radiation is great enough, then not all the snow from the preceding winter melts and, if these conditions persist over many years, continental-scale glaciers eventually form.

The periodic nature of the Milankovitch effect may be the primary cause of glacial–interglacial cycles. According to this theory, the first two processes—the arrangement of continents and the amount of reflected radiation—make Earth cold enough, and the third—the Milankovitch effect—causes the climate to teeter-totter between glacial and interglacial periods.

history, including the present time. There has been some climatic variation. For example, the "Little Ice Age" occurred between about AD 1200 and 1700. However, in general, the Holocene has been a relatively warm interglacial period.

The Holocene is sometimes called the "Human Age." This is somewhat inaccurate because *Homo sapiens* had evolved and occupied many regions of the globe well before the start of the Holocene. Yet the Holocene has witnessed all of humanity's recorded history, including the rise and fall of all historic civilizations. Humanity has had a great impact upon the Holocene environment. All organisms influence their environments to some degree, but few have ever changed Earth as much, or as fast, as we are doing in modern times.

▪ 21.8 Earth History in a Capsule

Our Earth has a long and exciting history—4.5 billion years' worth! To unravel this history we look at the rock record. Each event in Earth's long history is recorded in its rock layers. The order of geologic events is deciphered by relative dating, and absolute age is determined by radiometric dating. The geologic time scale provides a chronological listing of different time intervals according to the fossil record. When major changes in life occurred, a new era and period began.

Human Geologic Force

Although the "human age" amounts to only a brief 0.002% of geologic time, we are almost certainly the most clever and adaptable organisms to have evolved on the planet. All life forms alter their environments. Humans do it more, as we manipulate our environment to meet our needs. We have but to look at the irrigation systems of Mesopotamia, the cultivation of the Nile, the plowing of the prairies in the Great Plains, the invention of machines to further utilize the land, and the dams and locks on the Mississippi, Missouri, and Colorado rivers to illustrate the human role in geologic changes. Will the human activities that adversely affect our planetary life-support system—the large-scale burning of fossil fuels, the deterioration of the ozone layer, the destruction of the Amazon Rain Forest, and of particular concern, global warming—ultimately lead to a new period of extinctions? Perhaps our own. Who knows? Because we have the capacity to affect geologic change, it is imperative that we take care of our terrestrial home. It's the only one we have!

Throughout geologic time, life has changed, landforms have changed, and climate has changed. Change in one area influences change in another area. Each change is recorded in the rocks. As mentioned earlier, the Earth scientist is like a crime scene investigator. The "crime scene" is Earth, and the rock record provides us with clues to unravel *what* happened and *when* it happened. Now we need to look deeper to get an understanding of why some of these Earth-moving events happened. And that brings us to the topic of our next chapter—plate tectonics. We have learned about how Earth's surface has changed over time; now we need to understand the mechanisms that are responsible for these changes. Plate tectonics ties it all together. Earth is a big puzzle, and we understand the picture by fitting in one piece at a time.

SUMMARY OF TERMS

Relative dating The ordering of rocks in sequence by their comparative ages.

Original horizontality A relative dating principle stating that layers of sediment are deposited evenly, with each new layer laid down nearly horizontally over the older sediment.

Superposition A relative dating principle stating that in an undeformed sequence of sedimentary rocks, each bed or layer is older than the one above and younger than the one below.

Cross-cutting relationships A relative dating principle stating that where an igneous intrusion or fault cuts through other rocks, the intrusion or fault is younger than the rock it cuts.

Inclusions A relative dating principle stating that any inclusion (pieces of one rock type contained within another) is older than the rock containing it.

Lateral continuity A relative dating principle stating that sedimentary layers are deposited in all directions over large areas until some sort of obstruction, or barrier, limits their deposition.

Faunal succession A relative dating principle stating that fossil organisms succeed one another in a definite, irreversible, and determinable order.

Unconformity A break or gap in the geologic record, caused by erosion of preexisting rock or by an interruption in the sequence of deposition.

Angular unconformity An unconformity in which older tilted strata are overlaid by horizontal younger beds.

Radiometric dating A method for calculating the age of geologic materials based on the nuclear decay of naturally occurring radioactive isotopes.

Precambrian time The time of hidden life, which began about 4.5 billion years ago when Earth formed, lasted until about 543 million years ago (beginning of the Paleozoic), and makes up almost 90% of Earth's history.

Paleozoic era The time of ancient life, from 543 million years ago to 245 million years ago.

Pangaea The late-Paleozoic supercontinent made up of Gondwanaland (ancestral South America, Africa, Australia, Antarctica, and India) and Laurasia (ancestral North America, Europe, and Siberia/Asia).

Mesozoic era The time of middle life, from 245 million years ago to about 65 million years ago.

Cenozoic era The time of recent life; it began 65 million years ago and is still ongoing.

REVIEW QUESTIONS

21.1 The Rock Record—Relative Dating

1. What is relative dating?
2. What six principles are used in relative dating? Describe each one.
3. A granitic dike is found across a sandstone layer. What can be said about the relative ages of the dike and the sandstone? What principle applies here?
4. Why don't all rock formations show a continuous sequence from the beginning of time to the present?
5. How are fossils used in determining geologic time?
6. In a sequence of sedimentary rock layers, the oldest layer is on the bottom and the youngest layer is at the top. What relative dating principle applies here?
7. Explain how fossils of fishes and other marine animals occur at high elevations, such as the Himalayas.

21.2 Radiometric Dating

8. What is meant by radioactive half-life?
9. What are the half-lives of uranium-238, potassium-40, and carbon-14?
10. What isotope is preferred in dating very old rocks?
11. What isotope is commonly used for dating sediments or organic material from the Pleistocene?

21.3 Geologic Time

12. Which of the geologic time units spans the greatest length of time?
13. How old is Earth?

21.4 Precambrian Time

14. What key developments in life occurred during Precambrian time?
15. What evidence do we have of Precambrian life?

21.5 The Paleozoic Era

16. The Paleozoic era experienced several fluctuations in sea level. What effect did this have on life forms?
17. Name the periods of the Paleozoic era.
18. For what type of life form is the Silurian period best known?
19. The Devonian period is known as "the age of fishes." What were some of the Devonian life forms?
20. Why is the evolution of internal nostrils in the lobe-finned fishes considered an important step in the evolution of life on Earth?
21. Why do many geologists consider the lobe-finned fishes especially significant?
22. During what time period were most coal deposits laid down? Why was this period unique?
23. In what area of the United States do we find rich coal deposits?
24. What group evolved from the amphibians with the arrival of the amniote egg?

21.6 The Mesozoic Era

25. By what informal name is the Mesozoic era known?
26. What is the most likely cause of the Cretaceous extinction that wiped out the dinosaurs?
27. What effect did the breakup of Pangaea have on sea level?
28. How does the element iridium relate to the time of the extinction of the dinosaurs?

21.7 The Cenozoic Era

29. Which epochs make up the Tertiary period? The Quaternary period?
30. What important life forms evolved during the Cenozoic era?

EXERCISES

● BEGINNER   ■ INTERMEDIATE   ◆ EXPERT

1. ● Suppose you see a sequence of sedimentary rock layers overlaid by a basalt flow. A fault displaces the bedding of the sedimentary rock but does not intersect the basalt flow. Relate the fault to the ages of the two rock types in the formation.
2. ● If a sedimentary rock contains inclusions of metamorphic rock, which rock is older? Defend your answer.
3. ◆ How old are the oldest rocks on Earth? About how old are the oldest rocks on the Moon?
4. ◆ Which isotopes are most appropriate for dating rocks from the following ages? (a) early Precambrian time (b) the Mesozoic era (c) the late Pleistocene epoch

5. ■ Has the amount of uranium in Earth increased over geologic time? Has the amount of lead in Earth increased?
6. ◆ Granitic pebbles within a sedimentary rock have a radiometric age of 300 million years. What can you say about the age of the sedimentary rock? Nearby, a dike having a radiometric age of 200 million years intrudes an outcrop of the same sedimentary rock. What can you say about the age of the sedimentary rock?
7. ● Before the discovery of radioactivity, how did geologists estimate the age of rock layers?
8. ■ In dating a mineral, what is meant by "resetting the mineral's time clock"?

9. ◆ If we divide a number by 2, and then divide the result by 2, and so on indefinitely, the answer will never be zero. Why then is carbon dating useful only for materials that are no older than about 50,000 years? (Hint: What is the half-life of carbon-14?)

10. ● Geologists often refer to the early Paleozoic era as the *Cambrian explosion*. What do you think is meant by this term?

11. ■ Two isolated rock outcrops that share similar characteristics—sequence of layers and fossil assemblage—are separated by 50 km. Could these two outcrops be related? What principle helps confirm their relationship?

12. ■ Two isolated rock outcrops share a few similar characteristics—sequence of layers, mineralogy, and certain fossil assemblages—but not all layers are present. The rocks are separated by 300 km. Could the two outcrops be related? How can you explain the missing layers?

13. ● Name the eons in Precambrian time. Describe the eons, using no more than three words for each eon.

14. ◆ Suppose that in an undeformed sequence of rocks, you find a trilobite embedded in a black shale layer at the bottom of the formation and fossil leaves embedded in a green shale at the top of the formation. From your observation, what can you say about the ages of the formation?

15. ◆ In a sequence of sedimentary rock layers, the youngest layer is found at the bottom and the oldest layer at the top. What does this type of layering signify?

16. ● What is the difference between a nonconformity and an angular unconformity?

17. ■ How did the Precambrian atmosphere become nitrogen enriched?

18. ■ What factors are believed to have contributed to the generation of free oxygen during the late Precambrian? In what way did the increase in oxygen affect our planet?

19. ■ Before entering the atmosphere, where did much of the newly released oxygen end up during the Precambrian?

20. ● Why are Paleozoic marine sedimentary rocks such as limestone and dolomite found widely distributed in the continental interiors?

21. ■ Throughout the Paleozoic, sea level was variable; sometimes it was high and other times it was low. What was the primary cause of this variation in sea level?

22. ■ What was the coldest period of the Paleozoic? What may have contributed to this cooling period?

23. ■ During the Silurian, many continental landmasses were covered by shallow seas. Yet this time period marks the emergence of land plants. How can this be?

24. ● Coal beds form from the accumulation of plant material that has become trapped in swamp floors. Yet coal deposits are found on the continent of Antarctica, where no swamps or vegetation exists. What is your explanation?

25. ■ A radiometric date is determined from mica that has been removed from a rock. What does the date signify if the mica is found in granite? What does the date signify if the mica is found in schist?

26. ■ Most scientists think the iridium-rich sediments found straddling the Cretaceous/Tertiary boundary can be explained by a meteorite impact. Why are high concentrations of iridium significant?

27. ● During Earth's long history, life has emerged and life has perished. Briefly discuss the emergence of life and the extinction of life for each era.

28. ● In what ways can sea level be lowered? What effect might this have on existing life forms?

29. ■ What can cause a rise in sea level? Is this likely to happen in the future? Why or why not?

30. ■ What general assumption must be made to understand the processes that occurred throughout Earth's history?

31. ◆ What are some potential worldwide consequences that can occur when all landmasses are joined together, as in the formation of Pangaea?

32. ■ If fine muds were laid down at a rate of 1 cm/1000 yr, how long would it take to accumulate a sequence 1 km thick?

33. ◆ Why does sea level rise when the rate of seafloor spreading increases?

34. ● If sea level were to rise today, what land areas would be most affected? What life forms would be in danger of extinction?

35. ◆ What circumstances are likely to lead to the formation of continental-scale glaciers?

36. ◆ How does basaltic lava in a rift zone separate two landmasses?

37. ■ What is a likely cause of glacial–interglacial cycles? Are we currently in an ice age? Explain.

38. ■ What impact did the breakup of Pangaea have on sea level?

39. ● How have recent humans affected geological processes?

40. ◆ What geological event most likely resulted in the bending of the Hawaiian Island/Emperor Seamount chain?

41. ■ How did Pleistocene glaciations affect the land surface?

42. ■ How did the San Andreas Fault form?

43. ● How did the loss of coastlines affect shallow marine life?

44. ● What event allowed the evolution of many mammals in the early Cenozoic era?

45. ■ How was the Gulf of California formed?

PROBLEM

● **BEGINNER**   ■ **INTERMEDIATE**   ◆ **EXPERT**

1. ■ Refer to the figure. Using the principles of relative dating, determine the relative ages of the rock bodies and other lettered features. Start with this question: What was there first?

Sequence of events

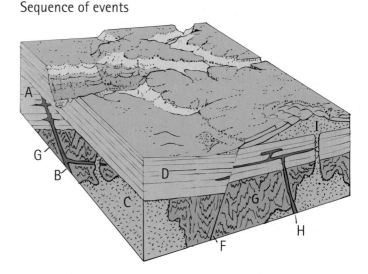

READINESS ASSURANCE TEST (RAT)

If you have a good handle on this chapter, if you really do, then you should be able to score at least 7 out of 10 on this RAT. If you score less than 7, you need to study further before moving on.

Choose the BEST answer to each of the following.

1. The principle of superposition is that each
   (a) sedimentary layer is older than the layer above.
   (b) sedimentary layer is younger than the layer above.
   (c) new layer of sediment is laid down nearly horizontally.
   (d) new layer of sediment is laid down accordingly.

2. The principle that different forms of animals throughout Earth's past occurred in a definite order is called the principle of
   (a) fossil assemblage.
   (b) faunal succession.
   (c) conformable fossils.
   (d) fossil determination.

3. The time it takes for 50% of a radioactive substance to decay is known as
   (a) radiometric dating.
   (b) carbon-14.
   (c) the proportion of atoms remaining.
   (d) the half-life.

4. Development of Earth's oceans was likely due to
   (a) water-rich meteors bombarding Earth's surface.
   (b) volcanic outgassing in Precambrian time.
   (c) slow convection in the mantle.
   (d) volcanic outgassing in the early Paleozoic.

5. The buildup of free oxygen was crucial to the emergence of life on Earth because it led to the formation of
   (a) O_2 for plants.
   (b) ozone, O_3, which helped screen Earth from harmful incoming UV radiation.
   (c) ozone, O_3, which primitive organisms could breathe.
   (d) the oceans, where life emerged.

6. The Paleozoic experienced several fluctuations in sea level. When sea level rises,
   (a) shallow seas cover the continents.
   (b) more water is tied up in glaciers, making the climate colder.
   (c) the climate turns warmer and swamps form.
   (d) ocean basins become shallow.

7. The most important event during the Cambrian period was the
   (a) emergence of the fishes.
   (b) ability of organisms to form an outer skeleton.
   (c) emergence of the trilobite.
   (d) ability of organisms to develop lungs.

8. The formation of the supercontinent of Pangaea
   (a) resulted from the collision of all major landmasses.
   (b) produced widespread mountain building in the Himalayas.
   (c) resulted in extensive volcanic activity and flood basalts.
   (d) all of these.
9. Glaciation during the Cenozoic resulted in
   (a) lowering of sea level worldwide.
   (b) carving Earth's surface (for example, the Swiss Alps).
   (c) land bridge connections between various continents.
   (d) all of these.

10. The birth of the San Andreas Fault corresponds to
    (a) the collision of the Pacific Plate and the Hawaiian Islands.
    (b) the collision of the Pacific Ridge System and North America.
    (c) the movement of the Pacific Plate over a turbulent zone.
    (d) the collision of India and Eurasia.

Answers to RAT
1a, 2b, 3d, 4b, 5b, 6a, 7b, 8a, 9d, 10b

EXPLORING FURTHER

Gould, Stephen Jay, Wonderful Life—The Burgess Shale and the Nature of History. New York: Norton, 1989.

www.scotese.com/earth.html
This is the Web site of the PALEOMAP Project, winner of *Scientific American*'s Science and Technology Web Award two years running. The Earth History section displays a series of full-color paleogeographic maps of past, present and future Earth. Go to the home page for animations, resource suggestions for software, and teaching tools.

www.ucmp.berkeley.edu/help/timeform.html
The Geologic Time Machine page of the UC Berkeley Museum of Paleontology Web site provides an easy-to-use timeline, full of links to detailed descriptions and photographs.

CHAPTER 21 ONLINE RESOURCES

The Physical Science Place

Interactive Figures
- 21.6, 21.8, 21.21

Tutorial
- Formation of an Angular Unconformity

Quiz

Flashcards

Links

plate tectonics and earth's interior

22

22.1 SEISMIC WAVES

22.2 EARTH'S INTERNAL LAYERS

22.3 CONTINENTAL DRIFT—AN IDEA BEFORE ITS TIME

22.4 ACCEPTANCE OF CONTINENTAL DRIFT

22.5 THE THEORY OF PLATE TECTONICS

22.6 CONTINENTAL EVIDENCE FOR PLATE TECTONICS

22.7 THE THEORY THAT EXPLAINS THE GEOSPHERE

■ If it were possible to dig a hole straight through Earth, what would we find in its interior? Because digging such a hole is impossible, what tools and techniques can we use to explore Earth's insides? Investigation begins with the rocks on the surface, which tell us a great deal about Earth's interior. Earthquakes and volcanic eruptions are also a link to the inner workings of our planet. Observations and careful measurements of their behavior can offer additional clues. All these features and processes are external expressions of Earth's internal processes.

FIGURE 22.1

Cross section of Earth's interior showing the four major layers and their approximate thickness.

FIGURE 22.2

Diagram of a seismograph. When Earth moves, the support unit attached to the ground also moves, but because of inertia, the mass at the end of the pendulum tends to stay in place. A pen attached to the mass marks the relative displacement on the slowly rotating drum beneath. In this way, the seismograph records ground movement.

FIGURE 22.3

INTERACTIVE FIGURE

Block diagrams show the effects of seismic waves. The yellow portion on the left side of each diagram represents the undisturbed area. (a) Primary body waves alternately compress and expand the material through which they travel—as shown by the different spacing between the vertical lines— similar to the action of a spring. (b) Secondary body waves cause the material through which they travel to oscillate up and down and from side to side. (c) Love surface waves whip back and forth like secondary body waves, but only in the horizontal direction. (d) Rayleigh surface waves have a rolling, up-and-down motion, similar to ocean waves.

■ 22.1 Seismic Waves

Earthquakes, besides being fearsome and destructive events, provide a key to understanding Earth's internal structure. An **earthquake** is the shaking or trembling of the ground that happens when rock under Earth's surface moves or breaks. These internal movements generate waves that travel through Earth's interior and across Earth's surface—*seismic waves*. The speed at which seismic waves travel and the paths that they take provide scientists a view into Earth's interior. What has been discovered is a layered planet. The major layers of Earth consist of the *crust, mantle, outer core,* and *inner core* (Figure 22.1).

Recall from Chapter 10 that a wave's speed depends on the medium through which it travels. We learned that sound waves generated by clicking two submerged rocks together travel faster through water than through air. And sound waves travel even faster through a solid. Just like sound waves, the speed of seismic waves depends on the elasticity and density of the material through which they travel. The greater the elasticity and density, the greater the wave speed. So measuring the speeds of seismic waves provides clues about Earth's composition.

Energy released during an earthquake travels in the form of seismic waves and radiates in all directions within Earth's interior. As the energy travels to Earth's surface, the ground shakes and moves. This ground movement is recorded on a *seismograph* (Figure 22.2). By combining seismograph records—*seismograms* from many different earthquakes, Earth's interior is revealed. So just as an X-ray reveals the interior of your body, a seismogram reveals Earth's interior.

There are two types of seismic waves: **body waves,** which travel through Earth's interior, and **surface waves,** which travel along Earth's surface (Figure 22.3). Body waves are further classified as either **primary waves** (P-waves) or **secondary waves** (S-waves). Primary waves, like sound waves, are longitudinal—they compress and expand the rock as they move through it. Like vibrations in a bell, primary waves move out in all directions from their source. Primary waves are the fastest of all seismic waves and so are the first to register on a seismograph. Because both solids and fluids can compress and expand, P-waves can travel through any type of material—solid granite, magma, water, or air. Secondary waves, like the waves produced on a vibrating violin string, are transverse—they vibrate the particles of their medium up and down and from side to side, perpendicular to the direction of wave travel. Because S-waves travel more slowly than P-waves, they are the second waves to register on a seismograph. S-waves cannot move through fluids—they travel only through solids.

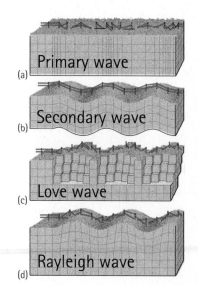

There are also two types of surface waves: *Rayleigh waves* and *Love waves*. Rayleigh waves have an up-and-down rolling motion, and Love waves have a side-to-side, whiplike motion. Both types of surface waves travel more slowly than P-waves and S-waves, and therefore they are the last to register on a seismograph.

In Earth's interior, seismic waves are reflected by the "surfaces" between differing materials. And when seismic waves pass into a different material, their wave speed changes, causing the wave to refract (Chapter 10). Geoscientists study the reflection, refraction, and speeds of the various types of seismic waves to piece together a story about Earth's interior. Seismic-wave research has revealed the architecture of Earth's internal layers (Figure 22.4).

22.2 Earth's Internal Layers

Near the beginning of the 20th century, Irish geologist Richard Oldham was examining records of a massive earthquake in India when he discovered that its S-waves traveled some distance through Earth and then stopped. He also observed that the P-waves traveled as far as the S-waves into Earth but then refracted at an angle and lost speed. Because S-waves cannot travel through liquid, but P-waves can (at a reduced speed), Oldham deduced that the earthquake waves had encountered an internal boundary—he had discovered Earth's core. The year was 1906.

Three years later, Croatian seismologist Andrija Mohorovičić (pronounced "moho-rovu-chick") analyzed seismic readings from a recent earthquake. He recognized a sharp increase in the speed of seismic waves at another boundary, one that lay at a shallower depth below Earth's surface. Knowing that wave speed depends on the properties of the material through which the wave passes, Mohorovičić concluded that the increase in speed was due to a density change within Earth—wave speed increased as it passed from a lower-density solid to a higher-density solid. Mohorovičić's seismographic data had literally drawn a map of the upper boundary of Earth's *mantle*, a layer of denser rock underlying the less dense crust. This boundary, known as the **Mohorovičić discontinuity** (called the "Moho" for short), separates Earth's crust from rocks of different composition in the mantle below.

In 1913, Beno Gutenberg reinforced Oldham's earlier findings by observing that both P-waves and S-waves are strongly influenced by a pronounced boundary approximately 2900 km deep—the core–mantle boundary. Seismic observations revealed that when P-waves reach this depth, they are reflected and refracted so strongly that the boundary actually casts a P-wave shadow over part of Earth (Figure 22.5). The shadow is a region where no waves are detected. The wave shadow develops between 105° and 140° from the surface location of the earthquake (the *epicenter*) and has no direct penetration of seismic waves. Because the boundary is so distinct, it marks an important change in the density of the materials present in Earth's interior. Both the overall density of Earth and the speed with which seismic waves travel through the core suggest that the core is composed of iron, a material that is much denser than the silicate rocks that make up the mantle. Furthermore, the sharp boundary between the mantle and core casts an S-wave shadow that is even more extensive than the P-wave shadow—S-waves are unable to pass through the core. Knowing that S-waves travel only through solids, English seismologist Sir Harold Jeffreys confirmed in 1926 that the core, or part of it, must be liquid.

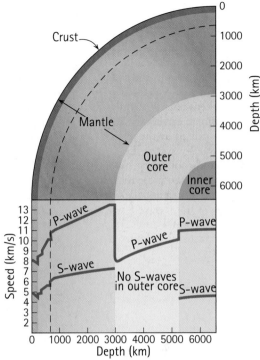

FIGURE 22.4

Cross section of Earth's internal layers, showing the increases and decreases of P-wave and S-wave velocity in the different layers.

The Physical Science Place
The Mantle and Crust

fyi

- How does density relate to elasticity? Elasticity is related to how rigid and springy a material is. It is a measure of a solid's ability to recover its shape once a deforming force is removed from it. For example, steel has a high elasticity, while fresh bread has a low elasticity. (A rubber band, interestingly, is called elastic because it returns to its original shape when released.) The density of rock beneath Earth's surface increases because it is compressed by the weight of material above it. The more the rock is compressed, the more rigid and elastic it becomes. So there is a strong connection between density and elasticity within Earth's interior.

FIGURE 22.5

Cutaway and cross-sectional diagrams showing the change in wave paths at the major internal boundaries and the P-wave shadow. The P-wave shadow between 105° and 140° from an earthquake's epicenter is caused by the refraction of the P-waves at the core–mantle boundary. Note that any location more than 105° from an earthquake's epicenter does not receive S-waves because the liquid outer core does not transmit S-waves.

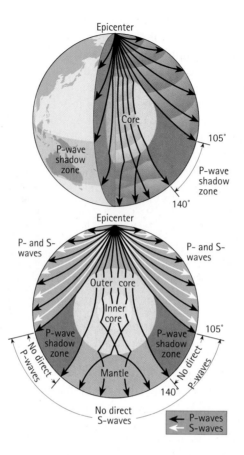

Taken together, the discoveries of Oldham, Mohorovičić, Gutenberg, and Jeffreys indicate that Earth consists of three layers of materials of different composition: the *crust*, *mantle*, and *core*. Each layer is a concentric sphere, so that Earth's overall structure resembles that of a boiled egg.

This simple picture of Earth's layers was refined in 1936 by Inge Lehmann, a Danish seismologist. Her research showed that P-waves refract not only at the core–mantle boundary, but also at a certain depth within the core, where they gain speed. This change in wave speed indicates that the inner region of the core must be solid. So the core was found to have two parts—a liquid outer core of molten iron and a solid-iron inner core. Adding Lehmann's work to earlier findings shows the complete and current picture of Earth's internal layered structure.

Do you suppose these layers in Earth's interior influence the geologic changes our planet experiences? The answer is yes, as you will now see.

CHECK POINT

What evidence supports the theory that Earth's inner core is solid and its outer core is liquid?

Was this your answer?

The differences between how P-waves and S-waves move through Earth's interior. As both waves encounter the boundary at 2900 km below the surface, a very pronounced wave shadow develops. P-waves are both reflected and refracted at that boundary, but S-waves are only reflected. S-waves cannot travel through liquids, implying that the outer core is liquid. As P-waves move through the outer core, there is a depth at which their speed suddenly increases. Knowing that waves travel faster in solids, we infer the existence of a solid inner core.

THE CORE

Earth's **core** is composed mainly of iron and smaller amounts of nickel. In the inner core, the iron and nickel are solid. Although the inner core is indeed very hot, intense pressure from the weight of the rest of Earth prevents the material of the inner core from melting (just as a pressure cooker prevents high-temperature water from boiling, as discussed in Chapter 7).

Because less weight is exerted on the outer core, the pressure is lower there, resulting in a liquid phase of iron and nickel. The molten outer core flows at the

■ Much that we know about Earth's interior was learned as a result of the Cold War between the United States and the former Soviet Union. In the 1960s, when testing of nuclear weapons was very common, underground nuclear explosions were found to produce seismic waves. Both countries installed sensitive seismographic stations to monitor their opponent's activities. It was the seismograms of this network of stations that revealed details of the unseen structure of our planet.

rate of several kilometers per year. This flow is evident far outside Earth's surface. The flowing molten outer core produces a flowing electric charge—an electric current. This electric current powers Earth's magnetic field. This magnetic field is not stable but has changed throughout geologic time. Recall from Chapter 9 that there have been times when Earth's magnetic field has diminished to zero, only to build up again with the poles reversed. These magnetic-pole reversals probably result from changes in the direction of fluid flow in the molten outer core of Earth.

CHECK POINT

Iron's normal melting point is 1535°C, yet Earth's inner core temperature is at least 5000°C. Why doesn't the solid inner core melt?

Was this your answer?
The intense pressure from the weight of Earth above crushes atoms together so tightly that even high temperature cannot budge them. Because of the pressure, melting cannot occur and the inner core stays solid.

THE MANTLE

Surrounding the core of the planet is the **mantle,** a rocky layer some 2900 km thick. Uniformly composed of hot, iron-rich silicate rocks, the mantle behaves like an elastic solid. Mantle rocks actually flow, even though they are solid—they behave in a *plastic* manner. This is similar to the behavior of Silly Putty®: it behaves like a solid under sudden stress (it breaks), but it behaves like a fluid when stress is applied slowly (it flows). Flow in the mantle occurs as *convection currents* that form *convection cells* (Figure 22.6). In other words, hot material in the mantle rises, cools, and then sinks (Chapter 7).

Although it is of relatively uniform bulk composition, the mantle is divided into two portions based on their physical properties. The lower mantle extends from the outer core to a depth of about 700 km (the dashed line in Figure 22.4). The lower mantle is completely solid because pressure in this region is too great for melting to occur.

The upper mantle, which extends from the 700-km depth upward to the crust–mantle boundary, has two zones (Figure 22.7). The lower part of the upper mantle, called the **asthenosphere,** is solid but contains very small amounts of liquid derived from the partial melting of mantle rocks. The asthenosphere is especially plastic, and it flows more easily than the lower mantle. Many geologists believe that convection cells in the asthenosphere do not extend below 700 km and are separate from convection cells in the lower mantle. But it is equally likely that the convection cells are not separated, extending through the asthenosphere and deep into the mantle, perhaps as deep as the core–mantle boundary. In any case, the constant flowing movements in the asthenosphere greatly affect the surface features of our planet.

Now we come to the rigid outer layer of Earth, which includes the crust and the uppermost part of the mantle. This region is the **lithosphere.** The mantle itself ends at the Moho, but the lithosphere includes the entire crust (Figure 22.7). So the upper mantle includes the asthenosphere and part of the lithosphere, and the lithosphere is part mantle and part crust. The reason for the strange divisions is the overlap between composition and physical properties, as revealed by seismic waves. The lithosphere behaves as a single unit, but there is a marked change in composition partway through the lithosphere.

- The density of rocks at Earth's surface is 2.7- 3.0 g/cm³, whereas the average density of Earth as a whole is 5.5 g/cm³. Thus, surface rocks are not representative of the planet's interior. To account for Earth's high average density, the density of the core must be at least 10 g/cm³. This and other reasons suggest that the core is composed of iron and smaller amounts of nickel, the most abundant of the heavier elements.

Link To Section 7.2

The discovery of magnetic pole reversals helped explain the motion of the continents.

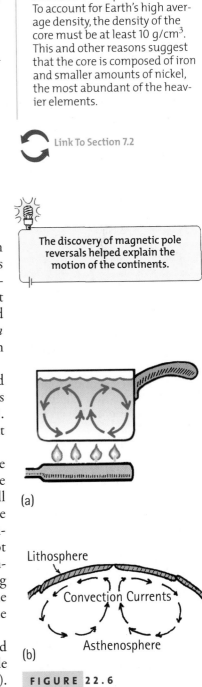

(a)

(b)

FIGURE 22.6

(a) A familiar example of convection is seen when water is heated in a pan. (b) A simple model showing convection currents in the asthenosphere.

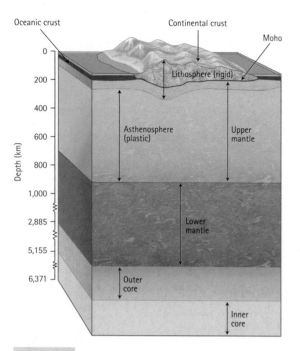

Oceanic crust

Continental crust

Moho

Lithosphere (rigid)

Asthenosphere (plastic)

Upper mantle

Lower mantle

Outer core

Inner core

Depth (km)

0
200
400
600
800
1,000
2,885
5,155
6,371

FIGURE 22.7

The core, mantle, and crust each have a different composition. The mantle starts at the core–mantle boundary and ends at the base of the crust. The lower mantle is essentially a single unit from a depth of 2900 km up to about 700 km. Although Earth's upper mantle is of a similar composition, it is divided into two distinct units. The lower portion of the upper mantle is the plastic asthenosphere. The top portion of the upper mantle, and the entire crust, together form the rigid lithosphere.

Link To Section 5.4

Link To Section 4.2

Unlike the asthenosphere, the lithosphere is rigid and brittle and does not flow appreciably. The lithosphere is, in a sense, riding on top of the asthenosphere like a raft on a pond. The lithosphere moves along with the motions of the material beneath it in the asthenosphere. Motions in the mantle are not uniform, and, because of this (as we shall see later in this chapter), the brittle lithosphere is broken into many individual pieces, or plates.

Mantle convection currents move at a leisurely pace, taking hundreds of millions of years to complete one loop. Even so, heat-driven motion in Earth's interior shapes and reshapes many of our surface features. The lithospheric plates are always in motion. Their movement causes earthquakes, volcanic activity, and the deformation of large masses of rock that create mountains.

THE CRUSTAL SURFACE

The uppermost portion of the lithosphere is the **crust.** The crust is subdivided into two categories: *continental crust* and *oceanic crust.* The density, composition, and thickness of the crust vary markedly from the deep ocean basins to the lofty continental plateaus. The crust of the ocean basins is compact. It's only about 10 km thick, and it is composed of dense basaltic rocks. Continental crust is generally between 20 and 60 km thick, and it is composed of granitic rocks, which are less dense than basaltic rocks. The lower density keeps most of the continental crust above sea level.

If continental crust is so much thicker than oceanic crust, why are the ocean basins underwater and the continents high and dry? The answer is found in their density differences and buoyancies (Chapter 5). The less-dense continental crust always sits higher than the more-dense oceanic crust, even if the continental crust has more mass. This is the principle of **isostasy.** Either type of crust reaches an equilibrium height with respect to the mantle when the upward-acting buoyant force, provided by the mantle, equals the weight of the crust.

The concept of isostasy can be made clear with an analogy. Imagine that Earth's crust is a cargo ship and that the mantle is the ocean. The ship establishes its vertical position in the water when the net force on it is zero. This happens when the gravitational force pulling the ship downward (its weight) equals the buoyant force pushing it upward (Figure 22.8). Note that the heavily loaded and denser ship floats lower, with more of it submerged, than the same ship when empty. Likewise, the crust's vertical position in the mantle rises and falls according to variations in density. Dense oceanic crust sits lower in the mantle than less-dense continental crust.

On the planetary scale, consider two blocks of crust of equal mass, one composed of oceanic crust and the other of continental crust (Figure 22.9). The mantle exerts the same buoyant force on both blocks, but the continental block floats higher because it is less dense. If we added more crust to the continental block, its upper surface would be higher than before, but it would also sink farther into the mantle because of the additional weight. So the higher the crust "floats" above the mantle, the deeper its "root" must be to support it.

Taking these ideas a bit further, blocks of crust achieve a state of equilibrium with each other. The weight of mantle material displaced by a block of floating crust equals the weight of the block. But where does the displaced mantle material go? Because pressure in the mantle from the overlying crust must be the same from

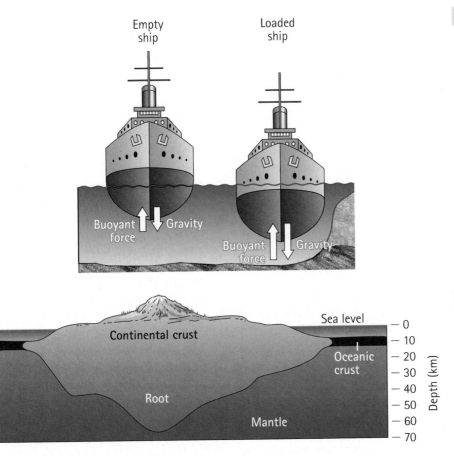

FIGURE 22.8

Isostasy: The vertical position of the crust is stable when the gravitational and buoyant forces balance. Denser oceanic crust therefore has a lower vertical position than the less-dense continental crust, just as the loaded ship sits lower in the water than the unloaded ship.

FIGURE 22.9

Continental crust is thicker and less dense than oceanic crust. As such, continental crust floats higher on the mantle than oceanic crust. To achieve isostatic balance, the higher the crust, the deeper the roots.

place to place at equal depths, the displaced material simply flows away from whatever is pressing down on it until the pressure in the mantle is equalized. This means that pressure in the mantle underneath the thick (but less dense) continental crust is about the same as the pressure in the mantle beneath the thinner (but denser) oceanic crust. To achieve this state of balance, the vertical positions of the different blocks of crust adjust until the pressure in the mantle is balanced.

CHECK POINT

1. How does Earth's crust behave like a ship floating in water?
2. Why is Earth's crust thicker beneath a mountain?

Were these your answers?

1. The crust's vertical position, as well as the ship's, is determined by the balance of the buoyant and gravitational forces acting on it.
2. Just as most of an iceberg is below sea level, likewise for mountains. Mountains sink until the upward buoyant force balances the downward gravitational force.

22.3 Continental Drift—An Idea Before Its Time

Have you ever noticed on a map that the western shoreline of Africa and the eastern shoreline of South America fit together like pieces of a jigsaw puzzle (Figure 22.10)? One person who took this observation seriously was German naturalist Alfred Wegener (1880–1930), shown in Figure 22.11. Wegener prepared

How old were you when you first noticed that the shoreline margins of South America and Africa fit together like a jigsaw puzzle?

The jigsaw-puzzle fit between continents is even better at the continental shelves than at the shorelines of the continents.

Continental shelf

Continent

fyi

- Scientists of the early 20th century believed that oceans and continents were geographically fixed. They regarded the surface of the planet as a static skin spread over a molten, gradually cooling interior. They also believed that the cooling of the planet resulted in its contraction, which caused the outer skin to contort and wrinkle into mountains and valleys.

Alfred Wegener (1880–1930) was a brilliant interdisciplinary scientist. In 1915 he published his hypothesis of continental drift, which eventually led to the discipline of plate tectonics. His interests included not only meteorology and climatology, but also astronomy, geology, geophysics, oceanography, and paleontology. Throughout his life, Wegener had a fascination with exploring the Arctic, and he was fortunate to survive several Arctic adventures. In 1930 at age 50, however, his luck ran out. Wegener died while crossing an ice sheet on an expedition of Greenland. His body still remains as part of the Greenland glacier. The life of Alfred Wegener—a productive life indeed!

a detailed hypothesis to explain this observation. His hypothesis, known as **continental drift,** stated that the world's continents were once joined together as a single supercontinent that he called *Pangaea.*

Wegener supported his continental drift hypothesis with impressive geological, biological, and climatological evidence. He proposed that the geological boundary of each continent lay not at its shoreline, but at the edge of its *continental shelf* (the gently sloping platform between the shoreline and the steep slope that leads to the deep ocean floor). When Wegener fit South America and Africa together along their continental shelves, the fit was even better than along their shorelines. And when he matched up the continental shelves, he found similar rock types on both sides of the Atlantic. Plus, in many cases, mountain chains of the same age were found to be continuous across the oceans—both in North America and Europe and in South America and Africa (Figure 22.12). In addition, fossils of nearly identical land-dwelling animals are found in South America and Africa but nowhere else. This was a strange finding because today, animals and plants of these regions are notable for their striking differences. And fossils of nearly identical trees are found in South America, India, Australia, and Antarctica.*

Even stronger evidence for a supercontinent comes from paleoclimatic (ancient climate) data. More than 300 million years ago (in the mid-Paleozoic era), a huge continental ice sheet covered parts of South America, southern Africa, India, and southern Australia (Figure 22.13 and Chapter 21). The ice sheet left evidence of its existence in thousands of well-preserved glacial striations that reveal the directions of ice flow. If these continents were in their present positions, the ice sheet would have had to cover the entire Southern Hemisphere, and, in some places, would have had to cross the equator! If the ice sheet were that extensive, the world climate would have been very cold. But there is no evidence of glaciation in the Northern Hemisphere at that time. In fact, the time of glaciation in the Southern Hemisphere was a time of subtropical climate in the Northern

* One fossil plant assemblage that offers especially strong support for Wegener's idea is the *Glossopteris* flora, which was named after the dominant gymnosperm tree found in the prehistoric southern temperate forests of South America, India, Australia, and Antarctica. Because the seeds from these trees were too large to be distributed by wind, the wide distribution of this flora supports Wegener's hypothesis that the continents were once joined together.

Hemisphere. To account for this enigmatic distribution of paleoclimates, Wegener proposed that Pangaea had been in existence 300 million years ago, with South Africa located over the South Pole. This reconstruction brings all the glaciated regions into close proximity near the South Pole and places the modern northern continents nearer the tropics.

Wegener described continental drift in his book *The Origin of Continents and Oceans*, published in 1915. Although he used evidence from many different scientific disciplines, his well-founded hypothesis was ridiculed by the scientific community. Antagonists complained that Wegener failed to provide a suitable driving force to account for the continental movements. (Wegener wrongly proposed that the tidal influence of the Moon could produce the needed force. He also proposed that the continents broke through Earth's crust like icebreakers cutting through ice.) Without a convincing explanation for his hypothesis, it was dismissed. Only later, in the light of newfound discoveries, did the scientific community accept Wegener's concept.

FIGURE 22.12

Wegener likened the fossil and rock matches to finding two pieces of torn newspaper with matching contours and lines of type. If the edges and the lines of type fit together, the two pieces of newspaper must have originally been one.

CHECKPOINT

1. What evidence might lead someone with no understanding of science to suspect that the continents were once connected?
2. Scientists of Wegener's day rejected continental drift because they couldn't imagine how massive, rocky continents could grind through the solid rock of the ocean floor. What did Wegener's contemporaries evidently not know about the mantle?

Were these your answers?

1. The most obvious evidence is the matching of the edges of the African and South American continents, which can be seen on any world map or globe.
2. Wegener's contemporaries assumed that if the continents were to move, they would have to push through solid rock. They did not know that the mantle has a "plastic-like" layer—the asthenosphere, over which "floating" continents can readily slide.

22.4 Acceptance of Continental Drift

One of the first key discoveries in support of continental drift came about through studies of Earth's magnetic field. We know from Chapter 9 that Earth is like a huge magnet, with its magnetic north and south poles near the geographic poles. Because certain minerals align themselves with the magnetic field when a rock is formed, rocks have a preserved imprint of changes in Earth's magnetism over geologic time. This magnetism from the geologic past is known as **paleomagnetism.** In the 1950s, a plot of the positions of the magnetic north pole through time revealed that over the past 500 million years, the position of the pole had apparently wandered extensively throughout the world (Figure 22.14). It seemed that either the magnetic poles had migrated through time or the continents had drifted. Because the apparent path of polar movement varied from continent to continent, it was more plausible that the continents had moved. Thus, the hypothesis of continental drift was revived, but a mechanism to explain how the movement occurred was still lacking.

Just as Wegener's hypothesis was supported by evidence in different disciplines, the mechanism to explain continental drift was assembled from several different disciplines. Piece by piece, it all came together in the 1960s, as a result of seafloor exploration.

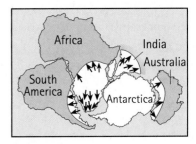

FIGURE 22.13

Paleoclimatic evidence supports Wegener's hypothesis of continental drift. Glacial striations in rock outcrops in South America, Africa, Antarctica, India, and Australia provide clues that these continents were once positioned together. The arrows depict the direction of glacial movement.

FIGURE 22.14

The path of the magnetic north pole during the last 500 million years. (The unit *m.y.a.* stands for "millions of years ago.") The lower red line is derived from evidence collected in Europe, and the upper red line is derived from evidence collected in North America. One would expect that these two lines would overlie each other. Thus, either the magnetic pole wanders erratically, or the continents have moved. But how could the pole be in more than one place at the same time?

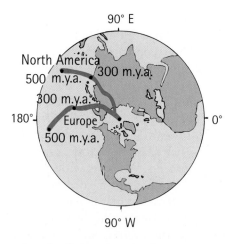

fyi

- *Paleo-* means "old" or "ancient." As a combining prefix, *paleo-* is used to describe things that occurred in the past. For example, *paleoclimate* is a way to describe ancient climates, *paleomagnetism* describes ancient magnetic data, and *paleontology* is the study of life in ancient geologic time.

fyi

- Measurements of the ocean floor began during World Wars I and II as echo-sounding devices—primitive sonar systems—began to measure ocean depth by recording the time it took for a sound signal (commonly called a "ping") from the ship to bounce off the ocean floor and return. The returned signals revealed that the ocean floor was much more rugged than previously thought.

FIGURE 22.15

Harry Hess (1906–1969). Hess's hypothesis of seafloor spreading helped establish a mechanism for Wegener's hypothesis of continental drift.

A key player who helped solve the puzzle of continental drift was Harry Hess (Figure 22.15), a geology professor from Princeton University. During World War II, Hess served in the U.S. Navy as a captain in charge of an attack transport ship. The ship was equipped with a *fathometer*, an innovative depth sounder that could map underwater topography. The main purpose of the device was to help the ship maneuver near shore during beach landings. But Hess, a scientist as well as a sailor, continued using the fathometer in the open sea to collect data about the deep ocean bottom. Through his wartime scientific surveying, Hess collected profiles of the ocean floor across the North Pacific Ocean. The depth sounder enabled him to make surveys at greater depths than had been previously accomplished. His findings expanded on and substantiated other recent discoveries and emerging ideas.

With the improved technology of the 1950s, oceanographers could map the ocean floor in greater detail. Huge mountain ranges running down the middle of the Atlantic, Pacific, and Indian oceans were discovered (Figure 22.16). The Mid-Atlantic Ridge, for example, was found to wind down the center of the Atlantic Ocean basin and parallel the American, European, and African coastlines. The ridge stretches 19,312 km, and its highest peaks emerge above sea level to form oceanic islands, such as Iceland and the Azores (Figure 22.17). In the center of the ridge and all along its length is a deep canyon—a volcanic *rift zone.** Another ocean floor feature was the discovery of deep *ocean trenches* (long, deep troughs in the seafloor) near some continental landmasses, particularly around the edges of the Pacific. So it was revealed that some of the deepest parts of the ocean are actually near some of the continents, and that some of the shallowest waters are in the middle of the oceans, at the midocean ridges.

As the topography of the ocean floor was revealed in the Atlantic Ocean basin, a different kind of mapping was being done in the Pacific basin—the intensity of Earth's magnetic field. The magnetic surveys of the ocean floor revealed a curious pattern of stripes—alternating strong and weak magnetic fields. The zebralike pattern was found to run parallel to the coastlines. Similar patterns of stripes, of alternating magnetic strength, were seen in other parts of the ocean.

With the discovery of the midocean rifts, Hess was inspired to look back at his data from years before. In 1960, Hess proposed that the seafloor is not permanent but is

---

* Recall the rift zones that formed in the breakup of Pangaea during the Mesozoic era. Volcanic activity is associated with rift zones.

FIGURE 22.16

Oceanographic cartographers Marie Tharp and Bruce Heezen co-created the first detailed map of the ocean floors on a global scale. Based on sonar readings, the map reveals enormous mountain ranges (brown regions) in the middle of the oceans and deep ocean trenches near some continental landmasses.

FIGURE 22.17

The Mid-Atlantic Ridge runs down the center of the Atlantic Ocean. Its highest peaks emerge above the water in several places, creating oceanic islands such as Iceland. This photo shows the exposed rift valley on Iceland.

Link To Section 9.3

constantly being renewed. He hypothesized that ocean ridges are located above upwelling convection cells in the mantle. As material from the mantle oozes upward, new lithosphere is formed. Old lithosphere is simultaneously destroyed in the deep ocean trenches near the edges of some continents. Thus, in a conveyor-belt fashion, new lithosphere forms at a spreading center and older lithosphere moves out from the ridge crest, eventually to be recycled back into the mantle at a deep ocean trench (Figure 22.18). Hess called his hypothesis **seafloor spreading.**

In 1962, magnetic surveys of the Atlantic basin began, with results quite similar to those found in the Pacific basin. Two English geologists, Vine and Matthews, puzzled over Hess's hypothesis of seafloor spreading and their new magnetic data. They proposed that as new basalt is extruded at an oceanic ridge, it is magnetized according to the existing magnetic field. So the magnetic surveys of the ocean's floor actually revealed not magnetic intensity but magnetic direction. The alternating stripes paralleling either side of the rift areas indicate periods of normal and reversed polarity (Figure 22.19). As in a very slow magnetic tape recording, the magnetic history of Earth is recorded in the spreading ocean floors. Because the dates of pole reversal can be determined, the magnetic

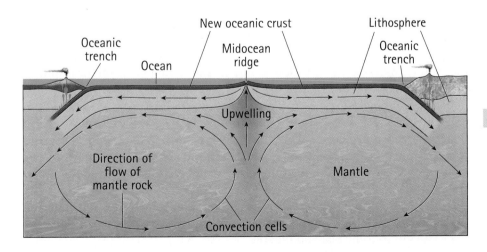

FIGURE 22.18

In conveyor-belt fashion, new lithosphere is formed at the midocean ridges (also called "spreading centers") as old lithosphere is recycled back into the asthenosphere at a deep ocean trench.

Normal polarity
Reversed polarity

FIGURE 22.19

As new material is extruded at an oceanic ridge (spreading center), it is magnetized according to the existing magnetic field. Magnetic surveys show alternating strips of normal and reversed polarity paralleling both sides of the rift area. Like a very slow magnetic tape recording, the magnetic history of Earth is thus recorded in the spreading ocean floors.

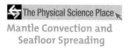
The Physical Science Place
Mantle Convection and Seafloor Spreading

fyi

■ Three essential bits of information are contained in the preserved magnetic record: (1) the polarity of Earth's magnetic field at the time the rock was formed, (2) the direction to the magnetic pole from the rock's location at the time the rock was formed, and (3) the magnetic latitude of the rock's location at the time the rock was formed. Once the magnetic latitude of a rock and the direction of the magnetic poles are known, the position of the magnetic pole at the time of formation can be determined.

If the seafloors spread, continents must move.

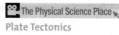
The Physical Science Place
Plate Tectonics

pattern of the spreading seafloor documents both the age of the seafloor and the rate at which it spreads.

The collaborative efforts of Tharp, Heezen, Hess, Vine, and Matthews (and others) convinced the scientific community, once and for all, that seafloor spreading and continental drift must in fact occur. Hess's hypothesis of seafloor spreading provided the mechanism to explain continental drift. The time was right for the revolutionary concepts to follow. The tide of scientific opinion had indeed switched in favor of a mobile Earth.

■ 22.5 The Theory of Plate Tectonics

Plate tectonics describes the motions of Earth's lithosphere that create ocean basins, mountain ranges, earthquake belts, and other large-scale features of Earth's surface. The theory of plate tectonics states that Earth's outer shell, the lithosphere, is divided into eight relatively large plates and a number of smaller ones (Figure 22.20). These lithospheric plates ride atop the relatively plastic asthenosphere below. So Wegener's hypothesis of continental drift was on the right track. The continents really do move—they move because they are embedded within the drifting tectonic plates.

Ultimately, lithospheric plates move in response to heat flow from Earth's interior. Recall from Chapter 7 that heat naturally moves from warmer regions to cooler regions. Inside Earth, heat moves from the core and mantle to the cooler crust. The core conducts an enormous amount of heat to the mantle. In the mantle, most of the heat flow is due to convection, and the lithospheric plates are the upper parts of the convection cells.

Earth's plates move in different directions at different speeds, ranging from 2 cm to about 15 cm per year. Plates carrying continents, such as the North American Plate, generally move slower, while oceanic plates, such as the Pacific Plate, tend to move much faster. Plates pull apart, crash, merge, and separate from one another over geologic time. Because of these interactions between plates, the edges of plates, the *plate boundaries*, are regions of intense geologic activity (Figure 22.21). While interiors of plates are relatively quiet, most earthquakes, volcanic eruptions, and mountain building occur where plates meet. There are three types of tectonic plate boundaries:

FIGURE 22.20

The lithosphere is divided into eight large plates and a number of smaller ones.

(a) Divergent boundaries—*where plates move away from each other*
(b) Convergent boundaries—*where plates move toward each other*
(c) Transform-fault boundaries—*where plates slide past each other*

DIVERGENT PLATE BOUNDARIES

Heat-driven convection cells in the mantle operate in loops, with symmetry between adjacent convection cells. Sites where the upward-moving parts of two adjacent convection cells reach the surface are places where plates spread apart. Tension is the dominant force where plates move away from each other. These spreading centers are **divergent plate boundaries** (Figure 22.21a). At the divergent boundary, the asthenosphere is very near the surface and the lithosphere is very thin.*

Mid-ocean ridges mark the locations of most divergent plate boundaries. At the ridge crests, gravity gives plates a small push away from divergent boundaries.

FIGURE 22.21

Plate boundaries are regions of intense geologic activity. They are also the sites of lithospheric formation and destruction. Named for the movement they accommodate, the three types of plate boundaries are (a) divergent, (b) convergent, and (c) transform-fault boundaries.

---

* Theoretically, the lithosphere has zero thickness at the exact location of the divergent boundary and the asthenosphere is at the surface.

Think of the ocean floor's topography—the highest elevations are at the mid-ocean ridges. The ridge area is high compared to the surrounding seafloor because of the lift it gets from rising mantle rock—rock that is convecting upward from the asthenosphere below. Because of the elevation difference between the ridge and the adjacent seafloor, gravity causes the plates to slide down and outward from the mid-ocean ridge like cookies sliding off a tilted cookie sheet.

Magma generated at a divergent boundary is from the partial melting of mantle rock brought upward with rising convection currents. Melting occurs because pressure is reduced on this rock as it nears the surface. Basaltic lava erupts where the plates diverge, partially filling the rift between the diverging plates. When cooled, the basalt becomes new oceanic crust. As the two plates continue moving apart, mantle rock beneath the new crust in the uppermost asthenosphere cools and hardens. Melded to the fresh crust above, new lithosphere is formed, which slowly moves away from the spreading center, cools, contracts, and becomes denser. Oceanic lithosphere is thin and young near the spreading ridge. At increasing distances, lithosphere is progressively thicker and older—equally on both sides of the ridge.

The Mid-Atlantic Ridge is the divergent boundary between the North American and Eurasian Plates in the North Atlantic, and the South American and African Plates in the South Atlantic. The rate of spreading at the Mid-Atlantic Ridge ranges between 1 cm and 6 cm per year. Although this spreading may seem slow, through geologic time the effect has been tremendous. Over the past 190 million years, seafloor spreading has transformed a tiny waterway through Africa, Europe, and the Americas into the vast Atlantic Ocean of today!

Spreading centers are not restricted to the ocean floors but also develop on land. Hot rock in Earth's interior rising beneath continental landmasses generates tension in Earth's crust, causing it to stretch and bend upward (upwarping).

The Physical Science Place
Plate Boundaries and
Plate Tectonics

fyi

- The spreading rate along the Mid-Atlantic Ridge averages about 2.5 cm/yr, or 25 km in a million years. This rate may seem slow, but 2.5 cm/yr over the past 100–200 million years adds up! From its infancy in the Mesozoic, the Atlantic Ocean has grown from a tiny inlet of water to the vast ocean that exists today.

FIGURING PHYSICAL SCIENCE

▪ Calculating the Age of the Atlantic Ocean

If you can estimate the rate of seafloor spreading and you know the present width of an ocean, you can calculate the ocean basin's age. Between the United States and Africa, the Atlantic Ocean is currently about 4,830 km or 4.8×10^8 cm wide. Let's assume that the rate of seafloor spreading in the Atlantic has been a constant 2.5 cm/yr over geologic time. We can then apply the familiar equation that relates speed, time, and distance.

$$\text{Time} = \text{distance/speed}$$
$$= (4.8 \times 10^8 \text{ cm})/(2.5 \text{ cm/yr})$$

$$= 1.92 \times 10^8 \text{ yr}$$
$$\sim 190 \text{ million years}$$

Based on these estimates, the age of the Atlantic Ocean is about 190 million years.

SAMPLE PROBLEM

The Red Sea is presently a narrow body of water located over a divergent plate boundary. The plates began to diverge apart from one another during the Oligocene period, 30 million years ago. Knowing the age and the width, what is the average rate of spreading?

SOLUTION:

If we take the current width of the Red Sea to be 300 km or 3.0×10^7 cm, and

the time of spreading to be 30 million years, the speed at which the seafloor spreads is then

$$\text{Speed} = \text{distance/time}$$
$$= (3.0 \times 10^7 \text{ cm})/(3.0 \times 10^7 \text{ yr})$$
$$= 1 \text{ cm/yr}$$

At this rate, it will take about 400 million years for the Red Sea to be as wide as the Atlantic Ocean. But Earth is dynamic and ever changing—and spreading rates can change. For example, the Ethiopian rift zone spread by about 8 m in 2006! At the same time, 2.5 km³ of magma intruded into the crust (enough to fill 2000 football stadiums)! Earth in action!

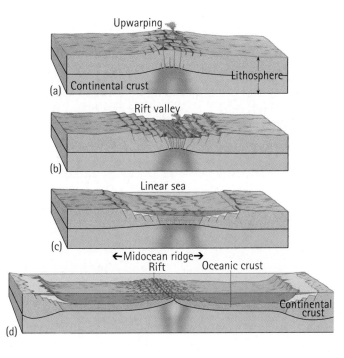

FIGURE 22.22

Formation of a rift valley and its transformation into an ocean basin. (a) Rising magma uplifts continental crust, causing the surface to crack. (b) Rift valley forms as crust is pulled apart. Africa's Great Rift Valley is in this stage today. (The two sides of the valley move away from each other because they are located above mantle convection cells that have the same circulation pattern as the cells in Figure 22.18.) (c) Water from the ocean drains in as the rift drops below sea level, forming a linear sea, so called because it is usually long and narrow. (d) Over millions of years, the rift continues to widen and becomes an ocean basin.

Gaps in the crust are produced, and large slabs of rock slide and sink down into these gaps. The large down-dropped valleys generated by this process are called either **rifts** or *rift valleys* (Figure 22.22). The Great Rift Valley of East Africa is an excellent example of such a feature; if the spreading continues, it may be the beginning of a new ocean basin.

CONVERGENT PLATE BOUNDARIES

Convergent plate boundaries, as the name implies, are where plates come together, or converge. Convergent boundaries are areas of compressive stress and, depending on the nature of the plate interactions, the recycling, or destruction of lithosphere. These regions of plate collisions are also regions of great mountain building. The type of convergence—or "slow collision"—that takes place depends on the type of lithosphere that is involved. The three kinds of convergent plate boundaries are:

- Oceanic–oceanic convergence (Figure 22.23a)

- Oceanic–continental convergence (Figure 22.23b)

- Continental–continental convergence (Figure 22.23c)

In *oceanic–oceanic convergence*, both plates have an oceanic leading edge. When two such plates meet, the older (and therefore cooler and denser) oceanic plate slides beneath the younger, less-dense oceanic plate. The process in which one plate bends and descends beneath the other is called **subduction,** and the area where this occurs is called a *subduction zone.* At Earth's surface, subduction zones are marked by deep ocean trenches that run parallel to the edges of convergent boundaries. The Marianas Trench, for example, is where the Pacific Plate and the slower-moving Philippine Plate collide. Dropping 11,000 m (7 mi) below sea level, the Marianas Trench is the deepest location on Earth's crust. If the world's highest mountain, Mt. Everest, were sunk to the bottom of the Marianas Trench, there would still be more than a mile of ocean above it!

fyi

- In East Africa, spreading processes have already torn Saudi Arabia away from the rest of the African continent, forming the Red Sea. So East Africa may be the site of Earth's next major ocean! Geologists believe that if spreading continues, the edge of the present-day African continent will separate completely, allowing the Indian Ocean to flood the area and making the easternmost corner of Africa (the Horn of Africa) a large island.

The three types of convergent margins: (a) oceanic–oceanic, (b) oceanic–continental, and (c) continental–continental.

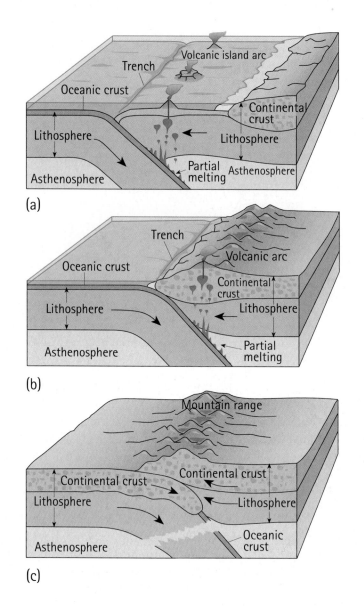

(a)

(b)

(c)

Subduction is an integral part of mantle convection. Each downgoing plate, or slab, controls the downward part of a convection cell. *Slab-pull*, in which gravity pulls the oldest (and therefore the coolest and densest) edge of the subducting plate into Earth's interior, is the dominant force driving plate tectonics. The rest of the subducting plate is pulled trenchward by the sinking leading edge. The longer the subduction zone, the more weight there is to pull on the rest of the downgoing plate. So, longer subduction zones mean faster plate movement. You can visualize a plate being pulled into the asthenosphere as being like a tablecloth being pulled slowly off a table (Figure 22.24).

The process of subduction induces magma generation and the formation of volcanoes on the seafloor. The descending, sediment-covered plate is saturated with seawater. As the wet plate heats up on its way down, fluids (mostly water vapor and carbon dioxide) buoyantly rise and interact with the wedge of mantle rock between the two plates (Figure 22.23a). The rising fluid lowers the melting point of this rock, causing partial melting and the formation of basaltic magma.

The magma rises and reservoirs of basaltic magma form in the crust. When magma reaches the ocean floor, undersea volcanoes form, which often start out with basaltic lava. But as crystallization of unerupted basaltic magma occurs, so does silica enrichment (Chapter 20)—the magma composition changes. Some

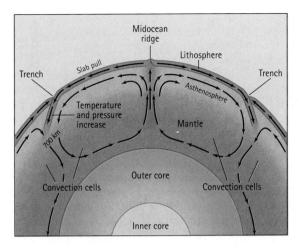

FIGURE 22.24

Simplified view of convection cells within the mantle, showing slab pull. The descending part of the plate—the "slab"—heats up, but generally does not melt, as temperature and pressure rise deeper in the mantle. The descending slab is a location of intense metamorphism.

partial melting of the basaltic oceanic crust may occur as well, which contributes to further silica enrichment of the magma. The volcanoes then begin erupting andesitic magma, which allows them to grow significantly higher. The volcanoes eventually break the surface of the ocean as a series of islands called an *island arc*. The size and elevation of the islands in an island arc increase over time because of continued volcanic activity. Such island arcs have formed the Aleutian Islands, the Marianas Islands, and the Tonga island group in the South Pacific Ocean, as well as the island-arc systems of the Alaskan Peninsula, the Philippines, and Japan.

The second type of plate collision is *oceanic–continental convergence* (Figure 22.23b). In this case, a plate with a continental leading edge slowly collides with a plate that has an oceanic leading edge. Being denser, the oceanic plate subducts beneath the less-dense continental plate. A deep-ocean trench forms offshore where the converging plates meet. As with oceanic–oceanic convergence, magma is generated. But rising through the thicker continental crust takes longer, allowing more time for crystallization and silica enrichment. And as magma rises through the overlying continental plate, varying amounts of continental crust are assimilated into the magma, further enriching it in silica. Together, these enrichment processes produce magma ranging from andesitic to granitic/rhyolitic. Most of the erupted lava is andesitic, and such eruptions can be quite intense and violent. The majority of the granitic magma does not erupt, but solidifies underground to form intrusive plutonic rock—granite. But in the later stages of such a volcano's life, eruptions of rhyolite can occur.

In western South America, the Andes Mountains were formed by such eruptions at an oceanic–continental convergent plate boundary. But there is another process that also contributes to the ongoing growth of the Andes. As subduction of the Nazca Plate beneath the South American Plate continues, marine sediments accumulated on the Nazca Plate are brought into the subduction zone's trench. These sediments are scraped off onto the granitic roots of the Andes. The scraped material becomes permanently attached to the South American Plate, adding thickness and buoyancy to the mountains and helping the Andes rise upward more rapidly than they are being weathered and eroded. Remnants of the original volcanic chain are the exposed batholiths and metamorphic rocks that flank the Andes on the western coast of South America.

In the western United States, examples of such volcanic activity are found in the Sierra Nevada, an ancient volcanic range, and the Cascade Range, which is currently active. The Sierra Nevada were produced by subduction of the ancient Farallon Plate beneath the North American Plate. The Sierra Nevada batholith is a remnant of the original volcanic range, while the California Coast Range has remnants of the sediments that accumulated in the trench. The Cascade Range,

Link To Section 20.7

FIGURE 22.25

Earthquakes at a subduction zone get deeper and deeper in the direction of subduction.

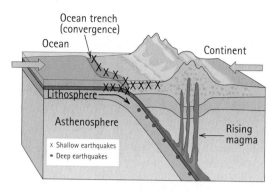

fyi

- The main difference between magma generation at convergent and divergent boundaries is the distance between the site of magma generation and Earth's surface. Magma generation at divergent boundaries is very near the surface, so basaltic magma moves upward unimpeded. At convergent boundaries, magma generation occurs deeper in the mantle, and the rising magma is impeded when it encounters the overlying lithosphere. These two factors increase the travel time to the surface for the mantle-derived basalt. More time means more significant crystallization can occur, resulting in the development of a variety of igneous rocks.

FIGURE 22.26

The continent–continent collision of India with Asia produced—and is still producing—the Himalayas.

produced from the subduction of the Juan de Fuca Plate (a piece of the old Farallon Plate) beneath the North American Plate, includes the volcanoes Mt. Rainier, Mt. Shasta, and Mt. St. Helens. The 1980 eruption of Mt. St. Helens gives testimony that the Cascade Range is still quite active.

As you may expect, the deep-ocean trenches that mark active subduction zones at oceanic–oceanic and oceanic–continental convergent boundaries are areas of intense earthquakes. Earthquakes occur along the subduction zones as the subducted plate grinds against the overriding plate. Earthquakes become steadily deeper and deeper in the direction of subduction (Figure 22.25).

The third type of plate collision occurs when both converging plates have continental crust at their leading edges—*continental–continental convergence*. In this case, blocks of colliding crust consist of the same type of buoyant granitic rock. Because the blocks of rock have similar density, neither sinks below the other when they collide—there is no subduction. Instead, convergence between two continental plates is more like a head-on collision (Figure 22.23c). Compression causes the plates to break and fold up on each other, making the crust very thick. Intensely compressed and metamorphosed rock defines the zone where continental plates meet. In contrast with convergence involving two oceanic plates or one continental and one oceanic plate, volcanic activity is not a characteristic feature of continental–continental collisions—but earthquakes are.

The collision between continental plates has produced some of the most famous mountain ranges, one majestic example being the snow-capped Himalayas, the highest mountain range in the world. This chain of towering peaks is still being thrust upward as India continues crunching up against Asia (Figure 22.26). The European Alps were formed in a similar fashion when part of the African Plate collided with the Eurasian Plate some 40 million years ago. Relentless pressure between the two plates continues, and it is slowly closing up the Mediterranean Sea. In America, the Appalachian Mountains were produced from a continental–continental collision that ultimately resulted in the formation of the supercontinent Pangaea.

FIGURE 22.27

Most transform boundaries occur in ocean basins where they offset oceanic ridges—for example, the Mid-Atlantic Ridge.

TRANSFORM PLATE BOUNDARIES

Transform plate boundaries are locations where two plates are neither colliding nor pulling apart but are rather sliding horizontally past one another. The fracture zone that forms a transform plate boundary is called a *transform fault*. Most transform faults are found in the ocean basin and connect offsets in the midocean ridges. Look at the Mid-Atlantic Ridge in Figure 22.27 and notice how it is broken up into segments. The offset ridge segments are connected by transform faults that "transform" the motion from one ridge segment to another. Between ridge segments, lithosphere coming from one ridge is moving in the opposite direction of the lithosphere coming from the other ridge (Figure 22.28). Can you see that the two slabs of lithosphere are on different plates? Now look at Figure 22.28 in the area of the inactive fracture zone. The sections of lithosphere on opposite sides of the fracture zone are part of the same plate; both sides are moving in the same direction. Along the transform fault, lithosphere is moving in opposite directions. Can you now see that the transform fault is indeed a plate boundary? Can you also see that fracture zones are former transform faults?

Because there is no tension or compression between the plates, there is no creation or destruction of the lithosphere. A transform plate boundary is a zone of horizontal accommodation of plate movement, with neither side of the boundary moving up or down relative to the other. As we will discuss in the following section, transform plate boundaries have the same type of motion as strike-slip faults and are called transform faults.

Although most transform plate boundaries are short and located within the ocean basins, a few are quite long, such as the San Andreas Fault in California (Figure 22.29). The San Andreas Fault stretches for 1500 km from Cape Mendocino in northern California to the East Pacific Rise in the Gulf of California. The Pacific Plate is moving northwest at a rate of about 5.0 cm/yr relative to the North American Plate. The San Andreas Fault accommodates about 70% of this motion, or about 3.5 cm/yr. The rest of the motion occurs along other faults (such as the Hayward Fault). Grinding and crushing take place as the two plates move past each other. When sections of the plates become locked together, stress

fyi

- The San Francisco earthquake of 1906 registered 8.2 on the Richter scale. This large earthquake did quite a bit of damage. Interestingly, most of the damage was due to the fires that burned the city afterward. The reason—all the water mains had burst and broken during the earthquake, so there was no water to put out the fires!

FIGURE 22.28

Transform faults allow two plates to slide past one another at places where two ridge segments are offset.

(a) The San Andreas Fault is a transform plate boundary famous for its earthquakes. The slice of California moving northwesterly lies on the Pacific Plate, while the rest of California sits on the North American Plate. (b) In this photo of the San Andreas Fault, notice the long valley created by many years of rock grinding along the fault.

(a)   (b)

builds up until it is relieved in the form of an earthquake. On April 18, 1906, the Pacific Plate lurched about 6 m northward over a 434-km stretch of the fault, releasing the built-up stress and resulting in the catastrophic San Francisco earthquake.

■ 22.6 Continental Evidence for Plate Tectonics

Convection in Earth's mantle and slab-pull cause the lithospheric plates to move slowly, but constantly. The interaction between plate boundaries creates stress, which produces strain in the rock. There are three types of stress and three resulting types of strain:

Compressional stress occurs when slabs of rock are pushed together. The resulting stain is shortening of the rock bodies.

Tensional stress occurs when slabs of rock are pulled apart. The resulting strain is extension of the rock bodies.

Shear stress occurs when slabs of rock are both pulled and pushed. The resulting strain is the sliding of one slab of rock past the other in opposite directions, without any noticeable shortening or extension.

Rocks respond to these stresses in three different ways. Rock can undergo *elastic deformation*, *brittle deformation*, or *plastic deformation*. Elastic doesn't necessarily mean "stretchy." A stressed elastic material simply returns to its original shape when the stress is removed. In a stretched rubber band, elastic deformation is observed after it is released—it returns to its original shape. If the rubber band breaks when stretched, it means that the *elastic limit* was exceeded. But the entire rubber band doesn't fall to pieces, does it? When the elastic limit is exceeded, brittle deformation happens in one or two places. Then the rubber band snaps back (and usually stings your fingers!). The remnants of the rubber underwent *elastic rebound*. Because rock is an elastic material, rocks also return

■ Rocks respond differently to stress. Some rocks are strong and some are weak; some rocks are prone to break and others are more likely to flow plastically. Increased pressure, temperature, and water content greatly affect a rock's response to stress. High pressure compresses mineral grains against one another, making the rock stronger, more difficult to break, and more prone to plastic flow. High temperature causes increased vibration of molecules in the rock, weakening rock strength and enhancing plastic flow. Wet rocks are generally weaker than dry rocks— water lubricates mineral grains and enhances microscopic slips that contribute to plastic flow. So rock deep below the surface is prone to plastic flow, but rock at the surface is more apt to fracture and break.

to their original shape after stress is removed—unless the elastic limit is exceeded.

When stress exceeds the elastic limit of the rock, the rock, or at least the part that failed, permanently loses its original form—the rock either breaks or flows. Rock that breaks undergoes *brittle deformation*; rock that flows has *plastic deformation*. The response of the rock to stress depends on temperature, pressure, and the composition of the rock. Brittle deformation occurs near the surface where temperature and pressure are low; it produces faults and fractured rocks. Plastic deformation typically occurs deep below the surface where temperature and pressure conditions are high; it causes rock to fold and flow. The geologic structures we see at Earth's surface, such as folds, faults, and related mountains, are examples of strain from tectonic stresses that exceeded the strength of the rock.

FOLDS

Deep in the crust under conditions of elevated temperature and pressure, compressive stress that pushes rock together causes buckling and folding. This is similar to the wrinkles you might find in a throw rug when you push one end of the rug toward the other end. Of course, to wrinkle sections of rock requires strong forces, which come from the movement of lithospheric plates.

Recall that sediments are deposited horizontally layer by layer, with the bottom layer deposited first. Thus the bottom layer is the oldest in the sequence of deposited layers, and the top layer is the youngest. As these originally flat-lying sedimentary rock layers are subjected to compressive stress, they tilt and become folded. Rock can be up-folded or down-folded. Each fold has an axis, with rock layers on one side of the axis a mirror image of the rock layers on the other side. You can imagine the axis as a plane extending downward into Earth, as Figure 22.30 shows. When the layers tilt in toward the fold axis, the fold is called a **syncline.** The rocks at the center, or core, of a syncline are the youngest, and as you move horizontally away from the axis, the rocks get older and older. If the fold layers tilt away from the axis, the fold is called an **anticline.** The rocks in the core of an anticline are oldest, and as you move horizontally away from the axis, the rocks get younger. Another way to think about this concept is that anticlines are pushed upward into an arch, and synclines are pushed downward into a sag.

The Physical Science Place
Folds and Faults in
   Earth's Crust
The San Andreas Fault
   Story

FIGURE 22.30

Anticline and syncline folds. Layer 1 is the oldest rock, and layer 6 is the youngest. The limbs of an anticline (an up-fold) tilt away from the axis of the fold (a marble would roll away from the axis), and the rock layers are oldest at the core of the fold. The limbs of a syncline (a down-fold) tilt toward the axis of the fold (a marble would roll toward the axis), and the rocks are youngest at the core.

CHECK POINT

Why are rocks at the core of a syncline younger than those farther out from the core, while the opposite is true for an anticline?

Was this your answer?

Think of the rug example. Assume the top surface of the rug is younger than the lower surface. When you push the rug, it can (1) fold upward or (2) fold downward. In the first case, the bottom surface makes up the core—an anticline. In the second case, the top surface makes up the core—a syncline. Makes sense!

FAULTS

When stress is stronger than rock and conditions are not hot enough or pressure not high enough for plastic deformation, brittle deformation occurs, and rock can fracture into two parts. If one part moves relative to the other part, the fracture

Footwall   Hanging wall

Footwall

FIGURE 22.31

The terms *footwall* and *hanging wall* were commonly used by miners because one could hang a lamp on a hanging wall and stand on a footwall.

FIGURE 22.32

A reverse fault. In a zone of compressional faulting, the hanging wall is pushed up relative to the footwall. (a) A reverse fault before erosion; (b) the same reverse fault after erosion.

FIGURE 22.33

A normal fault. In a zone of tensional faulting, the hanging wall drops down relative to the footwall, forming a normal fault. (a) A normal fault before erosion; (b) the same normal fault after erosion.

is called a **fault.** Movement along a fault can occur rapidly in the form of an earthquake, or slowly over time.

Faults are classified based on the relative direction of displacement (movement). Look at Figure 22.31 and note the oblique line in the top drawing; this line represents a fault. Imagine that you could pull the block diagram apart at the fault, as shown in the lower drawing. The half containing the fault surface where someone could stand is the *footwall* block. The fault surface of the other half is inclined and would make standing impossible; this is the *hanging wall* block. These terms were coined by miners because one could hang a lamp on a hanging wall, and one could stand on a footwall. Movement on this type of fault is mostly up-and-down motion—the hanging wall and footwall move vertically along the fault plane.

For a fault resulting from compressional forces, the hanging wall is pushed upward along the fault plane relative to the footwall, as Figure 22.32 shows. This type of fault is called a *reverse fault*. The Rocky Mountain foreland (east of the highest peaks), the Canadian Rockies, and the Appalachian Mountains, to name a few, were formed in part by reverse faulting.

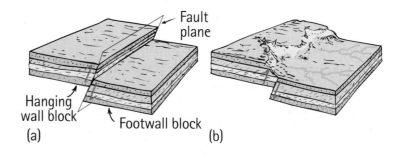

Stress in rock also occurs because of tension. Tensional forces, which pull at rocks, are the opposite of compressional forces, which push. Tension causes the hanging wall to drop downward along the fault plane relative to the adjacent footwall, producing a *normal fault* (Figure 22.33). Virtually the entire state of Nevada, eastern California, southern Oregon, southern Idaho, and western Utah are greatly affected by normal faulting.

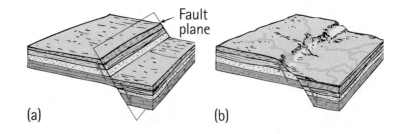

Faults that exhibit horizontal movement, in which blocks of rock slip past one another with very little vertical displacement, are called *strike-slip faults* (Figure 22.34). Some of the world's most famous faults, such as the San Andreas Fault in California, are strike-slip faults. Sticking and slipping of rock blocks along the San Andreas Fault (really a zone of faults) generates many of the earthquakes California is noted for, including the great San Francisco quake of 1906.

1. Reverse faults are the result of compressional forces. What happens to Earth's crust in a zone of reverse faulting?
2. Normal faults result from tensional stress. What surface feature may we expect to find in a zone of normal faulting?

Were these your answers?
1. Compressional stress pushes rocks together. So in a zone of reverse faulting we would find shortening of the crust. Look at Figure 22.32; can you see how compression shortens the crust?
2. Tensional stress pulls the crust apart. So in a zone of normal faulting we would find extension of the crust. Look at Figure 22.33; can you see how tension extends and elongates the crust?

FIGURE 22.34

The relative movement of a strike-slip fault is horizontal. The rock bodies are neither shortened nor extended.

EARTHQUAKES

Energy released during an earthquake travels in the form of seismic waves and radiates in all directions within Earth's interior. The point of origin where the rock initially slips is called the earthquake's *focus* (Figure 22.35). As energy travels to Earth's surface, the ground shakes and moves.

Most earthquakes are related to Earth's tectonic movement. Interactions at plate boundaries create stress, which produces strain in the rock. Strain begins at depth as elastic deformation. With continued stress, rocks at depth store up elastic energy. When the buildup of stress exceeds the rock's elastic limit, the rock breaks suddenly and slips into a new position—a fault forms. This same scenario can also occur along an existing fault. Faults have a certain amount of strength to resist slippage—fault surfaces are not smooth. Rocks on opposite sides of the fault become interlocked. When the strength of the fault is exceeded, it breaks suddenly.

Link To Section 10.2

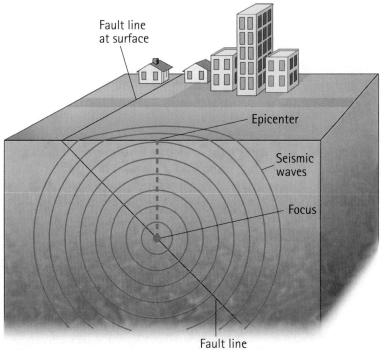

FIGURE 22.35

The actual underground location where fault rupture occurs is the *focus*. Directly overhead at the ground surface is the *epicenter*. Seismic waves radiate out from the focus in all directions.

Earthquake Measurement—The Mercalli and Richter Scales

Every year, hundreds of thousands of earthquakes occur. Although most are small and go undetected, the danger of large earthquakes certainly exists. Earthquake-prone regions experience large earthquakes about every 50–100 years.

The *Mercalli scale* measures earthquake intensity in terms of the effects the quake has on the local environment. The scale ranges from an intensity of I (barely detectable) to an intensity of XII (total destruction). The Mercalli intensity at any location depends on (1) how far that location is from where the earthquake occurred and (2) the nature of the subsurface materials at the location (for example, is the subsurface material solid rock or unconsolidated sediments?).

The Mercalli scale is a valuable yardstick. However, because it is based purely on observation, the Mercalli scale cannot precisely measure quake size. For this reason, seismologists developed a more precise way to estimate the energy released in an earthquake. The *Richter scale*, which is a magnitude scale, measures quake severity in terms of the amount of energy released and the amount of ground shaking at a standard distance from the location of the quake.

The Richter magnitude scale is based on the maximum seismic-wave amplitude of an earthquake, as recorded by a seismograph. The scale is logarithmic. This means that each increase of one unit on the scale is equivalent to a tenfold increase in the amplitude of ground shaking. So, compared with a magnitude 5 earthquake, a magnitude 6 earthquake produces ground motions 10 times as great, and a magnitude 6 earthquake is 100 times as great as a magnitude 4 earthquake.

But how does Richter magnitude relate to the energy released by an earthquake? The energy released by an earthquake is measured not only by the amplitude of seismic waves but also by their frequencies and wavelengths—and seismic waves occur over a wide range of amplitudes, frequencies, and wavelengths. Through careful analysis of many aspects of earthquakes and faults, scientists find that the energy released by an earthquake increases by about 30 times for each increase in Richter magnitude. For example, the 1964 Anchorage, Alaska, earthquake, which had a magnitude of 9.2, released about 30 times as much energy as, and produced 10 times as much ground shaking as, the 1923 Tokyo earthquake, which had a magnitude of 8.2. The magnitude 9.2 Anchorage quake released 900 times as much energy as, and produced 100 times as much ground shaking as, the 1995 magnitude 7.2 quake in Kobe, Japan.

The Richter scale was developed for California earthquakes, which usually have a shallow focus and are moderate in size. The scale breaks down at magnitudes of about 6.5 or greater. Other magnitude scales have been developed, such as the *moment magnitude scale*, which can more accurately represent larger earthquakes. Such scales take into account the length of fault rupture and the area over which the rupture occurred. For earthquakes less than 6.5, both scales yield equivalent magnitudes.

- Earthquakes cause the ground to shake and rupture. As the ground shakes, so do buildings on top of the land. It is often said that earthquakes don't kill people, but falling buildings do. Earthquakes can cause general property damage, lack of basic necessities, collapse of buildings, loss of life, higher insurance premiums, disease, landslides and avalanches, road and bridge damage, and fires generated by broken gas and electric lines.

For both cases, the sudden rupturing releases stored elastic energy. Rocks on both sides of the fault rebound to their original shapes. This snapping back into shape is the earthquake. The released energy travels away from the initial break site as seismic waves. At the point on Earth's surface directly above the focus—called the *epicenter*—strain often occurs as brittle deformation of the crust.

Because earthquakes generally result from movement along faults, the focus and epicenter of an earthquake usually coincide with known fault zones. Most earthquakes occur in just a few narrow zones (Figure 22.36). For example, the area of the Pacific Rim is a zone of great seismic activity, as is the area of the Mid-Atlantic Ridge. Comparing Figure 22.36 to Figure 22.20, can you see that most earthquakes occur at plate boundaries?

Even so, earthquakes occur all over the world, and not always along plate boundaries. Remember, Earth is very old, and plate boundaries can change over geologic time. So faults and related earthquakes can also be found *intraplate*—far from present plate boundaries—in areas that were once closer to plate boundaries. Because of their geologic history, these intraplate areas are zones of weakness. One such area is the New Madrid seismic zone in the central Mississippi Valley. In the winter of 1811–1812, a series of strong earthquakes along this zone permanently changed the course of the Mississippi River! With an estimated magnitude of 8.0 on the Richter scale, the earthquakes were so strong they caused church bells to

Modified Mercalli and Richter Scales

Mercalli Intensity	Richter Magnitude	Observation
I	1 to 2	Felt by very few people; barely noticeable.
II	2 to 3	Felt by a few people, especially on upper floors of buildings.
III	3 to 4	Felt indoors, especially on upper floors, but may not be recognized as an earthquake.
IV	4	Felt by many indoors, a few outdoors. May feel like a heavy truck passing by.
V	4 to 5	Felt by nearly everyone. Many awakened. Disturbances of trees, poles, and other tall objects.
VI	5 to 6	Felt by everyone. Some heavy furniture moved; a few instances of fallen plaster or damaged chimneys. Damage to buildings slight.
VII	6	Slight to moderate damage in well-built structures; considerable damage in poorly built structures.
VIII	6 to 7	Slight damage in specially designed structures; considerable in ordinary substantial buildings, with partial collapse; great damage in poorly built structures (fall of chimneys, factory stacks, columns, monuments, walls).
IX	7	Considerable damage in specially designed structures. Buildings shifted off foundations. Ground conspicuously cracked. Landslides.
X	7 to 8	Considerable damage. Some structures destroyed. Most masonry and frame structures destroyed with foundations. Ground badly cracked.
XI	8	Total damage. Few, if any, masonry structures remain standing. Bridges destroyed. Wide cracks in ground.
XII	8 or greater	Total damage. Waves seen on ground surfaces. Objects thrown up in air.

Key
- Shallow
- Intermediate
- Deep

fyi

- The area of the Pacific Rim that encircles much of the Pacific Ocean is often referred to as the "Ring of Fire." About 80% of the world's big earthquakes occur in the Ring of Fire, and because subduction is related to volcanic activity, about 75% of the world's volcanoes are located here as well.

FIGURE 22.36

Most earthquakes occur in just a few narrow zones.

LINK TO Physics

▪ Wave Motion—Tsunami

A tsunami is a wave, or series of waves, generated in a large body of water by any type of powerful disturbance that vertically displaces the water column. Earthquakes, landslides, explosions, and even meteorites can generate a tsunami.* When an earthquake causes the disturbance, a tsunami is sometimes referred to as a seismic sea wave.

Most tsunami** arise when an earthquake occurs on an underwater reverse fault. Rapid movement of the seafloor—generally upward—quickly thrusts the water above the uplifted area upward. The huge displaced mass of water then drops back down to sea level and a large wave is generated—a tsunami (Figure 22.37).

Tsunami are unlike most other ocean waves (Chapter 24). For example, wind-driven ocean waves have wavelengths usually less than 150 m, whereas tsunami have very long wavelengths—100 km or more. The depth over which the energy travels in a water wave is equal to one-half of the wave's wavelength, so most ocean waves affect only the uppermost part of the water column. But one-half of a tsunami's wavelength—about 50 km—is far deeper than the average ocean depth—about 4 km. So, the energy within a tsunami is spread over the entire water column, not just the uppermost layer. This is an enormous amount of energy. In the open ocean, tsunami are barely visible because of their huge wavelength. The ocean rises perhaps 1 m higher than usual, but that 1-m rise is spread over 100 km! And tsunami are fast—they travel through the open ocean at 800 km/h!

The true power of a tsunami becomes apparent as the wave approaches shore. Wave speed depends on water depth, so as the tsunami enters shallow water, the wave slows down. But the energy that was spread over 4 km is now

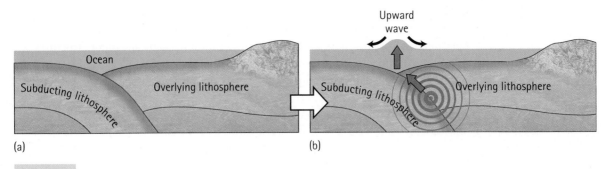

FIGURE 22.37

Most tsunami are generated by earthquakes in subduction zones. The edge of the overriding plate is pulled downward by the downgoing slab (a). The buildup of elastic energy is released suddenly, causing the earthquake. When the edge of the bent plate snaps back into place, the entire water column is flung upward (b) and then drops back down, causing the tsunami.

fyi

▪ In lands bordering the Pacific Ocean's "Ring of Fire," effective tsunami warning stations have been in existence since the 1960s. Unfortunately, the warning system did not extend into the Indian Ocean. After the catastrophe of the 2004 Sumatran tsunami, a warning system for the Indian Ocean was put in place. There is even the possibility of creating a unified global tsunami warning system, to include the Atlantic Ocean and the Caribbean.

ring in Boston, Massachusetts, 1000 mi away! Fortunately, because the region was sparsely settled at the time, the loss to human life and property was minimal.

Earthquakes vary in their size and destructiveness. At divergent plate boundaries, earthquakes are generally mild and shallow. At transform plate boundaries, earthquake size ranges from mild to moderate. And at convergent plate boundaries, earthquakes tend to be moderate to very strong. Out of all the plate boundaries, the strongest earthquakes occur along convergent plate boundaries where subduction occurs.

Devastating earthquakes can occur with all three types of faults: reverse, normal, or strike-slip. The 1906 San Francisco earthquake registered approximately 7.8 on the *Richter scale* (a measure of ground motion caused by earthquakes) and resulted in 700 deaths and extensive fire damage. The 1989 Loma Prieta earthquake near Santa Cruz, California, registered 7.1 on the Richter scale, caused 62 deaths, and did more than $6 billion in damage. Strike-slip faulting was involved in both of these earthquakes.

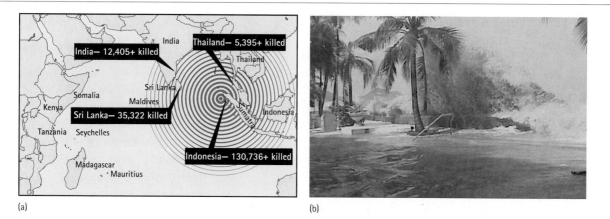

(a)   (b)

FIGURE 22.38

The Indonesian tsunami of 2004 was generated in response to a strong subduction earthquake off the coast of Sumatra. The tsunami struck the surrounding coastlines with catastrophic force; tens of thousands of people were killed, and many livelihoods were destroyed.

squeezed into the space of a few tens of meters, and then even less. Where does this energy go? Up! Wave height can grow from a mere meter to up to 30 m!

And tsunami don't "break" like ordinary ocean waves. As viewed from shore, a tsunami can appear as a very rapidly rising tide—a fast moving wall of water. But if a wave trough gets to shore first instead of a crest, it appears as a rapidly receding, very low tide. Unfortunately, this can cause people to investigate the odd sudden appear-

ance of exposed seafloor, complete with flopping fish that have been temporarily stranded. The wave crest quickly follows, drowning unsuspecting onlookers. In many ways, a tsunami on land is like an enormous sledgehammer delivering an extreme blow.

The 2004 Sumatra tsunami was generated by the subduction of the India Plate beneath the overriding Burma Plate. Because of its size (9.2 on the Richter scale) and huge release of energy, the Sumatran quake is termed

a *megathrust* earthquake. The energy released by this quake is comparable to the amount of energy released by a bomb made up of 100 billion tons of TNT (if it were possible to make such a bomb). Although the quake itself caused severe damage and casualties, the tsunami in the quake's aftermath was even more devastating. Coastal areas throughout the Indian Ocean basin were destroyed. This Indonesian tsunami killed about 184,000 people in 12 countries (Figure 22.38).

* The meteorite impact at the Chicxulub crater during the Cretaceous-Tertiary boundary would have certainly generated a large tsunami! Evidence of such an impact—meteorite as well as tsunami—has been found in rock layers in the southwestern United States.

** *Tsunami* is a Japanese word meaning "harbor wave." The plural and singular forms of the word are the same.

Still larger and more catastrophic earthquakes have occurred along reverse faults, usually those at subduction zones. One such earthquake, the 1964 earthquake in Anchorage, Alaska, registered 9.2 on the Richter scale. The quake and its aftermath caused 131 deaths and $300 million in damage.* Some of the greatest tragedies of recent times include the August 1999 earthquake in Turkey (7.6 on the Richter scale, 17,000 deaths), the January 2001 earthquake in El Salvador (7.7 on the Richter scale, 844 deaths), the January 2001 earthquake in India (7.6 on the Richter scale, 20,000 deaths), and the December 2004 earthquake in Sumatra (9.2 on the Richter scale, 184,000 deaths). Look to Figure 22.36 to see how these devastating earthquakes align with plate boundaries.

fyi

- Earthquakes in Indonesia are very common; in 2006 and 2007 this region experienced four earthquakes registering more than 6.0 on the Richter scale. A very shaky area, indeed.

* The death toll was largely due to great seismic sea waves, or tsunami. A tsunami is generated from the displacement of water as a result of an earthquake, a submarine landslide, or an underwater volcanic eruption.

TABLE 22.1   SOME NOTABLE EARTHQUAKES

Year	Location	Magnitude	Estimated Deaths	Comments
1811	New Madrid, MO	8.0	few	
1906	San Francisco, CA	7.8	700	Fires caused extensive damage
1923	Tokyo, Japan	8.2	150,000	Fire caused extensive destruction
1960	Southern Chile	9.5	5,700	The largest earthquake ever recorded
1964	Anchorage, AK	9.2	131	More than $300 million in damage
1970	Peru	7.9	66,000	Great rockslide
1971	San Fernando, CA	6.5	65	More than $5 billion in damage
1975	Liaoning, China	7.5	few	First major earthquake to be predicted
1976	Tangshan, China	7.6	500,000	
1985	Mexico City, Mexico	8.0	9,000	
1989	Loma Prieta, CA	7.1	62	More than $6 billion in damage
1994	Northridge, CA	6.7	57	More than $25 billion in damage
1995	Kobe, Japan	7.2	5,500	Between $95 billion and $147 billion in damage
1999	Izmit, Turkey	7.6	17,000	
2001	India	7.6	20.000	
2001	El Salvador	7.7	1,000	
2003	Bam, Iran	6.6	>30,000	
2004	Sumatra, Indonesia	9.2	184,000	Damage in >12 countries, mostly due to tsunami
2007	Sumatra, Indonesia	6.4	70	

Understanding earthquakes is obviously of major importance to society. Unfortunately, earthquakes and fault movement occur with little or no warning and thus are very difficult to predict. The best that can be done is to calculate the probability that earthquakes will occur within a given timeframe. Such probability estimates are based on the past occurrences of earthquakes in the target area. Table 22.1 lists some of the world's most notable earthquakes according to their impact on society.

■ 22.7 The Theory That Explains the Geosphere

Before the theory of plate tectonics was proposed, the underlying causes of processes such as mountain building, folding, and faulting were poorly understood. Plate tectonics explains the where and why of many geologic phenomena. Indeed, it can be thought of as a unifying theory because it links causes and effects.

Why are the Appalachian Mountains located where they are? What about the Sierra Nevada? The Rocky Mountains? The Alps? The plate tectonics model gives an answer: all large mountain-building events take place near convergent plate boundaries.

We can also relate the formation of the three different rock types to the plate tectonics theory. Although we are simplifying the following discussion of rock formation for the sake of clarity, all rocks are tied to plate interaction in one manner or another. First, consider the most common igneous volcanic rock. The formation of huge volumes of basalt is linked to divergent plate boundaries, where mantle rock partially melts to form new basaltic oceanic crust.

At subduction zones, the downgoing slabs are heated, releasing fluids that cause the overlying mantle wedge to partially melt. Basaltic magma is generated, which begins to crystallize. Recall from Chapter 20 that as crystallization proceeds, the remaining liquid contains more silica than the parent magma. Thus, andesite forms from basaltic magma, and we find belts of composite cone volcanoes, like those in the Andes and the Cascade Range. And how is granite formed? The large volumes of andesitic magma don't all erupt at once, but rather accumulate in Earth's crust. Some of the surrounding continental crust is partially melted and assimilated into the magma, enriching its silica content. And as cooling magma bodies undergo crystallization, the remaining liquid is further enriched in silica. When this silica-enriched magma solidifies, it forms granite. The Sierra Nevada is a largely granitic mountain range in California that formed in such a manner. The large batholiths of granitic rocks are the "roots," or solidified magma bodies, of a once-extensive volcanic belt formed as a result of subduction. Lastly, the intense heat and pressure caused by subduction and continental collisions result in extensive metamorphism of preexisting rocks—and we find the occurrence of regional metamorphism.

Many sedimentary rocks are linked directly to plate motions. As mountains grow by virtue of plate collisions, they also begin to weather and erode. The sediments produced are transported downslope, where they accumulate, layer upon layer, eventually becoming sedimentary rock.

Lastly, virtually all earthquake and volcanic activity can be tied directly to plate tectonics. These energetic responses to plate interactions are almost always found where plates interact; earthquakes are found at all types of plate boundaries, and volcanoes are concentrated where plates either collide or pull apart.

So the tectonic interaction between lithospheric plates, which occurs mostly at their boundaries, explains the origin of mountain chains, the formation and destruction of the ocean floors, the three types of rocks found on Earth, and the global distribution of earthquakes and volcanoes. The internal motions that change Earth's surface do so in a cycle. Using the principle of uniformitarianism—the present is the key to the past—Earth scientists can use the discoveries of the last 50 years to provide a framework for interpreting Earth history. Onward to Earth's surface!

SUMMARY OF TERMS

Earthquake The shaking or trembling of the ground that happens when rock under Earth's surface moves or breaks.

Body wave A type of seismic wave that travels through Earth's interior.

Surface wave A type of seismic wave that travels along Earth's surface.

Primary wave (P-wave) A longitudinal body wave that compresses and expands the material through which it moves;

it travels through solids, liquids, and gases and is the fastest seismic wave.

Secondary wave (S-wave) A transverse body wave that vibrates side to side or up and down through the material through which it moves; it cannot travel through liquids and so does not travel through Earth's outer core.

Mohorovičić discontinuity (Moho) The crust–mantle boundary, marking a depth at which the speed of P-waves traveling toward Earth's center increases.

Core The central portion of Earth's interior, divided into an outer liquid core and an inner solid core.

Mantle The middle layer in Earth's interior, between the crust and the core.

Asthenosphere A subdivision of the upper mantle situated below the lithosphere, a zone of plastic, easily deformed rock.

Lithosphere The entire crust plus the portion of the mantle above the asthenosphere.

Crust Earth's outermost layer.

Isostasy The process by which oceanic and continental crust come into vertical equilibrium, with respect to the mantle—the dense oceanic crust sits lower in the mantle than the less-dense continental crust.

Continental drift A hypothesis by Alfred Wegener that the world's continents are mobile and have moved to their present positions as the ancient supercontinent Pangaea broke apart.

Paleomagnetism The natural, ancient magnetization in a rock that can be used to determine the polarity of Earth's magnetic field and the rock's location of formation.

Seafloor spreading The moving apart of two oceanic plates at a rift in the seafloor.

Plate tectonics The theory that Earth's lithosphere is broken into pieces (plates) that move over the asthenosphere; boundaries between plates are where most earthquakes and volcanoes occur and where lithosphere is created and recycled.

Divergent plate boundary A plate boundary where lithospheric plates move away from one another—a spreading center; an area of tensional stress where new lithospheric crust is formed.

Rift (rift valley) A long, narrow gap that forms as a result of two plates diverging.

Convergent plate boundary A plate boundary where tectonic plates move toward one another; an area of compressive stress where lithosphere is recycled into the mantle, or shortened by folding and faulting.

Subduction The process in which one tectonic plate bends and descends beneath another plate at a convergent boundary.

Transform plate boundary A plate boundary where two plates are sliding horizontally past each other, without appreciable vertical movement.

Syncline A fold in rock with relatively young rocks at its core; rock age increases with horizontal distance from the fold core.

Anticline A fold in rock with relatively old rocks at its core; rock age decreases with horizontal distance from the fold core.

Fault A fracture along which visible displacement of one side relative to the other can be detected.

REVIEW QUESTIONS

22.1 Seismic Waves

1. How do P-waves travel through Earth's interior? How do S-waves travel through Earth's interior?
2. Can S-waves travel through liquids?
3. Name the two types of surface waves and describe the motion of each.

22.2 Earth's Internal Layers

4. What was Andrija Mohorovičić's major contribution to Earth science?
5. How did seismic waves contribute to the discovery of Earth's core?
6. What is the evidence that Earth's inner core is solid?
7. What is the evidence that Earth's outer core is liquid?
8. Describe the asthenosphere and the lithosphere. In what ways are they different from each other?
9. How does continental crust differ from oceanic crust?
10. Why does continental crust stand higher on the mantle than oceanic crust?

22.3 Continental Drift—An Idea Before Its Time

11. What key evidence did Alfred Wegener use to support his idea of continental drift?

12. Wegener proposed that the world's continents had at one time all been joined together into one supercontinent. What was the name of this supercontinent?

22.4 Acceptance of Continental Drift

13. What role did paleomagnetism play in supporting continental drift?
14. Where are the deepest parts of the ocean?
15. What major discovery at the bottom of the ocean did H. H. Hess make?
16. How is the ocean floor similar to a gigantic slow-moving tape recorder?
17. In what way does seafloor spreading support continental drift?

22.5 The Theory of Plate Tectonics

18. Name and describe the three types of plate boundaries.
19. What are two reasons that lithospheric plates move?
20. What is a rift?
21. What kind of boundary separates the South American Plate from the African Plate?

22. Describe the three types of plate collisions that occur at convergent boundaries.
23. What is a transform fault?

22.6 Continental Evidence for Plate Tectonics

24. Are folded rocks primarily the result of compressional or tensional forces?
25. Distinguish between anticlines and synclines.

26. What is the difference between reverse faults and normal faults?
27. Which kind of fault forms primarily from tension in Earth's crust? Primarily from compression?
28. What is elastic deformation? What happens when stress exceeds a rock's elastic limit?
29. Where are most of the world's volcanoes formed?
30. What is the source of a tsunami's huge amount of energy?

EXERCISES

● BEGINNER   ■ INTERMEDIATE   ◆ EXPERT

Please do not be intimidated by the large number of exercises and problems in this and other meatier chapters. If your course work is to cover many chapters, your instructor will likely assign only a few exercises and/or problems from each.

1. ● Compare the relative speeds of primary and secondary seismic waves. Which forms of matter can each travel through?
2. ● Explain how seismic waves indicate whether regions within Earth are solid or liquid.
3. ● How do seismic waves indicate layering of materials in Earth's interior?
4. ◆ What is the evidence that Earth's central core is solid?
5. ■ Speculate on why the lower part of the lithosphere is rigid and the asthenosphere is plastic, even though they are both part of the mantle.
6. ■ Even though the inner and outer cores are both predominantly composed of iron and nickel, the inner core is solid and the outer core is liquid. Why?
7. ■ What does the wave shadow, which develops between 105° and 140° from the origin of an earthquake, tell us about Earth's composition?
8. ● If Earth's mantle is composed of rock, how can we say that the crust floats on the mantle?
9. ■ Why is Earth's crust thicker beneath a mountain range?
10. ■ Which extends farther into the mantle, the continental crust or the oceanic crust? Why?
11. ◆ How does erosion and wearing away of a mountain affect the depth to which the crust extends into the lithosphere?
12. ■ Describe how the different paths of polar wandering helped establish that continents move over geologic time.
13. ● Why are most earthquakes generated near plate boundaries?
14. ● What is the driving force for mountain building in the Andes?
15. ● At what type of plate boundary were the Appalachian Mountains produced?
16. ◆ Why do mountains tend to form in long narrow ranges?

17. ■ What kind of plate boundary separates the North American Plate from the Pacific Plate?
18. ◆ Relate the formation of metamorphic rocks to plate tectonics. Would you expect to find metamorphic rocks at all three types of plate boundaries? Why or why not?
19. ■ Why does granite frequently form at oceanic–continental convergent boundaries but infrequently at oceanic–oceanic convergent boundaries?
20. ◆ Cite one line of evidence that suggests that subduction once occurred off the coast of California.
21. ■ Distinguish between continental drift and plate tectonics.
22. ◆ Are the present ocean basins a permanent feature on our planet? Discuss why or why not.
23. ◆ Are the present continents a permanent feature on our planet? Discuss why or why not.
24. ■ Why are the most ancient rocks found on the continents, not on the ocean floor?
25. ● Upon crystallization, certain minerals (the most important being magnetite) align themselves in the direction of the surrounding magnetic field, providing a magnetic fossil imprint. How does the seafloor's magnetic record support the theory of continental drift?
26. ■ How are the theories of seafloor spreading and continental drift supported by paleomagnetic data?
27. ● What kind of boundaries are associated with seafloor spreading centers?
28. ■ What is meant by magnetic pole reversals? What useful information do they tell us about Earth's history?
29. ● What does Earth's crust have in common with a conveyor belt?
30. ● Earthquakes, the result of sudden motion in Earth caused by abrupt release of slowly accumulated stress, cause rocks and faults to rupture and new faults to form. Relate such faulting to horizontal sliding motion between two plates. Where does this type of movement occur?
31. ● What type of fault is associated with the 1964 earthquake in Alaska?
32. ◆ The Mercalli scale measures earthquake intensity. The Richter scale measures earthquake magnitude. Which scale is the more precise measurement? Why?

33. ■ What is a very likely cause for the existence of Earth's magnetic field?

34. ■ Lithospheric rock is continuously created and destroyed. Where does this creation and destruction take place? Do the rates of the two processes balance each other?

35. ● Subduction is the process of one lithospheric plate descending beneath another. Why does the oceanic portion of the lithosphere undergo subduction while the continental portion does not?

36. ■ What geologic features are explained by plate tectonics?

37. ● In 1960, a large tsunami struck the Hawaiian Islands without warning, devastating the coastal town of Hilo, Hawaii. Since that time, a tsunami warning station has been established for the coastal areas of the Pacific. Why do you think these stations are located around the Pacific Rim?

38. ■ How did the Himalaya Mountains originate? How did the Andes Mountains originate?

39. ● What is the direct source of energy responsible for earthquakes in southern California?

40. ■ Describe how the presence of faults and folds supports the idea that lithospheric plates are in motion.

41. ● Reverse faults are created by compressional forces. Where in the United States do we find evidence of reverse faults?

42. ● Normal faults are created by tensional forces. Where in the United States do we find evidence of normal faults?

43. ● Strike-slip faults show horizontal motion. Where in the United States do we find strike-slip faulting?

44. ♦ If you found folded beds of sedimentary rock in the field, what detail would you need to know in order to tell if the fold was an anticline or a syncline?

45. ■ Does the fact that the mantle is beneath the crust necessarily mean that the mantle is denser than the crust? Explain.

46. ♦ Relate the generation of magma to plate boundaries. What type of magma is dominant at each of the types of boundaries? What are the mechanisms that drive magma generation?

47. ● Where is Earth's longest mountain range located?

48. ■ How old is the Atlantic Ocean thought to be? For how many years has lava been extruding at the Mid-Atlantic Ridge?

49. ■ What type of lava erupts at divergent boundaries? What types erupt at convergent boundaries?

50. ♦ What clues do we use to recognize the boundaries between ancient plates no longer in existence?

PROBLEMS

● BEGINNER   ■ INTERMEDIATE   ♦ EXPERT

1. ♦ The weight of ocean floor bearing down on the lithosphere is increased by the weight of ocean water. Relative to the weight of the 10-km-thick basaltic ocean crust (density 3 g/cm^3), how much weight does a 3-km-deep ocean (density 1 g/cm^3) contribute? Express your answer as a percentage of the crust's weight.

2. ■ The Richter scale is logarithmic, meaning that each increase of 1 on the Richter scale corresponds to an increase by a factor of 10 in the amplitude of the seismic waves recorded on a seismograph. An earthquake that measures 8 on the Richter scale has how many times more ground-shaking effect than a quake that measures 6 on the Richter scale?

3. ■ If you know the rate of movement along a fault, the amount of offset over a period of time can be calculated. The basic relationship is

$$\text{Rate} = \frac{\text{distance}}{\text{time}}$$

Movement along the San Andreas Fault is about 3.5 cm/yr. If a fence was built across the fault in 1990, how far apart will the two sections of the now-broken fence be in 2010?

4. ■ The San Andreas Fault separates the northwest-moving Pacific Plate, on which Los Angeles sits, from the North American Plate, on which San Francisco sits. If the plates slide past one another at a rate of 3.5 cm/yr, how long will it take the two cities to form one large city? (The distance between Los Angeles and San Francisco is 600 km.)

ACTIVE EXPLORATIONS

Look for a very old window, and note the lens effect in the bottom part of the glass. Glass has both solid and liquid properties; in fact, it is often thought of as a very viscous liquid. Over many years, its downward flow due to gravity is evidenced by the increased thickness near the bottom of the pane. Tie this observation of "plastic" behavior into our discussion of plate tectonics. Which parts of Earth behave plastically? Which parts are rigid? What do these ideas have to do with plate tectonics?

READINESS ASSURANCE TEST (RAT)

If you have a good handle on this chapter, if you really do, then you should be able to score 7 out of 10 on this RAT. If you score less than 7, you need to study further before moving on.

Choose the BEST answer to each of the following.

1. At divergent boundaries, basaltic magma is generated by the
   (a) crystallization of mantle magma.
   (b) partial melting of continental crust.
   (c) partial melting of mantle rock.
   (d) addition of water to mantle rock.
2. The hypothesis of continental drift is *not* supported by
   (a) seafloor spreading.
   (b) paleomagnetism.
   (c) isostasy.
   (d) glacial striations.
3. Which of the following statements is false?
   (a) The mantle includes part of the crust.
   (b) The lithosphere includes the entire crust.
   (c) The mantle includes part of the lithosphere.
   (d) The mantle includes the entire asthenosphere.
4. Seismic waves increase in speed when
   (a) they pass through a liquid.
   (b) rocks become denser and less rigid.
   (c) they form a wave shadow.
   (d) the elasticity of the rock increases.
5. Which part of Earth's crust is most dense?
   (a) the oceanic crust
   (b) the continental crust
   (c) the asthenosphere
   (d) the lithosphere

6. The source of a tsunami is usually
   (a) diverging tectonic plates.
   (b) an earthquake.
   (c) converging tectonic plates.
   (d) volcanism.
7. Which of the following types of seismic waves are the fastest?
   (a) Love waves
   (b) S-waves
   (c) P-waves
   (d) Rayleigh waves
8. Earthquakes are caused by
   (a) the friction between diverging plates.
   (b) the sudden release of energy that is stored elastically in deformed rocks.
   (c) the expansion of Earth's crust.
   (d) the combined motion of tectonic plates.
9. Seafloor spreading provided a driving force for continental drift because
   (a) the youngest seafloor is found near the continents.
   (b) seafloor spreading pushes continents apart.
   (c) mantle convection causes slippage.
   (d) subduction creates the youngest seafloor.
10. Lithosphere is created at _____ boundaries and destroyed at _____ boundaries.
   (a) convergent; divergent
   (b) divergent; convergent
   (c) divergent; transform
   (d) convergent; transform

Answers to RAT
1c, 2c, 3a, 4d, 5a, 6b, 7c, 8b, 9b, 10b

EXPLORING FURTHER

McPhee, John. *Assembling California.* New York: Farrar, Straus & Giroux, 1993. *Roadside Geology Series.* [Many titles covering various parts of North America.] Missoula, Montana: Mountain Press.

http://pubs.usgs.gov/publications/text/dynamic.html
Sponsored by the U.S. Geological Survey, this Dynamic Earth Web page explains plate tectonics, volcanoes, and earthquakes clearly with drawings, maps, and photographs. It is a great resource for historical information and scientific theory.

http://scign.jpl.nasa.gov/learn/plate1.htm
The Structure of the Earth section of the Southern California Integrated GPS Network Education Module home page provides an education module and activities for high school and college undergraduate students. This interesting and interactive site also covers plate tectonics and earthquakes.

http://www.tinynet.com/faults.html
This site describes four different types of faults. It also includes an extensive list of links to interesting geological sites, especially earthquake-related sites.

http://earth.leeds.ac.uk/faults/
This link is a good resource for detailed information about the different types of faults. It includes a large picture gallery of various faults. Click the Learn Structure icon to browse additional sections of the site.

http://www.ucmp.berkeley.edu/geology/tectonics.html
The Geology section of the UC Berkeley Museum of Paleontology Web site includes history and descriptions of plate tectonics and an impressive selection of plate tectonics animations.

http://greenwood.cr.usgs.gov/pub/open-file-reports/
ofr-99-0132/

Sponsored by the U.S. Department of the Interior and the U.S. Geological Survey, this site is a good resource for detailed instructions on how to build a model of the outer layers of Earth. The model can be used to develop a better understanding of the main features of plate tectonics.

CHAPTER 22 ONLINE RESOURCES

The Physical Science Place

Interactive Figure

- 22.3

Tutorials

- Mantle Convection and Seafloor Spreading
- Plate Boundaries and Plate Tectonics

Videos

- The Mantle and Crust
- Plate Tectonics
- Folds and Faults in Earth's Crust
- The San Andreas Fault Story

Quiz

Flashcards

Links

shaping earth's surface

23

23.1 THE HYDROLOGIC CYCLE

23.2 GROUNDWATER

23.3 THE WORK OF GROUNDWATER

23.4 SURFACE WATER AND DRAINAGE SYSTEMS

23.5 THE WORK OF SURFACE WATER

23.6 GLACIERS AND GLACIATION

23.7 THE WORK OF GLACIERS

23.8 THE WORK OF AIR

■ Imagine viewing Earth from space and surveying its entire surface. What would you see? The first thing we would notice is that most of Earth's surface—about 71%—is covered by water. The remaining 29% is taken up by continental landmasses. Erosion, transportation, and deposition of sediment shape and sculpt the surface of these landmasses.

The hydrosphere and the atmosphere both play roles in this reworking of the surface—water, wind, and ice are the agents that do most of the work. We approach our investigation of Earth's surface by first exploring how these agents flow. In turn, we then look at their effects on Earth's surface.

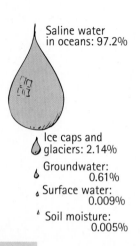

Saline water
in oceans: 97.2%

Ice caps and
glaciers: 2.14%

Groundwater:
0.61%

Surface water:
0.009%

Soil moisture:
0.005%

FIGURE 23.1

Distribution of Earth's water supply.

The Physical Science Place
The Hydrologic Cycle

The Physical Science Place
The Hydrologic Cycle Activity

23.1 The Hydrologic Cycle

Slightly more than 97% of Earth's water is in the oceans, and a little more than 2% is frozen in polar icecaps and glaciers. The remainder, less than 1%, consists of water vapor in the atmosphere, water in the ground, and water in rivers and lakes.

Water on Earth is constantly circulating, driven by the heat of the Sun and the force of gravity. As the Sun's energy evaporates ocean water, a cycle begins (Figure 23.2). Evaporation moves water molecules from Earth's surface to become part of the atmosphere. The resulting moist air may be transported great distances by wind. Some of the water molecules condense to form clouds and then precipitate as rain or snow. If precipitation falls on the ocean, the cycle is complete—from ocean back to ocean.

Completion of the cycle is more complex when precipitation falls on land, for water may drain to streams, then to rivers, and then journey back into the ocean. Or it may percolate into the ground, or evaporate back into the atmosphere before reaching the ocean. Also, water falling on land may become part of a snowpack or glacier. Although snow or ice may lock water up for many years, such water eventually melts or evaporates and returns to the cycle. This natural circulation of water—from the oceans to the air, then to the ground, then to the oceans, and then back to the atmosphere—is called the **hydrologic cycle.***

Water vapor plays an interesting role in the hydrologic cycle. The total amount of water vapor in the atmosphere remains relatively constant, even though clouds are continually forming. This can occur only if evaporation and precipitation balance each other. Because most of Earth's surface area is ocean, it makes sense that evaporation and precipitation are greatest over the oceans. In fact, 85% of the atmosphere's water vapor is water evaporated from the ocean, and 75% of the atmosphere's water vapor is precipitated back into the oceans. On the continents, precipitation exceeds evaporation. Of the atmosphere's water vapor, 15% is water evaporated from the continents and 25% is water precipitated back to the land. Balance is maintained between the amount of water taken up into the atmosphere (85% from oceans and 15% from continents) and

FIGURE 23.2

The hydrologic cycle. Water evaporated at Earth's surface enters the atmosphere as water vapor, condenses into clouds, precipitates as rain or snow, and falls back to the surface, only to evaporate again and go through the cycle yet another time.

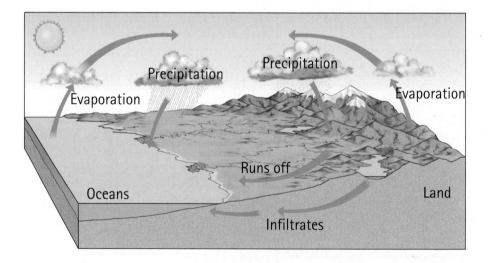

* This key concept is another conservation principle. Recall that in Chapter 3 we learned about conservation of momentum and energy; in Chapter 9, we learned about the conservation of electric charge; and in Chapter 13, we learned about the conservation of nucleons in nuclear reactions. So now we learn that the amount of water on Earth is conserved. A lack of it in one place means an abundance someplace else, which in most instances is the ocean.

TABLE 23.1   WATER RESOURCE RESIDENCE TIMES

Location	Average Residence Time
Atmosphere	1–2 weeks
Ocean	
Shallow depths	100–150 years
Deep depths	30,000–40,000 years
Continents	
Rivers	2–3 weeks
Lakes	10–100 years
Shallow groundwater	up to 100s of years
Deep groundwater	up to 1000s of years
Glaciers	10,000–20,000 years

fyi

The time for a certain parcel of water to complete the hydrologic cycle is its *residence time*—the average length of time a water molecule spends in a particular region. Water in polar ice and glaciers has a long residence time. For all practical purposes, the thousands-of-years residence time of deep groundwater means that if it is withdrawn, it will not be replenished. Water is truly a precious natural resource.

the amount precipitated out of the atmosphere (75% to the oceans and 25% to the continents).

The rain or snow that falls on the continents is Earth's only natural supply of fresh water. More than three-quarters of Earth's fresh water is locked up in the polar ice caps and glaciers. Surprisingly, most of the freely flowing fresh water is not in lakes and rivers, but rather beneath Earth's surface. As rain falls and sinks into the ground, it percolates downward. Some of the percolating water fills the open pore spaces between sediment grains. This water is now called **groundwater.**

CHECK POINT

1. What percentage of Earth's water supply is fresh water?
2. The volume of water evaporated from all of Earth's land surface is 60,000 km³/yr but the volume precipitated over the land surface is 96,000 km³/yr. The volume that precipitates each year is 36,000 km³ more than the volume that evaporates—so why isn't all the land flooded?

Were these your answers?

1. Less than 3%, as you can see by adding the freshwater values in Figure 23.1: 2.14% (glaciers) + 0.61% (groundwater) + 0.009% (surface water) + 0.005% (soil moisture) = 2.764%.
2. The excess water works its way back to the oceans. Excess water to the oceans does not cause sea level to rise because evaporation (85%) exceeds precipitation (75%) by 10%. Balance is maintained between the amount of water evaporated and precipitated over the oceans (85% − 75% = 10%) and the amount precipitated and evaporated over land (25% − 15% = 10%).

23.2 Groundwater

The liquid water in lakes, ponds, rivers, streams, springs, and puddles is the only fresh water that meets our eye, but all these water sources together hold only about 1.5% of Earth's non-ice fresh water. The other 98.5% resides in porous regions beneath Earth's surface.

Water beneath the ground exists as groundwater and *soil moisture*. Groundwater occurs in the *saturated zone*, the underground region where water has

FIGURE 23.3

The unsaturated zone is above the saturated zone. Water in the unsaturated zone does not completely fill the open pore spaces. This is soil moisture. Water in the saturated zone completely fills all open pore spaces. This is groundwater.

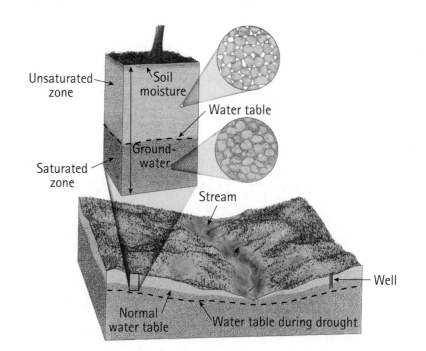

fyi

- As you will find out, water, wind, and ice are important agents that sculpt Earth's surface. Gravity, however, also plays a major role in altering the landscape. *Mass wasting* refers to the downslope movement of rock and soil under the direct influence of gravity. The combined effects of mass wasting and running water produce stream valleys—the running water cuts downward and mass wasting widens the valley. A prime example is the stream valley we call the Grand Canyon! Water helps set the stage for mass wasting to occur, but gravity is *the* controlling force.

completely filled all open pore spaces (Figure 23.3). Above the saturated zone is the *unsaturated zone*, where soil moisture resides. Pore spaces in the unsaturated zone are not completely filled with water—they contain a significant amount of air. As with water in a swimming pool, the pressure in groundwater increases with depth. Just as we can pump water from a swimming pool, we can pump groundwater from the ground. However, the presence of air in pore spaces prevents us from withdrawing water from the unsaturated zone.

Have you ever noticed how during a rainstorm sandy ground soaks up rain like a sponge? The water literally disappears into the ground. The type of surface material influences the amount of water that goes into the ground. Some soils, such as sand, soak up water easily. Other soils, such as clay, do not. Rocky surfaces with little or no soil are the poorest absorbers of water, with water penetration occurring mostly through fractures in the rock. Rain that does not soak into the ground becomes *runoff*, which then finds its way to bodies of water, such as lakes and rivers, or evaporates.

The amount of water that can be contained underground at a given location depends on the porosity of the soil or rock at that location. **Porosity** is the volume of open space in a soil, sediment, or rock sample compared to the total volume of solids plus voids. Porosity depends on the size and shape of the soil or sediment particles and on how tightly these particles are packed. For example, a soil composed of rounded particles of similar size has a higher porosity than a soil composed of rounded particles of various sizes. This is because the smaller particles fill up the spaces between the larger particles, thereby reducing the overall porosity of the soil.

Porosity is a measure of the volume of open space underground. As such, it represents the maximum *amount* of underground water at a given location. But the porosity does not tell us how groundwater *moves*. The **hydraulic conductivity**—a measure of permeability—tells us the degree to which geologic material can transmit water. If the pores are extremely small and poorly connected (as is the case with flattened clay particles), water may barely move at all. Think of it this way: it's a lot easier to sip soda through a large straw than through one of those very small straws intended for stirring coffee. Likewise, it is difficult for

FIGURING PHYSICAL SCIENCE

▪ Porosity

Porosity tells us the ratio of open space to total volume of soil, sediment, or rock sample:

$$\text{porosity} = \frac{\text{volume of open space}}{\text{volume of open space} + \text{volume of solids}}$$

SAMPLE PROBLEM

The volume of solids in a sediment sample is 975 cm^3 and the volume of open space is 325 cm^3. What is the porosity?

SOLUTION:

$$\text{Porosity} = \frac{325 \text{ cm}^3}{325 \text{ cm}^3 + 975 \text{ cm}^3} = 0.25$$

So the volume of open space is only one-fourth of the total volume.

---

water to move through the pores of clay. The hydraulic conductivity—or permeability—of clay is almost zero, even though the porosity of most clays is very high. In contrast, sand and gravel have large, open, well-connected pore spaces, and water moves freely from one pore space to the next. Thus, sand and gravel are highly porous and highly permeable (Figure 23.4).

CHECK POINT

Why is a sandy soil better for water flow than a clay soil?

Was this your answer?
Water flows easily through sandy soil because it is generally composed of rounded particles with pore spaces that are large and well connected; thus sandy soil is very permeable. We say the same when saying it has a high hydraulic conductivity. Water doesn't flow very easily through clay soils because clay is composed of flattened particles with small, poorly connected pore spaces between them; thus clay soils have a low hydraulic conductivity. Water flow is easier in soils with higher hydraulic conductivities.

THE WATER TABLE

When digging a hole in the ground, we find that wetness of the soil varies with depth. Just below the surface we encounter the unsaturated zone, where pore spaces are partially filled with water (Figure 23.3). As we descend farther, we enter the saturated zone, where pore spaces are completely filled with water. If our hole is entirely within the unsaturated zone, it does not fill with water. If we dig our hole deeper into the saturated zone, it partially fills with water. The upper boundary of the saturated zone is called the **water table** (Figure 23.3). The level of the water in our hole is the same level as the water table. In fact, the level of water in our hole *is* the water table at that location.

The depth of the water table beneath Earth's surface varies with precipitation and climate. It ranges from zero in marshes and swamps to hundreds of meters in some parts of the deserts. The water table also tends to rise and fall with the surface topography (Figure 23.5). At lakes and perennial streams (streams that flow all year), the water table is above the land surface.

(a)

(b)

FIGURE 23.4

Porosity and permeability. (a) The sediment particles in clay are small, flat, and tightly packed, with small, poorly connected pore spaces. Thus, clays have high porosity but low permeability (and low hydraulic conductivity). (b) Sediment particles in sand or gravel are relatively uniform in size and shape, with large and well-connected pore spaces. This allows water to flow freely. So sands and gravels can have both high porosity and high permeability (and high hydraulic conductivity).

The water table roughly parallels the ground surface. In times of drought, the water table falls, reducing stream flow and drying up wells. The water table also falls if the rate at which water is pumped out of a well exceeds the rate at which the groundwater is replaced.

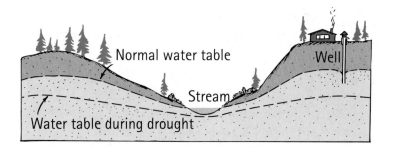

Want to see the water table? Most ponds and lakes are simply a place where the land surface dips below the water table.

AQUIFERS AND SPRINGS

Any water-bearing underground region through which groundwater can flow is an *aquifer*. These reservoirs underlie the land surface in many places and contain an enormous amount of water. More than half the land area in the United States is underlain by aquifers. One such aquifer is the Ogallala aquifer, which stretches from South Dakota to Texas and from Colorado to Arkansas!

Up to this point, we have been discussing *unconfined* aquifers. In unconfined aquifers, the soil or sediment above the water table is permeable, which allows the *recharge* of water—the process of water soaking into the ground—directly to the aquifer. All aquifers are at least partially unconfined, as illustrated in Figure 23.6.

Another type of aquifer is a *confined* aquifer. An aquifer is considered confined if it is sandwiched between continuous low-permeability layers (Figure 23.6). Confining layers* occur in sedimentary deposits that have alternating sand and clay layers, or alternating sandstone and shale layers. In confined aquifers, recharge does not come from directly above—water does not penetrate the confining layer. Natural recharge to confined aquifers comes only from unconfined portions of the aquifer at higher elevation.

As we learned in Chapter 5, pressure in water depends on the height of water above. Water anywhere in the confined portion of an aquifer is below the level of the water table in the recharge area. So groundwater in the confined aquifer—under pressure from water above—flows out through openings at

An artesian system is formed when groundwater in an aquifer confined between layers of low-permeability rock rises to the surface through any opening that taps the aquifer. Water flows freely if the water table height in the recharge area is greater than the height of the opening (flowing artesian well and artesian spring). If the height of the opening is greater than the height of the water table in the recharge area, the water does not flow (nonflowing artesian well).

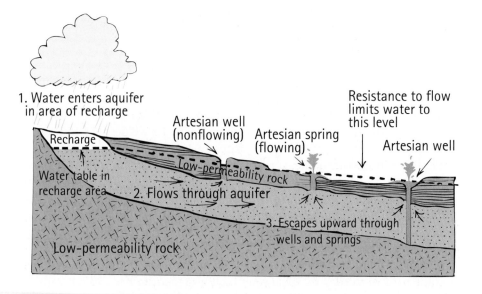

* Geologists call these layers *aquitards*.

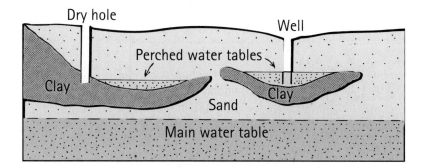

FIGURE 23.7

A perched water table is separated from the main water table by a low-permeability layer, clay in this case.

lower elevations. This is an **artesian system.** If the opening is natural and water flows out of the ground, it is an *artesian spring*. If the opening has been drilled, it is an *artesian well*. When first tapped, some artesian wells blast water tens of meters in the air!

Discontinuous low-permeability layers in an unconfined aquifer can intercept downward-percolating water above the water table. When this happens, a *perched* water table is created (Figure 23.7).

Where the water table intersects the land surface, groundwater emerges from an aquifer as a spring, stream, or lake (Figure 23.8). Springs can generally be found where the water table (or a perched water table) intersects the surface abruptly, such as on a hillside or coastal cliff. Because water tends to leak out of the ground through cracks and breaks in a rock, springs are often associated with faults. In fact, field geologists can often locate faults by looking for springs.

Link To Section 5.2

CHECK POINT
■

1. What is an aquifer?
2. What principal condition is required for an artesian system to occur?

Were these your answers?
1. An aquifer is a body of rock or sediment through which groundwater moves easily.
2. The principal condition required for an artesian system to occur is the presence of confining layers. Confining layers allow groundwater in confined aquifers to be under higher pressure than groundwater in unconfined aquifers. This allows water to rise above the top of the confined aquifer at natural or human-made openings.

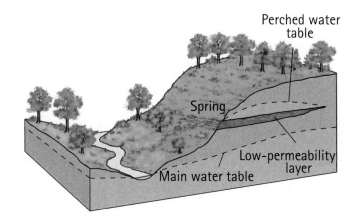

FIGURE 23.8

When the water table intersects the land surface, groundwater is released. From the perched water table, water is released via a spring; from the main water table, water is released by or into a stream.

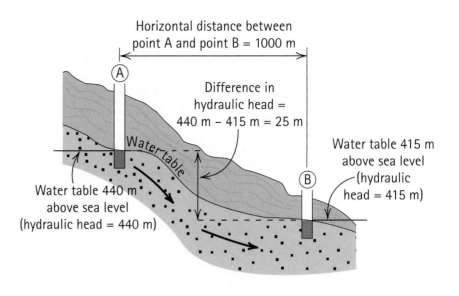

FIGURE 23.9

The hydraulic gradient is the difference in hydraulic head between any two locations divided by the horizontal distance between the locations. In this example, we have (440 m − 415 m)/1000 m.

GROUNDWATER MOVEMENT

The elevation of a water table above a particular location, usually sea level, is called the *hydraulic head*. This is the same elevation to which water in an unconfined aquifer rises in a well (Figure 23.9). Recall from Chapter 5 that liquid pressure is directly proportional to the depth of the liquid. Hence, the higher the hydraulic head above a particular location, the greater the water pressure at that location. The downward slope of a water table is the *hydraulic gradient*. It can be expressed like any slope: "rise over run," or in this case, the difference in hydraulic head between two points divided by the horizontal distance between those points.

Groundwater flows in response to pressure differences—it flows from high head to low head. Hence, groundwater flow through an aquifer is directly proportional to the hydraulic gradient. This can be expressed as

Groundwater flow rate $\sim$ hydraulic gradient

Flow rate also depends on the hydraulic conductivity of the soil and the cross-sectional area of the aquifer. The cross-sectional area of an aquifer by definition is always perpendicular to the flow direction. (Similarly for a water pipe, where the cross-sectional area is a circle defined by the diameter of the pipe). Hydraulic conductivity values are high for gravel and lower for fine sand or silt. When the hydraulic conductivity and cross-sectional area are introduced, the proportion can be expressed as the exact equation

Groundwater flow rate = hydraulic conductivity × cross-sectional area × hydraulic gradient

This relationship was first recognized by the French engineer Henri Darcy in 1856 and is aptly termed *Darcy's law*.

Topography plays an equally strong role in groundwater flow because it creates the hydraulic gradient. Groundwater moves from regions where the water table is high to regions where the water table is low, essentially flowing "downhill" underground (Figure 23.10).

The speed of groundwater movement is generally very slow compared to the flow speed in rivers and streams. The more permeable the aquifer, the faster the flow; the greater the hydraulic gradient, the faster the flow. The speed and route

fyi

- Groundwater flow, heat, and electric current all move in response to pressure differences. It's nice when different concepts connect in much the same way.

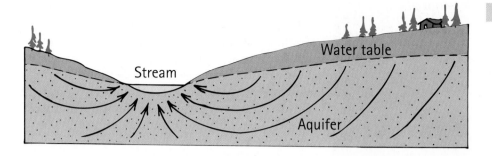

FIGURE 23.10

Groundwater flows from a high-hydraulic-head area, such as beneath a hill, to a low-hydraulic-head area, such as beneath a stream valley. The curved arrows indicate flow, which show that the stream is fed from below.

of groundwater flow can be measured by introducing dye into a well and noting the time it takes to travel to the next well. In most aquifers, groundwater speed is only a few centimeters per day, sufficient to keep underground reservoirs full.

■ 23.3 The Work of Groundwater

flowing groundwater—no matter how slow—can cause large changes in landscapes. These changes can occur because of humans, but more often than not they occur without any human interference.

Link To Section 18.4

LAND SUBSIDENCE

Most wells are drilled so that groundwater can be pumped from the ground. In areas where groundwater withdrawal has been extreme, the land surface is

---

FIGURING PHYSICAL SCIENCE

■ Darcy's Law

Knowing the dimensions and hydraulic conductivity of an aquifer, and the hydraulic head at the beginning and end of the aquifer, we can use Darcy's law to calculate the ground water flow rate. Again, Darcy's law is stated as follows:

Groundwater flow rate = hydraulic conductivity
$\times$ cross-sectional area $\times$ hydraulic gradient

Consider an aquifer with a length of 10 km that is between two lakes. Measured perpendicular to the length, a vertical slice of the aquifer has a cross-sectional area of 200,000 m². The elevation of the water surface at Lake A is 215 m above sea level. The elevation of the water surface at Lake B is 210 m above sea level. The hydraulic conductivity of the aquifer is 10 m/day.

SAMPLE PROBLEM 1

What is the hydraulic gradient in the aquifer?

SAMPLE PROBLEM 2

What is the groundwater flow rate in the aquifer?

SOLUTION TO PROBLEM 1:

The lake surfaces give us two locations where we know hydraulic head. First, we need to get the distances and hydraulic heads in the same units:

$$10 \text{ km} \times \frac{1{,}000 \text{ m}}{1 \text{ km}} = 10{,}000 \text{ m}$$

Then

$$\text{hydraulic gradient} = \frac{215 \text{ m} - 210 \text{ m}}{10{,}000 \text{ m}}$$
$$= 0.0005 \text{ (units cancel out)}$$

SOLUTION TO PROBLEM 2:

Groundwater flow rate = 10 m/day $\times$ 200,000 m²
$\times$ 0.0005 = 1000 m³/day

FIGURE 23.11

The Leaning Tower of Pisa. Construction began in about 1173 and was suspended when builders realized the foundation was inadequate. Work was later resumed, however, and the 60-m tower was completed 200 years later. Deviation from the vertical is about 4.6 m. The tower's foundation has been stabilized by groundwater withdrawal management, so now the tower should remain stable for years to come.

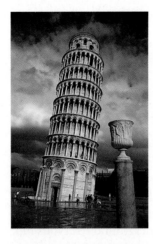

lowered—it *subsides*. The problem of land subsidence is most noticeable where the subsurface is composed of a thick sequence of poorly consolidated sediments rather than rock. Such sequences generally have many layers of easily compressed, water-bearing clays sandwiched between a series of sandy aquifers. Recall that clay has a very low hydraulic conductivity. As water is pumped from the aquifers, water slowly leaks out of the clay layers to replenish the aquifers, which usually continue to be pumped. As the clays lose water, they shrink, causing the land surface to subside.

Probably the most well-known example of land subsidence is the Leaning Tower of Pisa in Italy, built on the unconsolidated sediments from the Arno River. Over the years, the withdrawal of groundwater has resulted in subsidence and the increased tilt of the tower (Figure 23.11). In the United States, large amounts of groundwater have been pumped for irrigation in the San Joaquin Valley of California. This process has caused the water table to drop 75 m in 20 years, lowering the land surface by as much as 9 m (Figure 23.12). Because water for irrigation is now provided by canals, the sandy aquifers are slowly recharging (refilling with water), but most of the land subsidence caused by the compaction of the clay layers cannot be reversed.

FIGURE 23.12

The land surface in California's San Joaquin Valley lowered more than 9 m (30 ft) over a 50-year period because of the withdrawal of groundwater and the resulting compaction of sediments.

CHECK POINT

■ Why is land subsidence most evident in regions where the underlying geology is a series of clay layers sandwiched between sandy aquifers?

Was this your answer?
Clay layers lose water and compact as water is pumped from the adjacent aquifers. Compaction causes the land to subside.

CARBONATE DISSOLUTION

Groundwater is stored in vast limestone deposits that underlie millions of square kilometers of Earth's surface. But the groundwater slowly "eats away" at the limestone in which it is housed. Recall from Chapter 20 that limestone is made of the mineral calcite ($CaCO_3$). Rainwater chemically reacts with carbon dioxide in the air and soil to produce carbonic acid. When it seeps downward into limestone, the slightly acidic groundwater partially dissolves the rock. As groundwater steadily dissolves the limestone, it creates unusual erosional features, such as sinkholes and caverns. It is in limestone that we find the only true underground rivers—in other rocks and soils, underground water is found only in pore spaces, not in large, open channels.

CAVERNS AND CAVES The dissolving action of underground water has carved out magnificent caverns and caves (a cavern is simply a large cave). Groundwater flow in limestone aquifers occurs mostly through fractures in the rock, rather than through pores. Rainwater (enriched in carbonic acid) soaking into the limestone flows downward through the fractures toward the water table, dissolving rock as it goes. As groundwater flows toward its outlet, say a stream, the slightly acidic water dissolves the surrounding limestone, expanding the fractures and forming underground channels and caves (Figure 23.13a). The stream water is also acidic, and it dissolves and deepens the stream channel while groundwater continues to expand the caves. As the water level in the stream and

(a)

(b)

(c)

FIGURE 23.13

The formation of a cave begins with a layer of carbonate rock, mildly acidic groundwater, and an enormous span of time. (a) Groundwater makes its way toward a stream. (b) As the stream valley deepens because of erosion, the water table is lowered. The carbonate rock is eaten away as acidified water erodes and enlarges the existing fractures into small caves. (c) Further deepening of the stream valley causes the water table to drop even lower; water in the cave seeps downward, leaving an empty cave above a lowered groundwater level.

FIGURE 23.14

Cave dripstone formations.

the level of the water table drop, water drains from the first set of channels and caves. Groundwater continues to dissolve the limestone, forming a new, lower level of channels and caves (Figures 23.13b and 23.13c).

Water dripping from the cave ceiling, now rich in dissolved calcium carbonate, creates icicle-shaped stalactites as water evaporates and the calcium carbonate precipitates. Some of the water solution drips off the end of the stalactites to build corresponding cone-shaped stalagmites on the floor (Figure 23.14).

SINKHOLES Sinkholes are funnel-shaped cavities in the ground that are open to the sky. They are formed in much the same way as caves. Groundwater dissolves limestone and eventually the surface collapses in on itself. Some sinkholes are caves whose roofs have collapsed. Some sinkholes are formed by drought conditions or excessive groundwater pumping.

KARST REGIONS When sinkholes, caves, and caverns define the land surface, the terrain is called *karst topography*, named after the Karst region of Yugoslavia, where weathering and erosion of limestone characterizes the landscape. The pattern of streams in this type of landscape is very irregular; streams and rivers disappear into the ground and reappear as springs. Some karst areas appear as soft, rolling hills with large depressions that dot the landscape; the depressions are old sinkholes now covered with vegetation (Figure 23.15). In general, karst areas have sharp, rugged surfaces and thin to nonexistent soils as a result of high runoff and dissolution of surface material.

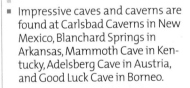

fyi

- Impressive caves and caverns are found at Carlsbad Caverns in New Mexico, Blanchard Springs in Arkansas, Mammoth Cave in Kentucky, Adelsberg Cave in Austria, and Good Luck Cave in Borneo.

■ 23.4 Surface Water and Drainage Systems

Streams—by which we mean all flowing surface water, from the Mississippi River to the shallowest woodland creek—are dynamic systems that affect both the surface of the land and the people who live on that land. Streams have many benefits to offer: they provide energy, irrigation, and a means of transportation.

FIGURE 23.15

Karst topography covered by vegetation makes up the rolling hills in south central Kentucky.

FIGURE 23.16

The karst landscape of China has been an inspiration to classical Chinese brush artists for centuries.

fyi

- Karst regions can be found throughout the world: in the Mediterranean basin; in sections of the Alps and the Pyrenees; in southern China; and in the United States in Kentucky, Missouri, Florida, and Tennessee.

FIGURE 23.17

Laminar flow is slow and steady, with no mixing of sediment in the channel. Turbulent flow is fast and jumbled, stirring up everything in the flow.

FIGURE 23.18

The long profile of a stream. At a stream's headwaters, the gradient is high, the channels are narrow and shallow, and the stream flow is rapid. As the stream progresses downslope, the gradient decreases, the channel widens, and discharge increases.

Streams also carve out and alter the landscape. The Grand Canyon is testimony to the mighty erosive power of the Colorado River. For millions of years, the river has been carving out the canyon walls, cutting deeper and deeper into the rock as it makes its way to the ocean. Yet surface water plays another important yet contrasting role as it shapes the landscape—it deposits sediments. In this way, surface water is both a destroyer and creator of sediments and sedimentary rocks.

Landscape evolution—progressive changes to Earth's surface—is driven by the characteristics of surface water flow. The manner of flow has a large effect on how water alters the landscape. The flow characteristics of moving water are of two types—turbulent and laminar (Figure 23.17). When water moves erratically downstream, stirring everything with which it comes in contact, the flow is **turbulent**. When water flows steadily downstream with no mixing of sediment, the flow is **laminar**. In general, slow, shallow flows tend to be laminar, and faster-moving flows tend to be turbulent. Whether a flow is laminar or turbulent depends on the nature and geometry of the stream channel and the speed of the flow.

STREAM FLOW GEOMETRY

Streams come in a variety of forms—straight or curved, fast or slow. At their headwaters (the stream origin), stream channels are narrow and water flows quickly through deeply incised, V-shaped mountain valleys. Farther downstream, channels widen so that water flows into and along broad, low valleys.

Three variables influence the speed of water in a stream—*stream gradient, stream discharge,* and *channel geometry.* The **gradient** is the vertical drop in the elevation of the stream channel divided by the horizontal distance for that drop. If we look at a long profile of a stream (Figure 23.18), we see that the gradient is steep near the stream's headwaters and gentler, almost horizontal, near its mouth. Because of gravity, stream speed tends to be greater where the stream gradient is steep. Downstream, discharge and channel geometry also influence stream speed.

Discharge is the volume of water that passes a given location in a channel in a certain amount of time. It is directly proportional to the cross-sectional area of the channel—the width times the depth—and to the *average* stream speed:

Discharge = cross-sectional area × average stream speed

Or put another way,

$$\text{Average stream speed} = \frac{\text{discharge}}{\text{cross-sectional area}}$$

Stream speed is usually not constant along the length of a stream. In the headwaters of a stream, the gradient is high, so stream speed is also high. In fact, these upland sections of a stream are often called "rapids." As a stream progresses downslope, the gradient gradually decreases, the channel typically widens, and, because tributary streams feed into it, discharge increases.

Because stream speed depends on three variables, downslope changes in stream speed are often counterintuitive. Consider a stream in which discharge doubles downslope, but the channel stays the same size and shape. Looking at the equation for stream speed, we can see that the stream speed also doubles. Now consider a stream in which discharge doubles downslope and the cross-sectional area of the channel also doubles. When friction is insignificant, the speed of this stream does not change, even though discharge has increased. The cross-sectional area of the stream channel increases by the same percentage as the discharge, compensating for the additional water without causing the stream speed to increase. Can you see that stream speed downslope can actually *decrease*?

The shape of the channel also affects stream speed. Now consider two streams that have the same cross-sectional area. Water flowing in the channel touches the channel bottom and sides. Friction between the water and the channel slows stream speed. The cross-sectional shape of a channel determines the amount of water in contact with the channel. The greater the contact area, the greater the friction (Figure 23.19). If the stream channel is rounded and deep, as opposed to flat-bottomed and relatively shallow, the stream speed is faster because less water is in contact with the channel.

fyi
- Squeeze the end of a garden hose and you'll see that water speeds up when the passage of flow becomes narrower. Likewise for the flow of streams.

The Physical Science Place
Shaping Planetary Surfaces

Link To Section 5.9

CHECK POINT

1. Consider a stream in which discharge doubles downslope, but the channel stays the same size and shape. Now look at the equation for stream speed. What happens to stream speed?
2. Now consider a stream in which discharge doubles and the cross-sectional area of the channel also doubles. Assume that no water enters from side tributaries and friction is insignificant. What happens to stream speed?
3. If the gradient of a stream decreases and the cross-sectional area stays the same, does the discharge stay the same? Assume that no water enters from side tributaries and friction is insignificant.

Were these your answers?
1. Stream speed doubles.
2. Cross-sectional area and discharge of the stream channel increase by the same percentage, so stream speed does not increase.
3. No, because the decreased gradient slows the stream speed. The discharge decreases.

(a) Rounded, deep channel

(b) Wide, shallow channel

FIGURE 23.19

(a) In a rounded, deep channel, the speed of water flow is relatively high because less water is in contact with the channel (there is less friction). (b) Wide, shallow channels tend to have slower flows because more water is in contact with the channel (there is more friction).

DRAINAGE BASINS AND NETWORKS

A stream is one small segment of a much larger system called a *drainage basin*. A drainage basin is defined as the total area that contributes water to a given stream. A drainage basin can cover a vast area or be as small as 1 km². Drainage basins are separated from one another by *divides*, lines tracing out the highest ground between streams. Under most circumstances, the separation is complete—rain that falls on one side of a divide cannot flow to an adjacent basin. A divide can be either very long, if it separates two enormous drainage basins, or a mere ridge separating two small gullies. The *Continental Divide*, a continuous line running north to south down the length of North America, separates the Pacific basin on the west from the Atlantic basin on the east. Water west of the divide eventually flows to the Pacific Ocean, and water east of it flows to the Atlantic Ocean (Figure 23.20).

As mentioned, streams merge with other streams as they flow downhill, becoming larger and larger. The entire assembly of streams draining a region is called a *drainage network*, which can be characterized by the branching pattern formed by its streams (Figure 23.21). Because streams erode the land surface and hence erode the rocks and rock material on the land, drainage patterns are greatly influenced by the rock type and rock material eroded.

CHECK POINT

Distinguish between a drainage basin and a drainage network.

Was this your answer?
A drainage basin is the total area that contributes water to a stream. It includes all the streams as well. A drainage network involves only the streams that drain water from the basins. So a drainage network is part of a drainage basin.

FIGURE 23.20

The Continental Divide in North America separates the Pacific basin on the west from the Atlantic basin on the east.

FIGURE 23.21

Different drainage patterns develop according to surface material and surface structure: (a) dendritic, (b) radial, (c) rectangular, (d) trellis.

23.5 The Work of Surface Water

flowing surface water shapes and sculpts Earth's surface in different ways depending on the *geologic environment*. Some environments are characterized by erosion, others by deposition. Some are quiet and placid, others are turbulent and wild.

EROSION AND TRANSPORT OF SEDIMENT

We learned in Chapter 20 that weathering and erosion create and move sediment. Erosion by water is the most common way clastic sediments are carried away from the places in which they formed. Surface water erodes sediment and rocks, transporting them downstream and eventually depositing them in another place. In this way, surface water reshapes our landscape.

Flowing water erodes stream channels in several different ways. First, stream water contains many dissolved substances that *chemically weather* and erode the rocks they encounter. Another powerful mechanism for erosion is *hydraulic action*—the sheer force of running water. Swiftly flowing streams and streams at flood stage have strong erosive power as they break up and loosen great quantities of sediment and rock. The most powerful type of erosion, however, is *abrasion*. Abrasion occurs when sediments and particles actually scour a channel, much like sandpaper scraping on wood. When powered by turbulently spiraling water, rock particles rotate like drill bits and carve out deep potholes (Figure 23.22). The faster the current, the greater the turbulence, and the greater the erosion.

Erosion is only the beginning of the story of how surface water alters Earth's surface. Streams carry more than just water—they transport great amounts of sediment from one location to another. In general, laminar flows can lift and carry only the very smallest and lightest particles. A turbulent flow, however, depending on its speed, can move and carry a range of particle sizes—from the smallest particles of clay to large pebbles and cobbles. A turbulent current gathers and moves particles downstream mainly by lifting them into the flow or by rolling and sliding them along the channel bottom. The smaller, finer particles are easily lifted into the flow, and they remain suspended to make the water murky.

As we would expect, the faster the current, the larger the particles that can be carried. Also, the larger the volume of water, the greater the volume of sediment that can be carried. So streams that have a higher discharge can carry larger volumes of sediment, and streams in which the water is moving fast can carry larger

FIGURE 23.22

When powered by turbulent circular currents of water, rock particles rotate like drill bits and carve out deep potholes.

- Channel dimension, gradient, discharge, sediment load, and velocity all work together as they influence stream flow. As such, a river is a system of interdependence in which any change in one variable influences change in the entire system.

sizes of sediment. The continuous abrasion of sediment in the stream channel breaks up the sediments and thus contributes to the overall decrease in particle size that we see as we move downstream. At the river's mouth, only finer particles of sand, silt, and clay remain. As we shall soon see, these tiny particles are deposited to form a delta when the stream loses speed as it enters the sea.

CHECK POINT

Which is more effective in transporting sediment, laminar or turbulent flow? Defend your answer.

Was this your answer?

Turbulent flow, because the water's motion is irregular and sediments have a greater tendency to remain in suspension. Turbulent flow carries these sediments because of the energy of its churning water. In laminar flow, water moves steadily in a straight-line path with no mixing of sediment in the channel.

EROSIONAL AND DEPOSITIONAL ENVIRONMENTS

Eventually, particles that are being transported by surface water drop out of suspension—they are deposited. This happens when the water loses energy and slows. As a river gradually loses energy, larger particles are deposited first and then smaller ones, causing most surface water deposits to be well sorted.

The most dominant feature of deposited sediments is the way the particles of sediment are laid down, layer upon horizontal layer. These layers are referred to as *beds*. Varying in both thickness and area, each bed represents one episode of deposition. For example, flooding in a particular year might produce a layer of sediment next to a river. A flood any time after that produces an overlying layer. Some beds represent deposition over millennia.

The deposition and erosion of sediments occur in many different environments, including oceans and shorelines, rivers and streams, deserts, and deltas. Each environment in which erosion, transportation, and deposition occur has its own specific characteristics.

STREAM VALLEYS AND FLOODPLAINS

As rainfall hits the ground, it loosens soil and washes it away. As more and more rain falls and the ground continues to lose soil, gullies form. Once water and soil particles funnel into such a gully, a stream channel is created. This erosive action may be extremely rapid, as in the erosion of unconsolidated sediments, or very slow, as in the erosion of solid rock. Water's erosive power enables a stream to widen and deepen its channel, to transport sediment away, and, in time, to create a valley. In high mountain areas, the erosive action of a stream cuts down into the underlying rock to form a narrow, V-shaped valley. Because the valley is narrow, the stream channel dominates the whole valley bottom. Fast-moving rapids and beautiful waterfalls are characteristic of V-shaped mountain stream valleys (Figure 23.23).

When a fast-flowing mountain stream leaves its narrow valley, it abruptly emerges onto a broad, relatively flat plain. The speed of the flow drops suddenly and the stream dumps its load of sediment. Streams of this type often flow no farther. The sediment deposits are generally fan-shaped and grow outward as additional sediment is deposited (Figure 23.24). The steep upper slopes of such deposits are dominated by boulders, cobbles, and gravels, while the base area and the adjacent plain are composed of sand, silt, and mud.

For streams that continue on, stream speed plays an important role in both erosion and deposition. In addition to the variations discussed above, water

FIGURE 23.23

At a stream's headwaters, high gradients contribute to fast-moving rapids. When there is an abrupt change in gradient, we see a beautiful cascading waterfall.

FIGURE 23.24

A fan-shaped clastic sedimentary deposit in Death Valley, California.

speed also varies within a channel from one bank across to the other. Flow speed is slower along the streambed, where water is in contact with the channel, creating friction; flow speed is greater near the water surface. In a large stream that is flowing in a straight channel, the maximum flow speed can be found midchannel (Figure 23.25b). In a stream running through a bending, looping channel, the maximum flow speed is found toward the outside of each bend (Figures 23.25a and 23.25c).

As a stream flows downhill and its gradient becomes gentler and its speed slows, the focus of its energy changes from eroding downward (deepening the channel) to eroding laterally in a side-to-side motion. As a result of this lateral action, the stream develops a more sinuous, *meandering* form (Figure 23.26). As the stream bends and curves, the flow speed in the channel shifts so that the maximum speed is always toward the outside of each bend (Figure 23.25). Rapidly moving water is very effective in eroding material from the outside of the bend, creating a steep bank called a **cut bank.** Material eroded from the cut bank is transported downstream, where it may eventually be deposited in areas where the stream speed decreases. Sandy **point bars** form on the insides of bends by this process.

Continuing their downhill path, streams meander back and forth across a river valley, depositing sediments as they go (Figure 23.27). The meandering

FIGURE 23.25

In a stream that bends (a and c), maximum flow speed is toward the outside of each bend and slightly below the surface. In a straight-channel stream (b), maximum speed is midchannel and near the water's surface. Erosion of the stream channel occurs where stream speed is greatest (cut bank); deposition occurs where stream flow slows (point bar).

FIGURE 23.26

The evolution of a stream valley and development of a floodplain. (a) At the headwaters, the V-shaped stream valley is characterized by steep gradients and fast-moving water that cuts down into the stream channel. Features in this area include cascading rapids and waterfalls. (b) Downstream, with reduced gradient, the stream focuses its erosive action in a side-to-side sinuous manner, thereby widening the stream valley. (c) Farther downstream, meandering increases and further widens the stream valley to form a large floodplain.

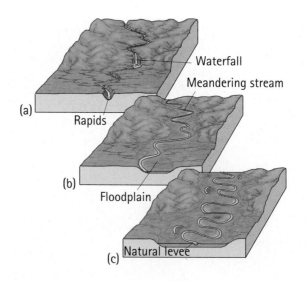

fyi

- You have probably heard the term "100-year flood." Does this term mean that a 100-year flood occurs once every 100 years? Not exactly. In fact, as you probably know, an uncommonly big flood can occur in any year. "One-hundred-year flood" is a statistical term, indicating a 1-in-100 chance that a flood this size will occur during any year. Perhaps a better term would be "1-in-100-chance flood."

movement creates a wide belt of almost flat land—a **floodplain.** As the name implies, this section of the river valley becomes flooded with water and sediments when a river overflows its banks. In a flood, as discharge and flow speed increase, so does the stream's ability to carry sediment. Thus, when a stream overflows its banks, sediment-rich water spills out onto the floodplain. The speed of the water quickly decreases as it spreads out over the wide, flat, floodplain, and a progression of large to small particles is deposited. As expected, larger, coarse-grained sediments are deposited along the edges of the channel and smaller, fine-grained sediments are deposited farther away from the stream channel on the floodplain. The larger particles deposited close to the stream channel form *natural levees* that help confine future floodwaters (Figure 23.28). The widening of the valley, as shown in Figure 23.27, occurs because sediment deposited by the stream, especially during floods, progressively fills the valley.

CHECK POINT

Floodplains are often prime agricultural areas. Why would people want to work and live in areas so prone to flooding?

Was this your answer?
People live and work in floodplain areas because such plains are next to rivers that provide easy access to water, food, and a means of transportation. Also, because of periodic flooding, floodplain soils are often extremely fertile and thus serve as prime farmland. As for the factor of danger—don't most people associate danger with anyone but themselves?

FIGURE 23.27

Cross section of a river valley. A floodplain is created when a river overflows its banks. Sands and gravels settle out first and act as natural levees to confine the river. Because the finer silt and clay particles can flow as a suspended load, they move beyond the levees and settle on the floodplain.

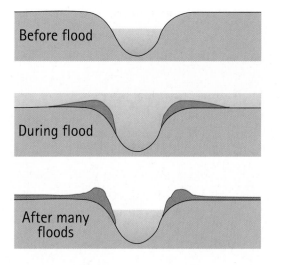

Before flood

During flood

After many floods

FIGURE 23.28

In a flood, increased discharge and flow speed help a stream carry not only a large sediment load but also larger particles. Larger, coarse-grained sediment deposited close to the stream channel forms natural levees that confine the stream between flood stages. Successive floods increase the height of the levees and may even raise the overall elevation of the channel bed so that it is higher than the surrounding floodplain.

DELTAS: THE END OF THE LINE FOR A RIVER

As a stream flows into a standing body of water, such as a sea, bay, or lake, the moving water gradually loses its forward momentum. With reduced energy, stream speed slows and the stream loses its ability to carry sediment. These changes cause the stream to dump its sediment load. In this way, the mouth of the stream and the area immediately offshore become filled with sediment. The dumped sediment forms a fan-shaped deposit called a **delta.** Sediment is deposited in order of decreasing weight, with heavy, coarse particles deposited first, at and near the shoreline. Light, fine particles are deposited farther offshore. With the continual addition of incoming sediment, the delta progressively builds itself outward as an extension of land into the body of water.

Deltas begin to form underwater, but the addition of incoming sediment eventually causes the delta to emerge as new land. As the main stream channel becomes choked with sediment, more energy is required for water to push through the accumulated sediment than to go around it. So new, smaller channels form off the main channel like branches on a tree. These *distributaries* allow water to flow unimpeded to the standing body of water. As the delta continues to extend outward, the distributaries also become clogged, as sediment continues to arrive from upstream. The first set of distributaries thus form distributaries of their own, taking on the appearance of branching fingers (Figure 23.29). When the fingers become clogged, the branching process repeats. As streams continue to flow to the standing water and as successive sediment layers are deposited one on top of the other, the delta continues to build and expand outward (Figure 23.30). Thus, delta environments are areas where new land is continuously created.

fyi

- Some of the world's greatest rivers have huge deltas at their mouths. Millions of years ago, the mouth of the Mississippi River was where Cairo, Illinois, is today. Since that time, the delta has extended 1600 km south to the city of New Orleans. Less than 5000 years ago, the site of New Orleans was underwater in the Gulf of Mexico!

What is the ultimate destination of all water flow and, hence, the eventual site of deposition of most sediments? Water flows eventually to the ocean, and sediments settle to the ocean floor.

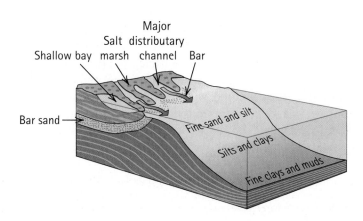

Major
Salt distributary
Shallow bay   marsh   channel   Bar

Bar sand

Fine sand and silt

Silts and clays

Fine clays and muds

FIGURE 23.29

Deltas are areas of land generation. As streams flow to the sea, they carry sediment. These sediments are deposited in order of decreasing weight, with heavy, coarse particles settling at or near the shoreline and light, fine particles settling farther offshore. Layer upon layer, the depositional platform called a delta takes form.

FIGURE 23.30

Satellite image of the Mississippi Delta. Note how smaller streams are formed branching off from the main river.

fyi

- Glaciers surge when the muddy till below the glacier is warmed by Earth's internal heat. Warmth is retained because insulation against the cold atmosphere is provided by the thick overlying ice. Melting occurs with meltwater seeping into the till, which softens and moves easily under the weight of the overlying ice.

FIGURE 23.31

When a deck of cards is pushed from one side, the individual playing cards slide past one another, thus shifting the whole deck.

■ 23.6 Glaciers and Glaciation

The mightiest rivers on Earth are frozen solid and normally flow a sluggish few centimeters per day. These great icy rivers are called **glaciers.** Glaciers covered significant portions of Earth several times in the distant past. Glaciation is still at work in many regions of the world, its agents being small alpine glaciers in mountainous areas, large alpine ice fields, and the huge Arctic and Antarctic continental ice sheets.

GLACIER FORMATION AND MOVEMENT

The ice of a glacier is formed from recrystallized snow. After snowflakes fall, their accumulation slowly changes the individual flakes to rounded lumps of icy material. As more snow falls, the pressure exerted on the bottom layers of icy snow compacts and recrystallizes it into glacial ice.

This ice does not become a true glacier, however, until it moves under its own weight. When the ice mass reaches a critical thickness of approximately 50 m, the pressure exerted by the overlying material causes the base of the ice to deform *plastically* and flow downslope. This plastic deformation can be likened to what happens to a deck of playing cards. When the deck is pushed from one end, as in Figure 23.31, individual cards slide past one another, shifting the entire deck. Plastic deformation in a glacier is greatest at the base of the ice, where pressure is greatest.

Plastic flow from the slippage of ice crystals is not the only component of glacial movement. The melting point of ice decreases as pressure increases. When melted ice—*meltwater*—forms at the base of the glacier, the process called *basal sliding* comes into play.* This second mechanism of glacial movement results in the sliding of the entire glacier downslope, with the meltwater acting as a lubricant. The net speed of the glacial ice increases from the base up, reaching its greatest values at the glacier's surface (Figure 23.32). Overlying layers are carried along in piggyback fashion as plastic deformation and basal sliding occurs in the lowest layers.

---

* Meltwater may result from the pressure of the overlying ice (the melting point of ice decreases as pressure increases), from the internal heat of Earth, or from the generation of heat from frictional drag as the glacier moves. Whatever the causes of its formation, meltwater contributes to the movement of a glacier.

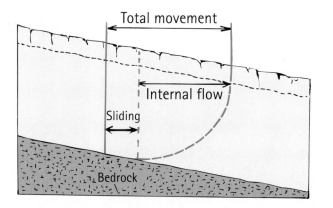

FIGURE 23.32

Cross section of a glacier. Glacial movement has two components: internal flow and sliding resulting from lubrication by meltwater. Movement is slowest at the base because of frictional drag and fastest at the surface. The upper parts of the glacier are carried along in piggyback fashion by plastic flow within the ice.

The uppermost portion of the glacier, carried along both by basal sliding and by internal plastic deformation, behaves like a rigid, brittle mass that may fracture. Huge, gaping cracks called *crevasses* may develop in this surface ice (Figure 23.33). These can extend to great depths and can therefore be quite dangerous for people attempting to cross a glacier.

Average glacier speed varies from glacier to glacier, and it can range from only a few centimeters to a few hundred centimeters per day. Such slow speeds are measured by placing a line of markers across the ice and recording their changes in position over a period of time, ranging from days to years. Ice moves fastest in the center and more slowly at the edges because of frictional drag (Figure 23.34). Some glaciers experience surges, or periods of much more rapid movement. These surges are probably caused by periodic melting of the base and sudden redistribution of mass. The flow rate in these relatively brief surges can be 100 times the normal rate. Viewed from above, flow bands of rock debris and ice normally have a parallel pattern, but during a surge the flow bands become intricately folded (Figure 23.35).

GLACIAL MASS BALANCE

From season to season, and over longer periods of time, the mass of a glacier changes. Typically, a glacier grows in the winter as snow accumulates on its surface. The amount of snow added and the process of adding snow to a glacier is aptly termed **accumulation.**

As ice accumulates and begins to flow downhill, it may move to an altitude where temperatures are warmer. When the ice begins to melt, the glacier loses some of its mass. A glacier may also lose mass as it moves downslope to a shoreline, where ice may break off, or *calve*, to form icebergs that float away to sea. Melting and calving are the two primary mechanisms by which glaciers lose mass. Although less noticeable, glaciers may also lose mass as the ice *sublimates* to water vapor. By whatever means, the total amount of ice lost and the process of losing ice is called **ablation** (Figure 23.36).

FIGURE 23.33

Crevasses can extend to very great depths and can be treacherous to cross.

Link To Section 7.6

FIGURE 23.34

Top view of a glacier. Movement is fastest at the center and gradually decreases along the edges because of friction.

FIGURE 23.35

Glacial flows: (a) normal flow; (b) surge flow.

(a)   (b)

fyi

- With global warming, ablation is exceeding accumulation in more and more places. As glaciers calve into the sea, or as ice sheets on land melt, sea level will rise.

When accumulation equals ablation, the size of the glacier remains constant. For example, in a mountain glacier, accumulation occurs with winter snowfall in the farther-back, higher-elevation parts of the glacier, and ablation occurs in the lower portions, where spring and summer melting is greatest. When accumulation rate and ablation rate are equal, the melting of the lower portions is offset by the downslope flow of ice from higher portions. As a result the location of the front edge of the glacier does not change. When accumulation exceeds ablation, the glacier advances—it grows. When ablation exceeds accumulation, the glacier retreats—it shrinks. Naturally, in all these cases, the ice of the glacier always flows downslope.

CHECK POINT

Under what conditions does the front of a glacier remain at the same location from year to year?

Was this your answer?

The front of a glacier remains at the same location when the rate of growth (accumulation) equals the rate of shrinking (ablation). In the spring, as ice at the glacier's front melts away, the glacier retreats up-slope. At the same time, the increased mass from the prior winter's accumulation causes the glacier to move forward. When the rate at which this forward movement matches the rate of melting, the location of the front edge doesn't change.

FIGURE 23.36

Accumulation on a glacier takes place at high elevations as snow falls on the glacier and turns to ice. Ablation takes place at lower elevations as ice melts or calves into icebergs or is lost through sublimation.

■ 23.7 The Work of Glaciers

Like flowing water in streams, glaciers can erode as well as deposit sediment. Both processes produce characteristic landforms, which yield clues about long-gone glaciers.

GLACIAL EROSION AND EROSIONAL LANDFORMS

Glaciers are powerful agents of erosion. Glaciation has created the beautiful landscapes of Tibet, Nepal, and Bhutan in Asia; the Alps of Switzerland; the fjords of Norway; and Yosemite Valley and the Great Lakes in North America. In many ways, a glacier is like a plow, as it scrapes and plucks up rock and sediment. It is also like a sled, as it carries its heavy load to distant places. As it moves across Earth's surface, a glacier loosens and lifts up blocks of rock, incorporating them into the ice. The large rock fragments carried at the bottom of a glacier scrape the underlying bedrock and leave long, parallel scratches (like sled tracks) aligned in the direction of ice flow (Figure 23.37). These scratches are called *striations*.

The two main types of glaciers, *alpine* and *continental*, have different erosional effects and produce dissimilar landforms. Alpine glaciers develop in mountainous areas and are often confined to individual valleys, while continental glaciers cover much larger areas. Alpine glaciers occur in most high mountain chains in the world, such as the Cascades, the Rockies, the Andes, and the Himalayas. The erosional features of alpine glaciation are depicted in Figure 23.38.

FIGURE 23.37

Striations mark the presence of a former glacier.

Link To Section 21.5

fyi

■ As you may recall, Alfred Wegener used glacial striations to reconstruct formerly joined landmasses. Evidence from glacial striations showed that glaciers moved from Africa toward the Atlantic Ocean and from the Atlantic Ocean onto South America. Wegener suggested that such glaciation is likely only if the Atlantic Ocean were missing and the continents joined.

FIGURE 23.38

The many erosional features of alpine glaciation. (a) The Matterhorn—named for its characteristic "horn" feature. (b) Hanging valleys are a spectacular feature found in areas that have been shaped by alpine glacial erosion. Bridalveil Falls in Yosemite National Park spills out of a hanging valley into the larger valley that was once occupied by the main glacier.

(a)   (b)

Small asymmetrical hills called *roches moutonnées* show the direction of continental glacial movement. On the side of the hill facing the approaching glacier, the slope is smooth and gentle. On the side facing away from the approaching glacier, the slope is rough and steep with a plucked appearance.

Roche moutonnée

Geologist Bob Abrams observes the grandeur of the Juneau Ice Field in Alaska.

Geologic features are best viewed from an airplane. Next time you fly in an airplane, request a window seat and enjoy the geology below.

Continental glaciers spread over the land surface, smoothing and rounding the underlying topography. Although striations are produced by both alpine and continental glaciers, they have played a larger role in the study of ancient continental glaciers. Because a continental glacier scours very large tracts of land, it tends to leave behind few obvious valleys (making it difficult to determine the glacier's direction of flow). By mapping striations on land once covered by continental glaciers, geologists can decipher the flow direction of the ice. Flow direction is also indicated by small, asymmetrical hills (Figure 23.39) that are known by the French term *roches moutonnées* (singular, *roche moutonnée*). On the "upstream" side of the ice flow, the hill's slope is smooth and striated from the abrasion of ice on bedrock. On the "downstream" side, the slope is rough and steep because the moving ice plucked rock fragments away from cracks in the bedrock.

GLACIAL SEDIMENTATION AND DEPOSITIONAL LANDFORMS

As a glacier advances across the land, it acquires and transports great quantities of debris. When the glacier retreats, this debris is left behind as it is melted out of the ice. Because a glacier abrades and picks up everything in its path, glacial deposits are characteristically composed of unsorted, angular rock fragments in a variety of shapes and sizes. Glacial deposits are collectively called **drift,** a term that dates back to the 19th century, when it was conjectured that all such debris had been "drifted in" by the great biblical flood.

Some glacial environments contain elements of several other depositional environments. This is due to the many processes at work, driven by water, wind, and ice. For example, glacial meltwater forms streams that may terminate as deltas in lakes, bays, or seas. Glacial deposits have some unique features, however. When glacial ice melts, it drops a poorly sorted, heterogeneous load of boulders, pebbles, sand, and clay. A wide range of particle sizes is the hallmark that differentiates glacial sediment from the much-better-sorted material deposited by streams and winds.

Drift is deposited in two main ways. When glacial sediment is released into meltwater, it is carried and deposited like any other waterborne sediment; thus, it is well sorted. Material deposited directly by melting ice forms an unsorted mixture of clay and bouldery rock debris. Many of the old stone walls and fences of New England are found in areas where the surface material is glacial till. Settlers who tried to farm this land had to remove all the larger boulders before they could plow, and they piled them along the edges of their fields. Oftentimes, large boulders that drastically differ from the local bedrock are found in glacial deposits. The large boulders provide proof of a glacier's ability to transport heavy loads for long distances. If a bedrock outcrop that matches the rock type of the out-of-place boulder can be found, then the distance and direction of glacial transport can be estimated.

The most common landform created by glaciers is the *moraine,* a ridge-shaped landform that marks the boundaries of ice flow. Of all the different types of moraines, probably the most important is the *terminal moraine,* as it marks the farthest point of a glacier's advance (Figure 23.40).

FIGURE 23.40

Glacial depositional land-forms. Of special importance is the terminal moraine, which marks the farthest point of a glacier's advance.

Another distinctive landform consisting of glacial sediments is the *drumlin*, an elongated hill shaped like the back of a whale. Formed by continental glaciation and lined up in the direction of ice flow, drumlins have a steep, blunt end in the direction from which the ice came and a tapered gentle slope on the downstream side (Figure 23.41). Perhaps the most famous drumlin in the United States is Bunker Hill in Massachusetts.

Many of the world's lakes, small and large, are the products of glacial action. Glaciers deepened valleys and deposited sediments that acted as dams, blocking stream drainage within some valleys and creating lakes. The Finger Lakes in upstate New York, the "10,000 Lakes" of Minnesota, and the Great Lakes of North America are all products of glacial action.

Link To Section 21.7

FIGURE 23.41

Topographic map showing numerous oval-shaped drumlins in upstate New York. Drumlins are steep and blunt on the side that faced the approaching glacier but tapered and gently sloping on the downflow side. Looking at the map, can you tell the direction of continental ice flow?

fyi

- A unique feature of sand dune formation is cross-bedding, which is found on the leeward side of a dune. The direction of cross-bedding indicates the direction of the wind (or water current) that deposited the sediments. A particularly great place to see ancient cross-bedding is in Zion National Park in Utah. Cross-bedding is also a common feature in river deltas and certain stream-channel deposits.

fyi

- Even though desert environments lack moisture, water is still the main cause of erosion and transportation of sediments. Scarce as water is in the desert, when a heavy rain falls, rainwater does not have time to soak into the ground and instead causes powerful flash floods. These flash floods transport and then deposit great quantities of debris and sediment at the bases of mountain slopes and on the floors of wide valleys and basins.

CHECK POINT

What land surface forms can be used to determine the direction of ice flow?

Was this your answer?
The direction of ice flow can be determined from striations (long, parallel scratches aligned in the direction of ice flow), *roches moutonnées* (small, asymmetrical hills), and *drumlins* (elongated hills shaped like the back of a whale).

23.8 The Work of Air

Water is the dominant agent of change altering our natural landscape, but air plays a role too. If you've ever been in a windstorm or at the beach on a windy day, you may have felt the sandblasting effect of the wind. Once in the air, particles of sediment can be carried great distances by the wind. Red dust from the Sahara is found on glaciers in the Swiss Alps and on islands in the Caribbean Ocean. Fine grains of quartz from central Asia blow onto the Hawaiian Islands.

In the desert, winds move over surfaces of dry sand, picking up the small, more easily transported particles but leaving the large, harder-to-move particles behind. The small particles bounce across the desert floor, knocking more particles into the air, to form *ripple marks*, which are actually tiny sand dunes (Figure 23.42). Ripple marks can also be formed by the movement of sand grains in water currents, as seen in shallow streams or under the waves at beaches.

Sand dunes begin to form where airflow is blocked by an obstacle, such as a rock or a clump of vegetation (Figure 23.42). As wind sweeps over and around the obstacle, the wind speed slows, causing sand grains to fall out of the air in the wind shadow (Figure 23.43). The falling sand forms a mound and blocks the flow of air even more. With more sand and more wind, the mound grows into a

FIGURE 23.42

Generated by blowing winds, ripple marks are narrow ridges of sand separated by wider troughs. They are small, elongated sand dunes. Large sand dunes can be seen in the background of the photograph.

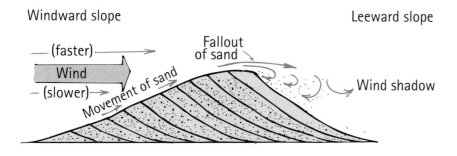

Windward slope

— (faster) →

Wind

— (slower) →

Movement of sand

Fallout of sand

Leeward slope

Wind shadow

FIGURE 23.43

Formation of a sand dune. When airflow is obstructed, wind speed drops and, as a result, sand grains settle in the wind shadow. With more wind, more sand settles, and a dune is formed. As the dune grows, sand grains on the windward slope move up and over the crest to fall on the leeward slope, which slowly causes movement of the whole dune downwind.

dune, which, with continued growth, begins to "move" downwind. The dune moves because grains on the windward slope are blown up and over the crest of the dune, falling on the leeward slope. In this way, wind removes sand from the back of the dune and redeposits it on the front of the dune. Over time, this continuous process moves the entire dune.

SUMMARY OF TERMS

Hydrologic cycle The natural circulation of all states of water from ocean to atmosphere to land, back to ocean.

Groundwater Underground water in the saturated zone.

Porosity The volume of open space in rock or sediment compared to the total volume of solids plus open space.

Hydraulic conductivity A measure of the ability of a porous rock or sediment to transmit fluid.

Water table The upper boundary of the saturated zone, below which every pore space is completely filled with water.

Artesian system A system in which confined groundwater under pressure can rise above the upper boundary of an aquifer.

Turbulent flow Water flowing rapidly and erratically in a jumbled manner, stirring up everything it touches.

Laminar flow Water flowing smoothly and fairly slowly in straight lines with no mixing of sediment.

Gradient The vertical drop in the elevation of a stream channel divided by the horizontal distance for that drop; the steepness of the slope.

Discharge The volume of water that passes a given location in a stream channel in a certain amount of time.

Cut bank A steep bank on the outside bend of a river's channel. An area of erosion.

Point bar A sandy, gentle bank on the inside bend of a river's channel. An area of deposition.

Floodplain A wide plain of almost flat land on either side of a stream channel. Submerged during flood stage, the plain is built up by sediments discharged during floods.

Delta An accumulation of sediments, commonly forming a triangular or fan-shaped plain, deposited where a stream flows into a standing body of water.

Glacier A large mass of ice formed by the compaction and recrystallization of snow, moving downslope under its own weight.

Accumulation The amount of snow added, and the process of adding snow, to a glacier.

Ablation The amount of ice lost, and the process of losing ice, from a glacier.

Drift A general term for all glacial deposits.

Sand dune A landform created when airflow is blocked by an obstacle, slowing air speed and therefore promoting the deposition of airborne sand.

REVIEW QUESTIONS

23.1 The Hydrologic Cycle

1. Where does most of Earth's precipitation occur?
2. As water is precipitated onto the land, where does it go? Where does most of the water on land end up?

23.2 Groundwater

3. Distinguish between *porosity* and *hydraulic conductivity*.
4. If a hole is dug in the unsaturated zone, does it fill with water? Why or why not?

5. Compare and contrast the unsaturated zone with the saturated zone.
6. What types of soil allow the greatest amount of rainfall to soak in?
7. What is an artesian system, and how is it formed?

23.3 The Work of Groundwater

8. Describe at least one consequence of overpumping groundwater.
9. How does rainwater become acidic? How does this affect limestone?
10. How do stalactites and stalagmites form?
11. Name three erosional features caused by groundwater in carbonate rocks.
12. What is the difference between a cave and a cavern?

23.4 Surface Water and Drainage Systems

13. What is meant by *stream gradient*, and how does it affect stream velocity?
14. What happens to stream speed when the discharge of a stream increases? What happens to discharge when the speed of a stream increases?
15. How does the shape of a stream channel affect flow?

23.5 The Work of Surface Water

16. What is the greater transporter of sediment, a laminar flow or a turbulent flow? Why?
17. Name three ways in which the movement of water erodes a stream channel. Which one creates potholes?

18. What factors are responsible for the formation of a stream valley?
19. Under what conditions do curvy, meandering rivers form along a floodplain?
20. What types of streams and stream valleys do we generally find in high mountainous regions?
21. Streams carry more than just water—they carry great quantities of sediment from one place to another. What is the size range of the particles that can be carried by a fast-moving stream?
22. What is a delta?

23.6 Glaciers and Glaciation

23. How is a glacier formed?
24. In what two main ways do glaciers flow?
25. Under what conditions does a glacier front advance? How about retreat?

23.7 The Work of Glaciers

26. What is the most common landform created by glaciers?
27. What erosional features are likely found in an area of alpine glaciation? See Figure 23.38.
28. What land features are formed from glacial deposits?

23.8 The Work of Air

29. How do sand dunes migrate?
30. How are ripple marks formed?

EXERCISES

● BEGINNER ■ INTERMEDIATE ◆ EXPERT

Again, please do not be intimidated by the large number of exercises and problems in this and other meatier chapters. If your course work is to cover many chapters, your instructor will likely assign only a few exercises and/or problems from each.

1. ● What percentage of Earth's supply of water is fresh water, and where is most of it located?
2. ■ Where does most rainfall on Earth finally end up before becoming rain again?
3. ● In a confined aquifer, the water in a well can rise above the top of the aquifer. What is this system called?
4. ■ In an unconfined aquifer, how high can water rise in a well that is not pumped?
5. ■ How is the local hydrologic cycle affected by the practice of drawing drinking water from a river and then returning sewage to the same river?
6. ● Is water in the unsaturated zone called groundwater? Why or why not?
7. ◆ In an aquifer, if the water table next to a stream is lower than the water level in the stream, does groundwater flow into the stream or does stream water flow into the ground? Explain.

8. ◆ As runoff into streams increases, which variables of stream flow (as discussed in the text) increase?
9. ● What is meant by channel geometry?
10. ● What happens to stream speed if the discharge in a stream doubles while the channel remains the same size and shape?
11. ● What happens to stream speed if discharge doubles and the cross-sectional area of the stream channel also doubles?
12. ● What three variables influence the speed of stream flow?
13. ● How does "frictional drag" play a role in the external movement of a glacier? How does this drag affect the internal movement?
14. ◆ As a stream progresses downslope, the gradient gradually decreases and the stream channel typically widens. What factors increase the stream's discharge? What factors decrease its discharge?
15. ● In the formation of a river delta, why are larger particles deposited first, followed by smaller particles farther out? Defend your answer.
16. ◆ What causes the formation of distributaries off the main channel of a river delta?

17. ● What is a sinkhole? What factors contribute to its formation?
18. ■ Which of the three agents of transportation—wind, water, or ice—transports the largest boulders? Why?
19. ■ Which of the three agents of transportation is limited to transporting small rocks? Why?
20. ◆ Carbonate rocks form mainly in marine environments. Why do we find abundant carbonate deposits on continental land?
21. ● Are underground rivers ever found in nature? Defend your answer.
22. ◆ Do you think a stream in which the flow is laminar can become turbulent without increasing the volume of water in the stream? Defend your answer.
23. ● Describe the formation of stalactites.
24. ■ Why is surface water both a creator and a destroyer of sediments and sedimentary rocks?
25. ■ Why do point bars form on the inside bends of meandering streams?

26. ■ How is a *roche moutonnée* different from a drumlin?
27. ● What well-known landscapes have been carved by glaciers?
28. ■ How do deposits from glacial ice differ from rocks deposited by rivers?
29. ■ Removal of groundwater can cause subsidence. If removal of groundwater is stopped, will the land likely rise again to its original level? Defend your answer.
30. ◆ Must a stream's speed increase in order for it to carry more sediment? Explain.
31. ● What distinguishes a huge block of ice from a glacier?
32. ■ Why do crevasses form on the surfaces of glaciers?
33. ■ Does all the ice in a glacier move at the same speed? Explain.
34. ■ What is the significance of the large, out-of-place boulders that are sometimes found in glacial deposits?
35. ● How are sand dunes formed?

PROBLEMS

● BEGINNER   ■ INTERMEDIATE   ◆ EXPERT

1. ■ We know from Figure 23.1 that 97.2% of Earth's water is in the oceans. The remaining 2.8% is Earth's freshwater supply. Of this freshwater supply, what percentage is found in the polar ice caps? In groundwater? In streams, lakes, and rivers?
2. ■ A particular stream widens as it progresses downstream. Using your answers for parts (a) and (b), briefly describe the changes in discharge.
(a) If the cross-sectional area of the stream is 1 m² and the stream speed is 0.5 m/s, what is the stream's discharge?
(b) If the cross-sectional area of the stream increases to 2 m² and the stream speed remains 0.5 m/s, what is the stream's discharge?

Refer to the Figuring Physical Science box on page 607 for Problems 3, 4, and 5.

3. ■ A pumping well was drilled and completed in a sand aquifer. Before the pump was turned on, the hydraulic gradient and the flow rate were measured to be 0.0001 and 1 m³/day, respectively. With the pump turned on, the gradient becomes 10 times larger. By how much does the flow rate increase?
4. ◆ Darcy's law gives us the *volume flow rate*—volume per time (for example, cubic meters per day, m³/day). Sometimes it is not possible to know the value of the cross-sectional area for a Darcy's-law calculation. This is especially true if we use a well to estimate flow in an aquifer. The size of a well is much smaller than the cross-sectional area of an entire aquifer, so we can't use the cross-sectional

area of the aquifer to calculate the flow rate. Another way to express volume is on a *per unit area* basis. For example, if we have a 1 m³ cube, we know its base is 1 m²—the cross-sectional area of the cube—and its height is 1 m. If we fill the cube with water, we can say we have 1 m of water per unit area. Darcy's law can be rearranged to calculate the volume flow rate per unit area. This is called the *specific discharge*, which has units of length per time (for example, meters per day, m/day). Completing parts (a) through (c) will illustrate how specific discharge relates to the volume flow rate. (Hint: Assume that the units for hydraulic conductivity are m/day and the units for cross-sectional area are m².)
(a) Suppose 1 m³ of water is pumped from a well into an empty cylindrical tank. If the water level is 2 m above the base, what is the cross-sectional area of the tank?
(b) If it takes a half a day to pump the 1 m³ of water into the tank, what is the flow rate in terms of both volume per time and specific discharge?
(c) Write Darcy's law so that it calculates the specific discharge.

Please make sure that you understand Problem 4 before doing Problem 5.

5. ◆ The hydraulic head at point A is 209 m. At point B, which is 300 m from point A, the hydraulic head is 210 m. The aquifer is composed of sand with a hydraulic conductivity of 150 m/day. Groundwater flows directly from point B to point A. What is the specific discharge?

READINESS ASSURANCE TEST (RAT)

If you have a good handle on this chapter, if you really do, then you should be able to score at least 7 out of 10 on this RAT. If you score less than 7, you need to study further before moving on.

Choose the BEST answer to each of the following.

1. The amount of discharge in a stream usually _____ downstream.
   (a) becomes turbulent
   (b) meanders
   (c) decreases
   (d) increases
2. Most of Earth's fresh water is found
   (a) in lakes.
   (b) in ice caps and glaciers.
   (c) in rivers.
   (d) underground.
3. The work of surface water does all of the following except
   (a) erosion.
   (b) deposition.
   (c) land subsidence.
   (d) delta formation.
4. The maximum amount of water a particular soil can hold is determined by its
   (a) porosity.
   (b) permeability.
   (c) degree of saturation.
   (d) amount of recharge.
5. Precipitation that does not soak into the ground or evaporate becomes
   (a) groundwater.
   (b) the water table.
   (c) soil moisture.
   (d) runoff.
6. Sand dunes form as wind
   (a) disperses sand.
   (b) blows sand from the back to the front of the dune.
   (c) blows sand from the front to the back of the dune.
   (d) interrupts the normal sequence of deposition.
7. Deltas form as
   (a) periodic flooding clogs stream channels.
   (b) erosion clogs stream channels.
   (c) stream gradient decreases.
   (d) streams enter a standing body of water.
8. What factors affect stream speed?
   (a) discharge
   (b) channel length
   (c) stream gradient
   (d) both (b) and (c)
   (e) both (a) and (c)
9. Snow converts to glacial ice when subjected to
   (a) decreasing temperature.
   (b) pressure.
   (c) rain.
   (d) basal sliding.
10. Underground water in the saturated zone is called
    (a) groundwater.
    (b) soil moisture.
    (c) the water table.
    (d) an artesian system.

Answers to RAT
1d 2b 3c 4a 5d 6b 7d 8e 9b 10a

EXPLORING FURTHER

http://water.usgs.gov/education.html
   This educational resource page is full of links to excellent sites about water and the environment. Check out the Water Science for Schools section for water information full of pictures, data, and map resources along with an interactive center to test your water knowledge!

http://www.geocities.com/monte7dco/
   This site is a good resource for learning characteristics of world landforms. It contains a descriptive picture glossary of landforms, along with an elementary-level lesson plan and links to additional plans.

http://www.enviroliteracy.org/article.php/244.html
   The home page for the Environmental Literacy Council is full of basic information about wind, water, erosion, ecosystems, and climate. It is text intensive with many resource links and suggestions.

CHAPTER 23 ONLINE RESOURCES

The Physical Science Place

Tutorials
- The Hydrologic Cycle Activity
- Shaping Planetary Surfaces

Video
- The Hydrologic Cycle

Quiz

Flashcards

Links

the oceans, atmosphere, and climatic effects

24

24.1 EARTH'S ATMOSPHERE AND OCEANS

24.2 COMPONENTS OF EARTH'S OCEANS

24.3 OCEAN WAVES, TIDES, AND SHORELINES

24.4 COMPONENTS OF EARTH'S ATMOSPHERE

24.5 SOLAR ENERGY

24.6 DRIVING FORCES OF AIR MOTION

24.7 GLOBAL CIRCULATION PATTERNS

■ The view of Earth from space shows that our planet is colored distinct shades of blue and silver. The blue is due to the water in the oceans and the silver to clouds in the atmosphere. How did our beautiful, swirling atmosphere and oceans come to be? What is the atmosphere made of? How do the atmosphere and oceans interact? So many questions come to mind.

We begin to answer these questions by learning about the evolution of Earth's atmosphere and oceans. We then explore important features of these two fluid shells as we investigate the transfer of heat between them—and how this heat transfer affects Earth's climate. We conclude with a look at the mechanisms that influence atmospheric and oceanic circulation patterns and the impact these patterns have on climate.

24.1 Earth's Atmosphere and Oceans

Link To Section 5.6

Seventy-one percent of Earth's surface is covered by water (Figure 24.1). The remaining 29% is land, most of which is located in the Northern Hemisphere (Figure 24.2). Although the many oceans are named for their various locations, they are really one big continuous ocean.

FIGURE 24.1

Most of Earth's surface is covered by water. We can divide Earth into (a) an ocean-dominated hemisphere and (b) a land-dominated hemisphere.

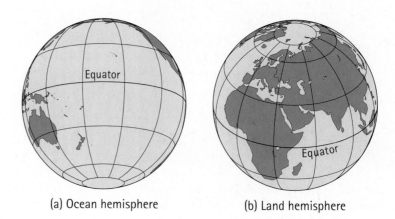

(a) Ocean hemisphere   (b) Land hemisphere

As we learned in Section 23.1 the oceans are the reservoir from which water evaporates into the atmosphere, only to precipitate later as rain and snow. The oceans play a major role in moderating Earth's temperature and climate. Recall from Chapter 6 that water has a high specific heat capacity—it is slow to heat up or cool down. Therefore, water transfers large amounts of thermal energy to its surroundings when it cools, and it absorbs large amounts of thermal energy from its surroundings when it warms. This property of water accounts for the moderate temperatures on lands bordering the oceans. The moderating influence of the oceans can be seen when we look at seasonal temperature variations for two cities at the same latitude: coastal San Francisco, California, and continental Wichita, Kansas (Figure 24.3). Whereas temperatures in San Francisco tend to have small seasonal variations, temperatures in Wichita show strong seasonal fluctuations—cold winters and hot summers. The oceans do a great job of moderating climate, both by making summers cooler and by making winters warmer.

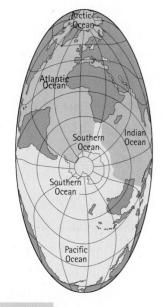

EVOLUTION OF EARTH'S ATMOSPHERE AND OCEANS

Earth probably had an atmosphere—a blanket of gases surrounding it—before the Sun was fully formed. This primitive atmosphere was likely composed only of hydrogen and helium, the two most abundant gases in the universe, along with trace amounts of ammonia and methane. The early atmosphere contained no free oxygen. Then, when the temperature and pressure in the contracting center of the still-forming Sun became high enough to ignite thermonuclear reactions, our Sun was born. The release of energy from the Sun's formation likely produced a strong outflow of charged particles—an outflow strong enough to sweep away Earth's earliest atmosphere.

But then a new atmosphere formed. The first stage in its formation probably occurred when gases trapped in Earth's hot interior escaped through volcanoes and fissures at Earth's surface. The gases spewed out by these early eruptions

FIGURE 24.2

When a map is centered over Antarctica, the expanse of the world ocean can be seen. In terms of size and volume, the Pacific Ocean accounts for more than half of the world ocean and is thus the largest ocean. In fact, the Atlantic and Indian Oceans combined would easily fit into the space occupied by the Pacific.

FIGURE 24.3

Comparison of seasonal temperature ranges for coastal San Francisco, California, and continental Wichita, Kansas.

Station: San Francisco, California
Latitude/longtitude: 37°37' N, 122°23' W
Average annual temperature: 14°C (57.2°F)
Total annual precipitation: 47.5 cm (18.7 in.)
Elevation: 5 m (16.4 ft)
Population: 750,000
Annual temperature range: 9°C (16.2°F)

Station: Wichita, Kansas
Latitude/longtitude: 37°39' N, 97°25' W
Average annual temperature: 13.7°C (56.6°F)
Total annual precipitation: 72.2 cm (28.4 in.)
Elevation: 402.6 m (1321 ft)
Population: 350,000
Annual temperature range: 27°C (48.6°F)

were probably much like the gases found in the volcanic eruptions of today—about 90%–95% water vapor, the rest mostly carbon dioxide. The early atmosphere still had no free oxygen and could not support the type of life we have today.

As we learned in Chapter 21, free oxygen did not occur in Earth's atmosphere until primitive bacteria known as cyanobacteria, algae-like microorganisms, appeared. Like the green plants that followed, these organisms used *photosynthesis* to convert carbon dioxide and water to carbohydrates and free oxygen:

$$CO_2 + H_2O + light \rightarrow CH_2O + O_2$$

After free oxygen (O_2) began accumulating, chemical reactions produced an ozone (O_3) layer in the upper atmosphere. Because the ozone layer acts as a filter to reduce the amount of ultraviolet radiation reaching Earth's surface, the surface was then much more able to support life.

As Earth cooled, the rich supply of water vapor in the atmosphere condensed to form the oceans. Cometary debris from interplanetary space also contributed some water to the oceans. These oceans, essential to the evolution of life and ultimately to the development of the present global environment, have remained for the rest of Earth's history.

Link To Section 11.3

Link To Section 21.4

Link To Section 6.7

CHECK POINT

1. Why are the hottest climates on Earth typically found in the interior sections of continents?
2. Did the ozone layer exist before cyanobacteria evolved?

Were these your answers?

1. The large specific heat capacity of water tends to keep coastal areas from experiencing extreme temperatures. Therefore, very hot climates are usually some distance from the ocean.
2. No. The formation of ozone, O_3, was preceded by the introduction of free oxygen, which came from photosynthesizing cyanobacteria.

■ 24.2 Components of Earth's Oceans

The world ocean is the dominant feature of our planet; some philosophy types have even wondered why our planet isn't called Ocean instead of Earth! If we could drain the water from Earth's oceans, we'd see enormous mountain ranges in the middle of the ocean basins and deep trenches bordering many of the continents, not the flat, featureless ocean bottom that had been imagined until the discoveries of the mid-20th century.

The ocean is an immense body of salty water. Its salty nature contributes to its variable density, which contributes to the movement of ocean currents. Waves move through the ocean in response to atmospheric movements, but the ocean also responds to extraterrestrial influences—the Moon and the Sun, which create tides. With so much of Earth's surface covered by oceans, the oceans hold a wealth of scientific treasures. But because it is so vast, we still have much to discover—we know more about the surface of the Moon than we know about the ocean floor!

THE OCEAN FLOOR

The features of the ocean floor are very pronounced. In fact, land rises, on average, about 840 m above sea level, while the ocean bottom drops, on average, about 3800 m below sea level. If we compare the height of Mt. Everest in the Himalayas, at a majestic 8848 m above sea level, with the depth of the Marianas Trench in the Pacific Ocean, an astounding 11,035 m below sea level, we see that the oceans are much deeper than land mountains are high.

The ocean floor consists of continental margins, deep-ocean basins, mid-ocean ridges, and deep trenches. Away from the continental margins, the topography of the ocean floor varies greatly. The midocean ridges that encircle the globe are tall and variable, the sediment-covered ocean bottom is relatively flat, and seafloor trenches near continental margins can be very deep (Figure 24.4).

FIGURE 24.4

Map of the ocean floor showing variation in topography. (a) Atlantic profile. (b) Pacific profile.

(a)

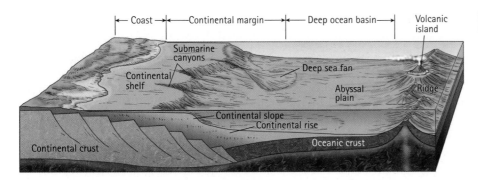

FIGURE 24.5

Profile of the continental margin going from land to the deep-ocean bottom.

The **continental margin** marks the boundary between continent and ocean. As Figure 24.5 shows, the continental margin consists of a *continental shelf* (the submerged upper, landward portion of the margin), a *continental slope* (the breakpoint where the shelf steepens as it descends to depths of 2–3 km), and a *continental rise* (the area from the base of the slope seaward to the deep-ocean floor).

Closest to the continent, the *continental shelf* is a gently sloping, underwater surface extending from the shoreline toward the ocean basin—it is an underwater extension of the continental crust. But because it is submerged, the continental shelf is also considered part of the ocean floor. The amount of a continent's edge that is considered continental shelf changes over geologic time, because sea level rises and falls. When sea level is lower than at present, such as it was during the last glacial period, the continental shelf is narrower, because less of the continental edge is submerged. On the other hand, when sea level is higher, as it would be if the ice caps on Antarctica and Greenland melted, the continental shelf would be wider because some of the continental land would be submerged. No matter what sea level is, the continental shelf varies widely from one geographic location to another. It is practically nonexistent in some locations, but extends seaward to a width of up to 1,500 km in others.

The *continental slope* is the sloping outer edge of a continent; it marks the boundary between continental crust and oceanic crust. The continental slope is steeper along mountainous coastlines that border convergent plate boundaries and more gently sloped along coastlines that lack plate boundaries (such as the eastern edge of North America and the Indian subcontinent).

The *continental rise* is the transition zone between the continental margin and the deep-ocean floor. It is basically a wedge of continental sediment that has accumulated at the base of the continental slope. Strong currents and gravity transport sediment to the continental rise from the shelf and slope, mostly because of undersea "landslides." Turbulent currents of sediment arising from fyi

- Not all midocean ridges are truly "midocean." The East Pacific Rise, for example, is much closer to the eastern margin of the Pacific Ocean than it is to the middle. In any case, midocean ridges are where seafloor spreading occurs. Consequently, they mark divergent plate boundaries wherever they are located.

Link To Section 18.1

fyi

- The evidence of volcanic activity on the ocean bottom is impressive—more than 20,000 volcanic peaks have been found in the Pacific Ocean alone!

these landslides chisel and carve out steep submarine canyons. As sediment collects at the base of the canyon, fan-shaped deposits called *deep-sea fans* form. Over time, the deep-sea fans grow and merge to form the continental rise. Not all continental margins have a continental rise. For example, along active convergent plate boundaries the continental slope leads directly into a deep-ocean trench.

The deep-ocean basins account for about 30% of Earth's surface. Deep-ocean basins are generally 3–5 km deep and are characterized by abyssal plains, ocean trenches, midocean ridges, and seamounts. Abyssal plains begin where the continental margin ends. They are some of the deepest and flattest regions on Earth—because thick accumulations of fine-grained sediment bury the underlying uneven oceanic crust.

Sticking up through the accumulated sediment of the abyssal plain are a variety of volcanic peaks—seamounts and small abyssal hills. Abyssal plains end where the midocean ridge begins. Midocean ridges form a continuous, underwater mountain chain that extends through all the oceans like the seams on a baseball. The average elevation at the crests of the ridge system is about 2.5 km above the surrounding seafloor. And a few islands, such as Iceland, occur where the midocean ridge rises above the water surface. The ridge system varies in width along its length, but overall it averages about 1000 km. The entire ridge system is volcanic and is composed of basalt; this is where oceanic crust is made. Along the crest, seafloor spreading creates a central rift zone, where tension from diverging plates causes normal faults to form and small earthquakes to rumble (see Chapter 22).

In ocean basins that border convergent plate boundaries, ocean trenches mark the deepest places on Earth—sometimes exceeding 10,000 m (more than 6 mi) in depth—much deeper than the height of most of Earth's surface mountains. Trenches are places of great tectonic activity. They occur at subduction zones, where lithosphere is forced down into the asthenosphere. This action causes earthquakes—sometimes quite large—and creates volcanoes that are capable of violent eruptions.

SEAWATER

Seawater is a complex solution of dissolved minerals, dissolved gases, and decomposing biological material. Interestingly, just about every natural compound can be found at some concentration in the ocean. Water, "the universal solvent," has a strong ability to dissolve salts (Chapter 18), so salts make up the majority of dissolved material in the sea. Overall, the composition of seawater is surprisingly simple, because only a few elements and compounds are present in abundance. Chloride, sodium, sulfate, magnesium, calcium, and potassium make up more than 99% of the salts in the sea (Table 24.1).

TABLE 24.1   PRINCIPAL ELEMENTS IN SEA SALTS

Element	Chemical Symbol	Percentage by Weight
Chloride	Cl^-	55.07
Sodium	Na^+	30.62
Sulfate	SO_4^-	7.72
Magnesium	Mg^{2+}	3.68
Calcium	Ca^{2+}	1.17
Potassium	K^+	1.10
Total		99.36

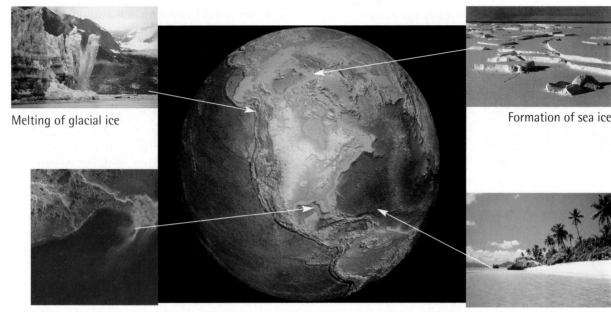

Melting of glacial ice

Formation of sea ice

Runoff

Evaporation

FIGURE 24.6

Salinity increases as the supply of fresh water decreases. Factors that increase salinity include formation of sea ice and evaporation. Salinity decreases as the supply of fresh water increases. Factors that decrease salinity include the runoff from streams and rivers, precipitation, and the melting of glacial ice, sea ice, and icebergs.

The amount of dissolved salts in seawater is measured as **salinity**—the mass of salts dissolved in 1000 g of seawater. On average, the salinity of seawater is about 35 g per 1000 g of seawater, or 35 parts per thousand. (Scientists note this with the symbol ‰.) So the salinity of the ocean is written 35‰.

Salinity varies from one part of the ocean to another, but the overall composition of seawater is the same from place to place—a mixture of 96.5% water and 3.5% salt. Salinity variation is, of course, influenced by factors that increase or decrease supplies of fresh water (Figure 24.6). Fresh water enters the ocean from three sources: runoff from streams and rivers, precipitation, and melting of ice. Fresh water leaves the ocean by two methods: evaporation and the formation of sea ice. Evaporation increases salinity because only pure water vapor leaves the seawater solution—the salts are left behind. And when sea ice forms, only the water molecules freeze. Salts are once again left behind in solution. There are slight regional variations in salinity as well. The salinity of oceans in the dry subtropics where evaporation is high can reach 37‰. And heavy precipitation dilutes the oceans in equatorial locations to as low as 33‰. Overall balance is maintained when evaporation is offset by precipitation/runoff and ice formation is offset by ice melting.

CHECK POINT

If you want to make up a batch of saltwater with the same salinity as seawater, how many grams of salt would it take to make 1 kg of seawater? How many grams of water?

Was this your answer?
The salinity of saltwater equals 35‰. So measure out 35 g of salt, then pour in 965 g of water to make 1 kg of seawater.

■ 24.3 Ocean Waves, Tides, and Shorelines

Ocean waves come in a variety of sizes and shapes, from tiny ripples to the gigantic waves powered by hurricanes. Water waves, like all other waves, begin with some kind of disturbance. The most common disturbance that

fyi

■ Marine life influences the composition of seawater by removing salts, dissolved gases, and other solutes. Shellfish, tiny foraminifera (marine protozoa), and various crustaceans (such as crabs and shrimp) remove calcium salts to build their bodies; diatoms (microscopic marine algae) and some types of microscopic animals draw heavily on the ocean's dissolved silica to form their shells. Some animals concentrate elements present in seawater in minute, almost undetectable amounts. For example, lobsters extract copper and cobalt; certain seaweeds concentrate iodine; and sea cucumbers extract vanadium!

(a)

(b)

FIGURE 24.7

Ocean waves have characteristics of simple sine waves.

When waves approach the beach head-on, so that the line of wave crests is parallel to the shore, rip currents form instead of longshore currents.

FIGURE 24.8

INTERACTIVE FIGURE

Movement of water particles with the passage of a wave. The particles move in a circular orbit. Circular motion is greatest at the surface and gradually decreases with depth. At depths greater than half a wavelength, circular motion is negligible.

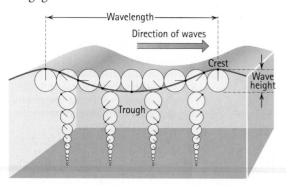

causes ocean waves is wind. Blow on a bowl filled with water and you'll see a succession of small ripples moving across the water's surface. The generation of waves in the ocean is similar. As wind speed increases, the ripples grow to full-sized waves. When stronger winds blow, larger waves are created. As waves travel away from their origin, they develop into regular patterns of smooth, rounded waves called *swell*—the mature undulations of the open ocean.

Recall from our study of waves in Chapter 10 that wave motion can be described in terms of a sine curve (Figure 24.7) and that it is the *disturbance* that is carried by a wave, not the material through which the wave is moving. The waveform travels across the ocean, while the water making up the wave remains, for the most part, in one place.

Waves on the ocean surface are *orbital* waves. As a water wave passes a given point, the water particles at that point move in circular paths. This circular motion can be seen by observing the behavior of a floating piece of wood on the ocean surface. The wood sways to and fro while bobbing up and down, actually tracing a circle during each wave cycle. This circular motion occurs near the water surface, and it decreases gradually with depth (Figure 24.8). At a depth of about one-half the wavelength, the circular component of the wave is negligible. For this reason, we can say, with reasonable accuracy, that wind-driven ocean waves occur mainly at the surface.

When a wave approaches shore, where water depth decreases, the circular motion is interrupted by the ocean bottom. As the water depth gets shallower, the lowermost circular paths are flattened—becoming elliptical—by dragging along the ocean bottom, and the wave slows down. This transition occurs at a depth equal to one-half the wave's wavelength. The wave period remains unchanged because the swell from deeper waters continues to advance toward shore. As a result, incoming waves gain on slower leading waves and the distance between waves decreases. This bunching-up of waves in a narrower zone produces higher, steeper waves. When wave height steepens to the point where the water can no longer support itself, the wave overturns and tumbles forward, breaking shoreward with a crash. The turbulent water created by the crash is called *surf*. The *surf zone* is the area of active water between the line of breakers and shore (Figure 24.9).

WAVE REFRACTION

As waves enter shallow water, their forward direction changes when they approach the shore at an angle. The portion of the wave closest to shore begins touching the ocean bottom at a depth of one-half the wave's wavelength. This near-shore part of the wave slows and lags behind portions of the wave still in the deeper water away from shore. As the next portion of the incoming wave touches bottom, it too slows. Thus, in a continuous fashion, the line of the wave crest bends as it moves into shallower water, in a sense pivoting around the slowest portion of the wave, to become more parallel to the coastline (Figure 24.10). This is wave refraction. Additionally, the oblique approach of waves causes a *longshore current* that flows parallel to shore.

Wave refraction also has a significant impact on irregular shorelines, making them straighter and more regular. The effect is greatest at shorelines that have protruding headlands and small bays. Refraction causes wave energy to be unevenly distributed (Figure 24.11). Wave energy is concentrated at headland areas where shorelines project into the water, because the wave encounters shallow water at such places first. The extra energy focused there increases the erosion rate. At adjacent bays,

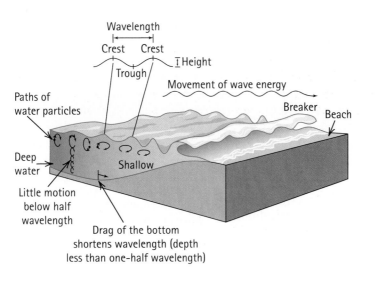

FIGURE 24.9

INTERACTIVE FIGURE

Waves change form as they travel from deep water through shallow water to shore. In deep water, orbital motion is circular. In shallow water, orbital motion becomes elliptical as a result of contact with the bottom. This change decreases the wave speed. As incoming waves continue to advance, the distance between waves decreases, causing wave height to increase. When waves reach a critical height, they break and crash into the surf zone.

the shoreline is further inland and wave energy is diluted. The gentler bay water facilitates sediment deposition. Headlands are worn back landward, while bays accumulate sediment seaward.

Link To Sections 10.2, 10.5

CHECK POINT

■

1. Why does a resting surfer tend to stay at the same spot in an area of regular ocean swell?
2. What causes a wave to slow as it approaches the shoreline?

Were these your answers?
1. Think of the bobbing piece of wood. The water under the resting surfer moves in a circular path, moving backward as much as it moves forward. So unless the surfer paddles to the shore or farther out to sea, the surfer basically bobs up and down in one place.
2. A wave slows when the bottommost part of its circular motion encounters the ocean bottom and flattens.

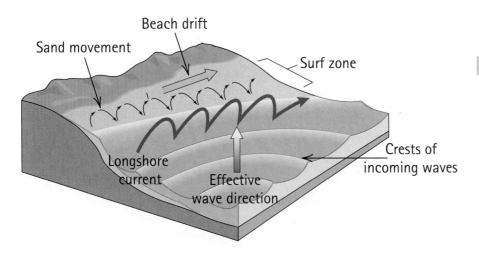

FIGURE 24.10

When waves approach a shoreline, they refract (bend) so that the crests of the approaching waves become more parallel to the shore as they move into shallower water. Because the overall direction of wave movement is oblique to the shore, a longshore current forms, which causes water and sand to move parallel to the shoreline.

FIGURE 24.11

On irregular coastlines, wave energy is concentrated as it converges on headlands and diffused as it diverges in coves and bays.

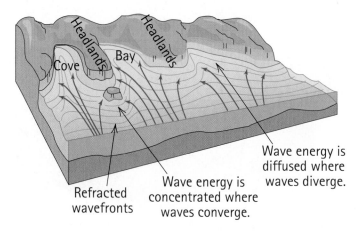

Refracted wavefronts

Wave energy is concentrated where waves converge.

Wave energy is diffused where waves diverge.

fyi

- About 280 barrier islands ring the Atlantic and Gulf coasts. They form a barricade between the coast and the open ocean. The lagoons that separate the narrow islands from the shore are relatively quiet waters. Small boats often use these lagoons as a "freeway" between Florida and New York, thus avoiding the potentially rough waters of the open Atlantic.

fyi

- Beach sand varies according to the area eroded. On tropical beaches, such as Florida or Hawaii, the beach is made up of organic material—sand-sized fragments from coral reefs and carbonate platforms. In contrast, the sand on many beaches of the continents is mostly inorganic—composed of silicate minerals eroded from continental rock.

THE WORK OF OCEAN WAVES

The power of oceans is apparent to anyone who visits the coast. Under normal conditions, some coastlines are quite placid, while others continually endure the ravages of large waves and rough surf. During large storms, all coastlines are at the mercy of the mighty ocean.

ALONG THE COAST

Winds blowing across the ocean surface generate waves, and as the waves enter shallow water near land, they become higher and steeper until they finally collapse, or break. In the surf zone, wave activity moves sediment back and forth, both shoreward and seaward. Because the amount of surf at a shoreline varies with time and because the rocks at any shoreline have different degrees of resistance to erosion, surf can form many different erosional features. Weakly consolidated sedimentary rocks and highly fractured igneous and metamorphic rocks erode fastest. Resilient, well-consolidated sedimentary rocks and unfractured igneous and metamorphic rocks erode more slowly.

Along shorelines consisting of hard rock, the pounding surf cuts into and notches the base of the land. As erosion proceeds, the notches deepen and the rocks above begin to jut out over the empty space at the base. As the overhanging rocks fall into the surf, the cliff progressively retreats. In time, waves cut into the cliff to form a relatively flat surface known as a *wave-cut platform* (Figure 24.12).

Rock particles eroded from the coast must sooner or later be deposited. Much of the material is deposited in the most common and well-known shoreline depositional environment—the *beach*. Sandy beaches are the result of turbulent motion in the surf zone. Beaches tend to be elongated by longshore currents that form when waves approach the shoreline at an oblique angle. For example, waves approaching a north–south coastline from the northwest would cause a

FIGURE 24.12

Characteristic coastal erosional landforms. Waves cut into cliffs to form platforms and terraces. Some notched cliffs erode to form sea caves. Sea arches can form if two sea caves, usually on opposite sides of a headland, become connected. When an arch collapses, a sea stack is left behind. In time, wave action also erodes away the sea stack.

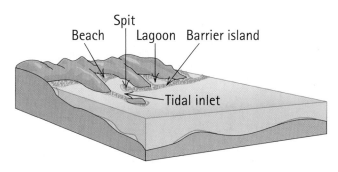

Spit
Beach   Lagoon   Barrier island

Tidal inlet

FIGURE 24.13

Characteristic coastal depositional landforms.

longshore current southward down the beach (Figure 24.10). These currents move sand down the length of the coast. Where the currents deposit sand, we have the formation of *spits*. Spits begin as submerged ridges of sand. As sand accumulates, the spit rises above the surface and projects from the coast into open water as a continuation of the beach, frequently as a fingerlike piece of land (Figure 24.13).

When sand ridges form offshore and parallel to the coast, they can eventually grow into *barrier islands*. Barrier islands form where ridges of sand break the surface of the water for a long-enough period of time that vegetation begins to take hold. During large storms, surf washes over the lowlands, making inlets into the lagoon area between the barrier island and the shore (Figure 24.13). The lagoon area is a quieter environment. It has finer-grained silts and muds that feature cross-bedding and small ripple marks caused by the oscillating motion of the lagoon water. On shore, smooth stones, rounded pebbles, and/or sand make up the beaches.

fyi

- Ancient coral reefs are composed of alternating layers of porous material and impermeable muds. This layered quality gives ancient coral reefs the potential to act as traps for oil and gas, making their discovery economically important.

LINK TO Biology

▪ Coral Reefs, an Offshore Environment

Coral reefs are composed of actively growing, individual coral organisms that secrete calcium carbonate exoskeletons as they grow. When we look at a piece of coral or a coral reef, we see the calcium carbonate exoskeleton, not the soft coral organism. Most corals, but not all of them, form colonies, and the colony-forming corals are the major reef builders. Reefs grow outward with time, but they can also grow upward as new corals cement themselves to the exoskeletons of dead coral below.

Many types of coral reefs can survive only in shallow water because a major food source for these corals is photosynthetic algae, which need bright light to live. The coral and algae live in a *symbiotic* relationship. The coral provides protection for the algae, and the algae provide oxygen and nutrients for the coral. These types of corals require warm, clear, and relatively sediment-free water to flourish.

Carbonate platforms are much larger than coral reefs, but their existence is also due to organisms. Carbonate platforms are the graveyards of calcium-secreting organisms and are formed in shallow waters close to or attached to continents. They account for the largest portion of carbonate sediment produced in the ocean.

Both coral reefs and carbonate platforms can be partially destroyed by constant, pummeling wave action as the reefs grow and approach the water surface. This destruction is natural. But coral reefs are also destroyed by less natural processes; thus, they indicate ocean health. We know that warm climates favor carbonate deposition because carbonates dissolve more easily in cold water than in warm water. But coral organisms are living creatures and when the water gets too warm, the symbiotic algae leaves the coral or the coral expels the algae. With nutrients cut off, the coral begins to lose its color—called *coral bleaching*—and slowly begins to die.* Scientists view coral bleaching as an indicator of global warming. Aside from warmer ocean temperatures, the symbiotic relationship between coral and algae may also be stressed by increased pollution; elevated ultraviolet (UV) radiation levels; changes in salinity, diseases, or predators; or a combination of these factors.

* The beautiful colors associated with coral reefs are the result of the relationship with the symbiotic algae. Corals by themselves are colorless.

- Earth is not a rigid solid, but is somewhat deformable (remember plastic flow in the mantle?). Because of this, we have Earth tides, though less pronounced than ocean tides. Twice each day the solid surface of Earth rises and falls as much as 25 cm (for comparison, the ocean rises and falls about 1 m each day). There are also atmospheric tides, which affect the intensity of cosmic rays that reach Earth's surface. These rays, regulated even more strongly by Earth's magnetic field, induce subtle changes in living things. Do any of your friends act strangely when there is a full Moon?

CHECK POINT

1. How are wave-cut platforms different from a beach?
2. Barrier islands provide a first line of defense against high-energy storm waves of the open ocean. Are barrier islands permanent features of the coast?

Were these your answers?

1. A wave-cut platform is formed by erosion; a beach is formed by deposition. Wave-cut platforms, terraces, sea arches, and sea stacks are erosional features. Beaches, spits, and barrier islands are depositional features.
2. Barrier islands are permanent features in the sense that they will always form. But they are impermanent in that they are vulnerable to the forces of nature. Barrier islands are made of transported sand—unconsolidated material susceptible to the erosive action of waves and high winds that constantly embattle the coastline.

TIDES

Along any coastline, it is easy to notice that the level of the ocean rises and falls cyclically. This daily change in ocean elevation is known as *tides*. Seafaring people have always known that there is a connection between ocean tides and the Moon. But until Isaac Newton, no one could offer a satisfactory explanation of why there are two high tides per day. Newton showed that ocean tides are caused by the *differences* in the gravitational pull exerted by the Moon on opposite sides of Earth. Gravitational force between the Moon and Earth is stronger on the side of Earth closest to the Moon, and it is weaker on the side of Earth farthest from the Moon. This is simply because gravitational force is weaker with increased distance.

To understand why tides are produced by the difference in the Moon's gravitational pulls on opposite sides of Earth, imagine that you have a big spherical ball of gelatin. If you exert the same force in the same direction on every part of the ball, it remains spherical. But if you pull harder on one side than on the other, there is a difference in forces and the ball becomes elongated (Figure 24.14).

That's what happens to this big ball we're living on. The side closer to the Moon is pulled with greater force toward the Moon than the far side is—and thus Earth is somewhat football-shaped. Surprisingly, the Moon does not simply circle Earth, in the strictest sense. And Earth actually accelerates toward the Moon because a force acts on it—where there is a net force there is acceleration. The force acting on Earth is a *centripetal* acceleration, for Earth circles the center of mass of the Earth–Moon system (a point within Earth about three-quarters of the way from the center to the surface). Both Earth and the Moon undergo centripetal acceleration as they circle each other about the Earth–Moon center

FIGURE 24.14

A ball of gelatin remains spherical when all parts of it are pulled equally in the same direction. When one side is pulled more than the other, however, its shape becomes elongated.

Link To Section 4.1

FIGURE 24.15

Two tidal bulges remain relatively fixed with respect to the Moon while Earth spins daily beneath them.

of mass. This makes both Earth and the Moon slightly elongated. Elongation of Earth is mainly in its oceans, which bulge equally on opposite sides.

Earth spins once per day, so a fixed point on Earth passes beneath both of these bulges each day. Any part of Earth that passes beneath one of the bulges has a high tide. Ideally, this produces two high tides per day, separated by almost 12.5 hours. But why aren't the high tides separated by 12 hours? It turns out that while Earth spins on its axis, the Moon moves in its orbit around Earth, such that the Moon appears at the same position in our sky every 24 hours and 50 minutes. So the second high tide follows the first by 12 hours and 25 minutes, and so on. That is why tides do not occur at the same time every day.

The Sun also contributes to ocean tides, although it is less than half as effective as the Moon in raising tides—even though its pull on Earth is 180 times as great as the pull of the Moon. So why doesn't the Sun cause tides 180 times as great as lunar tides? The answer has to do with a key word: *difference*. Because of the great distance of the Sun, the difference in its gravitational pull on opposite sides of Earth is very small (Figure 24.16). The percentage difference in the Sun's pull across Earth is only about 0.017%, compared to 6.7% across Earth by the Moon. That the Sun has any impact at all on Earth's tides is only because the pull of the Sun is 180 times as strong as the Moon's. Even so, Sun tides aren't even half as high as Moon tides (180 × 0.017% = 3%, nearly half of 6.7%).

Nevertheless, the Sun still plays a role in ocean tides on Earth. When the Sun, Earth, and Moon are aligned, tides due to the Sun and Moon coincide. We then have higher-than-average high tides and lower-than-average low tides—**spring tides** (Figure 24.17). (Spring tides have nothing to do with the spring season.) Spring tides occur at the times of a new or full Moon.

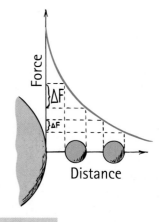

FIGURE 24.16

A plot of gravity versus distance (not to scale). The greater the distance from the Sun, the smaller the force F, which varies as $1/d^2$, and the smaller the difference in gravitational pulls on opposite sides of a planet, Δ_F, which varies as $1/d^3$, and hence, the smaller the tides.

FIGURE 24.17

When the Sun, Earth, and Moon are aligned with one another, as during a full or new Moon, spring tides occur.

When the Moon is halfway between a new Moon and a full Moon, the pulls of the Moon and Sun are perpendicular to each other. As a result, the solar and lunar tides do not overlap. Then high tides are not as high and low tides are not as low. These are called **neap tides** (Figure 24.18).

FIGURE 24.18

When the Sun and the Moon are about 90° apart (at the time of the half Moon), neap tides occur.

Midocean, the variation in water level—the range of the tide—is generally less than a meter. At shore, the range varies; it is greatest in some Alaskan fjords and is most notable in the basin of the Bay of Fundy, between New Brunswick and Nova Scotia in eastern Canada, where tidal differences sometimes exceed 15 m (Figure 24.19). This is largely due to the ocean floor, which funnels

FIGURE 24.19

Low and high tide at Hall's Harbour, Nova Scotia.

shoreward in a **V**-shape. The tide often comes in faster than a person can run. Don't dig clams near the water's edge at low tide in the Bay of Fundy! *

> **CHECK POINT**
>
> We know that both the Moon and the Sun produce our ocean tides. And we know the Moon plays the greater role because it is closer. Does its closeness mean it pulls on Earth's oceans with more gravitational force than the Sun does?
>
> **Was this your answer?**
> No, the Sun's pull is much stronger. The difference in distance is the key to tidal forces. If the Moon were closer to Earth, the tides on both Earth and the Moon would increase. Too close could catastrophically tear the Moon into pieces—the likely cause of the planetary rings of Saturn and other planets.

24.4 Components of Earth's Atmosphere

If gravity did not exist, gas molecules in Earth's atmosphere would fly off into outer space. Gases are compressible, which allows the invisible force of gravity to squeeze and hold a great number of gas molecules close to Earth's surface (where gravity is strongest). Thus the density of air molecules is greatest at Earth's surface and gradually decreases with height.

Because air has weight, it exerts pressure on Earth's surface. This pressure is known as *atmospheric pressure* or, simply, *air pressure*. The more weight, the more pressure. Like the atmosphere's density, air pressure also decreases with increasing height above Earth's surface. The higher you go, the lower the air pressure. Interestingly, the weight of the air on the ocean's surface keeps the ocean from boiling away. Recall from Chapter 7 that water boils at 0°C when no air pressure acts on it. So fish as well as birds appreciate the existence of the atmosphere.

***

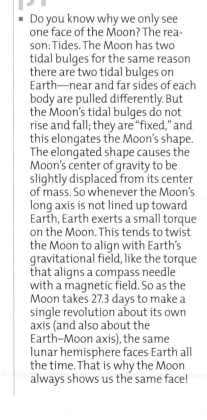

fyi

- Do you know why we only see one face of the Moon? The reason: Tides. The Moon has two tidal bulges for the same reason there are two tidal bulges on Earth—near and far sides of each body are pulled differently. But the Moon's tidal bulges do not rise and fall; they are "fixed," and this elongates the Moon's shape. The elongated shape causes the Moon's center of gravity to be slightly displaced from its center of mass. So whenever the Moon's long axis is not lined up toward Earth, Earth exerts a small torque on the Moon. This tends to twist the Moon to align with Earth's gravitational field, like the torque that aligns a compass needle with a magnetic field. So as the Moon takes 27.3 days to make a single revolution about its own axis (and also about the Earth–Moon axis), the same lunar hemisphere faces Earth all the time. That is why the Moon always shows us the same face!

* Our treatment of tides is quite simplified here, for tides are actually more complicated. Interfering landmasses and friction with the ocean floor, for example, complicate tidal motions. In many places, the tides break up into smaller "basins of circulation," where a tidal bulge travels like a circulating wave. These waves move around as if in a small basin of water that is tilted. For this reason the high tide may be hours away from an overhead Moon.

TABLE 24.2   COMPOSITION OF THE ATMOSPHERE

Permanent Gases			Variable Gases		
Gas	Symbol	Percentage by Volume	Gas	Symbol	Percentage by Volume
Nitrogen	N_2	78%	Water vapor	H_2O	0–4
Oxygen	O_2	21%	Carbon dioxide	CO_2	0.038**
Argon	Ar	0.9%	Ozone	O_3	0.000004*
Neon	Ne	0.0018%	Carbon monoxide	CO	0.00002*
Helium	He	0.0005%	Sulfur dioxide	SO_2	0.000001*
Methane	CH_4	0.0001%	Nitrogen dioxide	NO_2	0.000001*
Hydrogen	H_2	0.00005%	Particles (dust, pollen)		0.00001*

* Average value in polluted air.
** The amount of CO_2 in the atmosphere is increasing, as we will see later in this chapter and in Chapter 25.

■ Volcanic eruptions release a great deal of carbon dioxide, yet it is a minor constituent of Earth's atmosphere. That's because most carbon dioxide is gobbled up by the ocean, where it dissolves and ends up as calcium carbonate.

Table 24.2 shows that Earth's atmosphere is a mixture of various gases—primarily nitrogen and oxygen, with small percentages of water vapor, argon, and carbon dioxide and trace amounts of other elements and compounds.

⟳ Link To Sections 7.2, 7.7

VERTICAL STRUCTURE OF THE ATMOSPHERE

If you have ever gone mountain climbing, you have probably noticed that the air grows cooler and thinner with increasing elevation. At sea level, the air is generally warmer and denser. The greater density near Earth's surface is due to gravity. The density of the air, like the density of a deep pile of feathers, is greatest at the bottom and least at the top. More than half of the atmosphere's mass lies below an altitude of 5.6 km, and about 99% lies below an altitude of 30 km. Unlike a pile of feathers, however, the atmosphere doesn't have a distinct top. It gradually thins to the near vacuum of outer space.

Vertical Structure of the Atmosphere

The atmosphere is divided into layers, each distinct in its characteristics (Figure 24.20). The lowest layer, the **troposphere**, is where weather occurs. The troposphere extends to a height of 16 km over the equatorial regions and 8 km over the polar regions. Even though the troposphere is the thinnest atmospheric layer, it contains 90% of the atmosphere's mass and almost all of its water vapor

⟳ Link To Section 5.6

FIGURING PHYSICAL SCIENCE

■ Dense as Air

Knowing the density of air (1.25 kg/m³), it's a straightforward calculation to find the mass of air for any given volume—simply multiply air's density by the volume. The volume of an average-sized room is assumed to be 4.00 m × 4.00 m × 3.00 m = 48.0 cubic meters. Thus the mass of the air in the room is

$$\frac{1.25 \text{ kg}}{\text{m}^3} \times 48.0 \text{ m}^3 = 60.0 \text{ kg}$$

If you're curious to know how many pounds this is, multiply by the conversion factor 2.20 lb/1 kg.

$$60.0 \text{ kg} \times \frac{2.20 \text{ lb}}{\text{kg}} = 132 \text{ lb}$$

Because we know the density of air (1.25 kg/m³), it's a straightforward calculation to find the mass of air for any given volume.

SAMPLE PROBLEM

What is the mass in kilograms of the air in a classroom that has a volume of 796 m³?

SOLUTION:

Each cubic meter of air has a mass of 1.25 kg, so

$$796 \text{ m}^3 \times \frac{1.23 \text{ kg}}{\text{m}^3} = 995 \text{ kg}$$

which is as much as the combined mass of 17 students with a mass of about 60 kg each.

Link To Sections 9.5, 11.5

FIGURE 24.20

The atmospheric layers. The average temperature of the atmosphere varies in a zigzag pattern with altitude.

and clouds. Temperature in the troposphere decreases steadily (at about 6.5°C/km) with increasing altitude. At the top of the troposphere, temperature averages a freezing –50°C.

Above the troposphere is the **stratosphere**, which reaches a height of 50 km above the ground. Ozone molecules (O_3) form in the stratosphere and absorb UV radiation from the Sun. The absorption of UV radiation by the ozone layer causes temperature to rise from about –50°C at the bottom to about 0°C at the top.

Above the stratosphere, the **mesosphere** extends upward to about 80 km. The gases that make up the mesosphere absorb very little of the Sun's radiation. As a result, the temperature decreases again from about 0°C at the bottom of the layer to about –90°C at the top.

The situation is just the opposite in the layer above the mesosphere, the **thermosphere**. Here, temperature generally increases with altitude. This layer contains little air, but what air there is readily absorbs solar radiation. For this reason, temperatures are high, ranging from 500°C to 1500°C depending on solar activity. Because of low air density, however, this extreme temperature has little significance. In fact, it would actually be quite chilly if you could visit the thermosphere.

The **ionosphere** is an ion-rich region within the thermosphere and uppermost mesosphere. The ions are produced by the interaction of high-frequency solar radiation with atoms of atmospheric gases. Incoming solar rays strip electrons from nitrogen and oxygen atoms, producing a large concentration of free electrons and positively charged ions in this layer. The degree of ionization in the ionosphere depends on air density and on the amount of solar radiation. Ionization is greatest in the upper part of the ionosphere, where air density is low and solar radiation is high.

Ions in the ionosphere cast a faint glow that prevents moonless nights from becoming stark black. Near Earth's magnetic poles, fiery light displays called *auroras* occur as the solar wind (high-speed charged particles ejected by the Sun) strikes and excites molecules of atmospheric gases in the ionosphere (Figure 24.21). These auroral displays are particularly spectacular during times of solar flares—storms or eruptions of hot gases on the Sun.

Finally, above 500 km, in the **exosphere**, the thinning atmosphere gradually yields to the radiation belts and magnetic fields of interplanetary space.

FIGURE 24.21

The *aurora borealis* over Alaska is created by solar-charged particles that strike the upper atmosphere and light up the sky (just as similar particles on a smaller scale light up a fluorescent lamp).

■ 24.5 Solar Energy

Earth's equatorial regions are warmer than the polar regions. But why? Surface temperatures on Earth depend on the energy each part of Earth receives from the Sun each day. And how much energy is received depends on the angle at which the Sun's rays strike Earth's surface. The different angles at which sunlight strikes Earth can be mimicked by holding a flashlight vertically over a table and shining the light directly down on, and perpendicular to, the flat surface (Figure 24.22a). The light produces a bright circle. Now tip the flashlight at various angles and notice that the circle elongates into ellipses, spreading the same amount of energy over a greater area and, therefore, decreasing the intensity of the light. The same is true of sunlight on Earth's surface. High noon in equatorial regions is like the vertically held flashlight; high noon at higher latitudes is like the flashlight held at an angle.

THE SEASONS

The northern United States and Canada, both temperate regions, have distinct summer and winter seasons. These seasons change because the angle at which the Sun's rays strike these locations varies over the course of a year. Figure 24.23 shows how the tilt of Earth causes the variation in the rays' angle. Can you see that the rays are more perpendicular to the ground in the Northern Hemisphere when Earth's axis is tilted *toward* the Sun? When the Sun's rays are closest to perpendicular at any spot on Earth, that region experiences summer. Six months later, the rays fall upon the same region more obliquely, and we have winter. In between are fall and spring.

It is interesting to note that, because Earth follows an elliptical path around the Sun, Earth is farthest from the Sun when the Northern Hemisphere experiences summer. The angle of the Sun's rays, not the distance from the Sun, is most responsible for Earth's surface temperatures.

Another effect of the tilted rays is the length of daylight each day. Can you see in Figure 24.23 that a location in summer has more daylight per daily rotation of Earth than the same location when Earth is on the opposite side of the Sun in winter? If you have trouble visualizing this, look at the high latitudes near the poles. Consider the special latitude where daylight lasts nearly 24 hours during the summer solstice (around June 21) and night lasts about 24 hours at the winter solstice (around December 21). This latitude is called the Arctic Circle in the Northern Hemisphere and the Antarctic Circle in the Southern

FIGURE 24.22

(a) When the flashlight is held directly above at a right angle to the surface, the beam of light produces a bright circle. (b) When the light shines at an angle, the beam is dispersed over a larger area and is therefore less intense.

The Physical Science Place

Seasons

Link To Section 7.5

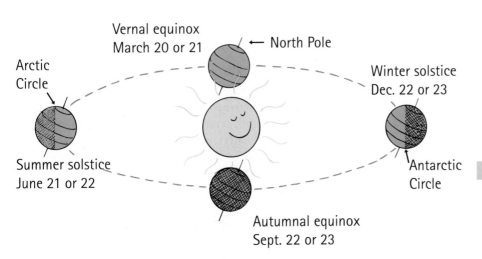

Vernal equinox
March 20 or 21

← North Pole

Arctic Circle

Winter solstice
Dec. 22 or 23

Summer solstice
June 21 or 22

Antarctic Circle

Autumnal equinox
Sept. 22 or 23

FIGURE 24.23

The tilt of Earth and the corresponding different spreading of solar radiation produce the yearly cycle of seasons.

FIGURE 24.24

The hot Sun emits visible short waves, and the cool Earth re-emits infrared long waves. Radiation emitted from Earth is called *terrestrial radiation*. Not all of the re-emitted infrared long waves are bounced back to space. Some are absorbed and trapped in the atmosphere by greenhouse gases. In this way, the greenhouse gases act like a blanket that warms the lower atmosphere.

Hemisphere. During the summer solstice, the North Pole leans towards the Sun and the South Pole leans away from the Sun. (Summer and winter are reversed, of course, in the two hemispheres.)

Halfway between the peaks of the winter and summer solstice, around mid-September and mid-March, the hours of daylight and night are of equal length. These are called the *equinoxes* (Latin for "equal nights"). The equal hours of day and night during the equinoxes are not restricted to high latitudes but occur all over the world.

CHECK POINT

1. Why are daylight hours fewer in winter months?
2. A friend says, "The tilt of Earth's axis, not Earth's distance from the Sun, causes Earth's seasons." Do you agree or disagree? Explain.

Were these your answers?

1. Earth is tilted on its axis, like a top leaning in one direction all the time. As Earth revolves around the Sun, the Northern Hemisphere is tilted toward the Sun in the summer and away from the Sun during the winter. When the hemisphere is tilted away, the Sun is closer to the horizon. Hence, the Sun rises later and sets earlier, resulting in shorter days.
2. Agree. The tilt of Earth's axis affects the angle at which solar radiation strikes a given location. When a location experiences winter, the angle at which the Sun's rays strike Earth's surface is farthest from perpendicular. In summer, the rays strike more directly.

TERRESTRIAL RADIATION

Solar radiation covers a wide spectrum of wavelengths, mostly in the visible short-wavelength part of the spectrum. Earth absorbs some of this energy and in turn reradiates part of it back to space. As we learned in Chapter 7, this is *terrestrial radiation*, emitted from Earth's surface (Figure 24.24). Terrestrial radiation is emitted in the infrared long-wavelength part of the spectrum.

Interestingly, terrestrial radiation rather than solar radiation directly warms the lower atmosphere, which explains why air close to the ground is so much warmer than air at higher elevations. The temperature of Earth's surface depends on the amount of solar radiation coming in compared with the amount of terrestrial radiation going out. In direct sunlight, the net effect is warming, because Earth's surface absorbs more energy from the Sun than it emits. At night, the net effect is cooling because Earth's surface emits more energy than it absorbs. Cloud cover blocks both incoming solar radiation and outgoing terrestrial radiation. Hence, cloudy days are much cooler than sunny days, and cloudy nights are generally warmer than clear nights.

THE GREENHOUSE EFFECT AND GLOBAL WARMING

Incoming short-wavelength solar radiation easily penetrates the atmosphere to reach and warm Earth's surface. The warmed surface then reradiates this energy back as long-wavelength terrestrial radiation. But not all of the outgoing long-wavelength terrestrial radiation can penetrate the atmosphere to escape into space. Much like the panes of glass in a greenhouse, atmospheric gases (mainly water vapor and carbon dioxide) trap long-wavelength terrestrial radiation, thereby warming the lower atmosphere. This warming of the lower atmosphere is called the **greenhouse effect**. The atmosphere acts like an insulating blanket, keeping Earth's surface warmer than it would be without an atmosphere.

In and of itself, the greenhouse effect is not harmful. In fact, it is essential to life on Earth as we know it. Without the greenhouse effect, Earth's surface would be a frigid –18°C! Like Earth's atmosphere, the origin of greenhouse gases is linked to the release of gases in volcanic eruptions. Of all the greenhouse gases, water vapor by far plays the largest role in confining Earth's heat. As part of Earth's natural hydrologic cycle, water vapor levels have remained relatively constant and abundant throughout time. And the concentrations of other heat-trapping greenhouse gases had been fairly stable over the course of modern human evolution (about 100,000 years) until the dawn of civilization about 10,000 years ago.

What now concerns scientists are increases in the concentration of greenhouse gases other than water vapor. Gases such as carbon dioxide (CO_2), methane (CH_4), nitrous oxide (N_2O), ozone (O_3), and chlorofluorocarbons (CFCs) are on the rise because of human activities. These gases are potent contributors to the greenhouse effect. As their concentrations increase, the atmosphere's ability to absorb and trap reradiated terrestrial heat energy also increases.

Climate changes have happened over the course of geologic time, but the present environmental concern is that increased levels of atmospheric carbon dioxide and other greenhouse gases may make Earth too warm, causing rapid changes to Earth's natural systems that would negatively affect humans. Since the Industrial Revolution of the 1800s, for example, atmospheric CO_2 levels have been steadily increasing (Figure 24.25). The CO_2 increase has been linked to the warming of Earth's surface by about 0.7°C since the late 1800s. Many long-standing, famous alpine glaciers are quickly receding, and some have disappeared altogether. Most scientists believe that further warming will occur if CO_2 emissions are not held in check.

The effects of warming Earth's surface are not fully known. One concern is that warming will cause the ice caps on Antarctica and Greenland to melt. This would raise sea level, resulting in the flooding of low-lying coastal lands. Warming would also change rainfall patterns, seriously affecting agriculture. The grain-growing regions of North America and Asia might shift northward as local climates warm and growing seasons lengthen. On the other hand, deserts in continental interiors might spread to cover much larger areas. We don't know. We do know that Earth has experienced warmer and colder periods in times past and that global-scale climatic changes may have contributed to many of the extinctions discussed in Chapter 21. Ongoing research is focused on understanding the impacts of global warming.

If planet Earth had *no* greenhouse effect, its average temperature would be –18°C. Brrrrr!

Link To Section 11.2

Link To Section 18.4

fyi

■ Past global temperatures are determined by measuring the ratio between hydrogen and its heavier isotope, deuterium, in polar ice cores. When global temperatures are high, the ocean is warmer, and larger fractions of water containing deuterium evaporate from the ocean and fall as snow. A high deuterium-to-hydrogen ratio in preserved ice cores therefore indicates a warmer climate.

FIGURE 24.25

Since the Industrial Revolution of the 1880s, atmospheric CO_2 levels and global temperatures have increased. Do you think there is a correlation?

CHECK
■ POINT

1. What does it mean to say the greenhouse effect is like a "one-way valve"?
2. Which gas in the atmosphere is the greatest contributor to the greenhouse effect?
3. What is the primary "natural" contributor to greenhouse gases in Earth's atmosphere?

Were these your answers?

1. The transparent material—atmosphere for Earth and glass for the florist's greenhouse—passes only incoming short waves and blocks outgoing long waves. In other words, radiation travels only one way.
2. Water vapor.
3. Volcanic eruptions. Interestingly, the 1991 volcanic eruption of Mount Pinatubo spewed more chlorine into the atmosphere than the combined leakage of CFCs over a century.

■ 24.6 Driving Forces of Air Motion

Link To Section 5.5

The underlying driving force for air movement is the unequal heating of Earth's surface by the Sun. It is often said that "warm air rises." Why is this so? The fundamental answer is that warm air is less dense than cool air. As we learned in Chapter 5, a less-dense fluid rises buoyantly when surrounded by a denser fluid. This is why hot-air balloons rise—the air inside the balloon is heated, making it less dense compared to the surrounding air.

But why is warm air less dense than cool air? Air molecules within an air parcel * are constantly colliding and bouncing off each other. When temperatures are cool, the molecules move more slowly and the distance they travel after each bounce is small. When the air parcel is heated, the molecules move faster and bounce farther after each collision. Thus, warm air requires more space—a larger volume. Because the mass of the air parcel stays the same, density is lower when volume is increased.

Link To Section 6.5

To envision this, imagine that the air parcel is surrounded by a very thin, stretchable container. The air molecules need to do work to cause the parcel to expand—they expend thermal energy (Chapter 6). Because the air parcel loses thermal energy, it cools as it expands. According to Boyle's law (Chapter 5), our air parcel can achieve this expansion only if the pressure around it is reduced. Air pressure decreases with altitude, so air that rises always expands and cools. The air parcel continues to rise until it becomes the same temperature and density as the surrounding air.

Conversely, sinking air heats up as it descends because the environment does work on a sinking air parcel (Chapter 6). The energy expended by the environment is transferred to the air parcel. The temperature of the air parcel goes up because its internal thermal energy has increased. A bicycle pump heats up when used to fill a bicycle tire for the same reason—work is done on the air inside the pump.

THE TEMPERATURE–PRESSURE RELATIONSHIP

When an air parcel is warmed, it rises. Cooler, denser air from higher altitudes then sinks to occupy the region left vacant by the rising warm air. Because the rising, warm air parcel cools on its way up, it too eventually sinks to replace other rising, warm air. Such motion constitutes a convection cycle and thermal circulation of air—in other words, a *convection current*.

As convection currents stir the atmosphere, the result is *wind*—defined as air with an average horizontal motion. Wind is generated in response to pressure

---

* Consider an air parcel to be a distinct "package" of air that is free to expand and contract—but does not mix with the surrounding air.

differences in the atmosphere, which, as we will see, are largely the result of temperature differences. A difference in pressure between two different locations is called a *pressure gradient*, and the force that causes air to move is the **pressure-gradient force**. The pressure-gradient force drives air from areas of high pressure toward areas of lower pressure.

To see how temperature affects pressure, consider the two equivalent air columns shown in Figure 24.26a. The two air columns are equivalent because they contain the same number of air molecules, distributed uniformly. To aid visualization, we make three simplifying assumptions: (1) there is no change in air density with height within a column (the air density of an entire column can change, however), (2) air cannot enter or leave either column (mass is constant, so surface air pressure does not change), and (3) the width of each column remains constant. So each column has a fixed amount of air with uniform density, which can move only up or down.

When the cities are at the same temperature, the columns are the same height, as shown in Figure 24.26a. At any elevation—say, the elevation marked X in the drawing—the air pressure in each column is the same, because the number of air molecules above X is the same in both cases. This is true for any elevation. Thus, we have no pressure difference between the two cities at any elevation because we have no temperature difference.

Now suppose that the temperature in city 1 drops while the temperature in city 2 rises. In this case, we get the situation shown in Figure 24.26b. The cooling air over city 1 contracts and becomes denser, and the warming air over city 2 expands upward (remember, column width is fixed) and becomes less dense. We still have the same number of air molecules over each city, but there is a cool, short air column over city 1 and a warmer, much taller air column over city 2. The air pressure at ground level stays the same in each city, because equal

Link To Section 3.10

Link To Section 7.2

(a)

(b)

FIGURE 24.26

(a) Air columns of the same temperature over two cities. Note that the air pressure at any elevation, such as elevation X shown here, is the same for both cities. (b) When city 1 is cold and city 2 is warm, the air column over city 2 is taller because of expansion, which means that for any given elevation above the surface, pressure is greater in the taller column. This produces a pressure gradient between the two cities. So wind blows from city 2 to city 1.

fyi

- Of course, air density is not really constant in a tall vertical column of atmosphere. But this simplification greatly helps in understanding the connection between temperature and pressure and its effect on air movement.

The Physical Science Place
The Coriolis Effect

FIGURE 24.27

If Earth were simply a nonrotating sphere, air circulation would be in a single Northern Hemisphere cell and a single Southern Hemisphere cell. In each cell, heated air would rise at the equator and move toward the polar regions, where it would cool, sink, and be drawn back to the warmer regions of the equator.

numbers of air molecules are pressing down on both. But the warming and cooling of the respective air columns causes pressure changes in the air aloft.*

Now consider elevation Y, halfway up air column 1. Because Y is halfway up column 1, the top half of air column 1 contains half the total number of molecules (again, assuming constant density with height). Elevation Y is less than halfway up in air column 2, because column 2 is taller. Thus, the part of column 2 above elevation Y contains more than half of column 2's air molecules. Because more air molecules are bearing down at elevation Y in air column 2 than in air column 1, the air pressure at Y must be greater in column 2 than it is in column 1. In other words, a difference in temperature has led to a difference in pressure aloft. So the number of air molecules above a particular elevation (i.e., the mass of the air above) governs the atmospheric pressure at that elevation. We come to a very important concept: warm air aloft is associated with high atmospheric pressure aloft, and cold air aloft is associated with low atmospheric pressure aloft.

So what does this pressure difference at elevations of a kilometer or more have to do with wind at Earth's surface? Well, differences in pressure cause air to move and, hence, wind to blow. Let us now allow the air between city 1 and city 2 to mingle. Because air moves from areas of high pressure to areas of low pressure, the wind aloft blows from city 2 toward city 1. As air aloft moves from column 2, air density in city 2 decreases and the surface pressure in city 2 drops. Meanwhile, as air accumulates in cooler column 1, the surface pressure in city 1 rises. Thus, we find that cold days are associated with high surface pressure, and warm days are associated with low surface pressure. Now that the surface pressure is higher in city 1, surface winds blow from city 1 to city 2. This, in turn, changes the pressure distribution again, as air moves away from city 1. A dynamic system indeed!

LARGE-SCALE AIR MOVEMENT

Globally, equatorial regions receive maximum radiant energy from the Sun and, as a result, have higher average temperatures than other regions of the world. As air heated by the hot ground (or ocean) at the equator rises, it moves toward the poles, cooling gradually in the upper atmosphere. This cooled air then sinks and is drawn back to the warmer regions near the equator. If Earth were a nonrotating sphere, the effect would be one simple single-cell circulation pattern in the Northern Hemisphere and another in the Southern Hemisphere, as shown in Figure 24.27.

But Earth rotates, which greatly affects the path of moving air. Think of Earth as a large merry-go-round rotating in a counterclockwise direction (the same direction Earth spins, as viewed from above the North Pole). Imagine that you and a friend are playing catch on this merry-go-round. When you throw the ball to your friend, the circular movement of the merry-go-round affects the direction in which your friend sees the ball travel. Although the ball travels in a straight-line path, it appears to curve to the right, as shown in Figure 24.28.

FIGURE 24.28

(a) On the nonrotating merry-go-round, a thrown ball travels in a straight line. (b) On the counterclockwise-rotating merry-go-round, the ball also moves in a straight line. However, because the merry-go-round is rotating, the ball appears to deflect to the right of its intended path.

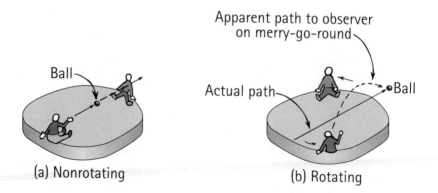

Apparent path to observer on merry-go-round

Ball

Actual path

Ball

(a) Nonrotating

(b) Rotating

---

* *Air aloft* is defined as air higher than 1 km above the ground.

(The ball travels straight but your friend never catches it, because the movement of the merry-go-round causes her position to change.) This apparent curving is similar to what happens on Earth. As Earth spins, all free-moving objects—air and water, aircraft and ballistic missiles, and even snowballs, to a small extent—appear to deviate from their straight-line paths as Earth rotates under them. This deflection due to the rotation of Earth is called the **Coriolis force.**[*]

A significant result of the Coriolis force is the deflection of winds toward the right in the Northern Hemisphere and toward the left in the Southern Hemisphere (Figure 24.29). The impact of the Coriolis force varies according to the speed of the wind. The faster the wind, the greater the deflection. Latitude also influences the degree of deflection. Deflection is greatest at the poles and decreases to zero at the equator. As Figure 24.30 shows, the Coriolis force has a significant impact on atmospheric motion—and airplanes—in the midlatitudes.

Air moving close to Earth's surface also encounters a frictional force. The rougher the surface, the greater the friction, and so the greater the drag on the wind. Because surface friction reduces wind speed, it also reduces the effect of the Coriolis force. This causes winds in the Northern Hemisphere to spiral out clockwise from a high-pressure region and spiral counterclockwise into a low-pressure region (top part of Figure 24.29c). In the Southern Hemisphere, these circulation patterns are reversed (bottom part of Figure 24.29c).

■ 24.7 Global Circulation Patterns

Cell-like circulation patterns are responsible for the redistribution of heat across Earth's surface and for global winds (Figure 24.31). At the equator, warmed air flows straight up with very little horizontal movement, resulting in a vast low-pressure zone at the surface. This rising motion creates a narrow, windless realm of air that is still, hot, and stagnant. Seamen of long ago cursed the equatorial seas as their ships floated listlessly for lack of wind, and they referred to the area as the *doldrums*. When the moist air from the doldrums rises, it cools and releases torrents of rain. Over land areas, such frequent rains give rise to the tropical rain forests that characterize the equatorial region.

The air of the sweltering doldrums rises to the boundary between the troposphere and stratosphere, where it divides and spreads out to the north and the south. (Relatively little wind crosses the equator into the neighboring hemisphere.) Because of the Coriolis force, the wind is deflected and travels along longer paths poleward than if it traveled straight toward the poles. By the time the wind has reached about 30°N and 30°S latitude, the air has cooled enough that it descends toward the surface. The descending air is compressed because air pressure increases with decreasing altitude. The compression warms the air and lowers its relative humidity.[**] A resulting high-pressure zone girdles Earth, creating a belt of hot, dry air at the surface and little rain.

On land, these high-pressure zones account for the world's great deserts—the Sahara in Africa, the Arabian Desert in the Middle East, the Mojave Desert in the United States, and the Great Victoria Desert in Australia. At sea, the hot,

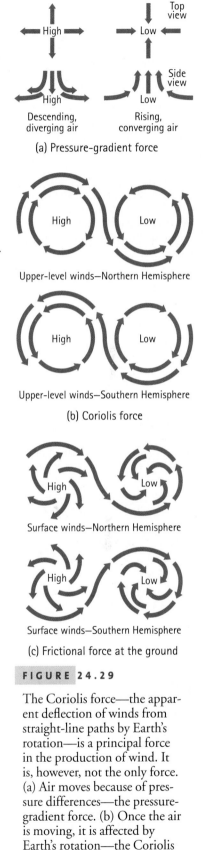

(a) Pressure-gradient force

(b) Coriolis force

(c) Frictional force at the ground

FIGURE 24.29

The Coriolis force—the apparent deflection of winds from straight-line paths by Earth's rotation—is a principal force in the production of wind. It is, however, not the only force. (a) Air moves because of pressure differences—the pressure-gradient force. (b) Once the air is moving, it is affected by Earth's rotation—the Coriolis force. (c) As air moves close to the ground, it slows because of frictional force.

---

[*] The Coriolis force is not a true force in the strictest sense, and is sometimes called the *Coriolis effect*. On the merry-go-round, for example, the curved path is perceived only by someone on the merry-go-round. Someone in a tree directly over the merry-go-round would see the ball move in a straight line. But because we are all on Earth, and not suspended above it, the Coriolis effect truly causes wind to change direction, as if directed by a force. So because the wind is truly deflected, the cause is referred to as the Coriolis force.

[**] Condensation, precipitation formation, and relative humidity are covered in Chapter 25.

Latitude influences the apparent deflection resulting from the Coriolis force. A free-moving object heading east (or west) appears to deviate from its straight-line path as Earth rotates beneath it. Deflection is greatest at the poles and decreases to zero at the equator. The Coriolis force is evident only for large-scale motions such as atmospheric winds or oceanic currents, in which one part has a greater tangential speed than the other about Earth's axis.

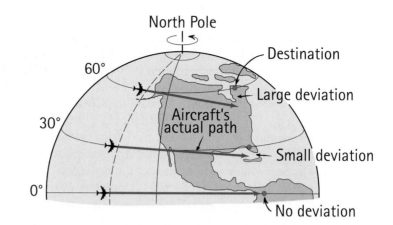

fyi
━━━━━━━━━━━━━

- Air currents are named for the direction the wind is coming from. Thus, the westerlies blow from the west—but air moves to the east.

Link To Section 7.8

Global circulation of the atmosphere results from a combination of two main factors: unequal heating of Earth's surface (which sets up convection cells) and Earth's rotation. The atmosphere has six cell-like circulation patterns; prevailing winds blow in the directions indicated by the arrows. The major prevailing winds are the *westerlies*, the *easterlies*, and the *trade winds*.

descending air produces very weak winds. According to legend, early sailing ships were frequently stalled at these latitudes, both north and south. As food and water supplies dwindled, horses on board were either eaten or cast overboard to conserve fresh water and reduce the load of the ship. As a result, this region is now known as the *horse latitudes*.

The thermal convection cycle that starts at the equator is completed when air flowing southward from the horse latitudes in the Northern Hemisphere and northward in the Southern Hemisphere is deflected westward by the Coriolis force to produce the *trade winds*. Air that flows northward from the horse latitudes in the Northern Hemisphere and southward in the Southern Hemisphere is deflected eastward to produce the prevailing *westerlies*.

Near the poles, frigid air continuously sinks, pushing the surface air outward. The Coriolis force is quite evident in the polar regions, as the wind deflects to the west to create the *polar easterlies* (Figure 24.31). The cool, dry polar air meets the warm, moist air of the westerlies at latitudes 60°N and 60°S. This boundary, called the *polar front*, is a zone of low pressure where contrasting air masses converge, often resulting in storms.

The middle latitudes are noted for their unpredictable weather. Although the winds tend to be from the west, they are often quite changeable, as the temperature and pressure differences between the subtropical and polar air masses at the polar front produce powerful winds. As air moves from regions of high pressure, where air is denser, toward regions of low pressure, the result is a whirlwind effect (as we will learn in Chapter 25, a *midlatitude cyclone*).

Irregularities in Earth's surface also influence wind behavior. Mountains, valleys, deserts, forests, and great bodies of water all play a part in determining which way the wind blows.

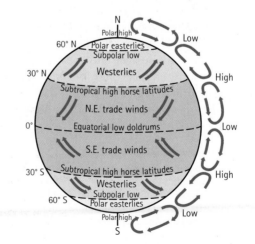

UPPER ATMOSPHERIC CIRCULATION

In the upper troposphere, "rivers" of rapidly moving air meander around Earth at altitudes of 9–14 km. These high-speed winds are the *jet streams*. With wind speeds averaging between 95 and 190 km/h, the jet streams play a critical role in the global transfer of heat from the equator to the poles.

The two most important jet streams, the *polar jet stream* and the *subtropical jet stream*, form in both the Northern and Southern Hemispheres. The formation of polar jet streams is due to a temperature gradient at the polar front—at about 60°N and 60°S latitude—where cool polar air meets warm tropical air. This temperature gradient causes a steep pressure gradient that increases the wind speed. During the winter, the polar jet stream is strong and migrates to lower latitudes, bringing strong winter storms and blizzards to the United States. In summer, the polar jet stream is weaker and migrates to higher latitudes.

The subtropical jet stream is generated by warm air carried poleward from the equator, which produces a sharp temperature gradient along the subtropical front—about 30°N and 30°S latitude. Once again, a pressure gradient caused by the temperature gradient generates strong winds.

The subtropical jet stream above Southeast Asia, India, and Africa merits special mention (Figure 24.32). The formation of this jet stream is related to the warming of the air above the Tibetan highlands. During the summer, the air above the continental highlands is warmer than the air above the ocean to the south. The warmer air rises, drawing in cooler, moist air from over the ocean. Thus, temperature and pressure gradients generate strong onshore winds that contribute to the region's *monsoon* climate.* During winter, the winds change direction to produce a dry season.

This cycle of winds characterizes the climates of much of Southeast Asia. The predictable rain-bearing summer wind from the sea that moves over the heated land is called the *summer monsoon*; the prevailing dry wind from land to sea in winter is called the *winter monsoon*.

(a)

(b)

FIGURE 24.32

Winds over Southeast Asia. (a) During the summer months, air over the oceans is cooler than the air over land. The summer monsoon brings heavy rains as the winds blow from sea to land. (b) During the winter months, air over continents is cooler than air over oceans. The winter monsoon generally has clear skies and winds that blow from land to sea.

CHECK POINT

1. What are the underlying causes of the trade winds, jet streams, and monsoons?
2. In the middle latitudes, airlines schedule shorter flight times for planes traveling west to east and longer flight times for planes traveling east to west. Why are eastbound planes faster?

Were these your answers?

1. Simply enough, the unequal heating of Earth's surface coupled with Earth's rotation.
2. The upper-level westerly winds of the jet stream account for faster-moving eastbound aircraft. As the jet stream moves from west to east, it carries along everything in its path. To save time and fuel, airline pilots seek the jet stream when traveling west to east and avoid it when traveling east to west.

OCEANIC CIRCULATION

The forces that drive the winds also affect the movement of seawater. In the open ocean, the movement of seawater results from two types of currents: wind-driven surface currents and density-driven deep-water currents. The density of

fyi

Air circulation affects ocean surface currents, as was found during a severe storm in 1990. Five cargo containers of athletic shoes were washed overboard from freighters en route from South Korea to the Pacific Northwest. Thousands of sneakers, hiking boots, and other shoes were picked up along beaches from British Columbia to Oregon and as far into the mid-Pacific as Hawaii. Although months at sea left many shoes mismatched, most shoes were still wearable after washing. Beachcombers formed "swap meets" to search for mates of found shoes!

* *Monsoon* is derived from the Arabic word *mausim*, which means "season." So monsoon is taken to mean "seasonal wind."

FIGURE 24.33

The ocean's vertical structure. In the surface zone, water is well mixed as it moves vertically in response to temperature and density changes, and as it moves horizontally in response to wind. Water in the transition zone moves along density surfaces. Water in the deep zone is density driven, as it circulates from cold polar regions to warmer equatorial regions.

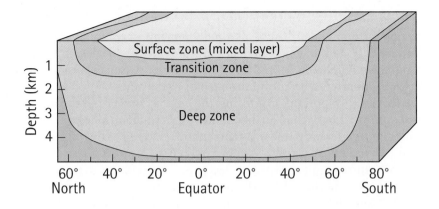

fyi

- There is some controversy over density-driven deep-ocean currents. There may not be enough potential energy in the system to drive such currents. Mixing by wind and tidal dissipation in the deep ocean may also contribute to driving the deep flow. In a sense, all ocean currents may ultimately be wind-driven!

fyi

- Where do the ocean's salts originate? One source is the chemical weathering of continental rocks. As the rocks are weathered, elements such as sodium, calcium, and potassium dissolve into water. The water carrying these chemicals usually flows downstream and makes its way to the ocean. Another source of elements is Earth's interior. For eons, volcanic eruptions have delivered huge quantities of chlorine, as well as water vapor and other gases, to the ocean's waters. Chlorine plus sodium? Sodium chloride—common table salt. Hence, the ocean's salty taste.

seawater is controlled by two factors—temperature and salinity. Near coastlines, water movement is affected not only by surface and deep-water currents but also by tides and the presence of coastal boundaries.

Like the atmosphere, the ocean can be divided into several vertical layers—the *surface zone*, a *transition zone*, and the *deep zone*. Scuba divers notice an increase in water pressure when swimming to lower depths. The deeper the descent, the greater the water pressure. The pressure is simply the weight of the water above pushing down on you (Chapter 5). Another factor that generally changes as you descend is temperature. Deeper waters are colder. So in addition to variations in salinity, seawater also varies in temperature and pressure. Because cold water is denser than warm water—for the same reasons that cold air is denser than warm air—cold seawater sinks below warm seawater. The salinity of the water also affects density: the greater the salinity, the greater the density. These variations are best illustrated when we look at the ocean's vertical structure (Figure 24.33).

CHECK POINT

Would you expect the pressure 100 m deep in high-salinity water to be the same as it is 100 m deep in lower-salinity water?

Was this your answer?

No. Higher salinity means higher density. A volume of denser water weighs more than an equal volume of less-dense water. Because pressure is weight per unit area, pressure at 100-m depth is greater if the water is more saline.

SURFACE CURRENTS

As winds blow across the ocean, frictional forces set surface waters into motion. If distances are short, the surface waters move in the same direction as the wind. For longer distances, however, other factors come into play.

One such factor is the deflective Coriolis force, which causes water currents to deflect to the right (in the Northern Hemisphere), so that they move in a different direction than the wind. This phenomenon, known as *Ekman transport* (Figure 24.34), causes the surface currents to deflect up to 45° from the wind direction! One consequence of Ekman transport is the *Ekman spiral*, a phenomenon that causes the deflection due to the Coriolis force to increase continuously with depth, so that the net direction of water transport over 100 m of depth can be up to 90° from the wind direction (Figure 24.34).

Figure 24.35 shows the global air circulation pattern between 40°N and 40°S latitudes. Focusing for the moment on the Northern Hemisphere, note that

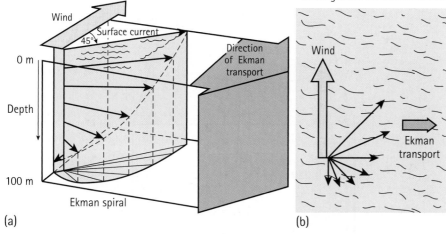

(a)

(b)

FIGURE 24.34

Two views of Ekman transport; part (a) shows velocity decrease with depth; part (b) shows a birds-eye view of the surface water direction and the overall direction of net water transport. With no outside force, water moves in the same direction as the wind. But the Coriolis force deflects surface water about 45° to the right of the wind direction (Northern Hemisphere). As you move down in depth, deflection increases to the right but water moves at a slower speed, causing the net direction of water transport to be between about 70° and 90° from the wind direction.

south of 30°N, the trade winds blow from northwest to southeast, while north of 30°N, the westerlies blow from west to east. Because of Ekman transport, ocean surface currents south of 30°N flow to the northwest, and north of 30°N to the southeast (Figure 24.35). As you can see from Figure 24.35, Ekman-influenced currents flow toward each other, or converge, at about 30°N. Such convergence causes a mounding of water (Figure 24.36). The ocean surface has highs and lows just like the land surface, except the "topography" of the ocean surface is much more subdued than land topography.

Just as on land, water in such mounds wants to flow "downhill" because of the pressure-gradient force. But once again the Coriolis force comes into play, causing this downhill flow to be deflected to the right (in the Northern Hemisphere). Now veering to the right, the Coriolis force acts again to deflect water further to the right, toward the peak of the mound. So two forces operate on the water, and they attempt to drive the water in opposite directions. The balance between the pressure-gradient force and the Coriolis force causes water to flow around the central, mounded area in a circular whirl pattern called a **gyre**

fyi

- Don't believe stories about swirling water in your sink being affected by the Coriolis force. Any effect due to the difference in speeds of one part of the sink compared with the other are minuscule, and they are masked by thermal motions in the water.

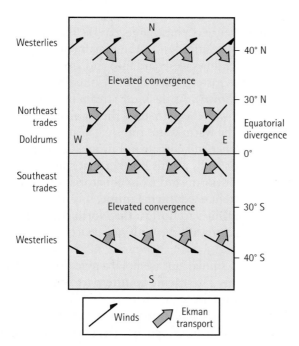

FIGURE 24.35

Ekman transport driven by prevailing winds causes currents to converge around 30°N and 30°S latitudes. The convergence causes the mounding of surface water in this area (see Figure 24.36).

FIGURE 24.36

Mounding of water in the Northern Hemisphere due to Ekman transport. Water initially flows "downhill" but is deflected to the right, as shown by the arrow coming from point A. Now flowing around the mound, geostrophic flow develops as the Coriolis force and the pressure-gradient force balance each other.

FIGURE 24.37

The gyre in the North Atlantic is composed of the Gulf Stream, the North Atlantic Current, the Canary Current, and the North Equatorial Current.

FIGURE 24.38

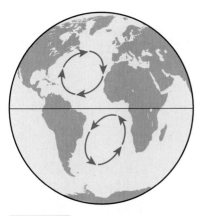

Separate gyres form north and south of the equator. Flow in the Northern Hemisphere is clockwise; flow in the Southern Hemisphere is counterclockwise.

(Figure 24.37). The circular motion is clockwise in the Northern Hemisphere and counterclockwise in the Southern Hemisphere. When the pressure-gradient force and the Coriolis force are in perfect balance, such flow is called *geostrophic* flow—water continues flowing around the mound instead of down it. Gyres dominate the overall flow of surface currents in each ocean basin. As Figure 24.38 shows, gyres don't cross the equator. There are separate gyres north and south of the equator in each ocean basin.

The gyre-generating, convergent flow around 30°N and 30°S latitude also causes *downwelling* that drives some surface water to flow downward. The opposite effect occurs at the equator, where the southeast trade winds slightly cross 0° latitude. Here, Ekman transport causes water to diverge at the equator, which causes *upwelling* of deeper, colder, nutrient-rich water (Figure 24.39). Such areas of upwelling are full of life. Ekman transport also causes upwelling or downwelling along coasts, depending on which way the dominant winds blow. Two such areas of upwelling are the coasts of Peru and California (Figure 24.40).

An important consequence of these large gyres is the transport of heat from equatorial regions to higher latitudes. In the North Atlantic Ocean, for example, warm equatorial water flows westward into and around the Gulf of Mexico, then northward along the eastern coast of the United States. This warm-water current is called the *Gulf Stream*. As the Gulf Stream flows northward along the North American coast, flow in the gyre steers the warm current eastward toward Europe (Figure 24.41). The warm water transfers heat to the westward-moving air, which influences weather and climate in Europe. Great Britain and Norway benefit from this heat, for lands at this northern latitude might well be colder were the heat not brought northward from equatorial regions and the Gulf of Mexico. As the warm current encounters Europe, it is turned southward toward the equator, where it continues to flow in the North Atlantic gyre. Eventually, this water moves westward toward the Gulf of Mexico and once again becomes part of the Gulf Stream.

Each ocean basin has a similar surface flow: a gyre flow pattern with a warm western boundary current. The Pacific counterpart of the Gulf Stream is the warm, northward-flowing current known as the *Kuroshio*. In the Southern Hemisphere, surface oceanic circulation (with the exception of the Antarctica Circumpolar Current) is similar, except that the gyres move counterclockwise.

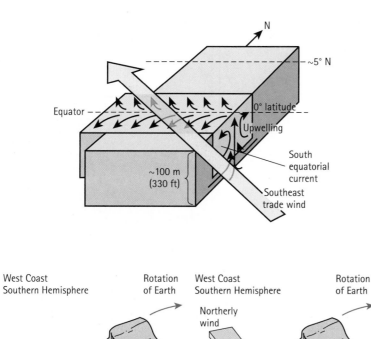

FIGURE 24.39

Ekman transport causes divergence of surface water at the equator, which causes upwelling of deep, cold, nutrient-rich water, allowing life to thrive there. The southeast trade winds, blowing from the Southern Hemisphere, actually cross the equator slightly. South of the equator, Ekman transport causes surface water to veer to the left, or southwest. North of the equator, Ekman transport causes surface water to veer to the right, or northwest. The surface water diverges, causing upwelling.

FIGURE 24.40

Ekman transport causes coastal upwelling (a) or downwelling (b), depending on the dominant wind direction.

Warm western boundary currents occur in all three Southern Hemisphere ocean basins.

DEEP-WATER CURRENTS

Surface waters are driven by winds, but deeper waters are driven by gravity—deep water flows because dense water sinks. Deep water moves more slowly than surface water, but vastly more water is transported by deep-water currents than by surface-water currents. Deep-water flow is like a huge global conveyor belt, moving water and heat all over the planet (Figure 24.42).

The densest water forms in the Weddell and Ross Seas adjacent to Antarctica. To understand how this water forms, we need to look at what happens when seawater begins to freeze. Seawater does not freeze easily. When it does, however, only the water freezes and the salt is left behind. Thus, the seawater that does not freeze experiences an increase in salinity, which, in turn, causes an increase in density.

Dense bottom water also forms in the Arctic Ocean, but the topography of the seafloor prevents most of it from escaping southward. Some of the deep water does escape into the North Atlantic, however. So dense bottom water is formed in the high latitudes and sinks to the ocean floor. Because dense water continues to form at the surface, cold, salty water more or less continuously sinks to the ocean bottom. The sinking water pushes still-deeper water out of the way, causing the bottom water to flow outward along the ocean floor. This is the start of the great conveyor belt.

FIGURE 24.41

The reddish areas in this false-color satellite image indicate warm water. The Gulf Stream transports huge amounts of heat northward and then eastward toward Europe.

FIGURE 24.42

Deep-ocean currents act like a conveyor belt, transporting cold water from the North Atlantic passed the equator and on to the Antarctic. From the Antarctic, water flows eastward and then northward into the Pacific Ocean and the Indian Ocean.

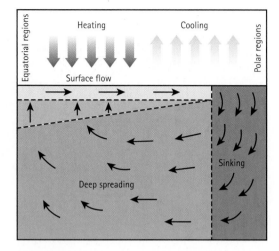

FIGURE 24.43

Cold, dense water sinks in the polar regions and then flows toward the equator. Eventually it is forced upward, where it warms, becomes part of the surface flow, and returns to the polar regions. Darker colors indicate colder water.

LINK TO Weather

■ The El Niño Condition

When weather is measured over time, an average weather pattern can be seen. Consistent behavior of weather over time is *climate*. During some periods, the *average* weather pattern departs from its norm. These variations from the average are expected and have recognizable short-term changes in the weather. A prime example of such a disruption is the El Niño condition.

Under normal conditions, weather patterns in the Pacific are controlled by warm, high-pressure systems located in both hemispheres near the equator. These high-pressure systems cause the trade winds to blow westward along the equator, dragging the warm equatorial surface waters along with them. As warm surface waters move westward, deeper, colder waters to the east rise upward to occupy the space left vacant by the warm surface water. The upwelling cold waters, rich in nutrients, attract a variety of sea life. Upwelling of

these cold waters has been especially important to the fishing industry along the coast of South America, where people earn their living catching anchovies that come to feed in the nutrient-rich waters.

Fishing is not always good, however. Each year in October the trade winds slacken, reversing the normal westward flow of warm tropical surface waters. As the warm surface waters drift eastward, upwelling decreases and so does the fishing industry. People along the South American coast refer to this occurrence as El Niño because it begins each year around the traditional December celebration of Christ's birth (Christ is *El Niño*, "The Child," in Spanish). Under normal conditions, the trade winds pick up again in early spring, the surface waters are again blown westward across the ocean, and everything returns to normal.

In some years, however, the trade winds fail to strengthen and the warm surface waters remain off the coast of South America for a year or longer. During these abnormal conditions,

upwelling of cold water ceases, and South American fishing industries fail. Although a small El Niño occurs each year, this extended El Niño is referred to as the *El Niño condition*.

The El Niño condition influences climate on both sides of the tropical Pacific Ocean. Under normal conditions, upwelling cold water on the eastern side of the Pacific coincides with dry cool air, high pressures, and clear skies. On the western side of the Pacific, surface waters—warmed and fueled by their long journey across the ocean—warm the surrounding air. As the warm moist air rises, low pressures and storms develop on the warm western side of the Pacific.

During an extended El Niño condition, the pattern is reversed. Warm water, rising warm moist air, low pressures, and storms are found on the eastern side of the Pacific rather than the western side. This exchange of pressure systems and weather patterns between east and west during an El Niño condition is sometimes referred to as the *southern oscillation*.

The conveyor belt continues, as shown in Figure 24.42. But also note that some of the deep water makes it back to the ocean surface. A combination of deep-water mixing by ocean-floor tidal stirring and upwelling due to favorable winds brings the deep waters slowly back to the surface. Figure 24.43 shows the cycle from cold, sinking water to warmer surface flow. Eventually, all of the deep water flows to the ocean surface, and vice versa, although the process may take hundreds or thousands of years.

SUMMARY OF TERMS

Continental margin The boundary between the continents and the ocean; it consists of a *continental shelf*, a *continental slope*, and a *continental rise*.

Salinity The mass of salts dissolved in 1000 g of seawater.

Spring tide A high or low tide that occurs when the Sun, Earth, and Moon are aligned so that the tides due to the Sun and Moon coincide, making the tides higher or lower than average. Occurs during the full Moon or new Moon.

Neap tide A tide that occurs when the Moon is midway between new and full, in either direction. The pulls of the Moon and Sun are perpendicular to one another, so the solar and lunar tides do not overlap. This makes high tides not as high and low tides not as low.

Troposphere The atmospheric layer closest to Earth's surface, 16 km high over the equator and 8 km high over the poles, containing 90% of the atmosphere's mass and essentially all of its water vapor and clouds.

Stratosphere The second atmospheric layer above Earth's surface, extending from the top of the troposphere up to 50 km. This is where stratospheric ozone forms.

Mesosphere The third atmospheric layer above Earth's surface, extending from the top of the stratosphere to 80 km.

Thermosphere The fourth atmospheric layer above Earth's surface, extending from the top of the mesosphere to 500 km.

Ionosphere An electrified region within the thermosphere and uppermost mesosphere where fairly large concentrations of ions and free electrons exist.

Exosphere The fifth atmospheric layer above Earth's surface, extending from the thermosphere upward and out into interplanetary space.

Greenhouse effect Warming caused by short-wavelength radiant energy from the Sun that easily enters the atmosphere and is absorbed by Earth. This energy is then reradiated at longer wavelengths that cannot easily escape Earth's atmosphere.

Pressure-gradient force The force that moves air from a region of high-pressure air to an adjacent region of low-pressure air.

Coriolis force The apparent deflection from a straight-line path observed in any body moving near Earth's surface, caused by Earth's rotation.

Gyre A circular or spiral whirl pattern, usually referring to very large current systems in the open ocean.

REVIEW QUESTIONS

24.1 Earth's Atmosphere and Oceans

1. Why do we have moderate temperatures on lands bordering the oceans?
2. What were the main components of Earth's first atmosphere? What happened to this atmosphere?
3. Earth's present atmosphere likely developed from gases that escaped from its interior during volcanic eruptions. What three principal atmospheric gases were produced by these eruptions?
4. Explain the importance of photosynthesis in the evolution of the atmosphere.

24.2 Components of Earth's Oceans

5. What are four features we may find between two continental margins (the ocean floor)?
6. Salinity of the ocean varies from one place to another. What two factors lead to an increase in salinity? What two factors lead to a decrease in salinity?

24.3 Ocean Waves, Tides, and Shorelines

7. Why do waves become taller as they enter shallow water?
8. What is wave refraction? Why does it occur in ocean waves?
9. Why is a barrier island's lagoon usually a quiet environment?
10. Why are all tides greatest at the time of a full or new Moon?
11. When would the highest high tides occur: during a spring tide or during a neap tide?

24.4 Components of Earth's Atmosphere

12. What elements make up today's atmosphere?
13. Why doesn't gravity flatten the atmosphere against Earth's surface?
14. In which atmospheric layer does all our weather occur?
15. Does temperature increase or decrease as one moves upward in the troposphere? As one moves upward in the stratosphere?

24.5 Solar Energy

16. What does the angle at which sunlight strikes Earth have to do with the temperate and polar regions?
17. What does Earth's tilt have to do with the change of seasons?
18. Why are the hours of daylight equal all around the world on the two equinoxes?
19. How does radiation emitted from Earth differ from that emitted by the Sun?
20. How is the atmosphere near Earth's surface heated from below?

24.6 Driving Forces of Air Motion

21. What is the underlying cause of air motion?
22. What causes pressure differences to arise, and hence causes the wind to blow?

23. In what direction does Earth spin—west to east or east to west?
24. What does the Coriolis force do to winds? To ocean currents?
25. How does the Coriolis force determine the general path of air circulation?

24.7 Global Circulation Patterns

26. Why are most of the world's deserts found in the area known as the horse latitudes?
27. What are the trade winds?
28. Why are eastbound aircraft flights usually faster than westbound ones?
29. What factors set surface ocean currents into motion?
30. Explain the circulation pattern of the Gulf Stream.

EXERCISES

● BEGINNER ■ INTERMEDIATE ◆ EXPERT

Please do not be intimidated by the large number of exercises and problems in this and other meatier chapters. If your course work is to cover many chapters, your instructor will likely assign only a few exercises and/or problems from each.

1. ■ If a gas fills all the space available to it, why doesn't the atmosphere go off into space?
2. ■ If there were no water on Earth's surface, would weather occur? Defend your answer.
3. ■ How do the wavelengths of radiant energy vary with the temperature of the radiating source? How does this affect solar and terrestrial radiation?
4. ● How is global warming affected by the relative transparencies of the atmosphere to long- and short-wavelength electromagnetic radiation?
5. ■ Explain why your ears pop when you ascend to higher altitudes.
6. ◆ How does the density of air in a deep mine compare with the density of air at sea level? Defend your answer.
7. ● Earth is closest to the Sun in January, but January is cold in the Northern Hemisphere. Why?
8. ■ How do the total number of hours of sunlight in a year compare for equatorial regions and polar regions of Earth? Why are polar regions so much colder?
9. ● If the composition of the upper atmosphere were changed so that it permitted a greater amount of terrestrial radiation to escape, what effect would this have on Earth's average temperature? How about if the atmosphere reduced the amount of terrestrial radiation that could escape?
10. ■ As humans consume more energy, the average temperature of Earth's surface tends to rise. Yet no matter how much energy is consumed, the temperature doesn't rise without limit. By what process is an indefinite rise prevented? Explain.
11. ● As the world's population increases, the amount of carbon dioxide emissions from fossil fuel combustion also increases. Yet the amount of carbon dioxide emitted

is greater than the amount found in the atmosphere. Where is the likely repository of excess atmospheric carbon dioxide?
12. ■ Why is it important that mountain climbers wear sunglasses and use sunblock even when the temperature is below freezing?
13. ■ If Earth were not spinning, in what direction would the surface winds blow where you live? In what direction does it actually blow on Earth at 15°S latitude and why?
14. ◆ What is the relationship between global atmospheric circulation and ocean currents? Relate oceanic gyres to patterns of subtropical high pressure.
15. ● Relate the jet stream to upper-air circulation. How does this circulation pattern relate to airline schedules from New York to San Francisco and the return trip to New York?
16. ■ What are the jet streams and how do they form?
17. ■ Why are temperature fluctuations greater over land than over water? Explain.
18. ◆ Because seawater does not freeze easily, sea ice never gets very thick. So from where do large icebergs originate?
19. ● The Mediterranean Sea is highly salty. What can you say about the relative rates of evaporation and precipitation over the Mediterranean?
20. ● What role does the Sun play in the circulation of ocean currents?
21. ● How does the ocean influence weather on land?
22. ■ What happens to the salinity of seawater when evaporation at the ocean surface exceeds precipitation? When precipitation exceeds evaporation? Defend your answer.
23. ◆ Why is there more concern about melting of polar ice caps than there is about melting icebergs?
24. ■ Water denser than surrounding water sinks. With respect to the densities of deeper water, how far does it sink?
25. ■ As a volume of seawater freezes, the salinity of the surrounding water increases. Explain.

26. ● Why are headland areas prime areas for erosion?

27. ◆ Suppose a breakwater is built offshore and perpendicular to the shore. How will this structure affect the longshore current and its transport of sand? Defend your answer.

28. ● Carbonate rocks are mainly formed in marine environments. Why do we find abundant carbonate deposits on continental land?

29. ● Why is the sand of some beaches composed of small pieces of seashells?

30. ◆ The oceans are composed of salt water, yet evaporation over the ocean surface produces clouds that precipitate fresh water. Why no salt?

31. ■ As waves approach shallow water, those with longer wavelengths slow down before those with shorter wavelengths do. Why?

32. ■ What effect does the formation of sea ice in polar regions have on the density of seawater? Explain.

33. ● Which receive more solar energy over the course of a year—tropical regions or temperate regions? How does this affect ocean salinity?

34. ● In tropical regions, solar energy exceeds terrestrial radiation. What effect does this have on the salinity of oceans in tropical regions?

35. ◆ Why do the temperate zones have unpredictable weather?

36. ◆ What happens to the water level in a glass of water when a floating ice cube in the glass melts? Similarly, what happens to the water level in the Great Lakes when floating chunks of ice melt?

37. ◆ Most people today know that the ocean tides are caused principally by the gravitational influence of the Moon. And most people therefore think that the gravitational pull of the Moon on Earth is greater than the gravitational pull of the Sun on Earth. What do you think?

38. ● Would ocean tides exist if the gravitational pull of the Moon (and Sun) were somehow equal on all parts of the world? Explain.

39. ■ Why aren't high ocean tides exactly 12 hours apart?

40. ● With respect to spring and neap ocean tides, when are the lowest tides? That is, when is it best for digging clams?

41. ◆ When the ocean tide is unusually high, is the following low tide unusually low? Defend your answer in terms of "conservation of water." (If you slosh water in a tub so it is extra deep at one end, is the other end extra shallow?)

42. ◆ How does the Coriolis force influence the movement of surface waters?

43. ■ What is the characteristic climate of the doldrums and why does it occur?

44. ● Because our atmosphere developed as a result of volcanic eruptions, why aren't there higher traces of atmospheric carbon dioxide, one of the principal volcanic gases?

45. ● Why is the thermosphere so much hotter than the mesosphere?

46. ● What is the source of the ions that give the ionosphere its name?

47. ● If it is winter and January in Chicago, what are the corresponding season and month in Sydney, Australia?

48. ● What causes the fiery displays of light called the auroras?

49. ◆ Explain why most of the bottom water of the oceans forms in the North Atlantic and near Antarctica.

50. ● How does the density of seawater vary with changes in temperature? How does density change with salinity?

PROBLEMS

● BEGINNER   ■ INTERMEDIATE   ◆ EXPERT

Refer to the Figuring Physical Science box on page 643 to answer the following two problems.

1. ● What is the mass in kilograms of the air in an "empty" nonpressurized scuba tank that has an internal volume of 0.0100 m^3?

2. ● What is the mass in kilograms of the air in a scuba tank that has an internal volume of 0.0100 m^3 and is pressurized so that the density of the air in the tank is 240 kg/m^3?

READINESS ASSURANCE TEST (RAT)

If you have a good handle on this chapter, if you really do, then you should be able to score at least 7 out of 10 on this RAT. If you score less than 7, you need to study further before moving on.

Choose the BEST answer to each of the following.

1. Earth's lower atmosphere is kept warm by
   (a) solar radiation.
   (b) terrestrial radiation.
   (c) shortwave radiation.
   (d) the ozone layer.

2. When compared to lands far from the oceans, lands that border the oceans tend to have
   (a) extreme seasonal variations.
   (b) small seasonal variations.
   (c) wet and cold winters.
   (d) dry hot summers.

3. Which pulls with the greater force on Earth's oceans?
   (a) the Sun
   (b) the Moon
   (c) both the Sun and the Moon; they have equal force
   (d) there is no force

4. Going *from* continental land *toward* the deep ocean basin, the continental margin consists of
(a) the slope, the shelf, and the rise.
(b) the rise, the slope, and the shelf.
(c) the shelf, the slope, and the rise.
(d) the shelf, the abyssal plain, and the midocean ridge.

5. Ocean waves refract because of
(a) wave interference.
(b) ocean swell.
(c) shallow water.
(d) longshore current.

6. A factor that increases ocean salinity is
(a) runoff from streams and rivers.
(b) formation of sea ice.
(c) precipitation.
(d) glacial melting.

7. The wind blows in response to
(a) frictional drag.
(b) Earth's rotation.
(c) pressure differences.
(d) moisture differences.

8. Planet Earth experiences changes of the seasons because of
(a) incoming solar radiation.
(b) Earth's rotation around the Sun.
(c) the Coriolis force.
(d) the tilt of Earth's axis.

9. The most significant result of the Coriolis force is
(a) the deflection of air and water currents.
(b) the creation of Ekman transport.
(c) the reflection of air currents and global air circulation.
(d) the creation of the jet streams.

10. The ultimate cause of ocean surface currents is
(a) divergence in equatorial regions.
(b) the gradient between the doldrums and the horse latitudes.
(c) density contrasts.
(d) frictional drag by prevailing winds.

Answers to RAT

1b, 2b, 3a, 4c, 5c, 6b, 7c, 8d, 9a, 10d

EXPLORING FURTHER

http://www.noaa.gov/
The National Oceanic and Atmospheric Administration home page is the place to find up-to-the-minute information about climate and weather. Full of articles, satellite photographs, maps, and links, this site should not be missed!

http://www.ncar.ucar.edu/ncar/
The home page for the National Center for Atmospheric Research is an excellent resource for detailed information about climate and weather, including climate modeling and weather prediction.

CHAPTER 24 ONLINE RESOURCES

The Physical Science Place

Interactive Figures

- 24.8, 24.9

Tutorials

- Vertical Structure of the Atmosphere
- Seasons
- The Coriolis Effect

Quiz

Flashcards

Links

driving forces of weather

25.1 ATMOSPHERIC MOISTURE

25.2 WEATHER VARIABLES

25.3 CLOUD DEVELOPMENT

25.4 AIR MASSES, FRONTS, AND STORMS

25.5 VIOLENT WEATHER

25.6 THE WEATHER—THE NUMBER ONE
TOPIC OF CONVERSATION

■ Surrounding our Earth is a life-giving shell of air—the atmosphere. Its mixture of gases provides us with air to breathe and protection from harmful ultraviolet radiation. In Chapter 24, we learned about the atmosphere's vertical structure and how most of the atmosphere's mass resides in the lowest layer—the troposphere. This is where Earth's weather occurs. And that brings us to our present topic—the weather. The factors that influence the weather are atmospheric moisture, temperature, air pressure, and the geographic arrangement of land and water features. We begin our discussion of weather by looking first at atmospheric moisture and at how the amount of moisture in the air influences atmospheric stability. Then we discuss the development of different air masses and some resulting weather patterns. We conclude with a look at the violent weather forces that greatly affect our planet's surface.

■ 25.1 Atmospheric Moisture

Link To Sections 18.4, 23.1

Link To Section 7.5

Water is certainly vital to life on Earth. But think for a moment about water's immense role in the physical processes of Earth—water shapes Earth's surface and governs its weather. And no matter how "dry" air may feel at times, it is never completely dry—some concentration of water vapor is always present.

But instead of saying "concentration of water vapor," we use the much shorter term **humidity**. Specifically, humidity is the mass of water vapor per volume of air, just as concentration is the mass of solute per volume of solution. It is important to note, however, that when you hear TV weather forecasters describing humidity, they are actually talking about **relative humidity**. Relative humidity depends on temperature—it describes the amount of water vapor currently in the air compared to the maximum amount of water vapor that could be in the air at that specific temperature. The maximum amount varies depending on the temperature.

$$\text{Relative humidity} = \frac{\text{water-vapor content}}{\text{water-vapor capacity}} \times 100\%$$

For example, a relative humidity of 50% means that the water-vapor content in the air is half the amount of the air's maximum capacity at that temperature.

Think of water-vapor capacity in the same way that you think of mineral solubility—the solubility of salt in water, for example. At a given temperature, the concentration of salt dissolved in water has an upper limit—the salt's *solubility* (Chapters 16 and 20). The solubility concentration cannot be exceeded—add more salt and it does not dissolve.

The same is true for the concentration of water vapor in air. At a given temperature, the humidity has an upper limit that cannot be exceeded, just as each mineral has a solubility limit. The humidity limit is called the **saturation vapor pressure**.* When air reaches its saturation vapor pressure, the humidity (or vapor pressure) cannot increase, and the relative humidity is 100%. The air is said to be *saturated*.

Warm air has a higher saturation vapor pressure than cold air (Table 25.1). Thus, when saturated air at 10°C is warmed to 20°C—and no water vapor is added—the air is no longer saturated and the relative humidity drops below 100%—to about 52% in this case. Even though the *relative* humidity dropped, the *actual* humidity—the concentration of water vapor in the air—stays the same. It is just that the warm air has a higher saturation vapor pressure and can, in a sense, "hold" more water vapor.

The saturation vapor pressure varies with temperature because there are two competing processes—the evaporation rate and the condensation rate (Chapter 7). The evaporation rate depends on temperature, but the condensation rate does not—it depends only on the humidity (Figure 25.1). When the evaporation rate equals the condensation rate—at any temperature—air is saturated, and the relative humidity is 100%.

---

* *Vapor pressure* measures the same thing as humidity—the amount of water vapor in air. Humidity and vapor pressure describe the same thing, but they have different units—grams per cubic meter for humidity and millibars (or millimeters of mercury) for vapor pressure.

TABLE 25.1   SATURATION VAPOR PRESSURE VS. TEMPERATURE

Temperature (°C)	Temperature (°F)	Saturation Vapor Pressure (millibars, mb)
0	32	6.1
5	41	8.7
10	50	12.3
15	59	17.0
20	68	23.4
25	77	31.7
30	86	42.5
35	95	56.3

Interestingly, evaporation and condensation are always happening—neither ever stops. When the evaporation rate is greater than the condensation rate, the relative humidity is always less than 100%. The greater the difference between the evaporation rate and the condensation rate, the lower the relative humidity, and the drier the air feels.

On the other hand, if the condensation rate is greater than the evaporation rate, the air is saturated *and* there is an excess of water vapor. In such a case, the excess water vapor condenses to form liquid water, because the evaporation rate is not fast enough to turn the freshly created liquid water back to vapor before more water condenses. Because slower-moving molecules characterize lower air temperatures, saturation and condensation are more likely to occur in cool air than in warm air (Figure 25.2).

In the example given on page 664, at 10°C the evaporation rate was the same as the condensation rate, so the relative humidity was 100%. When the temperature increased from 10°C to 20°C the condensation rate did not change, but the evaporation rate increased. As the temperature, evaporation rate, and saturation vapor pressure all increased, the air could accommodate more water vapor.

Saturation Vapor Pressure

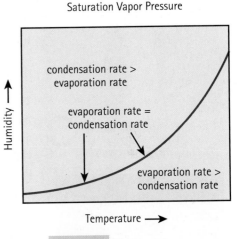

FIGURE 25.1

The curve in this figure is the saturation vapor pressure as a function of temperature and humidity (not relative humidity). On the curve, relative humidity is 100% and the evaporation rate equals the condensation rate. Below the curve, the relative humidity is less than 100% and the evaporation rate is greater than the condensation rate. Above the curve, the condensation rate is greater than the evaporation rate, and precipitation forms. This curve helps explain why hot summer days can feel very muggy. Your skin cannot evaporate water so easily. From the curve, can you see that warm air has a higher saturation vapor pressure than cold air?

Fast-moving H₂O molecules rebound upon collision

Slow-moving H₂O molecules condense upon collision

FIGURE 25.2

Condensation of water molecules.

fyi

- In rainy weather, when your car windshield fogs up, make sure the air conditioner is on when in defrost mode, even when blowing heated air. What causes window fogging is the humidity in the car caused by rain, by wet clothes, and by the breath of the passengers. Because the air from the air conditioner is very dry, it clears a foggy windshield very nicely in a very short time.

Link To Section 26.3

FIGURE 25.3

San Francisco is well known for its summer fog.

The relative humidity also dropped, because no water vapor was added—the air was no longer saturated at the higher temperature.

What would happen if we started with 20°C air and a relative humidity of 60%, then lowered the temperature to 10°C? As the temperature decreased, the evaporation rate would slow down, but water vapor would continue to condense at the same rate as it did when the air was 20°C. The saturation vapor pressure and the evaporation rate would both go down, so the relative humidity would go up. In fact, when the air temperature reaches 12°C, the evaporation rate and the condensation rate would be equal and the relative humidity would be 100%—the air would be saturated. When the temperature drops below 12°C, condensation rules and liquid water droplets form—rain.

Evaporation and condensation are essentially opposites of each other. They are phase-change processes between the liquid and gaseous states of water. And as we know, liquid water changes to a solid—it freezes—at 0°C. Water can also change phase from solid to gas—and from gas to solid—without becoming liquid in between. When ice changes phase directly to water vapor, the process is called *sublimation* (Chapter 7). The opposite process—water vapor changing directly to ice—is called *deposition*.*

TEMPERATURE CHANGES AND CONDENSATION

As an air parcel rises, it expands. The expansion occurs because the air moves to a region of lower air pressure. As we learned in Chapter 7, air cools when it expands. This seems like a contradiction, because warm air rises. But a rising air parcel must do work on the surrounding environment in order to expand. And doing that work uses energy, which is lost in the form of heat. As the air cools, water molecules move slower and condensation outpaces evaporation. Liquid water forms on microscopic particles of dust, smoke, and salt—*cloud condensation nuclei*—and this creates a cloud. As the sizes of the cloud droplets grow, they fall to Earth and we have rain. Rain is but one form of precipitation. Other familiar forms of precipitation are mist, hail, snow, and sleet.

Water vapor does not need to be high in a cloud to form precipitation—condensation can occur in air close to the ground, as well. When condensation occurs at or near Earth's surface, we call it *dew, fog,* or *frost*. On cool, clear nights, objects near the ground cool down more rapidly than the surrounding air. As the air cools, the saturation vapor pressure drops and the air cannot accommodate as much water vapor as when the air was warmer. When the temperature drops below a certain threshold, called the *dew point temperature*, the air becomes saturated—relative humidity is 100%—and condensation dominates. In this case, cloud condensation nuclei are not needed. Water from the now-saturated air condenses on any available surface—a twig, a blade of grass, the windshield of a car, and so on. We often call this type of condensation *early-morning dew*, because it occurs when daily temperatures are the coldest, just before sunrise. When the dew point is at or below water's freezing point, we have frost.** When a large mass of air cools and reaches its dew point, we get a cloud near the ground—fog.

CHECK POINT

What is the major difference between fog and a cloud?

Was this your answer?
Altitude.

---

* The term *deposition* in meteorology refers to the conversion of water vapor to a solid. This is different from the way we use this term in Chapters 20 and 23, where deposition refers to the laying down of sediment.

** Interestingly, frost is not frozen dew. Frost forms directly from water vapor as a result of *deposition*.

FIGURING PHYSICAL SCIENCE

▪Humidity

Humidity is the mass of water vapor per volume of air. Relative humidity is the ratio of the air's water vapor content compared with the air's water vapor capacity at a certain temperature.

For these three problems, consider a small, experimental air mass at 30°C that weighs 90 N.

SAMPLE PROBLEM 1

If the air density is 1.25 kg/m³, what is the volume of the air mass?

SOLUTION:

Newton's second law tells us that

$$a = \frac{F}{m}$$

Rearrange the equation and get

$$\text{Mass} = \frac{\text{force}}{\text{acceleration due to gravity}} = \frac{90\text{ N}}{10\text{ m/s}^2} = 9\text{ kg}$$

The volume of 9 kg can be found in this way:

$$9\text{ kg} \times \frac{1\text{ m}^3}{1.25\text{ kg}} = 7.2\text{ m}^3$$

SAMPLE PROBLEM 2

If there is 0.13 kg of water vapor in the air mass, what is the humidity of the air mass?

SOLUTION:

$$\text{Humidity} = \frac{\text{mass of water}}{\text{volume of air}} = \frac{0.13\text{ kg}}{7.2\text{ m}^3} = 0.018\text{ kg/m}^3$$

SAMPLE PROBLEM 3

At 30°C, the maximum amount of water vapor in the air mass is 30 g/m³ of water vapor. What is the relative humidity of the air mass?

SOLUTION:

First, we must convert the units:

$$\frac{30\text{ g}}{\text{m}^3} \times \frac{1\text{ kg}}{1000\text{ g}} = 0.03\text{ kg/m}^3$$

Then

Relative humidity

$$= \frac{\text{amount of water vapor in air}}{\text{maximum amount of water vapor in air at 30°C}}$$

$$= \frac{0.018\text{ kg/m}^3}{0.03\text{ kg/m}^3} \times 100\% = 60\%$$

▪ 25.2 Weather Variables

Air pressure, temperature, and density are three key variables that control how air behaves, and hence, they control the weather. To understand and predict the weather, we must understand all three. First, consider air pressure. Air is a mixture of molecules that move randomly and collide with one another like billiard balls on a pool table. When a molecule bumps into something, it exerts a small push on whatever it hits. Such pushes by countless molecules produce *air pressure*.

The faster the air molecules move, the greater their kinetic energy. The greater the kinetic energy, the greater the impact of molecular collisions—and the greater the air pressure. All else being equal, air composed of fast-moving molecules—warm air—exerts more air pressure on its surroundings than cooler air.

Another factor that affects air pressure is density. The denser the air, the more molecules present and hence the greater the number of molecular collisions. And more collisions means greater air pressure. Air becomes denser when it is compressed, and it becomes less dense when it expands. The changes in density occur because the volume of a given mass of air is made smaller by compression and larger by expansion.

↔ The Physical Science Place ↘
Surface Temperature of Planets

↻ Link To Section 6.5

ADIABATIC PROCESSES IN AIR

The concept of *heat exchange* shows us that air pressure, temperature, and density are interrelated. When heat is added to an air mass, the air temperature

The Physical Science Place ↘
Adiabatic Process

FIGURE 25.4

The temperature of a parcel of dry air that expands adiabatically changes by about 10°C for each kilometer of elevation.

Link To Section 5.5

increases, the air pressure increases, or both increase. Heat can be added to air by solar radiation, by moisture condensation, or by contact with warm ground. When heat is subtracted from an air mass, the temperature or the pressure of the air falls. Heat can be subtracted from air by radiation to space, by the evaporation of rain falling through dry air,* or by contact with cold surfaces.

But air can change temperature without the loss or gain of heat. When heat transfer is zero, or nearly so, we have an **adiabatic** process. Adiabatic processes occur when air is expanded or compressed. For adiabatic temperature changes to occur, expansion or compression must be rapid enough that there is insufficient time for heat exchange to occur while the volume change is occurring. To illustrate how bodies of air behave, imagine a body of air enclosed in a very thin plastic garment bag—an *air parcel*. Like a free-floating balloon, the parcel can expand and contract freely without heat transfer to or from air outside the parcel.

Recall from Chapter 24 that air pressure decreases with increasing height. For example, as an air parcel flows up the side of a mountain, the pressure exerted on it decreases, allowing the air parcel to expand—and cool without any heat exchange. Such changes in temperature, caused by changes in pressure, are described by the *ideal gas law*:**

$$\text{Pressure} \sim \text{density} \times \text{temperature}$$

Assuming relatively small changes in air density, which is generally the case in the lower atmosphere, the ideal gas law states that temperature goes down when pressure goes down. And we know that air pressure goes down when air expands. Conversely, temperature goes up when pressure goes up—air pressure goes up when air is compressed. Can you now see more clearly that the temperature of an air parcel decreases when lifted? And that air temperature increases when an air parcel descends? According to the ideal gas law, such temperature changes do not require the loss or gain of heat—they are adiabatic.

With adiabatic expansion, the temperature of a dry air parcel[†] decreases about 10°C for each kilometer it rises (Figure 25.4). This rate of cooling for dry air is called the *dry adiabatic lapse rate*. Air flowing up and over tall mountains or rising in thunderstorms may change elevation by several kilometers. Thus, when a dry air parcel at ground level at a comfortable 25°C rises 6 km, its temperature drops to a frigid –35°C! On the other hand, if air at 6 km elevation, at a typical temperature of –20°C, descends to the ground, its temperature rises to a whopping 40°C!

Adiabatic processes are not restricted to dry air. As rising air cools, its ability to accommodate water vapor decreases, increasing the relative humidity of the rising air. If the air cools to its dew point, the relative humidity climbs to 100%, water vapor condenses, and clouds form. Because condensation releases heat (Chapter 7), the air parcel is warmed. The heat counteracts to some extent the cooling due to expansion, making the air cool at a lesser rate—the *moist adiabatic lapse rate*. Although the moist adiabatic lapse rate varies according to the temperature and the moisture content of air, on average a saturated air parcel cools by about 6°C for every kilometer it rises.

A dramatic example of adiabatic warming is the *Chinook*—a dry wind that blows down from the Rocky Mountains across the Great Plains (Figure 25.5). Cold

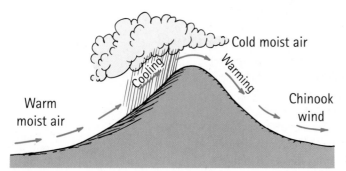

FIGURE 25.5

Chinooks—which are warm, dry winds—occur when high-altitude air descends and is adiabatically warmed.

* As we learned in Chapter 7, evaporation removes heat from the surrounding environment.
** The ideal gas law is related to Boyle's law, which was discussed in Chapter 5.
† A dry air parcel is an air parcel that is not saturated—its relative humidity is less than 100%.

air moving down a mountain slope is compressed as it moves to lower elevations (where air pressure is greater than at higher elevations), and it becomes much warmer. The effect of expansion or compression of gases is quite impressive.*

ATMOSPHERIC STABILITY

Let's consider a parcel of air descending from aloft. Air pressure at lower elevations is greater than air pressure aloft, so the air parcel is compressed as it descends, causing it to warm up. If the descending air parcel's temperature becomes warmer—and therefore less dense—than air at the lower elevation, the parcel will rise back to the elevation from which it came. Analogously, think of what happens when you try to submerge a flotation device, which is less dense than water, in a swimming pool. The flotation device pops right back to the surface. The flotation device and the descending air parcel in this example are both *stable*, with respect to their environments. They "want" to return to their original positions.

Now consider an air parcel that is forced upward. Such an air parcel expands as it rises, becoming cooler. If the temperature of the rising air becomes cooler than the surrounding air, it becomes denser and sinks back to the elevation where it started. This is also stable air.

On the other hand, if the rising air stays warmer than the surrounding air, it continues to rise instead of returning to its starting position. This is *unstable air*. Eventually, the air parcel expands and cools sufficiently to match the surrounding air. When the temperatures match, the air parcel rises no farther, but it does not sink back to its starting position. So unstable air rises, but stable air does not. Hot air balloons use this strategy to rise or descend to a desired elevation. Heat is added to the air in the balloon to make it rise, and the air inside the balloon is allowed to cool to make it descend.

The surrounding air plays an important role in determining whether an air parcel is stable or unstable. Under normal conditions, air temperature decreases with altitude. The *environmental lapse rate* describes the manner in which these temperature changes occur with altitude. A lower environmental lapse rate indicates that temperatures decrease less for each meter of increased altitude than with a higher environmental lapse rate. The environmental lapse rate varies from place to place and from day to day—it can even vary over the course of a day! The average environmental lapse rate is a decrease of 6.5°C for each kilometer rise in elevation.

A rising parcel of air continues to rise as long as it is warmer and less dense than the surrounding air. Air that is cooler and denser than its surroundings does the opposite—it sinks. Under some conditions, large parcels of cold air sink and remain at low elevations. This results in air above that is warmer. When upper regions of the atmosphere are warmer than lower regions, which is opposite of what normally occurs, we have a **temperature inversion**. In such cases, most rising air parcels can't pass through the upper layer of warmer air, because the rising air is cooler and denser. On a small scale, evidence of a temperature inversion is commonly seen over a cold lake when visible gases and small particles such as smoke spread out in a flat layer above the lake rather than rising and dissipating higher in the atmosphere (Figure 25.6).

fyi

- To see that expanding air cools, blow on your hand with your mouth wide open. Notice the warmth of the air. Then repeat this, but pucker your lips so that your mouth opening is very small. This time, your breath expands as it leaves your mouth. Aha! Cool!

Link To Section 7.2

FIGURE 25.6

The layer of campfire smoke over the lake indicates a temperature inversion. The air above the smoke is warmer than the smoke, and the air below is cooler.

* When you're flying at high altitudes where the outside air temperature is typically −35°C, you're quite comfortable in the warm cabin—but not because of heaters. The process of compressing outside air to maintain a cabin pressure that is nearly the same as the air pressure at sea level normally heats the air to a roasting 55°C (131°F). So air conditioners must be used to extract heat from the pressurized air.

Smog in Los Angeles is trapped by the mountains and a temperature inversion caused by warm air from the Mojave Desert overlying cool air from the Pacific Ocean.

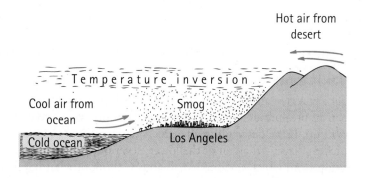

fyi

- Stable air that is forced to rise spreads out horizontally. When clouds develop in stable air, they too spread out into thin horizontal layers having flat tops and bottoms. Unstable air favors upward movement. When unstable rising air is moist, billowy and towering clouds develop.

The smog of Los Angeles is trapped by such an inversion, caused by cold air from the ocean being capped by a layer of hot air moving westward over the mountains from the hot Mojave Desert. The west-facing side of the mountains helps confine the trapped air (Figure 25.7). The Rocky Mountains on the western edge of Denver play a similar role in trapping smog beneath a temperature inversion.

CHECK POINT

1. If a parcel of dry air initially at 0°C expands adiabatically while flowing upward alongside a mountain, what is its temperature when it has risen 2 km? When it has risen 5 km?
2. What happens to the air temperature in a valley when dry, cold air blowing across the mountains descends into the valley?
3. When air is stable it resists vertical movement. When air is unstable, it wants to rise to a stable level—where its air temperature equals the temperature of its surroundings. In the example of a temperature inversion, is the air stable or unstable?

Were these your answers?

1. The air cools at the dry adiabatic lapse rate of 10°C for each kilometer it rises. When the parcel rises to an elevation of 2 km, its temperature is −20°C. At an elevation of 5 km, its temperature is −50°C.
2. The air is adiabatically compressed, and so its temperature increases. Residents of some valley towns in the Rocky Mountains, such as Salida, Colorado, benefit from this adiabatic compression and enjoy "banana belt" weather in midwinter.
3. A temperature inversion occurs when warmer air overlies cooler air. Air in this situation is very stable—the denser, cooler air cannot rise above the less dense, warmer air, and so it resists vertical movement.

■ 25.3 Cloud Development

When an unstable air parcel rises, the relative humidity increases as the air cools at the dry adiabatic lapse rate. At a certain elevation, the relative humidity reaches 100% and the air parcel is saturated. This key elevation is the **lifting condensation level**.

As the air parcel continues to rise, air now cools at the moist adiabatic rate. But more important, the condensation rate exceeds the evaporation rate above the lifting condensation level—cloud droplets begin to form. The lifting condensation level marks the base of cloud formation. The air parcel continues to

Link To Section 11.2

TABLE 25.2   THE FOUR MAJOR CLOUD GROUPS

1. High clouds (above 6000 m)	3. Low clouds (below 2000 m)
Cirrus	Stratus
Cirrostratus	Stratocumulus
Cirrocumulus	Nimbostratus
2. Middle clouds (2000–6000 m)	4. Clouds having vertical development
Altostratus	Cumulus
Altocumulus	Cumulonimbus

The altitude range of the major cloud groups varies somewhat with season and latitude. Also, some clouds extend vertically in more than one altitude range.

rise until it has cooled enough to match the temperature of the surrounding air. At this elevation, the *equilibrium level*, the air parcel stops rising and the condensation rate equals the evaporation rate—the air parcel is now stable. The equilibrium level marks the upper limit of cloud formation.

The elevation of the equilibrium level depends on the environmental lapse rate. Because the environmental lapse rate describes temperature changes of the surrounding air, the elevation at which a rising air parcel stabilizes can be quite variable.

To summarize, a rising air parcel cools at the dry adiabatic lapse rate until it reaches saturation. After saturation, the moist adiabatic lapse rate controls how thick the cloud becomes. Therefore, the height of the cloud base and how thick the cloud becomes depends on three variables: the environmental lapse rate, the dry adiabatic lapse rate, and the moist adiabatic lapse rate.

Clouds are classified according to their altitude and shape. There are ten principal cloud forms, each of which belongs to one of four major groups (Table 25.2).

> Clouds don't float! They are buoyed up by an invisible conveyor belt of air. Clouds are always moving up.

> Link To Section 7.2

CHECK POINT

To what does the environmental lapse rate refer?

Was this your answer?
The environmental lapse rate refers to the change of air temperature with altitude. It describes the temperature changes of the air that a rising air parcel passes through. The environmental lapse rate varies with local conditions. On average, the rate is about 6.5°C for each kilometer rise in elevation.

HIGH CLOUDS

High clouds are clouds that form at altitudes above 6000 m. High clouds (other than cirrus clouds) are denoted by the prefix *cirro-*. The air at this elevation is quite cold and dry, and so clouds this high are made up almost entirely of ice crystals.

The most common high clouds are thin, wispy *cirrus* clouds. Cirrus clouds are blown by high winds into their well-known wispy shapes, such as the classic "mare's tail" or "artist's brush." Cirrus clouds usually indicate fair weather, but they may also indicate approaching rain.

Cirrocumulus clouds are the familiar rounded white puffs. They are found in patches, and they seldom cover more than a small portion of the sky. Small ripples and a wavy appearance make cirrocumulus clouds look like the markings

fyi

- Montana is called "Big Sky Country" because of the high lifting condensation level. The dry climate means that cloud bases are usually high in the sky—hence the sky looks "bigger."

FIGURE 25.8

The four cloud groups.
(a) High clouds: cirrus, cirrostratus, cirrocumulus.
(b) Middle clouds:
altostratus, altocumulus.
(c) Low clouds: stratus,
stratocumulus, nimbostratus.
(d) Clouds with vertical
development: cumulus,
cumulonimbus.

(a)

(b)

(c)

(d)

Link To Section 11.5

on the body of a mackerel. Hence, cirrocumulus clouds are often said to make
up a *mackerel sky*.

Cirrostratus clouds are thin and sheetlike, and they often cover the entire sky.
The ice crystals in these clouds refract light and produce a halo around the Sun
or the Moon. When cirrostratus clouds thicken, they give the sky a white, glary
appearance—an indication of coming rain or snow.

MIDDLE CLOUDS

Middle clouds form at altitudes between 2000 and 6000 m. Middle clouds are
denoted by the prefix *alto-*. These clouds are composed of water droplets and,
when temperature allows, ice crystals.

Altostratus clouds are gray to blue-gray, and they often cover the sky for hundreds of square kilometers. Altostratus clouds are often so thick that they diffuse
incoming sunlight to the extent that objects on the ground don't produce shadows. Altostratus clouds often form before a storm. Look at the ground the next
time you're going on a picnic. If you don't see your shadow, cancel your plans!

Altocumulus clouds appear as gray, puffy masses in parallel waves or bands. The
individual puffs are much larger than those found in cirrocumulus clouds, and
the color is also much darker. The appearance of altocumulus clouds on a warm,
humid summer morning often indicates thunderstorms by late afternoon.

LOW CLOUDS

Low clouds ranging from the surface up to 2000 m are called *stratus* clouds.
They are almost always made up of water droplets, but in cold weather they may
also contain ice crystals and snow. Stratus clouds are uniformly gray, and they

often cover the whole sky. They are very common in winter, and they account for the sky's "hazy shade of winter." They resemble a high fog that doesn't touch the ground. Although stratus clouds are not directly associated with falling precipitation, they sometimes generate a light drizzle or mist.

Stratocumulus clouds either form a low, lumpy layer that grows in horizontal rows or patches or, with weak rising motion, appear as rounded masses. Their color is generally light to dark gray. To tell the difference between altocumulus clouds and stratocumulus clouds, hold your hand at arm's length and point toward the cloud in question. An altocumulus cloud commonly appears to be the size of your thumbnail; a stratocumulus cloud appears to be about the size of your fist. Precipitation of rain or snow is not usually produced by stratocumulus clouds.

Nimbostratus clouds are dark and foreboding. They are a wet-looking cloud layer associated with light to moderate rain or snow.

CLOUDS WITH VERTICAL DEVELOPMENT

Cumulus clouds are the most familiar of the many cloud types. Cumulus clouds resemble pieces of floating cotton, with sharp outlines and flat bases. They are white to light gray, and they generally occur about 1000 m above the ground. The tops of cumulus clouds are often in the form of rising towers, showing the upward limit of the rising air. These are the clouds childhood daydreams are made of. Did you ever see castles or the shapes of animals in the clouds?

When cumulus clouds turn dark and are accompanied by precipitation, they are referred to as *cumulonimbus* clouds. In this case, they indicate a coming storm. As we shall see, cumulonimbus clouds often become *thunderheads*.

PRECIPITATION FORMATION

Several things have to happen for precipitation to form. Each progressive step toward precipitation is part of the *collision-coalescence process*. The first requirement is the presence of dust—the condensation nuclei discussed earlier in this chapter.

Water vapor is less dense than air. But once cloud droplets form, the droplets are considerably denser than the air. The gravitational force pulling the droplets downward is enough to make them fall. So why don't all the water droplets in a cloud fall to the ground? The answer involves *updrafts*—the upward movement of air. A typical cumulus cloud has an updraft speed of at least 1 m/s, which is faster than the droplets can fall. So the droplets are "floated up" by the upward-rising air. Without updrafts, the droplets drift so slowly out of the bottom of the cloud and evaporate so quickly that they have no chance of reaching the ground. They are replaced by new droplets forming above.

In the collision-coalescence process, tiny droplets coalesce to form a range of droplet sizes. Early on, updrafts are stronger than the downward motion of the droplets, and all droplets are repeatedly blown upward—rain does not fall. As the droplets grow, they eventually fall at the same rate as the updraft, becoming more or less stationary. But the droplets are repeatedly bombarded from below by smaller droplets rising with the updraft. Such bombardment is like standing still on a beach when a strong wind blows and being constantly pelted by tiny sand grains blowing in the wind. This is when significant droplet growth occurs. Eventually, the stationary drops of water grow larger and become huge compared to typical cloud droplets—they become raindrops. Because raindrops fall faster than the updraft can push them upward, precipitation forms.

This process requires sufficient vertical development of the cloud; otherwise, there are not enough droplet collisions for individual droplets to grow big enough to reach the stationary point. So the presence of thicker clouds means a

The suffix *-nimbus* means "rain-producing."

fyi

- Cumulus clouds are denser than the surrounding air. So why don't they fall? The answer is, they do fall! They fall at the same speed at which the air is rising, and therefore they remain fixed in elevation. Without updrafts, there would be no cumulus clouds.

TABLE 25.3   CLASSIFICATION OF AIR MASSES AND THEIR CHARACTERISTICS

Typical Source Region	Classification	Symbol	Characteristics
Arctic	maritime arctic	mA	cool, moist, unstable
Greenland	continental arctic	cA	cold, dry, stable
North Atlantic and Pacific Oceans	maritime polar	mP	cool, moist, unstable
Canada, Siberia	continental polar	cP	cold, dry, stable
Caribbean Sea, Gulf of Mexico	maritime tropical	mT	warm, moist; usually unstable
Mexico, southwestern United States	continental tropical	cT	hot, dry, stable aloft; unstable at surface

higher chance of rain—and a higher chance that a hard rain will fall. Thicker clouds give droplets more time and space to coalesce into drops that are heavy enough to fall.

Raindrops form because the condensation rate exceeds the evaporation rate. But raindrops shrink as they fall because once they are out of the cloud, the evaporation rate exceeds the condensation rate. Sometimes, enough evaporation occurs that raindrops disappear before reaching the ground. Such precipitation is called *virga*.

■ 25.4 Air Masses, Fronts, and Storms

Link To Section 7.3

An air mass is a volume of air much larger than the parcels of air we've discussed so far. Various distinct air masses cover large portions of Earth's surface. Each has its own characteristics. An air mass formed over water in the tropics is different from one formed over land in the polar regions. Air masses are divided into six general categories, according to the type of land or water they form over and the latitude at which their formation occurs (Table 25.3 and Figure 25.9). The type of surface over which air-mass formation occurs is designated by a lowercase letter (m for maritime, c for continental). The source region in which an air mass forms is designated by a capital letter (A for arctic, P for polar, T for tropical.)

Continental polar (cP) and continental arctic (cA) air masses generally produce very cold, dry weather in winter and cool, pleasant weather in summer. Maritime polar (mP) and maritime arctic (mA) air masses, picking up moisture as they travel across the oceans, generally bring cool, moist weather to a region. Continental tropical (cT) air masses are generally responsible for the hot, dry weather of summer, and warm, humid conditions are due to maritime tropical (mT) air masses.

So we see that different types of air masses have their own characteristics. When two different air masses meet, a variety of weather conditions can develop.

FIGURE 25.9

Typical source regions of air masses for North America.

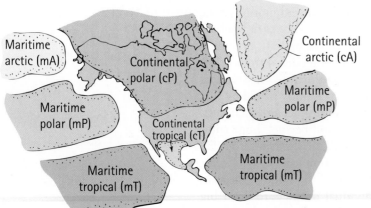

ATMOSPHERIC LIFTING

Clouds are great indicators of weather. For clouds to form, air must be lifted. The three principal lifting mechanisms in the atmosphere are convectional lifting, orographic lifting, and frontal lifting.

CONVECTIONAL LIFTING

Earth's surface is heated unequally. Some areas are better absorbers of solar radiation than others, so they heat up more quickly. The air that touches these surface "hot spots" becomes warmer than the surrounding air, and so it rises, expands, and cools. This rising of air is accompanied by the sinking of cooler air from above. The circulatory motion produces **convectional lifting.**

If cooling occurs close to the air's saturation temperature, the condensing moisture can form a cumulus cloud. Air within the cumulus cloud moves in a cycle: warm air rises, cool air descends. Because descending cool air inhibits the expansion of warm air beneath it, small cumulus clouds usually have a great amount of blue sky between them (Figure 25.10).

Cumulus clouds often remain in the places in which they formed, dissipating and reforming many times. As they grow, they shade the ground beneath from the Sun. This slows surface heating and inhibits the upward convection of warm air. Without a continuous supply of rising air, a cumulus cloud begins to dissipate. Once the cloud is gone, the ground reheats, allowing the air above it to warm and rise. Thus, convectional lifting begins again, and another cumulus cloud begins to form at the same location.

OROGRAPHIC LIFTING

An air mass that is pushed upward over an obstacle, such as a mountain range, undergoes **orographic lifting**—the rising air cools. If the air is humid, clouds form. The types of clouds that form depend on the air's stability and moisture content. If the air is stable, a layer of stratus clouds may form. If the air is unstable, cumulus clouds may form. As the air mass moves down the other side of the mountain (the leeward slope), it warms adiabatically. This descending air is dry and warm because most of its moisture was removed in the form of clouds and precipitation on the windward (upslope) side of the mountain. The descending air is warm because condensation releases heat. Because the dry leeward (downslope) sides of mountain ranges are sheltered from rain and moisture, they are often referred to as being in the *rain shadow* (Figure 25.11).

- As any glider pilot will attest, there's no way all air can rise. Some has to come back down. Where air rises and water vapor condenses we see clouds; where it descends we see blue sky between the clouds.

FIGURE 25.10

Cumulus clouds are often found as individual towering white clouds separated from each other by expanses of blue sky.

The Physical Science Place
Rain Shadow Activity

- Just as a "regular" shadow is a place where very little light falls because some obstacle blocks the light, a rain shadow is a place where little rain falls because some obstacle (such as a mountain) blocks precipitation.

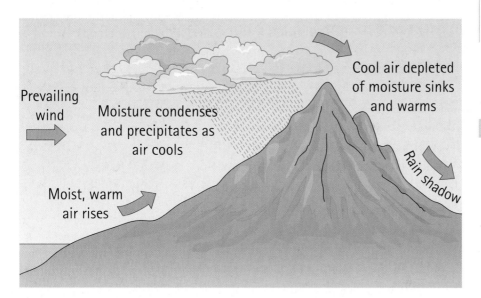

Prevailing wind

Moisture condenses and precipitates as air cools

Moist, warm air rises

Cool air depleted of moisture sinks and warms

Rain shadow

FIGURE 25.11

A mountain range may produce a rain shadow on its leeward slope. As warm, moist air rises on the windward slope, the air cools and precipitation develops. By the time it reaches the leeward slope, the air is depleted of moisture, so that the leeward side is dry. It lies in a rain shadow.

FIGURE 25.12

INTERACTIVE FIGURE

A cold front occurs when a cold air mass moves into a warm air mass. The cold air forces the warm air upward, where it condenses to form clouds. If the warmer air is moist and unstable, heavy rainfall and gusty winds develop.

The Physical Science Place

Cold Front Activity
Warm Front Activity

FRONTAL LIFTING

In weather reports, we often hear about fronts. A **front** is the contact zone between two different air masses. When two air masses make contact, differences in temperature, moisture, and pressure can cause one air mass to ride over the other. When this occurs, we have **frontal lifting**. If a cold air mass moves into an area occupied by a nonmoving warm air mass, the contact zone between them is called a *cold front*; and if warm air moves into an area occupied by a nonmoving mass of cold air, the zone of contact is called a *warm front*. Fronts are usually accompanied by wind, clouds, rain, and storms. An *occluded front* forms when a cold front overtakes a warm front or vice versa. If neither air mass is moving, the contact zone is called a *stationary front*.

Meteorologists and other observers of the sky can often tell when a cold front is approaching by observing high cirrus clouds, a shift in wind direction, a drop in temperature, and a drop in air pressure. As cold air moves into a warm air mass, forming a cold front, the warm air is forced upward (Figure 25.12). As it rises, it cools, and water vapor condenses into a series of cumulonimbus or nimbostratus clouds. The advancing wall of clouds at the front develops into thunderstorms with heavy showers and gusty winds. After the front passes, the air cools and sinks, pressure rises, and rain ceases. Except for a few fair-weather cumulus clouds, the skies become clear, and we have the calm after the storm.

When warm air moves into a cold air mass, forming a warm front, the less-dense warmer air gradually rides up and over the colder, denser air (Figure 25.13). The approach of a warm front, although less obvious and more gradual than the approach of a cold front, is also indicated by cirrus clouds. Ahead of

FIGURE 25.13

A warm front occurs when a warm air mass moves into a cold air mass. The less-dense warmer air rides up and over the colder, denser air, resulting in widespread cloudiness and light to moderate precipitation that can cover great areas.

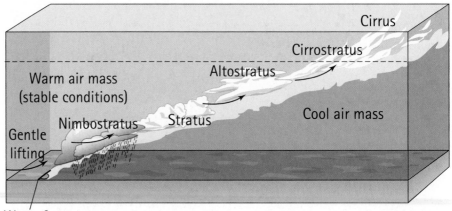

the front, the cirrus clouds descend and thicken into altocumulus and altostratus clouds that turn the sky an overcast gray. Moving still closer to the front, light to moderate rain or snow develops, and winds become brisk. At the front, air gradually warms, and the rain or snow turns to drizzle. Behind the front, the air is warm and the clouds scatter.

When a cold front and a warm front merge, the result is an occluded front. Several different mechanisms lead to occluded front formation. For simplicity, Figure 25.14 shows the steps leading to the formation of an occluded front when a cold front catches up with a slower-moving warm front. At first, the two fronts do not overlap; the northern ends of each front join at the center of a low-pressure area (Figure 25.14a). At this stage, warm air is in contact with the ground, between the two fronts. When the cold front first overtakes the warm front, the two fronts meet at the ground, such that the warm air above each front no longer touches the ground (Figure 25.14b). As the cold front continues to invade the warm front, the warm front itself no longer touches the ground (Figure 25.14c). The advancing cold front wedges itself under the warm front, and the intersection between the two fronts is above the ground. As you would imagine, a wide area of rainy weather accompanies occluded fronts.

fyi

■ No matter which air mass is advancing, the warmer, less dense air is always forced aloft. Warm air wants to rise. Cool air wants to sink. So the cooler, denser air acts as a wedge where lifting takes place.

FIGURE 25.14

Steps in the formation of an occluded front for the case when a cold front overtakes a warm front. The right side shows representative vertical cross sections for map views to the left. The location of each cross section is shown on its respective map.

fyi

- So why do we call the contact zone between air masses a front? The word *front* is a military term used to describe the boundary between two different armies. During World War I, Norwegian meteorologists adopted *front* as a way to describe the boundary line between two "warring" air masses.

When two different air masses are not strong enough to overtake the other, the boundary between them becomes a *stationary front*. A stationary front is like a stalemate between fronts and, as such, can remain over an area for several days. Eventually the stalemate ends and the front either dissipates or, depending on conditions aloft, changes into a cold or warm front.

MIDLATITUDE CYCLONES

Midlatitude cyclones are storm systems that typically sweep across the United States (and other places) from west to east. The formation of midlatitude cyclones is closely tied to the interactions between air masses at the polar front. The term *cyclone* refers to counterclockwise rotation, in the Northern Hemisphere, around a low-pressure center. *Anticyclonic* flow rotates clockwise around a high-pressure center.

The "typical" midlatitude cyclone progresses through six distinct steps, starting with *cyclogenesis*—the birth of the cyclone (Figure 25.15). This initial stage occurs at the polar front at a location with the characteristics of a stationary front, where both air masses are moving parallel to the front (Figure 25.15a).

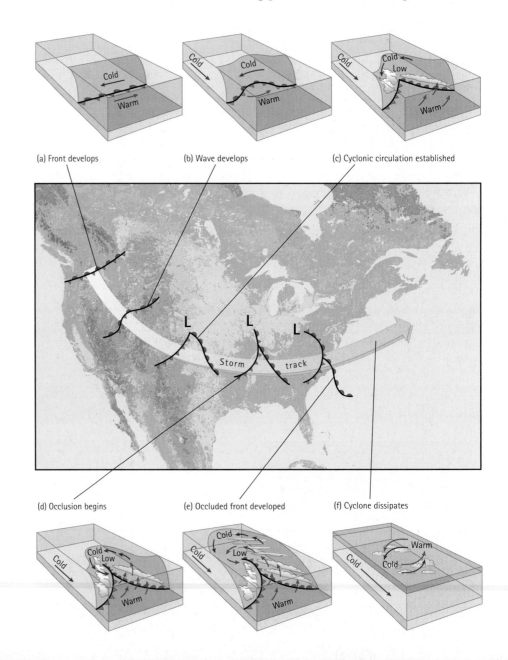

(a) Front develops   (b) Wave develops   (c) Cyclonic circulation established

(d) Occlusion begins   (e) Occluded front developed   (f) Cyclone dissipates

FIGURE 25.15

Steps in the life of a typical midlatitude cyclone.

Cyclogenesis happens when the linear front becomes disturbed and develops a wavelike curve, with warm air jutting into the cold air mass (Figure 25.15b). Usually, the wave tracks eastward with the westerlies.

If the wave doesn't die out, warm air continues to move poleward as cold air pushes toward the equator. At this stage two distinct fronts develop: a warm front on the leading edge of the system, trailed by a cold front. The fronts terminate at the center of what has become a roughly circular area of *low pressure* (Figure 25.15c). The pressure-gradient force drives air inward toward the center of the low, but the Coriolis force deflects the wind to the right. The deflected wind sets up a counterclockwise rotation around the central low. A pronounced temperature difference exists between the warmer east side of the system and the colder west side.

Under most conditions, the cold front moves eastward faster than the warm front. Cold air begins to lift the warm front, forming an occluded front (Figure 25.15d). As the occlusion continues to grow (Figure 25.15e), more and more warm air is displaced upward, until there is no significant temperature difference horizontally across the storm. But where did the warm air go? Figure 25.15f shows that all the warm air was forced up and over the steadily advancing cold front. Without the horizontal temperature difference, the storm is without an energy source. The counterclockwise cyclonic flow ceases, and clear, cold weather prevails.

This is the *Norwegian cyclone model*, and it does a great job of describing the beginning, middle, and end of a midlatitude cyclone. It was developed in the early 20th century, when meteorologists were limited to surface data. Data for the upper-level processes are needed to understand why midlatitude cyclones form and progress. The modern view of midlatitude cyclones invokes the idea of three major conveyor belts of air: a warm conveyor belt, a cold conveyor belt, and a dry conveyor belt (Figure 25.16).

The warm conveyor blows northward from the Gulf of Mexico, and hence carries significant moisture with it. Airflow originates near the surface but undergoes frontal lifting when it encounters the warm front. The warm air then cools adiabatically, leading to condensation and precipitation.

The cold conveyor belt blows in low from the east, north of the warm front. The cold air flows beneath the warm conveyor, picking up moisture from the evaporation of raindrops that fall through it from the warmer air above. When the cold conveyor belt approaches the center of the cyclone, the cold air is lifted, cooling even further. The rising air becomes saturated, producing more precipitation. Now at upper levels, the cold conveyor becomes incorporated into the westerlies.

The dry conveyor belt originates to the west of the storm, in the upper troposphere. It is a dry, cold flow that maintains the supply of cold air behind the cold front. After being caught and turned northward by the cyclonic flow, the dry conveyor belt also becomes incorporated into the upper-level westerlies.

The conveyor belt model supplements the original Norwegian polar-front model by illuminating the driving forces that feed a midlatitude cyclone.

FIGURE 25.16

The conveyor belt model of midlatitude cyclones, showing the relationship of the warm, cold, and dry conveyor belts to fronts observed at the surface.

■ 25.5 Violent Weather

The three types of lifting just discussed bring about many different weather conditions. Weather resulting from air masses in contact depends on the conditions of their source regions. Weather changes can occur slowly or very quickly. The most rapid changes, and the most violent ones, occur with three major types of storms: thunderstorms, tornadoes, and hurricanes.

FIGURE 25.17

The mature stage of a thunderstorm cloud appears as a towering cumulonimbus cloud that reaches up to about 12 km. Strong horizontal winds and icy temperatures flatten and distend the cloud's crown into a characteristic anvil shape.

Link To Section 8.8

THUNDERSTORMS

A thunderstorm begins with humid, unstable air rising, cooling, and condensing into a single cumulus cloud. This cloud builds and grows upward as long as it is fed by an updraft of rising warm air from below. Cloud droplets grow larger and heavier within the cloud until they eventually begin to fall as rain. The falling rain drags some of the cool air along with it, creating a downdraft—the chilled air is colder and denser than the air around it. Together, the rising warm updraft and the sinking, chilled downdraft make the cloud into a storm cell. This is the mature stage, at which the thunderstorm cloud appears as a lonely giant—dark and brooding in the sky. It typically has a base several kilometers in diameter, and it can tower to altitudes up to 12 km. At such high altitudes, horizontal winds and lower temperatures flatten and stretch the thunderhead crown into a characteristic anvil shape (Figure 25.17). After the thunderstorm dissipates, it leaves behind the cirrus anvil as a reminder of its once-mighty presence.

At any given time, about 1800 thunderstorms are in progress in Earth's atmosphere. Wherever thunderstorms occur, there is lightning and thunder. As water droplets in the cloud bump into and rub against one another, the cloud becomes electrically charged. Rather than being uniformly distributed throughout the cloud, the electrical charge separates—it is usually positively charged where there are ice crystals (at the colder cloud top) and negatively charged where the cloud is warmer (at the cloud bottom). As electrical stresses between the oppositely charged regions build up, the charge becomes great enough that electrical energy is released and passed to other points of opposite charge, which quite often means the ground. The electrical energy flowing from cloud to ground is lightning (Figure 25.18). As lightning heats up the air, the air expands and we hear lightning's noisy companion—thunder. Lightning strikes Earth roughly 100 times every second, with some bolts having an electric potential of as much as 100 million volts. Lightning claims about 100 human victims per year in the United States alone.

TORNADOES

A revolving object, such as a whirling ball on a string, speeds up when pulled toward its axis of revolution, thus conserving its angular momentum. Similarly, winds slowly rotating over a large area speed up when the radius of rotation

FIGURE 25.18

Time exposure of cloud-to-ground lightning during an intense thunderstorm. The blue, green, and red images of the flash were produced by a diffraction grating held over the camera lens when the photo was taken.

FIGURE 25.19

Like a gigantic vacuum cleaner, the strong wind of a tornado can pick up and obliterate everything in its path.

decreases. This increase in speed can produce a *tornado*, which is a funnel-shaped cloud that extends downward from a large cumulonimbus cloud. Produced as an extension of a powerful thunderstorm, the funnel cloud is called a tornado only after it touches the ground. The winds of a tornado travel in a counterclockwise direction (clockwise in the Southern Hemisphere) at wind speeds as low as 65 km/h but up to 450 km/h depending on the tornado's strength.

As a tornado moves across the land, at speeds from 45 to 95 km/h, it follows a path controlled by its parent thundercloud. The tornado can bounce and skip as it rises briefly from the ground and then touches back down again. A tornado acts like a gigantic vacuum cleaner, picking up everything in its path. It wreaks havoc not only by suction but also by the battering power of its whirling winds. In its wake it leaves a trail of flying dirt and debris (Figure 25.19).

Tornadoes occur in many parts of the world. In the flat central plains of the United States, a tornado zone extends from northern Texas through Oklahoma, Kansas, and Missouri. In this area, more than 300 tornadoes touch down each year. Hence the name for this area: Tornado Alley. Tornadoes are so frequent in this part of the country that many homes are built with underground storm shelters. The power of a tornado is terrifying and devastating.

HURRICANES

In the steamy tropics, where the Sun warms the oceans, heat transfer to the atmosphere by evaporation and conduction is so thorough that air and water temperatures are about equal. The high humidity in this part of the world favors the development of cumulus clouds and afternoon thunderstorms. Most of the individual storms are not severe, but thunderstorms sometimes become organized and begin to behave as a single system called a *tropical disturbance*. Warm, rising air creates a central low-pressure area that continuously draws more air toward the center of the disturbance. The inward-flowing air spirals because of the Coriolis force (Chapter 24). If this spiraling storm isn't broken apart by upper level winds, it can develop into a *tropical depression,* (wind speed less than 60 km/h), so called because of a central area of low pressure. If the storm intensifies, it progresses into a *tropical storm*, with increased wind speeds above 60 km/h. If favorable conditions continue, a more violent storm develops— a hurricane—with wind speeds above 120 km/h and up to nearly 300 km/h.

Hurricanes gain energy from the heat released by the condensation of water (Chapter 7). The condensation produces the vast amounts of rain that are typical of such storms. The heat warms the surrounding upper-level air, causing it to rise. As the upper-level air rises, surface air is sucked upward, intensifying the low-pressure center—the eye of the hurricane. Horizontal airflow spirals coun-

fyi

■ Hurricane Katrina formed on August 23, 2005, near the Bahamas. It struck the Gulf Coast of the United States six days later, wreaking havoc. New Orleans, Louisiana, and coastal Mississippi were hit particularly hard. Hurricane Katrina, however, was such a large storm that it devastated areas of the Gulf Coast up to 160 km from the eye of storm. At least 1,836 people were killed as a result of the hurricane and the floods that followed. It was the deadliest U.S. hurricane since the Okeechobee Hurricane of 1928. It was a costly storm, too. More than $100 billion was needed to repair storm damage and fund cleanup operations. Some areas still haven't recovered. It was the costliest natural disaster in U.S. history.

FIGURE 25.20

This satellite image of 2005's Hurricane Katrina shows the characteristic appearance of a hurricane. Bands of cumulonimbus clouds spiral around the low-pressure eye of the storm.

fyi

- We hear weather forecasters talk about short-, medium-, and long-range forecasting. What do these types of forecasting mean? In general, short-range forecasting predicts the next two days with considerable detail about temperature, wind, and weather. Medium-range forecasting predicts weather for the third to seventh days in less detail. Forecasting beyond seven days is considered long-range, and the predictions in such forecasts are in terms of conditions that are expected to be above or below normal.

Link To Section 12.5

fyi

- Swirling storms and disturbances caused by solar activity produce "space weather" between the Sun and Earth. Solar flares, coronal mass ejections, and magnetic storms affect not only Earth satellites but Earth's surface environment as well. Communications system failures, power blackouts, and brownouts are often attributed to space weather. As our use of space grows, so must our ability to predict its weather.

terclockwise (in the Northern Hemisphere) around the eye. The spiral bands of cumulonimbus clouds give the hurricane its familiar appearance (Figure 25.20).

The airflow that sets up in the eye forms a positive feedback loop. Condensation releases heat, which draws moist air upward from the ocean surface. The moist air cools and more condensation occurs, which releases more heat, which draws more warm, moist air upward. This cycle—essentially a natural heat engine—continues unless strong, upper-level winds from outside the storm disrupt the upward flow pattern, or the hurricane moves over land. Once over land, the hurricane is deprived of its energy source.

■ 25.6 The Weather—The Number One Topic of Conversation

Meteorologists have the important job of forecasting hurricanes and other storms. Weather forecasting is partly a matter of determining air-mass characteristics, how and why the characteristics might change, and in what direction air masses might move. In the case of hurricanes and tornadoes, such predictions are lifesaving. Meteorologists have a long and remarkable record of saving human lives and reducing property loss.

Weather forecasting involves great quantities of data from all over the world. Before the 1960s, most of these data were assembled, analyzed, and plotted by hand on weather maps and charts. This took thousands of calculations, a large workforce, and long hours. Now, with computers, great quantities of data from around the world can be processed in a matter of minutes. Computers do more than just plot and analyze data—they can help predict the weather as well. The computer draws maps of projected weather conditions, which the weather forecaster uses as a guide for predicting weather. Even so, the many variables involved often impede the making of accurate predictions, so don't count on an absence of rain on your parade!

Weather Maps

The weather forecaster's primary tool is the weather map, or chart. A weather map is essentially a representation of the frontal systems and the high-pressure and low-pressure systems that overlie the areas outlined in the map. Symbols on such a map are a shorthand notation to represent data gathered from various observation stations. These symbols are called *weather codes*.

This shorthand notation compiles 18 categories of data into a very small area called a *station model*. The circle at the center describes the overall appearance of the sky. Jutting from the circle is a wind arrow, its tail in the direction from which the wind comes and its feathers indicating wind speed. The other 15 weather elements are in standard position around the circle.

A weather map is covered with lines—*isobars*—that connect points of equal pressure. As air moves from a high-pressure region to a low-pressure region, it rises and cools, and the moisture in it condenses into clouds. In the vicinity of the low (L on a map), we see an extensive cloud cover. In the vicinity of the high (H on a map), we see clear skies. In a high-pressure region, air sinks and warms adiabatically. Because sinking air does not produce clouds, we find clear skies and fair weather. The heavy lines on a weather map represent fronts. Because fronts generally mean a change in the weather, they are of great importance on weather maps.

Weather maps show atmospheric conditions. As warm air rises, it expands and chills. As it chills, the water-vapor molecules condense to form clouds. Because air moves from high-pressure regions to low-pressure regions, low-pressure zones are accompanied by cloud cover. In a high-pressure zone, air generally sinks. Because sinking air does not usually produce clouds, we find clear skies and fair weather.

SUMMARY OF TERMS

Humidity A measure of the concentration or amount of water vapor in the air—the mass of water vapor per volume of air.

Relative humidity The amount of water vapor in the air at a given temperature expressed as a percentage of the maximum amount of water vapor the air can accommodate at that temperature.

Saturation vapor pressure The maximum amount of moisture the air can accommodate at a given temperature; the upper limit for humidity.

Adiabatic A term that describes temperature change in the absence of heat transfer—expanding air cools and compressing air warms.

Temperature inversion A condition in which the upper regions of the troposphere are warmer than the lower regions.

Lifting condensation level The height at which rising air cooling at the dry adiabatic lapse rate becomes saturated and condensation begins.

Convectional lifting An air-circulation pattern in which air warmed by the ground rises while cooler air aloft sinks.

Orographic lifting The lifting of an air mass over a topographic barrier such as a mountain.

Front The contact zone between two different air masses.

Frontal lifting The lifting of one air mass by another as two air masses converge.

Midlatitude cyclone West- to east-traveling storm with a central low-pressure area about which counterclockwise flow develops (in the Northern Hemisphere), and from which usually extends a cold front and a warm front; generally forms at the polar front.

REVIEW QUESTIONS

25.1 Atmospheric Moisture

1. Distinguish between humidity and relative humidity.
2. Why does relative humidity increase at night?
3. As air temperature decreases, does relative humidity increase, decrease, or stay the same?
4. What does the saturation point have to do with the dew point?
5. What happens to the water vapor in saturated air as the air cools?
6. What factors are responsible for condensation?
7. Does condensation occur more readily at high temperatures or low temperatures? Explain.
8. When water vapor condenses to liquid water, is heat absorbed or released?
9. Distinguish between condensation and precipitation.

25.2 Weather Variables

10. Explain why warm air rises and cools as it expands.
11. When a parcel of air rises, does it become warmer, become cooler, or remain at the same temperature?
12. What is an adiabatic process?
13. Name at least two ways that thermal energy in air can be increased.
14. Name at least two ways that thermal energy in air can be decreased.
15. What is a temperature inversion? Cite at least one location where these inversions often occur.
16. What happens to the air pressure and temperature of an air parcel as it flows up the side of a mountain?

25.3 Cloud Development

17. Explain how clouds form.

18. Name the cloud form associated with (a) the hazy shade of winter, (b) mackerel sky, (c) floating cotton, and (d) snowfall.
19. Name the cloud group to which each of the following cloud types belong: (a) altocumulus, (b) cirrostratus, (c) nimbostratus, and (d) cumulus clouds.
20. Rain or snow is most likely to be produced by which of the following cloud forms?
    a. cirrostratus
    b. nimbostratus
    c. altocumulus
    d. stratocumulus
21. Are clouds that have vertical development characteristic of stable air, stationary air, unstable air, or dry air?
22. Which type of clouds are most likely to become a thunderhead?

25.4 Air Masses, Fronts, and Storms

23. Explain how convectional lifting plays a role in the formation of cumulus clouds.
24. Explain how a convection cycle is generated.
25. Does a rain shadow occur on the windward side of a mountain range or on the leeward side? Explain.
26. Differentiate between a cold front and a warm front.
27. What are the three main atmospheric lifting mechanisms?
28. Under what conditions does orographic precipitation occur?

25.5 Violent Weather

29. How do downdrafts form in thunderstorms?
30. Briefly describe how thunder and lightning develop.

EXERCISES

Please do not be intimidated by the large number of exercises and problems in this and other meatier chapters. If your course work is to cover many chapters, your instructor will likely assign only a few exercises and/or problems from each.

1. ● Distinguish between weather and climate.
2. ■ Which produces precipitation: a rising moist air mass, a descending moist air mass, or both?
3. ■ Why do clouds tend to form above mountain peaks?
4. ■ Why does warm, moist air blowing over cold water result in fog?
5. ● Why does dew form on the ground during clear, calm summer nights?
6. ● Why does a July day in the Gulf of Mexico generally feel appreciably hotter than a July day in Arizona, even when temperatures are the same?
7. ■ Would you expect a glass of water to evaporate more quickly on a windy, warm, dry summer day or a calm, cold, dry, winter day? Defend your answer.
8. ●Why does the surface temperature of the ground increase on a clear, calm night as low cloud cover moves overhead?
9. ◆ During a summer visit to Cancún, Mexico, you stay in an air-conditioned room. Getting ready to leave your room for the beach, you put on your sunglasses. The minute you step outside your sunglasses fog up. Why?
10. ◆ After a day of skiing in the Rocky Mountains, you decide to go indoors to get a warm cup of cocoa. As you enter the ski lodge, your eyeglasses fog up. Why?
11. ● Can the temperature of an air mass change if no heat is added or subtracted? Explain.
12. ■ Why must an air mass rise in order to produce precipitation?
13. ● As an air mass moves first upslope and then downslope over a mountain, what happens to the air's temperature and moisture content?
14. ● Why are saturation and condensation more likely to occur on a cold day than on a warm day?
15. ● The sky is overcast, and it is raining. What type of cloud is above you, nimbostratus or cumulonimbus?
16. ■ Distinguish between dew and frost.
17. ■ What accounts for the large spaces of blue sky between cumulus clouds?
18. ● Why don't cumulus clouds form in an area where there is no air convection?
19. ■ Which cloud form is associated with a stable air mass?
20. ● What cloud form is associated with thunderstorms?
21. ■ What is the relationship between saturation vapor pressure and humidity?
22. ■ When the condensation rate is greater than the evaporation rate, what happens to the air?
23. ● Cite at least four types of information needed to predict the weather.
24. ● The accuracy of weather forecasts depends on great quantities of data and thousands of calculations. If the number of data points were decreased, would accuracy also decrease?
25. ■ What is the difference between rainfall that accompanies the passage of a warm front and rainfall that accompanies the passage of a cold front?
26. ■ When does an adiabatic process happen in the atmosphere?
27. ◆ How can a layer of altostratus clouds change into altocumulus clouds?
28. ◆ Antarctica is covered by glaciers and large ice sheets. Is the snowfall in Antarctica therefore heavy or light? Why?
29. ● How do fronts cause clouds and precipitation?
30. ■ Explain why freezing rain is more commonly associated with warm fronts than with cold fronts.
31. ■ What steps need to happen before precipitation occurs?.
32. ● In simplest terms, what is an occluded front?
33. ● How does a rain-shadow desert form?
34. ■ Sinking air warms, and yet the downdrafts in a thunderstorm are cold. Why?
35. ◆ Tornadoes form in the regions of a strong updraft, and yet they descend from the base of a cloud. Explain.
36. ■ Why are hurricanes more likely to occur on the East Coast of the United States than on the West Coast?
37. ● Why are clouds that form over water more efficient in producing precipitation than clouds that form over land?
38. ■ What is the source of the enormous amount of energy released by a hurricane?
39. ● What part of the United States has the highest frequency of tornadoes?
40. ● In which atmospheric layer does all our weather occur?

PROBLEMS

Refer to the Figuring Physical Science box on page 667 for the following two problems.

1. ◆ At 50°F the maximum amount of water vapor in air is 9 g/m^3. If the relative humidity is 40%, what is the mass of water vapor in 1 m^3 of air?

2. ◆ The relative humidity of an air parcel is 50% and the pressure is 1000 millibars (mb). If the pressure increases to 1053 mb without changing the temperature or the water vapor content of the air, what is the relative humidity? (Hint: Use Boyle's law from Chapter 5.)

READINESS ASSURANCE TEST (RAT)

If you have a good handle on this chapter, if you really do, then you should be able to score at least 7 out of 10 on this RAT. If you score less than 7, you need to study further before moving on.

Choose the BEST answer to each of the following.

1. Air that contains the maximum amount of water vapor for the temperature of the air mass is considered to
   (a) have a relative humidity of 100%.
   (b) be saturated.
   (c) have an evaporation rate equal to its condensation rate.
   (d) all of these

2. In most midlatitude cyclones, the warm front
   (a) is behind and to the west of the cold front.
   (b) is ahead and to the east of the cold front.
   (c) terminates in the high-pressure center.
   (d) convolutes back to a stationary front.

3. An adiabatic process has occurred when
   (a) air is warmed by solar radiation.
   (b) heat is subtracted by evaporation.
   (c) air expands and warms.
   (d) air expands and cools.

4. When air sinks,
   (a) it compresses and warms.
   (b) it reaches its equilibrium level and then begins to sink.
   (c) it expands and cools.
   (d) it forms into clouds.

5. When upper regions of the atmosphere are warmer than lower regions, we have
   (a) convective lifting.
   (b) a temperature inversion.
   (c) absolute instability.
   (d) an adiabatic process.

6. A key factor needed for precipitation to occur is
   (a) updraft motion in relatively thick clouds.
   (b) a condensation rate that exceeds the evaporation rate.
   (c) condensation nuclei.
   (d) all of these

7. For clouds to form, air must be lifted. The principal lifting mechanisms are
   (a) convectional, orographic, and frontal lifting.
   (b) continental, orogeny, and occluded lifting.
   (c) conversational, orthodontic, and face lifting.
   (d) stationary, occluded, and contact lifting.

8. As air temperature decreases, relative humidity
   (a) increases.
   (b) decreases.
   (c) stays the same.
   (d) none of these

9. In the Northern Hemisphere, tornadoes and hurricanes rotate in a _____ direction because of the _____.
   (a) clockwise, Coriolis force
   (b) counterclockwise, Coriolis force
   (c) clockwise, cyclogenesis force
   (d) counterclockwise, cyclogenesis force

10. When air is saturated,
    (a) the condensation rate is greater than the evaporation rate.
    (b) the condensation rate is less than the evaporation rate.
    (c) the condensation rate equals the evaporation rate.
    (d) the condensation rate depends on temperature.

Answers to RAT

1d, 2b, 3d, 4a, 5b, 6d, 7a, 8a, 9b, 10c

EXPLORING FURTHER

http://www.nws.noaa.gov
> The National Weather Service home page provides detailed weather information and forecasts. View maps and pictures and discover historical weather trends. Check out the Education Page for great links to activities and teaching tools.

http://lwf.ncdc.noaa.gov/oa/ncdc.html
> If you are looking for an active archive of worldwide weather data, check out this site. There is extensive information here, but the site requires multiple link clicks before you can access data.

CHAPTER 25 ONLINE RESOURCES

 The Physical Science Place

Interactive Figure
- 25.12

Tutorials
- Surface Temperature of Planets
- Rain Shadow Activity
- Cold Front Activity
- Warm Front Activity

Video
- Adiabatic Process

Quiz

Flashcards

Links

We began this book with the *physics* of the everyday world, then progressed to the submicroscopic world of *chemistry*. Then we explored *Earth science*, big by comparison. And now we conclude our explorations by studying the arena of the very big—the solar system, stars, galaxies, and the universe itself. We'll learn, for example, how the Moon, seen here in stages of an eclipse, becomes red due to sunsets all over the world. To *astronomy*!

part four
astronomy

the solar system

26.1 THE SOLAR SYSTEM AND ITS FORMATION

26.2 THE SUN

26.3 THE INNER PLANETS

26.4 THE OUTER PLANETS

26.5 EARTH'S MOON

26.6 FAILED PLANET FORMATION

■ Five billion years ago, there was no Sun. Instead, the region of the galaxy that would become our solar system was a gently swirling diffuse cloud of gas and dust. How then did this cloud evolve into our present solar system? How does the Sun produce so much energy? How are the planets similar, and how are they different? How did our Moon form and how does it go through phases? Why do we see only one side of the Moon? What are solar and lunar eclipses and why are they rare? What are meteors, asteroids, and comets? How frequently do they collide with our planet, and why does a comet's tail always point away from the Sun?

For thousands of years, people have gazed into the night sky and pondered questions such as these. Countless generations have wondered about our place in the universe. Only recently have we begun to understand it. We begin with a tour of the universe with a focus on our solar system, which on a cosmic scale is Earth's own backyard.

The Physical Science Place

Formation of the Solar System

The Physical Science Place

History of the Solar System
Orbits in the Solar System

Link To Section 4.8

The ancients could tell the difference between planets and stars because of the difference in their movements in the sky. The stars remain relatively fixed in their patterns in the sky, but the planets wander. The term *planet* is derived from the Greek for "wandering star."

Link To Section 21.4

■ 26.1 The Solar System and Its Formation

Our solar system is the collection of objects gravitationally bound to the Sun. Along with the Sun itself, the solar system contains at least eight **planets**, which are large orbiting bodies massive enough for their gravity to make them spherical but small enough to avoid having nuclear fusion in their cores. The planets also have successfully cleared all debris from their orbital paths, which all lie roughly in the same plane. This plane, called the **ecliptic**, is defined as the plane of Earth's orbit. Within the solar system are also numerous moons (objects orbiting planets), asteroids (small, rocky bodies), comets (small, icy bodies), and a collection of miniature planets known as dwarf planets that orbit on the outer edges of the solar system. The most well-known dwarf planet is Pluto, which was downgraded from planet status in 2006. The planets and all these other objects are quite small compared to the Sun. Figure 26.1 shows the sizes of the planets relative to the Sun, which contains a whopping 99.86% of the solar system's mass.

The vast distances between the Sun and the objects orbiting it can be grasped by imagining the Sun reduced to the size of a large beach ball 1 m in diameter. The closest planet, Mercury, would be an apple seed located about 40 m (130 ft) away. The next closest planet, Venus, would be the size of a pea about 80 m (255 ft) distant. Earth, also about the size of a pea, would be about 110 m away, which is greater than the length of a football field. The next planet, Mars, which is only a bit larger than Mercury, would be an apple seed almost two football field lengths away from our solar beach ball.

These first four planets—Mercury, Venus, Earth, and Mars—are called the **inner planets** because of their relatively close proximity to the Sun. All the inner planets are solid rocky planets. The **outer planets** are larger gaseous planets located much farther away. The first outer planet is Jupiter, which on the scale mentioned would be the size of a softball more than half a kilometer away. The second outer planet, Saturn, famous for its extensive ring system, would be the size of a baseball more than a kilometer away. Planets Uranus and Neptune would both be about the size of Ping-Pong balls located 2 and 3 km away, respectively. We see that solar system objects are mere specks in the vastness of the space about the Sun.

Because of these vast interplanetary distances, astronomers use the *astronomical unit* to measure them. One **astronomical unit (AU)** is about 1.5×10^8 km (about 9.3×10^7 mi) or the distance from Earth to the Sun. Table 26.1 gives the distances of planets from the Sun in AU. The data in Table 26.1 also show the division of the planets into two groups with similar properties. The inner planets—Mercury, Venus, Earth, and Mars—are solid and relatively small and dense. They are called

FIGURE 26.1

This illustration shows the order and relative sizes of planets. Moving away from the Sun, we have in order: Mercury, Venus, Earth, Mars, Jupiter, Saturn, Uranus, and Neptune. The planets range greatly in size, but the Sun dwarfs them all—containing more than 99% of the mass in the solar system. Note: distances are not to scale in this illustration.

the *terrestrial planets*. The outer planets are large, have many rings and satellites, and are composed primarily of hydrogen and helium gas. These are called the *jovian planets* because their large sizes and gaseous compositions resemble Jupiter.

TABLE 26.1   PLANETARY DATA

	Mean Distance from Sun (Earth distances, AU)	Orbital Period (years)	Diameter (km)	Average Mass (Earth = 1)	(kg)	Density (Earth = 1)	(g/cm³)	Inclination to Ecliptic
Sun			1,392,000	109.1	1.99×10^{30}	3.3×10^5	1.41	
Terrestrial								
Mercury	0.39	0.24	4,880	0.38	3.3×10^{23}	0.06	5.4	7.0°
Venus	0.72	0.62	12,100	0.95	4.9×10^{24}	0.81	5.2	3.4°
Earth	1.00	1.00	12,760	1.00	6.0×10^{24}	1.00	5.5	0.0°
Mars	1.52	1.88	6,800	0.53	6.4×10^{23}	0.11	3.9	1.9°
Jovian								
Jupiter	5.20	11.86	142,800	11.19	1.90×10^{27}	317.73	1.3	1.3°
Saturn	9.54	29.46	120,700	9.44	5.7×10^{26}	95.15	0.7	2.5°
Uranus	19.18	84.0	50,800	3.98	8.7×10^{25}	14.65	1.3	0.8°
Neptune	30.06	164.79	49,600	3.81	1.0×10^{26}	17.23	1.7	1.8°
Dwarf Planets								
Pluto	39.44	247.70	2,300	0.18	1.3×10^{22}	0.002	1.9	17°
Eris	67.67	557	2,400	0.19	1.6×10^{22}	0.002	1.9	44°

FIGURING PHYSICAL SCIENCE

▪ The Scale of the Solar System

Astronomical distances are mind-boggling. Try the following problems to better appreciate the sizes of bodies in the solar system and distances between them. (Use the distance formula: distance = speed × time and data from Table 26.1.)

SAMPLE PROBLEM 1

The distance between Earth and the Moon is 384,401 km. How many Earth diameters would fit between Earth and the Moon?

SOLUTION:

$$\frac{38,401 \text{ km}}{12,760 \text{ km}} \approx 30$$

About 30 Earth-sized planets would fit in the distance between Earth and the Moon.

SAMPLE PROBLEM 2

How long would it take to drive from Earth to the Moon if you drive at 55 mi/h? State your answer in hours and in years.

SOLUTION:

$$d = st \rightarrow t = \frac{d}{s}$$

Convert all units to metric.
(55 mi/h) (1km/0.62 mi) ≈ 89 km/h. Then t = 384,410 km/89 km/h = 4319 h (1 day/24h) (1y/365 days) ≈ 0.5 y. It would take about one-half year to drive to the Moon traveling at freeway speeds.

SAMPLE PROBLEM 3

If you could fly to the Sun on a jet that moves at 1000 km/h, how long would it take? State your answer in years.

SOLUTION:

$t = \frac{d}{s} = 1.5 \times 10^8$ km/1 ×
10^3 km/h = $(1.5 \times 10^5$ h)(1 day/24 h)

(1 yr/365 days) = 17 yr. So if you could fly to the Sun in a jet without becoming vaporized, it would take about 17 years.

SAMPLE PROBLEM 4

The diameter of the Sun is 1,390,000 km. What is its diameter in AU? How much greater is the mean distance between Earth and the Sun compared to the diameter of the Sun?

SOLUTION:

$(1.39 \times 10^6$ km)
$(1 \text{ AU}/1.5 \times 10^8$ km) ≈ 0.01 AU

$$\frac{1 \text{ AU}}{0.01 \text{ AU}} = 100$$

The diameter of the Sun is approximately equal to 0.01 AU, so the distance between Earth and the Sun is about 100 times the diameter of the Sun. Big as the Sun is, the solar system is mostly empty space.

FIGURE 26.2

This photograph, provided by the Hubble Telescope, shows the Orion Nebula. The Orion Nebula, like the nebula from which our solar system formed, is an interstellar cloud of gas and dust and the birthplace of stars.

Link To Section 4.1

FIGURE 26.3

(a) The nebula from which the solar system formed was originally a large, diffuse cloud that rotated quite slowly. The cloud began to collapse under the influence of gravity. (b) As the cloud collapsed, it heated up as gravitational potential energy converted to heat. It spun faster by the conservation of angular momentum. (c) The cloud flattened into a disk as a result of its fast rotation. A spinning, flattened disk was produced whose mass was concentrated at its hot center.

NEBULAR THEORY

The solar system is a highly ordered group of objects, with clear patterns and trends among planets. These regularities would be unlikely to arise from coincidence. Instead, they are the results of the physical processes that created the solar system. Thus, the observed regularities in the solar system are good clues as to how it formed.

Any theory of solar system formation must be able to explain two major regularities: (1) *the orderly motions among large bodies of the solar system* and (2) *the division of planets into two main types*—terrestrial *and* Jovian. Further, a viable theory of solar system formation must explain other known features of the solar system, including the existence of asteroids, comets, and moons.

The modern scientific theory that satisfies these requirements is called the **nebular theory**, which holds that the Sun and planets formed together from a cloud of gas and dust, a *nebula* (Latin for "cloud"). According to the theory, the solar system began to condense from the cloud of gas and dust about 5 billion years ago. The cloud would have been very diffuse and large, with a diameter thousands of times larger than Pluto's orbit. A nebula currently visible within the constellation of Orion is shown in Figure 26.2.

Within the nebula from which our solar system formed, the gravitational pull on particles exceeded the tendency of gas to disperse and fill all available space. Once the gravitational collapse of a cloud begins, gravity ensures that it continues. The universal law of gravity (Chapter 4) is an inverse-square law: the strength of the gravitational force increases dramatically as the particles become closer. The cloud maintained a constant mass as it shrank, gravitational forces grew ever stronger, and the cloud took on a spherical shape.

This nebula must also have had a slight net rotation, possibly due to the rotation of the galaxy itself. Then over the millions of years during which the solar nebula collapsed, it heated up, spun faster, and flattened into a disk shape. As a result, the nebula transformed from a large, diffuse, spherical cloud to a much smaller spinning disk with a hot center, as shown in Figure 26.3.

As the nebula shrank under the influence of gravity, much heat was released upon the collision of particles. Also, as the nebula shrank, it spun faster and faster (because of the *conservation of angular momentum* discussed in Chapter 3). As any spinning object contracts, the speed of its spin increases such that angular momentum is conserved. A familiar example is an ice skater whose spin rate increases when her extended arms are pulled inward. A nebula does the same.

What happens to the shape of a sphere as it spins faster and faster? The answer is, it flattens. A familiar example is the chef who turns a ball of pizza dough into a disk by spinning it on his hands. Even planet Earth is a slightly "flattened" sphere because of its daily spin. Saturn, with a greater spin, noticeably departs from a purely spherical shape. So the initially spherical nebula progressed to a spinning disk, the center of which became the *protosun*.

The formation of the spinning disk explains the orderly motions of our solar system today. All planets orbit the Sun in nearly the same plane because they

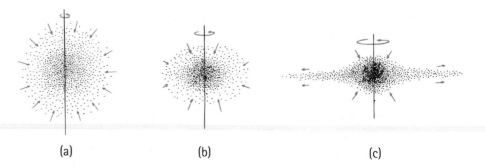

(a)   (b)   (c)

formed from that same flat nebular disk. The direction in which the disk was spinning became the direction of the Sun's rotation and the orbits of planets. It was also the preferred direction of rotation for planets—which is why most planets rotate in the same direction today.

CHECK POINT

As a nebula contracts, its rate of spin increases. What rule of nature is at play here?

Was this your answer?
The rule that applies to all spinning bodies is the *conservation of angular momentum*. That rule and the conservation of energy are dominant players in the universe.

The hot, central portion of the solar nebula, the protosun, was a clump of gas and dust that became the Sun when thermonuclear fusion ignited within it. The surrounding disk was the source of material that became the planets. In the spinning disk, matter collected in some regions more densely than in others. Perhaps small particles of gas and dust stuck together via gravity or electrostatic attraction. Because of their extra mass, these clumps exerted a stronger gravitational force on one another than on neighboring regions of the disk, and so they pulled in even more material to them. This led to the accretion of the nebular disk into small objects called *planetesimals*, which ranged in size from boulders to objects several kilometers in diameter. Planetesimals grew larger through countless collisions until they gravitationally dominated surrounding matter and finally became full-grown planets.

Link To Section 13.7

The planetesimals grew into planets at about the same time that the protosun was commencing thermonuclear fusion. (We'll return to thermonuclear fusion in the Sun in Section 26.2.) Once thermonuclear fusion occurred and the Sun began to radiate energy, the nebular disk warmed, with the inner portions reaching higher temperatures than the outer portions. As a result, the inner and outer planets developed differently. The inner planets formed from materials that remained solid at high temperatures; hence, the inner planets are rocky. The outer planets, by contrast, consist mainly of hydrogen and helium gas that coalesced in the cold regions of the solar system far from the Sun. Thus, we can see that the nebular theory accounts for the formation of the planets and the neat division of them into two groups.

26.2 The Sun

The Physical Science Place
The Sun

The Sun produces energy from the thermonuclear fusion of hydrogen to helium. Each second, approximately 657 million tons of hydrogen is fused to 653 million tons of helium. The 4 million tons of mass lost is discharged as radiant energy. This conversion of hydrogen to helium in the Sun has been going on since it formed nearly 5 billion years ago, and it is expected to continue at this rate for another 5 billion years. A tiny fraction of the Sun's energy reaches Earth and is converted by photosynthesizing organisms to chemical energy stored in large molecules. These energy-rich molecules are the primary energy source for almost all the organisms of this planet. The Sun, Earth's nearest star, is the solar system's power supply.

FIGURE 26.4

Never directly look at the Sun! Instead, you can get a nice view of the Sun by focusing the image of the Sun from a pair of binoculars onto a white surface. If the Sun is eclipsed by the Moon, which is a rare event, the Sun is seen as a crescent. More commonly, the Sun's image may reveal sunspots.

Solar energy is generated deep within the core of the Sun. The solar core constitutes about 10% of the Sun's total volume. It is very hot—more than 15,000,000 K. The core is also very dense, with more than 12 times the density of solid lead. Pressure in the core is 340 billion times Earth's atmospheric pressure! Because of these intense conditions, the hydrogen, helium, and minute quantities of other elements exist in the plasma state. (Plasma, recall, is the phase of matter beyond gas, consisting of ions and electrons rather than atoms—electrons have been stripped from atoms by high energies.) The nuclei of this plasma move fast enough to undergo nuclear fusion, as discussed in Section 13.7. The energy released from this nuclear fusion rises to the surface, where it causes gases to emit a broad spectrum of electromagnetic radiation, from infrared to X-rays.

The Sun's surface is a layer of glowing 5800 K plasma, which is much cooler than the Sun's core but hot enough to generate lots of light. This layer, called the *photosphere* (sphere of light), is about 500 km deep. Within the photosphere are relatively cool regions that appear as **sunspots** when viewed from Earth. Sunspots are cooler and darker than the rest of the photosphere and are caused by magnetic fields that impede hot gases from rising to the surface. As shown in Figure 26.4, sunspots can be seen by focusing the image of the Sun from a telescope or pair of binoculars onto a flat white surface. Sunspots are typically twice the size of Earth and move around because of the Sun's rotation, and last about a week or so. Often, they cluster in groups as shown in Figure 26.5.

FIGURE 26.5

Sunspots on the solar surface are relatively cool regions. We say relatively cool because they are hotter than 4000 K. They look dark only in contrast with their 5800 K surroundings.

The Sun spins slowly on its axis. Because the Sun is a fluid rather than a solid, different latitudes of the Sun spin at different rates. Equatorial regions spin once in 25 days, but higher latitudes take up to 36 days to make a complete rotation. This differential spin means the surface near the equator pulls ahead of the surface farther north or south. The Sun's differential spin wraps and distorts the solar magnetic field, which bursts out to form the sunspots mentioned earlier. A reversal of magnetic poles occurs every 11 years, and the number of sunspots also reaches a maximum every 11 years (currently). The complete cycle of solar activity is 22 years.

Above the Sun's photosphere is a transparent 10,000-km-thick shell called the *chromosphere* (sphere of color), seen during an eclipse as a pinkish glow surrounding the eclipsed Sun. The chromosphere is hotter than the photosphere, reaching temperatures of about 10,000 K. Its beautiful pink color, as shown in Figure 26.6, arises from the emission of light from hydrogen atoms.

Beyond the chromosphere are streamers and filaments of outward-moving, high-temperature plasmas curved by the Sun's magnetic field. This outermost region of the Sun's atmosphere is the *corona*, which

FIGURE 26.6

The pink chromosphere becomes visible when the Moon blocks most of the light from the photosphere during a solar eclipse.

extends out several million kilometers (Figure 26.7). The temperature of the corona is amazingly high—on the order of 1 million K—and it is where most of the Sun's powerful X-rays are generated. Because the corona is not very dense, its brightness is not as intense as the Sun's surface, which makes the corona safe to observe during (and only during) a total solar eclipse. High-speed protons and electrons are cast outward from the corona to generate the *solar wind*, which powers the aurora on Earth and produces the tails of comets.

FIGURE 26.7

The pearly white solar corona is visible only during a solar eclipse. Notice how this exceptional photo of the corona also captures some of the pink of the chromosphere as well as the face of the new Moon, which is faintly illuminated by light reflected from the full Earth. In other words, if you were standing on the Moon when this photo was taken, you would see your faint shadow cast by the fully lit Earth shining above you.

CHECK POINT

1. Was the Sun more massive 1000 years ago than it is today? Defend your answer.
2. Of the photosphere, chromosphere, and corona, which is thinnest? Which is hottest? Which is between the other two?

Were these your answers?

1. Yes, although slightly compared with its great mass. The Sun loses mass as hydrogen nuclei combine to make helium nuclei.
2. The photosphere is the thinnest. The corona is the hottest. The chromosphere is the pinkish layer above the photosphere and below the vast corona.

26.3 The Inner Planets

Compared with the outer planets, the four planets nearest the Sun are close together. These are Mercury, Venus, Earth, and Mars. These rocky planets each have a mineral-containing solid crust.

fyi

- We have seven days in a week because ancient Europeans decided to name days after the seven wandering celestial objects they could observe. The English day names were derived from the language of the Teutonic tribes who lived in the region that is now Germany. In Teutonic, the Sun is *Sun* (Sunday), the Moon is *Moon* (Monday), Mars is *Tiw* (Tuesday), Mercury is *Woden* (Wednesday), Jupiter is *Thor* (Thursday), Venus is *Fria* (Friday), and Saturn is *Saturn* (Saturday).

MERCURY

Mercury (Figure 26.8) is somewhat larger than Earth's Moon and similar in appearance. It is the closest planet to the Sun. Because of this closeness it is the fastest planet, circling the Sun in only 88 Earth days—which thus equals one Mercury "year." Mercury spins about its axis only three times for each two revolutions about the Sun. This makes its daytime very long and very hot, with temperatures as high as 430°C.

Because of Mercury's small size and weak gravitational field, it

FIGURE 26.8

This is a mosaic image of Mercury compiled from data obtained by the *Mariner 10* spacecraft on a flyby mission during 1974–1975. Areas for which data are missing are blank. Mercury is heavily cratered from the impacts of many meteorites. Mercury is a small planet, with only about 5% of the volume and mass of Earth.

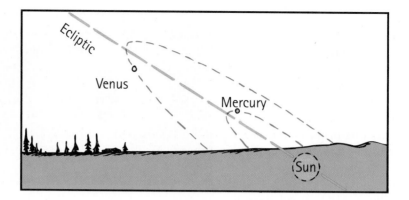

FIGURE 26.9

Because the orbits of Mercury and Venus lie inside the orbit of Earth, they are always near the Sun in our sky. Near sunset (or sunrise) they are visible as "evening stars" or "morning stars."

FIGURE 26.10

This photo of Venus, taken by NASA's Pioneer Venus Orbiter with cameras sensitive to ultraviolet light, reveals many of the planet's cloud features.

fyi

- Ancient American cultures ran their lives using three calendars. Their secular calendar, which told them when to plant seeds and so forth, followed the 365-day orbit of Earth. Their religious calendar centered on the roughly 260-day orbit of Venus, which gave them a calendar with 20-day weeks and 13-day months. Both of these calendars were cyclical and couldn't account for succeeding years. For that purpose they developed the "long count" calendar, which, interestingly enough, employed the concept of zero. They did this centuries before the concept of zero was recognized by accountants in India.

holds very little atmosphere. Mercury's atmosphere is only about a trillionth as dense as Earth's atmosphere—it's a better vacuum than laboratories on Earth can produce. So without a blanket of atmosphere, and because there are no winds to transfer heat from one region to another, nighttime on Mercury is very cold, about –170°C. Mercury is a fairly bright object in the nighttime sky and is best seen as an evening "star" during March and April or as a morning star during September and October. It is seen near the Sun at sunup or sunset.

VENUS

Venus, the second planet from the Sun, is frequently the first starlike object to appear after the Sun sets, so it is often called the evening "star," as illustrated in Figure 26.9. Compared with other planets, Venus most closely resembles Earth in size, density, and distance from the Sun. However, as shown in Figure 26.10, Venus has a very dense atmosphere and opaque cloud cover that generate high surface temperatures (470°C)—too hot for oceans. The atmosphere of Venus is about 96% CO_2. Remember from Chapter 24 that carbon dioxide is a "greenhouse gas." By this we mean that CO_2 blocks the escape of infrared radiation from Earth's surface and contributes to the warming of our planet. The thick blanket of CO_2 surrounding Venus effectively traps heat near the Venusian surface. This and Venus's proximity to the Sun make Venus the hottest planet in the solar system.

Another difference between Venus and Earth is in how the two planets spin about their axes. Venus takes 243 Earth days to make one full spin and only 225 Earth days to make one revolution around the Sun. Venus spins in a direction opposite to the direction of Earth's spin. So on Venus, the Sun rises in the west and sets in the east. But because the cloud cover is so intense, a Venusian sunrise or sunset is never visible from its surface.

The slow spin of Venus means that the atmosphere is not disturbed by the Coriolis effect described in Chapter 24. As a result, there is very little wind and weather on the surface of Venus. Instead, the stifling hot dense air sits still through its long days and nights.

In recent years, 17 probes have landed on the surface of Venus. There have been 18 flyby spacecraft (notably *Pioneer Venus* in 1978 and *Magellan* in 1993). From spacecraft data, scientists have been able to account for why the atmospheres of Venus and Earth are so different. According to the generally accepted model, when the two planets first formed they had similar amounts of water. Venus, however, is a bit closer to the Sun than Earth is. It also rotates much more slowly. These two factors combined to make the Sun-facing side of Venus significantly warmer than Earth. With greater warmth, more of Venus's water evaporated into its atmosphere. Like carbon dioxide, water vapor is a powerful greenhouse gas. So more water vapor in the atmosphere caused even more warming, which caused more of the oceans to evaporate, causing further warming—runaway global warming! Venus's early oceans contained massive amounts of dissolved carbonates, just as on Earth. As the oceans evaporated, these carbonates transformed into carbon dioxide and moved into the atmosphere, increasing the intensity of the greenhouse effect. Water vapor in the early Venusian atmosphere was subject to the Sun's ultraviolet rays, which broke the water down into hydrogen and oxygen. The hydrogen escaped into space while the oxygen chemically reacted with minerals on the surface. In the end, the planet's supply of water was forever lost. All that remains we see today as a thick atmosphere of heat-trapping carbon dioxide.

CHECK POINT

As the Sun gets older it also gets hotter. How will this affect the amount of water vapor in our atmosphere?

Was this your answer?

Initially, the additional heat from the Sun will increase the rate at which Earth's oceans evaporate. More water vapor in the atmosphere will then enhance the greenhouse effect, causing even warmer temperatures, which will provide for even more evaporation. As the oceans disappear, carbon dioxide levels will also rise dramatically, ensuring that the greenhouse effect keeps the water vaporized. Ultraviolet light from the Sun, however, will zap this water from the atmosphere. So in the end, as the Sun gets hotter, the amount of water vapor in the atmosphere will drop to near nothing. We will share the same fate as our sister planet Venus. You can breathe a bit easy because the time frame for this is about a billion years.

fyi

■ Not all of the Venusian water was lost to space. Some instead reacted with volcano-generated sulfur dioxide to form sulfuric acid, which now laces the upper levels of the Venusian atmosphere. That which stayed behind to react with sulfur dioxide, however, was predominately water containing the heavier isotope of hydrogen, known as deuterium. This has been confirmed with direct measurements from space probes, which show an unusually high proportion of deuterium. This, in turn, supports the theory that Venus lost its water because of a runaway greenhouse effect—if such didn't happen, then the proportion of deuterium in the atmosphere would remain similar to that currently found on Earth.

Link To Section 18.4

EARTH

Our home planet Earth resides within the Sun's *habitable zone*, which is a region not too close and not too far from the Sun so that water can exist predominately in the liquid phase, as shown in Figure 26.11. Earth has an abundant supply of liquid water covering about 70% of Earth's surface, which makes our planet the blue planet (Figure 26.12).

Earth's oceans support the *carbon dioxide cycle*, which acts as a thermostat to keep global temperatures from reaching harsh extremes. For example, if Earth froze over completely, carbon dioxide released by volcanoes would no longer be absorbed by the oceans. Instead the carbon dioxide would build up in the atmosphere, which would warm the atmosphere and hence melt the frozen Earth. Conversely, if Earth became hotter and hotter, more water would evaporate. This would lead to more precipitation, which would remove carbon dioxide from the atmosphere. With less carbon dioxide in the atmosphere, the greenhouse effect would be minimized and Earth would cool. So not only are we a nice distance from the Sun, but our atmosphere contains just enough water vapor and carbon dioxide to keep temperatures favorable for life. Furthermore, our relatively high daily spin rate allows only a brief and small lowering of temperature on the nighttime side of Earth. So temperature extremes of day and night are also kept moderate.

FIGURE 26.11

Earth resides on the inner side of the Sun's habitable zone, which is where conditions are favorable for life as we know it.

Solar System

FIGURE 26.12

Earth, the blue planet. This famous photo was taken by *Apollo 17* astronauts as they returned from the last manned mission to the Moon in 1972. It is the only existing photo of the full Earth from an appreciable distance. Can you see that it was taken in the summer months of the Southern Hemisphere?

Link To Section 25.1

Link To Section 14.1

For more information on the geology, weather, and history of planet Earth, review Chapters 20–25, which discuss Earth science.

CHECK POINT

If Venus was also once protected by the carbon dioxide cycle, what went wrong?

Was this your answer?
The carbon dioxide cycle requires water. The Venusian carbon dioxide cycle, therefore, broke down as the Venusian water was split apart by the Sun's ultraviolet rays.

MARS

Mars captures our fancy as another world, perhaps even as a world with life because it resides on the outer fringes of the habitable zone. Mars is a little more than half Earth's size; its mass is about one-ninth that of Earth; and it has a core, a mantle, a crust, and a thin, nearly cloudless atmosphere. It has polar ice caps and seasons that are nearly twice as long as Earth's because Mars takes nearly two Earth years to orbit the Sun. Mars and Earth spin on their axes at about the same rate, which means the lengths of their days are about the same. When Mars is closest to Earth, a situation that occurs once every 15 to 17 years, its bright, ruddy color outshines the brightest stars.

The Martian atmosphere is about 95% carbon dioxide, with only about 0.15% oxygen. Yet because the Martian atmosphere is relatively thin, it doesn't trap heat via the greenhouse effect as much as Earth's and Venus's atmospheres do. So the temperatures on Mars are generally colder than on Earth, ranging from about 30°C in the day at the equator to a frigid −130°C at night. If you visit Mars, never mind your raincoat, for there is far too little water vapor in the atmosphere for rain. Even the ice at the planet's poles consists primarily of carbon dioxide. And don't give a second thought to waterproof footwear, for the very low atmospheric pressure won't permit the existence of any puddles or lakes.

Conditions on Mars indicate that water was once abundant in its distant past. Channels on the Martian surface that appear to have been carved by water are seen by orbiting spacecraft. Landings on Mars, however, show it to be a very dry and windy place. Because the Martian atmosphere has a very low density, unequal heating produces Martian winds that are about 10 times as fast as winds on Earth.

In 2004, spacecraft orbiting Mars detected signs of the organic compound methane, CH_4, within the atmosphere. This is unusual because methane

FIGURE 26.13

(a) NASA's Mars Exploration Rover, *Spirit*, with cameras mounted on the white mast, took panoramic photos of the Martian surface. (b) *Spirit* took photographs in June 2004 for this composite, true-color image of the region named Columbia Hills on Mars. The vehicle later traveled to the hills to analyze their composition.

(a)

(b)

decomposes fairly rapidly, which tells us that this compound is currently being produced. The likely source is ongoing volcanic activity, which could potentially melt underground ice into liquid water. Indeed, scientists have found evidence of the leakage of underground liquid water onto the surface occurring since we started surveying Mars from space. Once on the surface, this water would evaporate, freeze, and sublime away. Underground pools of volcanically warmed liquid water, however, may harbor microscopic life forms.

Mars has two small moons—Phobos, the inner one, and Deimos, the outer. Both are potato-shaped and have cratered surfaces. They are likely captured asteroids. Phobos orbits in the same easterly direction in which Mars spins (like our Moon), at a distance of only 6000 km in a period of only 7.5 hours. From Mars it appears about half the size of our Moon. Deimos is about half the size of Phobos, and it orbits Mars in 30.3 hours at a distance of 20,000 km from the Martian surface.

26.4 The Outer Planets

The outer planets—Jupiter, Saturn, Uranus, and Neptune—are gigantic, gaseous, low-density worlds. They each formed from rocky and metallic cores that were much more massive than the terrestrial planets. The gravitational forces of these cores were strong enough to sweep up gases of the early planetary nebula, primarily hydrogen and helium. The cores continued to collect gases until the Sun ignited and the solar wind blew away all remaining interplanetary gases. The core of Jupiter was the first to develop, and hence it had the longest time to collect gas before solar ignition. This is why Jupiter is the largest of the outer planets. Another commonality is that they all have ring systems, Saturn's being the most prominent. We will explore the outer planets in the order of their distance from the Sun.

JUPITER

Jupiter is the largest of all the planets. Its yellow light in our night sky is brighter than that of any star. Jupiter spins rapidly about its axis in about 10 hours, a speed that flattens it so that its equatorial diameter is about 6% greater than its polar diameter. As with the Sun, all parts do not rotate in unison. Equatorial regions complete a full revolution several minutes before nearby regions in higher and lower latitudes. The atmospheric pressure at Jupiter's rocky surface is more than a million times the atmospheric pressure of Earth. Jupiter's atmosphere is about 82% hydrogen, 17% helium, and 1% methane, ammonia, and other molecules.

Jupiter's average diameter is about 11 times Earth's, which means Jupiter's volume is more than 1000 times Earth's. Jupiter's mass is greater than the combined masses of all the other planets. Because of its low density, however—about one-fourth of Earth's—Jupiter's mass is barely more than 300 times Earth's. Investigations of Jupiter tell us that its core is a solid sphere about 15 times as massive as the entire Earth, and it is composed of iron, nickel, and other minerals.

More than half of Jupiter's volume is an ocean of liquid hydrogen. Beneath the hydrogen ocean lies an inner layer of hydrogen compressed into a liquid metallic state. In it are abundant conduction electrons that flow to produce Jupiter's enormous magnetic field. The strong magnetic field about the planet captures high-energy particles and produces radiation belts 400 million times as

Why are planets round? All parts of a forming planet pull close together by mutual gravitation. No "corners" form because they're simply pulled in. So gravity is the cause of the spherical shapes of planets and other celestial bodies.

fyi

- Jupiter is near the point at which the addition of more matter would cause the size of this planet to contract. This is analogous to a stack of pillows. Start stacking pillows and the stack gets taller. Eventually, however, a point is reached at which the weight of upper pillows pushes down on lower pillows such that the column of pillows gets shorter. Interestingly, Jupiter is *larger* than the smallest stars, which, though smaller than Jupiter, are about 80 times as massive.

FIGURE 26.14

Jupiter, with its moons Io (orange dot over planet) and Europa (white dot to right of planet), as seen from the *Voyager 1* spacecraft in February 1979. The Great Red Spot (lower left), larger than Earth, is a cyclonic weather pattern of high winds and turbulence.

Link To Section 9.2

FIGURE 26.15

This artist's rendering shows aurorae (pink) in the upper atmosphere of Jupiter. Thunderclouds are seen below the aurorae, and the nearest major moon, Io, is seen at center left. Aurorae, like the northern lights on Earth, are caused by charged particles from the solar wind exciting gas molecules in the upper atmosphere. The gas molecules emit light as they return to an unexcited energy state.

energetic as Earth's Van Allen radiation belts. Radiation levels surrounding Jupiter are the highest ever recorded in space.

If you're planning to visit Jupiter, choose one of its moons instead. It has at least 63 of them, in addition to a faint ring. The four largest moons were discovered by Galileo in 1610; Io and Europa are about the size of our Moon, and Ganymede and Callisto are about as large as Mercury (Figure 26.16). Jupiter's moon Io has more volcanic activity than any other body in the solar system. Perhaps most intriguing of all, however, is Europa, whose surface is made of frozen water. As shown in Figure 26.17, deep beneath this ice is likely an ocean of water kept warm by the strong tidal forces from nearby Jupiter. If life were to be found anywhere in this solar system besides Earth, it would likely be on the floor of Europa's ocean adjacent to volcanic thermal vents. Such extraterrestrial life forms may be similar to the bizarre forms of life recently discovered adjacent to deep thermal vents on Earth's ocean floor. Alternatively, they may be single-cell organisms, such as bacteria. Then again, there may be nothing. So is life rare or common in this universe? The answer may be waiting for us in our own galactic backyard.

SATURN

Saturn is one of the most remarkable objects in the sky, with its rings clearly visible through a small telescope. It is quite bright—brighter than all but two stars—and it is second only to Jupiter in mass and size. Saturn is twice as far from Earth as Jupiter is. Its diameter, not counting its ring system, is nearly 10 times that of Earth, and its mass is nearly 100 times Earth's. It is composed primarily of hydrogen and helium, and it has the lowest density of any planet, only 0.7 times the density of water. These characteristics mean that Saturn would easily float in a bathtub, if the bathtub were large enough. Its low density and its 10.2-hour rapid spin produce more polar flattening than can be seen in the other planets. Notice its oblong nature in Figure 26.18. Like Jupiter, Saturn radiates about twice as much heat energy as it receives from the Sun.

FIGURE 26.16

Jupiter's four largest moons were discovered in 1610 by Galileo, who was the first to point the recently invented telescope toward the heavens. He noted the changing positions of these moons and concluded that they were orbiting Jupiter, which was a violation of the then widely held belief that all heavenly objects orbited Earth. His discovery was revolutionary. In his honor, these four moons are known as the Galilean moons.

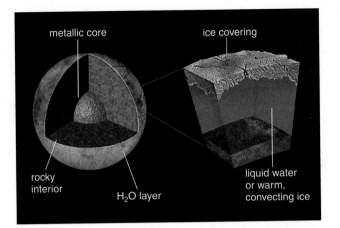

metallic core

ice covering

rocky
interior

H_2O layer

liquid water
or warm,
convecting ice

FIGURE 26.17

A model of the interior of
Europa with a zoomed-in view
of its ice-capped ocean, which,
according to magnetic measurements, likely covers the
entire sphere.

Saturn's rings, likely only a few kilometers thick, lie in a plane coincident
with Saturn's equator. Four concentric rings have been known for many years,
and spacecraft missions have detected others. The rings are composed of chunks
of frozen water and rocks, believed to be the material of a moon that never
formed or the remnants of a moon torn apart by tidal forces. All the rocks and
bits of matter that make up the rings pursue independent orbits about Saturn.
The inner parts of the ring travel faster than the outer parts, just as any satellite
near a planet travels faster than a more distant satellite.

Saturn has some 47 moons beyond its rings. The largest is Titan, 1.6 times
as large as our Moon and even larger than Mercury. It spins once every 16
days and has a methane atmosphere (not likely from any biological source)
with atmospheric pressure that is likely greater than Earth's. Its surface temperature is cold, roughly −170°C. A space probe built by NASA and the
European Space Agency landed on Titan in 2005. Remarkably, photos
revealed a landscape similar to Earth despite the fact that the materials are
completely different (Figure 26.19). Lakes and streams are filled with not
water but liquid methane. Rocks are made of ice. Instead of lava, Titan has a
flowing slush of ice and liquid ammonia. No life is expected to be found on
this moon because of the intensely cold temperatures. Titan, however, holds
an intriguing soup of organic molecules whose chemistry may provide a clue
to what Earth was like during the time before life arose here.

FIGURE 26.18

Saturn surrounded by its
famous rings, which are
composed of rocks and ice.

FIGURE 26.19

INTERACTIVE FIGURE

Images from Saturn's largest
moon, Titan, taken by the
Cassini spacecraft and its space
probe, the *Huygens*, which
successfully descended to the
surface.

1 km

FIGURE 26.20

This infrared image of Uranus was gathered by the 10-m Keck Telescope in Hawaii on July 11–12, 2004. Astronomers believe that Uranus tilts on its rotational axis as a result of a collision it had with a large body early in the solar system's history. The bright white and blue spots in the southern hemisphere of Uranus are clouds. Methane in the upper atmosphere absorbs red light, giving Uranus its blue-green color.

Recent studies of Galileo's notebooks show that Galileo saw Neptune in December 1612 and again in January 1613. He was interested in Jupiter at the time, and so he merely plotted Neptune as a background star.

FIGURE 26.21

Cyclonic disturbances on Neptune in 1989 produced a great dark spot, which was even larger than Earth and similar to Jupiter's Great Red Spot. The spot has now disappeared. The gray horizon in the foreground of this computer-generated montage is a close-up of Neptune's moon Triton, which has a composition and size similar to that of Pluto.

URANUS

Uranus (pronounced "YUR-uh-nus," accent on the first syllable) is twice as far from Earth as Saturn is, and it can barely be seen with the naked eye. Uranus was unknown to ancient astronomers and not discovered as a planet until 1781. The *Voyager 2* spacecraft first visited this planet in 1986. It has a diameter four times that of Earth and a density slightly greater than that of water. So if you could place Uranus in a giant bathtub, it would sink. The most unusual feature of Uranus is its tilt. Its axis is tilted 98° to the perpendicular of its orbital plane, so it lies on its side (Figure 26.20). Unlike Jupiter and Saturn, it appears to have no appreciable internal source of heat. Uranus is a cold place.

Uranus has at least 27 moons, in addition to a complicated faint ring system. Recall from Chapter 4 that perturbations in the planet Uranus led to the discovery in 1846 of a farther planet, Neptune.

NEPTUNE

Neptune has a diameter about 3.9 times that of Earth, a mass 17 times as great, and a mean density about a third of Earth's. Its atmosphere is mainly hydrogen and helium, with some methane and ammonia, which makes Neptune bluer than Uranus (Figure 26.21). Like Jupiter and Saturn, it emits about 2.5 times as much heat energy as it receives from the Sun. This is perplexing because calculations show that, like Uranus, Neptune should have already lost all of its original heat. One possible explanation is that, unlike Uranus, Neptune is still contracting.

The *Voyager 2* spacecraft flew by Neptune in 1989. It showed that Neptune has at least 13 moons in addition to a ring system. The largest moon, Triton, orbits Neptune in 5.9 days in a direction opposite to the planet's eastward spin. This suggests that Triton is a captured object. Triton's diameter is three-quarters the size of our Moon's diameter, and yet Triton is twice as massive as Earth's Moon. It has bright polar caps and geysers of liquid nitrogen.

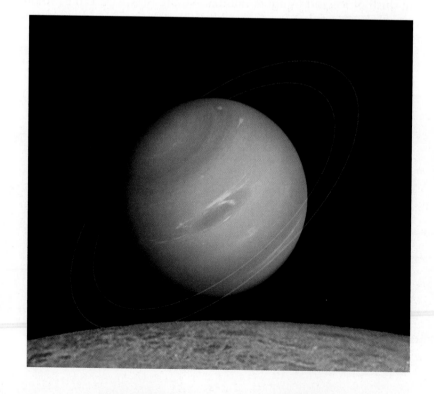

■ 26.5 Earth's Moon

Earth's Moon is puzzling. It is close to the size of Mercury, which is a planet and *not* a moon. The composition of Earth's Moon is nearly the same as Earth's mantle. Furthermore, the Moon possesses a rather small iron core. To explain these and a multitude of other facts about the Moon, scientists have pieced together the following probable scenario for its origin.

During the early history of the solar system, the young Earth had a Mercury-sized companion that accreted within an orbit close to that of Earth. Normally, if the companion's orbit were a bit closer to the Sun, then it would orbit faster than Earth and move ahead of Earth. Upon passing through a special point, known as the *Lagrangian point*, however, the object would find that the gravitational pull of Earth is strong enough to hold it back so that it orbited with Earth in unison. So early Earth may have had a twin that paraded with Earth around the Sun much like two horses running side by side on a circular track.

Eventually, a random event, such as the passing of an asteroid or comet, caused our companion to sway from the Lagrangian point and fall toward and collide with Earth. The collision would have been massively spectacular, spewing debris everywhere while turning Earth fully molten. Hitting askew, the impact sent Earth into a wild spin rotating once every five hours. The debris soon collected as a ring around Earth, and then, within about 1000 years, the ring coalesced into the Moon. This scenario is known as the *giant impact theory* of the origin of Earth's Moon. It explains why the Moon is so large (we started out as twin planets), why its composition is similar to Earth (it formed from our mantle and our mantle formed from it), why it has such a small iron core

> We see the aftermath of meteoroid bombardment on the Moon because it wears no makeup. Similar bombardment on Earth has been long erased by erosion.

FIGURE 26.22

Three steps to the formation of Earth's Moon. A Mercury-sized object collides with Earth, which turns molten. Debris collects in a ring that accretes into the Moon, which is quite close to the rapidly rotating Earth. Over the next billion years tidal forces slow the rate of Earth's rotation while also causing the Moon to move farther away.

A Mercury-sized planetesimal crashes into the young Earth, shattering both the planetesimal and our planet.

Hours later, our planet is completely molten and rotating very rapidly. Debris splashed out from Earth's outer layers is now in Earth orbit. Some debris rains back down on Earth, while some will gradually accrete to become the Moon.

Less than a thousand years later, the Moon's accretion is rapidly nearing its end, and relatively little debris still remains in Earth orbit.

FIGURE 26.23

Earth and the Moon as photographed in 1977 from the *Voyager 1* spacecraft on its way to Jupiter and Saturn.

■ During the time of the dinosaurs, a day was only about 19 hours. Today our days are about 24 hours. Billions of years from now, as Earth continues to slow down, a day will last about 47 days. At that time, Earth and the Moon will be gravity locked such that the Moon always appears in one location in the sky. To see the Moon, you will need to be on the Moon side of the planet, where perhaps real estate prices will be higher because of the view. One huge problem, however, is that by then, our Sun will have already gone through its dying phases, through which Earth will be subject to the fate of Venus, where it is forever cloudy. Nothing is permanent.

Most planets wobble significantly as they spin about their axes. The Moon, however, helps keep Earth's wobble to a minimum. As a result, our weather patterns are fairly consistent through the ages, which makes our planet even more favorable for the development of life. Thank you, Moon!

FIGURE 26.24

Edwin E. Aldrin, Jr., one of the three *Apollo 11* astronauts, stands on the dusty lunar surface. To date, 12 people have stood on the Moon.

(Earth's iron core had already differentiated and was not sent up with the debris), and much more. This impact theory is still the subject of much research and is thus being continually refined. Although it was developed only within the past couple decades, scientists are excited by its explanatory powers.

From a distance, Earth and the Moon still resemble a twin planet system, as you can see in Figure 26.23. Compared to Earth, however, the Moon is relatively small, with a diameter of about the distance from San Francisco to New York City. It once had a molten surface, but it cooled too rapidly for the establishment of moving crustal plates, like those of Earth. In its early history, it was intensely bombarded by meteoroids (as was Earth). A little more than 3 billion years ago, meteoroid bombardment and volcanic activity filled basins with lava to produce its present surface. It has undergone very little change since then. Its igneous crust is thicker than Earth's. The Moon is too small with too little gravitational pull to have an atmosphere, and so, without weather, the only eroding agents have been meteoroid impacts.

THE PHASES OF THE MOON

Sunshine always illuminates half of the Moon's surface. The Moon shows us different amounts of its sunlit half as it circles Earth each month. These changes are the **Moon phases** (Figure 26.25). The Moon cycle begins with the **new Moon**. In this phase, its dark side faces us and we see darkness. This occurs when the Moon is between Earth and the Sun (position 1 in Figure 26.26).

During the next seven days, we see more and more of the Moon's sunlit side (position 2 in Figure 26.26). The Moon is going though its waxing crescent phase (*waxing* means "increasing"). At the first quarter, the angle between the Sun, the Moon, and Earth is 90°. At this time, we see half the sunlit part of the Moon (position 3 in Figure 26.26).

During the next week, we see more and more of the sunlit part. The Moon is going through its waxing gibbous phase (position 4 in Figure 26.26). (*Gibbous* means "more than half.") We see a **full Moon** when the sunlit side of the Moon faces us squarely (position 5 in Figure 26.26). At this time, the Sun, Earth, and the Moon are lined up, with Earth in between. To view this full Moon you need to be on the nighttime side of Earth, at sunset when the full Moon rises from the east, or at sunrise when it sets in the west.

The cycle reverses during the following two weeks, as we see less and less of the sunlit side while the Moon continues in its orbit. This movement produces

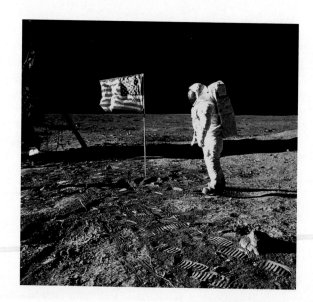

CHAPTER 26   THE SOLAR SYSTEM   **705**

the waning gibbous, last quarter, and waning crescent phases. (*Waning* means "shrinking.") The time for one complete cycle is about 29.5 days.*

FIGURE 26.25

The Moon in its various phases.

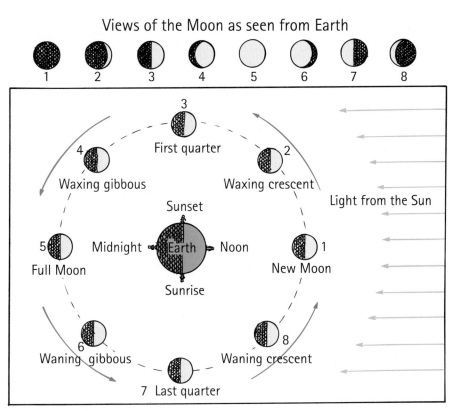

FIGURE 26.26

Sunlight always illuminates half of the Moon. As the Moon orbits Earth, we see varying amounts of its sunlit side. One lunar phase cycle takes 29.5 days.

CHECK POINT

1. Can a full Moon be seen at noon? Can a new Moon be seen at midnight?
2. Astronomers prefer to view the stars when the Moon is absent from the night sky. When, and how often, is the Moon absent from the night sky?

Were these your answers?
1. Figure 26.26 shows that at noontime, you would be on the wrong side of Earth to view the full Moon. Likewise, at midnight, the new Moon would be absent. The new Moon is in the sky in the daytime, not at night.
2. At the time of the new Moon and during the week on either side of the new Moon, the night sky does not show the Moon. Unless an astronomer wishes to study the Moon, these dark nights are the best time for viewing other objects. Astronomers usually view the night skies during two-week periods every two weeks.

If someone shone a flashlight on a ball in a dark room, you could tell where the flashlight was by looking at the illumination on the ball. The Moon is similarly lit by the Sun.

* The Moon actually orbits Earth once every 27.3 days relative to the stars. The 29.5-day cycle is relative to the Sun and is due to the motion of the Earth–Moon system as it revolves about the Sun.

The Moon spins about its own polar axis just as often as it circles Earth. So as the Moon circles Earth, it spins so that the same side (shown in yellow) always faces Earth. In each of the four successive positions shown here, the Moon has spun one-quarter of a turn.

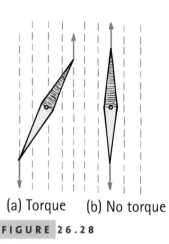

(a) Torque   (b) No torque

(a) When the compass needle is not aligned with the magnetic field (dashed lines), the forces represented by the blue arrows at either end produce a pair of torques that rotate the needle. (b) When the needle is aligned with the magnetic field, the forces no longer produce torques.

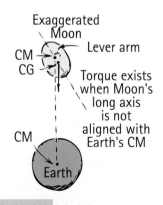

When the long axis of the Moon is not aligned with Earth's gravitational field, Earth exerts a torque that rotates the Moon into alignment. (CM: center of mass; CG: center of gravity)

WHY ONE SIDE ALWAYS FACES US

The first images of the back side of the Moon were taken by the unmanned Russian spacecraft *Lunik 3* in 1959. The first human witnesses of the Moon's back were *Apollo 8* astronauts, who orbited the Moon in 1968. From Earth, we see only a single lunar side. The familiar facial features of the "man in the Moon" are always turned toward us on Earth. Does this mean that the Moon doesn't spin about its axis as Earth does daily? No, but, relative to the stars, the Moon in fact does spin, although quite slowly—about once every 27 days. This monthly rate of spin matches the rate at which the Moon revolves about Earth. This explains why the same side of the Moon always faces Earth (Figure 26.27). This matching of monthly spin rate and orbital revolution rate is not a coincidence. After you answer the following Checkpoint, we'll explore why.

CHECKPOINT

A friend says that the Moon does not spin about its axis, and evidence for a nonspinning Moon is the fact that its same side always faces Earth. What do you say?

Was this your answer?

Place a quarter and a penny on a table. Pretend the quarter is Earth and the penny the Moon. Keeping the quarter fixed, revolve the penny around it in such a way that Lincoln's head is always pointed to the center of the quarter. Ask your friend to count how many rotations the penny makes in one revolution (orbit) around the quarter. He or she will see that it rotates once with each revolution. The key concept is that the Moon takes the same amount of time to complete one rotation as it does to revolve around Earth.

Think of a compass needle that lines up with a magnetic field. This lineup is caused by a *torque*—a "turning force with leverage" (like that produced by the weight of a child at the end of a seesaw). The compass needle on the left in Figure 26.28 rotates because of a pair of torques. The needle rotates counterclockwise until it aligns with the magnetic field. In a similar manner, the Moon aligns with Earth's gravitational field.

We know from the law of universal gravitation that gravity weakens with the inverse square of distance, so the side of the Moon nearer to Earth is gravitationally pulled more than the farther side. This stretches the Moon out to a football shape. (The Moon does the same to Earth and gives us tides). If its long axis doesn't line up with Earth's gravitational field, a torque acts on it as shown in Figure 26.29. Like a compass in a magnetic field, it turns into alignment. So the Moon lines up with Earth in its monthly orbit. One hemisphere always faces us. It's interesting to note that for many moons orbiting other planets, a single hemisphere faces the planet. We say these moons are "tidally locked."

FIGURE 26.30

A solar eclipse occurs when the Moon passes in front of the Sun as seen from Earth. The Moon's shadow has two portions: a dark, central umbra surrounded by the lighter penumbra. A total eclipse is seen from within the umbra and may last several minutes.

ECLIPSES

Although the Sun is 400 times as large in diameter as the Moon, it is also 400 times as far away. So from Earth, both the Sun and Moon measure the same angle (0.5°) and appear to be the same size in the sky. This coincidence allows us to see solar eclipses.

Both Earth and the Moon cast shadows when sunlight shines on them. When the path of either of these bodies crosses into the shadow cast by the other, an eclipse occurs. A **solar eclipse** occurs when the Moon's shadow falls on Earth. Because of the large size of the Sun, the rays taper to provide an umbra and a surrounding penumbra, as shown in Figures 26.30 and 26.31.

FIGURE 26.31

Geometry of a solar eclipse. During a solar eclipse, the Moon is directly between the Sun and Earth and the Moon's shadow is cast on Earth. Because of the small size of the Moon and tapering of the solar rays, a solar eclipse occurs only on a small area of Earth.

An observer in the umbra part of the shadow experiences darkness during the day—a total eclipse, *totality*. Totality begins when the Sun disappears behind the Moon, and ends when the Sun reappears on the other edge of the Moon. The average time of totality at any location is about 2 or 3 minutes, with a maximum no longer than 7.5 minutes. The eclipse time in any location is brief because of the Moon's motion. During totality, what appears in the sky is an eerie black disk surrounded by the pearly white streams of the corona, as was shown in Figure 26.7. It is an experience one can never forget. With binoculars, the features of the Moon can be seen because they are lit by the sunlight reflected from Earth. Pink flares from the chromosphere may also appear. But great caution is advised when viewing the totality, which must be a totality of 100%. The moment the first edges of the photosphere appear, which is the moment you now have 99.99% totality, is the very moment when you can seriously damage your eyes if you continue to look.* At that moment you have entered the penumbra, where the eclipse is partial. An ideal way to view the partial solar eclipse is to focus the light of the eclipse onto a white surface, as was shown in Figure 26.4. Alternatively, you can view the crescent Sun under the shade of a tree, which casts pinhole images of the Sun onto the ground, as shown in Figure 26.32. Check the map shown in Figure 26.33 to see if a solar eclipse is coming to your area soon. Many solar eclipse enthusiasts travel the world to view this inspiring natural phenomenon.

FIGURE 26.32

The crescent-shaped spots of sunlight are images of the partially eclipsed Sun formed as the sunlight passes through the leaves, which overlap to create image-forming pinholes. On a normal noneclipsed sunny day, these spots beneath a tree are round because the Sun is round. Look for round "Sun balls" on your next sunny day outside.

* People are cautioned not to look at the Sun at the time of a solar eclipse because the brightness and the ultraviolet light of direct sunlight damage the eyes. This good advice is often misunderstood by those who then think that sunlight is more damaging at this special time. However, staring at the Sun when it is high in the sky is harmful whether or not an eclipse occurs. In fact, staring at the bare Sun is more harmful than when part of the Moon blocks it. The reason for special caution at the time of an eclipse is simply that more people are interested in looking at the Sun during this time.

FIGURE 26.33

This map shows the paths of total solar eclipses from 2006 through 2030. More details about these and other future solar eclipses can be found on NASA's eclipse Web site, http://sunearth.gsfc.nasa.gov/eclipse/eclipse.html.

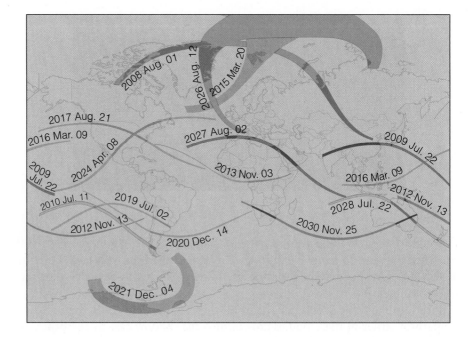

The corona of the Sun is about as bright as the full Moon to those in the path of totality.

FIGURE 26.34

A lunar eclipse occurs when Earth is directly between the Moon and the Sun and Earth's shadow is cast on the Moon.

FIGURE 26.35

A fully eclipsed Moon is not completely dark in the shadow of Earth but is quite visible. This is because Earth's atmosphere acts as a lens and refracts light into the shadow region—sufficient light to faintly illuminate the Moon.

The alignment of Earth, the Moon, and the Sun also produces a **lunar eclipse** when the Moon passes into the shadow of Earth, as shown in Figure 26.34. Usually a lunar eclipse precedes or follows a solar eclipse by two weeks. Just as all solar eclipses involve a new Moon, all lunar eclipses involve a full Moon. They may be partial or total. All observers on the dark side of Earth see a lunar eclipse at the same time. Interestingly enough, when the Moon is fully eclipsed, it is still visible as is shown and discussed in Figure 26.35.

Why are eclipses relatively rare events? This has to do with the different orbital planes of Earth and the Moon. Earth revolves around the Sun in a flat planar orbit. The Moon similarly revolves about Earth in a flat planar orbit. But the planes are slightly tipped with respect to each other—a 5.2° tilt, as shown in Figure 26.36. If the planes weren't tipped, eclipses would occur monthly. Because of the tip, eclipses occur only when the Moon intersects the Earth–Sun plane at the time of a three-body alignment (Figure 26.37). This occurs about two times per year, which is why there are at least two solar eclipses per year (visible only from certain locations on Earth). Sometimes there are as many as seven solar and lunar eclipses in a year.

CHECK POINT

1. Does a solar eclipse occur at the time of a full Moon or a new Moon?
2. Does a lunar eclipse occur at the time of a full Moon or a new Moon?

Were these your answers?

1. A solar eclipse occurs at the time of a new Moon, when the Moon is directly in front of the Sun. Then the shadow of the Moon falls on part of Earth.
2. A lunar eclipse occurs at the time of a full Moon, when the Moon and Sun are on opposite sides of Earth. Then the shadow of Earth falls on the full Moon.

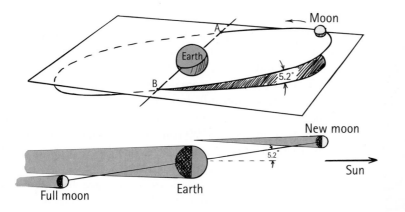

FIGURE 26.36

The Moon orbits Earth in a plane tipped 5.2° relative to the plane of Earth's orbit around the Sun. A solar or lunar eclipse occurs only when the Moon intersects the Earth–Sun plane (points A and B) at the precise time of a three-body alignment.

26.6 Failed Planet Formation

In three regions of our solar system, we find the remains of material that failed to collect into planets. These regions are the asteroid belt, the Kuiper belt, and the Oort Cloud.

THE ASTEROID BELT AND METEORS

The **asteroid belt** is a collection of rocks located between the orbits of Mars and Jupiter. More than 150,000 asteroids have been cataloged so far, but many more no doubt have yet to be discovered. They come in all shapes and sizes, but the largest asteroid, Ceres, is just under a thousand kilometers in diameter. Although Ceres is large enough to be fairly round, most asteroids are shaped more like a potato, as shown in Figure 26.38.

Evidence suggests that when the solar system was forming, the asteroid belt held much more mass than it does today. Massive Jupiter likely disrupted the orbits of this material, sending it off in many directions, including toward the inner planets and out of the solar system. The two moons of Mars, for example, are thought to be former asteroids. What remains of the asteroid belt is small. If all the presently remaining asteroids were scrunched together, they would make a sphere less than half the size of our Moon.

Jupiter also causes the collisions of asteroids with asteroids, which then break apart into smaller fragments. So rather than building into a planet, this material is slowly ground down and pushed off course. Asteroid fragments known as

Meteorites fall all over our planet, but the easiest place to find them is on the icy white surfaces found in polar regions. Are you looking to collect your own meteorites? Head south to Antarctica!

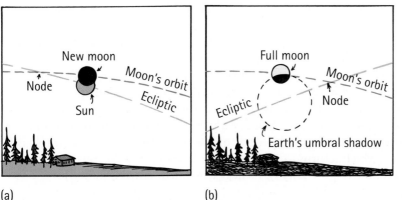

(a)   (b)

FIGURE 26.37

A total eclipse can occur only when the Moon's orbit intersects with the plane of Earth's orbit, which is the ecliptic. A solar eclipse occurs during the day as the new Moon passes in front of the Sun. A lunar eclipse happens only at night when the full Moon passes through Earth's shadow.

FIGURE 26.38

The asteroid Eros is about 40 km long, and like other small objects in the solar system, it is not spherical.

FIGURE 26.39

A meteor is produced when a meteoroid, usually about 80 km high, enters Earth's atmosphere. Most are sand-sized grains, which are seen as "falling" or "shooting" stars.

FIGURE 26.40

The Barringer Crater in Arizona, made 25,000 years ago by an iron meteorite with a diameter of about 50 m. The crater extends 1.2 km across and reaches 200 m deep.

meteoroids frequently find their way to Earth, where they are heated white-hot by friction with the atmosphere. As they descend with a fiery glow, they are called **meteors** (Figure 26.39).

If the meteoroid is large enough, it may survive to reach the surface, where it is called a **meteorite**. Most meteoroids, meteors, and meteorites are from asteroids, but many are also from comets, as we discuss later. Fortunately, smaller meteorites hit us more frequently than larger ones do. About 200 tons of small meteorites strike Earth every day. Every 10,000 years or so we are hit with a meteorite big enough to create a large crater, such as the one shown in Figure 26.40. Every 100 million years or so we are hit with one big enough—about 10 km in diameter—to cause a mass extinction, as occurred 65 million years ago at the end of the Cretaceous period discussed in Chapter 21. Accordingly, one of NASA's goals is to detect up to 90% of all large, near-Earth objects. If we can detect a dangerous space fragment early enough, we can take actions to alter its orbital path sufficiently to avoid impending disaster.

THE KUIPER BELT AND DWARF PLANETS

Beyond Neptune at a distance from about 30 to 50 AU is a region known as the **Kuiper belt** (pronounced "KI-pur," "KI" as in *kite*, rhymes with *hyper*). The Kuiper belt is occupied by many rocky, ice-covered objects. The most well-known Kuiper belt object is Pluto, which until recently was classified as a planet. Since its discovery in 1930, however, astronomers knew that Pluto was quite different from all the other known planets. For example, Pluto orbits at an angle to the plane of the solar system—the ecliptic. Also, Pluto is quite small, being only one-seventh as massive as our Moon. Then, starting in the 1990s, astronomers began discovering many more Kuiper belt objects as large as or larger than Pluto. So in 2006 these Pluto-sized Kuiper belt objects were officially classified as **dwarf planets**. The main reason they do not meet full planet status is that they have yet to accrete all the material in their orbital paths. In the outer edges of our solar system, however, matter is simply too sparse for that to happen. Interestingly, if the Kuiper belt were more dense with material, then these dwarf planets could have served as cores for additional jovian planets. But that never happened, and so the Kuiper belt is another zone of failed planet formation.

Space probes have yet to visit any of the dwarf planets of the Kuiper belt. Pluto and its moon, Charon, however, are due to be visited by the *New Horizons* spacecraft in 2015. We may have already had a preview, however, when the *Voyager 2* spacecraft took pictures of Neptune's moon Triton. Astronomers now suspect that Triton is a Kuiper belt dwarf planet pulled off course and captured into orbit around Neptune.

The larger Kuiper belt objects, such as Pluto, have a fair amount of inertia and so are not so easily thrown off course. Lighter Kuiper belt objects, however, are thrown off course quite frequently. Sometimes they are thrown toward the Sun, where the added heat and solar wind cause the ice and other volatile materials to be ejected, always in a direction away from the Sun. We see these objects as **comets,** which are characterized by their long and sometimes quite brilliant tails. The cover of this textbook shows the unusually bright comet McNaught, which passed close to the Sun in early 2007. Comets that come from the Kuiper belt tend to have orbital periods of less than 200 years. An example is Comet Halley, which returns to the inner solar system every 76 years—once in an average lifetime (Figure 26.43). Its next scheduled return is in 2061.

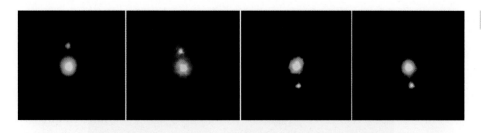

FIGURE 26.41

Theses infrared images of Pluto being orbited by its moon, Charon, are fuzzy because of the small size of these bodies and their great distance from Earth.

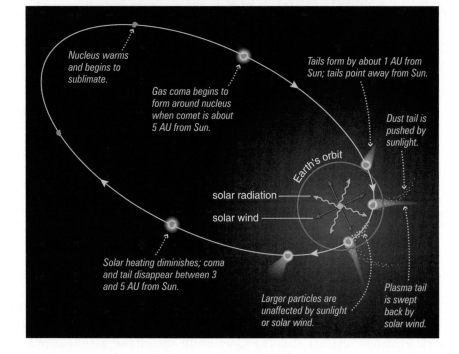

FIGURE 26.42

The comet warms as it gets closer to the Sun and initially develops a coma, which is a halo of gases surrounding the comet nucleus. From this coma arises the tail, which is blown outward by the solar wind. Note how the tail always extends away from the Sun. Most comets never make this journey and instead remain perpetually frozen within the outer reaches of our solar system.

Comets apparently reside in at least two regions. The first is the Kuiper belt, which lies roughly within the same plane of the solar system. The second region lies much farther out and surrounds our entire solar system—like a cloud.

THE OORT CLOUD AND COMETS

As the jovian planets grew, their gravitational pulls became stronger, which made them even more effective at pulling in additional interplanetary debris. Not all debris, however, was pulled fully into the jovian planets. In many cases, a

FIGURE 26.43

Observations of Comet Halley have been recorded for thousands of years. Although it usually provides a brilliant display, its last visit in 1986 was not so spectacular when viewed from Earth. We were ready, though, with space probes that flew close enough to Halley to capture dramatic images of its nucleus.

FIGURE 26.44

There are two major reservoirs for comets: the Kuiper belt and the Oort cloud.

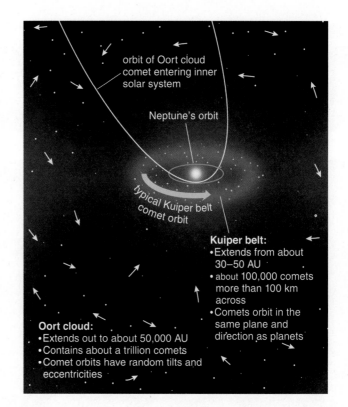

orbit of Oort cloud comet entering inner solar system

Neptune's orbit

typical Kuiper belt comet orbit

Kuiper belt:
• Extends from about 30–50 AU
• about 100,000 comets more than 100 km across
• Comets orbit in the same plane and direction as planets

Oort cloud:
• Extends out to about 50,000 AU
• Contains about a trillion comets
• Comet orbits have random tilts and eccentricities

Most comets usually last only a couple of orbits before they break up. But if the solar system is billions of years old, shouldn't they be depleted by now? This very question led to the idea of the Oort cloud, which provides a continual supply of new comets, replacing those that are destroyed.

Link To Section 21.5

chunk of rock or ice just missing a planet was instead whipped around the planet and then flung violently outward in any direction. Over billions of years this created a sphere of far-out objects just barely held to our solar system. We refer to this collection of far-out objects as the **Oort cloud** (*Oort* rhymes with *court*). Evidence suggests that the Oort cloud consists of a trillion objects extending as far out as 50,000 AU, which brings the cloud about a quarter of the way to the nearest star. A few of these objects periodically fall toward and then around the Sun, where they appear as comets. The orbital periods of comets originating from the Oort cloud is on the order of thousands or even millions of years. They come from nearly any angle.

Whether the comet comes from the Kuiper belt or the Oort cloud, it still has the potential for colliding with a planet. In 1994, Comet Shoemaker-Levy collided spectacularly with Jupiter, as shown in Figure 26.45. Also, the large meteorite that collided with Earth 65 million years ago causing the mass extinction of the dinosaurs may have been a comet.

FIGURE 26.45

Comet Shoemaker-Levy was already broken up into a string of objects just before it collided with Jupiter in 1994. The image on the left (a) shows an infrared view of the collision, which produced much heat as well as visible scars (the black dots), as shown in the photograph on the right (b).

(a)

(b)

TABLE 26.2   METEOR SHOWER DATA

Shower Name	Radiant*	Dates	Peak Dates	Meteors per Hour
Quadrantids	Pegasus	Jan. 1–6	Jan. 3	60
Eta Aquarids	Aquarius	May 1–10	May 6	35
Perseids	Perseus	Jul. 23–Aug. 20	Aug. 12	75
Orionids	Orion	Oct. 16–27	Oct. 22	25
Geminids	Gemini	Dec. 7–15	Dec. 13	75

* Meteors appear to radiate from a certain region of the sky, appropriately called a *radiant*. Radiants refer to constellations. See Chapter 27 for more on where the various constellations are located in the night sky.

FIGURE 26.46

When Earth crosses the orbit of a comet, we see a meteor shower.

The tail of the comet leaves behind a shower of particles. Each year Earth passes through the remnants of comet tails that create annual meteor showers, as indicated in Table 26.2. Meteor showers are beautiful to watch. Just go outside, look up at the sky, and with good eyesight every minute or so you will see a shooting star. Each streak is a tiny chip of a comet, once so very far away, that has fallen into Earth's neighborhood (Figure 26.46).

CHECK POINT

Of the asteroid belt, the Kuiper belt, and the Oort cloud:
1. Which is closest to the Sun?
2. Which generates comets?
3. Which gives us the most meteorites?
4. Which gives us the brightest meteor showers?
5. Which consists of fragments that never coalesced into planets?

Were these your answers?
1. the asteroid belt
2. the Kuiper belt and the Oort cloud
3. the asteroid belt
4. the Kuiper belt and the Oort cloud
5. all of them

SUMMARY OF TERMS

Planets The major bodies orbiting the Sun that are massive enough for their gravity to make them spherical and small enough to avoid having nuclear fusion in their cores. They also have successfully cleared all debris from their orbital paths.

Ecliptic The plane of Earth's orbit around the Sun. All major objects of the solar system orbit roughly within this same plane.

Inner planets The four planets orbiting within 2 AU of the Sun: Mercury, Venus, Earth, and Mars—all rocky planets and known as *terrestrial* planets.

Outer planets The four planets orbiting beyond 2 AU of the Sun: Jupiter, Saturn, Uranus, and Neptune—all gaseous planets and known as *jovian* planets.

Astronomical unit (AU) The average distance between Earth and the Sun, which is about 1.5×10^8 km (about 9.3×10^7 mi).

Nebular theory The idea that the Sun and planets formed together from a cloud of gas and dust, a *nebula*.

Sunspots Temporary, relatively cool and dark regions on the Sun's surface.

Moon phases The cycles of change of the "face" of the Moon, changing from *new*, to *waxing*, to *full*, to *waning*, and back to *new*.

New Moon The phase of the Moon when darkness covers the side facing Earth.

Full Moon The phase of the Moon when its sunlit side faces Earth.

Solar eclipse The phenomenon whereby the shadow of the Moon falls on Earth, producing a region of darkness in the daytime.

Lunar eclipse The phenomenon whereby the shadow of Earth falls on the Moon, producing relative darkness of the full Moon.

Asteroid belt A region between the orbits of Mars and Jupiter containing small, rocky, planetlike fragments that orbit the Sun. These fragments are called *asteroids*, which in Latin means "small star."

Meteoroid A small rock in interplanetary space, which can include a fragment of an asteroid or comet.

Meteor A meteoroid burning in Earth's atmosphere; a "shooting star."

Meteorite A meteoroid, or a part of a meteoroid, that has survived passage through Earth's atmosphere to reach the ground.

Kuiper belt The disk-shaped region of the sky beyond Neptune populated by many icy bodies and a source of short-period comets.

Dwarf planet A relatively large icy body, such as Pluto, originating within the Kuiper belt.

Comet A body composed of ice and dust that orbits the Sun, usually in a very eccentric orbit, and that casts a luminous tail produced by solar radiation pressure when close to the Sun.

Oort cloud The region beyond the Kuiper belt that is populated by trillions of icy bodies and is a source of long-period comets.

REVIEW QUESTIONS

26.1 The Solar System and Its Formation

1. How many known planets are in our solar system?
2. What dwarf planet was downgraded from planetary status in 2006?
3. How are the outer planets different from the inner planets aside from their location?
4. Why does a nebula spin faster as it contracts?
5. According to the nebular theory, did the planets start forming before or after the Sun ignited?

26.2 The Sun

6. What happens to the amount of the Sun's mass as it "burns"?
7. What are sunspots?
8. What is the solar wind?
9. How does the rotation of the Sun differ from the rotation of a solid body?
10. What is the age of the Sun?

26.3 The Inner Planets

11. Why are days on Mercury very hot and nights very cold?
12. What two planets are evening or morning "stars"?
13. Why is Earth called "the blue planet"?
14. What gas makes up most of the Martian atmosphere?
15. What evidence tells us that Mars was once wetter than it is now?

26.4 The Outer Planets

16. What surface feature do Jupiter and the Sun have in common?
17. Which move faster, Saturn's inner rings or the outer ones?
18. How tilted is Uranus's axis?
19. Why is Neptune bluer than Uranus?

26.5 Earth's Moon

20. Why does the Moon have no atmosphere?
21. Where is the Sun located when you view a full Moon?
22. Where are the Sun and the Moon located at the time of a new Moon?
23. Why don't eclipses occur monthly, or nearly monthly?
24. How does the Moon's rate of rotation about its own axis compare with its rate of revolution around Earth?

26.6 Failed Planet Formation

25. Between the orbits of what two planets is the asteroid belt located?
26. What is the difference between a meteor and a meteorite?
27. What is the Kuiper belt?
28. What is the Oort cloud, and what is it noted for?
29. What is a falling star?
30. What causes comet tails to point away from the Sun?

EXERCISES

● BEGINNER ■ INTERMEDIATE ◆ EXPERT

1. ■ According to the nebular theory, what happens to a nebula as it contracts under the force of gravity?
2. ● What happens to the shape of a nebula as it contracts and spins faster?
3. ■ What energy processes make the Sun shine? In what sense can it be said that gravity is the prime source of solar energy?
4. ■ A TV screen is normally light gray when not illuminated. How is the blackness of sunspots similar to the blackness in images on a TV screen?
5. ■ When a contracting ball of hot gas spins into a disk shape, it cools. Why?
6. ◆ If Earth didn't spin on its axis, but still revolved around the Sun, how long would a day be on Earth?

7. ♦ If Earth didn't spin on its axis, but still revolved around the Sun, would the Sun set on the eastern or western horizon or not at all?

8. ● The greenhouse effect is very pronounced on Venus but doesn't exist on Mercury. Why?

9. ● Where are elements heavier than hydrogen and helium formed?

10. ● What is the cause of winds on Mars (and also on almost every other planet)?

11. ■ Why is there so little wind on the surface of Venus?

12. ♦ If Venus were somehow transported into the habitable zone, would conditions once again become favorable for life?

13. ● What is the major difference between the terrestrial and jovian planets?

14. ● What does Jupiter have in common with the Sun that the terrestrial planets don't? What differentiates Jupiter from a star?

15. ■ When it comes to celestial bodies, such as planets and stars, why doesn't a larger size necessarily mean a larger mass?

16. ■ Why are the seasons on Uranus different from the seasons on any other planet?

17. ■ What were the similar historical circumstances that link the names of Neptune and Pluto with the elements neptunium and plutonium?

18. ● Earth rotates much faster than Venus. How does the giant impact theory of the formation of the Moon account for this fact?

19. ● Why are many craters evident on the surface of the Moon but not on the surface of Earth?

20. ■ Why is there no atmosphere on the Moon? Defend your answer.

21. ■ Is the fact that we see only one side of the Moon evidence that the Moon spins or that it doesn't? Defend your answer.

(a)   **(b)**

22. ♦ Photo (a) shows the Moon partially lit by the Sun. Photo (b) shows a Ping-Pong ball in sunlight. Compare the positions of the Sun in the sky when each photo was taken. Do the photos support or refute the claim that they were taken on the same day? Defend your answer.

23. ♦ We always see the same face of the Moon because the rotation of the Moon on its axis matches the rate at which it revolves around Earth. Does it follow that an observer on the Moon always sees the same face of Earth?

24. ■ If we never see the back side of the Moon, would an observer on the back side of the Moon ever see Earth?

25. ● In what alignment of the Sun, the Moon, and Earth does a solar eclipse occur?

26. ● In what alignment of the Sun, the Moon, and Earth does a lunar eclipse occur?

27. ● What does the Moon have in common with a compass needle?

28. ♦ If you were on the Moon and you looked up and saw a full Earth, would it be nighttime or daytime on the Moon?

29. ♦ If you were on the Moon and you looked up and saw a new Earth, would it be nighttime or daytime on the Moon?

30. ♦ Earth takes 365.25 days to revolve around the Sun. If Earth took this same amount of time to spin on its axis, what might we note about the Sun's position in the sky?

31. ● Do star astronomers make stellar observations during the full Moon part of the month or during the new Moon part of the month? Does it make a difference?

32. ● Nearly everybody has witnessed a lunar eclipse, but relatively few people have seen a solar eclipse. Why?

33. ■ Because of Earth's shadow, a partially eclipsed Moon looks like a cookie with a bite taken out of it. Explain with a sketch how the curvature of the bite indicates the size of Earth relative to the size of the Moon. How does the tapering of the Sun's rays affect the curvature of the bite?

34. ■ In what sense is Pluto a comet?

35. ■ Smaller chunks of asteroids are sent hurling toward Earth much more frequently than larger chunks of asteroids are. Why?

36. ● Why are meteorites more easily found on Antarctica than on other continents?

37. ● A meteor is visible only once, but a comet may be visible at regular intervals throughout its lifetime. Why?

38. ● What would be the consequence of a comet's tail sweeping across Earth?

39. ■ Chances are about 50-50 that in any night sky is at least one visible comet that has not been discovered. This keeps amateur astronomers busy looking night after night, because the discoverer of a comet gets the honor of having it named for him or her. With this high probability of comets in the sky, why aren't more found?

40. ■ In terms of the conservation of energy, describe why comets eventually burn out.

PROBLEMS

● **BEGINNER**   ■ **INTERMEDIATE**   ♦ **EXPERT**

1. ● Knowing that the speed of light is 300,000 km/s, show that sunlight takes about 8 min to reach Earth.

2. ■ How many days does sunlight take to travel the 50,000 AU from the Sun to the outer reaches of the Oort cloud?

3. ■ The light-year is a standard unit of distance used by astronomers. It is the distance light travels in one Earth year. In units of light-years, what is the approximate diameter of our solar system, including the outer reaches of the Oort cloud? (Assume that one light-year equals 63,000 AU.)

4. ■ The nearest star to our Sun is Alpha Centauri, which is about 4.4 light-years away. Assume that it too has an Oort cloud about 1.6 light-years in diameter. Show that there is enough space between us and it to fit about 1.75 solar systems.

5. ♦ If the Sun were the size of a beach ball, Earth would be the size of a green pea 110 m away. Show that the nearest star, Alpha Centauri (4.4 light-years away), would be about 30,000 km distant. (Hint: Find the distance to Alpha Centauri in AU.)

ACTIVE EXPLORATIONS

Ping-Pong Moon

Find a Ping-Pong ball on the next clear day when the Moon is out. Hold the Ping-Pong ball with your arm stretched out toward the Moon so that the Ping-Pong ball overlaps the Moon. Look carefully at how the Ping-Pong ball is lit by the Sun. Notice that this is the same way that the Moon is lit by the Sun! For an example, look at the photographs accompanying Exercise 22. To see the different phases that the Moon would have if it were elsewhere in the sky, move your Ping-Pong ball around. Note that as you bring the Ping-Pong ball closer to the Sun, the crescent on the Ping-Pong ball gets thinner. The same thing happens with the Sun. This activity is a great way to experience the roundness of the Moon.

Lunar Phases

Simulate the lunar phases. Insert a pencil into a Styrofoam ball. This represents the Moon. Position a lamp (representing the Sun) in another room near the doorway. Hold the ball in front of and slightly above yourself. Slowly turn yourself around, keeping the ball in front of you as you move. Observe the patterns of light and shadow on the ball. Relate this to the phases of the Moon.

Lunar Motion

When viewed from the North Pole, Earth spins counterclockwise, or toward the east. This means that the stars appear to move in the opposite direction, which is toward the west. This is just like when you're sitting in a train that begins moving eastward. The only way you know that you're moving eastward is because things outside your window give the appearance of moving westward. Just as Earth spins counterclockwise, the Moon revolves around us also counterclockwise, though not as fast as we spin. Look at where the Moon is located one night, say 11:00 PM. Look for the Moon the next night at the same time and you'll see that it has moved eastward (a counterclockwise direction) from where it was the previous night.

Reckoning Latitude

The crescent Moon always points to the Sun. You can use this fact to estimate your latitude. Down by the equator (0° latitude), the setting crescent Moon lies flat to the horizon, while up close to the North Pole (90° latitude) the crescent Moon stands on its end. (Small deviations from this arise when the Moon lies outside the ecliptic.) Next time you see the crescent Moon close to the horizon, look carefully at its angle and try to estimate your latitude. This method is most accurate when the Moon is passing through the ecliptic, that is, when the Moon falls in line with the planets.

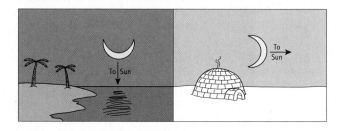

Space Shows

The dates of meteor showers are rather predictable, as shown in Table 26.2. How intense the shower might be, however, is still somewhat of a guess. So keep your eye to the nighttime sky for these meteor showers. While doing so, review in your mind the distinctions among meteoroids, meteors, meteorites, and comets. Also notable in our nighttime skies are lunar eclipses, which can seen by anyone on the nighttime side of Earth during the eclipse (assuming skies are clear). Here are the dates for upcoming total lunar eclipses reaching North America.

Date	Type
Feb. 21, 2008	Total
Feb. 9, 2009	Partial
Jul. 7, 2009	Partial
Aug. 6, 2009	Partial
Jun. 26, 2010	Partial
Dec. 21, 2010	Total
Dec. 10, 2011	Total
Jun. 4, 2012	Partial

READINESS ASSURANCE TEST (RAT)

If you have a good handle on this chapter, if you really do, then you should be able to score at least 7 out of 10 on this RAT. If you score less than 7, you need to study further before moving on.

Choose the BEST answer to each of the following.

1. The Sun contains what percentage of the solar system's mass?
   (a) about 35%
   (b) 85%
   (c) the percentage varies over time
   (d) more than 99%
2. The solar system is like an atom in that both
   (a) are governed principally through the electric force.
   (b) consist of a central body surrounded by objects moving in elliptical paths.
   (c) are composed of plasma.
   (d) are mainly empty space.
3. The nebular theory is based on the observation that the solar system
   (a) is highly ordered, indicating that it formed in a stepwise manner from physical processes.
   (b) has a structure much like an atom.
   (c) is highly complex and appears to have been built by chaotic processes.
   (d) appears to be very old.
4. When a contracting hot ball of gas spins into a disk shape, it cools faster because of
   (a) increased radiation transfer.
   (b) increased surface area.
   (c) decreased insulation.
   (d) increased convection currents.
   (e) eddy currents.
5. Each second, the burning Sun's mass
   (a) increases.
   (b) remains unchanged.
   (c) decreases.

6. Compared to your weight on Earth, your weight on Jupiter would be about
   (a) 3000 times as much.
   (b) half as much.
   (c) 3 times as much.
   (d) 300 times as much.
   (e) 100 times as much.
7. When the Moon assumes its characteristic thin crescent shape, the position of the Sun is
   (a) almost directly in back of the Moon.
   (b) almost directly behind Earth, so that Earth is between the Sun and the Moon.
   (c) at right angles to the line between the Moon and Earth.
8. When the Sun passes between the Moon and Earth, we have
   (a) a lunar eclipse.
   (b) a solar eclipse.
   (c) met our end.
9. Asteroids orbit
   (a) the Moon.
   (b) Earth.
   (c) the Sun.
   (d) all of these
   (e) none of these
10. With each pass of a comet about the Sun, the comet's mass
    (a) remains virtually unchanged.
    (b) increases.
    (c) is appreciably reduced.

Answers

1d, 2d, 3a, 4b, 5c, 6c, 7a, 8c, 9c, 10c

EXPLORING FURTHER

http://www.astro.uva.nl/demo/od95/
   Earth's Sun is a typical star, it's close, and it's well understood. Take a virtual tour of the Sun via this short and engaging solar science course with movies, diagrams, and first-rate images—and explanations. Be sure to check out the movies.

http://solar-center.stanford.edu/
   This award-winning solar site is a bright spot on the Web. For scientific background, go to About the Sun to link to information about solar physics. Solar Art & Literature and Solar Folklore will nicely round out your visit.

http://cannon.sfsu.edu/~gmarcy/cswa/history/history.html
   Women played a "starring" role in astronomy through history, from early observers who cataloged the stars to modern astronauts and research scientists. Click on the names of significant female astronomers, read about their accomplishments, and learn something about their interesting lives. This site includes wonderful archival photos.

http://www.solarviews.com/eng/history.htm
   This is a vast site befitting a vast subject—the history of space exploration. Click on Contents and see an outline

and a huge array of topic choices. The image index helps you quickly access graphics.

http://sse.jpl.nasa.gov/index.cfm
NASA's Solar System Exploration Web site includes amazing photos of the planets, their moons, and other solar system objects.

CHAPTER 26 ONLINE RESOURCES

Interactive Figure

- 26.19

Tutorials

- Formation of the Solar System
- The Sun

Videos

- History of the Solar System
- Orbits in the Solar System

Quiz

Flashcards

Links

stars and galaxies

27.1 OBSERVING THE NIGHT SKY

27.2 THE BRIGHTNESS AND COLOR OF STARS

27.3 THE HERTZSPRUNG–RUSSELL DIAGRAM

27.4 THE LIFE CYCLES OF STARS

27.5 BLACK HOLES

27.6 GALAXIES

■ The roots of astronomy reach back to prehistoric times when humans began to observe star patterns in the night sky. Though ancient people developed ways to measure the position and periodic movement of the stars, they knew nothing about the stars themselves. Today we know that Earth orbits a star—our Sun. We know that all the stars we see in the nighttime sky are much farther away from us than our fellow planets. On a moonless night we may guess we see many thousands or even millions of stars. But the unaided eye sees at most about 3000 stars, horizon to horizon. Many more stars become visible with a telescope, especially when the telescope is pointed toward a cloudlike band of light that stretches north to south. The ancient Greeks called this diffuse band of light the Milky Way.

Today we know the Milky Way to be a vast conglomeration of more than 100 billion stars. Our Sun and all the stars in our nighttime sky are located in the outer edges of this conglomeration. When viewed from afar, as shown in the chapter-opener illustration, all of these stars appear as a great swirl of stars known as a *galaxy*. The band of light identified by the ancient Greeks is the edge-on view of this galaxy. When modern telescopes peer away from the Milky Way, other galaxies come into view. How many? Again, the answer is more than 100 billion. So there are more than 100 billion galaxies each containing, on average, more than 100 billion stars. The total number of stars in our observable universe, therefore, is more than 100 billion × 100 billion, which is greater than the total number of grains of sands on all the beaches found on Earth. Stars are numerous.

In this chapter we explore the nature of stars—how they form, how they die, and how they are organized within galaxies. We will explore how there are many different types of stars, just as there are many different types of galaxies. We begin by taking a closer look at the stars visible to our unaided eyes. Compared to the size of the entire Milky Way Galaxy, these thousands of stars are like our immediate neighborhood.

Ursa Major

Big Dipper

FIGURE 27.1

The constellation Ursa Major, the Great Bear. The seven stars in the tail and back of Ursa Major form the Big Dipper.

The grouping of stars into constellations tells us about the thinking of astronomers of earlier times, but it tells us nothing about the stars themselves.

■ 27.1 Observing the Night Sky

Early astronomers divided the night sky into groups of stars, called *constellations*, such as the group of seven stars we now call the Big Dipper. The names of the constellations today carry over mainly from the names assigned to them by early Greek, Babylonian, and Egyptian astronomers. The Greeks, for example, included the stars of the Big Dipper in a larger group of stars that outlined a bear. The large constellation, Ursa Major (the Great Bear) is illustrated in Figure 27.1. The groupings of stars and the significance given to them has varied from culture to culture. To some cultures, the constellations stimulated storytelling and the making of great myths; to other cultures, the constellations honored great heroes, such as Hercules and Orion; to yet others, they served as navigational aids for travelers and sailors. To many cultures, including the African Bushmen and Masai, the constellations provided a guide for planting and harvesting crops because they were seen to move in the sky in concert with the seasons. Charts of this periodic movement became some of the first calendars. We can see in Figure 27.2 why the background of stars varies throughout the year.

The stars are at different distances from Earth. However, because all the stars are so far away, they appear equally remote. This illusion led the ancient Greeks and others to conceive of the stars as being attached to a gigantic sphere surrounding Earth, called the **celestial sphere**. Though we know it is imaginary, the celestial sphere is still a useful construction for visualizing the motions of the stars (Figure 27.3).

The stars appear to turn around an imaginary north–south axis once every 24 hours. This is the *diurnal motion* of the stars. Diurnal motion is easy to visualize as a rotation of the celestial sphere from east to west. This motion is a consequence of the daily counterclockwise rotation of Earth on its axis. When

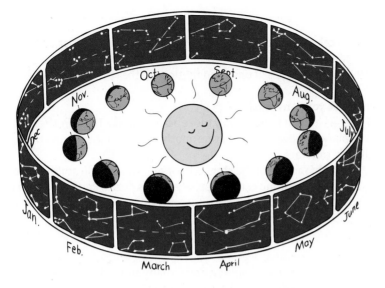

FIGURE 27.2

The night side of Earth always faces away from the Sun. As Earth circles the Sun, different parts of the universe are seen in the nighttime sky. Here the circle, representing 1 year, is divided into 12 parts—the monthly constellations. The stars in the nighttime sky change in a yearly cycle.

we speak of the diurnal motion of the stars, we are referring to the motions of celestial objects as a whole; this motion does not change the relative positions of objects. Figure 27.4 shows the diurnal motion of the stars making up the Big Dipper. Time-exposure photographs show that the Big Dipper appears to move in circles around the North Star (Figure 27.5). The North Star appears stationary as the celestial sphere rotates because it lies very close to the projection of Earth's rotational axis.

In addition to the diurnal motion of the sky, there is *intrinsic* motion of certain bodies that change their positions with respect to the stars. The Sun, the Moon, and planets, called "wanderers" by ancient astronomers, appear to migrate across the fixed backdrop of the celestial sphere. Interestingly, the stars themselves have intrinsic motion. They are so far away, however, that this motion is not apparent on the time scale of a human life. As shown in Figure 27.6, over thousands of years, the intrinsic movement of stars results in new patterns of stars. In other words, the constellations we see today are quite different from the ones that appeared to our earliest ancestors.

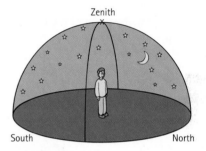

FIGURE 27.3

The celestial sphere is an imaginary sphere to which the stars are attached. We see no more than half of the celestial sphere at any given time. The point directly over our heads at any time is called the *zenith*.

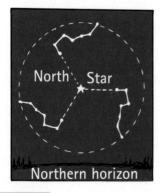

FIGURE 27.4

The pair of stars in the end of the Big Dipper's bowl point to the North Star. Earth rotates about its axis and therefore about the North Star, so over a 24-hour period the Big Dipper (and other surrounding star groups) makes a complete revolution.

FIGURE 27.5

A time exposure of the northern night sky.

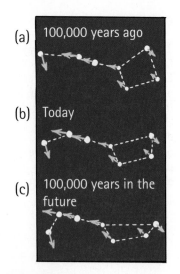

(a) 100,000 years ago

(b) Today

(c) 100,000 years in the future

FIGURE 27.6

The present pattern of the Big Dipper is temporary. Here we can see its pattern (a) 100,000 years ago; (b) as it appears at present; and (c) as it will appear in the future, about 100,000 years from now.

86 ly
69 ly   53 ly
360 ly
93 ly   116 ly   62 ly

FIGURE 27.7

Interestingly, the seven stars of the Big Dipper are at varying distances from Earth. Note their varying distances in light-years.

CHECK POINT

1. Which celestial bodies appear fixed relative to one another, and which celestial bodies appear to move relative to the others?
2. What are two types of observed motions of the stars in the sky?

Were these your answers?

1. The stars appear fixed as they move across the sky. The Sun, the Moon, and planets move relative to one another as they move across the backdrop of the stars.
2. One type of motion of the stars is their nightly rotation as if they were painted on a rotating celestial sphere; this is due to Earth's rotation on its own axis. Stars also appear to undergo a yearly cycle around the Sun because of Earth's revolution about the Sun.

Some stars on the celestial sphere are actually much farther away than others from Earth. Astronomers measure the vast distances between Earth and the stars using *light-years*. One **light-year** is the distance that light travels in 1 year, nearly 10 trillion km. For perspective, the diameter of Neptune's orbit is about 0.001 light-year. The distance from the Sun to the outer edges of the Oort cloud (the full radius of our solar system) is about 0.8 light-year. The star closest to our Sun, Proxima Centauri, is about 4.2 light-years away. The diameter of the Milky Way Galaxy is about 100,000 light-years. The next closest major galaxy, the Andromeda Galaxy, is about 2.3 million light-years distant. Figure 27.7 shows the distances to the seven stars making up the Big Dipper in light-years.

The speed of light (as we know from Chapter 11) is 3×10^8 m/s. Although this is very fast, it nevertheless takes light appreciable time to travel large distances. And so when you see the light emitted by a very distant object, you are actually seeing the light it emitted long ago—you are looking back in time. Consider the example of Supernova 1987a (a supernova is the explosion of a star, as you will learn more about in Section 27.4). This supernova occurred in a galaxy 190,000 light-years from Earth. Although we witnessed the supernova in 1987, the light from this explosion took 190,000 years to reach our planet, so the explosion actually occurred 190,000 years earlier. "News" of the supernova took 190,000 years to reach Earth!

27.2 The Brightness and Color of Stars

All stars have much in common with the Sun. All are born from clouds of interstellar dust with roughly the same chemical composition as the Sun (Chapter 26). About three-fourths of the interstellar material from which any star forms is hydrogen; one-fourth is helium; and no more than 2% of the material from which a star forms consists of heavier chemical elements. Stars shine brilliantly for millions or billions of years because of the nuclear fusion reactions that occur in their cores. And all stars, the Sun included, ultimately exhaust their nuclear fuel and die.

Yet not all stars are the same. If you look into the night sky, you will see that stars differ in two very visible ways: brightness and color. Brightness relates to how much energy a star produces, while its color indicates its surface temperature. However, although a star's brightness is related to its energy output, its brightness also depends on how far away it is from Earth. Recall from earlier chapters the inverse-square law: the intensity of light diminishes as the reciprocal of the square of the distance from the source. For example, the stars Betelgeuse and Procyon appear equally bright even though Betelgeuse emits about 5000 times as much light as Procyon. The reason? Procyon is much closer to Earth than Betelgeuse is.

To avoid confusing brightness with energy output, astronomers clearly distinguish between apparent brightness and the more important property, *luminosity*. Apparent brightness is the brightness of a star as it appears to our eyes. Luminosity, on the other hand, is the total amount of light energy that a star emits into space. Luminosity is usually expressed relative to the Sun's luminosity, which is noted L_{Sun}. For example, the luminosity of Betelgeuse is 38,000 L_{Sun}. This indicates that Betelgeuse is a very luminous star emitting about 38,000 times as much light into space as the Sun. On the other hand, Proxima Centauri is quite dim, with a luminosity of 0.00006 L_{Sun}. Astronomers have measured the luminosity of many stars and found that stars vary greatly in this respect. The Sun is somewhere in the middle of the luminosity range. The most luminous stars are about a million times as luminous as the Sun, while the dimmest stars produce about 1/10,000 as much light as the Sun.

Link To Section 12.6

Besides apparent brightness, a star's color is another property that varies widely among stars. Figure 27.8, a photograph of stars taken with the Hubble Telescope, shows this—stars come in every color of the rainbow. A star's color directly tells you about its surface temperature—for example, a blue star is hotter than a yellow star, and a yellow star is hotter than a red star. In fact, astronomers use color to measure the temperatures of stars. Why is it that a star's color corresponds to its temperature?

Link To Section 11.5

RADIATION CURVES OF STARS

As you learned in Chapters 7 and 11, all objects with a temperature above absolute zero emit energy in the form of electromagnetic radiation. The peak frequency $\bar{f}$ of the radiation is directly proportional to the absolute temperature T of the emitter:

$$\bar{f} \sim T$$

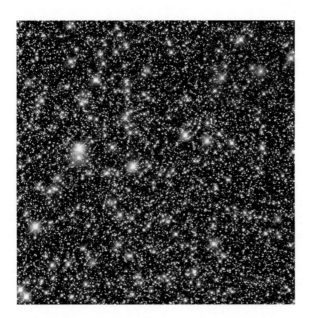

FIGURE 27.8

Most of the stars in this photograph are approximately the same distance—2000 light-years—from the center of the Milky Way Galaxy. A star's color indicates its surface temperature—a blue star is hotter than a yellow star, and a yellow star is hotter than a red star. This photo was taken by the Hubble Telescope.

FIGURE 27.9

FIGURE 27.9

These idealized radiation curves for stars of different surface temperatures show two important facts: (1) hotter stars emit radiation with higher average frequency than cooler stars, and (2) hotter stars emit more radiation per unit surface area at every frequency than cooler stars.

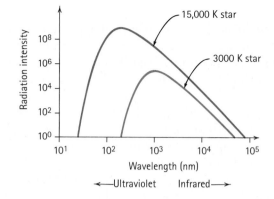

fyi

■ Interestingly, the Earth's atmosphere is transparent to a narrow band of light centered upon the Sun's peak frequency. Creatures here on Earth's surface evolved to be sensitive to these most abundant frequencies, which we now perceive as visible light. Within the spectrum of visible light we are most sensitive to a greenish-yellow, which is why many emergency vehicles are commonly painted greenish-yellow.

Stars have different colors because they emit different frequencies of electromagnetic waves in the visible range. Our eyes sense different frequencies of visible radiation as different colors. Figure 27.9 shows the radiation curves, which are graphs of the frequency of emitted radiation versus temperature of the emitter, for two stars with different temperatures. The radiation curves show that the hotter a star is, the shorter the wavelength of its peak frequency and the bluer it looks. So the blue stars in the night sky have higher temperatures than the red ones. The Sun, for example, with its approximately 5800 K temperature, emits most strongly in the middle of the visible spectrum and so appears yellow. Betelgeuse, on the other hand, appears red because of its cooler surface temperature (about 3400 K). Betelgeuse emits more red light than blue light.

Notice also from Figure 27.9 that the hotter a star is, the more radiant energy it emits. Thus we see that hot blue stars are more luminous than cooler red stars of the same size.

Link To Section 7.3

CHECK POINT

The temperature of Sirius is about 9400 K. What color is this star—and why?

Was this your answer?
Sirius has a slightly blue color. It emits more blue light than red light because of its high surface temperature.

■ 27.3 The Hertzsprung–Russell Diagram

When you compare the luminosity of stars to their temperature, interesting patterns emerge. Early in the 20th century, Danish astronomer Ejnar Hertzsprung and American astronomer Henry Norris Russell did just this. They produced a diagram known as the **Hertzsprung–Russell diagram**, or **H–R diagram**, which is of key importance in astronomy (Figure 27.10). The H–R diagram is a plot of the luminosity versus surface temperature of stars. Luminous stars are near the top of the diagram, and dim stars are toward the bottom. Hot bluish stars are toward the left side of the diagram and cool reddish stars are toward the right side.

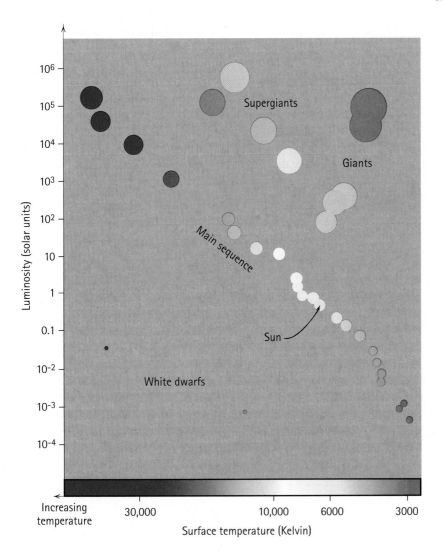

FIGURE 27.10

The H–R diagram shows a star's surface temperature on the horizontal axis and its luminosity on the vertical axis. The giant and supergiant stars shown here are not drawn to scale. The red supergiant, Antares, for example, is so large that, if drawn to scale, it would reach the ceiling of your classroom. Interestingly, although the radius of Antares is 700 times that of our Sun, its mass is only about 15 times greater. So, although Antares is much larger, it is also much less dense.

Because the giants and supergiants are so luminous, they are easy to see in the night sky even if they are not close to Earth. You can often identify them by their reddish color.

The H–R diagram shows several distinct regions of stars. Most stars are plotted on the band that stretches diagonally across the diagram. This band is called the **main sequence**. Stars on the main sequence, including our Sun, generate energy by fusing hydrogen to helium. As we would expect, the hottest main-sequence stars are the brightest and bluest stars and the coolest main-sequence stars are the most dim and red stars. Take a moment to locate the Sun on the H–R diagram. Can you see that the Sun is a roughly average main-sequence star in terms of its luminosity and temperature?

Toward the upper right of the diagram is a distinct group of stars—the **giant stars**. These stars clearly do not follow the pattern of the hydrogen-burning main-sequence stars. Because these stars are red, we know they must have low surface temperatures. If they were main-sequence stars, the giants would be dim. Yet notice how high the giants are on the luminosity scale—they are very bright. The fact that the giants are both much cooler and much brighter than the Sun tells us that these stars must also be much larger than the Sun. (Hence the name *giant*.) Above the giants on the H–R diagram are a few rare stars, the *supergiants*. The supergiants are even larger and brighter than the giants. As you will see in the next section, the giants and supergiants are stars nearing the end of their lives because the hydrogen fuel in their cores is running out.

fyi

- The H–R diagram is to astrophysicists what the periodic table is to chemists—an extremely important tool. A star's position on the H–R diagram can reveal its age. The age of our galaxy can be estimated by looking at the positions of our oldest stars and their white-dwarf remnants.

Toward the lower left are some stars that are so dim they cannot be seen with the unaided eye. The surfaces of these stars can be hotter than the Sun, which makes them blue or white in color. Yet their luminosities are quite low—on the order of 0.1 L_{Sun} to 0.0001 L_{Sun}. To be so hot and radiate so little light, these stars must be very small—they are called the **white dwarfs**. White dwarfs are typically the size of Earth or even smaller, yet they have mass comparable to the Sun. The density (or mass per volume) of a white dwarf is thus extremely high—about a million g/mL. For comparison, gold has a density of about 19 g/mL, while the average density of Earth is about 5.4 g/mL. As you will learn in the next section, white dwarfs are dead stars, the remnants of stars that have exhausted their nuclear fuel.

CHECK POINT

1. What characteristic do all main-sequence stars share?
2. Giants have cool surface temperatures yet are highly luminous. Does this mean that the frequency of light emitted by a giant does not depend on its surface temperature as described by Figure 27.9?

Were these your answers?

1. All main-sequence stars generate energy by the nuclear fusion of hydrogen to helium.
2. No, radiation curves hold for a giant star as for any other radiating body. Giants do have a relatively low energy output per unit surface area; they are highly luminous only because they are very large.

Astrology

There is more than one way to view the cosmos and its processes—astronomy is one and astrology is another. Astrology is a belief system that began more than 2000 years ago in Babylonia. Astrology has survived nearly unchanged since the second century AD, when some revisions were made by Egyptians and Greeks who believed that their gods moved heavenly bodies to influence the lives of people on Earth. Astrology today holds that the position of Earth in its orbit around the Sun at the time of birth, combined with the relative positions of the planets, has some influence over one's personal life. The stars and planets are said to affect such personal things as one's character, marriage, friendships, wealth, and death. Could the force of gravity exerted by these celestial bodies be a legitimate factor in human affairs? After all, the ocean tides are the result of the Moon's and Sun's positions, and the gravitational pulls between the planets perturb one another's orbits. Because slight variations in gravity produce these effects, might not slight variations in the planetary positions at the time of birth affect a newborn? If the influence of stars and planets is gravitational, then credit must also be given to the effect of the gravitational pull between the newborn and Earth itself. This pull is enormously greater than the combined pull of all the planets, even when lined up in a row (as occasionally happens). The gravitational influence of the hospital building on the newborn far exceeds that of the distant planets. So planetary gravitation cannot be an underlying agent for astrology.

Astrology is not a science, because it doesn't change with new information as science does, nor are its predictions borne out by fact. Rather, its predictions depend on coincidence and also on the tendency of many people to seek external explanations for their fates or personal behaviors. Astrological beliefs are built on anecdotal evidence that is neither reproducible nor testable. Astrology means different things to different people, but in any case, it is far outside the realm of science. It is a pseudoscience lying within the realm of superstition.

■ 27.4 The Life Cycles of Stars

In Chapter 26, we discussed the nebular theory, which explains how the Sun formed from an expansive, low-density cloud of gas and dust called a *nebula* (Figure 27.11). Other stars form in the same way. That is, over time, a nebula flattens, heats, and spins more rapidly as it gravitationally contracts. The center of the nebula becomes dense enough to trap infrared radiation so that this energy is no longer radiated away. The hot central bulge of a nebula is called a *protostar*.

Mutual gravitation between the gaseous particles in a protostar results in an overall contraction of this huge ball of gas, and its density increases still further as matter is crunched together, with an accompanying rise in pressure and temperature. When the central temperature reaches about 10 million K, hydrogen nuclei begin fusing to form helium nuclei. This thermonuclear reaction, converting hydrogen to helium, releases an enormous amount of radiant and thermal energy, as discussed in Chapter 26. The ignition of nuclear fuel marks the change from protostar to star. Outward-moving radiant energy and the gas accompanying it exert an outward pressure called *thermal pressure* on the contracting matter. When nuclear fusion occurs fast enough, thermal pressure becomes strong enough to halt the gravitational contraction. At this point, outward thermal pressure balances inward gravitational pressure, and the star's size stabilizes.

CHECK POINT

What do the processes of thermonuclear fusion and gravitational contraction have to do with the physical size of a star?

Was this your answer?
The size of a star is the result of these two continually occurring processes. Energy from thermonuclear fusion tends to blow the star outward like an ongoing hydrogen bomb explosion, and gravitation tends to contract its matter in an ongoing implosion. The outward thermonuclear expansion and inward gravitational contraction produce an equilibrium that accounts for the star's size.

Though all stars are born in the same way from contracting nebulae, they do not all progress through their lives in the same way. A star's mass determines the stages a star will go through from birth to death. There are limits on the mass that a star can attain. A star with a mass less than 0.08 times the mass of the Sun (0.08 M_{Sun}) never reaches the 10 million K threshold needed for sustained fusion of hydrogen. On the other hand, stars with masses above 100 M_{Sun} would undergo fusion at such a furious rate that gravity could not resist thermal pressure and the star would explode. So stars exist within the limits of about a tenth of the mass of the Sun and 100 times the solar mass.*

Most stars have masses on the order of that of the Sun. Such stars inhabit a central place on the main sequence of the H–R diagram. If you plot the life-cycle stages of average stars on a H–R diagram, they trace a curve similar to the one for our Sun, which is shown in Figure 27.12. The Sun was born about 4.5 billion years ago at position 1, when the fusion of hydrogen ignited. The Sun will spend most of its lifetime—some 10 billion years—on the main sequence, with thermal pressure keeping gravity at bay. Speaking more generally, a star's

*One solar mass, 1 M_{Sun}, is a unit of mass equivalent to that of the Sun: 2×10^{33} kg.

The Physical Science Place
Lives of Stars

The Physical Science Place
Stellar Evolution

FIGURE 27.11

This image of the Trifid Nebula was obtained by the Spitzer Space Telescope. This nebula is located 5400 light-years from Earth in the constellation Sagittarius. Within each of the four red dust clouds are developing stars.

 Link To Section 13.7

FIGURE 27.12

The stages of the Sun's life cycle are plotted on this H–R diagram.

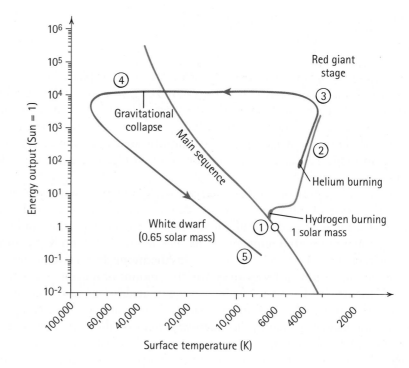

hydrogen-burning lifetime lasts for a period of a few million to 50 billion years, depending on its mass.

More-massive stars have shorter lives than less-massive stars. This may sound counterintuitive, because if they have more mass, they have more fuel to burn longer, right? High-mass stars, however, are more luminous than low-mass stars, meaning that they burn their hydrogen fusion fuel at a faster rate. Massive stars *must* be more luminous than small-mass stars so that the outward pressure of their nuclear fusion can offset the greater gravitational force of their contraction. Massive stars start out with more hydrogen fuel than small-mass stars, but they consume their fuel so much faster that they die billions of years younger than smaller stars.

No star lasts forever. In the old age of an average-mass star like our Sun, as the supply of hydrogen fuel diminishes, gravity overwhelms thermal pressure and the star pulls inward. Temperature rises as the burned-out hydrogen core contracts because of gravity. At a certain point, the temperature becomes high enough in the core to launch *helium burning*—the fusion of helium to carbon. The star then has a structure consisting of concentric shells. Helium fuses to carbon at the star's center while hydrogen fuses to helium in a surrounding shell. Energy output soars, moving the star off the main sequence.

With such intensified nuclear fusion within a star, the outward force of thermal pressure wins out over the inward force of gravity. The star balloons to become a giant (position 2). When our Sun reaches the giant stage about 5 billion years from now, its swelling and increased energy output will escalate Earth's temperatures. Earth will be stripped of its atmosphere and the oceans boiled dry. Ouch!

As fusion continues, carbon will continue to accumulate in the Sun's core, but temperatures will never become hot enough to allow the carbon to undergo fusion. Instead, carbon "ash" accumulates inside the star and fusion gradually tapers off. Then gravity predominates and the star contracts, which boosts its temperature. With higher temperatures, the color of the shrinking Sun will shift from red to blue, and its position will shift to the left in the H–R diagram.

A star's life cycle depends on its mass. The lowest-mass stars are brown dwarfs, dim but long-lived stars. Medium-mass stars progress from main-sequence stars to red giants or supergiants, then to white dwarfs. Very massive stars have short lives and die in massive explosions called *supernovae*.

Link To Section 4.1

When our Sun turns into a red giant billions of years from now, its diameter will encompass the orbit of Venus.

CHECK POINT

Why does a star shrink when its core runs out of nuclear fuel?

Was this your answer?

Outward thermonuclear expansion and inward gravitational contraction produce an equilibrium that accounts for the star's size. As the heat from the inner thermonuclear reactions begins to die down, gravity predominates and the star shrinks. Upon shrinking, matter becomes compressed, which is an additional source of heat to ignite further nuclear fusion. For a star the size of our Sun, heat from compression is hot enough to fuse elements to carbon. The fusion of heavier elements in our Sun, however, is not possible.

Astronomers have found evidence suggesting that the carbon within the center of many white dwarfs crystallizes into diamond. They expect that when our Sun transforms into a white dwarf 5 billion years from now, its ember core will crystallize as well, leaving a planet-sized diamond at the center of our solar system.

Our fuel-exhausted Sun will continue to shrink until the electrons within the Sun are so squeezed that they resist any further compression. Interestingly, the reason they resist further compression has to do with the ideas of the quantum hypothesis introduced in Chapter 12. Briefly, each subatomic particle has its own *quantum state* and no two subatomic particles can share the same quantum state. The Sun will shrink, but only up to the point at which electrons resist trespassing into the quantum states of their neighboring electrons. Having spent all of its nuclear fuel, our dead Sun, now quite small, will no longer be producing energy.

FIGURE 27.13

The Cat's Eye planetary nebula, seen here with the Hubble Space Telescope, measures about 1.2 light-years across, which is about a thousand times the diameter of Neptune's orbit. This planetary nebula is about 3000 light-years away, which places it within our galaxy. Clearly visible are the hot gases exploding away from the central Sun-sized star, which is in the process of transforming into a white dwarf.

As our Sun goes through this final collapse, the layers of plasma and gas surrounding the core will be ejected in a brilliant display, forming what is called a **planetary nebula** (Figure 27.13). Despite its name, a planetary nebula has nothing to do with planets. The name is derived from the fact that the planetary nebula looks like a nebula from which planets could form. The planetary nebula, however, will disperse within a million years, leaving the Sun's cooling carbon core behind as a white dwarf. White dwarfs have the mass of a star but the volume of a planet, and are thus far more dense than anything on Earth. Because the nuclear fires of a white dwarf have burned out, it is not actually a star anymore, but is more accurately called a stellar remnant. In any case, a white dwarf cools for eons in space until it becomes too cold to radiate visible light (Figure 27.14).

NOVAE AND SUPERNOVAE

There is another possible fate for a white dwarf, if it is part of a *binary star*. A binary star is a double star—a system of two stars that revolve about a common center, just as Earth and the Moon revolve about each other. If a white dwarf is a binary and close enough to its partner, the white dwarf may gravitationally pull hydrogen from its companion star. It then deposits this material on its own surface as a very dense hydrogen layer. Continued compacting increases the temperature of this layer, which ignites to embroil the white dwarf's surface in a thermonuclear blast that we see as a **nova**, which appears in the nighttime sky as a new star (*nova* is Latin for "new"). A nova is an event, not a stellar object. After a while, a nova subsides until enough matter accumulates to repeat the event. A given nova flares up at irregular intervals that may range from decades to hundreds of thousands of years.

Although low- and medium-mass stars become white dwarfs, the fate of stars more than about 10 M_{Sun} is quite different. When such a massive star contracts

FIGURE 27.14

A white dwarf, shown here, is the final stage in the evolution of low- and medium-mass stars. After a star has used all its nuclear fuel, its outer layers escape into space, leaving the dense core behind as a white dwarf. The strong gravitational field of a white dwarf causes it to attract matter from surrounding space to form an accretion disk. The disk is heated by friction where it meets the star, causing it to glow brightly.

after its giant or supergiant phase, more heat is generated than in the contraction of a small star. Such a star does not shrink to become a white dwarf. Instead, carbon nuclei in its core fuse and liberate energy while synthesizing heavier elements, such as neon and magnesium. Thermal pressure halts further gravitational contraction until all the carbon is fused. Then the core of the star contracts again to produce even greater temperatures, and a new fusion series produces even heavier elements. The fusion cycles repeat until the element iron is formed.

Fusion of elements with atomic numbers greater than those of iron consumes energy rather than liberating energy. (The reason for this, as you may recall from Chapter 13, is that the average mass per nucleon is lower for iron than any other element.) Once nuclei transform to iron, the fusion process stops. Thermal expansion that pushes against gravity, therefore, also stops. Gravity thus predominates and the entire star begins its final contraction.

Recall that with a dying medium-sized star, such as our Sun, contraction continues until gravity is counteracted by the resistance of electrons. With a supermassive supergiant, however, the gravitational forces are strong enough to overcome this resistance. The electrons, however, do not merge into one another. Instead, they combine with protons to form neutrons. What happens next is an astounding event called a **supernova**. Within minutes, the supergiant's iron core, about the size of Earth, collapses into a ball of neutrons only several kilometers in diameter. Massive amounts of energy are released—enough to outshine an entire galaxy. During this brief time of abundant energy, the heavy elements beyond iron are synthesized, as protons and neutrons outside the core mash into other nuclei to produce such elements as silver, gold, and uranium. These heavy elements are less abundant than the lighter elements because of the brief time available for synthesizing them.

Most of the energy during the collapse of the iron core is released in the form of *neutrinos*—nearly massless subatomic particles that rarely interact with matter. Concentrations of neutrinos released from the collapse of the iron core are great enough to blow the outer shells of the star outward at speeds in excess of 10,000 km/s, which is fast enough to travel 1 AU in about four hours. Over time, this supernova wind of heavy elements spreads to far reaches of the galaxy where the elements are taken in by nebulae destined to become new stars. The gold and platinum we wear for jewelry here on Earth, as well as the bulk of Earth itself, is the dust from supernovae that exploded many years before our solar system came to be.

Link To Section 13.6

Link To Section 12.1

CHECK POINT

A star can undergo a nova more than once. Can a star also go through multiple supernovae? Why or why not?

Was this your answer?
A nova is a thermonuclear explosion that occurs when a white dwarf collects sufficient mass from a very close neighboring star. As long as the neighboring star provides mass, this explosion can be repeated multiple times. A supernova is such an energetic release of energy that it is an end-all event occurring never more than once for a particular supergiant star.

A supernova flares up to millions of times its former brightness. In 1054 AD, Chinese astronomers recorded their observation of a star so bright that it could be seen by day as well as by night. This was a supernova (a "super new star"), its glowing plasma remnants now making up the spectacular Crab Nebula, shown in Figure 27.15. A less spectacular but more recent supernova was witnessed in 1987. The progress of this supernova, shown in Figure 27.16, has been very carefully monitored by modern scientific equipment.

FIGURE 27.15

The Crab Nebula is the remnant of a supernova explosion
first observed on Earth in 1054
AD. The explosion took place
within our galaxy at a distance
of about 7000 light-years from
Earth. Had the blast occurred
within 50 light-years, most life
on Earth would have likely
gone extinct. Are there any
nemesis stars right now within
this limit ready to supernova?
Good question. Check the
Web for information about
Betelgeuse.

A neutron star is a kilometer-sized atomic nucleus!

The superdense neutron core that remains after the supernova is called a
neutron star. In accord with the law of conservation of angular momentum, these
tiny bodies, with densities hundreds of millions times greater than those of white
dwarfs, can spin at fantastic speeds. Neutron stars provide an explanation for the
existence of *pulsars*. **Pulsars**, which are neutron stars, are rapidly varying sources
of low-frequency radio emissions. As a pulsar
spins, the beam of radiation it emits sweeps
across the sky. If the beam sweeps over Earth,
we detect its pulses. Of the approximately
300 known pulsars, only a few have been
found emitting X-rays or visible light. One is
in the center of the Crab Nebula (Figure
27.17). It has one of the highest rotational
speeds of any pulsar studied, rotating more
than 30 times per second. This is a relatively
young pulsar, for it is theorized that X-rays
and visible light are emitted only during a
pulsar's early history.

FIGURE 27.16

This image of the 1987A supernova was captured by the Hubble
Space Telescope about 20 years
after the initial explosion was
sighted. Note the development of
the ring systems, which continue
to expand outward. This supernova occurred safely outside the
Milky Way Galaxy some 160,000
light-years away within a nearby
smaller galaxy called the Large
Magellanic Cloud.

We saw earlier that a medium-sized star, such as our Sun, can collapse no further than a white dwarf because the force of gravity is not strong enough to overcome the resistance of electrons, which refuse to trespass into the quantum states
of neighboring electrons. Similarly, a neutron star stops collapsing because neutrons, like electrons, resist trespassing
into their neighboring neutrons. For
a dying star, however, the bigger they
are, the harder they fall. When the
collapsing star is the biggest of the
big, gravitational forces can be strong
enough to overcome even the resistance of neutrons. The collapse continues beyond the stage of a neutron
star, and the star disappears altogether from the observable universe.
What is left is a *black hole*.

(a)   (b)

FIGURE 27.17

The pulsar in the Crab Nebula
rotates like a searchlight, beam-
ing visible light and X-rays
toward Earth about 30 times
per second, blinking on and
off: (a) pulsar on, (b) pulsar off.

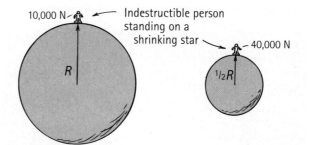

FIGURE **27.18**

If a star collapses to half its radius with no mass change, gravitation at its surface increases fourfold (in accordance with the inverse-square law). If the star collapses to one-tenth its radius, gravitation at its surface increases a hundredfold.

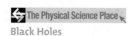

The Physical Science Place
Black Holes

fyi

- Contrary to stories about black holes, they're nonaggressive and don't reach out and swallow objects at a distance. Their gravitational fields are no stronger than the original fields about the stars before their collapse—except at distances smaller than the radius of the original star. Except when they are too close, black holes shouldn't worry future astronauts.

■ 27.5 Black Holes

A **black hole** is the remains of a supergiant star that has collapsed into itself. Upon this collapse, the force of gravity at the surface increases dramatically. Consider this from the perspective of Newton's law of gravity. According to this law, as discussed in Section 4.1, the force of gravity depends on the inverse square of the distance. If a star collapses to a tenth of its original size, the distance between the surface and the center of the star is one-tenth as much. The inverse square of one-tenth ($1/0.1^2$) equals 100. The weight at the surface, therefore, is 100 times as much, as illustrated in Figure 27.18. So the gravitational force at the surface of a collapsing star increases because the star is getting smaller.

As the force of gravity increases, so does the *escape speed*. Recall from Section 4.9 that escape speed is the speed a moving object needs to fly away without ever falling back. For planet Earth, the escape speed is 11.2 km/s. This means that an object shot outward at 11.2 km/s (about 25,000 mi/h) will never fall back to Earth. The escape speed from the surface of our Sun is 618 km/s. For a supergiant star that has collapsed past the neutron star stage, the escape speed shoots up exponentially to the speed of light, which is 300,000 km/s.

In the early 20th century, Einstein proposed that light and matter are related and thus both affected by gravity. We don't normally witness light being affected by gravity because light moves so fast, but with careful observations it is quite measurable. Starlight grazing the eclipsed Sun, for example, is seen to bend inward as the light passes through the Sun's strong gravitational field. So light is pulled downward by gravity. Sunlight can leave our Sun because the speed of light is much greater than the escape velocity. If a star such as our Sun, however, were to collapse to a radius of 3 km, the escape velocity from its surface would exceed the speed of light, and nothing—not even light—could escape. The Sun would be invisible. It would be a black hole.

The Sun, in fact, has too little mass to experience such a collapse, but when some stars with core masses many times greater than the mass of the Sun reach the end of their nuclear resources, they undergo collapse; their collapse continues until the stars reach infinite densities. Gravitation near the surfaces of these shrunken stars is so enormous that light cannot escape from them. They have crushed themselves out of visible existence.

A black hole has the same amount of mass after its collapse as before its collapse, so the gravitational field in regions at and beyond the star's original radius is no different in either case. An orbiting planet would keep on orbiting as though nothing happened. But closer distances near the vicinity of a black hole, beneath the star's original radius, are nothing less than the collapse of space itself, with a surrounding warp into which anything that passes too close—light, dust, or a spaceship—is drawn (Figure 27.19). Astronauts in a powerful spaceship could enter the fringes of this warp and still escape. Below a certain distance, however, they could not, and they would disappear from the observable universe.

FIGURE **27.19**

A rendering of a black hole stealing matter from a companion star.

If the Sun somehow suddenly collapsed to a black hole, what change would occur in the orbital speed of Earth?

Was this your answer?

None. This is best understood classically; nothing in Newton's law of gravitation, $F = G\frac{mM}{d^2}$, changes. The fact that the Sun is compressed doesn't change its mass, M, or its distance, d, from Earth. Because Earth's mass, M, and G don't change either, the force, F, holding Earth in its orbit does not change.

BLACK HOLE GEOMETRY

If we shine a beam of light toward, but slightly away from, a black hole at precisely the right distance, as shown in Figure 27.20, we can direct the light into *circular orbit* about the hole. This region above the black hole is called the *photon sphere*. The photon sphere is rather unstable because the slightest variation in the interaction of a light beam with the gravitational field will send the light beam either spiraling into the hole or back off into space. All beams of light that happen to be incident at this critical distance, however, are captured in the sphere.

An indestructible astronaut with a powerful enough spaceship could venture into the photon sphere of a black hole and come out again. While inside the photon sphere, she could still send beams of light back into the outside universe as shown in Figure 27.21. If she directed her flashlight sideways and toward the black hole, the light would quickly spiral into the black hole, but light directed vertically and at angles close to the vertical would still escape. As she drew closer and closer to the black hole, however, she would need to shine the light beams closer and closer to the vertical for escape. Moving closer still, our astronaut would find a particular distance where *no* light can escape. No matter in what direction the flashlight pointed, all the beams would be deflected into the black hole. Our unfortunate astronaut would have passed within the **event horizon**, the boundary where no light within can escape. Once inside the event horizon, she could no longer communicate with the outside universe; neither light waves, radio waves, nor any matter could escape from inside the event horizon. Our astronaut would have performed her last experiment in the universe as we conceive it.

The event horizon surrounding a black hole is often called the *surface* of the black hole, the diameter of which depends on the mass of the hole. For example, a black hole resulting from the collapse of a star 10 times as massive as the Sun has an event-horizon diameter of about 30 km. Estimated radii of event horizons for black holes of various masses are shown in Table 27.1. The event horizon is not a physical surface. Falling objects pass right through it. The event horizon is simply the boundary of no return.

When a collapsing star contracts within its own event horizon, the star still has substantial size. No known forces, however, can stop the continued contraction,

Light rays

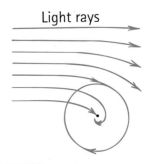

FIGURE 27.20

Light rays deflected by the gravitational field around a black hole. Light passing far away is bent only slightly; light passing closer can be captured into circular orbit; light passing closer still is drawn into the hole.

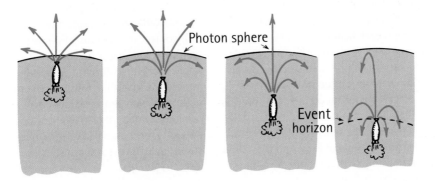

FIGURE 27.21

Just beneath the photon sphere, an astronaut can still shine light to the outside. But as she gets closer to the black hole, only light directed nearer to the vertical gets out, until finally even vertically directed light is trapped. This occurs at a distance called the event horizon.

TABLE 27.1   ESTIMATED RADII OF EVENT HORIZONS FOR NONROTATING BLACK HOLES OF VARIOUS MASSES

Mass of Black Hole	Radius of Event Horizon
1 Earth mass	0.8 cm
1 Jupiter mass	2.8 m
1 solar mass	3 km
2 solar masses	6 km
3 solar masses	9 km
5 solar masses	15 km
10 solar masses	30 km
50 solar masses	148 km
100 solar masses	296 km
1000 solar masses	2961 km

and the star quickly shrinks in size until finally it is crushed, presumably to the size of a pinhead, then to the size of a microbe, and finally to a realm of size smaller than ever measured by humans. At this point, according to theory, what remains has infinite density. This point is the **black-hole singularity**.

Falling into a Black Hole

Imagine yourself exploring a black hole on some futuristic scientific mission. Your spacecraft is cruising within a safe orbit around the black hole. Your first experiment is to launch a clock-bearing probe toward the black hole. The clock consists of a large LED display of blue lights. Through a telescope you watch as the probe descends. Remarkably, the closer the probe gets to the black hole, the slower the clock appears to run. Furthermore, the light coming from the clock shifts from blue to a lower-frequency red. As the probe gets closer still, the clock runs even slower. Soon you can't make out the clock at all because it has shifted to the infrared. So you switch to your infrared telescope and see that as the probe gets closer to the black hole its clock slows to a creep. Furthermore, the probe seems to be taking an unusually long time to descend. Eventually, the light from the clock is visible only with your microwave telescope, followed by your radio telescope as the frequency of the light from the clock gets lower and lower. Ultimately, just as the clock disappears completely, emitting no light whatsoever, you note that the

clock has frozen in time. To get to this point, however, would take, from your point of view, forever.

But you don't have forever, and you soon grow tired of watching the ultra-slow clock as it creeps ever so slowly toward the black hole. So you decide to move on to the second experiment, for which you have volunteered to place yourself within a second probe equipped with a blue clock and an array of telescopes. As you descend toward the black hole you note that your clock runs perfectly normally without changing color. The clock on the mother ship, however, is running rather fast. Furthermore, its color is shifting toward ultraviolet and beyond. Your old shipmates are moving rather fast as well. As you descend even deeper, their peculiar speed grows even faster. Soon they grow impatient waiting for you and they leave you behind for dead. Before you know it, the remaining visible stars quickly pass through their life spans, their light coming to you in a flash of ultrahigh frequencies but through a narrowing field of view. And then there is nothing. At that moment you pass

through the event horizon, which is a mathematical boundary, not a physical one. Singularity is still kilometers beneath you, but you are caught within its unrelenting grip. The universe you left behind has run through infinite time and exists no longer.

Unfortunately, such a fall through the event horizon of a regular-sized black hole would not be survivable. As you come closer, the gravitational pull on your feet would far exceed that on your head. As a result, your body would be stretched. You would be "ripped" of the opportunity to experience what it would be like inside the event horizon. Furthermore, as you continued to fall toward the black-hole singularity, your atoms would be compressed to an infinitely small size, which you would not survive. What would happen next is only conjecture. Perhaps your mass would explode like a Big Bang into another universe. Perhaps what happens to the mass that falls into the singularity is even stranger than we're capable of imagining. Maybe one day we crafty humans will come to understand such processes. If our species lasts that long!

We suspect that all dead stars with core masses greater than about five solar masses may be black holes. Locating these invisible star cores, however, is very difficult. One way to find them is to look for a binary system in which a single luminous star appears to orbit about an invisible companion, as was illustrated in Figure 27.19. If they are closely situated, matter ejected by the normal companion and accelerating into the neighboring black hole should emit X-rays. The first convincing candidate for a black hole, the X-ray star Cygnus X-1, was discovered in 1971. Many additional black hole candidates have since been found, which suggests that black holes are common. Studies of the center of our galaxy strongly suggest the presence of a black hole some 6 billion km in diameter, which is as large as our solar system! The origin of this mega black hole is likely related to the formation of the galaxy itself.

> As described in the box on page 734, tidal forces would rip you apart before falling into a regular-sized black hole. For a mega-sized black hole, like the one at the center of our galaxy, the tidal forces would be negligible—your spaceship would survive passage through the event horizon.

CHECK POINT

What determines whether a star becomes a white dwarf, a neutron star, or a black hole?

Was this your answer?
The mass of a star is the principal factor that determines its fate. Stars that are about as massive as the Sun, and those that are less massive, evolve to become white dwarfs; stars with masses of 10 M_{Sun} or greater evolve to become neutron stars; the most massive stars ultimately become black holes.

27.6 Galaxies

Look up into the clear nighttime sky away from the city lights and you will see plenty of stars. In between the stars you'll also see plenty of black. Before the early 20th century, the abundance of black in the night sky led many people to conclude that the universe consisted of an island of millions of stars nestled within a vast sea of emptiness. In addition to stars, however, are the cloudlike nebulae, some of them with a distinct spiral-shaped structure. As early as the 1750s, the German philosopher Immanuel Kant proposed that these spiral clouds were other islands of stars called *galaxies*. But without powerful telescopes, there was no way to tell whether that was true.

The debate about whether the universe consisted of one or many islands of stars was settled by the American astronomer Edwin Hubble. In 1927, working with the newly built largest telescope in the world, Hubble made out individual stars within the Andromeda spiral nebula (Figure 27.22). Some of these stars he noticed to be *Cepheids,* which are stars that change their luminosity over short periods of time. By measuring the rate at which they changed luminosity he estimated their distance, which he found to be much farther away than any star within our own galaxy. Spiral nebulae were not simply clouds—they were neighboring islands of stars within a vast emptiness that potentially extended forever.

But Hubble took his research a step further and discovered something even more amazing. He knew that the color of light emitted by a star or galaxy receding away from us shifts to the red because of the Doppler effect (Section 10.8). The degree of redshift could be measured quantitatively by focusing on the line spectrum of hydrogen (Section 12.6). The greater the shift in the lines of hydrogen's spectrum, the faster the receding speed. His research team measured both the distances and redshifts of numerous galaxies and discovered that the farther the galaxy, the greater the redshift. This meant that the galaxies were not static islands. Rather, they were receding from us in every direction, which meant that the universe itself was expanding.

fyi

Galaxies are cataloged by two systems. The first catalog is based on the work of Charles Messier, who in 1781 published a list of heavenly structures, such as galaxies, relatively easy to observe with small telescopes. The Andromeda Galaxy, for example, is the 31st entry of this catalog and is thus listed as M31. A "New General Catalog" was begun in 1888 that was subsequently used to identify all structures, including the many more that became visible with the advent of more powerful telescopes. Under this system, the Andromeda Galaxy is cataloged as NGC 224. You can use these catalog numbers in your Internet search engine to learn more about these objects, including their location in the nighttime sky.

Link To Section 10.8

Link To Section 28.1

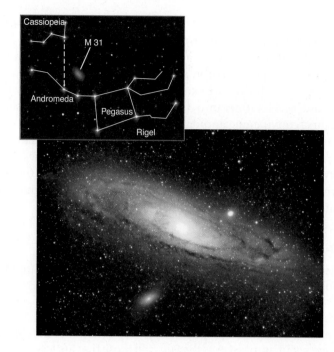

FIGURE 27.22

Hubble showed that the great spiral nebula within the Andromeda constellation was not just a swirling cloud of gas, but a neighboring galaxy of stars, which is now called the Andromeda Galaxy and catalogued as M31. You can see the Andromeda Galaxy for yourself by looking between the constellations of Cassiopeia and Pegasus in the winter nighttime sky. The galaxy appears huge, covering an area six times that of the full Moon. It is, of course, much dimmer than the Moon. Best viewing comes with a good pair of binoculars far away from city lights.

If distant galaxies were all moving away from one another, that could only mean that they were once much closer together. Running the cosmic movie backward would inevitably lead to a moment when all the galaxies were together within a single point. The universe as we know it, therefore, had a beginning. This moment has come to be known as the **Big Bang**, which we will discuss in more detail in Chapter 28. For the remainder of this chapter, however, we will simply describe the different kinds of galaxies and how they are organized within the observable universe.

A **galaxy** consists of a large assemblage of stars, interstellar gas, and dust. Galaxies are the breeding grounds of stars. Our own star, the Sun, is an ordinary star

FIGURE 27.23

A wide-angle photograph of the Milky Way, which appears as a north–south cloudlike band of light. The dark lanes and blotches are interstellar gas and dust obscuring the light from the galactic center. If it weren't for this dust, the Milky Way would be a much more spectacular nighttime display. This photograph also shows Comet Hyakutake, which appeared in 1996.

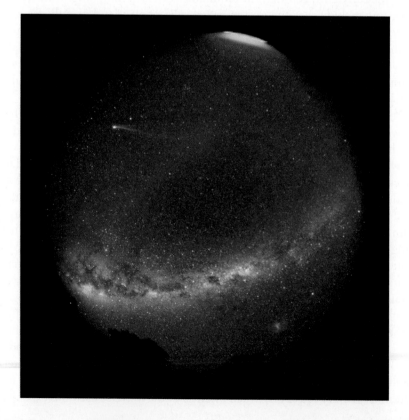

among more than 100 billion others in an ordinary galaxy known as the Milky Way Galaxy (Figure 27.23). With unaided eyes, we see the Milky Way as a faint band of light that stretches across the sky. The early Greeks called it the "milky circle" and the Romans called it the "milky road" or "milky way." The latter name has stuck.

The masses of galaxies range from about a millionth the mass of our galaxy to some 50 times more. Galaxies are calculated to have much more mass than can be seen with the telescope. A small proportion of the invisible mass is simply matter that has grown so cold that it doesn't emit enough light for us to see. The bulk of the invisible mass, however, is likely an unknown form of matter, called *dark matter*, that does not absorb or emit light. It does, however, possess mass and so its gravitational effects are quite measurable. In the next chapter we describe how dark matter likely played a key role in the formation and distribution of galaxies.

ELLIPTICAL, SPIRAL, AND IRREGULAR GALAXIES

The millions of galaxies visible on time-exposure photographs can be separated into three main classes—elliptical, spiral, and irregular. *Elliptical galaxies* are the most common galaxies in the universe. They are spherical in shape with the stars more crowded toward the center. Most contain little gas and dust, which makes them easy to see through. They also tend to be yellow in color, which tells us that they consist primarily of older stars—older stars are yellow, while hot young stars tend to be blue. Most ellipticals are small, consisting of fewer than a billion stars (Figure 27.24). An exception is the giant elliptical galaxy M87 (Figure 27.25). The largest ellipticals are about 5 times as large as our galaxy, and the smallest are 1/100 as large.

Spiral galaxies, such as the Andromeda Galaxy, shown in Figure 27.22, are perhaps the most beautiful arrangements of stars. Some spirals, such as the Sombrero Galaxy of Figure 27.26, have a spheroid central hub. Others, as shown in Figure 27.27, have a hub shaped like a bar. The Milky Way Galaxy is thought to look much like the NGC 6744 spiral galaxy, which is an intermediate between a barred and unbarred spiral (Figure 27.28).

Irregular galaxies are normally small and faint and are difficult to detect. They tend to contain large clouds of gas and dust mixed with both young (blue) and

FIGURE 27.24

This small elliptical galaxy, Leo I, found within the constellation Leo, is only about 2500 light-years in diameter. For comparison, the diameter of our Milky Way Galaxy is about 100,000 light-years.

fyi

- The Andromeda Galaxy is our closest spiral neighbor, being only some 2.5 million light-years away. It contains many more stars than the Milky Way, which makes it more luminescent. Also, its diameter is about 220,000 light-years, compared to the Milky Way's 100,000 light-years. Thus, our view of the Andromeda is likely more spectacular than the Andromeda's view of us.

FIGURE 27.25

The giant elliptical galaxy M87, one of the most luminous galaxies in the sky, is located near the center of the Virgo cluster, some 50 million light-years from Earth. It is about 120,000 light-years across and about 40 times as massive as our own galaxy, the Milky Way.

FIGURE 27.26

The Sombrero Galaxy, cataloged as M104, is about 80,000 light-years in diameter and about 32 million light-years from Earth. At its center is one of the most supermassive black holes measured in any nearby galaxy.

FIGURE 27.27

The beautiful barred spiral galaxy NGC 1300 is about 100,000 light-years across and some 70 million light-years away.

old (yellow) stars. The irregular galaxy first described by the navigator on Magellan's voyage around the world in 1521 is our nearest neighboring galaxy—the Magellanic Clouds. This galaxy consists of two "clouds," called the Large Magellanic Cloud (LMC) and the Small Magellanic Cloud (SMC), both of which are slowly being pulled into the Milky Way. The LMC is dotted with hot young stars with a combined mass of some 20 billion solar masses, and the SMC contains stars with a combined mass of about 2 billion solar masses (Figure 27.29). Some irregular galaxies, such as NGC 4038, shown in Figure 27.30, are the aftermaths of galactic collisions.

CHECK POINT
■ Is it possible for one type of galaxy to turn into another?

Was this your answer?
Yes, and this occurs as two symmetrically shaped galaxies collide to form an asymmetrically shaped irregular galaxy.

ACTIVE GALAXIES

By galactic standards, our Milky Way Galaxy is a pretty mellow place. Peering away from our galactic home, however, astronomers find galaxies, called *active galaxies*, that are brimming with energy. One type of active galaxy is the **starburst galaxy**, in which stars are forming at an unusually high rate. A starburst galaxy can produce more than 100 new stars per year. By comparison, our Milky Way produces an average of about 1 new star per year. A starburst's high rate of star formation is often the result of some violent disturbance, such as a collision between two galaxies. The irregular galaxy shown in Figure 27.30 is an example of a starburst galaxy. Another example is the Cigar Galaxy, M82, which is being deformed by the tidal forces from its much larger neighbor, M81 (Figure 27.31). A starburst tends to die down once the disturbance is removed or after the

FIGURE 27.28

The NGC 6744 galaxy is an intermediate between a barred and unbarred spiral galaxy. Studies of the Milky Way suggest that it too is an intermediate spiral. In other words, this is what we may look like from afar.

FIGURE 27.29

(a) The Large Magellanic Cloud and (b) the neighboring Small Magellanic Cloud are a pair of irregular galaxies. The Magellanic Clouds are our closest galactic neighbors, about 150,000 light-years distant. They likely orbit the Milky Way.

(a)   (b)

FIGURE 27.30

Shown in black and white is the ground-telescope view of an irregular galaxy resulting from the collision of two galaxies. Note the remnant arms that suggest two former spiral galaxies. The inset shows a close-up color view taken by the Hubble Space Telescope. Evident is the rapid formation of new stars (blue) occurring as the mass of the two galaxies combined.

starburst galaxy consumes of all its interstellar fuel. Many elliptical galaxies are thought to be former starburst galaxies because of their low abundance of interstellar dust and gases.

Other active galaxies are active by virtue of their galactic core, which hosts a black hole more massive than millions or even billions of Suns. The event horizons of these black holes are about as large as our solar system! Most large galaxies, including the Milky Way, have such black holes in their centers. What sets the "active galaxies" apart is the large amount of matter that continually falls into their supermassive black holes. Before falling into the black hole, the doomed mass forms a rapidly spinning disk, called an *accretion disk*, around the equator of the black hole. Charged particles in this hyperspinning disk create a narrow yet ultrastrong magnetic field that rises from the black hole's poles. Rather than falling into the black hole, some of the charged particles, such as electrons, are accelerated outward through these magnetic fields to nearly the speed of light. This results in two extremely long streams of particles, called *jets*, extending thousands of light-years away from the galactic center, which is called an **active galactic nucleus (AGN)**.

A relatively close AGN is found within the large elliptical galaxy M87, which was shown in Figure 27.25. High-resolution images of this galaxy, as shown in

FIGURE 27.31

The Cigar Galaxy, M82, is a spiral galaxy tilted away from us so that we see it from an edge-on view. Tidal forces from the nearby M81 galaxy disturb the distribution of matter within M82, which clumps, allowing for the formation of many new stars, as evidenced by M82's remarkable blue color. The red gases above and below the galactic plane are primarily hydrogen being pushed out by abundant stellar wind.

FIGURE 27.32

Material falling into the supermassive black hole in the center of M87 generates powerful jets that shoot out at near light speed.

Figure 27.32, reveal a jet of material streaming away from the center of this galaxy to a distance of about 7000 light-years. Interestingly, the jet is angled toward us. This plus the great speed of the jet (99.5% of the speed of light) helps make the jet appear more luminous. The opposite "counterjet" receding away from us at such great speeds is invisible for reasons having to do with Einstein's special theory of relativity.

Nearby active galactic nuclei, such as that of M87, give us a clue as to the possible nature of the most energetic galaxies of all—the *quasars*. Starting in the 1960s, astronomers began discovering extremely energetic bodies hundreds of times more luminous than our own galaxy, yet farther away than any observed object. Because they looked like radio emitting stars, they were dubbed "quasistellar radio sources," which was shortened to "quasars." Because all quasars are so very far away, they occurred a very long time ago—up to 13 billion years ago, which was close to the beginning of the universe. As we look to quasars, therefore, we are peering into the early lives of galaxies (Figure 27.33). During this galactic youth, much material was still falling into the supermassive black holes found within their galactic cores. The dynamics of this process allow for an efficient conversion of mass into energy, which would provide for colossal jets of highly energetic particles and light. When one of these ancient jets faces our direction, the result is a brilliant display of energy we call a **quasar**.

CHECK POINT

■ Are there any quasars found within the Milky Way Galaxy?

Was this your answer?

No. A quasar is the active galactic nucleus of a galaxy as it appeared towards the beginning of the universe. All quasars are billions of light-years away from our galaxy.

CLUSTERS AND SUPERCLUSTERS

Galaxies are not the largest structures in the universe; they tend to cluster into distinguishable groups. Our Milky Way Galaxy, for example, is part of a cluster of local galaxies that include two other major spiral galaxies, namely the Andromeda Galaxy and the Triangulum Galaxy. Also included are more than a

FIGURE 27.33

Each disk in this deep-space image taken by the Hubble Space Telescope is a galaxy. The quasar shown in the center is billions of light-years behind this cluster of galaxies. Interestingly, the cluster's gravity has bent the light from the quasar like a lens so that multiple images of the quasar are also seen.

Multiple Lensed Images

Quasar

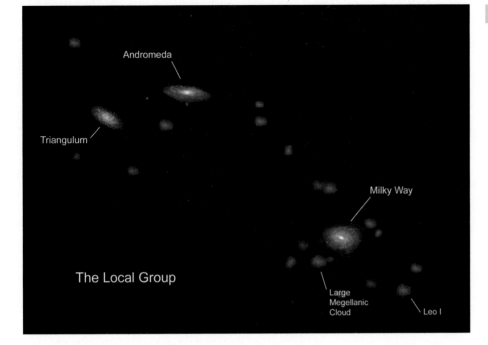

FIGURE 27.34

This two-dimensional composition shows the approximate relative distances between the members of our Local Group of galaxies. These galaxies are all moving toward each other and will one day collide into a larger supergalaxy.

dozen smaller elliptical galaxies, such as the Leo I galaxy shown in Figure 27.24, and a few irregular galaxies, such as the Large Magellanic Cloud. Altogether this cluster of galaxies is called the **Local Group**. Their approximate distributions are shown in Figure 27.34. If drawn correctly to scale, Andromeda is only about 20 Milky Way diameters away from the Milky Way. The Triangulum Galaxy, so named because it completes a triangle between the spirals, is even closer to Andromeda, but farther away from us.

Our Local Group of galaxies is also under the gravitational influence of neighboring galactic clusters. Our cluster plus all these other clusters makes for what is called a *supercluster*, which is a cluster of galactic clusters. Our Local Group is actually a rather minor component of our **Local Supercluster**, as is illustrated in Figure 27.35.

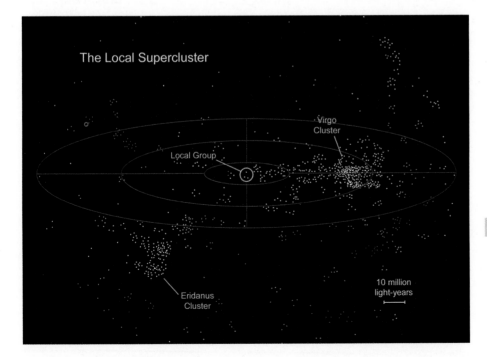

FIGURE 27.35

A supercluster is a cluster of galactic clusters. Each dot represents a galaxy. Note that our Local Group is found midway between two much larger clusters, the Virgo and Eridanus clusters.

Each cloud represents a super-cluster. Note that the super-clusters are strung together as though on the surface of a foam.

Our Local Supercluster is tied in with an elaborate network of many other superclusters, as shown in Figure 27.36. Together, these superclusters appear as though they reside on the surface of a foam inside which are large voids of empty space. Zooming out farther we find that the network of superclusters extends to the edges of the *observable universe*, as illustrated in Figure 27.37. By "observable universe" we mean all that we are able to see given the fact that the universe is only about 14 billion years old. The light coming from any object farther than 14 billion light-years has not had sufficient time to reach us.

So our observable universe is huge. Utterly huge. How large, then might be the entire universe—the whole she-bang? We don't know, and maybe never will.

The network of superclusters extends to the edges of the observable universe, which is no farther than 14 billion light-years away. This illustration, however, shows a hypothetical bird's-eye view of this observable universe fully matured to the present moment, which, because of cosmic expansion, would place those most distant objects now some 42 billion light-years away.

But that doesn't stop cosmologists and mathematicians from developing models proposing possible answers. One such model suggests that if the observable universe were the size of a proton, the entire universe would be about as large as planet Earth. Consider how many protons fit within the dimensions of Earth. This would be the number of observable universes within the entire universe. The number is so enormous that if you could travel $10^{10^{118}}$ m in any direction, you would have a large probability of coming across another observable universe that looks very much like the one you left. Continuing this speculation, you'd look for somebody like yourself who is reading a book exactly like the one you are reading now. What if this person is actually a future you? Besides living very, very far away, the only measurable difference is that he or she has already finished reading this paragraph. Scientists call this the "many worlds" model, in which each observable universe is static, representing one possible arrangement of matter. We don't move through time. Rather, we jump from one observable universe to the next, which gives the appearance of moving through time. Welcome. You have just jumped to a new observable universe. The old one you left six sentences ago is now ever so distant.

FIGURING PHYSICAL SCIENCE

■ The Scale of the Galaxy

SAMPLE PROBLEM 1

Earth is about 0.000016 light-year from the Sun and about 4.2 light-years from the next nearest star, which is Proxima Centauri. Compared to our distance from the Sun, how many times as far are we from Proxima Centauri?

SOLUTION:

Divide our distance from Proxima Centauri by our distance to the Sun:

$$\frac{4.2 \text{ light-years}}{0.000016 \text{ light-year}} = 262{,}500$$

So the nearest star is about 260,000 times as far from us as the Sun is.

SAMPLE PROBLEM 2

Our distance from the center of the Milky Way Galaxy is about 26,000 light-years. How many times as far is this compared to our distance from Proxima Centauri?

SOLUTION:

Divide the distance to the center of the galaxy by the distance to Proxima Centauri:

$$\frac{26{,}000 \text{ light-years}}{4.2 \text{ light-years}} = 6190$$

So the center of our galaxy is about 6200 times as far away as the nearest star.

SAMPLE PROBLEM 3

The Milky Way and Andromeda Galaxies are about 2,300,000 light-years apart. The diameter of the Milky Way Galaxy is about 100,000 light-years. How many Milky Way diameters distant is the Andromeda Galaxy?

SOLUTION:

Divide the distance to the Andromeda Galaxy by the diameter of the Milky Way Galaxy:

$$\frac{2{,}300{,}00 \text{ light-years}}{100{,}000 \text{ light-years}} = 23$$

So the Andromeda Galaxy is about 23 Milky Way Galaxy diameters away.

SAMPLE PROBLEM 4

The Andromeda Galaxy is moving toward the Milky Way Galaxy at a rate of about 300,000 mi/h. In how many years will these two galaxies collide?

SOLUTION:

Convert 300,000 mi/h to light-years per year. First convert the 300,000 mi to light-years:

$(300{,}000 \text{ mi})(1.61 \text{ km/1 mi}) \times$
$(1 \text{ light-year}/10{,}000{,}000{,}000{,}000 \text{ km})$
$\qquad = 4.83 \times 10^{-8} \text{ light-year}$

Second, convert hours to years:
$(1 \text{ h})(1 \text{ day}/24 \text{ h})(1 \text{ yr}/365.25 \text{ days})$
$\qquad = 1.14 \times 10^{-4} \text{ year}$

Put the two converted values together:

$300{,}000 \text{ mi/h}$
$\quad = 4.83 \times 10^{-8} \text{ light-years}/1.14 \times 10^{-10} \text{yr}$
$\quad = 4.23 \times 10^{-10} \text{ light-years/yr}$

Use the equation for speed from Chapter 1:

$$\text{speed} = \text{distance/time}$$

$4.23 \times 10^{-4} \text{ light-years/yr}$
$\qquad = 2{,}300{,}000 \text{ light-years}/x \text{ yr}$

Solve for x:

$$x = 5{,}400{,}000{,}000 \text{ yr}$$

So in roughly 5.4 billion years, the Andromeda and Milky Way Galaxies will collide. By this time our Sun will have exhausted most all of its nuclear fuel, so this is not something we Earthlings will be around to witness. Collisions between galaxies, however, are fairly common, and astronomers have photographed many such occurrences now in progress.

Polls show that about half of American adults do not know that it takes one year for Earth to go around the Sun. Many of us, therefore, are still struggling with scientific ideas of 400 years ago. Your knowledge is much more, and being aware of the amazing possibilities that science continues to reveal puts you in a very privileged minority. Celebrate that!

CHECK POINT

■ Which is greater, the number of stars in our galaxy or the number of galaxies in the universe?

Was this your answer?

There are far more galaxies within the entire universe than there are stars in our galaxy. Recall from the beginning of this chapter that astronomers estimate that there are about 100 billion stars in our galaxy and about 100 billion galaxies in our observable universe. If true, that means there are about 10^{22} stars in our observable universe, which is about the same number of water molecules in a drop of water. As large as the observable universe is large is as small as the fundamental building blocks of our body are small. As humans we are nicely situated between these two extremes.

SUMMARY OF TERMS

Celestial sphere An imaginary sphere surrounding Earth to which the stars are attached.

Light-year The distance light travels in one year.

Hertzsprung–Russell diagram (H–R diagram) A plot of intrinsic brightness versus surface temperature for stars. When so plotted, stars' positions take the form of a main sequence for average stars, with exotic stars above or below the main sequence.

Main sequence The diagonal band of stars on an H–R diagram; such stars generate energy by fusing hydrogen to helium.

Giant stars Cool giant stars above main-sequence stars on the H–R diagram.

White dwarf A dying star that has collapsed to the size of Earth and is slowly cooling off; located at the lower left on the H–R diagram.

Planetary nebula An expanding shell of gas ejected from a low-mass star during the latter stages of its evolution.

Nova An event wherein a white dwarf suddenly brightens and appears as a "new" star.

Supernova The explosion of a massive star caused by gravitational collapse with the emission of enormous quantities of matter.

Neutron star A small, extremely dense star composed of tightly packed neutrons formed by the welding of protons and electrons.

Pulsar A celestial object (most likely a neutron star) that spins rapidly, sending out short, precisely timed bursts of electromagnetic radiation.

Black hole The remains of a giant star that has collapsed on itself, so dense and with a gravitational field so intense that light cannot escape from it.

Event horizon The boundary region of a black hole from which no radiation may escape. Any events within the event horizon are invisible to distant observers.

Black-hole singularity The object of zero radius into which the matter of a black hole is compressed.

Big Bang The primordial explosion of space at the beginning of time.

Galaxy A large assemblage of stars, interstellar gas, and dust, usually categorized by its shape: elliptical, spiral, or irregular.

Starburst galaxy A galaxy in which stars are forming at an unusually high rate.

Active galactic nucleus (AGN) A supermassive black hole at the center of a galaxy into which matter is falling at a high rate, thereby releasing astronomical amounts of energy.

Quasar A distant galaxy with an active galactic nucleus emitting a beam of radiation pointed in our direction, which makes the quasar appear much brighter than the galaxy in which it resides.

Local Group Our immediate cluster of galaxies, including the Milky Way, Andromeda, and Triangulum spiral galaxies plus a few dozen smaller elliptical and irregular galaxies.

Local Supercluster A cluster of galactic clusters in which our Local Group resides.

REVIEW QUESTIONS

27.1 Observing the Night Sky

1. What are constellations?
2. Why does an observer at a given location see one set of constellations in the winter and a different set of constellations in the summer?
3. Why do the stars appear to turn on an imaginary north–south axis once every 24 hours?
4. Is the light-year a measurement of time or distance?

27.2 The Brightness and Color of Stars

5. Which is hotter, a red star or a blue star?
6. What is the difference between apparent brightness and luminosity?

27.3 The Hertzsprung–Russell Diagram

7. What is an H–R diagram?
8. Where are most stars plotted on an H–R diagram?
9. Where does our Sun reside on an H–R diagram?
10. Which is larger, a red star or a yellow star?

27.4 The Life Cycles of Stars

11. What process changes a protostar to a full-fledged star?
12. What are the outward forces that act on a star?
13. What are the inward forces that act on a star?
14. When will our Sun reach the red-giant stage?
15. Is the lifetime of a high-mass star longer or shorter than that of a low-mass star?

16. What is the relationship between the heavy elements that we find on Earth today and supernovae?
17. What is the relationship between a supernova and a neutron star?
18. What is the relationship between a neutron star and a pulsar?

27.5 Black Holes

19. What is the relationship between a supergiant star and a black hole?
20. Why do we not think the Sun will eventually become a black hole?
21. How does the mass of a star before its collapse compare with the mass of the black hole that it becomes?
22. If black holes are invisible, what is the evidence for their existence?
23. Is a black hole's event horizon a physical or mathematical boundary?

27.6 Galaxies

24. What type of galaxy is the Milky Way?
25. What are the consequences of galaxies colliding?
26. What is a starburst galaxy?
27. How does the brightness of a quasar compare with that of a large galaxy?
28. How many spiral galaxies are within the Local Group?
29. Is the Local Group a relatively small or large cluster of galaxies?
30. What are three galactic clusters found within our Local Supercluster?

EXERCISES

● BEGINNER   ■ INTERMEDIATE   ◆ EXPERT

1. ● The 19th-century author and social commentator Thomas Carlyle wrote, "Why did not somebody teach me the constellations and make me at home in the starry heavens, which are always overhead and which I don't half know to this day?" What besides the names of the constellations did Thomas Carlyle not know?
2. ■ Is any star bright enough for us to see on a sunny day?
3. ■ On the Moon, stars other than the Sun can be seen during the daytime. Why?
4. ● Which figure in the chapter best shows that a constellation seen in the background of a solar eclipse will be seen six months later in the night sky?
5. ■ We see the constellations as distinct groups of stars. Discuss why they would look entirely different from some other location in the universe, far distant from Earth.
6. ■ The Big Dipper is sometimes right side up (can hold water), and at other times upside down (cannot hold water). What length of time is required for the Dipper to change from one position to the other?

7. ● Why does the Big Dipper change its position in the night sky over the course of the evening, while Polaris remains relatively fixed in its position?
8. ■ If you traveled to the North Pole, where in the sky would you find Polaris, the North Star?
9. ● What is the relationship between a planetary nebula and a white dwarf?
10. ● What do the outward and inward forces acting on a star have to do with its size?
11. ● What is the relationship between a white dwarf and a nova?
12. ● What event marks the birth of a star?
13. ● When does a star die?
14. ■ What does the color of a star tell you about the star?
15. ● What color are the largest stars
16. ● What is expected to happen to the Sun in its old age?
17. ■ When can a burnt-out collapsing star rekindle itself?
18. ● In what sense are we all made of star dust?

19. ● How is the gold in a wedding ring evidence of ancient stars that ran their life cycles long before the solar system came into being?

20. ■ Would you expect metals to be more abundant in old stars or new stars? Defend your answer.

21. ◆ What is the evidence that our Sun is a relatively young star in the universe?

22. ● Why is there a lower limit on the mass of a star? (What can't happen in a low-mass accumulation of hydrogen atoms and other interstellar material?)

23. ● What keeps a main-sequence star from collapsing?

24. ● How does a protostar differ from a star?

25. ● How does the energy of a protostar differ from the energy that powers a star?

26. ● Why do nuclear fusion reactions not occur on the outer layers of stars?

27. ■ Why are massive stars generally shorter lived than low-mass stars?

28. ◆ Why are supermassive stars relatively rare?

29. ■ What does the spin rate of a star have to do with whether it has a system of planets?

30. ■ With respect to stellar evolution, what is meant by the statement, "The bigger they are, the harder they fall"?

31. ■ Why will the Sun not be able to fuse carbon nuclei in its core?

32. ■ Some stars contain fewer heavy elements than our Sun contains. What does this indicate about the age of such stars relative to the age of our Sun?

33. ● Which has the highest surface temperature: a red star, a white star, or a blue star?

34. ◆ In terms of the life cycle of the Sun, explain why life on Earth cannot last forever.

35. ◆ Elements heavier than iron are created in stars; are they formed in the same way as elements lighter than iron? Explain.

36. ■ If you fell into a black hole, you'd likely die as a result of tidal forces. Explain.

37. ■ A black hole is no more massive than the star from which it collapsed. Why, then, is gravitation so intense near a black hole?

38. ◆ What happens to the radial distance of the event horizon as more and more mass falls into a black hole?

39. ◆ What is the difference between a black hole's photon sphere and its event horizon?

40. ◆ How close are you to a black hole as you pass through the event horizon?

41. ● Will the Sun become a supernova? A black hole? Defend your answer.

42. ● Are there galaxies other than the Milky Way that can be seen with the unaided eye? Discuss.

43. ■ From where does a quasar release its energy?

44. ■ Does the Milky Way Galaxy contain an active galactic nucleus?

45. ◆ How should Figure 27.37 be modified to represent that which is currently observable to us?

PROBLEMS

● BEGINNER ■ INTERMEDIATE ◆ EXPERT

1. ● Suppose star A is four times as luminous as star B. If these stars are both 500 light-years away from Earth, how do their apparent brightnesses compare? How do the apparent brightnesses of these stars compare if star A is twice as far away as star B?

2. ■ The brightest star in the sky, Sirius, is about 8 light-years from Earth. Show that if you could somehow travel there at jet-plane speed, 2000 km/h, the trip would take about 4.3 million years. (Note: 1 light-year equals 9.46×10^{12} km.)

3. ● If you traveled straight out from the core of our galaxy and then looked back, you would have a grand view of the Milky Way's spiral shape. If the distance from the core to the outer edges is 50,000 light-years, how much

surface area would you be looking at? Assume the galaxy to be a circle whose area can be found by the equation $area = \pi r^2$.

4. ● Assume the Milky Way contains 100 billion stars evenly distributed with none concentrated toward the center. What would be the surface area density of stars? Use the equation $surface\ area\ density = number/surface\ area$.

5. ■ Use the information in Problem 4 to figure out the average amount of space around a single star in units of AU. (Note: 1 light-year = 63,000 AU.)

6. ■ Using your answer to Problem 5, would it be possible for two galaxies with stars evenly distributed to pass right through each other?

ACTIVE EXPLORATIONS

Passing Stars

To observe the diurnal motion of the stars, go star watching tonight. Pick a star or constellation that lines up with a stationary landmark such as a tree or house. Then come back in an hour or so and you will see that the star has

moved away from the landmark but remains in place relative to the other stars. In what direction has that star moved? To the east? West? Southwest? Northwest? Where will it be when the Sun rises? Will it be in the same location in 24 hours?

Passing Constellations

To observe the revolutionary motion of Earth around the Sun, go star watching and make note of the stars directly above you. Sketch their pattern on a piece of paper and write down the date and time at which you make this observation. If these stars are not already a well-known constellation, make up a new constellation and name it something creative. After a month has passed, look for this same constellation at the same time of night. Why is it no longer directly above you? In what direction has your constellation shifted? Why?

The Great Andromeda Galaxy

The nighttime sky is full of more than just stars and planets. If you are so lucky to be in a location where the clear night sky is free from the lights of a city, you simply must try to find the Andromeda Galaxy. The best viewing times are during the winter months in the Northern Hemisphere. First look for a large rectangle, which is the body of the flying horse Pegasus. Viewing Andromeda is best when the Pegasus box is directly overhead. Second, look to the north (close to where the Big Dipper is often seen) to find the letter W, which is the Cassiopeia constellation. Near Pegasus you will see a V of stars that extends from the box to the left of Cassiopeia. These are Pegasus's rear legs, which overlap the Andromeda constellation. The Andromeda Galaxy lies just above the upper leg directly between the Cassiopeia W and the Pegasus box. Look carefully at Figure 27.22 to help you pinpoint the exact location. With the unaided eye, it looks like a small dim fuzz. You'll see it best by not looking at it directly. The fuzz takes on more of an oval shape when viewed through a good pair of binoculars. What you see with the binoculars is mostly the central core. The full galaxy, visible only with more powerful telescopes, is about six times the diameter of the Moon!

READINESS ASSURANCE TEST (RAT)

If you have a good handle on this chapter, if you really do, then you should be able to score at least 7 out of 10 on this RAT. If you score less than 7, you need to study further before moving on.

Choose the BEST answer to each of the following.

1. Summer and winter constellations are different because
   (a) of the spin of Earth about its polar axis.
   (b) the night sky faces in opposite directions in summer and winter.
   (c) of the tilt of Earth's polar axis.
   (d) the universe is symmetric and harmonious.
2. Polaris is always directly over
   (a) the North Pole.
   (b) any location north of the equator.
   (c) the equator.
   (d) the South Pole.
3. The star nearest Earth is
   (a) Alpha Centauri.
   (b) Polaris.
   (c) Mercury.
   (d) the Sun.
4. The property of a star that relates to the amount of energy it is producing is its
   (a) luminosity.
   (b) apparent brightness.
   (c) color.
   (d) volume.
   (e) mass.
5. The longest-living stars are those of
   (a) low mass.
   (b) high mass.
   (c) intermediate mass.
   (d) infinite mass.
6. We do not see stars in the daytime because
   (a) the Sun blocks them.
   (b) they simply don't exist in the daytime part of the sky.
   (c) skylight overwhelms starlight.
   (d) of the lack of contrast with moonlight.
   (e) the solar wind obscures them from view.
7. Metals are relatively more abundant in
   (a) old stars.
   (b) new stars.
   (c) neither in particular.
   (d) neutron stars.
8. After our Sun burns its supply of hydrogen, it will become a
   (a) white dwarf.
   (b) black dwarf.
   (c) black hole.
   (d) red giant.
   (e) blue giant.
9. A black hole is
   (a) an empty region of space with a huge gravitational field.
   (b) a small region that has the mass of many galaxies.
   (c) the remains of a giant collapsed star.
   (d) as large as its photon sphere.
10. Scientists estimate the age of our universe to be about
   (a) 5000 years.
   (b) 1 billion years.
   (c) 14 billion years.
   (d) 42 billion years.

Answers to RAT
1b, 2a, 3d, 4a, 5a, 6c, 7b, 8d, 9c, 10c

EXPLORING FURTHER

http://imagine.gsfc.nasa.gov/index.html

Visit Imagine the Universe for topics ranging from supernovae to black holes. Enjoy clear explanations, great visual effects, and quizzes to test your understanding.

http://www.windows.ucar.edu

This site contains much information about stars, galaxies, and all things astronomical. You will find the history of space missions, current space weather forecasts, and multimedia tours of our solar system and beyond.

http://www.smv.org/hastings/student2.htm

This step-by-step sequence of lessons is designed to introduce you to the fundamentals of stars and galaxies.

http://oposite.stsci.edu
http://hubblesite.org

(Note that the spelling of the web address has "oposite" spelled with a single letter p.) This is the home page for the Space Telescope Science Institute, which helps connect the general public with the amazing discoveries made through our space telescopes, such as the Hubble Space Telescope. Follow the link to the photo gallery, which brings you to the HubbleSite. The Amazing Space education link for K–12 education is also worth your time.

http://www.atlasoftheuniverse.com

The illustrations for Figures 27.35 through 27.37 were developed from graphics created by Richard Powell through this Web site, which he created as part of his graduate school studies in astronomy. Featured is a wonderful interactive tutorial on the scale of the universe.

CHAPTER 27 ONLINE RESOURCES

The Physical Science Place

Tutorials

- Stellar Evolution
- Black Holes

Video

- Lives of Stars

Quiz

Flashcards

Links

28

the structure of space and time

28.1 THE BIG BANG

28.2 COSMIC INFLATION

28.3 GENERAL RELATIVITY

28.4 DARK MATTER

28.5 DARK ENERGY

28.6 THE FATE OF THE UNIVERSE

■ We began this textbook with physics focusing on the mechanics of our everyday physical world. Physic concepts were then applied to the world of the very small—the world of atoms and molecules, the domain of chemistry. We then applied physics and chemistry to planet Earth with a focus on its land, oceans, and atmosphere. Beyond Earth, two astronomy chapters discussed our galactic back-yard, the solar system; our neighborhood of stars, the Milky Way Galaxy; and our neighborhood of galaxies, the Local Group. We left off with a discussion on how superclusters of galaxies extend throughout the universe. Note in the above photograph that each spot of light is not a star, but an entire galaxy! This photo shows one tiny portion of our observable universe, a tiny corner of the universe as a whole.

Now we conclude with discussions of two very broad-reaching and most fascinating topics: *cosmology* and *relativity*. **Cosmology** is the study of the overall structure and evolution of the universe. Through cosmology we attempt to answer such questions as "How did the universe come into being?" and "What might be its ultimate fate?" **Relativity**, first postulated by Einstein, is the study of how space, energy, and mass are related to time.

We start this final capstone chapter by considering how it all began with a bang—the *Big Bang*.

Link To Section 27.6

The Physical Science Place
Hubble's Law

The Physical Science Place

From the Big Bang to Galaxies

28.1 The Big Bang

Not so long ago it was commonly thought that our Milky Way Galaxy comprised the whole universe. Then in the early 1920s, the astronomer Edwin Hubble, using a newly built powerful telescope, discovered that the Andromeda "nebula" was, in fact, a separate galaxy farther away than the outermost stars of the Milky Way. This was a monumental discovery, but Hubble didn't stop there. As discussed in the previous chapter, Hubble continued with his work, identifying and measuring the distances to numerous other galaxies. What he discovered next was most astonishing—the galaxies are all receding from one another. Furthermore, the farther away the galaxy, the greater the velocity with which it appears to be receding.

Hubble's observations had two major implications. The first was that if you could run the movie backward, you would find a time when the whole universe was compressed to a single point. Perhaps the universe as we know it has not always existed. Perhaps there was a moment at which the universe itself was born. This beginning point from which all matter and energy within our universe arose is referred to as the **Big Bang**.

The second major implication of Hubble's observations is that the universe is not contained within a region of space. Rather, space is *in* the universe and this space is rapidly expanding. This is peculiar because you may first think that the Big Bang occurred within an already existing infinite space and that matter and energy flew outward from this Big Bang to occupy this space. If this were the case, however, the distribution of galaxies we observe today and their relative motions would be quite different. To reemphasize this often confused concept: when we talk about the expansion of the universe, we are referring to an expansion of the very structure of space itself. A useful analogy is a group of ants on a balloon that is expanding, as shown in Figure 28.1. As the balloon is inflated, every ant sees every other ant moving farther away. Likewise, in an expanding universe, any observer sees all other galaxies moving away. So the Big Bang marked not only the beginning of time, but the beginning of space.

So how was Hubble able to measure the distances to incredibly distant galaxies? Hubble needed the power of the newly constructed Mt. Wilson telescope, which allowed him to distinguish individual stars within neighboring galaxies (Figure 28.2). As Hubble studied these distant stars he discovered that some of them were of a certain type, called *Cepheids*, that regularly change how bright they are over a period of a few days.

At the time of Hubble's observations, astronomers could calculate the luminosity of a Cepheid from its periodic changes in brightness. So by measuring the periods of the distant Cepheids, Hubble was able to calculate their luminosities. Astronomers use the term *luminosity* to mean how much energy a star puts out per second. The more luminous a star, the brighter it appears. Note that the

FIGURE 28.1

INTERACTIVE FIGURE

Every ant on the expanding balloon sees all other ants moving farther away. Each ant may therefore think that it is at the center of the expansion. Not so! There is no center to the surface of the balloon, just as there are no edges.

fyi

■ The North Star, also known as Polaris, is a Cepheid variable with a period of about four days. Its change in apparent brightness is not discernable to the naked eye. Interestingly, the North Star is in orbit around two companion stars, which means that the North Star is actually a triple star. These three stars are too close together and too far away (about 430 light-years) to be discernable with the naked eye.

apparent *brightness* of a star is not its luminosity. Why? Because brightness diminishes with distance. The farther away you are from a luminous star, the dimmer it appears. Like gravity, light grows weaker via the inverse-square law— if you double your distance from the light source, you see that the light is $\left(\frac{1}{2}\right)^2 = \frac{1}{4}$ as bright. So by comparing a Cepheid's luminosity with its brightness and plugging this into the inverse-square law, Hubble was able to calculate the distance to any galaxy containing Cepheids.

CHECK POINT

- Star A is four times as bright as star B, yet the two stars have the same luminosity.
  1. How can this be?
  2. If star A is 100 light-years away, how far away is star B?

Were these your answers?
1. Light from a light source becomes dimmer with distance. Star A, therefore, is brighter because it is closer to us.
2. The inverse-square law tells us that the brightness of a star is related to the inverse distance to the star squared:

$$\text{Apparent brightness} \sim \frac{1^2}{d}$$

So star A is four times as bright if star B is twice as far away:

$$\text{Apparent brightness of star A} \sim \frac{1^2}{1} = 1$$

$$\text{Apparent brightness of star B} \sim \frac{1^2}{2} = 0.25$$

Notice that a brightness of 1 is four times as great as a brightness of 0.25. From these values, we see that star A has a distance factor of 1, while Star B has a distance factor of 2. Star B is twice as far, which is 200 light-years.

FIGURE 28.2

Edwin Hubble (1889–1953), astronomer and cosmologist, is shown here in 1923 using the 100-inch telescope at the Mt. Wilson Observatory, where he worked for most of his life. In 1929, he announced Hubble's law, which states that the farther apart galaxies are, the faster they move.

How was Hubble able to calculate the velocity with which these distant galaxies were receding? Recall the Doppler effect, discussed in Chapter 10. Recall that sound waves stretch out when the sound source recedes and compress when the sound source approaches. Light waves do the same. As a galaxy recedes, the wavelength of light reaching us is stretched out. This longer wavelength means a lower frequency, which for visible light means a shift toward the red end of the color spectrum. Hubble studied the spectra of light coming from distant galaxies and measured the degree to which that light was shifted toward the red. The greater the *redshift*, the greater the velocity of the receding galaxy. Hubble's great accomplishment was collecting both the distance and redshift data for many galaxies and then correlating these data on a graph. His graph showed a clear relationship—the farther away the galaxy, the greater the redshift (Figure 28.3).

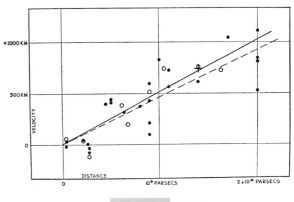

FIGURE 28.3

Hubble's original graph showing that more distant galaxies recede with greater velocities.

Galaxies recede from one another because the space between them is expanding. Astronomers thus follow an alternate explanation for why the spectra of galaxies show redshifts. As light waves travel through expanding space, the light waves themselves stretch out. This elongation of light waves due to the expansion of space is called the **cosmological redshift**. The farther away the galaxy,

Light from a distant galaxy is the light from glowing elements, which emit spectra of particular frequencies as discussed in Section 12.6. An examination of the full spectrum of a galaxy's light shows a pattern of peaks that are the sum of the spectra of all of the many glowing elements, primarily hydrogen and helium. If these peaks are shifted toward the red, then we know the galaxy must be receding away from us. By how much that frequency is shifted tells us how fast that galaxy is receding.

the longer its light has been traveling through expanding space, and hence the greater the cosmological redshift. The following Checkpoint illustrates this idea with an analogous rubber band.

CHECKPOINT

You draw three dots equally spaced along the length of a rubber band. Assume the distance between these dots is 5 mm. So the distance between the first and second dots is 5 mm, while the distance between the first and third dots is 10 mm. In 1 s you stretch the rubber band to 10 times its original length.
1. Show that the second dot recedes from the first dot with a speed of 45 mm/s. (Hint: speed = distance/time.)
2. Show that the third dot recedes from the first dot with a speed of 90 mm/s.
3. Why do farther galaxies appear to be receding from us at faster rates?

Were these your answers?
1. The original distance was 5 mm, which stretched to 50 mm. The change in distance, therefore, was 45 mm over a time of 1 s. So the speed equals 45 mm/s.
2. The difference between 100 mm and 10 mm is 90 mm over a time of 1 s, which equals 90 mm/s, or twice the speed of 45 mm/s.
3. For an expanding rubber band, we see that a dot twice as far away appears to travel twice as fast. The same goes with the universe: a galaxy twice as far away from us appears to be moving twice as fast. The reason for these receding velocities is the expansion of space itself.

Hubble suggested that a simple relationship exists between the distance of an object from Earth and its velocity of recession (because of the expansion of space). This simple relationship, which has been confirmed by numerous measurements over many decades, is known as **Hubble's law**:

$$\text{Hubble's law: } v = H \times d$$

where v is the velocity of a galaxy as deduced from its cosmological redshift, H is a constant known as Hubble's constant, and d is the distance of the galaxy from Earth. This law tells us, for example, that if one galaxy is twice as far away as another, the farther galaxy recedes twice as fast from us. Furthermore, if a galaxy were to move from where we are now to its present location with a velocity v, then the time of this trip would be the distance it travels divided by its velocity:

$$t = d/v$$

We use Hubble's law to substitute for v:

$$t = d/(H \times d)$$
$$= 1/H$$

When we enter the value of H into this equation, we have an estimate of the interval of time for the expansion. Said another way, we have determined the age of the universe. Plugging the currently accepted value of H into the equation indicates that the universe is nearly 14 billion years old. Wow!

Link To Section 10.10

Link To Section 1.10

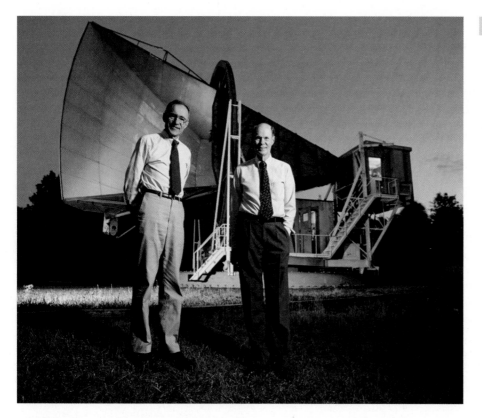

FIGURE 28.4

Arno Penzias and Robert
Wilson in front of the
microwave receiver they used
to detect the afterglow of the
Big Bang.

COSMIC BACKGROUND RADIATION

In addition to the expansion of the universe, a second line of evidence supporting the Big Bang theory is *cosmic background radiation*. In 1964, scientists Arno
Penzias and Robert W. Wilson, working at Bell Labs in New Jersey, used a
simple radio receiver to survey the heavens for radio signals (Figure 28.4). No
matter which way they directed their receiver, they detected microwaves with a
wavelength of 7.35 cm coming toward Earth. Penzias and Wilson were puzzled.
With no specific source of the radiation, where were the microwaves coming
from and why?

Remember that any object above absolute zero emits energy in the form of
electromagnetic radiation. The frequency of this radiation is proportional to the
absolute temperature of the emitter. Theorists at Princeton, working around the
same time as Penzias and Wilson, showed that if the universe began in a primordial explosion as described by the Big Bang, it would still be cooling off. Further, they showed that the temperature of the early universe would have cooled

An antenna-fed television (not
cable or satellite) tuned into a channel with no local station shows a
screen of static "snow." Interestingly, about 1% of this snow is due
to photons from the cosmic background radiation.

Link To Section 11.1

FIGURE 28.5

This all-sky map of the cosmic
background radiation taken by
the Wilkinson Microwave
Anisotropy Probe (WMAP)
satellite reveals an average temperature of about 2.73 K
everywhere. This is the cooled-
off remnants of the Big Bang.
The color shows minor temperature variation on the order
of ±0.0001 K.

Big Bang Helium

As the universe expands, it cools. The cosmic background microwave radiation tells us that it has cooled to an average of 2.73 K. Working backward from a present 2.73 K, scientists estimate a temperature in excess of 100 billion K within a few seconds after the Big Bang. At this extreme temperature, protons would be turning into neutrons and neutrons would be turning into protons. The rates of these transformations would be the about same, meaning the ratio of protons to neutrons in this super early universe would be about 1:1.

Over the next three minutes, the temperature dropped to less than 100 billion K, which favored the formation of protons.* Soon protons outnumbered neutrons. By the time the ratio of protons to neutrons was 7:1, the universe would have *cooled enough* to allow for nuclear fusion. We say "cooled enough" because the universe was still quite hot, but not as hot as before. At this point, the protons and neutrons would have begun fusing into deuterium nuclei (consisting of one proton and one neutron). Deuterium nuclei would have then fused to form helium. They would continue doing so until the deuterium

grew sparse. (Incidentally, this explains why deuterium is such a rare isotope today.) Continued cooling would prevent the further fusion of helium into heavier elements, such as carbon. By the time the three minutes of helium synthesis were over, the universe would be left with about 75% hydrogen and 25% helium, which is what we observe in the universe today. The Milky Way is actually about 28% helium. The 3% excess helium is likely from the fusion of hydrogen in the stars. No galaxy detected so far, however, has a helium abundance of less than 25%, just as predicted by the Big Bang.

* Neutrons are more massive than protons. The transformation of a proton into a neutron, therefore, requires the input of energy in accordance with $E = mc^2$. At temperatures below 100 billion K, there is insufficient energy to allow this transformation. The conversion of neutrons into protons, however, releases energy and so in accordance with the second law of thermodynamics is favored at lower temperatures.

by today to an average temperature of 2.73 K. A universe of this temperature would be expected to emit microwave radiation of just the frequency observed by Penzias and Wilson. Thus the influx of microwave radiation that initially puzzled Penzias and Wilson was found to be emitted by the cooling universe itself. This faint microwave radiation is now referred to as **cosmic background radiation** and is taken as strong evidence of the Big Bang (Figure 28.5).

THE ABUNDANCE OF HYDROGEN AND HELIUM

The Big Bang answers another cosmic mystery involving the element helium. Measurements show that matter in the universe is about 75% hydrogen and 25% helium. (Heavier elements such as those found on Earth are very minor contributors to the total amount of matter in the universe.) Hydrogen is the simplest of all elements, consisting of a single proton nucleus. It makes sense that hydrogen was the original element. Helium, however, is a more complex element containing a nucleus of two protons and two neutrons. We know helium is produced from the fusion of hydrogen in stars. But the number of stars is insufficient to account for all the observed helium—not more than 10% of the observed helium could have originated in stars. Most of the helium observed in the universe must have been created elsewhere. As described in the Big Bang Helium box, the Big Bang model predicts that the early universe would have been favorable to the formation of helium, but not the formation of other elements. A more detailed analysis shows that the amount of helium created just after the Big Bang should be about that which we observe in the universe today.

In summary, three major lines of evidence strongly support the Big Bang hypothesis. The first is the current expansion of space, which causes galaxies to recede from one another. The second is the discovery of cosmic background radiation, which is the Big Bang's afterglow. The third is the Big Bang's ability to explain the observed proportions of elements. Because of these and other similar lines of evidence, the Big Bang has come to be widely accepted by the scientific community as the most viable explanation for the beginning of our universe.

Where exactly did the Big Bang occur? Was it at some now far distant point from which we have long since traveled? The answer is an astounding NO! Rather, every point in the universe was present at the Big Bang. It's just that all of these points are now quite far away from one another. So if you want to point to the location of the Big Bang, just point your finger to the tip of your nose, or anywhere for that matter. You can't miss.

28.2 Cosmic Inflation

A *theory* is a comprehensive idea that can be used to explain a broad range of phenomena. As was discussed in the prologue of this textbook, theories are not carved in stone. Rather, they pass through stages of refinement. With each successive refinement, the theory grows stronger. In the early 1980s, the Big Bang theory gained such a refinement from the insights of physicists Alan Guth and Andrei Linde (Figure 28.6). The Big Bang theory had predicted that particles known as *magnetic monopoles* should be abundant in our current universe. Such monopoles, however, have never been detected despite countless attempts to find them. Guth realized that monopoles would become exceedingly rare if within an instant after the Big Bang, space expanded in a dramatic burst. (The monopoles would have been diluted beyond our ability to detect them.) This burst would have begun 10^{-38} seconds after the start of the Big Bang and ended a mere 10^{-36} seconds later. During this ultrabrief speck of time, the space within the universe inflated by a factor of 10^{30}. This moment of dramatic expansion is known as **cosmic inflation**. Soon after Guth first recognized the significance of cosmic inflation, Linde followed through with many important refinements on the theory. To the delight of Guth, Linde, and many others, the idea of cosmic inflation solved many other mysteries that had been plaguing the Big Bang theory.

One outstanding question was why the cosmic background radiation is so uniform in its temperature. Normally, two regions need to be in contact with each other in order to come to the same temperature. A cup of hot water and a cup of cold water, for example, won't blend to make a cup of warm water unless they are able to mix. In contrast, a constant expansion of space after the Big Bang with different regions so far apart would result in regions unable to mix their thermal energy with that of other regions, even after the lifetime of the universe. We would therefore expect some regions to remain very hot and others very cold. But studies of the cosmic background radiation clearly show a nearly uniform temperature. According to cosmic inflation, this uniform temperature was achieved in the moments before cosmic inflation. During these moments, all regions of the universe were still bound together. Being so close, they were able to intermingle and thus come to a homogenous temperature. So the Big Bang didn't explode right away. Instead, it held tight, expanding slowly for as long as it could before bursting like a bubble, which then calmed down almost immediately, but not before a dramatic inflation of space.

As indicated by the cosmic background radiation, the distribution of matter and energy within the expanded universe is remarkably uniform, but minor fluctuations still appear. According to cosmic inflation, the reason for these fluctuations can be traced to the realm of quantum mechanics, which tells us that physical attributes such as position and momentum become fuzzy on the scale of the very, very small—such as the size of subatomic particles. Indeed, the universe before inflation was very, very small—at one point even *smaller* than the size of subatomic particles. Assuming today's laws of quantum mechanics held true at the time of the Big Bang, then quantum fuzziness had to be a feature of the universe's birth (Figure 28.7). Upon inflation, ultrasmall quantum variations in position and momenta would have been magnified. The distribution of matter and energy in the post-inflation era would not have been perfectly uniform. The result would be a bit of clumpiness. Gravitational forces would then ensure

FIGURE 28.6

For their efforts in developing the theory of cosmic inflation, Alan Guth (left) and Andrei Linde (right) in 2004 together shared the prestigious Cosmology Prize of the Peter Gruber Foundation.

Link To Section 7.5

Link To Section 12.7

FIGURE 28.7

This image from NASA depicts the early history of the universe starting with the Big Bang. Quantum fluctuations were amplified by a sudden burst in size during a period known as inflation. Some 380,000 years later, particles settled such that the universe became transparent. It's from this moment that we see the cosmic background radiation afterglow. The first stars didn't form until about 400 million years later.

Link To Section 11.1

that these bits of clumpiness would be the seeds for further clumpiness, which would eventually give rise to such things as galaxies and superclusters of galaxies. Thus, when we look at the distribution of galaxies and superclusters of galaxies in the universe today, we are looking at the world of quantum mechanics magnified to a universal scale. So cosmic inflation explains not only the uniformity of matter and energy in the universe, but how minor fluctuations gave rise to the cosmic structure we see through our telescopes today. (See Figure 27.37 from the previous chapter.)

A third success of cosmic inflation has to do with the very shape of the universe. According to Einstein's *theory of general relativity*, which we discuss next, we live in three dimensions of space plus one dimension of time. Together these add up to a four-dimensional universe of **spacetime**. Mass has the effect of *curving* spacetime. With such curvature, parallel lines eventually meet or diverge. By analogy, consider the surface of planet Earth. Draw two parallel lines from the equator pointing exactly north–south. Because of the curvature of Earth, these parallel lines eventually merge at the poles. You'll see this happen with the longitudinal lines on any globe. Similarly, mass causes spacetime to curve in such a way that parallel lines eventually merge. Calculations of the total mass of the universe showed that the universe itself should be curved in very detectable ways. But this is not what astronomers observe. Instead, light travels in rather straight paths (unless the light passes through intense gravitational fields surrounding a star or black hole). We find that the universe is flat. Parallel lines remain parallel.

An explanation involves inflation. Consider an ant on a balloon. It may find that by walking straight it also walks in circles—it comes back to its starting point. Looking up, an unusually smart ant may see the curved horizon and conclude that its environment is highly curved. Take this balloon and inflate it to the size of the Sun. At this point the ant would walk on and on to new territories on what looked to the ant to be an infinite flat plane. Parallel lines would appear to remain parallel. Similarly, we humans on the surface of Earth drive our cars around as though the world is flat. But from afar, we know that Earth's surface is not flat—it is curved. Likewise for the universe: what we observe within our own corner of the universe appears to be flat, in a four-dimensional sort of way.

So is the universe as a whole curved? We do not know the answer to that question, but interestingly enough, most astrophysical data to date is consistent with a flat universe. This has major implications regarding the ultimate fate of the universe. But before delving into these implications, we need to explore the notion of spacetime, as well as the now firmly established idea that gravity is not a *force*, but a curvature of spacetime.

FIGURE 28.8

A two-dimensional analogy of four-dimensional warped spacetime. Spacetime near a star is curved in a way similar to the surface of a rubber sheet when a heavy ball rests on it.

28.3 General Relativity

In 1915, Einstein published his now famous **general theory of relativity**, which was a huge reworking of Newton's well-tested laws of gravity. According to Einstein, gravity is not a force exerted by one object on another through distance. Instead, gravity is the effect we witness when a large mass—such as a planet, star, or galaxy—causes a curvature in the shape of the spacetime in which it resides. The curvature of four-dimensional spacetime (three of distance and one of time) can be expressed mathematically, but is rather impossible for us to visualize. We can get a glimpse of this curvature by considering a simplified analogy in two dimensions: a heavy ball resting in the middle of a large rubber sheet. The more massive the ball, the more it warps the two-dimensional surface. A marble rolling past this warped surface would trace a bent path, as shown in Figure 28.8. If the marble were to roll closer to the ball, it might trace an elliptical curve around the ball. Rolled at the proper angle and speed (assuming no friction to slow it down), the marble could enter a perpetual orbit around the ball. In this analogy, we see here there is no "force" holding the marble to the ball. Rather, the marble is simply following the natural curvature of the rubber sheet.

In a similar manner, the Sun's mass curves the spacetime around it. Our planet is moving sideways along a path that follows this curvature. We are moving at just the proper speed so that the orbit is sustained. If we were to slow down, the steepness of the curvature would cause us to fall into the Sun. If we were to speed up sufficiently, we would escape this curvature and leave the Sun for good.

(a)

(b)

FIGURE 28.9

(a) Everything is weightless inside a nonaccelerating spaceship far away from gravitational influences. (b) When the spaceship accelerates, an occupant inside feels "gravity."

CHECK POINT

If the Sun suddenly became less massive, what would happen to our orbit?

Was this your answer?
With less mass, the Sun would have less of a warping effect on spacetime. If Earth maintained the same speed, it would be compelled to break away from the Sun. By Newton's laws, we would say that we escaped the Sun's now weaker gravitational pull. According to Einstein, however, we would say that spacetime became more flat, which allowed us to move along our merry way.

Einstein was led to this new theory of gravity by thinking about observers in accelerated motion. He imagined himself in a spaceship far away from gravitational influences (Figure 28.9). In such a spaceship at rest or in uniform motion relative to the distant stars, he and everything within the ship would float freely; there would be no "up" and no "down." But if rocket thrusters were activated to accelerate the ship, things would be different; phenomena similar to gravity would be observed. The wall adjacent to the rocket thrusters would push up against any occupants and become the floor, while the opposite wall would become the ceiling. Occupants in

FIGURE 28.10

To an observer inside the accelerating ship, a lead ball and a wood ball appear to fall together when released.

the ship would be able to stand on the floor and even jump up and down. If the acceleration of the spaceship were equal to g, the occupants could well be convinced the ship was not accelerating but was simply at rest on the surface of Earth.

Einstein concluded that gravity and motion through spacetime are related, a conclusion now called the **principle of equivalence**:

> **Local observations made in an accelerated frame of reference cannot be distinguished from observations made in a gravitational field.**

Link To Section 4.1

To examine this new "gravity" due to acceleration (as opposed to acceleration due to gravity, discussed in Chapter 4), Einstein considered the consequence of releasing two balls—say, one of wood and the other of lead. When released in a uniformly moving spaceship, the balls would continue to move with the spaceship at the moment of release. If the spaceship were moving at a *constant velocity* (zero acceleration), the balls would remain suspended in the same place, because both the spaceship and the balls would be moving by the same amount. But if the spaceship is accelerating, the floor moves upward faster than the balls, which are soon intercepted by the floor (Figure 28.10). Both balls, regardless of their masses, would meet the floor at the same time. Recall the legend of Galileo's demonstration at the Leaning Tower of Pisa. Occupants of the spaceship might be prone to attribute their observations to the force of gravity.

The two interpretations of the falling balls are equally valid, and Einstein incorporated this equivalence, or the impossibility of distinguishing between gravitation and acceleration, into the foundation of his general theory of relativity. The principle of equivalence states that observations made in an accelerated reference frame are indistinguishable from observations made in a Newtonian gravitational field. This equivalence would be interesting but not revolutionary if it applied only to mechanical phenomena, but Einstein went further and

FIGURE 28.11

(a) An outside observer sees a horizontally thrown ball travel in a straight line, and, because the spaceship is moving upward while the ball travels horizontally, the ball strikes the wall below a point opposite the window. (b) To an inside observer, the ball bends as if in a gravitational field.

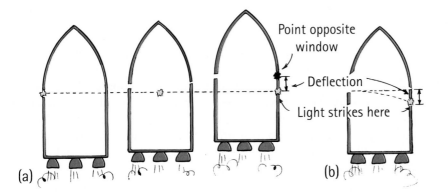

FIGURE 28.12

(a) An outside observer sees light travel horizontally in a straight line, but, like the ball in Figure 28.11, it strikes the wall slightly below a point opposite the window. (b) To an inside observer, the light also bends as if responding to a gravitational field.

stated that the principle holds for *all* natural phenomena; it holds for optical and all electromagnetic phenomena as well.

Consider a ball thrown sideways in a stationary spaceship in the absence of gravity. The ball follows a straight-line path relative to both an observer inside the spaceship and a stationary observer outside the spaceship. But if the spaceship is accelerating, the floor overtakes the ball and it hits the wall below a point opposite the window (Figure 28.11). An observer outside the spaceship still sees a straight-line path, but to an observer in the accelerating spaceship the path is curved; it is a parabola. The same result holds true for a beam of light (Figure 28.12). The only difference is the curvatures of both. If the ball were somehow thrown at the speed of light, both curvatures would be the same.

According to Newton, tossed balls curve because of a force of gravity. According to Einstein, both tossed balls and light curve when close to a planet or star not because of any force but because the spacetime in which they travel is curved (Figure 28.13).

FIGURE 28.13

The trajectory of a flashlight beam is identical to the trajectory of a baseball "thrown" at the speed of light. Both paths curve equally in a uniform gravitational field.

CHECK POINT

Whoa! We learned previously that the pull of gravity is an interaction between masses. And we learned that light has no mass. Now we say that light can be bent by gravity. Isn't this a contradiction?

Was this your answer?
There is no contradiction when the mass–energy equivalence (Chapter 13) is understood. It's true that light has no mass, but it is not "energyless." The fact that gravity pulls downward on light provides evidence that gravity pulls on the energy of light. Energy indeed is equivalent to mass!

Link To Section 13.6

General relativity, then, calls for a new geometry: a geometry not only of curved space but of curved time as well—a geometry of curved four-dimensional spacetime.* The mathematics of this geometry is too formidable to present here. The essence, however, is that the presence of mass produces the curvature or warping of spacetime. Instead of visualizing gravitational forces between masses, we abandon altogether the notion of force and instead think of masses responding in their motion to the curvature or warping of the spacetime they inhabit. It is the bumps, depressions, and warpings of geometrical spacetime that *are* the phenomena of gravity.

* Don't be discouraged if you cannot visualize four-dimensional spacetime, let alone the curving of spacetime. Einstein himself often told his friends, "Don't try. I can't do it either." Perhaps we are not too different from the great thinkers around Galileo who couldn't think of a moving Earth!

FIGURE 28.14

A precessing elliptical orbit when viewed from directly above is seen to advance forward in its orbit.

FIGURE 28.15

Starlight bends as it grazes the Sun. Point A shows the apparent position; point B shows the true position.

TESTS OF GENERAL RELATIVITY

Using equations for four-dimensional spacetime, Einstein recalculated the orbits of the planets about the Sun. Beyond the planets, space is almost flat, and objects travel along nearly straight-line paths. Near the Sun, planets and comets travel along curved paths because of the curvature of space. With only one minor exception, his theory gave almost exactly the same results as Newton's law of gravity. The exception was that Einstein's theory predicted that the elliptical orbits of the planets should slip forward with each revolution, a process known as *precession* (Figure 28.14). This precession would be very slight for distant planets and more pronounced for planets close to the Sun. Mercury is the only planet close enough to the Sun for the curvature of space to produce an effect on it that is not predicted by Newton's law.

Precession in the orbits of planets caused by the gravity of other planets was well known. Since the early 1800s, astronomers measured a precession of Mercury's orbit—about 574 seconds of arc per century. The gravity exerted by the other planets was found to account for the precession—except for 43 seconds of arc per century more than the calculated value. Even after all known corrections due to possible effects by other planets had been applied, the calculations of physicists and astronomers failed to account for the extra 43 seconds of arc. Either Venus was extra massive or some other, previously undiscovered planet was pulling on Mercury. And then came the explanation of Einstein, whose general-relativity equations, when applied to Mercury's orbit, predict the extra 43 seconds of arc per century!

As a second test of his theory, Einstein predicted that measurements of starlight passing close to the Sun would be deflected by an angle of 1.75 seconds of arc—large enough to be measured. This deflection of starlight can be observed during an eclipse of the Sun. (Measuring this deflection has become a standard practice at every total eclipse since the first measurements were made during the total eclipse of 1919.) A photograph taken of the darkened sky around the eclipsed Sun reveals the presence of the nearby bright stars. The positions of the stars then are compared with those in other photographs of the same area taken at other times during the night with the same telescope. In every instance, the deflection of starlight has supported Einstein's prediction (Figure 28.15).

CHECK POINT

Why do we not notice the bending of light by gravity in our everyday environment?

Was this your answer?
Only because light travels so fast; just as over a short distance, we do not notice the curved path of a high-speed bullet, we do not notice the curving of a light beam.

Einstein made a third prediction—that gravity causes a slowing down of time. He predicted, for example, that a minute of time on the surface of a massive planet should last longer compared to a minute of time on the surface of a less massive planet. Thus, a clock sitting on the more massive planet would be seen to be running slower than an identical clock sitting on the less massive planet. Gravity alters time! The stronger the gravity, the greater the effect.

Einstein suggested a way to measure this effect when he formulated the principle of equivalence. He knew that all atoms emit light at specific frequencies characteristic of the vibrational rate of electrons within the atom. Every atom is therefore a "clock," and a slowing down of atomic vibrations indicates the slowing down of such atomic clocks. An atom on the Sun, therefore, should emit

fyi

■ Technically, it's not mass that causes the curvature of spacetime, but the mass density. Earth's mass density is too small to provide noticeable effects on time. For a black hole, however, it's a different story. Recall from Chapter 27 that a clock falling toward a black hole's event horizon will appear to slow down as it gets closer and closer to the event horizon, which you would never see it reach.

light of a lower frequency (slower vibration) than light emitted by the same atom on Earth. Astronomers have found that the slowing down of time for atoms on our Sun is obscured by their thermal motion. The time-slowing effect of gravity, however, has since been observed and accurately measured in white dwarf stars, which have gravitational fields much stronger than that of our Sun. Furthermore, ultraprecise experiments here on Earth have shown that time runs slower at the bottom of a tall tower, which is closer to Earth's gravity, than at the top of the tower. The difference is exceedingly small, but the results of the experiment are reproducible and in strict accordance with Einstein's prediction (Figure 28.16).

Einstein's general theory of relativity was an elaboration of his **special theory of relativity**, which he had proposed ten years earlier in 1905. Through special relativity, Einstein showed how matter and energy are really two forms of the same thing as related by his now famous equation $E = mc^2$, introduced in Chapter 13. In Appendix E, as well as in the *Conceptual Physical Science Practice Book*, we present a fascinating aspect of special relativity that tells us how time changes with motion. For example, an astronaut taking a two-year round trip into space at high speeds could come back to find that Earth has aged 2000 years! For now, however, we turn to something even more mysterious, which is the potential presence of a form of matter completely invisible to our sight and sense of touch.

■ 28.4 Dark Matter

Matter can be defined as "that which occupies space." More specifically, we can say "that which occupies spacetime." Matter, however, more than simply occupies spacetime; its presence has the effect of curving spacetime. If this is our definition of matter, then evidence now suggests that the Big Bang generated matter in at least two different forms—one we can see and another that we can't.

The visible form of matter is the "ordinary matter" made of subatomic particles such as protons, neutrons, and electrons. As you learned in earlier chapters, these particles combine to make the atoms of the periodic table. Atoms then combine to make molecules, such as those that make our bodies. This matter we can touch. We interact with it directly. You, the planets, and the stars are made of this form of matter, which we will from here on refer to as **ordinary matter**.

The second form of matter generated from the Big Bang is quite unlike ordinary matter. This form of matter does not recognize the strong nuclear force, which means it cannot clump to form atomic nuclei. Neither does this second form of matter recognize the electromagnetic force, which makes it invisible to light as well as our sense of touch. As was explained in Chapter 12, the electromagnetic force is responsible for the repulsion between electrons. The reason you can't walk through a wall is because of the repulsions between the electrons in your body and the electrons in the wall. If the wall were made of this invisible matter you would be able to walk right through it. Of course, you wouldn't be able to see the wall either. This invisible form of matter that we cannot see or touch is known as **dark matter**.

So if dark matter is invisible to us, how do we know it's there? The answer is that this ghostly form of matter gives itself away by its gravitational effects. One of the first clues came to us as we were mapping the speeds at which stars orbit our galactic center. According to the laws of gravity, orbital speed is a function of the force of gravity between the orbiting object and the object being orbited—the greater the force of gravity, the greater the orbital speed. The inner planets of our solar system,

FIGURE 28.16

If you move from a distant point down to the surface of Earth, you move in the direction in which the gravitational force acts—toward a location where clocks run more slowly. A clock at Earth's surface runs slower than a clock located farther away.

Link To Section 12.1

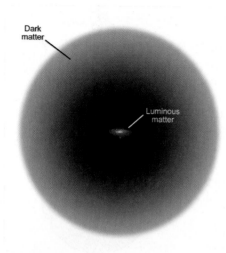

FIGURE 28.17

As large as a galaxy is, its diffuse halo of dark matter is much larger. This halo may measure up to 10 times the diameter of the luminous galaxy and be about six times as massive.

fyi

- An alternative theory to dark matter is Modified Newtonian Dynamics (MOND), proposed by the physicist Mordehai Milgrom in the early 1980s. According to MOND, Newton's equation $a = F/m$ fails when the force is exceedingly weak, perhaps because of quantum effects. A modified version of this equation was thus created to account for the observed orbital velocities of stars within galaxies. The theory is controversial but has yet to be fully refuted.

FIGURE 28.18

Images of dark matter, shown in blue, were created through the Hubble Space Telescope's Cosmic Evolution Survey. Of course, the dark matter is not blue. This graphic, however, shows dark matter's distribution over a narrow region of the sky back to about 6.5 billion years ago.

for example, orbit the Sun much faster than the outer planets because they are closer to the Sun and experience greater gravitational forces. Relative to our galaxy, we might expect the same trend—stars closest to the galactic center should have faster orbital speeds than stars farther out. Interestingly, that's not what we observe! Instead, stars closer to the galactic center and those farther out orbit with about the same speed. How can this be?

For a solar system, planets orbit as they do because most of the solar system's mass is concentrated within the central sun. For a galaxy such as the Milky Way or the Andromeda, it sure looks as though most of the mass is concentrated within the central bulge. The measured orbital speeds of stars, however, tell us that the bulk of the galaxy's mass lies outside the galaxy itself within a diffuse yet massive invisible halo many times the diameter of the visible galaxy, as shown in Figure 28.17. We know it's invisible because all of our telescopes see right through it! But something must be there affecting stellar orbital speeds.

Another bit of evidence for dark matter comes from measuring the speeds of galaxies as they orbit one another within clusters. The measured speeds tell us that the masses of these galaxies is many times greater than the total mass of all their stars.

Lastly, we know that the path of light is bent by gravity much as it is bent by an optical lens. A cluster of galaxies, therefore, can bend the light from an even farther cluster lying directly behind it—we say the foreground cluster behaves as a *gravitational lens*. Such a gravitational lensing effect was shown in Figure 27.33. The degree to which the light from the distant cluster bends is a function of the mass of the foreground cluster. Once again, the degree of light bending tells us that the mass of the closer cluster far exceeds that which we would expect based solely on the cluster's luminosity. So by carefully studying the bending of light from distant galaxies, we can build a map of the dark matter's distribution. Such a study using the Hubble Space Telescope is illustrated in Figure 28.18.

So the evidence for dark matter is strong. Our current problem, however, is trying to figure out exactly what dark matter is made of. It's clear that dark matter is not simply ordinary matter, such as expired stars, that have gotten so cold that they emit no light. Although numerous theories abound about the fundamental nature of dark matter, no dark matter particles have been detected. Until that happens, we are left with yet another fascinating mystery of our universe.

GALAXY FORMATION

We can speculate that when the universe formed, ordinary matter plus an even greater amount of dark matter was produced. Held together by gravity, the ordinary matter and dark matter would have been strewn outward in a clumpy fashion. Within a clump, ordinary and dark matter may initially have been uniformly mixed together. These two forms of matter differ significantly in that when ordinary matter collides with ordinary matter, energy is released as heat. With this loss of energy, the ordinary matter loses orbital speed and thus falls closer to the center of the clump. Over time, while all dark matter stayed distributed throughout the clump, the ordinary matter became concentrated at the center (Figure 28.19). This concentration of ordinary matter at the center of the clump allowed for the formation of stars. Also, as ordinary matter congregated toward the center, the rate of rotation would increase—angular momentum would have been conserved. If the original clump of ordinary and dark matter was just barely spinning, then the stars forming at the center would take on the form of an elliptical galaxy. If the original clump was spinning a bit faster, then the new stars would be spinning fast enough to flatten the galaxy, much like a rapidly spinning ball of pizza dough. The resulting disk would take on the form of a spiral galaxy. So from a clump of ordinary and dark matter, ordinary matter condensed to form a central galaxy. The dark matter remained diffuse, forming an invisible halo surrounding the newly formed galaxy.

FIGURE 28.19

Ordinary matter condensed out of a mixture of dark and ordinary matter.

28.5 Dark Energy

In the years just before Hubble's discovery of the expansion of the universe, Einstein was struggling to understand why gravity wasn't causing the universe to collapse in a Big Crunch. He was thinking of the universe as static, neither collapsing nor expanding. But in order for the universe to remain static against the inward curvature of gravity, there would need to be another fundamental outward curvature counteracting gravity. In other words, if gravity is the "pull inward," there should be a phenomenon that creates a "push outward." To allow for such a balance, he introduced into his equations the idea of what he called the *cosmological constant*. He had no proof for the existence of such a phenomenon. Rather, he just made it up to make his equations appear more natural.

About 10 years later, Hubble announced that the universe was *not* stable, but very dynamic and presently in a mode of expansion. Einstein later remarked that his own failure to predict a dynamic universe was the "greatest blunder of his career." Subsequent workings of Einstein's equations showed that a static universe would not be stable. Just the slightest push this way or that would cause it to either collapse or expand. Einstein abandoned his notion of a cosmological constant, which for the next 75 years remained a historical curiosity.

Then in the 1990s, some 40 years after Einstein's death, two teams of astronomers made a startling discovery. High-resolution data from very distant galaxies showed that space, beginning about 7.5 billion years ago, started to accelerate in its expansion. Galaxies are not simply coasting away from each other. Rather, some unknown form of energy is causing an increase in the rate at

Link To Section 4.1

The expansion of space started to accelerate about 7.5 billion years ago, which is shown on this diagram as a gradual widening just after the development of galaxies. The cause of this accelerated expansion has been given the name *dark energy*. But just because we can name something doesn't mean we understand it. Such is the case with dark energy.

fyi

- The space through which our planets orbit contains about 100 hydrogen atoms per liter. The space between the stars of our galaxy contains about 2 hydrogen atoms per liter. If you want really empty space, you'll need to travel to the vast voids that separate the superclusters as was shown in Figure 27.36. It's only from these regions that dark energy appears to be taking hold. The space within and between our local galaxies is too dense!

The Physical Science Place
Fate of the Universe

which galaxies are receding. Here was a phenomenon that acted as the opposite of gravity, causing an outward curvature in spacetime. Here was evidence of Einstein's once proposed cosmological constant!

At this writing, less than 10 years since the discovery of this phenomenon, there is much speculation about how these findings affect the fate of the universe. This form of unknown energy is generally described as **dark energy** (Figure 28.20). Current astronomical data suggest that dark energy, is in fact, Einstein's famed cosmological constant. Other possibilities, however, cannot yet be ruled out.

Some current models suggest that dark energy finds its source within the emptiness of spacetime. Matter has the effect of pulling spacetime together so that it contracts. The classic example is what happens upon the formation of a black hole—spacetime contracts to a point of zero volume (and infinite density). In the absence of matter—within a perfect vacuum—empty spacetime seethes with an energy that creates an opposite curvature, which allows spacetime to expand. As more empty spacetime is created, the *vacuum energy* becomes more predominant, which accelerates the formation of even more empty spacetime. Distant galaxies are thus seen to be accelerating from each other. It is as though gravity and dark energy are diametrically opposed. When gravity gains full rein, the result is infinite density—a black hole. When dark energy gains full rein, the result may be an infinite vacuum—an empty hole.

Many other intriguing models attempt to explain the nature of dark energy. Which model is best can be determined only after the collection of more evidence. Stay tuned for science news reports. In particular, the European Space Agency's *Planck Surveyor* may answer many of our current questions. However, this powerful space telescope and its many successors will, no doubt, also raise more questions than they answer. One thing is for sure: the universe holds no shortage of mysteries.

28.6 The Fate of the Universe

The universe is expanding. Matter, however, has the effect of reversing this expansion. Might enough matter exist within the universe to halt or even reverse this expansion? Before the discovery of dark energy and dark matter, astronomers had calculated that the ordinary matter in the universe was only 4%

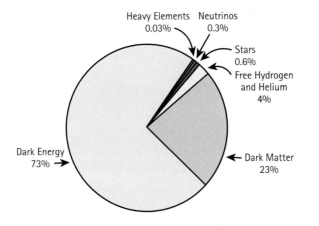

FIGURE 28.21

Ordinary matter, the stuff from which we and the galaxies we live in are made, makes up not more than 4% of the composition of the universe. The remainder is primarily dark matter (23%) and dark energy (73%), both of which we know very little about.

of the mass needed to halt the expansion. Dark matter was then discovered to be about six times as abundant as ordinary matter, making up about 23% of the mass needed to halt the expansion. If the remaining 100% − (4% + 23%) = 73% of matter could be accounted for, then the mass of the universe would be sufficient to one day halt the expansion. The geometric shape of such a universe would be flat. In other words, parallel lines would never touch. Because we observe the universe to be rather flat, astronomers puzzled over where this remaining 73% of matter might be.

Then came the discovery of dark energy. Recall that matter and energy are related by Einstein's equation $E = mc^2$. Both dark matter and dark energy, therefore, need to be included in tabulations of the total composition of the universe. Both would have an effect on the curvature of the observable universe. The abundance of dark energy then fits the bill for making up the remaining 73% of the composition of the observable universe, as shown in Figure 28.21. So it makes sense that our observable universe is flat, or quite close to it. There's a twist, however, in that dark energy causes the *expansion* of space, not contraction. The current thinking of most cosmologists, therefore, is that our flat universe is destined for an eternal expansion. This gives rise to a select number of possible scenarios for the fate of our universe.

In one scenario, called **heat death**, the universe will continue to expand, approaching absolute zero and a state of maximum entropy. After 10^{14} years, all stars will have exhausted all possible fuel. The universe will be fully dark. After 10^{16} years, planets and stars will be flung from their orbits because of random collisions, most then falling into supermassive black holes and the rest forming scattered stellar debris. After 10^{40} years, all protons and neutrons will have decayed, leaving behind gamma radiation and *leptons*, of which the electron is an example. From this time to about 10^{100} years in the future, supermassive black holes will be the dominant form of mass in the universe. But by the passing of 10^{150} years, they too will depart as they evaporate into photons and leptons. From then to perhaps 10^{1000} years in the future, the wavelengths of photons as well as all other remaining particles will be stretched to the lowest energy states possible. Entropy will have won supreme victory.

A second scenario, known as the **Big Rip**, recognizes that the influence of dark energy may grow stronger over time (Figure 28.22). As the universe expands exponentially, clusters of galaxies will be pulled farther apart, past the point of being visible to each other. Neighboring galaxies will then be pulled out of each other's sight. About 60 million years before the end, stars within galaxies will fly off in every direction. Some three months before the end, solar systems will disperse. In the last few minutes, stars and planets will become unbound. Then in the last instant, all atoms will be ripped apart, followed by their subatomic particles. The time frame for the Big Rip is estimated to be about 35 billion years after

Although *dark energy* and *dark matter* both begin with the word *dark*, they are uniquely different. One is a form of matter; the other is a form of energy. Both are still mysterious and may yet be disproven by new evidence or alternate explanations. By the next edition of this textbook, the picture may look quite different!

Link To Section 13.6

FIGURE 28.22

FIGURE 28.22

If the strength of dark energy remains constant, we can expect our universe to suffer heat death. If dark energy gains strength, then the ultimate demise of our universe may be the Big Rip.

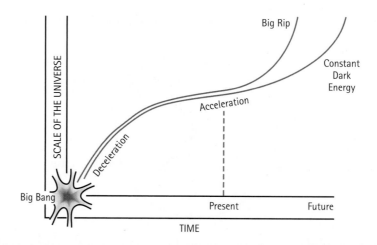

the Big Bang, which would about 21 billion years from now. And where will all the matter fly off to? Good question. Perhaps to subsequent Big Bangs?

Alan Guth, one of the early developers of cosmic inflation, supports the idea that the universe will not end everywhere at once. Some 14 billion years ago, a tiny patch of primordial material inflated to form our own observable universe. But inflation will keep on going for other patches of this primordial material and will continue to do so eternally. So while our region of the universe expands to infinity, other regions are just being born. This scenario, in which observable universes are spawned on a perpetual basis, is known as **eternal inflation**. With eternal inflation, the ultimate fate of our own observable universe is not the same as that of the universe as a whole.

It is interesting how the focus of our speculations has been narrowing down over our history. In the beginning, anything seemed possible. But as we opened our eyes and minds to the natural universe, we learned that some speculations were more worthy than others. We once viewed the constellations as heavenly gods. Of course, Earth was the center of the universe. Then we realized that Earth was just a planet orbiting the Sun, which itself was a medium-sized star among many. Our universe was the Milky Way Galaxy until Hubble pointed out otherwise. When galaxies were found to be receding, many hypothesized that gravity might be strong enough to pull them back together in a Big Crunch. This possibility has since been eliminated. Now we are at the point of wondering what might happen after eternal expansion. Our current speculations are just that—speculations. But they are highly refined speculations based on centuries of collected evidence. As we continue to look at the natural universe with ever more powerful telescopes, we can expect that our speculations will become ever more refined. This is the mind-opening art of science, which seeks to learn the nature of the universe for what it is—not for what we might wish it to be. Stay tuned.

SUMMARY OF TERMS

Cosmology The study of the overall structure and evolution of the universe.

Relativity The study of the relationships among space, energy, mass, and time.

Big Bang The primordial creation and expansion of space at the beginning of time.

Cosmological redshift The elongation of light waves due to the expansion of space.

Hubble's law The farther away a galaxy is from Earth, the more rapidly it is moving away from us: $v = H \times d$.

Cosmic background radiation The faint microwave radiation emanating from all directions that is the remnant heat of the Big Bang.

Cosmic inflation The theory that the universe experienced a sudden and brief burst in size immediately after the Big Bang.

Spacetime The continuum in which we live, consisting of three dimensions of space plus the fourth dimension of time.

General theory of relativity The theory first proposed by Einstein discussing the effects of gravity on spacetime.

Principle of equivalence Local observations made in an accelerated frame of reference cannot be distinguished from observations made in a Newtonian gravitational field.

Special theory of relativity The theory first proposed by Einstein discussing the effects of uniform motion on space, time, energy, and mass.

Ordinary matter Matter that responds to the strong nuclear, weak nuclear, electromagnetic, and gravitational forces. This is matter made of protons, neutrons, and electrons, which includes the atoms and molecules that make us and our immediate environment.

Dark matter Matter that responds only to weak nuclear and gravitational forces. This form of matter is invisible but reveals itself to us by its gravitational effects.

Dark energy An unknown form of energy that appears to be causing an acceleration of the expansion of space. It is thought to be associated with the energy exuded by a perfect vacuum.

Heat death A model for the end of the universe in which all matter and energy disperses to the point of maximum entropy.

Big Rip A model for the end of the universe in which dark energy grows stronger over time and causes all matter to rip apart.

Eternal inflation A model of the universe in which cosmic inflation is not a onetime event; rather, it progresses continuously, spawning an infinite number of observable universes in its wake.

REVIEW QUESTIONS

28.1 The Big Bang

1. Is the universe in space, or is space in the universe?
2. What is a Cepheid?
3. Which depends on distance, a star's brightness or its luminosity?
4. According to Hubble's law, what is the approximate age of the universe?
5. According to the cosmic background radiation, what is the average temperature of the universe today?

28.2 Cosmic Inflation

6. According to cosmic inflation theory, how long did it take for the universe to increase its size by a factor of 10^{30}?
7. At what point did the universe's temperature even out?
8. What did inflation do to the quantum fluctuations found within the early universe?
9. What would happen to the apparent curvature of a balloon if it could be inflated to the size of the Sun?
10. How many dimensions are there in spacetime?

28.3 General Relativity

11. In what year did Einstein publish his general theory of relativity?
12. Can an accelerated frame of reference be distinguished from a gravitational field?
13. You release a ball while standing against the floor of an accelerating spaceship. What happens to the ball?
14. What did general relativity predict about the orbit of Mercury?
15. What happens to starlight as it passes close to the Sun?

28.4 Dark Matter

16. What type of matter is visible?
17. If we can't see dark matter, how do we know it is there?
18. Is dark matter found mostly within or just outside a galaxy?
19. The closer a planet is to the Sun, the faster it orbits. Is it also true that the closer a star is to the center of the galaxy, the faster it orbits?
20. A huge cloud of ordinary and dark matter are uniformly mixed together. Over time, the ordinary matter becomes concentrated toward the center of this cloud. Why?

28.5 Dark Energy

21. Did Einstein first believe that the universe was static or dynamic?
22. What was Einstein's cosmological constant?
23. What did Einstein call "the greatest blunder of my career"?
24. According to recent evidence, how long ago did the expansion of the universe start accelerating?
25. What does WMAP stand for?

28.6 The Fate of the Universe

26. What is likely the major constituent of our universe?
27. Which is more abundant, dark or ordinary matter?
28. According to the heat death scenario, about how long will it take for the black holes of the universe to evaporate?
29. What does the Big Rip scenario assume about dark energy?
30. What scenario for the fate of the universe proposes that cosmic inflation is not a onetime event?

EXERCISES

1. ● What is the difference between cosmology and cosmetology?

2. ● When was most of the helium in the universe created?

3. ● What does the expansion of space do to light passing through it?

4. ● What is the relationship between dark energy and Einstein's cosmological constant?

5. ■ A police officer pulls you over for speeding. He tells you that his radar tracked you moving at a rate of 45 mi/h away from his parked police car. Were you really speeding away from him, or was the space between the two of you simply expanding?

6. ◆ Is space just the absence of matter?

7. ● The average temperature of the universe right now is about 2.73 K. Will this temperature likely go up or down over the next billion years?

8. ■ Mass can transform to energy and energy can transform into mass in accordance to Einstein's equation $E = mc^2$. So if the amount of mass–energy in the universe remains constant, what happens to the amount of space within the universe? What then is happening to the density of the universe?

9. ● If the initial universe had remained hotter for a longer period of time, would there likely be more or less helium?

10. ● If not from the Big Bang, then where did elements heavier than helium come from?

11. ● No galaxy found so far is made of less than 25% helium. If not from the stars, where did this helium come from?

12. ■ Can astronomers point their telescopes in the direction of where the Big Bang occurred?

13. ◆ A helium balloon here on Earth pops, releasing direct remnants of the Big Bang. True or false? Explain.

14. ● If we are made of stardust, what are stars made of?

15. ◆ Astronomers tell us that the average temperature of the universe is a rather homogenous 2.73 K(+/−0.0001K). But how can this be when we know that the temperatures of stars are ultrahot?

16. ● What are three lines of evidence supporting cosmic inflation?

17. ■ What if there was a symmetry to cosmic background radiation such that the pattern of temperature fluctuations in one direction was the same pattern seen in the exact opposite direction? What might you conclude about the curvature of the universe?

18. ● Which causes weight: gravity or acceleration?

19. ● If gravity is not a force, then what is it?

20. ● You toss a tennis ball up and down in front of you as you sit in a jet airplane accelerating down the runway for takeoff. Why is the tennis ball difficult to catch?

21. ● You toss a tennis ball up and down in front of you as you sit in a jet airplane cruising at a constant speed of 500 mi/h. Why is the tennis ball easy to catch despite the fact that the jet is moving so fast?

22. ■ You are free-floating in a spaceship in uniform motion deep in outer space. A ball is hovering in front of you. Suddenly the ball starts moving to the floor. What is happening to the spaceship? What happens to you?

23. ◆ Where does a clock run slower: at the front end or back end of an accelerating spaceship?

24. ■ Several billion years in the future our Sun will grow in size to become a red giant whose surface will extend to the present orbit of Earth. After this happens, will the slowing of time on the Sun's surface be more or less pronounced than it is today?

25. ◆ An astronaut is provided with "gravity" when the ship's engines are activated to accelerate the ship. This requires the use of fuel. Is there a way to accelerate and provide "gravity" without the sustained use of fuel? Explain. (Hint: Consider Appendix B.)

26. ● Under ultrasensitive conditions, should a person who worries about looking old live at the top or at the bottom of a tall apartment building?

27. ■ Under ultrasensitive conditions, should a person who wants to live a long life live at the top or at the bottom of a tall apartment building?

28. ■ Under ultrasensitive conditions, if you shine a beam of colored light to a friend above in a high tower, is the color of light your friend receives the same color you send? Explain.

29. ◆ Why does the gravitational attraction between the Sun and Mercury vary?

30. ◆ When do clocks move slowest on Mercury?

31. ● What might we assume about the distribution of dark matter if the planets in our solar system all orbited the Sun at about the same speed?

32. ● Early astronomers such as Kepler and Newton developed the laws of gravity based on the motion of the planets around the Sun. How might these laws have been different if our solar system were surrounded by a thick halo of dark matter?

33. ● What force allows dark matter to clump?

34. ■ If dark matter is affected by gravity, might there be lots of it surrounding us here on Earth's surface?

35. ● What is the relationship between dark matter and dark energy?

36. ◆ Why is dark energy not called the dark force?

37. ■ Is it dark matter or dark energy that keeps a galaxy from collapsing in on itself?

38. ■ If the universe were unchanging and there were an infinite number of stars, what effect might this have on the darkness of a clear night sky?

39. ■ If there are so many stars and galaxies, why do we see so much darkness in the clear night sky?

40. ◆ If we can't even predict the weather, how can we ever expect to predict the fate of the universe?

ACTIVE EXPLORATIONS

Your Inner Universe

Find a quiet and comfortable place to sit where you won't be disturbed. Place your hands on your legs and straighten your back. Imagine a string attached to the top of your head, lifting you upward. Gaze to the floor a couple of feet ahead of you. Relax your jaw so that your lips remain slightly open. Once you are comfortable, begin focusing on your exhaled breath. Simply notice your exhaled breath as it leaves your body and disperses into your surroundings. When you find yourself carried away by a thought, label the thought as "thinking" and let the thought drift away. New thoughts will keep popping in your mind, but return your attention to your exhaled breath. Try doing this for at least 10 minutes. Keep your experience simple and ordinary—don't get "trippy". Using this technique, you may come to the point of being able to distinguish your bare presence from your discursive thoughts.

Basic Goodness

If you tried the preceding activity, you will have found that it is not easy finding yourself amid your many thoughts. Nonetheless, consider it a worthy endeavor and healthy habit to get to know yourself in such a fashion. Once you feel comfortable with staying with your breath, you can practice welcoming in a particular thought, which becomes your focus while your breathing recedes to the background. A good thought to start with is the idea that your bare presence and the bare presence of every sentient being shines with a wisdom we might call "basic goodness." Underneath all the veneer, no matter what the flavor, there is basic goodness in everyone. Try doing this for at least 10 minutes.

Your Outer Universe

After you have given the contemplation of basic goodness a fair shake, a zillion other thoughts are also worthy of your contemplation. Within the context of this chapter, try contemplating the sheer size of the universe. Envision yourself in your chair, then mentally zoom out to see yourself in the room, then in the building, then on the planet. Zoom out stepwise so that you can experience each stage. Zoom out to the solar system circling within a galaxy that orbits neighboring galaxies within a supercluster that is one supercluster of billions of superclusters. Once you have zoomed fully out, focus on that sensation of bigness. When you lose the sensation, start over, moving as slowly or as fast as is comfortable.

READINESS ASSURANCE TEST (RAT)

If you have a good handle on this chapter, if you really do, then you should be able to score at least 7 out of 10 on this RAT. If you score less than 7, you need to study further before moving on.

Choose the BEST answer to each of the following.

1. Which of the following is not accepted evidence for the Big Bang
   (a) cosmic background radiation
   (b) the homogeneity of the temperature of the universe
   (c) the abundance of helium
   (d) dark energy

2. If the universe stopped expanding at this very moment, the cosmological redshift of distant galaxies would
   (a) promptly disappear.
   (b) gradually turn into a cosmological blueshift.
   (c) stop getting more redshifted than it already is.
   (d) not matter, because such a sudden halt to expansion would knock all galaxies and planets off their orbits and we would no longer be here to talk about it.

3. What percentage of galaxies were created during cosmic inflation?
   (a) 100%
   (b) about 70%
   (c) about 24%
   (d) 0%

4. What do cosmic inflation and dark energy have in common?
   (a) Both are responsible for the expansion of the universe.
   (b) Both are still occurring.
   (c) Both arise from the vast voids of space between superclusters.
   (d) Nothing.

5. Light bends in a gravitational field. Why is this bending not taken into consideration by surveyors who use laser beams as straight lines?
   (a) The bending of light occurs only in outer space.
   (b) Red light doesn't bend as much, which is why they use red lasers.
   (c) Laser light is an exception to this rule.
   (d) Light travels so fast that the curvature is not noticeable.

6. Time slows in a gravitational field. Would time slow in the artificial gravity produced in a rotating space habitat?
   (a) No, because there is no acceleration.
   (b) Yes, because of the principle of equivalence.
   (c) No, because the space station is spinning at a constant rate.
   (d) Yes, even though the gravity is artificial.

7. Should it be possible in principle for a photon to orbit a star?
   (a) Yes, if the photon is moving at the speed of light.
   (b) Yes, if the star's mass density is huge enough to make it a black hole.
   (c) No, because photons, by definition, always move in straight paths.
   (d) No, because stars already produce countless photons of their own.

8. Dark matter is
   (a) ordinary matter that is no longer emitting light.
   (b) altering the orbits of our planets.
   (c) attracted to ordinary matter.
   (d) repelled by ordinary matter.
9. Space in our local universe is
   (a) not empty.
   (b) seething with energy.
   (c) not well understood.
   (d) all of these

10. Which theory of the fate of the universe assumes that dark energy will grow stronger?
    (a) heat death
    (b) Big Rip
    (c) eternal inflation
    (d) all of these

Answers to RAT

1d, 2c, 3d, 4a, 5d, 6b, 7b, 8c, 9d, 10b

EXPLORING FURTHER

http://map.gsfc.nasa.gov
   NASA's home page for the Wilkinson Microwave Anisotropy Probe (WMAP), which has measured the cosmic background radiation in great detail (to be surpassed by the Planck Surveyor after 2008), includes images and discussions of the basic cosmology. Be sure to click on Universe, which will lead you to Cosmology 101—a launching pad to discussions of many of the topics discussed in this chapter and more.

http://en.wikipedia.org/wiki/Cosmic_inflation
   This Wikipedia page provides a broad overview of cosmic inflation, with numerous hyperlinks.

http://www.pbs.org/wgbh/nova/einstein/relativity/
   This description of Einstein and his theory of relativity was written by noted physicist Alan Lightman.

http://en.wikipedia.org/wiki/Modified_Newtonian_dynamics
http://www.astro.umd.edu/~ssm/mond/
   Visit these Web sites for more information regarding the Modified Newtonian Dynamics (MOND) model presented within this chapter as an FYI adjacent to the dark matter discussion, Section 28.4.

http://pauli.physics.lsa.umich.edu/w/arch/som/sto2001/Guth/real/n001.htm
   Watch Alan Guth present a seminar on the basic ideas of cosmic inflation and eternal inflation. His presentation gets rather technical, especially during the second half of this talk. After studying this chapter, however, you should be able to recognize many of the ideas.

http://space.mit.edu/home/tegmark/multiverse.html
   This is the home page of Max Tegmark, an MIT professor who specializes in the idea of the existence of parallel universes, which he classifies on four different levels.

CHAPTER 28 ONLINE RESOURCES

The Physical Science Place

Interactive Figure

- 28.1

Tutorials

- Hubble's Law
- Fate of the Universe

Video

- From the Big Bang to Galaxies

Two major systems of measurement prevail in the world today: the *United States Customary System* (USCS, formerly called the British system of units), used in the United States of America and in Burma, and the *Système International* (SI) (known also as the international system and as the metric system), used everywhere else. Each system has its own standards of length, mass, and time. The units of length, mass, and time are sometimes called the *fundamental units* because, once they are selected, other quantities can be measured in terms of them.

United States Customary System

Based on the British Imperial System, the USCS is familiar to everyone in the United States. It uses the foot as the unit of length, the pound as the unit of weight or force, and the second as the unit of time. The USCS is presently being replaced by the international system—rapidly in science and technology (all 1988 Department of Defense contracts) and some sports (track and swimming), but so slowly in other areas and in some specialties it seems the change may never come. For example, we will continue to buy seats on the 50-yard line. Camera film is in millimeters but computer disks are in inches.

For measuring time, there is no difference between the two systems except that in pure SI the only unit is the second (s, not sec) with prefixes; but in general, minute, hour, day, year, and so on, with two or more lettered abbreviations (h, not hr), are accepted in the USCS.

Systéme International

During the 1960 International Conference on Weights and Measures held in Paris, the SI units were defined and given status. Table A.1 shows SI units and their symbols. SI is based on the *metric system,* originated by French scientists after the French Revolution in 1791. The orderliness of this system makes it useful for scientific work, and it is used by scientists all over the world. The metric system branches into two systems of units. In one of these, the unit of length is the meter, the unit of mass is the kilogram, and the unit of time is the second. This is called the *meter-kilogram-second* (mks) system and is preferred in physics. The other branch is the *centimeter-gram-second* (cgs) system, which, because of its smaller values, is favored in chemistry. The cgs and mks units are related to each other as follows: 100 centimeters equal 1 meter; 1000 grams equal 1 kilogram. Table A.2 shows several units of length related to each other.

TABLE A.1 SI UNITS

Quantity	Unit	Symbol
Length	meter	m
Mass	kilogram	kg
Time	second	s
Force	newton	N
Energy	joule	J
Current	ampere	A
Temperature	kelvin	K

TABLE A.2   TABLE CONVERSIONS BETWEEN DIFFERENT UNITS OF LENGTH

Unit of Length	Kilometer	Meter	Centimeter	Inch	Foot	Mile
1 kilometer	= 1	1000	100,000	39,370	3280.84	0.62140
1 meter	= 0.00100	1	100	39.370	3.28084	6.21×10^{-4}
1 centimeter	= 1.0×10^{-5}	0.0100	1	0.39370	0.032808	6.21×10^{-6}
1 inch	= 2.54×10^{-5}	0.02540	2.5400	1	0.08333	1.58×10^{-5}
1 foot	= 3.05×10^{-4}	0.30480	30.480	12	1	1.89×10^{-4}
1 mile	= 1.60934	1609.34	160,934	63,360	5280	1

TABLE A.3   SOME PREFIXES

Prefix	Definition
micro-	One-millionth: a microsecond is one-millionth of a second
milli-	One-thousandth: a milligram is one-thousandth of a gram
centi-	One-hundredth: a centimeter is one-hundredth of a meter
kilo-	One thousand: a kilogram is 1000 grams
mega-	One million: a megahertz is 1 million hertz

One major advantage of the metric system is that it uses the decimal system, where all units are related to smaller or larger units by dividing or multiplying by 10. The prefixes shown in Table A.3 are commonly used to show the relationship among units.

▪ Meter

The standard of length of the metric system orginally was defined in terms of the distance from the north pole to the equator. This distance was thought at the time to be close to 10,000 kilometers. One ten-millionth of this, the meter, was carefully determined and marked off by means of scratches on a bar of platinum-iridium alloy. This bar is kept at the International Bureau of Weights and Measures in France. The standard meter in France has since been calibrated in terms of the wavelength of light—it is 1,650,763.73 times the wavelength of orange light emitted by the atoms of the gas krypton-86. The meter is now defined as being the length of the path traveled by light in a vacuum during a time interval of 1/299,792,458 of a second.

▪ Kilogram

FIGURE A.1

The standard kilogram.

The standard unit of mass, the kilogram, is a block of platinum-iridium alloy, preserved at the International Bureau of Weights and Measures located in France (Figure A.1). The kilogram equals 1000 grams. A gram is the mass of 1 cubic centimeter (cc) of water at a temperature of 4°C. (The standard pound is defined in terms of the standard kilogram; the mass of an object that weighs 1 pound is equal to 0.4536 kilogram.)

Second

The official unit of time for both the USCS and the SI is the second. Until 1956, it was defined in terms of the mean solar day, which was divided into 24 hours. Each hour was divided into 60 minutes and each minute into 60 seconds. Thus, there were 86,400 seconds per day, and the second was defined as 1/86,400 of the mean solar day. This proved unsatisfactory because the rate of rotation of the earth is gradually becoming slower. In 1956, the mean solar day of the year 1900 was chosen as the standard on which to base the second. In 1964, the second was officially defined as the time taken by a cesium-133 atom to make 9,192,631,770 vibrations.

Newton

One newton is the force required to accelerate 1 kilogram at 1 meter per second per second. This unit is named after Sir Isaac Newton.

Joule

One joule is equal to the amount of work done by a force of 1 newton acting over a distance of 1 meter. In 1948, the joule was adopted as the unit of energy by the International Conference on Weights and Measures. Therefore, the specific heat of water at 15°C is now given as 4185.5 joules per kilogram Celsius degree. This figure is always associated with the mechanical equivalent of heat—4.1855 joules per calorie.

Ampere

The ampere is defined as the intensity of the constant electric current that, when maintained in two parallel conductors of infinite length and negligible cross section and placed 1 meter apart in a vacuum, would produce between them a force equal to 2×10^{-7} newton per meter length. In our treatment of electric current in this text, we have used the not-so-official but easier-to-comprehend definition of the ampere as being the rate of flow of 1 coulomb of charge per second, where 1 coulomb is the charge of 6.25×10^{18} electrons.

Kelvin

The fundamental unit of temperature is named after the scientist William Thomson, Lord Kelvin. The kelvin is defined to be 1/273.15 the thermodynamic temperature of the triple point of water (the fixed point at which ice, liquid water, and water vapor coexist in equilibrium). This definition was adopted in 1968 when it was decided to change the name *degree Kelvin* (°K) to *kelvin* (K). The temperature of melting ice at atmospheric pressure is 273.15 K. The temperature at which the vapor pressure of pure water is equal to standard atmospheric pressure is 373.15 K (the temperature of boiling water at standard atmospheric pressure).

Unit square.

Area

The unit of area is a square that has a standard unit of length as a side. In the USCS, it is a square with sides that are each 1 foot in length, called 1 square foot and written $1\ ft^2$. In the international system, it is a square with sides that are 1 meter in length, which makes a unit of area of $1\ m^2$. In the cgs system it is $1\ cm^2$. The area of a given surface is specified by the number of square feet, square meters, or square centimeters that would fit into it. The area of a rectangle equals the base times the height. The area of a circle is equal to πr^2, where $\pi = 3.14$ and r is the radius of the circle. Formulas for the surface areas of other objects can be found in geometry textbooks.

Unit volume.

Volume

The volume of an object refers to the space it occupies. The unit of volume is the space taken up by a cube that has a standard unit of length for its edge. In the USCS, one unit of volume is the space occupied by a cube 1 foot on an edge and is called 1 cubic foot, written $1\ ft^3$. In the metric system it is the space occupied by a cube with sides of 1 meter (SI) or 1 centimeter (cgs). It is written $1\ m^3$ or $1\ cm^3$ (or cc). The volume of a given space is specified by the number of cubic feet, cubic meters, or cubic centimeters that will fill it.

In the USCS, volumes can also be measured in quarts, gallons, and cubic inches as well as in cubic feet. There are 1728 ($12 \times 12 \times 12$) cubic inches in $1\ ft^3$. A U.S. gallon is a volume of $231\ in^3$. Four quarts equal 1 gallon. In the SI volumes are also measured in liters. A liter is equal to $1000\ cm^3$.

Unit Conversion

Often in science, and especially in a laboratory setting, it is necessary to convert from one unit to another. To do so, you need only multiply the given quantity by the appropriate *conversion factor.*

All conversion factors can be written as ratios in which the numerator and denominator represent the equivalent quantity expressed in different units. Because any quantity divided by itself is equal to 1, all conversion factors are equal to 1. For example, the following two conversion factors are both derived from the relationship 100 centimeters = 1 meter:

$$\frac{100\ \text{centimeters}}{1\ \text{meter}} = 1 \qquad \frac{1\ \text{meter}}{100\ \text{centimeters}} = 1$$

Because all conversion factors are equal to 1, multiplying a quantity by a conversion factor does not change the value of the quantity. What does change are the units. Suppose you measured an item to be 60 centimeters in length. You can convert this measurement to meters by multiplying it by the conversion factor that allows you to cancel centimeters.

EXAMPLE

Convert 60 centimeters to meters.

ANSWER

$$(60 \; \text{centimeters}) \frac{(1 \; \text{meter})}{(100 \; \text{centimeters})} = 0.6 \; \text{meter}$$

quantity in centimeters ↑   conversion factor ↑   quantity in meters ↑

To derive a conversion factor, consult a table that presents unit equalities, such as Table A.2 or on the inside cover of this book. Then multiply the given quantity by the conversion factor, and *voilà,* the units are converted. Always be careful to write down your units. They are your ultimate guide, telling you what numbers go where and whether you are setting up the equation properly.

CHECK POINT

Multiply each physical quantity by the appropriate conversion factor to find its numerical value in the new unit indicated. You will need paper, pencil, a calculator, and a table of unit equalities.

a. 7320 grams to kilograms

b. 235 kilograms to pounds

c. 2.61 miles to kilometers

d. 100 calories to kilocalories

Were these your answers?

a. 7.32 kg

b. 518 lb

c. 4.20 km

d. 0.1 kcal

The Physical Science Place

APPENDIX A ONLINE RESOURCES

Tutorials

■ Significant Figures

When we describe the motion of something, we say how it moves relative to something else (Chapter 1). In other words, motion requires a reference frame (an observer, origin, and axes). We are free to choose this frame's location and to have it moving relative to another frame. When our frame of motion has zero acceleration, it is called an *inertial frame*. In an inertial frame, force causes an object to accelerate in accord with Newton's laws. When our frame of reference is accelerated, we observe fictitious forces and motions. Observations from a carousel, for example, are different when it is rotating and when it is at rest. Our description of motion and force depends on our "point of view."

We distinguish between *speed* and *velocity* (Chapter 1). Speed is how fast something moves, or the time rate of change of position (excluding direction): a *scalar* quantity. Velocity includes direction of motion: a *vector* quantity whose magnitude is speed. Objects moving at constant velocity move the same distance in the same time in the same direction.

Another distinction between speed and velocity has to do with the difference between distance and net distance, or *displacement*. Speed is *distance per duration* while velocity is *displacement per duration*. Displacement differs from distance. For example, a commuter who travels 10 kilometers to work and back travels 20 kilometers, but has "gone" nowhere. The distance traveled is 20 kilometers and the displacement is zero. Although the instantaneous speed and instantaneous velocity have the same value at the same instant, the average speed and average velocity can be very different. The average speed of this commuter's round-trip is 20 kilometers divided by the total commute time—a value greater than zero. But the average velocity is zero. In science, displacement is often more important than distance. (To avoid information overload, we have not treated this distinction in the text.)

Acceleration is the rate at which velocity changes. This can be a change in speed only, a change in direction only, or both. Negative acceleration is often called *deceleration*.

In Newtonian space and time, space has three dimensions—length, width, and height—each with two directions. We can go, stop, and return in any of them. Time has one dimension, with two directions—past and future. We cannot stop or return, only go. In Einsteinian space-time, these four dimensions merge (Appendix E).

Computing Velocity and Distance Traveled on an Inclined Plane

Recall from Chapter 1 Galileo's experiments with inclined planes. We considered a plane tilted such that the speed of a rolling ball increases at the rate of 2 meters per second each second—an acceleration of 2 m/s². So at the instant it starts moving its velocity is zero, and 1 second later it is rolling at 2 m/s, at the end of the next second 4 m/s, the end of the next second 6 m/s, and so on. The velocity of the ball at any instant is simply

FIGURE B.1

The ball rolls 1 m down the incline in 1 s and reaches a speed of 2 m/s. Its average speed, however, is 1 m/s. Do you see why?

Velocity = acceleration × time. Or, in shorthand notation $v = at$. (It is customary to omit the multiplication sign, ×, when expressing relationships in mathematical form. When two symbols are written together, such as the at in this case, it is understood that they are multiplied.)

How fast the ball rolls is one thing; how *far* it rolls is another. To understand the relationship between acceleration and distance traveled, we must first investigate the relationship between instantaneous velocity and *average velocity*. If the ball shown in Figure B.1 starts from rest, it will roll a distance of 1 meter in the first second. What will be its average speed? The answer is 1 m/s (it covered 1 meter in the interval of 1 second). But we have seen that the *instantaneous velocity* at the end of the first second is 2 m/s. Since the acceleration is uniform, the average in any time interval is found the same way we usually find the average of any two numbers: add them and divide by 2. (Be careful not to do this when acceleration is not uniform!) So if we add the initial speed (zero in this case) and the final speed of 2 m/s and then divide by 2, we get 1 m/s for the average velocity.

In each succeeding second we see the ball roll a longer distance down the same slope in Figure B.2. Note the distance covered in the second time interval is 3 meters. This is because the average speed of the ball in this interval is 3 m/s. In the next 1-second interval the average speed is 5 m/s, so the distance covered is 5 meters. It is interesting to see that successive increments of distance increase as a *sequence of odd numbers.* Nature clearly follows mathematical rules!

Investigate Figure B.2 carefully and note the *total* distance covered as the ball accelerates down the plane. The distances go from zero to 1 meter in 1 second, zero to 4 meters in 2 seconds, zero to 9 meters in 3 seconds, zero to 16 meters in 4 seconds, and so on in succeeding seconds. The sequence for *total distances* covered is of the *squares of the time.* We'll investigate the relationship between distance traveled and the square of the time for constant acceleration more closely in the case of free fall.

FIGURE B.2

If the ball covers 1 m during its first second, then in each successive second it will cover the odd-numbered sequence of 3, 5, 7, 9 m, and so on. Note that the total distance covered increases as the square of the total time.

CHECK POINT

During the span of the second time interval, the ball begins at 2 m/s and ends at 4 m/s. What is the *average speed* of the ball during this 1-s interval? What is its *acceleration?*

Were these your answers?

$$\text{Average speed} = \frac{\text{beginning} + \text{final speed}}{2} = \frac{2\,\text{m/s} + 4\,\text{m/s}}{2} = 3\,\text{m/s}$$

$$\text{Acceleration} = \frac{\text{change in velocity}}{\text{time interval}} = \frac{4\,\text{m/s} - 2\,\text{m/s}}{1\,\text{s}} = \frac{2\,\text{m/s}}{1\,\text{s}} = 2\,\text{m/s}^2$$

■ Computing Distance When Acceleration Is Constant

How far will an object released from rest fall in a given time? To answer this question, let us consider the case in which it falls freely for 3 seconds, starting at rest. Neglecting air resistance, the object will have a constant acceleration of about 10 meters per second each second (actually more like 9.8 m/s², but we want to make the numbers easier to follow).

$$\text{Velocity at the } \textit{beginning} \; = \; 0 \text{ m/s}$$

$$\text{Velocity at the } \textit{end} \text{ of 3 seconds} \; = \; (10 \times 3) \text{ m/s}$$

$$\textit{Average} \text{ velocity} \; = \; \frac{1}{2} \text{ the sum of these two speeds}$$

$$= \; \frac{1}{2} \times (0 + 10 \times 3) \text{ m/s}$$

$$= \; \frac{1}{2} \times 10 \times 3 \; = \; 15 \text{ m/s}$$

$$\text{Distance traveled} \; = \; \text{average velocity} \times \text{time}$$

$$= \; (\frac{1}{2} \times 10 \times 3) \times 3$$

$$= \; \frac{1}{2} \times 10 \times 3^2 \; = \; 45 \text{ m}$$

We can see from the meanings of these numbers that

$$\text{Distance traveled} \; = \; \frac{1}{2} \times \text{acceleration} \times \text{square of time}$$

This equation is true for an object falling not only for 3 seconds but for any length of time, as long as the acceleration is constant. If we let d stand for the distance traveled, a for the acceleration, and t for the time, the rule may be written, in shorthand notation,

$$d = \frac{1}{2} at^2$$

This relationship was first deduced by Galileo. He reasoned that if an object falls for, say, twice the time, it will fall with *twice the average speed.* Since it falls for *twice* the time at *twice* the average speed, it will fall *four* times as far. Similarly, if an object falls for *three* times the time, it will have an average speed *three* times as great and will fall *nine* times as far. Galileo reasoned that the total distance fallen should be proportional to the *square* of the time.

In the case of objects in free fall, it is customary to use the letter g to represent the acceleration instead of the letter a (g because acceleration is due to *gravity*). While the value of g varies slightly in different parts of the world, it is approximately equal to 9.8 m/s² (32 ft/s²). If we use g for the acceleration of a freely falling object (negligible air resistance), the equations for falling objects starting from a rest position become

$$v = gt$$

$$d = \frac{1}{2} gt^2$$

Much of the difficulty in learning physics, like learning any discipline, has to do with learning the language—the many terms and definitions. Speed is somewhat different from velocity, and acceleration is vastly different from speed or velocity. Please be patient with yourself as you find learning the similarities and the differences among physics concepts is not an easy task.

FIGURE B.3

When Chelcie Liu releases both balls simultaneously, he asks, "Which will reach the end of the equal-length tracks first?" (Hint: On which track is the average speed of the ball greater? Then, double hint: Which wins, the fast ball or the slow ball?)

CHECK POINT

1. An auto starting from rest has a constant acceleration of 4 m/s². How far will it go in 5 s?

2. How far will an object released from rest fall in 1 s? In this case the acceleration is $g = 9.8$ m/s².

3. If it takes 4 s for an object to freely fall to the water when released from the Golden Gate Bridge, how high is the bridge?

Were these your answers?

1. Distance $= \dfrac{1}{2} \times 4 \times 5^2 = 50$ m

2. Distance $= \dfrac{1}{2} \times 9.8 \times 1^2 = 4.9$ m

3. Distance $= \dfrac{1}{2} \times 9.8 \times 4^2 = 78.4$ m

Notice that the units of measurement when multiplied give the proper units of meters for distance:

$$d = \frac{1}{2} \times 9.8 \text{ m} \times 16 = 78.4 \text{ m}$$

FIGURE B.4

When a phonograph record turns, a ladybug farther from the center travels a longer path in the same time and has a greater tangential speed.

FIGURE B.5

The entire disk rotates at the same rotational speed, but ladybugs at different distances from the center travel at different tangential speeds. A ladybug twice as far from the center moves twice as fast.

Circular Motion

Linear speed is what we have been calling simply *speed*—the distance traveled in meters or kilometers per unit of time. A point on the perimeter of a merry-go-round or turntable moves a greater distance in one complete rotation than a point nearer the center. Moving a greater distance in the same time means a greater speed. The speed of something moving along a circular path is **tangential speed**, because the direction of motion is tangent to the circle.

Rotational speed (sometimes called angular speed) refers to the number of rotations or revolutions per unit of time. All parts of the rigid merry-go-round turn about the axis of rotation *in the same amount of time.* All parts share the same rate of rotation, or *number of rotations or revolutions per unit of time.* It is common to express rotational rates in revolutions per minute (rpm).* Phonograph records that were common a few years ago rotate at 33 1/3 rpm. A ladybug sitting anywhere on the surface of the record revolves at 33 1/3 rpm.

Tangential speed is *directly proportional* to rotational speed (at a fixed radial distance). Unlike rotational speed, tangential speed depends on the distance from the axis (Figure B.5). Something at the center of a rotating platform has no tangential speed at all, and merely rotates. But, approaching the edge of the platform, tangential speed increases. Tangential speed is directly proportional to the distance from the axis (for a given rotational speed). Twice as far from the rotational axis, the speed is twice as great. Three times as far from the rotational axis, there is three times as much tangential

---

* Physics types usually describe rotational speed in terms of the number of "radians" turned in a unit of time, for which they use the symbol ω (the Greek letter *omega*). There's a little more than 6 radians in a full rotation (2π radians, to be exact).

speed. When a row of people locked arm in arm at the skating rink makes a turn, the motion of "tail-end Charlie" is evidence of this greater speed. So tangential speed is directly proportional both to rotational speed and to radial distance.*

CHECK POINT

On a rotating platform similar to the disk shown in Figure B.5, if you sit halfway between the rotating axis and the outer edge and have a rotational speed of 20 rpm and a tangential speed of 2 m/s, what will be the rotational and tangential speeds of your friend who sits at the outer edge?

Was this your answer?
Since the rotating platform is rigid, all parts have the same rotational speed, so your friend also rotates at 20 rpm. Tangential speed is a different story; since she is twice as far from the axis of rotation, she moves twice as fast—4 m/s.

■ Torque

Whereas force causes changes in speed, *torque* causes changes in rotation. To understand torque (rhymes with *dork*), hold the end of a meterstick horizontally with your hand. If you dangle a weight from the meterstick near your hand, you can feel the meterstick twist. Now if you slide the weight farther from your hand, the twist you feel is greater, although the weight is the same. The force acting on your hand is the same. What's different is the torque.

FIGURE B.6

If you move the weight away from your hand, you will feel the difference between force and torque.

$$\text{Torque} = \text{lever arm} \times \text{force}$$

Lever arm is the distance between the point of application of the force and the axis of rotation. It is the shortest distance between the applied force and the rotational axis. Torques are intuitively familiar to youngsters playing on a seesaw. Kids can balance a seesaw even when their weights are unequal. Weight alone doesn't produce rotation. Torque does, and children soon learn that the distance they sit from the pivot point is every bit as important as weight (Figure B.7). When the torques are equal, making the net torque zero, no rotation is produced.

Recall the equilibrium rule in Chapter 1—that the sum of the forces acting on a body or any system must equal zero for mechanical equilibrium. That is, $\Sigma \boldsymbol{F} = 0$. We now see an additional condition. The *net torque* on a body or on a system must also be zero for mechanical equilibrium. Anything in mechanical equilibrium doesn't accelerate—neither linearly nor rotationally.

Suppose that the seesaw is arranged so that the half-as-heavy girl is suspended from a 4-meter rope hanging from her end of the seesaw (Figure B.8). She is now 5 meters from the fulcrum, and the seesaw is still balanced. We see that the lever-arm distance is 3 meters, not 5 meters. The lever arm about any axis of rotation is the perpendicular distance from the axis to the line along which the force acts. This will always be the shortest distance between the axis of rotation and the line along which the force acts.

FIGURE B.7

No rotation is produced when the torques balance each other.

FIGURE B.8

The lever arm is still 3 m.

---

* When customary units are used for tangential speed *v*, rotational speed *ω*, and radial distance *r*, the direct proportion of *v* to both *r* and *ω* becomes the exact equation $v = r\omega$. So the tangential speed will be directly proportional to *r* when all parts of a system simultaneously have the same *ω*, as for a wheel, disk, or rigid wand. (The direct proportionality of *v* to *r* is not valid for the planets because planets don't all have the same *ω*.)

FIGURE B.9

Although the magnitudes of the force are the same in each case, the torques are different.

This is why the stubborn bolt shown in Figure B.9 is turned more easily when the applied force is perpendicular to the handle, rather than at an oblique angle, as shown in the first figure. In the first figure, the lever arm is shown by the dashed line and is less than the length of the wrench handle. In the second figure, the lever arm is equal to the length of the wrench handle. In the third figure, the lever arm is extended with a pipe to provide more leverage and a greater torque.

CHECK POINT

1. If a pipe effectively extends a wrench handle to three times its length, by how much will the torque increase for the same applied force?

2. Consider the balanced seesaw in Figure B.7. Suppose the girl on the left suddenly gains 50 N, such as by being handed a bag of apples. Where should she sit in order to balance, assuming the heavier boy remains in place?

Were these your answers?

1. Three times more leverage for the same force gives three times more torque. (This method of increasing torque sometimes results in shearing off the bolt!)

2. She should sit half-a-meter closer to the center. Then her lever arm is 2.5 m. This checks: 300 N × 2.5 m = 500 N × 1.5 m.

Angular Momentum

Things that rotate, whether a cylinder rolling down an incline or an acrobat doing a somersault, keep on rotating until something stops them. A rotating object has an "inertia of rotation." Recall, from Chapter 3, that all moving objects have "inertia of motion" or *momentum*—the product of mass and velocity. This kind of momentum is **linear momentum**. Similarly, the "inertia of rotation" of rotating objects is called **angular momentum**.

For the case of an object that is small compared with the radial distance to its axis of rotation, like a tetherball swinging from a long string or a planet orbiting around the sun, the angular momentum can be expressed as the magnitude of its linear momentum, mv, multiplied by the radial distance, r, (Figure B.10).* In shorthand notation, angular momentum = mvr. Like linear momentum, angular momentum is a vector quantity and has direction as well as magnitude. In this appendix, we won't treat the vector nature of angular momentum (or even of torque, which also is a vector).

Just as an external net force is required to change the linear momentum of an object, an external net torque is required to change the angular momentum

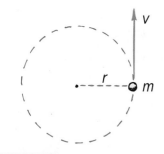

FIGURE B.10

A small object of mass m whirling in a circular path of radius r with a speed v has angular momentum mvr.

* For rotating bodies that are large compared with radial distance—for example, a planet rotating about its own axis—the concept of *rotational inertia* must be introduced. Then angular momentum is rotational inertia × rotational speed. See any of Hewitt's *Conceptual Physics* textbooks for more information.

of an object. We can state a rotational version of Newton's first law (the law of inertia):

> **An object or system of objects will maintain its angular momentum unless acted upon by an unbalanced external torque.**

We see application of this rule when we look at a spinning top. If friction is low and torque also low, the top tends to remain spinning. The earth and planets spin in torque-free regions, and once they are spinning, they remain so.

Conservation of Angular Momentum

Just as the linear momentum of any system is conserved if no net forces are acting on the system, angular momentum is conserved if no net torque acts on the system. In the absence of an unbalanced external torque, the angular momentum of that system is constant. This means that its angular momentum at any one time will be the same as at any other time.

Conservation of angular momentum is shown in Figure B.11. The man stands on a low-friction turntable with weights extended. To simplify, consider only the weights in his hands. When he is slowly turning with his arms extended, much of the angular momentum is due to the distance between the weights and the rotational axis. When he pulls the weights inward, the distance is considerably reduced. What is the result? His rotational speed increases!* This example is best appreciated by the turning person, who feels changes in rotational speed that seem to be mysterious. But it's straight physics! This procedure is used by a figure skater who starts to whirl with her arms and perhaps a leg extended and then draws her arms and leg in to obtain a greater rotational speed. Whenever a rotating body contracts, its rotational speed increases.

The law of angular momentum conservation is seen in the motions of the planets and the shape of the galaxies. When a slowly rotating ball of gas in space gravitationally contracts, the result is an increase in its rate of rotation. The conservation of angular momentum is far-reaching.

FIGURE B.11

Conservation of angular momentum. When the man pulls his arms and the whirling weights inward, he decreases the radial distance between the weights and the axis of rotation, and the rotational speed increases correspondingly.

---

* When a direction is assigned to rotational speed, we call it *rotational velocity* (often called *angular velocity*). By convention, the rotational velocity vector and the angular momentum vector have the same direction and lie along the axis of rotation.

APPENDIX C

VECTORS

Vectors and Scalars

A *vector* quantity is a directed quantity—one that must be specified not only by magnitude (size) but by direction as well. Recall from Chapter 1 that velocity is a vector quantity. Other examples are force, acceleration, and momentum. In contrast, a *scalar* quantity can be specified by magnitude alone. Some examples of scalar quantities are speed, time, temperature, and energy.

Vector quantities may be represented by arrows. The length of the arrow tells you the magnitude of the vector quantity, and the arrowhead tells you the direction of the vector quantity. Such an arrow drawn to scale and pointing appropriately is called a *vector*.

FIGURE **C.1**

Adding Vectors

Vectors that add together are called *component vectors.* The sum of component vectors is called a *resultant.*

To add two vectors, make a parallelogram with two component vectors acting as two of the adjacent sides (Figure C.2). (Here our parallelogram is a rectangle.) Then draw a diagonal from the origin of the vector pair; this is the resultant (Figure C.3).

Caution: Do not try to mix vectors! We cannot add apples and oranges, so velocity vectors combine only with velocity vectors, force vectors combine only with force vectors, and acceleration vectors combine only with acceleration vectors—each on its own vector diagram. If you ever show different kinds of vectors on the same diagram, use different colors or some other method of distinguishing the different kinds of vectors.

FIGURE **C.2**

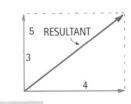

FIGURE **C.3**

Finding Components of Vectors

Recall from Chapter 2 that to find a pair of perpendicular components for a vector, first draw a dashed line through the tail of the vector (in the direction of one of the desired components). Second, draw another dashed line through the tail end of the vector at right angles to the first dashed line. Third, make a rectangle whose diagonal is the given vector. Draw in the two components. Here we let **F** stand for "total force," **U** stand for "upward force," and **S** stand for "sideways force."

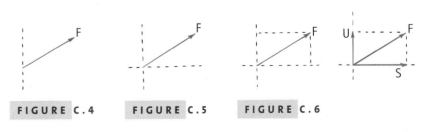

FIGURE **C.4**   FIGURE **C.5**   FIGURE **C.6**

EXAMPLES

1. Ernie Brown pushes a lawnmower and applies a force that pushes it forward and also against the ground. In Figure C.7, **F** represents the force applied by the man. We can separate this force into two components.
The vector **D** represents the downward component, and **S** is the sideways component, the force that moves the lawnmower forward. If we know the magnitude and direction of the

FIGURE C.7

vector **F**, we can estimate the magnitude of the components from the vector diagram.

2. Would it be easier to push or pull a wheelbarrow over a step? Figure C.8 shows the force at the wheel's center. When you push a wheelbarrow, part of the force is directed downward, which makes it harder to get over the step. When you pull, however, part of the pulling force is directed upward, which helps to lift the wheel over the step. Note that the vector diagram suggests that pushing the wheelbarrow may not get it over the step at all. Do you see that the height of the step, the radius of the wheel, and the angle of the applied force determine whether the wheelbarrow can be pushed over the step? We see how vectors help us analyze a situation so that we can see just what the problem is!

FIGURE C.8

3. If we consider the components of the weight of an object rolling down an incline, we can see why its speed depends on the angle. Note that the steeper the incline, the greater the component **S** becomes and the faster the object rolls. When the incline is vertical, **S** becomes equal to the weight, and the object attains maximum acceleration, 9.8 m/s^2. There are two more force vectors that are not shown: the normal force **N**, which is equal and oppositely directed to **D**, and the friction force **f**, acting at the barrel-plane contact.

FIGURE C.9

4. When moving air strikes the underside of an airplane wing, the force of air impact against the wing may be represented by a single vector perpendicular to the plane of the wing (Figure C.10). We represent the force vector as acting midway along the lower wing surface, where the dot is, and pointing above the wing to show the direction of the resulting wind impact force. This force can be broken up into two components, one sideways and the other up. The upward component, **U**, is called *lift*. The sideways component, **S**, is called *drag*. If the aircraft is to fly at constant velocity at constant altitude, then lift must

FIGURE C.10

equal the weight of the aircraft and the thrust of the plane's engines must equal drag. The magnitude of lift (and drag) can be altered by changing the speed of the airplane or by changing the angle (called *angle of attack*) between the wing and the horizontal.

5. Consider the satellite moving clockwise in Figure C.11. Everywhere in its orbital path, gravitational force **F** pulls it toward the center of the host planet. At position A we see **F** separated into two components: **f**, which is tangent to the path of the projectile, and **f′**, which is perpendicular to the path. The relative magnitudes of these components in comparison to the magnitude of **F** can be seen in the imaginary rectangle they compose: **f** and **f′** are the sides, and **F** is the diagonal. We see that component **f** is along the orbital path but against the direction of motion of the satellite. This force component reduces the speed of the satellite. The other component, **f′**, changes the direction of the satellite's motion and pulls it away from its tendency to go in a straight line. So the path of the satellite curves. The satellite loses speed until it reaches position B. At this farthest point from the planet (apogee), the gravitational force is somewhat weaker but perpendicular to the satellite's motion, and component **f** has reduced to zero. Component **f′**, on the other hand, has increased and is now fully merged to become **F**. Speed at this point is not enough for circular orbit, and the satellite begins to fall toward the planet. It picks up speed because the component **f** reappears and is in the direction of motion as shown in position C. The satellite picks up speed until it whips around to position D (perigee), where once again the direction of motion is

FIGURE C.11

perpendicular to the gravitational force, **f**′ blends to full **F**, and **f** is nonexistent. The speed is in excess of that needed for circular orbit at this distance, and it overshoots to repeat the cycle. Its loss in speed in going from D to B equals its gain in speed from B to D. Kepler discovered that planetary paths are elliptical, but never knew why. Do you?

6. Refer to the Polaroids held by Ludmila back in Chapter 11, in Figure 11.58. In the first picture (a), we see that light is transmitted through the pair of Polaroids because their axes are aligned. The emerging light can be represented as a vector aligned with the polarization axes of the Polaroids.

When the Polaroids are crossed (b), no light emerges because light passing through the first Polaroid is perpendicular to the polarization axes of the second Polaroid, with no components along its axis. In the third picture (c), we see that light is transmitted when a third Polaroid is sandwiched at an angle between the crossed Polaroids. The explanation for this is shown in Figure C.12.

FIGURE C.12

Sailboats

*S*ailors have always known that a sailboat can sail downwind, in the direction of the wind. Sailors have not always known, however, that a sailboat can sail upwind, against the wind. One reason for this has to do with a feature that is common only to recent sailboats—a fin-like keel that extends deep beneath the bottom of the boat to ensure that the boat will knife through the water only in a forward (or backward) direction. Without a keel, a sailboat could be blown sideways.

Figure C.13 shows a sailboat sailing directly downwind. The force of wind impact against the sail accelerates the boat. Even if the drag of the water and all other resistance forces are negligible, the maximum speed of the boat is the wind speed. This is because the wind will not make impact against the sail if the boat is moving as fast as the wind. The wind would have no speed relative to the boat and the sail would simply sag. With no force, there is no acceleration. The force vector in Figure C.13 *decreases* as the boat travels faster. The force vector is maximum when the boat is at rest and the full impact of the wind fills the sail, and is minimum when the boat travels as fast as the wind. If the boat is somehow propelled to a speed faster than the wind (by way of a motor, for example), then air resistance against the front side of the sail will produce an oppositely directed force vector. This will slow the boat down. Hence the boat when driven only by the wind cannot exceed wind speed.

If the sail is oriented at an angle, as shown in Figure C.14, the boat will move forward, but with less acceleration. There are two reasons for this:

1. The force on the sail is less because the sail does not intercept as much wind in this angular position.

2. The direction of the wind impact force on the sail is not in the direction of the boat's motion, but is perpendicular to the surface of the sail. Generally speaking, whenever any fluid (liquid or gas) interacts with a smooth surface, the force of interaction is perpendicular to the smooth surface.* The boat does not move in the same direction as the perpendicular force on the sail, but is constrained to move in a forward (or backward) direction by its keel.

FIGURE C.13

FIGURE C.14

* You can do a simple exercise to see that this is so. Try bouncing a coin off another on a smooth surface, as shown. Note that the struck coin moves at right angles (perpendicular) to the contact edge. Note also that it makes no difference whether the projected coin moves along path A or path B. See your instructor for a more rigorous explanation, which involves momentum conservation.

FIGURE C.15

FIGURE C.16

FIGURE C.17

We can better understand the motion of the boat by resolving the force of wind impact, **F**, into perpendicular components. The important component is that which is parallel to the keel, which we label **K**, and the other component is perpendicular to the keel, which we label **T**. It is the component **K**, as shown in Figure C.15, that is responsible for the forward motion of the boat. Component **T** is a useless force that tends to tip the boat over and move it sideways. This component force is offset by the deep keel. Again, maximum speed of the boat can be no greater than wind speed.

Many sailboats sailing in directions other than exactly downwind (Figure C.16) with their sails properly oriented can exceed wind speed. In the case of a sailboat cutting across the wind, the wind may continue to make impact with the sail even after the boat exceeds wind speed. A surfer, in a similar way, exceeds the velocity of the propelling wave by angling his surfboard across the wave. Greater angles to the propelling medium (wind for the boat, water wave for the surfboard) result in greater speeds. A sailcraft can sail faster cutting across the wind than it can sailing downwind.

As strange as it may seem, maximum speed for most sailcraft is attained by cutting into (against) the wind, that is, by angling the sailcraft in a direction upwind! Although a sailboat cannot sail directly upwind, it can reach a destination upwind by angling back and forth in a zigzag fashion. This is called *tacking*. Suppose the boat and sail are as shown in Figure C.17. Component **K** will push the boat along in a forward direction, angling into the wind. In the position shown, the boat can sail faster than the speed of the wind. This is because as the boat travels faster, the impact of wind is increased. This is similar to running in a rain that comes down at an angle. When you run into the direction of the downpour, the drops strike you harder and more frequently, but when you run away from the direction of the downpour, the drops don't strike you as hard or as frequently. In the same way, a boat sailing upwind experiences greater wind impact force, while a boat sailing downwind experiences a decreased wind impact force. In any case the boat reaches its terminal speed when opposing forces cancel the force of wind impact. The opposing forces consist mainly of water resistance against the hull of the boat. The hulls of racing boats are shaped to minimize this resistive force, which is the principal deterrent to high speeds.

Iceboats (sailcraft equipped with runners for traveling on ice) encounter no water resistance and can travel at several times the speed of the wind when they tack upwind. Although ice friction is nearly absent, an iceboat does not accelerate without limits. The terminal velocity of a sailcraft is determined not only by opposing friction forces but also by the change in relative wind direction. When the boat's orientation and speed are such that the wind seems to shift in direction, so the wind moves parallel to the sail rather than into it, forward acceleration ceases—at least in the case of a flat sail. In practice, sails are curved and produce an airfoil that is as important to sailcraft as it is to aircraft, as discussed in Chapter 6.

The Physical Science Place

APPENDIX C ONLINE RESOURCES

Tutorials
- Vectors

APPENDIX D

EXPONENTIAL GROWTH AND DOUBLING TIME*

One of the most important things we seem unable to perceive is the process of exponential growth. We think we understand how compound interest works, but we can't get it through our heads that a fine piece of tissue paper folded upon itself 50 times (if that were possible) would be more than 20 million kilometers thick. If we could, we could "see" why our income buys only half of what it did 4 years ago, why the price of everything has doubled in the same time, why populations and pollution proliferate out of control.**

When a quantity such as money in the bank, population, or the rate of consumption of a resource steadily grows at a fixed percent per year, we say the growth is exponential. Money in the bank may grow at 4 percent per year; electric power generating capacity in the United States grew at about 7 percent per year for the first three-quarters of the 20th century. The important thing about exponential growth is that the time required for the growing quantity to double in size (increase by 100 percent) is also constant. For example, if the population of a growing city takes 12 years to double from 10,000 to 20,000 inhabitants and its growth remains steady, in the next 12 years the population will double to 40,000, and in the next 10 years to 80,000, and so on.

There is an important relationship between the percent growth rate and its *doubling time,* the time it takes to double a quantity:†

$$\text{Doubling time} = \frac{69.3}{\text{percent growth per unit time}} \approx \frac{70}{\%}$$

So to estimate the doubling time for a steadily growing quantity, we simply divide the number 70 by the percentage growth rate. For example, the 7 percent growth rate of electric power generating capacity in the United States means that in the past the capacity had doubled every 10 years [70%/(7%/year) = 10 years]. A 2 percent growth rate for world population means the population of the world doubles every 35 years [70%/(2%/year) = 35 years]. A city planning commission that accepts what seems like a modest 3.5 percent growth rate may not realize that this means that doubling will occur in 70/3.5 or 20 years; that's double capacity for such things as water supply, sewage-treatment plants, and other municipal services every 20 years.

What happens when you put steady growth in a finite environment? Consider the growth of bacteria that grow by division, so that one bacterium becomes two, the two divide to become four, the four divide to become eight, and so on. Suppose the division time for a certain strain of bacteria is 1 minute.

---

* This appendix is drawn from material by University of Colorado physics professor Albert A. Bartlett, who strongly asserts, "The greatest shortcoming of the human race is man's inability to understand the exponential function." See Professor Bartlett's still-timely article, "Forgotten Fundamentals in the Energy Crisis" (*American Journal of Physics,* September 1978) or his revised version (*Journal of Geological Education,* January 1980).

** K. C. Cole, *Sympathetic Vibrations* (New York: Morrow, 1984).

† For exponential decay we speak about half-life, the time required for a quantity to reduce to half its value. This case is treated in Chapter 16.

FIGURE D.1

An exponential curve. Notice that each of the successive equal time intervals noted on the horizontal scale corresponds to a doubling of the quantity indicated on the vertical scale. Such an interval is called the doubling time.

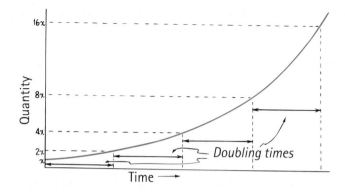

This is then steady growth—the number of bacteria grows exponentially with a doubling time of 1 minute. Further, suppose that one bacterium is put in a bottle at 11:00 A.M. and that growth continues steadily until the bottle becomes full of bacteria at 12 noon. Consider seriously the following question.

CHECK POINT

When was the bottle half-full?

Was this your answer?
11:59 A.M.; the bacteria will double in number every minute!

It is startling to note that at 2 minutes before noon the bottle was only 1/4 full. Table D.1 summarizes the amount of space left in the bottle in the last few minutes before noon. If you were an average bacterium in the bottle, at which time would you first realize that you were running out of space? For example, would you sense there was a serious problem at 11:55 A.M., when the bottle was only 3% filled, (1/32), and had 97% of open space (just yearning for development)? The point here is that there isn't much time between the moment that the effects of growth become noticeable and the time when they become overwhelming.

Suppose that at 11:58 A.M. some farsighted bacteria see that they are running out of space and launch a full-scale search for new bottles. Luckily, at 11:59 A.M. they discover three new empty bottles, three times as much space as they had ever known. This quadruples the total resource space ever known to the bacteria, for they now have a total of four bottles, whereas before the discovery they had only one. Further suppose that, thanks to their technological proficiency, they are able to migrate to their new habitats without difficulty. Surely, it seems to most of the bacteria that their problem is solved—and just in time.

FIGURE D.2

TABLE D.1   THE LAST MINUTES IN THE BOTTLE

Time	Part Full (%)	Part Empty
11:54 A.M.	1/64 (1.5%)	63/64
11:55 A.M.	1/32 (3%)	31/32
11:56 A.M.	1/16 (6%)	15/16
11:57 A.M.	1/8 (12%)	7/8
11:58 A.M.	1/4 (25%)	3/4
11:59 A.M.	1/2 (50%)	1/2
12:00 noon	full (100%)	none

CHECK
■ **POINT**

If the bacteria growth continues at the unchanged rate, what time will it be when the three new bottles are filled to capacity?

Was this your answer?

12:02 P.M.!

We see from Table D.2 that quadrupling the resource extends the life of the resource by only two doubling times. In our example the resource is space—but it could as well be coal, oil, uranium, or any nonrenewable resource.

Continued growth and continued doubling lead to enormous numbers. In two doubling times, a quantity will double twice ($2^2 = 4$; quadruple) in size; in three doubling times, its size will increase eightfold ($2^3 = 8$); in four doubling times, it will increase sixteenfold ($2^4 = 16$); and so on.

FIGURE D.3

A single grain of wheat placed on the first square of the chessboard is doubled on the second square, this number is doubled on the third, and so on, presumably for all 64 squares. Note that each square contains one more grain than all the preceding squares combined. Does enough wheat exist in the world to fill all 64 squares in this manner?

TABLE D.2	EFFECTS OF THE DISCOVERY OF THREE NEW BOTTLES
Time	**Effect**
11:58 A.M.	Bottle 1 is 1/4 full
11:59 A.M.	Bottle 1 is 1/2 full
12:00 noon	Bottle 1 is full
12:01 P.M.	Bottles 1 and 2 are both full
12:02 P.M.	Bottles 1, 2, 3, and 4 are all full

This is best illustrated by the story of the court mathematician in India who years ago invented the game of chess for his king. The king was so pleased with the game that he offered to repay the mathematician, whose request seemed modest enough. The mathematician requested a single grain of wheat on the first square of the chessboard, two grains on the second square, four on the third square, and so on, doubling the number of grains on each succeeding square until all squares had been used. At this rate there would be 2^{63} grains of wheat on the 64th square. The king soon saw that he could not fill this "modest" request, which amounted to more wheat than had been harvested in the entire history of the Earth!

It is interesting and important to note that the number or grains on any square is one grain more than the total of all grains on the preceding squares. This is true anywhere on the board. Note from Table D.3 that when eight grains are placed on the fourth square, the eight is one more than the total of seven grains that were already on the board. Or the 32 grains placed on the sixth square is one more than the total of 31 grains that were already on the board. We see that in one doubling time we use more than all that had been used in all the preceding growth!

So if we speak of doubling energy consumption in the next however many years, bear in mind that this means in these years we will consume more energy than has heretofore been consumed during the entire preceding period of steady growth. And if power generation continues to use predominantly fossil fuels, then except for some improvements in efficiency, we would burn up in the next doubling time a greater amount of coal, oil, and natural gas than has already been consumed by previous power generation, and except for improvements in pollution control, we can expect to discharge even more toxic wastes into the environment than the millions upon millions of tons already discharged over all the previous years of industrial civilization. We would also expect more human-made calories of heat to be absorbed by Earth's

TABLE D.3	FILLING THE SQUARES ON THE CHESSBOARD	
Square Number	Grains on Square	Total Grains Thus Far
1	1	1
2	2	3
3	4	7
4	8	15
5	16	31
6	32	63
7	64	127
.	.	.
.	.	.
.	.	.
64	2^{63}	$2^{64} - 1$

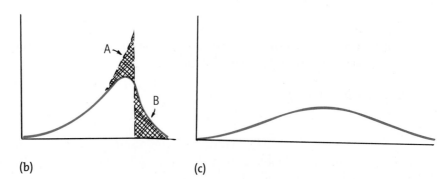

(a)   (b)   (c)

FIGURE D.4

(a) If the exponential rate of consumption for a nonrenewable resource continues until it is depleted, consumption falls abruptly to zero. The shaded area under this curve represents the total supply of the resource. (b) In practice, the rate of consumption levels off and then falls less abruptly to zero. Note that the cross-hatched area A is equal to the crosshatched area B. Why? (c) At lower consumption rates, the same resource lasts a longer time.

FIGURE D.5

A curve showing the rate of consumption of a renewable resource such as agricultural or forest products, where a steady rate of production and consumption can be maintained for a long period, provided this production, is not dependent upon the use of a nonrenewable resource that is warning in supply.

ecosystem than have been absorbed in the entire past! At the previous 7 percent annual growth rate in energy production, all this would occur in one doubling time of a single decade. If over the coming years the annual growth rate remains at half this value, 3.5 percent, then all this would take place in a doubling time of two decades. Clearly this cannot continue!

The consumption of a nonrenewable resource cannot grow exponentially for an indefinite period, because the resource is finite and its supply finally expires. The most drastic way this could happen is shown in Figure D.4 (a), where the rate of consumption, such as barrels of oil per year, is plotted against time, say in years. In such a graph the area under the curve represents the supply of the resource. We see that when the supply is exhausted, the consumption ceases altogether. This sudden change is rarely the case, for the rate of extracting the supply falls as it becomes more scarce. This is shown in Figure D.4(b). Note that the area under the curve is equal to the area under the curve in (a). Why? Because the total supply is the same in both cases. The principal difference is the time taken to finally extinguish the supply. History shows that the rate of production of a nonrenewable resource rises and falls in a nearly symmetric manner, as shown in (c). The time during which production rates rise is approximately equal to the time during which these rates fall to zero or near zero.

Production rates for all nonrenewable resources decrease sooner or later. Only production rates for renewable resources, such as agriculture or forest products, can be maintained at steady levels for long periods of time (Figure D.5), provided such production does not depend on waning nonrenewable resources such as petroleum. Much of today's agriculture is so petroleum-dependent that it can be said that modern agriculture is simply the process whereby land is used to convert petroleum into food. The implications of petroleum scarcity go far beyond rationing of gasoline for cars or fuel oil for home heating.

The consequences of unchecked exponential growth are staggering. It is important to ask: Is growth really good? In answering this question, bear in mind that human growth is an early phase of life that continues normally through adolescence. Physical growth stops when physical maturity is reached. What do we say of growth that continues in the period of physical maturity? We say that such growth is obesity—or worse, cancer.

QUESTIONS TO PONDER

1. According to a French riddle, a lily pond starts with a single leaf. Each day the number of leaves doubles, until the pond is completely covered by leaves on the 30th day. On what day was the pond half covered? One-quarter covered?

2. In an economy that has a steady inflation rate of 7 percent per year, in how many years does a dollar lose half its value?

3. At a steady inflation rate of 7 percent, what will be the price every 10 years for the next 50 years for a theater ticket that now costs $20? For a coat that now costs $200? For a car that now costs $20,000? For a home that now costs $200,000?

4. If the sewage treatment plant of a city is just adequate for the city's current population, how many sewage treatment plants will be necessary 42 years later if the city grows steadily at 5 percent annually?

5. If world population doubles in 40 years and world food production also doubles in 40 years, how many people then will be starving each year compared to now?

6. Suppose you get a prospective employer to agree to hire your services for wages of a single penny for the first day, 2 pennies for the second day, and double each day thereafter providing the employer keeps to the agreement for a month. What will be your total wages for the month?

7. In the preceding exercise, how will your wages for only the 30th day compare to your total wages for the previous 29 days?

8. If fusion power were harnessed today, the abundant energy resulting would probably sustain and even further encourage our present appetite for continued growth and in a relatively few doubling times produce an appreciable fraction of the solar power input to the earth. Make an argument that the current delay in harnessing fusion is a blessing for the human race.

APPENDIX E

SPECIAL RELATIVITY

You're sitting inside your car at a traffic light. Suddenly, the car beside you starts moving backwards. A moment later you realize it's you who is moving forward—your foot had come off the brake and you're about to hit the car in front of you! Sound feasible? Here on Earth we rely on a background of trees or other objects to tell whether or not we're moving. In the absence of these external references, our motion is impossible to gauge. Similarly, imagine you are resting inside a spaceship in deep outerspace. A second spaceship whizzes past. Which spaceship is moving? The astronaut in the second spaceship may argue that she was the one at rest while your spaceship went whizzing past.

Einstein thought about these ideas and added his conclusion: there is no experiment you can perform to decide who is moving and who is not. There is no such thing as absolute rest and all motion is relative. Einstein realized that observers can never detect their uniform motion except *relative* to other objects. From this realization, Einstein postulated:

> **All laws of nature are the same in all uniformly moving reference frames.**

By "uniformly moving," Einstein meant constant velocity and no acceleration. For example, the physical laws within a uniformly moving space ship (constant velocity; zero acceleration) are the same as those in a stationary laboratory. An important physical law is that the speed of light has a constant value of 299,792 km/s. Any measurement of the speed of light in either place, therefore, would show the same value:

> **The speed of light in free space will have the same value to all observers, regardless of the motion of the source or the motion of the observer. The speed of light is a constant.**

If you move away from a tossed baseball, when you catch it, you'll catch it at a slower speed. Do this for light, however, and you have a different story. Pretend we're in a high-speed rocket moving away from a light source at nearly the speed of light. Good old common sense tells us that the light catches up to us and passes us slower than if we weren't moving—just like a ball. But according to Einstein, the speed of light remains constant, no matter what your motion (Figure E.1). This has been confirmed many times over by many repeated experiments.

So if the light doesn't slow down as you recede away from the light source, something else must be happening. That something else is the stretching of time. Light has farther to travel as you recede away from the source, right? In other words, the space through which light must travel to reach you is greater. In order to keep the speed of light constant, the duration of a unit of time also becomes greater.

Speed is distance divided by time. We can generically discuss the speed of light (denoted by the letter c) as the amount of space through which light travels divided by an amount of time it takes to travel through that space:

FIGURE E.1

The speed of light is measured to be the same in all frames of reference.

fyi

- Einstein was 26 in 1905 when he published three major papers that provided the blueprint for much of today's physics. One was on the quantum theory of light and the photoelectric effect, the second on the explanation of Brownian motion, and the third on special relativity. He won the Nobel prize for his quantum explanation of the photoelectric effect—not for relativity.

$$\frac{\text{SPACE}}{\text{TIME}} = \frac{\text{SPACE}}{\text{TIME}} = C$$

So as the space gets larger, the time gets larger as well. The ratio of the two, which is the speed of light, is the same.

But think carefully about this. Who notices a change in time? From our point of view in the fast moving spaceship, one second continues to feel like one second. From the point of view of our friends back on Earth, one second also continues to feel like one second. It's only when the two quantities of time are compared that the difference becomes apparent. For the duration of our high-speed space trip, our friends on Earth may have witnessed a full week. Upon our return, however, we may have experienced only a few days. So, one's time depends upon one's frame of reference. Different frames of reference moving at different speeds experience different durations of time!

Time Dilation

To explore this idea further, pretend you are Einstein at the turn of the twentieth century riding in a trolleycar, which provided the high-speed travel back then. Suppose that the trolleycar is stationed beneath a huge clock displayed in a village square. You see this clock move to the future at a rate of 60 seconds per minute. At exactly 12 noon the trolley suddenly zips off at the speed of light. You are then traveling alongside the light that carries the information "12 noon." You look to this light, which tells you it's 12 noon in the village square. You look a moment later and it's still 12 noon in the village square. Time in the village square is frozen!

When the trolleycar was not moving, you saw the village square clock move into the future at the rate of 60 seconds per minute; when you moved at the speed of light, you saw seconds on the clock taking infinite time. These are the two extremes. What's in between? What happens for speeds that are less than the speed of light? A little thought will show that the clock will be seen to run somewhere between the rate 60 seconds per minute and the rate of 60 seconds per an infinity of time. From your high-speed (but less than *c*) moving frame of reference, the clock and all events in the reference frame of the clock will be seen in slow motion. Time will be stretched. How much depends on speed. The faster you go, the more time stretches. This is **time dilation**. Time dilation has nothing to do with the mechanics of clocks, but with the nature of time itself.

Nothing is unusual about a traveling clock itself; it is simply ticking to the rhythm of a different time. The faster a clock travels, the slower it runs as viewed by an observer not traveling with the clock. If it were possible for an observer to watch a clock zip by at the speed of light, the clock would not appear to be

If you were to speed away from a clock at the speed of light, the time on the clock would appear frozen.

The graph shows how 1 second on a stationary clock is stretched out, as measured on a moving clock. Note that the stretching becomes significant only at speeds near the speed of light.

running at all. This observer would measure the interval between ticks to be infinite. Time would be frozen and the clock would be ageless!

Time dilation has been confirmed in the laboratory with atomic particle accelerators. The lifetimes of fast-moving radioactive particles increase as the speed goes up, and the amount of increase is just what Einstein's equation predicts.

Time dilation has been confirmed also for not-so-fast motion. In 1971, to test Einstein's theory, four cesium-beam atomic clocks were twice flown on regularly scheduled commercial jet flights around the Earth, once eastward and once westward. The clocks indicated different times after their round trips. Relative to the atomic time scale of the U.S. Naval Observatory, the observed time differences, in billionths of a second, were in accord with relativistic prediction.

FIGURE E.4

When we see the rocket at rest, we see it traveling at the maximum rate in time: 24 hours per day. If we see the rocket traveling at the maximum rate through space (the speed of light), we see its time standing still.

CHECK POINT

1. If you are moving in a spaceship at a high speed relative to the Earth, would you notice a difference in your pulse rate? In the pulse rate of the people back on Earth?

2. Does time dilation mean that time really passes more slowly in moving systems or that it only seems to pass more slowly?

Were these your answers?

1. There would be no relative speed between you and your own pulse, which share the same frame of reference, so you would notice no relativistic effects in your own pulse. There would, however, be a relativistic effect between you and people back on Earth. You would find their pulse rate slower than normal (and, likewise, they would find your pulse rate slower than normal). Relativity effects are always attributed to the other guy.

2. The slowing of time in moving systems is not merely an illusion resulting from motion. Time really does pass more slowly in a moving system compared to one at relative rest. (This is dramatically shown in "The Twin Trip," in the Conceptual Physical Science Practice Book.)

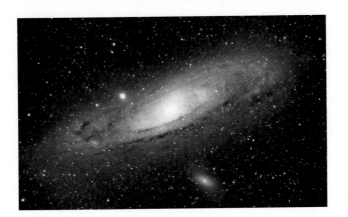

FIGURE E.5

From the Earth frame of reference, light takes 25,000 years to travel from the center of the galaxy to our solar system. From the frame of reference of a high-speed spaceship, the trip takes less time. From the frame of reference of light itself, the trip takes no time. There is no time in a speed-of-light frame of reference.

Different Views of Spacetime

So how can two people within the same universe experience different times? At the heart of our confusion is the fact that we are three-dimensional creatures living within four-dimensional spacetime. Consider the following scenario. Two off-duty astronauts are sitting together in a lounge admiring the coffee table between them. They come to an agreement on the shape of the

FIGURE E.6

Two-dimensional photographs are limited in their ability to capture the reality of a three-dimensional object. Why would a movie camera be more effective?

We often find ourselves in a different space—you're reading this book, while your friend is out shopping. Now you know it's also possible to find ourselves in a different time. But we are always together within this same universe, which is the same spacetime.

FIGURE E.7

Einstein was more than a great scientist; he was a man of unpretentious disposition with a deep concern for the welfare of his fellow beings. The choice of Einstein as the person of the century by *Time* magazine at the end of the 1900s was most appropriate—and noncontroversial.

coffee table. They recognize this coffee table as a three-dimensional object. They then decide to take pictures of the coffee table from many different angles. They print out these pictures on two-dimensional flat sheets of paper, which they lay out on the coffee table to examine. Which *one* picture should they choose to best represent the coffee table? It is quickly obvious to them that each picture of the coffee table looks different. One photo taken from directly above might make the table look short, while another taken from the side might make the table look long. No single two-dimensional photo, however, can accurately describe the whole three-dimensional object.

Similarly, spacetime is four-dimensional, which is something we three-dimensional creatures can look at from different perspectives. Two space traveling astronauts in different ships moving at different speeds will have different perspectives. When they get back together to compare their experiences, which are their three-dimensional "snapshots," they can well expect to have seen different angles of the same thing. One astronaut may have aged one year while the other aged two years. But all that while, they were both enjoying the same universe, which is the same spacetime. Just as there are many ways to experience a coffee table, there are also many ways we can experience spacetime.

We can tie special relativity together with general relativity, discussed in Section 28.3, by asking a simple question: What do you have to do in order to change your motion? Review Chapter 1 and you'll recall that in order to change your motion you must accelerate. As you accelerate within a spaceship, however, you experience things as though you are within a gravitational field (see Figure 28.9). According to general relativity, gravity, or anything that simulates gravity such as acceleration, has the effect of slowing down time. So for the space traveler, the change from one time reference to another occurs as he or she is accelerating. Furthermore, to return home, the space traveler will have to decelerate. During this deceleration, time alters back to the frame of reference the traveler once left.

It is important to note the *relativistic* nature of time both in special relativity and in general relativity. In both theories, there is no way that you can extend the duration of your own experience. Others moving at different speeds or in different gravitational fields may attribute a great longevity to you, but your longevity is seen from *their* frame of reference, never from your own. Changes in time and other relativistic effects are always attributed to "the other guy." That's relativity.

Space Travel

One of the old arguments advanced against the possibility of human interstellar travel was that the human life span is too short. It was argued, for example, that the nearest star (after the sun), Alpha Centauri, is four light-years away, and that a round trip, even at the speed of light, would require at least eight years. And even a speed-of-light voyage to the center of our galaxy, which is 25,000 light-years distant, would require a 25,000-year lifetime. But these arguments fail to take into account time dilation. Time for a person on Earth and time for a person in a high-speed rocketship are not the same.

A person's heart beats to the rhythm of the region of spacetime it is in, which can be very different for an observer who stands outside the person's frame of reference. For example, consider a team of astronauts traveling at 99% the speed of light (0.99c) on a round trip journey to the star Procyon (10.4 light-years distant). It would take light itself 20.8 years to make the same round trip. Because of time dilation, it would seem to the astronauts that only three years had gone by. This is what all their clocks would tell them—and, biologically, they would be only three years older. The space officials greeting them on their return, however, would be 21 years older!

At higher speeds, the results are even more impressive. At a rocket speed of 0.9999c, travelers could travel about 70 light-years in a single year of their own time. Traveling for a single year at 0.99999c would bring them to a destination some 220 light-years distant.

Present technology does not permit such journeys. Spaceships traveling at relativistic speeds would encounter a dangerous hail of interstellar particles. If, somehow, a way were devised to solve this problem, there would be the problem of energy and fuel. Spaceships traveling at relativistic speeds would require billions of times the energy used to put a space shuttle into orbit. The practicalities of such space journeys are enormously prohibitive. The only alternative is for us to stop thinking about moving through space at high speeds and instead focus on ways we might cause the space behind us to expand and the space in front of us to contract. As we study dark energy (Chapter 28), perhaps we will find the mechanism of its action and be able to apply that to space ships. Perhaps a form of exotic matter, or perhaps an ultravacuum on the back of our spaceship (and the reverse of such on the front of our ship) would allow us to remain stationary while traveling vast distances at superluminal speeds. We could call this the "warp drive" and we could use it to boldly go where no one has gone before. Might you be interested in such a trek? You've got the basics. We encourage you to continue with your learning about our physical universe. We live in a fascinating place during a fascinating time.

fyi

- Special relativity also tells us that the mass of an object increases with speed. The effect is negligible at normal speeds, but at relativistic speeds it becomes significant. Notably, to push the object faster and faster becomes more and more difficult. Upon reaching the speed of light, the object would have acquired an infinite mass, which would have required an infinite amount fuel. So no rocket ship can ever be pushed to the speed of light. The speed of light is the speed limit for any form of traveling matter.

FIGURE E.8

The best science fiction builds from a solid foundation of that which is known to a speculation of that which may one day be possible, including peace, joy, compassion for fellow sentient beings, and a devout respect for the environment that sustains us.

Aberrations   Limitations on the formation of perfect images, which are inherent, to some degree, in all optical systems.

Ablation   The amount of ice lost, and the process of losing ice, from a glacier.

Absolute zero   The theoretical temperature at which a substance possesses no thermal heat and the temperature at which particles of a substance have their minimum kinetic energy.

Acceleration   The rate at which velocity changes with time; the change in velocity may be in magnitude, or in direction, or in both. It is usually measured in meters per second squared.

Accumulation   The amount of snow added, and the process of adding snow, to a glacier.

Acid   A substance that donates hydrogen ions.

Acidic solution   A solution in which the hydronium-ion concentration is higher than the hydroxide-ion concentration.

Activation energy   The minimum energy required in order for a chemical reaction to proceed.

Active galactic nucleus   A supermassive black hole at the center of a galaxy into which matter is falling at a high rate, thereby releasing astronomical amounts of energy.

Adiabatic   A term that describes temperature change in the absence of heat transfer—expanding air cools, and compressing air warms.

Addition polymer   A polymer formed by the joining together of monomer units with no atoms being lost as the polymer forms.

Additive primary colors   The three colors—red, blue, and green—that, when added in certain proportions, will produce any color in the spectrum.

Air resistance   The force of friction acting on an object due to its motion through air.

Alcohol   An organic molecule that contains a hydroxyl group bonded to a saturated carbon.

Aldehyde   An organic molecule containing a carbonyl group, the carbon of which is bonded either to one carbon atom and one hydrogen atom or to two hydrogen atoms.

Alpha particle   The nucleus of a helium atom, which consists of two neutrons and two protons, ejected by certain radioactive elements.

Alternating current (ac)   Electric current that repeatedly reverses its direction; the electric charges vibrate about relatively fixed points. In the United States, the vibrational rate is 60 Hz.

Amide   An organic molecule containing a carbonyl group, the carbon of which is bonded to a nitrogen atom.

Amine   An organic molecule containing a nitrogen atom bonded to one or more saturated carbon atoms.

Amphoteric   A description of a substance that can behave either as an acid or as a base.

Amplitude   For a wave or vibration, the maximum displacement on either side of the equilibrium (midpoint) position.

Angular unconformity   An unconformity in which older, tilted strata are overlain by younger, horizontal beds.

Anode   The electrode where oxidation occurs.

Anticline   A fold in rock with relatively old rocks at the core; rock age decreases with horizontal distance from the fold core.

Applied research   A branch of scientific research that focuses on developing applications built upon the principles discovered through basic research.

Archimedes' principle   An immersed body is buoyed up by a force equal to the weight of the fluid it displaces.

Aromatic compound   Any organic molecule containing a benzene ring.

Artesian system   A system in which confined groundwater under pressure can rise above the upper boundary of an aquifer.

Asteroid   A small rocky planet-like fragment that orbits the Sun. Tens of thousands of these objects make up an asteroid belt between the orbits of Mars and Jupiter.

Asthenosphere   A subdivision of the upper mantle situated below the lithosphere, a zone of plastic, easily deformed rock.

Astronomical unit (AU)   The average distance between Earth and the Sun, which is about 1.5×10^8 kilometers (about 9.3×10^7 mi).

Atmospheric pressure   The pressure exerted against bodies immersed in the atmosphere resulting from the weight of air pressing down from above. At sea level, atmospheric pressure is about 101 kPa.

Atom   The smallest particle of an element that has all of the element's chemical properties.

Atomic mass   The mass of an element's atoms listed in the periodic table as an average value based on the relative abundance of the element's isotopes.

Atomic mass unit (amu)   The standard unit of atomic mass, which is equal to one-twelfth the mass of the common atom of carbon, arbitrarily given the value of exactly 12.

Atomic nucleus   The core of an atom, consisting of two basic subatomic particles—protons and neutrons.

Atomic number   The number that designates the identity of an element, which is the number of protons in the nucleus of an atom; in a neutral atom, the atomic number is also the number of electrons in the atom.

Atomic orbital   A region of space in which an electron in an atom has a 90% chance of being located.

Atomic spectrum   The pattern of frequencies of electromagnetic radiation emitted by the atoms of an element, considered to be the element's "fingerprint."

Atomic symbol   An abbreviation for an element or atom.

Avogadro's number   The number of particles—6.02×10^{23}—contained in 1 mole of anything.

Barometer   Any device that measures atmospheric pressure.

Base   A substance that accepts hydrogen ions.

Basic research   A branch of scientific research that focuses on a greater understanding of how the natural world operates.

Basic solution   A solution in which the hydroxide-ion concentration is higher than the hydronium-ion concentration.

Beats   A series of alternate reinforcements and cancellations produced by the interference of two waves of slightly different frequency, heard as a throbbing effect in sound waves.

Bernoulli's principle   The pressure in a fluid moving steadily, without friction or an input of outside energy; decreases when the fluid velocity increases.

Beta particle   An electron (or positron) emitted during the radioactive decay of certain nuclei.

Big Bang   The primordial explosion of space at the beginning of time.

Big rip   A model for the end of the universe in which dark energy grows stronger over time and causes all matter to rip apart.

Binary star   Pairs of stars that orbit about a common center of mass.

Black hole   The remains of a giant star that has collapsed upon itself, so dense and gravitational field so intense that light itself cannot escape.

Black hole singularity   The object of zero radius into which the matter of a black hole is comprised.

Body wave   A type of seismic wave that travels through Earth's interior.

Boiling   A rapid state of evaporation that takes place within the liquid as well as at its surface. As with evaporation, cooling of the liquid results.

Bond energy   The amount of energy that is either absorbed as a chemical bond breaks or is released as a chemical bond forms.

Bow wave   The V-shaped wave made by an object moving across a liquid surface at a speed greater than the wave speed.

Boyle's law   The product of pressure and volume is a constant for a given mass of confined gas regardless of changes either in pressure or in volume individually, so long as temperature remains unchanged:

$$P_1 V_1 = P_2 V_2$$

Buffer solution   A solution that resists large changes in pH, made either from a weak acid and one of its salts or from a weak base and one of its salts.

Buoyant force   The net upward force that a fluid exerts on an immersed object.

Carbonyl group   A carbon atom double-bonded to an oxygen atom; found in ketones, aldehydes, amides, carboxylic acids, and esters.

Carboxylic acid   An organic molecule containing a carbonyl group, the carbon of which is bonded to a hydroxyl group.

Catalyst   Any substance that increases the rate of a chemical reaction without itself being consumed by the reaction.

Cathode   The electrode where reduction occurs.

Celestial sphere   An imaginary sphere surrounding Earth to which the stars are attached.

Cenozoic era   The time of recent life; it began 65 million years ago and is still ongoing.

Chain reaction   A self-sustaining reaction in which the products of one reaction event stimulate further reaction events.

Chemical bond   The force of attraction between two atoms that holds them together.

Chemical change   A change in which the atoms of one or more substances are rearranged into one or more new substances.

Chemical equation   A representation of a chemical reaction in which reactants are drawn before an arrow that points to the products.

Chemical formula   A notation used to indicate the composition of a compound, consisting of the atomic symbols for the different elements of the compound and numerical subscripts indicating the ratio in which the atoms combine.

Chemical property   A property that characterizes the ability of a substance to undergo a change that transforms it into a different substance.

Chemical reaction   Synonymous with chemical change.

Chemical sediments   Sediments that form by the precipitation of minerals from water on Earth's surface.

Cleavage   The tendency of a mineral to break along planes of weakness.

Combustion   An exothermic oxidation–reduction reaction between a nonmetallic material and molecular oxygen.

Comet   A body composed of ice and dust that orbits the Sun, usually in a very eccentric orbit, and which casts a luminous tail produced by solar radiation pressure when it is close to the Sun.

Complementary colors   Any two colors that, when added, will produce white light.

Compound   A material in which atoms of different elements are bonded to one another.

Compression   Condensed region of the medium through which a longitudinal wave travels.

Concentration   A quantitative measure of the amount of solute in a solution.

Conceptual model   A representation of a system that helps in making predictions about how the system behaves.

Condensation   The change of phase from gas to liquid; the opposite of evaporation. Warming of the liquid results.

Condensation polymer   A polymer formed by the joining together of monomer units accompanied by the loss of small molecules, such as water.

Conduction   The transfer of thermal energy by molecular and electronic collisions within a substance (especially a solid).

Conductor   Any material having free charged particles that easily flow through it when an electric force acts on them.

Conformation   One of the possible spatial orientations of a molecule.

Conservation of energy   In the absence of external work input or output, the energy of a system remains unchanged. Energy cannot be created or destroyed.

Conservation of energy and machines   The work output of any machine cannot exceed the work input. In an ideal machine, where no energy is transformed into heat:

$$\text{work}_{\text{input}} = \text{work}_{\text{output}}$$

$$(Fd)_{\text{input}} = (Fd)_{\text{output}}$$

Conservation of momentum   In the absence of an external force, the momentum of a system remains unchanged. Hence, the momentum before an event involving only internal forces is equal to the momentum after the event:

$$mv_{(\text{before event})} = mv_{(\text{after event})}$$

Continental drift   A hypothesis by Alfred Wegener that the world's continents are mobile and have moved to their present positions as the ancient supercontinent Pangaea broke a apart.

Continental margin   The boundary between continental land and deep ocean basins, consisting of continental shelf, continental slope, and continental rise.

Convection   The transfer of thermal energy in a gas or liquid by means of currents in the heated fluid. The fluid flows, carrying energy with it.

Convectional lifting   An air-circulation pattern in which air warmed by the ground rises while cooler air aloft sinks.

Convergent plate boundary   A plate boundary where tectonic plates move toward one another; an area of compressive stress where lithosphere is recycled into the mantle, or shortened by folding and faulting.

Converging lens   A lens that is thicker in the middle than at the edges and that refracts parallel rays passing through it to a focus.

Core   The central portion of Earth's interior, divided into an outer liquid core and an inner solid core.

Coriolis force   The apparent deflection from a straight-line path observed in any body moving near Earth's surface, caused by Earth's rotation.

Corrosion   The deterioration of a metal, typically caused by atmospheric oxygen.

Cosmic background radiation   The faint microwave radiation eminating from all directions that is the remnant heat of the Big Bang.

Cosmic inflation   The theory that the universe experienced a sudden and brief burst in size immediately after the Big Bang.

Cosmological red shift   The elongation of light waves due to the expansion of space.

Cosmology   The study of the overall structure and evolution of the universe.

Coulomb   The SI unit of electrical charge. One coulomb (C) is equal in magnitude to the total charge of 6.25×10^{18} electrons.

Coulomb's law   The relationship among electrical force, charge, and distance. If the charges are alike in sign, the force is repelling; if the charges are unlike, the force is attractive:

$$F = k\frac{q_1 q_2}{d^2}$$

Covalent bond   A chemical bond in which atoms are held together by their mutual attraction for two or more electrons they share.

Covalent compound   An element or chemical compound in which atoms are held together by covalent bonds.

Critical angle   The minimum angle of incidence inside a medium at which a light ray is totally reflected.

Critical mass   The minimum mass of fissionable material in a reactor or nuclear bomb that will sustain a chain reaction.

Cross-cutting relationships   A relative dating principle stating that where an igneous intrusion or fault cuts through other rocks, the intrusion or fault is younger than the rock it cuts.

Crust   Earth's outermost layer.

Crystal form   The outward expression of the orderly internal arrangement of atoms in a crystal.

Crystallization   The growth of a solid from a material whose constituent atoms can come together in the proper chemical proportions and geometric arrangements.

Cut bank   A steep bank on the outside bend of a river's channel. An area of erosion.

Dark energy   An unknown form of energy that appears to be causing an accleration of the expansion of space. Thought to be associated with the energy exuded by a perfect vacuum.

Dark matter   Matter that responds only to weak nuclear and gravitational forces. This form of matter is invisible but reveals itself to us by its gravitational effects.

Delta   An accumulation of sediments, commonly forming a triangular or fan-shaped plain, deposited where a stream flows into a body of water.

Density   The ratio between the mass of a substance and its volume.

Deposition   The stage of sedimentary rock formation in which eroded particles come to rest.

Diffraction   The bending of light that passes around an obstacle or through a narrow slit, causing the light to spread and to produce light and dark fringes.

Dipole   A separation of charge that occurs in a chemical bond because of differences in the electronegativities of the bonded atoms.

Direct current (dc)   An electric current flowing in one direction only.

Discharge   The volume of water that passes a given location in a stream channel in a certain amount of time.

Dissolving   The process of mixing a solute in a solvent to produce a homogeneous mixture.

Distillation   A purifying process in which a vaporized substance is collected by exposing it to cooler temperatures over a receiving flask, which collects the condensed purified liquid.

Divergent plate boundary   A plate boundary where lithospheric plates move away from one another—a spreading center; an area of tensional stress where new lithospheric crust is formed.

Diverging lens   A lens that is thinner in the middle than at the edges, causing parallel rays passing through it to diverge as if from a point.

Doppler effect   The change in frequency of wave motion resulting from motion of the sender or the receiver.

Drift   A general term for glacial deposits.

Dwarf planet   Relatively large icy bodies, such as Pluto, originating within the Kuiper belt.

Earthquake   The shaking or trembling of the ground that happens when rock under Earth's surface moves or breaks.

Ecliptic   The plane of Earth's orbit around the Sun. All major objects of the solar system orbit roughly within this same plane.

Efficiency   The percentage of the work put into a machine that is converted into useful work output. (More generally, efficiency is useful energy output divided by total energy input.)

Elastic collision   A collision in which colliding objects rebound without lasting deformation or the generation of heat.

Electric current   The flow of electric charge that transports energy from one place to another. It is measured in amperes, where 1 A is the flow of 6.25×10^{18} electrons per second, or 1 coulomb per second.

Electric field   Defined as force per unit charge, it can be considered to be an energetic "aura" surrounding charged objects. About a charged point, the field decreases with distance according to the inverse-square law, like a gravitational field. Between oppositely charged parallel plates, the electric field is uniform.

Electric potential The electric potential energy per amount of charge, measured in volts, and often called *voltage*:

$$\text{Voltage} = \frac{\text{electric energy}}{\text{amount of charge}}$$

Electric potential energy The energy a charge possesses by virtue of its location in an electric field.

Electric power The rate of energy transfer, or the rate of doing work; the amount of energy per unit time, which can be measured by the product of current and voltage:

$$\text{Power} = \text{current} \times \text{voltage}$$

It is measured in watts (or kilowatts), where $1 \text{ A} \times 1 \text{ V} = 1 \text{ W}$.

Electrical resistance The property of a material that resists the flow of an electric current through it. It is measured in ohms (Ω).

Electrically polarized Term applied to an atom or molecule in which the charges are aligned so that one side has a slight excess of positive charge and the other side a slight excess of negative charge.

Electrochemistry The branch of chemistry concerned with the relationship between electrical energy and chemical change.

Electrode Any material that conducts electrons into or out of a medium in which electrochemical reactions are occurring.

Electrolysis The use of electrical energy to produce chemical change.

Electromagnet A magnet whose field is produced by an electric current. It is usually in the form of a wire coil with a piece of iron inside the coil.

Electromagnetic induction The induction of voltage when a magnetic field changes with time. If the magnetic field within a closed loop changes in any way, a voltage is induced in the loop. This is a statement of Faraday's law. (The induction of voltage is actually the result of a more fundamental phenomenon: the induction of an electric *field*, as defined in Maxwell's counterpart to Faraday's law.)

Electromagnetic spectrum The range of electromagnetic waves that extends in frequency from radio waves to gamma rays.

Electromagnetic wave An energy-carrying wave emitted by vibrating electrical charges (often electrons) and composed of oscillating electric and magnetic fields that regenerate one another.

Electron A negatively charged particle in an atom.

Electron-dot structure A shorthand notation of the shell model of the atom in which valence electrons are shown around an atomic symbol.

Electronegativity The ability of an atom to attract a bonding pair of electrons to itself when bonded to another atom.

Electrostatics The study of electric charge at rest (not *in motion,* as in electric currents).

Element Any material that is made up of only one type of atom.

Elemental formula A notation that uses the atomic symbol and (sometimes) a numerical subscript to denote how atoms of the element are bonded together.

Ellipse The oval path followed by a satellite. The sum of the distances from any point on the path to two points called foci is a constant. When the foci are together at one point, the ellipse is a circle. As the foci get farther apart, the ellipse gets more "eccentric."

Elliptical galaxy A galaxy that is round or elliptical in outline. It has little gas and dust, no disk or spiral arms, and few hot and bright stars.

Endothermic A term that describes a chemical reaction in which there is a net absorption of energy.

Energy The property of a system that enables it to do work.

Entropy The measure of energy dispersal of a system. Whenever energy freely transforms from one form to another, the direction of transformation is toward a state of greater disorder and, therefore, toward one of greater entropy.

Equilibrium rule The vector sum of forces acting on a nonaccelerating object equals zero: $\Sigma F = 0$.

Erosion The wearing away of rocks, and the processes by which rock particles are transported by water, wind, or ice.

Escape speed The speed that a projectile, a space probe, or a similar object must reach in order to escape the gravitational influence of Earth or of another celestial body to which it is attracted.

Ester An organic molecule containing a carbonyl group, the carbon of which is bonded to one carbon atom and one oxygen atom bonded to another carbon atom.

Eternal inflation A model of the universe in which cosmic inflation is not a one time event, rather it progresses continuously spawning an infinite number of observable universes in its wake.

Ether An organic molecule containing an oxygen atom bonded to two carbon atoms.

Evaporation The change of phase at the surface of a liquid as it passes to the gaseous phase.

Event horizon   The boundary region of a black hole from which no radiation may escape. Any events within the event horizon are invisible to distant observers.

Exosphere   The fifth atmospheric layer above Earth's surface, extending from the thermosphere upward and out into interplanetary space.

Exothermic   A term that describes a chemical reaction in which there is a net release of energy.

Fact   A phenomenon about which competent observers can agree.

Faraday's law   The induced voltage in a coil is proportional to the number of loops, multiplied by the rate at which the magnetic field changes within those loops.

Fault   A fracture along which visible displacement can be detected on one side relative to the other.

Faunal succession   A relative dating principle stating that fossil organisms succeed one another in a definite, irreversible, determinable order.

First law of thermodynamics   A restatement of the law of energy conservation, usually as it applies to systems involving changes in temperature: Whenever heat flows into or out of a system, the gain or loss of thermal energy equals the amount of heat transferred.

Floodplain   A wide plain of almost flat land on either side of a stream channel. Submerged during flood stage, the plain is built up by sediments deposited during floods.

Fold   A series of ripples in the crust that result from compressional deformation of the lithosphere.

Force   Simply stated, a push or a pull.

Force pair   The action-and-reaction pair of forces that constitute an interaction.

Force vector   An arrow drawn to scale so that its length represents the magnitude of a force and its direction represents the direction of the force.

Forced vibration   The setting up of vibrations in an object by a vibrating force.

Formula mass   The sum of the atomic masses of the atoms in a chemical compound or element.

Fracture   A break that does not occur along a plane of weakness.

Frame of reference   A vantage point (usually a set of coordinate axes) with respect to which position and motion may be described.

Free fall   Motion under the influence of gravitational pull only.

Freezing   The process of changing state from liquid to solid, as from water to ice.

Frequency   For a vibrating body or medium, the number of vibrations per unit time. For a wave, the number of crests that pass a particular point per unit time.

Friction   The resistive force that opposes the motion or attempted motion of an object through a fluid or past another object with which it is in contact.

Front   The contact zone between two different air masses.

Frontal lifting   The lifting of one air mass by another as two air masses converge.

Full Moon   The phase of the Moon when its sunlit side is the side facing Earth.

Functional group   A specific combination of atoms that behaves as a unit in an organic molecule.

Fundamental frequency   The lowest frequency of vibration, or the first harmonic. In a string, the vibration makes a single segment.

Galaxy   A large assemblage of stars, interstellar gas, and dust, usually categorized by its shape: elliptical, spiral, or irregular.

Galaxy cluster   Pertains to a group of more than one galaxy.

Galaxy super cluster   A group of an enormous number of galaxies.

Gamma ray   High-frequency electromagnetic radiation emitted by the nuclei of radioactive atoms.

General theory of relativity   The second of Einstein's theories of relativity, which discusses the effects of gravity on space and time.

Generator   An electromagnetic induction device that produces electric current by rotating a coil within a stationary magnetic field.

Geodesic   The shortest distance between two points in various models of space.

Giant stars   Cool giant stars above main-sequence stars on the H–R diagram.

Glacier   A large mass of ice formed by the compaction and recrystallization of snow, moving downslope under its own weight.

Gradient   The vertical drop in the elevation of a stream channel divided by the horizontal distance for that drop; the steepness of the slope.

Gravitational red shift   The lengthening of the waves of electromagnetic radiation due to escape from a gravitational field.

Gravitational wave   The transport of energy by the motion of waves in a gravitational field.

Greenhouse effect   Warming caused by short-wavelength radiant energy from the Sun that easily enters the atmosphere and is absorbed by Earth. This energy is then reradiated at longer wavelengths that cannot easily escape Earth's atmosphere.

Groundwater   Underground water in the saturated zone.

Group   A vertical column in the periodic table, also known as a family of elements.

Gyre   Circular or spiral whirl pattern, usually referring to very large current systems in the open ocean.

Half-life   The time required for half the atoms in a sample of a radioactive isotope to decay.

Half reaction   One portion of an oxidation–reduction reaction, represented by an equation showing electrons as either reactants or products.

Hang time   The time that one's feet are off the ground during a vertical jump.

Hard water   Water containing large amounts of calcium and magnesium ions.

Harmonic   A partial tone that is an integer multiple of the fundamental frequency. The vibration that begins with the fundamental vibrating frequency is the first harmonic, twice the fundamental is the second harmonic, and so on in sequence.

Heat   The thermal energy that flows from a substance of higher temperature to a substance of lower temperature, commonly measured in calories or joules.

Heat death   A model for the end of the universe in which all matter and energy disperses to the point of maximum entropy.

Heat of fusion   The amount of energy needed to change any substance from solid to liquid (and vice versa). For water, this is 334 J/g (or 80 cal/g).

Heat of vaporization   The amount of energy required to change any substance from liquid to gas (and vice versa). For water, this is 2256 J/g (or 540 cal/g).

Hertz   The SI unit of frequency. One hertz (Hz) equals one vibration per second.

Heteroatom   Any atom other than carbon or hydrogen in an organic molecule.

Heterogeneous mixture   A mixture in which the various components can be seen as individual substances.

Homogeneous mixture   A mixture in which the components are so finely mixed that the composition is the same throughout.

H-R diagram   (Hertzsprung-Russell diagram) A plot of intrinsic brightness versus surface temperature of stars. When so plotted, stars' positions take the form of a main sequence for average stars, with exotic stars above or below the main sequence.

Hubble's law   The farther away a galaxy is from Earth, the more rapidly it is moving away from us: $v = H \times d$

Humidity   A measure of the concentration or amount of water vapor in the air—the mass of water vapor per volume of air.

Hydraulic conductivity   A measure of the ability of a porous rock or sediment to transmit fluid.

Hydrocarbon   A chemical compound containing only carbon and hydrogen atoms.

Hydrogen bond   A strong dipole–dipole attraction between a slightly positive hydrogen atom on one molecule and a pair of nonbonding electrons on another molecule.

Hydrologic cycle   The natural circulation of all states of water from ocean to atmosphere to land, back to ocean.

Hydronium ion   A water molecule after accepting a hydrogen ion.

Hydroxide ion   A water molecule after losing a hydrogen ion.

Hypothesis   An educated guess or reasonable explanation. When the hypothesis makes a prediction that can be tested by experiment, it qualifies as a scientific hypothesis.

Igneous rocks   Rocks formed by the cooling and crystallization of hot, molten rock material called magma (or lava).

Impulse   The product of the force acting on an object and the time during which it acts.

Impure   In chemistry, this term refers to a material that is a mixture of more than one element or compound.

Inclusions   A relative dating principle stating that any inclusion (pieces of one rock type contained within another) is older than the rock containing it.

Induced dipole   A dipole temporarily created in an otherwise nonpolar molecule, induced by a neighboring charge.

Inelastic collision   A collision in which the colliding objects become distorted, generate heat, and possibly stick together.

Inertia   The property of things to resist changes in motion.

Infrasonic   Describes a sound of a frequency too low to be heard by the normal human ear—below 20 hertz.

Inner planets   The four planets orbiting within 2 AU of the Sun including Mercury, Venus, Earth and Mars. These planets are all rocky planets, known as the *terrestrial* planets.

Insoluble   Not capable of dissolving to any appreciable extent in a given solvent.

Insulator   Any material without free charged particles and through which current does not easily flow.

Interaction   Mutual action between objects in which each one exerts an equal and opposite force on the other.

Interference   The result of superposing different waves of the same wavelength. Constructive interference results from crest-to-crest reinforcement; destructive interference results from crest-to-trough cancellation. The interference of selected wavelengths of light produces colors known as *interference colors.*

Interference pattern   The pattern formed by superposition of different sets of waves, which produces mutual reinforcement in some places and cancellation in others.

Inverse-square law   A law relating the intensity of an effect to the inverse square of the distance from the cause. Gravity follows an inverse-square law, as do the effects of electric, magnetic, light, sound, and radiation phenomena:

$$\text{Intensity} \sim \frac{1}{\text{distance}^2}$$

Ion   An electrically charged particle created when an atom either loses or gains one or more electrons.

Ionic bond   A chemical bond in which an attractive electric force holds ions of opposite charge together.

Ionic compound   Any chemical compound containing ions.

Ionosphere   An electrified region within the thermosphere and uppermost mesosphere where fairly large concentrations of ions and free electrons exist.

Irregular galaxy   A galaxy with a chaotic appearance and with large clouds of gas and dust, but without spiral arms.

Isostasy   The process by which oceanic and continental crust come into vertical equilibrium, with respect to the mantle—the dense oceanic crust sits lower in the mantle than the less-dense continental crust.

Isotopes   Different forms of an element whose atoms contain the same number of protons but different numbers of neutrons.

Ketone   An organic molecule containing a carbonyl group, the carbon of which is bonded to two carbon atoms.

Kilogram   The unit of mass. One kilogram (kg) is the mass of 1 liter (L) of water at 4°C.

Kinetic energy   Energy of motion, described by the relationship:

$$\text{Kinetic energy} = 1/2\ mv^2$$

Kuiper Belt (pronounced KI-pur)   The disk-shaped region of the sky beyond Neptune populated by many icy bodies and a source of short-period comets.

Laminar flow   Water flowing smoothly and fairly slowly in straight lines with no mixing of sediment.

Lateral continuity   A relative dating principle stating that sedimentary layers are deposited in all directions over large areas until some sort of obstruction, or barrier, limits their deposition.

Lava   Molten magma that moves upward from inside Earth and flows onto the surface. The term *lava* refers both to the molten rock itself and to solid rocks that form from it.

Law   A general hypothesis or statement about the relationship of natural quantities that has been tested over and over again and has not been contradicted. Also known as a *principle.*

Law of mass conservation   Matter is neither created nor destroyed during a chemical reaction—atoms merely rearrange, without any apparent loss or gain of mass, to form new molecules.

Law of reflection   The angle of incidence equals the angle of reflection. The incident and reflected rays lie in a plane that is normal to the reflecting surface.

Law of universal gravitation   Every mass in the universe attracts every other mass with a force that for two masses is directly proportional to the product of their masses and inversely proportional to the square of the distance separating them:

$$F = G\frac{m_1 m_2}{d^2}$$

Length contraction   The contraction of objects in their direction of motion as a result of speed.

Lifting condensation level   The height at which rising air cooling at the dry adiabatic lapse rate becomes saturated and condensation begins.

Light year   The distance light travels in one year.

Lithosphere   The entire crust plus the portion of the mantle above the asthenosphere.

Local group   Our immediate cluster of galaxies including the Milky Way, Andromeda, and Trangulum spiral galaxies plus a few dozen smaller elliptical and irregular galaxies.

Local supercluster   A cluster of galactic clusters in which our local group resides.

Longitudinal wave   A wave in which the medium vibrates in a direction parallel (longitudinal) with the direction in which the wave travels. Sound consists of longitudinal waves.

Lunar eclipse   The phenomenon whereby the shadow of the Earth falls upon the Moon producing relative darkness of the full Moon.

Magma   Molten rock in Earth's interior.

Magnetic domains   Clustered regions of aligned magnetic atoms. When these regions themselves are aligned with one another, the substance containing them is a magnet.

Magnetic field The region of magnetic influence around a magnetic pole or a moving charged particle.

Magnetic force (1) Between magnets, it is the attraction of unlike magnetic poles for each other and the repulsion between like magnetic poles. (2) Between a magnetic field and a moving charge, it is a deflecting force due to the motion of the charge: the deflecting force is perpendicular to the velocity of the charge and perpendicular to the magnetic field lines. This force is greatest when the charge moves perpendicular to the field lines and is smallest (zero) when it moves parallel to the field lines.

Main sequence The diagonal band of stars on an H–R diagram; such stars generate energy by fusing hydrogen to helium.

Mantle The middle layer in Earth's interior, between the crust and the core.

Mass The quantity of matter in an object. More specifically, it is the measure of the inertia or sluggishness that an object exhibits in response to any effort made to start it, stop it, deflect it, or change its state of motion in any way.

Mass-energy equivalence The relationship between mass and energy as given by the equation $E = mc^2$.

Mass number The total number of nucleons in an atomic nucleus.

Maxwell's counterpart to Faraday's law A magnetic field is induced in any region of space in which an electric field is changing with time. The magnitude of the induced magnetic field is proportional to the rate at which the electric field changes. The direction of the induced magnetic field is at right angles to the changing electric field.

Mechanical deformation Metamorphism caused by stress, such as increased pressure.

Mesosphere The third atmospheric layer above Earth's surface, extending from the top of the stratosphere to 80 km.

Mesozoic era The time of middle life, from 248 million years ago to about 65 million years ago.

Metamorphic rocks Rocks formed from preexisting rocks that have been changed or transformed by high temperature, high pressure, or both.

Metamorphism The changes in rock that happen as physical and chemical conditions change.

Meteor The streak of light produced by a meteoroid burning in Earth's atmosphere; a "shooting star."

Meteorite A meteoroid or part of a meteoroid that has survived passage through Earth's atmosphere to reach the ground.

Meteoroid A small rock in interplanetary space.

Midlatitude cyclone West-to-east traveling storm with a central low-pressure area about which counterclockwise flow develops (in the Northern Hemisphere), and from which usually extends a cold front and a warm front; generally forms at the polar front.

Milky Way The name of the galaxy to which we belong. Our cosmic home.

Mineral A naturally formed, inorganic crystalline solid composed of an ordered arrangement of atoms with a specific chemical composition.

Mixture A combination of two or more substances in which each substance retains its properties.

Mohorovic discontinuity (Moho) The crust-mantle boundary, marking the depth at which the speed of P-waves traveling toward Earth's center increases.

Mohs scale of hardness A ranking of a mineral's hardness, which is its resistance to scratching.

Molarity A unit of concentration equal to the number of moles of a solute per liter of solution.

Molar mass The mass of 1 mole of a substance.

Mole The amount of any pure substance that contains as many atoms, molecules, ions, or other elementary units as the number of atoms in 12 grams of carbon-12. This is equal to 6.02×10^{23} particles.

Molecule A submicroscopic particle consisting of a group of atoms. Also, a group of atoms held tightly together by covalent bonds.

Momentum The product of the mass of an object and its velocity.

Monomers The small molecular units from which a polymer is formed.

Moon phases The cycles of change of the "face" of the Moon, changing from *new* to *waxing*, to *full*, to *waning*, and back to *new*.

Natural frequency A frequency at which an elastic object naturally tends to vibrate, so that minimum energy is required to produce a forced vibration or to continue vibration at that frequency.

Neap tide A tide that occurs when the Moon is midway between new and full, in either direction. The pulls of the Moon and Sun are perpendicular to one another, so the solar and lunar tides do not overlap. This makes high tides not as high and low tides not as low.

Nebular theory The idea that the Sun and planets formed together from a cloud of gas and dust, a *nebula*.

Net force The combination of all forces that act on an object.

Neutral solution   A solution in which the hydronium-ion concentration is equal to the hydroxide-ion concentration.

Neutralization   A reaction in which an acid and base combine to form a salt.

Neutron   An electrically neutral subatomic particle in an atomic nucleus.

Neutron star   A small, highly dense star composed of tightly packed neutrons formed by the welding of protons and electrons.

New Moon   The phase of the Moon when darkness covers the side facing Earth.

Newton   The scientific unit of force.

Newton's first law of motion   Every object continues in a state of rest, or in a state of motion in a straight line at a constant speed, unless it is compelled to change that state by forces exerted upon it.

Newton's law of cooling   The rate of loss of thermal energy from an object is proportional to the temperature difference between the object and its surroundings.

Newton's second law of motion   The acceleration produced by a net force on an object is directly proportional to the net force, is in the same direction as the net force, and is inversely proportional to the mass of the object.

Newton's third law of motion   Whenever one object exerts a force on a second object, the second object exerts an equal and opposite force on the first object.

Nonbonding pairs   Two paired valence electrons that tend not to participate in a chemical bond.

Nonpolar   Said of a chemical bond that has no dipole.

Nonsilicate   A mineral that does not contain silica (silicon + oxygen).

Nova   An event wherein a white dwarf suddenly brightens and appears as a "new" star.

Nuclear fission   The splitting of the nucleus of a heavy atom, such as uranium-235, into two main parts, accompanied by the release of much energy.

Nuclear fusion   The combining of nuclei of light atoms to form heavier nuclei, with the release of much energy.

Nucleon   A nuclear particle; a proton or a neutron in an atomic nucleus.

Ohm's law   The statement that the current in a circuit varies in direct proportion to the potential difference or voltage and inversely with the resistance. A potential difference of 1 V across a resistance of 1 Ω produces a current of 1 A:

$$\text{Current} = \frac{\text{voltage}}{\text{resistance}}$$

Oort Cloud   The region beyond the Kuiper Belt populated by trillions of icy bodies and is a source of long-period comets.

Ordinary matter   Matter that responds to the strong nuclear, weak nuclear, electromagnetic, and gravitational forces. This is matter made of protons, neutrons, and electrons, which includes the atoms and molecules that make us and our immediate environment.

Ore   A mineral deposit containing valuable metals that can be economically extracted from the ground to yield a profit.

Organic chemistry   The study of carbon-containing compounds.

Original horizontality   A relative dating principle stating that layers of sediment are deposited evenly, with each new layer laid down almost horizontally over the older sediment.

Orographic lifting   The lifting of an air mass over a topographic barrier such as a mountain.

Osmosis   The diffusion of a water or some other fluid through a semipermeable membrane, from a solution with a low concentration of solutes to a solution with a higher concentration of solutes.

Outer planets   The four planets orbiting beyond 2 AU of the Sun including Jupiter, Saturn, Uranus, and Neptune—all gaseous and known as the *jovian* planets.

Oxidation   The process whereby a reactant loses one or more electrons.

Paleomagnetism   The natural ancient magnetization in a rock that can be used to determine the intensity and direction of Earth's magnetic field at the time of the rock's formation.

Paleozoic era   The time of ancient life, from 543 million years ago to 248 million years ago.

Pangaea   The late-Paleozoic supercontinent made up of Gondwanaland (ancestral South America, Africa, Australia, Antarctica, and India) and Laurasia (ancestral North America, Europe, and Siberia/Asia).

Parabola   The curved path followed by a projectile near Earth under the influence of gravity only.

Parallel circuit   An electric circuit with two or more devices connected in such a way that the same voltage acts across each one, and any single one completes the circuit independently of all the others.

Partial melting   The incomplete melting of rocks, resulting in magmas of different compositions.

Partial tone   One of the frequencies present in a complex tone. When a partial tone is an integer multiple of the lowest frequency, it is a harmonic.

Period   The time required for a vibration or a wave to make a complete cycle; equal to 1/frequency.

Period   A horizontal row in the periodic table.

Periodic table   A chart in which all known elements are listed in order of atomic number.

pH   A measure of the acidity of a solution, equal to the negative of the base-10 logarithm of the hydronium-ion concentration.

Phenol   An organic molecule in which a hydroxyl group is bonded to a benzene ring.

Photoelectric effect   The emission of electrons from a metal surface when light shines on it.

Physical change   A change in which a substance changes one or more of its physical properties without transforming it into a new substance.

Physical model   A representation of an object on some convenient scale.

Physical property   Any physical attribute of a substance, such as color, density, or hardness.

Planets   The major bodies orbiting the Sun, massive enough for their gravity to make them spherical, but small enough to avoid nuclear fusion in their cores.

Plate tectonics   The theory that Earth's lithosphere is broken into pieces (plates) that move over the asthenosphere; boundaries between plates are where most earthquakes and volcanoes occur and where lithosphere is created and recycled.

Planetary nebula   An expanding shell of gas ejected from a low-mass star during the latter stages of its evolution.

Pluton   A very large intrusive body formed below Earth's surface.

Plutonic rock   Intrusive igneous rock formed from magma that cools beneath Earth's surface. Granite is a plutonic rock.

Point bar   A sandy, gentle bank on the inside bend of a river's channel. An area of deposition.

Polar   Said of a chemical bond that has a dipole.

Polarization   The alignment of the transverse electric vectors that make up electromagnetic radiation. Such waves of aligned vibrations are said to be *polarized*.

Polyatomic ion   An ionically charged molecule.

Polymer   A long, organic molecule made of many repeating units.

Polymorphs   Two or more minerals that contain the same elements in the same proportions but have different crystal structures.

Porosity   The volume of open space in rock or sediment compared with the total volume of solids plus open space.

Postulates of the special theory of relativity   (1) All laws of nature are the same in all uniformly moving frames of reference. (2) The speed of light in free space has the same measured value regardless of the motion of the source or the motion of the observer; that is, the speed of light is a constant.

Potential difference   The difference in potential between two points, measured in volts, and often called *voltage difference.*

Potential energy   The stored energy that a body possesses because of its position.

Power   The time rate of work.

Precambrian time   The time of hidden life, which began about 4.5 billion years ago when Earth formed, lasted until about 543 million years ago (beginning of the Paleozoic), and makes up almost 90% of Earth's history.

Precipitate   A solute that has come out of solution.

Pressure   The ratio of force to the area over which that force is distributed:

$$\text{Pressure} = \frac{\text{force}}{\text{area}}$$

Liquid pressure = weight density × depth.

Pressure-gradient force   The force that moves air from a region of high-pressure to an adjacent region of low-pressure air.

Primary wave (P-wave)   A longitudinal body wave that compresses and expands the material through which it moves; it travels through solids, liquids, and gases and is the fastest seismic wave.

Principle of equivalence   Local observations made in an accelerated fram of reference cannot be distinguished from observations made in a Newtonian gravitational field.

Principal quantum number, (n)   An integer that specifies the quantized energy level of an atomic orbital.

Principle of flotation   A floating object displaces a weight of fluid equal to its own weight.

Probability cloud   The pattern of electron positions plotted over time to show the likelihood of an electron's being at a given position at a given time.

Products   The new materials formed in a chemical reaction.

Projectile   Any object that moves through the air or through space under the influence of gravity.

Proton   A positively charged particle in an atomic nucleus.

Protostar   The aggregation of matter that goes into and precedes the formation of a star.

Pseudoscience   A theory or practice that mimics science but is considered to be without scientific foundation.

Pulsar   Likely a neutron star that rapidly spins, sending short precisely timed bursts of electromagnetic radiation.

Pure   Having a uniform of homogeneous composition, or being without impurities. In chemistry, the term is used to denote a material that consists of a single element or compound.

Quality   The characteristic timbre of a musical sound, which is governed by the number and relative intensities of partial tones.

Quantum   The smallest elemental unit of a quantity. For example, one quantum of electromagnetic energy is called a photon.

Quantum hypothesis   The idea that the smallest elemental units of energy, such as light, are contained in discrete packets called quanta.

Quasar   A distant galaxy with an active galactic nucleus emitting a beam of radiation pointed in our direction, which makes the quasar appear much brighter than the galaxy in which it resides.

Radiation   The transfer of energy by means of electromagnetic waves.

Radioactivity   The process whereby unstable atomic nuclei break down and emit radiation.

Radiometric dating   A method for calculating the age of geologic materials based on the nuclear decay of naturally occurring radioactive isotopes.

Rarefaction   Rarefied region, or region of lessened pressure, of the medium through which a longitudinal wave travels.

Reactants   The reacting substances in a chemical reaction.

Reaction rate   A measure of how quickly the concentration of products in a chemical reaction increases or the concentration of reactants decreases.

Real image   An image formed by light rays that converge at the location of the image. A real image can be displayed on a screen.

Recrystallization   A process that occurs when rocks are subjected to high temperatures and pressures and go through a change in minerals; often accompanied by the loss of H_2O or CO_2.

Red giant   Cool giant stars above main sequence stars on the H-R diagram.

Reduction   The process whereby a reactant gains one or more electrons.

Reflection   The return of light rays from a surface in such a way that the angle at which a given ray is returned is equal to the angle at which it strikes the surface. When the reflecting surface is irregular, light is returned in irregular directions; this is *diffuse reflection*.

Refraction   The bending of an oblique ray of light when it passes from one transparent medium to another. This is caused by a difference in the speed of light in the transparent media. When the change in medium is abrupt (say, from air to water), the bending is abrupt; when the change in medium is gradual (say, from cool air to warm air), the bending is gradual, which accounts for mirages.

Refraction   The bending of a wave, either through a nonuniform medium or from one medium to another, caused by differences in wave speed.

Relationship of impulse and momentum   Impulse is equal to the change in the momentum of the object upon which the impulse acts. In symbol notation, $Ft = \Delta mv$.

Relative dating   The ordering of rocks in sequence by their comparative ages.

Relativity   The study of the relationships among space, energy, mass, and time.

Relative humidity   The amount of water vapor in the air at a given temperature expressed as a percentage of the maximum amount of water vapor the air can accommodate at that temperature.

Resonance   The response of a body when a forcing frequency matches its natural frequency.

Resultant   The net result of a combination of two or more vectors.

Reverberation   Re-echoed sound.

Reverse osmosis   A technique for purifying water by forcing it through a semipermeable membrane.

Rift (rift valley)   A long, narrow gap that forms as a result of two plates diverging.

Rock   An aggregate of minerals. Some rocks are aggregates of fossil shell fragments, solid organic matter, or any combination of these components.

Rock cycle   A sequence of events involving the formation, destruction, alteration, and reformation of rocks as a result of the generation and movement of magma; the weathering, erosion, transportation, and deposition of sediment; and the metamorphism of preexisting rocks.

Salinity   The mass of salts dissolved in 1000 g of seawater.

Salt   An ionic compound formed from the reaction between an acid and a base.

Sand dune   Landform created when air flow is blocked by an obstacle, slowing the air speed and therefore promoting the deposition of airborne sand.

Satellite   A projectile or small celestial body that orbits a larger celestial body.

Saturated hydrocarbon   A hydrocarbon containing no multiple covalent bonds, with each carbon atom bonded to four other atoms.

Saturated solution   A solution containing the maximum amount of solute that will dissolve in its solvent.

Saturation vapor pressure   The maximum amount of moisture the air can accommodate at a given temperature; the upper limit for humidity.

Science   The collective findings of humans about nature, and a process of gathering and organizing knowledge about nature.

Scientific method   An orderly method for gaining, organizing, and applying new knowledge.

Seafloor spreading   The moving apart of two oceanic plates at a rift in the seafloor.

Second law of thermodynamics   Heat never spontaneously flows from a cold substance to a hot substance. Also, all systems tend to become more and more disordered as time goes by.

Secondary wave (S-wave)   A transverse body wave that vibrates side to side or up and down through the material through which it moves; it cannot travel through liquids and so does not travel through Earth's outer core.

Sedimentary rocks   Rocks formed from the accumulation of weathered material (sediments) that have been eroded by water, wind, or ice.

Sedimentation   The stage of sedimentary rock formation in which deposited sediments accumulate and change (lithify) into sedimentary rock through the processes of compaction and, usually, cementation.

Semipermeable membrane   A membrane that allows only the passage of molecules small enough to fit through its submicroscopic pores.

Series circuit   An electric circuit with devices connected in such a way that the same electric current flows through each of them.

Shell   A set of overlapping atomic orbitals of similar energy levels; in other words, a region of space in which electrons of similar energy levels in an atom have a 90% chance of being located.

Shock wave   The cone-shaped wave made by an object moving at supersonic speed through a fluid.

Silicate   A mineral that contains both silicon and oxygen and (usually) other elements in its chemical composition; silicates are the largest and most common rock-forming mineral group.

Sine curve   A wave form traced by simple harmonic motion, which can be made visible on a moving conveyor belt by a pendulum swinging at right angles above the moving belt.

Solar eclipse   The phenomenon whereby the shadow of the Moon falls upon Earth producing a region of darkness in the daytime.

Solubility   A measure of the ease with which a mineral can be dissolved. Low-solubility minerals are difficult to dissolve. High-solubility minerals are easier to dissolve.

Soluble   Capable of dissolving to an appreciable extent in a given solvent.

Solute   Any component in a solution that is not the solvent.

Solution   A homogeneous mixture in which all components are in the same phase.

Solvent   The component in a solution that is present in the largest amount.

Sonic boom   The loud sound resulting from a shock wave.

Space-time   The four-dimensional continuum in which all events take place and all things exist: Three dimensions are the coordinates of space and the fourth is of time.

Special theory of relativity   The first of Einstein's theories of relativity, which discusses the effects of uniform motion on space, time, energy, and mass.

Specific gravity   The ratio between the weight of a substance and the weight of an equal volume of water.

Specific heat capacity   The quantity of heat per unit mass required to raise the temperature of a substance by 1°C.

Spectroscope   A device that uses a prism or a diffraction grating to separate light into its component colors.

Speed   The distance traveled per time.

Spiral galaxy   A disk-shaped galaxy with hot bright stars, and spiral arms. Our Milky Way is a spiral galaxy.

Spring tide   A high or low tide that occurs when the Sun, Earth, and Moon are aligned so that the tides due to the Sun and Moon coincide, making the high tides higher or lower than average. Occurs during the full Moon or new Moon.

Standing wave   A stationary wave pattern formed in a medium when two sets of identical waves pass through the medium in opposite directions.

Starburst galaxy   A galaxy in which stars are forming at an unusually high rate.

Stratosphere   The second atmospheric layer above Earth's surface, extending from the top of the troposphere up to 50 km. This is where stratospheric ozone forms.

Structural isomers   Molecules that have the same molecular formula but different chemical structures.

Subduction   The process in which one tectonic plate bends and descends beneath another plate at a convergent boundary.

Sublimation   The change of phase directly from solid to gas, bypassing the liquid phase.

Submicroscopic   Refers to the realm of atoms and molecules, which is a realm so small that we are unable to observe it directly with optical microscopes.

Subtractive primary colors The three colors of absorbing pigments—magenta, yellow, and cyan—that, when mixed in certain proportions, will reflect any color in the spectrum.

Sunspots Temporary, relatively cool and dark regions on the Sun's surface.

Superconductor Any material with zero electrical resistance, wherein electrons flow without losing energy and without generating heat.

Supernova An exploding massive star caused by gravitational collapse with the emission of enormous quantities of matter.

Superposition A relative dating principle stating that, in an undeformed sequence of sedimentary rocks, each bed or layer is older than the one above and younger than the one below.

Support force The force that supports an object against gravity, often called the *normal force.*

Surface wave A type of seismic wave that travels along Earth's surface.

Suspension A homogeneous mixture in which the various components are in different phases.

Syncline A fold in rock with relatively young rocks at its core; rock age increases with horizontal distance from the fold core.

Technology The means of solving practical problems by applying the findings of science.

Tectonic plates Sections into which Earth's crust is broken up; they move in response to heat flow and convection in Earth's interior.

Temperature A measure of the hotness or coldness of substances, related to the average kinetic energy per molecule in a substance; measured in degrees Celsius, or in degrees Fahrenheit, or in kelvins.

Temperature inversion A condition in which the upper regions of the troposphere are warmer than the lower regions.

Terminal speed The speed at which the acceleration of a falling object terminates when air resistance balances its weight.

Terrestrial radiation The radiant energy emitted by Earth.

Theory A synthesis of a large body of information that encompasses well-tested hypotheses about certain aspects of the natural world.

Thermal energy (*internal energy*) The total energy (kinetic plus potential) of the submicroscopic particles that make up a substance.

Thermodynamics The study of heat and its transformation to different forms of energy.

Thermonuclear fusion Nuclear fusion produced by high temperature.

Thermonuclear reaction The fusion reaction brought about by high temperatures.

Thermosphere The fourth atmospheric layer above Earth's surface, extending from the top of the mesosphere to 500 km.

Third law of thermodynamics No system can reach absolute zero.

Time dilation The slowing of time as a result of speed.

Total internal reflection The total reflection of light traveling within a medium that strikes the boundary of another medium at an angle at, or greater than, the critical angle.

Transform plate boundary A plate boundary where two plates are sliding horizontally past each other, without appreciable vertical movement.

Transformer A device for transferring electric power from one coil of wire to another by means of electromagnetic induction.

Transmutation The conversion of an atomic nucleus of one element into an atomic nucleus of another element through a loss or gain in the number of protons.

Transparent The term applied to materials through which light can pass in straight lines.

Transverse wave A wave in which the medium vibrates in a direction perpendicular (transverse) to the direction in which the wave travels. Light consists of transverse waves.

Troposphere The atmospheric layer closest to Earth's surface, 16 km high over the equator and 8 km high over the poles, containing 90% of the atmosphere's mass and essentially all of its water vapor and clouds.

Turbulent flow Water flowing rapidly and erratically in a jumbled manner, stirring up everything it touches.

Ultrasonic Describes a sound of a frequency too high to be heard by the normal human ear—above 20,000 hertz.

Unconformity A break or gap in the geologic record, caused by erosion of preexisting rock or by an interruption in the sequence of deposition.

Unsaturated hydrocarbon A hydrocarbon containing at least one multiple covalent bond.

Unsaturated solution A solution that is capable of dissolving additional solute.

Valence electron An electron that is located in the outermost occupied shell of an atom and can participate in chemical bonding.

Valence shell The outermost occupied shell of an atom.

Vector component   Parts into which a vector can be separated and that act in different directions from the vector.

Vector quantity   A quantity that specifies direction as well as magnitude.

Velocity   The speed of an object with specification of its direction of motion.

Velocity vector   An arrow drawn to scale so that its length represents the magnitude of a velocity and its direction represents the direction of motion.

Virtual image   An image formed by light rays that do not converge at the location of the image. Mirrors, converging lenses used as magnifying glasses, and diverging lenses all produce virtual images.

Volcanic rocks   Extrusive igneous rocks formed by the eruption of molten rock at Earth's surface. Basalt is a volcanic rock.

Volcano   A central vent through which lava, gases, and ash erupt and flow.

Water table   The upper boundary of the saturated zone, below which every pore space is filled with water.

Wave speed   The speed with which waves pass a particular point:

Wave speed = frequency × wavelength

Wavelength   The distance between successive crests, troughs, or identical parts of a wave.

Weathering   Disintegration and/or decomposition of rock at or near Earth's surface.

Weight   Simply stated, the force due to gravity on an object. More specifically, the gravitational force with which a body presses against a supporting surface.

Weightlessness   A condition encountered in freefall wherein a support force is lacking.

White dwarf   Dying star that has collapsed to the size of Earth and is slowly cooling off; located at the lower left of the H-R diagram.

Work   The product of the force and the distance through which the force moves:

$$W = Fd$$

Work-Energy Theorum   The work done on an object equals the change in kinetic energy of the object:

$$\text{Work} = \Delta KE$$

INDEX

Ablation, 619
Absolute zero, 143
Absorption of radiant energy, 164–165
Acceleration, 27–31, 41–44
Accumulation, 619
Acid rain, 448–452
Acidic solution, 445–448
    acidity, 447–448
    pH scale, 447–448
Acidity, 447–448
Acids, 438–442
    strengths, 442–445
Acoustical imaging, dolphins and, 238
Action
    identification, 46–48
    on masses, 48–50
Active galactic nucleus, 739
Active galaxies, 738–740
Addition polymers, 490–494
Additive primary colors, 268
Adiabatic processes, atmospheric, 667–669
Advances in science, history of, 2
AGN. See Active galactic nucleus
Air, in hydrologic cycle, 624–625
Air masses, 674–679
    atmospheric lifting, 675
    convectional lifting, 675
    frontal lifting, 676–678
    midlatitude cyclones, 678–679
    orographic lifting, 675
Air motion, 648–651
    forces, 648–651
    large-scale air movement, 650
    temperature-pressure relation-ship, 648–650
Air resistance, 24
Alcohols, 480–484
Aldehyde, 486
Alkaloids, 484–485
Alloy, 362
Alpha particle, 312–314
Alternating current, 193
Amide, 487
Amines, 484–485
Ampere, 192
Amplitude, 232
Annual radiation exposure, 315
Anode, 456
Anticlines, 585
Aquifers, 604–605
Archimedes' principle, 120–122

Aristotle, 16
Aromatic compound, 477
Art, science and, 7–8
Artesian system, 605
Artificial atomic transmutation, 321
Asteroid belt, 709–710
Asthenosphere, 569
Astrology, 726
Astronomical unit, 690
Astronomy, 9–10, 687–770
Atmosphere, 630–632, 642–644
    air motion, 648–651
        large-scale air movement, 650
        temperature-pressure rela-tionship, 648–650
    composition of, 643
    evolution of, 630–632
    global circulation patterns, 651–659
        deep-water currents, 657–659
        oceanic circulation, 653–654
        surface currents, 654–657
        upper atmospheric circula-tion, 653
    vertical structure of, 643–644
Atmospheric lifting, 675
Atmospheric moisture, 664–668
    adiabatic processes, 667–669
    atmospheric stability, 669–670
    humidity, 664
    relative humidity, 664
    saturation vapor pressure, 664–665
    temperature changes, 666
    temperature inversion, 669
Atmospheric pressure, 125–128
Atomic half-life, transmutation, 318–321
    artificial transmutation, 321
    natural transmutation, 319–320
Atomic mass, 290–291
Atomic nucleus, 316–318
    strong nuclear force, 316
Atomic number, 288
Atomic symbol, 288
Atoms, 283–502
    atomic mass, 290–291
    atomic number, 288
    atomic symbol, 288
    bonding, 353–380
        covalent bonds, 362–365

dipoles, 372–373
electron-dot structures, 354–355
induced dipoles, 373–376
    molecules, 358
ionic bonds, 359–361
ions, 372–373
    formation of, 356–358
metallic bonds, 361–362
molecular attractions, 371–376
molecular polarity, 368–371
polar covalent bonds, 365–368
conceptual model, 296–298
electron waves, 302–304
electrons, 286
elements, 287–288
isotopes, 290–291
molecules, counting by mass, 414–418
    Avogadro's number, 416
    formula mass, 415
    molar mass, 417
neutrons, 288–291
nucleons, 289
periodic table, 288, 292
    groups, 293–296
    periods, 293–296
physical model, 296–298
protons, 288–291
quantum hypothesis, 300–302
    principal quantum number n, 301
    quantum, 300
    quantum hypothesis, 300
shell model, 304–305
spectroscope identification, 298–300
    atomic spectrum, 299
subatomic particles, 289
valence electrons, 305
Attitude, scientific, 3–5
    fact, 3
    law, 3
    theory, 4
AU. See Astronomical unit
Average speed, 26
Avogadro's number, 416

Barometers, 127–128
Bases, 438–445
    strengths, 442–445
Basic solutions, 445–452
    acidity, 447–448

Batteries, 455–457
    anode, 456
    cathode, 456
    electrode, 455
Beats, 242–243
Bernoulli's principle, 131–133
    applications of, 132–134
Beta particle, 312–314
Big Bang, 736, 750–754
Big Rip, 765
Black holes, 732–735
    geometry, 733–735
Black-holes, singularity, 734
Body, effect of electric currents on, 195
Body waves, 566
Boiling, 171–173
Boiling points, polar, 371
Bonding, atoms, 353–380
    covalent bonds, 362–365
    dipoles, 372–373
    electron-dot structures, 354–355
    induced dipoles, 373–376
        molecules, 358
    ionic bonds, 359–361
    ions, 372–373
        formation of, 356–358
    metallic bonds, 361–362
    molecular attractions, 371–376
    molecular polarity, 368–371
    polar covalent bonds, 365–368
Bottled water, 402–403
Bow waves, 245–247
Boyle's law, 124–125
Breeder reactor, 326–327
Buoyancy in gas, 130–131
Buoyancy in liquid, 119–120
Buoyant force, 119

Cambrian period, 549–550
Carbon-14 dating, 322
Carbonate dissolution, 608–609
Carboniferous period, 551
Carbonyl compounds, 485–489
Carbonyl group, 485
Carboxylic acid, 487
Catalysts, 423–425
Cathode, 456
Caverns, 608–609
Caves, 608–609
Cenozoic era, 556–559

Cenozoic life, 558–559
Chain reaction, 324
Change of phase, 168–171, 174–175
    energy, 174–175
    heat transfer, 168–171
Chemical bond, 342
Chemical changes, 342–345
Chemical classification, matter, 383–386
    heterogeneous mixtures, 385
    homogeneous mixtures, 385
    impure, 384
    pure, 384
    solution, 385
    suspension, 386
Chemical equations, 412–414
    law of mass conservation, 412
    products, 412
    reactants, 412
Chemical formulas, 346
Chemical properties, 341
Chemical reactions, 343, 411–435, 437–469
    acid-base reaction, 441–442
    acid rain, 448–452
    acidic solution, 445–448
        acidity, 447–448
        pH scale, 447–448
    acids, 438–442
        strengths, 442–445
    atoms, molecules, counting by mass, 414–418
    bases, 438–442
        strengths, 442–445
    basic solutions, 445–452
        acidity, 447–448
    catalysts, 423–425
    chemical equations, 412–414
        law of mass conservation, 412
        products, 412
        reactants, 412
    combustion, 461–463
    corrosion, 461–463
    electrolysis, 459–461
    electrons
        energy, 454–459
            batteries, 455–457
                anode, 456
                cathode, 456
                electrode, 455
                electrochemistry, 454
                fuel cells, 457–459
        gaining, 452–453
            half reaction, 452
            oxidation, 452
            reduction, 452
        losing, 452–453
            half reaction, 452
            oxidation, 452
            reduction, 452
    energy, 425–429
        bond energies, 426
        endothermic reaction, 428–429
        exothermic reaction, 426–428

entropy, 429–430
    hydronium ion, 439
    hydroxide ion, 440
    neutral solution, 445–448
    neutralization, 441
    protons, 438–442
    reaction rates, 418–423
        activation energy, 422
    salt, 441–442
Chemical sediments, 516
Chemical sedimentary rock, 527–528
Chemistry, 9–10, 283–502
    applied research, 339
    basic research, 338
    compounds, 346
    compounds, naming, 347–348
    elemental formula, 345
    elements, 345–347
    submicroscopic world, 340–341
        molecules, 340
        submicroscopic, 340
Circular orbits, 103–104
Circulation patterns
    atmosphere, 651–659
        deep-water currents, 657–659
        oceanic circulation, 653–654
        surface currents, 654–657
        upper atmospheric circulation, 653
    global, 651–659
Clouds, 271
    cloud groups, 671
    development, 670–674
    high clouds, 671–672
    lifting condensation level, 670
    low clouds, 672–673
    middle clouds, 672
    precipitation formation, 673–674
    vertical development, 673
Clusters, 740–744
Collisions, 66–67
    elastic, 66
    inelastic, 66
Color, 265–271, 512
    of stars, 722–724
Colored lights, mixing, 267–268
Colored pigments, mixing, 269
Colors
    complementary, 268–269
    primary, 268
Combustion, 461–463
Comets, 710–713
Complementary colors, 268–269
Components of, 632–635
Compounds, 345–347
    naming, 347–348
Compression, 235
Concentration, 388
Conceptual model, atom, 296–298
Condensation, 170–171, 666

Condensation polymers, 490–491, 494–497
Conduction, 160
Conductor, 190
Configuration, 473
Conformations, 474
Conservation laws, 67
Conservation of charge, 183–184
Conservation of energy, 74–75
    machines, 76
Conservation of momentum, 65–68
    law of, 65
Continental drift, 571–576
    paleomagnetism, 573
    seafloor spreading, 575
Continental margin, 633
Control of nuclear fusion, 331–332
Convection, 161–162, 569–570
Convectional lifting, 675
Convergent plate boundaries, 579–582
Core,
    Earth, 568–569
Coriolis force, 651
Corrosion, 461–463
Cosmic background radiation, 753–754
Cosmic inflation, 755–757
Cosmological redshift, 751
Cosmology, 749
Coulomb, 185
Coulomb's law, 184–186
Covalent bonds, 362–365
    polar, 365–368
Covalent compound, 363
Cretaceous extinction, 555–556
Critical mass, 324
Cross-cutting relationships, rock dating, 541
Crystallization, minerals, 514
    in magma, 515–516
    in water-solutions, 516–517
Current electricity, 181–208

Dark energy, 763–764
Dark matter, 761–763
Dating of rock, 540–544
    cross-cutting relationships, 541
    faunal succession, 542
    inclusions, 541
    lateral continuity, 541
    original horizontality, 541
    superposition, 541
Deep-water currents, 657–659
Deltas, 617–618
Density, 116–117
Depositional environments, 614
Depositional landforms, 622–624
Desalination, 399–402
Detergents, 394–396
Devonian period, 551
Diffuse reflection, 262
Dipoles, 366, 372–373
    induced, 373–376

Direct current, 193
Dispersion, 272–274
Dissolving, 388
Distance, gravity, inverse-square law, 90–91
Distillation, 383
Divergent plate boundaries, 577–579
Dolphins, 238
Doppler effect, 244–245
Drainage basins, 612–613
    networks, 612–613
Drainage systems, 609–613
Drift, 622
Dwarf planets, 710–711
Dynamic equilibrium, 23–24

Earth, 697
    asthenosphere, 569
    core, 568–569
    crustal surface, 570–571
    folds, 585
    internal layers, 567–571
    internal mantle, 569–570
    isostasy, 570
    lithosphere, 569
    mantle, 569–570
    Mohorovičić discontinuity, 567
    movement of, 39
    plate tectonics, 567–571
Earth science, 9–10, 123, 503–686
Earthquakes, 566, 587–592
    measurement, 588
    Mercalli scale, 588
    Richter scale, 588
    Tsunami, 590
Eclipses, 707–709
Ecliptic, 690
Efficiency, 77–78
    sources of energy, 78–79
Elastic collision, 66
Electric charge, 182–184
Electric circuits, 197–200
Electric current, 191–193, 213–215
Electric field, 186–188
Electric meters, 216–217
Electric motors, 217–218
Electric potential, 189–190
Electric power, 200–202
Electric resistance, 193–194
Electric shock, 195–197
Electrically polarized, 186
Electricity, 181–208
    alternating current, 193
    ampere, 192
    body, effect of electric currents on, 195
    conductor, 190
    conservation of charge, 183–184
    coulomb, 185
    Coulomb's law, 184–186
    direct current, 193
    electric charge, 182–184
    electric circuits, 197–200

electric current, 191–193, 213–215
electric field, 186–188
electric potential, 189–190
electric power, 200–202
electric resistance, 193–194
electric shock, 195–197
electrically polarized, 186
electronics technology, 184
ionized bracelets, 185
lightening, 680
magnetic fields, 213–215
microwave oven, 187
Ohm's law, 194–197
overloading, 199–200
parallel circuits, 198–200
potential difference, 190
safety fuses, 200
series circuits, 197–198
sparks, 184
superconductors, 194
voltage sources, 190–191
volts, 193
Electrochemistry, 454
Electrode, 455
Electrolysis, 459–461
Electromagnet induction, 209–230
Electromagnetic induction, 218–220
Electromagnetic spectrum, 256–257
Electromagnetic wave, 256
Electromagnets, 214–215
Electron-dot structures, 354–355
Electron waves, 302–304
Electronegativity, 366
Electronics technology, 184
Electrons, 286
  energy of, 454–459
    batteries, 455–457
      anode, 456
      cathode, 456
      electrode, 455
    electrochemistry, 454
    fuel cells, 457–459
  gaining, 452–453
  half reaction, 452
  losing, 452–453
  oxidation, 452
  reduction, 452
Elemental formula, 345
Elements, 287–288, 345–347
Ellipse, 104
Elliptical galaxies, 737–738
Elliptical orbits, 104–105
Emission of radiant energy, 163–164
Endothermic reaction, 428–429
Energy, 59–86, 174–175, 425–429
  bond energies, 426
  change of phase, 174–175
  conservation of energy, 74–75
  endothermic reaction, 428–429
  exothermic reaction, 426–428

kinetic energy, 70
  momentum, compared, 73–74
law of conservation of energy, 74
potential energy, 69–70
radiant, reflection of, 165–166
sources of, 78–79
work-energy theorem, 71–74
Entropy, 146–147, 429–430
Equilibrium, dynamic, 23–24
Equilibrium rule, 22
Erosion, 525, 613–614
Erosional environments, 614
Erosional landforms, 621–622
Escape speed, 106–108
Esters, 489
Ethers, 480–484
Evaporation, 169
Exosphere, 644
Exothermic reaction, 426–428
Expansion, thermal, 150–151
Expansion of water, 151–153
Eye, 265

Fact, 3
Failed planet formation, 709–713
Faraday's law, 219–200
  Maxwell's counterpart to, 223
Fate of universe, 764–766
Faults, 585–587
Faunal succession, rock dating, 542
Field induction, 223
First law of thermodynamics, 145
Floodplains, 614–617
Floor of ocean, 632–634
Flotation, 122
Fluid mechanics, 115–140
  Archimedes' principle, 120–122
  atmospheric pressure, 125–128
  barometers, 127–128
  Bernoulli's principle, 131–133
    applications of, 132–134
  Boyle's law, 124–125
  buoyancy in gas, 130–131
  buoyancy in liquid, 119–120
  density, 116–117
  flotation, 122
  Pascal's principle, 129–130
  pressure, 117–119
    in gas, 124–125
    in liquid, 118–119
Foliated metamorphic rock, 531–532
Force, 17
Force of friction, 24–25
  air resistance, 24
  friction, 24
Force pair, 45
Forced vibrations, 239–240
Forces, 45–46
Formation of rock, 514–517
Fossils, 528
Fourier analysis, 249
Free fall, 29, 41
  acceleration, 41–44

Free-fall velocity, 29
Freezing, 173–174
Frequency, 232
Friction, 24
Front, 676–678
Frontal lifting, 676–678
Fuel cells, 457–459
Full Moon, 704
Functional groups, 478–480
Fundamental frequency, 248
Fusion, heat of, 174

G-free fall acceleration, 41–43
G-non-free fall acceleration, 43–44
Galaxies, 719–748, 763
  active, 738–740
  active galactic nucleus, 739
  elliptical, 737–738
  formation, 763
  irregular, 737–738
  spiral, 737–738
  starburst, 738
Galilei, Galileo. *See* Galileo
Galileo, 17
  concept of inertia, 17–18
  force, 17
  inertia, 17–18
Gamma rays, 312–314
Gaps in rock record, 543
  angular unconformity, 543
  unconformities, 543
Gas
  buoyancy in, 130–131
  pressure in, 124–125
Gases, 393–394
General theory of relativity, 757–761
Generators, 220–221
Geologic time, 546
Giant stars, 725
Glacial erosion, 621–622
Glacial sedimentation, 622–624
Glaciation, 618–620
Glacier mass balance, 619–620
Glaciers, 618–624
  formation, 618–619
  movement, 618–619
Global circulation patterns, 651–659
  deep-water currents, 657–659
  oceanic circulation, 653–654
  surface currents, 654–657
  upper atmospheric circulation, 653
Global warming, 167–168, 646–648
Grams, moles, converting between, 416–418
Gravitation, universal, 93–94
Gravity, 87–114
  distance, inverse-square law, 90–91
  law of universal gravitation, 88
  universal gravitational constant, 89–90
  universal law of, 88–90

Greenhouse effect, 167–168, 646–648
Groundwater, 601–607
  aquifers, 604–605
  movement, 606–607
  springs, 604–605
  water table, 603–604
Gyre, 655

Half reaction, electrons, 452
Hard water, softening, 396–397
Harmonics, 248
Heat, 144
  death, 765
  of fusion, 174
  quantity of, 145
  specific capacity, 147–150
    of water, 148
  thermal energy, 144
  of vaporization, 174
Heat capacity, specific, 147–150
  of water, 148
Heat death, 765
Heat of fusion, 174
Heat of vaporization, 174
Heat transfer, 159–179
  absorption of radiant energy, 164–165
  boiling, 171–173
  change of phase, 168–171, 174–175
  condensation, 170–171
  conduction, 160
  convection, 161–162
  emission of radiant energy, 163–164
  energy, 174–175
  evaporation, 169
  freezing, 173–174
  global warming, 167–168
  greenhouse effect, 167–168
  heat of fusion, 174
  heat of vaporization, 174
  melting, 173–174
  Newton's law of cooling, 166–167
  radiation, 163–166
  reflection of radiant energy, 165–166
  sublimation, 169
  terrestrial radiation, 164
Helium, 754
Hertz, 232
Hertzsprung-Russell diagram, 724–726
Heteroatom, 478
Heterogeneous mixture, 385
High clouds, 671–672
Homogeneous mixtures, 385
Humidity, 664
Hurricanes, 681–682
Hydraulic conductivity, 602–603
Hydrocarbon, unsaturated, 477
Hydrocarbons, 472–476
  saturated, 477
  unsaturated, 476–478
Hydrogen, 754
Hydrogen bond, 372

Hydrologic cycle, 600–601
Hydronium ion, 439
Hydroxide ion, 440
Hypothesis, 3

Igneous rock, 517–523
    lava, 521
    plutonic rock, 523
    volcanic, 521
Imaging
    acoustical, dolphins and, 238
    magnetic resonance, 218
Impulse, 60–62
Impure matter, chemical classifi-
        cation, 384
Inclusions, rock dating, 541
Induced dipoles, 373–376
    molecules, 358
Induction
    electromagnetic, 218–220
    field, 223
Inelastic collision, 66
Inertia, 17–18, 38
    mass-measure, 18–20
        mass, 18
        volume, 19
        weight, 18
Inner planets, 690, 695–699
    Earth, 697
    Mars, 698–699
    Mercury, 695–696
    Venus, 696–697
Insolubility, 392
Instantaneous speed, 25–26
Interaction, 45
Interactions, 45–46
Inverse-square law, 90–91
Ionic bonds, 359–361
Ionic compounds, 359
Ionized bracelets, 185
Ionosphere, 644
Ions, 358, 372–373
    formation of, 356–358
    polyatomic, 358
Irregular galaxies, 737–738
Isomers, structural, 473
Isostasy, 570
Isotopes, 290–291
    radiometric dating, 545

Jupiter, 699–700

Karst regions, 609
Ketone, 486
Kilogram, 20
Kinetic energy, 70
    momentum, compared, 73–74
Kuiper belt, 710–711

Land subsidence, 607–608
Laminar flow, 610
Large-scale air movement, 650
Lateral continuity, rock dating,
        541
Lateral inhibition, 260
Lava, 521
Law of acceleration, 52
Law of action-reaction, 52

Law of conservation of energy, 74
Law of conservation of momen-
        tum, 65
Law of cooling, 166–167
Law of gravity, 88–90
Law of inertia, 52
Law of mass conservation, 412
Law of reflection, 260–262
Law of thermodynamics,
        145–146
Law of universal gravitation, 88
Laws of thermodynamics,
        145–146
    first law of, 145
    second law of, 146
    third law of, 146
Layers, Earth, 600–601
    core, 568–569
    crustal surface, 570–571
    internal, 567–571
    mantle, 569–570
Lever, 76
Life cycles of stars, 727–731
Lifting condensation level, 670
Light, 255–282
    color, 265–271
    dispersion, 272–274
    electromagnetic spectrum,
        256–257
    opaque materials, 257–260
    polarization, 269–271,
        274–277
        colored lights, mixing,
            267–268
        colored pigments, mixing,
            269
        complementary colors,
            268–269
        diffuse reflection, 262
        law of reflection, 260–262
        rainbows, 272–274
        selective reflection, 266
        selective transmission,
            266–267
    reflection, 260–262
    refraction, 262–265
    transparent, 257–260
Light-year, 722
Limitations of science, 6
Liquid
    buoyancy in, 119–120
    pressure in, 118–119
Lithosphere, 569
Longitudinal waves, 234–235
Loudspeakers, 236
Low clouds, 672–673
Lunar eclipse, 708

Machine, 76
Machines, 76–77
    conservation of energy, 76
Magma, 514, 520–521
    generation of, 518–520
Magnetic domains, 212–213
Magnetic fields, 211–215
Magnetic force, 210
    on current-carrying wires, 216
    on moving charges, 215–218

Magnetic poles, 210–211
Magnetic resonance imaging, 218
Magnetic therapy, 224
Magnetism, 209–230
    alternating current, 220–221
    electric currents, 213–215
    electric meters, 216–217
    electric motors, 217–218
    electromagnet, 214
    electromagnetic induction,
        218–220
    electromagnets, 214–215
    Faraday's law, 219–200
        Maxwell's counterpart to,
            223
    field induction, 223
    generators, 220–221
    magnetic domains, 212–213
    magnetic fields, 211–215
    magnetic force, 210
        on current-carrying wires,
            216
        on moving charges,
            215–218
    magnetic poles, 210–211
    magnetic resonance imaging,
        218
    magnetic therapy, 202
    power production, 221–222
    superconducting electromag-
        nets, 215
    transformer, 222
    transformer-boosting,
        222–223
    voltage, 222–223
Mantle, Earth, 569–570
Mars, 698–699
Mass, 18
Mass-energy equivalence,
        327–329
Masses, reaction on, 48–50
Mathematics, 2–3
Melting, 173–174
Mercury, 695–696
Mesosphere, 644
Mesozoic era, 554–556
Metallic bonds, 361–362
Metamorphic rock, 518,
        528–532
    classifying, 531–532
    foliated, 531–532
    metamorphism, 528
    types of, 530–531
    nonfoliated, 532
    recrystallization, 528
Meteor shower data, 713
Meteorite, 710
Meteoroids, 710
Meteors, 709–710
Methods, scientific, 3
Microwaves, 187
Middle clouds, 672
Midlatitude cyclones, 678–679
Minerals, 503–538
    chemical sediments, 516
    classification, 513–514
    cleavage, 511–512
    color, 512

crystal form, 509–510
crystallization, 514
    in magma, 515–516
    in water-solutions, 516–517
density, 512–513
formation, 514–517
fracture, 511–512
hardness, 510–511
    Mohs scale, 511
magma, 514
nonsilicates, 513
polymorphs, 510
properties, 509–513
rock-forming, 513–514
silicates, 513
solubility, 517
Mixing colored lights, 267–268
Mixing colored pigments, 269
Mixtures, 381–410
    chemical classification, matter,
        383–386
        heterogeneous mixture, 385
        homogeneous mixtures, 385
        impure, 384
        pure, 384
        solution, 385
        suspension, 386
    detergents, 394–396
    hard water, softening,
        396–397
    physical separation, 383–384
    soaps, 394–396
    solubility, 391–394
        gases, 393–394
        insoluble, 392
        soluble, 391
        temperatures changes,
            392–394
    solutions, 386–390
        concentration, 388
        dissolving, 388
        molarity, 389
        mole, 389
        saturated solution, 388
        solutes, 387
        solvent, 387
        unsaturated solution, 388
    wastewater treatment,
        403–405
    water purification, 397–403
        bottled water, 402–403
        desalination, 399–402
        osmosis, 401
        reverse osmosis, 401
        semipermeable membrane,
            400
Mohorovičić discontinuity, 567
Mohs scale, mineral hardness,
        511
Moisture, atmospheric, 664–668
    adiabatic processes, 667–669
    atmospheric stability, 669–670
    humidity, 664
    relative humidity, 664
    saturation vapor pressure, 664
        vs. temperature, 665
    temperature changes, 666
    temperature inversion, 669

Molar mass, 417
Molarity, 389
Mole, 389
Molecular attractions, 371–376
Molecular polarity, 368–371
Molecules, 340, 353–380
Moles, grams, converting
    between, 416–418
Momentum, 59–86
  collisions, 66–67
  conservation laws, 67
  conservation of, 65–68
  decreasing over long time,
    61–62
  decreasing over short time,
    62–64
  elastic collision, 66
  impulse, 60–62
  increasing, 61
  inelastic collision, 66
  kinetic energy, compared,
    73–74
  law of conservation of momen-
    tum, 65
Monomers, 489
Moon, 703–709
  eclipses, 707–709
  full Moon, 704
  lunar eclipse, 708
  new Moon, 704
  phases of, 704–705
  solar eclipse, 707
Motion, 15–36
  acceleration, 27–31
  Aristotle, 16
  dynamic equilibrium, 23–24
  equilibrium rule, 22
  force of friction, 24–25
    air resistance, 24
    friction, 24
  free fall, 29
  Galileo
    force, 17
    inertia, 17–18
  inertia, mass-measure, 18–20
    mass, 18
    volume, 19
    weight, 18
  kilogram, 20
  motion, relative nature of, 27
  net force, 20–21
    vector, 20
    vector quantity, 20
  relative nature of, 27
  speed, 25–27
    average speed, 26
    instantaneous speed, 25–26
    in units, 26
  support force, 23
  velocity, 27
    free-fall, 29
MRI. *See* Magnetic resonance
    imaging

Naming compounds, 347–348
Nanotechnology, 224
Natural atomic transmutation,
    319–320

Natural frequency, 239
Nature of science, 1–11
Neap tides, 641
Nebula, planetary, 729
Nebular theory, 692
Neptune, 702
Net force, 20–21
  vector, 20
Neutral solution, 445–448
Neutralization, 441
Neutron star, 731
Neutrons, 288–291
New Moon, 704
Newton, Isaac
  law of acceleration, 52
  law of action-reaction, 52
  law of cooling, 166–167
  law of inertia, 52
  laws of motion, 37–58
  summary of laws, 52
Non-free fall acceleration, 43–44
Nonbonding pairs, 355
Nonfoliated metamorphic rock,
    532
Nonpolar, 366
Nonpolar substances, boiling
    points, 371
Nonsilicate minerals, 513
Novae, 729–731
Nuclear fission, 323–327
  breeder reactor, 326–327
  chain reaction, 324
  critical mass, 324
Nuclear fusion, 330–332
  controlling fusion, 331–332
  thermonuclear fusion, 330
Nucleons, 289

Oceanic circulation, 653–654
Oceans, 629–659
  along coast, 638–640
  as basic solutions, 448–452
  components of, 632–635
  evolution of, 630–632
  ocean floor, 632–634
  seawater, 634–635
  shorelines, 635–642
  tides, 635–642
  wave refraction, 636–638
  waves, 635–642
    along coast, 638–640
    refraction, 636–638
    tides, 640–642
Ohm's law, 194–197
Oort cloud, 711–713
Opaque materials, 257–260
Orbits
  circular, 103–104
  elliptical, 104–105
Ordinary matter, 761
Ordovician period, 550
Organic chemistry, 472
Organic compounds, 471–502
  alcohols, 480–484
  alkaloids, 484–485
  amines, 484–485
  carbonyl compounds,
    485–489

ethers, 480–484
  functional groups, 478–480
  hydrocarbons, 472–476
  phenols, 480–484
  polymers, 489–497
    addition polymers,
      492–494
    condensation polymers,
      494–497
    unsaturated hydrocarbons,
      476–478
Organic molecules, functional
    groups, 479
Original horizontality, rock
    dating, 541
Orographic lifting, 675
Osmosis, 401
Outer planets, 690, 699–702
  Jupiter, 699–700
  Neptune, 702
  Saturn, 700–701
  Uranus, 702
Overloading, 199–200
Oxidation, 452
Ozone, 549, 631

Paleomagnetism, 573
Paleozoic era, 549–553
Pangaea, 551
Parabola, 95
Parallel circuits, 198–200
Partial melting, 520
Partial tones, 248
Pascal's principle, 129–130
Periodic table, 288, 292
  groups, 293–296
  periods, 293–296
Permian period, 552–553
pH scale, 447–448
Phases of Moon, 704–705
Phenols, 480–484
Physical model, 297
  atom, 296–298
Physical separation, mixtures,
    383–384
Physics, 9–10
  historical perspective, 122
Pigments
  colored, mixing, 269
  mixing, 269
Pitch, 235
Planet, failed formation,
    709–713
Planetary data, 691
Planetary nebula, 729
Planets, 690, 695–702
  Earth, 697
  Jupiter, 699–700
  Mars, 698–699
  Mercury, 695–696
  Neptune, 702
  Saturn, 700–701
  Uranus, 702
  Venus, 696–697
Plate tectonics, 565–598
  continental drift, 571–576
  convergent plate boundaries,
    579–582

divergent plate boundaries,
    577–579
  earthquakes, 587–592
  faults, 585–587
  folds, 585
  layers, Earth, 600–601
  seismic waves, 566–567
  transform plate boundaries,
    583–584
Plutonic rock, 523
Polar, 366
Polar covalent bonds, 365–368
Polar substances, boiling points,
    nonpolar substances,
    371
Polarization, 269–271, 274–277
  colored lights, mixing,
    267–268
  colored pigments, mixing, 269
  complementary colors,
    268–269
  diffuse reflection, 262
  law of reflection, 260–262
  rainbows, 272–274
  selective reflection, 266
  selective transmission,
    266–267
Poles, magnetic, 210–211
Polyatomic ions, 358
Polymers, 489–497
  addition polymers, 492–494
  condensation polymers,
    494–497
Porosity, 602
Potential difference, 190
Potential energy, 69–70
Power, 75
  lever, 76
  machines, 76–77
    conservation of energy, 76
  production of, 221–222
Precambrian time, 546–549
Precipitate, 393
Precipitation formation,
    673–674
Pressure, 117–119
  atmospheric, 125–128, 642
  in gas, 124–125
  in liquid, 118–119
  saturation vapor, 664–665
  temperature, relationship,
    648–650
Pressure-gradient force, 649
Primary waves, 566
Principal quantum number n,
    301
Principle of equivalence, 758
Principle of flotation, 122
Projectiles, 94
  fast-moving, 101
  launched at angle, 96–98
  launched horizontally, 95–96
  motion of, 94–101
  parabola, 95
  satellites, 101–102
Protons, 288–291, 438–442
Pseudoscience, 6, 185
Pulsars, 731

Pure matter, chemical classification, 384
Purification of water, 397–403
   bottled water, 402–403
   desalination, 399–402
   osmosis, 401
   reverse osmosis, 401
   semipermeable membrane, 400

Quality, 248
Quantity of heat, 145
Quantum hypothesis, 300–302
Quasar, 740

Rad, 314
Radiant energy
   absorption of, 164–165
   emission of, 163–164
   reflection of, 165–166
Radiation, 163–166
   cosmic background, 753–754
   dosage, 314–315
   terrestrial, 164
Radiation curves, stars, 723–724
Radioactive tracers, 316
Radioactivity, 312–316
   alpha particle, 312–314
   annual radiation exposure, 315
   beta particle, 312–314
   gamma rays, 312–314
   rad, 314
   radiation dosage, 314–315
   radioactive tracers, 316
   rem, 314
Radiometric dating, 321–323, 544–545
   carbon-14 dating, 322
   isotopes, 545
Rain, acid, 448–452
Rainbows, 272–274
Rarefaction, 235
Reaction
   identification, 46–48
   on masses, 48–50
Reaction rates, 418–423
   activation energy, 422
Reduction, 452
Reflection, 237–239, 260–262
   selective, 266
Reflection of radiant energy, 165–166
Refraction, 237–239, 262–265
   waves, 636–638
Relative dating, rock dating, 541
Relative humidity, 664
Relative rock dating, 541
   cross-cutting relationships, 541
   faunal succession, 542
   inclusions, 541
   lateral continuity, 541
   original horizontality, 541
   superposition, 541
Relativity, 750
   general theory of, 757–761
      tests of, 760–761
Religion, science and, 7–8
Rem, 314

Resistance, electrical, 193
Resonance, 239–240
Reverberation, 237
Reverse osmosis, 401
Rock, 503–538
   deposition, 525
   erosion, 525
   formation, 514–517
   igneous, 517–523
      lava, 521
      plutonic rock, 523
      volcanic, 521
   magma, 520–521
      generation of, 518–520
   metamorphic, 518, 528–532
      classifying, 531–532
      foliated, 531–532
      metamorphism, 528
         types of, 530–531
      nonfoliated, 532
      recrystallization, 528
   partial melting, 520
   plutonic rock, 523
   relative dating, 540–544
   rock cycle, 532–533
   rock-forming minerals, 513–514
   rock record, 539–564
   sedimentary, 518, 523–528
      chemical, 527–528
      classifying, 526–527
      formation of, 524–526
   sedimentation, 525
   types, 517–518
   weathering, 524
Rock cycle, 532–533
Rock-forming minerals, 513–514

Safety fuses, 200
Salinity, 635
Salt, acid-base reaction, 441–442
Sand dunes, 624
Satellites, 101–102
   circular orbits, 103–104
Saturated hydrocarbons, 477
Saturated solution, 388
Saturation vapor pressure, 664–665
Saturn, 700–701
Science, 1, 7–8
   astronomy, 9–10
   chemistry, 9–10
   conceptual physical, 2–3
   earth science, 9–10
   history of advances in, 2
   hypothesis, 3
   limitations of, 6
   mathematics, 2–3
   nature of, 1–11
   physics, 9–10
   pseudoscience, 6, 185
   society and, 6
   technology and, 8–9
Scientific attitude, 3–5
   fact, 3
   law, 3
   theory, 4
Scientific methods, 3

Sea salts, elements of, 634
Seafloor spreading, 554, 575
Seasons, 645–646
Seawater, 634–635
Second law of thermodynamics, 146
Secondary waves, 566
Sediment transport, 613–614
Sedimentary rock, 518, 523–528
   chemical, 527–528
   classifying, 526–527
   formation of, 524–526
Sedimentation, 525
Seismic waves, 566–567
   body waves, 566
   earthquake, 566
   primary waves, 566
   secondary waves, 566
   surface waves, 566
Selective reflection, 266
Selective transmission, 266–267
Semipermeable membrane, 400
Series circuits, 197–198
Shell model, atom, 304–305
Shock wave, 246
Shorelines, 635–642
Silicates, 513
Silurian period, 550–551
Sinkholes, 609
Sky, color, 269–270
Soaps, 394–396
Solar eclipse, 707
Solar energy, 645–648
   global warming, 646–648
   greenhouse effect, 646–648
   seasons, 645–646
   terrestrial radiation, 646
Solar system, 689–718
   asteroid belt, 709–710
   astronomical unit, 690
   comets, 710–713
   dwarf planets, 710–711
   ecliptic, 690
   failed planet formation, 709–713
   inner planets, 690, 695–699
      Earth, 697
      Mars, 698–699
      Mercury, 695–696
      Venus, 696–697
   Kuiper belt, 710–711
   meteor shower data, 713
   meteorite, 710
   meteoroids, 710
   meteors, 709–710
   Moon, 703–709
      eclipses, 707–709
      full Moon, 704
      lunar eclipse, 708
      new Moon, 704
      phases of, 704–705
      solar eclipse, 707
   nebular theory, 692
   Oort cloud, 711–713
   outer planets, 690, 699–702
      Jupiter, 699–700
      Neptune, 702
      Saturn, 700–701

   Uranus, 702
   planetary data, 691
   planets, 690
   Sun, 693–695
   sunspots, 694
Solubility, 391–394
   gases, 393–394
   insoluble, 392
   minerals, 517
   soluble, 391
   temperatures changes, 392–394
Soluble, 391
Solutes, 387
Solutions, 385–390
   concentration, 388
   dissolving, 388
   molarity, 389
   mole, 389
   saturated solution, 388
   solutes, 387
   solvent, 387
   unsaturated solution, 388
Solvent, 387
Sonic boom, 245–247
Sound
   acoustical imaging, dolphins and, 238
   bow waves, 245–247
   Doppler effect, 244–245
   forced vibrations, 239–240
   Fourier analysis, 249
   interference, 240–244
      beats, 242–243
      standing waves, 243
   musical sounds, 247–248
      fundamental frequency, 248
      harmonics, 248
      partial tones, 248
      quality, 248
   natural frequency, 239
   reflection, 237–239
   refraction, 237–239
   resonance, 239–240
   reverberation, 237
   shock wave, 246
   sonic boom, 245–247
   speed of, 236
Sound waves, 235–237
   loudspeakers, 236
   pitch, 235
   speed of sound, 236
Sources of energy, 78–79
Space/time structure, 749–770
   Big Bang, 750–754
   Big Rip, 765
   cosmic background radiation, 753–754
   cosmic inflation, 755–757
   cosmological redshift, 751
   cosmology, 749
   dark energy, 763–764
   dark matter, 761–763
   fate of universe, 764–766
   galaxy formation, 763
   general theory of relativity, 757–761
   heat death, 765

helium, 754
hydrogen, 754
ordinary matter, 761
principle of equivalence, 758
relativity, 750
special theory of relativity, 761
tests of general relativity, 760–761
Sparks, 184
Special theory of relativity, 761
Specific heat capacity, 147–150
of water, 148
Spectroscope identification, atom, 298–300
atomic spectrum, 299
Speed, 25–27
average speed, 26
instantaneous speed, 25–26
terminal, 43
in units, 26
Speed of sound, 236
Spiral galaxies, 737–738
Springs, 604–605
Standing waves, 243
Starburst galaxies, 738
Starburst galaxy, 738
Stars, 719–748
brightness, 722–724
color of, 722–724
Hertzsprung-Russell diagram, 725
life cycles of, 727–731
neutron star, 731
radiation curves, 723–724
starburst galaxy, 738
Static electricity, 181–208
Stratosphere, 644
Stream flow geometry, 610–611
discharge, 610
gradient, 610
Stream valleys, 614–617
cut bank, 615
point bars, 615
Strong nuclear force, 316
Structural isomers, 473
Structure of space, time, 749–770
Subatomic particles, 289
Subduction, 579–582
Sublimation, 169, 620
Submicroscopic life, 340–341
molecules, 340
Subtractive primary colors, 269
Sun, 693–695
Sunsets, color, 270–271
Sunspots, 694
Superclusters, 740–744
Superconducting electromagnets, 215
Superconductors, 194
Supernovae, 729–731
Superposition, rock dating, 541
Support force, 23
Surface currents, 654–657
Surface water, 609–617
laminar, 610
turbulent, 610

Surface waves, 566
Suspension, 386
Synclines, 585
Technology, science and, 8–9
Tectonics. *See* Plate tectonics
Temperature, 142–143, 666
atmospheric, 666
solubility, 392–394
Temperature inversion, 669
Temperature-pressure relationship, atmosphere, 648–650
Terminal speed, 43
Terminal velocity, 43
Terrestrial radiation, 164, 646
Tests of general relativity, 760–761
Theory, 4
Thermal energy, 141–158
absolute zero, 143
heat, 144
quantity of heat, 145
temperature, 142–143
Thermal expansion, 150–151
Thermodynamics, 141–158
laws of, 145–146
first law of, 145
second law of, 146
third law of, 146
Thermonuclear fusion, 330
Thermosphere, 644
Third law of thermodynamics, 146
Thunderstorms, 680
Tides, 635–642
Time/space structure, 749–770
Big Bang, 750–754
Big Rip, 765
cosmic background radiation, 753–754
cosmic inflation, 755–757
cosmological redshift, 751
cosmology, 749
dark energy, 763–764
dark matter, 761–763
fate of universe, 764–766
galaxy formation, 763
general theory of relativity, 757–761
heat death, 765
helium, 754
hydrogen, 754
ordinary matter, 761
principle of equivalence, 758
relativity, 750
special theory of relativity, 761
tests of general relativity, 760–761
Tornadoes, 681
Transfer of heat, 159–179
absorption of radiant energy, 164–165
boiling, 171–173
change of phase, 168–171, 174–175
condensation, 170–171
conduction, 160

convection, 161–162
emission of radiant energy, 163–164
energy, 174–175
evaporation, 169
freezing, 173–174
global warming, 167–168
greenhouse effect, 167–168
heat of fusion, 174
heat of vaporization, 174
melting, 173–174
Newton's law of cooling, 166–167
radiation, 163–166
reflection of radiant energy, 165–166
sublimation, 169
terrestrial radiation, 164
Transform plate boundaries, 583–584
Transformer, 222
Transformer-boosting, 222–223
Transmission, selective, 266–267
Transparency, 257–260
Transparent materials, 257–260
Transverse waves, 234–235
Troposphere, 643
Turbulent flow, 610
Types of rock, 517–518

Universal gravitation, 93–94
Universal gravitational constant, 89–90
Universal law of gravity, 88–90
Unsaturated hydrocarbons, 476–478
Unsaturated solution, 388
Upper atmospheric circulation, 653
Uranus, 702

Valence electrons, 305
Valence shell, 354
Vaporization, heat of, 174
Vector, 20
Vector quantity, 20
Velocity, 25–27
free-fall, 29
terminal, 43
Venus, 696–697
Vertical development, 673
Vertical structure, atmosphere, 643–644
Vibrations, 232–233
Violent weather, 679–682
Volcanic rock, 521
Volcanoes, 521–523
Voltage sources, 190–191
Volts, 193
Volume, 19

Wastewater treatment, 403–405
Water
bottled, 402–403
hard, softening, 396–397
wastewater treatment, 403–405

Water purification, 397–403
bottled water, 402–403
desalination, 399–402
osmosis, 401
reverse osmosis, 401
semipermeable membrane, 400
Water resource residence times, 601
Water table, 603–604
Wavelength, 232
Waves, 231–254, 635–642
amplitude, 232
coastal, 638–640
compression, 235
frequency, 232
hertz, 232
longitudinal waves, 234–235
motion, 233–234
rarefaction, 235
refraction, 636–638
sound waves, 235–237
acoustical imaging, dolphins and, 238
bow waves, 245–247
Doppler effect, 244–245
forced vibrations, 239–240
Fourier analysis, 249
interference, 240–244
beats, 242–243
standing waves, 243
loudspeakers, 236
musical sounds, 247–248
fundamental frequency, 248
harmonics, 248
partial tones, 248
quality, 248
natural frequency, 239
pitch, 235
reflection, 237–239
refraction, 237–239
resonance, 239–240
reverberation, 237
shock wave, 246
sonic boom, 245–247
speed of sound, 236
speed, 233–234
tides, 640–642
transverse waves, 234–235
vibrations, 232–233
wavelength, 232
Weather, 663–686
air masses, 674–679
atmospheric lifting, 675
convectional lifting, 675
frontal lifting, 676–678
midlatitude cyclones, 678–679
orographic lifting, 675
atmospheric moisture, 664–668
adiabatic processes, 667–669
atmospheric stability, 669–670
humidity, 664
relative humidity, 664

Weather (*Continued*)
  saturation vapor pressure, 664
    vs. temperature, 665
    temperature changes, 666
    temperature inversion, 669
  cloud development, 670–674
    cloud groups, 671
    high clouds, 671–672

lifting condensation level, 670
low clouds, 672–673
middle clouds, 672
precipitation formation, 673–674
vertical development, 673
hurricanes, 681–682

maps, 683
thunderstorms, 680
tornadoes, 681
violent, 679–682
weather variables, 667–670

Weight, 18, 92–93
Weightlessness, 92–93

White dwarfs, 726
Work, energy and, 68–71
Work-energy theorem, 71–74

Zero, absolute, 143

Eon	Era	Period	Subperiod	Epoch	Ma
Phanerozoic	Cenozoic	Quaternary		Holocene	0.01
				Pleistocene	1.8
		Tertiary		Pliocene	5.3
				Miocene	23.8
				Oligocene	33.7
				Eocene	54.8
				Paleocene	65
	Mesozoic	Cretaceous			144
		Jurassic (first bird)			206
		Triassic			248
	Paleozoic	Permian (first reptiles)			290
		Carboniferous	Pennsylvanian		323
			Mississippian		354
		Devonian (first amphibians)			417
		Silurian (first insect fossils)			443
		Ordovician (first vertebrate fossils)			490
		Cambrian (first plant fossils)			543
Precambrian Time	Proterozoic				2500
	Archean				3800
	Hadean				4500